MECHANISMS OF ATMOSPHERIC OXIDATION OF THE ALKANES

MECHANISMS OF ATMOSPHERIC OXIDATION OF THE ALKANES

JACK G. CALVERT, RICHARD G. DERWENT,
JOHN J. ORLANDO, GEOFFREY S. TYNDALL,
AND TIMOTHY J. WALLINGTON

OXFORD
UNIVERSITY PRESS

2008

Oxford University Press, Inc., publishes works that further
Oxford University's objective of excellence
in research, scholarship, and education.

Oxford New York
Auckland Cape Town Dar es Salaam Hong Kong Karachi
Kuala Lumpur Madrid Melbourne Mexico City Nairobi
New Delhi Shanghai Taipei Toronto

With offices in
Argentina Austria Brazil Chile Czech Republic France Greece
Guatemala Hungary Italy Japan Poland Portugal Singapore
South Korea Switzerland Thailand Turkey Ukraine Vietnam

Published by Oxford University Press, Inc.
198 Madison Avenue, New York, New York 10016
www.oup.com

Library of Congress Cataloging-in-Publication Data
Mechanisms of atmospheric oxidation of the alkenes / Jack G. Calvert . . . [et al.].
p. cm.
Includes bibliographical references and index.
ISBN 978-0-19-536581-8
1. Alkanes – Oxidation. 2. Atmospheric chemistry. I. Calvert, Jack G. (Jack George), 1923– II. Title.
QC879.7.M429 2008
551.51'1—dc22 2007047420

9 8 7 6 5 4 3 2 1

Printed in the United States of America
on acid-free paper

Acknowledgments

The authors thank several agencies and their members for the support of this research effort. Timothy Belian, former Executive Director, and current Executive Director, Brent K. Bailey, of the Coordinating Research Council (3650 Mansell Road, Suite 140, Alpharetta, GA 30022), and the members of the Coordinating Research Council's Atmospheric Impacts Committee who gave the authors their support and monitored this study: Rory S. MacArthur (Chevron Energy Tech. Co., Richmond, CA); James C. Ball (Ford Motor Co., Dearborn, MI); Daniel C. Baker [Shell Global Solutions (U.S.), Houston, TX]; Steven H. Cadle (General Motors R&D Center, Warren, MI); David P. Chock (Ford Research and Advanced Engineering, Dearborn, MI); Alan M. Dunker, General Motors R&D Center, Warren, MI); Eric Foote, DaimlerChrysler, Chelsea, MI); John German (American Honda Motor Company, Ann Arbor, MI); Alan J. Krol (BP America, Naperville, IL); Mani Natarajan (Marathon Petroleum Co. LLC, Findlay, OH); Charles H. Schleyer (Exxon Mobil Research and Engineering, Paulsboro, NJ); James J. Simnick, (BP, Global Fuels Technology, Naperville, IL); Ken Wright (ConocoPhillips Petroleum Co., Bartlesville, OK).

The authors acknowledge their research organizations and educational institutions for the support of their participation in this research: National Center for Atmospheric Research (NCAR), Atmospheric Chemistry Division, Boulder, CO; Ford Motor Company, Physical and Environmental Sciences Department of the Scientific Research Laboratories, Dearborn, MI; Oak Ridge National Laboratories, Environmental Sciences Division, Oak Ridge, TN; the United Kingdom Department for the Environment, Food and Rural Affairs, Air and Environmental Quality Division, London, England, for support under contract EPG 1/2/200.

The authors thank their many coworkers who provided them with important new information and other assistance: Geert Moortgat and Hannelore Keller-Rudek (Max-Planck Institute, Mainz, Germany) for providing reprints and spectral data from the files of Mainz Spectral Atlas; James Burkholder (Aeronomy Laboratory, National Oceanic and Atmospheric Administration, Boulder, CO) for providing original spectral data on the halogen-substituted acetones; William B. DeMore (NASA-retired) for providing original data from relative rate experiments for reactions of OH radicals with several haloalkanes; Mike Hurley (Ford Motor Co.) for helpful discussions on the use of SARs to describe the reactivity of haloalkanes towards OH radicals; Ian Smith for providing original kinetic data for reactions of haloalkyl peroxy radicals with NO; Georges LeBras (CRNS) and Robert Lesclaux (University of Bordeaux-retired) for providing reprints; Ole John Nielsen (University of Copenhagen) for helpful discussions; Nathan Soderborg (Ford Motor Co.) for advice on linear regressions; library staff of NCAR (Gale Grey, Leslie Forehand) and Oak Ridge National Laboratories (Larkee Moore, Mary Jacqueline) for help in obtaining reprints of the published data referenced in this study; U.S. Department of Energy, Office of Science, Biological and Environmental Research Program for Jack Calvert's Visiting Scientist Appointment in the Environmental Sciences Division, Oak Ridge National Laboratories, Stephen G. Hildebrand and Gary K. Jacobs for hosting this appointment, and Thomas Edward Stanford and Linda W. Armstrong for their valuable assistance. Michael Jenkin and Bill Carter are thanked for their careful reading of the descriptions of the MCM and SAPRC chemical mechanisms in chapter VIII. Roger Atkinson and Harvey Jeffries are acknowledged for their longstanding contributions to the understanding of the atmospheric chemistry of the VOCs and modeling ozone formation, respectively, without which chapters I and VIII would have been impossible to write authoritatively. The authors are most grateful to Dr. Steven M. Japar for his helpful scientific review of the original manuscript and Lois Mott (CRC) for her insightful editing of the original manuscript.

Contents

About the Authors

Jack G. Calvert studied at the University of California, Los Angeles where he received his B.S. (Chemistry) in 1944 and his Ph.D. in 1949 with Professor F. E. Blacet. He continued postdoctoral studies in photochemistry with Dr. E. W. R. Steacie of the National Research Council, Canada (1949–1950). He joined the faculty of The Ohio State University in 1950 and served as Kimberly Professor of Chemistry until 1981. He studied with Drs. R. A. Cox and S. A. Penkett at the Harwell Laboratory of the United Kingdom's Atomic Energy Research Establishment (1978). He was appointed Senior Scientist at the National Center for Atmospheric Research (Boulder, Colorado) in 1981 where he led the Atmospheric Kinetics and Photochemistry Group of the Atmospheric Chemistry Division until his retirement when he was appointed Emeritus Senior Scientist at NCAR (1993–present). Since 2002 he has been a Visiting Scientist in the Environmental Sciences Division of the Oak Ridge National Laboratory. Professor Calvert has authored or coauthored approximately 300 publications in the scientific literature that relate to various aspects of photochemistry, reaction kinetics, and atmospheric chemistry. His honors that he has received include: The Distinguished Research Award, The Ohio State University (1981); Simon Guggenheim Memorial Fellowship (1978); American Chemical Society, Columbus Section Award (1981); American Chemical Society Award for Creative Advances in Environmental Science and Technology (1982); Chambers Award of the Air Pollution Control Association (1986). He has been a member and chair of many national and international scientific committees organized to solve recognized problems in atmospheric chemistry. Calvert is coauthor of the two preceding books on mechanisms of atmospheric oxidation of the alkenes and the aromatic hydrocarbons published by Oxford University Press.

Richard G. Derwent was born in Darlington, England in 1947. He studied at Queens' College, University of Cambridge where he received the B.A., M.A., and Ph.D. degrees under Professor B.A. Thrush. Dr. Derwent has spent much of his research career studying atmospheric chemistry. Initially, this work was carried out in the Air Pollution Division, Warren Spring Laboratory, Stevenage, where he set up the first U.K. monitoring networks for ozone, NO_x, carbon monoxide and hydrocarbons. He then spent a period of 16 years (1974–1990) at the Harwell Laboratory of the United Kingdom's Atomic Energy Research Establishment with Drs. R. A. Cox and S. A. Penkett. Here he built the first European models of stratospheric ozone depletion, tropospheric ozone build-up, acid rain and ground level photochemical smog formation. During the 1990–1993 period he was research manager in the Department of Environment and Her Majesty's Inspectorate of Pollution. During the next ten years (1993–2003) he worked at the United Kingdom Meteorological Office, Bracknell, where he modeled the processes of acid rain, photochemical ozone formation and the build up of greenhouse gases. Since 2003 he has been a self-employed consultant. In recognition of his contribution to atmospheric chemistry research he was appointed as an Officer of the Order of the British Empire (OBE) by Her Majesty Queen Elizabeth II in 2000. He received the Fitzroy Prize from the Royal Meteorological Society in 1997; and the Royal Society of Chemistry Award and Medal for Environmental Chemistry, in 1993. He is author or joint author of over 350 published research papers, reports, conference papers and memoranda related to acid rain, urban pollution, photochemical smog and global atmospheric chemistry. He has served on many national and international committees related to atmospheric modeling and air quality.

John J. Orlando was born in Timmins, Ontario, Canada in 1960. He received both his B.Sc. (1982) and Ph.D. degrees (1987) in Chemistry from McMaster University in Hamilton, Canada with Professors D. R. Smith and J. Reid. He then spent two years as a postdoctoral fellow at the National Oceanic and Atmospheric Administration's Aeronomy Laboratory under the supervision of Drs. C. J. Howard and A. R. Ravishankara, where he worked on problems related to the effects of halogens on stratospheric ozone. Dr. Orlando is presently a Scientist III in the Atmospheric Chemistry Division of the National Center for Atmospheric Research (NCAR) where he has been employed since 1989 as an active member of the laboratory kinetics group. His research at NCAR has focused largely on the determination of the mechanisms of the atmospheric oxidation of carbon-containing species, on halogen chemistry of relevance to the stratosphere, and on the atmospheric chemistry of nitrogen oxide species. He is author or coauthor of over 100 scientific publications related to atmospheric chemistry. Dr. Orlando was the recipient of the American Meteorological Society

Special Award (2001), the Velux Foundation Visiting Professor Fellowship at the University of Copenhagen (2002), and the NCAR Special Achievement Award in 2003. Dr. Orlando was coeditor and coauthor of the book *Atmospheric Chemistry and Global Change* published by Oxford University Press in 1998.

Geoffrey S. Tyndall was born in Mansfield, U.K., in 1955. His undergraduate and graduate studies were conducted in the Churchill College, Cambridge University, where he obtained a B. A. in chemistry in 1978, and a Ph.D. in physical chemistry in 1982 with Professor B. A. Thrush. His postdoctoral work was carried out at the Max-Planck-Institut für Chemie, in Mainz, Germany, with Dr. G. K. Moortgat (1983–1986), and at the NOAA Aeronomy laboratory, with Dr. A. R. Ravishankara (1986–1988). Dr. Tyndall joined the scientific staff at NCAR in 1988, where he is currently a Senior Scientist in the Atmospheric Chemistry Division and leader of the Laboratory Kinetics Group. His research over the years has focused on many aspects of atmospheric photochemistry, including the study of the kinetics and spectroscopy of HO_2 and organic peroxy radicals, the study of atmospheric sulfur and nitrogen chemistry, and studies of the atmospheric oxidation mechanisms of anthropogenic and biogenic hydrocarbons. He was the recipient of the American Meteorological Society Special Award (2001), and the NCAR/ACD Achievement Award (2003). Dr. Tyndall is author and coauthor of over 100 scientific publications related to atmospheric chemistry and was a coauthor and coeditor of the Oxford University Press book *Atmospheric Chemistry and Global Change* published in 1998.

Timothy J. Wallington was born and educated in England. He received B.A. (1981), M.A. (1982), and Ph.D. (1983) degrees from Corpus Christi College, Oxford University where he studied with Professor R. P. Wayne and Dr. R. A. Cox. He has carried out extensive research on various aspects of atmospheric chemistry and the kinetics and mechanisms of many different transient atmospheric species. His postgraduate research studies were made at the University of California, Los Angeles (1984–1986) with Professor J. N. Pitts and Dr. R. A. Atkinson. He was Guest Scientist at the U.S. National Bureau of Standards (1986–1987) with Dr. M. J. Kurylo. He joined the Research staff at the Ford Motor Company in 1987 where he is currently a Technical Leader in the Physical and Environmental Sciences Department. Dr. Wallington has studied the atmospheric chemistry of vehicle and manufacturing emissions and their contribution to local, regional, and global air pollution and global climate change. He has helped develop policy and strategy to address global environmental issues associated with transportation. He is coauthor of over 300 peer reviewed scientific publications dealing with various aspects of air pollution chemistry. He is the recipient of eleven Ford Research Publication Awards (1991–2006), the Ford Motor Company Technical Achievement Award (1995), the Henry Ford Technology Award (1996), the Humboldt Research Fellowship, Universität Wuppertal (1998–1999), with Professor K. H. Becker, and the American Chemical Society Award in Industrial Chemistry (2008). Dr. Wallington was coauthor on the two previous books on the mechanisms of atmospheric oxidation of the alkenes and the aromatic hydrocarbons that were published by Oxford University Press.

MECHANISMS OF ATMOSPHERIC OXIDATION OF THE ALKANES

I

Importance of Alkanes in Atmospheric Chemistry of the Urban, Regional, and Global Scales

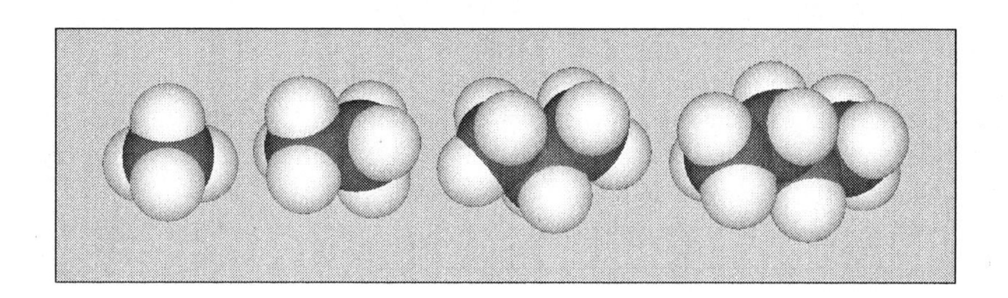

I-A. Introduction to the Alkanes

The alkanes are a class of organic compounds that make up a major component of the trace-gas composition of the atmosphere. They take part in atmospheric chemical reactions on a range of scales from urban to global. Their participation in these atmospheric chemistry processes makes a substantial contribution to photochemical ozone formation on the urban and regional scales, controls the build-up of methane and ozone (the second and third most important greenhouse gases after carbon dioxide), and drives stratospheric ozone layer depletion (for halogen-substituted alkanes only).

The objective of this review is to assemble the information required to describe quantitatively the participation of alkanes in photochemical ozone formation, global warming, and stratospheric chemistry for use in atmospheric models. Atmospheric models are the main tools for integrating our knowledge of the atmospheric chemistry of the alkanes with that of the other important processes such as emissions, transport, dispersion, and deposition. Here we assemble and evaluate the available information in the peer-reviewed literature on the kinetics and mechanisms of the chemical reactions under tropospheric and stratospheric conditions of the alkanes emitted into the atmosphere and their first-generation products.

As there are a large number of alkanes emitted into the atmosphere by human activities and by natural processes (see the discussion of section I-B), it is helpful at the outset to divide them into their major subdivisions:

- Straight- and branched-chain hydrocarbons,
- Cyclic hydrocarbons,
- Halogen-substituted hydrocarbons.

Straight- and branched-chain hydrocarbons contain a linear framework of carbon–carbon bonds with the general structural formulae C_nH_{2n+2}. This homologous series starts with methane, CH_4, a gas under normal atmospheric conditions of temperature and pressure, and stretches through liquids and ultimately to waxy solids; see section I-B. Because of the large numbers of straight- and branched chain alkanes present in the atmosphere, it is useful to break this subdivision further into low-molecular-mass alkanes and high-molecular-mass alkanes. This further subdivision is somewhat arbitrary, but we will find it convenient to draw a distinction somewhere between hexane and octane. The emissions of low-molecular-mass $<C_6$—C_8 are dominated by road transport emissions and the leakage and evaporation of liquid fuels (Carter, 2000). The emissions of high-molecular-mass alkanes $>C_6$—C_8 are dominated by the evaporation of hydrocarbon solvents such as white and mineral spirits (Carter, 2000) and by the emissions of fuel and lubricants from vehicles with diesel engines (Cautreels and van Cauwenberghe, 1978).

Cyclic hydrocarbons (naphthenes) have skeletal frameworks with a single ring containing three or more carbon atoms and have the general structural formulae C_nH_{2n}. This homologous series begins with cyclopropane, C_3H_6, and, in the atmosphere, extends up to cyclooctane. The bulk of the cyclic hydrocarbons emitted into the atmosphere appear to be derivatives of cyclohexane with varying numbers of straight- or branched-chain alkyl substituents (Carter, 2000). If the ring contains an atom other than carbon, then the compound would be classed as a heterocyclic compound. Such compounds are relatively rare, though European emission inventories of organic compounds do contain heterocyclic compounds such as dioxane, dioxolone, thiophene, and dimethylfuran (Passant, 2002). The emissions of cyclic hydrocarbons are dominated by emissions from the evaporation of hydrocarbon solvents such as white spirits (Passant, 2002).

Halogenated hydrocarbons or haloalkanes are straight-chain or cyclic hydrocarbons in which one or more hydrogen atoms has been replaced with a fluorine, chlorine, bromine, or iodine atom. As a class, they are largely man-made with a few important exceptions such as methyl chloride (CH_3Cl) (Khalil and Rasmussen, 1999). They have been developed since the 1930s for a few specialist applications as refrigeration and air conditioning fluids, aerosol propellants, foam blowing agents, and solvents (McCarthy et al., 1977). Generally speaking, only the chlorine- and bromine-substituted alkanes take part in the depletion of the stratospheric ozone layer and only the fluorine-substituted alkanes are major greenhouse gases. Emissions of the iodoalkanes are dominated by natural marine sources, and they are relatively rare trace atmospheric constituents.

I-B. Structure and Properties of the Alkanes

Molecules of alkane hydrocarbons contain carbon and hydrogen atoms joined by single, covalent bonds between the atoms. The backbone of the alkane molecules is made up of carbon atoms linked together in continuous or branched chains as in butane and 2-methylpropane, respectively; see figure I-B-1. Each carbon atom is attached to four

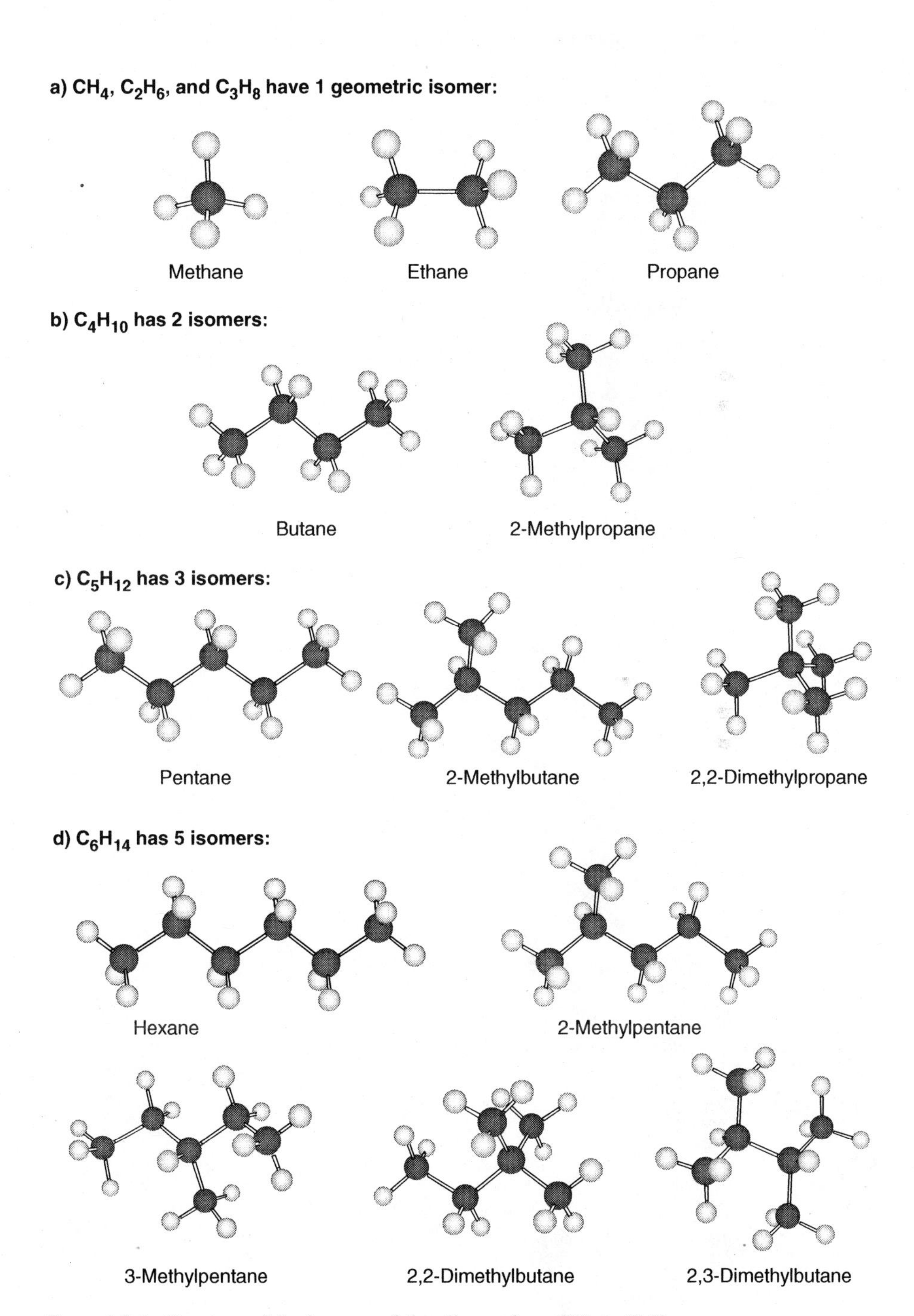

Figure 1-B-1. Structure of the isomers of the alkanes from CH_4 to C_7H_{16}.

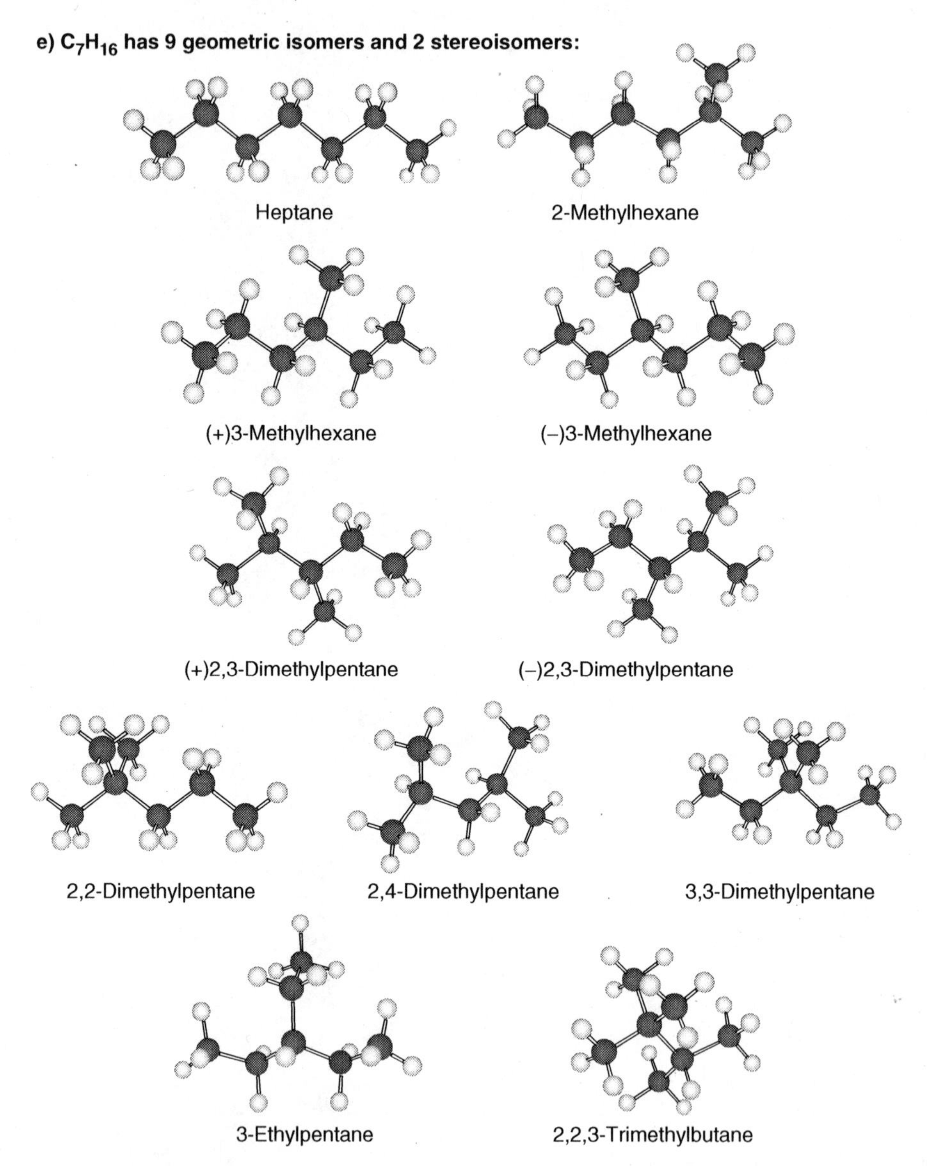

Figure I-B-1. *(continued)*

other atoms in a tetrahedral arrangement with H—C—H and C—C—C angles of about 110°. The open-chain alkanes have the general formula C_nH_{2n+2}. Rotation of CH_3— or other alkyl groups about C—C bonds can occur freely at ordinary temperatures. For the usual gas phase or liquid phase conditions, groups attached to the carbon atoms do not exchange positions on the skeletal chain without dissociation and destruction of the molecule. Migration of alkyl groups and molecule reformations are effected through the use of specific catalysts under careful control of temperature and other variables.

There are several different kinds of C—H bonds in the alkanes, with different chemical reactivity toward OH radicals and other reactive species. The three H-atoms in CH_3- groups are called *primary H-atoms*; the two H-atoms in C—CH_2—C groups are *secondary H-atoms*, while a single H-atom attached to a C-atom that is also attached to three other C-atoms is called a *tertiary H-atom*. Reactivity toward abstraction of an H-atom by OH or other reactive species increases in the order: primary-H $<$ secondary-H $<$ tertiary-H.

A special characteristic of the alkanes is the vastness of the number of different structural patterns, or isomers, that can exist. Consider the different alkane isomers of chemical formulae CH_4, C_2H_6, C_3H_8, C_4H_{10}, C_5H_{12}, C_6H_{14}, and C_7H_{16} shown in figure I-B-1. There is only one possible geometric arrangement of the CH_4, C_2H_6, and C_3H_8 alkanes, while C_4H_{10} has two possible isomers, C_5H_{12} has three, C_6H_{14} has five, and C_7H_{16} has nine. Of the nine different geometric isomers of C_7H_{16}, 3-methylhexane and 2,3-dimethylpentane contain carbon atoms on which four chemically different groups are attached. With such molecules, the mirror image of each is not superimposable on the parent molecule; indeed, these are different molecules. Although these optical isomers or stereoisomers have many identical properties (boiling points, melting points, vapor pressures, etc.), they differ in the direction in which they rotate the plane of polarized light. However, there are no significant differences in chemical reactivity between alkane stereoisomers.

Mathematical studies have determined accurately the number of possible geometric and stereoisomers that can exist (e.g., see: Davies and Freyd, 1989; Bytautas and Klein, 1998). As the number of C-atoms increases, there is a surprising increase in the number of isomers; see figure I-B-2. $C_{10}H_{22}$ has 75 geometric isomers, $C_{15}H_{32}$ has 4347, $C_{20}H_{42}$ has 366,319, while $C_{40}H_{82}$ has 62,481,801,147,341, and $C_{80}H_{162}$ is expected to have approximately 10^{28} isomers. In view of the extremely complex mixture of hydrocarbons that fuels contain, one can appreciate the great difficulty in separation and identification that one has in analyses of air samples that contain traces of these uncounted components of fuels together with their many oxidation products.

The physical as well as the chemical properties of the alkanes vary widely with the size of the molecule. The vapor pressure of an alkane is one of its important physical properties that determine its usefulness in different fuels and solvents. At a temperature in the range 280–320 K, the vapor pressure of the liquid state of the straight-chain alkanes decreases over 8 to 10 orders of magnitude as one progresses from CH_4 to n-$C_{20}H_{42}$; see figure I-B-3. The effect of branching of the alkane structure on the vapor pressure at 300 K can be seen in figure I-B-4. For alkanes of equal carbon number, the more highly branched alkanes have a somewhat higher vapor pressure. The cycloalkanes (C_nH_{2n}) which contain the carbon atoms joined in a ring, have somewhat lower vapor pressures than the open- or branched-chain alkanes of the same carbon number. The normal boiling points (1 atmosphere pressure) increase from –161.5°C for CH_4 to 440.8°C for n-$C_{29}H_{60}$; see figure I-B-5. As one compares straight-chain alkanes to branched alkanes with the same number of carbon atoms per molecule, a small decrease in boiling points is observed for the 2-methylalkanes and a further decrease for 2,2-dimethylalkanes, reflecting the increased vapor pressures at a given

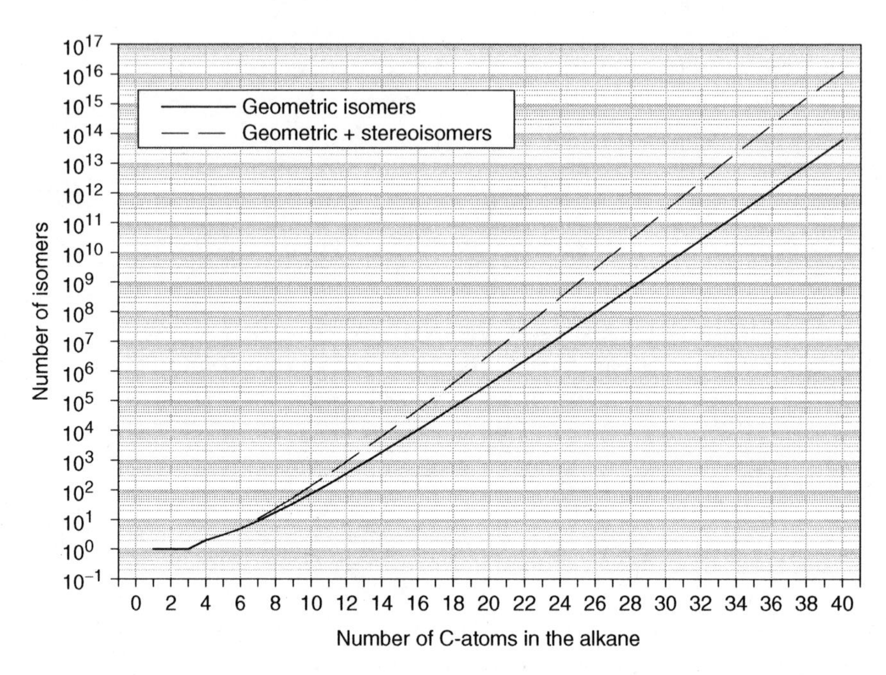

Figure I-B-2. The number of geometric isomers and stereoisomers of alkanes of general formula C_nH_{2n+2} from CH_4 to $C_{40}H_{82}$ (Davies and Freyd, 1989; Bytautas and Klein, 1998).

temperature. In general, the melting points of the alkanes rise with carbon number, with solids being the stable form at 298 K for alkanes greater than about C_{18}.

Figure I-B-6 shows the approximate carbon number in hydrocarbons that are commonly used in various fuels. The alkanes are an important component of liquid fuels and lubricants in use today. These are usually used in combination with somewhat smaller amounts of aromatic hydrocarbons, alkenes, and certain fuel additives that enhance the desired qualities of the product. The composition of the fuel is determined by the specific requirements of the engines and compliance with the regulations of the country, state, or community.

I-C. General Survey of the Atmospheric Chemistry of the Alkanes

The pathways for initiating the oxidation reactions of volatile organic compounds (VOCs) in the lower atmosphere, alkanes included, generally involve reactions with ozone (O_3), hydroxy radicals (OH), nitrate radicals (NO_3), oxygen atoms [$O(^3P)$], and photolysis at wavelengths longer than about 290 nm. A qualitative comparison of the reaction rates of the different major classes of VOCs via these channels in urban atmospheres (Atkinson, 1995) is shown in table I-C-1. From this table, it is apparent that, for the alkenes, reactions with O_3, OH, NO_3, and $O(^3P)$ are potentially important. As a result, the alkenes react rapidly and decay away quickly following their emission into the urban atmosphere. For the aromatic hydrocarbons, reactions

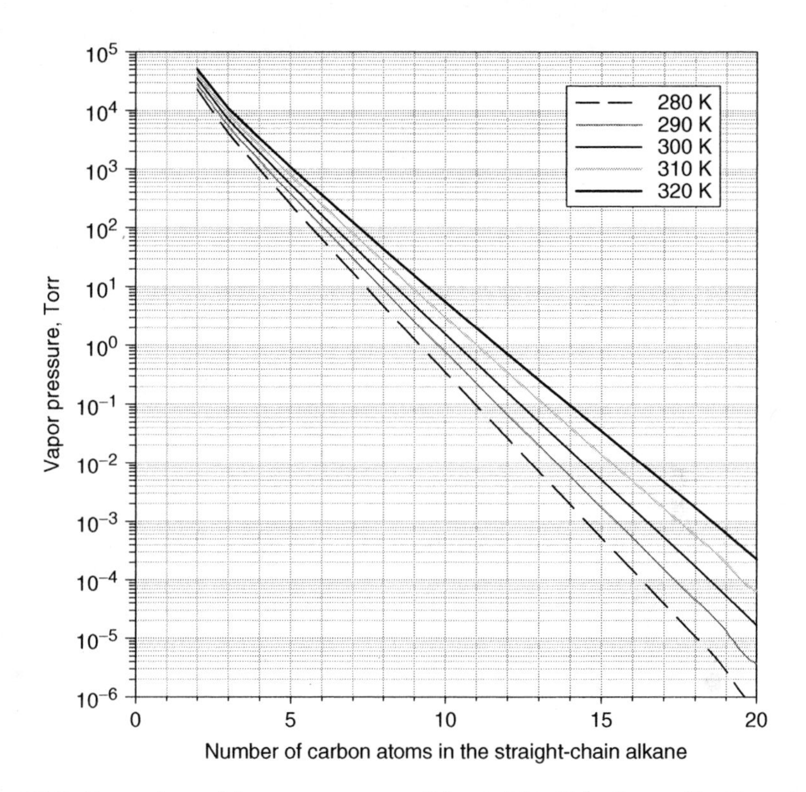

Figure I-B-3. Comparison of the vapor pressures of the straight-chain alkanes (liquid state) C_2H_6 to n-$C_{20}H_{42}$ at temperatures from 280 to 320 K. For the higher alkanes, the data represent metastable liquid droplets cooled below their freezing points; these data are necessarily approximate, since extrapolations of the liquid-phase vapor pressure data from the range of measurements are required to reach the temperatures shown here. Data are derived from compilations in the *CRC Handbook of Chemistry and Physics* (Lide, 2001–2002).

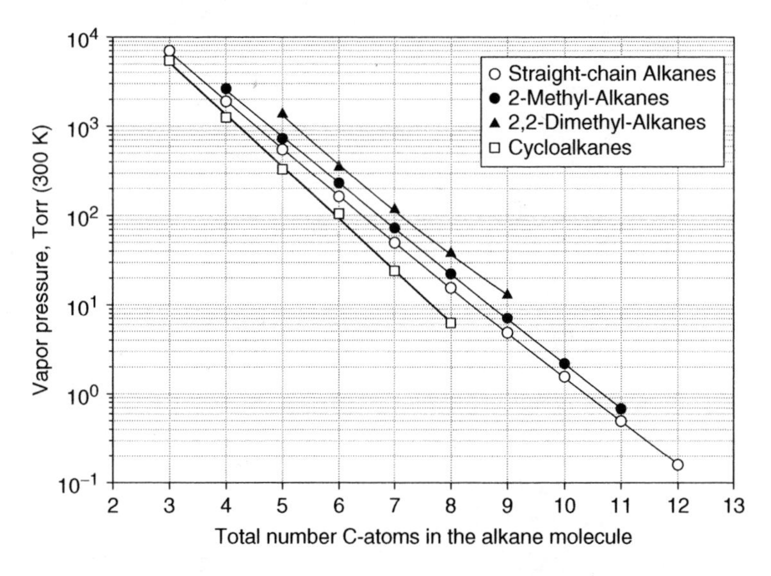

Figure I-B-4. Comparison of the vapor pressures of the straight-chain alkanes, 2-methylalkanes, 2,2-dimethylalkanes, and cycloalkanes of carbon numbers C_3–C_{12}. Data are derived from compilations in the *CRC Handbook of Chemistry and Physics* (Lide, 2001–2002).

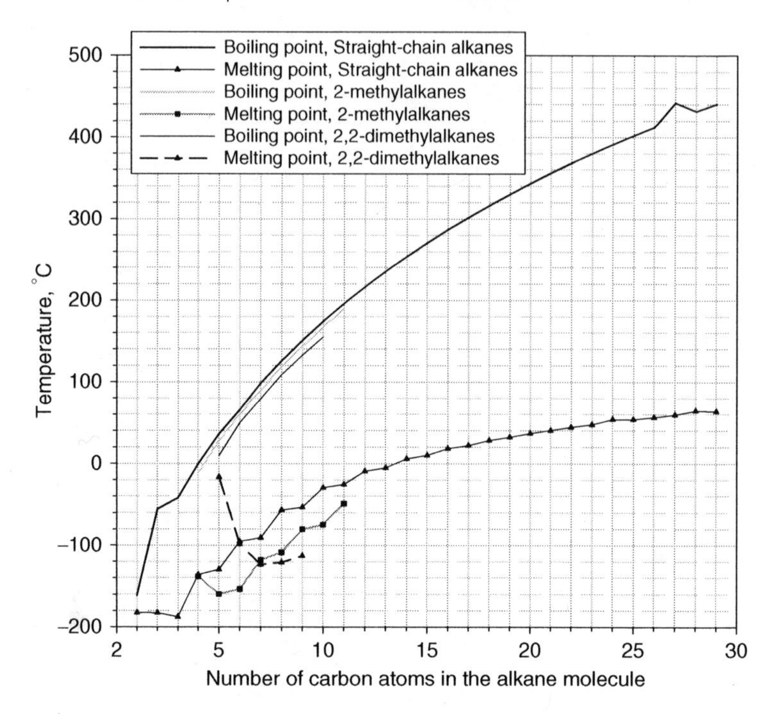

Figure I-B-5. Comparison of the normal boiling points (1 atmosphere pressure) and melting points of the straight-chain alkanes (CH_4 to $C_{29}H_{60}$), 2-methylalkanes (C_4H_{10} to $C_{11}H_{24}$), and 2,2-dimethylalkanes (C_5H_{12} to $C_{10}H_{22}$). Data are derived from compilations in the *CRC Handbook of Chemistry and Physics* (Lide, 2001–2002). Note that the reported boiling point of $C_{27}H_{56}$ appears to be anomalous.

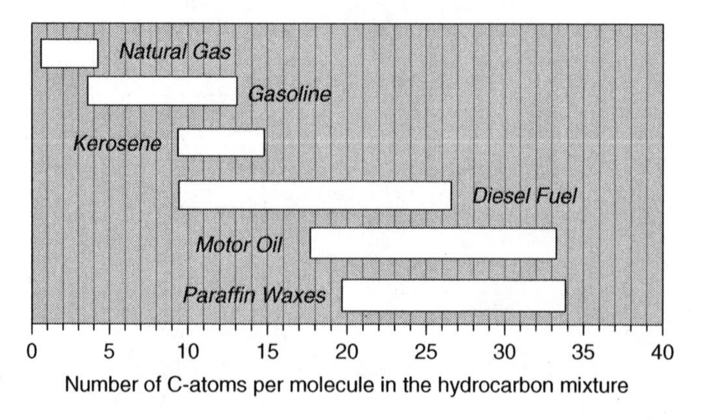

Figure I-B-6. Approximate range of carbon numbers of hydrocarbons often included in various fuels, lubricants, and waxes.

with OH are always important and, as with the alkenes, reactions are generally rapid. Consequently, alkenes and aromatic hydrocarbons contribute significantly to urban-scale atmospheric chemistry processes because of their fast reactions with OH radicals.

Table I-C-1. Qualitative assessment of the reaction rates for VOCs with the main tropospheric reaction initiators in urban atmospheres, following Atkinson (1994)

VOCs	OH	O_3	NO_3	$O(^3P)$	Photolysis
Alkanes	Slow to medium	Extremely slow			
Alkenes	Medium to fast	Fast	Medium to fast	Fast	
Aromatics	Slow to fast	None to medium[a]	None to fast[b]		
Oxygenated VOCs	Slow to fast		Slow to medium		Slow to fast

[a] Only for aromatic alkenes.
[b] Only for aromatic alkenes and phenols.

In contrast, table I-C-1 shows that the alkanes are much less reactive on the urban scale than the alkenes and aromatic compounds. The reactions of the alkanes with the main tropospheric reaction initiators are much slower than those of the alkenes and aromatic hydrocarbons. The reactions with OH radicals are important whereas reactions of O_3, NO_3, and photolysis are relatively unimportant. Because of the slow to medium reactivities exhibited by the alkanes, little of the alkanes is oxidized in urban areas close to its points of emission, while the remainder is transported over large distances to chemical environments that are characteristically different to the urban environments represented in table I-C-1. It will therefore be necessary to consider a much wider range of reaction initiators (e.g., Cl in the marine boundary layer, $O(^1D)$ and Cl in the stratosphere) for the alkanes, compared with the limited range in table I-C-1 which is appropriate for alkenes and aromatic compounds (Calvert et al., 2000, 2002).

Reaction of OH dominates the different reaction pathways in table I-C-1 for the individual alkanes. Of course, the importance depends on the relative concentrations of the different initiators such as OH and their reaction rate coefficients, and under special atmospheric conditions all the initiators but ozone can react with the alkanes to some extent; see chapter III. The concentrations of the initiators are highly variable in space and time because of variations in levels of the photochemically labile precursor species that produce them, because of variations in the concentrations of sink species, and because of variations in solar actinic irradiance. Many of the reaction rate coefficients of the alkane reactions are highly temperature dependent, and this can lead to further variations in the relative importance of the different oxidation reactions.

Because most of the alkanes derived from human activities are emitted in urban areas, urban photochemical oxidation driven by OH radicals is the first major atmospheric chemical system that involves the alkanes. The most reactive straight- or branched-chain alkanes such as decane would undergo oxidation with a conversion rate in excess of 10% per hour. Decane contributes strongly to urban-scale photochemical ozone formation, driven by its reactions with OH and the subsequent reactions of its degradation products. Butane and the cyclic hydrocarbons may achieve conversion rates of 1–5% per hour. These alkanes barely have time to react on the urban scale before they are transported out into the rural areas surrounding the source regions. However, over several days almost complete oxidation would occur. Because their reactions and those of their degradation products with OH are potent sources of ozone, low-molecular-mass alkanes such as butane and the cyclic alkanes are important sources of ozone on the multi-day or long-range transport scale. Since multi-day ozone formation has always been recognized as an important facet of the trans-boundary transport of ozone in Europe, much attention has been paid to controlling emissions of low-molecular-mass alkanes.

For many alkanes, the conversion rates in urban areas are significantly less than 5% per hour, and hence they are transported within the atmospheric boundary layer into the environment downwind of the major source regions. Exchange occurs with the free troposphere region above the atmospheric boundary layer within a time-scale of days. Unreactive alkanes will become well mixed around latitude circles within a time-scale of approximately 50 days, vertically up to the tropopause on a time-scale of approximately 30 days, and between the hemispheres on a time-scale of approximately 360 days.

The atmospheric chemistry processes occurring on the urban and regional scales adjacent to the source regions and within the atmospheric boundary layer may be thought of as acting as a chemical filter. Those alkanes with large values for their OH reaction rate coefficients will be rapidly oxidized, and their spatial patterns will be highly localized around the source regions. For those increasingly unreactive alkanes, transport distances extend so that spatial patterns become increasingly hemispheric and ultimately global in scale. The atmospheric boundary chemical filter effectively removes the reactive cyclic hydrocarbons, together with the high-molecular-mass straight- and branched-chain alkanes. Propane, ethane, and methane survive and are increasingly well mixed on the hemispheric and global scales.

Most all the haloalkanes undergo little oxidation within the atmospheric boundary layer on the urban and regional scales. Consequently, they tend to have well characterized hemispheric and global distributions. For those alkanes, and haloalkanes in particular, with atmospheric lifetimes much longer than months, transport and dispersion processes will take them increasingly up to the tropopause. For those with atmospheric lifetimes longer than a year or so, transport processes can take them through the rising branches of the tropical Hadley cells and into the tropical lower stratosphere (Newell et al., 1972). Once in the lower stratosphere, because of the extremely low temperatures, OH oxidation processes tend to be inhibited and transport processes can take them to higher altitudes in the stratosphere. Here the alkanes and haloalkanes are exposed to a profoundly different set of oxidizing agents, chemical reaction initiators, and to solar radiation with a markedly different range of wavelengths to those in the troposphere (Molina and Rowland, 1974). Oxidation or photolysis of any chlorine- or bromine-substituted alkanes releases reactive chlorine or bromine into the stratosphere which catalyses the destruction of the stratospheric ozone layer.

For some haloalkanes, principally the man-made perfluorocarbons such CF_4 and C_2F_6, there are no identified and significant oxidation or photolytic destruction mechanisms operating in the troposphere and stratosphere. The only known removal process are incineration by passing through combustion systems such as electricity-generating power stations and photolysis by Lyman-α radiation (121.6 nm) in the mesosphere

(Ravishankara et al., 1993). Perfluorocarbons have atmospheric lifetimes measured in thousands of years and so these compounds are essentially immortal on human time-scales. Consequently, they will steadily accumulate in the atmosphere and become essentially permanent atmospheric contaminants. Since many of the perfluorocarbons are intensely radiatively active (Weston, 1996; Hurley et al., 2005), they will contribute a permanent, man-made greenhouse effect.

I-D. The Role of Alkanes in Urban Photochemical Ozone Formation

Photochemical ozone formation is a widespread air quality problem associated with large urban and industrial conurbations typified by Los Angeles in the United States of America. Photochemical smog consists of a mixture of strong oxidizing agents such as ozone and visibility-reducing aerosol particles (Haagen-Smit, 1952), produced by intense sunlight-driven chemical processes over time-scales of hours and distances of tens of kilometers. This intense photochemical activity is driven by the urban and industrial emissions of the main ozone precursors, the VOCs and oxides of nitrogen, NO and NO_2, that are often designated as NO_x.

The VOCs play a dual role in urban photochemical ozone formation by firstly reacting with OH radicals and so stimulating the conversion of OH radicals into hydroperoxy (HO_2) and organic peroxy radicals (RO_2). Secondly, this increased supply of peroxy radicals stimulates the conversion of nitric oxide NO into nitrogen dioxide NO_2:

$$RO_2 + NO \rightarrow RO + NO_2 \quad (1)$$

$$HO_2 + NO \rightarrow OH + NO_2 \quad (2)$$

and this results in O_3 formation:

$$NO_2 + h\nu \rightarrow NO + O(^3P) \quad (3)$$

$$O(^3P) + O_2(+M) \rightarrow O_3(+M) \quad (4)$$

This is the mechanism identified by Blacet (1952) and quantified by Leighton (1961). It has largely remained unchanged over the last 50 years.

Because of the short lifetime of all the radical species involved, the O_3 formation occurs where the hydroxyl radical attack on the VOCs occurs. OH radical oxidation of the VOCs is thus the rate-determining step for O_3 formation and is a competitive process involving all the VOC species. This competition is driven by the product of the concentrations of the VOC species [VOC_i] and the rate coefficients for the OH + VOC_i reactions, $k_{OH+VOCi}$, that is by the $k_{OH+VOCi} \times$ [VOC_i] terms. The magnitude of the subsequent ozone formation following the attack on a specific VOC species depends on the fate and behavior of the initial OH reaction products. In the presence of NO_x, these initial reaction products can convert NO into NO_2 in reactions (1) and (2), and hence form O_3 in reactions (3) and (4). The RO radicals formed in (1) can act as a source of HO_2 radicals. For example, 2-$C_4H_9O_2$ forms following OH attack on the secondary H-atoms in butane, and this forms the2-C_4H_9O radical in reaction (1). This 2-butoxy radical reacts rapidly to form HO_2in reaction (5):

$$CH_3CH(O)CH_2CH_3 + O_2 \rightarrow HO_2 + CH_3C(O)CH_2CH_3 \qquad (5)$$

This reaction is in competition with the alkoxy radical decomposition (see chapter VI for full details). These HO_2 radicals can convert NO into NO_2in reaction (2), which again can form O_3 in reactions (3) and (4). These reactions also produce first-generation reaction products such as aldehydes and ketones; for example, methyl ethyl ketone formation in reaction (5). Since these first-generation products often have significant reactivities with OH, or are photochemically active, the urban VOC composition changes rapidly away from that in the emission sources, and this change is accompanied by the formation of ozone.

In table I-D-1, we show how the competition for hydroxyl radicals occurs between the different classes of VOCs by constructing the individual $k_{OH+VOCi} \times$ [VOC_i] terms using the surface mixing ratios of the individual VOC_i species in a typical Los Angeles atmosphere. Because of their higher emission rates and lower reactivity, alkanes account for 42% of the surface mixing ratios of observed VOCs in the Los Angeles atmosphere, whereas alkenes account for 7% and aromatics 19%. Despite the much higher mixing ratios for the alkanes compared with the alkenes and aromatics, alkanes account for less than 15% of the OH loss rate coefficient compared with 27% and 18% for the alkenes and aromatics, respectively. This relatively small contribution from the alkanes to the total OH loss rate coefficient can be attributed to their lower OH reaction rate coefficients than those of the alkenes and aromatics.

To translate the OH loss rate into an ozone formation rate, we need to understand how many NO to NO_2 interconversion steps occur in the degradation of the initial reaction products ultimately through to carbon dioxide and water. Furthermore, we need to understand how much of this subsequent chemistry occurs within the urban area itself and how much in the downwind environment. This understanding has been especially developed for the Los Angeles conurbation and has been assembled into the Maximum Incremental Reactivity (MIR) scale (Carter, 1994). The MIR scale quantifies the mass of ozone formed per mass of VOC reacted at the point of maximum ozone under the conditions of maximum ozone productivity. By weighting the surface mixing ratios of the individual VOCs with their respective MIR values, an appreciation can be gained of the contribution made by each VOC to ozone formation in the Los Angeles atmosphere; see table I-D-1.

The degradation products of the OH + alkane reactions appear to be efficient at forming ozone so that, although alkanes only account for less than 15% of the OH loss, they account for 18% of the MIR-weighted reactivity. The low-molecular-mass alkanes form aldehydes as major degradation products, and these act as major ozone sources on urban time-scales. The higher molecular-mass alkanes form organic nitrogen species as their first reaction products a significant fraction of the time, and this impedes ozone formation on the urban scale. These latter alkanes therefore contribute somewhat to the OH reactivity but almost nothing to the MIR-weighted reactivity. Alkenes, in

Table I-D-1. Typical composition of Los Angeles morning air in October 1995 and contribution to reactivity using OH reactivity and the Carter (1994) maximum incremental reactivity (MIR) factors[b]

VOC Species	Average ppbC	Percentage of OH Reactivity	Percentage of MIR Reactivity
Alkanes (42% of non-methane VOCs)	211.2	14.6	18.4
2-Methylbutane	28.69	2.1	3.0
Propane	27.50	1.0	1.0
Ethane	20.23	0.2	0.4
Butane	16.58	1.0	1.3
Pentane	15.39	1.2	1.3
2-Methylpentane	12.26	1.1	1.5
Methylcyclopentane	8.73	1.0	1.9
Hexane	8.16	0.7	0.6
2-Methylpropane	8.08	0.4	0.7
2,2,4-Trimethylpentane	7.86	0.3	0.6
3-Methylpentane	7.41	0.6	0.9
3-Methylhexane	6.72	0.7	0.7
Methylcyclohexane	6.28	0.9	0.9
Heptane	5.02	0.5	0.3
2-Methylhexane	4.94	0.5	0.4
2,3-Dimethylpentane	4.72	0.3	0.6
2,3-Dimethylbutane	3.27	0.3	0.3
Cyclohexane	3.14	0.4	0.3
2,4-Dimethylpentane	2.81	0.2	0.4
2-Methylheptane	2.29	0.2	0.2
2,3,4-Trimethylpentane	2.23	0.2	0.3
Octane	2.09	0.2	0.1
Cyclopentane	1.95	0.2	0.4
3-Methylheptane	1.94	0.2	0.2
Nonane	1.62	0.2	0.0
2,2-Dimethylbutane	1.24	0.0	0.0
Alkenes (7% of non-methane VOCs)	34.2	27.3	18.8
Ethene	14.91	6.2	7.7
Propene	5.88	5.0	4.1
2-Methylpropene & 1-butene[a]	4.97	6.2	1.8
2-Methyl-2-butene	1.48	2.1	0.9
trans-2-Hexene	0.80	0.9	0.5
2-Methyl-1,3-butadiene	0.68	1.3	0.4
cis-2-Pentene	0.66	0.8	0.5
trans-2-Butene	0.60	0.9	0.5
cis-2-Butene	0.56	0.8	0.5
Cyclopentene	0.38	0.5	0.2
3-Methyl-1-butene	0.35	0.2	0.2
2-Methyl-1-pentene	0.32	0.3	0.0
4-Methyl-1-pentene	0.29	0.2	0.1
cis-2-Hexene	0.23	0.2	0.1
Aromatics (19% of non-methane VOCs)	96.4	17.9	29.9
Toluene	33.02	2.7	6.1
m- & *p*-Xylene[a]	19.92	4.6	9.4
1,2,4-Trimethylbenzene	12.12	4.2	7.1

(continued)

Table I-D-1. (*continued*)

VOC Species	Average ppbC	Percentage of OH Reactivity	Percentage of MIR Reactivity
Aromatics (*continued*)			
Benzene	9.73	0.2	0.3
o-Xylene	7.14	1.2	3.0
Ethylbenzene	5.41	0.5	1.0
1,3,5-Trimethylbenzene	3.47	2.1	2.3
Styrene	3.06	2.2	0.4
1-Propylbenzene	1.60	0.1	0.2
2-Propylbenzene	0.94	0.0	0.1
Other VOCs (33% of non-methane VOCs)	165.2	27.0	26.8
Ethyne	17.35	0.7	0.6
Methyl 2-methyl-2-propyl ether	20.46	1.2	1.1
Other identified	73.86	14.6	14.6
Unidentified	53.57	10.6	10.6
Methane	2000	1.7	2.3
Carbon monoxide	500	11.6	3.8

[a] The chromatographic methods employed were unable to separate the isomers in question.
[b] Taken from Calvert et al. (2002).

contrast, account for 27% of the OH loss and only 19% of the MIR-weighted reactivity, presumably in part because major fractions of the first carbonyl products of the branched chain alkenes are ketones, which are relatively unreactive on urban time-scales. A large fraction of the aromatic compounds generates strongly sunlight-absorbing dialdehydes as first carbonyl products and so dominates the MIR-weighted reactivity. Overall, it is clear from table I-D-1 that, despite the much higher emission rates and hence surface mixing ratios for the alkanes than for the alkenes and aromatics, alkanes account for only 18% of the urban-scale photochemical ozone formation in the Los Angeles conurbation. This relatively small contribution from the alkanes to urban-scale ozone formation can be attributed to their low OH reaction rate coefficients.

Because of the unique and authoritative work of Carter (1994), the MIR scale gives clear and unmistakable insights into the contributions made by each individual VOC species, the alkanes included, in urban-scale photochemical ozone formation in the Los Angeles air basin. However, because of the complex interplay between the different atmospheric chemical processes, it is not straightforward to generalize the MIR scale from Los Angeles to other urban areas. This is a particularly important issue for the alkanes. The main issue here is the detailed NO_x environments in which the VOCs generally, and the alkanes in particular, are processed. Photochemical ozone formation is inhibited at both high and low NO_x levels. At high NO_x levels inhibition results because NO_2 competes with the individual VOC species for OH radicals to form HNO_3, an important removal process for NO_x. The rate coefficient for the reaction of OH radical with NO_2, $OH + NO_2 \rightarrow HONO_2$, is significantly larger than those for most alkanes but significantly smaller than those for alkenes and aromatic compounds. At low NO_x levels, ozone production is hindered because $HO_2/RO_2{+}NO$ reactions become inefficient relative to HO_2/HO_2 and HO_2/RO_2 reactions. The longer lifetime alkanes tend to be processed in urban environments at lower NO_x levels than the shorter lifetime alkenes and aromatic compounds and hence produce less ozone per ppb of VOC reacted. Such differences between the different classes of VOC species will, in principle, be operating in all urban areas, but the details have only been quantified in detail for the Los Angeles conurbation within the MIR scale.

The alkane fraction of the VOC composition of the ambient air in Los Angeles has been strongly contaminated by the exhaust emissions of gasoline-engined motor vehicles and by the evaporation of the fuel required to power them. Fuel can evaporate from the gasoline tanks onboard vehicles while they are driven, or parked, and while they are being refueled at the gasoline pump. Gasoline stations themselves can emit gasoline vapors from spillages and during storage tank refilling by tankers (Rubin et al., 2006). Gasoline tankers, regional storage facilities, and oil refineries also emit vapors into the ambient atmospheres of urban and industrial centers (CONCAWE, 1990). Alkanes are therefore a major component of emissions from the marketing and distribution of motor fuels. It is hardly surprising then that the average VOC composition over a large number of cities in the USA is relatively uniform and, as in Los Angeles, exhibits a major influence from motor vehicles and their fuels.

Table I-D-2 compares the alkane composition of ambient air over North American and European cities on the basis of their average surface mixing ratios. Relative to propylene and toluene, important alkenes, and aromatic compounds from motor vehicle exhausts, the ambient air over cities in North America contains more pentanes and butanes. This difference reflects the warmer summer conditions found in North America and the resulting influence on the formulation of gasoline composition and hence the impact on motor vehicle evaporative emissions.

Increasingly, both exhaust and evaporative emissions from motor vehicles are being controlled in North America, Europe, and in the rest of the world, through the implementation of exhaust gas catalyst and evaporative canister technologies (Commission of the European Communities, 1991; Duffy et al., 1999). Furthermore, gasoline vapor recovery technologies are being implemented throughout the gasoline marketing and distribution chain from oil refinery to the gasoline station (CONCAWE, 1990). Figure I-D-1 shows the running 90-day mean concentrations of a number of VOC species monitored over the period from 1998 to 2004 at a curbside location in the center of London, using data reported in Dollard et al. (2007). The figure shows the declining trends in the concentrations of two VOC species largely derived from motor vehicle exhausts: ethylene and propylene, and of two alkanes largely derived from gasoline evaporation: butane and 2-methylbutane. The impact of the European-wide implementation of motor vehicle exhaust and evaporative emission controls during the 1990s is clearly apparent from this figure.

A completely different $C_2 - C_5$ alkane composition is found in the urban atmospheres of the major central and south American cities, including Mexico City, Mexico (Blake and Rowland, 1995), Santiago, Chile (Chen et al., 2001), and Porto Alegre, Brazil (Grosjean et al., 1998). In these cities, liquefied petroleum gas (LPG) is the major domestic heating and cooking fuel. This is evident in the atmospheric composition of the low-molecular-mass alkanes with the predominance of propane and butane, compared with that in the major cities of North America and Europe. This same predominance of

Table I-D-2. Typical VOC composition in USA and European cities in ppb

VOC	USA[a]	UK[b]	Italy[c]	Greece[d]
Ethane	11.7	5.2	4.8	
Propane	7.8	3.0	2.2	1.2
Butane	10.1	4.1	3.2	2.1
2-Methylpropane	3.7	1.9	2.9	1.1
Pentane	4.4	0.7	3.8	4.2
2-Methylbutane	9.1	2.6	4.0	11.7
Hexane	2.2	0.3	0.4	1.6
2-Methylpentane	2.5	1.0		3.3
3-Methylpentane	1.8			2.3
2,3-Dimethylbutane	0.6			0.6
Heptane	0.7	0.1	0.6	2.4
2-Methylhexane	1.0			1.8
3-Methylhexane	0.8			3.5
2,4-Dimethylpentane	0.3			
Octane	0.3			0.6
3-Methylheptane	0.3			0.8
2,2,4-Trimethylpentane	0.8			
2,3,4-Trimethylpentane	0.3			
Nonane	0.2			1.7
Decane	0.3			3.1
Cyclopentane	0.4			
Methylcyclopentane	1.1			0.7
Cyclohexane	0.3			0.3
Methylcyclohexane	0.5			1.8
Ethylene	10.7	3.0	8.5	
Propylene	2.6	1.6	1.5	3.9
2-Methylpropene & 1-Butene	1.5	0.2	0.2	0.9
cis-2-Butene		0.2	0.1	0.3
trans-2-Butene	0.6	0.2	0.6	0.4
cis-2-Pentene	0.7	0.1	0.4	
trans-2-Pentene	0.6	0.2	0.1	
2-Methyl-1-butene	0.5		1.2	0.4
Benzene	2.1	1.2	0.8	5.0
Toluene	4.6	2.2	3.9	14.3
Ethylbenzene	0.7	0.4	1.2	2.7
o-Xylene	0.9	0.5	1.2	3.7
m- & *p*-Xylene	2.2	1.3	2.4	12.1
m-Ethyltoluene	0.5			
1,2,3-Trimethylbenzene	0.4			3.3
1,3,5-Trimethylbenzene	0.3		2.0	9.2
1,2-Dimethyl-3-ethylbenzene	0.3			
1,4-Diethylbenzene	0.2			
Acetylene	6.5	4.7	1.4	

[a] Taken from Parrish et al. (1998) as the median concentrations over 39 cities.
[b] Taken from Derwent et al. (2003) as averages for London, Cardiff, Birmingham, Leeds, and Edinburgh.
[c] Taken from Latella et al. (2005) for Milan, Italy.
[d] Taken from Moschonas and Glavas (1996) for Athens, Greece.

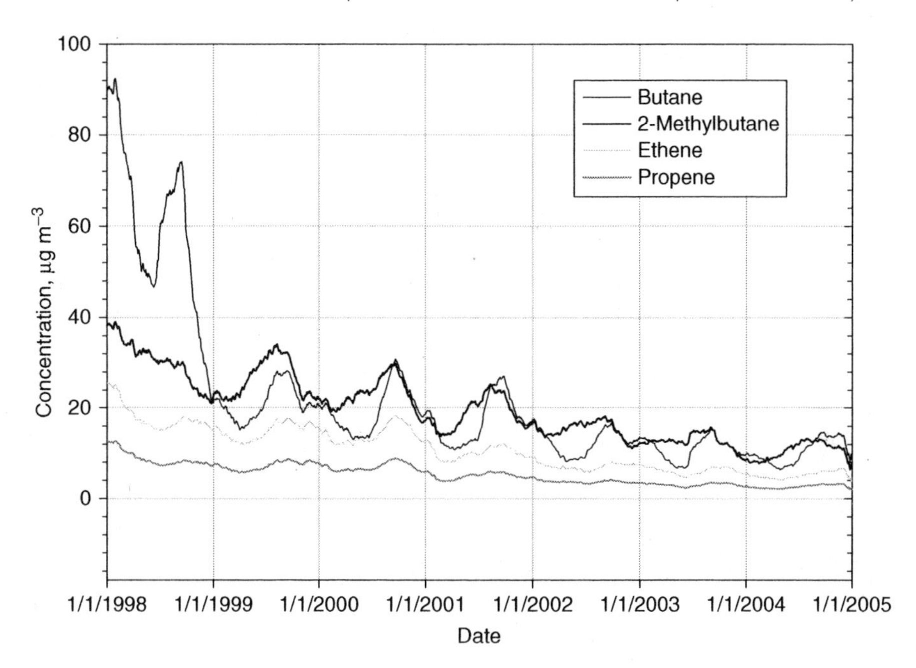

Figure I-D-1. Running 90-day mean VOC levels at a curbside location in London from 1998 to end of 2004.

propane and butane has been reported for some Asian cities including Karachi, Pakistan (Barletta et al., 2002) and Bombay, India (Mohan-Rao et al., 1997).

Air masses passing over urban and industrial centers pick up substantial loadings of VOCs and NO_X. Under conditions of strong sunlight, rapid processing of the emitted VOCs occurs, driven by OH radicals. The subsequent reactions of the primary reaction products in the presence of NO_X generate ozone and other oxidants. On the timescales of hours, many of the highly reactive VOCs react away. For the high-molecular-mass straight- and branched-chain alkanes, close to 90% will have reacted by the time the air masses leave the urban centers. These reactions will have produced ozone and first-generation oxygen- and nitrogen-containing organic compounds, and these products will be transported into the downwind environment. For the low-molecular-mass straight- and branched-chain alkanes, less than 50% will have reacted. Hence a mixture of unreacted alkanes and some first-generation reaction products pass into the downwind environment. At this point, the urban scale gives way to the regional scale and the unreacted alkanes contribute to atmospheric processes on the regional scale.

I-E. The Role of Alkanes in Regional-Scale Photochemical Ozone Formation

The boundary between the urban and regional scales is usually drawn at about 50 kilometers, with the urban scale covering 1 to 50 km and the regional scale from

50 to 2500 km. Travel times are several hours on the urban scale and several days on the regional scale. Because of the different geographical distributions of population in North America and Europe, photochemical ozone has been seen differently on the two continents.

Photochemical episodes have often been considered as being an urban-scale phenomenon in North America and regional-scale within Europe. In North America there is a tendency for the major population and industrial centers to be considered as separate air basins. Under these conditions, the output from one population center has little impact as the input to another. In Europe, particularly northwest Europe, the closer proximity of the major population centers has meant that ozone concentrations appear to rise in concert and to show elevated levels that are similar over large areas (Guicherit and van Dop, 1977). In Europe, the size of the countries is comparable to the characteristic length of the regional scale, so that transboundary fluxes of ozone are important policy issues (Derwent et al., 1991). There are, of course, exceptions. Ozone transport is an important feature in the non-attainment of ozone air quality standards extending over a broad area from Texas to the Atlantic Ocean coast and from the Gulf of Mexico up through the northeast corridor of the USA from Baltimore to Maine and perhaps beyond (Meagher et al., 1998). The large cities in the Mediterranean region, such as Athens, act as large regional sources of ozone without any substantial input from transboundary transport (Klemm et al., 1998; Kalabokas et al., 2000).

The VOC composition on the regional scale differs from that on the urban scale because of the rapid depletion of highly reactive VOCs by reaction with OH radicals on the urban time-scales. Table I-E-1 compares the VOC composition between urban and rural locations in the United Kingdom and France. Relative to propylene and toluene, the low-molecular-mass alkanes such as ethane and propane show higher mixing ratios at the rural site than at the urban site. With progressive remoteness, alkanes become relatively more predominant than alkenes and aromatics.

Table I-E-1. Urban and rural VOC composition in northwestern Europe (ppb)

VOC	UK Cities[a]	Harwell	Porspoder[b]
Ethane	5.22	2.13	1.64
Propane	3.03	1.27	0.73
Butane	4.08	0.95	0.34
2-Methylpropane	1.88	0.37	0.26
Pentane	0.74	0.22	0.11
2-Methylbutane	2.62	0.67	0.23
Hexane	0.30	0.07	0.04
2-Methylpentane	0.99	0.21	0.04
Heptane	0.14	0.05	0.05
Ethene	3.04	0.93	0.49
Propene	1.55	0.62	0.21
Benzene	1.15	0.38	0.18
Toluene	2.21	0.60	0.34
Ethyne	4.74	2.79	0.58

[a] Data for UK cities and the rural site Harwell, Oxfordshire, United Kingdom were adapted from Derwent et al. (2003).
[b] Data for the rural site Porspoder, Brittany, France were adapted from Boudries et al. (1994).

In assessing the contribution made by individual VOC species to photochemical ozone formation on the regional scale, a number of different factors have to be taken into account. The spatial distribution of VOC and NO_x emissions and the species distribution of the former exert an important influence. The spatial scales of the OH-driven oxidation of each VOC species is determined by its OH reaction rate coefficient. However, the rate of ozone production is not simply related to that of VOC oxidation because of the limited availability of NO_x which is determined by the relative reaction rates of OH with NO_2 and the VOC species. Furthermore, each VOC species will have a different carry-over of unreacted parent VOC and first generation reaction products and show a different propensity to form organic nitrogen compounds. These nitrogen compounds are predominately alkyl nitrates ($RONO_2$) and acylperoxy nitrates ($RCOO_2NO_2$) that can act as temporary reservoirs of NO_x because they can decompose on the regional scale and resupply enough NO_x to support continued ozone production.

The Master Chemical Mechanism (Jenkin et al., 1997; 2003; Saunders et al., 2003) attempts to take into account quantitatively these many facets of regional ozone formation. (The full mechanism is available via the University of Leeds website: http://mcm.leeds.ac.uk/MCM.) The version v3.0 contains 12,692 reactions involving 4342 chemical species and has been used to describe photochemical ozone formation on a single air parcel trajectory that travels across Europe from Austria across to the United Kingdom during a period of 5 days. Ozone builds up in the air parcel and on the final day reaches 86 ppb, maximum hourly mean concentration. The model was rerun setting the emissions of each VOC species to zero, in turn, to determine the contribution that VOC makes to the final ozone concentration (Derwent et al., 2003); see table I-E-2. Of the 36.8 ppb attributable to the emitted VOCs, the alkanes accounted for just over one-third, 14.2 ppb, the aromatics about one quarter, with the remainder shared between the alkenes and the oxygenates. Butane made the largest single contribution of any VOC species, accounting for 12% of the attributable photochemical ozone formation. Alkanes play a large part in regional-scale ozone formation across Europe because transboundary time-scales exceed the reaction time-scales of many of the emitted straight- and branched-chain alkanes.

In principle, if the regional timescale is long enough, all the VOCs emitted will have reacted, and the integrated rate of VOC oxidation will be determined solely by emission. Under these conditions, reactivity drops out of consideration. These statements deal with VOC oxidation and serve only as a guide to photochemical ozone formation. There are two important considerations that intervene with the result that regional-scale photochemical ozone formation is only partially determined by VOC emissions. Reactivity is an important issue still because the duration of photochemical episodes is generally only 2–5 days so that VOC oxidation does not often go through to completion. A further important consideration is the lifetime of ozone in the atmospheric boundary layer. Dry deposition is an important loss process for ozone on the regional scale, limiting its lifetime to 2–5 days and its transport distances to 500–2500 km. Consequently, even if photochemical processing could be extended to ensure that the oxidation of the VOCs was complete, apparent ozone formation would be reduced because of dry deposition of any ozone formed with the longer travel times.

Table I-E-2. Contribution to regional-scale ozone formation over Europe from a range of VOC species[a,b]

VOCs	Contribution to Peak Hourly Ozone (ppb)
Alkanes (total)	14.23
Ethane	0.32
Propane	0.68
Butane	4.31
2-Methylpropane	0.84
Pentane	1.78
2-Methylbutane	1.91
Hexane	1.76
2-Methylpentane	0.18
3-Methylpentane	0.11
Heptane	0.76
2-Methylhexane	0.16
3-Methylhexane	0.14
Octane	0.47
Nonane	0.21
Decane	0.34
Undecane	0.13
Dodecane	0.04
Cyclohexane	0.09
Alkenes (total)	6.47
Ethene	3.13
Propene	1.83
1-Butene	0.31
cis-2-Butene	0.05
trans-2-Butene	0.02
2-Methylpropene	0.38
1,3-Butadiene	0.34
1-Pentene	0.21
cis-2-Pentene	0.06
trans-2-Pentene	0.04
3-Methyl-1-butene	0.03
2-Methyl-1,3-butadiene	0.02
1-Hexene	0.03
α−Pinene	0.03
Aromatic hydrocarbons (total)	9.57
Benzene	0.45
Toluene	2.49
Ethylbenzene	0.67
m-Xylene	1.68
o-Xylene	0.96
p-Xylene	0.80
1,2,4-Trimethylbenzene	1.50
1,3,5-Trimethylbenzene	0.50
1,2,3-Trimethylbenzene	0.31
m-Ethyltoluene	0.08
o-Ethyltoluene	0.02
p-Ethyltoluene	0.04
1-Propylbenzene	0.07

Table I-E-2. (*continued*)

VOCs	Contribution to Peak Hourly Ozone (ppb)
Oxygenates (total)	5.63
Ethanol	2.74
Formaldehyde	0.65
2-Butanone	0.36
Methanol	0.33
Ethyl acetate	0.24
Acetaldehyde	0.23
Acetone	0.17
2-Butanol	0.15
1-Butyl acetate	0.14
2-Propanol	0.13
Propanal	0.07
2-Butoxyethanol	0.07
Dimethyl ether	0.06
3,3-Dimethyl-2-butanone	0.06
Pentanal	0.05
Benzaldehyde	0.05
Acetic acid	0.05
1,2-Dihydroxypropane	0.04
2-Methylpropanol	0.04
Other VOCs	0.92
Total	36.82

[a] This is the contribution from each emitted VOC species to the peak ozone concentration of 87.8 ppb found in an air parcel that has traveled for 5 days across Europe from Austria to the United Kingdom.
[b] Data taken from Derwent et al. (2003).

To integrate all the details of regional-scale ozone formation and transboundary air pollution transport into a single reactivity scale requires a highly detailed treatment of VOC oxidation and the formation of degradation products including organic nitrogen compounds, together with the spatial distributions of the emissions of ozone precursors and ozone dry deposition. In Europe, the Master Chemical Mechanism has been used to compile the Photochemical Ozone Creation Potential (POCP) scale, addressing regional-scale ozone formation over 5 days travel time for a large number of individual VOC species (Derwent et al., 1998). While the scale has been specifically developed to cover the case of an air parcel crossing Europe en route from Austria to the United Kingdom, it should be appropriate for much of northwest Europe.

POCPs are determined by using a photochemical trajectory model to describe ozone formation in a single air parcel traveling across Europe from Austria to the United Kingdom over a period of 5 days. The instantaneous emissions of each VOC, in turn, at each point along the trajectory path are increased by a given mass emission and the increment in the average ozone concentration along the 5-day trajectory is calculated. These average ozone increments are calculated relative to the increment for ethylene and expressed as the POCP index with ethylene = 100. Table I-E-3 presents POCPs for a large number of VOCs, together with the fraction of the total mass emission

Table I-E-3. Percentage of total VOC emissions, POCP, and POCP-weighted emissions for a large number of VOC species in northwestern Europe[a]

VOCs	Percent of Emissions[b]	POCP[c]	Percent × [POCP]
All VOCs (total)	100		2640.3
Alkanes (total)	42.61		1264.9
Ethane	5.049	7.7	38.8
Propane	3.883	13.8	53.5
Butane	8.922	31.4	280.2
2-Methylpropane	1.921	27.9	53.6
Pentane	3.531	39.7	140.3
2-Methylbutane	3.251	33.7	109.4
2,2-Dimethylpropane	0.015	17.9	0.3
Hexane	2.619	40.1	104.9
2-Methylpentane	0.356	40.7	14.5
3-Methylpentane	0.229	43.3	9.9
2,2-Dimethylbutane	0.044	21.8	1.0
2,3-Dimethylbutane	0.068	50.0	3.4
Heptane	2.004	34.9	70.0
2-Methylhexane	0.356	32.4	11.5
3-Methylhexane	0.280	41.7	11.7
Octane	1.699	34.3	58.3
2-Methylheptane	0.099	34.0	3.4
3-Methylheptane	0.162	36.9	6.0
Nonane	0.603	34.0	20.5
2-Methyloctane	0.085	34.3	2.9
3-Methyloctane	0.078	33.7	2.6
4-Methyloctane	0.084	36.5	3.1
3,4-Dimethylheptane	0.062	35.6	2.2
Decane	0.958	35.6	34.1
2-Methylnonane	0.172	34.9	6.0
3-Methylnonane	0.199	39.1	7.8
4-Methylnonane	0.136	34.9	4.8
2,5-Dimethyloctane	0.054	38.1	2.1
2,6-Dimethyloctane	0.160	36.2	5.8
2-Methyl-3-ethylheptane	0.194	34.3	6.7
2,2-Dimethyl-3,3-dimethylhexane	0.051	19.2	1.0
Undecane	0.482	35.6	17.1
2-Methyldecane	0.141	33.7	4.7
3-Methyldecane	0.158	35.6	5.6
4-Methyldecane	0.264	35.9	9.5
5-Methyldecane	0.071	34.6	2.5
Dodecane	0.255	33.0	8.4
Tridecane	0.311	42.0	13.0
Tetradecane	1.213	45.5	55.2
Dichloromethane	0.973	2.6	2.5
Chloroethane	0.112	10.9	1.2
Methylcyclopentane	0.054	48.7	2.6
Cyclohexane	0.083	28.2	2.3
Methylcyclohexane	0.188	65.1	12.2
Ethylcyclohexane	0.072	62.5	4.5
1-Propylcyclohexane	0.165	59.9	9.9
1,2,3-Trimethylcyclohexane	0.062	57.1	3.6
2-Propylcyclohexane	0.074	59.9	4.4
1-Butylcyclohexane	0.240	59.3	14.2
1-Methylpropylcyclohexane	0.072	59.3	4.3

Table I-E-3. *(continued)*

VOCs	Percent of Emissions[b]	POCP[c]	Percent × [POCP]
Alkanes (*continued*)			
1-Methyl-3-propylcyclohexane	0.132	59.6	7.9
1-Methyl-4-propylcyclohexane	0.127	55.8	7.1
Pentylcyclohexane	0.033	55.8	1.8
Hexylcyclohexane	0.003	55.8	0.2
Alkenes (total)	9.67		761.1
Ethene	2.867	100.0	286.7
Propene	1.240	117.0	145.0
1-Butene	0.218	104.2	22.7
cis-2-Butene	0.226	112.8	25.5
trans-2-Butene	0.226	116.3	26.3
2-Methylpropene	0.771	62.5	48.2
1,3-Butadiene	0.360	88.8	31.9
cis-2-Pentene	0.235	109.0	25.6
trans-2-Pentene	0.235	111.2	26.1
1-Pentene	0.221	94.6	20.9
2-Methyl-1-butene	0.016	75.3	1.2
3-Methyl-1-butene	0.020	73.1	1.4
2-Methyl-2-butene	0.026	81.7	2.1
2-Methyl-1,3-butadiene	0.002	114.4	0.2
1-Hexene	0.022	88.5	2.0
cis-2-Hexene	0.073	104.5	7.6
trans-2-Hexene	0.073	102.2	7.4
α−Pinene	0.027	67.9	1.8
β−Pinene	0.023	33.3	0.8
Limonene	0.219	71.5	15.6
Tetrachloroethene	0.514	1.3	0.7
Trichloroethene	1.967	28.5	56.1
1,2-Dichloroethane	0.094	54.5	5.1
Aromatics (total)	15.53		1073.5
Benzene	1.100	9.6	10.6
Toluene	3.643	43.9	160.0
o-Xylene	1.020	77.9	79.4
m-Xylene	2.030	85.6	173.7
p-Xylene	0.879	72.1	63.4
Ethylbenzene	1.274	45.5	58.0
Propylbenzene	0.149	37.5	5.6
1-Methylethylbenzene	0.043	31.7	1.4
1,2,3-Trimethylbenzene	0.404	105.4	42.6
1,2,4-Trimethylbenzene	1.431	109.9	157.3
1,3,5-Trimethylbenzene	0.534	106.7	56.9
o-Ethyltoluene	0.128	72.8	9.3
m-Ethyltoluene	0.257	78.2	20.1
p-Ethyltoluene	0.180	63.5	11.4
3,5-Dimethylethylbenzene	0.254	104.5	26.5
3,5-Diethyltoluene	1.695	98.1	166.3
1,2,3,5-Tetramethylbenzene	0.095	103.8	9.9
1,2,4,5-Tetramethylbenzene	0.066	100.3	6.6
1-Methyl-4-1-methylethylbenzene	0.136	75.0	10.2
1-Methyl-3-1-methylethylbenzene	0.040	87.8	3.5
Styrene	0.174	4.8	0.8

(continued)

Table I-E-3. (*continued*)

VOCs	Percent of Emissions[b]	POCP[c]	Percent × [POCP]
Oxygenates (total)	21.058		627.5
Formaldehyde	1.692	45.8	77.6
Acetaldehyde	0.371	54.8	20.3
Propanal	0.114	71.8	8.2
2-Methylpropanal	0.036	49.7	1.8
Butanal	0.042	69.9	2.9
Pentanal	0.053	70.5	3.7
3-Methylbutanal	0.004	40.7	0.2
Benzaldehyde	0.136	−18.6	−2.5
2-Methylbenzaldehyde	0.054	−27.9	−1.5
3-Methylbenzaldehyde	0.073	−18.3	−1.3
4-Methylbenzaldehyde	0.027	5.1	0.1
Methanol	2.035	13.1	26.7
Ethanol	7.500	34.3	257.2
1-Propanol	0.289	48.1	13.9
2-Propanol	0.855	18.3	15.6
1-Butanol	0.446	51.9	23.2
2-Methylpropanol	0.116	35.6	4.1
1-Methylpropanol	0.023	40.4	0.9
2-Methyl-2-propanol	0.000	1.6	0.0
3-Methyl-1-butanol	0.003	43.9	0.1
Phenol	0.036	−5.4	−0.2
o-Cresol	0.009	19.2	0.2
2,5-Xylenol	0.007	55.4	0.4
2,4-Xylenol	0.005	53.8	0.3
2,3-Xylenol	0.005	34.0	0.2
Cyclohexanol	0.002	44.6	0.1
4-Hydroxy-4-methyl-2-pentanone	0.109	28.5	3.1
Acetone	1.660	6.4	10.6
2-Butanone	1.081	32.1	34.6
4-Methyl-2-pentanone	0.485	51.6	25.0
Cyclohexanone	0.059	28.8	1.7
Methyl formate	0.023	2.9	0.1
Methyl acetate	0.374	7.1	2.6
Ethyl acetate	0.768	19.2	14.8
2-Propyl acetate	0.113	21.5	2.4
1-Butyl acetate	0.951	25.6	24.4
1-Methylpropyl acetate	0.000	0.3	0.0
1-Propyl acetate	0.073	24.4	1.8
Formic acid	0.004	2.6	0.0
Ethanoic acid	0.042	9.0	0.4
Propanoic acid	0.001	13.1	0.0
Dimethyl ether	0.198	17.9	3.5
Diethyl ether	0.004	46.2	0.2
Di-2-propyl ether	0.045	43.9	2.0
1,2-Dihydroxyethane	0.088	33.0	2.9
1,2-Dihydroxypropane	0.096	39.4	3.8
2-Butoxyethanol	0.402	44.6	17.9
1-Methoxy-2-propanol	0.284	34.3	9.7
2-Methoxyethanol	0.056	29.2	1.6
2-Ethoxyethanol	0.041	36.5	1.5
2-Propenal	0.120	53.8	6.4
2-Butenal	0.037	92.3	3.4

Table I-E-3. (*continued*)

VOCs	Percent of Emissions[b]	POCP[c]	Percent × [POCP]
Oxygenates (*continued*)			
Ethanedial	0.007	21.8	0.1
2-Oxopropanal	0.005	100.6	0.5
Other VOCs (total)	1.02		13.3
Ethyne	0.93	7	6.6
Propyne	0.09	73	6.7

[a] Calculated with MCMv3.1 using 12,871 reactions and 4414 species.

[b] VOC speciation in percentage by mass taken for the United Kingdom and assumed the constant across Europe; taken from Passant (2002).

[c] POCP is an index expressed relative to ethylene = 100 that gives the ozone increment in ppb averaged over a 5-day period from an incremental increase in mass emissions of a VOC relative to that from the same incremental increase in ethylene emissions (Derwent et al. 1998).

of VOCs in the UK that is accounted for by that VOC species. It is assumed that the VOC speciation for the United Kingdom is equally applicable across northwest Europe. In view of the similarity of the motor vehicles, their fuels, and the solvents used, this is probably a reasonable working assumption. POCP values cover the range from 116 for *trans*-2-butene, the most reactive VOC species considered here, to –28 for 2-methylbenzaldehyde, the least reactive VOC species. The latter VOC species exhibits a negative POCP because its degradation mechanism includes the formation of organic nitrogen-containing compounds that decrease the availability of NO_x and hence decrease the ozone-formation potential of air masses under the condition of long-range transboundary transport. POCPs for the alkanes cover the range from 65 for methylcyclohexane to 8 for ethane and are generally about one-half to one-third of the values for the alkenes and aromatic compounds.

Table I-E-3 also contains the product of the fraction of the total VOC emissions accounted for by that VOC with its POCP value, to generate a figure that represents the POCP-weighted mass emissions for that species. Using this reactivity scale, it is apparent that alkanes contribute significantly to both mass emissions (42.6%) and POCP-weighted emissions (47.9%). Alkane oxidation appears to be a dominant source of POCP-weighted emissions on the regional scale in Europe, accounting for slightly more than the aromatics (40.7%) and significantly more than the alkenes (28.8%). This situation appears to be in marked contrast with that found on the urban scale in Los Angeles, where alkanes contribute only weakly to urban ozone formation. This difference reflects the longer time-scale for reaction with OH radicals on the regional scale than on the urban scale and hence the importance of long-range transport.

I-F. The Role of Alkanes in Global Tropospheric Chemistry

I-F-1. Spatial Distributions of Alkanes in the Troposphere and the Role of OH Reactions

The fate and behavior of the alkanes in the troposphere is largely determined by their reactivity with OH radicals. OH radicals are the main oxidant for VOCs and have been likened to the 'detergent' in the tropospheric 'washing machine'. The OH radical oxidation processes remove over half a billion metric tons of methane per year and the majority of the rest of the alkanes. Only perfluoroalkanes and some chloroalkanes and bromoalkanes survive the tropospheric OH sink because of their negligible OH reactivity. Broadly speaking, this lower limit of reactivity is delineated by OH + alkane rate coefficients that are much less than about 10^{-15} cm^3 molecule^{-1} s^{-1}. Alkanes with such low OH reactivities behave as well-mixed tropospheric constituents with little in the way of mixing ratio gradients between the poles and between the surface and the tropopause.

The global atmospheric lifetime (in years) characterizes the time taken to turn over the global atmospheric burden. It is defined as the atmospheric burden B (in Tg, say) divided by the global mean removal rate R (in Tg/yr) for a trace gas in steady state, that is with unchanging burden (Prather, 2002). A corollary of this definition is that, when in steady state, the atmospheric burden B of a trace gas equals the product of its atmospheric lifetime τ and the emission rate E. That is to say:

$$B = E \times \tau = R \times \tau$$

For the alkanes with tropospheric OH oxidation sinks only and OH + alkane reaction rate coefficients, k, then the removal rate coefficient becomes k [OH], and

$$\tau = 1/k[\mathrm{OH}]$$

Because both k and [OH] vary throughout the troposphere, the global removal rates of alkanes with OH radicals must be obtained by integration over the three dimensions of the troposphere and over time. This spatial averaging can be carried out with a three-dimensional chemistry-transport model.

To this end, the global Lagrangian chemistry-transport model STOCHEM (Collins et al., 1997) has been used to examine the spatial distributions of a synthetic alkane which is emitted by man-made activities with a source strength of 1 Tg yr^{-1} and a variable temperature-independent reaction rate with OH radicals. STOCHEM describes the chemistry of methane, CO, NO_x, SO_2, DMS, and 12 non-methane VOCs using 160 chemical reactions to describe the full diurnal, seasonal, and spatial variations in the distribution of tropospheric OH. Table I-F-1 presents the atmospheric lifetimes of the synthetic alkane calculated in each model experiment carried out with different OH + alkane reaction rate coefficients. Atmospheric lifetimes for the synthetic alkane varied between 2 and 1060 days for temperature-independent OH + alkane reaction rate coefficients in the range from 1.0×10^{-11} and 1.0×10^{-14} cm^3 molecule^{-1} s^{-1}. Averaging over the seven model experiments, the synthetic alkane experienced a mean tropospheric OH radical number density of $1.1 \pm 0.03 \times 10^6$ molecule cm^{-3}.

Figures I-F-1a to I-F-1g present the annual mean surface spatial distributions of the synthetic alkane, averaged over an entire year under steady-state conditions, for different OH + alkane reaction rate coefficients. For those synthetic alkanes

Table I-F-1. Atmospheric lifetimes for a synthetic alkane calculated with a global Lagrangian 3-D chemistry-transport model for different rate coefficients for the reaction between OH and the alkane

OH + Alkane Rate Coefficient, k (cm^3 $molecule^{-1}$ s^{-1})	Atmospheric Lifetime (days)
1.0×10^{-11}	2
3.0×10^{-12}	5
1.0×10^{-12}	13
3.0×10^{-13}	40
1.0×10^{-13}	110
3.0×10^{-14}	360
1.0×10^{-14}	1060

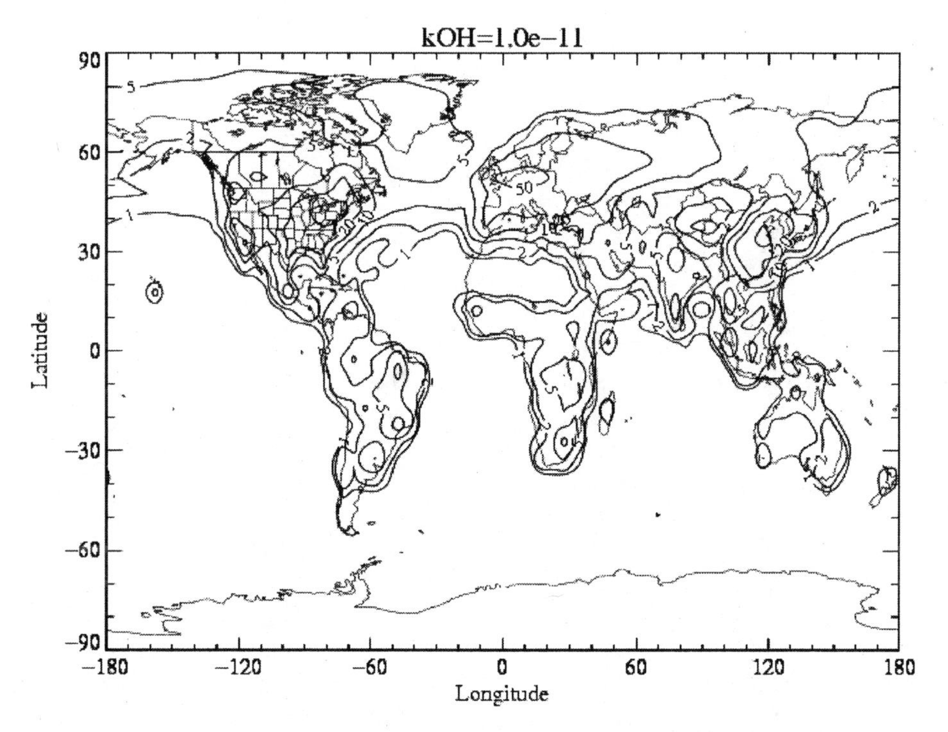

Figure I-F-1a. Annual mean surface distribution of a synthetic alkane with a man-made source strength of 1 Tg yr^{-1} and an OH reaction rate coefficient of 1.0×10^{-11} cm^3 $molecule^{-1}$ s^{-1}.

with OH + alkane reaction rate coefficients above 3.0×10^{-12} cm^3 $molecule^{-1}$ s^{-1}, (figures I-F-1a and I-F-1b), the surface spatial distributions of the alkane are entirely focused around the source regions. Far away from sources, mixing ratios become vanishingly small, consistent with the short lifetime of the alkane in the range

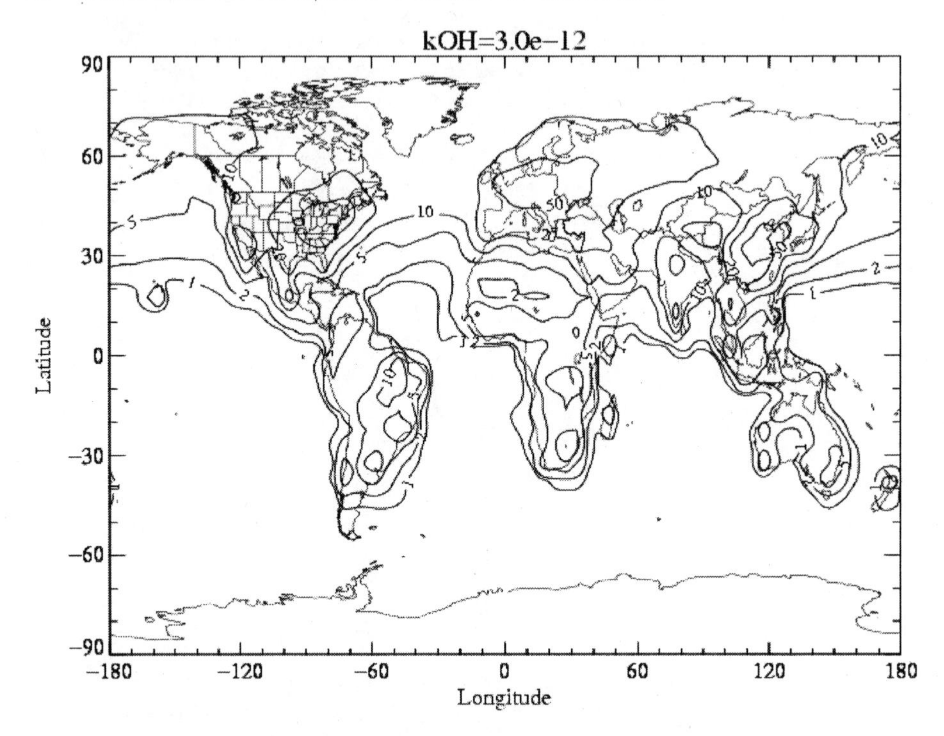

Figure I-F-1b. Annual mean surface distribution of a synthetic alkane with a man-made source strength of 1 Tg yr^{-1} and an OH reaction rate coefficient of 3.0×10^{-12} cm^3 $molecule^{-1}$ s^{-1}.

2–5 days. Southern-hemisphere mixing ratios are dominated by the presence of southern-hemisphere sources rather than by transport from the northern hemisphere. Under these conditions, oxidation rates are highly variable in space and time, being heavily influenced by photochemical activity occurring close to the surface in the vicinity of the individual source regions.

For those synthetic alkanes with OH + alkane reaction rate coefficients above 3.0×10^{-13} cm^3 $molecule^{-1}$ s^{-1} (figures I-F-1c and I-F-1d), spatial distributions become much less focused around source regions and intercontinental transport becomes apparent, particularly in the northern hemisphere. A continuous band of elevated mixing ratios joins the major continents together, and some evidence of transport into the southern hemisphere is apparent, consistent with atmospheric lifetimes in the range 13–40 days.

For alkanes with OH + alkane reaction rate coefficients above 3.0×10^{-14} cm^3 $molecule^{-1}$ s^{-1} (figures I-F-1e and I-F-1f), spatial gradients become small across the northern hemisphere. Significant interhemispheric transport into the southern hemisphere occurs so that the influence of local southern-hemisphere sources becomes less apparent for the alkanes with atmospheric lifetimes in the range 110–360 days. For the longest lived alkane, see figure I-F-1g, with a lifetime of 1060 days, the spatial distribution becomes essentially well-mixed with small spatial gradients across the northern and southern hemispheres.

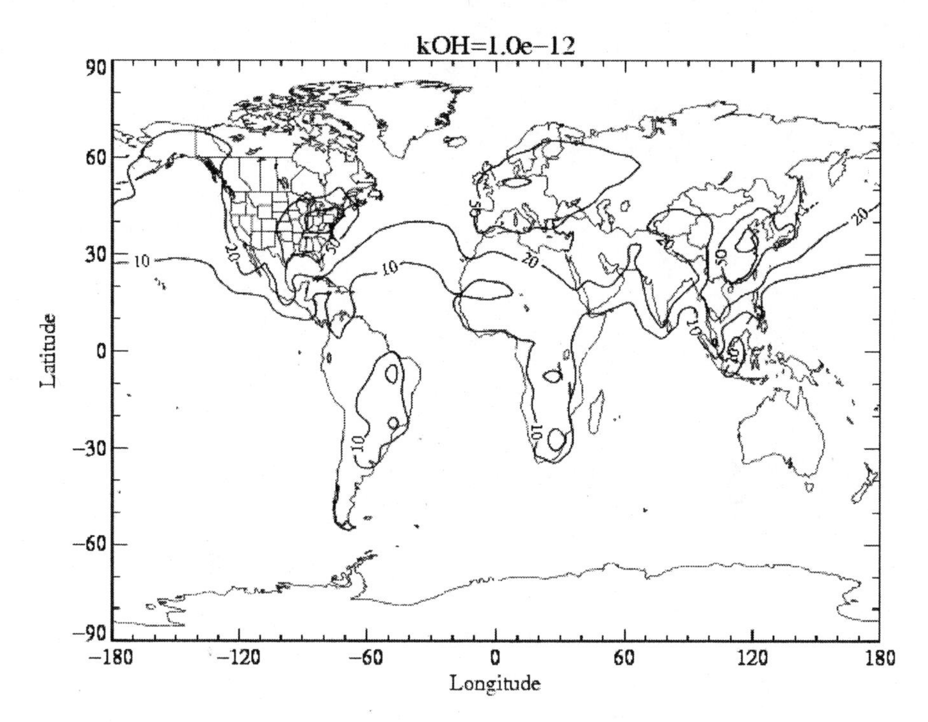

Figure I-F-1c. Annual mean surface distribution of a synthetic alkane with a man-made source strength of 1 Tg yr^{-1} and an OH reaction rate coefficient of 1.0×10^{-12} cm^3 $molecule^{-1}$ s^{-1}.

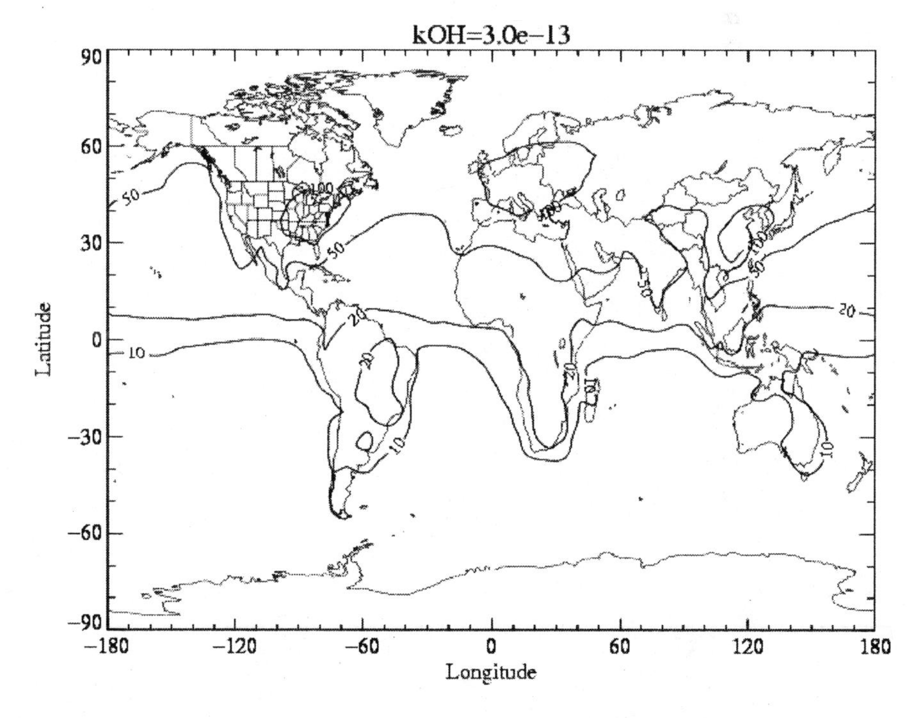

Figure I-F-1d. Annual mean surface distribution of a synthetic alkane with a man-made source strength of 1 Tg yr^{-1} and an OH reaction rate coefficient of 3.0×10^{-13} cm^3 $molecule^{-1}$ s^{-1}.

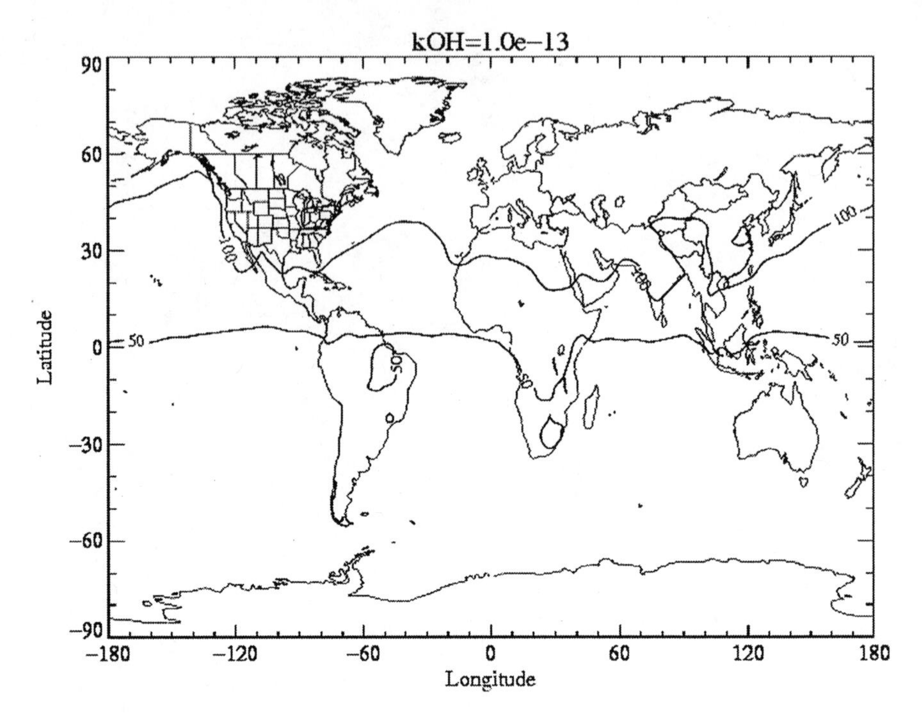

Figure I-F-1e. Annual mean surface distribution of a synthetic alkane with a man-made source strength of 1 Tg yr^{-1} and an OH reaction rate coefficient of 1.0×10^{-13} cm^3 $molecule^{-1}$ s^{-1}.

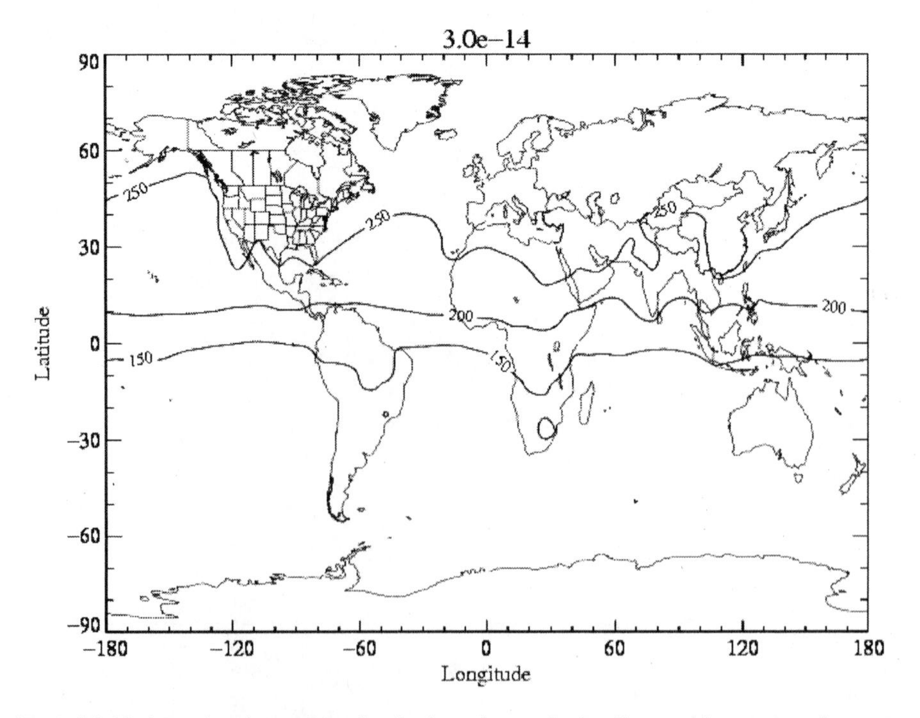

Figure I-F-1f. Annual mean surface distribution of a synthetic alkane with a man-made source strength of 1 Tg yr^{-1} and an OH reaction rate coefficient of 3.0×10^{-14} cm^3 $molecule^{-1}$ s^{-1}.

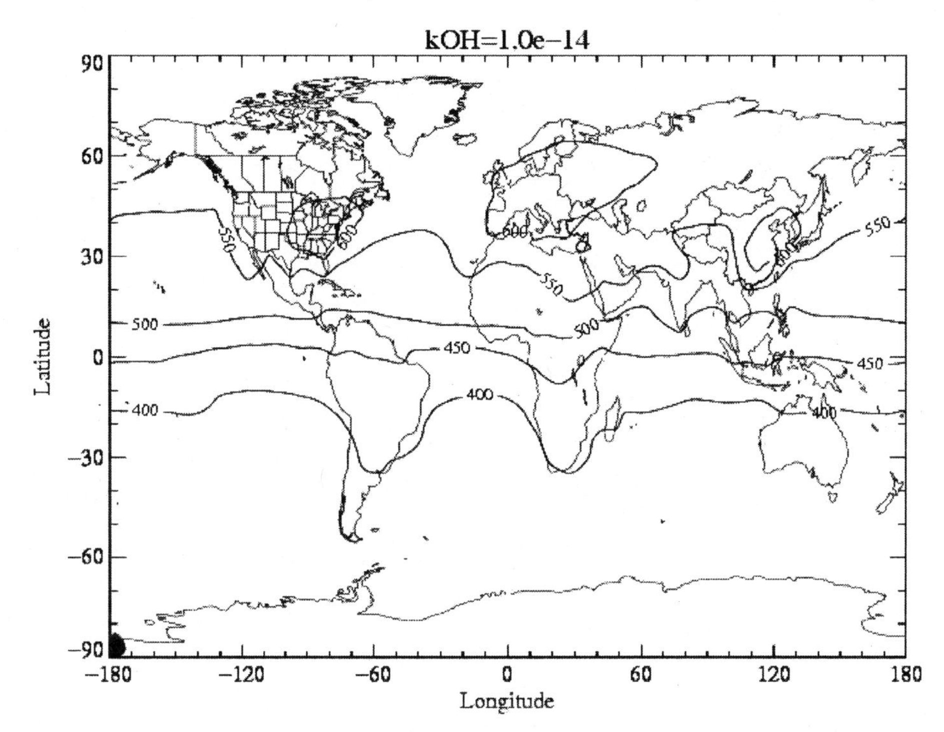

Figure I-F-1g. Annual mean surface distribution of a synthetic alkane with a man-made source strength of 1 Tg yr^{-1} and an OH reaction rate coefficient of 1.0×10^{-14} cm^3 $molecule^{-1}$ s^{-1}.

Figure I-F-2 presents the annual mean distributions of the surface mixing ratios of the synthetic alkanes around the latitude circle at 40°–45°N normalized by the mixing ratio found over the North American continent. Three peaks are found in all the curves corresponding to the intense man-made emissions over North America, Europe, and Asia. Three minima are apparent, corresponding to the Atlantic Ocean, Mongolia, and the Pacific Ocean. The depth of the minima increases as the atmospheric lifetime of the synthetic alkane decreases. The ratio of the minimum to the peak mixing ratio is equal to or greater than 1/e = 0.3678 for those alkanes with atmospheric lifetimes in excess of about 50 days, an estimate of the characteristic time-scale for transport around a latitude circle.

By way of comparison, figure I-F-3 presents the annual mean altitude mixing ratio profiles of the synthetic alkanes above 40°–45°N, 5°–10°E. Mixing ratios fall off quickly with altitude for the alkanes emitted at the earth's surface. The fall-off in mixing ratio is more than 1/e = 0.3678 for those alkanes with atmospheric lifetimes less than about 30 days, an estimate of the characteristic time-scale for vertical transport in the troposphere.

The characteristic time-scale for inter-hemispheric exchange between the northern and southern hemispheres from figure I-F-4 appears to be 360 days. This plot shows the annual mean surface mixing ratios along the 5°–10°E meridian, normalized by the mixing ratio found over northwest Europe.

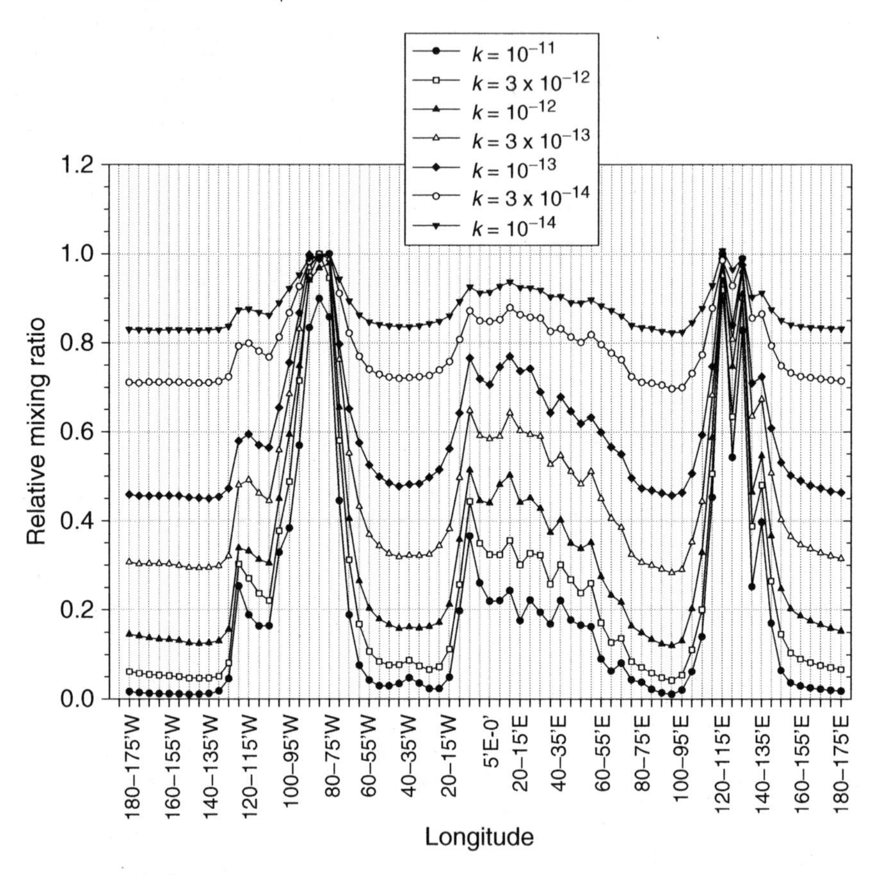

Figure I-F-2. Annual mean surface distribution around a latitude circle at 40°–45°N of synthetic alkanes with man-made source strengths of 1 Tg yr^{-1} and OH reaction rate coefficients in the range between 10^{-11} and 10^{-14} cm^3 $molecule^{-1}$ s^{-1}.

I-F-2. Atmospheric Lifetimes of the Haloalkanes

The atmospheric lifetime τ of a haloalkane, that has only tropospheric OH radical oxidation sinks, can be expressed as:

$$\tau = 1/k[OH]$$

Consequently, it is a straightforward matter to use the results of the synthetic alkane calculations from section I-F-1 above to calculate atmospheric lifetimes for those haloalkanes for which rate coefficient data are available for their reaction with hydroxyl radicals. Table I-F-2 presents the atmospheric lifetime due to hydroxyl radical attack using the reaction rate coefficient data from table VI-A-76. The estimated atmospheric lifetimes with respect to reaction with OH radicals span the range from 250 years for HFC-236fa to 2 days for hexylbromide.

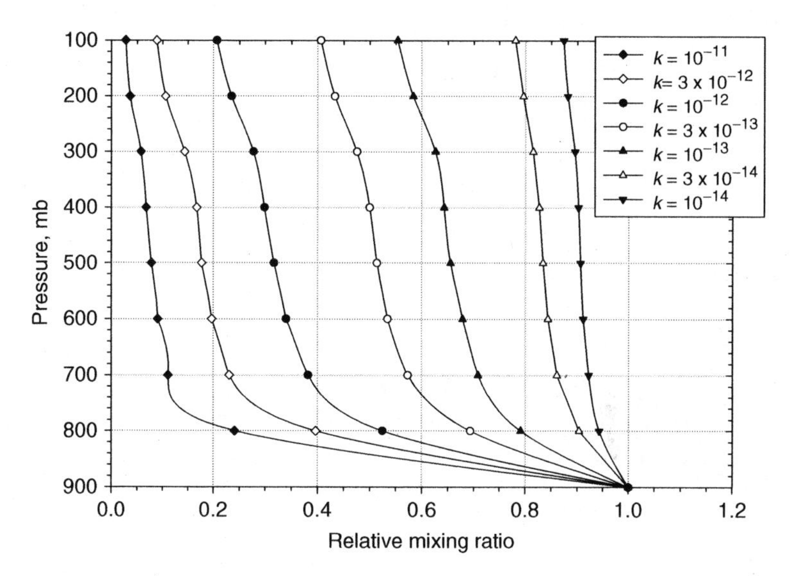

Figure I-F-3. Annual mean altitude distributions above 40°–45°N and 5°–10°E of synthetic alkanes with man-made source strengths of 1 Tg yr^{-1} and OH reaction rate coefficients in the range between 10^{-11} and 10^{-14} cm^3 $molecule^{-1}$ s^{-1}.

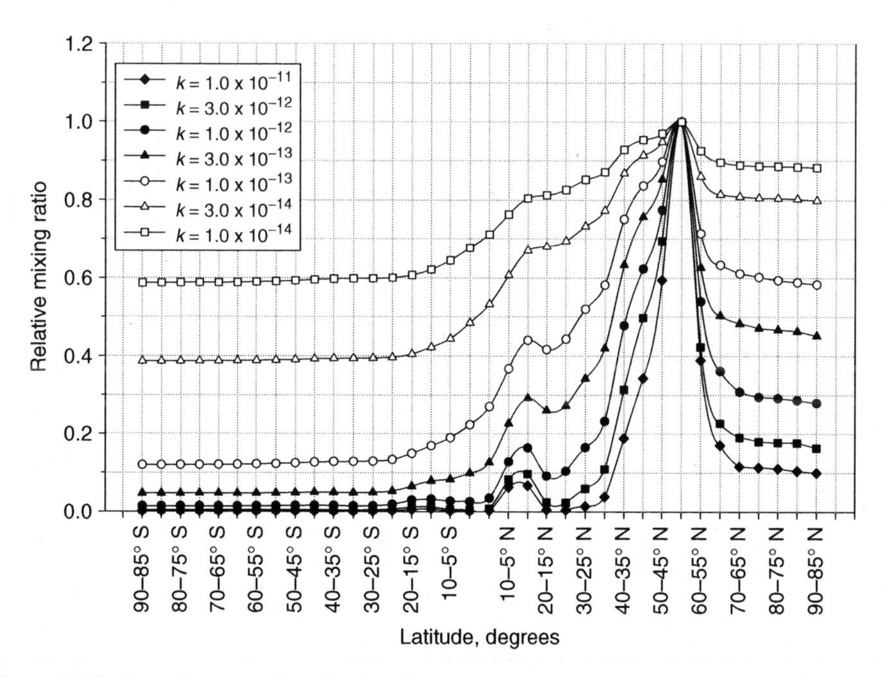

Figure I-F-4. Annual mean surface distribution around the 5°–10°E meridian of synthetic alkanes with man-made source strengths of 1 Tg yr^{-1} and OH reaction rate coefficients in the range between 10^{-11} and 10^{-14} cm^3 $molecule^{-1}$ s^{-1}.

Table I-F-2. Atmospheric lifetimes due to hydroxyl radical OH attack for a range of haloalkanes[a]

Haloalkane	Atmospheric Lifetime (yr)[b]
CH_3F (HFC-41)	2.5
CH_2F_2 (HFC-32)	4.9
CHF_3 (HFC-23)	241
CH_3CH_2F (HFC-161)	0.2
CH_2FCH_2F (HFC-152)	0.4
CH_3CHF_2 (HFC-152a)	1.3
CH_2FCHF_2 (HFC-143)	3.6
CH_3CF_3 (HFC-143a)	51.8
CHF_2CHF_2 (HFC-134)	9.5
CF_3CH_2F (HFC-134a)	14.1
CHF_2CF_3 (HFC-125)	30.7
$CF_3CH_2CH_3$ (HFC-263fb)	0.9
CF_3CHFCH_2F (HFC-245eb)	2.7
$CHF_2CHFCHF_2$ (HFC-245ea)	2.8
$CF_3CH_2CHF_2$ (HFC-245fa)	7.2
$CF_3CH_2CF_3$ (HFC-236fa)	250
$CF_3CHFCHF_2$ (HFC-236ea)	11.3
$CF_3CF_2CH_2F$ (HFC-236cb)	6.7
CF_3CHFCF_3 (HFC-227ea)	41.0
$CHF_2CF_2CH_2F$ (HFC-245ca)	8.1
$CF_3CH_2CF_2CH_3$(HFC-365mfc)	8.0
$CF_3CH_2CH_2CF_3$(HFC-356mff)	7.1
$CF_3CF_2CH_2CH_2F$ (HFC-356mcf)	1.1
$CHF_2CF_2CF_2CHF_2$(HFC-338pcc)	12.8
$CF_3CH_2CF_2CH_2CF_3$(HFC-458mfcf)	23.6
$CF_3CHFCHFCF_2CF_3$(HFC-43-10mee)	16.7
$CF_3CF_2CH_2CH_2CF_2CF_3$(HFC-55-10mcff)	3.5
CH_3Cl	1.3
CH_2Cl_2	0.4
$CHCl_3$	0.5
CH_3CH_2Cl	0.1
CH_2ClCH_2Cl	0.2
CH_3CHCl_2	0.1
$CH_2ClCHCl_2$	0.2
CH_3CCl_3	5.4
CCl_3CCl_2H	12.4
$CH_3CH_2CH_2Cl$	0.1
$CH_3CHClCH_3$	0.1
$CH_2ClCH_2CH_2Cl$	3.9
$CH_3CH_2CH_2CH_2Cl$	0.03
$(CH_3)_3CCl$	7.5
$CH_3(CH_2)_3CH_2Cl$	0.0
$CH_3(CH_2)_4CH_2Cl$	0.0
CH_3Br	1.6
CH_2Br_2	0.3
$CHBr_3$	0.2
CH_3CH_2Br	0.2
CH_2BrCH_2Br	0.2
$CH_3CH_2CH_2Br$	0.031
$CH_3CHBrCH_3$	0.036
$CH_3CH_2CH_2CH_2Br$	0.013

Table I-F-2. *(continued)*

Haloalkane	Atmospheric Lifetime (yr)[b]
$CH_3(CH_2)_3CH_2Br$	0.008
$CH_3(CH_2)_4CH_2Br$	0.005
CH_3I	0.6
CH_3CH_2I	0.037
$CH_3CH_2CH_2I$	0.020
CH_3CHICH_3	0.024
CH_2FCl (HCFC-31)	1.2
CHF_2Cl (HCFC-22)	12.5
$CHFCl_2$ (HCFC-21)	1.6
CH_3CF_2Cl (HCFC-142b)	18.4
CH_3CFCl_2 (HCFC-141b)	9.3
$CF_3CF_2CCl_2H$ (HCFC-225ca)	2.0
CF_3CCl_2H (HCFC-123)	1.1
CH_2ClCF_2Cl (HCFC-132b)	3.1
$CHCl_2CF_2Cl$ (HCFC-122)	0.8
$CHFClCFCl_2$ (HCFC-122a)	2.6
$CH_3CF_2CFCl_2$ (HCFC-243cc)	25.4
$CHFClCF_2Cl$ (HCFC-123a)	3.9
CH_2ClCF_3 (HCFC-133a)	3.9
$CHFClCF_3$ (HCFC-124)	5.5
CF_2ClCF_2CHFCl (HCFC-225cb)	5.5
$CF_3CH_2CFCl_2$ (HCFC-234fb)	30.3
CHF_2Br (Halon-1201)	5.3
CF_3CH_2Br (Halon-2301)	3.2
CF_3CHFBr (Halon-2401)	2.8
CF_3I	3.0
CH_2ClBr	0.4
$CHCl_2Br$	0.2
$ClCH_2CHBrCH_2Br$	0.1
$CF_3CHClBr$ (Halon-2311)	0.8
$CF_2BrCHFCl$	3.5

[a] OH + haloalkane rate coefficient data taken from table VI-A-76.

[b] Set so that the atmospheric lifetime due to OH attack on methane is 9.6 years (IPCC, 2001) using the OH + CH_4 rate coefficient of 2.45×10^{-12} exp(-1775/T) based on DeMore et al. (1997).

I-F-3. Other Tropospheric Removal Mechanisms for the Alkanes

Much of this discussion of the atmospheric lifetimes, inter-hemispheric gradients, and spatial distributions of the alkanes has centered around their reactivity with tropospheric OH radicals and their OH reaction rate coefficients. This has been possible because of the overriding importance of OH oxidation as a removal process for alkanes in particular and VOCs in general.

The presence of the stratospheric ozone layer above the troposphere protects the troposphere from much of the damaging ultraviolet radiation from the sun. The shortest wavelengths reaching the troposphere are generally about 285 nm. While solar radiation of longer wavelength than 285 nm drives much of tropospheric chemistry through the photolysis of ozone, formaldehyde, and nitrogen dioxide, only bromoalkanes and iodoalkanes are significantly photochemically labile at these wavelengths. For some compounds, particularly the halons, tropospheric photolysis, rather than OH oxidation, controls the atmospheric lifetimes. For most iodoalkanes, photolysis appears to dominate over OH reactivity.

Because of the low reactivity of the shorter chain alkanes, they can readily undergo long-range transport and can participate in atmospheric chemistry processes in regions that are characteristically different from those in which they were emitted. Examples include the chemistry of the remote marine boundary layer and of the polar sunrise.

In the marine boundary layer, strong acids such as sulfuric and nitric acids can displace hydrogen chloride from sea-salt particles (Savoie and Prospero, 1982):

$$HNO_3 + NaCl = NaNO_3 + HCl$$

Nitryl chloride can also be formed by reactions with N_2O_5 at nighttime (Finlayson-Pitts et al., 1989):

$$N_2O_5 + NaCl = NaNO_3 + ClNO_2$$

The subsequent reactions of HCl and $ClNO_2$ generate a small steady state of chlorine atoms in the daylight marine boundary layer (Singh and Kasting, 1988). Because of the different patterns of reactivity with alkanes between chlorine atoms and OH radicals, local oxidation rates may be different in the marine boundary layer from those in the troposphere, in general. This is particularly the case for ethane, propane, and tetrachloroethylene (Singh et al., 1996).

Polar sunrise chemistry produces the almost complete destruction of ozone (Barrie et al., 1988) and gaseous elemental mercury (Schroeder et al., 1998) within a short timescale in the polar boundary layer, shortly after sunrise in the arctic regions of Canada and at coastal locations in the Antarctic (Temme et al., 2003). The enhanced chemical processing required to explain the ozone and mercury depletion events appears to involve elevated levels of reactive bromine compounds and, to a smaller extent, chlorine compounds (Calvert and Lindberg, 2003). The presence of iodine-containing compounds can in theory enhance the ozone and mercury depletion (Calvert and Lindberg, 2004a,b).

Because of the markedly different reactivities of the different alkanes with bromine and chlorine atoms compared with OH radicals, local alkane oxidation rates may differ in the polar sunrise from those in the troposphere, in general. The elevated levels of reactive chlorine produce marked changes in the local concentrations of C_2—C_5 alkanes whereas those of reactive bromine have little influence on the VOCs, except for the marked destruction of acetylene (Jobson et al., 1994). While the reactions of bromine atoms with alkanes are of negligible importance, the bromine atom is believed to be the main reactant leading to ozone destruction in polar sunrise episodes (e.g., see Calvert and Lindberg, 2003).

I-G. Alkanes and the Greenhouse Effect

For the majority of the alkanes with OH rate coefficients of $<10^{-12}$ cm^3 molecule^{-1} s^{-1} and with large man-made sources, well-defined, semi-uniform tropospheric distributions will result. If the alkane has infrared absorption bands that overlap with the wavelength distribution of the outgoing terrestrial radiation, the alkane may act as a radiatively active trace gas; such a gas is commonly called a greenhouse gas (Highwood et al., 1999). Furthermore, if its absorption bands overlap with the so-called atmospheric window region, the alkane may be intensely radiatively active.

It is commonly found that the haloalkanes that contain C—F bonds have strong infrared absorptions in the atmospheric window region, and hence they behave as strong greenhouse gases (IPCC, 2005). In contrast, haloalkanes containing C—Cl and C—Br are not as efficient absorbers in the infrared region. The process by which alkanes interfere with the outgoing terrestrial radiation and lead to a warming of the surface–atmosphere system is termed direct radiative forcing (IPCC, 1990). Radiative forcing increases with increasing tropospheric burden and can be characterized through the radiative efficiency term. This is defined as the change in radiative flux at the top of an atmospheric column caused by the introduction of 1 ppb of the trace gas into that atmospheric column. Radiative efficiency is an instantaneous property of the atmosphere that depends on the location and strength of the infrared absorption bands of the trace gas, the spectral intensity distribution of the outgoing terrestrial radiation, and the spatial distribution of the trace gas. The radiative efficiencies of a number of alkanes are presented in table I-G-1.

A more useful concept for policy-making is the global warming potential (GWP) (IPCC, 1990). The global warming potential is the time-integrated radiative forcing resulting from the emission of 1Tg of the trace gas compared with 1Tg of CO_2, integrated over a time horizon of, for example, 100 years. The GWP concept thus integrates into one index, the concepts of persistence and radiative efficiency, and expresses their product on a basis relative to carbon dioxide. Table I-G-1 presents the GWPs of a wide range of trace gases and shows the importance of those haloalkanes containing C—F bonds as greenhouse gases. The perfluoroalkanes appear to show the highest GWPs of all the trace gases in table I-G-1. By multiplying the current emissions with their respective GWPs, we can derive an estimate for the contribution to the greenhouse effect from current emissions; see table I-G-2. This contribution is heavily dominated by the contributions from the alkanes as a class.

In addition to their direct radiative forcing, alkanes also contribute to radiative forcing and climate change because their emissions can change the global distributions of methane and ozone (Collins et al., 2002), the second and third most important greenhouse gases after CO_2. The reactions of alkanes with OH radicals act as the major sink for some alkanes. Depending on the chemistry of the degradation reactions and the fates and behavior of the first-generation reaction products, the oxidation of the alkanes can change the methane and ozone distributions in the troposphere. If the peroxy radicals formed in the initial oxidation of the alkanes by OH radicals can convert NO into NO_2, they will form additional tropospheric ozone. Furthermore, if the reaction of OH with the alkanes does not efficiently recycle the OH radical lost in

Table I-G-1. Radiative efficiencies[a] and global warming potentials[b] (GWP) of the alkanes

Alkane	Chemical Formula	Radiative Efficiency[a] ($W\ m^{-2}\ ppb^{-1}$)	GWP for 100 yr Time Horizon[b]
Alkanes controlled by the Montreal Protocol			
CFC-11	CCl_3F	0.25	4680
CFC-12	CCl_2F_2	0.32	10720
CFC-113	CCl_2FCClF_2	0.30	6030
CFC-114	$CClF_2CClF_2$	0.31	9880
CFC-115	$CClF_2CF_3$	0.18	7250
Halon-1301	$CBrF_3$	0.32	7030
Halon-1211	$CBrClF_2$	0.30	1860
Halon-2402	$CBrF_2CBrF_2$	0.33	1620
Tetrachloromethane	CCl_4	0.13	1380
Bromomethane	CH_3Br	0.01	5
1,1,1-Trichloroethane	CH_3CCl_3	0.06	144
HCFC-22	$CHClF_2$	0.20	1780
HCFC-123	$CHCl_2CF_3$	0.14	76
HCFC-124	$CHClFCF_3$	0.22	599
HCFC-141b	CH_3CCl_2F	0.14	713
HCFC-142b	CH_3CClF_2	0.20	2270
HCFC-225ca	$CHCl_2CF_2CF_3$	0.20	120
HCFC-225cb	$CHClFCF_2CClF_2$	0.32	586
Alkanes controlled by the Kyoto Protocol			
Methane	CH_4	3.7×10^{-4}	23
HFC-23	CHF_3	0.19	14310
HFC-32	CH_2F_2	0.11	670
HFC-125	CHF_2CF_3	0.23	3450
HFC-134a	CH_2FCF_3	0.16	1410
HFC-143a	CH_3CF_3	0.13	4400
HFC-152a	CH_3CHF_2	0.09	122
HFC-227ea	CF_3CHFCF_3	0.25	3140
HFC-236fa	$CF_3CH_2CF_3$	0.28	9500
HFC-245fa	$CHF_2CH_2CF_3$	0.28	1020
HFC-365mfc	$CH_3CF_2CH_2CF_3$	0.21	782
HFC-43-10mee	$CF_3CHFCHFCF_2CF_3$	0.40	1610
PFC-14	CF_4	0.08	5820
PFC-116	C_2F_6	0.26	12010
PFC-218	C_3F_8	0.26	8690
PFC-318	$c\text{-}C_4F_8$	0.32	10090
PFC-3-1-10	C_4F_{10}	0.33	8710
PFC-5-1-14	C_6F_{14}	0.49	9140
Other alkanes			
Ethane	CH_3CH_3	0.0032	
Propane	$CH_3CH_2CH_3$	0.0031	
Butane	$CH_3CH_2CH_2CH_3$	0.0047	
2-Methylpropane	$(CH_3)_2CHCH_3$	0.0047	
Pentane	$CH_3(CH_2)_3CH_3$	0.0046	
Dichloromethane	CH_2Cl_2	0.03	10
Chloromethane	CH_3Cl	0.01	16

[a] Radiative efficiencies and GWP data taken from IPCC (2005).

[b] Global warming potential (GWP) of a substance is defined as the ratio of the time-integrated radiative forcing from the instantaneous emission of 1 kg of that substance relative to that of 1 kg of a reference gas such as carbon dioxide.

Table I-G-2. The GWP-weighted current European emissions of a number of alkanes[c]

Alkane	European Emissions[a] ($Gg\ yr^{-1}$)	GWP-Weighted Emissions[b] ($Tg\ yr^{-1}$)
Methane	15, 000	345
CFC-11	5.3	25
CFC-12	4.3	46
CFC-113	0.8	5
CFC-115	0.5	4
Halon-1301	0.8	6
Halon-1211	0.7	1.3
HCFC-124	0.6	0.4
HCFC-141b	5.8	4
HCFC-142b	1.6	4
HCFC-22	14	25
HCFC-123	0.2	0.02
HFC-125	2.3	8
HFC-134a	14.1	20
HFC-152a	2.5	0.3
HFC-365	1.0	0.8
Bromomethane	2.1	0.01
Chloromethane	42	0.7
1,1,1-Trichloroethane	1.1	0.2
Tetrachloromethane	2.1	3
Dichloromethane	55	0.6

[a] Current European emissions are those required in a sophisticated Lagrangian dispersion model to support the observations of elevated alkane levels at the Mace Head atmospheric baseline station during the 2003–2004 period (Manning et al., 2003).

[b] GWPs refer to the 100-year time horizon.

[c] For reference, values for carbon dioxide emissions in Europe are of the order of 5,000,000 $Gg\ yr^{-1}$ and GWP-weighted emissions, 5000 $Tg\ yr^{-1}$.

Table I-G-3. Indirect GWPs for some simple organic compounds[a]

Organic Compound	GWP for 100 yr Time Horizon
Ethane	8.4
Propane	6.3
Butane	7.0
Ethene	6.8
Propene	4.9
Toluene	6.0

[a] Taken from Collins et al. (2002), for organic compounds of man-made origin.

the initial attack, oxidation of the alkanes will lead to a depletion of tropospheric OH and hence increase the build-up of methane in the troposphere. In this way, alkanes can contribute an indirect radiative forcing by perturbing the tropospheric distributions of methane and ozone. Table I-G-3 presents the indirect GWPs for a few simple organic compounds, including some alkanes.

The presence of the stratospheric ozone layer exerts a major warming influence on the stratosphere because of the absorption of solar ultraviolet and visible radiation by ozone. Increases in the ozone content of the stratosphere due to man-made activities will therefore act as a positive radiative forcing agent and will lead to climate warming. Conversely then, stratospheric ozone layer depletion acts as a negative radiative forcing agent and as a climate cooling process (IPCC, 2005).

For the simple alkanes, the direct and indirect radiative forcings involving tropospheric methane and ozone are of a comparable magnitude and are of the same sign. Alkanes are therefore global warming agents (Collins et al., 2002). For the bromoalkanes and halons, direct radiative forcing is weak and indirect forcing through ozone layer depletion dominates, and so they are global cooling agents (IPCC, 2005). For the chlorofluorocarbons (CFCs) and hydrochlorofluorocarbons (HCFCs), direct and indirect radiative forcings are of similar magnitudes but of opposite signs, leading to a cancellation and minimal climate effects (IPCC, 2005). The hydrofluorocarbons (HFCs) and perfluorocarbons (PFCs) have enormous positive direct radiative forcing impacts and are an important class of radiatively active trace gases (IPCC, 2005). Because of their long atmospheric lifetimes, they will continue to make a permanent contribution to radiative forcing and climate change over the next century and beyond.

I-H. The Role of Alkanes in Stratospheric Chemistry and Ozone Layer Depletion

Those alkanes and haloalkanes with long enough atmospheric lifetimes survive the tropospheric OH oxidation sink and are transported up through the rising branches of the tropical Hadley cells into the lower stratosphere. Transport processes then move them both polewards and upwards over the time-scales of several years. The lower stratosphere is intensely cold and dry, and hence oxidation of alkanes and haloalkanes by OH radicals is severely inhibited in the lower stratosphere.

As the alkanes and haloalkanes are transported up into the middle stratosphere, the wavelength distribution of the solar ultraviolet radiation steadily shifts towards the vacuum ultraviolet. This is because both ozone and oxygen number densities decline with altitude. As a result, solar actinic irradiances steadily increase in the 190–240 nm and 240–285 nm wavelength ranges, for a given solar zenith angle. Ultimately, at some altitude in the stratosphere, solar photolysis time-scales and transport time-scales become comparable and the haloalkane mixing ratio profiles begin to decline steeply with increasing altitude. If the haloalkane contains chlorine and bromine substituents, these will be released in the primary photolysis step or the subsequent reactions of the primary photolysis products. Once the reactive chlorine or bromine atoms are released, they can take part in stratospheric ozone depletion reactions with serious global consequences (WMO, 1989).

Generally speaking, the higher the altitude in the stratosphere at which the photolysis of the haloalkane takes place, the longer its atmospheric lifetime and the more efficient are the reactive chlorine and bromine products at destroying ozone. These factors are quantified using the ozone depletion potential, the ODP factor, originally introduced

Table I-H-1. Ozone depletion potentials (ODPs) for a series of haloalkanes

Alkane	Ozone Depletion Potential[a]
CFC-11	1.0
CFC-12	1.0
CFC-113	1.0
CFC-114	0.94
CFC-115	0.44
Halon-1301	12
Halon-1211	6.0
Halon-2402	< 8.6
Tetrachloromethane	0.73
1,1,1-Trichloroethane	0.12
HCFC-22	0.05
HCFC-123	0.02
HCFC-124	0.02
HCFC-141b	0.12
HCFC-142b	0.07
HCFC-225ca	0.02
HCFC-225cb	0.03
Chloromethane	0.02
Bromomethane	0.38
HFC-134a	$< 1.5 \times 10^{-5}$
HFC-23	$< 4 \times 10^{-4}$
HFC-125	$< 3 \times 10^{-5}$

[a] Ozone depletion potentials taken from WMO (2003). Ozone depletion potential (ODP) is the ratio of integrated perturbations to total ozone for a differential mass emission of a particular compound relative to an equal emission of CFC-11.

by Wuebbles (1983). Up-to-date assessments of ODPs are given in table I-H-1. Policy actions under the Montreal Protocol and its Amendments have used ODP values to gauge the severity and urgency of emissions reductions for the different CFCs, HCFCs, and other ozone-depleting substances.

The implementation of the Montreal Protocol and its Amendments has illuminated many examples of how atmospheric fate and behavior are strongly influenced by the tropospheric OH oxidation sink. Methylchloroform (1,1,1-trichloroethane) production and usage were phased out under the Montreal Protocol at the end of 1995. As a result of its short atmospheric lifetime of 5 years due to OH oxidation, its global burden has declined rapidly as its atmospheric release declined due to phase-out (WMO, 2003). Hydrochlorofluorocarbons (HCFCs) have been useful as temporary replacements for CFCs in essential refrigeration and air-conditioning applications (IPCC, 2005). Because they contain C—H bonds, they are subject to oxidation in the troposphere, dramatically reducing their ODPs below those of the CFCs they have replaced.

Methane itself plays an important role in the atmospheric chemical processes that bring about stratospheric ozone layer depletion. Much of the methane that enters the lower stratosphere is removed by oxidation with OH radicals, as it is, indeed, in the troposphere. Because of the large activation energy of the OH + methane reaction rate coefficient, methane has to be transported into the middle stratosphere before this reaction becomes significantly rapid for oxidation and transport times-scales to become comparable. At this level in the stratosphere, methane reacts with chlorine atoms converting reactive odd chlorine into the temporary reservoir species HCl (Nicolet, 1977). The main sink for HCl is transport to the tropopause and below, followed by rain-out in the troposphere. So the reaction of methane with chlorine atoms represents the main chain termination step for stratospheric ozone layer depletion. It is through the trace gas methane that stratospheric ozone layer depletion and global warming become intimately linked as long term global environmental problems.

In addition to its reactions with OH and Cl atoms, methane also reacts with electronically excited oxygen atoms $O(^1D)$ (Nicolet, 1977), as do many other haloalkanes in the stratosphere. In the troposphere, $O(^1D)$ atoms react only with water vapor in addition to being deactivated to $O(^3P)$ by collisions with oxygen and nitrogen molecules. Because of the relative dryness of the stratosphere compared with the troposphere and the lower atmospheric pressures there, $O(^1D)$ exhibits a much wider reactivity with the haloalkanes. For some of the long-lived haloalkanes, the $O(^1D)$ reactions may be the only atmospheric sink of any significance.

The number densities of oxygen $O(^3P)$ atoms rise steadily through the stratosphere and mesosphere as oxygen molecules are increasingly photolysed and dissociated. However, there appear to be no reactions between $O(^3P)$ and the alkanes that are of any significance to atmospheric chemistry in these regions.

Photolysis by short-wavelength solar radiation down to the wavelength of Lyman-α radiation at 121.6 nm becomes increasingly important in the upper stratosphere and mesosphere (Ravishankara et al., 1993) and represents an important destruction process for water vapor, carbon dioxide, and some haloalkanes. For some PFCs there is virtually no photochemical destruction at these extreme wavelengths (Ravishankara et al., 1993) and all the man-made emissions ever emitted will remain in the atmosphere as permanent minor constituents.

II

Reactions of Alkanes with the Hydroxyl Radical (OH)

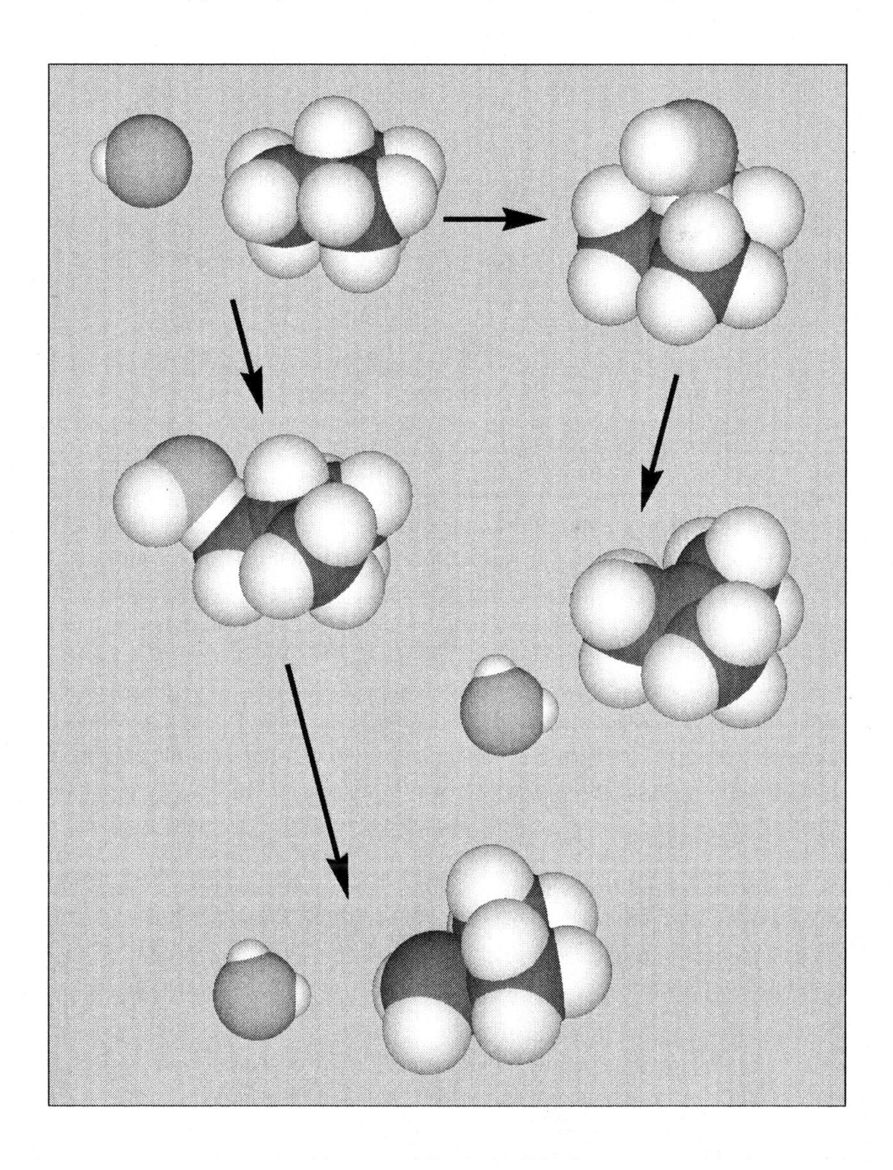

II-A. Introduction

This chapter contains a review of the kinetics and mechanisms of the reactions of alkanes with the hydroxyl radical. Reaction with OH is the major atmospheric fate of all alkanes. Since the alkanes are fully saturated, the reaction invariably involves abstraction of a hydrogen atom to form an alkyl radical, which adds molecular oxygen to form an alkyl peroxy radical.

$$OH + RH \rightarrow H_2O + R$$

$$R + O_2 + M \rightarrow RO_2 + M$$

The reactions of these alkyl peroxy radicals, and the alkoxy radicals subsequently formed from them, define the ultimate product yields; these will be described in later chapters.

The H—OH bond strength in water is approximately 500 kJ mol^{-1}, and C—H bond strengths in alkanes range from 438 kJ mol^{-1} in methane to 388 kJ mol^{-1} for the tertiary hydrogen atom in 2-methylpropane; hence hydrogen abstraction by OH is always thermodyamically possible. In general, the rate coefficients for hydrogen abstraction by hydroxyl from alkanes increase with the size of the alkane, as a result of the larger number of hydrogen atoms available. As will be seen later, subtle effects lead to differences in the abstraction rate at different sites, as a result of different bond strengths and electronic or steric effects. For substituted alkanes, such as the haloalkanes, these differences become larger, and dominate the site specificity of the reactions. Figure II-A-1 shows a plot of the recommended rate coefficient at 298 K for the *n*-alkanes as a function of the carbon number. It can be seen that the rate coefficients increase more-or-less linearly throughout the series, reflecting the addition of an extra (secondary) $—CH_2—$ group for each molecule. As will be seen later, the more heavily substituted a carbon atom is with alkyl groups, the more rapid hydrogen abstraction will be at that carbon atom. In general, the reactivity of a given alkane tends to be dominated by the number of $—CH_2—$ groups present, and this in turn influences the product distribution observed.

The variation of rate coefficients for elementary reactions as a function of temperature is usually parameterized in terms of the Arrhenius expression

$$k = A\exp(-E_a/RT) = A\exp(-B/T) \qquad \text{(I)}$$

where A is the pre-exponential factor (related to the change in entropy in the transition state) and E_a is the activation energy. The parameter B, which is equal to E_a/R, is often reported in kinetics studies, and can be referred to as the activation temperature. Thus, a plot of $\ln(k)$ versus $1/T$ (referred to as an Arrhenius plot) would be straight, with a slope equal to $-E_a/R$. In reality, Arrhenius plots curve, even for simple reactions in which only one reaction site is present. The wide range of temperature over which OH–alkane reactions have been studied gives a unique insight into the extent of curvature. Some curvature is always to be expected. The relative velocity between two particles varies as

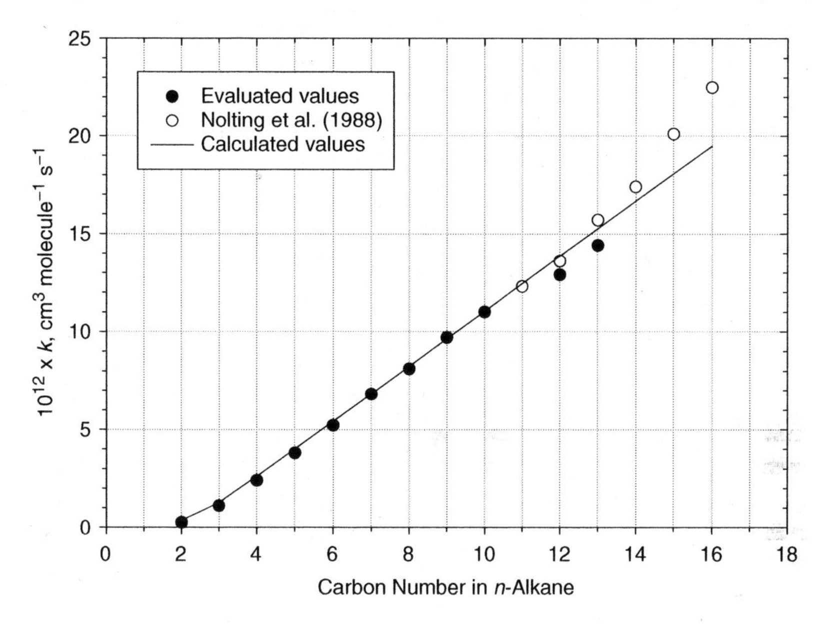

Figure II-A-1. Plot of recommended rate coefficients for the reactions of OH with the *n*-alkanes versus carbon number.

the square root of the temperature, leading to curvature even under the simple transition state theory. In addition, the *A*-factor is predicted to increase with temperature, since the increase in temperature results in the opening up of low-frequency vibrational modes, which changes the partition function. Finally, tunneling of hydrogen atoms can lead to a reduction in the apparent activation energy at low temperatures. All of these factors can contribute to curvature in Arrhenius plots, even for molecules in which there is only one type of hydrogen atom to be abstracted. Furthermore, most alkanes contain a mixture of primary, secondary, and tertiary hydrogen atoms, each with its own rate of abstraction; each behaves independently within a given molecule. As will be described later, primary hydrogen atoms have a higher activation energy for abstraction than secondary or tertiary hydrogen atoms, and so increased curvature is often observed at higher temperatures as the relative contribution of the primary hydrogen atoms increases. As an example, figure II-A-2 shows a composite of currently available rate coefficients for the reaction of OH with 2-methylpropane (isobutane) over a wide range of temperature. The solid line is a fit made by Tully et al. (1986b) to their own data measured between 293 and 864 K, extrapolated down to 213 K. By using a series of isotopically labeled molecules, Tully et al. were able to separate the rate coefficients for abstraction at primary and tertiary sites. These partial rate coefficients are shown as the dashed and dotted curves. As can be seen, considerable curvature is introduced by the rapid increase in abstraction at the primary site, although both group rate coefficients show a mild curvature.

It is important to take correct account of curvature in the Arrhenius plots, particularly in extrapolating data to lower temperatures. In many cases, little curvature is found if only data measured at and below 298 K are considered, as pointed out by DeMore

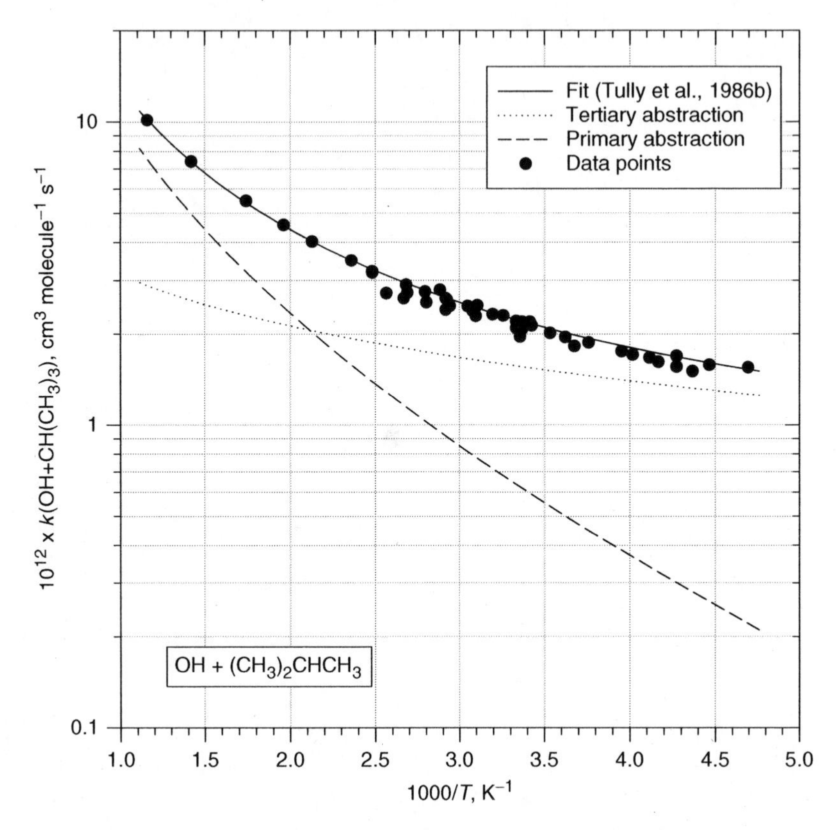

Figure II-A-2. Arrhenius plot of the rate coefficients for the reaction of OH with 2-methylpropane.

and Bayes (1999). This can be a result of scatter between various methods, and also insufficient precision of a single measurement technique. In such situations, relative rate data can often provide valuable information, since systematic errors tend to cancel out or are constant over a fairly wide temperature range, enabling trends in data to be identified clearly.

It has been found empirically by Atkinson (1986a, b, 2003, and references therein) that a modified Arrhenius expression can often describe the combined effects of all these factors:

$$k = CT^n \exp(-D/T) \qquad \text{(II)}$$

In this case, the local activation energy at temperature T (the local slope of a plot of ln(k) versus $1/T$) is given by $DR + nRT$, and the activation temperature is thus $D + nT$. The exponent n varies between 1 and 3 for the systems that have been studied extensively. In general, there are insufficient measurements available below about 250 K to adequately describe the atmospheric chemistry of the alkanes, or to evaluate the extent of curvature in the Arrhenius plots. To reduce the number of free parameters in the fits, n is usually set to 2.

Donahue et al. (1998) have suggested that the T^2 dependence introduces too much curvature, especially at low temperature. By considering the functional form of the transition-state representation of an abstraction reaction they suggested the following expression:

$$k = \frac{B\exp(-E_a/RT)}{T \prod_{i=1-3} (1 - \exp(-144\nu_i/RT))} \tag{III}$$

Expression (III) makes a more realistic attempt to represent the "switching on" of low-frequency modes in the partition function. The values of the vibrational frequencies ν_i are not known, but Donahue et al. suggested using values of 300 cm^{-1} (degenerate) and 500 cm^{-1} to represent loose C—H—O and H—O—H bending modes in the transition state. Setting all three values to 350 cm^{-1} does not noticeably change the form of the function. A further advantage of this formulation is that the activation energy in the exponent of the numerator is the "real" one, corresponding to the barrier height for abstraction. At very high temperature, the expression acquires a $T^2\exp(-D/T)$ dependence, similar to Atkinson's 3-parameter fit. A comparison of expression (III) with the 3-parameter modified Arrhenius expression (II) shows a close correspondence, although at low temperature expression (III) shows somewhat less curvature than the modified Arrhenius expression. Thus expression (III) may give a more realistic dependence of the rate coefficient, but the T^2–Arrhenius expression is more convenient to use. Since a major purpose of this book is to provide data for atmospheric chemists and their models, and Arrhenius data are not expected to curve very much in the atmospheric temperature regime, we have decided to limit the evaluation to data below about 500 K, and to use a simple Arrhenius expression if this fits the data sufficiently well. This approach has the added advantages of simplifying the transfer of relative rate data from one standard to another, and should also give a better extrapolation to low temperatures when no or few measurements are available. Of course, if curvature in a temperature dependence were clearly visible (or if a strong theoretical reason existed) it would be most appropriate to use the "curved" expression to place relative rate data on an absolute basis. Specific cases where a 3-parameter fit is used to describe the data will be discussed in more detail in the relevant sections.

The remainder of the chapter summarizes the available kinetic data for reactions of OH radicals with alkanes. While the book is primarily aimed at atmospheric scientists, it is sometimes informative to consider measurements made at higher temperatures, since these can shed light on the temperature dependence of the data, which may not be visible simply from a consideration of measurements made under atmospheric conditions. Since alkanes are one of the major constituents of gasoline, diesel oil, etc., kinetic measurements have been made for a number of alkanes ranging from lower stratospheric temperatures up to 1000 degrees, making these some of the most extensively studied bimolecular reactions. This leads to unique possibilities for theoretical interpretation of the data. However, often the data sets are not continuous, and scant data exist between ambient temperatures and combustion temperatures. In these cases, care must be taken in interpolating (and, more importantly, extrapolating to low temperature) the combined data sets.

A useful method of estimating the rate coefficients for OH abstraction reactions is the structure–additivity relationship (SAR). In this approach, rather than considering abstraction of individual hydrogen atoms by OH, groups such as methyl $—CH_3$, methylene $>CH_2$, and methylidene $(R)_3C—H$ are considered (Kwok and Atkinson, 1995, and references therein). Rate coefficients for abstraction from each group are provided, and these rate coefficients are modified according to the neighboring group, which can enhance or decrease the rate coefficient. These group rate coefficients can be used to predict the overall rate coefficient for abstraction and, in principle, the relative fractions of abstraction at each site in a molecule (although they are usually less successful at doing this). Structure–additivity relationships are discussed in more detail in section II-E.

A comprehensive review of the kinetic data for alkanes with OH was given by Atkinson (2003), and the interested reader should consult that review for more details, along with tabulations of the rate coefficients. Where new data have become available since 2003, we have updated the tables and refit the data. The tables and figures in the present work omit much of the earlier, less reliable data that is to be found in Atkinson (2003). We have also taken advantage of the recommendations of the IUPAC data evaluation panel (Atkinson et al., 2005b), who give simple, 2-parameter representations of Atkinson's 2003 evaluation for a number of smaller alkanes. We cover in this chapter the linear and branched alkanes containing up to eight carbon atoms (with isotopic substitution where available), the longer *n*-alkanes, and a number of cycloalkanes. In the following sections of this chapter, first the experimental data on the reaction kinetics will be presented, followed by a discussion of theoretical and empirical ways to estimate the rate coefficients. The measurement techniques that were employed in the various rate coefficient studies are designated by abbreviations in the text and in table. These are defined in table II-A-1.

II-B. Rate Coefficients for the Reactions of OH with the Alkanes

II-B-1. Reaction of OH with Methane (CH_4)

Methane is the simplest alkane, and also the most abundant in the atmosphere. This is a result of its large atmospheric source strength, coupled with a fairly slow reaction rate with OH. As a result, methane is fairly well mixed in the troposphere. Local gradients occur, as do interhemispheric ones; these mostly reflect the differing source strengths. Methane has both anthropogenic and biogenic sources, and it has been shown that these can have different isotopic signatures. Consequently, the kinetic isotope effects on the different isotopologs (forms of a molecule containing one or more isotopic substitutions) become an important tool in defining source apportionment in the atmosphere, and also in determining the relative extent of attack by OH and Cl atoms in the lower stratosphere. Since the kinetic isotope effects for methane tend to be large, and also geophysically relevant, they will be reviewed along with the rate coefficients for methane. For the larger alkanes, rate coefficients for molecules that are fully deuterated (or deuterated fully at one site) will be considered with the kinetic data, while data

Table II-A-1. Abbreviations for the experimental techniques used in the text and in tables of rate constant determinations

Abbreviation	Experimental Technique
API	Atmospheric pressure ionization
CIMS	Chemical ionization mass spectroscopy
DF	Discharge flow tube
ECD	Electron capture detection
EPR	Electron paramagnetic resonance spectroscopy
FID	Flame ionization detection
FP	Flash photolysis
FTIR	Fourier transform infrared
GC	Gas chromatography
HPLC	High-pressure liquid chromatography
IR abs	Infrared absorption
IR em	Infrared emission
LIF	Laser-induced fluorescence
MM abs	Molecular modulation absorption
MMS	Molecular modulation spectroscopy
MS	Mass spectrometry
PR	Pulse radiolysis
PTR	Proton transfer reaction
RF	Resonance fluorescence
RR	Relative rate measurement
SF	Stopped flow reactor
SPME	Solid phase microextraction fiber technique
UV abs	Ultraviolet absorption
VLPR	Very low pressure reactor
VUV abs	Vacuum ultraviolet absorption

pertaining to isotopes present at natural abundance are collected together at the end of the chapter, along with secondary isotope effects.

Direct measurements of the rate coefficient for methane with OH have been made from 190 K to 1500 K, while relative rate measurements exist up to 1840 K. The reaction is quite slow, with a room temperature value of 6.4×10^{-15} cm^3 molecule^{-1} s^{-1} and a reasonably large activation temperature $E/R \sim 1690$ K. Hence, the rate coefficient varies by four orders of magnitude over the range of measurements. At stratospheric temperatures the reaction is very slow, and some of the earlier rate coefficient measurements were influenced by the presence of minor impurities which caused the measurements to be overestimated. However, there is now a number of studies which give a consistent picture of the variation of the rate coefficient with temperature below room temperature.

A complete tabulation of the available data is given in the review by Atkinson (2003). The measurements of Vaghjiani and Ravishankara (1991), Finlayson-Pitts et al. (1992), Dunlop and Tully (1993a), Mellouki et al. (1994), Gierczak et al. (1997), and Bonard et al. (2002) form a consistent data set between 190 K and 400 K. Over the temperature range 190 K to 1500 K curvature of the Arrhenius plot is observed, and Atkinson (2003) recommends a rate coefficient given by $k = 1.85 \times 10^{-20} T^{2.82} \exp(-987/T)$ cm^3 molecule^{-1} s^{-1}. Over the atmospheric range of temperatures, an Arrhenius expression of $k = 1.85 \times 10^{-12} \exp(-1690/T)$ cm^3 molecule^{-1} s^{-1} can be used (Atkinson et al., 2005b) without too much deviation from the measured values. The estimated uncertainty at 298 K is ±15%.

II-B-1a. Reactions of OH with Isotopic Variants of Methane

The rate coefficient for the reaction of OH with $^{13}CH_4$ (carbon-13 substituted methane) has been measured four times. Secondary kinetic isotope effects such as this, where the reaction does not occur at the isotopically substituted site, are usually small (less than 1% change from $^{12}CH_4$ to $^{13}CH_4$) and are often expressed in per mil difference according to expression (IV):

$$\varepsilon = ((k_{12}/k_{13}) - 1) \times 1000 \tag{IV}$$

The first measurement of the kinetic isotope effect was by Rust and Stevens (1980), who found $\varepsilon = 3$ per mil at 298 K. Later, Davidson et al. (1987) measured $\varepsilon = 10 \pm 7$ per mil, taking advantage of a much higher conversion (which makes the technique more sensitive). Cantrell et al. (1990b) extended the study of Davidson et al., finding $\varepsilon = 5.4 \pm 0.9$ per mil, independent of temperature between 273 and 353 K. However, in a recent study, Saueressig et al. (2001) found $\varepsilon = 3.9 \pm 0.4$ per mil, intermediate between the values of Rust and Stevens and Cantrell et al. The measurements of Cantrell et al. (1990b) and Saueressig et al. (2001) fall just outside the overlap of their stated uncertainties. The data of Saueressig seem less scattered, but Cantrell's data cover a wider range of conversion. A weighted mean of these two sets of data gives a recommended secondary kinetic isotope effect of $\varepsilon = 4.4 \pm 1.0$ per mil, independent of temperature for atmospheric purposes.

The reactions of OH with all the deuterium-substituted methanes have also been studied. These primary kinetic isotope effects are much larger than the secondary effect just described, since they involve a large change in relative mass of the atoms being abstracted. The reaction of CH_3D (methane-d_1) with OH radicals has been studied by Gordon and Mulac (1975), DeMore (1993b), Gierczak et al. (1997), and Saueressig et al. (2001). The studies of Gordon and Mulac (at 416 K) and Gierczak et al. (249–422 K) were absolute measurements, while those of DeMore and Saueressig et al. used the relative-rate technique; both were made relative to CH_4. The results are in quite good agreement, with the derived rate coefficients of DeMore (from 273 to 353 K) slightly higher than those of Gierczak et al. (i.e., a smaller kinetic isotope effect). The room-temperature value of Saueressig et al. agrees very well with that of Gierczak et al. A fit to the rate coefficients of DeMore, Gierczak et al. and Saueressig et al. (Atkinson, 2003) gives $k = 5.19 \times 10^{-18} T^2 \exp(-1332/T)\,\text{cm}^3\,\text{molecule}^{-1}\,\text{s}^{-1}$; the data of Gierczak et al. can be fit also by a simple 2-parameter Arrhenius expression $k = 3.11 \times 10^{-12} \exp(-1910/T)\,\text{cm}^3\,\text{molecule}^{-1}\,\text{s}^{-1}$. The ratio $k(CH_4)/k(CH_3D) = 1.25$ at 298 K.

Rate coefficients for the reactions of CH_2D_2 and CHD_3 with OH have been made by Gordon and Mulac at 416 K and by Gierczak et al. between 270 and 354 K. The values of Gierczak are recommended for atmospheric purposes: $k(CH_2D_2) = 2.3 \times 10^{-12} \exp(-1930/T)$; $k(CHD_3) = 1.46 \times 10^{-12} \exp(-1970/T)\,\text{cm}^3\,\text{molecule}^{-1}\,\text{s}^{-1}$. It should be noted that the rate coefficients were measured over a limited temperature range and should not be extrapolated to very low temperature.

The reaction of OH with CD_4 has been measured by Gordon and Mulac (1975), Dunlop and Tully (1993a), and Gierczak et al. (1997). The single data point of Gordon and Mulac lies considerably lower than in the other two studies, which are in excellent agreement over the region of overlapping temperatures. A fit to the complete data set of Dunlop and Tully and Gierczak et al. leads to the 3-parameter expression $k = 5.7 \times 10^{-18}\ T^2 \exp(-1882/T)\,cm^3\ molecule^{-1}\,s^{-1}$, while for atmospheric purposes Gierczak et al. recommend $k = 1.0 \times 10^{-12} \exp(-2100/T)\,cm^3\ molecule^{-1}\,s^{-1}$. The value of the rate coefficient at 298 K is $8.7 \times 10^{-16}\ cm^3\ molecule^{-1}\,s^{-1}$. Gierczak et al. pointed out that the rate coefficients for the family of substituted methanes could be expressed as linear combinations of the rate coefficients for CH_4 and CD_4; hence it appears that the abstractions from the C—H and C—D bonds can be treated independently. The relative rate coefficients for OH reaction with the various methane isotopologs at 298 K are: $k(CH_4) : k(CH_3D) : k(CH_2D_2) : k(CHD_3) : k(CD_4)$ :: 1.00 : 0.80 : 0.55 : 0.31 : 0.14. The activation energy increases down the series, and the A-factor decreases, consistent with the fact that the abstraction is dominated by the number of C—H bonds in the molecule.

II-B-2. Reaction of OH with Ethane (CH_3CH_3)

The reaction of OH with ethane is one of the best studied gas phase reactions, since it has been used as a test for many experimental systems and also used as a reference reaction in relative rate experiments. A number of absolute studies have defined the rate coefficient over the atmospheric temperature range, and these are used as the basis of the recommendation. The studies of Smith et al. (1984), Stachnik et al. (1986), Wallington et al. (1987b), Talukdar et al. (1994), Donahue et al. (1998), and Clarke et al. (1998) between 180 and 360 K are all in excellent agreement. When taken together with the data of Tully et al. (1986a) at higher temperatures it is clear that the Arrhenius plot curves, and Atkinson (2003) recommends a rate coefficient of $1.49 \times 10^{-17} T^2 \exp(-499/T)\,cm^3\ molecule^{-1}\,s^{-1}$, with a room temperature value of $2.48 \times 10^{-13}\,cm^3\,molecule^{-1}\,s^{-1}$. However, over the range of atmospheric interest little or no curvature is observed, and we have fitted the above data sets from 180 to 400 K to derive the simple Arrhenius expression $k = 8.61 \times 10^{-12} \exp(-1047/T)\,cm^3\,molecule^{-1}\ s^{-1}$. That expression has been used in this work to normalize relative rate coefficients for larger alkanes. The data used and the fits are shown in figure II-B-1, and table II-B-1. The estimated uncertainty at 298 K is $\pm 15\%$.

Tully and coworkers also measured rate coefficients for the deuterium-substituted ethanes CH_3CD_3, and C_2D_6, showing that the CH_3 and CD_3 groups can be treated independently. They derived a kinetic isotope effect k_{CH3}/k_{CD3} of 4.61 at room temperature, which was subsequently found to be characteristic of all primary hydrogen atoms in larger alkanes. This can be compared to the kinetic isotope effect of roughly 7 between CH_4 and CD_4 discussed in the previous section.

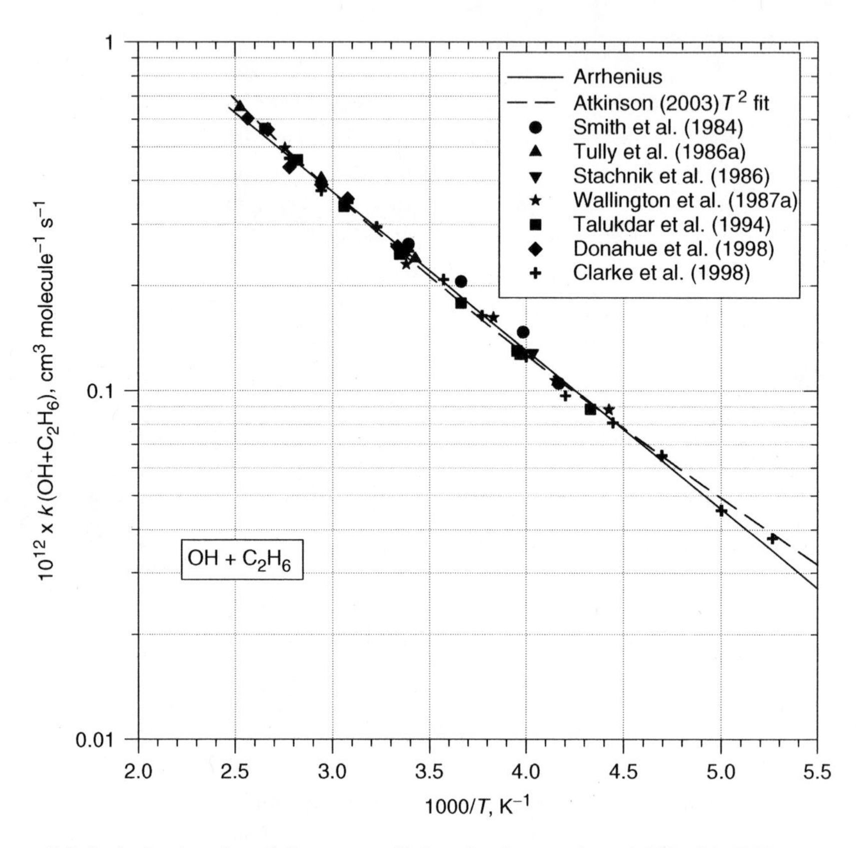

Figure II-B-1. Arrhenius plot of the rate coefficient for the reaction of OH with C_2H_6.

Table II-B-1. Rate coefficients $k = Ae^{-B/T}$ (cm^3 $molecule^{-1}$ s^{-1}) for the reaction of OH with C_2H_6 (selected data used in fit)

$10^{12} \times A$	B(K)	$10^{12} \times k$	T(K)	Technique[a]	Reference	Temperature Range (K)
18.0	1240 ± 110	0.105 ± 0.004	240	FP–RF	Smith et al. (1984)	240–295
		0.147 ± 0.006	251			
		0.205 ± 0.009	273			
		0.263 ± 0.010	295			
$k(T)$[b]		0.239 ± 0.010	292	PLP–RF	Tully et al. (1986a)	292–396
		0.407 ± 0.017	340			
		0.651 ± 0.027	396			
		0.128 ± 0.010	248	PLP–RA	Stachnik et al. (1986)	248–297
		0.250 ± 0.007	297			
8.4	1050 ± 100	0.088 ± 0.013	226	FP–RF	Wallington et al. (1987b)	226–363
		0.107 ± 0.010	241			
		0.162 ± 0.018	261			

Table II-B-1. (*continued*)

$10^{12} \times A$	B(K)	$10^{12} \times k$	T(K)	Technique[a]	Reference	Temperature Range (K)
		0.230 ± 0.026	296			
		0.497 ± 0.055	363			
10.3	1108 ± 40	0.088 ± 0.002	231	PLP–LIF	Talukdar et al. (1994)	231–377
		0.127 ± 0.007	252			
		0.130 ± 0.002	253			
		0.178 ± 0.004	273			
		0.246 ± 0.003	299			
		0.338 ± 0.004	327			
		0.459 ± 0.005	355			
		0.564 ± 0.008	377			
9.7	1086	0.259 ± 0.008	300	DF–LIF	Donahue et al. (1998)	300–390
		0.355 ± 0.011	325			
		0.390 ± 0.012	340			
		0.438 ± 0.023	360			
		0.561 ± 0.017	375			
		0.604 ± 0.018	390			
7.9	1027	0.027 ± 0.003	180	DF–LIF	Clarke et al. (1998)	180–360
		0.038 ± 0.001	190			
		0.045 ± 0.003	200			
		0.065 ± 0.001	213			
		0.081 ± 0.001	225			
		0.097 ± 0.002	238			
		0.125 ± 0.001	250			
		0.164 ± 0.002	265			
		0.208 ± 0.002	280			
		0.252 ± 0.002	295			
		0.295 ± 0.005	310			
		0.346 ± 0.004	325			
		0.374 ± 0.004	340			
		0.464 ± 0.016	360			

[a] Abbreviations used for the techniques are defined in table II-A-1.
[b] $k(T) = 8.51 \times 10^{-18} T^{2.06} \exp(-430/T)$

II-B-3. Reaction of OH with Propane ($CH_3CH_2CH_3$)

The reaction of OH with propane has also been studied extensively. The preferred rate coefficients are shown in figure II-B-2 and table II-B-2. The absolute measurements of Mellouki et al. (1994), Talukdar et al. (1994), Donahue et al. (1998), Clarke et al. (1998), and Kozlov et al. (2003a) define the rate coefficient between 190 and 376 K, while the measurements of Droege and Tully (1986a) can be used to extend the data up to 850 K. The relative rate data of DeMore and Bayes (1999), normalized using the rate coefficient for ethane given above, are in excellent agreement with the absolute data.

Atkinson (2003) recommends a 3-parameter fit of $1.65 \times 10^{-17} T^2 \exp(-87/T)$ cm^3 molecule^{-1} s^{-1}, leading to a room temperature value of 1.1×10^{-12} cm^3 molecule^{-1} s^{-1}. However, we have fitted the data using the simple Arrhenius expression

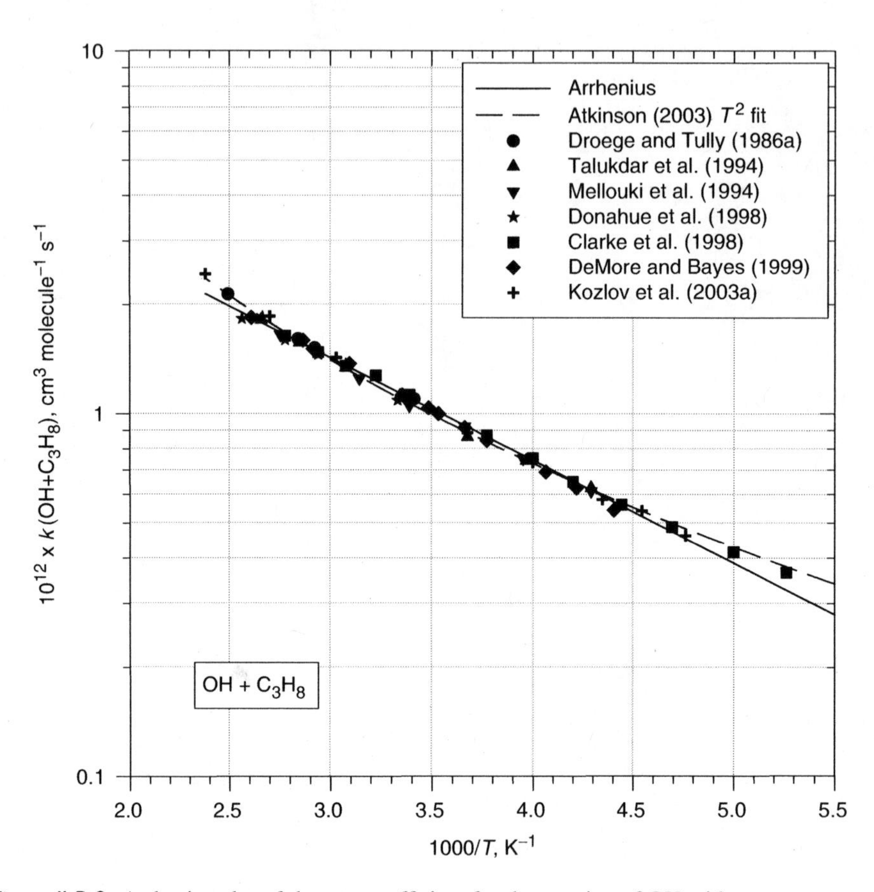

Figure II-B-2. Arrhenius plot of the rate coefficient for the reaction of OH with propane.

Table II-B-2. Rate coefficients $k = A\ e^{-B/T}$ ($cm^3\ molecule^{-1}\ s^{-1}$) for the reaction of OH with C_3H_8 (selected data used in fit)

$10^{12} \times A$	B(K)	$10^{12} \times k$	T(K)	Technique[a]	Reference	Temperature Range (K)
$k(T)$[b]		1.10 ± 0.04	293	PLP–RF	Droege and Tully (1986a)	210–420
		1.52 ± 0.06	342			
		1.61 ± 0.07	352			
		2.14 ± 0.09	401			
10.1	657 ± 46	0.62 ± 0.02	233	PLP–LIF	Talukdar et al. (1994)	233–376
		0.74 ± 0.02	252			
		0.86 ± 0.02	272			
		1.12 ± 0.04	299			
		1.34 ± 0.06	325			
		1.57 ± 0.05	351			
		1.82 ± 0.07	376			

Table II-B-2. *(continued)*

$10^{12} \times A$	B(K)	$10^{12} \times k$	T(K)	Technique[a]	Reference	Temperature Range (K)
9.81	650 ± 30	0.61 ± 0.61	233	PLP–LIF	Mellouki et al. (1994)	233–363
		0.75 ± 0.04	253			
		0.92 ± 0.04	273			
		1.05 ± 0.04	295			
		1.25 ± 0.04	318			
		1.51 ± 0.02	343			
		1.65 ± 0.07	363			
11.2	693	1.09 ± 0.03	300	DF–LIF	Donahue et al. (1998)	300–390
		1.37 ± 0.04	325			
		1.46 ± 0.04	340			
		1.60 ± 0.09	360			
		1.85 ± 0.06	375			
		1.83 ± 0.10	390			
9.24	624	0.36 ± 0.02	190	DF–LIF	Clarke et al. (1998)	190–360
		0.41 ± 0.02	200			
		0.49 ± 0.01	213			
		0.56 ± 0.01	225			
		0.65 ± 0.01	238			
		0.75 ± 0.01	250			
		0.87 ± 0.02	265			
		1.13 ± 0.02	295			
		1.28 ± 0.02	310			
		1.36 ± 0.02	325			
		1.48 ± 0.02	340			
		1.64 ± 0.03	360			
11.4	689	0.54	227	RR(C_2H_6)[c]	DeMore and Bayes (1999)	227–383
		0.62	237			
		0.69	246			
		0.84	265			
		0.91	273			
		1.00	283			
		1.04	287			
		1.13	298			
		1.37	323			
		1.51	343			
		1.60	349			
		1.84	383			
$k(T)$[d]		0.46 ± 0.02	210	FP–RF	Kozlov et al. (2003a)	210–420
		0.54 ± 0.01	220			
		0.58 ± 0.01	230			
		0.73 ± 0.02	250			
		0.88 ± 0.02	272			
		1.13 ± 0.02	298			
		1.43 ± 0.03	330			
		1.86 ± 0.05	370			
		2.43 ± 0.04	420			

[a] Abbreviations used for the techniques are defined in table II-A-1.

[b] $k(T) = 1.04 \times 10^{-16} T^{1.72} \exp(-145/T)$.

[c] Placed on an absolute basis using $k(C_2H_6) = 8.61 \times 10^{-12} \exp(-1047/T)$ cm^3 molecule^{-1} s^{-1}.

[d] $k(T) = 5.81 \times 10^{-17} T^{1.83} \exp(-167/T)$.

$1.0 \times 10^{-11} \exp(-655/T) \mathrm{cm}^3$ molecule^{-1} s^{-1} for atmospheric purposes, which is essentially identical to the expression given by Talukdar et al. (1994). The estimated uncertainty at 298 K is ±15%.

Droege and Tully (1986a) studied a series of deuterated propanes, $CH_3CD_2CH_3$, $CH_3CH_2CD_3$, $CH_3CD_2CD_3$, $CD_3CH_2CD_3$, and $CD_3CD_2CD_3$, and were able to derive partial rate coefficients for abstraction at the various carbon atoms, along with kinetic isotope effects for primary and secondary hydrogen atoms. They showed that reaction of C_3H_8 with OH at 298 K should proceed by reaction according to 72% at the CH_2 group and 28% at the CH_3 groups. They also derived a primary kinetic isotope effect of $k_{CH2}/k_{CD2} = 2.62$ at 298 K for the secondary hydrogen atoms.

II-B-4. Reaction of OH with Butane ($CH_3CH_2CH_2CH_3$)

The reaction of OH with butane has been well studied. This is in part because of the importance of the reaction and in part because it has been used as a reference reaction for many relative rate studies, since it is relatively fast, and butane is easy to handle. The reaction has been studied at temperatures down to 230 K using both absolute methods (flash photolysis and discharge flow) and the relative rate technique, where its rate coefficient has been measured relative to the well-known reactions OH + ethane and OH + propane. The data are in good agreement over the range of temperatures studied. The tendency for the absolute data measured by Droege and Tully (1986b), Talukdar et al. (1994), and Donahue et al. (1998) is to show a slight upward curvature, consistent with the 3-parameter fit. However, if the temperature-dependent relative rate data of DeMore and Bayes (1999) are calculated using 3-parameter expressions for the reference reactions, they show a pronounced upward curvature at low temperature. Using the 2-parameter fit for propane given by Talukdar et al. leads to a more reasonable shape for the DeMore data at low temperature, although the relative rate data still tend to be systematically lower than the absolute data. The data at 298 ± 2 K are in good agreement, with a mean of $2.36 \times 10^{-12} \mathrm{cm}^3$ molecule^{-1} s^{-1}. We have taken the data of Talukdar et al., Donahue et al., Droege and Tully (below 400 K), and DeMore and Bayes scaled to the 2-parameter fit for propane given above to construct an Arrhenius plot, leading to the expression $k = 1.40 \times 10^{-11} \exp(-520/T) \mathrm{cm}^3$ molecule^{-1} s^{-1}. The data and fits are summarized in figure II-B-3 and table II-B-3. The estimated uncertainty at 298 K is ±15%.

Droege and Tully (1986b) also measured rate coefficients for the reaction of OH with fully deuterated butane, C_4D_{10}. By assuming that the H/D isotope effect for the primary hydrogen atoms was the same as in ethane and 2,2-dimethylpropane, they were able to calculate the kinetic isotope effect at the secondary carbons, and hence to determine the sites of attack as a function of temperature. The temperature-dependent values for overall abstraction at the primary (k_p) and secondary (k_s) sites were:

$$2k_p = 6.86 \times 10^{-17} T^{1.73} \exp(-380/T) \mathrm{cm}^3 \mathrm{molecule}^{-1} \mathrm{s}^{-1}$$

$$2k_s = 1.20 \times 10^{-16} T^{1.64} \exp(125/T) \mathrm{cm}^3 \mathrm{molecule}^{-1} \mathrm{s}^{-1}$$

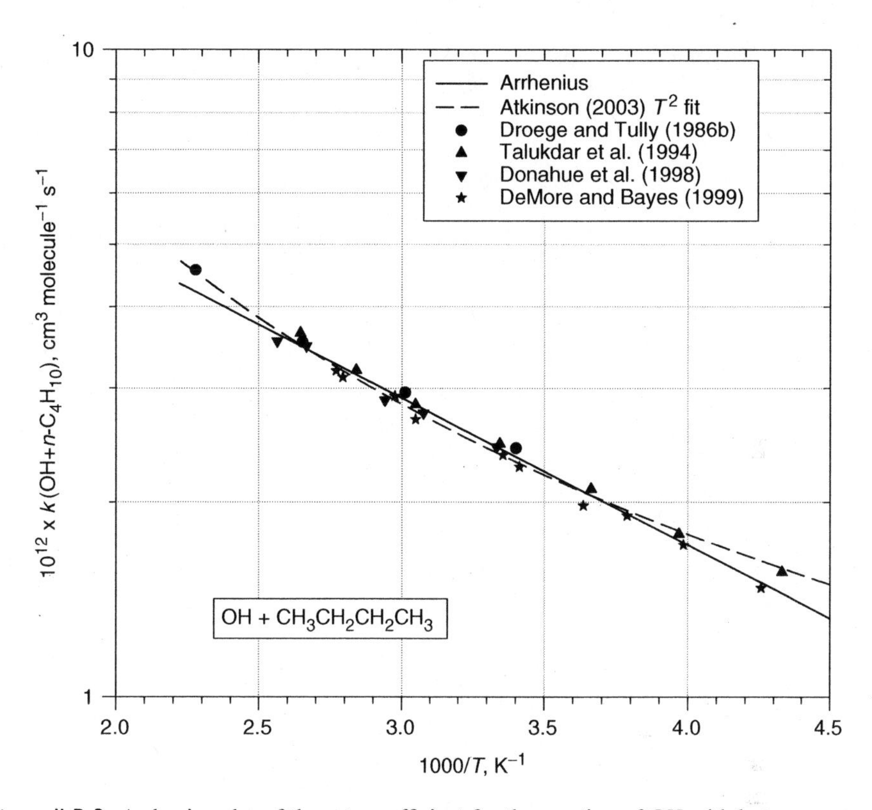

Figure II-B-3. Arrhenius plot of the rate coefficient for the reaction of OH with butane.

Table II-B-3. Rate coefficients $k = A\ e^{-B/T}$ ($cm^3 molecule^{-1}\ s^{-1}$) for the reaction of OH with butane (selected data used in fit)

$10^{12} \times A$	B(K)	$10^{12} \times k$	T(K)	Technique[a]	Reference	Temperature Range (K)
$k(T)$[b]		2.42 ± 0.10	294	PLP–RF	Droege and Tully (1986b)	294–439
		2.95 ± 0.12	332			
		3.53 ± 0.15	377			
		4.56 ± 0.19	439			
11.8	470 ± 40	1.56 ± 0.02	231	PLP–LIF	Talukdar et al. (1994)	231–378
		1.79 ± 0.02	252			
		2.10 ± 0.02	273			
		2.46 ± 0.02	299			
		2.83 ± 0.06	328			
		3.20 ± 0.03	352			
		3.65 ± 0.02	378			
13.6	514	2.43 ± 0.07	300	DF–LIF	Donahue et al. (1998)	300–390
		2.74 ± 0.08	325			
		2.87 ± 0.09	340			

(continued)

Table II-B-3. *(continued)*

$10^{12} \times A$	B(K)	$10^{12} \times k$	T(K)	Technique[a]	Reference	Temperature Range (K)
		3.48 ± 0.10	375			
		3.54 ± 0.11	390			
13.1	513	1.47	235	RR (C_3H_8)[c]	DeMore and Bayes (1999)	235–361
		1.72	251			
		1.91	264			
		1.98	275			
		2.26	293			
		2.36	298			
		2.68	328			
		2.91	336			
		3.12	358			
		3.19	361			

[a] Abbreviations used for the techniques are defined in table II-A-1.
[b] $k(T) = 2.34 \times 10^{-17} T^{1.95} \exp(134/T)$.
[c] Placed on an absolute basis using $k(C_3H_8) = 1.0 \times 10^{-11} \exp(-655/T)$ cm^3 molecule^{-1}s^{-1}.

At room temperature, 87% of the attack was found to be at the secondary site, leading to 2-butyl $CH_3CH{\cdot}CH_2CH_3$ radicals, while at lower temperatures this fraction would be even higher. Thus oxidation of butane under atmospheric conditions is expected to produce predominantly 2-butyl peroxy radicals.

II-B-5. Reaction of OH with 2-Methylpropane [Isobutane, $CH_3CH(CH_3)CH_3$]

The kinetics of the reaction of OH with 2-methylpropane (isobutane) have been measured between 213 and 864 K using both absolute and relative rate methods. Measurements around room temperature are tightly constrained, giving a rate coefficient of 2.12×10^{-12} cm^3 molecule^{-1} s^{-1}. The temperature-dependent data of Tully et al. (1986b), Talukdar et al. (1994), Donahue et al. (1998), and Wilson et al. (2006), along with the measurements at single elevated temperatures by Hucknall et al. (1975), Baldwin and Walker (1979), and Bott and Cohen (1989) all form a consistent data set. The recommended expression (Atkinson, 2003) for the rate coefficient is $k = 1.17 \times 10^{-17} T^2 \exp(213/T)$ cm^3 molecule^{-1} s^{-1}. An Arrhenius fit to the data of Talukdar et al. Donahue et al. and Wilson et al. between 213 K and 390 K leads to the recommendation of $7.0 \times 10^{-12} \exp(-350/T)$ cm^3 molecule^{-1} s^{-1} for atmospheric modeling purposes. The estimated uncertainty at 298 K is ±15%. The preferred data and fits are shown in figure II-B-4 and table II-B-4.

Tully et al. (1986b) also measured rate coefficients for the various symmetrically deuterated isotopologs of 2-methylpropane: $CD(CH_3)_3$, $CH(CD_3)_3$, and $CD(CD_3)_3$ between 293 and 864 K. The series of rate coefficients is internally consistent and shows that group additivity applies to the two groups of hydrogen atoms. The kinetic isotope effect at the tertiary site ranges between 1.9 at 293 K and 1.1 at 864 K. By combining the kinetic isotope effects for the different types of hydrogen with the rate

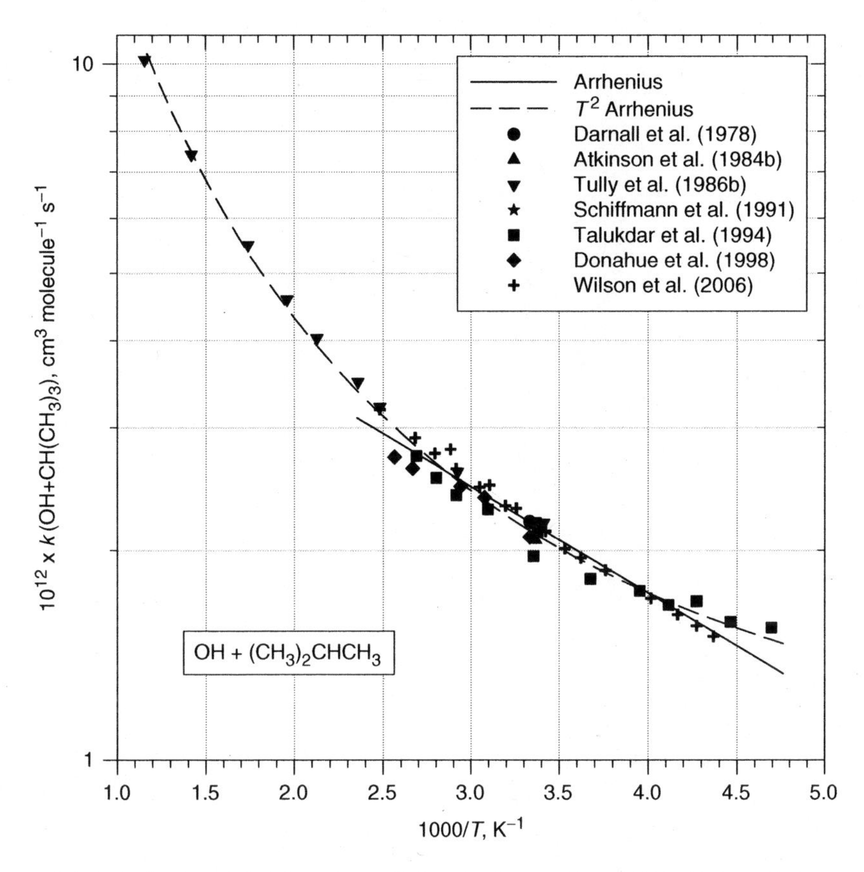

Figure II-B-4. Arrhenius plot of the rate coefficient for the reaction of OH with 2-methylpropane.

Table II-B-4. Rate coefficients $k = A\ e^{-B/T}$ (cm^3 molecule^{-1} s^{-1}) for the reaction of OH with 2-methylpropane (selected data used in fit)

$10^{12} \times A$	B(K)	$10^{12} \times k$	T(K)	Technique[a]	Reference	Temperature Range (K)
		2.20 ± 0.05	300	RR (n-C_4H_{10})[b]	Darnall et al. (1978)	
		2.07 ± 0.05	297	RR (n-C_4H_{10})[b]	Atkinson et al. (1984b)	
$k(T)$[c]		2.19 ± 0.11	293	PLP–RF	Tully et al. (1986b)	293–864
		2.59 ± 0.15	342			
		3.21 ± 0.16	403			
		3.49 ± 0.17	424			
		4.03 ± 0.20	470			
		4.58 ± 0.23	510			
		5.49 ± 0.27	574			
		7.40 ± 0.37	705			
		10.13 ± 0.51	864			

(continued)

Table II-B-4. (*continued*)

$10^{12} \times A$	B(K)	$10^{12} \times k$	T(K)	Technique[a]	Reference	Temperature Range (K)
		2.11 ± 0.09	298	PLP–IR	Schiffmann et al. (1991)	
5.72	293 ± 40	1.55 ± 0.18	213	PLP–LIF	Talukdar et al. (1994)	213–372
		1.58 ± 0.09	224			
		1.69 ± 0.06	234			
		1.67 ± 0.06	243			
		1.75 ± 0.06	253			
		1.82 ± 0.11	272			
		2.13 ± 0.10	296			
		2.19 ± 0.04	297			
		1.96 ± 0.09	298			
		2.29 ± 0.07	323			
		2.40 ± 0.10	343			
		2.54 ± 0.03	357			
		2.73 ± 0.06	372			
6.24	321	2.09 ± 0.06	300	DF–LIF	Donahue et al. (1998)	300–390
		2.38 ± 0.07	325			
		2.47 ± 0.09	340			
		2.62 ± 0.08	375			
		2.72 ± 0.14	390			
8.39	397	1.51	229	RR (n-C_4H_{10})[b]	Wilson et al. (2006)	229–403
		1.56	234			
		1.62	240			
		1.71	249			
		1.87	266			
		2.13	292			
		2.30	307			
		2.48	322			
		2.79	347			
		1.95	276			
		2.01	283			
		2.16	298			
		2.32	313			
		2.46	328			
		2.62	343			
		2.75	358			
		2.90	373			
		3.18	403			

[a] Abbreviations used for the techniques are defined in table II-A-1.
[b] Placed on an absolute basis using $k(n\text{-}C_4H_{10}) = 1.4 \times 10^{-11} \exp(-520/T)\,\text{cm}^3\,\text{molecule}^{-1}\,\text{s}^{-1}$.
[c] $k(T) = 4.31 \times 10^{-17} T^{1.8} \exp(175/T)$.

coefficients for the deuterium-substituted molecules, rate coefficients could be determined for the individual sites. The value for abstraction at the tertiary site was derived to be $k_{tert} = 1.56 \times 10^{-12}\,\text{cm}^3\,\text{molecule}^{-1}\,\text{s}^{-1}$ at 298 K. The results obtained are not entirely consistent with what would be predicted from structure–additivity relationships (SARs). The rate coefficient for $CH(CD_3)_3$ should provide an upper bound to the group contribution for a tertiary >CH— in alkanes, since the deuterated methyl groups are relatively unreactive. However, the SAR gives 1.94×10^{-12} compared to 1.70×10^{-12} cm^3 molecule^{-1} s^{-1} for the measured total rate coefficient for OH + 2-methylpropane. The usefulness of using SARs to predict the site of attack will be explored in more detail later in the section on estimating rate coefficients.

II-B-6. Reaction of OH with Pentane ($CH_3CH_2CH_2CH_2CH_3$)

The reaction of OH with pentane has been the subject of a number of studies. Four of these (Harris and Kerr, 1988; Talukdar et al., 1994; Donahue et al., 1998; DeMore and Bayes, 1999) covered a range of temperatures relevant for atmospheric studies, whereas the single point of Baldwin and Walker (1979) was obtained at 753 K. The temperature-dependent data include both absolute and relative studies. The data for this reaction appear more scattered than those for butane, and, in addition, the relative rate data of DeMore and Bayes (1999) and Harris and Kerr (1988) lie systematically below the absolute studies of Donahue et al. (1998) and Talukdar et al. (1994). The relative rate studies at room temperature by Atkinson et al. (1982c), Behnke et al. (1987), and Nolting et al. (1988) tend to be lower than the absolute determinations. The reasons for this are unknown. Talukdar et al. (1994) performed detailed purity checks on their sample of pentane and showed that the contributions due to other identified alkanes should be negligible. The mean of the most reasonable (i.e., within 10% of the mean) room temperature data is $(3.92 \pm 0.17) \times 10^{-12}$ cm^3 molecule^{-1} s^{-1} ($\pm\sigma$). When constructing the Arrhenius curve, as with butane, the relative rate data of DeMore and Bayes (1999) show strong curvature at low temperature if normalized using the modified Arrhenius expressions of Atkinson (2003), and in fact they agree with Talukdar's data of at 233 K (see figure 9 of Atkinson, 2003). However, if linear Arrhenius expressions are used, the resulting points lie below, but parallel to, the data of Talukdar et al. The difference in values calculated using the two approaches is 17% at 233 K.

We have used the Arrhenius expressions for propane and for butane, derived above, to place the relative rate data for pentane and hexane on an absolute scale. The pentane data used are shown in figure II-B-5 and in table II-B-5. A fit to the data from Talukdar et al. (1994), Donahue et al. (1998), and DeMore and Bayes (1999) returns the expression $k = 1.81 \times 10^{-11} \exp(-452/T)$ with a value at 298 K of 3.96×10^{-12} cm^3 molecule^{-1} s^{-1}, which is within 2% of the mean of the measured values at that temperature. The estimated uncertainty at 298 K is ±20%.

Structure additivity relationships (Kwok and Atkinson, 1995) predict that approximately 57% of the attack should occur at the 2-sites ($CH_3\underline{CH_2}CH_2\underline{CH_2}CH_3$), with 35% at the 3-site ($CH_3CH_2\underline{CH_2}CH_2CH_3$). The distribution of 2-pentyl and 3-pentyl nitrates corresponds to the SAR (6% and 4%) if one assumes that the yields of all secondary nitrates are the same for a given carbon number (Arey et al., 2001). A detailed product study by Atkinson et al. (1995) led to the conclusion that the attack is more like 45–50% at each site. However, in a more recent paper (Reisen et al., 2005), direct measurements of 5-hydroxypentan-2-one (a product of alkoxy radical isomerization) were more in line with the prediction from the SAR. Clearly, more work is required

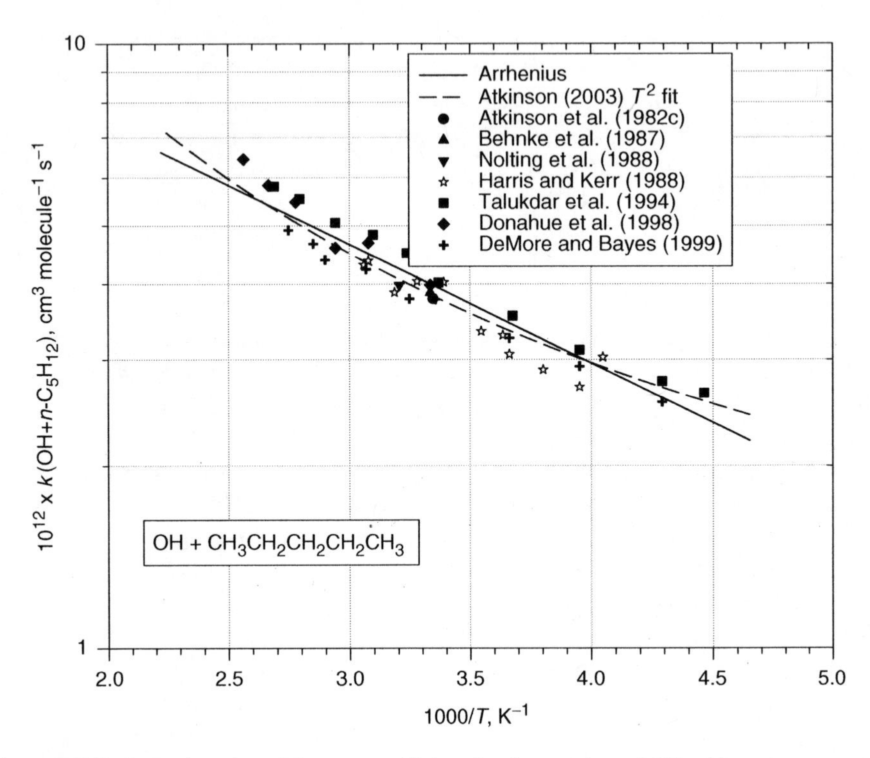

Figure II-B-5. Arrhenius plot of the rate coefficient for the reaction of OH with pentane.

Table II-B-5. Rate coefficients $k = A\ e^{-B/T}$ (cm^3 $molecule^{-1}$ s^{-1}) for the reaction of OH with pentane (selected data used in fit)

$10^{12} \times A$	B(K)	$10^{12} \times k$	T(K)	Technique[a]	Reference	Temperature Range (K)
		3.78 ± 0.05	299	RR (n-C_6H_{14})[b]	Atkinson et al. (1982c)	
		3.9	300	RR (n-C_4H_{10})[c]	Behnke et al. (1987)	
		3.97 ± 0.15	312	RR (n-C_7H_{16})[d]	Nolting et al. (1988)	
18.1	470	3.03	247	RR (n-C_4H_{10})[c]	Harris and Kerr (1988)	247–327
		2.70	253			
		2.89	263			
		3.06	273			
		3.29	275			
		3.34	282			
		4.03	295			
		4.04	305			
		3.87	314			
		4.37	325			
		4.30	327			
15.0	392 ± 40	2.64 ± 0.06	224	PLP–LIF	Talukdar et al. (1994)	224–372
		2.76 ± 0.14	233			

Table II-B-5. (*continued*)

$10^{12} \times A$	B(K)	$10^{12} \times k$	T(K)	Technique[a]	Reference	Temperature Range (K)
		3.11 ± 0.09	253			
		3.54 ± 0.11	272			
		4.02 ± 0.15	297			
		4.50 ± 0.12	309			
		4.83 ± 0.08	323			
		5.06 ± 0.12	340			
		5.54 ± 0.26	358			
		5.81 ± 0.16	372			
29.7	609	3.98 ± 0.13	300	DF–LIF	Donahue et al. (1998)	300–390
		4.68 ± 0.33	325			
		4.59 ± 0.18	340			
		5.47 ± 0.16	360			
		5.83 ± 0.44	375			
		6.44 ± 0.37	390			
15.0	415	3.78	308	RR (C_3H_8)[e]	DeMore and Bayes (1999)	233–364
		4.39	345			
		2.55	233	RR (n-C_4H_{10})[c]		
		2.92	253			
		3.26	273			
		3.77	298			
		4.23	326			
		4.67	351			
		4.91	364			

[a] Abbreviations used for the techniques are defined in table II-A-1.
[b] Placed on an absolute basis using $k(n\text{-}C_6H_{14}) = 1.98 \times 10^{-11} \exp(-394/T)\,cm^3\,molecule^{-1}\,s^{-1}$.
[c] Placed on an absolute basis using $k(n\text{-}C_4H_{10}) = 1.4 \times 10^{-11} \exp(-520/T)\,cm^3\,molecule^{-1}\,s^{-1}$.
[d] Placed on an absolute basis using $k(n\text{-}C_7H_{16}) = 2.76 \times 10^{-11} \exp(-430/T)\,cm^3\,molecule^{-1}\,s^{-1}$.
[e] Placed on an absolute basis using $k(C_3H_8) = 1.0 \times 10^{-11} \exp(-655/T)\,cm^3\,molecule^{-1}\,s^{-1}$.

to identify the relative rates of attack at each carbon atom, and to include this in predictive rules.

II-B-7. Reaction of OH with 2-Methylbutane [Isopentane, $CH_3CH(CH_3)CH_2CH_3$]

Isopentane (2-methylbutane) is often the major non-aromatic emission (in terms of ppbC) related to gasoline in urban regions. Hence, it is surprising that it has only been studied using the relative rate technique, with only one study as a function of temperature. The studies at ambient temperature of Lloyd et al. (1976a), Darnall et al. (1978), and Atkinson et al. (1984b) all used butane as the reference compound, whereas Cox et al. (1980b) used ethene. Wilson et al. (2006) studied the reaction as a function of temperature using butane and 2-methylpropane as references. Table II-B-6 lists all the measured rate coefficients, normalized to the expressions for the reference reactions derived in this work. As can be seen from figure II-B-6, the later studies are all in excellent agreement, and an Arrhenius fit gives $k = 1.01 \times 10^{-11} \exp(-296/T)\,cm^3\,molecule^{-1}\,s^{-1}$, with a value of $(3.75 \pm 0.5) \times 10^{-12}\,cm^3\,molecule^{-1}\,s^{-1}$ at 298 K.

Table II-B-6. Rate coefficients $k = A\ e^{-B/T}$ (cm^3 $molecule^{-1}$ s^{-1}) for the reaction of OH with 2-methylbutane

$10^{12} \times A$	B(K)	$10^{12} \times k$	T(K)	Technique[a]	Reference	Temperature Range (K)
		2.70 ± 0.54	305 ± 2	RR (n-C_4H_{10})[b]	Lloyd et al. (1976a)	
		3.30 ± 0.07	300 ± 1	RR (n-C_4H_{10})[b]	Darnall et al. (1978)	
		3.7	300	RR (C_2H_6)[c]	Cox et al. (1980b)	
		3.60 ± 0.10	297 ± 2	RR (n-C_4H_{10})[b]	Atkinson et al. (1984b)	
10.2	296	3.61	296	RR (n-C_4H_{10})[b]	Wilson et al. (2006)	222–407
		4.07	323			
		4.65	364			
		5.23	407			
		2.65	213			
		2.80	222			
		3.16	254			
		3.34	266			
		3.65	288			
		3.89	304			
		4.67	347			
		5.37	382			
		3.43	283	RR ($CH(CH_3)_3$)[d]		
		3.70	310			
		3.91	323			
		4.09	338			
		4.29	352			
		4.52	370			
		4.64	381			
		4.91	399			
		3.15	242			
		3.36	264			
		3.61	292			
		3.76	313			

[a] Abbreviations used for the techniques are defined in table II-A-1.
[b] Placed on an absolute basis using $k(n\text{-}C_4H_{10}) = 1.4 \times 10^{-11} \exp(-520/T)\ cm^3\ molecule^{-1}\ s^{-1}$.
[c] Placed on an absolute basis using $k(C_2H_6) = 8.61 \times 10^{-12} \exp(-1047/T)\ cm^3\ molecule^{-1}\ s^{-1}$.
[d] Placed on an absolute basis using $k(CH(CH_3)_3) = 7.0 \times 10^{-12} \exp(-350/T)\ cm^3\ molecule^{-1}\ s^{-1}$.

However, further studies, especially as a function of temperature, are desirable. The estimated uncertainty at 298 K is ±20%. It is expected from the SAR that abstraction from the secondary and tertiary sites should dominate, with the tertiary attack being favored by a factor of about 2.

II-B-8. Reaction of OH with 2,2-Dimethylpropane [Neopentane, $C(CH_3)_4$]

The reaction of OH with 2,2-dimethylpropane (neopentane) is relatively slow, since all the hydrogen atoms are primary. Of the seven studies, those of Greiner (1970) and Tully et al. (1985, recalculated in 1986a) were performed over extended temperature ranges,

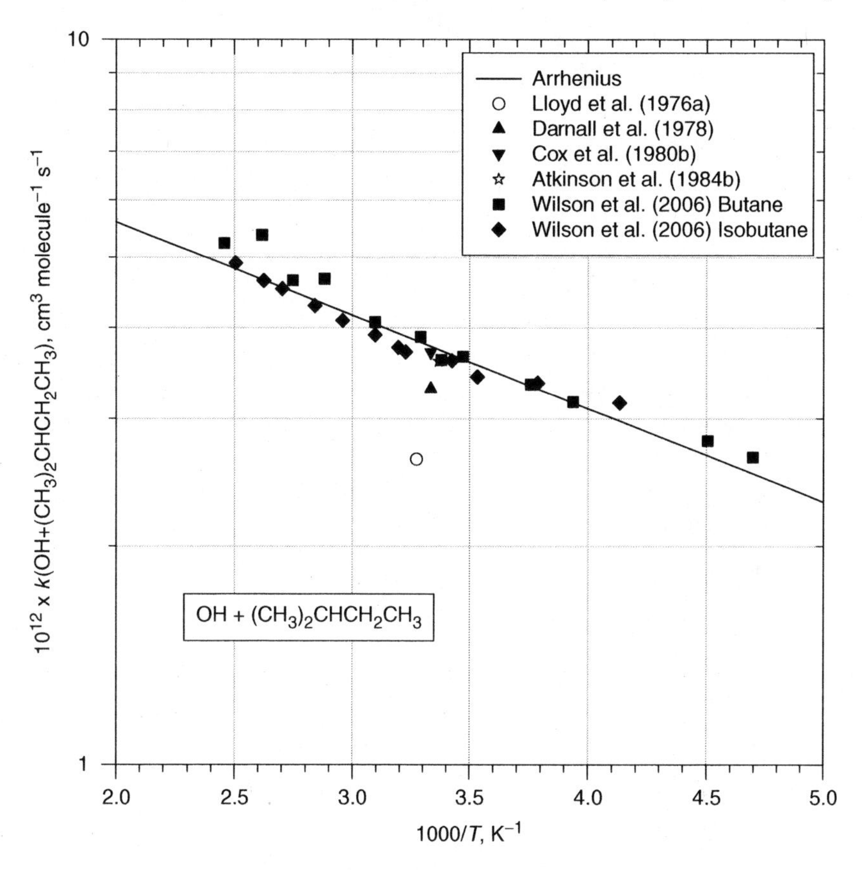

Figure II-B-6. Arrhenius plot of the rate coefficient for the reaction of OH with 2-methylbutane.

while the relative rate study of Baker et al. (1976) was made at a single temperature of 753 K. There are no measurements below 287 K. The unweighted mean of the values near room temperature is $(8.2 \pm 1.6) \times 10^{-13}$ cm^3 molecule^{-1} s^{-1} (2σ), with individual measurements varying between 7.1×10^{-13} and 9.1×10^{-13} cm^3 molecule^{-1} s^{-1}. Interestingly, the relative rate measurement of Atkinson et al. (1982e) is the lowest reported. The overall rate coefficient seems rather high, relative to the expected abstraction rate from methyl groups in other molecules (SARs give $k \approx 7 \times 10^{-13}$); this is especially true compared to 2,2,3,3-tetramethylbutane ($k = 9.7 \times 10^{-13}$). The temperature dependencies measured by Greiner et al. (1970) and Tully et al. (1985, 1986a) are in very good agreement over the region of overlap, and combining these data with the room-temperature value leads to a recommendation of $1.9 \times 10^{-17} T^2 \exp(-207/T)$. The estimated uncertainty at 298 K is ±15%. Since the curvature in the Arrhenius plot is slight, an activation temperature of ~800 K could be used with caution below room temperature.

Tully and coworkers (1985) also measured the kinetics for the fully deuterated species between 290 and 903 K. The kinetic isotope effect was identical to that found for ethane over the entire range, being 4.6 at room temperature. Thus, there is good evidence that the kinetic isotope effect for a given type of hydrogen atom does not depend strongly on the identity of the neighboring group.

II-B-9. Reaction of OH with Hexane [$CH_3(CH_2)_4CH_3$]

The reaction of OH with hexane has been studied using both absolute and relative methods. The room temperature values from 15 studies are in excellent agreement, with a mean of $(5.48 \pm 0.35) \times 10^{-12}$ cm^3 molecule^{-1} s^{-1}. The only studies of the temperature dependence of the reaction are by Donahue et al. (1998), who used discharge flow–LIF, and by DeMore and Bayes (1999), who used the relative rate technique. The values from these studies agree within a few percent at all temperatures. No studies have been conducted below 292 K. The data from DeMore and Bayes have been put on an absolute basis using the rate coefficients recommended here, and this leads to a slightly more coherent data base than that shown in figure 11 of Atkinson (2003). A fit to the data shown in figure II-B-7 and table II-B-7 gives an Arrhenius expression $k = 1.98 \times 10^{-11} \exp(-394/T)$, and a value at 298 K of 5.27×10^{-12} cm^3 molecule^{-1} s^{-1}. The estimated uncertainty at 298 K is ±20%. Greater than 90% of the abstraction will occur at the 3-positions ($CH_3CH_2\underline{CH_2}\underline{CH_2}CH_2CH_3$) or 2-positions ($CH_3\underline{CH_2}CH_2CH_2\underline{CH_2}CH_3$. Product distributions have been measured by Eberhard et al. (1995), Arey et al. (2001), and Reisen et al. (2005), and these indicate

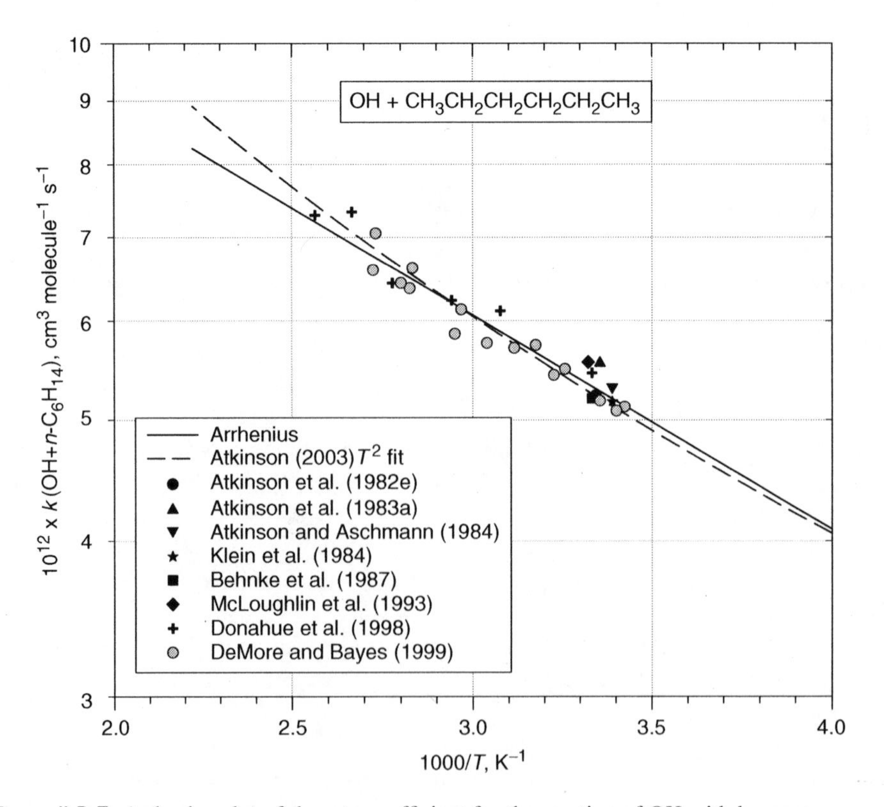

Figure II-B-7. Arrhenius plot of the rate coefficient for the reaction of OH with hexane.

Table II-B-7. Rate coefficients $k = A\ e^{-B/T}$ (cm^3 $molecule^{-1}$ s^{-1}) for the reaction of OH with hexane (selected data used in fit)

$10^{12} \times A$	B(K)	$10^{12} \times k$	T(K)	Technique[a]	Reference	Temperature Range (K)
		5.23 ± 0.09	299	RR (n-C_4H_{10})[b]	Atkinson et al. (1982e)	
		5.55 ± 0.20	298	RR (C_3H_6)[c]	Atkinson et al. (1983a)	
		5.31 ± 0.46	295	RR (C_3H_6)[c]	Atkinson and Aschmann (1984)	
		5.17 ± 0.52	295	RR (n-C_4H_{10})[b]	Klein et al. (1984)	
		5.2	300	RR (n-C_4H_{10})[b]	Behnke et al. (1987)	
		5.37 ± 0.20	301	RR (n-C_5H_{12})[d]	McLoughlin et al. (1993)	
19.5	384	5.45 ± 0.16	300	DF–LIF	Donahue et al. (1998)	300–390
		6.11 ± 0.18	325			
		6.23 ± 0.19	340			
		6.43 ± 0.45	360			
		7.33 ± 0.62	375			
		7.29 ± 0.22	390			
20.3	406	5.12	292	RR (C_3H_8)[e]	DeMore and Bayes (1999)	292–367
		5.18	298			
		5.49	307			
		5.74	315			
		5.71	321			
		6.13	337			
		5.86	339			
		6.61	353			
		6.43	357			
		7.05	366			
		5.08	294	RR (n-C_4H_{10})[b]		
		5.43	310			
		5.76	329			
		6.37	354			
		6.59	367			

[a] Abbreviations used for the techniques are defined in table II-A-1.
[b] Placed on an absolute basis using $k(n\text{-}C_4H_{10}) = 1.4 \times 10^{-11} \exp(-520/T)\,cm^3\ molecule^{-1}\ s^{-1}$.
[c] Placed on an absolute basis using $k(C_3H_6) = 2.6 \times 10^{-11}\ cm^3\ molecule^{-1}\ s^{-1}$.
[d] Placed on an absolute basis using $k(n\text{-}C_5H_{12}) = 1.81 \times 10^{-11} \exp(-452/T)\,cm^3\ molecule^{-1}\ s^{-1}$.
[e] Placed on an absolute basis using $k(C_3H_8) = 1.0 \times 10^{-11} \exp(-655/T)\,cm^3\ molecule^{-1}\ s^{-1}$.

that attack at the 3-position is slightly favored over the 2-position, in agreement with the SAR.

II-B-10. Reaction of OH with 2-Methylpentane [Isohexane, $CH_3CH(CH_3)CH_2CH_2CH_3$]

The rate coefficient of this reaction has been measured at room temperature by Lloyd et al. (1976a), Cox et al. (1980b), and Atkinson et al. (1984b), and as a function of temperature by Wilson et al. (2006). All the studies used the relative

Table II-B-8. Rate coefficients $k = Ae^{-B/T}$ (cm^3 $molecule^{-1}$ s^{-1}) for the reaction of OH with 2-methylpentane

$10^{12} \times A$	B(K)	$10^{12} \times k$	T(K)	Technique[a]	Reference	Temperature Range (K)
		4.34 ± 0.87	305	RR (n-C_4H_{10})[b]	Lloyd et al. (1976a)	
		5.3	300	RR (C_2H_4)[c]	Cox et al. (1980b)	
		5.15 ± 0.22	297 ± 2	RR (n-C_4H_{10})[b]	Atkinson et al. (1984b)	
16.2	320	4.95	283	RR (c-C_6H_{12})[d]	Wilson et al. (2006)	283–387
		5.28	292			
		5.45	312			
		5.94	332			
		6.37	355			
		7.02	387	RR (n-C_7H_{16})[e]		
		5.34	285			285–398
		5.77	293			
		5.72	302			
		6.06	303			
		6.09	317			
		5.91	327			
		6.23	333			
		6.47	347			
		6.81	359			
		6.62	372			
		6.96	377			
		7.57	398			

[a] Abbreviations used for the techniques are defined in table II-A-1.
[b] Placed on an absolute basis using $k(n\text{-}C_4H_{10}) = 1.4 \times 10^{-11} \exp(-520/T)$ cm^3 $molecule^{-1}$ s^{-1}.
[c] Placed on an absolute basis using $k(C_2H_4) = 8.52 \times 10^{-12}$ cm^3 $molecule^{-1}$ s^{-1} at 1 bar and 298 K.
[d] Placed on an absolute basis using $k(\text{cyclohexane}) = 3.0 \times 10^{-11} \exp(-440/T)$ cm^3 $molecule^{-1}$ s^{-1}.
[e] Placed on an absolute basis using $k(n\text{-}C_7H_{16}) = 2.76 \times 10^{-11} \exp(-430/T)$ cm^3 $molecule^{-1}$ s^{-1}.

rate technique, and data are given in table II-B-8. As can be seen from figure II-B-8, the data measured by Wilson et al. relative to heptane are more scattered than those relative to cyclohexane. The kinetic data for cyclohexane are deemed to be more reliable, and so the measurements relative to cyclohexane are preferred. They are used along with those of Cox et al. and Atkinson et al. to derive the Arrhenius expression $k = 1.77 \times 10^{-11} \exp(-362/T)$ cm^3 $molecule^{-1}$ s^{-1}. The value at 298 K is 5.26×10^{-12} cm^3 $molecule^{-1}$ s^{-1}. The estimated uncertainty at 298 K is ±25%. Further work is required to define the temperature dependence of this reaction.

II-B-11. Reaction of OH with 3-Methylpentane [$CH_3CH_2CH(CH_3)CH_2CH_3$]

This reaction has been studied only three times, all using the relative rate method. The two room-temperature studies by Lloyd et al. (1976a) and Atkinson et al. (1984b) are in fair agreement, differing by about 10%. The data are given in table II-B-9. As shown in figure II-B-9, the data of Wilson et al. (2006) made relative to cyclohexane

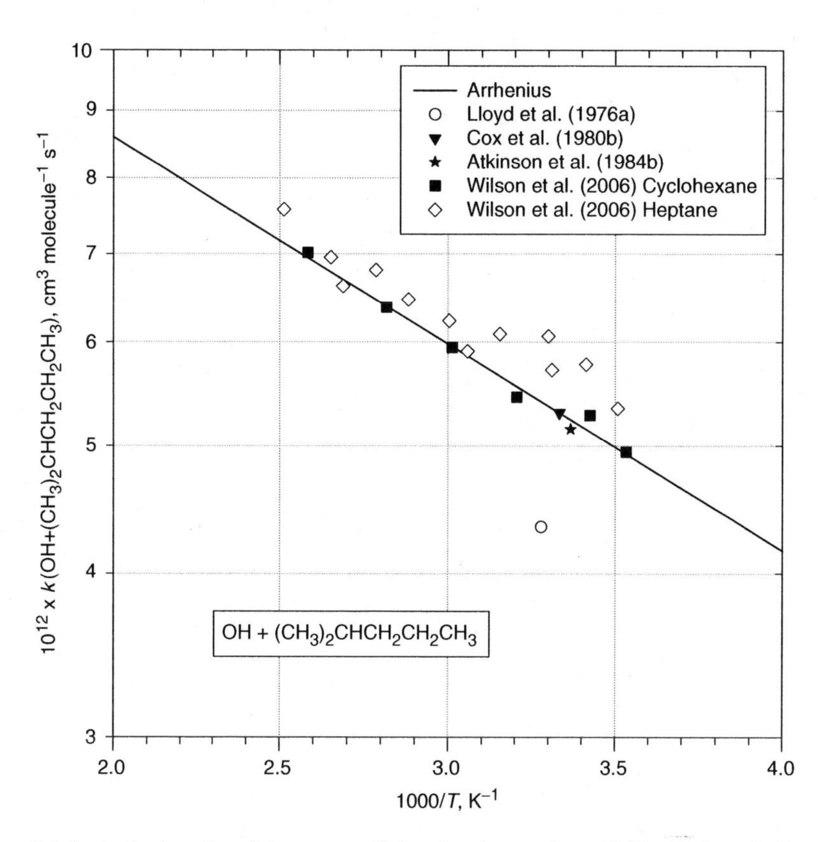

Figure II-B-8. Arrhenius plot of the rate coefficient for the reaction of OH with 2-methylpentane.

Table II-B-9. Rate coefficients $k = Ae^{-B/T}$ ($cm^3 molecule^{-1} s^{-1}$) for the reaction of OH with 3-methylpentane

$10^{12} \times A$	B(K)	$10^{12} \times k$	T(K)	Technique[a]	Reference	Temperature Range (K)
		5.9 ± 1.2	305 ± 2	RR (n-C_4H_{10})[b]	Lloyd et al. (1976a)	
		5.24 ± 0.10	297 ± 2	RR (n-C_4H_{10})[b]	Atkinson et al. (1984b)	
17.2	336	5.22	284	RR (c-C_6H_{12})[c]	Wilson et al. (2006)	284–400
		5.79	303			
		6.12	323			
		6.36	342			
		6.93	370			
		7.47	400			
17.7	309	6.07	284	RR (n-C_7H_{16})[d]		284–381
		6.21	292			
		6.39	312			
		6.87	331			
		7.34	353			
		8.04	381			

[a] Abbreviations used for the techniques are defined in table II-A-1.
[b] Placed on an absolute basis using $k(n\text{-}C_4H_{10}) = 1.4 \times 10^{-11} \exp(-520/T)$ cm^3 molecule^{-1} s^{-1}.
[c] Placed on an absolute basis using k(cyclohexane) $= 3.0 \times 10^{-11} \exp(-440/T)$ cm^3 molecule^{-1} s^{-1}.
[d] Placed on an absolute basis using $k(n\text{-}C_7H_{16}) = 2.76 \times 10^{-11} \exp(-430/T)$ cm^3 molecule^{-1} s^{-1}.

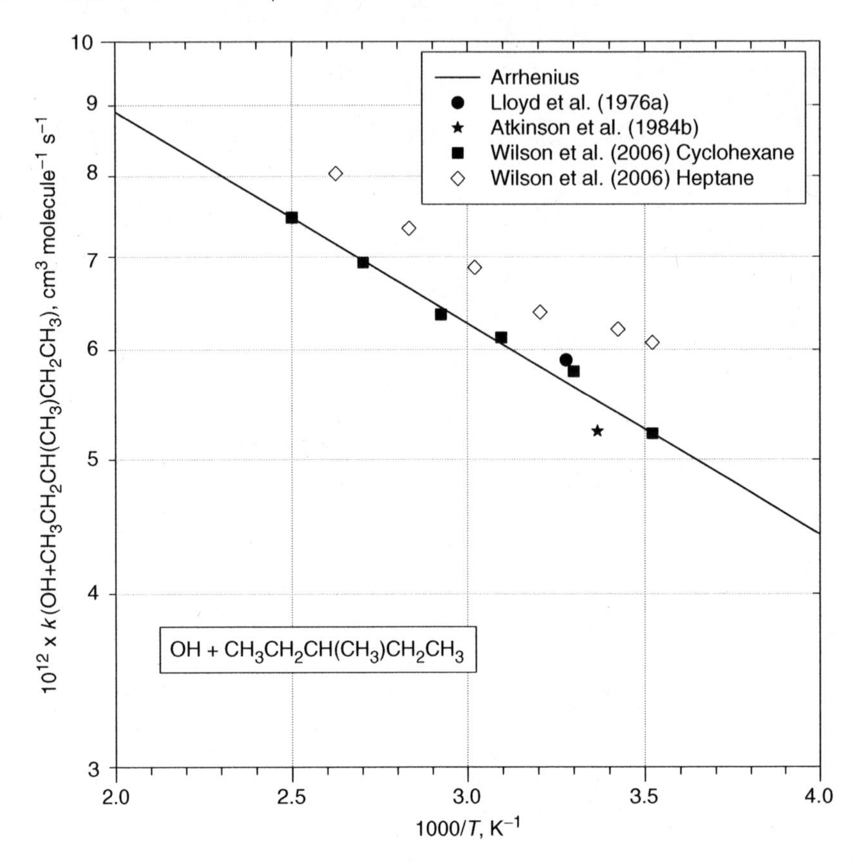

Figure II-B-9. Arrhenius plot of the rate coefficient for the reaction of OH with 3-methylpentane.

agree with the earlier studies, while those made relative to heptane lie 15–20% higher. The data relative to cyclohexane are deemed to be more reliable, since the rate coefficients for that reaction are better known. An Arrhenius fit to the data of Lloyd et al. Atkinson et al. and Wilson et al. relative to cyclohexane yields $k = 1.80 \times 10^{-11} \exp(-351/T)\,\mathrm{cm^3\,molecule^{-1}\,s^{-1}}$, with a value of $5.53 \times 10^{-12}\,\mathrm{cm^3\,molecule^{-1}\,s^{-1}}$ at 298 K. The estimated uncertainty at 298 K is ±25%. Further work is required to define the temperature dependence of this reaction.

II-B-12. Reaction of OH with 2,2-Dimethylbutane [$CH_3C(CH_3)_2CH_2CH_3$]

The reaction of OH with 2,2-dimethylbutane has been studied, using the relative rate technique, by Atkinson et al. (1984b) relative to butane at 297 K, and by Harris and Kerr (1988) relative to pentane between 245 and 328 K. The rate coefficients at 298 K agree within their stated uncertainties. Over this narrow temperature range, the rate coefficients show Arrhenius behavior. We recommend Atkinson's (2003) expression for the rate coefficient $k = 3.37 \times 10^{-11} \exp(-809/T)\,\mathrm{cm^3\,molecule^{-1}\,s^{-1}}$, with a value at 298 K of 2.23×10^{-12} cm^3 molecule^{-1} s^{-1}. The estimated uncertainty at 298 K is ±25%. The rate coefficient for this reaction predicted by the SAR is 1.8×10^{-12} cm^3 molecule^{-1} s^{-1}, which is more than 20% lower than the measured value.

II-B-13. Reaction of OH with 2,3-Dimethylbutane [$CH_3CH(CH_3)CH(CH_3)_2$]

The reaction of OH radicals with 2,3-dimethylbutane has been studied seven times, including a high-temperature study by Bott and Cohen (1991). A selection of the data is shown in table II-B-10 and figure II-B-10. Greiner (1970) studied the reaction by flash photolysis at five temperatures, but found no systematic dependence on temperature.

Table II-B-10. Rate coefficients $k = A\ e^{-B/T}$ (cm^3 molecule^{-1} s^{-1}) for the reaction of OH with 2,3-dimethylbutane (selected data used in fit)

$10^{12} \times A$	B(K)	$10^{12} \times k$	T(K)	Technique[a]	Reference	Temperature Range (K)
		5.79 ± 0.03	299 ± 2	RR (c-C_6H_{12})[b]	Atkinson et al. (1982e)	
		5.87	247	RR (n-C_4H_{10})[c]	Harris and Kerr (1988)	247–327
		4.88	253			
		5.35	263			
		5.20	273			
		5.25	275			
		5.35	282			
		5.59	295			
		5.69	305			
		4.77	314			
		5.82	325			
		5.60	327			
		5.55	247	RR (n-C_5H_{12})[d]		
		5.19	253			
		5.65	263			
		5.46	273			
		5.63	275			
		5.68	282			
		5.44	295			
		5.96	305			
		5.45	314			
		5.99	325			
		5.91	327			
9.1	112	220	5.65	RR (n-C_6H_{14})[e]	Wilson et al. (2006)	220–407
		227	5.55			
		231	5.56			
		239	5.72			
		253	5.74			
		261	5.86			
		285	5.86			
		293	6.19			

(continued)

Table II-B-10. (*continued*)

$10^{12} \times A$	B(K)	$10^{12} \times k$	T(K)	Technique[a]	Reference	Temperature Range (K)
		299	5.99			
		311	6.49			
		324	6.31			
		348	6.57			
		372	6.80			
		407	7.09			
12.6	202	245	5.58	RR (n-C_7H_{16})[f]		
		251	5.58			
		266	5.84			
		274	5.92			
		288	6.24			
		300	6.66			
		322	6.77			
		333	6.74			
		346	6.91			
		371	7.41			
		398	7.52			
15.4	284	302	6.06	RR (c-C_6H_{12})[b]		
		328	6.28			
		355	6.83			
		378	7.34			
		318	6.30			
		342	6.78			
		403	7.57			

[a] Abbreviations used for the techniques are defined in table II-A-1.
[b] Placed on an absolute basis using k(cyclohexane) = $3.0 \times 10^{-11} \exp(-440/T)$ cm^3 molecule^{-1} s^{-1}.
[c] Placed on an absolute basis using $k(n\text{-}C_4H_{10}) = 1.4 \times 10^{-11} \exp(-520/T)$ cm^3 molecule^{-1} s^{-1}.
[d] Placed on an absolute basis using $k(n\text{-}C_5H_{12}) = 1.81 \times 10^{-11} \exp(-452/T)$ cm^3 molecule^{-1} s^{-1}.
[e] Placed on an absolute basis using $k(n\text{-}C_6H_{14}) = 1.98 \times 10^{-11} \exp(-394/T)$ cm^3 molecule^{-1} s^{-1}.
[f] Placed on an absolute basis using $k(n\text{-}C_7H_{16}) = 2.76 \times 10^{-11} \exp(-430/T)$ cm^3 molecule^{-1} s^{-1}.

Harris and Kerr (1988) used the relative rate technique between 247 and 327 K, and also found a fairly weak dependence on temperature. Measurements at room temperature scatter between 4.0×10^{-12} cm^3 molecule^{-1} s^{-1} (Cox et al., 1980b) and 7.5×10^{-12} cm^3 molecule^{-1} s^{-1} (Greiner, 1970). Wilson et al. (2006) measured the rate coefficient as a function of temperature relative to those of hexane, cyclohexane, and heptane. Although their measurements extended down to 220 K, measurements of the rate coefficients for the reference compounds hexane and heptane are available only to 292 K, and so the full data set has not been fitted in figure II-B-10. We recommend an Arrhenius expression $k = 1.25 \times 10^{-11} \exp(-212/T)$ cm^3 molecule^{-1} s^{-1}, corresponding to a value at 298 K of 6.14×10^{-12} cm^3 molecule^{-1} s^{-1}. The estimated uncertainty at 298 K is ±25%. The available data base is clearly scattered, and more studies are desirable. The rate coefficient seems unusually large compared to those of structurally related molecules such as 2,3-dimethylpentane and 2,4-dimethylpentane.

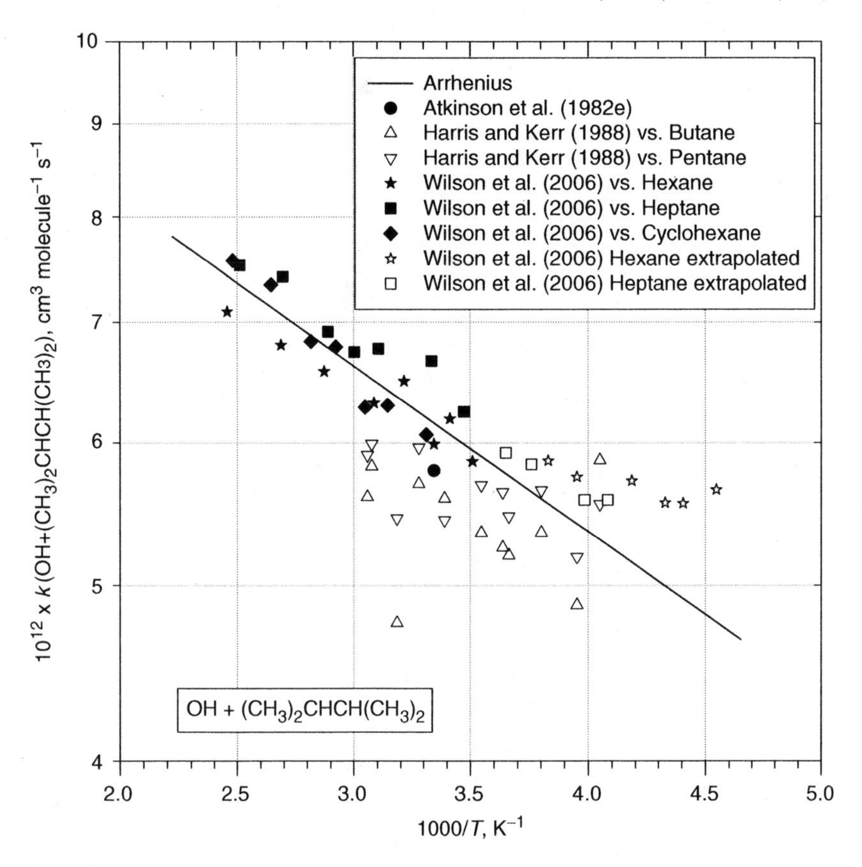

Figure II-B-10. Arrhenius plot of the rate coefficient for the reaction of OH with 2,3-dimethylbutane.

II-B-14. Reaction of OH with Heptane [$CH_3(CH_2)_5CH_3$]

The data base for this reaction is small. Several measurements have been made at atmospheric temperatures using the relative rate technique (Atkinson et al., 1982c; Klöpffer et al., 1986; Behnke et al., 1987, 1988; Ferrari et al., 1996; Wilson et al., 2006); the single absolute measurement is that of Koffend and Cohen (1996) at 1100 K. The measurement of Klöpffer et al. is considerably higher than the others, and is not considered further here. The remaining data at atmospherically relevant temperatures are listed in table II-B-11. Wilson et al. (2006) made measurements as a function of temperature relative to hexane, cyclopentane, and cyclohexane. As in the case of 2,3-dimethylbutane, the measurements were made beyond the range where the reference reactions have been measured, although these data points fit the extrapolated Arrhenius expression quite well. The recommended Arrhenius fit to the data shown in figure II-B-11 is $k = 2.76 \times 10^{-11} \exp(-430/T)$ cm^3 molecule^{-1} s^{-1} with a value at 298 K of 6.50×10^{-12} cm^3 molecule^{-1} s^{-1}. The estimated uncertainty at 298 K is ±20%.

Table II-B-11. Rate coefficients $k = A\ e^{-B/T}$ (cm^3 $molecule^{-1}$ s^{-1}) for the reaction of OH with heptane (selected data used in fit)

$10^{12} \times A$	B(K)	$10^{12} \times k$	T(K)	Technique[a]	Reference	Temperature Range (K)
		6.68 ± 0.11	299 ± 2	RR (n-C_6H_{14})[b]	Atkinson et al. (1982c)	
		6.60	300 ± 3	RR (n-C_4H_{10})[c]	Behnke et al. (1987)	
		6.78 ± 0.08	300	RR (n-C_8H_{18})[d]	Behnke et al. (1988)	
		6.97 ± 0.29	295 ± 2	RR (n-C_8H_{18})[d]	Ferrari et al. (1996)	
		4.74	241	RR (n-C_6H_{14})[b]	Wilson et al. (2006)	241–406
		5.02	250			
		5.36	261			
		5.75	271			
		6.80	299			
		7.19	314			
		6.90	324			
		7.58	326			
		8.65	362			
		8.06	373			
		9.45	389			
		6.21	279	RR (c-C_5H_{10})[e]		
		6.23	288			
		6.18	302			
		6.77	310			
		6.97	317			
		8.14	342			
		8.10	359			
		9.86	406			
		5.83	284	RR (c-C_6H_{12})[f]		
		6.13	293			
		6.33	296			
		6.42	305			
		6.93	322			
		7.84	331			
		8.08	338			
		8.17	346			
		7.98	355			
		8.36	368			
		9.47	379			
		8.80	384			
		9.10	397			

[a] Abbreviations used for the techniques are defined in table II-A-1.
[b] Placed on an absolute basis using $k(n\text{-}C_6H_{14}) = 1.98 \times 10^{-11}\ \exp(-394/T)\ cm^3\ molecule^{-1}\ s^{-1}$.
[c] Placed on an absolute basis using $k(n\text{-}C_4H_{10}) = 1.4 \times 10^{-11}\ \exp(-520/T)\ cm^3\ molecule^{-1}\ s^{-1}$.
[d] Placed on an absolute basis using $k(n\text{-}C_8H_{18}) = 3.43 \times 10^{-11}\ \exp(-441/T)\ cm^3\ molecule^{-1}\ s^{-1}$.
[e] Placed on an absolute basis using $k(\text{cyclopentane}) = 2.2 \times 10^{-11}\ \exp(-450/T)\ cm^3\ molecule^{-1}\ s^{-1}$.
[f] Placed on an absolute basis using $k(\text{cyclohexane}) = 3.0 \times 10^{-11}\ \exp(-440/T)\ cm^3\ molecule^{-1}\ s^{-1}$.

II-B-15. Reaction of OH with 2,3-Dimethylpentane [$CH_3CH(CH_3)CH(CH_3)CH_2CH_3$]

The single study of this reaction was made by Wilson et al. (2006), who measured the rate coefficient as a function of temperature relative to the reactions of OH with hexane and cyclohexane (see table II-B-12). Data using the two references are in good

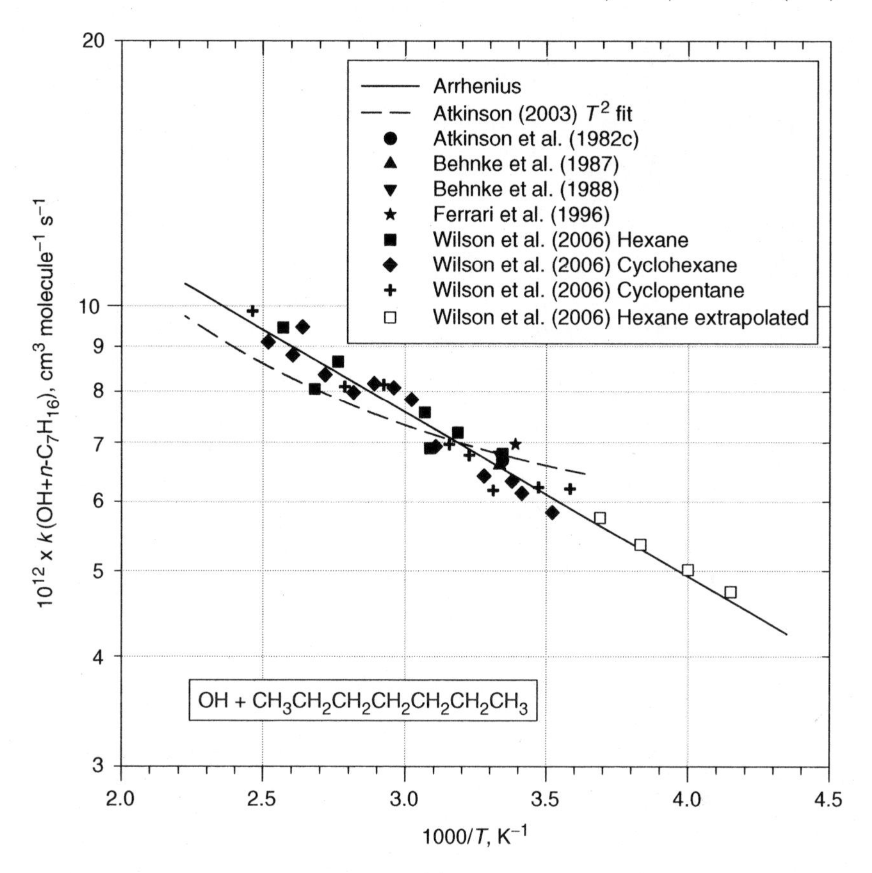

Figure II-B-11. Arrhenius plot of the rate coefficient for the reaction of OH with heptane.

agreement (figure II-B-12). After normalization to the expressions derived in the present work, an Arrhenius expression $k = 1.53 \times 10^{-11} \exp(-252/T)\,\text{cm}^3\ \text{molecule}^{-1}\ \text{s}^{-1}$ is derived, leading to a value at 298 K of $6.58 \times 10^{-12}\ \text{cm}^3\ \text{molecule}^{-1}\ \text{s}^{-1}$. The estimated uncertainty at 298 K is $\pm 40\%$.

II-B-16. Reaction of OH with 2,4-Dimethylpentane [$CH_3CH(CH_3)CH_2CH(CH_3)_2$]

This reaction has been studied by Atkinson et al. (1984b) at 298 K and by Wilson et al. (2006) as a function of temperature. The data are given in table II-B-13. The measurements of Wilson et al. at 297 $\pm$ 3 K are some 25% higher than the measurement of Atkinson et al. (figure II-B-13), but are in reasonable agreement with predictions using the SAR of Kwok and Atkinson (1995). After renormalization of the data of Wilson et al. an unweighted fit to all the data gives $k = 2.7 \times 10^{-11} \exp(-460/T)\,\text{cm}^3\ \text{molecule}^{-1}\ \text{s}^{-1}$. The rate coefficient at 298 K is $5.7 \times 10^{-12}\ \text{cm}^3\ \text{molecule}^{-1}\ \text{s}^{-1}$. However, the data of Wilson et al. were measured

Table II-B-12. Rate coefficients $k = A\ e^{-B/T}$ (cm^3 $molecule^{-1}$ s^{-1}) for the reaction of OH with 2,3-dimethylpentane

$10^{12} \times A$	B(K)	$10^{12} \times k$	T(K)	Technique[a]	Reference	Temperature Range (K)
14.0	226	6.42	287	RR (n-C_6H_{14})[b]	Wilson et al. (2006)	287–403
		6.68	299			
		6.34	308			
		6.82	311			
		7.00	323			
		6.76	333			
		7.81	344			
		7.60	365			
		7.67	376			
		7.82	403			
17.5	287	6.49	292	RR (c-C_6H_{12})[c]		292–395
		7.05	314			
		7.40	332			
		7.55	342			
		8.01	369			
		8.43	395			

[a] Abbreviations used for the techniques are defined in table II-A-1.
[b] Placed on an absolute basis using $k(n\text{-}C_6H_{14}) = 1.98 \times 10^{-11} \exp(-394/T)\, cm^3\, molecule^{-1}\, s^{-1}$.
[c] Placed on an absolute basis using $k(\text{cyclohexane}) = 3.0 \times 10^{-11} \exp(-440/T)\, cm^3\, molecule^{-1}\, s^{-1}$.

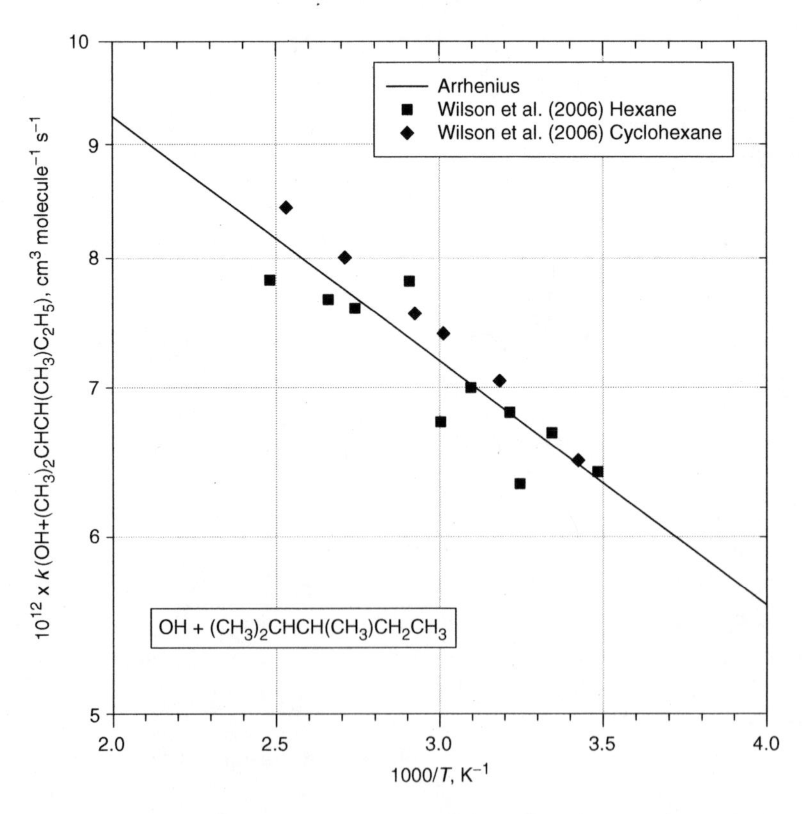

Figure II-B-12. Arrhenius plot of the rate coefficient for the reaction of OH with 2,3-dimethylpentane.

Table II-B-13. Rate coefficients $k = A\ e^{-B/T}$ ($cm^3\ molecule^{-1}\ s^{-1}$) for the reaction of OH with 2,4-dimethylpentane

$10^{12} \times k$	T(K)	Technique[a]	Reference	Temperature Range (K)
4.77 ± 0.10	297 ± 2	RR (n-C_4H_{10})[b]	Atkinson et al. (1984b)	
5.27	283	RR (n-C_7H_{16})[c]	Wilson et al. (2006)	283–388
6.30	294			
6.62	312			
7.98	331			
8.09	354			
9.22	388			
4.78	272	RR (2,3-dimethylbutane)[d]		272–410
6.21	299			
7.22	320			
6.56	334			
6.61	348			
7.11	365			
7.67	386			
8.44	410			

[a] Abbreviations used for the techniques are defined in table II-A-1.
[b] Placed on an absolute basis using $k(n\text{-}C_4H_{10}) = 1.4 \times 10^{-11} \exp(-520/T)\ cm^3\ molecule^{-1}\ s^{-1}$.
[c] Placed on an absolute basis using $k(n\text{-}C_7H_{16}) = 2.76 \times 10^{-11} \exp(-430/T)\ cm^3\ molecule^{-1}\ s^{-1}$.
[d] Placed on an absolute basis using k(2,3-dimethylbutane) $= 1.25 \times 10^{-11} \exp(-212/T)\ cm^3\ molecule^{-1}\ s^{-1}$.

relative to reference compounds for which the rate coefficients are not well defined, and so a large uncertainty of ±40% should be assigned.

II-B-17. Reaction of OH with 2,2,3-Trimethylbutane [$CH_3C(CH_3)_2CH(CH_3)_2$]

This reaction has been studied a number of times, over a wide temperature range. Greiner (1970) made absolute measurements at five temperatures between 296 and 497 K, whereas Baldwin et al. (1981) made a single determination using the relative rate method (relative to H_2) at 753 K. Harris and Kerr (1988) made relative rate measurements between 243 and 324 K using three different reference compounds, and Atkinson et al. (1984b) used butane as reference compound at 297 K. The measurements of Harris and Kerr are somewhat scattered and show only a weak temperature dependence, essentially zero within the measurement uncertainties (typically 10%) at each temperature. Atkinson (2003) recommends a rate coefficient at 298 K of $3.81 \times 10^{-12}\ cm^3\ molecule^{-1}\ s^{-1}$, with an uncertainty of ±25%. No recommendation for the temperature dependence can be made.

II-B-18. Reaction of OH with Octane [$CH_3(CH_2)_6CH_3$]

The data base for this reaction is a little more complete than that for heptane. In addition to three relative rate studies near 298 K (Atkinson et al., 1982c; Behnke et al., 1987;

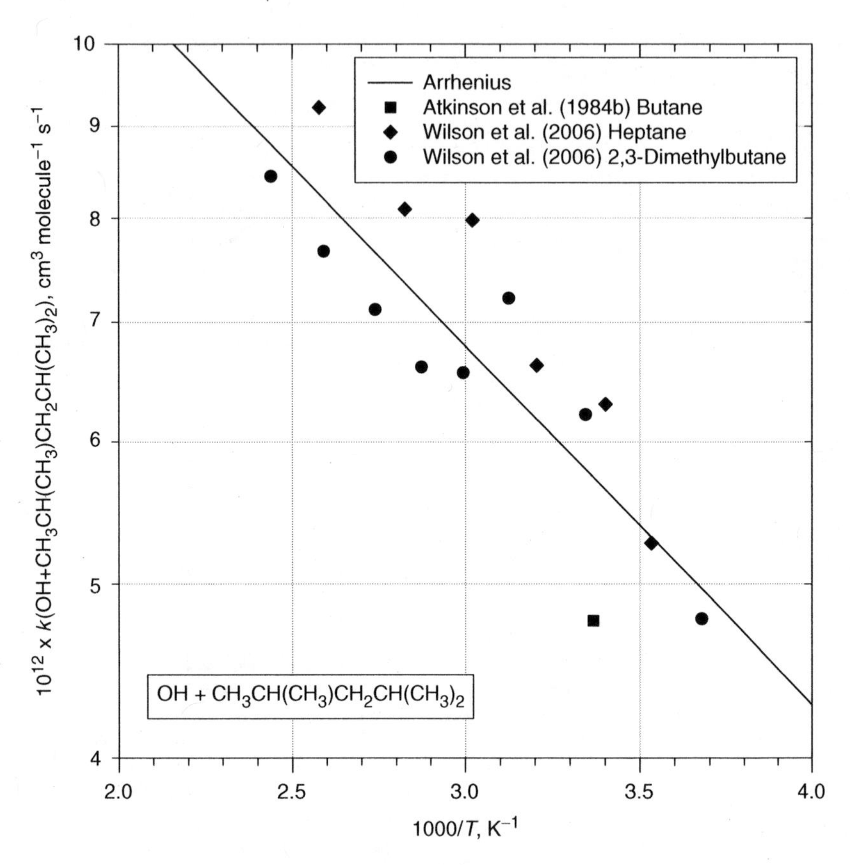

Figure II-B-13. Arrhenius plot of the rate coefficient for the reaction of OH with 2,4-dimethylpentane.

Nolting et al., 1988) which are in excellent agreement, and a relative rate study by Wilson et al. (2006) between 284 and 393 K, there is a direct study by Greiner (1970) between 296 and 497 K, and a high-temperature study by Koffend and Cohen (1996). The data are given in table II-B-14. An Arrhenius fit (figure II-B-14) to the relative rate data of Atkinson et al., Behnke et al. and Wilson et al. gives $k = 3.43 \times 10^{-11} \exp(-441/T)$ cm^3 molecule^{-1} s^{-1}, with a value at 298 K of 7.81×10^{-12} cm^3 molecule^{-1} s^{-1}. The estimated uncertainty at 298 K is ± 20%.

II-B-19. Reaction of OH with 2,2,4-Trimethylpentane [Isooctane, $(CH_3)_3CCH_2CH(CH_3)_2$]

The reaction of 2,2,4-trimethylpentane (isooctane) with OH radicals has been studied from 298 to 493 K by Greiner (1970), at 1186 K by Bott and Cohen (1991), and at 298 K by Atkinson et al. (1984b). The data at 298 K from the studies of Atkinson et al. and Greiner et al. are in overall agreement, and a rate coefficient at 298 K of 3.34×10^{-12} cm^3 molecule^{-1} s^{-1} is recommended. The estimated

Table II-B-14. Rate coefficients $k = A\,e^{-B/T}$ (cm^3 $molecule^{-1}$ s^{-1}) for the reaction of OH with octane

$10^{12} \times A$	B(K)	$10^{12} \times k$	T(K)	Technique[a]	Reference	Temperature Range (K)
29.5	364 ± 61	8.42 ± 1.25	296	FP–UV abs	Greiner (1970)	296–497
		12.0 ± 0.7	371			
		10.8 ± 0.5	371			
		14.3 ± 0.4	497			
		8.25 ± 0.11	299 ± 2	RR (n-C_6H_{14})[b]	Atkinson et al. (1982c)	
		8.03	300 ± 3	RR (n-C_4H_{10})[c]	Behnke et al. (1987)	
		8.16 ± 0.28	312	RR (n-C_7H_{16})[d]	Nolting et al. (1988)	
		7.38	284	RR (n-C_7H_{16})[d]	Wilson et al. (2006)	284–393
		7.56	295			
		7.87	304			
		8.88	323			
		9.31	344			
		10.1	362			
		10.8	384			
		7.36	296	RR (c-C_6H_{12})[e]		
		8.43	314			
		9.08	334			
		10.2	353			
		11.0	373			
		10.8	393			

[a] Abbreviations used for the techniques are defined in table II-A-1.
[b] Placed on an absolute basis using $k(n\text{-}C_6H_{14}) = 1.98 \times 10^{-11} \exp(-394/T)$ cm^3 $molecule^{-1}$ s^{-1}.
[c] Placed on an absolute basis using $k(n\text{-}C_4H_{10}) = 1.4 \times 10^{-11} \exp(-520/T)$ cm^3 $molecule^{-1}$ s^{-1}.
[d] Placed on an absolute basis using $k(n\text{-}C_7H_{16}) = 2.76 \times 10^{-11} \exp(-430/T)$ cm^3 $molecule^{-1}$ s^{-1}
[e] Placed on an absolute basis using k(cyclohexane) $= 3.0 \times 10^{-11} \exp(-440/T)$ cm^3 $molecule^{-1}$ s^{-1}.

uncertainty at 298 K is ±25%. No recommendation is given for the temperature dependence. The value for the rate coefficient predicted by the structure–additivity relationship is almost 40% higher than the measured value. Aschmann et al. (2002) studied the products from the OH-initiated oxidation, and identified products resulting from abstraction of a hydrogen atom at each of the four reaction sites in the molecule. They note that measured products only account for about half of the product yields, and suggest that abstraction at the tertiary site is much lower than expected, due to steric hindrance. Acetone product was attributed to decomposition of the *tert*-butoxy radical ($(CH_3)_3CO\cdot$) formed from the $(CH_3)_3CCH(O\cdot)CH(CH_3)_2$ alkoxy radical. However, in the presence of 10 ppm NO, *tert*-butoxy will preferentially react with NO to form *tert*-butyl nitrite, so that the observed acetone should really be attributed to the dominant alkoxy radical $(CH_3)_3CCH_2C(O\cdot)(CH_3)_2$ formed at the tertiary site. This would increase the overall product yield, but the measured yields of hydroxycarbonyls still seem unrealistically low. Considering the commercial importance of isooctane in automotive fuels, the limited kinetic and mechanistic data base for its reaction with OH radicals is surprising. Further work is needed.

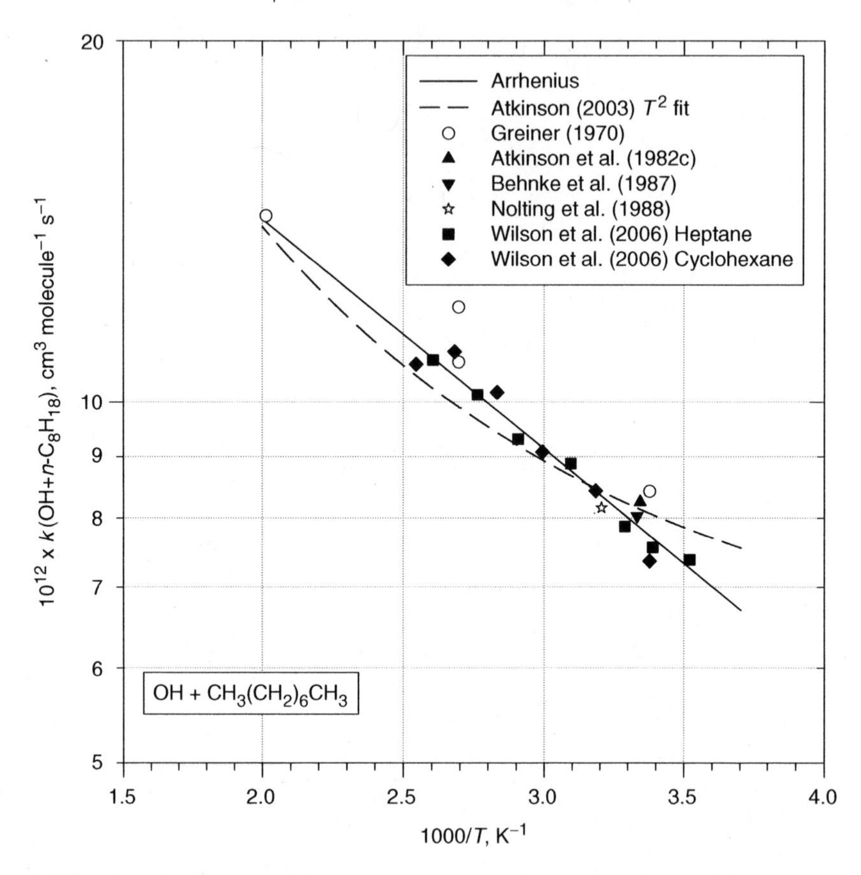

Figure II-B-14. Arrhenius plot of the rate coefficient for the reaction of OH with octane.

II-B-20. Reaction of OH with 2,3,4-Trimethylpentane [$CH_3CH(CH_3)CH(CH_3)CH(CH_3)_2$]

The reaction of OH with 2,3,4-trimethyl pentane has been studied three times, by Harris and Kerr (1988), Aschmann et al. (2004), and Wilson et al. (2006). The data of Harris and Kerr, which were obtained over a limited temperature range (243–313 K), are somewhat scattered, with no trend observed with temperature. The data of Wilson et al. cover the range 244–373, and also show a very weak dependence on temperature. A rate coefficient $k = 6.6 \times 10^{-12}$ cm^3 molecule^{-1} s^{-1} is recommended, with no recommendation for the temperature dependence. The estimated uncertainty at 298 K is ±25%. The SAR of Kwok and Atkinson (1995) predicts a rate coefficient at 298 K of 8.54×10^{-12} cm^3 molecule^{-1} s^{-1}, somewhat higher than the measured value. Aschmann et al. (2004) performed a detailed product study and found evidence for much less abstraction at the 2- and 4-sites than predicted by the SAR (~20% observed versus 56% calculated). Aschmann et al. attribute this discrepancy (and the overall lower rate coefficient) to steric hindrance at the tertiary sites. Such effects are not accounted for in the SAR.

II-B-21. Reaction of OH with 2,2,3,3-Tetramethylbutane [$(CH_3)_3CC(CH_3)_3$]

This reaction has been the subject of five studies, which are in very good agreement. Absolute studies were made by Greiner (1970) from 294 to 495 K, by Tully et al. (1985) from 290 to 738 K, and by Bott and Cohen (1991) at 1180 K. Relative rate studies were made by Baldwin and Walker (1979) at 753 K, and by Atkinson et al. (1984b) at 297 K. The values measured at room temperature agree within ±10%, and the temperature trend is also well constrained by the data at elevated temperature. Atkinson recommends a value at 298 K of 9.72×10^{-13} cm^3 molecule^{-1} s^{-1} and for the temperature dependence $k = 1.99 \times 10^{-17} T^2 \exp(-178/T)$ over the range 290–1180 K. The estimated uncertainty at 298 K is ±20%. For temperatures below 298 K an activation temperature of ~750 ± 25 K would be appropriate (based on the 3-parameter fit, evaluated at 273 K, and the presence of only primary hydrogen atoms in the molecule).

As a consistency check, we note that the rate coefficient for 2,2,3,3-tetramethylbutane, which contains six methyl groups at which reaction may occur, should be 1.5 times that of 2,2-dimethylpropane, which contains four methyl groups. The measured room temperature values are 9.7×10^{-13} and 8.2×10^{-13} cm^3 molecule^{-1} s^{-1} respectively, which differ by a factor of only 1.18. The values predicted by the SAR (e.g., Kwok and Atkinson, 1995) are 1.0×10^{-12} and 6.7×10^{-13} cm^3 molecule^{-1} s^{-1}, which agrees slightly better for the larger molecule. It should be noted that the value at 297 K measured by Atkinson et al. (1982e) for 2,2-dimethyl propane, 7.05×10^{-13} cm^3 molecule^{-1} s^{-1}, is much closer to the SAR, and gives a ratio of rate coefficients closer to that expected.

II-B-22. Reaction of OH with Nonane [$CH_3(CH_2)_7CH_3$]

The reaction of OH with nonane has been the subject of a number of studies at room temperature, all using the relative rate technique. The rate coefficient reported by Coeur et al. (1998, 1999) is somewhat smaller than the other values and has a large uncertainty. The mean of the other values (Atkinson et al., 1982c; Behnke et al., 1987, 1988; Nolting et al., 1988; Ferrari et al., 1996; Kramp and Paulson, 1998) is $(9.75 \pm 0.30) \times 10^{-12}$ cm^3 molecule^{-1} s^{-1}. The estimated uncertainty at 298 K is ±20%. Other than a high temperature measurement by Koffend and Cohen (1996), no systematic data exist as a function of temperature.

II-B-23. Reaction of OH with Decane [$CH_3(CH_2)_8CH_3$]

Four measurements of this rate coefficient have been made using the relative rate technique at or close to room temperature (Atkinson et al., 1982c; Nolting et al., 1988; Behnke et al., 1988; Aschmann et al., 2001); these are all in fairly good agreement. Li (2004) also made a measurement relative to that of octane, using a fast flow–relative rate technique; this is in broad agreement with Atkinson's relative rate measurements. Atkinson (2003) recommends a rate coefficient at 298 K of 1.1×10^{-11} cm^3 molecule^{-1} s^{-1} with an estimated overall uncertainty of ±20%.

No recommendation for the temperature dependence is made. Aschmann et al. (2001) performed a product analysis, and found that the extent of nitrate formation from the 3-, 4-, and 5-sites was very similar, suggesting that the OH attack is roughly the same at each, in accord with the SAR.

II-B-24. Reaction of OH with Higher *n*-Alkanes

The only rate coefficients for alkanes beyond decane are those measured by Zetzsch's group (Nolting et al., 1988; Behnke et al., 1988) using the relative rate technique. These data are not entirely consistent. As was shown earlier (figure II-A-1), a plot of k versus the carbon number (n) shows a steady increase of k with n, and an increment of about 1.4×10^{-12} cm^3 molecule^{-1} s^{-1} per additional carbon atom beyond propane. The data of Nolting et al. initially show this same trend, but then tend upward with larger rate coefficients. The largest alkane measured by Behnke et al. was tridecane, and this lies slightly below the trend line but within experimental uncertainty. There does not seem to be a good reason why the reactivity of these higher hydrocarbons should increase above C_{12}, and it is suggested that there is some unidentified error above tridecane in the data of Nolting et al. In this case, the preferred rate coefficient is obtained either by the SAR or by the linear extrapolation of the data for $n < 13$, as described above.

II-C. Reactions of OH with Cycloalkanes

The cycloalkanes are a further class of fully saturated compounds containing only C—C and C—H bonds, but including rings of carbon groups. The general formula for the simple cycloalkanes is C_nH_{2n}, as for the alkenes. Examples are cyclopropane (*cyclo*-C_3H_6), cyclobutane (*cyclo*-C_4H_8), cyclopentane (*cyclo*-C_5H_{10}), etc. These simple cycloalkanes consist solely of methylene groups —CH_2— joined in a ring. Structures of some cycloalkanes are depicted in Appendix A. The smaller members are known to be strained, with bond angles less than 109°; this affects the thermochemistry of the compounds. Normally this is expressed as a strain energy, which is the difference between the heat of formation of the cyclic compounds and the heat of formation estimated for the corresponding unstrained compound on the basis of group additivity. The strain energies associated with the C_3, C_4, C_5, C_6, and C_7 rings are 115, 110, 26, 0, and 27 kJ mol^{-1}, respectively. The presence of strain in the molecules affects the C—H bond strengths, since the radicals formed will also be more or less strained than those derived from open chain compounds. For example, the C—H bond strength in cyclopropane is 422 kJ mol^{-1}, compared to 401 kcal mol^{-1} for secondary hydrogen atoms in unstrained alkanes. These factors together influence the kinetics of the cycloalkanes with OH.

II-C-1. Reaction of OH with Cyclopropane (*cyclo*-C_3H_6)

To date the reaction of OH with cyclopropane has been studied five times from around 200 K to 500 K by Jolly et al. (1985), Dóbé et al. (1991, 1992), Clarke et al. (1998), DeMore and Bayes (1999), and Wilson et al. (2001). The data (presented in table II-C-1) are quite scattered, especially at lower temperatures. Room-temperature values range from 6.2×10^{-14} to 1.1×10^{-13} cm^3 molecule^{-1} s^{-1}. The majority of the studies seem to suggest that the data of Dóbé et al. are too high at 298 K, but that they agree more closely at higher temperatures, possibly as a result of impurities in the sample. The activation energies derived from the individual studies are also

Table II-C-1. Rate coefficients $k = A\ e^{-B/T}$ (cm^3 molecule^{-1} s^{-1}) for the reaction of OH with cyclopropane

$10^{12} \times A$	B(K)	$10^{12} \times k$	T(K)	Technique[a]	Reference	Temperature Range (K)
		0.062 ± 0.014	298	FP–RA	Jolly et al. (1985)	
4.4	1143	0.111 ± 0.024	298	PLP–RF	Dóbé et al. (1992)	298–492
		0.133 ± 0.008	325			
		0.115 ± 0.022	337			
		0.191 ± 0.042	362			
		0.236 ± 0.016	388			
		0.365 ± 0.058	467			
		0.412 ± 0.034	476			
		0.462 ± 0.062	492			
0.81	723	0.021 ± 0.004	200	DF–LIF	Clarke et al. (1998)	200–360
		0.030 ± 0.004	213			
		0.035 ± 0.001	225			
		0.038 ± 0.001	238			
		0.043 ± 0.001	250			
		0.052 ± 0.002	265			
		0.059 ± 0.001	280			
		0.066 ± 0.001	295			
		0.075 ± 0.002	310			
		0.090 ± 0.003	325			
		0.101 ± 0.001	340			
		0.116 ± 0.008	360			
4.4	1208	0.056	276	RR (C_2H_6)[b]	DeMore and Bayes (1999)	276–421
		0.077	298			
		0.079	300			
		0.097	316			
		0.138	348			
		0.162	363			
		0.186	383			
		0.222	403			
		0.253	421			
5.9	1290	0.078	298	RR (C_2H_6)[b]	Wilson et al. (2001)	298–459
		0.109	323			
		0.181	373			
		0.238	401			
		0.269	416			
		0.325	445			
		0.356	459			

[a] Abbreviations used for the techniques are defined in table II-A-1.
[b] Placed on an absolute basis using $k(C_2H_6) = 8.61 \times 10^{-12} \exp(-1047/T)$ cm^3 molecule^{-1} s^{-1}.

quite different; the data of Clarke et al. (200–360 K) show a much shallower temperature dependence (~750/T) than the others (Dóbé et al., DeMore and Bayes, and Wilson et al.), which are in the range (1000–1250)/T. Clearly, the measurement of such slow reactions is problematic (cf. the earlier discussion on methane), and additional, careful measurements are required, especially below room temperature. The existing data, when combined, indicate very high curvature over the whole range of measurements; this result does not seem to fit with any expected theory. The data of Clarke et al. (1998) lead to a very low A-factor, $\sim 8 \times 10^{-13}$ cm^3 molecule^{-1} s^{-1} at 298 K, whereas the steeper data of DeMore and Bayes (1999) and Wilson et al. (2001) suggest an A-factor of $\sim 5 \times 10^{-12}$ cm^3 molecule^{-1} s^{-1}. Clarke et al. noted that OH was regenerated in their discharge-flow experiments. An unweighted fit to the data of Jolly et al., DeMore and Bayes, and Wilson et al. (figure II-C-1) returns a simple Arrhenius expression of $6.0 \times 10^{-12} \exp(-1300/T)$ cm^3 molecule^{-1} s^{-1}, which is clearly not ideal, but agrees with all of the data above 250 K to around 20%. The rate coefficient at 298 K is $k = 7.4 \times 10^{-14}$ cm^3 molecule^{-1} s^{-1}, with an estimated overall uncertainty of ±30%.

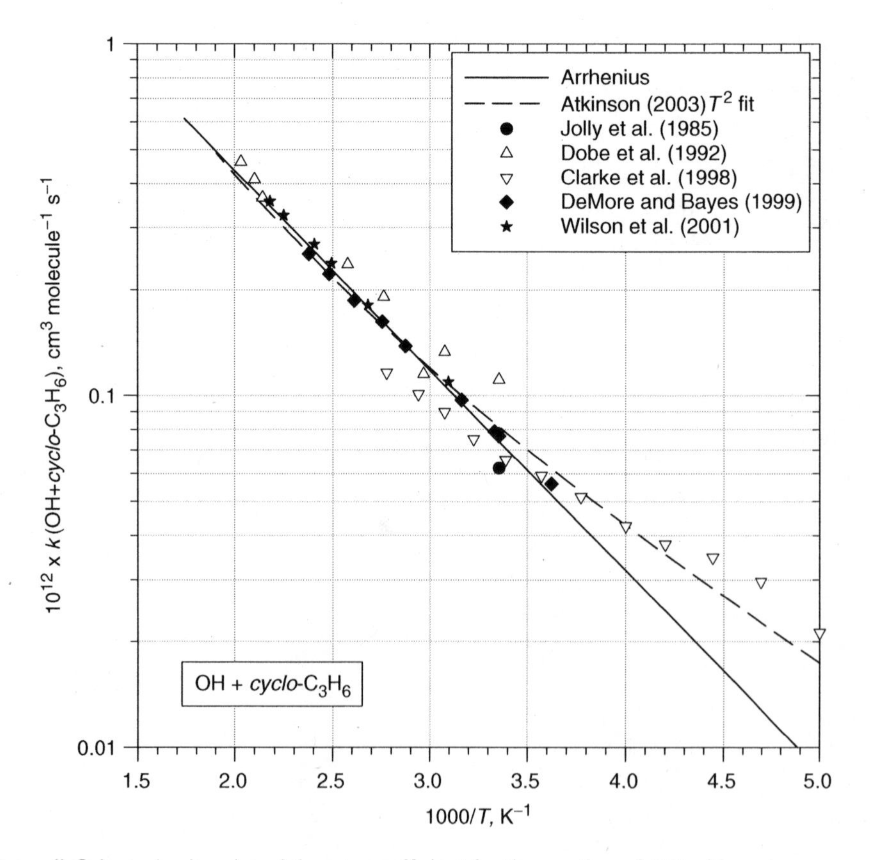

Figure II-C-1. Arrhenius plot of the rate coefficient for the reaction of OH with cyclopropane.

The *A*-factor per —CH_2— group in cyclopropane is lower than for the other cycloalkanes (and *n*-alkanes), by around a factor of 2–3. However, the overall rate coefficient is lower by around two orders of magnitude. Hence, the majority of the decrease in rate coefficient for cyclopropane relative to the other cycloalkanes comes from the larger activation energy, not the *A*-factor. Some efforts have been made to explain this in terms of the increase in C—H bond strength in the smaller, more strained cycloalkanes. However, Donahue et al. (1998) have argued in favor of a frontier orbital effect that involves the overlap of the unpaired electron of the hydroxyl radical with the highest occupied orbitals in the cyclopropane. They suggest that the maximum electron density in cyclopropane is in the plane of the ring, rather than at the hydrogen atoms, leading to the increased activation energy for H-abstraction.

Clearly, more work is required on this reaction. A careful relative rate study at low temperatures which could resolve the temperature dependence would be particularly useful in defining the activation energy.

II-C-2. Reaction of OH with Isopropylcyclopropane (*iso*-C_3H_7-*cyclo*-C_3H_5)

There is a solitary measurement of this rate coefficient. Atkinson and Aschmann (1988a) measured $k = 2.61 \times 10^{-12}$ cm^3 molecule^{-1} s^{-1} at 296 K relative to butane. The estimated uncertainty at 298 K is ±25%. Since most of the reactivity is associated with the isopropyl group, this measurement does not help to resolve the discrepancy in the cyclopropane rate coefficient (which is approximately a factor of 35 smaller). Interestingly, the rate coefficient is approximately 25% larger than that for 2-methylpropane; this is surprising, since the cyclopropyl group is not expected to be much more reactive than a methyl group, which possibly indicates a large activation of the tertiary hydrogen atom.

II-C-3. Reaction of OH with Cyclobutane (*cyclo*-C_4H_8)

The reaction of OH with cyclobutane has been the subject of three studies (Gorse and Volman, 1974; Dóbé et al., 1992; DeMore and Bayes, 1999). The agreement among them is poor. The early measurement of Gorse and Volman, made relative to CO at 298, lies well below the later studies and is not considered further here. In contrast to the data for cyclopropane, the absolute results of Dóbé et al. lie 15–20% lower than those of DeMore and Bayes, but with a similar temperature dependence. The data from these studies are given in table II-C-2.

There is no strong evidence for curvature in either data set, and no reason to discount either study. Hence, we recommend (figure II-C-2) a simple Arrhenius fit through the data of Dóbé et al. (1992) and DeMore and Bayes (1999), leading to $k = 1.11 \times 10^{-11} \exp(-510/T)$ cm^3 molecule^{-1} s^{-1} that is valid over the range of measurements (272–469 K) with an uncertainty of ±25%. The corresponding rate coefficient at 298 K is 2.0×10^{-12} cm^3 molecule^{-1} s^{-1}. The estimated uncertainty at 298 K is ±25%.

Table II-C-2. Rate coefficients $k = A\ e^{-B/T}$ (cm^3 $molecule^{-1}$ s^{-1}) for the reaction of OH with cyclobutane

$10^{12} \times A$	B(K)	$10^{12} \times k$	T(K)	Technique[a]	Reference	Temperature Range (K)
15.9	661	1.75 ± 0.12	298	PLP–RF	Dóbé et al. (1992)	298–469
		2.22 ± 0.20	327			
		2.36 ± 0.14	360			
		2.89 ± 0.30	392			
		3.49 ± 0.28	429			
		3.98 ± 0.40	469			
13.1	546	1.80	272	RR (C_3H_8)[b]	DeMore and Bayes (1999)	272–366
		2.00	288			
		2.02	293			
		1.97	298			
		2.18	303			
		2.30	309			
		2.64	343			
		3.00	366			

[a] Abbreviations used for the techniques are defined in table II-A-1.
[b] Placed on an absolute basis using $k(C_3H_8) = 1.0 \times 10^{-11} \exp(-655/T)\ cm^3\ molecule^{-1}\ s^{-1}$.

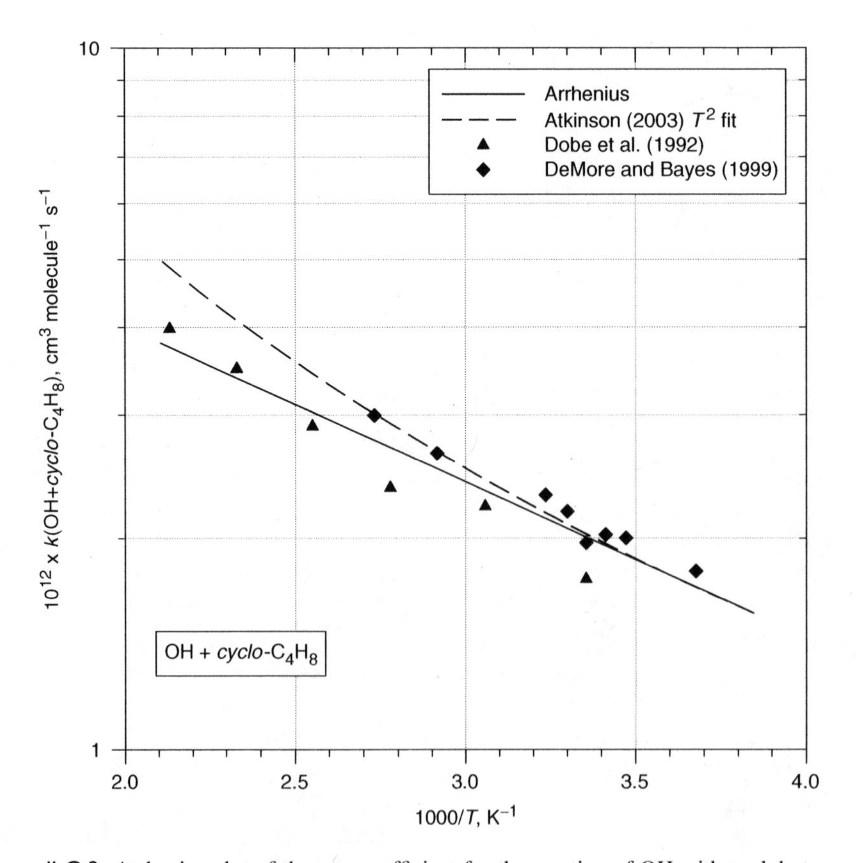

Figure II-C-2. Arrhenius plot of the rate coefficient for the reaction of OH with cyclobutane.

II-C-4. Reaction of OH with Cyclopentane (*cyclo*-C_5H_{10})

The reaction of OH with cyclopentane has been studied quite extensively, and the data are in reasonably good agreement. The temperature-dependent data of Droege and Tully (1987), DeMore and Bayes (1999), and Wilson et al. (2006) are in particularly good accord, and these fit well with the room-temperature values of Atkinson et al. (1982e) and Kramp and Paulson (1998) (table II-C-3). The relative rate data of Kramp and

Table II-C-3. Rate coefficients $k = A\ e^{-B/T}$ (cm^3 $molecule^{-1}$ s^{-1}) for the reaction of OH with cyclopentane (selected data used in fit)

$10^{12} \times A$	B(K)	$10^{12} \times k$	T(K)	Technique[a]	Reference	Temperature Range (K)
		4.93 ± 0.05	299	RR (c-C_6H_{12})[b]	Atkinson et al. (1982e)	
		5.18 ± 0.38	298	FP–RA	Jolly et al. (1985)	
$k(T)$[c]		5.02 ± 0.22	295	PLP–RF	Droege and Tully (1987)	295–491
		6.12 ± 0.27	344			
		7.23 ± 0.32	403			
		9.45 ± 0.41	491			
		4.85 ± 0.15	296 ± 2	RR[d]	Kramp and Paulson (1998)	
18.8	352	5.89 ± 0.26	300	DF–LIF	Donahue et al. (1998)	300–390
		6.35 ± 0.59	325			
		6.18 ± 0.19	340			
		7.64 ± 0.23	360			
		7.25 ± 0.38	375			
		7.53 ± 0.45	390			
21.0	434	4.37	273	RR (n-C_4H_{10})[e]	DeMore and Bayes (1999)	273–423
		4.79	292			
		5.06	310			
		5.38	326			
		6.13	348			
		6.10	354			
		6.59	373			
		7.17	398			
		7.52	423			
21.4	441	2.59	209	RR (n-C_4H_{10})[e]	Wilson et al. (2006)	209–407
		3.18	228			
		3.78	253			
		4.14	267			
		5.21	310			
		6.11	345			
		7.17	386			
		4.46	288			
		5.41	323			
		6.36	367			
		7.16	407			

[a] Abbreviations used for the techniques are defined in table II-A-1.
[b] Placed on an absolute basis using k(cyclohexane) $= 3.0 \times 10^{-11} \exp(-440/T)\,cm^3\ molecule^{-1}\ s^{-1}$.
[c] $k(T) = 6.04 \times 10^{-16}\ T^{1.52} \exp(111/T)$.
[d] Mean of values measured relative to nonane, propene, and *trans*-2-butene.
[e] Placed on an absolute basis using $k(n\text{-}C_4H_{10}) = 1.4 \times 10^{-11} \exp(-520/T)\,cm^3\ molecule^{-1}\ s^{-1}$.

Paulson, which used a number of reference compounds, represent a good check on the consistency of the OH data base. The rate coefficient for cyclopentane was determined relative to four reference compounds. Three of these determinations (nonane, propene, and *trans*-2-butene) are in perfect agreement, while that measured relative to 1,3-butadiene is some 10% lower, possibly as a result of the large difference in rate coefficients for the two species. Hence the average of the first three determinations is used in our fit to the data. The measurements of Donahue et al. (1998) lie 15–20% higher than those of the other groups and show a weaker dependence on temperature, as was the case for cyclopropane.

Wilson et al. (2006) extended the temperature range down to 209 K using butane and cyclohexane as reference compounds. We have taken their data relative to butane, for which the rate coefficient is well known, along with the data of Atkinson et al. (1982e), Jolly et al. (1985), Droege and Tully (1987), Kramp and Paulson (1998), and DeMore and Bayes (1999) measured relative to butane, to derive an Arrhenius expression for the rate coefficient (figure II-C-3). At 298 K a rate coefficient of 4.9×10^{-12} cm^3 molecule^{-1} s^{-1} is recommended, with an Arrhenius expression of

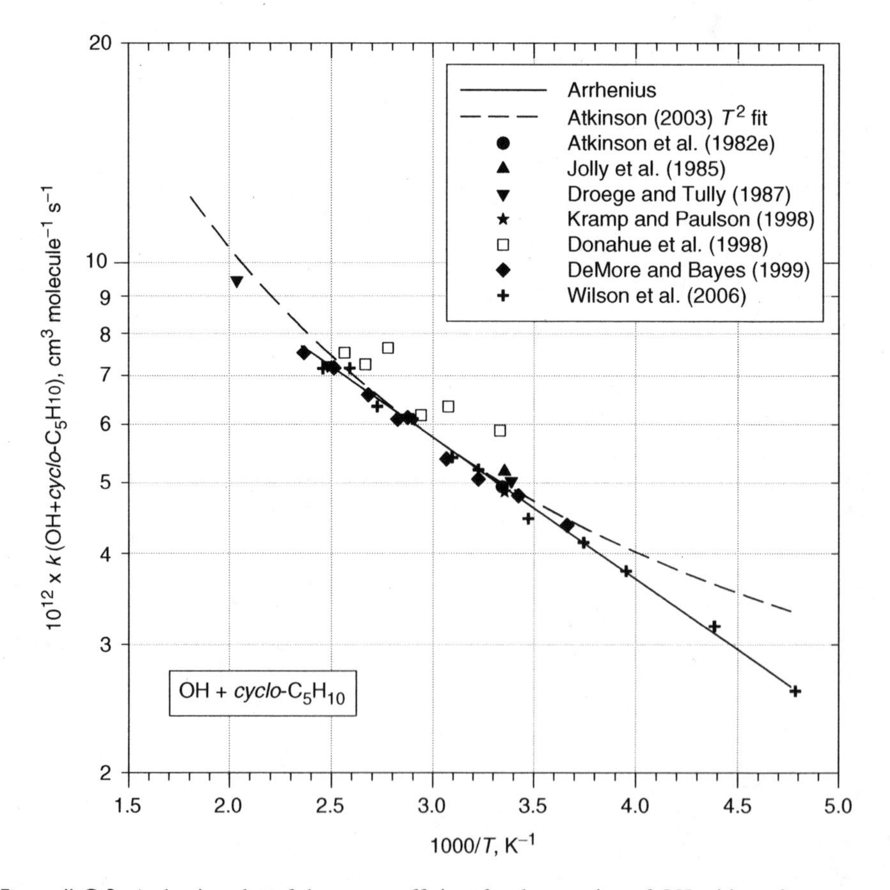

Figure II-C-3. Arrhenius plot of the rate coefficient for the reaction of OH with cyclopentane.

$2.2 \times 10^{-11}\exp(-450/T)$ for atmospheric purposes. The estimated uncertainty at 298 K is ±20%.

Droege and Tully (1987) also studied the reaction of OH with fully deuterated cyclopentane, *cyclo*-C_5D_{10}. The observed kinetic isotope effect $k_H/k_D = 1.16 \times \exp(254/T)$ for abstraction from the CH_2 groups is essentially identical to that derived for the methylene groups in propane and butane, further indicating that the KIE is independent of the chemical environment for a given type of hydrogen atom.

II-C-5. Reaction of OH with Cyclohexane (*cyclo*-C_6H_{12})

The reaction of OH with cyclohexane has been studied between 296 and ~500 K using both relative and absolute methods (table II-C-4). The overall agreement between

Table II-C-4. Rate coefficients $k = A\,e^{-B/T}$ (cm^3 $molecule^{-1}$ s^{-1}) for the reaction of OH with cyclohexane

$10^{12} \times A$	B(K)	$10^{12} \times k$	T(K)	Technique[a]	Reference	Temperature Range (K)
		6.95 ± 0.05	299 ± 2	RR (n-C_4H_{10})[b]	Atkinson et al. (1982e)	
		6.86 ± 0.10	300 ± 3	RR (n-C_4H_{10})[b]	Tuazon et al. (1983)	
k(T)[c]		7.14 ± 0.31	292	PLP–RF	Droege and Tully (1987)	292–491
		8.49 ± 0.37	342			
		10.1 ± 0.44	401			
		12.9 ± 0.56	491			
		7.10 ± 0.20	296	RR[d]	Kramp and Paulson (1998)	
28.2	408	7.66 ± 0.23	300	DF–LIF	Donahue et al. (1998)	300–390
		7.68 ± 0.23	325			
		7.84 ± 0.24	340			
		9.96 ± 0.34	360			
		9.06 ± 0.27	375			
		10.3 ± 0.6	390			
30.7	442	6.92	298	RR (n-C_4H_{10})[b]	DeMore and Bayes (1999)	298–363
		8.00	326			
		8.67	350			
		9.04	363			
27.6	422	4.20	225	RR (n-C_4H_{10})[b]	Wilson et al. (2006)	222–408
		5.17	253			
		5.67	267			
		6.68	304			
		8.27	344			
		9.12	386			
		5.84	288			
		7.12	323			
		8.26	369			
		9.31	408			

(*continued*)

Table II-C-4. *(continued)*

$10^{12} \times A$	B(K)	$10^{12} \times k$	T(K)	Technique[a]	Reference	Temperature Range (K)
		4.05	222	RR (c-C_5H_{10})[e]		
		4.88	240			
		5.78	264			
		6.69	292			
		7.46	317			
		6.17	280			
		6.47	294			
		7.06	308			
		7.67	323			
		8.12	337			
		8.61	353			
		8.98	367			
		9.37	382			
		9.73	395			

[a] Abbreviations used for the techniques are defined in table II-A-1.
[b] Placed on an absolute basis using $k(n\text{-}C_4H_{10}) = 1.4 \times 10^{-11} \exp(-520/T)\,\text{cm}^3\,\text{molecule}^{-1}\,\text{s}^{-1}$.
[c] $k(T) = 1.09 \times 10^{-15} T^{1.47} \exp(125/T)$
[d] Mean of values measured relative to toluene, *trans*-2-butene, and 1,3-butdiene.
[e] Placed on an absolute basis using $k(\text{cyclopentane}) = 2.2 \times 10^{-11} \exp(-450/T)\,\text{cm}^3\,\text{molecule}^{-1}\,\text{s}^{-1}$.

the different studies is quite good, although some scatter exists, particularly at room temperature (figure II-C-4). The room-temperature values vary from $5.24 \times 10^{-12}\,\text{cm}^3\,\text{molecule}^{-1}\,\text{s}^{-1}$ (Nielsen et al., 1986) to $8.6 \times 10^{-12}\,\text{cm}^3\,\text{molecule}^{-1}\,\text{s}^{-1}$ (Bourmada et al., 1987), both of which were absolute measurements. Most of the room-temperature values lie between 6.5×10^{-12} and $7.5 \times 10^{-12}\,\text{cm}^3\,\text{molecule}^{-1}\,\text{s}^{-1}$. Temperature-dependent rate coefficients were reported by Greiner (1970), Droege and Tully (1987), Donahue et al. (1998), and DeMore and Bayes (1999). The data of DeMore and Bayes (298–368 K) are in very good agreement with those of Droege and Tully (292–491 K) in the region of overlap. The data of Donahue et al. are fairly scattered, as was the case for cyclopentane. Kramp and Paulson again provided data relative to a number of different reference compounds, and their measurements fall between 6.76×10^{-12} and $7.33 \times 10^{-12}\,\text{cm}^3\,\text{molecule}^{-1}\,\text{s}^{-1}$, or $\pm 4\%$ of the mean.

The recommended rate coefficient, taken from the evaluation of Atkinson (2003), is $k = 3.26 \times 10^{-17} T^2 \exp(262/T)\,\text{cm}^3\,\text{molecule}^{-1}\,\text{s}^{-1}$, giving a value at 298 K of $6.97 \times 10^{-12}\,\text{cm}^3\,\text{molecule}^{-1}\,\text{s}^{-1}$. We have fit a subset of the data, shown in Figure II-C-4, to an Arrhenius expression to give $k = 3.0 \times 10^{-11} \exp(-440/T)\,\text{cm}^3\,\text{molecule}^{-1}\,\text{s}^{-1}$. The estimated uncertainty at 298 K is $\pm 20\%$. The value of the rate coefficient at 298 K is somewhat lower than what would be expected from the SAR. Since cyclohexane is an unstrained ring, the rate coefficient is expected to be that for 6 —CH_2— groups, or $8.5 \times 10^{-12}\,\text{cm}^3\,\text{molecule}^{-1}\,\text{s}^{-1}$. This difference, of the order of 20%, is surprising, since it is usually assumed that cyclohexane is the model for unstrained cycloalkanes, and the SARs are built on the assumption that 6-membered rings can be represented as unstrained CH_2 groups. Droege and Tully (1987) also reported rate coefficients for

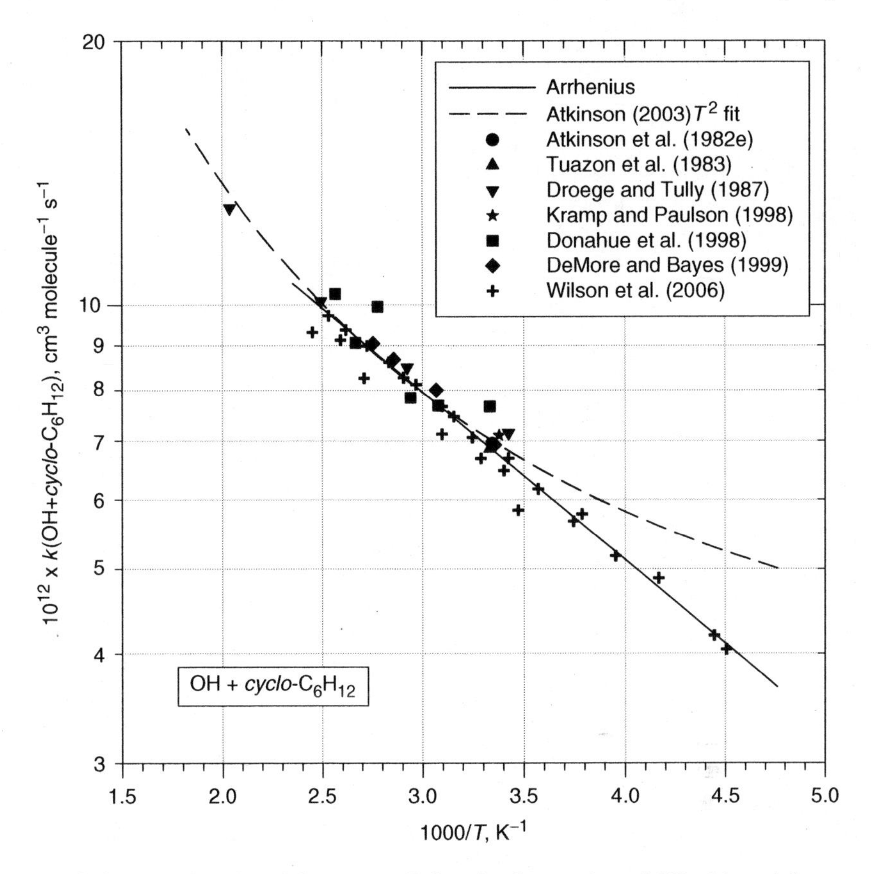

Figure II-C-4. Arrhenius plot of the rate coefficient for the reaction of OH with cyclohexane.

cyclo-C_6D_{12} with OH, finding kinetic isotope effects that were indistinguishable from those for cyclopentane.

II-C-6. Reaction of OH with Methylcyclopentane (*cyclo*-C_5H_9—CH_3)

This reaction has been studied only once, by Anderson et al. (2004). They used a relative rate method, and found $k = 8.6 \times 10^{-12}$ cm^3 molecule^{-1} s^{-1} at 298 K. The value seems rather high, relative to the well studied reaction OH + cyclopentane and considering structure–activity relationships, which predict a rate coefficient less than 6×10^{-12} cm^3 molecule^{-1} s^{-1}. The estimated uncertainty at 298 K is ±30%.

II-C-7. Reaction of OH with Cycloheptane (*cyclo*-C_7H_{14})

This reaction has been studied by Jolly et al. (1985), Donahue et al. (1998), and Wilson et al. (2006). The first two studies were absolute determinations (table II-C-5). The room temperature value of Jolly et al. is higher than the others, although it

Table II-C-5. Rate coefficients $k = A\ e^{-B/T}$ (cm^3 $molecule^{-1}$ s^{-1}) for the reaction of OH with cycloheptane

$10^{12} \times A$	B(K)	$10^{12} \times k$	T(K)	Technique[a]	Reference	Temperature Range (K)
		13.1 ± 2.1	298 ± 2	FP–RA	Jolly et al. (1985)	
41.2	372	12.0 ± 0.3	300	DF–LIF	Donahue et al. (1998)	300–390
		13.4 ± 0.4	325			
		13.3 ± 0.4	340			
		14.9 ± 0.4	375			
		16.4 ± 1.4	390			
38.5	386	9.50	274	RR (n-C_6H_{14})[b]	Wilson et al. (2006)	274–408
		10.5	300			
		11.6	326			
		12.5	351			
		13.2	373			
		14.0	404			
		10.3	291	RR ($cyclo$-C_6H_{12})[c]		
		11.2	312			
		12.5	336			
		13.6	361			
		15.1	386			
		15.3	408			

[a] Abbreviations used for the techniques are defined in table II-A-1.
[b] Placed on an absolute basis using $k(n\text{-}C_6H_{14}) = 1.98 \times 10^{-11} \exp(-394/T)$ cm^3 $molecule^{-1}$ s^{-1}.
[c] Placed on an absolute basis using k(cyclohexane) $= 3.0 \times 10^{-11} \exp(-440/T)$ cm^3 $molecule^{-1}$ s^{-1}.

agrees within uncertainties with the measurement of Donahue et al. On the basis of an unweighted fit to the temperature dependent data of Donahue et al. (1998) and Wilson et al. (2006) shown in figure II-C-5, we recommend $k = 4.0 \times 10^{-11} \exp(-388/T)$ cm^3 $molecule^{-1}$ s^{-1}, giving a rate coefficient at 298 K of 1.1×10^{-11} cm^3 $molecule^{-1}$ s^{-1}. The estimated uncertainty at 298 K is ±25%.

II-C-8. Reaction of OH with Methylcyclohexane ($cyclo$-C_6H_{11}-CH_3)

This reaction has been studied only once, in a relative rate study by Atkinson et al. (1984b). The rate coefficient obtained, $k = 9.64 \times 10^{-12}$ cm^3 $molecule^{-1}$ s^{-1}, is some 40% larger than that for unsubstituted cyclohexane. Interestingly, the predicted value for this rate coefficient is only 5–6% higher than the measured value, in contrast with the situation for unsubstituted cyclohexane. The estimated uncertainty in the rate coefficient at 298 K is 30%.

II-C-9. Reaction of OH with Cyclooctane ($cyclo$-C_8H_{16})

This reaction has been studied only once, by Donahue et al. (1998), between 300 and 390 K. A simple Arrhenius fit to the data gives $k = 5.2 \times 10^{-11} \exp(-408/T)$, with a 298 K rate coefficient of 1.3×10^{-11} cm^3 $molecule^{-1}$ s^{-1}. The activation energy is

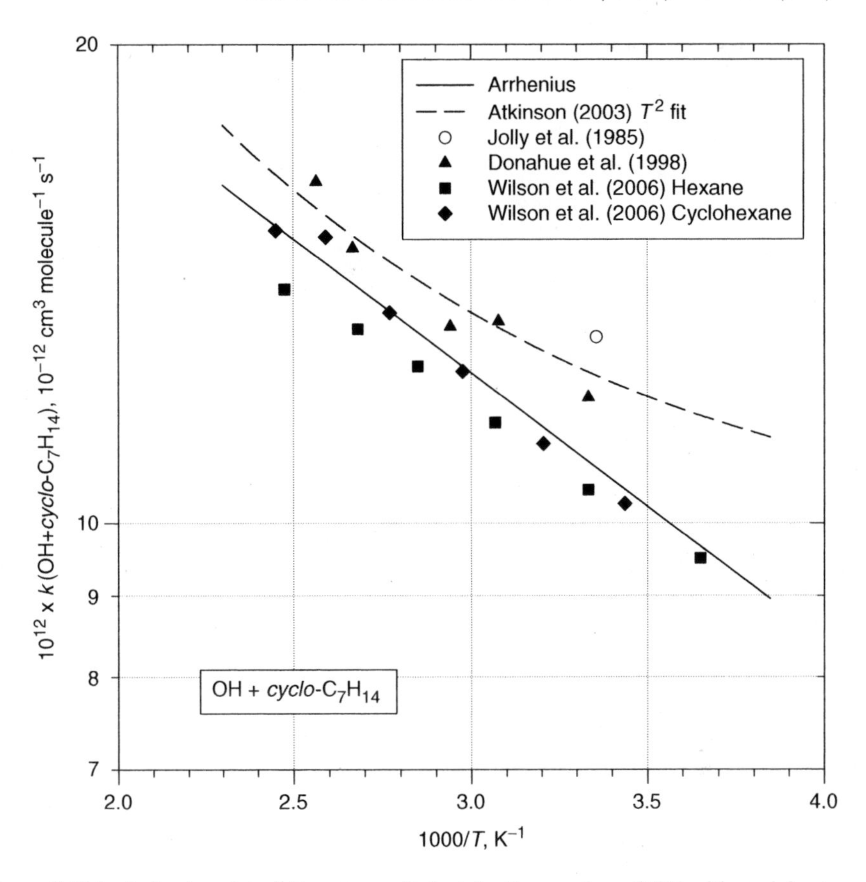

Figure II-C-5. Arrhenius plot of the rate coefficient for the reaction of OH with cycloheptane.

indistinguishable from that of cycloheptane, and the *A*-factor is consistent with the greater number of hydrogen atoms in the larger molecule. The estimated uncertainty is correspondingly higher, ±30% at 298 K, in view of the paucity of studies.

II-C-10. Reaction of OH with Larger Cycloalkanes and Polycycloalkanes

Atkinson and coworkers have measured the rate coefficients at 298 K for OH radicals with a number of larger cycloalkanes containing 3-, 4-, 5- and 6-membered rings (Atkinson et al., 1983b, Atkinson and Aschmann, 1992). The rate coefficients are collected in table II-C-6. These rate coefficients have been used to derive the factors for strained rings used in the SAR calculations. Atkinson showed that to estimate the rate coefficients using SAR, only the ring(s) in which a particular carbon atom is located affect(s) the rate coefficient for abstraction at that carbon atom. The compound *trans*-pinane is the hydrogenated form of both α- and β-pinene, two of the most prevalent biogenic hydrocarbons. *trans*-Pinane has a rate coefficient for OH attack of 1.34×10^{-11} cm^3 $molecule^{-1}$ s^{-1}. This suggests that abstraction could be

Table II-C-6. Rate coefficients $k = A\ e^{-B/T}$ (cm^3 $molecule^{-1}$ s^{-1}) for the reaction of OH with some larger cycloalkanes at 296 ± 2 K

Molecule	$10^{12} \times k$	Technique[h]	Reference
Bicyclo[2.2.1]heptane[a]	5.1	RR (n-C_6H_{14})[i]	Atkinson et al. (1983b)
Bicyclo[2.2.2]octane	13.7		
Bicyclo[3.3.0]octane	10.3		
cis-Bicyclo[4.3.0]nonane	16.0		
trans-Bicyclo[4.3.0]nonane	16.5		
cis-Bicyclo[4.4.0]decane[b]	18.6		
trans-Bicyclo[4.4.0]decane[c]	19.0		
Tricyclo[5.2.1.0^{2,6}]decane	10.6		
Tricyclo[3.3.1.1^{3,7}]decane[d]	21.5		
(1R,2R)-2,6,6-Trimethylbicyclo[3.1.1]heptane[e]	12.4	RR (n-C_4H_{10})[j]	Atkinson and Aschmann (1992)
1,7,7-Trimethyltricyclo[2.2.1.0^{2,6}]heptane[f]	2.7		
Quadricyclo[2.2.1.0^{2,6}0^{3,5}]heptane[g]	1.7		

[a] Norbornane.
[b] *cis*-Decalin.
[c] *trans*-Decalin.
[d] Adamantane.
[e] *trans*-Pinane.
[f] Tricyclene.
[g] Quadricyclane.
[h] Abbreviations used for the techniques are defined in table II-A-1.
[i] Placed on an absolute basis using $k(n\text{-}C_6H_{14}) = 1.98 \times 10^{-11} \exp(-394/T)$ cm^3 $molecule^{-1}$ s^{-1}.
[j] Placed on an absolute basis using $k(n\text{-}C_4H_{10}) = 1.4 \times \exp(-520/T)$ cm^3 $molecule^{-1}$ s^{-1}.

a non-negligible pathway in the reactions of α-pinene and β-pinene (Peeters et al., 2001), whose overall rate coefficients with OH are 5.4×10^{-11} cm^3 $molecule^{-1}$ s^{-1} and 7.9×10^{-11} cm^3 $molecule^{-1}$ s^{-1}, respectively.

II-C-11. General Comments on the Kinetics of OH Radicals with Cycloalkanes

There are still a number of unanswered questions concerning the reactions of OH with cycloalkanes. Clearly, they have not been studied as extensively as the open-chain alkanes. The trends in reactivity for cycloalkanes are not well understood at room-temperature, as the larger cycloalkanes seem to be more reactive than expected. The temperature-dependent rate coefficients for cyclobutane, cyclopentane, cyclohexane, cycloheptane and cyclooctane are shown as a composite in figure II-C-6. The activation temperatures are very similar (400–500 K) and the A-factors increase with the number of CH_2 groups. However, the A-factors do not scale linearly with the number of carbon atoms, and the larger members of the series (cycloheptane and cyclooctane) appear to be too reactive compared to cyclohexane. More work is required both to study the temperature dependences of these reactions and to understand them from a theoretical standpoint.

As shown by Dóbé et al. (1992) and Jolly et al. (1985), a good correlation is found between the logarithm of the rate constant and the bond strength, or, by implication,

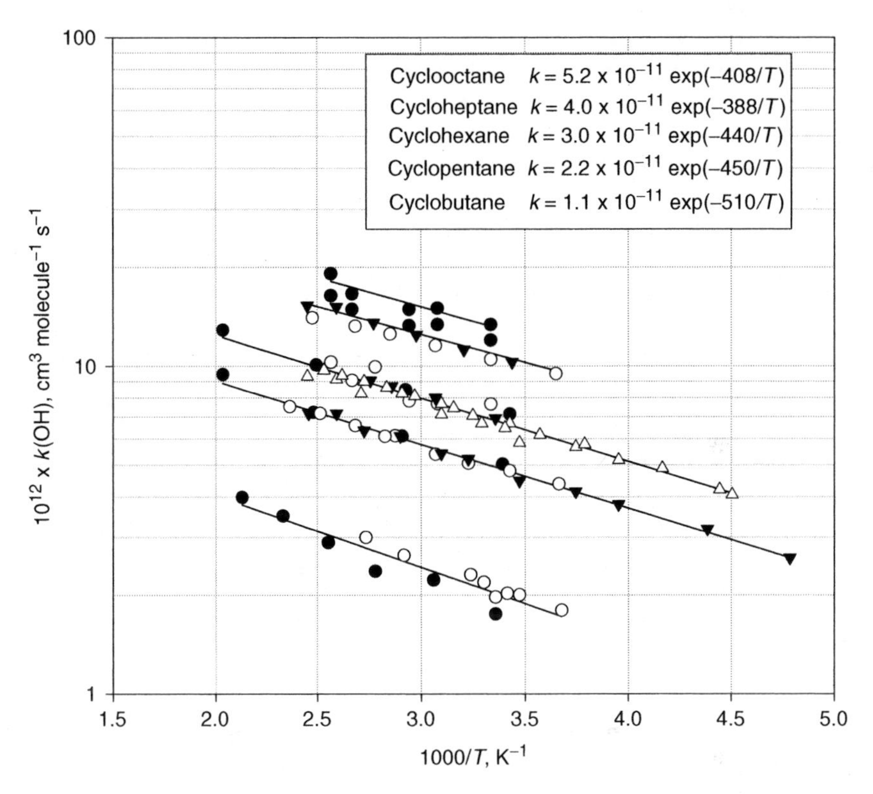

Figure II-C-6. Comparison of the Arrhenius plots of the rate coefficients for reactions of OH with the cyclic alkanes.

the activation energy and the bond strength. This strictly assumes that the *A*-factor does not change with the size of the ring. However, the *A*-factor per $—CH_2—$ group does appear to increase from around 2×10^{-12} cm^3 molecule^{-1} s^{-1} for cyclopropane to 6×10^{-12} cm^3 molecule^{-1} s^{-1} for cycloheptane and cyclooctane. Hence, the reactivity trends cannot be explained by changes in activation energy alone (as figure II-C-6 shows, the activation energies are in fact not very different after cyclopropane). The reasons for the observed trends in the *A*-factors of the cycloalkanes are not at all apparent. An alternative approach to explaining the reactivity suggested by Donahue et al. (1998) relates the activation energy to the energetics of the orbital overlap between the OH radical and the cycloalkane. This approach offers some insight into the mechanism, but still does not account well for the reactivity of cyclopropane.

II-D. Reactions of OH with Isotopically Substituted Alkanes at Natural Abundance

Except for the secondary kinetic isotope effect in methane, all of the rate data discussed thus far for OH reactions with alkanes pertains to pure compounds which have deuterium substitution at one (or more) sites. In the atmosphere, however, the isotopic labeling differs in that typically only one atom in a given molecule is labeled. This is because the natural abundances of deuterium (1.15×10^{-4}) and ^{13}C (1.07×10^{-2}) are very low, so statistically only one substitution per molecule is expected. There has been a number of measurements of both the primary (i.e., k_H/k_D) and secondary (i.e., $^{12}C/^{13}C$) kinetic isotope effects for a series of alkanes made by Rudolph's group (Iannone et al., 2004; Anderson et al., 2004); these are summarized in table II-D-1.

Iannone et al. (2004) measured the hydrogen atom isotope effects for the deuterated species at natural isotopic abundance. They found kinetic isotope effects ε ranging between 97.3 per mil, or roughly a 10% difference in rate coefficients, for isobutane, to 29 per mil for decane. The measured values are smaller than those derived by Tully and coworkers, since they include the statistical probability of a particular hydrogen atom being deuterated. Iannone et al. point out that the trend for the measured isotope effect falls off with the number of hydrogen atoms (N_H) as KIE $\varepsilon = 640/N_H$ per mil. For butane, Tully found that the kinetic isotope effect for secondary hydrogen atoms is a factor of 2.62. Hence, for larger *n*-alkanes, which consist mostly of secondary sites, a dependence of $[1-(1/2.62)] \times 1000$ is expected, or $\varepsilon = 618/N_H$ per mil, in excellent agreement with that observed. 2-Methylpropane has a large KIE, since it is dominated by primary hydrogen atoms for which the KIE is larger, a factor of 4.6 in ethane (Tully et al., 1986a).

Anderson et al. (2004) measured secondary carbon kinetic isotope effects in a series of alkanes. The values, expressed in per mil differences, are also listed in table II-D-1. They also noted a decrease in the KIE with the size of the molecule, which could be expressed as KIE $\varepsilon = 16.1/N_C$. Anderson et al. also extended the treatment to provide structure–additivity rules for KIE, analogous to those for estimating rate coefficients (see next section).

Table II-D-1. Kinetic isotope effect values for reactions of non-methane hydrocarbons with OH radicals

Molecule	Hydrogen KIE per mil[a]	Carbon KIE per mil[b]
Ethane		8.57 ± 1.95
Propane		5.46 ± 0.35
Butane	51.6 ± 2.1	5.16 ± 0.67
2-Methylpropane	97.3 ± 12.5	8.45 ± 1.49
Pentane	65.9 ± 7.0	2.85 ± 0.79
2-Methylbutane		2.91 ± 0.43
Hexane	52.8 ± 5.0	2.20 ± 0.07
Heptane	38.9 ± 7.8	1.96 ± 0.26
Octane	33.4 ± 3.1	2.13 ± 0.39
Nonane	29.6 ± 1.6	
Decane	29.0 ± 5.3	
Cyclopentane	63.2 ± 5.9	1.84 ± 0.13
Cyclohexane	79.5 ± 9.6	4.46 ± 0.51
Methylcyclopentane		1.77 ± 0.53

[a] From Iannone et al. (2004).
[b] From Anderson et al. (2004).

II-E. Estimation and Prediction of Rate Coefficients of OH with Alkanes

The most common way to estimate rate coefficients for OH with alkanes (and other organic species) is through a structure activity relationship (SAR). This approach is described by Atkinson et al. (1982c, 1984b) for *n*-alkanes and for branched alkanes, and has been modified and updated a number of times since those studies then to include substituent groups (Atkinson, 1986a, b, 1987; Kwok and Atkinson, 1995; Bethel et al., 2001); the SAR method treats the rate of abstraction from a group within a molecule (rather than a singular hydrogen atom). From a simultaneous fit of all the available data for the alkanes as a function of temperature, Kwok and Atkinson obtained the following group rate coefficients at 298 K:

$$k(\text{—CH}_3) = 1.67 \times 10^{-13}; \quad k(\text{—CH}_2\text{—}) = 9.34 \times 10^{-13};$$

$$k(>\text{CH—}) = 1.94 \times 10^{-12}\ \text{cm}^3\text{molecule}^{-1}\text{s}^{-1}.$$

The expressions for the temperature-dependent parameters are given in table II-E-1. The effects of neighboring groups *X* are taken into account by multiplying by an appropriate factor *F(X)*, also derived by simultaneously fitting a large number of rate coefficients. The factors for cycloalkanes are then derived, based on the above coefficients for the open chain alkanes. Substituent factors are given in table II-E-2.

When evaluating the effectiveness of the fit, one could consider either the mean deviation of the fitted rate coefficients from the measured ones (actually, Kwok and Atkinson minimized the relative deviation, not the absolute deviation), or the performance of the fit in reproducing certain reactivity trends that are observed. For example, we have noted in the foregoing discussion that the rate coefficients for compounds containing tertiary groups are consistently not well represented by this fit. As an example, the data measured by Tully for 2-methylpropane give a value for abstraction from the tertiary group of 1.70×10^{-12} cm^3 molecule^{-1} s^{-1}, whereas the value from the SAR is 1.94×10^{-12} cm^3 molecule^{-1} s^{-1}. A further constraint would seem to be from the series of *n*-alkanes shown in figure II-E-1. The slope of this plot represents the incremental rate coefficient when a CH_2 group is added, and this should equate to $k(CH_2)F^2$.

Table II-E-1. Group values for abstraction reactions of organic compounds

Group	$10^{12} \times k(298)^a$	$10^{18} \times C^{a,b}$	$D(K)^b$	$10^{12} \times k(298)^a$
		Kwok and Atkinson (1995)		This Work[c]
$-CH_3$	0.136	4.49	320	0.136
$>CH_2$	0.934	4.50	−253	0.78
$-CH<$	1.94	2.12	696	1.37
$-OH$	0.14	2.1	85	

[a] Units are cm^3 molecule^{-1} s^{-1}.
[b] The values of $k(T)$ can be calculated from $k = CT^2 \exp(-D/T)$.
[c] From the "phenomenological" fit to data (see text).

Table II-E-2. Substituent factors F(X) for OH abstraction reactions of alkanes and oxygenates at 298 K.

Group	Kwok and Atkinson (1995)[a]	This Work[b]
$—CH_3$	1.00	1.00
$>CH_2$	1.23	1.35
$—CH<$	1.23	1.35
$>C<$	1.23	1.35
$=O$	8.7	
$—CHO$	0.75	
$>C=O$	0.75	
$—CH_2C(O)—$	3.9	
$>CHC(O)—$	3.9	
$—C(R_2)C(O)—$	3.9	
$—OH$	3.5 (2.9)[c]	
$—C(OH)<$	2.6[c]	
$—CH_2ONO_2$	0.20	
$>CHONO_2$	0.20	
$—CR_2ONO_2$	0.20	
$—ONO_2$	0.04	
3-Membered ring	0.02	
4-Membered ring	0.28	
5-Membered ring	0.64	

[a] The values given by Kwok and Atkinson (1995) are weakly temperature dependent.
[b] From the "phenomenological" fit to alkane data (see text).
[c] Updated by Bethel et al. (2001).

We have performed a "phenomenological" fit to the data at 298 K, again minimizing the relative deviation, to try to represent some of these trends more explicitly. These results are also given in the table II-E-1. The main change is a reduction in the value for $k(>CH—)$ from 1.94 to 1.37. The mean deviation for the alkanes containing tertiary groups is about a factor of 2 less than that using Atkinson's parameters. Of course, none of these fits is unique or definitive. Harris and Kerr (1988) argued that doing a phenomenologically constrained fit was not appropriate, and that a global fit better accounted for scatter in the data. As can be seen in figure II-E-1, the slopes of the plots of fitted against measured rate coefficients are indistinguishable for the two formulations.

We have noted in the preceding discussion that apparent discrepancies exist between the relative efficiency of different sites of attack for the SAR compared to what is observed. Unfortunately, not enough data are available to make a definitive statement about this. Most of what is known is from the deuterium substitution experiments of Tully's group, which considered smaller, symmetrical molecules. The use of the "phenomenological" fits is an attempt to take advantage of this information. However, since we do not have data for the larger molecules, this may not be justified. As the molecules become larger and more branched, other, more subtle effects may come into play. The product data that are available for larger, branched alkanes do not seem to be in particularly good agreement with the SAR, but there are insufficient experimental data to be able to draw a definitive conclusion at present.

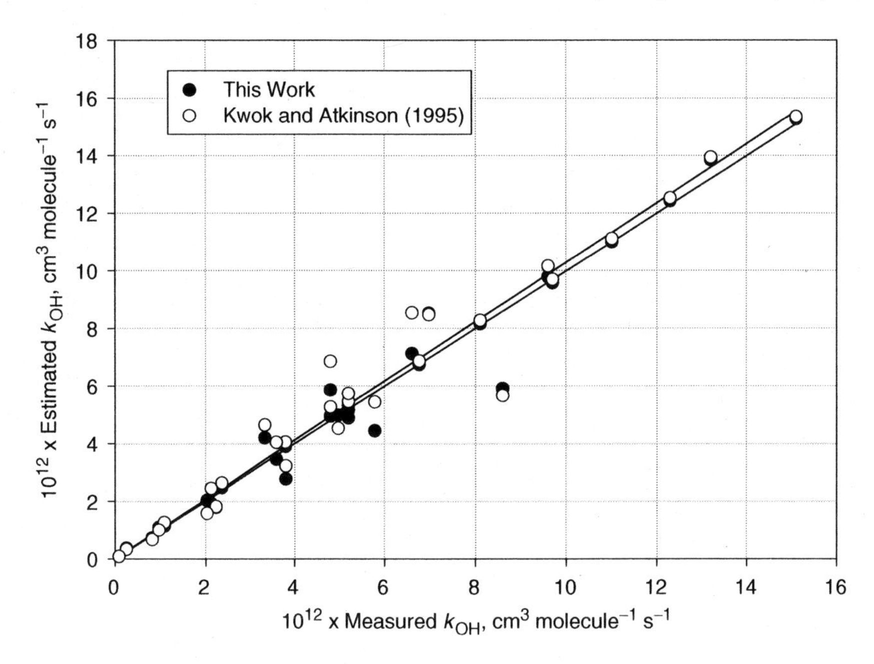

Figure II-E-1. Plot of the measured rate coefficients against those estimated using the SARs of Kwok and Atkinson (1995) and those derived in this work.

A more generalized method of estimating rate coefficients is from so-called Evans–Polanyi plots. These take advantage of the observation that reactivity is often largely controlled by a single molecular parameter among a series of similar molecules. This could be the bond strength of the bond being broken, the ionization potential of the molecule, or the electron affinity of the radical produced. Since the properties of the attacking radical obviously remain constant, some reactivity trends are expected for the substrates. The plots are most useful when correlating rate coefficients for attack by different radicals. For example, plots of $\log(k_O(^3P)$ versus $\log(k_{OH})$ are often found to be linear for a large set of molecules, where $(k_O(^3P))$ represents the rate coefficient for attack by $O(^3P)$, and k_{OH} that for attack by OH (figure III-F-I). Other plots, such as $\log(k)$ versus ionization potential or electron affinity, are usually more scattered, partly because of the lack of accurate values of the molecular parameters, and partly because when plotting one rate coefficient against another subtle internal effects tend to cancel out. As a result of the Arrhenius formulation of the rate coefficient, a plot of $\log(k)$ versus some other parameter is essentially mapping out some variation in the activation energy as a function of molecular structure. Since OH is an electrophilic reagent, some correlation with the ionization potential of the other reactant is expected.

Donahue (2003, and references therein) has approached this problem by considering the reactions in terms of a curve crossing between covalent and ionic forms of the reactants and products. For example, the reaction $RH + OH \rightarrow R + H_2O$ is visualized in terms of potential curves joining $RH + OH$ to $R^- + H_2O^+$ and $R + H_2O$ to

RH^{+}+ OH^{-}. The reaction barrier is caused by an avoided crossing of these curves, and its height can be related to the ionization potentials and electron affinities of the participating species. This gives a simplified but physically meaningful interpretation of the dependence of rate coefficients on electronic properties.

DeMore (2005) has suggested an alternative approach to the estimation of rate coefficients. The use of Evans–Polanyi type correlations assumes implicitly that the *A*-factors of these reactions simply scale with the number of hydrogen atoms in the molecule, *n*(H). By considering a large body of data on abstraction reactions in both alkanes and substituted organics, DeMore reached the conclusion that there was actually a correlation between the rate coefficient at 298 K and the *A*-factor, and that as the reactions become slower, the *A*-factor becomes smaller (see the discussion on cyclopropane earlier in this chapter). This effect derives from the fact that for a slow reaction the transition state is tighter, and so the *A*-factor is reduced. DeMore has obtained correlations between k/n(H) and E/R and k/n(H) and *A*, which enable a rate coefficient to be predicted as a function of temperature once the room-temperature rate coefficient is known.

II-F. Estimation of Rate Coefficients for Substituted Molecules

The concept of the structure–additivity relationship, described in section II-E, can be extended to molecules with electronegative substituents replacing carbon or hydrogen atoms. Since the oxidation products of alkanes are all substituted alkanes, it is also appropriate to consider these here. To accommodate these molecules, additional multiplicative factors *F(X)* are used to describe the effect of nearest neighbor substituents (Kwok and Atkinson, 1995, Bethel et al., 2001). Thus, for a molecule containing the grouping R—CH_2X (e.g., ethanol, where $R = CH_3$ and $X =$ OH), the rate of abstraction at the central carbon atom would be $k(\text{sec})F(R)F(X)$, where $F(R)$ would normally be 1.23 for an alkyl group (see above) and $F(\text{OH}) = 3.5$. Table II-E-2 also includes a representative set of $F(X)$ values for the types of molecules generated from alkanes (aldehydes, ketones, alcohols, nitrates). Substituent factors for halogens are derived in chapter 6.

The simple additivity approach does not always work well for substituted compounds. The presence of electronegative groups (>C=O, $-ONO_2$) in a molecule changes the electronic environment considerably, and the effects can sometimes extend two or more carbon atoms away. This effect can be seen clearly in the reactions of ketones with OH (Atkinson et al., 1982a; Wallington and Kurylo, 1987a; Le Calvé et al., 1998), where considerable activation of the carbon atom two sites away from the carbonyl group occurs (see chapter V–C). This activation can be accounted for in the SAR by the introduction of an *F* for $-CH_2C(O)R$. It should be pointed out that the origin of this activation may not reside in electronic effects, but rather in an enhanced steric factor as a result of a hydrogen bond between the carbonyl group and the OH radical. More complex oxygenates (not considered in this work) such as esters, require the introduction of additional groups which take into account longer range effects (Le Calvé et al., 1997, Szilágyi et al., 2004).

Neeb (2000) has considered a different approach in which the additivity is expressed in terms of abstraction from modified groups, rather than modifying the reactivity at an alkyl group by a substituent. Thus the rate coefficient for abstraction of the secondary hydrogen atoms in ethanol would be expressed as $k(RCH_2O{-}X)F(X)$, where $X = H$, and $F(H)$ in this case is equal to 1.0. While this treatment offers some flexibility in how the substituents are handled, especially for carbonyl species such as esters (e.g., Le Calvé et al., 1997, Szilágyi et al., 2004), the increased number of individual rate coefficients, and resulting loss of simplicity, makes the formulation less useful than the standard Atkinson approach (Kwok and Atkinson, 1995). Furthermore, the participation of hydrogen-bonded intermediates and pre-reactive complexes (see chapter V) limits the utility of SARs for the prediction of rate coefficients for reactions of OH radicals with highly polar molecules.

III

Kinetics and Mechanisms of Reactions of Cl, $O(^3P)$, NO_3, and O_3 with Alkanes

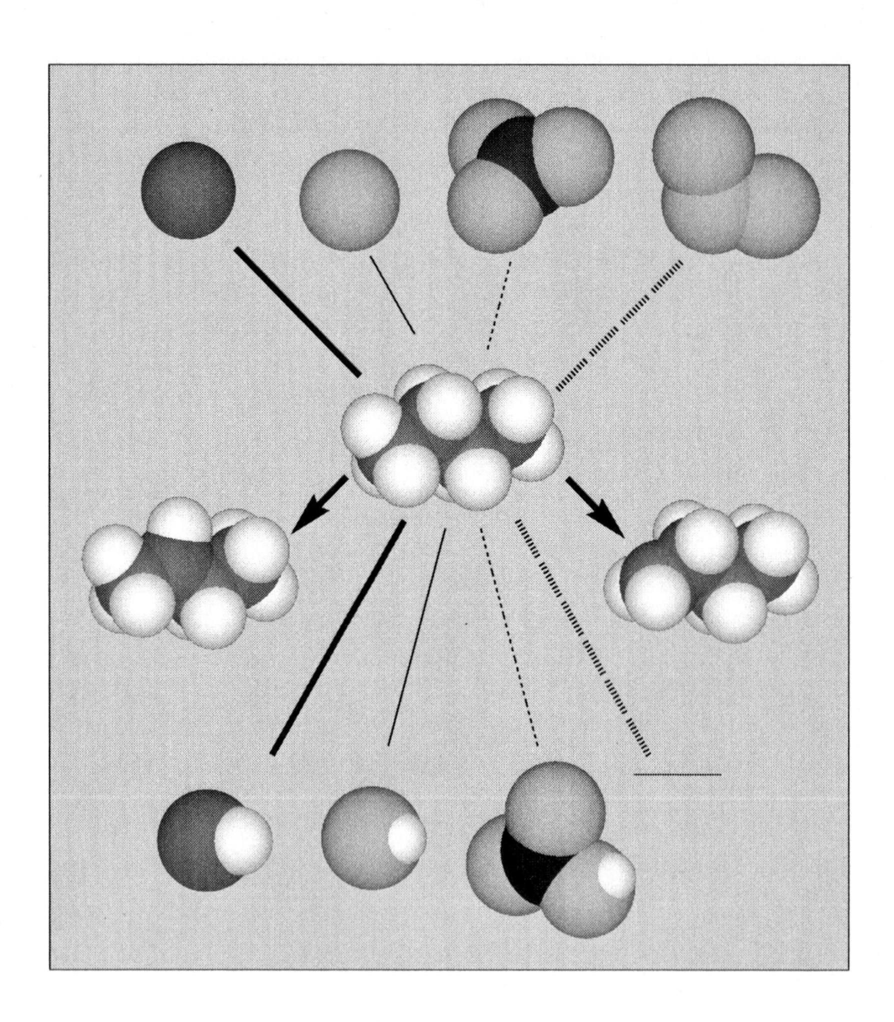

III-A. Introduction

In this chapter, kinetic data for the reactions of alkanes with Cl-atoms, $O(^3P)$, NO_3, and O_3 will be presented. As was noted in chapter I-C, reaction with OH is clearly the major loss process for the alkanes under most atmospheric conditions. However, reaction with

Cl, O(^3P), and NO_3 can play a role in certain circumstances. For example, although ambient concentrations of Cl-atoms have not been firmly established and tropospheric average concentrations are likely to be less than 10^3 molecule cm^{-3} (Rudolph et al., 1996), elevated concentrations (10^4 molecule cm^{-3} or more) may be present in the marine boundary layer (MBL) (Jobson et al., 1994; Wingenter et al., 1996; 2005), in coastal areas (Spicer et al., 1998), and/or in urban regions (Tanaka et al., 2003). Reaction of O(^3P) with the alkanes may play some role in polluted areas, especially in the early morning when photolysis of NO_2 [a source of O(^3P)] is maximized relative to O_3 photolysis (a major OH source). NO_3 is most abundant at night and, in fact, is the most important nighttime oxidant for the alkanes, although the rate at which this nighttime chemistry occurs cannot compete with daytime OH-initiated oxidation rates. As will be summarized in section III-E, the occurrence of any reaction between ozone and the alkanes has not been established, and these processes can certainly be ignored for any atmospheric conditions (Atkinson and Carter, 1984). Table III-A-1 presents atmospheric lifetimes for a representative set of alkanes against various oxidation pathways (OH, Cl, O(^3P), NO_3) for conditions found in various regions of the lower atmosphere.

The reactions of OH, Cl, O(^3P), and NO_3 with the alkanes almost certainly occur via abstraction in all cases, and thus it is not surprising that correlations can be found in the data sets; see section III-F for details. As shown in table III-A-2, reaction of OH with the alkanes is exothermic by at least 58 kJ mol^{-1} (methane) and by as much 97 kJ mol^{-1} (2-methylpropane). On the other hand, abstraction of a hydrogen from methane by Cl, O(^3P) or NO_3 is endothermic by 7–13 kJ mol^{-1}, while abstraction from other alkanes is only slightly exothermic. Nonetheless, for a given alkane, it is the Cl-atom rate coefficient that is generally largest, because these reactions tend to occur without an energy barrier (except in the special cases of reaction with methane and cyclopropane) and they occur with large, essentially gas-kinetic A-factors. In general, for a given alkane, the OH rate coefficient is 1–2 orders of magnitude lower than the Cl-atom rate coefficient but 20–1000 times higher than the O(^3P) rate coefficient and 4–5 orders of magnitude higher than the NO_3 rate coefficient.

In the sections that follow, the available data for reaction of the alkanes with Cl (section III-B), O(^3P) (section III-C), NO_3 (section III-D), and O_3 (section III-E) will be summarized and evaluated. These sections will also include a discussion of rate coefficient estimation methods when available. Finally, an examination of the correlations that exist between these data sets and the data set for reaction of OH with the alkanes will be summarized (section III-F). Note that, in any fits of experimental data to Arrhenius expressions, data are weighted by a factor of $1/k^2$ so as to minimize the percent deviation of the data from the fit.

III-B. Reactions of Cl-Atoms with Alkanes

Reactions of Cl-atoms with alkanes are believed to play a role in the atmospheric oxidation of alkanes in some cases and at some specific locations. The actual levels of Cl-atoms in the atmosphere are quite uncertain and vary significantly, both spatially and

Table III-A-1. Lifetimes (τ, hr) for reaction of a selected set of alkanes with various oxidants, under representative sets of atmospheric conditions; reactant concentrations, molecules cm^{-3}

Alkane	τ(OH) Urban, Daytime [OH] = 3×10^6	τ(Cl), Free Troposphere, Daytime [Cl] = 1×10^3	τ(Cl) MBL[a] Daytime [Cl] = 3×10^4	$\tau(O^3P)$ Urban, Morning $[O(^3P)] = 10^5$	$\tau(NO_3)$ Urban, Nighttime $[NO_3] = 3 \times 10^9$	$\tau(O_3)$ Urban, All Day $[O_3] = 2 \times 10^{12}$
Methane	1.47×10^4	2.8×10^6	9.3×10^4	4×10^8	$> 9.3 \times 10^4$	$> 1.4 \times 10^8$
Ethane	373	4708	157	2.8×10^6	$> 9.3 \times 10^3$	$> 1.4 \times 10^7$
Propane	85	1984	66	3.1×10^5	$> 1.9 \times 10^3$	$> 2.3 \times 10^7$
2-Methylpropane	44	1903	63	2.3×10^4	926	$> 7 \times 10^6$
2-Methylbutane	26	1263	42	2.1×10^4	578	
Hexane	18	868	29	4.6×10^4	770	
2,3-Dimethylbutane	16	1245	42	1.4×10^4	226	
3-Methylpentane	18	1056	35		463	
Octane	11	630	21	3.1×10^4	514	
2,2,4-Trimethylpentane	27	1124	37	3.1×10^4	1234	
Decane	8.4	527	18		356	

[a] Marine boundary layer.

Table III-A-2. Rate coefficients at 298 K (cm^3 $molecule^{-1}$ s^{-1}) for reaction of some simple alkanes with OH, Cl, $O(^3P)$, and NO_3 and the corresponding enthalpy changes ($kJ\ mol^{-1}$) for reaction at the weakest C—H bond. Enthalpy data are taken from the IUPAC (http://www.iupac-kinetic.ch.cam.ac.uk) and NIST chemical kinetics websites (http://kinetics.nist.gov)

	OH + Alkane		Cl + Alkane		$O(^3P)$ + Alkane		NO_3 +Alkane	
	k	ΔH_r	k	ΔH_r	k	ΔH_r	k	ΔH_r
Methane	6.3×10^{-15}	−58	1.0×10^{-13}	+7	7×10^{-18}	+9	$< 1 \times 10^{-18}$	+13
Ethane	2.4×10^{-13}	−74	5.7×10^{-11}	−9	1×10^{-15}	−7	$< 2 \times 10^{-17}$	−3
Propane	1.1×10^{-12}	−88	1.4×10^{-10}	−23	9×10^{-15}	−21	$< 5 \times 10^{-17}$	−17
2-Methylpropane	2.2×10^{-12}	−97	1.4×10^{-10}	−32	1.2×10^{-13}	−30	1×10^{-16}	−26

temporally, making their relative importance difficult to establish. Typically, concentrations $< 10^3$ atom cm^{-3} are estimated in the free troposphere (Rudolph et al., 1996). However, Cl-atom levels in the marine boundary layer (MBL) or in coastal areas are thought to be higher (10^4 atom cm^{-3} or higher) (Wingenter et al., 1996; Spicer et al., 1988; Wingenter et al., 2005), likely due to their liberation via the occurrence of heterogeneous reactions on sea salt aerosol. The contribution to hydrocarbon oxidation of elevated levels of Cl-atoms, [Cl] $\sim 10^4$ atom cm^{-3}, during polar MBL springtime ozone depletion events is now well documented (e.g., Jobson et al., 1994). Some recent isotope data suggest that the depletion of methane by Cl in the MBL may be of global signficance (Allan et al., 2005). There may also exist urban sources of inorganic Cl (Chang et al., 2002; Tanaka et al., 2003), in particular for those cities located in coastal areas where reactions of nitrogen oxide reservoirs with sea salt aerosol likely play a role (Finlayson-Pitts, 1993). Thus Cl-atom chemistry may contribute to hydrocarbon oxidation (and ozone production) here as well. Finally, Cl-atoms are present at levels of 10^3–10^4 atom cm^{-3} in the lower stratosphere, and are thought to contribute to the destruction of methane (and perhaps other longer-lived alkanes) in this region as well.

The data base for reactions of Cl-atoms with the alkanes is generally quite well established. Reactions involving the smaller alkanes (methane through the pentanes) have been studied using both absolute and relative rate techniques, whereas only relative rate data are available for the larger alkanes (mostly using butane as the reference compound). Similarly, the temperature dependencies of the smaller species have been studied in some detail. This is particularly the case for the methane reaction, which has a measurable activation energy owing to the endothermicity of the overall reaction. The available data suggest very little, if any, temperature dependence to the rate coefficient for reactions of Cl-atoms with alkanes containing three carbons or more. The available data for each of the alkanes are discussed in this section and are summarized in tables III-B-1 through III-B-10. Note that in compling the data from relative rate studies, the approach has been to build up the data base from smaller to larger compounds. That is, in any relative rate study the smaller of the two compounds under investigation is treated as the reference species, starting with H_2 and proceeding through methane, ethane, propane, etc.

III-B-1. Cl + Methane (CH_4)

The rate coefficient for the reaction of Cl-atoms with methane has been directly measured over a wide range of temperatures (188–1100 K) in experiments using either flow tubes (Clyne and Walker, 1973; Poulet et al., 1974; Leu and DeMore, 1976; Michael and Lee, 1977; Lin et al., 1978; Zahniser et al., 1978; Keyser, 1978; Schlyer et al., 1978; Sawerysyn et al., 1987; Beichert et al., 1995; Seeley et al., 1996; Wang and Keyser, 1999; Bryukov et al., 2002), flash photolysis (Davis et al., 1970; Watson et al., 1976; Manning and Kurylo, 1977; Whytock et al., 1977; Ravishankara and Wine, 1980; Clark et al., 1982; Matsumi et al., 1997; Pilgrim et al., 1997; Mellouki, 1998), or very-low pressure reactor systems (Baghal-Vayjooee et al., 1979; Heneghan et al., 1981; Dobis and Benson, 1987; Lazarou et al., 1992). The data are summarized in table III-B-1 and are plotted in figure III-B-1; note that, for clarity, a number of studies conducted only at ambient temperature are not included in the plot. Data obtained over the range of temperatures of interest for atmospheric chemistry (180–300 K) have been reviewed recently by Wang and Keyser (1999) and by the IUPAC panel (Atkinson et al., 2005a). These two sources report essentially identical Arrhenius expressions, $k = 6.6 \times 10^{-12} \exp(-1240/T)\,\mathrm{cm^3\ molecule^{-1}\ s^{-1}}$, which gives $k = 1.03 \times 10^{-13}\,\mathrm{cm^3\ molecule^{-1}\ s^{-1}}$ at 298 K. This expression is adopted here for the temperature range 180–300 K, with an estimated uncertainty of ±10% at 298 K. Note, however, that the NASA–JPL recommendation (Sander et al., 2005) for this reaction, $k = 9.6 \times 10^{-12} \exp(-1360/T)\,\mathrm{cm^3\ molecule^{-1}\ s^{-1}}$, yields rate coefficients that are similar at ambient temperatures, but 20% lower near 200 K than those obtained from the IUPAC expression.

At higher temperatures, the data show deviations from Arrhenius behavior (upward curvature). A least-squares fit to the modified Arrhenius expression, $k = A\,T^2 \exp(-B/T)$, is given in figure III-B-1 for all the data (Watson et al., 1976, corrected for the presence of ethane; Watson, 1977; Whytock et al., 1977, Manning and Kurylo, 1977; Keyser, 1978; Lin et al., 1978; Zahniser et al., 1978; Ravishankara and Wine, 1980; Heneghan et al., 1981; Lazarou et al., 1992; Seeley et al., 1996; Pilgrim et al., 1997; Wang and Keyser, 1999; Bryukov et al., 2002). This leads to $k = 1.5 \times 10^{-17}\,T^2 \exp(-766/T)\,\mathrm{cm^3\ molecule^{-1}\ s^{-1}}$, which adequately fits the entire data set from 180 to 1104 K. Bryukov et al. (2002) studied the Cl-atom reaction with methane between 295 and 1104 K; their modified Arrhenius expression, $k = 1.3 \times 10^{-19}\,T^{2.69} \exp(-497/T)\,\mathrm{cm^3\ molecule^{-1}\ s^{-1}}$, returns rate coefficients that are about 10% higher than our fit at room temperature and 35% higher at 1100 K. These deviations result largely because their rate coefficients at high temperature are larger than those reported by Pilgrim et al. (1997) (e.g., by about 50% at 800 K).

A few authors (Pritchard et al., 1954, 1955; Knox and Nelson, 1959) have presented relative rate data for the reaction of Cl-atoms with methane, using H_2 as the reference compound. These data, once converted to absolute values using the IUPAC-recommended expression for the Cl + H_2 reaction ($k_{H2} = 3.9 \times 10^{-11} \exp(-2310/T)\,\mathrm{cm^3\ molecule^{-1}\ s^{-1}}$), all lie below the recommendations just presented, although the deviations are typically only 15–30%.

Table III-B-1. Summary of measurements of the rate coefficient (k) for reaction of Cl-atoms with methane; $k = A \times T^n \times e^{-B/T}$ (cm^3 $molecule^{-1}$ s^{-1})

A	B (K)	n	$10^{13} \times k$	T (K)	Technique[a]	Reference	Temperature Range (K)
1.25×10^{-11}	1485				RR[b]	Pritchard et al. (1954,1955)	293–484
1.17×10^{-11}	1485				RR[b]	Knox and Nelson (1959)	193–593
			1.5 ± 0.1	298	FP–RF	Davis et al. (1970)	
$(5.08 \pm 0.50) \times 10^{-11}$	1792 ± 37		1.3 ± 0.1	300	DF–MS	Clyne and Walker (1973)	300–686
			2.3 ± 0.4	329			
			4.0 ± 0.6	374			
			8.4 ± 1.5	418			
			8.8 ± 1.5	449			
			11 ± 2	484			
			19 ± 3	530			
			25 ± 4	591			
			29 ± 5	654			
			49 ± 8	686			
			1.2 ± 0.3	295	DF–MS	Leu and DeMore (1976)	
7.94×10^{-12}	1260				FP–RF	Watson et al. (1976); Watson (1977)	218–401
5.44×10^{-19}	608	2.50	9.06 ± 0.73	500	FP–RF	Whytock et al. (1977)	200–500
			5.71 ± 0.41	447			
			3.77 ± 0.28	404			
			2.74 ± 0.15	371			
			2.04 ± 0.14	343			
			1.38 ± 0.05	318			
			1.13 ± 0.07	299			
			0.749 ± 0.059	276			
			0.571 ± 0.044	260			
			0.411 ± 0.026	245			
			0.324 ± 0.020	232			
			0.234 ± 0.018	220			
			0.191 ± 0.014	210			
			0.146 ± 0.009	200			

(continued)

Table III-B-1. *(continued)*

A	B (K)	n	$10^{13} \times k$	T (K)	Technique[a]	Reference	Temperature Range (K)
			1.08 ± 0.06	298	DF–RF	Michael and Lee (1977)	
$(7.93 \pm 1.53) \times 10^{-12}$	1273 ± 51		0.249 ± 0.018	218	FP–RF	Manning and Kurylo (1977)	218–323
			0.273 ± 0.14	224			
			0.349 ± 0.042	235			
			0.483 ± 0.040	250			
			0.456 ± 0.140	254			
			0.615 ± 0.040	263			
			0.735 ± 0.066	277			
			0.858 ± 0.056	285			
			1.04 ± 0.11	296			
			1.41 ± 0.10	313			
			1.67 ± 0.05	323			
$(7.4 \pm 2.0) \times 10^{-11}$	1291 ± 68		0.218 ± 0.008	220	DF–RF	Keyser (1978)	220–298
			0.351 ± 0.021	243			
$(1.65 \pm 0.32) \times 10^{-11}$	1530 ± 68		0.522 ± 0.16	263			298–423
			1.01 ± 0.02	298			
			1.27 ± 0.04	318.5			
			1.91 ± 0.06	343			
			2.89 ± 0.29	373.5			
			3.32 ± 0.13	398			
			4.57 ± 0.20	423			
$(1.05 \pm 0.25) \times 10^{-11}$	1420 ± 80		0.52 ± 0.03	268.4	DF–MS	Lin et al. (1978)	268–423
			0.96 ± 0.06	296.0			
			1.41 ± 0.09	335.0			
			2.19 ± 0.11	371.1			
			3.88 ± 0.34	423.1			
$(8.2 \pm 0.6) \times 10^{-12}$	1320 ± 20	0	9.10 ± 0.25	504	DF–RF	Zahniser et al. (1978)	200–300
			3.78 ± 0.25	407			
8.6×10^{-18}	795	2.11	2.29 ± 0.04	361			200–500
			1.03 ± 0.02	300			
			0.98 ± 0.034	298			

0.934 ± 0.033	296			
0.498 ± 0.023	254			
0.294 ± 0.004	235			
0.186 ± 0.007	215			
0.109 ± 0.003	200			
1.3 ± 0.2	298	DF–RF	Schlyer et al. (1978)	
0.93 ± 0.05	298	VLPR	Baghal-Vayjooee et al. (1979)	
2.86 ± 0.10	375	FP–RF	Ravishankara and Wine (1980)	221–375
1.96 ± 0.13	352			
1.64 ± 0.04	333			
1.23 ± 0.05	315			
0.980 ± 0.045	298			
0.756 ± 0.023	283			
0.631 ± 0.027	272			
0.514 ± 0.036	260			
0.446 ± 0.015	249			
0.402 ± 0.013	241			
0.328 ± 0.015	231			
0.269 ± 0.016	221			
0.342 ± 0.016	241			
0.268 ± 0.010	231			
0.214 ± 0.007	221			
0.310 ± 0.015	233	VLPR	Heneghan et al. (1981)	233–338
0.470 ± 0.025	253			
0.960 ± 0.050	298			
1.80 ± 0.09	338			
1.03 ± 0.22	298	FP–VUV abs.	Clark et al. (1982)	
0.917 ± 0.073	294	DF–MS	Sawerysyn et al. (1987)	
0.993 ± 0.013	298	VLPR	Dobis and Benson (1987)	

(continued)

Table III-B-1. *(continued)*

A	B (K)	n	$10^{13} \times k$	T (K)	Technique[a]	Reference	Temperature Range (K)
$(1.1 \pm 0.8) \times 10^{-11}$	1410 ± 150	0			VLPR	Lazarou et al. (1992)	263–303
			0.94 ± 0.04	298	DF–RF	Beichert et al. (1995)	
$(7.0 \pm 1.6) \times 10^{-12}$	1270 ± 60		0.842 ± 0.034	291	DF–RF	Seeley et al. (1996)	181–291
			0.952 ± 0.042	283			
			0.551 ± 0.014	260			
			0.360 ± 0.017	246			
			0.340 ± 0.016	241			
			0.179 ± 0.013	213			
			0.163 ± 0.010	209			
			0.109 ± 0.013	205			
			0.100 ± 0.004	194			
			0.112 ± 0.008	192			
			0.066 ± 0.003	181			
$(3.7^{+8.2}_{-2.5}) \times 10^{-13}$	385 ± 320	2.6 ± 0.7	0.93 ± 0.09	292	FP–IR abs.	Pilgrim et al. (1997)	
			1.8 ± 0.2	350			292–800
$(8.9 \pm 0.9) \times 10^{-13}$	660 ± 40	2	3.0 ± 0.2	400			
			6.5 ± 0.4	500			
			12.6 ± 0.9	600			
			18.6 ± 0.9	700			
			30 ± 2	800			

			1.07 ± 0.03	297	FP–RF	Mellouki (1998)	
$(6.5 \pm 0.9) \times 10^{-12}$	1235 ± 34		1.01 ± 0.06	298	DF–RF	Wang and Keyser (1999)	218–298
			0.756 ± 0.070	279			
			0.598 ± 0.043	261			
			0.369 ± 0.020	239			
			0.220 ± 0.028	218			
1.3×10^{-19}	497	2.69	1.04 ± 0.11	295	DF–RF	Bryukov et al. (2002)	295–1104
			1.084 ± 0.055	298			
			1.190 ± 0.067	300			
			2.18 ± 0.11	351			
			3.85 ± 0.13	399			
			8.06 ± 0.51	498			
			17.48 ± 0.36	606			
			31.2 ± 1.1	704			
			40.2 ± 1.8	777			
			55.7 ± 2.4	843			
			60.0 ± 2.4	909			
			105.6 ± 6.9	1012			
			122.8 ± 3.9	1104			

[a] Abbreviations used here and throughout the book to describe the experimental techniques employed are defined in table II-A-1.
[b] Placed on an absolute basis using $k_{H2} = 3.9 \times 10^{-11} \exp(-2310/T)$ cm^3 molecule^{-1} s^{-1} (Atkinson et al., 2005a).

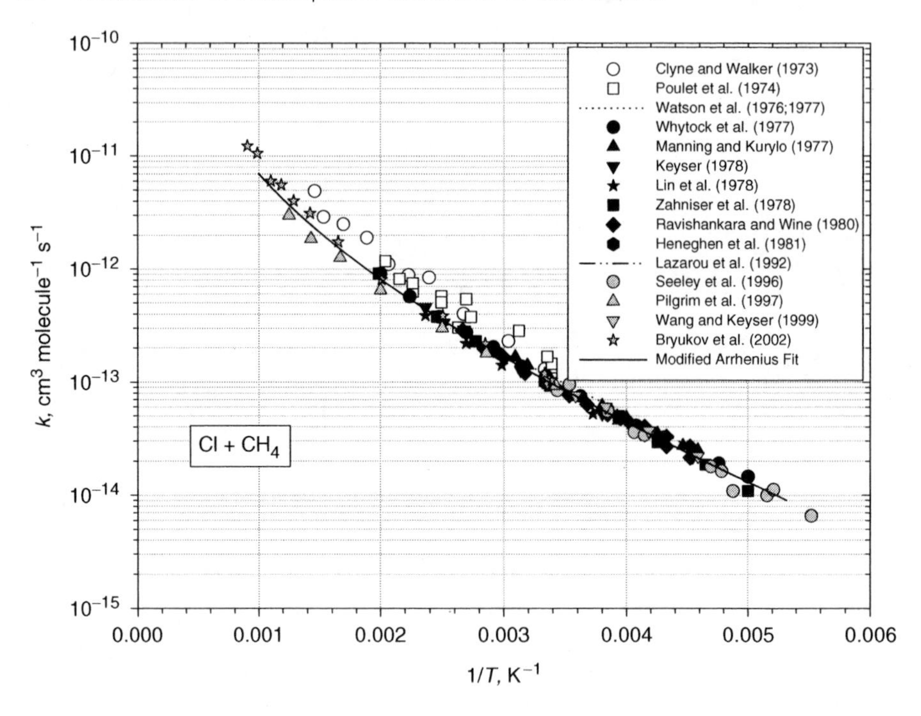

Figure III-B-1. Measurements of the rate coefficient for reaction of Cl-atoms with methane using absolute methods. For clarity, a number of ambient temperature data points are not shown.

III-B-2. Cl + Ethane (C_2H_6)

There are numerous absolute measurements of the rate coefficient for reaction of Cl-atoms with ethane, covering temperatures between about 180 and 1000 K. These data have been obtained using flow tubes (Schlyer et al., 1978; Lewis et al., 1980; Ray et al., 1980; Beichert et al., 1995; Bryukov et al., 2003; Hickson and Keyser, 2004), flash photolysis (Davis et al., 1970; Manning and Kurylo, 1977; Kaiser et al., 1992; Stickel et al., 1992; Pilgrim et al., 1997; Tyndall et al., 1997a; Mellouki, 1998; Hitsuda et al., 2001b), or a very low pressure reactor (Dobis and Benson, 1990; 1991). The data are given in table III-B-2 and plotted in figure III-B-2. Note that, for clarity, a number of determinations made only at ambient temperatures are not included in the plot. With the exception of the single data point from Schlyer et al. (1978), the data are in excellent agreement. The IUPAC panel (Atkinson et al., 2005a) recommends $k = 8.3 \times 10^{-11} \exp(-100/T)\,\text{cm}^3\ \text{molecule}^{-1}\ \text{s}^{-1}$ for the temperature range 220–600 K, with $k = (5.9 \pm 0.8) \times 10^{-11}\,\text{cm}^3\ \text{molecule}^{-1}\ \text{s}^{-1}$ at 298 K, and that recommendation is adopted here. The recommendation of the NASA–JPL panel (Sander et al., 2005), $k = 7.7 \times 10^{-11} \exp(-90/T)\,\text{cm}^3\ \text{molecule}^{-1}\ \text{s}^{-1}$ between 200 and 300 K, gives similar results (e.g., $k = 5.7 \times 10^{-11}\,\text{cm}^3\ \text{molecule}^{-1}\ \text{s}^{-1}$ at 298 K). Some curvature in the data set at high temperatures is evident in the data of Bryukov et al. (2003) and to a lesser extent in the Pilgrim et al. (1997) data set.

A number of relative rate studies involving reaction of Cl-atoms with ethane versus methane have been conducted (Pritchard et al., 1955; Knox and Nelson, 1959; Lee and

Table III-B-2. Summary of measurements of the rate coefficient (k) for reaction of Cl-atoms with ethane; $k = AT^n \exp(-B/T)\,cm^3\ molecule^{-1}\ s^{-1}$

A	B (K)	n	$10^{11} \times k$	T (K)	Technique[a]	Reference	Temperature Range (K)
7.00×10^{-17}	660	2			RR[b]	Pritchard et al. (1955)	349–563
5.78×10^{-17}	644	2			RR[b]	Knox and Nelson (1959)	193–593
			6.0 ± 1.5	298	FP–RF	Davis et al. (1970)	
$(7.29 \pm 1.23) \times 10^{-11}$	61 ± 44	0	5.45 ± 0.46	222	FP–RF	Manning and Kurylo (1977)	222–322
			5.48 ± 0.46	232			
			5.76 ± 0.24	243			
			5.90 ± 0.28	255			
			6.11 ± 0.36	267			
			5.74 ± 0.28	280			
			5.93 ± 0.44	296			
			5.92 ± 0.50	322			
			3.5	361	RR[b]	Lee and Rowland (1977)	
			3.7	298			
			6.4	243			
6.24×10^{-17}	672	2	5.24 ± 0.23	296	RR[b]	Lin et al. (1978)	198–296
			5.56 ± 0.22	273			
			5.85 ± 0.28	253			
			6.16 ± 0.34	233			
			6.31 ± 0.59	220			
			6.47 ± 0.33	198			
			4.0 ± 1.2	298	DF–RF	Schlyer et al. (1978)	
			5.95 ± 0.28	298	DF–MS	Ray et al. (1980)	
$(9.01 \pm 0.48) \times 10^{-11}$	133 ± 15	0	5.33 ± 0.22	220	DF–RF	Lewis et al. (1980)	220–604
			4.70 ± 0.21	223			
			5.03 ± 0.25	250			
			5.48 ± 0.30	298			
			6.83 ± 0.33	402			

(continued)

Table III-B-2. *(continued)*

A	B (K)	n	$10^{11} \times k$	T (K)	Technique[a]	Reference	Temperature Range (K)
			6.61 ± 0.33	410			
			7.04 ± 0.53	515			
			7.09 ± 0.55	604			
11.0×10^{-17}	532	2			RR[b]	Tschuikow-Roux et al. (1985a)	280–368
$(8.20 \pm 0.12) \times 10^{-11}$	86 ± 10		5.45 ± 0.07	203	VLPR–MS	Dobis and Benson (1991)	203–343
			5.77 ± 0.03	233			
			5.95 ± 0.06	268			
			6.10 ± 0.22	298			
			6.53 ± 0.17	343			
			7.05 ± 1.4	298	FP–RF	Kaiser et al. (1992)	
			5.9 ± 0.6	297	FP–IR abs.	Stickel et al. (1992)	
			5.53 ± 0.21	298	DF–RF	Beichert et al. (1995)	
			5.75 ± 0.45	298	FP–RF	Tyndall et al. (1997a)	
8.6×10^{-11}	135	0	5.5 ± 0.2	292	FP-IR abs.	Pilgrim et al. (1997)	292–600
			6.0 ± 0.3	400			
3.1×10^{-11}	−150	0.7	6.4 ± 0.1	450			292–800
			6.4 ± 0.2	500			
			6.7 ± 0.1	550			
			7.0 ± 0.3	600			
			7.1 ± 0.1	650			
			7.6 ± 0.2	700			
			8.1 ± 0.7	800			
			6.5 ± 0.3	297	FP–RF	Mellouki (1998)	
			5.85 ± 0.55	298	FP–LIF	Hitsuda et al. (2001b)	
4.91×10^{-12}	82	0.47	5.23 ± 0.20	299	DF–RF	Bryukov et al. (2003)	299–1002
			5.46 ± 0.56	299			
7.23×10^{-13}	−117	0.70	6.14 ± 0.26	346			203–1400
			6.94 ± 0.60	402			
			6.88 ± 0.40	426			
			7.66 ± 0.63	498			
			8.27 ± 0.46	604			

			9.44 ± 0.58	707			
			10.03 ± 0.26	758			
			9.93 ± 0.68	758			
			10.68 ± 0.63	855			
			11.63 ± 0.60	1002			
7.32×10^{-11}	74	0	5.15 ± 1.44	353	DF–RF	Hickson and Keyser (2004)	177–353
			5.11 ± 0.99	324			
			5.10 ± 0.96	296			
			5.36 ± 0.70	296			
			5.25 ± 0.48	276			
			4.90 ± 0.86	257			
			5.11 ± 0.54	237			
			5.21 ± 0.53	217			
			5.43 ± 0.50	197			
			4.73 ± 0.54	177			
			8.00	690	RR[b]	Kaiser et al. (2004)	252–690
			6.81	677			
			7.11	600			
			6.85	544			
			7.08	542			
			7.28	539			
			6.94	539			
			6.32	486			
			6.99	485			
			6.70	450			
			6.50	384			
			6.33	297			
			6.54	297			
			6.39	297			
			6.17	297			
			6.08	297			
			6.23	297			
			6.37	267			
			5.98	255			
			6.16	252			

[a] Abbreviations for the techniques are defined in table II-A-1.

[b] Placed on an absolute basis using $k_{Cl+CH4} = 1.5 \times 10^{-17} T^2 \exp(-766/T)$.

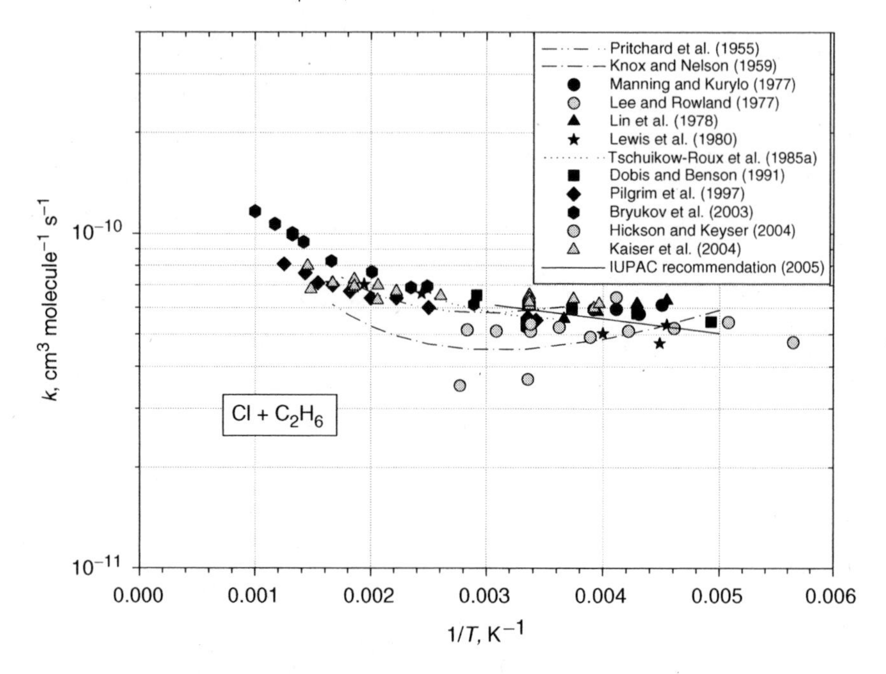

Figure III-B-2. Measurements of the rate coefficient for reaction of Cl-atoms with ethane. For clarity, a number of ambient temperature data points are not shown.

Rowland, 1977; Lin et al., 1978; Tschuikow-Roux et al., 1985a; Kaiser et al., 2004) and, as pointed out by Kaiser et al. (2004), this collective data set provides an independent means for testing the absolute measurements of methane and ethane. With the exception of the two highest temperature points of Lee and Rowland (1977), the relative rate data are in very good agreement. On the basis of the available relative rate data and pairs of recent absolute measurements from three laboratories (Pilgrim et al., 1997; Wang and Keyser, 1999; Bryukov et al., 2002, 2003; Hickson and Keyser, 2004), Kaiser et al. (2004) derived the expression $k_{C2H6}/k_{CH4} = 41 \times (T/298)^{-1.58} \times \exp(786/T)$. This expression retrieves rate coefficient ratios that are within 15% of the values obtained from the ratio of the IUPAC expression for ethane ($k = 8.3 \times 10^{-11} \exp(-100/T)$ cm^3 molecule^{-1} s^{-1}) and the modified Arrhenius expression derived above for methane ($k_{CH4} = 1.5 \times 10^{-17} T^2 \exp(-766/T)$ cm^3 molecule^{-1} s^{-1}) for the entire range from 180 to 750 K. Furthermore, the ratio of the ethane and methane Arrhenius expressions from the IUPAC panel yields essentially identical rate coefficient ratios to those of the Kaiser et al. (2004) expression over the range 180–300 K. The various relative rate studies, converted to absolute Cl + ethane data using $k_{CH4} = 1.5 \times 10^{-17} T^2 \exp(-766/T)$ cm^3 molecule^{-1} s^{-1}, are included in figure III-B-2.

III-B-3. Cl + Propane (C_3H_8)

A large body of absolute rate coefficient data (Lewis et al., 1980; Beichert et al., 1995; Pilgrim et al., 1997; Mellouki, 1998; Hitsuda et al., 2001a) and relative rate data (Pritchard et al., 1955; Knox and Nelson, 1959; Tschuikow-Roux et al., 1985b; Beichert et al., 1995; Tyndall et al., 1997a; Sarzynski and Sztuba, 2002) are available for this reaction. The results of these studies are generally in excellent agreement; see table III-B-3 and figure III-B-3.

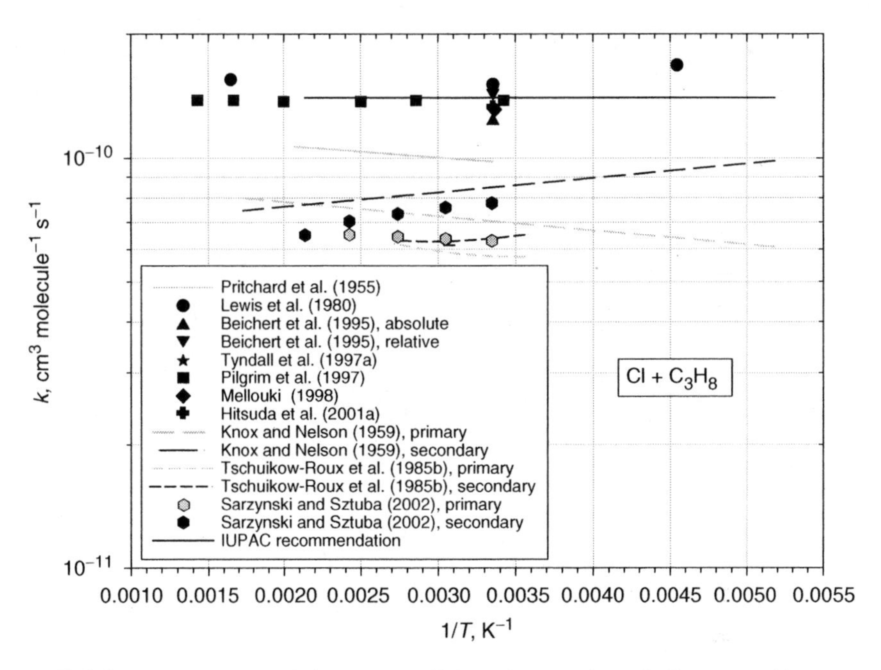

Figure III-B-3. Measurements of the rate coefficient for reaction of Cl-atoms with propane. (Dashed lines indicate data for reactions only at the $—CH_2$ group or the $—CH_3$ groups; see the legends.)

Table III-B-3. Summary of measurements of the rate coefficient (k) of the reaction of Cl-atoms with propane; $k = A\ \exp(-B/T)\ \mathrm{cm^3\ molecule^{-1}\ s^{-1}}$

A	B (K)	$10^{10}\times k$	T (K)	Technique[f]	Reference	Temperature Range (K)
1.22×10^{-10}	65			RR[c]	Pritchard et al. (1955)	298–484
9.2×10^{-11}	80			RR[c], P[a]	Knox and Nelson (1959)	193–593
6.5×10^{-11}	−80			RR[c], S[b]	Knox and Nelson (1959)	193–593
$(1.36 \pm 0.10) \times 10^{-11}$	-44 ± 25	1.68 ± 0.05	220	DF–RF	Lewis et al. (1980)	220–607
		1.51 ± 0.06	298			
		1.55 ± 0.12	607			
4.5×10^{-11}	−80			RR[d], P[a]	Tschuikow-Roux et al. (1985b)	281–368

(continued)

Table III-B-3. (*continued*)

A	B (K)	$10^{11} \times k$	T (K)	Technique[f]	Reference	Temperature Range (K)
3.2×10^{-11}	−200			RR[d], S[b]	Tschuikow-Roux et al., (1985b)	281–368
		1.23 ± 0.10	298	DF–RF	Beichert et al. (1995)	
		1.45 ± 0.15	298	RR[c]	Beichert et al. (1995)	
		1.5 ± 0.2	297	RR[c,e]	Tyndall et al. (1997a)	
1.38×10^{-10}	0	1.38 ± 0.02	292	FP–IR abs.	Pilgrim et al. (1997)	292–700
		1.38 ± 0.05	350			
		1.37 ± 0.03	400			
		1.37 ± 0.01	500			
		1.38 ± 0.02	600			
		1.38 ± 0.03	700			
		1.31 ± 0.03	297	FP–RF	Mellouki (1998)	
		1.33 ± 0.03	298	FP–LIF	Hitsuda et al. (2001a)	
7.0×10^{-11}	30	0.629 ± 0.017	298.6	RR[c], P[a]	Sarzynski and Sztuba (2002)	298–468
		0.636 ± 0.012	328	P[a]		
		0.644 ± 0.012	365	P[a]		
		0.651 ± 0.026	412.3	P[a]		
		0.650 ± 0.028	467.7	P[a]		
4.9×10^{-11}	−142	0.778 ± 0.014	298.6	S[b]		298–468
		0.759 ± 0.015	328	S[b]		
		0.732 ± 0.005	365	S[b]		
		0.703 ± 0.024	412.3	S[b]		
		0.650 ± 0.24	467.7	S[b]		

[a] P, Rate coefficients for abstraction from the primary sites in propane.
[b] S, Rate coefficients for abstraction from the secondary sites in propane.
[c] Placed on an absolute basis using $k_{C2H6} = 8.3 \times 10^{-11} \exp(-100/T)$ cm^3 molecule^{-1} s^{-1}.
[d] Placed on an absolute basis using $k_{CH4} = 6.6 \times 10^{-12} \exp(-1240/T)$ cm^3 molecule^{-1} s^{-1}.
[e] Placed on an absolute basis using $k_{C2H4} = 10.3 \times 10^{-11}$ cm^3 molecule^{-1} s^{-1} (Atkinson et al., 2005a)
[f] Abbreviations for the techniques are defined in table II-A-1.

Absolute determinations of this rate coefficient have been made using the discharge flow tube–resonance fluorescence technique (Lewis et al., 1980; Beichert et al., 1995), or via flash photolysis with either Cl-atom detection by resonance fluorescence (Mellouki, 1998), vacuum ultraviolet (VUV) laser-induced fluorescence (Hitsuda et al., 2001a), or HCl detection via time-resolved IR absorption (Pilgrim et al., 1997). Room-temperature values fall in the range $(1.23–1.58) \times 10^{-10}$ cm^3 molecule^{-1} s^{-1}. No measurable temperature dependence of the rate coefficient has been observed either by Pilgrim et al. (292–700 K) or by Lewis et al. (220–607 K).

Relative rate determinations have been made using either methane (Tschuikow-Roux et al., 1985b), ethane (Pritchard et al., 1955; Knox and Nelson, 1959; Beichert et al., 1995; Tyndall et al., 1997a; Sarzynski and Sztuba, 2002), or ethene (Tyndall et al., 1997a) as the reference compound. All data, once normalized to currently recommended reference rate coefficients, fall in the range (1.22–1.55) $\times 10^{-10}$ cm^3 molecule^{-1} s^{-1} at room temperature, in excellent agreement with the absolute rate determinations just discussed. Site-specific rate coefficients have been determined via quantification of the 1- and 2-chloropropane products of chain chlorination at room temperature by Tyndall et al. (1997a), and over ranges of temperature by Sarzynski and Sztuba (2002), Tschuikow-Roux et al. (1985b), and Knox and Nelson (1959).

An unweighted average of the data of Pritchard et al. (1955), Knox and Nelson (1959), Lewis et al. (1980), Tschuikow-Roux et al. (1985b), Beichert et al. (1995), Pilgrim et al. (1997), Tyndall et al. (1997a), Mellouki (1998), Hitsuda et al. (2001a), and Sarzynski and Sztuba (2002) at ambient temperatures leads to a value of $(1.4 \pm 0.2) \times 10^{-10}$ cm^3 molecule^{-1} s^{-1}. There is strong evidence (Pritchard et al., 1955; Knox and Nelson, 1959; Lewis et al., 1980; Tschuikow-Roux et al., 1985b; Pilgrim et al., 1997) that the rate coefficient is essentially independent of temperature between 200 and 600 K. The IUPAC panel recommends $k = (1.4 \pm 0.2) \times 10^{-10}$ cm^3 molecule^{-1} s^{-1}, independent of temperature, and that recommendation is adopted here.

Regarding the site-specific rate coefficients, we recommend the expressions of Sarzynski and Sztuba (2002), placed on an absolute basis using the IUPAC (Atkinson et al., 2005a) rate coefficient expression for ethane, $k(\text{primary}) = 7.0 \times 10^{-11} \exp(-30/T)$ cm^3 molecule^{-1} s^{-1} and $k(\text{secondary}) = 4.9 \times 10^{-11} \exp(142/T)$ cm^3 molecule^{-1} s^{-1}. These data suggest that about 56% of the reaction occurs at the secondary site at room temperature. These expressions are similar to those of Knox and Nelson (1959) and agree very well with the 298 K branching ratios reported by Tyndall et al. (1997a).

III-B-4. Cl + Butane (n-C_4H_{10})

There are a number of absolute and relative rate studies of this reaction that have been summarized in table III-B-4. The rate coefficient is seemingly very well established. In fact, the reaction has become the standard upon which much of the relative rate coefficient database is established (e.g., Atkinson and Aschmann, 1985; Hooshiyar and Niki, 1995; Aschmann and Atkinson, 1995a). Absolute determinations of this rate coefficient have been made using discharge flow tube–resonance fluorescence (Lewis et al., 1980; Beichert et al., 1995; Stutz et al., 1998), or flash photolysis with either: a) Cl-atom detection by resonance fluorescence (Tyndall et al., 1997a; Mellouki, 1998) or VUV laser-induced fluorescence (Hitsuda et al., 2001a), or b) HCl detection via IR absorption (Qian et al., 2002) or emission (Nesbitt and Leone, 1982). All values reported at room temperature are in excellent agreement, ranging between 1.80 and 2.26×10^{-10} cm^3 molecule^{-1} s^{-1}. In addition, Lewis et al. (1980) report very little temperature dependence to the rate coefficient between 298 and 598 K, $k = 2.14 \times 10^{-10} \exp(10/T)$ cm^3 molecule^{-1} s^{-1}, or $k = 2.25 \times 10^{-10}$ cm^3 molecule^{-1} s^{-1} independent of temperature.

Relative rate studies for the butane reaction have been conducted at room temperature using ethane (Knox and Nelson, 1959; Atkinson and Aschmann, 1985; Wallington et al., 1988c; Beichert et al., 1995; Hooshiyar and Niki, 1995; Tyndall et al., 1997a;

Table III-B-4. Summary of measurements of the rate coefficient (k) of the reaction of Cl-atoms with butane; $k = A\ \exp(-B/T)\ cm^3\ molecule^{-1}\ s^{-1}$

A	B (K)	$10^{10} \times k$	T (K)	Technique[e]	Reference	Temperature Range (K)
7.4×10^{-11}	−25			RR[c] P[a]	Knox and Nelson (1959)	193–593
7.8×10^{-11}	−285			RR[c] S[b]	Knox and Nelson (1959)	193–593
$(2.15 \pm 0.10) \times 10^{-10}$	-12 ± 26	2.25 ± 0.10	298	DF–RF	Lewis et al. (1980)	298–607
		2.17 ± 0.09	422			
		2.19 ± 0.03	598			
		1.8 ± 0.2	298	FP–IR em.	Nesbitt and Leone (1982)	
		1.83 ± 0.05	296	RR[c]	Atkinson and Aschmann (1985)	
		2.06 ± 0.08	296	RR[d]	Atkinson and Aschmann (1985)	
		1.89 ± 0.02	298	RR[c]	Atkinson and Aschmann (1987)	
		1.72 ± 0.13	295	RR[c]	Wallington et al. (1988c)	
		1.97 ± 0.05	295	RR[d]	Wallington et al. (1988c)	
		2.11 ± 0.18	298	DF–RF	Beichert et al. (1995)	
		2.27 ± 0.11	298	RR[c]	Beichert et al. (1995)	
		2.02 ± 0.16	296	RR[c]	Hooshiyar and Niki (1995)	
		2.13 ± 0.03	296	RR[d]	Hooshiyar and Niki (1995)	

2.17 ± 0.10	297	RR[c]	Tyndall et al. (1997a)	
2.15 ± 0.15	298	FP–RF	Tyndall et al. (1997a)	
2.25 ± 0.06	297	FP–RF	Mellouki (1998)	
2.0 ± 0.4	293	DF–RF	Stutz et al. (1998)	
2.4 ± 0.3	Room	RR[c]	Stutz et al. (1998)	
2.04 ± 0.12	298	FP–LIF	Hitsuda et al. (2001a)	
1.91 ± 0.10	298	FP–IR abs.	Qian et al. (2002)	
0.627 ± 0.041	295.4	RR[c], P[a]	Sarzynski and Sztuba (2002)	295–469
0.647 ± 0.049	325.5	P[a]		
0.644 ± 0.025	365.2	P[a]		
0.677 ± 0.020	411.7	P[a]		
0.651 ± 0.040	469.4	P[a]		
1.53 ± 0.08	295.4	S[b]		
1.47 ± 0.10	325.5	S[b]		
1.43 ± 0.08	365.2	S[b]		
1.42 ± 0.08	411.7	S[b]		
1.24 ± 0.08	469.4	S[b]		

[a] Rate coefficients for abstraction from the primary sites in butane.
[b] Rate coefficients for abstraction from the secondary sites in butane.
[c] Placed on an absolute basis using $k_{C2H6} = 8.3 \times 10^{-11} \exp(-100/T)$ cm^3 molecule^{-1} s^{-1}.
[d] Placed on an absolute basis using $k_{C3H8} = 1.4 \times 10^{-10}$ cm^3 molecule^{-1} s^{-1}.
[e] Abbreviations for the techniques are defined in table II-A-1.

Stutz et al., 1998; Sarzynski and Sztuba, 2002), propane (Wallington et al., 1988c; Hooshiyar and Niki, 1995), or ethene (Tyndall et al., 1997a) as the reference compound. When normalized to currently accepted values for the reference reaction rate coefficients, all data except those of Knox and Nelson fall in the range $(1.7–2.3) \times 10^{-10}$ cm^3 molecule^{-1} s^{-1}. The Knox and Nelson (1959) data extend over the range 193–593 K and show a slightly higher value at room temperature, $k = 2.8 \times 10^{-10}$ cm^3 molecule^{-1} s^{-1}, with a measurable negative temperature dependence. Site specific rate coefficients, obtained via the quantification of the 1- and 2-chlorobutane products of the chain oxidation, have been reported at room temperature by Tyndall et al. (1997a) and over a range of temperatures by Knox and Nelson (1959) and Sarzynski and Sztuba (2002).

An unweighted average of the room temperature data from the absolute and relative rate measurements of Lewis et al. (1980), Nesbitt and Leone (1982), Atkinson and Aschmann (1985), Beichert et al. (1995), Hooshiyar and Niki (1995), Tyndall et al. (1997a), Mellouki (1998), Hitsuda et al. (2001a), Qian et al. (2002), and Sarzynski and Sztuba (2002) leads to a value of $k = (2.1 \pm 0.3) \times 10^{-10}$ cm^3 molecule^{-1} s^{-1}, which is recommended here. There appears to be little or no temperature dependence to the overall rate coefficient (Lewis et al., 1980; Tyndall et al., 1997a; Sarzynski and Sztuba, 2002). The data of Sarzynski and Sztuba (2002) and Tyndall et al. (1997a) provide a consistent picture for the site-specific rate coefficients at room temperature, k(primary) $= (6.2 \pm 0.6) \times 10^{-11}$ cm^3 molecule^{-1} s^{-1}; k(secondary) $= (1.5 \pm 0.15) \times 10^{-10}$ cm^3 molecule^{-1} s^{-1}. For the temperature dependence of the site-specific rate coefficients, the data of Sarzynski and Sztuba are recommended as placed on an absolute basis using the IUPAC (Atkinson et al., 2005a) expression for the Cl + ethane reference reaction: k(primary) $= 7.2 \times 10^{-11} \exp(-40/T)$ cm^3 molecule^{-1} s^{-1}; k(secondary) $= 9.5 \times 10^{-11} \exp(140/T)$ cm^3 molecule^{-1} s^{-1}. The data of Knox and Nelson for k(primary) are 20–25% higher than this recommendation; their data for k(secondary) also give a higher value at 298 K than the recommendation and show a stronger negative temperature dependence.

III-B-5. Cl + 2-Methylpropane [Isobutane, $(CH_3)_2CHCH_3$]

A number of absolute and relative measurements of the rate coefficient for the reaction of Cl-atoms with 2-methylpropane (isobutane) have been reported; see table III-B-5. Most studies are in excellent agreement.

Absolute measurements have been made at room temperature via the techniques of discharge flow–resonance fluorescence (Lewis et al., 1980; Beichert et al., 1995) and flash photolysis (Hitsuda et al., 2001a; Smith et al., 2002b). The results of the four absolute measurements are all essentially identical, falling in the range $(1.40–1.49) \times 10^{-10}$ cm^3 molecule^{-1} s^{-1}.

A suite of room-temperature relative rate measurements has been reported, all using butane as the reference compound. Once a common reference rate coefficient is established (k(butane) $= 2.1 \times 10^{-10}$ cm^3 molecule^{-1} s^{-1}, see previous section), the data of Atkinson and Aschmann (1985) give: $(1.46 \pm 0.02) \times 10^{-10}$ cm^3 molecule^{-1} s^{-1}; those of Wallington et al. (1988c), $(1.41 \pm 0.08) \times 10^{-10}$ cm^3 molecule^{-1} s^{-1}; those

Table III-B-5. Summary of measurements of the rate coefficient (k) of the reaction of Cl-atoms with 2 methyl-propane; $k = A \exp(-B/T)\,cm^3\,molecule^{-1}\,s^{-1}$

A	B (K)	$10^{10} \times k$	T (K)	Technique[a]	Reference	Temperature Range (K)
1.36×10^{-10}	30			RR[b]	Pritchard et al. (1955)	298–484
1.08×10^{-10}	−10			RR[b], P[c]	Knox and Nelson (1959)	193–593
1.49×10^{-11}	−410			RR[b], T[d]	Knox and Nelson (1959)	193–593
7.3×10^{-11}	−20			RR[b], P[c]	Cadman et al. (1976)	252–367
2.5×10^{-11}	−174			RR[b], T[d]		
		1.46 ± 0.06	298	DF–RF	Lewis et al. (1980)	
		1.46 ± 0.03	296	RR[e]	Atkinson and Aschmann (1985)	
		1.41 ± 0.08	295	RR[e]	Wallington et al. (1988c)	
		1.40 ± 0.08	298	DF–RF	Beichert et al. (1995)	
		1.43 ± 0.06	298	RR[e]	Beichert et al. (1995)	
		1.40 ± 0.01	296	RR[e]	Hooshiyar and Niki (1995)	
		1.40 ± 0.09	298	FP–LIF	Hitsuda et al. (2001a)	
		1.49 ± 0.02	295	FP–IR abs.	Smith et al. (2002b)	
		1.01 ± 0.02	297.5	RR[b], P[c]	Sarzynski and Sztuba (2002)	295–469
		0.978 ± 0.037	327.1	P[c]		
		1.04 ± 0.14	364	P[c]		
		1.07 ± 0.03	411.6	P[c]		
		1.07 ± 0.07	466.6	P[c]		
		0.546 ± 0.036	297.5	T[d]		
		0.495 ± 0.061	327.1	T[d]		
		0.479 ± 0.063	364	T[d]		
		0.449 ± 0.085	411.6	T[d]		

[a] Abbreviations for the techniques are defined in table II-A-1.
[b] Placed on an absolute basis using $k_{C2H6} = 8.3 \times 10^{-11} \exp(-100/T)\,cm^3\,molecule^{-1}\,s^{-1}$.
[c] Rate coefficients for abstraction from the primary sites in 2-methylpropane.
[d] Rate coefficients for abstraction from the tertiary site in 2-methylpropane.
[e] Placed on an absolute basis using $k_{C4H10} = 2.1 \times 10^{-10}\,cm^3\,molecule^{-1}\,s^{-1}$.

of Hooshiyar and Niki (1995), $(1.40 \pm 0.01) \times 10^{-10}$ cm^3 molecule^{-1} s^{-1}; and those of Beichert et al. (1995) give k(butane) $= (1.43 \pm 0.08) \times 10^{-10}$ cm^3 molecule^{-1} s^{-1}. All are seen to be in excellent agreement.

More recently, Sarzynski and Sztuba (2002), using ethane as the reference compound, studied the site-specific reaction of Cl-atoms with 2-methylpropane. Using the IUPAC (Atkinson et al., 2005a) expression for the reference reaction rate coefficient, their data give $k = (1.52 \pm 0.05) \times 10^{-10}$ cm^3 molecule^{-1} s^{-1} for the total reaction rate coefficient, independent of temperature (295–412 K), in agreement with the absolute and relative rate data reported above. Furthermore, they report that the ratio for reaction at the primary and tertiary hydrogen atoms in 2-methylpropane is given by the expression k(primary)/k(tertiary) $= 4.48 \exp(-259/T)$, with k(primary)/k(tertiary) $=$ 1.88 at 298 K.

Some older relative rate studies of the reaction of Cl with 2-methylpropane are also available (Pritchard et al., 1955; Knox and Nelson, 1959; Cadman et al., 1976). These data are broadly consistent with the more modern data presented above. The Pritchard et al. (1955) study, using the IUPAC value for the reference reaction rate coefficient, yields $k = 1.25 \times 10^{-10}$ cm^3 molecule^{-1} s^{-1}, essentially independent of temperature over the range 298–484 K. The Knox and Nelson (1959) data for both the total and site-specific rate coefficients are in good agreement with the Sarzynski and Sztuba (2002) data at ambient temperatures, but their k(total) and k(tertiary) data both show strong negative temperature dependencies. The Cadman et al. (1976) total and site-specific rate coefficients are generally lower than the other measurements (by about 20%), although the temperature-dependencies of their site-specific rate coefficients agree quite well with those of Sarzynski and Sztuba (2002).

To summarize, a value of $k = 1.4 \times 10^{-10}$ cm^3 molecule^{-1} s^{-1} is recommended for this rate coefficient at 298 K, based on an unweighted average of the data of Lewis et al. (1980), Atkinson and Aschmann (1985), Wallington et al. (1988c), Beichert et al. (1995), Hooshiyar and Niki (1995), Hitsuda et al. (2001a), and Smith et al. (2002b). The uncertainty is estimated at ±10%. The rate coefficient is essentially independent of temperature. The branching ratio k(primary)/k(tertiary) $= 4.48 \exp(-259/T)$ reported by Sarzynski and Sztuba (2002) is adopted.

III-B-6. Cl + Pentane (n-C_5H_{12})

The three available relative rate determinations, all made at room temperature, are in substantive agreement; see table III-B-6. Using the value for the Cl + butane reference reaction recommended above, $k = 2.1 \times 10^{-10}$ cm^3 molecule^{-1} s^{-1}, leads to values of $(2.69 \pm 0.13) \times 10^{-10}$ (Atkinson and Aschmann, 1985), $(2.71 \pm 0.02) \times 10^{-10}$ (Hooshiyar and Niki, 1995), and $(2.90 \pm 0.15) \times 10^{-10}$ cm^3 molecule^{-1} s^{-1} (Wallington et al., 1988c). These data are in reasonable agreement with the flash photolysis–TDL absorption measurement of Qian et al. (2002), $k = (2.46 \pm 0.12) \times 10^{-10}$ cm^3 molecule^{-1} s^{-1}. An unweighted average of these four determinations leads to a value of $k = 2.7 \times 10^{-10}$ cm^3 molecule^{-1} s^{-1}, which is recommended here with an estimated uncertainty of ±15%.

Table III-B-6. Summary of measurements of the rate coefficient (k) of the reaction of Cl-atoms with pentane; $k = A\exp(-B/T)\,cm^3\,molecule^{-1}\,s^{-1}$

$10^{10} \times k$	T (K)	Technique[a]	Reference
2.69 ± 0.13	296	RR[b]	Atkinson and Aschmann (1985)
2.90 ± 0.15	295	RR[b]	Wallington et al. (1988c)
2.71 ± 0.02	296	RR[b]	Hooshiyar and Niki (1995)
2.46 ± 0.12	298	FP–IR abs.	Qian et al. (2002)

[a] Abbreviations for the techniques are defined in table II-A-1.
[b] Placed on an absolute basis using $k_{C4H10} = 2.1 \times 10^{-10}\,cm^3\,molecule^{-1}\,s^{-1}$.

III-B-7. Cl + 2-Methylbutane [Isopentane, $(CH_3)_2CHCH_2CH_3$]

There are two relative rate measurements available for this reaction (Atkinson and Aschmann, 1985; Hooshiyar and Niki, 1995), both obtained at ambient temperatures using the relative rate technique with butane as the reference compound. The data are summarized in table III-B-7. Upon normalization to the recommended rate coefficient for the reference reaction, values of $(2.16 \pm 0.08) \times 10^{-10}$ and $(2.12 \pm 0.02) \times 10^{-10}\,cm^3\,molecule^{-1}\,s^{-1}$ are obtained, respectively. These relative rate determinations are in agreement with the flash photolysis–TDL absorption measurement of Qian et al. (2002), who report $k = (1.94 \pm 0.10) \times 10^{-10}\,cm^3\,molecule^{-1}\,s^{-1}$. A value of $2.1 \times 10^{-10}\,cm^3\,molecule^{-1}\,s^{-1}$ is recommended, with an estimated uncertainty of ±15%.

Table III-B-7. Summary of measurements of the rate coefficient (k)for the reaction of Cl-atoms with 2-methylbutane (isopentane); $k = A\exp(-B/T)\,cm^3\,molecule^{-1}\,s^{-1}$

$10^{10} \times k$	T (K)	Technique[a]	Reference
2.16 ± 0.09	296	RR[b]	Atkinson and Aschmann (1985)
2.12 ± 0.02	296	RR[b]	Hooshiyar and Niki (1995)
1.94 ± 0.10	298	FP–IR abs.	Qian et al. (2002)

[a] Abbreviations for the techniques are defined in table II-A-1.
[b] Placed on an absolute basis using $k_{C4H10} = 2.1 \times 10^{-10}\,cm^3\,molecule^{-1}\,s^{-1}$.

III-B-8. Cl + 2,2-Dimethylpropane [Neopentane, $(CH_3)_4C$]

This rate coefficient has been measured by both absolute (Kambanis et al., 1995; Qian et al., 2002) and relative rate methods (Pritchard et al., 1955; Knox and Nelson, 1959; Atkinson and Aschmann, 1985; Rowley et al., 1992a); see table III-B-8 and figure III-B-4. All room-temperature measurements fall in the range $(1.0\text{–}1.2) \times 10^{-10}\,cm^3\,molecule^{-1}\,s^{-1}$. The data of Pritchard et al. (1955), Knox and Nelson (1959), Rowley et al. (1992a), and Kambanis et al. (1995) all suggest a very weak (or zero) temperature dependence to the rate coefficient. A value of

Table III-B-8. Summary of measurements of the rate coefficient (k) for the reaction of Cl-atoms with 2,2-dimethylpropane (neopentane); $k = A \exp(-B/T)\,cm^3\,molecule^{-1}\,s^{-1}$

A	B (K)	$10^{10} \times k$	T (K)	Technique[a]	Reference	Temperature Range (K)
0.85×10^{-10}	−50			RR[b]	Pritchard et al. (1955)	298–484
$(1.16 \pm 0.03) \times 10^{-10}$	−10			RR[c]	Knox and Nelson (1959)	193–593
		1.17 ± 0.03	298	RR[d]	Atkinson and Aschmann (1985)	
$(1.16 \pm 0.05) \times 10^{-10}$	0	1.18 ± 0.14	248	RR[e]	Rowley et al. (1992a)	248–366
		1.26 ± 0.11	273			
		1.05 ± 0.09	294			
		1.15 ± 0.06	326			
		0.89 ± 0.09	348			
		1.12 ± 0.07	366			
$(1.11 \pm 0.13) \times 10^{-10}$	0	1.14 ± 0.13	273	VLPR	Kambanis et al. (1995)	273–333
		1.13 ± 0.14	303			
		1.05 ± 0.09	333			
		1.01 ± 0.05	298	FP–IR abs.	Qian et al. (2002)	

[a] Abbreviations for the techniques are defined in table II-A-1.
[b] Placed on an absolute basis using $k_{C2H6} = 8.3 \times 10^{-11} \exp(-100/T)\,cm^3\,molecule^{-1}\,s^{-1}$.
[c] Placed on an absolute basis using k_{C3H8}(primary site) $= 7.0 \times 10^{-11} \exp(-30/T)\,cm^3\,molecule^{-1}\,s^{-1}$.
[d] Placed on an absolute basis using $k_{C4H10} = 2.1 \times 10^{-10}\,cm^3\,molecule^{-1}\,s^{-1}$.
[e] Placed on an absolute basis using $k_{CH3OH} = 5.7 \times 10^{-11}\,cm^3\,molecule^{-1}\,s^{-1}$.

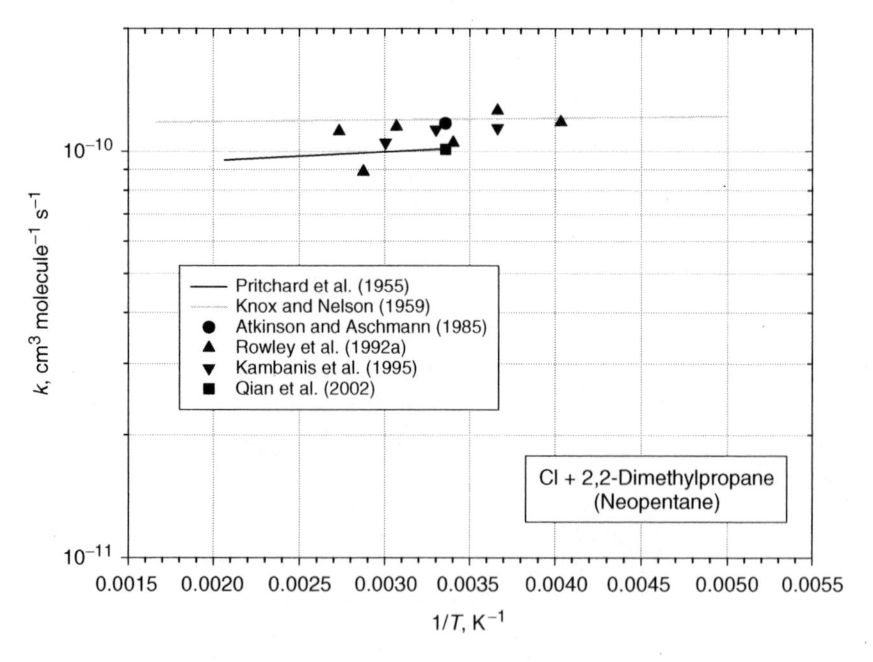

Figure III-B-4. Measurements of the rate coefficient for reaction of Cl-atoms with 2,2-dimethylpropane (neopentane).

$k = 1.1 \times 10^{-10}$ cm^3 molecule^{-1} s^{-1} is recommended, independent of temperature, with an estimated uncertainty of ±15%.

III-B-9. Cl + Hexane (n-C_6H_{14})

The three relative rate studies of this reaction (Atkinson and Aschmann, 1985; Wallington et al., 1988c; Hooshiyar and Niki, 1995), when normalized to a common value of 2.1×10^{-10} cm^3 molecule^{-1} s^{-1} for the butane reference rate coefficient, all yield $k = (3.2 - 3.3) \times 10^{-10}$ cm^3 molecule^{-1} s^{-1} near room temperature (see table III-B-9). These values are in good agreement with the recent flash photolysis–tunable diode laser absorption data of Qian et al. (2002), $k = (3.44 \pm 0.17) \times 10^{-10}$ cm^3 molecule^{-1} s^{-1}. A value of

Table III-B-9. Summary of measurements of the rate coefficient (k, cm^3 molecule^{-1} s^{-1}) for the reaction of Cl-atoms with C_6–C_{10} acylic alkanes at temperatures near 298 K

Alkane	$10^{10} \times k$	Technique[a,b]	Reference
Hexane	3.23 ± 0.06	RR	Atkinson and Aschmann (1985)
	3.22 ± 0.21	RR	Wallington et al. (1988c)
	3.30 ± 0.04	RR	Hooshiyar and Niki (1995)
	3.44 ± 0.17	FP–IR abs.	Qian et al. (2002)
2-Methylpentane	2.71 ± 0.09	RR	Aschmann and Atkinson (1995a)
	2.81 ± 0.09	RR	Hooshiyar and Niki (1995)
3-Methylpentane	2.69 ± 0.07	RR	Aschmann and Atkinson (1995a)
2,3-Dimethylbutane	2.21 ± 0.06	RR	Atkinson and Aschmann (1985)
	2.31 ± 0.08	RR	Wallington et al. (1989b)
	2.16 ± 0.06	RR	Hooshiyar and Niki (1995)
Heptane	3.64 ± 0.13	RR	Atkinson and Aschmann (1985)
	3.95 ± 0.06	RR	Hooshiyar and Niki (1995)
2,2,3-Trimethylbutane	1.93 ± 0.08	RR	Aschmann and Atkinson (1995a)
2,4-Dimethylpentane	2.77 ± 0.13	RR	Aschmann and Atkinson (1995a)
2-Methylhexane	3.38 ± 0.04	RR	Hooshiyar and Niki (1995)
Octane	4.39 ± 0.13	RR	Aschmann and Atkinson (1995a)
	4.43 ± 0.13	RR	Hooshiyar and Niki (1995)
2,2,3,3-Tetramethylbutane	1.69 ± 0.10	RR	Aschmann and Atkinson (1995a)
2,2,4-Trimethylpentane	2.50 ± 0.09	RR	Aschmann and Atkinson (1995a)
	2.44 ± 0.08	RR	Hooshiyar and Niki (1995)
Nonane	4.64 ± 0.13	RR	Aschmann and Atkinson (1995a)
Decane	5.27 ± 0.19	RR	Aschmann and Atkinson (1995a)

[a] Abbreviations for the techniques are defined in table II-A-1.
[b] All relative rate measurements placed on an absolute basis using $k_{C4H10} = 2.1 \times 10^{-10}$ cm^3 molecule^{-1} s^{-1}.

3.3×10^{-10} cm^3 molecule^{-1} s^{-1} is recommended, with an estimated uncertainty of ±15%.

III-B-10. Cl + 2-Methylpentane [$(CH_3)_2CHCH_2CH_2CH_3$]

Relative rate studies of this reaction (Aschmann and Atkinson, 1995a; Hooshiyar and Niki, 1995) yield $k = (2.7–2.8) \times 10^{-10}$ cm^3 molecule^{-1} s^{-1} near 298 K when an updated value of 2.1×10^{-10} cm^3 molecule^{-1} s^{-1} is used for the butane reference rate coefficient; see table III-B-9. A value of 2.8×10^{-10} cm^3 molecule^{-1} s^{-1} is recommended, with an estimated uncertainty of ±15%.

III-B-11. Cl + 3-Methylpentane [$CH_3CH_2CH(CH_3)CH_2CH_3$]

The relative rate study of Aschmann and Atkinson (1995a), using the updated value of 2.1×10^{-10} cm^3 molecule^{-1} s^{-1} for the butane reference rate coefficient, yields $k = (2.69 \pm 0.06) \times 10^{-10}$ cm^3 molecule^{-1} s^{-1} at 298 K (see table III-B-9). This value is recommended, but with an estimated uncertainty of ±20% to reflect the fact that only one study has been made.

III-B-12. Cl + 2,3-Dimethylbutane [$(CH_3)_2CHCH(CH_3)_2$]

Three relative rate studies (Atkinson and Aschmann, 1985; Wallington et al., 1989b; Hooshiyar and Niki, 1995) of the reaction of Cl with 2,3-dimethylbutane have been conducted; see table III-B-9. The data, all obtained using butane as the reference species and all conducted at ambient temperatures, are in excellent agreement, $k = (2.21 \pm 0.06) \times 10^{-10}$, $(2.31 \pm 0.08) \times 10^{-10}$, and $(2.16 \pm 0.06) \times 10^{-10}$ cm^3 molecule^{-1} s^{-1}, respectively, once standardization to a common butane rate coefficient (2.1×10^{-10} cm^3 molecule^{-1} s^{-1}) is carried out. A value of 2.2×10^{-10} cm^3 molecule^{-1} s^{-1} is recommended, with an estimated uncertainty of ±15%.

III-B-13. Cl + Heptane (n-C_7H_{16})

Two relative rate measurements of this reaction rate coefficient have been made at ambient temperatures (Atkinson and Aschmann, 1985; Hooshiyar and Niki, 1995), using butane as the reference compound (see table III-B-9). The data are in good accord. Using the updated value for the reference rate coefficient, 2.1×10^{-10} cm^3 molecule^{-1} s^{-1}, yields $(3.64 \pm 0.13) \times 10^{-10}$ cm^3 molecule^{-1} s^{-1} (Atkinson and Aschmann, 1985) and $(3.95 \pm 0.07) \times 10^{-10}$ cm^3 molecule^{-1} s^{-1} (Hooshiyar and Niki, 1995). A rate coefficient $k = 3.8 \times 10^{-10}$ cm^3 molecule^{-1} s^{-1} is recommended, with an estimated uncertainty of ±15%.

III-B-14. Cl + 2,2,3-Trimethylbutane [$(CH_3)_3CCH(CH_3)_2$]

Aschmann and Atkinson (1995a) studied this reaction using the relative rate technique, with butane as the reference compound (see table III-B-9). A value of

$k = (1.93 \pm 0.08) \times 10^{-10}\,\text{cm}^3$ molecule^{-1} s^{-1} is obtained at ambient temperatures using the updated rate coefficient ($2.1 \times 10^{-10}\,\text{cm}^3$ molecule^{-1} s^{-1}) for the reference reaction. This value is recommended, with the uncertainty increased to $\pm 20\%$ to reflect the fact that only one study has been made.

III-B-15. Cl + 2,4-Dimethylpentane [$(CH_3)_2CHCH_2CH(CH_3)CH_3$]

Aschmann and Atkinson (1995a) measured this rate coefficient relative to butane, and reported a rate coefficient ratio of 1.32 ± 0.07. Using the updated rate coefficient ($2.1 \times 10^{-10}\,\text{cm}^3$ molecule^{-1} s^{-1}) for the reference reaction yields a value of $k = (2.77 \pm 0.13) \times 10^{-10}\,\text{cm}^3$ molecule^{-1} s^{-1} at room temperature; see table III-B-9. This value is recommended, with an increased uncertainty ($\pm 20\%$) since the results from only one study are available.

III-B-16. Cl + 2-Methylhexane [$(CH_3)_2CHCH_2CH_2CH_2CH_3$]

The only study of this reaction is a relative rate measurement of Hooshiyar and Niki (1995), using butane as the reference compound; see table III-B-9. Using the updated reference reaction rate coefficient results in a 296 K value of $k = (3.38 \pm 0.04) \times 10^{-10}\,\text{cm}^3$ molecule^{-1} s^{-1} for the 2-methylhexane reaction. This value is recommended, with an uncertainty of $\pm 20\%$.

III-B-17. Cl + Octane (n-C_8H_{18})

The two relative rate measurements of this reaction rate coefficient (Hooshiyar and Niki, 1995; Aschmann and Atkinson, 1995a) are in very good agreement; see table III-B-9. Values of $(4.44 \pm 0.13) \times 10^{-10}\,\text{cm}^3$ molecule^{-1} s^{-1} and $(4.39 \pm 0.13) \times 10^{-10}\,\text{cm}^3$ molecule^{-1} s^{-1}, respectively, are obtained at ambient temperatures, using the currently recommended value for the butane reference reaction. A value of $k = 4.4 \times 10^{-10}\,\text{cm}^3$ molecule^{-1} s^{-1} is recommended, with an estimated uncertainty of $\pm 15\%$.

III-B-18. Cl + 2,2,3,3-Tetramethylbutane [$(CH_3)_3CC(CH_3)_3$]

The only measurement is the relative rate study of Aschmann and Atkinson (1995a); see table III-B-9. Using an updated value for the butane reference reaction yields $k = (1.69 \pm 0.10) \times 10^{-10}\,\text{cm}^3$ molecule^{-1} s^{-1} at room temperature. This value is adopted with an uncertainty of $\pm 20\%$.

III-B-19. Cl + 2,2,4-Trimethylpentane [$(CH_3)_3CCH_2CH(CH_3)_2$]

The relative rate determinations of Aschmann and Atkinson (1995a) and Hooshiyar and Niki (1995), both using butane as the reference compound, are in excellent agreement; see table III-B-9. Values of $k = (2.50 \pm 0.09) \times 10^{-10}\,\text{cm}^3$ molecule^{-1} s^{-1} and

$(2.44 \pm 0.09) \times 10^{-10}$ cm^3 molecule^{-1} s^{-1}, respectively, are obtained at room temperature using the updated reference reaction rate coefficient. The recommended value is thus 2.5×10^{-10} cm^3 molecule^{-1} s^{-1}, with an estimated uncertainty of $\pm 15\%$.

III-B-20. Cl + Nonane (n-C_9H_{20})

This reaction has been studied by Aschmann and Atkinson (1995a) using the relative rate technique, with butane as the reference compound; see table III-B-9. The value obtained at room temperature, using the updated reference rate coefficient, is $(4.64 \pm 0.13) \times 10^{-10}$ cm^3 molecule^{-1} s^{-1}. A value of 4.6×10^{-10} cm^3 molecule^{-1} s^{-1} is adopted with an uncertainty of $\pm 20\%$, to reflect the fact that only one study is available.

III-B-21. Cl + Decane (n-$C_{10}H_{22}$)

Aschmann and Atkinson (1995a) have studied this reaction using the relative rate technique, with butane as the reference compound; see table III-B-9. The value obtained, using the updated reference reaction rate coefficient, is $(5.27 \pm 0.19) \times 10^{-10}$ cm^3 molecule^{-1} s^{-1} at ambient temperatures. An uncertainty of $\pm 20\%$ is suggested, given that only one study of this reaction is available.

III-B-22. Cl + Cyclopropane ($cyclo$-C_3H_6)

This reaction has been studied by a variety of absolute and relative rate techniques; data are summarized in table III-B-10 and displayed in figure III-B-5. DeSain et al. (2003) reported a value of $k = (1.36 \pm 0.40) \times 10^{-13}$ cm^3 molecule^{-1} s^{-1} at ambient temperatures by monitoring the appearance profiles of HO_2 following reaction of Cl with cyclopropane in O_2 in a laser flash photolysis-IR absorption system:

$$Cl + cyclo\text{-}C_3H_6 \rightarrow cyclo\text{-}C_3H_5 + HCl$$

$$cyclo\text{-}C_3H_5 + O_2 \rightarrow HO_2, \text{ other products}$$

In a companion publication, Hurley et al. (2003) report $k = (1.15 \pm 0.17) \times 10^{-13}$ cm^3 molecule^{-1} s^{-1} at 298 K from relative rate studies. Hurley et al. (2003) also studied the temperature dependence of the Cl + cyclopropane rate coefficient by monitoring appearance profiles for HCl in a laser flash photolysis-IR absorption system; they report $k = (1.8 \pm 0.3) \times 10^{-10} \exp((-2150 \pm 110)/T)$ cm^3 molecule^{-1} s^{-1} over the temperature range 293–623 K, with k measured to be $(1.12 \pm 0.22) \times 10^{-13}$ cm^3 molecule^{-1} s^{-1} at 293 K. Using a very low pressure reactor–mass spectrometer system, Baghal-Vayjooee and Benson (1979) report a room-temperature value of $k = (1.21 \pm 0.05) \times 10^{-13}$ cm^3 molecule^{-1} s^{-1}. Finally, Knox and Nelson (1959) studied the reaction of Cl with cyclopropane over the range 300–500 K relative to the reaction of Cl with ethane. Using an updated value for the reference reaction rate coefficient yields $k = 4.6 \times 10^{-11} \exp(-1660/T)$, with $k = (1.8 \pm 0.2) \times 10^{-13}$ cm^3 molecule^{-1} s^{-1} at 298 K.

An unweighted average of the ambient temperature data presented here yields $k = (1.4 \pm 0.4) \times 10^{-13}$ cm^3 molecule^{-1} s^{-1}, which we recommend. Agreement between

Table III-B-10. Summary of measurements of the rate coefficients (k) for the reaction of Cl-atoms with cycloalkanes; $k = A \exp(-B/T)$ cm^3 molecule^{-1} s^{-1}

Cycloalkane	A	B (K)	k	T (K)	Technique[a]	Reference	Temperature Range (K)
Cyclopropane	4.6×10^{-11}	1660			RR[b]	Knox and Nelson (1959)	293–593
			$(1.21 \pm 0.05) \times 10^{-13}$	296	VLPR–MS	Baghal-Vayjooee and Benson (1979)	
			$(1.36 \pm 0.40) \times 10^{-13}$	296	FP–IR abs.	DeSain et al. (2003)	
			$(1.15 \pm 0.17) \times 10^{-13}$	295	RR[c,d]	Hurley et al. (2003)	
	$(1.8 \pm 0.3) \times 10^{-10}$	2150 ± 100	$(1.12 \pm 0.22) \times 10^{-13}$	293	FP–IR abs.	Hurley et al. (2003)	293–623
			$(5.30 \pm 0.54) \times 10^{-13}$	373			
			$(1.04 \pm 0.07) \times 10^{-12}$	423			
			$(1.88 \pm 0.09) \times 10^{-12}$	473			
			$(2.77 \pm 0.13) \times 10^{-12}$	523			
			$(4.09 \pm 0.19) \times 10^{-12}$	573			
			$(5.71 \pm 0.37) \times 10^{-12}$	623			
Cyclobutane	1.8×10^{-10}	−56			RR[e]	Knox and Nelson (1959)	193–593
Cyclopentane	2.0×10^{-10}	−110			RR[b]	Pritchard et al. (1955)	298–484
			$(3.05 \pm 0.09) \times 10^{-10}$	295	RR[f]	Wallington et al. (1989b)	
			$(2.10 \pm 0.51) \times 10^{-10}$	249			
	$(2.24 \pm 0.23) \times 10^{-10}$	0	$(2.03 \pm 0.38) \times 10^{-10}$	274	FP–UV abs.	Rowley et al. (1992b)	249–364
			$(1.88 \pm 0.51) \times 10^{-10}$	300			
			$(2.20 \pm 0.06) \times 10^{-10}$	324			
			$(2.34 \pm 0.13) \times 10^{-10}$	348			
			$(2.36 \pm 0.13) \times 10^{-10}$	364			

(*continued*)

Table III-B-10. *(continued)*

Cycloalkane	A	B (K)	k	T (K)	Technique[a]	Reference	Temperature Range (K)
	1.3×10^{-10}	−300			FP–IR,UV abs.	Crawford et al. (1997)	267–359
Cyclohexane			$(2.0 \pm 0.2) \times 10^{-10}$	298	FP–RF	Davis et al. (1970)	
			$(3.32 \pm 0.15) \times 10^{-10}$	296	RR[f]	Atkinson and Aschmann (1985)	
			$(3.37 \pm 0.14) \times 10^{-10}$	295	RR[f]	Wallington et al. (1988c)	
	$(2.34 \pm 0.24) \times 10^{-10}$	0	$(2.71 \pm 0.51) \times 10^{-10}$	249	FP-UV abs.	Rowley et al. (1992b)	249–364
			$(2.61 \pm 0.16) \times 10^{-10}$	258			
			$(2.48 \pm 0.33) \times 10^{-10}$	274			
			$(2.30 \pm 0.26) \times 10^{-10}$	300			
			$(2.35 \pm 0.15) \times 10^{-10}$	324			
			$(2.24 \pm 0.09) \times 10^{-10}$	368			
			$(3.33 \pm 0.13) \times 10^{-10}$	296	RR[f]	Aschmann and Atkinson (1995a)	
Methyl-cyclohexane			$(3.75 \pm 0.13) \times 10^{-10}$	296	RR[f]	Aschmann and Atkinson (1995a)	
cis-Bicyclo [4.4.0]decane			$(4.66 \pm 0.09) \times 10^{-10}$	296	RR[f]	Aschmann and Atkinson (1995a)	

[a] Abbreviations for the techniques are defined in table II-A-1.
[b] Placed on an absolute basis using $k_{C2H6} = 8.3 \times 10^{-11} \exp(-100/T)\,cm^3\,molecule^{-1}\,s^{-1}$.
[c] Placed on an absolute basis using $k_{CH4} = 1.0 \times 10^{-13}\,cm^3\,molecule^{-1}\,s^{-1}$.
[d] Placed on an absolute basis using $k_{CH3Cl} = 4.8 \times 10^{-13}\,cm^3\,molecule^{-1}\,s^{-1}$.
[e] Placed on an absolute basis using k_{C3H8}(primary site) $= 7.0 \times 10^{-11} \exp(-30/T)\,cm^3\,molecule^{-1}\,s^{-1}$.
[f] Placed on an absolute basis using $k_{C4H10} = 2.1 \times 10^{-10}\,cm^3\,molecule^{-1}\,s^{-1}$.

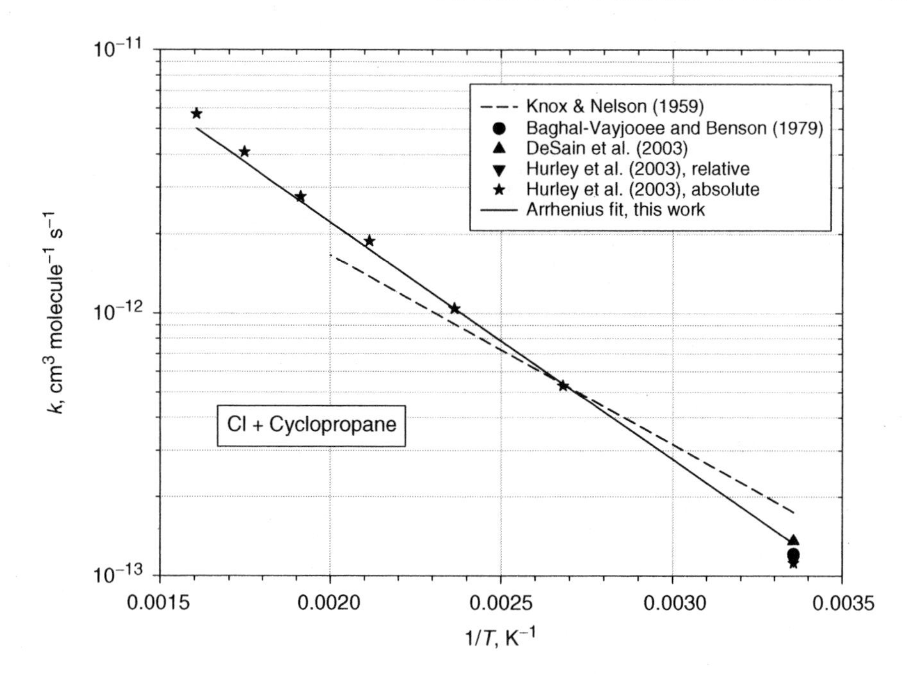

Figure III-B-5. Measurements of the rate coefficient for reaction of Cl with cyclopropane.

the two studies of the temperature dependence to the rate coefficient (Knox and Nelson, 1959; Hurley et al., 2003) is better than 40% over the range of overlap. A fit to the two data sets between 300 and 450 K, scaled to give the 298 K value of 1.4×10^{-13} cm^3 molecule^{-1} s^{-1}, yields $k = 1.4 \times 10^{-10} \exp(-2074/T)$ cm^3 molecule^{-1} s^{-1}.

The reaction of Cl with cyclopropane is anomalous in that it occurs with an activation energy of about 17 kJ mol^{-1}, and thus its rate coefficient is two orders of magnitude slower than might be expected on the basis of the rate coefficients for reaction of Cl with propane or cyclobutane. The existence of a significant barrier to the reaction suggests that hydrogen abstraction by Cl-atoms might be an endothermic process. Baghal-Vayjooee and Benson (1979) measured the equilibrium constant for the system Cl + *cyclo*-C_3H_6 $\leftrightarrow$ HCl + *cyclo*-C_3H_5 and reported that the H-atom abstraction by Cl was endothermic by about 14 kJ/mole. However, in a recent theoretical study of the system, Hurley et al. (2003) found the reaction to be slightly exothermic but to occur with a barrier height of about 17 kJ/mole, consistent with experiment.

III-B-23. Cl + Cyclobutane (*cyclo*-C_4H_8)

The only study available is the relative rate study of Knox and Nelson (1959) who reported a rate coefficient ratio (relative to Cl-atom abstraction from the CH_3 groups in propane) of 2.54 exp $(-86/T)$; see table III-B-10. Using the currently recommended rate coefficient for reaction of Cl with propane (see section III-B-3) yields k(Cl + cyclobutane) = $1.8 \times 10^{-10} \exp(56/T)$ cm^3 molecule^{-1} s^{-1}, with k = 2.2×10^{-10} cm^3 molecule^{-1} s^{-1} near 298 K. Further studies of this rate coefficient are required before a firm recommendation can be made.

III-B-24. Cl + Cyclopentane (*cyclo*-C_5H_{10})

The four measurements of the rate coefficient for reaction of Cl with cyclopentane are in rough agreement; see table III-B-10. Using the relative rate technique with butane as the reference, Wallington et al. (1989b) obtained a value of $(3.05 \pm 0.09) \times 10^{-10}$ cm^3 molecule^{-1} s^{-1}. Rowley et al. (1992b), from an analysis of initial HO_2 and cyclopentylperoxy radical yields following flash photolysis of Cl_2–CH_3OH–cyclopentane mixtures, found k(cyclopentane)/k(methanol) = 4.0; this yields k(cyclopentane) = $(2.24 \pm 0.23) \times 10^{-10}$ cm^3 molecule^{-1} s^{-1} independent of temperature (249–364 K). A similar study by Crawford et al. (1997) from 267 to 359 K found k(cyclopentane)/k(methanol) = $(2.3 \pm 0.2) \exp[(300 \pm 200)/T]$, yielding k(cyclopentane) = $1.3 \times 10^{-10} \exp(300/T)$ cm^3 molecule^{-1} s^{-1} and k(cyclopentane, 298 K) = 3.6×10^{-10} cm^3 molecule^{-1} s^{-1}. An earlier relative rate study (Pritchard et al., 1955), placed on an absolute basis using an updated value for the ethane reference reaction rate, yields k = $2.0 \times 10^{-10} \exp(110/T)$ cm^3 molecule^{-1} s^{-1} for T = 298–484 K, with k = 2.9×10^{-10} cm^3 molecule^{-1} s^{-1} at 298 K. The recommended value at 298 K, k = 2.9×10^{-10} cm^3 molecule^{-1} s^{-1}, is obtained from an unweighted average of the four available measurements, with an estimated uncertainty of ±20%. Although there is an indication of a negative temperature dependence in some of the data sets, we recommend k = 2.9×10^{-10} cm^3 molecule^{-1} s^{-1} independent of temperature, on the basis of data for other rapid Cl + alkane reactions.

III-B-25. Cl + Cyclohexane (*cyclo*-C_6H_{12})

Three relative rate studies (Atkinson and Aschmann, 1985; Wallington et al., 1988c; Aschmann and Atkinson, 1995a), all using butane as the reference species, are in excellent agreement, k = $(3.3\text{–}3.4) \times 10^{-10}$ cm^3 molecule^{-1} s^{-1}, once standardization to a common butane rate coefficient (2.1×10^{-10} cm^3 molecule^{-1} s^{-1}) is carried out; see table III-B-10. Values reported by Davis et al. (1970) using flash photolysis–resonance fluorescence in experiments at ambient temperatures, and Rowley et al. (1992b), who monitored initial HO_2 and cyclohexylperoxy yields following flash photolysis of Cl_2–CH_3OH–cyclohexane mixtures, are measurably lower than the relative rate studies, 2.0×10^{-10} and $(2.33 \pm 0.24) \times 10^{-10}$ cm^3 molecule^{-1} s^{-1}, respectively. The Rowley et al. (1992b) data suggest no significant temperature dependence to the cyclohexane reaction between 248 and 364 K. Given the consistency of the relative rate data, and the somewhat indirect nature of the Rowley et al. study, the value obtained in the relative rate studies, 3.3×10^{-10} cm^3 molecule^{-1} s^{-1}, is recommended at ambient temperatures with an estimated uncertainty of ±15%. As shown by Rowley et al. (1992b), there is very likely no temperature dependence to the rate coefficient.

III-B-26. Cl + Methylcyclohexane (*cyclo*-C_6H_{11}–CH_3)

As shown in table III-B-10, a value of $k = (3.75 \pm 0.13) \times 10^{-10}$ cm^3 molecule^{-1} s^{-1} is obtained at 296 K from the work of Aschmann and Atkinson (1995a), after updating to the currently recommended rate coefficient for the reference reaction of Cl with butane. An uncertainty of ±20% is adopted, given that only one study is available.

III-B-27. Cl + *cis*-Bicyclo[4.4.0]decane (*bicyclo*-$C_{10}H_{18}$)

This reaction was studied via the relative rate method by Aschmann and Atkinson (1995a). After updating to the currently recommended rate coefficient for the reference reaction (Cl + butane), a value of $k = (4.66 \pm 0.09) \times 10^{-10}$ cm^3 molecule^{-1} s^{-1} is obtained at 296 K; see table III-B-10. Given that only one study of this reaction is available, an uncertainty of ±20% is adopted.

III-B-28. Cl-Atom Reactions with Isotopic Variants of the Alkanes

There are now available a number of rate coefficient measurements for the reaction of Cl-atoms with partially or fully deuterated alkanes and with alkanes containing a ^{13}C atom. These data are summarized in this section and in tables III-B-11 to III-B-14.

III-B-28.1. Cl + Methane Isotopomers

Very precise measurements of the rate coefficient ratio k_{12CH4}/k_{13CH4} are available at room temperature: 1.066 ± 0.002 (Saueressig et al., 1995); 1.066 ± 0.002 (Crowley et al., 1999); 1.0621 ± 0.0004 (Tyler et al., 2000); and 1.058 ± 0.002 (Feilberg et al., 2005b). The data of Saueressig et al. (1995) and Tyler et al. (2000) both suggest a weak, negative temperature dependence to this ratio. Saueressig et al. (1995) report $k_{12CH4}/k_{13CH4} = 1.043\times \exp(6.455/T)$ for the range 223–297 K, while a fit to the data of Tyler et al. (2000) gives $k_{12CH4}/k_{13CH4} = 1.036\times \exp(7.43/T)$ for the temperature range 273–349 K; See table III-B-11.

There are data available regarding the rate coefficient for reaction of Cl-atoms with CH_3D, CH_2D_2, CHD_3, CD_4, and $^{13}CH_4$; see table III-B-12. The reaction of CH_3D with Cl has been studied relative to the reaction of Cl with CH_4 by Wallington and Hurley (1992a), Saueressig et al. (1996), Wallington and Hurley (an updated value quoted in Saueressig et al., 1996), Tyler et al. (2000), Boone et al. (2001), and Feilberg et al. (2005b). The Saueressig et al. (1996) study covers the range 223–298 K, the Tyler et al. (2000) data are in the range 273–349 K, while the other measurements were restricted to ambient temperatures. The data are all in excellent agreement, with values of k_{CH4}/k_{CH3D} ranging from 1.47 to 1.54 at ambient temperature and increasing slightly with decreasing temperature. An unweighted least-squares fit to the data (all normalized to $k_{CH4} = 6.6\times 10^{-12} \exp(-1240/T)$ cm^3 molecule^{-1} s^{-1}), gives $k_{CH3D} = 6.21\times 10^{-12} \exp(-1337/T)$ cm^3 molecule^{-1} s^{-1}. Equivalently, the rate coefficient ratio k_{CH4}/k_{CH3D} can be expressed as $1.063 \times \exp(97/T)$.

Table III-B-11. Measurements of the ^{13}C kinetic isotope effect in the reaction of Cl with methane (CH_4)

A (cm^3 molecule^{-1} s^{-1})	B (K)	$k(^{12}C)/k(^{13}C)$	T (K)	Technique[a]	Reference	Temperature Range (K)
1.043	6.455	1.075 ± 0.005	223	RR	Saueressig et al. (1995)	223–297
		1.069 ± 0.004	243			
		1.070 ± 0.004	263			
		1.066 ± 0.002	297			
		1.065 ± 0.003	297			
		1.066 ± 0.002	298	RR	Crowley et al. (1999)	
1.035	7.55 ± 1.64	1.0638 ± 0.0004	273	RR	Tyler et al. (2000)	273–349
		1.0621 ± 0.0004	299			
		1.0599 ± 0.0004	323			
		1.0575 ± 0.0002	349			
		1.06 ± 0.01	298	RR	Feilberg et al. (2005b)	

[a] Abbreviations for the techniques are defined in table II-A-1.

Table III-B-12. Measurements of the rate coefficient (k, cm^3 molecule^{-1} s^{-1}) for reaction of Cl-atoms with deuterated methanes[a]

Alkane	$10^{14} \times k$	T (K)	Technique[a]	Reference	Temperature Range (K)
CH_3D	7.57 ± 0.22	295	RR[a]	Wallington and Hurley (1992a)	
CH_3D	6.82 ± 0.19	298		Saueressig et al. (1996)	223–298
	6.69 ± 0.19	296			
	3.76 ± 0.12	263			
	2.53 ± 0.08	243			
	1.59 ± 0.06	223			
	6.73 ± 0.13	296			
CH_3D	7.00 ± 0.42	295	RR[a]	Wallington and Hurley, as quoted in Saueressig et al. (1996)	
CH_3D	4.65 ± 0.22	273		Tyler et al. (2000)	273–349
	7.08 ± 0.22	299			
	10.2 ± 2.2	323			
	14.0 ± 2.2	349			
CH_3D	6.7 ± 0.5	298	RR[a]	Boone et al. (2001)	
CH_3D	7.01 ± 0.14	298	RR[a]	Feilberg et al. (2005b)	
CH_2D_2	4.57 ± 0.22	295	RR[a]	Wallington and Hurley (1992a)	
CH_2D_2	7.0 ± 0.8	298	FP–LIF	Matsumi et al. (1997)	
CH_2D_2	4.3 ± 0.5	298	RR[a]	Boone et al. (2001)	
CH_2D_2	4.20 ± 0.09	298	RR[a]	Feilberg et al. (2005b)	
CHD_3	2.32 ± 0.05	295	RR[a]	Wallington and Hurley (1992a)	
CHD_3	2.0 ± 0.3	298	RR[a]	Boone et al. (2001)	

Table III-B-12. (*continued*)

Alkane	$10^{14} \times k$	T (K)	Technique[a]	Reference	Temperature Range (K)
CHD_3	2.19 ± 0.05	298	RR[a]	Feilberg et al. (2005b)	
CD_4	1.03 ± 0.06	303.7	RR[a]	Chiltz et al. (1963)	303.7–461.4
	1.03 ± 0.08	304.4			
	1.47 ± 0.14	321.9			
	2.40 ± 0.14	343.3			
	5.78 ± 0.46	400.4			
	8.00 ± 0.67	430.7			
	11.0 ± 1.4	451.6			
	12.3 ± 0.9	461.4			
CD_4	0.852	298	DF–MS (RR)[b]	Clyne and Walker (1973)	298–1018
	0.792	299			
	1.12	322			
	2.02	340			
	2.87	366			
	6.47	407			
	7.50	426			
	10.8	460			
	12.3	471			
	26.6	520			
	33.5	549			
	38.7	593			
	67.7	625			
	78.1	642			
	77.1	681			
	113	687			
	127	731			
	163	775			
	214	787			
	172	816			
	198	854			
	242	871			
	270	891			
	292	924			
	274	930			
	422	973			
	356	1000			
	419	1018			
CD_4	0.64	298	RR[a,d]	Strunin et al. (1975)	163–613
CD_4	0.61 ± 0.05	295	RR[a]	Wallington and Hurley (1992a)	
CD_4	0.82 ± 0.10	298	FP–LIF	Matsumi et al. (1997)	
CD_4	0.54 ± 0.04	298	RR[c]	Boone et al. (2001)	
CD_4	0.701 ± 0.014	298	RR[a]	Feilberg et al. (2005b)	

[a] Placed on an absolute basis using $k_{CH4} = 6.6 \times 10^{-12} \exp(-1240/T)$ cm^3 molecule^{-1} s^{-1}.
[b] Placed on an absolute basis using $k_{CH4} = 1.5 \times 10^{-17} T^2 \exp(-766/T)$ cm^3 molecule^{-1} s^{-1}.
[c] Placed on an absolute basis using $k_{13CO} = 3.2 \times 10^{-14}$ cm^3 molecule^{-1} s^{-1}.
[d] Authors report $k_{CH4}/k_{CD4} = 2.2 \exp(590/T)$ cm^3 molecule^{-1} s^{-1} for $163 < T < 300$ K; significant curvature (larger effective difference in activation energies) was observed at higher temperatures.
[e] Abbreviations for the techniques are defined in table II-A-1.

The data set for reaction of Cl with CH_2D_2 includes three relative rate studies (Wallington and Hurley, 1992a; Boone et al., 2001; Feilberg et al., 2005b) and a flash photolysis–laser induced fluorescence study (Matsumi et al., 1997). All data were obtained at ambient temperature. The relative rate measurements carried out relative to the Cl + CH_4 reaction are in good agreement, $k = (4.5 \pm 0.2) \times 10^{-14}$ (Wallington and Hurley, 1992a), $k = (4.2 \pm 0.5) \times 10^{-14}$ (Boone et al., 2001), and $k = (4.20 \pm 0.09) \times 10^{-14}$ cm^3 molecule^{-1} s^{-1} (Feilberg et al., 2005b). The flash photolysis value of Matsumi et al. (1997), $(7.0 \pm 0.8) \times 10^{-14}$ cm^3 molecule^{-1} s^{-1}, is not in keeping with the remainder of the data set. A value of 4.3×10^{-14} cm^3 molecule^{-1} s^{-1} is recommended on the basis of the relative rate studies, with an estimated uncertainty of ±15%.

Data are available for the reaction of Cl with CHD_3: $k = (2.0 \pm 0.3) \times 10^{-14}$ (Boone et al., 2001); $k = (2.4 \pm 0.1) \times 10^{-14}$ (Wallington and Hurley, 1992a); $k = (2.19 \pm 0.05) \times 10^{-14}$ (Feilberg et al., 2005) cm^3 molecule^{-1} s^{-1}. All measurements were made relative to the rate coefficient for reaction of Cl with methane. A value of $k = 2.2 \times 10^{-14}$ cm^3 molecule^{-1} s^{-1} is recommended, with an estimated uncertainty of ±15%.

The reaction of Cl with CD_4 has been studied using the relative rate approach (Chiltz et al., 1963; Clyne and Walker, 1973; Strunin et al., 1975; Wallington and Hurley, 1992a; Boone et al., 2001; Feilberg et al., 2005b) and directly via flash photolysis–laser induced fluorescence (Matsumi et al., 1997). Although data measured at ambient temperatures range from 5.4×10^{-15} cm^3 molecule^{-1} to 1.0×10^{-14} cm^3 molecule^{-1} s^{-1}, there is good agreement among the more recent relative rate measurements of Wallington and Hurley (1992a), Boone et al. (2001), and Feilberg et al. (2005b). These values are preferred and a room-temperature value of 6.2×10^{-15} cm^3 molecule^{-1} s^{-1} is recommended, with an estimated uncertainty of ±25%. Measurements made over ranges of temperature (Chiltz et al., 1963; Clyne and Walker, 1973; Strunin et al., 1975) indicate a large activation energy for the reaction, $E_a \sim 20$ kJ mol^{-1}.

III-B-28.2. Measurements of the Cl + C_2–C_8 Alkane Rate Coefficients Using Samples of Natural Abundance

Rudolph and co-workers have recently examined kinetic isotope effects (KIEs) for both D/H (Iannone et al., 2005) and $^{12}C/^{13}C$ (Anderson et al., 2007) in a number of C_2–C_8 alkanes. These measurements were made on samples of natural abundance using gas chromatographic–isotope ratio mass spectrometer techniques. Data are reported as values of $^{Cl}\varepsilon_{H/D}$ and $^{Cl}\varepsilon_{12/13}$ in per mille, with $^{Cl}\varepsilon_{H/D}(‰) = (^{Cl}k_H/^{Cl}k_D - 1) \times 1000$ and $^{Cl}\varepsilon_{12/13}(‰) = (^{Cl}k_{12}/^{Cl}k_{13} - 1) \times 1000$. KIEs obtained in these two studies (Iannone et al., 2005; Anderson et al., 2007) are summarized in figure III-B-6 and table III-B-13. The data show a monotonically decreasing KIE with number of carbons in the alkane, as expected from statistical considerations. The work of both Iannone et al. (2005) and Anderson et al. (2007) also provides, via the relative rate approach, rate coefficients for a number of Cl + alkane reactions; these data are in good agreement with the recommended data provided earlier in this chapter, as summarized in table III-B-13.

Table III-B-13. Measured KIEs and rate coefficients (cm^3 $molecule^{-1}$ s^{-1}) for reaction of Cl with some alkanes

Compound	${}^{Cl}\varepsilon_{H/D}$ (%)	${}^{Cl}\varepsilon_{12/13}$ (%)	$10^{10} \times k(Cl + alkane)$	Reference
Ethane		10.73 ± 0.20	0.48 ± 0.01	Anderson et al. (2007)
Propane		6.44 ± 0.14	1.59 ± 0.23	Anderson et al. (2007)
Butane	39.6 ± 2.7			Iannone et al. (2005)
		3.94 ± 0.01	2.09 ± 0.03	Anderson et al. (2007)
2-Methylpropane		6.18 ± 0.18	1.38 ± 0.05	Anderson et al. (2007)
Pentane	28.2 ± 0.9		2.84 ± 0.19[a]	Iannone et al. (2005)
		3.22 ± 0.17	2.62 ± 0.03	Anderson et al. (2007)
2-Methylbutane (isopentane)		1.79 ± 0.42	1.93 ± 0.07	Anderson et al. (2007)
Hexane	24.6 ± 1.0		3.39 ± 0.39[a]	Iannone et al. (2005)
		2.02 ± 0.40	3.20 ± 0.15	Anderson et al. (2007)
Heptane	24.0 ± 1.2		3.81 ± 0.23[b]	Iannone et al. (2005)
		2.06 ± 0.19	3.50 ± 0.13	Anderson et al. (2007)
Octane	17.9 ± 3.3		4.74 ± 0.23[b]	Iannone et al. (2005)
		1.54 ± 0.15	3.82 ± 0.18	Anderson et al. (2007)
Nonane	15.1 ± 0.7		5.10 ± 0.23[b]	Iannone et al. (2005)
Decane	14.9 ± 1.8		5.33 ± 0.23[b]	Iannone et al. (2005)
Cyclopentane		3.04 ± 0.09	3.05 ± 0.06	Anderson et al. (2007)
Cyclohexane		2.30 ± 0.09	3.83 ± 0.12	Anderson et al. (2007)
Methylcyclopentane		2.56 ± 0.25	2.82 ± 0.11	Anderson et al. (2007)

[a] Placed on an absolute basis using $k(Cl + butane) = 2.1 \times 10^{-10}$ cm^3 $molecule^{-1}$ s^{-1}.
[b] Placed on an absolute basis using $k(Cl + hexane) = 3.3 \times 10^{-10}$ cm^3 $molecule^{-1}$ s^{-1}.

Of particular note is the measurement of the rate coefficient for reaction of Cl with methylcyclopentane, $k = (2.82 \pm 0.11) \times 10^{-10}$ cm^3 $molecule^{-1}$ s^{-1} (Anderson et al., 2007), which represents the first measurement of this rate coefficient.

III-B-28.3. Studies of the Rate Coefficients for the Cl–Alkane Reaction with Fully Deuterated Alkanes

III-B-28-3.1. Cl + C_2D_6. The rate coefficient for reaction of Cl-atoms with fully deuterated ethane, C_2D_6, has been measured using both absolute (Parmar and Benson, 1989; Dobis et al., 1994; Hitsuda et al., 2001b) and relative rate methods (Chiltz et al., 1963; Tschuikow-Roux et al., 1985a; Wallington and Hurley, 1992a); see table III-B-14. The Hitsuda et al. (2001b) measurement is specific to ground-state Cl-atoms ($^2P_{3/2}$), whereas all other measurements involve thermalized Cl (mixture

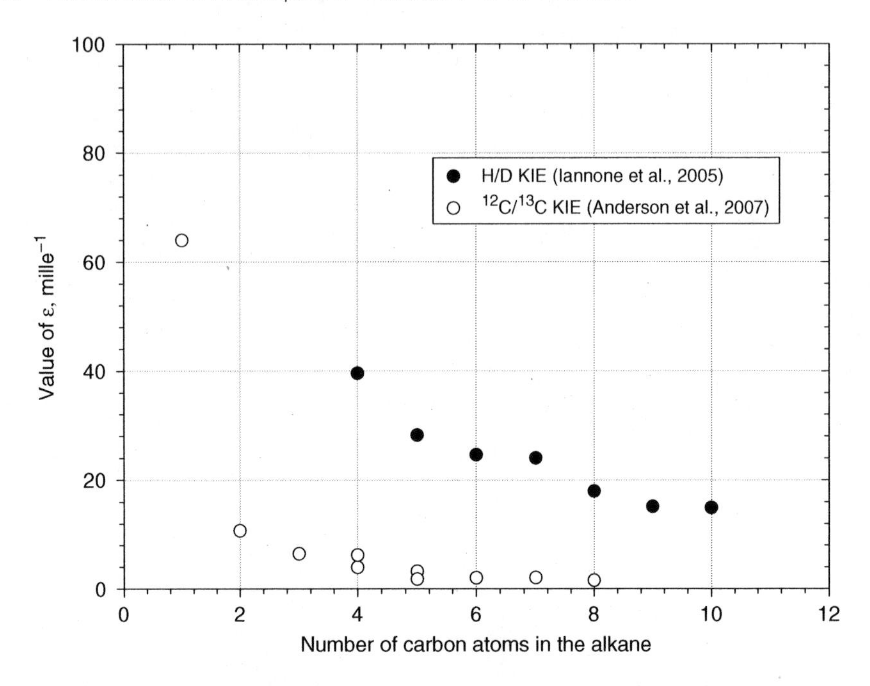

Figure III-B-6. Measured kinetic isotope effects for reaction of Cl with acyclic alkanes.

of $^2P_{3/2}$ and a small amount of $^2P_{1/2}$). With the exception of the early study of Parmar and Benson (1989), the entire data set is seen to be in very good agreement after conversion of the relative rate data to absolute values (using $k_{C2H6} = 8.3 \times 10^{-11} \exp(-100/T)\,\text{cm}^3\ \text{molecule}^{-1}\ \text{s}^{-1}$). Values measured at ambient temperatures range from 1.87 to $2.24 \times 10^{-11}\,\text{cm}^3\ \text{molecule}^{-1}\,\text{s}^{-1}$, and an unweighted average of the five determinations is recommended, $k_{C2D6} = (2.0 \pm 0.3) \times 10^{-11}\,\text{cm}^3$ molecule^{-1} s^{-1}. The data of Chiltz et al. (1963) and Tschuikow-Roux et al. (1985a) indicate a slight decrease in the ratio k_{C2H6}/k_{C2D6} with increasing temperature. A least-squares fit to the entire data set yields $k_{C2D6} = 5.8 \times 10^{-11} \times \exp(-310/T)\,\text{cm}^3$ molecule^{-1} s^{-1}.

Hitsuda et al. (2001b) also specifically studied the reaction of excited Cl-atoms ($^2P_{1/2}$) with C_2D_6, and showed that this reaction takes place almost exclusively via quenching of the Cl($^2P_{1/2}$) rather than via reaction to generate DCl.

III-B-28-3.2. Cl + C_3D_8. The reaction of Cl-atoms with fully deuterated propane, C_3D_8, has been studied by Hitsuda et al. (2001a) using a laser flash photolysis–laser induced fluorescence technique. Reaction of ground-state Cl($^2P_{3/2}$) with C_3D_8 was found to occur with a rate coefficient of $(8.6 \pm 0.5) \times 10^{-11}\,\text{cm}^3\ \text{molecule}^{-1}\ \text{s}^{-1}$. Collisions of excited Cl($^2P_{1/2}$) with C_3D_8 resulted almost exclusively in deactivation of the excited Cl($^2P_{1/2}$), rather than via reaction to produce DCl. The total rate coefficient for loss of Cl($^2P_{1/2}$) with C_3D_8 was reported to be $(3.1 \pm 0.2) \times 10^{-10}\,\text{cm}^3$ molecule^{-1} s^{-1}.

Table III-B-14. Summary of measurements of the rate coefficients for reaction of Cl-atoms with deuterated alkanes; $k = A \exp(-B/T)$ cm^3 molecule^{-1} s^{-1}

Alkane	A	B (K)	k	T (K)	Technique[a]	Reference	Temperature Range (K)
Ethane-d_6			2.23×10^{-11}	302.8	RR[b]	Chiltz et al. (1963)	303–433
			2.23×10^{-11}	310.2			
			2.54×10^{-11}	335.7			
			2.61×10^{-11}	367.3			
			2.47×10^{-11}	395.1			
			2.78×10^{-11}	431.9			
			2.99×10^{-11}	433			
	2.81×10^{-11}	93			RR[c]	Tschuikow-Roux et al. (1985a)	280–368
			$(8.3 \pm 0.7) \times 10^{-12}$	295	VLPR–MS	Parmar and Benson (1989)	
			$(1.92 \pm 0.14) \times 10^{-11}$	295	RR[b]	Wallington and Hurley (1992a)	
			$(2.11 \pm 0.05) \times 10^{-11}$	298	VLPR–MS	Dobis et al. (1994)	
			$(1.87 \pm 0.12) \times 10^{-11}$	298	FP–LIF	Hitsuda et al. (2001b)	
Propane-d_8			$(8.6 \pm 0.5) \times 10^{-11}$	298	FP–LIF	Hitsuda et al. (2001a)	
Butane-d_{10}			$(1.5 \pm 0.3) \times 10^{-10}$	293	DF–RF	Stutz et al. (1998)	
			$(1.7 \pm 0.2) \times 10^{-10}$	298	RR[b]	Stutz et al. (1998)	
Cyclohexane-d_{12}			$(2.77 \pm 0.09) \times 10^{-10}$	296	RR[d]	Aschmann and Atkinson (1995a)	

[a] Abbreviations for the techniques are defined in table II-A-1.
[b] Placed on an absolute basis using $k_{C2H6} = 8.3 \times 10^{-11} \exp(-100/T)$ cm^3 molecule^{-1} s^{-1}.
[c] Placed on an absolute basis using $k_{CH4} = 6.6 \times 10^{-12} \exp(-1240/T)$ cm^3 molecule^{-1} s^{-1}.
[d] Placed on an absolute basis using $k_{C4H10} = 2.1 \times 10^{-10}$ cm^3 molecule^{-1} s^{-1}.

III-B-28-3.3. $Cl + CD_3CD_2CD_2CD_3$. Stutz et al. (1998) have used a combination of discharge flow tube and relative rate techniques to determine the rate coefficient for reaction of Cl with fully deuterated butane (C_4D_{10}). The value obtained using the discharge flow system, $(1.5 \pm 0.3) \times 10^{-10}$ cm^3 molecule^{-1} s^{-1}, agrees very well with their relative rate measurements using ethane as the reference compound, $(1.7 \pm 0.2) \times 10^{-10}$ cm^3 molecule^{-1} s^{-1} at 1 Torr and $(1.6 \pm 0.2) \times 10^{-10}$ cm^3 molecule^{-1} s^{-1} at atmospheric pressure. The authors recommend a value of 1.6×10^{-10} cm^3 molecule^{-1} s^{-1}, which is adopted here, with an estimated uncertainty of ±15%.

III-B-28-3.4. Cl + Cyclohexane-d_{12}(cyclo-C_6D_{12}). Aschmann and Atkinson (1995a) have reported a rate coefficient for reaction of Cl with fully deuterated cyclohexane measured by the relative rate technique, with butane as the reference compound. Using an updated value for the reference reaction rate coefficient, a value of $(2.8 \pm 0.1) \times 10^{-10}$ cm^3 molecule^{-1} s^{-1} is obtained. The rate coefficient is about 18% smaller than that for the cyclohexane itself, and is recommended with an estimated uncertainty of ±20%.

III-B-29. Rate Coefficient Estimation Methods; Structure–Additivity Relations

Structure–additivity relationships (SARs) for reactions of Cl with alkanes at 298 K have been reported previously by Atkinson and Aschmann (1985), Aschmann and Atkinson (1995a), Hooshiyar and Niki (1995), Tyndall et al. (1997a), and Qian et al. (2002). Methodology and nomenclature are the same as for the OH structure–additivity relationships presented in chapter II.

Tyndall et al. (1997a) recommended the following parameterizations on the basis of an analysis of the data for twenty alkane species:

$$k(\text{—CH}_3) = 2.91 \times 10^{-11}\ \text{cm}^3\ \text{molecule}^{-1}\ \text{s}^{-1}$$
$$k(\text{—CH}_2\text{—}) = 9.14 \times 10^{-11}\ \text{cm}^3\ \text{molecule}^{-1}\ \text{s}^{-1}$$
$$k(\text{>CH—}) = 6.53 \times 10^{-11}\ \text{cm}^3\ \text{molecule}^{-1}\ \text{s}^{-1}$$
$$F(\text{—CH}_3) = 1.00;\ F(\text{—CH}_2\text{—}) = F(\text{>CH—}) = F(\text{>C<}) = 0.8.$$

In that work, the rate coefficient for reaction at a —CH_3 group was assumed to be unaffected by neighboring groups (i.e., k(—CH_3) was fixed at 2.91×10^{-11} cm^3 molecule^{-1} s^{-1}). A re-analysis was performed herein, using methodology identical to that in Tyndall et al. (1997a) and the data for twenty-four alkanes (the twenty used in Tyndall et al., plus 2-methylhexane, methylcyclohexane, cyclopentane, and cyclobutane). Note that the rate coefficient for reaction of Cl with cyclopropane is about 1000 times slower than predicted, so this compound was not included in the fit. Parameters obtained via minimization of the function $\Sigma((k_{calc} - k_{meas})/k_{meas})^2$ are only slightly different from those of Tyndall et al. (1997a):

$$k(\text{—CH}_3) = 2.84 \times 10^{-11}\ \text{cm}^3\ \text{molecule}^{-1}\ \text{s}^{-1}$$
$$k(\text{—CH}_2\text{—}) = 8.95 \times 10^{-11}\ \text{cm}^3\ \text{molecule}^{-1}\ \text{s}^{-1}$$

$k(>\text{CH}-) = 6.48 \times 10^{-11}\ \text{cm}^3\ \text{molecule}^{-1}\ \text{s}^{-1}$
$F(-\text{CH}_3) = 1.00;\ F(-\text{CH}_2-) = F(>\text{CH}-) = F(>\text{C}<) = 0.80.$

Exclusion of the cyclic compounds (cyclobutane, cyclopentane, cyclohexane, and methyl-cyclohexane) from the fit generated essentially the same parameters (less than 5% difference). Measured and predicted rate coefficients are given in table III-B-15 and the data are presented in graphical form in figure III-B-7. For completeness, parameterizations from other studies (Aschmann and Atkinson, 1995a; Hooshiyar and Niki, 1995; Qian et al., 2002) are presented in table III-B-16 for comparison.

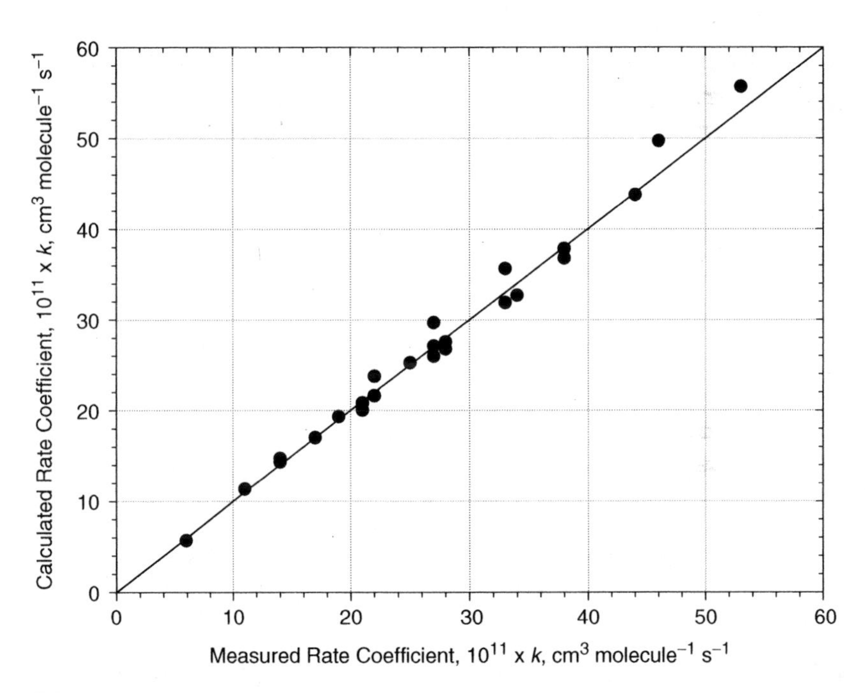

Figure III-B-7. Comparison of measured and calculated rate coefficients for reaction of Cl-atoms with the alkanes. Calculations are carried out using parameters described in the text. Solid line shows a 1:1 relationship between the variables.

III-C. Reaction of O(^3P) with Alkanes

III-C-1. Introduction and Mechanism

The reactions of O(^3P) with the alkanes are generally quite slow, as a result of the large activation barriers involved. For example, the rate coefficient for reaction of O(^3P) with 2-methylpropane is about $3 \times 10^{-14}\ \text{cm}^3\ \text{molecule}^{-1}\ \text{s}^{-1}$ at 298 K, with an activation energy of about 21 kJ/mole. Correlations between the O(^3P) reactions and those involving other atmospheric oxidants are discussed in detail in section III-F, but, in general, the O(^3P) reactions with smaller alkanes are about 20–1000 times

Table III-B-15. Comparison of measured and calculated rate coefficients (k, cm^3 $molecule^{-1}$ s^{-1}) for reaction of Cl-atoms with alkanes (~298 K)

Alkane	Recommended Rate Coefficient $10^{11} \times k$	Calculated Rate Coefficient $10^{11} \times k$
Ethane	5.9	5.6
Propane	14	15
Butane	21	20
2-Methylpropane	14	15
Pentane	27	26
2-Methylbutane	21	21
2,2-Dimethylpropane	11	11
Hexane	33	31
2-Methylpentane	28	27
3-Methylpentane	27	27
2,3-Dimethylbutane	22	22
Heptane	38	37
2,4-Dimethylpentane	28	28
2,2,3-Trimethylbutane	19	19
2-Methylhexane	34	32
Octane	44	43
2,2,4-Trimethylpentane	25	25
2,2,3,3-Tetramethylbutane	17	17
Nonane	46	49
Decane	53	54
Cyclohexane	33	34
Methylcyclohexane	38	36
Cyclopentane	27	29
Cyclobutane	22	23
Cyclopropane	0.014	(18)

Table III-B-16. Comparison of parameters for structure–additivity-based estimation of Cl-atom rate coefficients from different studies. Values of k are in cm^3 $molecule^{-1}$ s^{-1}, F values are unitless[a]

Reference	$k(-CH_3)$	$k(-CH_2-)$	$k(-CH-)$	$F(-CH_3)$	$F(-CH_2-) = F(>CH-) = F(>C<)$
This study	2.88×10^{-11}	8.95×10^{-11}	6.49×10^{-11}	1.00	0.8
Tyndall et al. (1997a)	2.91×10^{-11}	9.14×10^{-11}	6.53×10^{-11}	1.00	0.8
Aschmann and Atkinson (1995a)	3.32×10^{-11}	8.34×10^{-11}	6.09×10^{-11}	1.00	0.79
Hooshiyar and Niki (1995)	2.9×10^{-11}	8.2×10^{-11}	6.1×10^{-11}	1.00	0.8
Qian et al. (2002)	2.61×10^{-11}	8.40×10^{-11}	5.90×10^{-11}	1.00	0.85

[a] In this work and in Tyndall et al. (1997a), the value for $k(-CH_3)$ is assumed to be independent of the neighboring groups. In all other work, the $k(-CH_3)$ value is allowed to be influenced by the neighboring group.

slower than OH reactions but are considerably faster (hundreds of times) than the corresponding NO_3 reactions.

The data set for reaction of $O(^3P)$ with the alkanes was developed largely in the 1980s and earlier, for application to combustion chemistry. The importance of these reactions in combustion requires that a wide range of temperatures be considered, from 300 K to temperatures greater than 1000 K. A variety of methods have been employed over the years. Most recently (i.e., the 1980s and 1990s), the methods of choice have included flow tubes, flash photolysis systems, and shock tubes, with detection of $O(^3P)$ via resonance absorbance, resonance fluorescence, or vacuum ultraviolet (VUV) pulsed laser-induced fluoresence. The requirement that data be available over a large temperature range has led to extrapolations to high temperature based on transition state theory (Cohen and Westberg, 1986, 1991).

In general, there is a fair amount of scatter in the available data for the rate coefficients of $O(^3P)$ with the alkanes. Many complications are present in these studies with regard to secondary chemistry, and reanalysis of older studies is often difficult or impossible. A series of reviews and evaluations of the data base were put together in the late 1980s and early 1990s (Herron, 1988; Cohen and Westberg, 1991; Baulch et al., 1992), and these authors generally estimate the uncertainties in the rate coefficient data near 298 K to be a factor of two or more, even for some of the better studied systems (e.g., ethane).

There have been very few studies of the reactions of $O(^3P)$ with the alkanes since the time of these latest review articles. A notable exception is the work of Miyoshi et al. (1993, 1994), who studied a number of C_1–C_6 alkanes, predominantly above 850 K but in some cases at 300–400 K as well. As will be seen in this section, their work on model compounds such as neopentane, hexane, and cyclohexane provides a firmer foundation for the development of structure–additivity relationships.

Given the relative unimportance of the reactions of $O(^3P)$ with alkanes in the atmosphere, the lack of an abundance of recent studies, the difficulties in interpretation of some of the older experiments, and the availability of a number of thorough reviews (Herron, 1988; Cohen and Westberg, 1991; Baulch et al., 1992; Miyoshi et al., 1993, 1994), a full presentation and evaluation of all available data is not given here. Instead, what will be presented is a summary of the rate coefficients recommended near room temperature by these authors. The recommendations of the various reviews are also summarized in tables III-C-1 through III-C-4.

III-C-2. $O(^3P)$ + Methane (CH_4)

There is a general consensus (Cohen and Westberg, 1991; Baulch et al., 1992; Miyoshi et al., 1993) on the value of the rate coefficient for reaction of $O(^3P)$ with methane near 298 K, $k = 7 \times 10^{-18}$ cm^3 molecule^{-1} s^{-1}, with an estimated uncertainty of a factor of 2–3 (see table III-C-1). Herron (1988) does not make a recommendation below 400 K, but extrapolation of the expression given yields $k(298\,\text{K}) = 6 \times 10^{-18}$ cm^3 molecule^{-1} s^{-1}, consistent with the other reviews.

Table III-C-1. Summary of recommended rate coefficients (k) for reaction $O(^3P)$ with C_1–C_4 alkanes; $k = A \times T^n \times \exp(-B/T)$, $cm^3\ molecule^{-1}\ s^{-1}$

Alkane	A	B (K)	n	k at 298 K	Reference	Temperature Range (K)
Methane	1.15×10^{-15}	4270	1.56		Herron (1988)	400–2500
	8.1×10^{-18}	3820	2.2	7×10^{-18}	Cohen and Westberg (1991)	298–2500
	1.5×10^{-15}	4270	1.56	6.5×10^{-18}	Baulch et al. (1992)	300–2500
Ethane	1.8×10^{-10}	3950	0		Herron (1988)	400–1100
	1.7×10^{-15}	2940	2.4		Herron (1988)	1100–2000
	1.8×10^{-31}	140	6.5	1.1×10^{-15}	Cohen and Westberg (1991)	298–1300
	4.3×10^{-9}	3820	0		Cohen and Westberg (1991)	1300–2500
	1.66×10^{-15}	2920	1.5	4.7×10^{-16}	Baulch et al. (1992)	300–1200
Propane	1.7×10^{-10}	3170	0		Herron (1988)	400–1100
	8×10^{-22}	1280	3.5		Herron (1988)	1100–2000
	1.3×10^{-21}	1280	3.5	8.1×10^{-15}	Cohen and Westberg (1991)	298–2000
	8.5×10^{-19}	1588	2.56	8.9×10^{-15}	Miyoshi et al. (1993, 1994)	298–1260
Butane	7.8×10^{-23}	780	3.9	2.6×10^{-14}	Herron (1988)	300–2000
	7.8×10^{-23}	780	3.9	2.6×10^{-14}	Cohen and Westberg (1991)	250–2000
	4.1×10^{-21}	1167	3.43	2.5×10^{-14}	Miyoshi et al. (1993, 1994)	250–1140
2-Methylpropane	9.5×10^{-11}	2650	0		Herron (1988)	400–1000
	1.4×10^{-23}	80	3.9		Herron (1988)	1000–2000
	3.3×10^{-23}	80	3.9	1.2×10^{-13}	Cohen and Westberg (1991)	250–2000

III-C-3. $O(^3P)$ + Ethane (C_2H_6)

Divergent values for the rate coefficient for reaction of $O(^3P)$ with ethane near 298 K have been presented; see table III-C-1. Cohen and Westberg (1991) recommend $k = 1.1 \times 10^{-15}$ cm^3 molecule^{-1} s^{-1}, largely on the basis of the work of Mahmud et al. (1988), while Baulch et al. (1992) recommend 5×10^{-16} cm^3 molecule^{-1} s^{-1} apparently without any inclusion of the Mahmud et al. (1988) data in their evaluation. Uncertainties in these recommendations are about a factor of 2–2.5. A rate coefficient of approximately 1×10^{-15} cm^3 molecule^{-1} s^{-1}, with an uncertainty of a factor of 2.5, seems reasonable.

III-C-4. $O(^3P)$ + Propane (C_3H_8)

Cohen and Westberg (1991) and Miyoshi et al. (1993) both recommend a rate coefficient for reaction of $O(^3P)$ with propane at 298 K of $(7–9) \times 10^{-15}$ cm^3 molecule^{-1} s^{-1}. Cohen and Westberg (1991) suggest an uncertainty of a factor of 2. Calculations of Cohen and Westberg (1991) suggest that the vast majority of the reaction (>85%) leads to the formation of 2-propyl radical.

III-C-5. $O(^3P)$ + Butane (n-C_4H_{10})

Herron (1988), Cohen and Westberg (1991), and Miyoshi et al. (1993) all recommend a value of $k = 2.6 \times 10^{-14}$ cm^3 molecule^{-1} s^{-1} at ambient temperatures, with an uncertainty of about a factor of 1.5–2.5. At least 94% of the reaction is expected to generate *sec*-butyl radical, with the remainder leading to 1-butyl (Cohen and Westberg, 1991).

III-C-6. $O(^3P)$ + 2-Methylpropane [Isobutane, $(CH_3)_2CHCH_3$)]

Cohen and Westberg (1991) recommend $k = 1.2 \times 10^{-13}$ cm^3 molecule^{-1} s^{-1} near 298 K, with an uncertainty of a factor of 1.5. However, it should be noted that the recommendation of Herron (1988), which is based largely on the work of Michael et al. (1982) but does not extend below 400 K, clearly supports a substantially (by at least a factor of three) lower rate coefficient at ambient temperatures. The calculations of Cohen and Westberg suggest >98% of the reaction occurs via abstraction of the tertiary H-atom.

III-C-7. $O(^3P)$ + Pentane (n-C_5H_{12})

The reviews of Herron (1988), Cohen and Westberg (1991), and Miyoshi et al. (1993) all give a rate coefficient of $(6–7) \times 10^{-14}$ cm^3 molecule^{-1} s^{-1} for this reaction at 298 K. Uncertainty estimates range from a factor of 1.5 (Cohen and Westberg, 1991) to a factor of 3 (Herron, 1988).

III-C-8. $O(^3P)$ + 2-Methylbutane [Isopentane, $(CH_3)_2CHCH_2CH_3$]

Very little information is available regarding the rate coefficient for reaction of 2-methylbutane (isopentane) with $O(^3P)$. The only measurement is that of Herron and Huie (1969), $k = 1.33 \times 10^{-13}$ cm^3 molecule^{-1} s^{-1} at 307 K. An uncertainty of a factor of three is recommended (Herron, 1988).

III-C-9. $O(^3P)$ + 2,2,-Dimethylpropane [Neopentane, $(CH_3)_4C)$]

Cohen and Westberg (1991) recommend $k = 7.5 \times 10^{-16}$ cm^3 molecule^{-1} s^{-1} near 298 K, with an estimated uncertainty of a factor of two; their recommendation is based on a fit to data obtained by Michael et al. (1982), using transition state theory. The recommendations of Herron (1988) (400–1000 K), also based on the data of Michael et al. (1982), are consistent with those of Cohen and Westberg (1991). However, Miyoshi et al. (1994) suggest that these estimates are too low on the basis of their more recent high-temperature (880–1100 K) data for this reaction. They suggest a 298 K value of $k = 5.7 \times 10^{-15}$ cm^3 molecule^{-1} s^{-1}, consistent with an older study of the reaction by Herron and Huie (1969). Note that this higher recommendation of Miyoshi et al. (1994) forms the basis of the —CH_3 group rate coefficient used in the structure–reactivity calculations described in section III-C-21, $k(-CH_3) = 0.25 \times k(\text{neopentane}) = 1.4 \times 10^{-15}$ cm^3 molecule^{-1} s^{-1}.

III-C-10. $O(^3P)$ + Hexane (n-C_6H_{14})

This is one of the few reactions of $O(^3P)$ with alkanes that has been studied at ambient temperatures since the time of the latest reviews. Miyoshi et al. (1994) measured this rate coefficient using a pulsed laser photolysis–VUV, pulsed laser-induced fluorescence technique between 296 and 399 K. They report $k = 5.56 \times 10^{-11} \times \exp(-2020/T)$ cm^3 molecule^{-1} s^{-1}, with $k = (6.0 \pm 0.4) \times 10^{-14}$ cm^3 molecule^{-1} s^{-1} at 296 K.

The reviews of Herron (1988) and Cohen and Westberg (1991) both give $k = 9 \times 10^{-14}$ cm^3 molecule^{-1} s^{-1} for this rate coefficient near room temperature, with large uncertainties, a factor of 1.5 (Cohen and Westberg, 1991) or a factor of 3 (Herron, 1988). The recent data from Miyoshi et al. (1994) are preferred. Note also that the recent room-temperature data of Miyoshi et al. (1994) for reaction of $O(^3P)$ with hexane and cyclohexane provide a reasonable estimate for the —CH_2— group rate coefficient used in the structure–additivity calculations described in section III-C-21, $k(-CH_2-) \sim 1.4 \times 10^{-14}$ cm^3 molecule^{-1} s^{-1}.

III-C-11. $O(^3P)$ + 2,3-Dimethylbutane [$(CH_3)_2CHCH(CH_3)_2$]

Herron (1988) and Cohen and Westberg (1991) recommend $k = 2 \times 10^{-13}$ cm^3 molecule^{-1} s^{-1} for the rate coefficient for reaction of $O(^3P)$ with 2,3-dimethylbutane at 298 K, with estimated uncertainties of a factor of 1.5 (Cohen and Westberg, 1991) or 3 (Herron, 1988). Reaction at the tertiary H-atoms is expected to dominate (Cohen and Westberg, 1991).

III-C-12. $O(^3P)$ + Heptane (n-C_7H_{16})

Herron (1988) and Cohen and Westberg (1991) recommend $k = 1.3 \times 10^{-13}$ cm^3 molecule^{-1} s^{-1} for reaction of $O(^3P)$ with heptane at 298 K, with uncertainties of a factor of 1.5 (Cohen and Westberg, 1991) or a factor of 3 (Herron, 1988). Note, however, that the recent Miyoshi et al. (1994) data for reaction of $O(^3P)$ with hexane suggest a somewhat lower rate coefficient, $k \approx 7.6 \times 10^{-14}$ cm^3 molecule^{-1} s^{-1}; see structure–additivity calculations in section III-C-21. Rate coefficients for the reactions of $O(^3P)$ with the C_5–C_7alkanes are compared in table III-C-2.

III-C-13. $O(^3P)$ + Octane (n-C_8H_{18})

A value of $k = 1.7 \times 10^{-13}$ cm^3 molecule^{-1} s^{-1} at 298 K is recommended by both Herron (1988) and Cohen and Westberg (1991), with estimated uncertainties of a factor of 1.5 (Cohen and Westberg, 1991) or a factor of 3 (Herron, 1988). As with heptane, the recent data and analysis from Miyoshi et al. (1994) suggest a lower rate coefficient, $k \approx 9 \times 10^{-14}$ cm^3 molecule^{-1} s^{-1}; see structure–additivity calculations in section III-C-21.

III-C-14. $O(^3P)$ + 2,2,3,3-Tetramethylbutane[$(CH_3)_3CC(CH_3)_3$]

There is only one measurement of this rate coefficient (Herron and Huie, 1969), $k = 1.4 \times 10^{-14}$ cm^3 molecule^{-1} s^{-1} at 307 K, but it is believed (Cohen and Westberg, 1991) that this measurement is an overestimate. A rate coefficient of $k = 3.4 \times 10^{-15}$ cm^3 molecule^{-1} s^{-1} near 298 K is recommended by Cohen and Westberg (1991), with an uncertainty of a factor of 2.5. Herron (1988) adopts an earlier rate coefficient expression from Cohen and Westberg (1986), derived from transition-state theory, which generates rate coefficients that are about one-half those from Cohen and Westberg (1991). The updated expression from Cohen and Westberg (1991), with $k(298\text{ K}) = 3.4 \times 10^{-15}$ cm^3 molecule^{-1} s^{-1}, is adopted here.

III-C-15. $O(^3P)$ + 3-Methylheptane [$CH_3CH_2CH(CH_3)CH_2CH_2CH_2CH_3$]

The only measurement of this rate coefficient is that of Ford and Endow (1957), who reported a value of $k = 1.1 \times 10^{-13}$ cm^3 molecule^{-1} s^{-1} at ambient temperatures. The reaction has not been evaluated in any of the recent review articles. Recommended rate coefficients for the reactions of $O(^3P)$ with C_8 alkanes are summarized in table III-C-3.

Table III-C-2. Summary of recommended rate coefficients for reaction $O(^3P)$ with C_5–C_7 alkanes; $k = A \times T^n \times \exp(-B/T)$ cm^3 $molecule^{-1}$ s^{-1}

Alkane	A	B (K)	n	k at 298 K	Reference	Temperature Range (K)
Pentane	6.3×10^{-20}	1010	3.0	5.9×10^{-14}	Herron (1988)	300–2000
	6.3×10^{-20}	1010	3.0	5.9×10^{-14}	Cohen and Westberg (1991)	250–2000
	8.9×10^{-18}	1370	2.37	6.5×10^{-14}	Miyoshi et al. (1993, 1994)	250–1140
2-Methylbutane				1.3×10^{-13}(307 K)	Herron (1988)	307
2,2-Dimethylpropane	1.5×10^{-10}	3600	0		Herron (1988)	400–1000
	2.3×10^{-18}	2720	2.5		Herron (1988)	1000–2000
	4.2×10^{-18}	2720	2.5	7.5×10^{-16}	Cohen and Westberg (1991)	298–2000
Hexane	1.7×10^{-19}	1000	2.9	9.1×10^{-14}	Herron (1988)	300–2000
	1.7×10^{-19}	1000	2.9	9.1×10^{-14}	Cohen and Westberg (1991)	250–2000
				$(6.0 \pm 0.4) \times 10^{-14}$	Miyoshi et al. (1994)	296
2,3-Dimethylbutane	3.9×10^{-22}	140	3.6	2.0×10^{-13}	Herron (1988)	300–2000
	3.9×10^{-22}	140	3.6	2.0×10^{-13}	Cohen and Westberg (1991)	250–2000
Heptane	1.5×10^{-19}	880	2.9	1.3×10^{-13}	Herron (1988)	300–2000
	1.5×10^{-19}	880	2.9	1.3×10^{-13}	Cohen and Westberg (1991)	250–2000

Table III-C-3. Summary of recommended rate coefficients for reaction $O(^3P)$ with C_8 alkanes; $k = A \times T^n \times \exp(-B/T)$ cm^3 molecule^{-1} s^{-1}

Alkane	A	B (K)	n	k, 298 K	Reference	Temperature Range (K)
Octane	3.2×10^{-19}	840	2.8	1.7×10^{-13}	Herron (1988)	300–2000
	3.2×10^{-19}	840	2.8	1.7×10^{-13}	Cohen and Westberg (1991)	250–2000
2,2,3,3-Tetramethylbutane	5.3×10^{-18}	2570	2.5	1.5×10^{-15}	Herron (1988)	300–2000
	5.3×10^{-18}	2360	2.5	3.4×10^{-15}	Cohen and Westberg (1991)	250–2000
3-Methylheptane				1.1×10^{-13}	Ford and Endow (1957)	300
2,3,4-Trimethylpentane				0.5×10^{-13} (307 K)	Herron (1988)	307
2,2,4-Trimethylpentane				0.9×10^{-13} (307 K)	Herron (1988)	307

III-C-16. $O(^3P)$ + Some Dimethyl- and Trimethyl-Substituted Pentanes

Rate coefficients for these reactions near room temperature have been reviewed by Herron (1988) on the basis of measurements made by Herron and Huie (1969). Values of 1.1×10^{-13} cm^3 molecule^{-1} s^{-1} for 2,2-dimethylpentane, 1.7×10^{-13} cm^3 molecule^{-1} s^{-1} for 2,4-dimethylpentane, 5×10^{-14} cm^3 molecule^{-1} s^{-1} for 2,3,4-trimethylpentane, and 9×10^{-14} cm^3 molecule^{-1} s^{-1} for 2,2,4-trimethylpentane are reported (Herron, 1988). Uncertainties are estimated to be a factor of three in all cases.

III-C-17. $O(^3P)$ + 3,3-Diethylpentane [$CH_3CH_2C(C_2H_5)_2CH_2CH_3$]

This rate coefficient has been studied directly over the temperature range 298–773 K by Eichholtz et al. (1997), using a discharge flow–mass spectrometer system. They report an Arrhenius expression of $k = (2.0 \pm 0.3) \times 10^{-11} \exp((-2100 \pm 210)/T)$ cm^3 molecule^{-1} s^{-1}. This expression gives $k = 1.7 \times 10^{-14}$ cm^3 molecule^{-1} s^{-1} at ambient temperatures.

III-C-18. $O(^3P)$ + Cyclopentane (*cyclo*-C_5H_{10})

Herron (1988) and Cohen and Westberg (1991) both recommend $k = 1.2 \times 10^{-13}$ cm^3 molecule^{-1} s^{-1} for this rate coefficient at 298 K, with recommended uncertainties of about a factor of two or three. However, the recent work of Miyoshi et al. (1994) based upon hexane and cyclohexane rate data, suggests that a lower value, $k \approx 7 \times 10^{-14}$ cm^3 molecule^{-1} s^{-1}, might be more reasonable.

III-C-19. $O(^3P)$ + Cyclohexane (*cyclo*-C_6H_{12})

This rate coefficient has recently been studied at low temperature (296–400 K) by Miyoshi et al. (1994), using a laser flash photolysis–pulsed VUV laser-induced fluorescence method. They report $k = 3.6 \times 10^{-11} \exp[(-1800 \pm 144/T)]$ cm^3 molecule^{-1} s^{-1}, with $k = (8.6 \pm 0.5) \times 10^{-14}$ cm^3 molecule^{-1} s^{-1} at 296 K. The recommendations of Herron (1988) and of Cohen and Westberg (1991) are roughly a factor of two larger near room temperature, with uncertainties of a factor of two or three. The newer data of Miyoshi et al. (1994) are preferred. As noted in the discussion of hexane above, this new (lower) rate coefficient for cyclohexane forms the basis of the —CH_2— group rate coefficient used in the structure–additivity calculations described in section III-C-21, $k(—CH_2—) = 1/6 \times k(\text{cyclohexane}) = 1.4 \times 10^{-14}$ cm^3 molecule^{-1} s^{-1}.

III-C-20. $O(^3P)$ + Cycloheptane (*cyclo*-C_7H_{14})

A room-temperature value for the rate coefficient of $O(^3P)$ with cycloheptane of $k = 3 \times 10^{-13}$ cm^3 molecule^{-1} s^{-1} is recommended by both Herron (1988) and Cohen and Westberg (1991), with estimated uncertainties of about a factor of two (Cohen and Westberg, 1991) or three (Herron, 1988). However, a lower value

($\sim 1 \times 10^{-13}$ cm^3 molecule^1 s^{-1}) is suggested for cyclohexane by the recent work of Miyoshi et al. (1994). A summary of recommended rate coefficients for the reactions of O(^3P) with selected alkyl-substituted pentanes and some cycloalkanes is given in table III-C-4.

III-C-21. Rate Coefficient Estimation Methods for O(^3P) + Alkanes (SARs)

Miyoshi et al. (1994) have presented the following expressions for the temperature dependence (300–1100 K) of the reactivity of O(^3P) atoms at —CH_3, —CH_2—, and >CH— groups (all in units of cm^3 molecule^{-1} s^{-1}):

$k(\text{—CH}_3) = 6.9 \times 10^{-22}\, T^{3.469} \exp(-1556/T)$; $k = 1.43 \times 10^{-15}$ at 298 K
$k(\text{—CH}_2\text{—}) = 1.87 \times 10^{-21}\, T^{3.267} \exp(-832/T)$; $k = 1.39 \times 10^{-14}$ at 298 K
$k(\text{>CH—}) = 4.17 \times 10^{-16}\, T^{1.444} \exp(-821/T)$; $k = 9.91 \times 10^{-14}$ at 298 K

As noted above, these expressions are based on the data obtained by Miyoshi et al. (1993, 1994); the reactivity at —CH_3 groups is based on neopentane, and reactivity at —CH_2— groups is based on cyclohexane. In the absence of a more complete and accurate data set, a more complex analysis taking into account the effects on neighboring groups is not justified currently. In table III-C-5 a comparison is given between those rate coefficients recommentded by Herron (1988), Cohen and Westberg (1991), Baulch et al. (1992), and those recently measured by Miyoshi et al. (1994). Note in particular that the new data of Miyoshi et al. (1994) for reaction of O(^3P) with hexane and cyclohexane suggest a lower reactivity per —CH_2— group than previously believed. Thus the calculated rates for some of the alkanes (pentane, heptane, octane) and cycloalkanes are seen to be lower than the previous recommendations. Comparisons of rate coefficients calculated using the Miyoshi et al. (1994) expressions with those measured experimentally are shown in figure III-C-1.

III-D. Reactions of NO_3 Radicals with Alkanes

III-D-1. Introduction

Reactions of NO_3 radicals with alkanes have been fairly extensively studied at ambient temperatures, although little is known about the temperature dependence of these rate coefficients. The majority of studies were carried out during the late 1980s and early 1990s, mostly using relative rate techniques, although a variety of direct techniques (discharge flow, stopped flow, flash photolysis, modulated photolysis) were also used to study the reactions of some of the smaller alkanes. The reactions have been shown to be rather slow ($k < 10^{-18}$ cm^3 molecule^{-1} s^{-1} for methane; $k \approx$ (0.5–3) $\times 10^{-16}$ cm^3 molecule^{-1}s^{-1} for most C_4–C_{10} species). These rate coefficients are typically 4–5 orders of magnitude less than the corresponding OH rate coefficient (see rate coefficient correlations, section III-F). Since atmospheric levels of NO_3 rarely exceed 10^9 molecule cm^{-3}, even in polluted urban regions, and thus are only three orders of magnitude higher than OH, the NO_3 reactions are of little significance as

Table III-C-4. Summary of recommended rate coefficients (k) for reaction $O(^3P)$ with selected alkanes; $k = A\ T^n \exp(-B/T)$ cm^3 molecule^{-1} s^{-1}

Alkane	A	B (K)	n	k, 298 K	Reference	Temperature Range (K)
2,2-Dimethylpentane				1.1×10^{-13}(307 K)	Herron (1988)	307
2,4-Dimethylpentane				1.7×10^{-13}(307 K)	Herron (1988)	307
3,3-Diethylpentane	$(2.0 \pm 0.3) \times 10^{-11}$	2100 ± 210	0	1.7×10^{-14}	Eichholtz et al. (1997)	
Cyclopentane	4.8×10^{-18}	1390	2.6	1.2×10^{-13}	Herron (1988)	300–2000
	4.8×10^{-18}	1390	2.6	1.2×10^{-13}	Cohen and Westberg (1991)	298–2000
Cyclohexane	4.3×10^{-18}	1290	2.6	1.4×10^{-13}	Herron (1988)	300–2000
	4.3×10^{-18}	1240	2.6	1.4×10^{-13}	Cohen and Westberg (1991)	298–2000
				$(8.5 \pm 0.6) \times 10^{-14}$	Miyoshi et al. (1994)	297
Cycloheptane	3.5×10^{-18}	1090	2.6	2.4×10^{-13}	Herron (1988)	300–2000
	3.8×10^{-18}	1020	2.6	3.3×10^{-13}	Cohen and Westberg (1991)	298–2000

Table III-C-5. Comparison between recommended (Herron, 1988; Cohen and Westberg, 1991; Miyoshi et al., 1994) and calculated rate coefficients for reaction of $O(^3P)$-atoms with alkanes near 298 K. All data in units of cm^3 $molecule^{-1}$ s^{-1}

Alkane	Recommended Rate Coefficient[a] (see text for details regarding each alkane)	Measured Values (Miyoshi et al., 1994)	Calculated (using group reactivity expressions of Miyoshi et al., 1994)
Ethane	1×10^{-15}		3×10^{-15}
Propane	9×10^{-15}		1.75×10^{-14}
Butane	2.7×10^{-14}		3.2×10^{-14}
2-Methylpropane	1.2×10^{-13}		1.1×10^{-13}
Pentane	7×10^{-14}		4.7×10^{-14}
2-Methylbutane	1.3×10^{-13}		1.2×10^{-13}
2,2-Dimethlpropane	8×10^{-16}		6×10^{-15}
Hexane	9×10^{-14}	$(6.0 \pm 0.4) \times 10^{-14}$	***6×10^{-14}***[b]
2,3-Dimethylbutane	2×10^{-13}		2.1×10^{-13}
Heptane	1.2×10^{-13}		***7.6×10^{-14}***[b]
Octane	1.7×10^{-13}		***9×10^{-14}***[b]
2,2,3,3-Tetramethylbutane	3.4×10^{-15}		***9×10^{-15}***[b]
3-Methylheptane	1.1×10^{-13}		1.6×10^{-13}
2,3,4-Trimethylpentane	5×10^{-14}		3×10^{-13}
2,2,4-Trimethylpentane	9×10^{-14}		1.2×10^{-13}
2,4-Dimethylpentane	1.7×10^{-13}		2.2×10^{-13}
2,2-Dimethylpentane	1.1×10^{-13}		1.4×10^{-13}
3,3-Diethylpentane	1.7×10^{-13}		6.4×10^{-14}
Cyclopentane	1.2×10^{-13}		***7×10^{-14}***[b]
Cyclohexane	1.8×10^{-13}	$(8.6 \pm 0.5) \times 10^{-14}$	***9×10^{-14}***[b]
Cycloheptane	2.6×10^{-13}		***1.0×10^{-13}***[b]

[a] Prior to measurements of Miyoshi et al. (1993,1994).

[b] Data highlighted in bold italics indicate systems where updates are recommended on the basis of data and structure–additivity analysis of Miyoshi et al. (1993,1994).

sinks for the alkanes. Nonetheless, these reactions may play some role as sources of radicals and nitric acid in the nighttime chemistry.

The reactions of NO_3 with alkanes have been studied by both direct and relative rate methods. Direct measurements have been made using low-pressure discharge flow tube (Bagley et al., 1990), stopped flow (Boyd et al., 1991), flash photolysis (Wallington et al., 1986), and modulated photolysis techniques (Burrows et al., 1985), with NO_3 monitored by visible absorption in all cases. However, great care is required in these direct studies to account for secondary reactions. For example, as discussed in detail by Bagley et al. (1990), reaction of NO_3 with the parent alkane will be rapidly followed by further reaction of NO_3 with the initial alkyl radical product:

$$NO_3 + RH \rightarrow HONO_2 + R \quad \text{(III-D-1)}$$

$$NO_3 + R \rightarrow RONO_2 \quad \text{(III-D-2a)}$$

$$\rightarrow NO_2 + RO \quad \text{(III-D-2b)}$$

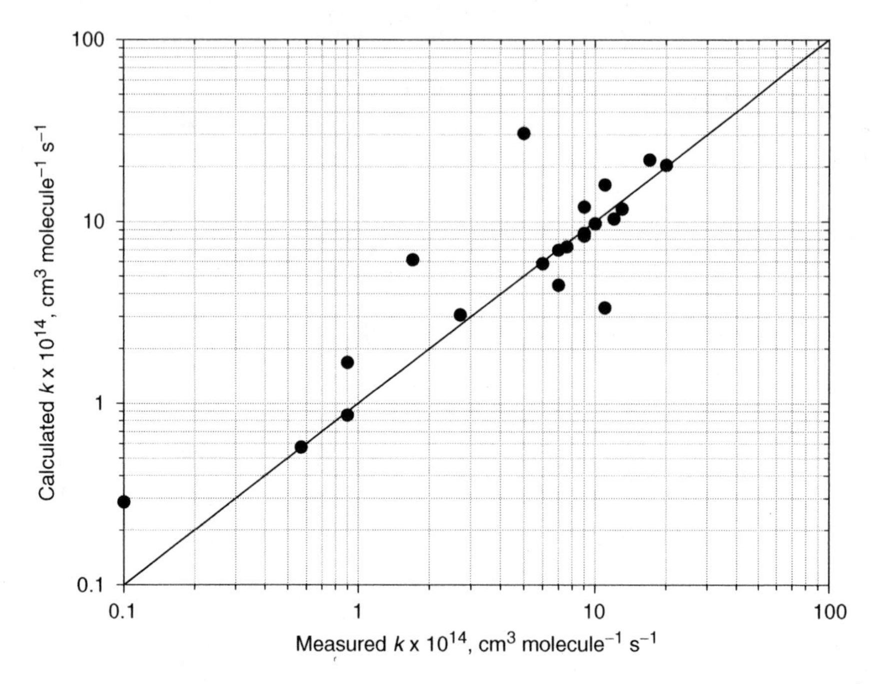

Figure III-C-1. Comparison of measured and calculated rate coefficients for reaction of $O(^3P)$ atoms with the alkanes. Calculations are made using parameters described in the text. The line is the 1:1 relation between the variables.

Thus, observed NO_3 decays must be corrected by at least a factor of two to account for the reactions III-D-2a and III-D-2b. Subsequent reaction of NO_3 with products of reaction III-D-2, for example alkoxy radicals, leads to further complications. In the presence of O_2, even higher stoichiometries (three–five) are calculated as a result of chain reactions involving RO_2 and RO radicals generated from reaction of the alkyl radicals with O_2.

A number of "traditional" relative rate studies have been carried out (e.g., Atkinson et al., 1984c, 1988; Aschmann and Atkinson, 1995b), in which the loss of two organics (the compound under study and a reference species) are simultaneously monitored in the presence of NO_3. Although few alkane + NO_3 rate coefficients were accurately known when measurements were first made in the mid-1980s, sufficient information is now available to allow the emergence of a consistent set of absolute rate coefficients. In particular, a set of relative rate measurements made by Atkinson et al. (1988) allows for the entire data set to be pinned to absolute measurements for reaction of NO_3 with *trans*-2-butene. (Note that reactions of NO_3 with alkenes occur via addition rather than abstraction, and thus they are less likely subject to secondary chemistry complications and thus can be considered more reliable.)

In a variant of the relative rate technique, rate coefficients for reactions of NO_3 with alkanes are essentially measured relative to the $NO_3 + NO_2 \leftrightarrow N_2O_5$ equilibrium constant, by monitoring the increase in the rate of N_2O_5 loss upon addition of an organic reactant (Atkinson et al., 1984c). Rate coefficients obtained in this manner are often interpreted as upper limits due to potential reaction of NO_3 with products of the reaction with the parent alkane.

Although no product studies exist, reaction of NO_3 with alkanes is generally accepted to occur via abstraction, e.g.,

$$NO_3 + RH \rightarrow HONO_2 + R \qquad \text{(III-D-1)}$$

a conclusion supported by the theoretical work of Bravo-Pérez et al. (2002). Since the H—ONO_2 bond strength is about 426 kJ mol^{-1}, hydrogen abstraction from methane is an endothermic process, ΔH_r (298 K) = 13 kJ mol^{-1}, while abstraction from all other alkanes is slightly exothermic (by about 3, 17, and 26 kJ mol^{-1} for primary, secondary, and tertiary hydrogens, respectively). Surprisingly, only one study of the temperature dependence of the NO_3+ alkane reactions exists (Bagley et al., 1990), and this work indicates typical *A*-factors of 10^{-12} cm^3 molecule^{-1} s^{-1} and activation barriers of 22–27 kJ mol^{-1} for C_4–C_5 alkanes. Structure–additivity relationships for abstraction at room temperature and as a function of temperature have been developed; see details in section III-D-16. In general, reactivity is found to increase in the expected manner, with compounds containing tertiary >CH— sites being more reactive than compounds containing secondary —CH_2— sites, which are in turn more reactive than primary —CH_3 sites. However, the theoretical study of Bravo-Pérez et al. (2002) indicates that the determination of overall rate coefficients and the site of NO_3 attack may be complicated by the presence of tunneling and subtle symmetry effects. For example, their calculations on 2-methylpropane suggest more reaction of NO_3 at the methyl groups than at the tertiary C—H site.

A summary of the available data for reactions of NO_3 with the alkanes is given in tables III-D-1 (methane, ethane, propane), III-D-2 (butane, 2-methylpropane), III-D-3 (C_5–C_6 alkanes), and III-D-4 (C_7–C_{10} alkanes). In cases where sufficient information is available, rate coefficient recommendations are also given.

III-D-2. NO_3 + Methane (CH_4)

The reaction of NO_3 with methane has been studied by a variety of direct and relative techniques (Burrows et al., 1985; Wallington et al., 1986; Cantrell et al., 1987; Boyd et al., 1991), and in all cases only an upper limit to the rate coefficient has been obtained. Neither Burrows et al. (1985), using a molecular modulation technique, Wallington et al. (1986), using a flash photolysis technique, nor Boyd et al. (1991), using a stopped-flow reactor technique, observed any decay of NO_3 that could be attributed to reaction with CH_4, leading to upper limits for *k* of 4×10^{-16}, 2×10^{-17}, and 8×10^{-19} cm^3 molecule^{-1}s^{-1}, respectively. Cantrell et al. (1987), using the NO_3+ $NO_2 \leftrightarrow N_2O_5$ equilibrium relative rate technique described above, reported $k < 1 \times 10^{-18}$ cm^3 molecule^{-1} s^{-1} from the lack of measurable CH_4 decay and $k < 4 \times 10^{-21}$ cm^3 molecule^{-1} s^{-1} from the lack of detection of the assumed end-products, CO and CO_2. Using updated equilibrium constant data (Atkinson et al., 2005a) results in a modest (20%) increase to these upper limits (1.2×10^{-18} cm^3 molecule^{-1} s^{-1} and 5×10^{-21} cm^3 molecule^{-1} s^{-1}, respectively).

Table III-D-1. Summary of rate coefficient measurements for reaction of NO_3 radicals with methane, ethane, and propane; $k = A \times \exp(-B/T)$ cm^3 $molecule^{-1}$ s^{-1}

Alkane	A	B (K)	k	T (K)	Technique[a]	Reference	Temperature Range (K)
Methane			$<4 \times 10^{-16}$	298	MM–Abs	Burrows et al. (1985)	
			$<2 \times 10^{-17}$	298	FP–Abs	Wallington et al. (1986)	
			$<1.2 \times 10^{-18}$	298	RR[b]	Cantrell et al. (1987)	
			$<5 \times 10^{-21}$	298	RR[b]	Cantrell et al. (1987)	
			$<8 \times 10^{-19}$	298	SF–Abs	Boyd et al. (1991)	
Ethane			$<4 \times 10^{-18}$	298	FP–Abs	Wallington et al. (1986)	
	5.7×10^{-12}	4428	$(3.0 \pm 0.5) \times 10^{-16}$	453	DF–Abs	Bagley et al. (1990)	453–553
			$(4.7 \pm 0.8) \times 10^{-16}$	473			
			$(1.4 \pm 0.3) \times 10^{-15}$	523			
			$(1.6 \pm 0.3) \times 10^{-15}$	553			
			$<2.7 \times 10^{-17}$	298	SF–Abs	Boyd et al. (1991)	
			$(>1.1 \times 10^{-17})$	298	SF–Abs	Boyd et al. (1991)	
Propane			$<4.8 \times 10^{-17}$	298	SF–Abs	Boyd et al. (1991)	
			$(>2.2 \times 10^{-17})$	298	SF–Abs	Boyd et al. (1991)	

[a] Abbreviations for the techniques are defined in table II-A-1.
[b] Placed on an absolute basis using $K_{eq} = 3.0 \times 10^{-11}$ cm^3 $molecule^{-1}$ for the $NO_3 + NO_2 \leftrightarrow N_2O_5$ equilibrium system (Atkinson et al., 2005a).

Table III-D-2. Summary of rate coefficient measurements for reaction of NO_3 radicals with C_4 compounds; $k = A \times \exp(-B/T)$ cm^3 $molecule^{-1}$ s^{-1}

Alkane	A	B (K)	k	T (K)	Technique[d]	Reference	Temperature Range (K)
Butane			$(6.7 \pm 1.8) \times 10^{-17}$	296	RR[a]	Atkinson et al. (1984c)	
			$<2 \times 10^{-17}$	298	FP–Abs.	Wallington et al. (1986)	
			$(4.5 \pm 0.6) \times 10^{-17}$	298	DF–Abs.	Bagley et al. (1990)	298–523
			$(1.44 \pm 0.12) \times 10^{-16}$	333			
			$(4.6 \pm 1.2) \times 10^{-16}$	373			
			$(11.2 \pm 1.2) \times 10^{-16}$	423			
			$(32 \pm 3) \times 10^{-16}$	473			
			$(90 \pm 4) \times 10^{-16}$	523			
	2.8×10^{-12}	3280				Atkinson et al. (2005a)	298–423
2-Methylpropane			$(9.8 \pm 2.1) \times 10^{-17}$	296	RR[b]	Atkinson et al. (1984c)	
			$(1.1 \pm 0.2) \times 10^{-16}$	298	DF–Abs	Bagley et al. (1990)	
			$(4.5 \pm 1.6) \times 10^{-16}$	348			
			$(8.0 \pm 0.8) \times 10^{-16}$	373			
			$(23 \pm 4) \times 10^{-16}$	423			
			$(54 \pm 12) \times 10^{-16}$	473			
			$(130 \pm 24) \times 10^{-16}$	523			
			$<6 \times 10^{-16}$	298	SF–Abs	Boyd et al. (1991)	
	3.9×10^{-12}	3150				This work[c]	298–423

[a] Placed on an absolute basis using $k(NO_3 + \text{heptane}) = 1.4 \times 10^{-16}$ cm^3 $molecule^{-1}$ s^{-1} at 296 K.
[b] Placed on an absolute basis using $k(NO_3 + \text{2,3-dimethylbutane}) = 4.1 \times 10^{-16}$ cm^3 $molecule^{-1}$ s^{-1} at 296 K.
[c] Derived from a fit of Bagley et al. (1991) data from 298 to 423 K; see text for details.
[d] Abbreviations for the techniques are defined in table II-A-1.

Table III-D-3. Summary of rate coefficient measurements for reaction of NO_3 radicals with C_5–C_6 compounds; $k = A \times \exp(-B/T)$ cm^3 $molecule^{-1}$ s^{-1}

Alkane	K	T (K)	Technique[e]	Reference	Temperature Range (K)
Pentane	$(8.3 \pm 1.8) \times 10^{-17}$	296	RR[a]	Atkinson et al. (1984c)	
2-Methylbutane	$(1.6 \pm 0.2) \times 10^{-16}$	298	DF–Abs	Bagley et al. (1990)	298–523
	$(3.9 \pm 1.4) \times 10^{-16}$	323			
	$(10.4 \pm 1.2) \times 10^{-16}$	373			
	$(25 \pm 4) \times 10^{-16}$	423			
	$(61 \pm 14) \times 10^{-16}$	473			
	$(129 \pm 37) \times 10^{-16}$	523			
	$<6 \times 10^{-16}$	298	SF–Abs	Boyd et al. (1991)	
	$(1.57 \pm 0.16) \times 10^{-16}$	296	RR[b]	Aschmann and Atkinson (1995b)	
Hexane	$(1.1 \pm 0.2) \times 10^{-16}$	296	RR[a]	Atkinson et al. (1984c)	
	$(1.47 \pm 0.30) \times 10^{-16}$	298	DF–Abs	Langer et al. (1993)	
2,3-Dimethylbutane	$(4.3 \pm 0.9) \times 10^{-16}$	296	RR[c]	Atkinson et al. (1984c)	
	$(4.1 \pm 0.2) \times 10^{-16}$	296	RR[a]	Atkinson et al. (1984c)	
	$(4.1 \pm 0.3) \times 10^{-16}$	296	RR[d]	Atkinson et al. (1988)	
Cyclohexane	$(1.35 \pm 0.25) \times 10^{-16}$[b]	296	RR[b]	Atkinson et al. (1984c)	
2-Methylpentane	$(1.72 \pm 0.18) \times 10^{-16}$[b]	296	RR[b]	Aschmann and Atkinson (1995b)	
3-Methylpentane	$(2.05 \pm 0.21) \times 10^{-16}$[b]	296	RR[b]	Aschmann and Atkinson (1995b)	

[a] Placed on an absolute basis using $k(NO_3 + \text{heptane}) = 1.4 \times 10^{-16}$ cm^3 $molecule^{-1}$ s^{-1} at 296 K.
[b] Placed on an absolute basis using $k(NO_3 + 2,3\text{-dimethylbutane}) = 4.1 \times 10^{-16}$ cm^3 $molecule^{-1}$ s^{-1} at 296 K.
[c] Placed on an absolute basis using $K_{eq} = 3.8 \times 10^{-11}$ cm^3 $molecule^{-1}$ for the $NO_3 + NO_2 \leftrightarrow N_2O_5$ equilibrium system at 296 K (IUPAC, 2005).
[d] Placed on an absolute basis using $k(NO_3 + trans\text{-2-butene}) = 3.89 \times 10^{-13}$ cm^3 $molecule^{-1}$ s^{-1} at 296 K.
[e] Abbreviations for the techniques are defined in table II-A-1.

Table III-D-4. Summary of rate coefficient (k, cm^3 molecule^{-1} s^{-1}) measurements for reaction of NO_3 radicals with C_7–C_{10} compounds at 296 K

Alkane	k	Technique[d]	Reference
2,2,4-Trimethylpentane	$(0.75 \pm 0.32) \times 10^{-16}$	RR[b]	Aschmann and Atkinson (1995b)
2,2,3,3-Tetramethylbutane	$<0.49 \times 10^{-16}$	RR[b]	Aschmann and Atkinson (1995b)
2,2,3-Trimethylbutane	$(2.23 \pm 0.33) \times 10^{-16}$	RR[b]	Aschmann and Atkinson (1995b)
2,4-Dimethylpentane	$(1.45 \pm 0.45) \times 10^{-16}$	RR[b]	Aschmann and Atkinson (1995b)
Heptane	$(1.10 \pm 0.30) \times 10^{-16}$	RR[c]	Atkinson et al. (1984c)
	$(1.37 \pm 0.14) \times 10^{-16}$	RR[a]	Atkinson et al. (1988)
Octane	$(1.86 \pm 0.24) \times 10^{-16}$	RR[a]	Atkinson et al. (1984c)
Nonane	$(2.46 \pm 0.31) \times 10^{-16}$	RR[a]	Atkinson et al. (1984c)
	$(1.93 \pm 0.53) \times 10^{-16}$	RR[b]	Aschmann and Atkinson (1995b)
Decane	$(2.61 \pm 0.61) \times 10^{-16}$	RR[b]	Aschmann and Atkinson (1995b)

[a] Placed on an absolute basis using $k(NO_3 + \text{heptane}) = 1.4 \times 10^{-16}$ cm^3 molecule^{-1} s^{-1} at 296 K.
[b] Placed on an absolute basis using $k(NO_3 + 2,3\text{-dimethylbutane}) = 4.1 \times 10^{-16}$ cm^3 molecule^{-1} s^{-1} at 296 K.
[c] Placed on an absolute basis using $K_{eq} = 3.8 \times 10^{-11}$ cm^3 molecule^{-1} for the $NO_3 + NO_2 \leftrightarrow N_2O_5$ equilibrium system at 296 K (Atkinson et al., 2005a).
[d] Abbreviations for the techniques are defined in table II-A-1.

Atkinson (1991) recommends $k < 1 \times 10^{-18}$ cm^3 molecule^{-1} s^{-1}, on the basis of the data available at the time and on the basis of correlations between OH and NO_3 rate coefficients with the alkanes. This conservative recommendation is followed here, but the lack of CO or CO_2 production in the experiments of Cantrell et al. (1987) provides strong evidence for a much smaller rate coefficient.

III-D-3. NO_3 + Ethane (C_2H_5)

Reaction of NO_3 with ethane has been studied by three direct methods. Wallington et al. (1986), using a flash photolysis–visible absorption method, reported $k < 4 \times 10^{-18}$ cm^3 molecule^{-1} s^{-1}. Bagley et al. (1990), using a discharge flow–visible absorption method, observed reaction of NO_3 with ethane at temperatures between 453 K and 553 K, and reported $k = (5.7 \pm 4.0) \times 10^{-12} \exp(-4426 \pm 337)/T)$ cm^3 molecule^{-1} s^{-1}; see figure III-D-1. Extrapolation of this Arrhenius expression to 298 K yields $k = 2.0 \times 10^{-18}$ cm^3 molecule^{-1} s^{-1}. The authors also suggest a 298 K upper limit of about 10^{-17} cm^3 molecule^{-1} s^{-1} on the basis of their observations. In the stopped-flow study of Boyd et al. (1991), decays of NO_3 were enhanced upon addition of ethane, and were consistent with $k < 2.7 \times 10^{-17}$ cm^3 molecule^{-1} s^{-1} at 298 K. Also, on the basis of modeling studies of their reaction system, the authors suggested a lower limit for k near 10^{-17} cm^3 molecule^{-1} s^{-1}. However, lower limits suggested by this group for reaction of NO_3 with 2-methylpropane seem high compared to relative rate data, and so should probably be treated with caution. On the basis of the data just summarized, the IUPAC panel (Atkinson et al., 2005a) recommends $k < 1 \times 10^{-17}$ cm^3 molecule^{-1} s^{-1} at 298 K, a recommendation that is adopted here.

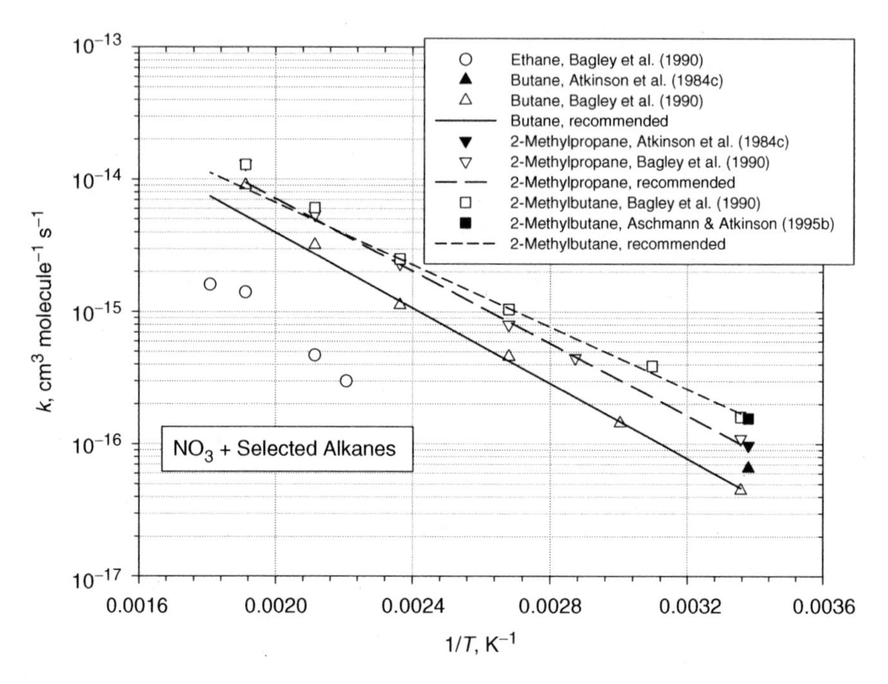

Figure III-D-1. Measurements of the rate coefficients for reaction of NO_3 with selected alkanes.

III-D-4. NO_3 + Propane (C_3H_8)

The only study of the reaction of NO_3 with propane is that of Boyd et al. (1991), using a stopped-flow reactor with NO_3 detection via visible absorption. This study leads to an upper limit of $k < 7 \times 10^{-17}\,cm^3\,molecule^{-1}\,s^{-1}$. The authors, from modeling studies of their reaction system, also suggested a lower limit of $k > 2 \times 10^{-17}\,cm^3\,molecule^{-1}\,s^{-1}$. However, we note that the lower limit these authors derived for reaction of NO_3 with 2-methylpropane is higher than other, seemingly reliable data for this reaction, and so should be treated with caution. The IUPAC panel (Atkinson et al., 2005a) recommends $k < 7 \times 10^{-17}\,cm^3\,molecule^{-1}\,s^{-1}$ at 298 K on the basis of the Boyd et al. (1991) study, a recommendation that is adopted here.

III-D-5. NO_3 + Butane (n-C_4H_{10})

The rate coefficient for this reaction has been measured by Atkinson et al. (1984c), Wallington et al. (1986), and Bagley et al. (1990). Atkinson et al. (1984c) monitored the simultaneous decay of a series of straight-chain alkanes in the presence of NO_3. Using the reaction of NO_3 with heptane as the reference reaction and an updated value for the reference reaction rate coefficient (Atkinson, 1991) leads to $k = (6.7 \pm 1.8) \times 10^{-17}\,cm^3\,molecule^{-1}\,s^{-1}$. Wallington et al. (1986), using flash photolysis–visible absorption, observed enhanced decay rates for NO_3 upon addition of butane, and report $k \leq 2 \times 10^{-17}\,cm^3\,molecule^{-1}\,s^{-1}$. (This was reported as an upper limit due to the possibility of the presence of reactive impurities.) Bagley et al. (1990) studied this reaction over the temperature range 298–523 K using a discharge flow–visible absorption apparatus. Assuming loss of two NO_3 radicals per butane reacted, they report values ranging from $(4.5 \pm 0.6) \times 10^{-17}$ cm^3 molecule^{-1} s^{-1} at 298 K to $(9.0 \pm 0.4) \times 10^{-16}$ cm^3 molecule^{-1} s^{-1} at 523 K. The authors noted upward curvature in their Arrhenius plots at high temperature, likely the result of further secondary chemistry (i.e., more than two NO_3 radicals consumed per butane reaction). In light of this curvature, the IUPAC panel (Atkinson et al., 2005a) recommends $k = 2.8 \times 10^{-12} \times \exp(-3280/T)$ cm^3 molecule^{-1} s^{-1} for $T = 298$–423 K, obtained from a least-squares fit to the Bagley et al. (1990) data over this limited temperature range. The 298 K value retrieved from this expression, $k = 4.6 \times 10^{-17}$ cm^3 molecule^{-1} s^{-1}, is roughly in accord with the updated value (Atkinson, 1991) obtained by relative rate measurements. However, this is inconsistent with the upper limit reported by Wallington et al. (1986). An uncertainty of ±25% is estimated for k at 298 K. Relevant data are displayed in figure III-D-1.

III-D-6. NO_3 + 2-Methylpropane [Isobutane, $(CH_3)_2CHCH_3$]

Both relative (Atkinson et al., 1984c) and absolute rate coefficient data (Bagley et al., 1990; Boyd et al., 1991) are available for reaction of NO_3 with 2-methylpropane. Room-temperature data from Atkinson et al. (1984c) and Bagley et al. (1990) are in excellent agreement. The Atkinson et al. (1984c) study, which used 2,3-dimethylbutane as the reference compound, leads to $k = (9.8 \pm 2.1) \times 10^{-17}$ cm^3 molecule^{-1} s^{-1} (Atkinson, 1991; Aschmann and Atkinson, 1995b), while Bagley et al. (1990) report a room-temperature value, $k = (1.1 \pm 0.2) \times 10^{-16}$ cm^3 molecule^{-1} s^{-1}, after correction by a factor of two for NO_3 reaction with the alkyl radical product of the reaction. Using a stopped-flow technique, Boyd et al. (1991) report an upper limit of $k = 6 \times 10^{-16}$ cm^3 molecule^{-1} s^{-1}, obtained by dividing the observed rate coefficient by a factor of two to account for the reaction of NO_3 with the alkyl radical products. In a modeling study of their system, however, these authors were unable to reconcile their observed NO_3 decays with the much lower rate coefficients of Atkinson et al. (1984c) and Bagley et al. (1990).

Bagley et al. (1990) also studied this reaction at elevated temperatures (298–523 K). An Arrhenius plot of the data (see figure III-D-1) showed slight upward curvature at the highest temperatures studied, likely the result of NO_3 consumption by additional secondary chemistry under these conditions. For consistency with butane, the expression recommended herein is obtained from a fit of the Bagley et al. (1990) data from 298–423 K, $k = 3.9 \times 10^{-12} \times \exp(-3150/T)$ cm^3 molecule^{-1} s^{-1}, yielding $k = 1.0 \times 10^{-16}$ cm^3 molecule^{-1} s^{-1} at 298 K. Uncertainties in k of ±20% are estimated.

III-D-7. NO_3 + Pentane (n-C_5H_{12})

The rate coefficient for reaction of NO_3 with pentane has been the subject of only one relative rate study. Atkinson et al. (1984c) determined

k(pentane)/k (heptane) = 0.59 ± 0.12 at 296 K. Using an updated value for the reference rate coefficient leads to $k(NO_3$+ pentane) = $(8.3 \pm 1.8) \times 10^{-17}$ cm^3 molecule^{-1} s^{-1}.

III-D-8. NO_3 + 2-Methylbutane [Isopentane, $(CH_3)_2CHCH_2CH_3$]

The rate coefficients of Bagley et al. (1990) and Aschmann and Atkinson (1995b), measured at ambient temperatures, are in excellent agreement, $k = (1.6 \pm 0.2) \times 10^{-16}$ and $(1.57 \pm 0.16) \times 10^{-16}$ cm^3 molecule^{-1} s^{-1}, respectively. Bagley et al. (1990) also studied this reaction at elevated temperatures. Data are displayed in figure III-D-1. As was the case with butane and 2-methylpropane, the data of Bagley et al. (1990) above 425 K were higher than an extrapolation of an Arrhenius fit to the 298–423 K data; this is likely indicative of the occurrence of secondary chemistry above 425 K. A fit of their data between 298 and 423 K yields $k = 1.5 \times 10^{-12} \exp(-2707/T)$. This expression, which yields a value of $k = 1.7 \times 10^{-16}$ cm^3 molecule^{-1} s^{-1} at 298 K, is recommended, with estimated uncertainties of ±15%.

III-D-9. NO_3 + Hexane (n-C_6H_{14})

The rate coefficient for reaction of NO_3 with hexane has been studied at room temperature by both relative (Atkinson et al., 1984c) and absolute techniques (Langer et al., 1993), and the two measurements are in reasonable accord. Atkinson et al. (1984c) found k(hexane)/k(heptane) = 0.77 ± 0.14, yielding $k(NO_3$+ hexane) = $(1.1 \pm 0.2) \times 10^{-16}$ cm^3 molecule^{-1} s^{-1} at 296 K. Langer et al. (1993), using a discharge-flow apparatus with NO_3 detection via visible absorption, report k(hexane) = $(1.47 \pm 0.30) \times 10^{-16}$ cm^3 molecule^{-1} s^{-1}. Although the authors assumed loss of two NO_3 per hexane molecule consumed, their resulting value should strictly be treated as an upper limit. The rate coefficient of Atkinson et al. (1984c), k = 1.1×10^{-16} cm^3 molecule^{-1} s^{-1}, is recommended, with the uncertainty increased to ±35 % to encompass the flow tube value.

III-D-10. NO_3 + 2,3-Dimethylbutane [$(CH_3)_2CHCH(CH_3)_2$]

The reaction of $NO_3$with 2,3-dimethylbutane, which is often used as a reference reaction in relative rate studies, has itself been the subject of two relative rate studies at ambient temperatures (Atkinson et al., 1984c, 1988). In the first study, the N_2O_5 equilibrium constant and the reaction of NO_3 with heptane were used as reference reactions. Using updated data for these reference reactions gives $k = (4.3 \pm 0.9) \times 10^{-16}$ cm^3 molecule^{-1} s^{-1} and $k = (4.1 \pm 0.2) \times 10^{-16}$ cm^3 molecule^{-1} s^{-1} for the reaction of NO_3 with 2,3-dimethylbutane at 296 K. In a series of relative rate measurements, which can be traced back to the reaction of NO_3 with *trans*-2-butene as the primary reference, Atkinson et al. (1988) report $k(NO_3$ + 2,3-dimethylbutane) = $(4.1 \pm 0.3) \times 10^{-16}$ cm^3 molecule^{-1} s^{-1}. The recommended rate coefficient is $k = 4.1 \times 10^{-16}$ cm^3 molecule^{-1} s^{-1} ($T \sim 296$ K) with an estimated uncertainty of ±15% (Atkinson, 1991; Aschmann and Atkinson, 1995b).

III-D-11. NO_3 + Cyclohexane (C_6H_{12})

The rate coefficient for reaction of NO_3 with cyclohexane has been the subject of one relative rate study. Atkinson et al. (1984c) determined *k*(cyclohexane)/*k*(2,3-dimethylbutane) = 0.33 ± 0.06 at 296 K. Using the recommended rate coefficient for the reference reaction gives $k(NO_3$ + cyclohexane) = 1.35×10^{-16} cm^3 molecule^{-1} s^{-1}, with an estimated uncertainty of ±25%.

III-D-12. NO_3 + Heptane (n-C_7H_{16})

This reaction has also often been employed as the standard in relative rate studies. Relative rate measurements, made against the N_2O_5 equilibrium constant (Atkinson et al., 1984c) and against *trans*-2-butene (Atkinson et al., 1988), are in reasonable agreement, $k = (1.10 \pm 0.30) \times 10^{-16}$ cm^3 molecule^{-1} s^{-1} and $k = (1.37 \pm 0.14) \times 10^{-16}$ cm^3 molecule^{-1} s^{-1}, respectively. Given the higher uncertainties involved in studies using the N_2O_5 equilbrium system, a value of $k = 1.4 \times 10^{-16}$ cm^3 molecule^{-1} s^{-1} is recommended for temperatures near 296 K, with an estimated uncertainty of ±20%.

III-D-13. NO_3 + Octane (n-C_8H_{18})

The rate coefficient for reaction of NO_3 with octane has been studied by Atkinson et al. (1984c), who found *k*(octane)/*k*(heptane) = 1.33 ± 0.16 at 296 K. Using an updated value for the reference rate coefficient leads to *k* (NO_3+octane) = 1.9×10^{-16} cm^3 molecule^{-1} s^{-1} (Atkinson, 1991; Aschmann and Atkinson, 1995b). Given that only one study of this reaction is available, an uncertainty of ±25% is suggested.

III-D-14. NO_3 + Nonane (n-C_9H_{20})

This rate coefficient has been determined in the relative rate studies of Atkinson et al. (1984c) and Aschmann and Atkinson (1995b), using heptane and 2,3-dimethylbutane as reference species, respectively. Data are in reasonable accord. An average of the two determinations, $k = 2.2 \times 10^{-16}$ cm^3 molecule^{-1} s^{-1}, is recommended at 296 K, with an estimated uncertainty of ±25%.

III-D-15. Reaction of NO_3 with Other Alkanes

A series of alkanes has been the subject of one relative rate study, that of Aschmann and Atkinson (1995b). The reference compound was 2,3-dimethylbutane, *k* (NO_3 + 2, 3-dimethylbutane) = 4.1×10^{-16} cm^3 molecule^{-1} s^{-1} (see section III-D-10). The alkanes studied, in addition to 2-methylbutane and nonane that have already been discussed, include 2,2,4-trimethylpentane; 2,2,3,3-tetramethylbutane; 2,2,3-trimethylbutane; 2,4-dimethylpentane; 2-methylpentane; 3-methylpentane; and decane. Rate coefficient ratios and absolute rate coefficients for these species at 296 K are summarized in tables III-D-3 and III-D-4; estimated uncertainties are ±25% in all cases.

III-D-16. NO_3–Alkane Rate Coefficient Estimation Methods (SARs)

Although the data set for reaction of NO_3 with the alkanes is not nearly as well established as for OH or Cl-atom reactions, there is sufficient information available to allow for reasonable structure–additivity relationships to be created. From their studies of reaction of NO_3 with some of the smaller alkanes, Bagley et al. (1990) showed that the following expressions for reaction at primary, secondary, and tertiary alkyl sites provided rate coefficient estimates in reasonable agreement with their observations:

k_{primary} (298 K) $= 1.1 \times 10^{-18}$ cm^3 molecule^{-1} s^{-1}; $E/R = 4400$
$k_{\text{secondary}}$ (298 K) $= 2.4 \times 10^{-17}$ cm^3 molecule^{-1} s^{-1}; $E/R = 3250$
k_{tertiary} (298 K) $= 1.1 \times 10^{-16}$ cm^3 molecule^{-1} s^{-1}; $E/R = 2960$

Note that in this work the authors are assuming that no substituent effects are operative, i.e., that in the notation of Atkinson and co-workers (e.g., Kwok and Atkinson, 1995) for OH structure–additivity relationships, $F(\text{—CH}_3) = F(\text{—CH}_2\text{—}) = F(\text{>C—}) = F(\text{>C<}) = 1$.

Aschmann and Atkinson (1995b) followed a procedure entirely analogous to that developed for OH structure–additivity estimations, and obtained the following parameters:

k_{primary} (298 K) $= 7 \times 10^{-19}$ cm^3 molecule^{-1} s^{-1}
$k_{\text{secondary}}$ (298 K) $= 1.22 \times 10^{-17}$ cm^3 molecule^{-1} s^{-1}
k_{tertiary} (298 K) $= 7.5 \times 10^{-17}$ cm^3 molecule^{-1} s^{-1}

with $F(\text{—CH}_3) = 1.00$ by definition, and $F(\text{—CH}_2\text{—}) = F(\text{>C—}) = F(\text{>C<}) = 1.67$.

We have conducted a similar exercise, using rate coefficient values recommended in this evaluation. With k_{primary} (298 K) fixed to a value of 7×10^{-19} cm^3 molecule^{-1} s^{-1} and $F(\text{—CH}_3) = 1.00$, the parameters obtained are as follows:

$k_{\text{secondary}}$ (298 K) $= 1.74 \times 10^{-17}$ cm^3 molecule^{-1} s^{-1}
k_{tertiary} (298 K) $= 11.7 \times 10^{-17}$ cm^3 molecule^{-1} s^{-1}
$F(\text{—CH}_2\text{—}) = F(\text{>C—}) = F(\text{>C<}) = 1.33$.

The three parameterizations just described provide the expected result: reaction at the weaker bonded tertiary C—H site is the most rapid, and reaction at the stronger bonded primary —CH_3 groups is slowest. The three sets of parameters provide reasonable estimates of the measured rate coefficients; see table III-D-5 and figure III-D-2. However, as noted by Aschmann and Atkinson (1995b), four branched species denoted with a superscript "a" in table III-D-5 show marked discrepancies between measured and predicted rate coefficients. Particularly difficult to reconcile is the large difference in rate coefficient between the structurally similar species 2,3-dimethylbutane (4.1×10^{-16} cm^3 molecule^{-1} s^{-1}) and 2,4-dimethylpentane (1.45×10^{-16} cm^3 molecule^{-1} s^{-1}); with an additional site of attack, the 2,4-dimethylpentane species will always have the higher calculated rate coefficient. Some of the explanation for these discrepancies may come from the theoretical work of Bravo-Pérez et al. (2002), who showed that tunneling and entropic factors might play a large role in determining the rate and site of attack. For example, their calculations show that there is a very large symmetry-related entropic

Table III-D-5. Recommended and estimated rate coefficients (k, cm^3 $molecule^{-1}$ s^{-1}) for reactions of NO_3 with the alkanes at 298 K

	$10^{16} \times k$			
Alkane	Recommended k	Estimated k (Bagley et al., 1990)	Estimated k (Aschmann and Atkinson, 1995b)	Estimated k (This Work)
Methane	<0.01			
Ethane	<0.1	0.02		
Propane	<0.7	0.26		
Butane	0.46	0.50	0.43	0.48
2-Methylpropane	1.0	1.13	0.79	1.2
Pentane	0.83	0.74	0.77	0.79
2-Methylbutane	1.7	1.37	1.49	1.83
Hexane	1.1	0.98	1.11	1.10
2,3-Dimethylbutane[a]	4.1	2.24	2.55	3.17
Cyclohexane	1.35	1.44	2.04	1.85
Heptane	1.4	1.22	1.45	1.41
Octane	1.9	1.46	1.79	1.72
Nonane	2.2	1.70	2.13	2.03
2,2,4-Trimethylpentane[a]	0.75	1.40	1.65	1.92
2,2,3,3-Tetramethylbutane	<0.49	(0.66)	(0.42)	(0.56)
2,2,3-Trimethylbutane[a]	2.2	1.15	1.31	1.61
2,4-Dimethylpentane[a]	1.45	2.48	2.89	3.48
2-Methylpentane	1.7	1.61	1.83	2.14
3-Methylpentane	2.05	1.61	2.53	2.58
Decane	2.6	1.94	2.47	2.34

[a] Rate coefficients for these alkanes show marked discrepancies between measured and predicted values.

enhancement of the rate of reaction at the $—CH_3$ groups in 2-methylpropane, such that reaction at the CH_3 groups is actually more rapid than reaction at the tertiary C—H position. This result is not captured by the parameterizations. Further theoretical and product studies will be required to determine the actual sites of attack of NO_3 on the alkanes, and to fully resolve this issue.

III-E. Reactions of Ozone with Alkanes

Reactions of ozone with the alkanes have been shown to be sufficiently slow to be considered entirely negligible in the atmosphere (Atkinson and Carter, 1984). The bulk of the available rate data, see table III-E-1, stems from the work of Schubert, Dillemuth and their co-workers (Schubert and Pease, 1956a; Schubert and Pease, 1956b; Dillemuth et al., 1960; Morrissey and Schubert, 1963). These investigators studied the reactions of ozone with the C_1–C_4 alkanes, using static mixtures of ozone in excess hydrocarbon, and they obtained rate data from the rate of decay of ozone, as measured by IR absorption. Room-temperature rate coefficients range from 1.4×10^{-24} cm^3 $molecule^{-1}$ s^{-1} for methane to 2×10^{-23} cm^3 $molecule^{-1}$ s^{-1} for 2-methylpropane, implying atmospheric lifetimes with respect to ozone reaction in excess of 1000 years for all alkanes studied.

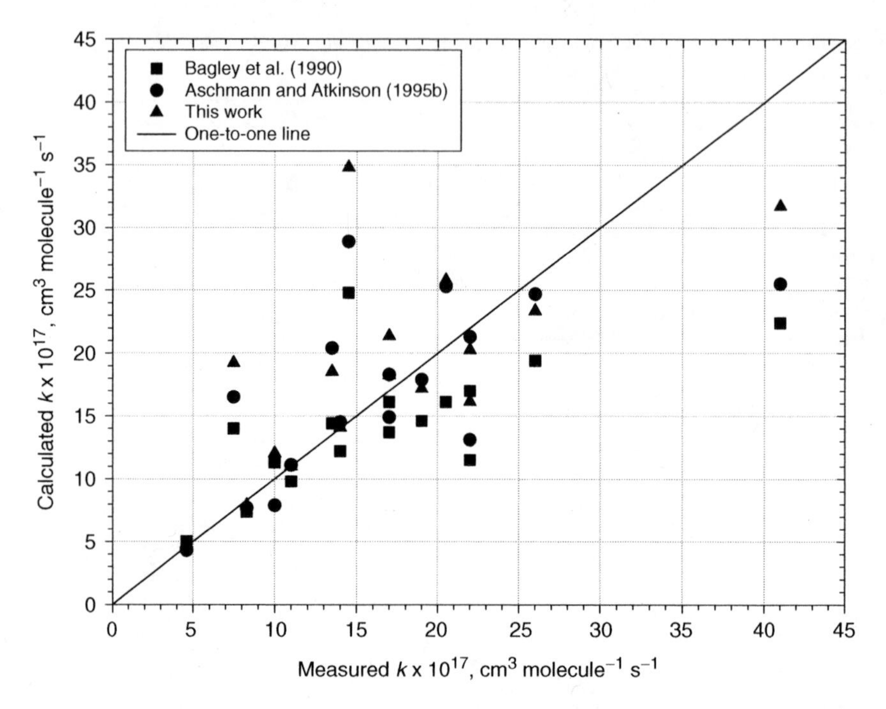

Figure III-D-2. Comparison of measured and calculated rate coefficients for reaction of NO_3 with the alkanes. Calculations were made using parameters described in the text.

Although these data are sufficient to eliminate the reactions of the alkanes with ozone from further consideration for atmospheric chemistry, there are a number of reasons (Atkinson and Carter, 1984) to believe that the data in table III-E-1 actually represent upper limits to the true rate coefficients. Firstly, these experiments would be very susceptible to low concentrations of reactive impurities, such as alkenes (note that reactions of ozone with C_2–C_4 alkenes are at least five orders of magnitude faster than the values obtained for the corresponding alkanes). Furthermore, the endothermicity of the most likely elementary reaction pathway, represented by reaction (III-E-1),

$$RH + O_3 \rightarrow R + HO + O_2 \quad \text{(III-E-1)}$$

ranges from about 63 kJ mol^{-1} for 2-methylpropane to about 117 kJ mol^{-1} for methane, higher than the measured activation energies; thus it is unlikely that the data obtained correspond to the initial elementary reaction (Atkinson and Carter, 1984). Even if more geometrically complex, but energetically more favorable pathways were operative, e.g.

$$CH_3CH_3 + O_3 \rightarrow CH_3O_2 + CH_3O$$

$$\rightarrow CH_3CH_2O + HO_2$$

$$\rightarrow CH_3CH_2O_2 + OH$$

Table III-E-1. Summary of rate coefficient measurements for reaction of ozone with alkanes; $k = A \exp(-B/T)$ cm^3 molecule^{-1} s^{-1}

Alkane	A	B (K)	k	T (K)	Technique[a]	Reference	Temperature Range (K)
Methane	1.2×10^{-13}	7500				Schubert and Pease (1956a)	313–323
	2.7×10^{-13}	7725			Static reactor	Dillemuth et al. (1960)	308–340
	2.4×10^{-14}	7000			Static reactor	Dillemuth et al. (1960)	308–340
			$<1.2 \times 10^{-21}$	298		Stedman and Niki (1973)	
Ethane	2.1×10^{-13}	7000			Static reactor	Morrissey and Schubert (1963)	273–333
	5.8×10^{-13}	7400			Static reactor	Morrissey and Schubert (1963)	273–333
Propane	5.1×10^{-15}	6090			Static reactor	Schubert and Pease (1956a)	304–323
	3.5×10^{-13}	7200			Static reactor	Morrissey and Schubert (1963)	273–333
	1.9×10^{-13}	7350			Static reactor	Morrissey and Schubert (1963)	273–333
Butane	1.4×10^{-15}	5590			Static reactor	Schubert and Pease (1956a)	298–323
2-Methylpropane	7.3×10^{-16}	5180			Static reactor	Schubert and Pease (1956a, 1956b)	298–323

these would likely result in ozone-consuming chain reactions and erroneously high rate coefficients. For example, in the ethane case shown above, HO_2 radicals might either be directly produced in the ethane–ozone reaction, or might result from the CH_3O_2–CH_3O_2 or CH_3O_2–$C_2H_5O_2$ reactions. The likely fate of HO_2 in these systems is reaction with ozone, a process that consumes excess ozone but also generates OH radicals, which themselves will propagate the peroxy radical chain chemistry.

III-F. Correlations between Rate Coefficients for Reactions of OH, Cl, $O(^3P)$, and NO_3 with the Alkanes

As alluded to in the introduction, the reaction of OH, Cl, $O(^3P)$, and NO_3 with the alkanes all occur via H-atom abstraction, and it is not surprising to find that correlations exist between the various data sets. Figure III-F-1 shows log–log (base 10) plots of the rate coefficients for reactions of the alkanes with Cl, $O(^3P)$, and NO_3 versus those for reaction of the alkanes with OH.

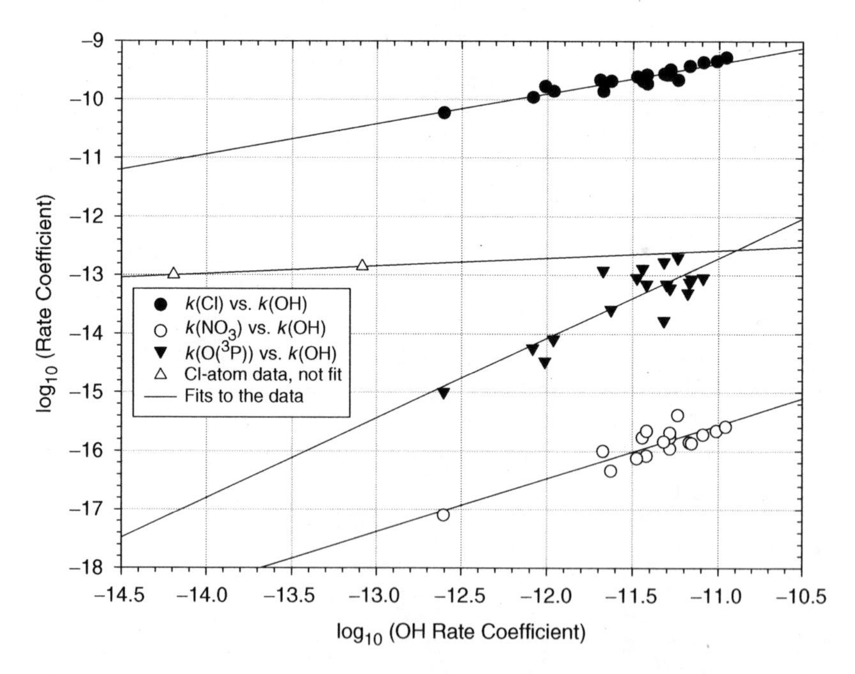

Figure III-F-1. Plots of logarithm of the rate coefficients (cm^3 $molecule^{-1}$ s^{-1}) for reaction of Cl, $O(^3P)$ and NO_3 with the alkanes versus those for reaction of OH with the corresponding alkane. Solid lines are unweighted least-squares fits to the data.

The correlation is strongest for the Cl versus OH rate coefficient data ($R^2 = 0.888$), likely reflecting the relatively high quality of the two data sets. The data for methane (for which the Cl-atom reaction is endothermic) and cyclopropane (which also occurs through a significant energy barrier) are clearly anomalous and are not included in the fit. An unweighted linear least-squares fit to the data yields: $\log_{10}[k(\text{Cl}+\text{alkane})] = 0.51 \times \log_{10}[k(\text{OH}+\text{alkane})] - 3.68$, where k is in units of cm^3 $molecule^{-1}$ s^{-1}.

The correlation between the $O(^3P)$ and OH rate coefficients shows some scatter, $R^2 = 0.739$, very likely the result of the large uncertainties (factors of two or more) in the $O(^3P)$ rate coefficients. An unweighted least-squares fit to the data yields, in units of cm^3 $molecule^{-1}$ s^{-1}, $\log_{10}k[(O^3P+\text{alkane})] = 1.36 \times \log_{10}[k(\text{OH}+\text{alkane})] + 2.29$. The fit using the updated values (highlighted in bold italics in table III-C-5) is much improved relative to the fit obtained using the pre-Miyoshi et al. (1993, 1994) data (lefthand column of table III-C-5).

There is a reasonable correlation between the NO_3 and OH rate coefficients with the alkanes. An unweighted least-squares fit to the data gives (in units of cm^3 $molecule^{-1}$ s^{-1}): $\log_{10}[k(NO_3 + \text{alkane})] = 0.910 \times \log_{10}[k(\text{OH} + \text{alkane})] - 5.55$, with $R^2 = 0.789$. This correlation is essentially identical to that reported by Wayne et al. (1991). The largest outlier in the data set is 2,3-dimethylbutane, for which the NO_3 rate coefficient seems about 2.5 times too high relative to the remainder of the data.

IV

Mechanisms and End-Products of the Atmospheric Oxidation of Alkanes

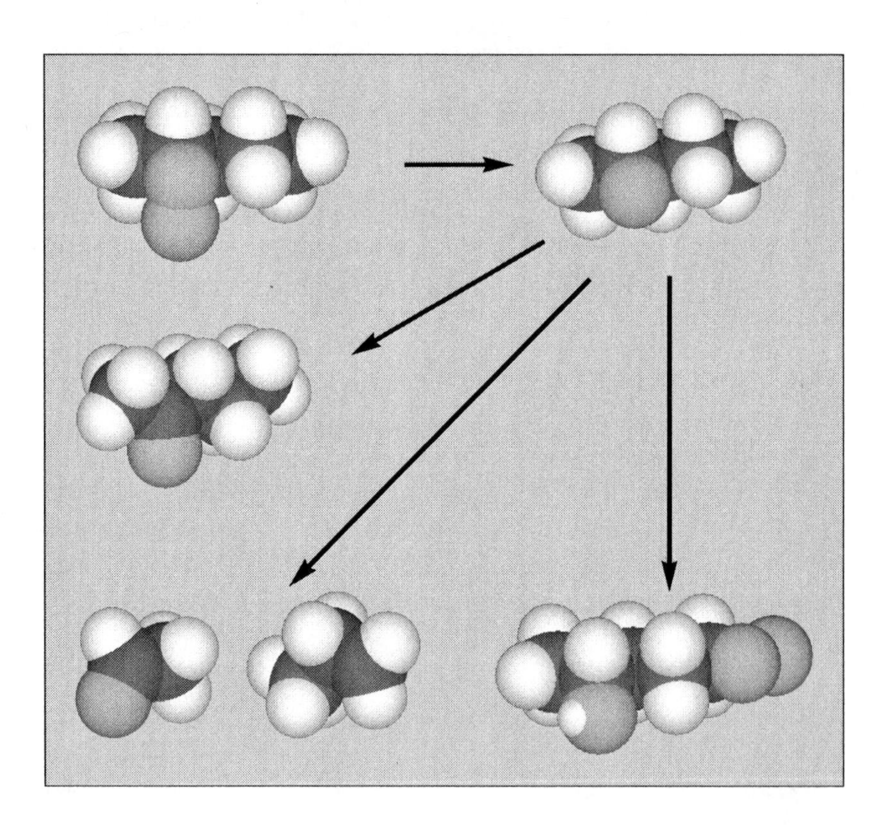

In the previous two chapters, the kinetics of the reactions of the alkanes with the major atmospheric oxidants [OH, Cl, NO_3, $O(^3P)$, and O_3] have been discussed. These initiation reactions are followed by a series of rapid reactions involving the alkylperoxy (RO_2) and alkoxy (RO) radicals, which lead, on a time-scale of minutes, to the first-generation end-products of the oxidation process. The chemistry of these radical species, and the end-products that result, are the topics of this chapter.

As we have seen, reactions of the alkanes with OH and other oxidants occur via abstraction, leading to the formation of alkyl radicals, as shown for ethane:

$$OH + CH_3CH_3 \rightarrow CH_3CH_2\bullet + H_2O$$

The sole atmospheric fate of the alkyl radicals is addition of O_2, e.g.,

$$CH_3CH_2\bullet + O_2 + M \rightarrow CH_3CH_2O_2\bullet + M$$

a process that occurs on a timescale of a few tens of nanoseconds in the lower atmosphere. Because of the rapidity and exclusivity of the O_2 addition, the two reactions shown above are often considered to occur simultaneously and the alkylperoxy radicals (such as $CH_3CH_2O_2$) can then be thought of as the product of the initial oxidation step.

The alkylperoxy radicals have atmospheric lifetimes on the order of minutes and, as detailed in section IV-A, reactions with NO, NO_2, HO_2, and organic peroxy radicals generally need to be considered. The reactions of the peroxy radicals with NO are of paramount importance, since they provide the only photochemical source of ozone in the troposphere, via photolysis of the major product of the reaction, NO_2, e.g.,

$$CH_3CH_2O_2\bullet + NO \rightarrow CH_3CH_2O\bullet + NO_2$$

$$NO_2 + h\nu \rightarrow NO + O(^3P)$$

$$O(^3P) + O_2 + M \rightarrow O_3 + M$$

A competing reaction channel, leading to organic nitrate formation, e.g.,

$$CH_3CH_2O_2\bullet + NO \rightarrow CH_3CH_2ONO_2$$

serves to "short-circuit" the ozone formation chemistry, and results in the loss of both HO_x and NO_x radicals from the system. Reactions of the peroxy radicals with HO_2 and with other organic peroxy radicals proceed in many cases via the formation of stable products, e.g.,

$$CH_3CH_2O_2\bullet + HO_2 \rightarrow CH_3CH_2OOH + O_2$$

$$CH_3CH_2O_2\bullet + CH_3O_2\bullet \rightarrow CH_3CH_2OH + CH_2O + O_2$$

$$\rightarrow CH_3CHO + CH_3OH + O_2$$

$$\rightarrow CH_3CH_2O\bullet + CH_3O\bullet + O_2$$

The first three of these reactions are sinks for HO_2 and RO_2 radicals and thus are limiters of ozone production. Reactions of peroxy radicals with NO_2 are of minimal importance in the lower troposphere, since the peroxynitrate products of these reactions are thermally unstable and dissociate rapidly (within a few seconds) back to reactants under these conditions, e.g.,

$$CH_3CH_2O_2\bullet + NO_2 + M \leftrightarrow CH_3CH_2O_2NO_2 + M$$

The peroxynitrates are, however, sufficiently stable to play a role as radical reservoir species in the colder upper troposphere.

In addition to stable end-products like the organic nitrates, peroxynitrates, hydroperoxides, alcohols, and carbonyls, the major by-products of peroxy radical chemistry are the alkoxy radicals, such as $CH_3CH_2O\bullet$ in the above example. The alkoxy radicals are very reactive under atmospheric conditions, with lifetimes usually less than 100 μs at the Earth's surface. The alkoxy radicals can potentially undergo a number of competing reactions, including reaction with O_2, unimolecular dissociation, or rearrangement processes; see section IV-B. This chemistry produces a complex array of end-products, including various carbonyls as well as multi-functional species such as hydroxycarbonyls and hydroxynitrates.

In section IV-C, the observed end-products of the oxidation of the alkanes (the carbonyl compounds, organic nitrates, hydroperoxides, alcohols, and multifunctional hydroxycarbonyls and hydroxynitrates) are presented. To the extent possible, the origin of these products will be rationalized from a consideration of the chemistry of the specific peroxy and alkoxy radicals involved in the oxidation process, as outlined in the earlier sections of this chapter.

IV-A. Peroxy Radical Chemistry

Alkylperoxy radicals react in the atmosphere with NO, NO_2, HO_2, and organic peroxy radicals (RO_2). The kinetics of these reactions are detailed in this section, building on previous reviews by Wallington et al. (1992c), Lightfoot et al. (1992), Wallington et al. (1997), and Tyndall et al. (2001).

The reaction of alkylperoxy radicals with NO is believed to occur via the formation of a short-lived peroxynitrite intermediate, which can decompose back to reactants, decompose to form NO_2 and alkoxy radical products, or rearrange to form an organic nitrate:

$$RO_2\bullet + NO \leftrightarrow (ROONO)^{\ddagger} \rightarrow RO\bullet + NO_2$$

$$\rightarrow RONO_2$$

Although the details of the nitrate formation process are still a matter of debate (e.g., see Barker et al., 2003; Lohr et al., 2003; Zhang et al., 2004a), it is generally accepted that the nitrate yield increases with: i) increasing pressure, ii) decreasing temperature, and iii) increasing size of the alkyl (R) group. Branching ratios for these reactions are dealt with in detail in section IV-C. Rate coefficients for these reactions appear to be essentially identical for all alkylperoxy radicals for which recommendations can be made, $k = (7 - 11) \times 10^{-12}$ cm^3 molecule^{-1}s^{-1}; see figure IV-A-1. The reactions occur with small negative activation energies ($\sim 2.5 - 3.3$ kJ mol^{-1}), consistent with the formation of the peroxynitrite complex.

Reactions of alkylperoxy radicals with NO_2 occur via addition,

$$RO_2\bullet + NO_2 + M \rightarrow RO_2NO_2 + M$$

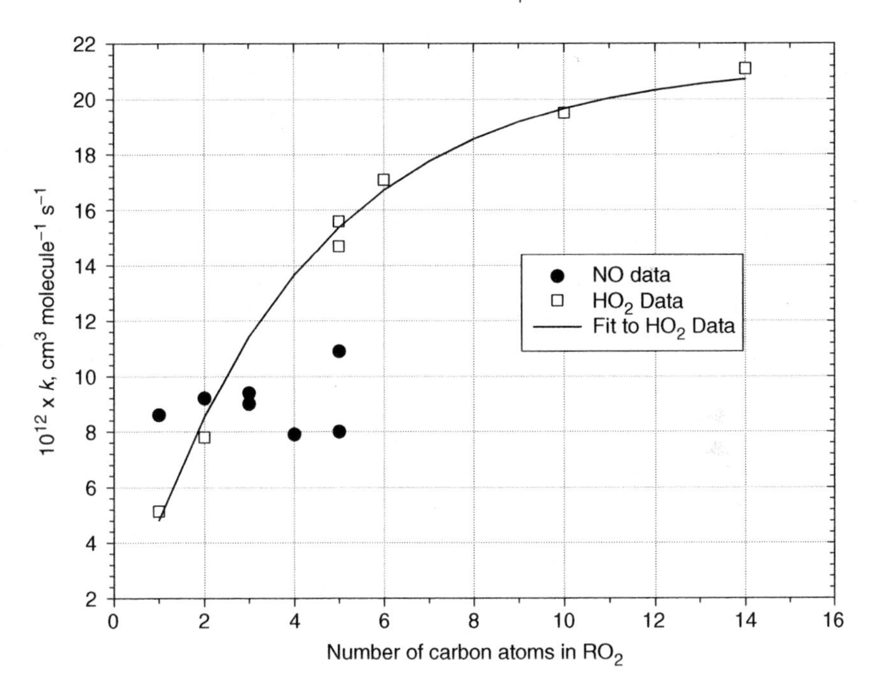

Figure IV-A-1. Rate coefficient data for reaction of alkylperoxy radicals with NO and HO_2, as a function of the number of carbon atoms in the peroxy radical. Rate coefficient values are those recommended in this work.

leading to the formation of thermally unstable peroxynitrates. These reactions are typical three-body processes, and their rate coefficients vary with pressure and temperature in the manner described by the Troe relation (Troe, 1969, 1983):

$$k = \frac{k_0[M]k_\infty}{k_\infty + k_0[M]} F_c^{[1+(\log_{10}(k_0[M]/k_\infty))^2]^{-1}}$$

where k_0 is the (third-order) rate coefficient at the low pressure limit, $[M]$ is the buffer gas density (molecule cm^{-3}), k_∞ is the (second-order) rate coefficient at the high pressure limit, and F_c is the broadening factor, which determines the shape of the fall-off curve. At least for the cases of methyl- and ethylperoxy, rate coefficients are in the fall-off region at atmospheric pressure.

In contrast to the NO reactions, rate coefficients for reactions of alkylperoxy radicals with HO_2 increase with the size of the alkyl fragment (Boyd et al., 2003); see figure IV-A-1. Boyd et al. (2003) have shown that the rate coefficient can be fitted to a function of the form $k = A\times(1-\exp(Bn))$, where A and B are fit parameters and n is the number of carbons in the alkyl fragment. A least-squares fit (see figure IV-A-1) to the rate coefficients recommended in this section yields: $k = 21.3\times[1-\exp(-0.256n)]$, with k in units of 10^{-12} cm^3 $molecule^{-1}s^{-1}$. All RO_2 + HO_2 reactions have large negative activation energies (E/R typically $\approx$ 800–1100 K), which suggests they proceed via the formation of a short-lived complex. As discussed in more detail in section IV-C, the reactions lead mainly (if not exclusively) to the formation of alkyl hydroperoxides, with the formation of a carbonyl and water a possible minor channel, e.g.,

$$CH_3O_2\bullet + HO_2 \rightarrow CH_3OOH + O_2 \quad \text{(major)}$$

$$\rightarrow CH_2O + H_2O + O_2 \quad \text{(minor)}$$

Reactions of alkylperoxy radicals with like or unlike alkyperoxy radicals are generally quite slow and occur via radical and molecular channels, e.g.,

$$RCH_2O_2\bullet + CH_3O_2\bullet \rightarrow RCH_2O\bullet + CH_3O\bullet + O_2$$

$$\rightarrow RCH_2OH + CH_2O + O_2$$

$$\rightarrow RCHO + CH_3OH + O_2$$

With the exception of the special case of the $CH_3O_2\bullet$–$CH_3O_2\bullet$ reaction, this class of reactions is not well characterized, particularly with regard to the product distributions. Since the methylperoxy radical is the most abundant atmospheric organic peroxy radical in the atmosphere, the only alkylperoxy–alkylperoxy reactions considered here are those involving this species.

IV-A-1. Methylperoxy Radical ($CH_3O_2\bullet$)

IV-A-1.1. $CH_3O_2\bullet + NO$

This reaction has been studied using molecular modulation (Anastasi et al., 1978; Cox and Tyndall, 1980), flash photolysis–UV absorption (Adachi and Basco, 1979a; Sander and Watson, 1980; Simonaitis and Heicklen, 1981; Zellner et al., 1986), flash photolysis-laser-induced fluorescence (Ravishankara et al., 1981), flash photolysis–mass spectrometry (Masaki et al., 1994; Xing et al., 2004), pulse radiolysis–UV absorption (Sehested et al., 1993b), flow tube–mass spectrometry (Plumb et al., 1979, 1981; Kenner et al., 1993; Villalta et al., 1995; Scholtens et al., 1999, Bacak et al., 2004), and relative rate methods (Simonaitis and Heicklen, 1979). The data are collected in table IV-A-1 and plotted in figure IV-A-2. The data points of Adachi and Basco (1979a) and Simonaitis and Heicklen (1979) are considerably smaller than the remainder of the data, and are not considered further. A least-squares fit to the remaining data gives $k = 1.96 \times 10^{-12} \exp(403/T)$ cm^3 molecule^{-1}s^{-1}, with $k = 7.6 \times 10^{-12}$ cm^3 molecule^{-1}s^{-1} at 298 K. The uncertainty at 298 K is estimated to be $\pm 15\%$. In a recent review of peroxy radical kinetics, Tyndall et al. (2001) recommend a weaker temperature dependence to the reaction, $k = 2.8 \times 10^{-12} \exp(300/T)$ cm^3 molecule^{-1}s^{-1}, largely on the basis of the work of Villalta et al. (1995). The expression obtained here, $k = 1.96 \times 10^{-12} \exp(403/T)$ cm^3 molecule^{-1}s^{-1}, gives rate coefficients that are 15% larger at 200 K and 15% smaller at 400 K than those recommended by Tyndall et al. (2001). Finally, we note that the rate coefficient of Helleis et al. (1993) for the reaction of CD_3O_2 with NO, $k = (8.6 \pm 1.0) \times 10^{-12}$ cm^3 molecule^{-1}s^{-1}, is consistent with the body of data regarding the CH_3O_2 + NO reaction.

Table IV-A-1. Rate coefficients (k) for reaction of CH_3O_2 with NO, with $k = A \times \exp(-B/T)$ $cm^3 molecule^{-1} s^{-1}$

A	B (K)	$10^{12} \times k$	T (K)	Technique[a]	Reference	Temperature Range (K)
		> 1	298	MMS–UV Abs.	Anastasi et al. (1978)	
		8.0 ± 2.0	295	DF–MS	Plumb et al. (1979)	
		3.0 ± 0.2	298	FP–UV Abs.	Adachi and Basco (1979a)	
		3.2 ± 1.8	296	RR [b]	Simonaitis and Heicklen (1979)	
		6.5 ± 2.0	298	MMS–UV Abs.	Cox and Tyndall (1980)	
		6.1 ± 0.7	298	FP–UV Abs.	Sander and Watson (1980)	
		6.3 ± 0.9	298			
		8.2 ± 1.1	298			
		8.9 ± 0.7	298			
$(8.1 \pm 1.6) \times 10^{-12}$	0	8.4 ± 1.5	240	FP–LIF	Ravishankara et al. (1981)	240–339
		8.6 ± 1.1	250			
		9.0 ± 1.1	270			
		7.8 ± 1.2	298			
		7.8 ± 1.4	339			
$(2.1 \pm 1.0) \times 10^{-12}$	−380 ± 250	13 ± 1.4	218	FP–UV Abs.	Simonaitis and Heicklen (1981)	218–365
		17 ± 2.2	218			
		7.7 ± 0.9	296			
		6.3 ± 1.0	365			
		8.6 ± 2.0	295	DF–MS	Plumb et al. (1981)	
		7± 2	298	FP–UV Abs.	Zellner et al. (1986)	
		9.1 ± 2.0	293	DF–MS	Kenner et al. (1993)	
		8.8 ± 1.4	298	PR–UV Abs.	Sehested et al.(1993b)	
		11.2 ± 1.5	298	FP–MS	Masaki et al. (1994)	

(continued)

Table IV-A-1. *(continued)*

A	B (K)	$10^{12} \times k$	T (K)	Technique[a]	Reference	Temperature Range (K)
$(2.8 \pm 0.5) \times 10^{-12}$	-285 ± 60	11.9	200	DF–MS	Villalta et al. (1995)	200–429
		9.9	223			
		8.9	248			
		7.4	273			
		7.4	298			
		6.3	329			
		5.9	372			
		5.5	394			
		5.9	410			
		5.8	429			
$(9.2^{+6.0}_{-3.9}) \times 10^{-13}$	-600 ± 140	6.5 ± 1.9	295	DF–MS	Scholtens et al. (1999)	203–295
		8.5 ± 1.3	295			
		8.3 ± 1.6	295			
		7.1 ± 1.1	295			
		7.3 ± 0.9	273			
		8.2 ± 2.6	263			
		12.2 ± 3.5	248			

		9.2 ± 3.0	243			
		8.5 ± 2.5	238			
		10.5 ± 3.6	233			
		15.4 ± 4.7	228			
		11.4 ± 3.8	223			
		13.5 ±6.2	218			
		15.8 ± 5.5	218			
		18.1 ± 4.0	213			
		15.6 ± 4.5	213			
		17.6 ± 5.9	208			
		18.3 ± 4.0	203			
		9.9 ± 2.1	298	FP–MS	Xing et al. (2004)	
$(1.75^{+0.28}_{-0.24}) \times 10^{-12}$	−435 ± 35	7.41 ± 0.22	298	DF–CIMS	Bacak et al. (2004)	193–300
		7.42 ± 0.17	298			
		9.97 ± 0.10	253			
		12.01 ± 0.04	233			
		12.82 ± 0.08	223			
		13.09 ±0.08	213			
		13.25 ± 0.07	203			
		17.67 ± 0.16	193			

[a] Abbreviations for the experimental techniques employed are defined in table II-A-1.
[b] Placed on an absolute basis using $k(CH_3O_2 + SO_2) = 8.2 \times 10^{-15} cm^3 molecule^{-1} s^{-1}$.

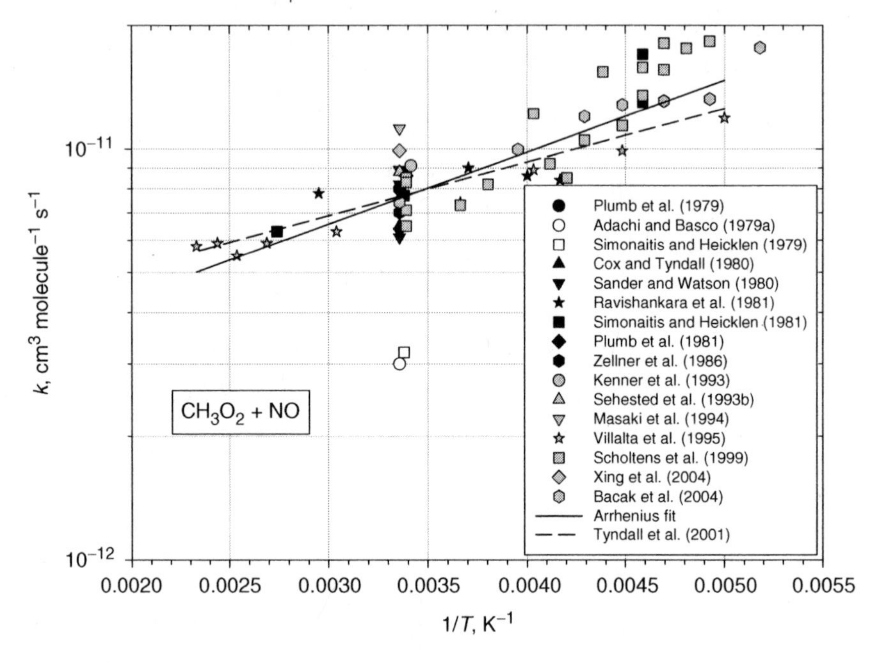

Figure IV-A-2. Measurements of the rate coefficient for reaction of CH_3O_2 with NO.

IV-A-1.2. CH_3O_2• + NO_2

The reaction has been the subject of six studies covering a range of temperatures, pressures and bath gases (Adachi and Basco, 1980; Cox and Tyndall, 1979; Ravishankara et al., 1980; Sander and Watson, 1980; Bridier et al., 1992; Wallington et al., 1999). In contrast to other studies, Adachi and Basco (1980) did not observe the dependence of the rate coefficient on pressure (50 and 580 Torr) expected for an association reaction, and these data are not considered further. The JPL panel (Sander et al., 2005) recommends the following Troe equation parameters which provide a good fit to the data at atmospheric pressure and below: $F_c = 0.6$; $k_0 = (1.5 \pm 0.8) \times 10^{-30}(T/300)^{-4.0}$cm^6 molecule^{-2}s^{-1}; and $k_\infty = (6.5 \pm 3.2) \times 10^{-12}(T/300)^{-2.0}$ cm^3 molecule^{-1}s^{-1}. These parameters are very similar to those recommended by Tyndall et al. (2001). However, the work of Wallington et al. (1999), which extends to total pressures of 14 atm SF_6, demonstrates that the true value for k_∞ is much higher than given by the fit parameters, $k_\infty = (1.81 \pm 0.25) \times 10^{-11}$cm^{3}molecule^{-1}s^{-1}. All the experimental measurements are presented in table IV-A-2.

IV-A-1.3. CH_3O_2• + HO_2

Rate coefficients have been obtained for this reaction using the techniques of molecular modulation–UV absorption (Cox and Tyndall, 1979, 1980; Jenkin et al., 1988; Moortgat et al., 1989), or flash photolysis–UV absorption (McAdam et al., 1987; Kurylo et al., 1987; Dagaut et al., 1988a; Lightfoot et al., 1990c, 1991). The original data are summarized in table IV-A-3. Tyndall et al. (2001) conducted a thorough review of this

Table IV-A-2. Rate coefficients (k) for reaction of CH_3O_2 with NO_2, with $k = A\times \exp(-B/T)$ cm^3 $molecule^{-1}s^{-1}$

$10^{12}\times k$	Pressure, Torr/ (Buffer Gas)	T (K)	Technique[a]	Reference	Temperature Range (K)
1.2 ± 0.3	50 / (Ar/CH_4)	298	MM–UV Abs	Cox and Tyndall	
1.6 ± 0.3	540 / N_2	298		(1979; 1980)	
1.53 ± 0.07	53–580 / Ar	298	FP–UV Abs	Adachi and Basco (1980)	
0.884 ± 0.064	50 / He	298	FP–UV Abs	Sander and Watson (1980)	
1.18 ± 0.16	100 / He	298			
1.74 ± 0.14	225 / He	298			
2.26 ± 0.27	350 / He	298			
2.53 ± 0.31	500 / He	298			
2.80 ± 0.35	700 / He	298			
1.15 ± 0.10	50 / N_2	298			
1.58 ± 0.15	100 / N_2	298			
2.22 ± 0.31	225 / N_2	298			
2.98 ± 0.23	350 / N_2	298			
3.67 ± 0.21	500 / N_2	298			
3.94 ± 0.17	700 / N_2	298			
1.28 ± 0.16	50 / SF_6	298			
1.99 ± 0.25	100 / SF_6	298			
3.07 ± 0.35	225 / SF_6	298			
3.90 ± 0.36	350 / SF_6	298			
4.21 ± 0.37	500 / SF_6	298			
4.81 ± 0.59	700 / SF_6	298			
1.36 ± 0.23	76 ± 3 / N_2	298	FP–UV Abs	Ravishankara et al. (1980)	253–353
1.90 ± 0.32	157 ± 2 / N_2	298			
2.60 ± 0.56	258 ± 3 / N_2	298			
2.82 ± 0.28	352 ± 3 / N_2	298			
3.36 ± 0.32	519 ± 2 / N_2	298			
4.12 ± 0.38	722 ± 1 / N_2	298			
1.19 ± 0.15	330 ± 3 / N_2	353			
1.35 ± 0.21	354 ± 2 / N_2	353			
1.71 ± 0.10	511 ± 2 / N_2	353			
1.93 ± 0.20	696 ± 2 / N_2	353			
2.50 ± 0.34	109 ± 1 / N_2	253			
3.85 ± 0.30	250 ± 2 / N_2	253			
5.1 ± 0.50	503 ± 3 / N_2	253			
5.8 ± 1.0	519 / N_2	253			
4.4 ± 0.4	760 / (N_2/O_2)	298	FP–UV Abs	Bridier et al. (1992)	298–373
2.84 ± 0.15		333			
2.25 ± 0.32		353			
2.09 ± 0.9		368			
1.4 ± 0.4		373			
3.7	380 / SF_6	298	PR–UV Abs	Wallington et al. (1999)	
9.96	3040 / SF_6				
11.2	4560 / SF_6				
11.6	6080 / SF_6				
13.0	7600 / SF_6				
11.5	7980 / SF_6				
13.2 ± 1.0	10640 / SF_6				

[a] Abbreviations for the experimental techniques employed are defined in table II-A-1.

Table IV-A-3. Rate coefficients (k, cm^3 molecule^{-1} s^{-1}) for reaction of CH_3O_2 with HO_2

$10^{12} \times k$	T (K)	Technique[a]	Reference	Temperature Range (K)
8.5 ± 1.5	274	MM–UV Abs	Cox and Tyndall (1979, 1980)	
6.5 ± 1.0	298			
3.5 ± 0.5	338			
1.3	298	FTIR	Kan et al. (1980)	
6.4 ± 1.0	298	FP–UV Abs	McAdam et al. (1987)	
2.9 ± 0.4	298	FP–UV Abs	Kurylo et al. (1987)	
6.8 ± 0.5	228	FP–UV Abs	Dagaut et al. (1988a)	228–380
5.5 ± 0.3	248			
4.1 ± 0.3	273			
2.4 ± 0.5	340			
2.1 ± 0.3	380			
5.4 ± 1.1	300	MM–UV Abs, IR Abs	Jenkin et al. (1988)	
6.8 ± 0.9	303			
4.8 ± 0.2	298	MM–UV Abs	Moortgat et al. (1989)	
10.37 ± 4.72	248	FP–UV Abs	Lightfoot et al. (1990c)	248–573
7.63 ± 1.70	273			
5.63 ± 1.02	298			
5.22 ± 1.24	323			
2.98 ± 0.84	368			
3.11 ± 0.48	373			
2.39 ± 0.36	473			
1.83 ± 0.38	573			
1.3 ± 0.13	600	FP–UV Abs	Lightfoot et al. (1991)	600–678
1.2 ± 0.10	616			
1.1 ± 0.15	648			
0.72 ± 0.25	678			
5.13 ± 0.55	298	FP–UV Abs	Boyd et al. (2003)	

[a] Abbreviations for the experimental techniques employed are defined in table II-A-1.

reaction, including a reanalysis of some of the data using updated UV absorption cross sections for the two reactants. Their recommendation, $k = 4.15 \times 10^{-13}\exp((750 \pm 150)/T)\,cm^3 molecule^{-1} s^{-1}$, with $k = 5.14 \times 10^{-12}\,cm^3 molecule^{-1} s^{-1}$ at 298 K, is adopted here. The uncertainty at 298 K (Tyndall et al., 2001) is estimated to be ±30%. The value recently reported by Boyd et al. (2003), $k = (5.13 \pm 0.55) \times 10^{-12} cm^3$ molecule$^{-1}s^{-1}$, is in excellent agreement with this recommendation.

IV-A-1.4. CH_3O_2• + CH_3O_2•

This reaction has been studied by numerous groups using either flash photolysis or molecular modulation to generate the CH_3O_2 radicals and UV absorption to monitor their temporal profiles (Parkes et al., 1973; Parkes, 1974; Hochanadel et al., 1977; Parkes, 1977; Anastasi et al., 1978; Cox and Tyndall, 1979; Kan et al., 1979; Sanhueza et al., 1979; Adachi et al., 1980; Sander and Watson, 1980, 1981; Kurylo and Wallington, 1987; McAdam et al., 1987; Jenkin et al., 1988;

Lightfoot et al., 1990a; Simon et al., 1990; Jenkin and Cox, 1991; Lightfoot et al., 1991; Bloss et al., 2002). Alcock and Mile (1975) also studied this reaction via end-product analysis following photolysis of azomethane/O_2/2,3-dimethylbutane mixtures.

As summarized in Tyndall et al. (2001), there are two issues that need to be considered when evaluating data obtained for this reaction using UV absorption. First, the reaction is second order in CH_3O_2 and therefore absolute CH_3O_2 concentrations are required to obtain the rate coefficient. Because essentially all measurements made to date used UV absorption to monitor CH_3O_2, reported rate coefficients are dependent on the UV absorption cross section used in the analysis; i.e., what is actually measured in the laboratory is the ratio of the observed rate coefficient, k_{obs}, to the absorption cross section, σ, at the monitoring wavelength. The second issue that needs to be dealt with is the fact that the CH_3O_2–CH_3O_2 reaction leads unavoidably to the production of HO_2, which itself reacts rapidly with CH_3O_2:

$$CH_3O_2\bullet + CH_3O_2\bullet \rightarrow CH_3OH + CH_2O + O_2$$

$$CH_3O_2\bullet + CH_3O_2\bullet \rightarrow CH_3O\bullet + CH_3O\bullet + O_2$$

$$CH_3O\bullet + O_2 \rightarrow CH_2O + HO_2$$

$$CH_3O_2\bullet + HO_2 \rightarrow CH_3OOH + O_2$$

It can be shown that the observed rate coefficient, k_{obs}, exceeds the true rate coefficient for the CH_3O_2–CH_3O_2 reaction by a factor $(1 + \alpha)$, where α is the branching fraction to the radical-forming channel. Values of k_{obs}/σ are summarized in table IV-A-4. Tyndall et al. (2001) have converted all available k_{obs}/σ data to actual k values, using their recommended UV absorption cross section and branching ratios; actual k values, using the Tyndall et al. cross sections and branching ratios, are also shown in table IV-A-4. Tyndall et al. (2001) report $k = 9.5 \times 10^{-14} \exp((390 \pm 100)/T)$ cm^3 molecule^{-1}s^{-1} with $k = 3.5 \times 10^{-13}$ cm^3 molecule^{-1}s^{-1} at 298 K, a recommendation that is adopted here. The estimated uncertainty at 298 K is ±20% (Tyndall et al., 2001).

IV-A-2. Ethylperoxy Radical ($CH_3CH_2O_2\bullet$)

IV-A-2.1. $C_2H_5O_2\bullet$ + NO

This rate coefficient has been measured a total of nine times, using flash photolysis–UV absorption (Adachi and Basco, 1979b; Maricq and Szente, 1996), pulse radiolysis–UV absorption (Sehested et al., 1993b), or discharge flow–mass spectroscopy (Plumb et al., 1982; Daële et al., 1995; Eberhard and Howard, 1996; Ranschaert et al., 2000; Xing et al., 2004; Bardwell et al., 2005). The data of Maricq and Szente (1996), Eberhard and Howard (1996), Ranschaert et al. (2000), and Bardwell et al. (2005) cover ranges of temperature, whereas the rest of the measurements were made at room temperature (~298 K). The data are summarized in table IV-A-5 and displayed in

Table IV-A-4. Rate coefficient data for reaction of CH_3O_2 with CH_3O_2. Data for k_{actual} are in units of cm^3 $molecule^{-1}s^{-1}$

$10^{-5}\times (k_{obs}/\sigma)$ cm s^{-1}	$10^{13}\times k_{actual}$	λ (nm)	T (K)	Technique[b]	Reference	Temperature Range (K)
0.75 ± 0.25	2.3	240	298	MM–UV Abs	Parkes et al. (1973)	
	7.8		373	End-product Analysis	Alcock and Mile (1975)	
1.31 ± 0.23	3.75	248	295	FP–UV Abs	Hochanadel et al. (1977)	
1.00 ± 0.18	3.13	238	288–298	MM–UV Abs	Parkes (1977)	
0.8 ± 0.1	2.5	240	298	MM–UV Abs	Anastasi et al.(1978)	298–325
0.6 ± 0.2	1.6	240	325			
1.33 ± 0.23	3.67	250	298	MM–UV Abs	Cox and Tyndall(1979)	
2.05 ± 0.25	3.71	265	298	FP–UV Abs	Kan et al. (1979)	
1.27 ± 0.10	3.24	253.7	298	FP–UV Abs	Sanhueza et al.(1979)	
1.08 ± 0.08	3.34	240	298	FP–UV Abs	Adachi et al. (1980)	
1.06 ± 0.07	3.15	245	298	FP–UV Abs	Sander and Watson (1980)	
1.31 ± 0.23	4.08	250	248	FP–UV Abs	Sander and Watson (1981)	248–417
1.40 ± 0.32	4.12		270			
1.22 ± 0.18	3.37		298			
1.18 ± 0.12	3.06		329			
1.02 ± 0.16	2.48		373			
0.95 ± 0.12	2.20		417			
1.35 ± 0.16	4.42	250	228	FP–UV Abs	Kurylo and Wallington (1987)	228–380
1.28 ± 0.20	3.98		248			
1.16 ± 0.14	3.39		273			
1.10 ± 0.14	3.04		298			
1.21 ± 0.09	3.08		340			
1.00 ± 0.12	2.41		380			
1.34 ± 0.23	3.70	250	298	FP–UV Abs	McAdam et al. (1987)	
1.11 ± 0.12	3.06	250	298	MM–UV Abs	Jenkin et al. (1988)	

1.27 ± 0.25	3.95	250	248	FP–UV Abs	Lightfoot et al. (1990a)	248–573
1.16 ± 0.15	3.39		273			
1.17 ± 0.36	3.23		298			
0.97 ± 0.13	2.37		368			
1.04 ± 0.10	2.53		373			
0.91 ± 0.08	2.17		388			
0.91 ± 0.09	2.09		423			
0.95 ± 0.04	2.11		473			
0.90 ± 0.07	1.94		523			
0.98 ± 0.06	2.08		573			
1.16 ± 0.12	3.19	250	300	MM–UV Abs	Simon et al. (1990)	
1.07	3.42	230	268	MM–UV Abs	Jenkin and Cox (1991)	268–350
1.23	3.93		268			
1.13	3.57		273			
1.11	3.46		278			
1.05	3.24		283			
1.07	3.26		288			
1.22	3.67		293			
1.01	3.01		298			
1.01	2.98		303			
1.01	2.92		313			
1.00	2.86		318			
0.98	2.73		333			
0.95	2.60		343			
1.04	2.81		350			
1.02	2.76		350			
	1.54		600	FP–UV Abs	Lightfoot et al. (1991)	
	1.82		623 ± 4			
	1.93		648 ± 8			
1.23	2.6 ± 0.7	260	298	FP–UV Abs	Bloss et al. (2002)	

[a] Values of k_{obs}/σ are converted to k_{actual} using CH_3O_2 UV absorption cross sections and CH_3O_2–CH_3O_2 branching ratios from Tyndall et al. (2001); see text for details.
[b] Abbreviations for the experimental techniques employed are defined in table II-A-1.

Table IV-A-5. Rate coefficients (k) for reaction of $C_2H_5O_2$ with NO, with $k = A \times \exp(-B/T)$ cm^3 molecule^{-1}s^{-1}

A	B (K)	$10^{12} \times k$	T (K)	Technique[a]	Reference	Temperature Range (K)
		2.7 ± 0.2	298	FP–UV Abs	Adachi and Basco (1979b)	
		8.9 ± 3.0	295	DF–MS	Plumb et al. (1982)	
		8.5 ± 1.2	298	PR–UV Abs	Sehested et al. (1993b)	
		8.2 ± 1.6	298	DF–LIF	Daële et al. (1995)	
$(2.9 \pm 0.5) \times 10^{-12}$	−350 ± 60	16.9 ± 1.1	207	DF–CIMS	Eberhard and Howard (1996)	220–355
		17.9 ± 1.1	208			
		14.0 ± 0.3	223			
		14.1 ± 1.1	244			
		11.9 ± 0.7	252			
		11.6 ± 0.3	261			
		8.46 ± 0.40	298			
		7.41 ± 0.82	298			
		9.16 ± 0.20	298			
		9.30 ± 0.44	298			
		8.56 ± 0.36	298			
		8.05 ± 0.30	298			
		9.06 ± 0.22	298			
		9.67 ± 0.46	298			
		9.20 ± 0.26	325			
		8.02 ± 0.36	355			
		6.77 ± 0.28	402			
		6.74 ± 0.92	402			
		6.62 ± 0.16	403			
		6.53 ± 0.44	403			

$(3.1^{+1.5}_{-1.0}) \times 10^{-12}$	–330 ± 110	15.7 ± 3.0	220	FP–IR and UV abs	Maricq and Szente (1996)	220–355
		12.8 ± 2.4	220			
		12.9 ± 1.4	228			
		12.0 ± 1.9	229			
		10.2 ± 1.9	232			
		11.6 ± 2.2	232			
		10.7 ± 2.0	232			
		12.3 ± 2.3	253			
		11.2 ± 2.2	253			
		11.2 ± 1.5	253			
		12.3 ± 2.5	254			
		15.7 ± 2.8	254			
		10.7 ± 2.0	271			
		14.0 ± 2.7	271			
		13.7 ± 2.6	271			
		12.3 ± 2.5	295			
		10.3 ± 2.5	295			
		11.8 ± 2.2	295			
		9.3 ± 1.8	295			
		8.1 ± 1.5	295			
		8.6 ± 1.6	295			
		11.2 ± 2.1	295			
		8.5 ± 1.1	295			
		7.3 ± 1.5	323			
		11.0 ± 2.2	323			
		9.2 ± 1.7	324			
		8.4 ± 1.6	324			
		7.4 ± 1.4	354			
		6.9 ± 1.2	354			
		6.9 ± 1.2	354			
		6.9 ± 1.0	355			

(continued)

Table IV-A-5. *(continued)*

A	B (K)	$10^{12} \times k$	T (K)	Technique[a]	Reference	Temperature Range (K)
$(3.8^{+2.1}_{-1.3}) \times 10^{-12}$	-290 ± 110	9.6 ± 1.1	299	DF–MS	Ranschaert et al. (2000)	213–299
		11.1 ± 0.7	299			
		9.6 ± 1.5	299			
		10.2 ± 1.0	273			
		13.8 ± 4.2	263			
		11.8 ± 1.6	263			
		11.0 ± 0.7	258			
		12.4 ± 0.7	253			
		13.1 ± 4.0	248			
		12.1 ± 2.2	248			
		12.2 ± 2.7	243			
		14.6 ± 3.1	238			
		13.3 ± 6.2	238			
		12.2 ± 1.7	233			
		15.4 ± 1.3	233			
		13.9 ± 1.3	228			
		15.0 ± 3.5	223			
		13.4 ± 1.2	213			
		11.0 ± 0.8	298	FP–MS	Xing et al. (2004)	
$(1.75^{+0.14}_{-0.13}) \times 10^{-12}$	-462 ± 19	8.32	298	DF–CIMS	Bardwell et al. (2005)	203–298
		10.5	258			
		11.8	243			
		13.8	223			
		17.3	203			
		8.09	298			
		9.38	273			
		10.6	263			
		12.1	233			

[a] Abbreviations for the experimental techniques employed are defined in table II-A-1.

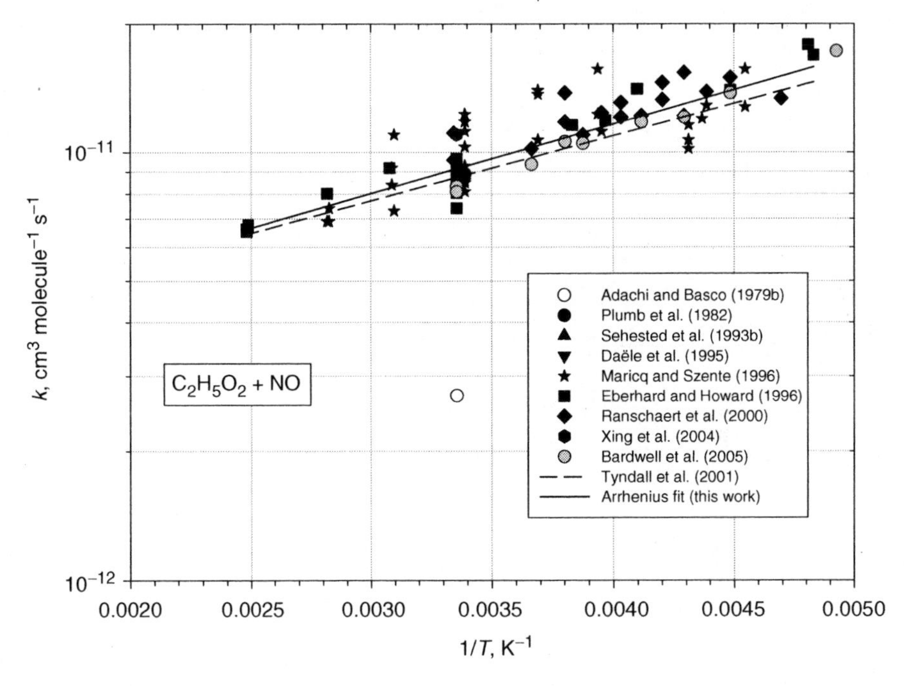

Figure IV-A-3. Measurements of the rate coefficient for reaction of $C_2H_5O_2$ with NO.

figure IV-A-3. With the exception of the single data point of Adachi and Basco (1979b), the data are in excellent agreement. A leastsquares fit to the remainder of the data yields $k = 2.62 \times 10^{-12} \exp(373/T)$ cm^3 molecule^{-1}s^{-1}, with $k = 9.2 \times 10^{-12}$ cm^3 molecule^{-1}s^{-1} at 298 K. The estimated uncertainty is $\pm15\%$. The expression returns rate coefficient values within 10% of those obtained from the expression recommended by Tyndall et al. (2001), $k = 2.7 \times 10^{-12} \exp(350 \pm 50/T)$ cm^3 molecule^{-1}s^{-1}, over the range 200–400 K.

IV-A-2.2. $C_2H_5O_2\bullet + NO_2$

This reaction has been studied by Adachi and Basco (1979c), using a flash photolysis–UV absorption system, Elfers et al. (1990), using a relative rate method, and Ranschaert et al. (2000), using a flow-tube-chemical ionization mass spectrometer system; see table IV-A-6. The data of Adachi and Basco (1979c) display no pressure dependence over the range 44–676 Torr, and are inconsistent with the Elfers et al. (1990), data and with expectations for the behavior of an addition reaction. Tyndall et al. (2001) have reanalyzed and extended the data of Elfers et al. (1990), using molecular parameters obtained from theoretical calculations and RRKM calculations fitted to the Elfers et al. data. The following Troe equation parameters were recommended: $k_0 = 4.87 \times 10^{-30}(T/298)^{-5.8}$cm^{6}molecule^{-2}s^{-1}, $k_\infty = 7.62 \times 10^{-12} \times (T/298)^{-0.9}$ cm^3 molecule^{-1}s^{-1}, and $F_c = 0.6$. The data of Ranschaert et al. (2000), obtained at 100 Torr total pressure between 213 and 299 K, fall within 3–17% of the rate coefficients retrieved from these parameters.

Table IV-A-6. Rate coefficients (k, cm^3 $molecule^{-1}s^{-1}$) for reaction of $C_2H_5O_2$ with NO_2

$10^{12} \times k$	Pressure Torr/ (Buffer Gas)	T (K)	Technique[a]	Reference	Temperature Range (K)
1.25 ± 0.07	44–676 / (O_2/Ar)	298	FP–UV Abs	Adachi and Basco (1979c)	
8.9	755 / (N_2/O_2)	255.6	RR[b]	Elfers et al. (1990)	
5.5	76 / (N_2/O_2)	253.8			
2.3	7.8 / (N_2/O_2)	253.9			
4.7	76.8 / (N_2/O_2)	264.7			
2.77 ± 0.28	100 / N_2	299	DF–CIMS	Ranschaert et al. (2000)	213–299
3.38 ± 0.20		297			
4.10 ± 0.39		283			
4.01 ± 0.63		283			
4.89 ± 0.57		273			
4.76 ± 0.54		273			
5.06 ± 2.10		263			
4.42 ± 0.56		263			
4.43 ± 0.88		253			
5.64 ± 0.46		253			
6.49 ± 1.23		243			
6.33 ± 1.97		243			
6.31 ± 1.48		233			
7.00 ± 0.54		233			
7.07 ± 0.52		228			
6.61 ± 0.83		223			
7.89 ± 0.52		223			
7.95 ± 0.33		213			
8.48 ± 0.79		213			

[a] Abbreviations for the experimental techniques employed are defined in table II-A-1.
[b] Measured relative to rate coefficient for reaction of ethylperoxy with NO, $k = 1.1 \times 10^{-11}$ at 253–265 K.

IV-A-2.3. $C_2H_5O_2\bullet + HO_2$

The rate coefficient for this reaction has been measured by Cattell et al. (1986) and Boyd et al. (2003) at room temperature, and by Dagaut et al. (1988b), Fenter et al. (1993), and Maricq and Szente (1994) over a range of temperatures, all using UV absorption to monitor the radical temporal profiles. Original data are summarized in table IV-A-7. The data from Cattell et al. (1986), Dagaut et al. (1988b), Fenter et al. (1993), and Maricq and Szente (1994) were reviewed and re-analyzed by Tyndall et al. (2001) using their recommended UV absorption cross sections. A rate coefficient $k = 7.4 \times 10^{-13}\exp((-700 \pm 200)/T)$ was recommended. This expression yields $k = 7.8 \times 10^{-12}$ cm^3 $molecule^{-1}s^{-1}$ at 298 K, with an estimated uncertainty of ±30% (Tyndall et al., 2001). A large excess of HO_2 over $C_2H_5O_2$ was used in the more recent measurement of Boyd et al. (2003), making the result less reliant on accurate knowledge of the $C_2H_5O_2$ UV absorption cross section. The rate coefficient obtained in that study, $k = (8.14 \pm 0.38) \times 10^{-12}cm^3$ $molecule^{-1}s^{-1}$, is in excellent agreement with the recommendation of Tyndall et al. (2001).

Table IV-A-7. Rate coefficients (k, cm^3 $molecule^{-1}s^{-1}$) for reaction of $C_2H_5O_2$ with HO_2

$10^{12} \times k$	T (K)	Technique[a]	Reference	Temperature Range (K)
6.3 ± 0.9	295	MM–UV/IR Abs	Cattell et al. (1986)	
7.3 ± 1.0	248	FP–UV Abs	Dagaut et al. (1988b)	248–380
6.0 ± 0.5	273			
5.4 ± 1.2	298			
3.4 ± 1.0	340			
3.1 ± 0.5	380			
24.5 ± 2.7	248	FP–UV Abs	Fenter et al.(1993)	248–480
15.8 ± 2.8	273			
10.4 ± 1.2	298			
8.7 ± 1.0	323			
7.6 ± 0.8	332			
4.6 ± 0.5	372			
3.1 ± 0.3	410			
1.9 ± 0.2	480			
19.8	210	FP–UV Abs	Maricq and Szente (1994)	210–363
16.5	233			
11.5	253			
9.5	273			
8.1	295			
6.6	323			
6.1	363			
8.14 ± 0.38	298	FP–UV Abs	Boyd et al. (2003)	

[a] Abbreviations for the experimental techniques employed are defined in table II-A-1.

IV-A-2.4. $C_2H_5O_2\bullet + CH_3O_2\bullet$

This reaction was studied by Villenave and Lesclaux (1996), using a flash photolysis–UV absorption system. The reported rate coefficient, see table IV-A-8, is $(2.0 \pm 0.5) \times 10^{-13} cm^3 molecule^{-1}s^{-1}$, with the quoted uncertainty representing statistical errors (1σ) only. This rate coefficient is recommended, with an uncertainty estimated (Villenave and Lesclaux, 1996) to be $\pm 40\%$.

IV-A-3. 1-Propylperoxy Radical (n-Propylperoxy, $CH_3CH_2CH_2O_2\bullet$)

IV-A-3.1. $CH_3CH_2CH_2O_2\bullet + NO$

This rate coefficient has been measured over a range of temperatures by Eberhard and Howard (1996), using a flow-tube–chemical ionization mass spectrometer system; see table IV-A-9 and figure IV-A-4. These authors report $k = (2.9 \pm 0.5) \times 10^{12} \exp((350 \pm 60)/T)$ cm^3 $molecule^{-1}s^{-1}$, with $k = (9.4 \pm 1.6) \times 10^{-12}$ cm^3 $molecule^{-1}s^{-1}$ at 298 K. These values are recommended with the uncertainty increased to ±25%, to account for the fact that measurements from only one group are available. Note also

Table IV-A-8. Rate coefficients (cm^3 $molecule^{-1} s^{-1}$) for RO_2–RO_2, RO_2–NO, and RO_2–NO_2 reactions for which only a single study has been reported

Reactants		k (cm^3 $molecule^{-1}$ s^{-1})	T (K)	Technique[a]	Reference
$C_2H_5O_2$	CH_3O_2	$(2.0 \pm 0.5) \times 10^{-13}$	298	FP–UV abs	Villenave and Lesclaux (1996)
2-$C_3H_7O_2$	CH_3O_2	1×10^{-12}	373	GC	Alcock and Mile (1975)
$(CH_3)_3CO_2$	CH_3O_2	$(1.0 \pm 0.5) \times 10^{-13}$	298	MM–UV abs	Parkes (1974)
$(CH_3)_3CCH_2O_2$	CH_3O_2	$(1.5 \pm 0.5) \times 10^{-12}$	298	FP–UV abs	Villenave and Lesclaux (1996)
cyclo-$C_6H_{11}O_2$	CH_3O_2	$(9.0 \pm 0.15) \times 10^{-14}$	298	FP–UV abs	Villenave and Lesclaux (1996)
$C_{10}H_{21}O_2$	HO_2	$(1.95 \pm 0.21) \times 10^{-11}$	298	FP–UV abs	Boyd et al. (2003)
$C_{14}H_{29}O_2$	HO_2	$(2.11 \pm 0.23) \times 10^{-11}$	298	FP–UV abs	Boyd et al. (2003)
2-$C_5H_{11}O_2$	NO	$(8.0 \pm 1.4) \times 10^{-12}$	297	DF–CIMS	Eberhard and Howard (1997)
$(CH_3)_3CCH_2O_2$	NO	$(4.7 \pm 0.4) \times 10^{-12}$	298	PR–UV abs	Sehested et al. (1993b)
$(CH_3)_3CC(CH_3)_2CH_2O_2$	NO	$(1.8 \pm 0.2) \times 10^{-12}$	298	PR–UV abs	Sehested et al. (1993b)
cyclo-$C_5H_9O_2$	NO	$(10.9 \pm 1.9) \times 10^{-12}$	297	DF–CIMS	Eberhard and Howard (1997)
cyclo-$C_6H_{11}O_2$	NO	$(6.7 \pm 0.9) \times 10^{-12}$	296	PR–UV abs	Platz et al. (1999)
2-$C_3H_7O_2$	NO_2	$(5.7 \pm 0.2) \times 10^{-12}$	298	FP–UV abs	Adachi and Basco (1982)
cyclo-$C_6H_{11}O_2$	NO_2	$(9.5 \pm 1.5) \times 10^{-12}$	296	PR–UV abs	Platz et al. (1999)

[a] Abbreviations for the experimental techniques employed are defined in table II-A-1.

Table IV-A-9. Rate coefficients (k) for reaction of 1-$C_3H_7O_2$ with NO, with $k = A \times \exp(-B/T)$ cm^3 $molecule^{-1}s^{-1}$

A	B (K)	$10^{12} \times k$	T (K)	Technique[a]	Reference	Temperature Range (K)
$(2.9 \pm 0.5) \times 10^{-12}$	-350 ± 60	17.5 ± 1.2	201	DF–CIMS	Eberhard and Howard (1996)	201–403
		15.9 ± 0.7	201			
		17.0 ± 0.9	201			
		13.3 ± 0.4	221			
		11.0 ± 0.6	259			
		9.16 ± 0.46	298			
		9.40 ± 0.64	298			
		9.28 ± 0.18	298			
		9.33 ± 0.72	298			
		8.33 ± 0.40	325			
		7.44 ± 0.64	355			
		7.06 ± 0.20	402			
		7.03 ± 0.46	402			

[a] Abbreviations for the experimental techniques employed are defined in table II-A-1.

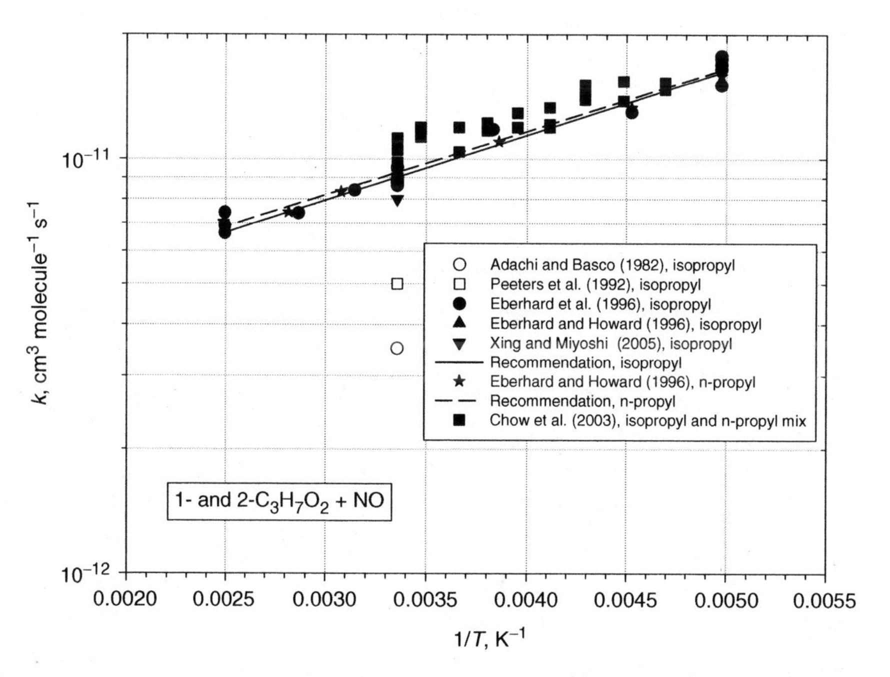

Figure IV-A-4. Measurements of the rate coefficient for reaction of 1-propylperoxy and 2-propylperoxy radicals with NO.

that Chow et al. (2003) measured the rate coefficient for reaction of NO with a mixture of 2-propylperoxy (~55 %) and 1-propylperoxy radicals (~45%), generated from the reaction of Cl with propane; these data are discussed in detail in the following section IV-A-4.1.

IV-A-4. 2-Propylperoxy Radical [Isopropylperoxy Radical, $(CH_3)_2CHO_2$]

IV-A-4.1. $(CH_3)_2CHO_2\bullet + NO$

This reaction has been the subject of six studies (Adachi and Basco, 1982; Peeters et al., 1992; Eberhard et al., 1996; Eberhard and Howard, 1997; Chow et al., 2003; Xing and Miyoshi, 2005). The studies of Eberhard et al. (1996) and Chow et al. (2003) covered a range of temperatures, 201–401 K and 213–298 K respectively, whereas the other studies were restricted to room temperature (~298 K). Note that in the study of Chow et al. (2003) peroxy radicals were generated from the reaction of Cl-atoms with propane, and this actually involves a mixture of 2-propylperoxy (~55 %) and 1-propylperoxy radicals (~45 %) (Tyndall et al., 1997a). The values of Adachi and Basco (1982) and Peeters et al. (1992) are considerably lower than the more recent data; see table IV-A-10 and figure IV-A-4. The IUPAC panel recommends the data of Eberhard and Howard (1997): $k = 2.7 \times 10^{-12} \exp(360/T)$ cm^3 molecule^{-1}s^{-1}, with $k = 9.0 \times 10^{-12}$ cm^3 molecule^{-1}s^{-1} at 298 K. That recommendation is adopted here, with an estimated uncertainty of ±15%. The agreement between the data of Chow et al. (2003) and the very similar rate coefficients for reaction of NO with 1-propylperoxy and 2-propylperoxy radicals provides a consistent picture regarding these two reaction rate coefficients.

IV-A-4.2. $(CH_3)_2CHO_2\bullet + NO_2$

This reaction was studied by Adachi and Basco (1982), using the flash photolysis–UV absorption method. They report $k = (5.7 \pm 0.2) \times 10^{-12}$ cm^3 molecule^{-1}s^{-1}, independent of pressure (55–400 Torr Ar buffer gas); see table IV-A-8. Given the problems associated with the Adachi and Basco data for the ethylperoxy + NO_2 reaction, noted previously, and the expectation of a pressure dependence to the rate coefficient for the $(CH_3)_2CHO_2 + NO_2$ reaction, no recommendation is given.

IV-A-4.3. $(CH_3)_2CHO_2\bullet + CH_3O_2\bullet$

The only data concerning this reaction are from Alcock and Mile (1975), who estimated $k = 1 \times 10^{-12}$ cm^3 molecule^{-1}s^{-1} at 373 K from an end-product analysis of 2,3-dimethylbutane oxidation (see table IV-A-8). In the absence of confirming data, no recommendation is made.

Table IV-A-10. Rate coefficients (k) for reaction of 2-$C_3H_7O_2$ with NO, with $k = A \times \exp(-B/T)$ cm^3 $molecule^{-1} s^{-1}$

A	B (K)	$10^{12} \times k$	T (K)	Technique[a]	Reference	Temperature Range (K)
		3.5 ± 0.3	298	FP–UV Abs	Adachi and Basco (1982)	
		5.0 ± 1.2	298	DF–MS	Peeters et al. (1992)	
		9.1 ± 1.5	298	DF–CIMS	Eberhard and Howard (1997)	
$(2.7 \pm 0.5) \times 10^{-12}$	−360	15.08 ± 1.08	201	DF–CIMS	Eberhard et al. (1996)	201–401
		17.75 ± 1.52	201			
		16.51 ± 0.48	201			
		16.93 ± 1.16	201			
		17.48 ± 1.20	201			
		12.96 ± 3.12	221			
		11.78 ± 0.66	261			
		8.63 ± 0.44	298			
		8.75 ± 1.48	298			
		8.63 ± 1.18	298			
		8.81 ± 1.26	298			
		8.87 ± 0.40	298			
		8.99 ± 0.64	298			
		9.55 ± 0.86	298			
		8.41 ± 0.34	318			
		7.40 ± 1.04	349			
		6.62 ± 0.52	401			
		6.89 ± 1.44	401			
		7.42 ± 0.72	401			

(continued)

Table IV-A-10. *(continued)*

A	B (K)	$10^{12} \times k$	T (K)	Technique[a]	Reference	Temperature Range (K)
$(4.3^{+1.0}_{-0.9}) \times 10^{-12}$	-268 ± 56	11.2 ± 1.9	298	DF–CIMS	Chow et al. (2003)	213–298
		10.5 ± 0.7	298			
		9.8 ± 1.6	298			
		11.4 ± 0.9	288			
		11.9 ± 0.9	288			
		11.3 ± 1.6	288			
		11.9 ± 0.8	273			
		10.4 ± 0.7	273			
		11.7 ± 1.4	263			
		12.2 ± 0.9	263			
		11.9 ± 0.7	253			
		12.9 ± 1.0	253			
		12.1 ± 1.4	243			
		11.9 ± 1.8	243			
		13.3 ± 1.9	243			
		13.3 ± 0.2	243			
		15.1 ± 1.2	233			
		14.2 ± 0.7	233			
		13.9 ± 0.5	233			
		13.8 ± 1.3	223			
		15.4 ± 1.1	223			
		14.7 ± 3.4	213			
		15.3 ± 1.3	213			
		8.0 ± 1.5	298	FP–MS	Xing and Miyoshi (2005)	

[a] Abbreviations for the experimental techniques employed are defined in table II-A-1.

IV-A-5. 2-Methyl-2-propylperoxy Radical [*tert*-Butylperoxy, $(CH_3)_3CO_2\bullet$]

IV-A-5.1. $(CH_3)_3CO_2\bullet + NO$

There are four measurements of this rate coefficient, made using a variety of techniques. The data are summarized in table IV-A-11. Anastasi et al. (1978), using molecular modulation spectroscopy, reported a lower limit of $k > 1 \times 10^{-12}$ cm^3 molecule^{-1}s^{-1}. Peeters et al. (1992), using a discharge flow–mass spectrometer system to detect NO_2, reported $k = (4.0 \pm 1.1) \times 10^{12}$ cm^3 molecule^{-1}s^{-1} at 298 K. This value is likely a lower limit, given that "late" NO_2 production will occur from reaction of the by-product CH_3O_2 with NO. Measurements made via detection of the peroxy radical itself have been reported by Eberhard and Howard (1997) and Xing and Miyoshi (2005), obtained using discharge flow systems with either chemical ionization or electron transfer mass spectrometric detection. The value of Eberhard and Howard (1997), $k = (7.9 \pm 1.3) \times 10^{-12}$cm^3 molecule^{-1}s^{-1} at 298 K, which was obtained in the most direct manner, is recommended, with an estimated uncertainty of $\pm 20\%$.

Table IV-A-11. Rate coefficients (k, cm^3 molecule^{-1}s^{-1}) for reaction of $(CH_3)_3CO_2$ with NO

$10^{12} \times k$	T (K)	Technique[a]	Reference
> 1	298	MM–UV Abs	Anastasi et al. (1978)
4.0 ± 1.1	298	DF–MS	Peeters et al. (1992)
7.9 ± 1.3	297	DF–CIMS	Eberhard and Howard (1997)
8.6 ± 1.4	298	FP–MS	Xing and Miyoshi (2005)

[a] Abbreviations for the experimental techniques employed are defined in table II-A-1.

IV-A-5.2. $(CH_3)_3CO_2\bullet + NO_2$

The only measurement of this rate coefficient is a lower limit reported by Anastasi et al. (1978), 5×10^{-13}cm^3 molecule^{-1}s^{-1}, measured at 298 K using molecular modulation spectrometry. No recommendation can be made at this time.

IV-A-5.3. $(CH_3)_3CO_2\bullet + CH_3O_2\bullet$

The only information regarding this reaction comes from the work of Parkes (1974), using a molecular modulation technique. A rate coefficient of $(1.0 \pm 0.5) \times 10^{-13}$ cm^3 molecule^{-1}s^{-1} is reported; see table IV-A-8. In the absence of confirming data, no recommendation can be made at this time.

IV-A-6. 2-Pentylperoxy Radical ($CH_3CH_2CH_2CH(O_2\bullet)CH_3$)

IV-A-6.1. $CH_3CH_2CH_2CH(O_2\bullet)CH_3 + NO$

The only measurement is that of Eberhard and Howard (1997); see table IV-A-8. Using a flow tube–chemical ionization mass spectrometric technique, these authors report a 297 K value of $k = (8.0 \pm 1.4) \times 10^{-12}$ cm^3 molecule^{-1}s^{-1}. This value is recommended, with the uncertainty increased to ±25% to reflect the fact that only one study has been conducted.

IV-A-7. 2,2-Dimethyl-1-propylperoxy Radical [Neopentylperoxy, $(CH_3)_3CCH_2O_2\bullet$]

IV-A-7.1. $(CH_3)_3CCH_2O_2\bullet + NO$

The only study of this reaction is that of Sehested et al. (1993b), using a pulsed radiolysis–UV absorption technique; see table IV-A-8. The reported value, $k = (4.7 \pm 0.4) \times 10^{-12}$ cm^3 molecule^{-1}s^{-1}, is approximately a factor of two lower than rate coefficients obtained for smaller radicals, possibly the result of complications in the analysis resulting from the formation of $(CH_3)_3CO_2$ radicals and their subsequent reaction with NO:

$$(CH_3)_3CCH_2O_2\bullet + NO \rightarrow (CH_3)_3CCH_2O\bullet + NO_2$$

$$(CH_3)_3CCH_2O\bullet \rightarrow (CH_3)_3C\bullet + CH_2O$$

$$(CH_3)_3C\bullet + O_2 \rightarrow (CH_3)_3CO_2\bullet$$

$$(CH_3)_3CO_2\bullet + NO \rightarrow (CH_3)_3CO\bullet + NO_2$$

No recommendation for this reaction rate coefficient is given.

IV-A-7.2. $(CH_3)_3CCH_2O_2\bullet + HO_2$

The reaction has been the subject of two flash photolysis–UV absorption studies (Rowley et al., 1992a; Boyd et al., 2003); see table IV-A-12. Room-temperature data from the two studies are in excellent agreement, $k = 1.4 \times 10^{-11}$ cm^3 molecule^{-1}s^{-1}. The only study of the temperature dependence is that of Rowley et al. (1992a), who reported $k = (1.43 \pm 0.46) \times 10^{-13} \exp((1380 \pm 100)/T)$ cm^3 molecule^{-1}s^{-1}. Our fit of

Table IV-A-12. Rate coefficients (k) for reaction of 2,2-dimethyl-1-propylperoxy radical with HO_2, with $k = A\times \exp(-B/T)$ cm^3 molecule^{-1}s^{-1}

A	B (K)	$10^{11}\times k$	T (K)	Technique[a]	Reference	Temperature Range (K)
$(1.43 \pm 0.46) \times 10^{-13}$	-1380 ± 100	3.22 ± 0.83	248	FP–UV Abs	Rowley et al. (1992a)	248–365
		2.29 ± 0.56	273			
		1.69 ± 0.23	293			
		1.42 ± 0.13	298			
		1.13 ± 0.13	326			
		0.89 ± 0.14	348			
		0.85 ± 0.29	348			
		0.61 ± 0.03	365			
		1.38 ± 0.23	298	FP–UV Abs	Boyd et al. (2003)	

[a] Abbreviations for the experimental techniques employed are defined in table II-A-1.

the Arrhenius expression to the combined data set yields $k = 2.39 \times 10^{-13}\exp(1228/T)$ cm^{3}molecule^{-1} s^{-1}; $k = 1.47 \times 10^{-11}$ cm^3 molecule^{-1}s^{-1} at 298 K. The two expressions yield data within 10% of each other over the temperature range covered in the Rowley et al. (1992a) study.

IV-A-7.3. $(CH_3)_3CCH_2O_2\bullet + CH_3O_2\bullet$

This reaction was studied by Villenave and Lesclaux (1996), using the flash photolysis/UV absorption technique; see table IV-A-8. They report a rate coefficient of $(1.5 \pm 0.5) \times 10^{-12}$ cm^3 molecule^{-1}s^{-1}, with the quoted uncertainty representing statistical errors (1σ) only. This rate coefficient is recommended, with an uncertainty of $\pm 50\%$ (Villenave and Lesclaux, 1996).

IV-A-8. Cyclopentylperoxy Radical, *cyclo*-$C_5H_9O_2\bullet$

IV-A-8.1. cyclo-$C_5H_9O_2\bullet + NO$

The only measurement is that of Eberhard and Howard (1997), who obtained $k = (10.9 \pm 1.9) \times 10^{-12}$ cm^3 molecule^{-1}s^{-1} at 297 K, using a flow tube–chemical ionization mass spectrometric technique; see table IV-A-8. This value is recommended, with the uncertainty increased to $\pm 25\%$ to reflect the fact that only one study has been conducted.

IV-A-8.2. cyclo-$C_5H_9O_2\bullet + HO_2$

There are two studies of this reaction (Rowley et al., 1992b; Crawford et al., 1997), as summarized in table IV-A-13. Rowley et al. (1992b) used a flash photolysis–UV absorption technique, whereas Crawford et al. (1997) used a flash photolysis system with UV absorption to monitor cyclopentylperoxy radicals and both UV and IR absorption to monitor HO_2 radicals. The UV spectra for cyclopentylperoxy radicals reported by Rowley et al. (1992b) and Crawford et al. (1997) are indistinguishable, and both used HO_2 UV absorption cross sections that are consistent with current recommendations (Tyndall et al., 2001). A least-squares fit of the Arrhenius expression to the combined data set gives $k = 3.65 \times 10^{-13}\exp(1106/T)$ cm^3 molecule^{-1}s^{-1}, with $k = 1.49 \times 10^{-11}$cm^3 molecule^{-1}s^{-1} at 298 K. However, Crawford et al. (1997) note that their four lowest temperature data points may be systematically lower than the remainder of the data set, possibly due to the low cyclopentane vapor pressure under these conditions. Refitting the data excluding these four low-temperature points yields $k = 1.37 \times 10^{-13}\exp(1427/T)$ cm^3 molecule^{-1}s^{-1}, with $k = 1.56 \times 10^{-11}$cm^3 molecule^{-1}s^{-1} at 298 K. This latter expression is recommended, with an uncertainty at 298 K of $\pm 25\%$.

IV-A-9. Cyclohexylperoxy Radical (*cyclo*-$C_6H_{11}O_2\bullet$)

IV-A-9.1. cyclo-$C_6H_{11}O_2\bullet + NO$

The only measurement is that of Platz et al. (1999), using a pulse radiolysis–UV absorption technique; see table IV-A-8. The value reported at 298 K is

Table IV-A-13. Rate coefficients (k) for reaction of cyclopentylperoxy radical with HO_2, with $k = A \times \exp(-B/T)$ cm^3 $molecule^{-1} s^{-1}$

A	B (K)	$10^{11} \times k$	T (K)	Technique[a]	Reference	Temperature Range (K)
$(2.1 \pm 1.3) \times 10^{-13}$	-1323 ± 185	3.73 ± 1.06	249	FP–UV Abs	Rowley et al. (1992b)	248–364
		2.85 ± 0.49	274			
		1.83 ± 0.57	300			
		2.07 ± 0.31	300			
		1.31 ± 0.13	324			
		1.02 ± 0.28	348			
		0.71 ± 0.17	364			
$(3.2^{+15}_{-1.1}) \times 10^{-13}$	-1150 ± 200	6.6 ± 1.4	214	FP–UV/IR Abs	Crawford et al. (1997)	214–359
		4.7 ± 1.3	214			
		4.3 ± 1.0	225			
		3.6 ± 0.8	231			
		4.0 ± 0.8	253			
		2.7 ± 0.4	267			
		3.0 ± 0.6	267			
		2.4 ± 0.4	268			
		2.9 ± 0.5	268			
		3.0 ± 1.0	271			
		1.6 ± 0.4	295			
		1.8 ± 0.4	295			
		1.6 ± 0.3	295			
		1.0 ± 0.2	324			
		1.2 ± 0.2	325			
		0.6 ± 0.1	356			
		0.8 ± 0.2	356			
		0.7 ± 0.2	357			
		0.7 ± 0.2	357			
		0.7 ± 0.2	358			
		0.8 ± 0.1	359			

[a] Abbreviations for the experimental techniques employed are defined in table II-A-1.

$k = (6.7 \pm 0.9) \times 10^{-12}$ cm^3 molecule^{-1}s^{-1}. However, the measurement is very likely influenced by the generation of peroxy radicals formed following ring-opening and the subsequent reaction of these species with NO, and so most likely represents a lower limit to the actual rate coefficient. No recommendation is given at this time.

IV-A-9.2. cyclo-$C_6H_{11}O_2$• + NO_2

The only study of this reaction is that of Platz et al. (1999), who used a pulse radiolysis–UV absorption system to monitor NO_2 loss in time-resolved fashion; see table IV-A-8. The reported rate coefficient is $(9.5 \pm 1.5) \times 10^{-12}$ cm^3 molecule^{-1}s^{-1}. This value is adopted, with an estimated uncertainty of ±30%.

IV-A-9.3. cyclo-$C_6H_{11}O_2$• + HO_2

This reaction has been the subject of two studies (Rowley et al., 1992b; Boyd et al., 2003), both conducted using the flash photolysis–UV absorption technique. In both studies, data were analyzed using HO_2UV absorption cross sections that are within 10% of current recommendations (Tyndall et al., 2001). The data are in very good agreement; see table IV-A-14. A least-squares fit of the Arrhenius expression to the data yields $k = 2.66 \times 10^{-13}\exp(1240/T)$cm^3 molecule^{-1}s^{-1}, with $k = 1.71 \times 10^{-11}$ cm^3 molecule^{-1}s^{-1} at 298 K. An uncertainty of ±25% is estimated at 298 K.

Table IV-A-14. Rate coefficients (k) for reaction of cyclohexylperoxy with HO_2, with $k = A \times \exp(-B/T)$ cm^3 molecule^{-1}s^{-1}

A	B (K)	$10^{11} \times k$	T (K)	Technique[a]	Reference	Temperature Range (K)
$(2.6 \pm 1.2) \times 10^{-13}$	-1245 ± 124	3.77 ± 0.76	249	FP–UV Abs	Rowley et al. (1992b)	248–364
		3.25 ± 0.38	258			
		2.41 ± 0.70	274			
		1.98 ± 0.34	300			
		1.68 ± 0.38	300			
		1.17 ± 0.18	324			
		0.78 ± 0.11	364			
		1.71 ± 0.17	298	FP–UV Abs	Boyd et al. (2003)	

[a] Abbreviations for the experimental techniques employed are defined in table II-A-1.

IV-A-9.4. cyclo-$C_6H_{11}O_2$• + CH_3O_2•

The only measurement is that of Villenave and Lesclaux (1996); see table IV-A-8. Using a flash photolysis–UV absorption system, a rate coefficient of $(9.0 \pm 0.15) \times 10^{-14}$ cm^3 molecule^{-1}s^{-1} was obtained at 298 K. The quoted uncertainty represents statistical errors only. This rate coefficient is recommended, with an estimated uncertainty of ±35% (Villenave and Lesclaux, 1996).

IV-A-10. 2,2,3,3-Tetramethyl-1-Butylperoxy Radicals [$(CH_3)_3CC(CH_3)_2CH_2O_2\bullet$]

IV-A-10.1. $(CH_3)_3CC(CH_3)_2CH_2O_2\bullet + NO$

The only study of this reaction is that of Sehested et al. (1993b), using a pulse radiolysis–UV absorption technique; see table IV-A-8. The reported value, $k = (1.8 \pm 0.2) \times 10^{-12}$ cm^3 molecule^{-1}s^{-1}, is considerably lower than that obtained for smaller radicals, possibly the result of interference from the formation of $(CH_3)_3CC(CH_3)_2O_2$ radicals and their subsequent reaction with NO:

$$(CH_3)_3CC(CH_3)_2CH_2O_2\bullet + NO \rightarrow (CH_3)_3CC(CH_3)_2CH_2O\bullet + NO_2$$

$$(CH_3)_3CC(CH_3)_2CH_2O\bullet \rightarrow (CH_3)_3CC\bullet(CH_3)_2 + CH_2O$$

$$(CH_3)_3CC\bullet(CH_3)_2 + O_2 \rightarrow (CH_3)_3CC(CH_3)_2O_2\bullet$$

$$(CH_3)_3CC(CH_3)_2O_2\bullet + NO \rightarrow (CH_3)_3CC(CH_3)_2O\bullet + NO_2$$

No recommendation for this reaction rate coefficient is given.

IV-A-11. Decylperoxy Radicals (n-$C_{10}H_{21}O_2\bullet$)

IV-A-11.1. $C_{10}H_{21}O_2\bullet + HO_2$

Boyd et al. (2003) measured the rate coefficient for this reaction using flash photolysis–UV absorption at 298 K; see table IV-A-8. The decylperoxy radicals were created from reaction of OH with decane and thus included a mixture of 1-, 2-, 3-, 4-, and 5-decylperoxy. The measurements were carried out with a large excess of HO_2 over $C_{10}H_{21}O_2$, and a rate coefficient $k = (19.5 \pm 2.1) \times 10^{-12}$ cm^3 molecule^{-1}s^{-1} was reported. This value is recommended, with an estimated uncertainty of $\pm 35\%$.

IV-A-12. Tetradecylperoxy Radicals (n-$C_{14}H_{29}O_2\bullet$)

IV-A-12.1. $C_{14}H_{29}O_2\bullet + HO_2$

Boyd et al. (2003) used flash photolysis–UV absorption to determine the rate coefficient for this reaction at 298 K; see table IV-A-8. The tetradecylperoxy radicals were generated from the reaction of OH with tetradecane, and thus included a mixture of 1-, 2-, 3-, 4-, 5-, 6-, and 7-tetradecylperoxy radicals. The measurements were carried out with a large excess of HO_2 over $C_{14}H_{29}O_2$, and a rate coefficient $k = (2.11 \pm 0.23) \times 10^{-11}$ cm^3 molecule^{-1}s^{-1} was reported. This value is recommended, with an estimated uncertainty of $\pm 35\%$.

IV-B. Alkoxy Radical Chemistry

IV-B-1. Introduction

In the standard hydrocarbon oxidation sequence, it is the chemistry of the alkoxy radicals that determines the bulk of the distribution of first-generation end-products.

In general, the atmospheric chemistry of alkoxy radicals is characterized by a competition between three classes of reactions (e.g., Atkinson, 1990; Atkinson and Carter, 1991; Atkinson, 1994, 1997a,b; Atkinson and Arey, 2003; Devolder, 2003; Orlando et al., 2003b): 1) reaction with O_2, which preserves the carbon chain of the parent alkane and results in the production of a carbonyl compound and HO_2; 2) unimolecular decomposition, which usually results in the formation of an alkyl radical and a carbonyl compound with a shortening of the carbon chain; and 3) unimolecular isomerization, which usually leads to multifunctional oxidation products (e.g., 1,4-hydroxycarbonyls and 1,4-hydroxynitrates) and a preservation of the carbon chain. These potentially competing pathways are illustrated in figure IV-B-1 for the 2-pentoxy radical (Orlando et al., 2003b).

Absolute rate coefficients for these reactions have been obtained for only a few of the smaller alkoxy radicals. For example, rate coefficients have only been firmly established over a range of temperatures for reaction of a subset of the C_1–C_6 alkoxy radicals with O_2, dissociation rate coefficients have only been directly measured for ethoxy, 2-propoxy, 2-butoxy, and 2-methyl-2-propylperoxy radicals (Balla et al., 1985; Blitz et al., 1999; Caralp et al., 1999; Devolder et al., 1999; Fittschen et al., 2000; Falgayrac et al., 2004), and no direct measurement of isomerization rates have been reported to date. A large portion of the data base describing the atmospheric behavior of alkoxy radicals has been built up primarily from two sources: 1) environmental chamber experiments, in which end-product distributions observed under atmospheric conditions have been used to infer relative rates of competing alkoxy radical reactions (e.g., Carter et al., 1976; Cox et al., 1981; Niki et al., 1981b; Eberhard et al., 1995; Aschmann et al., 1997; Orlando et al., 2000; Cassanelli et al., 2006); and 2) from theoretical methodologies that lend themselves well to the study of unimolecular processes (e.g., Somnitz and Zellner,

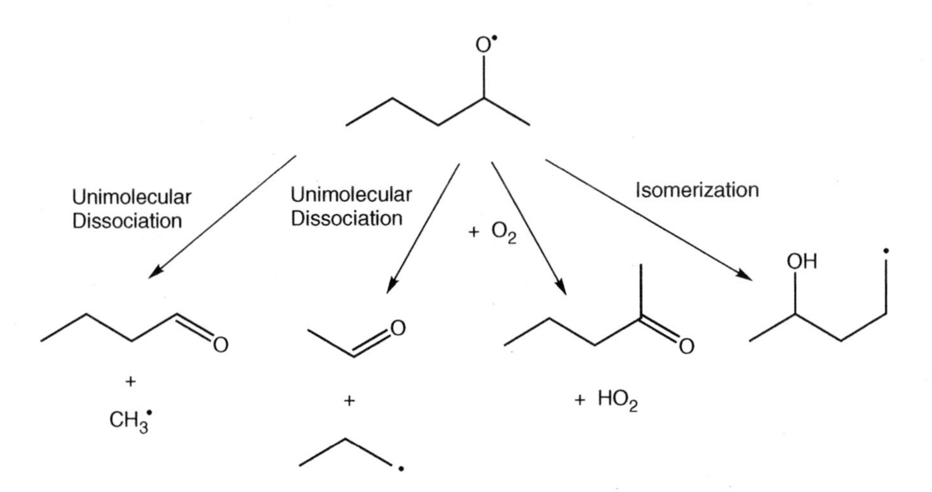

Figure IV-B-1. Possible reaction pathways for the 2-pentoxy radical under atmospheric conditions (Orlando et al., 2003b). The figure has been reprinted from Orlando et al. (2003b) with the permission of the American Chemical Society.

2000a,b,c; Méreau et al., 2000a,b; Fittschen et al., 2000; Lin and Ho, 2002; Méreau et al., 2003).

An overview of these three classes of competing alkoxy radical reactions (reaction with O_2, unimolecular decomposition, and isomerization) is given in this section. This is followed by a summary of the available information on the reaction rate coefficients for the individual alkoxy radicals. We note also that alkoxy radicals can react with both NO and NO_2:

$$RO\bullet + NO_x \rightarrow RONO_x \quad \text{(IV-B-1a)}$$

$$\rightarrow R'{=}O + HNO_x \quad \text{(IV-B-1b)}$$

These reactions have been shown to occur with rate coefficients of about $(2.0 - 4.5) \times 10^{-11}\,cm^3\,molecule^{-1}s^{-1}$ (e.g., Atkinson, 1997a), and they occur mainly via formation of the alkyl nitrite or nitrate (channel IV-B-1a). While these reactions will not be dealt with explicitly in this chapter, they can play a role in very polluted atmospheres (i.e., with $NO_x \geq 5$ ppm or so) and in environmental chamber experiments, and thus cannot be totally ignored.

IV-B-2. Modes of Alkoxy Radical Reaction in the Atmosphere

IV-B-2.1. Alkoxy Radical Reactions with O_2

Reaction with O_2 is an important atmospheric fate for many alkoxy radicals:

$$RR'CHO\bullet + O_2 \rightarrow RR'C{=}O + HO_2$$

Formally, the reaction occurs via O_2 abstraction of an α-hydrogen, resulting in the production of a carbonyl compound and the HO_2 radical (Zellner, 1987; Hartmann et al., 1990). Thus, the reaction is common to all primary and secondary alkoxy radicals, but cannot occur for tertiary radicals such as 2-methyl-2-propoxy radical [*tert*-butoxy, $(CH_3)_3CO\bullet$] and 2-methyl-2-butoxy radical [$CH_3CH_2C(CH_3)_2O\bullet$]. The reaction, however, does not appear to involve a simple abstraction process. Recent theoretical studies on the methoxy radical reaction with O_2 suggest that the reaction occurs instead via the formation of an alkoxy radical–O_2 complex ($RR'CHO \cdots O_2$) that is held together by a non-covalent bond, and which subsequently rearranges via H-atom transfer to products (Bofill et al., 1999; Setokuchi and Sato, 2002). Tunneling appears to play a major role in the H-atom transfer process, at least in the case of methoxy (Setokuchi and Sato, 2002).

Absolute rate coefficients for alkoxy radical reactions with O_2 have been well established over a range of temperatures for only the simplest of the alkoxy radicals (e.g., methoxy, ethoxy, 1- and 2-propoxy radicals). Direct studies of a few C_4–C_7 radicals exist, although insufficient data are available to make firm recommendations in most cases. Nonetheless, some general observations can be made and some trends established. As a result of the complexity of the mechanism involved,

A-factors and reaction rate coefficients are quite small. With the exception of methoxy $+O_2(k_{298K} = 2 \times 10^{-15} cm^3 \text{ molecule}^{-1} s^{-1})$, rate coefficients near 298 K for those radicals studied fall within a factor of two of 10^{-14} cm^3 $\text{molecule}^{-1} s^{-1}$. *A*-factors fall within an order of magnitude of 10^{-14} cm^3 $\text{molecule}^{-1} s^{-1}$ in all cases studied, and, as illustrated in figure IV-B-2, activation energies tend to decrease with increasing carbon number (from about 10 kJ mol^{-1} for the methoxy radical reaction to -4 kJ mol^{-1} for the corresponding 2-butoxy radical reaction). A notable exception is the cyclohexoxy radical, $k = 4.7 \times 10^{-12} \exp(-1720/T)$ cm^3 $\text{molecule}^{-1} s^{-1}$ (Zhang et al., 2004b), for which an anomalously high *A*-factor and activation energy have been reported; see figure IV-B-2.

Despite the relatively small rate coefficients for the alkoxy radical reactions with O_2, the large atmospheric abundance of O_2 results in a very short lifetime for the alkoxy radicals throughout the troposphere. For example, assuming for illustrative purposes $k_{RO+O2} \sim 10^{-14}$ cm^3 $\text{molecule}^{-1} s^{-1}$ independent of temperature, RO lifetimes of 20 μs at the Earth's surface and 140 μs at 16 km are obtained. Obviously, only those unimolecular processes that occur on these timescales will be able to compete with the O_2 reaction and be of any atmospheric relevance. Again, the one exception to this general discussion is the tertiary alkoxy radicals for which no O_2 reaction can occur, implying longer atmospheric lifetimes and the importance of slower unimolecular processes.

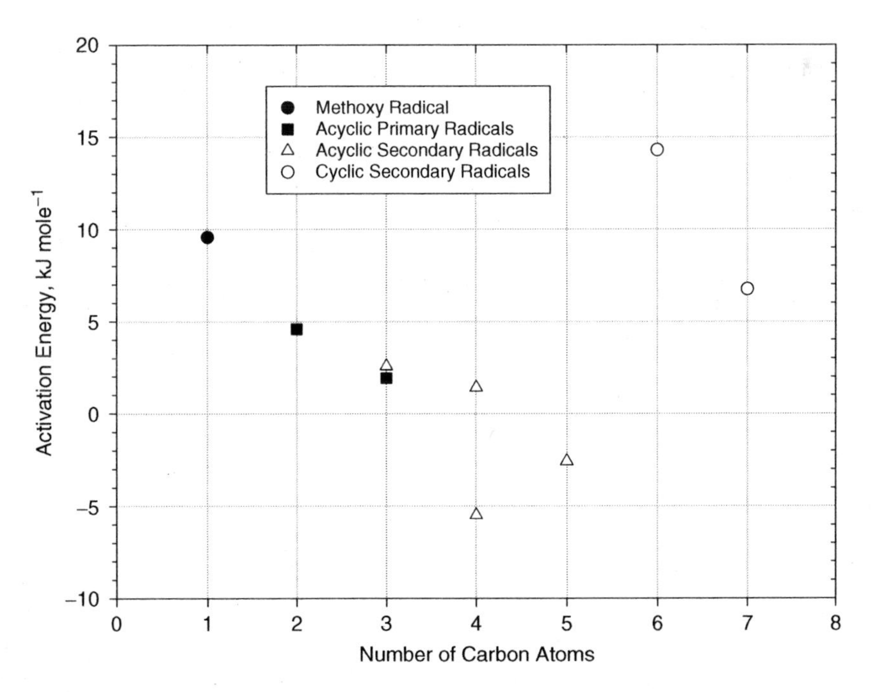

Figure IV-B-2. Plot of activation energy (kJ/mole) versus number of carbon atoms for reaction of alkoxy radicals with O_2. Data for methoxy, ethoxy, 1-propoxy, and 2-propoxy are from rate expressions recommended in this work; data for larger alkoxy radicals represent individual determinations.

IV-B-2.2. Unimolecular Decomposition Reactions of the Alkoxy Radicals

In general, alkoxy radicals can undergo unimolecular decomposition via three separate bond rupture processes, as illustrated below for the generic alkoxy radical RR′R″CO•:

$$RR'R''CO\bullet \rightarrow R\bullet + R'R''C{=}O$$
$$\rightarrow R'\bullet + RR''C{=}O$$
$$\rightarrow R''\bullet + R'RC{=}O$$

where R, R′, and R″ represent alkyl groups or H-atoms for alkoxy radicals derived from the alkanes. Carbon–carbon bond scissions are more favorable than carbon–hydrogen scissions, and so a shortening of the carbon chain almost always results. As a general rule of thumb, a more reactive set of products is generated from the bond dissociation processes than is generated from the O_2 reaction; i.e., more aldehydes from bond dissociation vs. more ketones from the O_2 reactions. Table IV-B-1 summarizes some of the more atmospherically relevant alkoxy radical dissociation processes. A more comprehensive summary is provided in table 1 of Orlando et al. (2003b).

Rate coefficients for alkoxy radical decomposition processes have been measured directly (via LIF detection of the alkoxy radical) in only a few cases, namely for

Table IV-B-1. Major alkoxy radical unimolecular decomposition reactions encountered in the atmospheric chemistry of the alkanes. Rate coefficients and activation energies are those given in Orlando et al. (2003b), except where indicated

Radical	Decomposition Products	Approximate Rate (298 K, 1 atm.)	E_a (kJ mol^{-1})	Competing Reactions
2-Butoxy	$CH_3CH_2\bullet + CH_3CHO$	2.5×10^4 s^{-1}	53	Reaction with O_2
2-Methyl-2-propoxy	$CH_3C(O)CH_3 + \bullet CH_3$	3000 s^{-1}	60	None
2-Methyl-1-propoxy	$CH_3CH(\bullet)CH_3 + CH_2O$	5.7×10^4 s^{-1}	52	Reaction with O_2
3-Pentoxy	$CH_3CH_2\bullet + CH_3CH_2CHO$	3×10^4 s^{-1}	54	Reaction with O_2
2,2-Dimethyl-1-propoxy	$(CH_3)_3C\bullet + CH_2O$	2×10^6 s^{-1}	40	Reaction with O_2 [b]
2-Methyl-2-butoxy	$CH_3CH_2\bullet + CH_3C(O)CH_3$	9.4×10^5 s^{-1}	44	None
2-Methyl-1-butoxy [a]	$CH_3CH_2CH(\bullet)CH_3 + CH_2O$	$\sim 10^5$ s^{-1}	~ 47	Isomerization (major) Reaction with O_2
3-Methyl-2-butoxy [a]	$CH_3CH(\bullet)CH_3 + CH_3CHO$	$\sim 10^7$ s^{-1}	~ 38	Reaction with O_2 [b]
Cyclopentoxy	$\bullet CH_2CH_2CH_2CH_2CHO$	$> 10^7$ s^{-1}	< 38	Reaction with O_2 [b]
Cyclohexoxy	$\bullet CH_2CH_2CH_2CH_2CH_2CHO$	5×10^4 s^{-1}	48	Reaction with O_2

[a] Rate coefficient and E_a estimated using parameterizations given in Orlando et al. (2003b).
[b] Minor process at 298 K.

ethoxy (Caralp et al., 1999), 2-propoxy (Balla et al., 1985; Devolder et al., 1999), 2-methyl-2-propylperoxy (Blitz et al., 1999; Fittschen et al., 2000), and 2-butoxy (Falgayrac et al,. 2004). A number of theoretical studies of these reactions where conducted in the year 2000, and have led to the production of a large body of internally consistent rate data for most alkoxy radicals containing 2–5 carbon atoms (Méreau et al., 2000b; Somnitz and Zellner, 2000a,b; Fittschen et al., 2000). Relative rate data have been obtained for a number of systems from end-product studies, usually in static environmental chambers, but more recently in slow-flow systems as well. A classic example is the 2-butoxy radical, for which decomposition and reaction with O_2 are competitive in 1 atm air at 298 K (Carter et al., 1979; Cox et al., 1981; Drew et al., 1985; Libuda et al., 2002; Meunier et al., 2003; Cassanelli et al., 2005):

$$CH_3CH_2CH(O\bullet)CH_3 \rightarrow CH_3CH_2\bullet + CH_3CHO \qquad \text{(IV-B-2)}$$

$$CH_3CH_2CH(O\bullet)CH_3 + O_2 \rightarrow CH_3CH_2C(O)CH_3 + HO_2 \qquad \text{(IV-B-3)}$$

Here, ratios of acetaldehyde and 2-butanone yields provide a measure of the rate coefficient ratio for the competing processes.

Given the mechanistic similarities of all the alkoxy radical decomposition reactions, it is not surprising to find that they are all characterized by similar A-factors. The body of recent theoretical and experimental data points to A-factors at the high-pressure limit that fall between about 10^{13} and 10^{14} s^{-1}, somewhat lower than earlier estimates (Baldwin et al., 1977; Atkinson, 1997b). Because the decomposition reactions are typically in the fall-off region at atmospheric pressure, effective A-factors at atmospheric pressures are somewhat lower, with estimates in the 10^{12} or 10^{13} range for a large body of two- to five-carbon alkoxy radicals (e.g., Somnitz and Zellner, 2000b; Falgayrac et al., 2004).

Alkoxy radical decomposition reactions are almost always endothermic, and possess energy barriers that exceed the endothermicity. Note that in the remainder of this chapter we will distinguish as best as possible between the energy barrier (activation energy at 0 K, designated E_0 or E_b) and the Arrhenius activation energy (E_a) for a given process. For typical atmospheric temperatures, the activation energy at infinite pressure exceeds the energy barrier by 3–4 kJ mol^{-1}, while effective activation energies near 1 atm are often intermediate between these two extremes.

As first discussed by Baldwin et al. (1977), the endothermicity for an alkoxy radical decomposition and the activation energy for the reaction are correlated, $E_a \approx A + B\times(\Delta H_r)$, so that a reasonable estimate of the rate coefficient for a given decomposition reaction can be obtained from a knowledge of the endothermicity. Choo and Benson (1981) later showed that the intercept of such plots was dependent on the nature of the alkyl radical leaving group, and could in fact be correlated with the ionization potential (IP) of the leaving group. Thus, structure–reactivity estimates of the following form are now found in the literature (Atkinson,1997b; Méreau et al., 2000b; Orlando et al., 2003b):

$$E_a = (A \times \text{IP} - B) + C \times (\Delta H_r) \qquad \text{(IV-B-A)}$$

Orlando et al. suggested $E_a(\text{kJ mol}^{-1}) = (10.0 \times \text{IP(ev)} - 49.4) + 0.58 \times \Delta H_r$ (kJ mol^{-1}) on the basis of available data involving CH_3 radical elimination.

Other structure-reactivity approaches have appeared in the recent literature. For example, Johnson et al. (2004b) showed that both the 298 K rate coefficient k_{298K} and the activation energy for alkoxy radical decomposition reactions correlated well with the mean of the ionization potential ($\text{IP}_{m,}$ in ev) of the two reaction products. Their analysis for unsubstituted (i.e., alkane-derived) alkoxy radicals leads to the following expressions:

$$\ln(k_{298K}, \text{cm}^3\text{molecule}^{-1}\text{s}^{-1}) = -2.57(\text{IP}_m)^2 + 43.76(\text{IP}_m) - 176.00$$

$$E_a\,(\text{kJ mol}^{-1}) = 14.62(\text{IP}_m) - 82.58$$

where an A-factor per available reaction pathway of 10^{13} s^{-1} has been assumed at atmospheric pressure for all reactions.

Méreau et al. (2000b) demonstrated that a reasonable estimate of the activation energy for alkoxy radical decomposition reactions could be obtained using two parameters, the reaction enthalpy and the number of hydrogens (n_H) on the carbon bonded to the alkoxy oxygen:

$$E_a(\text{kJ mol}^{-1}) = (45.81 - 22.34 \times n_H) + 1.2 \times \Delta H_r(\text{kJ mol}^{-1}) \qquad \text{(IV-B-B)}$$

They further combined equation (IV-B-B) with their own version of (IV-B-A) to obtain an equation based only on the readily available leaving-group IP and n_H parameters:

$$E_a(\text{kJ mol}^{-1}) = 10.46 \times \text{IP(ev)} + 8.79 \times n_H - 43.51 \qquad \text{(IV-B-C)}$$

Somnitz and Zellner (2000c) pointed out that decompositions of linear alkoxy radicals could be divided into three categories: those for which both products (the alkyl radical and the carbonyl species) contain one carbon atom (ethoxy radical, $E_b = 72.4$ kJ mol^{-1}); those for which one product contains two or more carbon atoms and the other one carbon (e.g., 1-propoxy, 1-butoxy, 2-propoxy, $E_b \sim 62$ kJ mol^{-1}); and those for which each product contains two or more carbon atoms (e.g., 2-butoxy, 3-pentoxy, $E_b = 50.2$ kJ mol^{-1}).

Peeters et al. (2004), using ethoxy radical decomposition as the standard ($E_b = 73.2$ kJ mol^{-1}), demonstrated that substitution of hydrogen atoms in ethoxy with alkyl radical fragments resulted in a lowering of the energy barrier by 8.8 kJ mol^{-1} for substitution at the α-carbon and 13 kJ mol^{-1} for substitution at the β-carbon. They extended this approach to include alkoxy radicals containing hydroxy and carbonyl substituents, and reported the following expression:

$$E_b(\text{kJ mol}^{-1}) = 73.2 - 8.8 \times N_\alpha(\text{alk}) - 13.0 \times N_\beta(\text{alk}) - 33.5 \times N_\alpha(\text{OH}) - 33.5 \times N_\beta(\text{OH}) - 50.2 \times N_\alpha({=}\text{O}) - 33.5 \times N_\beta({=}\text{O}) \qquad \text{(IV-B-D)}$$

Comparisons of recommended activation energies (Orlando et al., 2003b) with those obtained from the various structure–reactivity relationships show that all of the approaches listed above provide a reasonable means for activation energy (or energy barrier) estimation. However, scatter on the order of a few kJ mol^{-1} is present in most cases, and this translates to an uncertainty of a factor of 2 or more in the estimated rate coefficient near 298 K.

IV-B-2.3. Unimolecular Isomerization Reactions of the Alkoxy Radicals

Another important type of reaction that some alkoxy radicals can undergo is unimolecular isomerization via a 1,5-hydrogen shift, as shown here for the 1-butoxy radical (Carter et al., 1976; Baldwin et al., 1977; Atkinson, 1997b):

$$CH_3CH_2CH_2CH_2O\bullet \rightarrow \bullet CH_2CH_2CH_2CH_2OH \qquad \text{(IV-B-4)}$$

For the alkoxy radicals derived from the alkanes, these reactions are roughly thermoneutral or slightly exothermic (depending on the nature of the hydrogen being abstracted). They have reduced *A*-factors relative to the simple decomposition reactions just discussed, and in all cases occur through barriers of less than 42 kJ mol^{-1}. Although other isomerization processes (e.g., 1,4- or 1,6-hydrogen shifts) are also possible in theory, the 1,5-H shifts via a six-membered ring transition state are the most favored (Carter et al., 1976; Baldwin et al., 1977) and are the only processes sufficiently rapid to occur under atmospheric conditions. Thus, the reactions need be considered only for alkanes possessing carbon chains of four atoms or more. Although great strides have been made since 1995, the rate coefficients for these processes are still not well characterized. In fact, no direct measurement of an isomerization reaction rate coefficient had been made by the year 2005; rate data have been derived either from theoretical studies (Jungkamp et al., 1997; Lendvay and Viskolcz, 1998; Somnitz and Zellner, 2000a,b; Hack et al., 2001; Lin and Ho, 2002; Ferenac et al., 2003; Vereecken and Peeters, 2003; Méreau et al., 2003) or, in a relative sense, from end-product data analysis (Carter et al., 1976; Cox et al., 1981; Niki et al., 1981b; Eberhard et al., 1995; Atkinson et al., 1995; Heiss and Sahetchian, 1996; Geiger et al., 2002; Johnson et al., 2004a; Cassanelli et al., 2005, 2006). A summary of available data, updated from table 4 of Orlando et al. (2003b), is presented in table IV-B-2.

It is now firmly established (Atkinson et al., 1995; Kwok et al., 1996; Arey et al., 2001; Reisen et al., 2005) that alkoxy radical isomerization results in the eventual formation of a 1,4-hydroxycarbonyl and 1,4-hydroxynitrate compounds, via the reaction sequence outlined below for 1-butoxy (stable end-products are highlighted for clarity):

$$CH_3CH_2CH_2CH_2O\bullet \rightarrow \bullet CH_2CH_2CH_2CH_2OH$$

$$\bullet CH_2CH_2CH_2CH_2OH + O_2 \rightarrow \bullet O_2CH_2CH_2CH_2CH_2OH$$

$$\bullet O_2CH_2CH_2CH_2CH_2OH + NO \rightarrow \bullet OCH_2CH_2CH_2CH_2OH + NO_2$$

$$\rightarrow \mathbf{O_2NOCH_2CH_2CH_2CH_2OH}$$

Table IV-B-2. Summary of available data on isomerization of alkane–derived alkoxy radicals. For theoretical studies of activation energies–barrier heights, data have been reported as either barrier heights (activation energies at 0 K) (A), activation energies (kJ mol^{-1}) at 1 atm. (B), or activation energies at infinite pressure (C). Level of theory used is given in parentheses (SAR = structure–additivity relationship)

Radical	Isomerization Product	Approximate Rate near 298 K, 1 atm (s^{-1})	Activation Energy or Barrier Height (kJ mol^{-1})		Reference
1-Butoxy	$\bullet CH_2CH_2CH_2CH_2OH$	1.6×10^5			Carter et al. (1979)
		1.5×10^5			Cox et al. (1981)
		1.9×10^5			Niki et al. (1981b)
		1×10^5	40.6		Heiss and Sahetchian (1996)
		2×10^5	35.1 (SAR)		Atkinson (1997b)
		1.3×10^5	39.3 (b3lyp)	A	Jungkamp et al. (1997)
		1.2×10^5	38.5 (bac4)	A	Lendvay and Viskolcz (1998)
		1.1×10^5	41.4 (g2)	B	Somnitz and Zellner (2000b)
		1.7×10^5	36.8 (dft)	B	Méreau et al. (2003)
			41.4 (b3lyp)	A	Ferenac et al. (2003)
			41.0 (b3lyp)	A	Lin and Ho (2002)
			40.2 (g2/mp2)	A	Hack et al. (2001)
		1.4×10^5	34.3 (b3lyp)	B	Vereecken and Peeters (2003)
		1.3×10^5			Geiger et al. (2002)
		2.0×10^5	27.2 [a]		Cassanelli et al. (2005)
		2.0×10^5	29.3 [a]		Cassanelli et al. (2006)
2-Pentoxy	$\bullet CH_2CH_2CH_2CH(OH)CH_3$	2.5×10^5			Atkinson et al. (1995)
		2×10^5	35.1(SAR)		Atkinson (1997b)
		3.3×10^5	38.9	C	Méreau et al. (2003)
		5.0×10^5	36.8 (g2)	B	Somnitz and Zellner (2000b)
			36.1 (b3lyp)	A	Lin and Ho (2002)
		4.1×10^5	30.5[b]		Johnson et al. (2004a)
2-Methyl-1-butoxy	$\bullet CH_2CH_2CH(CH_3)CH_2OH$	2×10^5	35.1 (SAR)		Atkinson (1997b)
		2.3×10^5	40.6	C	Méreau et al. (2003)

3-Hexoxy	$CH_3CH_2CH(OH)CH_2CH_2CH_2\bullet$	2×10^5	35.1 (SAR)		Atkinson (1997b)
		1.7×10^5	40.2	C	Méreau et al. (2003)
		$(1.8–4.3) \times 10^5$			Eberhard et al. (1995)
2-Methyl-2-pentoxy	$\bullet CH_2CH_2CH_2C(OH)(CH_3)_2$	2×10^5	35.1 (SAR)		Atkinson (1997b)
		5.3×10^5	37.7	C	Méreau et al. (2003)
			33.6 (b3lyp)	A	Lin and Ho (2002)
1-Pentoxy	$CH_3CH(\bullet)CH_2CH_2CH_2OH$	2×10^6	28.5 (SAR)		Atkinson (1997b)
		2.0×10^6	32.2 (g2)	B	Somnitz and Zellner (2000b)
		2.2×10^6	33.1	C	Méreau et al. (2003)
		3.4×10^6	24.3[a]		Johnson et al. (2004a)
1-Hexoxy	$CH_3CH_2CH(\bullet)CH_2CH_2CH_2OH$	2×10^6	28.5 (SAR)		Atkinson (1997b)
		2.9×10^6	32.2	C	Méreau et al. (2003)
2-Hexoxy	$CH_3CH(\bullet)CH_2CH_2CH(OH)CH_3$	2×10^6	28.5 (SAR)		Atkinson (1997b)
		$(1.4–4.7) \times 10^6$			Eberhard et al. (1995)
		5×10^6	31.0	C	Méreau et al. (2003)
			25.0 (b3lyp)	A	Lin and Ho (2002)
2-Methyl-2-hexoxy	$CH_3CH(\bullet)CH_2CH_2C(OH)(CH_3)_2$	2×10^6	28.5 (SAR)		Atkinson (1997b)
		1.1×10^7	29.3	C	Méreau et al. (2003)
4-Methyl-1-pentoxy	$(CH_3)_2C\bullet CH_2CH_2CH_2OH$	5×10^6	23.0 (SAR)		Atkinson (1997b)
		1.1×10^8	21.0	C	Méreau et al. (2003)
5-Methyl-2-hexoxy	$(CH_3)_2C\bullet CH_2CH_2CH(OH)CH_3$	5×10^6	23.0 (SAR)		Atkinson (1997b)
		9.5×10^6	19.7[b]		Johnson et al. (2004a)
		5×10^8	15.5	C	Méreau et al. (2003)
			15.8(b3lyp)	A	Lin and Ho (2002)
2,5-Dimethyl-2-hexoxy	$(CH_3)_2C\bullet CH_2CH_2C(OH)(CH_3)_2$	5×10^6	23.0 (SAR)		Atkinson (1997b)
		4.4×10^8	16.3	C	Méreau et al. (2003)

[a] Determined from measurements of k_{isom} / k_{O2}, using $k_{O2} = 6 \times 10^{-14} \exp(-550/T)$ cm^3 molecule^{-1} s^{-1}.
[b] Determined from measurements of k_{isom} / k_{O2}, using $k_{O2} = 1.5 \times 10^{-14} \exp(-200/T)$ cm^3 molecule^{-1}s^{-1}.

$$\bullet OCH_2CH_2CH_2CH_2OH \rightarrow HOCH_2CH_2CH_2CH(\bullet)OH$$

$$HOCH_2CH_2CH_2CH(\bullet)OH + O_2 \rightarrow \mathbf{HOCH_2CH_2CH_2CHO + HO_2}$$

Note that in the standard case a second isomerization process, more rapid than the first, is thought to occur before the stable end-products are formed. The multifunctional nature of these end-products makes their detection and quantification difficult, a fact that has complicated the characterization of the isomerization reactions. In a series of experiments, Atkinson and co-workers (Atkinson et al., 1995; Kwok et al., 1996; Arey et al., 2001; Reisen et al., 2005) have employed an array of gas chromatographic and mass spectrometric techniques to first identify and later to quantify, and in some cases speciate, the 1,4-hydroxycarbonyl and 1,4-hydroxynitrate compounds formed in the OH-initiated oxidation of a number of C_4–C_8 straight- and branched-chain alkanes. The specifics of these data, which are extremely important both in terms of confirming the occurrence of the isomerization reactions and in assessing their overall impact on the alkane chemistry, will be discussed in more detail in section IV-C.

The earliest experimental data on the rates of the isomerization processes involved studies of the competition between isomerization of 2-pentoxy (Carter et al., 1976) and 1-butoxy radical (Carter et al., 1979; Cox et al., 1981; Niki et al., 1981b) and their reaction with O_2, e.g.,

$$CH_3CH_2CH_2CH_2O\bullet \rightarrow \bullet CH_2CH_2CH_2CH_2OH \qquad \text{(IV-B-4)}$$

$$CH_3CH_2CH_2CH_2O\bullet + O_2 \rightarrow CH_3CH_2CH_2CHO + HO_2 \qquad \text{(IV-B-5)}$$

Carter et al. (1976) noted a lower yield of 2-pentanone than 3-pentanone in the OH-initiated oxidation of pentane, consistent with a significant occurrence of isomerization for the 2-pentoxy radical. Note that 3-pentoxy cannot undergo isomerization via a 1,5-H shift. In the case of 1-butoxy, from observed yields of butanal versus O_2 partial pressure, $k_{IV\text{-}B\text{-}4}/k_{IV\text{-}B\text{-}5}$ values of about (1.5–2) $\times 10^{19}$ molecule cm^{-3} were established by a number of groups (Carter et al., 1979; Cox et al., 1981; Niki et al., 1981b). Similar studies were later conducted on 2-pentoxy and 2- and 3-hexoxy (Dobé et al., 1986; Atkinson et al., 1995; Eberhard et al,. 1995). From a combination of these room-temperature data and estimated *A*-factors (Baldwin et al., 1977), the following set of parameters was derived to describe the rate coefficients for the isomerization reactions at atmospheric pressure (Atkinson, 1997b):

$k(\text{primary}) = 1.6 \times 10^5 \text{s}^{-1}$ $k(\text{sec.}) = 1.6 \times 10^6 \text{s}^{-1}$ $k(\text{tert.}) = 4.0 \times 10^6 \text{s}^{-1}$

$A(\text{primary}) = 2.4 \times 10^{11} \text{s}^{-1}$ $A(\text{sec.}) = 1.6 \times 10^{11} \text{s}^{-1}$ $A(\text{tert.}) = 8.0 \times 10^{10} \text{s}^{-1}$

$E_a/R(\text{primary}) = 4240$ K $E_a/R(\text{sec.}) = 3430$ K $E_a/R(\text{tert.}) = 2745$ K

$F(CH_3) = 1.00$ $F(\text{alkyl}) = 1.27$ $F(OH) = 4.3$

Here, the k, A, and E_a/R values are the 298 K rate coefficients, A-factors and activation temperatures for isomerization involving the 1,5-transfer of a hydrogen from a $—CH_3$ (primary), $—CH_2—$ (secondary), or $>CH—$ (tertiary) group. The multiplicative F-factors account for the presence of the various substituents (methyl, alkyl, hydroxy) at the abstraction site.

A number of theoretical and experimental studies of the isomerization reactions have appeared since 1996 (Heiss and Sahetchian, 1996; Jungkamp et al., 1997; Lendvay and Viskolcz, 1998; Somnitz and Zellner, 2000a,b; Méreau et al., 2000a; Hack et al., 2001; Lin and Ho, 2002; Vereecken and Peeters, 2003; Méreau et al., 2003; Ferenac et al., 2003; Johnson et al., 2004a; Cassanelli et al., 2005, 2006). Most of these address the temperature dependence of the rate parameters; see table IV-B-2. For the 1-butoxy isomerization, activation energies or energy barriers derived from theory vary between 33 and 42 kJ mol^{-1} (Jungkamp et al., 1997; Lendvay and Viskolcz, 1998; Méreau et al., 2000a; Somnitz and Zellner, 2000a,b; Lin and Ho, 2002; Vereecken and Peeters, 2003; Ferenac et al., 2003), while estimated A-factors at 1 atm pressure fall between about 10^{11} and $10^{12} s^{-1}$. However, as discussed in detail by Vereecken and Peeters (2003), these studies neglect the contribution of different rotamers of both the 1-butoxy reactant and the transition state to isomerization and often neglect the effects of tunneling, and thus may be subject to significant but potentially canceling errors. The treatment of Vereecken and Peeters (2003), which seems the most complete, gives $k_{IV\text{-}B\text{-}4} = 1.4 \times 10^{11} \exp(-4100/T)$, in very good agreement with the Atkinson (1997b) recommendation. However, the end-product data of Cox and co-workers (Cassanelli et al., 2005, 2006), obtained using either a slow-flow system or an environmental chamber, yields somewhat different parameters ($A \approx 2 \times 10^{10} s^{-1}$, $E_a/R \approx 3520K$).

Studies of the isomerizations of larger and, in some cases, more substituted radicals have also been conducted since the year 2000 (Somnitz and Zellner, 2000a,b; Lin and Ho, 2002; Méreau et al., 2003; Johnson et al., 2004a). The data show the expected decrease in activation energy (or barrier heights) for isomerization involving H-transfer from a primary, secondary, or tertiary group, in support of the Atkinson parameterizations. For abstraction from a secondary site, e.g., 1-pentoxy, 1-hexoxy, and 2-hexoxy radical isomerizations, the data center on activation energies of about 25–33 kJ mol^{-1} and 298 K rate coefficients of (2–5) $\times 10^6 s^{-1}$, in fairly good agreement with the Atkinson parameterization (Somnitz and Zellner, 2000b; Lin and Ho, 2002; Johnson et al., 2004a; Méreau et al., 2003). With regard to isomerizations involving transfer of a tertiary hydrogen, the picture is less clear. Theoretical studies of Méreau et al. (2003) and Lin and Ho (2002) find much lower activation energies (or barrier heights), 15–21 kJ mol^{-1}, than are implied by the Atkinson parameterization, and consequently much faster rate coefficients [e.g., $> 10^8 s^{-1}$, compared with $5 \times 10^6 s^{-1}$ given by Atkinson (1997b)]. The end-product study of Johnson et al. (2004a) for 5-methyl-2-hexoxy isomerization gives an intermediate result, $E_a = 21$ kJ mol^{-1} with $k = 1 \times 10^7 s^{-1}$ near 300 K.

It is apparent that more systematic studies of these isomerization reactions will be required before firm conclusions can be made regarding their rates, or comprehensive structure–additivity relationships can be constructed. Méreau et al. (2003) suggested revisions to the Atkinson parameterization (1997b) that result in larger rate coefficients for all types of radicals. Johnson et al. (2004a) noted a correlation between the isomerization rate coefficient and the bond energy of the hydrogen being abstracted; although this was based on only three isomerization data points, data for the analogous reaction of alkoxy radicals with alkanes were also included in the development of the correlation. Finally, as first noted by Somnitz and Zellner (2000b), decreases in activation energy due to substitution on the carbon atom at the C—O• site are seen in some cases (Somnitz and Zellner, 2000b; Lin and Ho, 2002; Méreau et al, 2003), and these effects have yet to be captured in structure–additivity relationships.

Given that a great deal more data are now available regarding the Arrhenius parameters for the isomerization reactions, it is instructive to derive a structure–additivity relationship using the most current data. Figure IV-B-3 gives calculated activation energies, $E_a(\text{kJ mol}^{-1}) = 37.7 - 8.8 \times N_{\text{alk,abs}} - 2.0 \times N_{\text{alk,oxy}}$ for radicals of structure $HC(R)(R)CH_2CH_2C(R')(R')O\bullet$, where $N_{\text{alk,abs}}$ refers to the number of alkyl substituents (R-groups) at the abstraction site and $N_{\text{alk,oxy}}$ refers to the number of alkyl substituents (R′-groups) at the oxy radical site. The parameters (37.7, 8.8, 2.0) were obtained from a least-squares fit to the data presented in table IV-B-2. Most of the scatter is from the wide range of 1-butoxy data points (calculated E_a = 36 kJ mol^{-1}, "measured" values between 27 and 41 kJ mol^{-1}). Some of the scatter is due also to the interchangeable use of calculated barrier heights and activation energies at infinite pressure and E_a at 1 atm. In any event, it is clear that there is a separation of the data

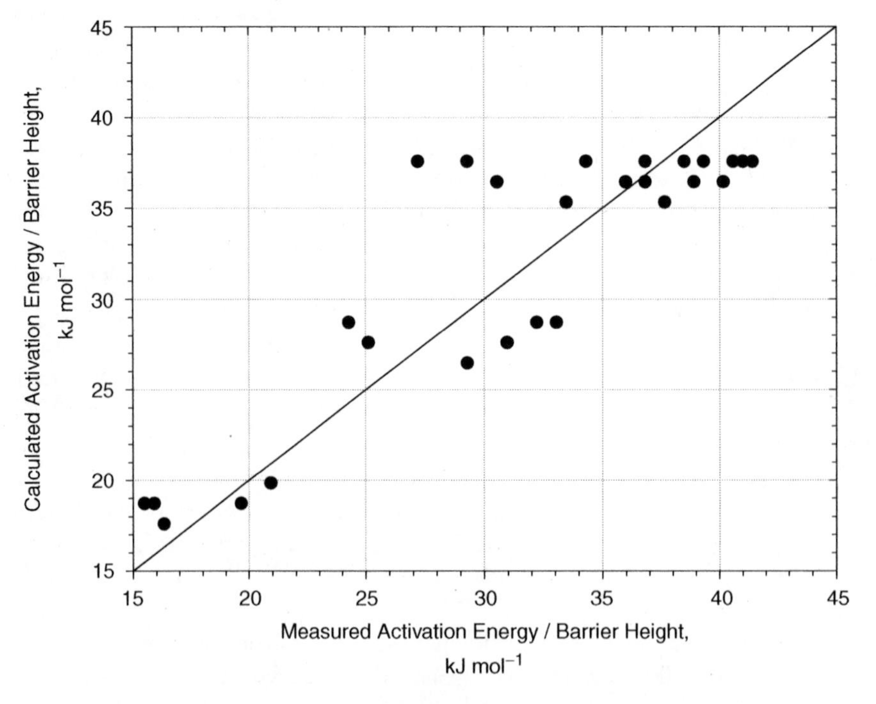

Figure IV-B-3. Comparison of calculated (using the structure–additivity method described in the text) versus measured activation energies (or energy barriers) for isomerization of alkoxy radicals.

into abstraction from CH_3 groups (activation energy 29–42 kJ mol^{-1}), CH_2 groups (25–33 kJ mol^{-1}), and CH groups (17–21 kJ mol^{-1}).

IV-B-2.4. Chemical Activation in the Chemistry of Alkoxy Radicals

The unimolecular decomposition of some alkoxy radicals is known to be influenced by a so-called "chemical activation" effect (e.g., Wallington et al., 1996b; Orlando et al., 1998, 2003b). The origin of the effect is as follows. Reaction of a typical alkylperoxy radical with NO to form an alkoxy radical and NO_2 is an exothermic process, $\Delta H_{rxn} \sim -63$ kJ mol^{-1}, and much of this excess energy is imparted to the alkoxy radical fragment. In cases where the energy barrier to unimolecular decomposition of the alkoxy radical is reasonably low, a significant fraction of the nascent alkoxy radicals can actually possess sufficient energy to overcome the barrier, allowing for their "prompt", non-thermal decomposition. The remainder of the alkoxy radicals are thermalized by collisions with the bath gas, and then undergo unimolecular reaction or reaction with O_2 on slower, thermal timescales.

While the chemical activation effect has been shown to be prevalent for many halogen- and oxygen-substituted alkoxy radicals (Orlando et al., 2003b), it was commonly believed that barriers to decomposition of the unsubstituted (i.e., alkane-derived) radicals were too high (E_a typically greater than 50 kJ mol^{-1}) for significant chemical activation to occur. However, there are some recent reports, both experimental (Libuda et al., 2002; Geiger et al., 2002; Cassanelli et al., 2005) and theoretical (Caralp and Forst, 2003), that suggest measurable occurrence of chemical activation in the chemistry of some unsubstituted alkoxy radicals. Libuda et al. (2002), in environmental chamber studies of 2-butoxy chemistry,

$$CH_3CH_2CH(O\bullet)CH_3 \rightarrow CH_3CH_2\bullet + CH_3CHO \qquad \text{(IV-B-2)}$$

$$CH_3CH_2CH(O\bullet)CH_3 + O_2 \rightarrow CH_3CH_2C(O)CH_3 + HO_2 \qquad \text{(IV-B-3)}$$

noted that the *apparent* rate coefficient ratio $k_{IV\text{-}B\text{-}2}/k_{IV\text{-}B\text{-}3}$ increased with increasing O_2 partial pressure. This apparent increase is exactly what would be expected if chemical activation were occurring but is not incorporated into the analysis; the decomposition reaction (IV-B-2) appears increasingly more rapid as more O_2 is present, as the O_2 reaction is unable to compete with prompt decomposition of the activated radicals. The authors estimated that prompt decomposition of 2-butoxy was occurring with a yield of 6–10% for temperatures from 280 to 313 K. Cassanelli et al. (2005) reached a similar conclusion for 2-butoxy, reporting ~5% prompt decomposition at 283 K and ~9% at 298 K. Similar results were reported by Geiger et al. (2002) in studies of the chemistry of 2-methyl-1-propoxy (*iso*-butoxy), 17% prompt decomposition, and 3-pentoxy, 14% prompt decomposition. Geiger et al. (2002) also estimated that prompt isomerization of the 1-butoxy radical occurred with a yield of about 21%. Theoretical studies of these same radicals (Caralp and Forst, 2003) led to similar estimates of the extent of the occurrence of prompt decomposition (2-butoxy, 8%; 2-methyl-1-propoxy, 12%; 3-pentoxy, 5%) or isomerization (1-butoxy, 18%).

While these studies clearly point to the likelihood of a measurable influence of chemical activation in the chemistry of some alkane-derived alkoxy radicals, more work will be required to fully quantify this effect for the entire suite of atmospherically relevant decomposition and isomerization processes. There are a number of competing factors that will play a role in determining the extent of activation for a particular process. Clearly, lower barrier processes will be more prone to activation than higher barrier processes. Also, there will be an anti-correlation between the size of the radical and the extent of chemical activation; for radicals containing perhaps six carbons or more, the RRKM lifetimes of the excited radicals or of the peroxynitrite intermediate involved in their production is likely to become sufficiently long that collisional deactivation will begin to compete effectively with prompt decomposition. Finally, the *A*-factor for the unimolecular process will have an effect. For example, collisional deactivation will likely be more competitive with the lower *A*-factor isomerization processes than with the higher *A*-factor dissociation processes. Interesting candidates for further study of chemical activation include the neopentoxy radical, where the activation energy for dissociation is estimated to be only about 42 kJ mol^{-1} (Orlando et al., 2003b), or radicals such as 2-hexoxy or 4-methyl-1-pentoxy, where activation energies to isomerization are likely ~30 kJ mol^{-1} (e.g., Méreau et al., 2003).

It is also important to point out that the occurrence of chemical activation, if verified, would have a non-negligible effect on some previously measured rate coefficient data for unimolecular alkoxy radical reactions. As will become apparent in the following discussion, alkoxy radicals (RO•) are prepared in the laboratory in three ways: via the photolysis of organic nitrites (RONO), via reaction of alkylperoxy radicals with like radicals ($RO_2 + RO_2$), or via reaction of peroxy radicals (RO_2) with NO. In general, peroxy radical–peroxy radical reactions are thermoneutral or even slightly endothermic and thus impart little energy to the alkoxy product. Hot alkoxy effects from organic nitrite photolysis also appear to be minimal, perhaps 10% or less (e.g., Johnson et al., 2004a; Cassanelli et al., 2005). Thus, the largest effects are likely to be seen in studies which involve alkoxy radical formation via the $RO_2 + NO$ reaction; for example, in the case of 2-butoxy radical alluded to earlier (Carter et al., 1979; Cox et al., 1981; Drew et al., 1985; Libuda et al., 2002; Meunier et al., 2003; Cassanelli et al., 2005). Here, it can be shown that the occurrence of about 10% prompt decomposition would mean that observed (effective) values of $k_{IV\text{-}B\text{-}2} / k_{IV\text{-}B\text{-}3}$ (which are typically about 4×10^{18} molecule cm^{-3}) derived from end-product data in air at 298 K would overestimate the true rate coefficient ratio by ~20%. The overestimation would be large for slower decomposition processes and smaller for faster decomposition processes, for the same extent of prompt decomposition.

At present, there is insufficient quantitative information available regarding the extent of occurrence of prompt decomposition or isomerization to justify the application of corrections to the previously measured rate coefficient data; in the sections that follow, data are presented as measured. Thus, rate coefficient ratios reported for species such as 1-butoxy (Carter et al., 1979; Cox et al., 1981), 2-butoxy (Carter et al., 1979; Cox et al., 1981; Drew et al., 1985; Meunier et al., 2003), 2- and 3-pentoxy (e.g., Atkinson et al., 1995), 2- and 3-hexoxy (e.g., Eberhard et al., 1995), derived from studies of the OH-initiated oxidation of the parent alkane in the presence of NO_x, may represent effective rate coefficient ratios in air rather than the true thermalized values.

A summary of the available rate data regarding the atmospherically significant reactions of the individual alkoxy radicals (with O_2, unimolecular dissociation and isomerization) is given in the following sections. The review will focus on the more recent (and direct) measurements of the alkoxy radical rate coefficients when available. However, we note that a number of ground-breaking studies of alkoxy radical chemistry were carried out during the 1970s and 1980s, particularly by Batt and co-workers, Heicklen and co-workers, and Benson and co-workers (e.g., Wiebe et al., 1973; Glasson, 1975; Mendenhall et al., 1975; Batt and McCulloch, 1976; Batt and Milne, 1976; Barker et al., 1977; Batt and Milne, 1977a,b; Batt et al., 1978; Batt, 1979; Batt and Robinson, 1979, 1982; Zabarnick and Heicklen, 1985a,b,c; Batt et al., 1989). Although these older studies were usually of an indirect nature, and often involved fairly complex mechanisms that are difficult to re-interpret, they were instrumental in providing, at the very least, a semi-quantitative picture of alkoxy radical chemistry and formed the knowledge base for alkoxy radical reactivity through the 1990s (e.g., see Atkinson, 1997b).

IV-B-3. Methoxy Radical ($CH_3O\bullet$)

The predominant fate of methoxy radical in the atmosphere is reaction with O_2:

$$CH_3O\bullet + O_2 \rightarrow CH_2O + HO_2$$

Thermal decomposition via H-atom elimination is expected to be exceedingly slow, $k < 1\ s^{-1}$, and will not occur under atmospheric conditions.

A number of direct studies of the rate coefficient for the O_2 reaction have been made at temperatures between 298 and 973 K using the laser flash-photolysis–laser-induced fluorescence method (Sanders et al., 1980a,b; Gutman et al., 1982; Lorenz et al., 1985; Wantuck et al., 1987). Data are summarized in table IV-B-3 and plotted in figure IV-B-4. While only an upper limit was reported by Sanders et al. (1980a,b), the latter three of these studies are in substantive agreement. On the basis of these three studies, we recommend $k = 7.37 \times 10^{-14} \exp(-1143/T)$ $cm^3\ molecule^{-1} s^{-1}$ over the range 298–610 K, with $k = 1.6 \times 10^{-15}\ cm^3\ molecule^{-1} s^{-1}$ at 298 K. The measurements of Wantuck et al. (1987), which extend from 298 K to 973 K, show substantial upward curvature in the rate coefficient above 600 K, a result that has been attributed to tunneling effects (Setokuchi and Sato, 2002).

There has also been a number of relative rate studies of the $CH_3O\bullet + O_2$ reaction (Wiebe et al., 1973; Glasson, 1975; Mendenhall et al., 1975; Alcock and Mile, 1977; Barker et al., 1977; Batt and Robinson, 1979; Cox et al., 1980a; Kirsch and Parkes, 1981). These generally involved the investigation of the reaction of methoxy with O_2 occurring in competition with its reaction with NO or NO_2. As summarized in Orlando et al. (2003b), these experiments are often rather indirect and involve an analysis of fairly complex chemistry. Nonetheless, the bulk of these data are in reasonable

Table IV-B-3. Rate coefficients for reaction of methoxy radicals with O_2, with $k = A \times \exp(-B/T)$

A (cm^3 molecule^{-1}s^{-1})	B (K)	k (cm^3 molecule^{-1} s^{-1})	T (K)	Technique [a]	Reference	Temperature Range (K)
		$< 1.7 \times 10^{-15}$	298	FP–LIF	Sanders et al. (1980a) Sanders et al. (1980b)	
1.0×10^{-13}	1308	4.6×10^{-15}	413	FP–LIF	Gutman et al. (1982)	413–628
		6.0×10^{-15}	475			
		1.1×10^{-14}	563			
		1.3×10^{-14}	608			
$(5.5 \pm 2.0) \times 10^{-14}$	1000	1.51×10^{-15}	298	FP–LIF [b]	Lorenz et al. (1985)	298–450
		2.29×10^{-15}	332			
		2.95×10^{-15}	380			
		4.17×10^{-15}	450			
		$(2.3 \pm 0.2) \times 10^{-15}$	298	FP–LIF	Wantuck et al. (1987)	298–973
		$(1.9 \pm 0.2) \times 10^{-15}$	298			
		$(4.0 \pm 0.2) \times 10^{-15}$	348			
		$(4.7 \pm 0.3) \times 10^{-15}$	423			
		$(4.4 \pm 0.1) \times 10^{-15}$	423			
		$(4.6 \pm 0.2) \times 10^{-15}$	423			
		$(7.2 \pm 0.6) \times 10^{-15}$	498			
		$(1.4 \pm 0.1) \times 10^{-14}$	573			
		$(1.2 \pm 0.3) \times 10^{-14}$	573			
		$(1.4 \pm 0.1) \times 10^{-14}$	573			
		$(3.0 \pm 0.1) \times 10^{-14}$	673			
		$(6.7 \pm 0.4) \times 10^{-14}$	773			
		$(1.4 \pm 0.1) \times 10^{-13}$	873			
		$(3.6 \pm 1.1) \times 10^{-13}$	973			

[a] Abbreviations for the experimental techniques employed are defined in table II-A-1.
[b] Data shown are estimated from graphs presented in the original reference.

agreement (a factor of two or better) with the direct studies summarized in the previous discussion (Orlando et al., 2003b).

Since reaction with O_2 is essentially the sole fate of the methoxy radical throughout the atmosphere, and this reaction is not rate-limiting in any way, the exact rate coefficient for the reaction is not critical for atmospheric modeling. Nonetheless, measurements of this rate coefficient would be worthwhile at temperatures below 298 K, a region yet to be studied. Confirmation of the curvature observed by Wantuck et al. (1987) above 600 K would also be valuable from a theoretical standpoint (Orlando et al., 2003b).

IV-B-4. Ethoxy Radical, ($C_2H_5O\bullet$)

Thermal decomposition of the ethoxy radical, which occurs via CH_3 elimination,

$$CH_3CH_2O\bullet \rightarrow CH_3\bullet + CH_2O$$

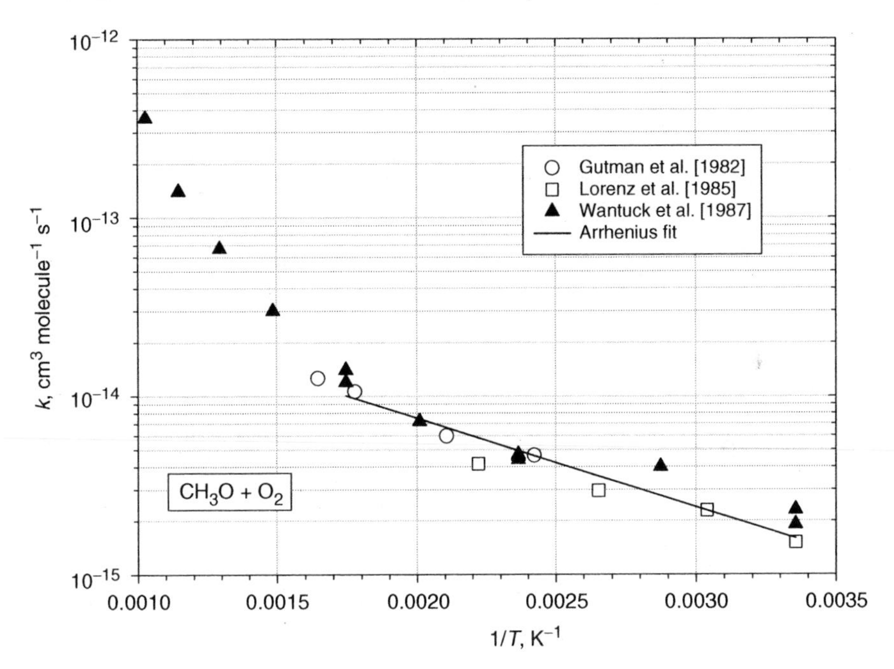

Figure IV-B-4. Direct measurements of the rate coefficient for reaction of methoxy radical with O_2. The figure has been adapted from Orlando et al. (2003b) with the permission of the American Chemical Society.

is a slow process. The reaction has been shown in a variety of experimental and theoretical studies to have an activation energy of 74.9–78.2 kJ mol^{-1}, with $k \sim 3$ s^{-1} at 298 K, 1 atm. pressure (Caralp et al., 1999; Méreau et al., 2000b; Somnitz and Zellner, 2000b; Fittschen et al., 2000).

Thus, the atmospheric chemistry of the ethoxy radical ($C_2H_5O\bullet$) is dominated by its reaction with O_2:

$$CH_3CH_2O\bullet + O_2 \rightarrow CH_3CHO + HO_2$$

Absolute studies of the O_2 reaction have been carried out by four groups (Gutman et al., 1982; Hartmann et al., 1990; Fittschen et al., 1999; Devolder, 2003); laser flash-photolysis–laser-induced fluorescence or discharge flow–laser-induced fluorescence methods were used. Collectively, the data set covers the range 245–411 K. Data are shown in table IV-B-4 and plotted in figure IV-B-5. A combined fitting of the three earliest data sets (Gutman et al., 1982; Hartmann et al., 1990; Fittschen et al., 1999), conducted by Orlando et al. (2003b), led to an Arrhenius expression of $k = \left(5.13^{+9.70}_{-3.32}\right) \times 10^{-14}$ exp[$-(550 \pm 350)/T$] cm^3 molecule^{-1} s^{-1} over the temperature range 280–420 K, with $k = 8.1 \times 10^{-15}$ cm^3 molecule^{-1} s^{-1} at 298 K. This expression is essentially identical to that recently reported by Devolder (2003), $k = 4.85 \times 10^{-14}\exp(-530/T)$ cm^3 molecule^{-1}s^{-1} for $T = 245$–400 K.

Table IV-B-4. Rate coefficients (k, cm^3 $molecule^{-1}$ s^{-1}) for reaction of ethoxy radicals with O_2, with $k = A \times \exp(-B/T)$

A ($cm^3 molecule^{-1} s^{-1}$)	B (K)	$10^{15} \times k$	T (K)	Technique [a]	Reference	Temperature Range (K)
		8.0	296	FP–LIF	Gutman et al. (1982)	296–353
		9.8	353			
$(7.1 \pm 0.7) \times 10^{-14}$	552 ± 64	10.8 ± 2.0	295	FP–LIF	Hartmann et al. (1990)	295–411
		13.0 ± 2.4	328			
		15.0 ± 3.0	348			
		16.1 ± 3.5	373			
		18.2 ± 1.4	411			
$(2.4 \pm 0.9) \times 10^{-14}$	325	8.5	286	FP–LIF	Fittschen et al. (1999)	295–354
		7.5	292			
		7.3	295			
		8.3	310			
		6.7	316			
		8.0	316			
		7.2	332			
		10.8	354			
		9.1	354			
		7.4	355			
		10.6	373			
		10.2	390			
4.85×10^{-14}	530			DF–LIF	Devolder (2003)	245–400

[a] Abbreviations for the experimental techniques employed are defined in table II-A-1.

A relative rate study by Zabarnick and Heicklen (1985a), conducted between 225 and 393 K using the reaction of ethoxy with NO as the reference reaction, provides data that are consistent with the absolute rate coefficients (Orlando et al., 2003b).

IV-B-5. 1-Propoxy Radical (n-Propoxy, n-$C_3H_7O\bullet$)

Thermal decomposition of the 1-propoxy radical via elimination of ethyl radical,

$$CH_3CH_2CH_2O\bullet \rightarrow CH_3CH_2\bullet + CH_2O$$

has been characterized by theoretical means (Méreau et al., 2000b; Somnitz and Zellner, 2000a,b; Fittschen et al., 2000). The reaction has an activation energy of $\sim 65.7 \pm 2.0$ kJ mol^{-1}, occurs with a rate coefficient of about 300 s^{-1} at 298 K and 1 atm. total pressure, and thus is too slow (by about two orders of magnitude) to compete under atmospheric conditions with the O_2 reaction:

$$CH_3CH_2CH_2O\bullet + O_2 \rightarrow CH_3CH_2CHO + HO_2$$

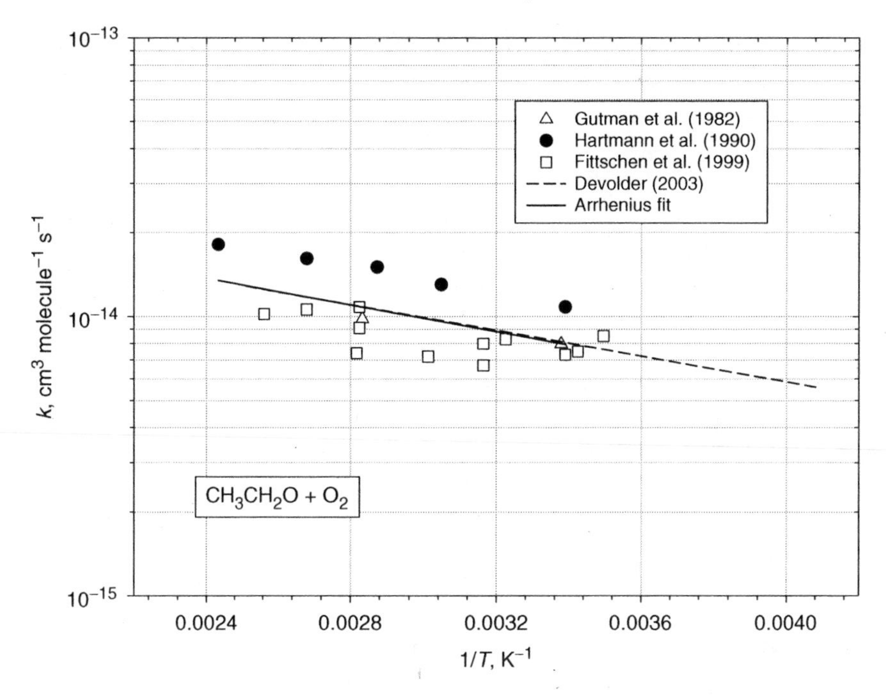

Figure IV-B-5. Direct measurements of the rate coefficient for reaction of ethoxy radical with O_2.

There are two direct mesurements of the O_2 reaction rate coefficient (Mund et al., 1998; Fittschen et al., 1999), both conducted using laser flash-photolysis–laser-induced fluorescence. These data are summarized in table IV-B-5 and shown in figure IV-B-6. Combining the data sets from these two studies (Orlando et al., 2003b) leads to $k = (2.38^{+1.03}_{-0.72} \times 10^{-14})\,\exp(-230 \pm 100/T)$ cm^3 molecule^{-1}s^{-1} over the range 220–400 K, with $k = 1.1 \times 10^{-14}$ cm^3 molecule^{-1}s^{-1} at 298 K. A relative rate study of k_{O2} by Zabarnick and Heicklen (1985b) is in reasonable accord with this recommendation (within a factor of 1.5–3).

Meunier et al. (2003), in an environmental chamber study, examined product distributions obtained from the photolysis of 1-propyl nitrite at different O_2 partial pressures. Their data lead to a rate coefficient ratio $k_{decomp}/k_{O2} = (3.8 \pm 0.4) \times 10^{16}$ molecule cm^{-3}, in reasonable agreement with the rates of the individual reactions.

IV-B-6. 2-Propoxy Radical (Isopropoxy Radical, $CH_3CH(O\bullet)CH_3$)

The thermal decomposition of 2-propoxy radical,

$$CH_3CH(O\bullet)CH_3 \rightarrow CH_3CHO + CH_3\bullet$$

has been studied using both direct experimental (Balla et al., 1985; Devolder et al., 1999) and theoretical approaches (Méreau et al., 2000b; Somnitz and Zellner, 2000a,b; Fittschen et al., 2000). Reported activation energies range from 63.6 to 68.6 kJ mol^{-1},

Table IV-B-5. Rate coefficients (k, cm^3 molecule^{-1} s^{-1}) for reaction of 1-propoxy radicals with O_2, with $k = A \times \exp(-B/T)$

A (cm^{3}molecule^{-1}s^{-1})	B (K)	$10^{14} \times k$	T (K)	Technique [a]	Reference	Temperature Range (K)
$(1.4 \pm 0.6) \times 10^{-14}$	108 ± 50	1.02	218	FP–LIF [b]	Mund et al. (1998)	218–315
		0.90	229		Mund et al. (1999)	
		0.89	240		(correction)	
		0.84	252			
		0.89	265			
		1.00	276			
		1.00	288			
		0.99	308			
		1.01	315			
$(2.5 \pm 0.5) \times 10^{-14}$	240	1.07	289	FP–LIF	Fittschen et al. (1999)	289–381
		1.09	308			
		1.08	308			
		1.31	333			
		1.32	354			
		1.20	366			
		1.65	381			

[a] Abbreviations for the experimental techniques employed are defined in table II-A-1.
[b] Data shown are estimated from graphs presented in the original reference.

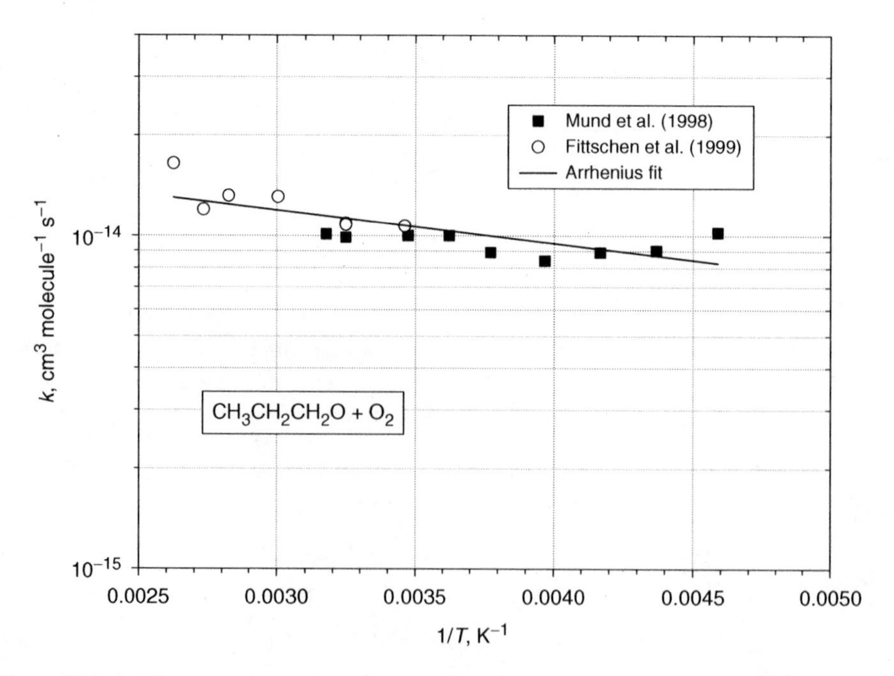

Figure IV-B-6. Direct measurements of the rate coefficient for reaction of 1-propoxy radical with O_2.

and measured k_{decomp} at 298 K and 1 atm. air vary from about 150 s^{-1} to about 600 s^{-1}. The dissociation reaction is thus about two orders of magnitude too slow to compete with the O_2 reaction:

$$CH_3CH(O\bullet)CH_3 + O_2 \rightarrow CH_3C(O)CH_3 + HO_2$$

Reaction of 2-propoxy with O_2 has been studied directly via the laser flash-photolysis/laser-induced fluorescence technique at temperatures between 200 and 400 K (Balla et al., 1985; Mund et al., 1998; Fittschen et al., 1999; Deng et al., 2000). Data are summarized in table IV-B-6 and shown in figure IV-B-7. On the basis of these, studies Orlando et al. (2003b) recommended $k_{O2} = (1.95^{+0.52}_{-0.40}) \times 10^{-14} \exp(-310 \pm 70/T)$ cm^3 molecule^{-1}s^{-1} between 200 and 400 K, with $k_{O2} = 6.8 \times 10^{-15}$ cm^3 molecule^{-1}s^{-1} at 298 K.

Table IV-B-6. Rate coefficients (k, cm^3 molecule^{-1} s^{-1}) for reaction of 2-propoxy radicals with O_2, with $k = A \times \exp(-B/T)$

A (cm^{3}molecule^{-1}s^{-1})	B (K)	$10^{15} \times k$	T (K)	Technique [a]	Reference	Temperature Range (K)
$(1.51 \pm 0.70) \times 10^{-14}$	196 ± 141	6.90 ± 0.37	296	FP–LIF	Balla et al. (1985)	295–384
		7.06 ± 0.56	295			
		8.22 ± 0.28	296			
		7.55 ± 0.30	294			
		8.65 ± 0.26	314			
		7.48 ± 0.55	330			
		8.60 ± 0.36	346			
		9.16 ± 0.92	367			
		8.27 ± 0.80	384			
$(1.0 \pm 0.3) \times 10^{-14}$	217 ± 48	4.79	212	FP–LIF[b]	Mund et al. (1998)	212–327
		5.25	233			
		5.75	253			
		5.50	263			
		6.32	277			
		6.24	290			
		6.31	300			
		6.32	327			
$(1.6 \pm 0.2) \times 10^{-14}$	265	5.9	288	FP–LIF	Fittschen et al. (1999)	288–364
		6.5	298			
		6.6	307			
		7.1	316			
		6.9	316			
		7.5	331			
		7.9	340			
		8.0	354			
		9.0	364			
		8.73 ± 2.46	296	FP–LIF	Deng et al. (2000)	296–330
		6.73 ± 1.92	314			
		7.45 ± 0.96	330			

[a] Abbreviations for the experimental techniques employed are defined in table II-A-1.
[b] Data shown are estimated from graphs presented in the original reference.

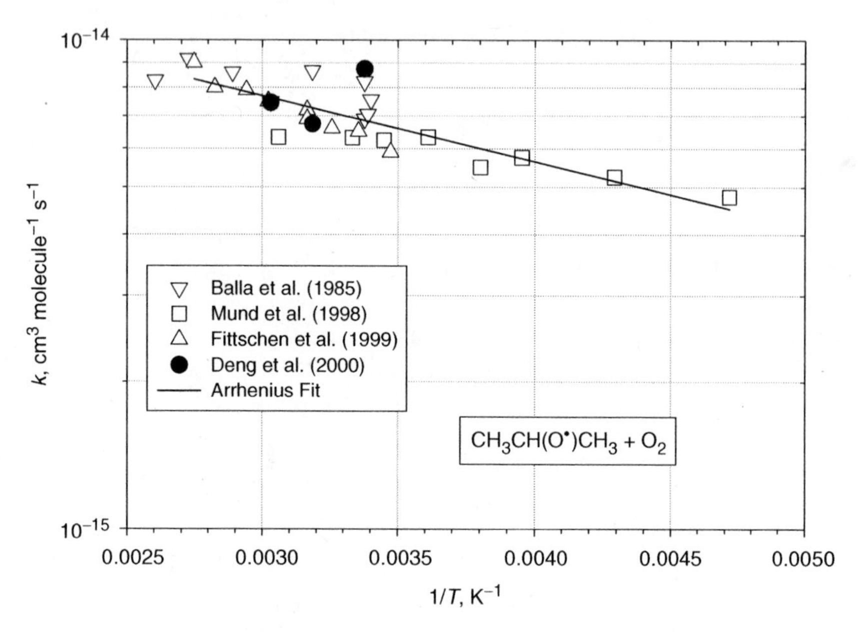

Figure IV-B-7. Direct measurements of the rate coefficient for reaction of 2-propoxy radical with O_2.

Meunier et al. (2003) report $k_{decomp}/k_{O2} = (2.9 \pm 0.3) \times 10^{16}$ molecule cm^{-3}, on the basis of end-product studies conducted from the photolysis of 2-propyl nitrite in an environmental chamber over a range of O_2 partial pressures; this result is in reasonable accord with the data for the individual reactions.

IV-B-7. 1-Butoxy Radical (*n*-Butoxy Radical, $CH_3CH_2CH_2CH_2O\bullet$)

As discussed in section IV-B-2.3, 1-butoxy is the simplest radical that can isomerize via a 1,5 hydrogen shift:

$$CH_3CH_2CH_2CH_2O\bullet \rightarrow \bullet CH_2CH_2CH_2CH_2OH$$

Thermal decomposition and reaction with O_2 are also feasible processes:

$$CH_3CH_2CH_2CH_2O\bullet \rightarrow CH_3CH_2CH_2\bullet + CH_2O$$

$$CH_3CH_2CH_2CH_2O\bullet + O_2 \rightarrow CH_3CH_2CH_2CHO + HO_2$$

Theoretical studies (Méreau et al., 2000b; Somnitz and Zellner, 2000a,b; Fittschen et al., 2000) have shown that the dissociation reaction is quite slow, with an activation energy of about 63–68 kJ mol^{-1}, and that it occurs with a first-order rate coefficient of approximately 100 s^{-1} at 298 K and 1 atm. air. Thus the reaction is unimportant throughout the atmosphere.

The reaction of 1-butoxy with O_2 is not as well characterized as is the case for the smaller alkoxy radicals. Hein et al. (1999) monitored NO_2 and OH time profiles following flash photolysis of 1-bromobutane/O_2/NO mixtures and determined $k = (1.4 \pm 0.7) \times 10^{-14}$ cm^3 molecule^{-1}s^{-1} at 293 K. Morabito and Heicklen (1987) determined k_{decomp} and k_{O2} relative to the reaction of 1-butoxy with NO. Using the currently recommended value for this reaction, a value of $k_{O2} = (1.3 \pm 0.1) \times 10^{-14}$ cm^3 molecule^{-1}s^{-1} is obtained at room temperature (Orlando et al., 2003b), in agreement with Hein et al. (1999). Morabito and Heicklen (1987) also reported a positive temperature dependence to k_{O2} with values ranging from 8.7×10^{-15} to 3.7×10^{-14}cm^3 molecule^{-1}s^{-1} for temperatures between 265 and 393 K. Given the indirect nature of the data available, a rate coefficient recommendation is not warranted currently.

As summarized in table IV-B-2 and section IV-B-2.3, a significant amount of effort, both theoretical and experimental, has been expended on the study of the 1-butoxy isomerization reaction (Carter et al., 1976, 1979; Cox et al, 1981; Niki et al., 1981b; Morabito and Heicklen, 1987; Heiss and Sahetchian, 1996; Jungkamp et al., 1997; Lendvay and Viskolcz, 1998; Méreau et al., 2000a; Somnitz and Zellner, 2000a,b; Hack et al., 2001; Geiger et al., 2002; Vereecken and Peeters, 2003; Ferenac et al., 2003). A consensus value of $k_{isom}/k_{O2} = (1.8 \pm 0.4) \times 10^{19}$ molecule^{-1} cm^{-1} has been reached, indicating that isomerization will account for $(78 \pm 5)\%$ of the 1-butoxy chemistry in 1 atm. air at 298 K, while O_2 reaction will account for the remainder. Because of the reasonably low barrier (Morabito and Heicklen, 1987; Heiss and Sahetchian, 1996; Jungkamp et al., 1997; Lendvay and Viskolcz, 1998; Somnitz and Zellner, 2000b; Hack et al., 2001; Lin and Ho, 2002; Vereecken and Peeters, 2003; Méreau et al., 2003; Ferenac et al., 2003; Cassanelli et al., 2005, 2006), the isomerization will remain competitive with the O_2 reaction at the reduced temperatures of the free and upper troposphere. For example, the experimental data of Cassanelli et al. (2005, 2006) imply that isomerization will account for about 30% of the 1-butoxy chemistry at 12 km altitude, $T = 215$ K.

IV-B-8. 2-Butoxy Radical (*sec*-Butoxy, $CH_3CH(O\bullet)CH_2CH_3$)

The chemistry of the 2-butoxy radical is characterized by a competition between reaction with O_2 and thermal decomposition:

$$CH_3CH_2CH(O\bullet)CH_3 \rightarrow CH_3CH_2\bullet + CH_3CHO$$

$$CH_3CH_2CH(O\bullet)CH_3 + O_2 \rightarrow CH_3CH_2C(O)CH_3 + HO_2$$

Thermal decomposition via methyl radical elimination has been shown to be insignificant via both experimental and theoretical methodologies (e.g., Drew et al., 1985; Fittschen et al., 2000; Mereau et al., 2000b; Somnitz and Zellner, 2000 a,b).

Reaction of 2-butoxy with O_2 has been studied directly via flash photolysis/LIF by three groups (Deng et al., 2000, 2001; Lotz and Zellner, unpublished data quoted in Orlando et al., 2003b; Falgayrac et al., 2004), and in a less direct fashion by Hein et al. (1998). The data are all in qualitative accord with room-temperature rate coefficients, falling in the range (6.5–13) $\times 10^{-15}$ cm^3 $molecule^{-1}s^{-1}$; see table IV-B-7. The data from Deng et al. (2001) [which supersedes their earlier study, Deng et al. (2000)] show a slight negative temperature dependence, $k = (1.2 \pm 1.0) \times 10^{-15} \exp[(550 \pm 190)/T]$ cm^3 $molecule^{-1}s^{-1}$, whereas those of Lotz and Zellner indicate the opposite temperature trend, $k = (2.3 \pm 1.0) \times 10^{-14} \exp[(-170 \pm 100)/T]$ cm^3 $molecule^{-1}s^{-1}$. Given the factor of $\sim$2 scatter in the room temperature data and the conflicting temperature dependences reported by the various authors, no firm recommendation for this rate coefficient can be given (Orlando et al., 2003b). Nonetheless, a value near 1.2×10^{-14} cm^3 $molecule^{-1}s^{-1}$, roughly independent of temperature, provides a reasonable ($\pm$40%) representation of the data obtained in the more direct studies (Deng et al., 2001; Lotz and Zellner, unpublished data quoted in Orlando et al, 2003b; Falgayrac et al., 2004).

Table IV-B-7. Rate coefficients (k, cm^3 $molecule^{-1}$ s^{-1}) for reaction of 2-butoxy radicals with O_2, with $k = A \times \exp(-B/T)$

A ($cm^3 molecule^{-1}s^{-1}$)	B (K)	$10^{15} \times k$	T (K)	Technique [a]	Reference	Temperature Range (K)
$(1.2 \pm 1.0) \times 10^{-15}$	-550 ± 190			FP–LIF	Deng et al. (2001)	295–384
$(2.3 \pm 1.0) \times 10^{-14}$	170 ± 100			FP–LIF	Lotz and Zellner (as quoted in Orlando et al. (2003b))	233–305
		9.0 ± 0.3 9.0 ± 0.2	291 295	FP–LIF	Falgayrac et al. (2004)	
		6.5	293	FP–UV Abs.	Hein et al. (1998)	

[a] Abbreviations for the experimental techniques employed are defined in table II-A-1.

The dissociation process has recently been measured via flash photolysis/LIF by Falgayrac et al. (2004) over a range of pressures between 291 and 348 K. These authors combined their measured data with a Rice, Ramsberger, Kassel, Marcus (RRKM) analysis to derive the following expression at 1 atm. total pressure: $k = 6.7 \times 10^{12} \exp(-5740/T)\,s^{-1}$, with $k = 2.9 \times 10^4\,s^{-1}$ at 298 K. The activation energies and 298 K rate coefficients reported by Falgayrac et al. (2004) are in good agreement with most recent theoretical studies (Fittschen et al., 2000; Méreau et al., 2000b; Somnitz and Zellner, 2000a,b). The lone exception is the study of Jungkamp et al. (1997) in which substantially lower barriers were estimated.

A number of end-product studies have been conducted over the years to examine the competition between 2-butoxy decomposition and its reaction with O_2 (Carter et al., 1979; Cox et al., 1981; Libuda et al., 2002; Meunier et al., 2003; Cassanelli et al., 2005). Rate coefficient ratios k_{decomp}/k_{O2} in the range (2.5–5.5) $\times 10^{18}$ molecule cm^{-3} have been reported at room temperature. An O_2 rate coefficient near 1.2×10^{-14} cm^3 $molecule^{-1}s^{-1}$ implies dissociation rate coefficients of about (2–5) $\times 10^4 s^{-1}$, in very good agreement with the direct measurement of Falgayrac et al. (2004) and the body of theoretical data (Fittschen et al., 2000; Méreau et al., 2000b; Somnitz and Zellner, 2000a,b). End-product studies over ranges of temperature give differences in activation energy between decomposition and O_2 reaction of about 48 kJ mol^{-1} (Libuda et al., 2002; Cassanelli et al., 2005), in reasonable accord with the available data for the individual reactions. Both of these groups also noted small contributions (5–10%) of chemically activated decomposition for 2-butoxy radicals generated from the reaction of NO with 2-butyl peroxy.

It is thus clear that, for 2-butoxy radicals, decomposition and reaction with O_2 occur with competitive rates under ambient conditions. Taking $k_{decomp} \sim 3 \times 10^4$ s^{-1} and $k_{O2} \sim 10^{-14}$ cm^3 $molecule^{-1}s^{-1}$ implies about 40% decomposition and 60% reaction with O_2 at 298 K, 1 atm. pressure. With a barrier to decomposition of about 50 kJ mol^{-1}, reaction with O_2 will preponderate in the upper troposphere.

IV-B-9. 2-Methyl-2-propoxy Radical [*tert*-Butoxy Radical, $(CH_3)_3CO\bullet$]

The 2-methyl-2-propoxy radical, $(CH_3)_3CO\bullet$, represents something of a special case; since it possesses no α-hydrogens, reaction with O_2 is not possible. Furthermore, it possesses no carbon chains of sufficient length to allow for a favorable isomerization reaction to occur. Thus the only viable atmospheric process for this reaction (other than combination with NO or NO_2) is thermal decomposition, which occurs via methyl radical elimination to generate acetone:

$$(CH_3)_3CO\bullet \rightarrow CH_3C(O)CH_3 + CH_3\bullet$$

This reaction has been studied directly, using LIF to detect the 2-methyl-2-propoxy radicals (Blitz et al., 1999; Fittschen et al., 2000), and has also been studied theoretically (Méreau et al., 2000b; Fittschen et al., 2000). Estimates of the activation energy for the decomposition process range from 56.9 to 61.9 kJ mol^{-1}, with rate coefficients at 298 K and 1 atm. air falling between 1500 and 3000 s^{-1}. The decomposition will be exceedingly slow in the upper troposphere, <1 s^{-1}.

IV-B-10. 2-Methyl-1-propoxy Radical [Isobutoxy Radical, $(CH_3)_2CHCH_20\bullet$]

There are two reaction pathways available to the 2-methyl-1-propoxy (isobutoxy) radical under atmospheric conditions, reaction with O_2 and unimolecular decomposition:

$$(CH_3)_2CHCH_2O\bullet \rightarrow CH_3CH(\bullet)CH_3 + CH_2O$$

$$(CH_3)_2CHCH_2O\bullet + O_2 \rightarrow (CH_3)_2CHCHO + HO_2$$

The only study of the 2-methyl-1-propoxy radical reaction with O_2 is the relative rate study of Zabarnick and Heicklen (1985c), who used the 2-methyl-1-propoxy reaction with NO as the reference reaction. Values for k_{O2} increased from 1.4×10^{-14} to 4.4×10^{-14} cm^3 $molecule^{-1}s^{-1}$ for temperatures from 265 to 393 K. Orlando et al. (2003b) made no recommendation for this rate coefficient due to the complexity involved, both in the chemical system used by Zabarnick and Heicklen (1985c) and in the analysis of the data.

The decomposition of 2-methyl-1-propoxy has been studied via theoretical methods (Méreau et al., 2000b; Fittschen et al., 2000) and more recently via end-product analysis (Geiger et al., 2002). Geiger et al. (2002) report a rate coefficient ratio $k_{decomp}/k_{O2} = (6.2 \pm 1.2) \times 10^{18}$ molecule cm^{-3}, implying about 55% decomposition and 45% reaction with O_2 in 1 atm. air at 298 K.

There is excellent agreement between the two reported theoretical studies, with activation energies at infinite pressure of 51.5 (Fittschen et al., 2000) and 51.9 kJ mol^{-1} (Méreau et al., 2000b) having been reported. Fittschen et al. (2000) calculate a decomposition rate coefficient of $5.7 \times 10^4 s^{-1}$ at 298 K, 1 atm., reasonably consistent with that derived from the combination of the Geiger et al. (2002) and Zabarnick and Heicklen (1985c) data. The theoretical data imply a significant diminution of the decomposition rate coefficient with altitude, such that reaction with O_2 will dominate the chemistry of 2-methyl-1-propoxy in the upper troposphere (k_{decomp} < 100 s^{-1} at 215 K).

IV-B-11. 1-Pentoxy Radical (*n*-Pentoxy Radical, $CH_3CH_2CH_2CH_2CH_2O\bullet$)

The 1-pentoxy radical can potentially undergo unimolecular decomposition, isomerization, or reaction with O_2:

$$CH_3CH_2CH_2CH_2CH_2O\bullet \rightarrow CH_3CH_2CH_2CH_2\bullet + CH_2O$$

$$CH_3CH_2CH_2CH_2CH_2O\bullet \rightarrow CH_3CH(\bullet)CH_2CH_2CH_2OH$$

$$CH_3CH_2CH_2CH_2CH_2O\bullet + O_2 \rightarrow CH_3CH_2CH_2CH_2CHO + HO_2$$

Theoretical studies of the 1-pentoxy thermal dissociation (Fittschen et al., 2000; Somnitz and Zellner, 2000 a,b) show this to be a very slow process, with an activation energy near 63 ± 2 kJ mol^{-1} and a room temperature rate coefficient of about 3 s^{-1}. The reaction is thus of no atmospheric significance.

Hein et al. (1999) examined the competition between reaction with O_2 and isomerization for the 1-pentoxy radical at 50 mbar by monitoring NO_2 and OH time profiles following flash photolysis of 1-bromopentane/O_2/NO mixtures. Their data demonstrated the dominance of the isomerization process for these conditions, but only limits could be placed on the reaction rates, $k_{isom} \geq 10^5 s^{-1}$ and $k_{O2} \leq 1 \times 10^{-13}$ cm^3 molecule^{-1}s^{-1}.

Theoretical studies of the isomerization process (Somnitz and Zellner, 2000 a,b; Méreau et al., 2003) provide rate coefficients of $\sim 2 \times 10^6$ at 298 K, 1 atm. air and activation energies (infinite pressure) of 33.1 – 33.7 kJ mol^{-1}. These data confirm the dominance of isomerization over the O_2 reaction in the lower atmosphere, and suggest that isomerization remains the major reaction in the upper troposphere (Méreau et al., 2003). A recent end-product study by Johnson et al. (2004a) confirms the rapid

298 K isomerization rate coefficient ($k = 3.4 \times 10^6$ s^{-1} at 298 K, 1 atm). The study, conducted over the range 243–283 K, finds a lower activation energy for isomerization, ~24 kJ mol^{-1}, again suggesting the continued dominance of the isomerization reaction in the upper troposphere.

Finally, the end-product study of Heimann et al. (2002), involving photolysis of 1-pentyl iodide, gives rate coefficient ratios for the reactions of the 1-pentoxy radical that differ significantly from the body of data reviewed here. Difficulties in identifying and quantifying the products of isomerization, coupled with apparently spurious sources of potential 1-pentoxy decomposition products (e.g., butanal), likely contributed to this discrepancy.

IV-B-12. 2-Pentoxy Radical [$CH_3CH_2CH_2CH(O\bullet)CH_3$]

The 2-pentoxy radical can potentially undergo unimolecular decomposition, isomerization, or reaction with O_2:

$$CH_3CH_2CH_2CH(O\bullet)CH_3 \rightarrow CH_3CH_2CH_2\bullet + CH_3CHO$$

$$CH_3CH_2CH_2CH(O\bullet)CH_3 \rightarrow \bullet CH_2CH_2CH_2CH(OH)CH_3$$

$$CH_3CH_2CH_2CH(O\bullet)CH_3 + O_2 \rightarrow CH_3CH_2CH_2C(O)CH_3 + HO_2$$

Surprisingly, there are no direct laboratory studies of any of these reaction processes. A value near 1×10^{-14} cm^3 molecule^{-1}s^{-1} at room temperature (with a small temperature dependence) can be assumed for the O_2 reaction on the basis of data for similar radicals (e.g., 2-butoxy and 3-pentoxy).

The decomposition reaction appears to play a minor role in the atmospheric chemistry of 2-pentoxy. End-product studies conducted by Dóbé et al. (1986) at temperatures between 363 and 413 K and by Heimann et al. (2002) at room temperature suggest $k_d \approx 10^4$ s^{-1} near 298 K. Theoretical studies of Fittschen et al. (2000), Méreau et al. (2000b), and Somnitz and Zellner (2000 a,b) all retrieved activation energies for decomposition between 54.8 and 55.6 kJ mol^{-1}, giving 298 K rate coefficients of $(1–2) \times 10^4$ s^{-1}.

The isomerization reaction, as characterized by three independent theoretical studies (Somnitz and Zellner, 2000a,b; Lin and Ho, 2002; Méreau et al., 2003), appears to have an activation energy (or reaction barrier) near 38 kJ mol^{-1} and a room temperature rate coefficient of $(3–5) \times 10^5$s^{-1}; thus this represents the major reaction for 2-pentoxy at 298 K. The dominance of the isomerization process is also evident from the end-product studies of Atkinson et al. (1995) and Johnson et al. (2004a), who report k_{isom}/k_{O2} ratios of $\sim 3 \times 10^{19}$ molecule cm^{-3} and $\sim 5 \times 10^{19}$ molecule cm^{-3}, respectively, near 298 K. Johnson et al. (2004a) also studied the temperature dependence of the competition between isomerization and reaction with O_2, and reported an A-factor for isomerization near 10^{11} s^{-1} and an activation energy of 31 kJ mol^{-1}. Both values are lower than those obtained in the theoretical studies, although similar room-temperature rate coefficients are retrieved. Extrapolation of the Johnson et al. (2004a) data suggests that isomerization will remain an important (but perhaps not dominant) fate for 2-pentoxy radical in the upper troposphere.

The substantially lower isomerization rate coefficient reported in the end-product study of Dóbé et al. (1986), $k_{isom} = 6 \times 10^3\ s^{-1}$, is believed to be in error due to difficulties in accurately quantifying the multifunctional products generated following the isomerization. The isomerization rate coefficient obtained in the end-product study of Heimann et al. (2002) also seems somewhat low, $k_{isom} = 10^5 s^{-1}$ at 298 K, perhaps for the same reason.

IV-B-13. 3-Pentoxy Radical [$CH_3CH_2CH(O\bullet)CH_2CH_3$]

The 3-pentoxy radical has two available reaction pathways, reaction with O_2 and unimolecular decomposition:

$$CH_3CH_2CH(O\bullet)CH_2CH_3 + O_2 \rightarrow CH_3CH_2C(O)CH_2CH_3 + HO_2$$

$$CH_3CH_2CH(O\bullet)CH_2CH_3 \rightarrow CH_3CH_2CHO + CH_3CH_2\bullet$$

The O_2 reaction has been studied by Deng et al. (2001), using laser flash-photolysis with LIF for direct detection of the 3-pentoxy radicals, and by Hein et al. (2000), who monitored NO_2 and OH time profiles following flash photolysis of 3-bromopentane/O_2/NO mixtures; see table IV-B-8. Deng et al. (2001) report $k = (4.1 \pm 1.2) \times 10^{-15} \exp[(310\pm70)/T]$ cm^3 molecule^{-1}s^{-1} over the range 220 to 285 K, which extrapolates to $k = (1.2\pm0.6)\times10^{-14}$ cm^3 molecule^{-1} s^{-1} at 298 K. This room-temperature value is $\sim$ 50% higher than the 298 K value reported by Hein et al. (2000), $(7.2 \pm 3.5) \times 10^{-15}$ cm^3 molecule^{-1} s^{-1}, although the data points are within the rather large combined uncertainties. Further studies of the reaction are required before a firm recommendation can be made.

Table IV-B-8. Rate coefficients for reaction of 3-pentoxy with O_2, with $k = A \times \exp(-B/T)$

A (cm^3 molecule^{-1}s^{-1})	B (K)	k (cm^3 molecule^{-1} s^{-1})	T (K)	Technique [a]	Reference	Temperature Range (K)
$(4.1 \pm 1.2) \times 10^{-15}$	-310 ± 70			FP–LIF	Deng et al. (2001)	220–285
		$(7.2 \pm 3.5) \times 10^{-15}$	298	FP–UV Abs	Hein et al. (2000)	

[a] Abbreviations for the experimental techniques employed are defined in table II-A-1.

The only direct measurement of the unimolecular decomposition of 3-pentoxy was made by Hein et al. (2000) at reduced pressure [$k = (5.0 \pm 2.5) \times 10^3 s^{-1}$ at 50 mbar]. Thus, information on the rate of this process at atmospheric pressure comes mainly from end-product studies (Atkinson et al., 1995; Geiger et al., 2002; Heimann et al., 2002; Meunier et al., 2003) and from theoretical treatments (Hein et al., 2000; Méreau et al., 2000b; Somnitz and Zellner, 2000a,b). The available product data are in reasonable agreement, with reported k_{decomp}/k_{O2} rate coefficient ratios falling in the range (3.1–4.1) $\times 10^{18}$ molecule cm^{-3} (Atkinson et al., 1995; Heimann et al., 2002; Geiger et al., 2002; Meunier et al., 2003). Using the k_{O2} value of Deng et al. (2001) yields k_{decomp} values of about 3 $\times 10^4$ s^{-1} at 298 K, 1 atm., in good agreement with the theoretical data, all of which lie between (1.6 and 3.4) $\times 10^4$ s^{-1}. It can then be concluded that unimolecular decomposition accounts for a significant (30 $\pm$ 15%) fraction of the chemistry of 3-pentoxy in 1 atm. air at 298 K. The rather large activation energy for the decomposition process, $\sim$ 54 kJ mol^{-1} (Méreau et al., 2000b; Somnitz and Zellner, 2000a,b; Hein et al., 2000), implies a decreasing importance of the decomposition reaction with altitude in the troposphere. For example, at 12 km and 215 K, $k_{decomp} \leq 10$ s^{-1} and reaction with O_2 will clearly dominate.

IV-B-14. 2,2-Dimethyl-1-propoxy [Neopentoxy Radical, $(CH_3)_3CCH_2O\bullet$]

The $(CH_3)_3CCH_2O\bullet$ radical can react with O_2 or undergo unimolecular decomposition:

$$(CH_3)_3CCH_2O\bullet + O_2 \rightarrow (CH_3)_3CCHO + HO_2$$

$$(CH_3)_3CCH_2O\bullet \rightarrow (CH_3)_3C\bullet + CH_2O$$

$$(CH_3)_3C\bullet + O_2 \rightarrow (CH_3)_3COO\bullet$$

Lightfoot et al. (1990b) conducted flash photolysis/UV absorbance measurements of Cl_2/neopentane/O_2/N_2 mixtures, and were able to retrieve k_{decomp}/k_{O2} values from an examination of the temperature (248–298 K) and O_2 partial pressure dependence of the 2-methyl-2-propylperoxy (neopentoxy) radical yield in this system. Using $k_{O2} = 10^{-14}$ cm^3 molecule^{-1}s^{-1} gives $k_{decomp} = 2 \times 10^6$ s^{-1} at 298 K, decreasing to $\sim 3 \times 10^5$ s^{-1} at 248 K. Wallington et al. (1992a) examined end-products of neopentane oxidation in an environmental chamber at 298 K. In the presence of NO, a unity (105 $\pm$ 6%) yield of CH_2O and $<$10% yield of $(CH_3)_3CCHO$ in 1 atm. air confirmed the rapidity of the dissociation process compared to O_2 reaction ($k_{decomp}/k_{O2} > 5 \times 10^{19}$ cm^3 molecule^{-1} s^{-1}; $k_{decomp} > 5 \times 10^5$ s^{-1}). In the absence of NO, and at O_2 partial pressures approaching 1 atm. Wallington et al. (1992a) saw evidence for some $(CH_3)_3CCHO$ production from the O_2 reaction, suggesting the onset of competition between decomposition and reaction with O_2 under these conditions.

The rapidity of the neopentoxy dissociation has been confirmed in theoretical studies of this reaction (Méreau et al., 2000b; Fittschen et al., 2000) in which activation energies at infinite pressure of 39.2 kJ mol^{-1} (Fittschen et al., 2000) and 40.2 kJ mol^{-1} (Méreau et al., 2000b) and a 298 K rate coefficient of $\sim 3 \times 10^6$ s^{-1} (Méreau et al., 2000b) have been reported.

The combined data sets indicate a dominance of decomposition over O_2 reaction in 1 atm air at 298 K by at least a factor of 20. The decomposition is likely to remain competitive with O_2 reaction in the upper troposphere, with $k_{decomp} \sim 10^3 - 10^4$ s^{-1} at 210–215 K suggested by the theoretical studies (Méreau et al., 2000b; Fittschen et al., 2000). The low activation barrier to neopentoxy radical decomposition makes this system prime for a chemical activation effect. A thorough study of neopentoxy chemistry, perhaps via end-product analyses over the range of tropospheric temperatures, would be instructive.

IV-B-15. 2-Methyl-2-butoxy Radical [$CH_3CH_2C(O\bullet)(CH_3)_2$]

The 2-methyl-2-butoxy radical is generated in substantial yield from the reaction of OH with 2-methylbutane (isopentane). The radical cannot react with O_2 (it possesses no α-hydrogen atoms) or isomerize, but can undergo two possible dissociation reactions:

$$CH_3CH_2C(O\bullet)(CH_3)_2 \rightarrow CH_3CH_2\bullet + CH_3C(O)CH_3$$

$$\rightarrow CH_3CH_2C(O)(CH_3) + \bullet CH_3$$

The relative rate of these two processes has been determined between 283 and 338 K (Johnson et al., 2005) via end-product analysis following the photolysis of 2-methyl-2-butyl nitrite in a slow-flow system. A weighted least-squares fit to their data yields rate coefficient ratios $k_{ethyl}/k_{methyl} = 0.31\ \exp(1520/T)$. Thus, ethyl radical elimination is favored by about a factor of 50 near room temperature, and by even larger amounts at lower temperatures. The theoretical calculations of Méreau et al. (2000b) are reasonably consistent with these data, giving $k_{ethyl}/k_{methyl} = 0.3\ \exp(1700/T)$ at infinite pressure, and $k_{ethyl}/k_{methyl} = 85$ at room temperature, 1 atm. total pressure.

IV-B-16. 2-Methyl-1-butoxy Radical [$CH_3CH_2CH(CH_3)CH_2O\bullet$]

The 2-methyl-1-butoxy radical is obtained in minor yield from the oxidation of 2-methylbutane (isopentane). The radical can isomerize, react with O_2, or decompose:

$$CH_3CH_2CH(CH_3)CH_2O\bullet \rightarrow \bullet CH_2CH_2CH(CH_3)CH_2OH$$

$$CH_3CH_2CH(CH_3)CH_2O\bullet + O_2 \rightarrow CH_3CH_2CH(CH_3)CHO + HO_2$$

$$CH_3CH_2CH(CH_3)CH_2O\bullet \rightarrow CH_2CH_2CH(\bullet)CH_3 + CH_2O$$

The only study of this radical is that of Méreau et al. (2003), who have estimated an activation energy for the isomerization process of 40.6 kJ mol^{-1}, with a 298 K rate coefficient of 2.3×10^5 s^{-1} in 1 atm air. The isomerization reaction is thus expected to be the most important under these conditions. The decomposition, by analogy to 2-methyl-1-propoxy radical, should occur with a 298 K rate coefficient of about 6×10^4 s^{-1} and with an activation energy of about 50 kJ mol^{-1}, and is thus of minor importance. Reaction with O_2 (expected first-order rate coefficient $\approx 5 \times 10^4$ s^{-1} at 298 K, 1 atm air) will be of some importance at the Earth's surface and will compete favorably with isomerization at the colder temperatures of the free and upper troposphere.

IV-B-17. 2-Hexoxy Radical [$CH_3CH_2CH_2CH_2CH(O\bullet)CH_3$]

The 2-hexoxy radical has available to it three possible reaction pathways, reaction with O_2, unimolecular decomposition, and isomerization:

$$CH_3CH_2CH_2CH_2CH(O\bullet)CH_3 + O_2 \rightarrow CH_3CH_2CH_2CH_2C(O)CH_3 + HO_2$$

$$CH_3CH_2CH_2CH_2CH(O\bullet)CH_3 \rightarrow CH_3CH_2CH_2CH_2\bullet + CH_3CHO$$

$$\rightarrow CH_3CH(\bullet)CH_2CH_2CH(OH)CH_3$$

No direct studies of the O_2 reaction or the unimolecular decomposition, either theoretical or experimental, are available, although rate coefficients of $\sim 10^{-14}$ cm^3 molecule^{-1}s^{-1} and $\sim 3 \times 10^4$s^{-1}, respectively, can be estimated at 298 K (Atkinson, 1997b; Orlando et al., 2003b). End-product data from Eberhard et al. (1995) [and, less quantitatively, Carter et al. (1976)] demonstrate the predominant occurrence of isomerization, $k_{isom} > 10^6$s^{-1}, in agreement with conclusions that can be drawn from theoretical studies of this process (Lin and Ho, 2002; Méreau et al., 2003). The reasonably low energy barrier to isomerization, 25–32 kJ mol^{-1}, suggests that isomerization will remain dominant throughout most of the troposphere.

IV-B-18. 3-Hexoxy Radical [$CH_3CH_2CH(O\bullet)CH_2CH_2CH_3$]

The chemistry of the 3-hexoxy radical should mimic that of 2-pentoxy, with reaction with O_2, decomposition, and isomerization all of potential importance:

$$CH_3CH_2CH(O\bullet)CH_2CH_2CH_3 + O_2 \rightarrow CH_3CH_2C(O)CH_2CH_2CH_3 + HO_2$$

$$CH_3CH_2CH(O\bullet)CH_2CH_2CH_3 \rightarrow CH_3CH_2CH_2\bullet + CH_3CH_2CHO$$

$$\rightarrow CH_3CH_2CH(OH)CH_2CH_2CH_2\bullet$$

The end-product data of Eberhard et al. (1995), the theoretical study of Méreau et al. (2003), and the structure–additivity analysis of Atkinson (1997b) all suggest that the isomerization reaction will be the most important reaction near 298 K, $k_{isom} \approx 2 \times 10^5$ s^{-1}. This can be compared to the expected first-order rate coefficient for reaction with O_2, 5×10^4s^{-1}, assuming $k_{O2} \approx 10^{-14}$ cm^3 molecule^{-1}s^{-1}. By analogy to 2-pentoxy, decomposition is estimated to occur with an activation energy of about 54 kJ mol^{-1} and a rate coefficient of $\sim (1 - 2) \times 10^4$ s^{-1} at 298 K, 1 atm. air, and thus will be of limited importance. Although further low-temperature studies are needed, the isomerization reaction [activation energy $\sim$ 40 kJ mol^{-1} (Méreau et al., 2003)] will likely remain competitive with the O_2 reaction in the upper troposphere.

IV-B-19. 1-Hexoxy Radical (*n*-Hexoxy Radical, $CH_3CH_2CH_2CH_2CH_2CH_2O\bullet$)

There is only one study of 1-hexoxy chemistry in the literature, a theoretical study of its isomerization reaction by Méreau et al. (2003):

$$CH_3CH_2CH_2CH_2CH_2CH_2O\bullet \rightarrow CH_3CH_2CH(\bullet)CH_2CH_2CH_2OH$$

These authors obtained an activation energy (infinite pressure) to the isomerization of 32.2 kJ mol^{-1}, a 298 K rate coefficient of 3×10^6 s^{-1}, and a rate coefficient in excess of 10^4 s^{-1} for conditions of the upper troposphere (~12 km, 215 K). The atmospheric behavior of 1-hexoxy is likely to be very similar to that of 1-pentoxy discussed in section IV-B-11, with isomerization being the predominant reaction in the lower troposphere and remaining the major reaction in the upper troposphere. Reaction with O_2 is also likely to contribute in the upper troposphere.

IV-B-20. 2-Methyl-2-pentoxy Radical [$CH_3CH_2CH_2C(O\bullet)(CH_3)_2$]

The 2-methyl-2-pentoxy radical will be generated in substantial yield from the oxidation of 2-methylpentane. There are two decomposition processes and one isomerization process available to this radical (Johnson et al., 2005):

$$\begin{aligned} CH_3CH_2CH_2C(O\bullet)(CH_3)_2 &\rightarrow CH_3CH_2CH_2\bullet + CH_3C(O)CH_3 \\ &\rightarrow CH_3CH_2CH_2C(O)CH_3 + \bullet CH_3 \\ &\rightarrow \bullet CH_2CH_2CH_2C(OH)(CH_3)_2 \end{aligned}$$

End-product studies have been carried out by Johnson et al. (2005) following photolysis of 2-methyl-2-pentyl nitrite between 283 and 345 K. Using an estimated rate coefficient for the isomerization process, $k_{isom} = 9.0 \times 10^{10} \exp(-3645/T)\,s^{-1}$, the authors report $k_{propyl} = 4.5 \times 10^{12} \exp(-5700/T)$ s^{-1} and $k_{methyl} = 7.1 \times 10^{12} \exp(-6375/T) s^{-1}$. Clearly, isomerization will be the major reaction pathway throughout the troposphere, accounting for approximately 95% of the fate of the 2-methoxy-2-pentoxy radical at 298 K, and $\gg$ 95% at the lower temperatures of the upper troposphere. The data also show that the dominant decomposition process is via elimination of the propyl radical, as is expected from the nature of the leaving groups involved.

IV-B-21. 4-Methyl-1-pentoxy Radical [$(CH_3)_2CHCH_2CH_2CH_2O\bullet$]

The 4-methyl-1-pentoxy radical is formed in minor yield in the oxidation of 2-methylpentane. The radical is of interest as it is the simplest alkoxy radical that can isomerize via a 1,5-H shift involving transfer of a tertiary H-atom. The radical can potentially undergo isomerization, decomposition, or reaction with O_2:

$$(CH_3)_2CHCH_2CH_2CH_2O\bullet \rightarrow (CH_3)_2C(\bullet)CH_2CH_2CH_2OH$$

$$\rightarrow (CH_3)_2CHCH_2CH_2\bullet + CH_2O$$

$$(CH_3)_2CHCH_2CH_2CH_2O\bullet + O_2 \rightarrow (CH_3)_2CHCH_2CH_2CHO + HO_2$$

The isomerization reaction has been studied theoretically by Méreau et al. (2003). These authors report an activation energy (at infinite pressure) of 20.9 kJ mol^{-1}, with a rate coefficient of 1.1×10^8 s^{-1} at 1 atm. pressure and 298 K. The Méreau data suggest that the isomerization will dominate throughout the troposphere. However, we note that the activation energy retrieved by Méreau et al. (2003) for isomerization of 5-methyl-2-hexoxy (also a tertiary H-atom transfer) is about 4 kJ mol^{-1} lower than the measured value of Johnson et al. (2004a). Thus there is the possibility that the Méreau rate coefficients for 4-methyl-1-pentoxy isomerization are somewhat overestimated. Nonetheless, there is little doubt that isomerization will dominate at 298 K, and will remain of major importance throughout the troposphere.

IV-B-22. 5-Methyl-2-hexoxy Radical [$(CH_3)_2CHCH_2CH_2CH(O\bullet)CH_3$]

Isomerization is expected to be the dominant reaction of the 5-methyl-2-hexoxy radical. This isomerization process has been studied both by end-product analysis (Johnson et al., 2004a) and theoretically (Lin and Ho, 2002; Méreau et al., 2003). Theoretically derived activation energies, 15.5 kJ mol^{-1} (Méreau et al., 2003), or reaction barriers, 15.9 kJ mol^{-1} (Lin and Ho, 2002), are lower than the activation energy reported by Johnson et al. (2004a), 19.7 kJ mol^{-1}. Thus the corresponding 298 K rate coefficients differ considerably (5×10^8 s^{-1} [Méreau et al., 2003] versus 9.5×10^6 s^{-1}[Johnson et al., 2004a]). However, even if the smaller rate coefficients and higher barrier from Johnson are correct, the isomerization reaction will dominate throughout the troposphere.

IV-B-23. 2-Methyl-2-hexoxy Radical [$CH_3CH_2CH_2CH_2C(O\bullet)(CH_3)_2$]

Isomerization of the 2-methyl-2-hexoxy radical

$$CH_3CH_2CH_2CH_2C(CH_3)_2O\bullet \rightarrow CH_3CH(\bullet)CH_2CH_2C(CH_3)_2OH$$

has been studied theoretically by Méreau et al. (2003). These authors found an activation energy (at infinite pressure) for this process of 29.3 kJ mol^{-1}, with $k = 1.1 \times 10^7$ s^{-1} at 298 K, 1 atm. By analogy to 2-methyl-2-butoxy, the decomposition reaction

$$CH_3CH_2CH_2CH_2C(CH_3)_2O\bullet \rightarrow CH_3CH_2CH_2CH_2\bullet + CH_3C(O)CH_3$$

is expected to be an order of magnitude slower than isomerization at 298 K. Because of the higher barrier to decomposition, isomerization will exceed decomposition by even greater amounts in the colder upper troposphere.

IV-B-24. 2,5-Dimethyl-2-hexoxy Radical $[(CH_3)_2CHCH_2CH_2C(O\bullet)(CH_3)_2]$

Isomerization is the only reaction of the 2,5-dimethyl-2-hexoxy radical that has been studied to date:

$$(CH_3)_2CHCH_2CH_2C(CH_3)_2O\bullet \rightarrow (CH_3)_2C(\bullet)CH_2CH_2C(CH_3)_2OH$$

Using theoretical methodologies, Méreau et al. (2003) found an activation energy for this process of 16.3 kJ mol^{-1}, with $k = 4.4 \times 10^8$ s^{-1} at 298 K, 1 atm, and $k = 9 \times 10^7$ s^{-1} at 220 K, 0.2 atm. While the magnitude of these rate coefficients may be overestimated (see section IV-B-22), it is nonetheless very likely that isomerization will be the dominant fate of 2,5-dimethyl-2-hexoxy throughout the troposphere.

IV-B-25. Cyclopentoxy Radical [*cyclo*-$C_5H_9O\bullet$]

A number of end-product studies (Takagi et al., 1981; Rowley et al., 1992b; Aschmann et al., 1997; Orlando et al., 2000) have shown clearly that the predominant atmospheric fate of the cyclopentoxy radical is ring-opening rather than reaction with O_2:

CH2 CH2 CHO• CH2 CH2 (ring) → •CH2H2CH2CH2CHO

+ O2 → C(=O) CH2 CH2 CH2 CH2 (ring) + HO2

Note that the ring-opening process is totally analogous to the decomposition reaction of acyclic alkoxy radicals. The data of Orlando et al. (2000) show that the decomposition reaction has an activation energy of no more than 42 kJ mol^{-1}, and that it remains rapid ($k > 10^6$s^{-1}) at temperatures as low as 230 K. Thus this reaction is expected to be the dominant process throughout the entire troposphere. The predominance of the dissociation process for cyclopentoxy is attributed (Takagi et al., 1981) to the ring strain in the system, which lowers the enthalpy (and hence the activation energy) to the ring-opening process.

IV-B-26. Cyclohexoxy Radical [*cyclo*-$C_6H_{11}O_2$]

The atmospheric chemistry of the cyclohexoxy radical involves competition between decomposition via ring-opening and reaction with O_2:

$$\textit{cyclo}\text{-}C_6H_{11}O^\bullet \longrightarrow {}^\bullet CH_2CH_2CH_2CH_2CH_2CHO$$

$$\textit{cyclo}\text{-}C_6H_{11}O^\bullet \xrightarrow{(+O_2)} \textit{cyclo}\text{-}C_6H_{10}{=}O + HO_2$$

The rate coefficient for reaction of cyclohexoxy with O_2 has been obtained by Zhang et al. (2004b) using the flash photolysis/laser-induced fluorescence technique, $k_{O2} = (5.8 \pm 2.3) \times 10^{-12}\exp[(-1720 \pm 100)/T]$. As shown in figure IV-B-2, these Arrhenius parameters are quite anomalous (higher *A*-factor and higher activation energy) when compared to reactions of acyclic alkoxy radical reactions with O_2. Zhang et al. (2004b) also used theoretical methodologies to estimate rate coefficients for the ring-opening process. For the more stable equatorial conformer, their calculations yielded $k_{decomp} = 2.4 \times 10^{13}\exp(-6070/T)$ s^{-1} at 1 atm, with $k_{decomp} = 3.4 \times 10^4$ s^{-1} at 298 K.

End-product studies from a number of groups (Takagi et al., 1981; Rowley et al., 1991; Aschmann et al., 1997; Platz et al., 1999; Orlando et al., 2000; Hanson et al., 2004) demonstrate that decomposition of cyclohexoxy is competitive with its O_2 reaction at 298 K. The study of Orlando et al. (2000), conducted over the range 273–298 K, provides the following rate coefficient ratio: $k_{decomp}/k_{O2} = (7.7 \pm 2.4) \times 10^{26} \exp[(-5550 \pm 1110)/T)]$. The body of available data implies that reaction with O_2 will increase in importance with altitude in the troposphere, contributing about 30% of the reactivity at 298 K, 1 atm. and essentially all of the reactivity in the upper troposphere.

IV-B-27. 4-Methyl-cyclohexoxy Radical [*cyclo*-CH_3-$C_6H_{10}O\bullet$]

The rate coefficient for reaction of *trans*-4-methyl-1-cyclohexoxy radical with O_2 has recently been obtained by Zhang et al. (2005) using flash photolysis/LIF. They report $k = (1.4^{+8}_{-1}) \times 10^{-13} \exp[(-810 \pm 400)/T]$ cm^3 $molecule^{-1}s^{-1}$ between 228 and 292 K, with $k = 9.2 \times 10^{-15}$ cm^3 $molecule^{-1}s^{-1}$ near room temperature. While the activation energy for this reaction is higher than for reaction of C_2–C_5 acyclic alkoxy radicals with O_2, the anomaly is not as large as for the cyclohexoxy reaction studied by the same research group (Zhang et al., 2004b); see figure IV-B-2.

Zhang et al. (2005) also obtained ring-opening Arrhenius parameters for *trans*-4-methyl-1-cyclohexoxy via theoretical methodologies, $k = 1.8 \times 10^{13} \exp(-6140/T)\ s^{-1}$; they showed that these calculations were consistent with the extrapolation of their observed first-order decays of 4-methyl-1-cyclohexoxy to zero O_2 partial pressure. Like the cyclohexoxy radical (section IV-B-26), decomposition via ring-opening and reaction with O_2 appear to be competing fates of the 4-methyl-1-cyclohexoxy radical in the lower troposphere. The relatively large barrier to ring-opening will lead to near-quantitative occurrence of the O_2 reaction in the upper troposphere.

A potentially interesting isomerization of the *cis*-4-methyl-1-cyclohexoxy radical, via a 1,6 H-shift involving the methyl group, has been shown theoretically by Zhang et al. (2005) to be very slow, $k < 1\ s^{-1}$ at room temperature.

IV-B-28. Other Five- and Six-Carbon Alkoxy Radicals

No experimental or theoretical studies are available regarding the chemistry of many of the alkoxy radicals formed in the oxidation of five- and six-carbon alkanes such as 2-methylbutane (except 2-methyl-2-butoxy and 2-methyl-1-butoxy), 2-methylpentane (except 2-methyl-2-pentoxy and 4-methyl-1-pentoxy), 3-methylpentane, 2,3-dimethylbutane, and 2,2-dimethylbutane. Table IV-B-9 presents a summary of the expected chemistry for a number of these radicals showing the hydrocarbon they are derived from and the products and 298 K rate coefficients for the relevant reactions. The estimates provided in the table derive largely from comparison with structurally similar radicals and from the structure–reactivity relationships discussed in section IV-B-2 and in Orlando et al. (2003b).

IV-C. Major Products of the Atmospheric Oxidation of the Alkanes

IV-C-1. Introduction

The first-generation products of the oxidation of the alkanes include a complex array of carbonyl species (aldehydes and ketones), hydroperoxides, alcohols, organic nitrates, and multifunctional species such as hydroxycarbonyls and hydroxynitrates. These products result from the chemistry of the alkylperoxy and alkoxy radicals, described in sections IV-A and IV-B. In general, the product distributions obtained will vary with location in the atmosphere as they are influenced by variables such as temperature, total pressure, O_2 partial pressure, and the mixing ratios of NO_x and HO_x radicals.

The first issue that must be considered when trying to understand the end-products of an oxidation process is the site of attack of the oxidant (usually OH) on the alkane; this determines the initial distribution of peroxy radicals formed. For example, in pentane there are three separate sites at which reaction can occur: the methyl group at the 1-position, the two equivalent CH_2 groups (2-position), and the central CH_2 group (the 3-position), leading to the formation of 1-, 2- and 3-pentylperoxy radicals (see figure IV-C-1). It is important to realize that the sites of attack (distribution of peroxy

Table IV-B-9. Major expected reaction pathways for alkoxy radicals derived from a number of five- and six-carbon alkanes. The radicals presented in the table have not been studied directly; data given are derived from structure–additivity relationships, or by comparison to structurally similar radicals

Parent Alkane	Alkoxy Radical Produced	Expected Major Reaction(s)	Products from Expected Reaction	Estimated Rate (s^{-1}) at 298 K, 1 atm Air
2-Methylbutane	3-Methyl-2-butoxy	Decomposition	$CH_3C{\bullet}HCH_3 + CH_3CHO$	5×10^6
		Reaction with O_2	$(CH_3)_2CHC(O)CH_3$	5×10^4
2-Methylbutane	3-Methyl-1-butoxy	Isomerization	${\bullet}CH_2CH(CH_3)CH_2CH_2OH$	2×10^5
		Reaction with O_2	$(CH_3)_2CHCH_2CHO + HO_2$	5×10^4
2-Methylpentane	2-Methyl-3-pentoxy	Decomposition	$CH_3CH_2CHO + CH_3C{\bullet}HCH_3$	5×10^6
		Reaction with O_2	$CH_3CH_2C(O)CH(CH_3)_2 + HO_2$	5×10^4
2-Methylpentane	4-Methyl-2-pentoxy	Isomerization	${\bullet}CH_2CH(CH_3)CH_2CH(OH)CH_3$	2×10^5
		Decomposition	$(CH_3)_2CHCH_2{\bullet} + CH_3CHO$	3×10^4
		Reaction with O_2	$(CH_3)_2CHCH_2C(O)CH_3 + HO_2$	5×10^4
2-Methylpentane	2-Methyl-1-pentoxy	Isomerization	$CH_3CH{\bullet}CH_2CH(CH_3)CH_2OH$	2×10^6
		Decomposition	$CH_3CH_2CH_2CH{\bullet}CH_3 + CH_2O$	6×10^4
		Reaction with O_2	$CH_3CH_2CH_2CH(CH_3)CHO + HO_2$	5×10^4
3-Methylpentane	3-Methyl-1-pentoxy	Isomerization	$CH_3CH{\bullet}CH(CH_3)CH_2CH_2OH$	2×10^6
		Reaction with O_2	$CH_3CH_2CH(CH_3)CH_2CHO + HO_2$	5×10^4
3-Methylpentane	3-Methyl-2-pentoxy	Decomposition	$CH_3CH_2C{\bullet}HCH_3 + CH_3CHO$	5×10^6
		Isomerization	${\bullet}CH_2CH_2CH(CH_3)CH(OH)CH_3$	2×10^5

(continued)

Table IV-B-9. *(continued)*

Parent Alkane	Alkoxy Radical Produced	Expected Major Reaction(s)	Products from Expected Reaction	Estimated Rate (s^{-1}) at 298 K, 1 atm Air
3-Methylpentane	3-Methyl-3-pentoxy	Decomposition	$CH_3CH_2\bullet + CH_3C(O)CH_2CH_3$	1×10^6
3-Methylpentane	2-Ethyl-1-butoxy	Isomerization	$\bullet CH_2CH_2CH(C_2H_5)CH_2OH$	2×10^5
		Decomposition	$CH_3CH_2CH\bullet CH_2CH_3 + CH_2O$	6×10^4
		Reaction with O_2	$CH_3CH_2CH(C_2H_5)CHO + HO_2$	5×10^4
2,3-Dimethylbutane	2,3-Dimethyl-1-butoxy	Isomerization	$\bullet CH_2CH(CH_3)CH(CH_3)CH_2OH$	2×10^5
		Decomposition	$(CH_3)_2CHCH\bullet CH_3 + CH_2O$	6×10^4
		Reaction with O_2	$(CH_3)_2CHCH(CH_3)CHO + HO_2$	5×10^4
2,3-Dimethylbutane	2,3-Dimethyl-2-butoxy	Decomposition	$CH_3CH\bullet CH_3 + CH_3C(O)CH_3$	$\geq 10^7$
2,2-Dimethylbutane	2,2-Dimethyl-1-butoxy	Decomposition	$CH_3CH_2C\bullet(CH_3)_2 + CH_2O$	2×10^6
		Isomerization	$\bullet CH_2CH_2C(CH_3)_2CH_2OH$	2×10^5
		Reaction with O_2	$CH_3CH_2C(CH_3)_2CHO + HO_2$	5×10^4
2,2-Dimethylbutane	3,3-Dimethyl-2-butoxy	Decomposition	$(CH_3)_3C\bullet + CH_3CHO$	$\geq 10^7$
		Reaction with O_2	$(CH_3)_3CC(O)CH_3 + HO_2$	5×10^4
2,2-Dimethylbutane	3,3-Dimethyl-1-butoxy	Isomerization	$\bullet CH_2C(CH_3)_2CH_2CH_2OH$	2×10^5
		Reaction with O_2	$(CH_3)_3CCH_2CHO + HO_2$	5×10^4

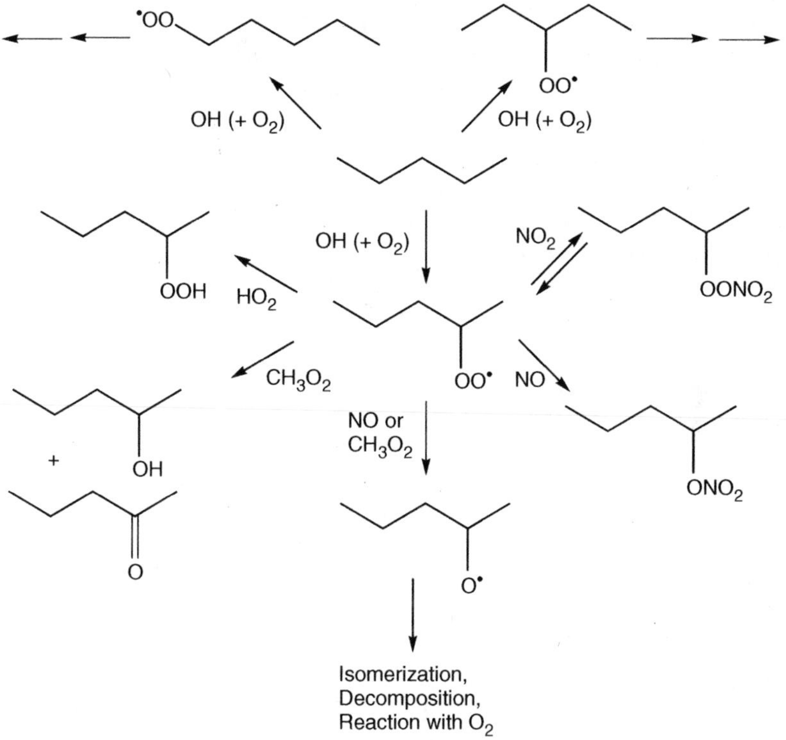

Figure IV-C-1. Partial oxidation scheme for pentane, showing the formation of peroxy radicals and the subsequent reactions of the 2-pentyl peroxy species.

radicals) vary with the oxidant involved [OH, Cl, NO_3, $O(^3P)$] and will even vary with temperature for a given oxidant.

The peroxy radicals formed react in the atmosphere with NO to form alkoxy radicals and organic nitrates, with NO_2 to form peroxy nitrates, with HO_2 to form hydroperoxides, and with CH_3O_2 and other peroxy radicals to form alkoxy radicals, carbonyls, and alcohols, as described in section IV-A. These competitive reaction channels are illustrated in figure IV-C-1 for 2-pentylperoxy. Further reactions of the alkoxy radicals lead to additional products, including carbonyls, multifunctional species, and smaller peroxy radicals. The relative rates of the competing peroxy radical reactions are obviously controlled by the relative abundance of the potential reactants (NO, NO_2, HO_2, CH_3O_2, other RO_2), and thus are dependent on location in the atmosphere. In general, NO_x mixing ratios are highest near anthropogenic sources (e.g., urban centers) and reactions with NO_x will dominate in these regions. Reactions with HO_2 and CH_3O_2 will play an increasingly important role in more pristine, low-NO_x regions, such as the marine boundary layer.

Before delving into the actual end-products observed for the individual alkanes, there are a number of common features to the atmospheric chemistry of alkylperoxy radicals that will be presented first (e.g., Atkinson, 1997a; Atkinson and Arey, 2003).

Product studies of the reactions of alkylperoxy radicals with HO_2 have been only conducted in a few specific cases: see table IV-C-1. Large (near 100%) yields of the hydroperoxides have most often been observed,

$$RO_2\bullet + HO_2 \rightarrow ROOH + O_2$$

and this channel is generally assumed to occur exclusively (Atkinson, 1997a; Tyndall et al., 2001). However, it should also be pointed out that a second channel in the reaction of CH_3O_2 with HO_2, involving the production of water, has also been suggested in the literature (Jenkin et al., 1988; Moortgat et al., 1989; Elrod et al., 2001):

$$CH_3O_2\bullet + HO_2 \rightarrow CH_3OOH + O_2$$

$$\rightarrow CH_2O + H_2O + O_2$$

Most recently, Elrod et al. (2001) presented evidence for the occurrence of the H_2O channel with a yield of 11% at 298 K and 31% at 218 K. Further studies of the branching ratios in this reaction and the reaction of larger alkylperoxy radicals with HO_2, particularly at low temperature, are required to confirm Elrod's observations.

Table IV-C-1. Summary of direct studies of the end-products observed in the reactions of various alkyl peroxy radicals with HO_2. All data were obtained at room temperature, unless otherwise noted

Peroxy Radical	Product Channel and Yield	Technique[a]	Reference
CH_3O_2	$CH_3OOH + O_2$, > 75% $CH_2O + H_2O + O_2$, < 25%		Jenkin et al. (1988)[b]
CH_3O_2	$CH_3OOH + O_2$, 92 ± 8%	Chamber–FTIR	Wallington and Japar (1990b)
CH_3O_2	$CH_3OOH + O_2$, 92 ± 5%	Chamber–FTIR	Wallington (1991)
CD_3O_2	$CD_3OOH + O_2$, 100 ± 4%	Chamber–FTIR	Wallington and Hurley (1992c)
CH_3O_2	$CH_2O + H_2O + O_2$, 11% at 298 K $CH_2O + H_2O + O_2$, 31% at 218 K	Flow tube–MS	Elrod et al (2001)
$C_2H_5O_2$	$CH_3CH_2OOH + O_2$, 100%	Chamber–FTIR	Wallington and Japar (1990a)
$C_2H_5O_2$	$CH_3CH_2OOH + O_2$, 104 ± 5% $CH_3CHO + H_2O + O_2$, 2 ± 4%	Chamber–FTIR (284, 298 & 312 K)	Spittler et al. (2000)
$C_2H_5O_2$	$CH_3CH_2OOH + O_2$, 93 ± 10%	Chamber–FTIR and HPLC	Hasson et al. (2004)
cyclo-$C_5H_9O_2$	*cyclo*-$C_5H_9OOH + O_2$, 96 ± 18%	Chamber–FTIR	Rowley et al. (1992b)
cyclo-$C_6H_{11}O_2$	*cyclo*-$C_6H_{11}OOH + O_2$, 99 ± 18%	Chamber–FTIR	Rowley et al. (1992b)
$(CH_3)_3CCH_2O_2$	$(CH_3)_3CCH_2OOH + O_2$, 92 ± 17%	Chamber–FTIR	Rowley et al. (1992a)

[a] Abbreviations for the experimental techniques employed are defined in table II-A-1.
[b] Data reanalyzed in Wallington (1991).

The most abundant alkylperoxy radical in the atmosphere is CH_3O_2, and thus reaction of alkylperoxy species with CH_3O_2 may be of some atmospheric significance. In general, these reactions can occur through three pathways, as illustrated below for the case of ethylperoxy:

$$CH_3CH_2O_2\bullet + CH_3O_2\bullet \rightarrow CH_3CH_2O\bullet + CH_3O\bullet + O_2$$

$$\rightarrow CH_3CHO + CH_3OH + O_2$$

$$\rightarrow CH_3CH_2OH + CH_2O + O_2$$

However, accurate product yields in these $RO_2 + CH_3O_2$ cross-reactions have yet to be established (except in the special case of the CH_3O_2–CH_3O_2reaction), and estimates are required for modeling purposes. A common estimation method (Madronich and Calvert, 1990; Villenave and Lesclaux, 1996) uses the average of the branching ratios to radical production in the self-reactions of the two radicals of interest to approximate the branching to radical production in the cross-reaction.

Another common feature to all alkylperoxy radicals is their addition reaction with NO_2 to form peroxynitrate species:

$$RO_2\bullet + NO_2 + M \leftrightarrow RO_2NO_2 + M$$

The peroxynitrates, however, are thermally unstable (e.g., Zabel et al., 1989; Zabel, 1995), and decompose back to RO_2 and NO_2 on timescales of ~ 0.3 s at 1 atm pressure and 298 K; thus their concentrations do not build up to any significant extent under these conditions. The peroxynitrate lifetimes are, however, considerably longer at low temperature (e.g., 2–3 hours at 12 km, 215 K), and their formation may be of some significance in the upper troposphere.

The reaction of alkylperoxy radicals with NO has been shown to proceed through two channels, the major channel giving the alkoxy radical and NO_2 and the minor channel giving an organic nitrate (e.g., Darnall et al., 1976; Takagi et al., 1981; Atkinson et al., 1982d, 1983c, 1984a, 1987; Harris and Kerr, 1989; Carter and Atkinson, 1989; Atkinson, 1994; Eberhard et al., 1995; Atkinson, 1997a; Aschmann et al., 1997; Platz et al., 1999; Scholtens et al., 1999; Ranschaert et al., 2000; Orlando et al., 2000; Arey et al., 2001; Aschmann et al., 2001; Chow et al., 2003; Espada et al., 2005):

$$RO_2\bullet + NO \rightarrow RO\bullet + NO_2 \qquad \text{(IV-C-1a)}$$

$$RO_2\bullet + NO \rightarrow RONO_2 \qquad \text{(IV-C-1b)}$$

Branching ratios for this class of reactions have been obtained for a wide range of peroxy radicals at room temperature and 1 atm. pressure, and in a few cases over ranges of temperature or pressure (see table IV-C-2 for a summary). It has been concluded (e.g., by Atkinson, 1997a; Arey et al., 2001; Atkinson and Arey, 2003) that the branching ratio to nitrate production 1) increases monotonically

Table IV-C-2. Summary of measured organic nitrate yields near atmospheric pressure from reaction of peroxy radicals with NO, along with those calculated for secondary peroxy radicals from the formula of Arey et al. (2001).

Peroxy Radical	Measured $k_{IV\text{-}C\text{-}1b}$ / ($k_{IV\text{-}C\text{-}1a} + k_{IV\text{-}C\text{-}1b}$)	Technique Used [a] and Conditions [d]	Calculated [c] $k_{IV\text{-}C\text{-}1b}$ / ($k_{IV\text{-}C\text{-}1a} + k_{IV\text{-}C\text{-}1b}$)	Reference for Measured Value
Methylperoxy	< 0.03	Flow tube-CIMS (100 Torr, 298 K)		Scholtens et al. (1999)
Ethylperoxy	0.006 at 298 K, 0.02 at 213 K	Flow tube–CIMS (100 Torr, 213–298 K)		Ranschaert et al. (2000)
	< 0.014	735 Torr, 299 ± 2 K		Atkinson et al. (1982d)
1-Propylperoxy	0.024	735 Torr, 299 ± 2 K		Atkinson et al. (1982d)
2-Propylperoxy	0.04	735 Torr, 299 ± 2 K	0.041, 298 K	Atkinson et al. (1982d)
	0.006, 298 K 0.020, 213 K	Flow tube–CIMS (100 Torr, 213–298 K)		Chow et al. (2003)
1-Butylperoxy	0.040	735 Torr, 299 ± 2 K		Atkinson et al. (1982d)
	0.077 ± 0.003	~ 1 atm., room temp.		Espada et al. (2005)
2-Butylperoxy	0.084	760 Torr, 298 ± 2 K	0.066	Arey et al. (2001) [b]
	0.070 ± 0.002	~ 1 atm., room temp.		Espada et al. (2005)
2-Pentylperoxy	0.106	760 Torr air, 298 ± 2 K	0.108	Arey et al. (2001) [b]
3-Pentylperoxy	0.126	760 Torr air, 298 ± 2 K	0.108	Arey et al. (2001) [b]
2-Pentylperoxy and 3-Pentylperpoxy (Combined)	0.150, 284 K 0.125, 300 K 0.074, 327 K 0.071, 337 K	740 Torr air, 284–337 K	0.108	Atkinson et al. (1983c)

Neopentylperoxy	0.0513 ± 0.0053	735 Torr, 298 ± 2 K		Atkinson et al. (1984a)
	0.0755 ± 0.0062, 282 K	733–741 Torr, 282–323 K		Atkinson et al. (1987)
	0.0507 ± 0.0056, 300 K			
	0.0326 ± 0.0034, 323 K			
2-Methyl-2-butylperoxy	0.0533 ± 0.0022	735 Torr, 298 ± 2 K		Atkinson et al. (1984a)
	0.0670 ± 0.0057, 282 K	740 Torr, 282–323 K		Atkinson et al. (1987)
	0.0560 ± 0.0017, 300 K			
	0.0364 ± 0.0024, 323 K			
3-Methyl-2-butylperoxy	0.109 ± 0.003	735 Torr, 298 ± 2 K	0.108	Atkinson et al. (1984a)
	0.196 ± 0.008, 282 K	740 Torr, 282–323 K	0.145	Atkinson et al. (1987)
	0.150 ± 0.004, 300 K		0.108	
	0.0912 ±0.0042, 323 K		0.069	
2-Hexylperoxy	0.140	760 Torr air, 298 ± 2 K	0.168	Arey et al. (2001) [b]
	0.24 ± 0.07	725 ± 5 Torr air, 297 ± 3 K		Eberhard et al. (1995)
3-Hexylperoxy	0.158	760 Torr air, 298 ± 2 K	0.168	Arey et al. (2001) [b]
	0.16 ± 0.04	725 ± 5 Torr air, 297 ± 3 K		Eberhard et al. (1995)
2-Methyl-2-pentylperoxy	0.0350 ± 0.0096			
3-Methyl-2-pentylperoxy and 4-methyl-2-pentyl-peroxy (combined)	0.165 ± 0.016	735 Torr, 298 ± 2 K	0.168	Atkinson et al. (1984a)
3-Methyl-2-pentylperoxy	0.140 ± 0.014	735 Torr, 298 ± 2 K	0.108	Atkinson et al. (1984a)
	0.225 ± 0.016, 281 K	740 Torr, 281–323 K	0.240	Atkinson et al. (1987)
	0.162 ± 0.009, 300 K		0.168	
	0.109 ± 0.008, 323 K		0.090	

(continued)

Table IV-C-2. *(continued)*

Peroxy Radical	Measured $k_{IV\text{-}C\text{-}1b}$ / ($k_{IV\text{-}C\text{-}1a} + k_{IV\text{-}C\text{-}1b}$)	Technique Used [a] and Conditions [d]	Calculated [c] $k_{IV\text{-}C\text{-}1b}$ / ($k_{IV\text{-}C\text{-}1a} + k_{IV\text{-}C\text{-}1b}$)	Reference for Measured Value
2-Heptylperoxy	0.177	760 Torr air, 298 ± 2 K	0.220	Arey et al. (2001) [b]
3-Heptylperoxy	0.202	760 Torr air, 298 ± 2 K	0.220	Arey et al. (2001) [b]
4-Heptylperoxy	0.175	760 Torr air, 298 ± 2 K	0.220	Arey et al. (2001)[b]
2-, 3-, and 4-heptylperoxy (combined)	0.308, 284 K 0.287, 300 K 0.199, 322 K 0.153, 341K	740 Torr air, 284–341 K	0.220	Atkinson et al. (1983c)
	0.484 ± 0.141, 253 K 0.401 ± 0.026, 263 K 0.385 ± 0.056, 273 K 0.303 ± 0.015, 283 K 0.251 ± 0.050, 299 K 0.220 ± 0.017, 300 K 0.216 ± 0.023, 303 K 0.230 ± 0.018, 314 K 0.185 ± 0.032, 325 K 0.196 ± 0.012, 325 K 0.162 ± 0.025, 325 K	1 atm. air, 253–325 K		Harris and Kerr (1989)
2-Octylperoxy	0.224	760 Torr air, 298 ± 2 K	0.253	Arey et al. (2001) [b]
3-Octylperoxy	0.238	760 Torr air, 298 ± 2 K	0.253	Arey et al. (2001) [b]
4-Octylperoxy	0.243	760 Torr air, 298 ± 2 K	0.253	Arey et al. (2001) [b]
2-Decylperoxy	0.184 ± 0.034	740 Torr air, 296 ± 2 K	0.284	Aschmann et al. (2001)

3-Decylperoxy	0.244 ± 0.044	740 Torr air, 296 ± 2 K	0.284	Aschmann et al. (2001)
4-Decylperoxy	0.252 ± 0.052	740 Torr air, 296 ± 2 K	0.284	Aschmann et al. (2001)
5-Decylperoxy	0.240 ± 0.040	740 Torr air, 296 ± 2 K	0.284	Aschmann et al. (2001)
cyclo-Pentylperoxy	0.045 ± 0.015	760 Torr air, room temp.	0.108	Takagi et al. (1981)
	< 0.17, 298 K	700 Torr, FTIR detection		Orlando et al. (2000)
	< 0.26, 230 K			
cyclo-Hexylperoxy	0.090 ± 0.044	760Torr air, room temp.	0.168	Takagi et al. (1981)
	0.160 ± 0.015	735 Torr, 298 ± 2 K		Atkinson et al. (1984a)
	0.165 ± 0.021	740 Torr, 298 ± 2 K		Aschmann et al. (1997)
	0.16 ± 0.04	FTIR detection, 700 Torr, 296 ± 2 K		Platz et al. (1999)
	0.15 ± 0.04, 298 K	FTIR detection, 700 Torr		Orlando et al. (2000)
	0.21 ± 0.06, 273 K		0.168, 298 K	
	0.18 ± 0.07, 261 K		0.286, 273 K	
	0.17 ± 0.04	≈ 1 atm., room temp.	0.349, 261K	Espada et al. (2005)
*Cyclo-H*eptylperoxy	0.050 ± 0.011	760 Torr air, room temp.	0.220	Takagi et al. (1981)

[a] Abbreviations for the experimental techniques employed are defined in table II-A-1.

[b] Earlier data from Atkinson and co-workers not shown.

[c] Calculated branching ratio (for secondary peroxy radicals only) using parameters described by Arey et al. (2001). Calculation is for 298 K, 1 atm. total pressure unless otherwise stated.

[d] Experiments conducted in environmental chambers with GC-based detection of nitrates, unless otherwise noted.

with the number of carbons in the R group; 2) increases with decreasing temperature; 3) increases with increasing total pressure; and 4) is significantly larger (by factors of 2–3) for secondary alkylperoxy radicals than for primary or tertiary ones. Some conflicting evidence has recently been presented, however, regarding this fourth conclusion. Espada et al. (2005) report branching fractions for nitrate formation for reaction of 1-butylperoxy and 2-butylperoxy with NO that are indistinguishable from each other, 7.7 ± 0.3% and 7.0 ± 0.2% respectively. Furthermore, Flocke et al. (2006) have an unpublished data set which shows that, for some C_5–C_7 branched hydrocarbons, branching ratios to formation of primary, secondary, and tertiary nitrates are all essentially equal for a given carbon number.

For secondary radicals, Atkinson and co-workers (Atkinson et al., 1987; Carter and Atkinson, 1989; Atkinson, 1997a; Arey et al., 2001) have shown that the rate coefficient ratio $k_{(IV.C.1b)}/k_{(IV.C.1a)}$ can be estimated as a function of temperature, pressure, and number of carbons from the following empirical expression:

$$k_{(IV\text{-}C\text{-}1b)}/k_{(IV\text{-}C\text{-}1a)} = [A/(1+(A/B))] \times F^{z}$$

where $A = Y_0^{300}\,[M]\,(T/300)^{-m_0}$; $B = Y_\infty^{300}(T/300)^{-m_\infty}$; $z = (1 + \{\log_{10}[A/B]\}^2)^{-1}$; $Y_0^{300} = \alpha \times \exp(\beta \times n)$; n is the number of carbons in the peroxy radical, and α and β are constants. Currently recommended parameters (Arey et al., 2001) are $Y_\infty^{300} = 0.43$, $\alpha = 2 \times 10^{-22}$, $\beta = 1.0$, $m_0 = 0$, $m_\infty = 8.0$, and $F = 0.41$. Branching ratios calculated from this formula for secondary radicals are given in table IV-C-2.

In the following sections, a summary of laboratory data regarding the end-products of the oxidation of the individual alkanes is presented. Some of the common features of the oxidation processes that we have described (formation of hydroperoxides from $RO_2 + HO_2$ reactions; formation of peroxynitrates from reactions of $RO_2 + NO_2$; formation of nitrates from reactions of $RO_2 + NO$) will not be dealt with explicitly for each hydrocarbon, except as needed to explain specific laboratory observations. Observed product distributions will be rationalized to the degree possible from a consideration of the initial site of attack of OH on the parent hydrocarbon and the chemistry of the resulting alkylperoxy and alkoxy radicals. For compounds with six carbons or less for which no end-product data are available, structure–additivity relationships will be used to provide likely end-products and reaction mechanisms.

IV-C-2. Products of Atmospheric Oxidation of Methane (CH_4)

The basics of the atmospheric oxidation of methane in the presence of NO_x were established by Cox et al. (1976b) via end-product analysis following photolysis of mixtures of HONO, methane, and NO_x in air. Formaldehyde was observed with near 100% yield, while methyl nitrite and methyl nitrate were observed as minor products. The proposed mechanism is now universally accepted, and involves stoichiometric production of CH_3O_2 radicals following the initial attack of OH on methane, and the subsequent conversion of CH_3O_2 to CH_3O via reaction with NO. The major fate of the CH_3O radical is reaction with O_2, with smaller amounts reacting in the chamber system with NO and NO_2:

$$OH + CH_4 \rightarrow CH_3 + H_2O$$

$$CH_3 + O_2 \rightarrow CH_3O_2\bullet$$

$$CH_3O_2\bullet + NO \rightarrow CH_3O + NO_2$$

$$CH_3O\bullet + O_2 \rightarrow CH_2O + HO_2$$

$$CH_3O\bullet + NO \rightarrow CH_3ONO$$

$$CH_3O\bullet + NO_2 \rightarrow CH_3ONO_2$$

There are a number of studies examining the end-products of methane oxidation (or, equivalently, of methyl radical oxidation) in the absence of NO_x (Alcock and Mile, 1975; Weaver et al., 1975; Parkes, 1977; Selby and Waddington, 1979; Kan et al., 1980; Niki et al., 1981a; Hanst and Gay, 1983; Horie et al., 1990; Tyndall et al., 1998; Qi et al., 1999). Observed primary products are CH_2O, CH_3OH, CH_3OOH, and possibly CH_3OOCH_3, with the secondary products CO, HCOOH, CO_2, $HOCH_2OOH$ also being reported in some studies (e.g., Hanst and Gay, 1983; Qi et al., 1999). The primary products arise from the CH_3O_2–CH_3O_2 and CH_3O_2–HO_2 reactions,

$$CH_3O_2\bullet + CH_3O_2\bullet \rightarrow CH_3O\bullet + CH_3O\bullet + O_2 \qquad \text{(IV.C.2a)}$$

$$\rightarrow CH_3OH + CH_2O + O_2 \qquad \text{(IV.C.2b)}$$

$$\rightarrow CH_3OOCH_3 + O_2 \qquad \text{(IV.C.2c)}$$

$$CH_3O\bullet + O_2 \rightarrow CH_2O + HO_2$$

$$CH_3O_2\bullet + HO_2 \rightarrow CH_3OOH$$

$$\rightarrow CH_2O + H_2O + O_2$$

with the subsequent oxidation of CH_2O accounting for the secondary products. Currently recommended branching ratios for the CH_3O_2–CH_3O_2 reaction are $k_{IV.C.2a}/k_{IV.C.2b} = 26.2\ \exp(-1130/T)$, $k_{IV.C.2c} = 0$, with $k_{IV.C.2a}/k = 0.37$ and with $k_{IV.C.2b}/k = 0.63$ at 298 K (Tyndall et al., 2001). As discussed in section IV-A, Tyndall et al. (2001) recommend 100% formation of CH_3OOH in the reaction of CH_3O_2 with HO_2, although the minor occurrence of the CH_2O channel cannot be completely ruled out (Elrod et al., 2001).

IV-C-3. Products of Atmospheric Oxidation of Ethane (C_2H_6)

Five end-product studies of ethane oxidation (or, equivalently, ethylperoxy radical oxidation) are available in the literature, all conducted in the absence of NO_x.

Niki et al. (1982) photolyzed mixtures of Cl_2, ethane, O_2, and N_2 at 700 Torr, and identified three major products: acetaldehyde, ethyl hydroperoxide, and ethanol, with relative yields of 1.0, $\sim$0.5, and 0.28, respectively. They also presented evidence for the possible formation of $C_2H_5OOC_2H_5$. These authors observed essentially the same product distributions from the photolysis of $(C_2H_5)_2N_2/O_2$ mixtures, with carbon mass balances of about 90%.

Anastasi et al. (1983) quantified the production of CH_3CHO, CH_3CH_2OH, and C_2H_5OOH following the continuous photolysis of $(C_2H_5)_2N_2$–O_2–N_2 mixtures.

Hanst and Gay (1983) also photolyzed Cl_2−ethane mixtures in air and observed acetaldehyde, ethyl hydroperoxide, peracetic acid, CO_2, and CO as products. Concentration vs. time profiles are consistent with acetaldehyde and ethyl hydroperoxide being primary products, while CO, CO_2, and peracetic acid appear to be secondary (see their figure 2). Mass balances at reasonably low ethane conversions (<60%) are very good, 90–100%, and acetaldehyde is clearly the major observed product, in agreement with the Niki et al. (1982) study.

Wallington et al. (1989a) observed acetaldehyde, ethyl hydroperoxide, and ethanol as products of the Cl-atom initiated-oxidation of ethane, with molar yields of 0.54 ± 0.14, 0.36 ± 0.10, and 0.10 ± 0.04, respectively. These data are entirely consistent with those reported by Niki et al. (1982).

Qi et al. (1999) also photolyzed Cl_2–ethane mixtures in air. Following the near-complete oxidation of the ethane, the presence of CO_2, CO, acetaldehyde, formic acid, ethyl hydroperoxide, peracetic acid, methyl hydroperoxide, and hydroxymethyl hydroperoxide was noted via a combination of HPLC and FTIR analysis.

The data can be explained from the quantitative formation of ethylperoxy radicals from either reaction of Cl with ethane or from $(C_2H_5)_2N_2$ photolysis, followed by their reaction with HO_2 to generate ethyl hydroperoxide, with other $C_2H_5O_2$ to generate ethanol and acetaldehyde or ethoxy radicals, and the subsequent reaction of ethoxy radical with O_2 to produce acetaldehyde:

$$CH_3CH_2O_2\bullet + HO_2 \rightarrow CH_3CH_2OOH + O_2$$

$$CH_3CH_2O_2\bullet + CH_3CH_2O_2\bullet \rightarrow CH_3CHO + CH_3CH_2OH + O_2$$

$$\rightarrow CH_3CH_2O\bullet + CH_3CH_2O\bullet + O_2$$

$$CH_3CH_2O\bullet + O_2 \rightarrow CH_3CHO + HO_2$$

Further oxidation of acetaldehyde in the Hanst and Gay (1983) and Qi et al. (1999) studies could lead to peracetic acid, acetic acid, methyl hydroperoxide, formaldehyde, methanol, hydroxymethyl hydroperoxide, CO, and CO_2, many of which were observed. On the basis of data from Niki et al. (1982), Anastasi et al. (1983), and Wallington et al. (1989a), the IUPAC panel (Atkinson et al., 2005a) estimates a 62% branching fraction for radical formation in the $C_2H_5O_2$–$C_2H_5O_2$ reaction.

In the presence of NO, near-quantitative conversion of ethylperoxy radicals to acetaldehyde, via the formation of ethoxy radical, would be expected, although this does not appear to have been demonstrated experimentally:

$$CH_3CH_2O_2\bullet + NO \rightarrow CH_3CH_2O\bullet + NO_2$$

$$\rightarrow CH_3CH_2ONO_2$$

$$CH_3CH_2O\bullet + O_2 \rightarrow CH_3CHO + HO_2$$

IV-C-4. Products of Atmospheric Oxidation of Propane (C_3H_8)

The products of the oxidation of propane have been studied by three groups, one conducted in the presence of NO_x (Cox et al., 1980b) and two in its absence (Hanst and Gay, 1983; Qi et al., 1999). In addition, Kirsch et al. (1979) studied a subset of this chemistry, via photolysis of *trans*-2,2′-azopropane in O_2/N_2 mixtures. Cox et al. (1980b) observed a substantial yield of acetone (56%), while the other two studies reported small yields of peracetic acid and organic hydroperoxide (Hanst and Gay, 1983) or peracetic acid, ethyl and methyl hydroperoxide, and hydroxymethyl hydroperoxide (Qi et al., 1999). Kirsch et al. (1979) quantified acetone, 2-propanol and 2-propyl hydroperoxide as end-products of 2-propylperoxy chemistry.

The observations can be rationalized as follows. Reaction of OH with propane leads to two peroxy radicals, 1-propylperoxy and 2-propylperoxy, with estimated yields of about 26% and 74%, respectively (Kwok and Atkinson, 1995). In the presence of NO_x, these peroxy radicals are converted to the corresponding alkoxy radicals, 1-propoxy and 2-propoxy, as well as small yields of 1- and 2-propyl nitrate. The alkoxy radicals then react with O_2 to give propanal (expected yield 26%) and acetone (expected yield 74%). In the absence of NO_x, one would expect the production of 1- and 2-propyl hydroperoxide from reaction of the propylperoxy radicals with HO_2. 1-Propanol, 2-propanol, propanal, and acetone are expected products of the propylperoxy–propylperoxy reaction and the reactions of propylperoxy with other alkylperoxy radicals. Kirsch et al. (1979) estimate a 58% branching to 2-propoxy formation in the 2-propylperoxy + 2-propylperoxy reaction. The observed peracetic acid, ethyl and methyl hydroperoxides, and hydroxymethyl hydroperoxide (Hanst and Gay, 1983; Qi et al., 1999) may have originated from the subsequent oxidation of propanal.

IV-C-5. Products of Atmospheric Oxidation of Butane (*n*-Butane, *n*-C_4H_{10})

End-products of butane oxidation have been studied directly, both in the presence of nitrogen oxides (Carter et al., 1979; Cox et al., 1981; Kwok et al., 1996; Cassanelli et al., 2005) and in their absence (Hanst and Gay, 1983). Furthermore, subsets of the overall butane chemistry have been studied, using various sources for the formation of either butyl or butoxy radicals, i.e., via the photolysis of 1- or 2-butyl nitrite (Niki et al., 1981b; Libuda et al., 2002; Meunier et al., 2003; Cassanelli et al., 2005, 2006), via the photolysis of 1- or 2-butyl iodide (Geiger et al., 2002; Libuda et al., 2002), or via the reaction of F-atoms with 2-butanol (Drew et al., 1985). Major end-products observed in the various studies are butanone, acetaldehyde, butanal, 4-hydroxybutanal, 2-butyl nitrate, ethyl nitrate, 2-butyl nitrite, ethyl nitrite, formaldehyde, and peroxyacetyl nitrate (PAN).

Oxidation of butane can occur at two sites, leading to the production of 2-butylperoxy and 1-butylperoxy radicals. Using the methods of Kwok and Atkinson (1995), yields of 87% and 13%, respectively, are estimated from the OH-initiated attack at room temperature:

$$OH + CH_3CH_2CH_2CH_3(+O_2) \rightarrow CH_2CH(O_2\bullet)CH_2CH_3 + H_2O \qquad (87\%)$$

$$\rightarrow CH_2CH_2CH_2CH_2O_2\bullet + H_2O \qquad (13\%)$$

The further oxidation of 2-butylperoxy in the presence of NO_x leads to two of the major end-products, butanone and acetaldehyde, as well as the minor products 2-butyl nitrate and ethyl nitrate (Carter et al., 1979; Cox et al., 1981; Kwok et al., 1996; Libuda et al., 2002; Meunier et al., 2003; Cassanelli et al., 2005):

$$CH_3CH_2CH(O_2\bullet)CH_3 + NO \rightarrow CH_3CH_2CH(O\bullet)CH_3 + NO_2$$

$$\rightarrow CH_3CH_2CH(ONO_2)CH_3$$

$$CH_3CH_2CH(O\bullet)CH_3 + O_2 \rightarrow CH_3CH_2C(O)CH_3 + HO_2$$

$$CH_3CH_2CH(O\bullet)CH_3 \rightarrow CH_3CH_2\bullet + CH_3CHO$$

$$CH_3CH_2 + O_2 \rightarrow CH_3CH_2O_2\bullet$$

$$CH_3CH_2O_2\bullet + NO \rightarrow CH_3CH_2O\bullet + NO_2$$

$$\rightarrow CH_3CH_2ONO_2$$

$$CH_3CH_2O\bullet + O_2 \rightarrow CH_3CHO + HO_2$$

The observation of such species as formaldehyde, peroxyacetyl nitrate (PAN), ethyl nitrate, ethyl nitrite, and methyl nitrate can be explained from the subsequent oxidation of the primary products (largely from the more reactive acetaldehyde).

The formation of 2-butanone vs. acetaldehyde is controlled by the competition between the decomposition of 2-butoxy radicals and their reaction with O_2. The recognition of the formation and subsequent decomposition of activated 2-butoxy radicals (Libuda et al., 2002; Cassanelli et al., 2005) complicates the analysis of some of the older data regarding this competition, but a central value of $k_{decomp}/k_{O2} = (2.3 \pm 0.5) \times 10^{18}$ molecule cm^{-3} can be extracted from the collective data set at 298 K. Thus, in 1 atm. air at 298 K, a 2-butanone yield of $68 \pm 5\%$ and an acetaldehyde yield of $64 \pm 10\%$ is obtained per 2-butoxy radical formed; note that two acetaldehydes are formed per 2-butoxy decomposition. Product studies as a function of temperature (Libuda et al., 2002; Cassanelli et al., 2005) are consistent with a barrier to 2-butoxy decomposition of 46–50 kJ mol^{-1}, and this barrier results in a sharp decline in the yield of acetaldehyde with decreasing temperature.

The oxidation of the 1-butylperoxy radical in the presence of NO_x leads to the other major primary products, butanal (Carter et al., 1979; Cox et al., 1981; Kwok et al., 1996a; Geiger et al., 2002; Cassanelli et al,. 2005), 1-butyl nitrate (Atkinson et al., 1982d; Espada et al., 2005) and the more recently identified 4-hydroxybutanal (Kwok et al., 1996):

$$CH_3CH_2CH_2CH_2O_2\bullet + NO \rightarrow CH_2CH_2CH_2CH_2O\bullet + NO_2$$

$$\rightarrow CH_2CH_2CH_2CH_2ONO_2$$

$$CH_3CH_2CH_2CH_2O\bullet \rightarrow \bullet CH_2CH_2CH_2CH_2OH$$

$$CH_3CH_2CH_2CH_2O\bullet + O_2 \rightarrow CH_3CH_2CH_2CHO + HO_2$$

$$\bullet CH_2CH_2CH_2CH_2OH + O_2 \rightarrow \bullet O_2CH_2CH_2CH_2CH_2OH$$

$$\bullet O_2CH_2CH_2CH_2CH_2OH + NO \rightarrow \bullet OCH_2CH_2CH_2CH_2OH + NO_2$$

$$\rightarrow O_2NOCH_2CH_2CH_2CH_2OH$$

$$\bullet OCH_2CH_2CH_2CH_2OH \rightarrow HOCH_2CH_2CH_2CH(\bullet)OH$$

$$HOCH_2CH_2CH_2CH(\bullet)OH + O_2 \rightarrow HOCH_2CH_2CH_2CHO + HO_2$$

Studies of the competition between 1-butoxy radical isomerization and its reaction with O_2, largely through an observation of the change in butanal yield with O_2 partial pressure, have led to a rate coefficient ratio of $k_{isom}/k_{O2} = (1.8 \pm 0.4) \times 10^{19}$ molecule cm^{-3} at 298 K and 1 atm. air (Carter et al., 1979; Cox et al., 1981; Niki et al., 1981b; Geiger et al., 2002). Therefore, the yield of butanal per 1-butoxy formed is about $22 \pm 5\%$ under these conditions, with the remainder forming largely 4-hydroxy-butanal via isomerization. Due to the existence of a ~29 kJ mol^{-1} barrier to the isomerization, the yield of butanal will increase with decreasing temperature, although isomerization is likely to remain of some importance in the upper troposphere (Cassanelli et al., 2005). The possibility of chemically activated isomerization, following formation of 1-butoxy from reaction of 1-butylperoxy with NO, must also be considered; yields of activated 1-butoxy radicals of about 20% have been reported (Geiger et al., 2002; Caralp and Forst, 2003).

Product studies of butane oxidation in the absence of NO_x are less extensive and less detailed than those conducted in the presence of NO_x. Hanst and Gay (1983) observed the formation of CO and CO_2, peracetic acid, butyl hydroperoxide(s), and unidentified carbonyl compounds. The butyl hydroperoxide species likely results from reaction of HO_2 with butylperoxy radicals, while CO, CO_2, and peracetic acid are among the products expected from the oxidation of acetaldehyde in the absence of NO_x.

Libuda et al. (2002) photolyzed 2-butyl iodide in air in the absence of NO_x, as part of a comprehensive study of 2-butoxy radicals. Although acetaldehyde and 2-butanone were apparently observed in the absence of NO_x, no yields were reported.

Thus, quantitative details of reaction of 1- or 2-butylperoxy with HO_2, or with CH_3O_2 or other peroxy radicals, remain unknown at this point. Possible reaction channels are:

$$CH_3CH_2CH_2CH_2O_2\bullet + HO_2 \rightarrow CH_3CH_2CH_2CH_2OOH + O_2$$

$$\rightarrow CH_3CH_2CH_2CHO + H_2O + O_2$$

$$CH_3CH_2CH(O_2\bullet)CH_3 + HO_2 \rightarrow CH_3CH_2CH(OOH)CH_3 + O_2$$

$$\rightarrow CH_3CH_2C(O)CH_3 + H_2O + O_2$$

$$CH_3CH_2CH_2CH_2O_2\bullet + CH_3O_2\bullet \rightarrow CH_2CH_2CH_2CH_2O\bullet + CH_3O\bullet + O_2$$

$$\rightarrow CH_3CH_2CH_2CHO + CH_3OH + O_2$$

$$\rightarrow CH_3CH_2CH_2CH_2OH + CH_2O + O_2$$

$$CH_3CH_2CH(O_2\bullet)CH_3 + CH_3O_2\bullet \rightarrow CH_3CH_2CH(O\bullet)CH_3 + CH_3O\bullet + O_2$$

$$\rightarrow CH_3CH_2CH(OH)CH_3 + CH_2O + O_2$$

$$\rightarrow CH_3CH_2C(O)CH_3 + CH_3OH + O_2$$

IV-C-6. Products of Atmospheric Oxidation of 2-Methylpropane [Isobutane, $(CH_3)_3CH$]

Reaction of OH with 2-methylpropane can occur at two sites, at the tertiary hydrogen or at any of the three equivalent methyl groups, leading to the formation of 2-methyl-2-propylperoxy radicals or 2-methyl-1-propylperoxy radicals:

$$OH + (CH_3)_3CH\ (+O_2) \rightarrow (CH_3)_3CO_2\bullet + H_2O$$

$$\rightarrow (CH_3)_2CHCH_2O_2\bullet + H_2O$$

Predicted yields using the structure–additivity formulation of Kwok and Atkinson (1995) are 79% for the 2-methyl-2-propylperoxy (*tert*-butylperoxy) and 21% for the 2-methyl-1-propylperoxy (*iso*-butylperoxy) species.

The chemistry of the 2-methyl-2-propylperoxy radical in the absence of NO_x was explored by Thomas and Calvert (1962) from the continuous photolysis of 2, 2′-azoisobutane. Products identified by FTIR spectroscopy were acetone, formaldehyde, *tert*-butyl hydroperoxide, and *tert*-butyl alcohol, which can arise from the reactions of $(CH_3)_3CO_2$ with HO_2, CH_3O_2, and other $(CH_3)_3CO_2$ radicals:

$$(CH_3)_3CO_2\bullet + (CH_3)_3CO_2\bullet \rightarrow 2(CH_3)_3CO\bullet + O_2$$

$$(CH_3)_3CO\bullet\ (+O_2) \rightarrow CH_3C(O)CH_3 + CH_3O_2\bullet$$

$$(CH_3)_3CO_2\bullet + CH_3O_2\bullet \rightarrow (CH_3)_3CO\bullet + CH_3O\bullet + O_2$$

$$\rightarrow (CH_3)_3COH + CH_2O + O_2$$

$$(CH_3)_3CO\bullet \rightarrow CH_3\bullet + CH_3C(O)CH_3$$

$$CH_3\bullet + O_2 \rightarrow CH_3O_2\bullet$$

$$CH_3O\bullet + O_2 \rightarrow CH_2O + HO_2$$

$$CH_3O_2\bullet + CH_3O_2\bullet \rightarrow CH_3O\bullet + CH_3O\bullet + O_2$$

$$\rightarrow CH_2O + CH_3OH + O_2$$

$$(CH_3)_3COO\bullet + HO_2 \rightarrow 2(CH_3)_3COOH + O_2$$

Although no direct studies are available in the literature, the chemistry of $(CH_3)_3CO_2$ radicals in the presence of NO_x should be straightforward, leading to high yields of acetone and formaldehyde:

$$(CH_3)_3COO\bullet + NO \rightarrow (CH_3)_3CO\bullet + NO_2$$

$$\rightarrow (CH_3)_3CONO_2$$

$$(CH_3)_3CO\bullet \rightarrow CH_3C(O)CH_3 + CH_3\bullet$$

$$CH_3\bullet\ (+2O_2, NO) \rightarrow\rightarrow CH_2O(+NO_2 + HO_2)$$

Note, however, that the decomposition of $(CH_3)_3CO$ (to acetone and methyl radical) is reasonably slow, $k \approx 3000\ s^{-1}$, and formation of $(CH_3)_3CONO$ and $(CH_3)_3CONO_2$ [via addition of NO or NO_2 to $(CH_3)_3CO$] is often observed in laboratory environments, e.g., Wallington et al. (1992a).

The chemistry of 2-methyl-1-propylperoxy radical in the presence of NO_x has been studied by Geiger et al. (2002):

$$(CH_3)_2CHCH_2OO\bullet + NO \rightarrow (CH_3)_2CHCH_2O\bullet + NO_2$$

$$\rightarrow (CH_3)_2CHCH_2ONO_2$$

$$(CH_3)_2CHCH_2O\bullet + O_2 \rightarrow (CH_3)_2CHCHO + HO_2$$

$$(CH_3)_2CHCH_2O\bullet \rightarrow (CH_3)_2CH\bullet + CH_2O$$

$$(CH_3)_2CH\bullet + O_2 + NO_x \rightarrow\rightarrow CH_3C(O)CH_3, (CH_3)_2CHONO_2$$

These authors report $k_{decomp}/k_{O2} = (6.2 \pm 1.2) \times 10^{18} cm^3\ molecule^{-1}$ for 2-methyl-1-propoxy radical, thus implying about 55% decomposition (giving mainly acetone and formaldehyde) and 45% reaction with O_2 (giving 2-methylpropanal) in 1 atm. air at 298 K. At lower temperatures of the free and upper troposphere, decomposition will be slower [see theoretical studies of 2-methyl-1-propylperoxy radical chemistry (Méreau et al., 2000b; Fittschen et al., 2000)], and increased yields of 2-methylpropanal would be expected.

IV-C-7. Products of Atmospheric Oxidation of Pentane (n-Pentane, n-C_5H_{12})

The OH-initiated oxidation of pentane in the presence of NO_x has been studied a number of times over the years (Carter et al., 1976; Cox et al., 1980b; Atkinson et al., 1995; Kwok et al., 1996; Arey et al., 2001; Reisen et al., 2005); the most extensive studies are by Atkinson and co-workers. Primary products observed and reported yields in air at 298 K are summarized in table IV-C-3. Additional information regarding the chemistry involved can be derived from end-product studies involving 1-pentoxy, 2-pentoxy, and 3-pentoxy radicals (Geiger et al., 2002; Meunier et al., 2003; Johnson et al., 2004a). Further studies in the absence of NO_x, summarized at the end of this section, have been carried out by Heimann et al. (2002).

Attack by OH on pentane can occur at three positions, with predicted yields at 298 K of 8%, 57%, and 35%, respectively, for the 1-, 2-, and 3-pentylperoxy species (Kwok and Atkinson, 1995):

$$OH + CH_3CH_2CH_2CH_2CH_3 \rightarrow \bullet CH_2CH_2CH_2CH_2CH_3 + H_2O \qquad (8\%)$$

$$\rightarrow CH_3CH(\bullet)CH_2CH_2CH_3 + H_2O \qquad (57\%)$$

$$\rightarrow CH_3CH_2CH(\bullet)CH_2CH_3 + H_2O \qquad (35\%)$$

The chemistry of the 3-pentylperoxy system in the presence of NO_x is reasonably straightforward, involving competing fates (reaction with O_2 and decomposition) of the 3-pentoxy radical:

$$CH_3CH_2CH(\bullet)CH_2CH_3 + O_2 \rightarrow CH_3CH_2CH(O_2\bullet)CH_2CH_3$$

$$CH_3CH_2CH(O_2\bullet)CH_2CH_3 + NO \rightarrow CH_3CH_2CH(O\bullet)CH_2CH_3 + NO_2$$

$$\rightarrow CH_3CH_2CH(ONO_2)CH_2CH_3$$

$$CH_3CH_2CH(O\bullet)CH_2CH_3 + O_2 \rightarrow CH_3CH_2C(O)CH_2CH_3 + HO_2$$

$$CH_3CH_2CH(O\bullet)CH_2CH_3 \rightarrow CH_3CH_2\bullet + CH_3CH_2CHO$$

$$CH_3CH_2\bullet + O_2 \rightarrow CH_3CH_2O_2\bullet$$

$$CH_3CH_2O_2\bullet + NO \rightarrow CH_3CH_2O\bullet + NO_2$$

$$\rightarrow CH_3CH_2ONO_2$$

$$CH_3CH_2O\bullet + O_2 \rightarrow CH_3CHO + HO_2$$

This chemistry is the origin of the 3-pentyl nitrate, 3-pentanone, acetaldehyde, and propanal which have been observed as primary products. The PAN and peroxypropionyl nitrate (PPN) species identified by Cox et al. (1980b) also arise

Table IV-C-3. Summary of available data regarding the end-products of pentane in the presence of NO_x. Reported product yields (%) in 1 atm. air at 298 K are given when available, *x*'s indicate observed products for which no yield information is readily retrievable

Primary Product	Carter et al. (1976)	Cox et al. (1980b) [a]	Atkinson et al. (1982d)	Atkinson et al. (1983c)	Atkinson et al. (1995)	Kwok et al. (1996)	Arey et al. (2001)	Reisen et al. (2005)
Pentanal	*x*							
2-Pentanone	*x*				5.4 ± 0.8			
3-Pentanone	*x*				25 ± 3			
Total pentanones						25		
1-Pentyl nitrate		*x*						
2-Pentyl nitrate		*x*	7.1 ± 0.9	7.4 ± 0.2	6.1 ± 1.2			
3-Pentyl nitrate		*x*	4.6 ± 0.6	5.2 ± 0.4	4.7 ± 0.6			
Acetaldehyde								
Propanal								
1,4-Hydroxy-carbonyls (total)					*x*	33	36	54 ± 20
5-Hydroxy-2-pentanone								49 ± 19
4-Hydroxy-pentanal								5.2
1,4-Hydroxy-nitrates (total)							2.6	

[a] Secondary products PAN, PPN, and CH_3ONO_2 are also observed. Specific isomer(s) of pentyl nitrate not identified.

primarily from this channel, via the oxidation of acetaldehyde and propanal. Combining the estimated 35% yield of 3-pentylperoxy from the initial OH reaction and the observed 25 ± 3% yield (Atkinson et al,. 1995) of 3-pentanone indicates that: 1) approximately 10% yields of acetaldehyde and propanal can be expected from this channel; and 2) that the reaction with O_2 is about 2.5 times more rapid than decomposition for 3-pentoxy at 298 K, consistent with the data presented in section IV-B.

The chemistry of the 2-pentylperoxy and 1-pentylperoxy systems is more complex, due to the dominance of the isomerization channels in the chemistry of the corresponding alkoxy radicals:

$$CH_3CH_2CH_2CH(O\bullet)CH_3 \rightarrow \bullet CH_2CH_2CH_2CH(OH)CH_3$$

$$CH_3CH_2CH_2CH(O\bullet)CH_3 + O_2 \rightarrow CH_3CH_2CH_2CH(O)CH_3 + HO_2$$

$$CH_3CH_2CH_2CH(O\bullet)CH_3 \rightarrow CH_3CH_2CH_2\bullet + CH_3CHO$$

$$\bullet CH_2CH_2CH_2CH(OH)CH_3 + O_2 \rightarrow \bullet OOCH_2CH_2CH_2CH(OH)CH_3$$

$$\bullet OOCH_2CH_2CH_2CH(OH)CH_3 + NO \rightarrow \bullet OCH_2CH_2CH_2CH(OH)CH_3 + NO_2$$

$$\rightarrow O_2NOCH_2CH_2CH_2CH(OH)CH_3$$

$$\bullet OCH_2CH_2CH_2CH(OH)CH_3\ (+O_2) \rightarrow HOCH_2CH_2CH_2C(O)CH_3(+HO_2)$$

$$CH_3CH_2CH_2\bullet\ (+O_2, NO_x) \rightarrow\rightarrow CH_3CH_2CHO;\ CH_3CH_2CH_2ONO_2$$

$$\bullet OCH_2CH_2CH_2CH_2CH_3\ (+O_2) \rightarrow HOCH_2CH_2CH_2CH(O_2\bullet)CH_3$$

$$HOCH_2CH_2CH_2CH(O_2\bullet)CH_3 + NO \rightarrow HOCH_2CH_2CH_2CH(O\bullet)CH_3 + NO_2$$

$$\rightarrow HOCH_2CH_2CH_2CH(ONO_2)CH_3$$

$$HOCH_2CH_2CH_2CH(O\bullet)CH_3\ (+O_2) \rightarrow O{=}CHCH_2CH_2CH(OH)CH_3(+HO_2)$$

Atkinson and co-workers (Carter et al, 1976; Atkinson et al., 1995; Kwok et al., 1996; Arey et al., 2001; Reisen et al., 2005) have expended a great deal of effort in first identifying and later quantifying and speciating the multifunctional products expected from the isomerization process. These methodologies, which have also been applied to hexane, heptane, and octane, will be described in some detail here. Initial evidence for the occurrence of isomerization was surmised from the absence of a large yield of 2-pentanone in the pentane/NO_x/O_2 system (Carter et al., 1976; Cox et al., 1980b). Subsequently, an increasingly sophisticated array of atmospheric pressure ionization–mass spectrometer (API–MS) and GC–FID/GC–MS techniques (with derivatization) have been applied to the study of these isomerizations (Atkinson et al., 1995; Kwok et al., 1996; Arey et al., 2001; Reisen et al., 2005). Initial experiments (Atkinson et al., 1995; Kwok et al., 1996) involved the use of the API–MS system in positive ion mode, with protonated water clusters acting as the main reagent ion.

Peaks corresponding to the 1,4-hydroxycarbonyl compounds (5-hydroxy-2-pentanone from 2-pentoxy; 4-hydroxy-pentanal from 1-pentoxy) were clearly observed from the oxidation of both pentane and its fully deuterated isomer. More recently, Arey et al. (2001) applied the API–MS system in the negative ion mode, with pentafluorobenzyl alcohol (PFBOH) or NO_2 added to the system to generate reagent ions such as NO_2^-, $[PFBOH{\cdot}O_2]^-$, and $[PFBOH{\cdot}NO_2]^-$. Using this technique, Arey et al. (2001) were able to identify clearly peaks due to the 1,4-hydroxycarbonyls and 1,4-hydroxynitrates from both pentane and pentane-d_{12}. They were also able to estimate the total (unspeciated) yields of these species; a hydroxycarbonyl yield of 36% and a hydroxynitrate yield of 2.6% were reported, with estimated uncertainties of about a factor of two. In a more recent study, Reisen et al. (2005) used an on-fiber derivatization [with O-(2,3,4,5,6-pentafluorobenzyl)-hydroxylamine] technique with solid-phase microextraction, coupled with GC–FID and GC–MS for the speciation and quantification of the hydroxycarbonyl species. Identification and quantification of 5-hydroxy-2-pentanone (49 ± 19% yield) and 4-hydroxypentanal (~5% yield) were reported.

The 2-pentoxy radical is expected to be formed with roughly a 55% yield (Kwok and Atkinson, 1995) and, from the discussion in section IV-B, isomerization should be its dominant fate at 298 K, 1 atm air (~90%), with reaction with O_2 contributing the bulk of the remaining reactivity. The observed yields of 2-pentanone, 5.4 ± 0.8% (Atkinson et al., 1995), and 5-hydroxy-2-pentanone, 49 ± 19% (Reisen et al., 2005), are entirely consistent with these expectations. Isomerization should be even more dominant in the case of 1-pentoxy. This fact, coupled with the small initial formation yield of 1-pentylperoxy, explains the reasonably small yields of 4-hydroxypentanal, 5.2% (Reisen et al., 2005), and the absence of detectable quantities of pentanal in the system.

Heimann et al. (2002) carried out end-product studies of the photolysis of 1-, 2-, and 3-iodopentane (and thus of 1-, 2-, and 3-pentylperoxy radicals) in air in the absence of NO_x; see table IV-C-4 for a summary.

Photolysis of 3-iodopentane led to the production of 3-pentanol, 3-pentanone, propanal, acetaldehyde, and ethanol. The product distribution was quantitatively consistent with a detailed model that included the pentylperoxy radical reactions with similar peroxy radicals and cross-reactions of the relevant peroxy radicals (3-pentylperoxy, ethylperoxy and HO_2). Competition between decomposition of 3-pentoxy and its reaction with O_2 was studied explicitly (by varying the O_2 partial pressure); the rate coefficient ratio obtained, $k_{decomp}/k_{O2} = (3.2 \pm 0.3) \times 10^{18}$ molecule cm^{-3}, is essentially identical to that reported by Atkinson et al. (1995), $k_{decomp}/k_{O2} = 3.3 \times 10^{18}$ molecule cm^{-3}.

Product distributions observed by Heimann et al. (2002) following the irradiation of 1-iodopentane and 2-iodopentane were more complex, owing to the preponderance of multifunctional products resulting from isomerization of the 1- and 2-pentoxy radicals. Although some GC peaks were not conclusively identified (presumably due to multifunctional species), and some quantitative discrepancies exist in the 1-pentyl peroxy case (see section IV-B-11), the observed products were reasonably consistent with detailed models of the reaction systems.

Table IV-C-4. Products and their yields relative to 1-, 2-, or 3-pentanol observed by Heimann et al. (2002) from the photolysis of iodopentanes in air

Product	Relative Yield	Likely Product Origin
a) From 1-iodopentane:		
1-Pentanol	1.0 ± 0.09	1-Pentylperoxy + RO_2
Pentanal	1.14 ± 0.06	1-Pentylperoxy + RO_2; 1-pentoxy + O_2
Butanal	0.102 ± 0.008	Chemistry of 1-butylperoxy, formed from 1-pentoxy decomposition
1-Butanol	0.102 ± 0.008	Chemistry of 1-butylperoxy, formed from 1-pentoxy decomposition
Pentan-1,4-diol	0.080 ± 0.025	Isomerization of 1-pentoxy, followed by rxn. of resulting peroxy radical with RO_2
Unidentified	0.91 ± 0.07	Likely from isomerization of 1-pentoxy
b) From 2-iodopentane:		
2-Pentanol	1.0 ± 0.11	2-Pentylperoxy + RO_2
2-Pentanone	1.47 ± 0.18	2-Pentylperoxy + RO_2; 2-pentoxy + O_2
Propanal	0.073 ± 0.012	Decomposition of 2-pentoxy, and subsequent chemistry of 1-propyl radical
Propanol	0.023 ± 0.004	Decomposition of 2-pentoxy, and subsequent chemistry of 1-propyl radical
Pentan-1,4-diol	0.015 ± 0.01	Oxidation of 2-pentanol
Acetaldehyde	0.16 ± 0.02	Decomposition of 2-pentoxy
Unidentified	1.07 ± 0.22	Isomerization of 2-pentoxy
c) From 3-iodopentane:		
3-Pentanol	1.00 ± 0.03	3-Pentylperoxy + RO_2
3-Pentanone	1.80 ± 0.04	3-Pentylperoxy + RO_2; 3-pentoxy + O_2
Propanal	0.49 ± 0.02	3-Pentoxy decomposition
Acetaldehyde	0.33 ± 0.04	3-Pentoxy decomposition, and subsequent chemistry of ethyl radical
Ethanol	0.10 ± 0.04	3-Pentoxy decomposition, and subsequent chemistry of ethyl radical

IV-C-8. Products of Atmospheric Oxidation of 2-Methylbutane [Isopentane, $(CH_3)_2CHCH_2CH_3$]

Cox et al. (1980b) studied the OH-initiated oxidation of 2-methylbutane in the presence of NO_x. Using a combination of GC–FID and GC–ECD techniques, they identified acetone, acetaldehyde, PAN, methyl nitrate, as well as propyl and pentyl nitrates as reaction products; no yield data were reported.

Although four unique sites of OH attack on 2-methylbutane exist, the chemistry will be dominated by the attack at the tertiary (—CH—) and secondary (—CH_2—) sites, with predicted yields of 59% and 28%, respectively (Kwok and Atkinson, 1995):

$$OH + (CH_3)_2CHCH_2CH_3 + O_2 \rightarrow (CH_3)_2C(O_2\bullet)CH_2CH_3 \qquad (59\%)$$

$$\rightarrow (CH_3)_2CHCH(O_2\bullet)CH_3 \qquad (28\%)$$

Subsequent steps in the oxidation of 2-methyl-2-butylperoxy, $(CH_3)_2C(O_2\bullet)CH_2CH_3$, as first outlined by Cox et al. (1980b), explain their observation of acetone, acetaldehyde, PAN, and methyl nitrate (the latter two species being by-products of acetaldehyde oxidation):

$$(CH_3)_2C(OO\bullet)CH_2CH_3 + NO \rightarrow (CH_3)_2C(O\bullet)CH_2CH_3 + NO_2$$

$$\rightarrow (CH_3)_2C(ONO_2)CH_2CH_3$$

$$(CH_3)_2C(O\bullet)CH_2CH_3 \rightarrow CH_3C(O)CH_3 + CH_3CH_2\bullet$$

$$CH_3CH_2\bullet + O_2\ (+NO_x, O_2) \rightarrow\rightarrow CH_3CHO, CH_3CH_2ONO_2$$

Oxidation of 3-methyl-2-butylperoxy radical likely occurs as follows:

$$(CH_3)_2CHCH(O_2\bullet)CH_3 + NO \rightarrow (CH_3)_2CHCH(O\bullet)CH_3 + NO_2$$

$$\rightarrow (CH_3)_2CHCH(ONO_2)CH_3$$

$$(CH_3)_2CHCH(O\bullet)CH_3 + O_2 \rightarrow (CH_3)_2CHC(O)CH_3 + HO_2$$

$$(CH_3)_2CHCH(O\bullet)CH_3 \rightarrow (CH_3)_2CH\bullet + CH_3CHO$$

$$(CH_3)_2CH\bullet\ (+O_2, NO_x) \rightarrow\rightarrow CH_3C(O)CH_3, (CH_3)_2CHONO_2$$

The structure–additivity rules of Peeters et al. (2004) predict a rather low energy barrier ($\sim$38 kJ mol^{-1}) for decomposition of 3-methyl-2-butoxy radical, $(CH_3)_2CHCH(O\bullet)CH_3$ (see table IV-B-11). Thus the decomposition reaction (leading to acetone and acetaldehyde) should dominate over the O_2 reaction at 298 K and should remain a major reaction pathway even in the upper troposphere.

IV-C-9. Products of Atmospheric Oxidation of 2,2-Dimethylpropane [Neopentane, $(CH_3)_3CCH_3$]

Aspects of the atmospheric oxidation of neopentane have been studied using both flash photolysis/UV absorption and environmental chamber/FTIR absorption techniques (Lightfoot et al., 1990b; Wallington et al., 1992a; Rowley et al., 1992a).

Reaction of neopentane with OH can lead to only one peroxy radical, $(CH_3)_3CCH_2O_2\bullet$. Reaction of this radical with NO will result in the production of either neopentyl nitrate [estimated yield 10% at 298 K, 1 atm. pressure (Atkinson et al., 1984a, 1987)] or the neopentoxy radical, $(CH_3)_3CCH_2O\bullet$:

$$(CH_3)_3CCH_2O_2\bullet + NO \rightarrow (CH_3)_3CCH_2O\bullet + NO_2$$

$$\rightarrow (CH_3)_3CCH_2ONO_2$$

Both Lightfoot et al. (1990b) and Wallington et al. (1992a) have examined the atmospheric fate of the neopentoxy radical, $(CH_3)_3CCH_2O\bullet$, which might decompose or react with O_2:

$$(CH_3)_3CCH_2O\bullet \rightarrow (CH_3)_3C\bullet + CH_2O$$

$$(CH_3)_3CCH_2O\bullet + O_2 \rightarrow (CH_3)_3CCHO + HO_2$$

Lightfoot et al. (1990b), from a consideration of the yields of $(CH_3)_3CO_2\bullet$ and HO_2 following the $(CH_3)_3CCH_2O_2\bullet$–$(CH_3)_3CCH_2O_2\bullet$ radical reaction in their flash photolysis system, estimated that decomposition of the $(CH_3)_3CCH_2O\bullet$ radical occurs with a rate coefficient of 2×10^6 s^{-1} at 298 K, decreasing to 3×10^5 s^{-1} at 248 K.

Environmental chamber data from Wallington et al. (1992a) confirm the rapid decomposition of neopentoxy. Using FTIR spectroscopy to examine products of the Cl-atom-initiated oxidation of neopentane in the presence of NO_x, these authors observed a 105 ± 6% yield of formaldehyde and no observable 2,2-dimethylpropanal (yield< 10%), consistent with the predominant occurrence of $(CH_3)_3CCH_2O$ decomposition. Both $(CH_3)_3CONO$ and $(CH_3)_3CONO_2$ were observed from subsequent reaction of the $(CH_3)_3CO$ radical in the presence of elevated NO_x concentrations:

$$(CH_3)_3C\bullet + O_2 \rightarrow (CH_3)_3CO_2\bullet$$

$$(CH_3)_3CO_2\bullet + NO \rightarrow (CH_3)_3CO\bullet + NO_2$$

$$\rightarrow (CH_3)_3CONO_2$$

$$(CH_3)_3CO\bullet + NO[NO_2] \rightarrow (CH_3)_3CONO, [(CH_3)_3CONO_2]$$

Wallington et al. (1992a) also examined the Cl-initiated oxidation of neopentane in the absence of NO_x at three O_2 partial pressures (10, 150, and 700 Torr). Yields of formaldehyde (47±4% yield), a hydroperoxide (presumably neopentyl hydroperoxide, 21±4%), acetone (14±3%), neopentanol (12±2%) and methanol (2.6±0.3%) were found to be independent of O_2, whereas the yield of neopentanal increased somewhat with O_2 (37 ± 3% at 10 and 150 Torr O_2, 49 ± 4% at 700 Torr O_2). The increased yield of neopentanal at high O_2 partial pressures might be indicative of the onset of some reaction of the neopentoxy radical with O_2 under these conditions.

Thus, the predominant fate of the $(CH_3)_3CCH_2O\bullet$ radical in 1 atm. air at 298 K is decomposition to $(CH_3)_3C\bullet$ and CH_2O. Subsequent products of the oxidation of the $(CH_3)_3C\bullet$ fragment in the atmosphere will include acetone, $(CH_3)_3CONO_2$, and formaldehyde. At sub-ambient temperatures, some reaction of the $(CH_3)_3CCH_2O$ with O_2 may occur to generate pivaldehyde $[(CH_3)_3CCHO]$. Again it is suggested that a more detailed study of neopentane oxidation would be valuable, with particular attention paid to the examination of low-temperature product yields and the possible occurrence of chemical activation in the chemistry of neopentoxy radical.

IV-C-10. Products of Atmospheric Oxidation of Hexane (n-Hexane, n-C_6H_{14})

The reaction of OH with hexane can occur at three unique sites, with calculated probabilities at 298 K of 52% for the 3-position, 42% for the 2-position, and 6% for the 1-position (Kwok and Atkinson, 1995). The expected reaction of each of the resultant alkoxy radicals (3-, 2-, and 1-hexoxy) is isomerization at 298 K, so the major products (in addition to the 1-, 2-, and 3-hexyl nitrate species) are expected to be an array of multifunctional six-carbon 1,4-hydroxycarbonyl and 1,4-hydroxynitrate compounds. The expected reaction sequence for 3-hexoxy radical, leading to 6-nitrooxy-3-hexanol and 6-hydroxy-3-hexanone, is illustrated below as an example:

$$CH_3CH_2CH(O\bullet)CH_2CH_2CH_3\,(+O_2) \rightarrow CH_3CH_2CH(OH)CH_2CH_2CH_2O_2\bullet$$

$$CH_3CH_2CH(OH)CH_2CH_2CH_2O_2\bullet + NO \rightarrow CH_3CH_2CH(OH)CH_2CH_2CH_2O\bullet + NO_2$$

$$\rightarrow CH_3CH_2CH(OH)CH_2CH_2CH_2ONO_2$$

$$CH_3CH_2CH(OH)CH_2CH_2CH_2O\bullet \rightarrow CH_3CH_2C(\bullet)(OH)CH_2CH_2CH_2OH$$

$$CH_3CH_2C(\bullet)(OH)CH_2CH_2CH_2OH + O_2 \rightarrow CH_3CH_2C(O)CH_2CH_2CH_2OH + HO_2$$

Carter et al. (1976) were the first to report on the end-products of hexane oxidation. Their observation of very low yields of simple hexanone/hexanal species was consistent with the expected preponderance of hexoxy radical isomerization. Similar conclusions were drawn by Cox et al. (1980b), who were unable to detect any hexanone isomers.

Eberhard et al. (1995) were the first to identify and quantify the formation of the 1,4-hydroxycarbonyl products in the hexane system. These authors studied the chemistry of 2-hexoxy and 3-hexoxy radicals directly (through photolysis of the corresponding nitrites in air in the presence of NO_x), and also detected end-products of the OH-initiated oxidation of hexane in the presence of NO_x. A combination of GC–ECD (for alkyl nitrite quantification), GC–MS (for qualitative product identification), and derivatization with 2,4-dinitrophenylhydrazine (DNPH) followed by HPLC separation (for quantitative determination of carbonyl compounds) was used. Although a great number of products were observed, some of which seem to be artifactual, the authors clearly showed that 5-hydroxy-2-hexanone was the major product of 2-hexoxy oxidation (yield $> 70\%$). For 3-hexoxy, the absence of large yields of 3-hexanone pointed to the occurrence of isomerization, although the expected hydroxycarbonyl species, 6-hydroxy-3-hexanone, was not identified or quantified.

Kwok et al. (1996), using their API–MS system in the positive ion mode, were able to identify the hydroxycarbonyl compounds from hexane and its fully deuterated analogue, although no quantitative information was obtained. Hexanone/hexanal isomers were formed in very small yield (3.3%, uncertainty of a factor of two), again consistent with the prevalence of isomerization of the hexoxy radicals generated during the oxidation process.

Arey et al. (2001) used a combination of GC–FID, GC–MS, and API–MS techniques to detect and quantify products of the OH-initiated oxidation of hexane. Yields of 2-hexyl and 3-hexyl nitrate were determined, using GC–FID, to be 5.9 ± 0.7 and $8.2 \pm 1.0\%$, respectively. When normalized to the expected yield of the relevant hexylperoxy radicals [on the basis of the estimation procedure of Kwok and Atkinson (1995)], a nitrate yield of $15 \pm 1.9\%$ per secondary peroxy radical was determined. Using the API–MS system in the negative ion mode, products with mass/charge ratio of 116 (C_6-hydroxycarbonyls) and 163 (C_6-hydroxynitrates) were identified. Studies of fully deuterated hexane gave rise to peaks corresponding to the analogous deuterated products. A total yield for the C_6 hydroxycarbonyl species of 53% and for the C_6 hydroxynitrates of 3.5% (estimated uncertainty of a factor of two) were determined. Small yields of C_4- and C_5-hydroxycarbonyls and C_5-hydroxynitrates were also reported.

Finally, Reisen et al. (2005) used their precoated SPME fiber technique, coupled with GC–FID and GC–MS detection systems, to identify and quantify the various 1,4-hydroxycarbonyl isomers formed. They found near-equal yields of 5-hydroxy-2-hexanone (24% yield) and 6-hydroxy-3-hexanone (26%), which are expected from OH attack at the 2- and 3-position, respectively. A 7% yield of 4-hydroxyhexanal, the product expected from OH reaction at the 1-position, was found. These yields are all about 50–100 % of the expected (Kwok and Atkinson, 1995) yield of the parent peroxy radical. Combining the yields of the hexyl nitrates [about 20% (Arey et al., 2001)] with those for the hydroxycarbonyls and hydroxynitrates accounts for about 80% of the hexane reaction products.

IV-C-11. Products of Atmospheric Oxidation of 2-Methylpentane [$(CH_3)_2CHCH_2CH_2CH_3$]

The chemistry of 2-methylpentane has not been studied in great detail. Cox et al. (1980b) studied its OH-initiated oxidation in the presence of NO_x, using a combination of GC–FID and GC–ECD. They identified acetone and propanal (both with small yields), PAN and PPN as secondary products, 1-propyl nitrate, and two isomers of hexyl nitrate.

Although there are five unique sites for OH attack to occur, the chemistry will be dominated by abstraction at the tertiary (>CH—) and the two secondary (—CH_2—) sites, with predicted yields of 42%, 25%, and 25%, respectively (Kwok and Atkinson, 1995):

$$OH + (CH_3)_2CHCH_2CH_2CH_3(+O_2) \rightarrow (CH_3)_2C(O_2\bullet)CH_2CH_2CH_3 \quad (42\%)$$

$$\rightarrow (CH_3)_2CHCH(O_2\bullet)CH_2CH_3 \quad (25\%)$$

$$\rightarrow (CH_3)_2CHCH_2CH(O_2\bullet)CH_3 \quad (25\%)$$

The chemistry of the corresponding alkoxy species, which will result from reaction of the peroxy radicals with NO, will generate acetone, propanal, 1- and 2-propyl nitrate, 5-hydroxy-2-methyl-2-pentanone, and 5-nitrooxy-4-methyl-2-pentanol, as illustrated in figure IV-C-2.

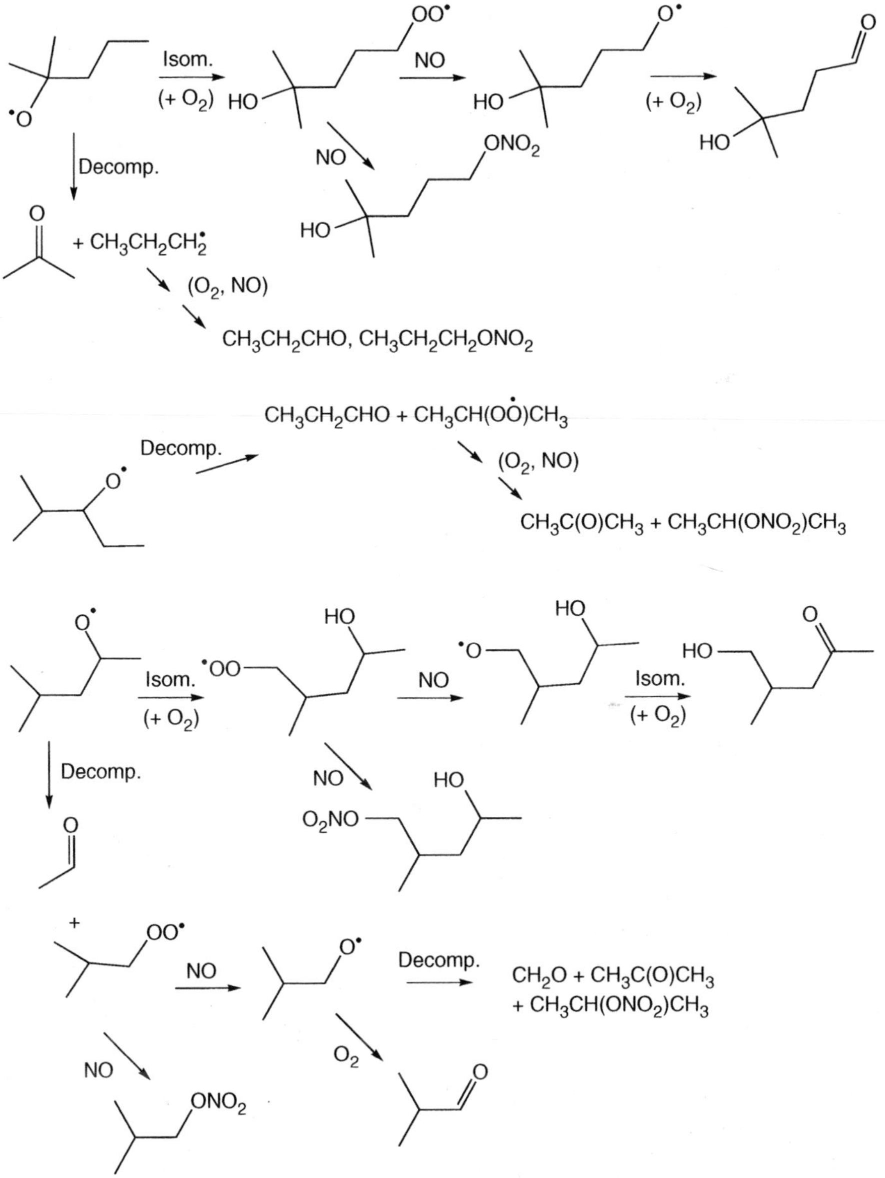

Figure IV-C-2. Reaction sequence for the alkoxy radicals formed from the atmospheric oxidation of 2-methylpentane.

Johnson et al. (2005) have shown that isomerization is the main fate of the $(CH_3)_2C(O\bullet)CH_2CH_2CH_3$ radical throughout the troposphere, with decomposition contributing no more than 6% to its chemistry. Thus a ~40% yield of 4-hydroxy-4-methylpentanal is anticipated. Decomposition will be the main fate of the 2-methyl-3-pentoxy radical, giving ~25% yields for both acetone and propanal. Isomerization is predicted (see table IV-B-11) to be the major fate of the $(CH_3)_2CHCH_2CH(O\bullet)CH_3$ radical at 298 K and atmospheric pressure, giving rise to a significant (~25%) yield of 5-hydroxy-4-methyl-2-pentanone. Reaction of this latter radical with O_2 will also likely play a role under conditions of the upper troposphere. The contribution of decomposition is expected to be minor throughout the troposphere.

IV-C-12. Products of Atmospheric Oxidation of 3-Methylpentane [$(CH_3CH_2CH(CH_3)CH_2CH_3)$]

There are no studies of the oxidation of 3-methylpentane in the open literature. However, a reasonable assessment of the likely end-products can be derived using structure–additivity relationships both for the site of initial OH attack and for the behavior of the relevant alkoxy radicals.

Reaction with OH is estimated (Kwok and Atkinson, 1995) to occur about 50% of the time at the tertiary site (3-position) and a total of about 40% of the time at the two equivalent secondary sites (2-position), leading to the production of 3-methyl-3-pentylperoxy and 3-methyl-2-pentylperoxy radicals. Reactions of these peroxy species are expected to occur in the usual way, leading to the production of nitrates and alkoxy radicals from reaction with NO, hydroperoxides from reaction with HO_2, and alkoxy radicals, alcohols, and ketones from reaction with CH_3O_2 or other alkylperoxy species. Expected reactions of the two alkoxy radical species (3-methyl-3-pentoxy and 3-methyl-2-pentoxy) are outlined in figure IV-C-3. Major products are thus likely to be 2-butanone and acetaldehyde, while smaller yields of ethyl nitrate, 2-butyl nitrate, 5-nitrooxy-3-methyl-2-pentanol, 4-oxo-3-methyl-pentanal, and 3-methyl-2-pentanone are also expected. The 3-methyl-2-pentoxy radical is of particular interest, since isomerization, decomposition, and reaction with O_2 are all plausible reactions. Structure–additivity rules (Peeters et al., 2004; Atkinson, 1997b) can be used to estimate activation energies of ~38 kJ mol^{-1} for the decomposition process and ~33 kJ mol^{-1} for the isomerization; see table IV-B-9. Because of the larger *A*-factor for decomposition, this process should dominate at 298 K, 1 atm. pressure. However, in the upper troposphere, measurable contributions from all three competing processes seem likely.

IV-C-13. Products of Atmospheric Oxidation of 2,3-Dimethylbutane [$(CH_3)_2CHCH(CH_3)_2$]

The oxidation of 2,3-dimethylbutane has been studied both in the presence of NO_x (Cox et al., 1980b) and in its absence (Heimann and Warneck, 1992). Cox et al. (1980b) observed acetone (150% molar yield), propyl and hexyl nitrates, and the secondary formation of PAN following the photolysis of HONO/2,3-dimethylbutane/NO/NO_2/air mixtures. Heimann and Warneck (1992) reported the formation of acetone (46.8%), 2-propanol (7.5%), 2,3-dimethylbutanal (8.7%), 2,3-dimethyl-1-butanol (3.4%), 2,3-dimethyl-2-butyl hydroperoxide (29.3%), and 2,3-dimethyl-2-butanol (2.2%), following the photolysis of H_2O_2/2,3-dimethylbutane/air mixtures.

The results of the two experiments can be rationalized as follows. Attack of OH on 2,3-dimethylbutane can occur at two sites, either at the two equivalent tertiary

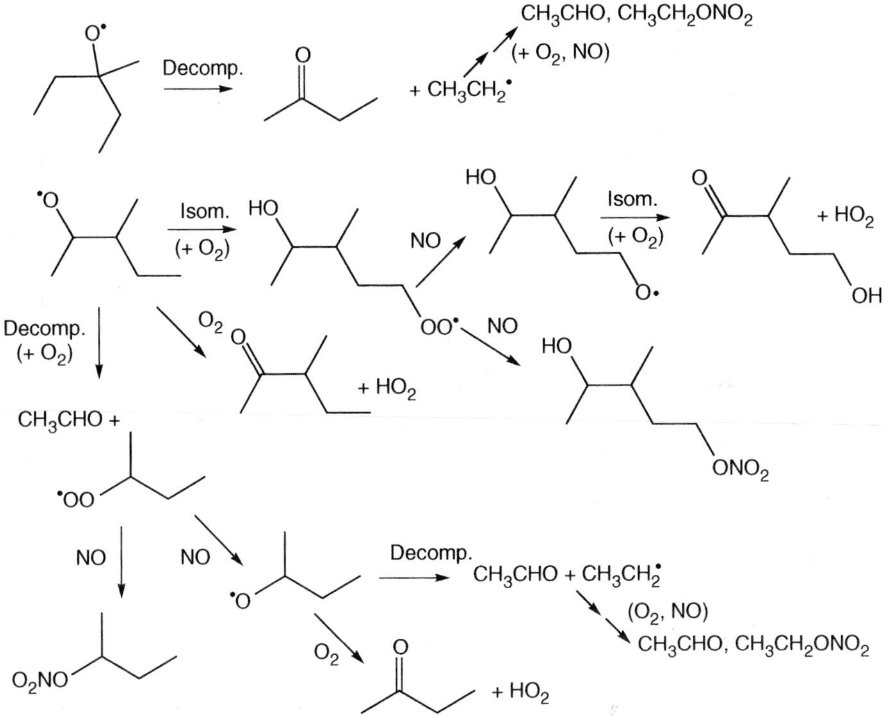

Figure IV-C-3. Sequence of reactions for the alkoxy radicals expected from the atmospheric oxidation of 3-methylpentane.

sites, leading to the formation of 2,3-dimethyl-2-butylperoxy radicals, or at the four equivalent methyl groups, leading to the formation of 2,3-dimethyl-1-butylperoxy radicals:

$$\mathrm{OH + (CH_3)_2CHCH(CH_3)_2 + O_2 \rightarrow (CH_3)_2CHC(O_2\bullet)(CH_3)_2 + H_2O}$$
$$\mathrm{\rightarrow (CH_3)_2CHCH(CH_3)CH_2O_2\bullet + H_2O}$$

From their observed product yields and a consideration of the likely mechanism involved in their formation, Heimann and Warneck (1992) estimated that 83% of the OH attack was occurring at the tertiary sites, in very good agreement with the estimation method of Kwok and Atkinson (1995) that predicts 78% reaction at the tertiary sites. In the presence of NO_x, the major product obtained from the subsequent chemistry of the 2,3-dimethyl-2-butylperoxy radicals is acetone, with smaller yields of 2,3-dimethyl-2-butyl nitrate and 2-propyl nitrate also expected:

$$\mathrm{(CH_3)_2CHC(O_2\bullet)(CH_3)_2 + NO \rightarrow (CH_3)_2CHC(O\bullet)(CH_3)_2 + NO_2}$$
$$\mathrm{\rightarrow (CH_3)_2CHC(ONO_2)(CH_3)_2}$$

$$(CH_3)_2CHC(O\bullet)(CH_3)_2\ (+O_2) \rightarrow CH_3C(O)CH_3 + (CH_3)_2CHO_2\bullet$$

$$(CH_3)_2CHO_2\bullet + NO \rightarrow (CH_3)_2CHO\bullet + NO_2$$

$$\rightarrow (CH_3)_2CHONO_2$$

$$(CH_3)_2CHO\bullet + O_2 \rightarrow CH_3C(O)CH_3 + HO_2$$

The high yield of 2,3-dimethyl-2-butylperoxy radicals and the production of nearly two molecules of acetone per 2,3-dimethyl-2-butylperoxy consumed is then consistent with the 150% yield of acetone observed by Cox et al. (1980b).

Subsequent chemistry of the 2,3-dimethyl-1-butylperoxy radical in the presence of NO_x has not been explicitly explored, but this can result in the formation of acetone, acetaldehyde, 3-methyl-2-butanone and formaldehyde (via decomposition of 2,3-dimethyl-1-butoxy), 2,3-dimethyl-butanal (via its reaction with O_2), or 2,3-dimethyl-4-hydroxy-butanal and 2,3-dimethyl-4-nitrooxy-butanol (via isomerization), with isomerization likely to be the major pathway at 298 K; see figure IV-C-4.

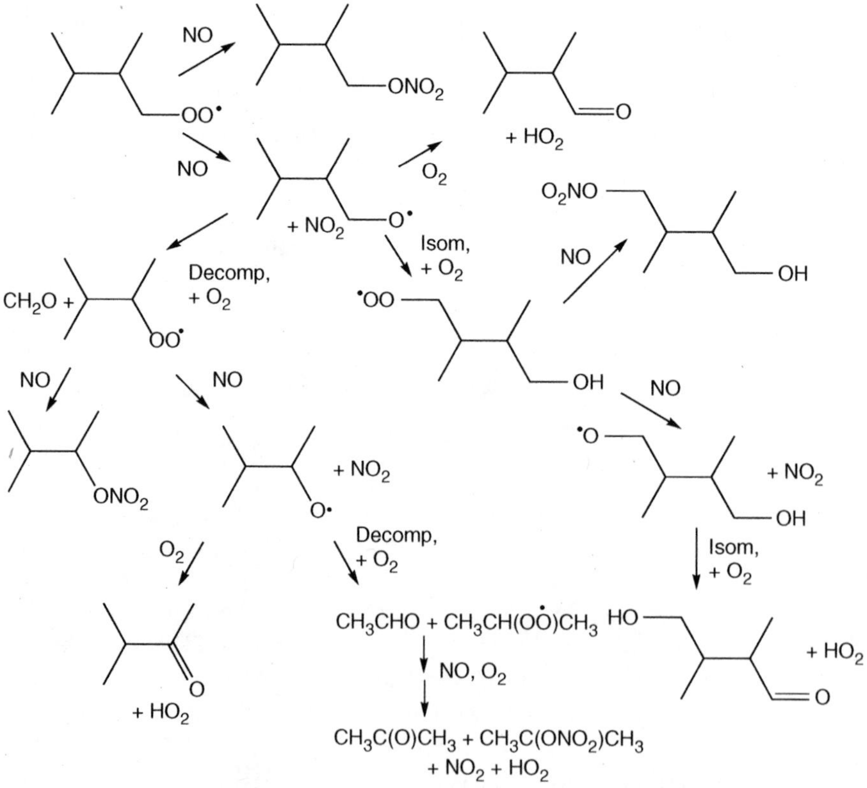

Figure IV-C-4. Sequence of reactions expected for the 2,3-dimethyl-1-butoxy radical formed in the atmospheric oxidation of 2,3-dimethylbutane.

The chemistry in the absence of NO_x is complicated by the occurrence of reactions between peroxy radicals of the same structure or in cross-reactions of the various peroxy radicals formed in the system, including the initially formed 2,3-dimethyl-2-butylperoxy and 2,3-dimethyl-1-butylperoxy species. Heimann and Warneck (1992) were able to adequately model their observed product distributions using reasonable estimates for the rate coefficients and branching ratios of these peroxy radical reactions.

IV-C-14. Products of Atmospheric Oxidation of Heptane (n-Heptane, n-C_7H_{16})

The OH-initiated oxidation of heptane in the presence of NO_x has been studied over the last decade by Atkinson and co-workers, using a combination of gas chromatographic and mass spectral techniques (Kwok et al., 1996; Arey et al., 2001; Reisen et al., 2005). The major products formed are C_7 1,4-hydroxycarbonyls (from the isomerization of the 1-,2-,3-, and 4-heptoxy radicals), C_7 1,4-hydroxynitrates (also end-products of heptoxy isomerization), heptyl nitrates (from reactions of the various heptylperoxy radicals with NO), and C_7 carbonyl compounds (from reactions of the heptoxy radicals with O_2). With four possible sites of attack of OH on heptane, four isomers of each class of compounds are expected.

Kwok et al. (1996), using API–MS in the positive ion mode, identified the formation of hydroxycarbonyls and heptanone compounds from both heptane and its fully deuterated analogue. The combined yield of the heptanone species was only 1.1% (with an estimated uncertainty of a factor of two). This low yield is in keeping with the prevalence of isomerization of the heptoxy radicals involved in the oxidation.

Arey et al. (2001) applied a combination of GC–FID, GC–MS and API–MS techniques to the study of heptane oxidation. An accounting for $70 \pm 28\%$ of the products was possible. Using GC–FID, yields of 2-, 3-, and 4-heptyl nitrate of (5.9 ± 0.6), (8.3 ± 0.9), and $(3.6 \pm 0.5)\%$, respectively, were obtained. When normalized to the calculated site of OH radical attack [using parameters of Kwok and Atkinson (1995)], this corresponds on average to a nitrate yield of $(18.7 \pm 2.3)\%$ per secondary peroxy radical formed. The API–MS studies of Arey et al. (2001) were run in negative ion mode, with either pentafluorobenzyl alcohol (PFBOH) or NO_2 added to the system to act as the reagent ion. Compounds with mass/charge ratios of 130 (the 1,4-hydroxycarbonyls) and 177 (the 1,4-hydroxynitrates) were readily identified under these conditions. Analogous studies of fully deuterated heptane gave the corresponding species at mass/charge ratios of 143 and 191. A total yield of 1,4-hydroxycarbonyl compounds of 46% (estimated uncertainty of a factor of two) was obtained. C_4, C_5, and C_6 hydroxycarbonyl compounds were also observed in small yield; these may result from decomposition of the heptoxy radicals. Small yields (about 5%) of the C_7 hydroxynitrate species were also reported, along with even smaller yields of C_5 and C_6 hydroxynitrate compounds.

Reisen et al. (2005) used solid-phase microextraction fibers precoated with O-(2,3,4,5,6-pentafluorobenzyl)hydroxylamine for on-fiber derivatization, coupled with GC–FID and GC–MS detection, to identify and quantify products of heptane oxidation. From an analysis of observed fragmentation patterns and retention times, these authors were able to identify and quantify specific hydroxycarbonyl isomers.

Yields of 5-hydroxy-2-heptanone (product of attack at the 2-position, 18% molar yield), 6-hydroxy-3-heptanone (product of attack at the 3-position, 16%), 1-hydroxy-4-heptanone (product of attack at the 4-position, 11%), and 4-hydroxyheptanal (product of attack at the 1-position, 6%) were reported. The yields of the hydroxyketones are roughly half of the calculated yield of the parent peroxy radical, based on the estimation procedure of Kwok and Atkinson (1995). Coupled with the ~20% yield of the various heptyl nitrates per peroxy radical and an additional ~5% yield of hydroxynitrate, an accounting can be made for about 75% of the products (with a large uncertainty).

IV-C-15. Products of Atmospheric Oxidation of Octane (*n*-Octane, *n*-C_8H_{18})

Atkinson and co-workers (Kwok et al., 1996; Arey et al., 2001; Reisen et al., 2005) have applied similar techniques to the study of the OH-initiated oxidation of octane as discussed above for the smaller straight-chain alkanes. As with the heptane system described in section IV-C-14, four sites of attack of OH on octane are possible. Given the expected predominance of isomerization for the four resultant alkoxy radicals, the major products formed are likely to be four C_8 isomers of each of the 1,4-hydroxycarbonyl, 1,4-hydroxynitrate, octyl nitrate, and octanone/octanal species.

The study of Kwok et al. (1996), using the API–MS system in positive-ion mode, showed the combined yield of octanone/octanal compounds to be exceedingly small, about 1–2%, consistent with isomerization being the major fate of all four alkoxy species formed. The formation of C_8 1,4-hydroxycarbonyl species was also noted from both octane and its fully-deuterated analogue, although these species were not speciated or quantified.

Arey et al. (2001) were able to quantify (55 ± 22)% of the products of octane oxidation, using a combination of GC–FID, GC–MS, and API–MS techniques. Yields of 2-, 3-, and 4-octyl nitrate of (6.2 ± 0.6), (8.1 ± 0.8), and (8.3 ± 0.2)%, respectively, were obtained, corresponding to an average nitrate yield of (23.5 ± 3.1)% per secondary peroxy radical formed. Using the API system, C_8-hydroxycarbonyls (molecular weight 144) and hydroxynitrates (191) were observed. Studies of fully deuterated octane gave the corresponding deuterated species. The total yield of the C_8-hydroxycarbonyl and hydroxynitrate compounds were found to be 27% and 5.4%, respectively, with an estimated uncertainty of a factor of two. Smaller yields of C_4-through C_7-hydroxycarbonyls and C_5- through C_7-hydroxynitrates were also in evidence.

Reisen et al. (2005), using the on-fiber derivatization technique discussed previously, were able to obtain some quantitative data regarding the yields of the individual hydroxycarbonyl isomers (estimated uncertainties ±30%). The 5-hydroxy-2-octanone and 6-hydroxy-3-octanone species (expected from attack at the 2- and 3-position) were not separable, but a combined yield for these two compounds of 38% was reported. Yields of the 7-hydroxy-4-octanone species, expected from attack at the 4-position, and 4-hydroxyoctanal, the product of attack at the 1-position, were 12% and 3%, respectively. These reported yields, which account for about 50% of the oxidation of octane, all fall about 20% to a factor of 2–3 below the expected yields of the parent peroxy radical (Kwok and Atkinson, 1995). When combined with the reported yields of the octyl nitrates and hydroxynitrates, one can account for about 80% of the overall oxidation of the octane.

IV-C-16. Products of Atmospheric Oxidation of 2,2,4-Trimethylpentane [$(CH_3)_3CCH_2CH(CH_3)_2$]

Aschmann et al. (2002) used a combination of GC–FID, GC–MS, and API–MS techniques to detect and quantify products obtained in the OH-initiated oxidation of 2,2,4-trimethylpentane in the presence of NO_x. Using GC–FID, the primary products acetone (54 ± 7%), 2-methyl propanal (26 ± 3%), and 4-hydroxy-4-methyl-2-pentanone (5.1 ± 0.6%) were quantified. Using the API–MS system in both positive- and negative-ion modes, a C_7-hydroxycarbonyl species [$HOCH_2C(CH_3)_2CH_2C(O)CH_3$, yield ~3%], a C_8-hydroxycarbonyl [probably $(CH_3)_2C(OH)CH_2C(CH_3)_2CHO$, yield ~ 3%], a C_8-carbonyl [probably $(CH_3)_3CCH_2CH(CH_3)CHO$], and C_4-,C_7-,and C_8-hydroxynitrates were identified.

The reaction of OH with 2,2,4-trimethylpentane can occur at four different positions, with reaction at the tertiary 4-position expected to be the most important (52%), followed by reaction at the secondary, 3-position (30%), at the three equivalent methyl groups attached to the 2-position (11%), and at the two equivalent methyl groups attached to the 4-position (7%) (Kwok and Atkinson, 1995; Aschmann et al., 2002). In the presence of NO_x, the four peroxy radicals thus generated will react to generate four isomers of octyl nitrate as well as four branched alkoxy radicals; see Aschmann et al. (2002).

Decomposition and isomerization should be competing fates of the most prominent alkoxy radical formed, $(CH_3)_3CCH_2C(O\bullet)(CH_3)_2$, as is summarized in figure IV-C-5 (Aschmann et al. 2002). Decomposition will lead mostly to the formation of two molecules each of acetone and formaldehyde, while isomerization will also generate 4-hydroxy-4-methyl-2-pentanone and a complex array of hydroxynitrate and hydroxycarbonyl compounds (Aschmann et al., 2002); see figure IV-C-5. Although there is qualitative agreement between observed and predicted products, the observed yields of acetone and 4-hydroxy-4-methyl-2-pentanone can account for about one-half of the reacted 2,2,4-trimethylpentane.

Attack at the $-CH_2-$ group in the 3-position leads to the production of the alkoxy radical $(CH_3)_3CCH(O\bullet)CH(CH_3)_2$. This radical has two decomposition channels available to it, with elimination of $(CH_3)_3C\bullet$ radical (and thus formation of acetone, formaldehyde, and 2-methyl propanal) likely favored:

$$(CH_3)_3CCH(O\bullet)CH(CH_3)_2 \rightarrow (CH_3)_3C\bullet + (CH_3)_2CHCHO$$

$$(CH_3)_3C\bullet\ (+O_2, NO_x) \rightarrow\rightarrow CH_3C(O)CH_3 + CH_2O$$

The observed 2-methyl propanal yield, 26 ± 3%, is consistent with the predicted occurrence of 30% of the OH-attack at this site.

Attack at the three equivalent CH_3 groups, which accounts for about 11% of the reaction, gives rise to a large yield of the C_8-hydroxycarbonyl species

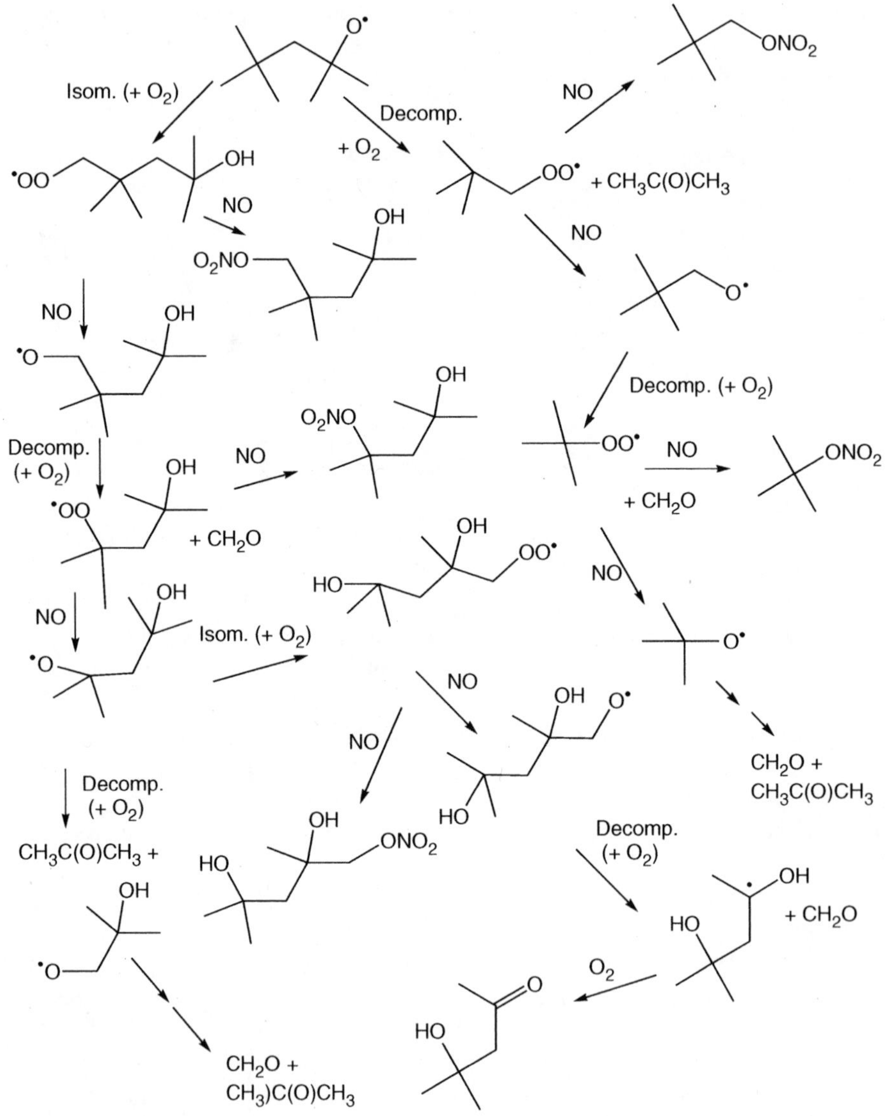

Figure IV-C-5. Sequence of reactions of alkoxy radicals formed in the atmospheric oxidation of 2,2,4-trimethylpentane.

$(CH_3)_2C(OH)CH_2C(CH_3)_2CHO$, as well as lesser amounts of acetone, formaldehyde, 2-methylpropanal, the C_7-hydroxycarbonyl $(CH_3)_2C(OH)CH_2CH(CH_3)CHO$, and the C_8-hydroxynitrate $HOCH_2C(CH_3)_2CH_2C(ONO_2)(CH_3)_2$ (Aschmann et al., 2002).

Finally, about 7% of the reaction occurs on the two equivalent CH_3 groups, which is predicted to give the C_8-carbonyl compound $(CH_3)_3CCH_2CH(CH_3)CHO$, a C_7-hydroxycarbonyl species $HOCH_2C(CH_3)_2CH_2C(O)CH_3$, and a C_7-hydroxynitrate species $CH_3CH(OH)CHC(CH_3)_2CH_2ONO_2$ (Aschmann et al., 2002).

IV-C-17. Products of Atmospheric Oxidation of 2,3,4-Trimethylpentane [$(CH_3)_2CHCH(CH_3)CH(CH_3)_2$]

The OH-initiated oxidation of 2,3,4-trimethylpentane has been studied under atmospheric conditions in the presence of NO_x by Aschmann et al. (2004), using a combination of GC–FID, GC–MS, and API–MS techniques. The following products, which accounted for 69 ± 6% of the carbon, were observed using GC-based techniques (molar yields in parentheses): acetaldehyde (47 ± 6%), acetone (76 ± 11%), 3-methyl-2-butanone (41 ± 5%), 3-methyl-2-butyl nitrate (1.6 ± 0.2%), 2-propyl nitrate (6.2 ± 0.8%), and (tentatively) an octyl nitrate. Using the API–MS system in both positive-and negative-ion modes, acetone, 3-methyl-2-butanone, a C_5- and C_8-hydroxynitrate, and a C_8-hydroxycarbonyl compound were also identified.

Although four sites of attack of OH on 2,3,4-trimethylpentane are possible, the chemistry will be dominated by reaction at the two equivalent tertiary sites (the 2- and 4-position), which are predicted to account for about 56% of the reaction, and attack at the tertiary 3-position, which is predicted to account for about 34% of the reaction (Kwok and Atkinson, 1995; Aschmann et al., 2004). Attack at the four equivalent methyl groups (the 1-position) is predicted to account for the bulk of the remaining reactivity (about 10%).

The major alkoxy radical formed is predicted to be the $(CH_3)_2C(O\bullet)CH(CH_3)CH(CH_3)_2$ species, which will undergo decomposition predominantly to give acetone, 3-methyl-2-butyl nitrate, and acetaldehyde; see figure IV-C-6. Isomerization of this species is expected to make a minor contribution. Abstraction of the tertiary hydrogen at the 3-position leads to the formation of the $(CH_3)_2CHC(O\bullet)(CH_3)CH(CH_3)_2$ radical, the predominant fate of which will be decomposition to give 3-methyl-2-butanone, acetone, and 2-propyl nitrate:

$$(CH_3)_2CHC(O\bullet)(CH_3)CH(CH_3)_2 \rightarrow (CH_3)_2CH\bullet + CH_3C(O)CH(CH_3)_2$$

$$(CH_3)_2CH\bullet + O_2 \rightarrow (CH_3)_2CHO_2\bullet$$

$$(CH_3)_2CHO_2\bullet + NO \rightarrow (CH_3)_2CHO\bullet + NO_2$$

$$(CH_3)_2CHO_2\bullet + NO \rightarrow (CH_3)_2CHONO_2$$

$$(CH_3)_2CHO\bullet + O_2 \rightarrow CH_3C(O)CH_3 + HO_2$$

Attack at any of the four equivalent methyl groups leads to the $(CH_3)_2CHCH(CH_3)CH(CH_3)CH_2O\bullet$ radical, the dominant fate of which will be isomerization. Expected end-products include the hydroxynitrate species $(CH_3)_2C(ONO_2)CH(CH_3)CH(CH_3)CH_2OH$ and $CH_3CH(ONO_2)CH(CH_3)CH_2OH$, the hydroxycarbonyl species $(CH_3)_2C(OH)CH(CH_3)CH(CH_3)CHO$, acetone, acetaldehyde, and formaldehyde (Aschmann et al., 2004).

Quantitative agreement between the measured and predicted product yields can be achieved if it is assumed that OH reaction occurs about 40% at the 2-position [somewhat less than the predicted 56% (Kwok and Atkinson, 1995)], 25–30% at the 3-position, and 10–15% at the 1-position (Aschmann et al., 2004).

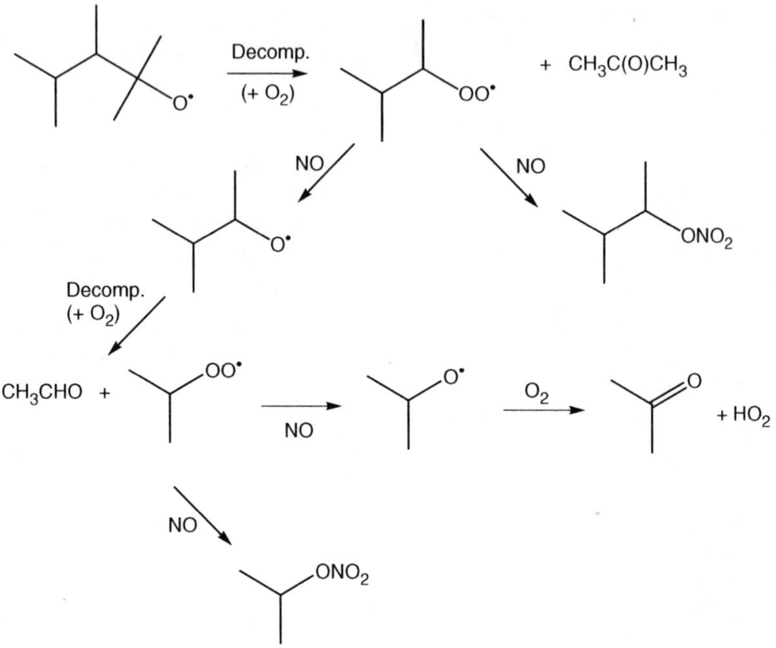

Figure IV-C-6. Sequence of reactions of the $(CH_3)_2C(O\bullet)CH(CH_3)CH(CH_3)_2$ alkoxy radical formed in the atmospheric oxidation of 2,3,4-trimethylpentane.

IV-C-18. Products of Atmospheric Oxidation of Nonane (*n*-Nonane, *n*-C_9H_{20})

To the best of our knowledge, the oxidation of nonane has not been explicitly studied under atmospheric conditions. Reaction of OH with nonane can occur at five distinct sites, with the most prominent sites of attack expected to be the 3- and 4-position (~29% apiece), followed by the 2-position (~24%), the 5-position (~15%), and the 1-position (~3%) (Kwok and Atkinson, 1995). The major fate of each of the resultant alkoxy radicals is expected to be isomerization. Thus, by analogy to hexane, heptane, and octane discussed in sections IV-C-10, IV-C-14, and IV-C-15, respectively, the major products obtained are likely to be various isomers of nonyl nitrate, and of nine-carbon 1,4-hydroxycarbonyls and 1,4-hydroxynitrates.

IV-C-19. Products of Atmospheric Oxidation of Decane (*n*-Decane, *n*-$C_{10}H_{22}$)

The mechanism of the atmospheric oxidation of decane in the presence of NO_x has been studied by Aschmann et al. (2001), using a combination of GC–FID, GC–MS and API–MS (positive and negative ion) techniques. Products observed included 2-, 3-, 4-, and 5-decyl nitrate, C_{10}-hydroxycarbonyl and hydroxynitrate species, and lesser yields of C_6- through C_9-hydroxycarbonyl compounds. Overall, about 50% of the reacted decane was accounted for, although this value is probably uncertain by about a factor of 2–3.

Reaction of OH can occur at five sites, the 1- through 5-positions, with calculated probabilities of 3.1, 20.7, 25.4, 25.4, and 25.4%, respectively (Kwok and Atkinson, 1995; Aschmann et al., 2001). Thus, the end-product distributions will be dominated by the chemistry of 2-, 3-, 4-, and 5-decylperoxy radicals, which are formed in nearly equal yields.

Reaction of these four major peroxy radicals with NO is expected to generate the corresponding nitrates and alkoxy radicals. The gas chromatographic analyses of Aschmann et al. (2001) indeed demonstrated the formation of 2-, 3-, 4-, and 5-decyl nitrates, with yields of 0.038 ± 0.007, 0.062 ± 0.011, 0.064 ± 0.013, and $0.061 \pm 0.010\%$, respectively. Using these measured yields (which were all corrected for secondary reaction with OH) and the calculated probabilities of reaction at the different sites (Kwok and Atkinson, 1995), an average nitrate yield of $23.3 \pm 4.0\%$ per secondary peroxy radical reacted was obtained for all four radicals. This value is in reasonable agreement with the predicted yield based on discussions in section IV-C-1.

The predominant fate of all four alkoxy radicals (2-, 3-, 4-,and 5-decoxy) is likely to be isomerization, leading to the formation of a unique 1,4-hydroxycarbonyl and a 1,4-hydroxynitrate compound from each radical. For example, isomerization of 3-decoxy results in the formation of 6-nitrooxy-3-decanol and 6-hydroxy-3-decanone; see figure IV-C-7. Using their API–MS system in the positive-ion mode, Aschmann et al. (2001) indeed observed peaks at masses consistent with the presence of ten-carbon hydroxycarbonyl and hydroxynitrate compounds.

Using the API–MS in negative ion mode, with NO_2^- acting as the reagent ion, Aschmann et al. (2001) saw evidence for the presence of C_6-through C_9-hydroxycarbonyls in addition to the C_{10} species. Estimated yields for the various hydroxycarbonyls, believed to be uncertain by a factor of 2–3, are 13% for the C_{10}-hydroxycarbonyls and 2% for each of the C_6- through C_9-hydroxycarbonyls. The estimated yield of C_{10}-hydroxynitrates is quite low, 2.4%. This may indicate either that nitrate yields from the reaction of γ-hydroxy-substituted peroxy radicals with NO are considerably lower than for reactions of the corresponding unsubstituted radicals, or that there is a systematic problem in the method used to calibrate the API–MS system for the 1,4-hydroxynitrate species.

IV-C-20. Products of Atmospheric Oxidation of 3,4-Diethylhexane [$CH_3CH_2CH(C_2H_5)CH(C_2H_5)CH_2CH_3$]

The OH-initiated oxidation of 3,4-diethylhexane, $(C_2H_5)_2CHCH(C_2H_5)_2$, was studied in the presence of NO_x by Aschmann et al. (2001), using GC–FID and GC–MS techniques for the quantification of carbonyls and nitrates, and API–MS for identification and semi-quantitative analysis of hydroxycarbonyls and hydroxynitrates. Using the GC-based techniques, acetaldehyde (40% molar yield), propanal ($37 \pm 6\%$), 3-pentanone ($40 \pm 4\%$), two C_{10} alkyl nitrates (total yield $17.7 \pm 3.2\%$), and 3-pentyl nitrate ($2.3 \pm 1.1\%$) were quantified. Using the API–MS system in positive ion mode, the authors were able to identify peaks due to 3-pentanone and propanal, as well as to C_{10}-hydroxycarbonyls and C_{10}-hydroxynitrates. In negative ion mode, C_5-, C_8-,

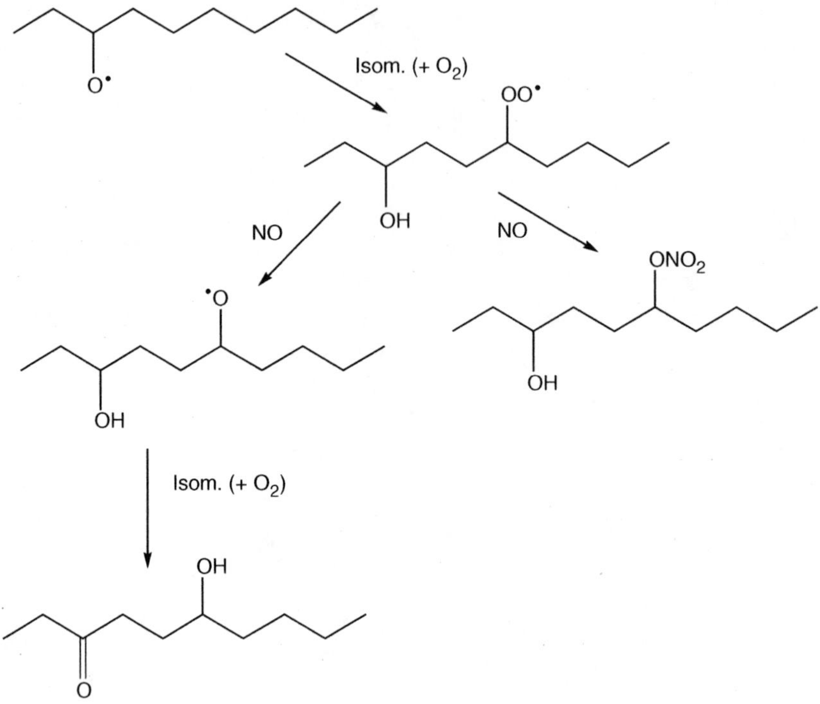

Figure IV-C-7. Reaction sequence of one of the alkoxy radicals, 3-decoxy, formed in the atmospheric oxidation of decane.

and C_{10}-hydroxynitrates (total yield about 1.7%), as well as C_5-, C_6-, C_8-, C_9-, and C_{10}-hydroxycarbonyls (total yield ~11%), were also identified. In total, an accounting could be made for about 70% of the reacted 3,4-diethylhexane.

Using the estimation method of Kwok and Atkinson (1995), the reaction of OH with 3,4-diethylhexane is predicted to occur with a 58% probability at the two equivalent >CH— groups in the 3-position, with a 37% probability at the four equivalent $-CH_2-$ groups in the 2-position, and with a 5% probability at the four equivalent $-CH_3$ groups in the 1-position. Thus, three C_{10}-alkyl nitrates are possible. The two species observed (Aschmann et al., 2001) are likely the secondary nitrate, $CH_3CH(ONO_2)CH(C_2H_5)CH(C_2H_5)_2$ (observed yield 14.5%), and the tertiary compound, $(C_2H_5)_2C(ONO_2)CH(C_2H_5)_2$ (observed yield 3.2%).

Attack at the tertiary site (the two equivalent >CH— groups in the 3-position) leads to the formation of the $(C_2H_5)_2C(O\bullet)CH(C_2H_5)_2$ radical in large yield. This radical is expected to react nearly quantitatively via decomposition to yield 3-pentanone and the 3-pentyl radical. Further oxidation of the 3-pentyl radical should then give a mixture of 3-pentanone, propanal, acetaldehyde, and 3-pentyl nitrate (Aschmann et al., 2001):

$$(C_2H_5)_2C(O\bullet)CH(C_2H_5)_2 \rightarrow (CH_3CH_2)_2CO + CH_3CH_2CH(\bullet)CH_2CH_3$$

$$CH_3CH_2CH(\bullet)CH_2CH_3 + O_2 \rightarrow CH_3CH_2CH(O_2\bullet)CH_2CH_3$$

$$CH_3CH_2CH(O_2\bullet)CH_2CH_3 + NO \rightarrow CH_3CH_2CH(O\bullet)CH_2CH_3 + NO_2$$

$$\rightarrow CH_3CH_2CH(ONO_2)CH_2CH_3$$

$$CH_3CH_2CH(O\bullet)CH_2CH_3 + O_2 \rightarrow (CH_3CH_2)_2CO + HO_2$$

$$CH_3CH_2CH(O\bullet)CH_2CH_3 \rightarrow CH_3CH_2\bullet + CH_3CH_2CHO$$

$$CH_3CH_2\bullet + O_2 \rightarrow CH_3CH_2O_2\bullet$$

$$CH_3CH_2O_2\bullet + NO \rightarrow CH_3CH_2O\bullet + NO_2$$

$$\rightarrow CH_3CH_2ONO_2$$

$$CH_3CH_2O\bullet + O_2 \rightarrow CH_3CHO + HO_2$$

Attack at the four equivalent —CH_2— groups in the 2-position gives the alkoxy radical $CH_3CH(O\bullet)CH(C_2H_5)CH(C_2H_5)_2$ (in addition to the secondary nitrate discussed previously). This alkoxy radical is expected to decompose largely to give acetaldehyde, a C_8-nitrate (4-ethyl-3-nitrooxy-hexane), propanal, and the 3-pentylperoxy radical:

$$CH_3CH(O\bullet)CH(C_2H_5)CH(C_2H_5)_2\ (+O_2) \rightarrow CH_3CHO + C_2H_5CH(O_2\bullet)CH(C_2H_5)_2$$

$$C_2H_5CH(O_2\bullet)CH(C_2H_5)_2 + NO \rightarrow C_2H_5CH(O\bullet)CH(C_2H_5)_2 + NO_2$$

$$\rightarrow C_2H_5CH(ONO_2)CH(C_2H_5)_2$$

$$C_2H_5CH(O\bullet)CH(C_2H_5)_2\ (+O_2) \rightarrow C_2H_5CHO + CH_3CH_2CH(O_2\bullet)CH_2CH_3$$

The 3-pentylperoxy radical reacts as shown to give additional 3-pentyl nitrate, 3-pentanone, propanal, and acetaldehyde (Aschmann et al., 2001). Isomerization of the $CH_3CH(O\bullet)CH(C_2H_5)CH(C_2H_5)_2$ radical is expected to be a minor channel, giving mostly C_{10}- and C_8-hydroxycarbonyl and hydroxynitrates species, as well as acetaldehyde and propanal.

The alkoxy radical formed via attack at the four equivalent CH_3— groups in the 1-position, $(C_2H_5)_2CHCH(C_2H_5)CH_2CH_2O\bullet$, is predicted to isomerize to yield a C_{10}-hydroxynitrate and the $(C_2H_5)_2C(O\bullet)CH(C_2H_5)CH_2CH_2OH$ species. This alkoxy radical is expected to isomerize in small part to give the hydroxyaldehyde, but will likely decompose readily to give 3-pentanone and $HOCH_2CH_2CH(\bullet)CH_2CH_3$. This resulting radical should eventually be oxidized to yield the corresponding C_5-hydroxynitrate, 1-hydroxy-3-pentanone, and propanal, with either glycolaldehyde or two formaldehyde molecules as the co-product (Aschmann et al., 2001).

Thus, essentially all of the observed products can be accounted for by the mechanism just presented. However, as pointed out by Aschmann et al. (2001), the yields of 3-pentanone and 3-pentyl nitrate are insufficient to support the supposition of 55–60% of the OH attack occurring at the tertiary C—H groups. Note that the measured yield of 3-pentanone is 40%, and about 1.4 molecules of 3-pentanone are expected per attack at the tertiary site. Aschmann et al. (2001) show much better agreement between observed and predicted product distributions if abstraction at the tertiary site accounts for only about 25–30% of the reaction.

IV-C-21. Products of Atmospheric Oxidation of Cyclopropane (*cyclo*-C_3H_6)

The mechanism of the oxidation of the Cl-atom-initiated oxidation of cyclopropane has been studied by DeSain et al. (2003). In a conventional environmental chamber study conducted in 700 Torr air, the only products observed via FTIR spectroscopy were CO and CO_2; yields were not reported. It is likely that both the CO and CO_2 are secondary products of the oxidation, since the reaction of Cl with cyclopropane is considerably slower than its reaction with many of the anticipated initial products (DeSain et al., 2003) (e.g., ethene, oxirane, acrolein, propanedial; see the following discussion).

DeSain et al. (2003) also carried out experiments at very high cyclopropane concentrations ($\sim$0.5 Torr), using ethyl chloride as a "tracer" to determine the amount of cyclopropane consumed. Oxirane and ethene were observed as primary products. The yield of oxirane was seen to decrease with total pressure (molar yield of $14 \pm 2\%$ at 6 Torr total pressure, $8 \pm 2\%$ at 10 Torr total pressure, $6 \pm 2\%$ at 50 Torr total pressure), while the yield of ethene increased with total pressure (molar yield of $11 \pm 3\%$ at 6 Torr total pressure, $15 \pm 1\%$ at 10 Torr total pressure, $30 \pm 6\%$ at 50 Torr total pressure). However, because these experiments were conducted at reduced total pressures, it is unclear whether the observed products are formed from a unimolecular rearrangement/decomposition of excited cyclopropylperoxy radicals (which may not occur at higher pressure) or from the chemistry of stabilized cyclopropylperoxy radicals. While the increased yield of ethene with total pressure may be indicative of its formation from the chemistry of stabilized cyclopropylperoxy radicals, further environmental chamber experiments at atmospheric pressure using very high cyclopropane concentrations ($\sim$0.5 Torr) in the presence of a tracer are required to conclusively identify and quantify the end-products of cyclopropane under atmospheric conditions. Note that propanedial would be a likely expected end-product of cyclopropylperoxy radicals, should its oxidation proceed in a traditional manner, e.g.,

$$cyclo\text{-}C_3H_5O_2\bullet + NO \rightarrow cyclo\text{-}C_3H_5O\bullet + NO_2$$

$$cyclo\text{-}C_3H_5O\bullet(+O_2) \rightarrow \bullet O_2CH_2CH_2CHO$$

$$\bullet O_2CH_2CH_2CHO + NO \rightarrow \bullet OCH_2CH_2CHO + NO_2$$

$$\bullet OCH_2CH_2CHO + O_2 \rightarrow OCHCH_2CHO + HO_2$$

IV-C-22. Products of Atmospheric Oxidation of Cyclobutane (*cyclo*-C_4H_8)

To the best of our knowledge, no studies of the OH-initiated oxidation of cyclobutane have been reported in the literature. The likely oxidation pathway (assuming formation of cyclobutylperoxy and its subsequent conversion to cyclobutoxy) has, however, been speculated upon by Carter and Atkinson (1985), and is detailed below.

Note that isomerization of the oxy radical formed subsequent to ring-opening occurs via transfer of an aldehydic H-atom. Major expected products thus include a four-carbon hydroxy-substituted acylperoxy nitrate species ($HOCH_2CH_2CH_2C(O)OONO_2$) and 3-hydroxy-propanal ($HOCH_2CH_2CHO$); see figure IV-C-8.

IV-C-23. Products of Atmospheric Oxidation of Cyclopentane [*cyclo*-C_5H_{10}]

The oxidation of cyclopentane has been studied using environmental chamber techniques (Takagi et al., 1981; Rowley et al., 1992c; Orlando et al., 2000), flash-photolysis/UV absorption (Rowley et al., 1992b,c), and a slow-flow tube equipped

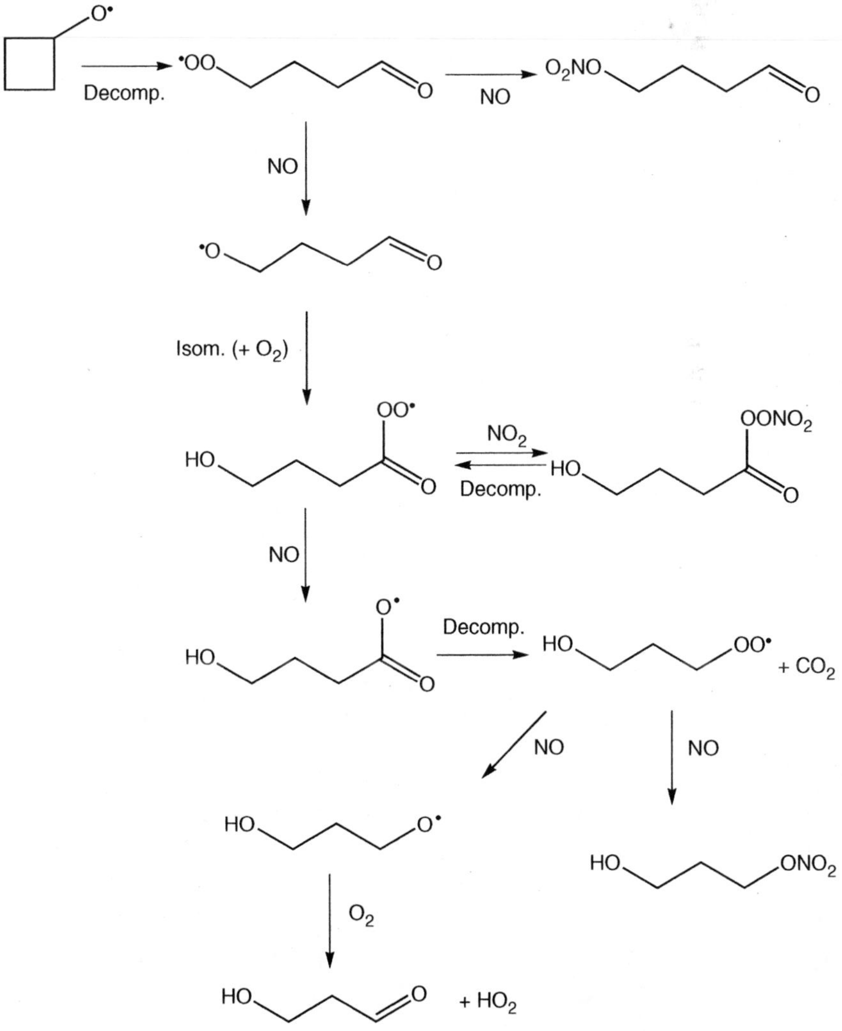

Figure IV-C-8. Expected sequence of reactions of the cyclobutoxy radical formed in the atmospheric oxidation of cyclobutane.

with a PTR–MS system (Hanson et al., 2004). Obviously, there is only one unique site of OH attack on cyclopentane, which leads to the formation of the cyclopentylperoxy radical. Reaction of this peroxy species with NO generates in part cyclopentyl nitrate, with a yield of $\leq$12% at room temperature (Takagi et al., 1981; Orlando et al., 2000). The data of Rowley et al. (1992b) show that reaction of the peroxy radical with HO_2 gives the corresponding cyclopentyl hydroperoxide quantitatively.

The predominant fate of the cyclopentyloxy species is ring-opening to form the $\bullet CH_2CH_2CH_2CH_2CHO$ radical. This ring-opening is facile compared with cyclohexane (see the following discussion), and is attributed to a lowering of the barrier due to the presence of ring strain in the cyclopentyl system (Takagi et al., 1981). The major end-product following ring-opening has been identified as 4-hydroxybutanal (Hanson et al., 2004). Minor end-products include 2-glutaric dialdehyde (Rowley et al., 1992c; Orlando et al., 2000; Hanson et al., 2004), as well as 2-hydroxypentanedial, 5-hydroxy-2-nitrooxypentanal, and 5-nitrooxypentanal (Hanson et al., 2004). Major reactions occurring pursuant to ring-opening are shown below:

$$\bullet O_2CH_2CH_2CH_2CH_2CHO + NO \rightarrow \bullet OCH_2CH_2CH_2CH_2CHO + NO_2$$

$$\rightarrow O_2NOCH_2CH_2CH_2CH_2CHO$$

$$\bullet OCH_2CH_2CH_2CH_2CHO + O_2 \rightarrow OCHCH_2CH_2CH_2CHO + HO_2$$

$$\bullet OCH_2CH_2CH_2CH_2CHO \rightarrow HOCH_2CH_2CH_2CH(\bullet)CHO$$

$$HOCH_2CH_2CH_2CH(\bullet)CHO + O_2 \rightarrow HOCH_2CH_2CH_2CH(O_2\bullet)CHO$$

$$HOCH_2CH_2CH_2CH(O_2\bullet)CHO + NO \rightarrow HOCH_2CH_2CH_2CH(O\bullet)CHO + NO_2$$

$$\rightarrow HOCH_2CH_2CH_2CH(ONO_2)CHO$$

$$HOCH_2CH_2CH_2CH(O\bullet)CHO\ (+O_2) \rightarrow OCHCH_2CH_2CH(OH)CHO + HO_2$$

$$HOCH_2CH_2CH_2CH(O\bullet)CHO\ (+O_2) \rightarrow HOCH_2CH_2CH_2CHO + CO + HO_2$$

Takagi et al. (1981) also detected CH_2O, HCOOH, and C_2H_2 in their study of cyclopentane oxidation, but it is likely that these products are artifacts.

IV-C-24. Products of Atmospheric Oxidation of Cyclohexane [*cyclo*-C_6H_{12}]

The oxidation of cyclohexane under atmospheric conditions has been studied by a number of groups, using environmental chamber and slow-flow tube systems (Takagi et al., 1981; Atkinson et al,. 1984a; Rowley et al., 1991, 1992b; Atkinson et al., 1992a; Aschmann et al., 1997; Platz et al., 1999; Orlando et al., 2000; Hanson et al., 2004).

Reaction of OH with cyclohexane will lead to the quantitative production of the cyclohexylperoxy radical. Production of cyclohexyl nitrate from reaction of the peroxy radical with NO has been observed by many groups (Takagi et al., 1981; Atkinson et al., 1992a; Aschmann et al., 1997; Platz et al., 1999; Orlando et al., 2000), and the yield at room temperature and 1 atm. total pressure is found to be $15 \pm 4\%$ from the average of the five determinations. Data from Orlando et al. (2000) show that the cyclohexyl nitrate yield does not change dramatically with temperature, with yields of only about 20% being measured at 273 and 261 K. This is in disagreement with the predicted rapid rise in the nitrate yield implied by the formulae in section IV-C-1. The data of Rowley et al. (1992b) show that reaction of the cyclohexylperoxy radical with HO_2 gives the hydroperoxide quantitatively.

The cyclohexoxy radical can react with O_2 to generate cyclohexanone, or decompose to give $\bullet CH_2CH_2CH_2CH_2CH_2CHO$. In 1 atm. air at 298 K, the cyclohexanone yield has been reported to be 25–30% (Takagi et al, 1981; Atkinson et al., 1992a; Aschmann et al., 1997; Platz et al., 1999; Orlando et al., 2000); thus, most of the chemistry proceeds via ring-opening under these conditions. Because of the reasonably high barrier to the ring-opening process, 48.1 ± 9.2 kJ mol^{-1} (Orlando et al., 2000), the cyclohexanone yield is expected to increase with decreasing temperature, and thus with increasing altitude in the troposphere. The ring-opening leads to a complex set of products (Aschmann et al., 1997; Hanson et al., 2004), since the dominant fate of the alkoxy species $\bullet OCH_2CH_2CH_2CH_2CH_2CHO$ is isomerization. Aschmann et al. (1997) used their API–MS system to identify a C_6-hydroxydicarbonyl, probably $OCHCH_2CH(OH)CH_2CH_2CHO$, and a C_6-carbonyl nitrate, probably $OCHCH_2CH_2CH_2CH_2CH_2ONO_2$, as major products of the ring-opening. Hanson et al. (2004), using PTR–MS in a slow-flow system, identified a C_6-hydroxycarbonyl nitrate (probably $HOCH_2CH_2CH_2CH(ONO_2)CH_2CHO$), a C_6-hydroxydicarbonyl (probably $OCHCH_2CH(OH)CH_2CH_2CHO$), and a C_6-dicarbonyl (probably $OCH(CH_2)_4CHO$). Major reactions expected to follow the ring-opening are shown in figure IV-C-9.

IV-C-25. Products of Atmospheric Oxidation of Cycloheptane (*cyclo*-C_7H_{14})

The only product study of the OH-initiated oxidation of cycloheptane available at the time of this writing is that of Takagi et al. (1981), conducted at 1 atm. pressure of air at ambient temperature. These authors reported a cycloheptyl nitrate yield of only 5%, well below the expected yield for a 7-carbon molecule. However, the yields of cyclohexyl nitrate and particularly cyclopentyl nitrate reported by this group seem low as well, leaving open the possibility of the existence of a systematic problem with the data. Takagi et al. (1981) observed only a very small yield of cycloheptanone, 3%, implying the predominance of ring-opening in the chemistry of the cycloheptoxy radical, analogous to the cyclopentoxy system discussed in section IV-C-23. Again, the different behavior of the cyclo-C_5 and C_7 systems vs. cyclohexoxy is suggested to result from the added ring strain in the C_5 and C_7 systems which lowers the barrier to the ring-opening process (Takagi et al., 1981; Aschmann et al., 1997). The end-products of ring-opening have not been determined experimentally, but a likely mechanism, adapted from Carter and Atkinson (1985), is given in figure IV-C-10.

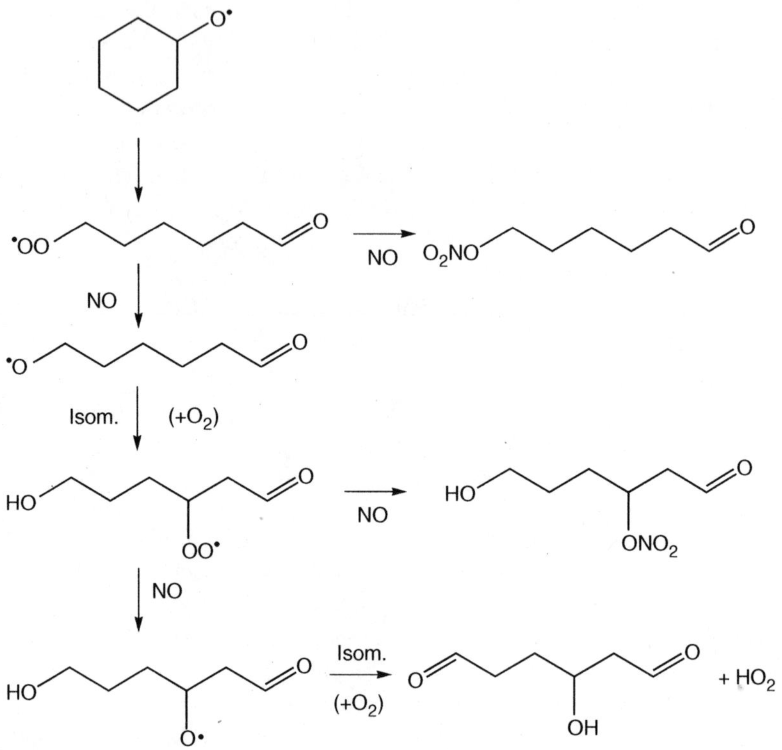

Figure IV-C-9. Sequence of reactions following ring-opening of the cyclohexoxy radical formed in the atmospheric oxidation of cyclohexane.

Takagi et al. (1981) also report the formation of CH_2O, HCOOH, and C_2H_2 in this system but, as was the case for their studies of cyclopentane and cyclohexane, these products are likely artifacts.

IV-C-26. Products of Atmospheric Oxidation of 1-Butylcyclohexane (*cyclo*-C_6H_{11}-1-C_4H_9)

Aschmann et al. (2001) studied the OH-initiated oxidation of 1-butylcyclohexane, using GC–FID and GC–MS to detect and quantify carbonyls and nitrates, and API–MS for detection and quantification of 1,4-hydroxycarbonyls and 1,4-hydroxynitrates. Products observed included butanal ($7.2 \pm 0.9\%$ molar yield); cyclohexanone ($4.9 \pm 0.7\%$); seven different C_{10} alkyl nitrates ($19 \pm 6\%$ total yield); hydroxycarbonyls, mostly containing ten carbons (37%); and ten-carbon hydroxynitrates and/or hydroxycarbonyl-nitrates (2.3%). Overall, about 65% (40–110%) of the oxidized 1-butylcyclohexane is accounted for by the observations.

Using the structure–reactivity scheme of Kwok and Atkinson (1995), the dominant site of OH attack on 1-butylcyclohexane is expected to be at the tertiary site (24%), with reaction at each of the eight CH_2 groups expected to contribute about 8–10% to the reactivity.

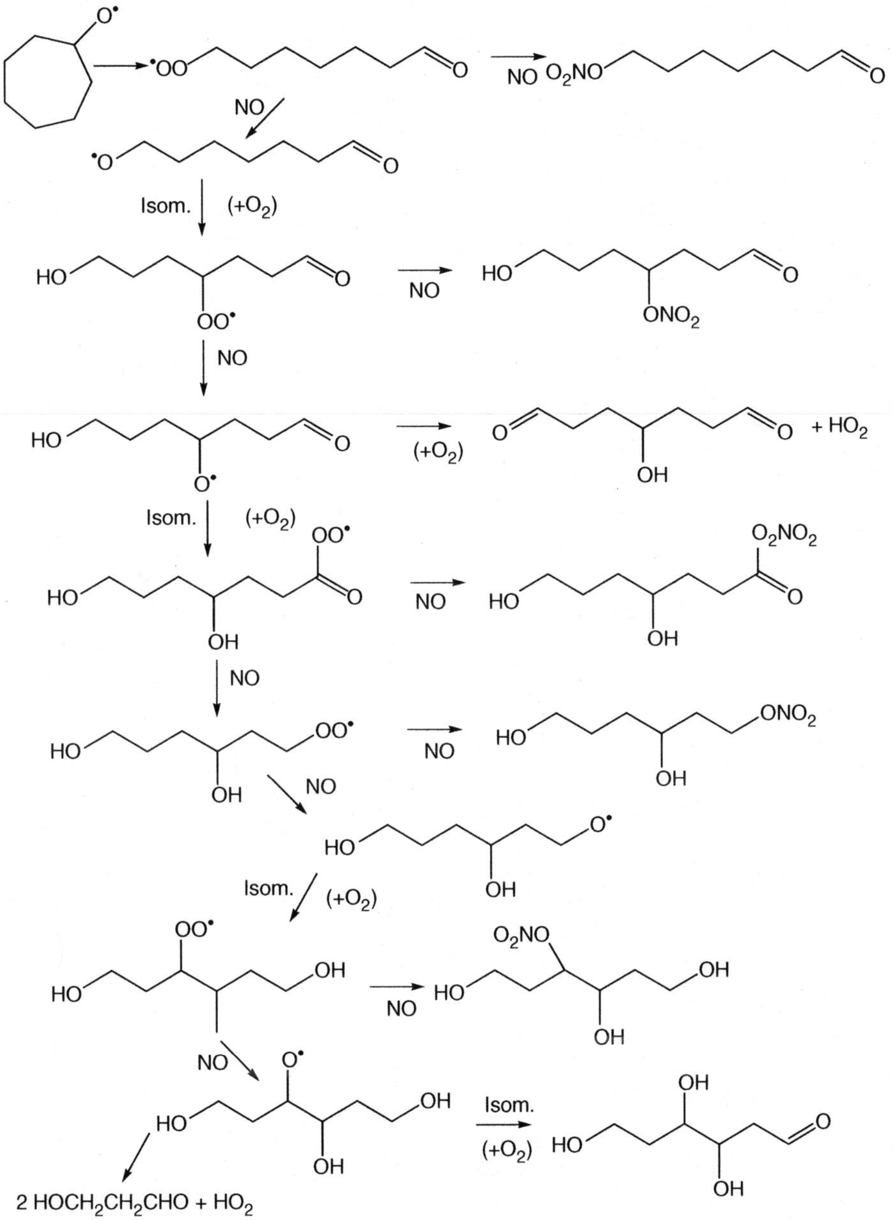

Figure IV-C-10. Sequence of expected reactions of the cycloheptoxy radical formed in the atmospheric oxidation of cycloheptane.

The dominant fates of the major expected alkoxy radicals, 1-butyl-1-cyclohexoxy, as well as of the *cyclo*-C_6H_{11}-$CH_2CH_2CH(O\bullet)CH_3$ and *cyclo*-C_6H_{11}-$CH_2CH(O\bullet)CH_2CH_3$ radicals, are predicted to be isomerization, generating a ten-carbon hydroxycarbonyl and hydroxynitrate as major products in each case. The *cyclo*-C_6H_{11}-$CH(O\bullet)CH_2CH_2CH_3$

radical, on the other hand, will decompose to give butanal and the cyclohexyl radical, which will in turn oxidize to give cyclohexyl nitrate, cyclohexanone, $OCHCH_2CH(OH)CH_2CH_2CHO$, and other multifunctional ring-opening products (see discussion of cyclohexane oxidation in section IV-C-23).

Attack by OH on the ring will also generate the 2-butyl-1-cyclohexoxy, 3-butyl-1-cyclohexoxy, and 4-butyl-1-cyclohexoxy radicals. By analogy with cyclohexoxy itself, each of these radicals may decompose (to generate ten-carbon hydroxydicarbonyls, carbonylnitrates, or hydroxycarbonylnitrates) or react with O_2 (to generate a butyl-substituted cyclohexanone). Additionally, the 2-butyl-1-cyclohexoxy and 3-butyl-1-cyclohexoxy species may isomerize by transfer of a hydrogen from the butyl chain, provided a favorable conformation of the six-membered transition state is attainable (Aschmann et al., 2001). These isomerizations would likely lead to the formation of hydroxybutyl-substituted cyclohexanones.

IV-C-27. Predicted Major Products of Some Additional Methyl-Substituted Alkanes

There are a number of seven- and eight-carbon alkanes for which no end-product studies have been reported in the literature. The expected major products for some of these species in the presence of NO_x, as derived from structure–additivity relationships, are summarized in table IV-C-5.

IV-C-28. Secondary Organic Aerosol Formation from Reaction of OH with the Straight-Chain Alkanes

While it is now well established that the oxidations of aromatic compounds and of biogenic (alkene) species are important atmospheric sources of secondary organic aerosol (SOA), there is an increasing awareness that the oxidation of larger alkanes may also make a measurable contribution as well. Lim and Ziemann (2005) measured SOA yields from the OH-initiated oxidation of the straight-chain alkanes (C_8–C_{15}) in the presence of aerosol seed particles. The data obtained, as well as similar data from other sources (Wang et al., 1992; Takekawa et al., 2003), are summarized in figure IV-C-11. The plot of SOA yield vs. carbon number is sigmoidal, with SOA yields being essentially zero for octane, increasing rapidly from undecane to tridecane, before leveling off near 50% for tetradecane and pentadecane. The abrupt increase in the SOA yield near C_{11} may be related, at least in part, to the onset of the participation of the first-generation 1,4-hydroxycarbonyls in aerosol nucleation and/or growth (Lim and Ziemann, 2005).

Lim and Ziemann (2005) also used a thermal desorption particle beam mass spectrometer system to determine SOA composition, either in real-time or via temperature-programmed thermal desorption. Compounds tentatively identified in the SOA produced from pentadecane included C_{15}-hydroxynitrates (first-generation products expected from isomerization of pentadecoxy radicals), as well as C_{15} carbonyl nitrates, dinitrates, and hydroxydinitrates (all expected products of the oxidation of the primary products pentadecyl nitrates, ketones, or hydroxyketones). Also observed was an array of compounds that arise from the OH-initiated oxidation of substituted dihydrofurans. These dihydrofurans are thought to exist in equilibrium with the linear

Table IV-C-5. Major end-products predicted to be formed in the atmospheric oxidation of a number of seven- and eight-carbon alkanes in the presence of NO_x near 298 K, as derived from structure–additivity relationships

Alkane	Peroxy Radical Formed, % yield[a]	Expected Products
2,2,3-Trimethylbutane	$(CH_3)_3CC(CH_3)(OO\bullet)CH_3$, 75	Acetone, formaldehyde, $(CH_3)_3CC(CH_3)(ONO_2)CH_3$, $(CH_3)_3CONO_2$
2,4-Dimethylpentane	$(CH_3)_2C(OO\bullet)CH_2CH(CH_3)_2$, 70	Acetone, formaldehyde, 2-methylpropanal, $(CH_3)_2C(ONO_2)CH_2CH(CH_3)_2$, $(CH_3)_2CHCH_2(ONO_2)$
	$(CH_3)_2CHCH(OO\bullet)CH(CH_3)_2$, 20	2-Methylpropanal, acetone, $(CH_3)_2CHCH(ONO_2)CH(CH_3)_2$
2-Methylhexane	$(CH_3)_2C(OO\bullet)CH_2CH_2CH_2CH_3$, 35	2-Methyl-2-hexyl nitrate, acetone, 1-butyl nitrate, butanal, 4-hydroxybutanal, 5-nitrooxy-2-methyl-2-hexanol, acetaldehyde, 4-nitrooxy-2-methyl-2-butanol, 3-hydroxy-3-methyl-butanal, 4-hydroxy-2-butanone, formaldehyde, 3,4-dihydroxy-3-methylbutanal
	$(CH_3)_2CHCH(OO\bullet)CH_2CH_2CH_3$, 20	2-Methyl-3-hexylnitrate, butanal, acetone
	$(CH_3)_2CHCH_2CH(OO\bullet)CH_2CH_3$, 20	5-Methyl-3-hexylnitrate, 6-nitrooxy-5-methyl-3-hexanol, 6-hydroxy-5-methyl-3-hexanone
	$(CH_3)_2CHCH_2CH_2CH(OO\bullet)CH_3$, 17	5-Methyl-2-hexyl nitrate, 5-nitrooxy-5-methyl-2-hexanol, 5-hydroxy-5-methyl-2-hexanone
3-Methylhexane	$CH_3CH_2C(CH_3)(OO\bullet)CH_2CH_2CH_3$, 41	2-Butanone, propanal, 3-methyl-3-hexyl nitrate
	$CH_3CH_2CH(CH_3)CH(OO\bullet)CH_2CH_3$, 20	4-Methyl-3-hexyl nitrate, propanal, 2-butyl nitrate, 2-butanone, acetaldehyde
	$CH_3CH(OO\bullet)CH(CH_3)CH_2CH_2CH_3$, 16	3-Methyl-2-hexyl nitrate, acetaldehyde, 2-pentyl nitrate, 2-pentanone, propanal, 5-nitrooxy-2-pentanol, 5-hydroxy-2-pentanone, 5-nitrooxy-3-methyl-2-hexanol, 5-hydroxy-3-methyl-2-hexanone
	$CH_3CH_2CH(CH_3)CH_2CH(OO\bullet)CH_3$, 16	4-Methyl-2-hexyl nitrate, 5-nitrooxy-4-methyl-2-hexanol, 5-hydroxy-4-methyl-2-hexanone
2,2-Dimethylhexane		Butanal, 2,2-dimethyl-3-hexylnitrate, $(CH_3)_3CONO_2$, acetone, formaldehyde
	$(CH_3)_3CCH(OO\bullet)CH_2CH_2CH_3$, 30 $(CH_3)_3CCH_2CH(OO\bullet)CH_2CH_3$, 30	5,5-Dimethyl-3-hexyl nitrate, 6-nitrooxy-5,5-dimethyl-3-hexanol, 6-hydroxy-5,5-dimethyl-3-hexanone
	$(CH_3)_3CCH_2CH_2CH(OO\bullet)CH_3$, 25	5,5-Dimethyl-2-hexanone, 5,5-dimethyl-2-hexylnitrate, acetaldehyde, 3,3-dimethyl-butylnitrate, 4-nitrooxy-3,3-dimethyl-1-butanol, 4-hydroxy-3,3-dimethyl-butanal
2,2,3,3-Tetramethylbutane	$(CH_3)_3CC(CH_3)_2CH_2O_2\bullet$, 100	Formaldehyde, acetone, $(CH_3)_3CC(CH_3)_2CH_2ONO_2$, $(CH_3)_3CC(CH_3)_2ONO_2$, $(CH_3)_3CONO_2$

[a] Percent yields determined using structure–additivity data of Kwork and Atkinson (1995) to determine the OH site of attack.

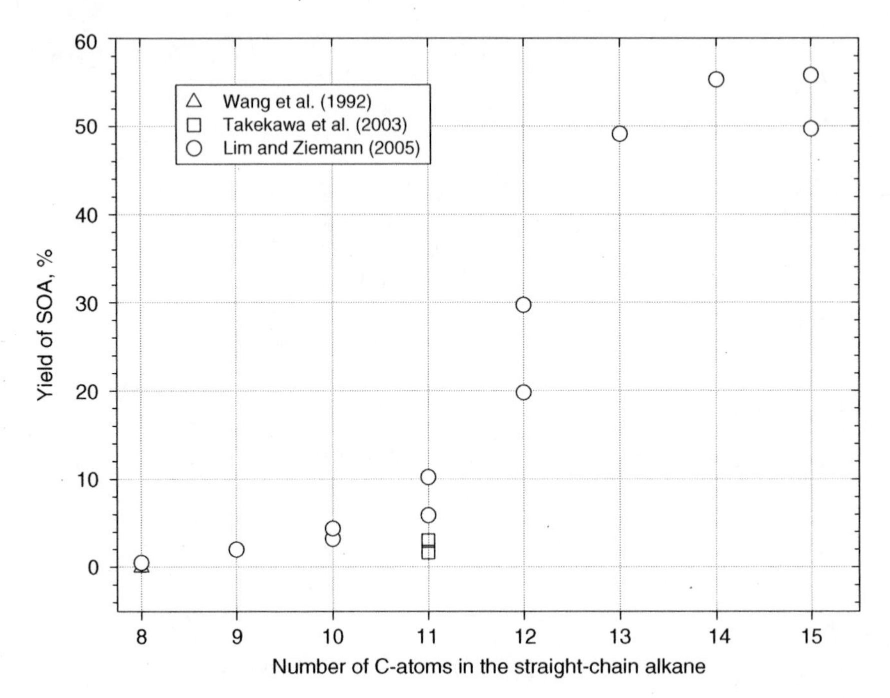

Figure IV-C-11. Secondary organic aerosol (SOA) yield from OH-initiated oxidation of straight-chain alkanes in the presence of NO_x. The figure has been adapted from Lim and Ziemann (2005) with the permission of the American Chemical Society.

1,4-hydroxycarbonyl species (e.g., Martin et al., 2002; Gong et al., 2005; Holt et al., 2005), as illustrated in the following reaction scheme (Lim and Ziemann, 2005), and thus can themselves be considered as first-generation products of the oxidation of long-chain alkanes:

R_2 O R_1 OH $\rightleftharpoons$ R_2 O OH R_1 $\rightleftharpoons$ R_2 O R_1 + H_2O

Substituted Dihydrofuran

On the basis of the Lim and Ziemann (2005) data described here, it is interesting to note that the potential for SOA formation from cetane (n-$C_{16}H_{34}$), a major component of diesel fuel, is quite high, with an anticipated SOA yield of about 55%.

V

Reactions of Products of Alkane Reactions

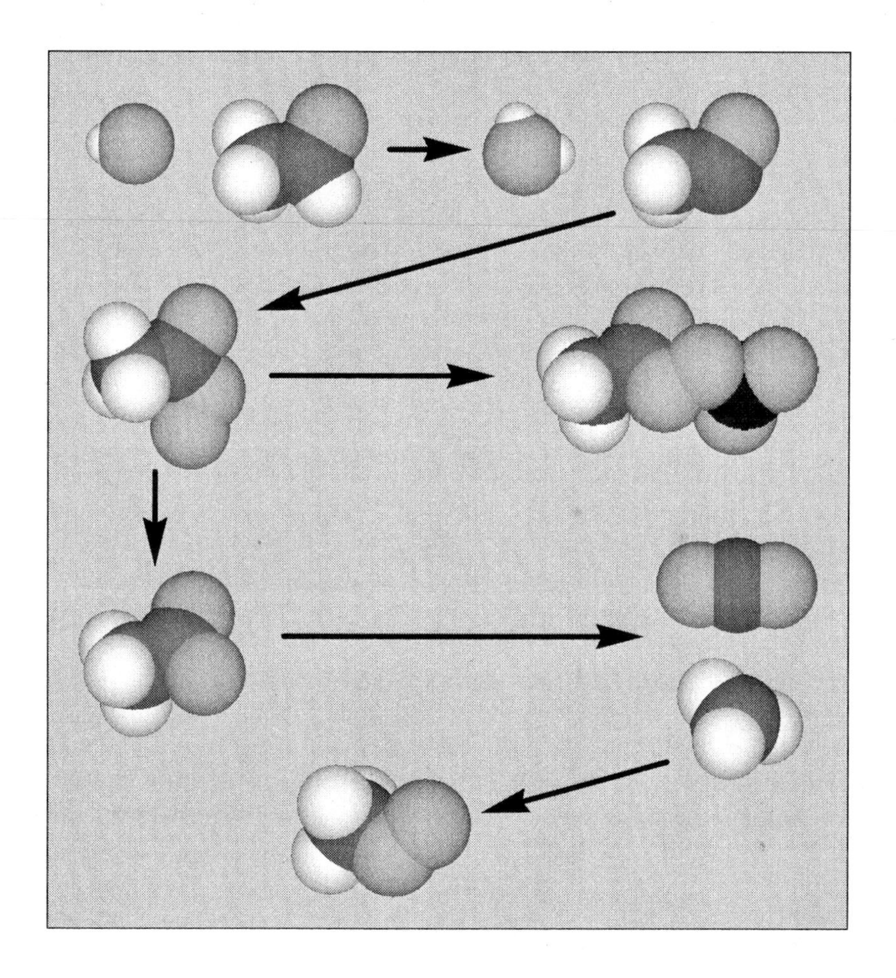

V-A. Introduction

The first-generation products of the reactions of OH radicals with alkanes are all oxygenated organics—alcohols, aldehydes, ketones, hydroperoxides, and nitrates. A number of these products are difunctional (following isomerization of the initially formed alkoxy radical), but the majority are monofunctional. The database for reactions of OH with oxygenates is not as extensive as that for the parent alkanes. In particular, the range of temperatures over which the reactions have been studied tends to be smaller, and usually does not extend to high temperatures. Increasingly, however, studies of the temperature dependences are appearing in the literature, and they show a reasonably consistent picture. The rate coefficients increase with increasing length of the carbon chain, as was the case for alkanes. There is strong evidence for activation of hydrogen atoms several carbons removed from the primary functional group, and many of the rate coefficients exhibit negative temperature dependences.

The following sections describe the kinetics of reactions of oxygenated organic compounds with ozone (O_3), hydroxy radicals (OH), and nitrate radicals (NO_3). Tables are given listing measurements at each temperature. The uncertainties in the tables are as given by the original author, except when multiple determinations have been reported at a given temperature (often the case at room temperature); in which case, the values have been averaged and assigned an uncertainty encompassing two standard deviations (again taken from the original paper, where given). In the case of relative rate experiments, we have recalculated the values using rate coefficients for the reference reactions derived in the present work, if appropriate. An overview of the kinetics and mechanisms of OH reactions with oxygenates is given by Mellouki et al. (2003), and a number of OH and NO_3 reactions have been evaluated by Atkinson et al. (2005b).

V-B. Reactions of Ozone with Oxygenated Products from Alkane Oxidation

A small number of studies have been carried out to investigate the reactions of ozone with oxygenated organics. In general, the reactions are very slow, and often too slow to measure. Considering the rapid reactions of most oxygenates with OH, the ozone reactions will be too slow to be of importance in the atmosphere. Furthermore, rate coefficients as small as the ones given here must always be treated with caution. Low levels of impurities can react with ozone, eventually producing OH radicals, and the ensuing chain reactions can maintain an apparent reaction. A summary of the available data follows.

The reaction of formaldehyde with ozone has been the subject of one published study. Braslavsky and Heicklen (1976) found a rate coefficient of 2.1×10^{-24} cm^3 molecule^{-1} s^{-1}. The reaction of acetaldehyde with ozone has been studied twice. Stedman and Niki (1973) reported a rate coefficient of 3.4×10^{-20} cm^3 molecule^{-1} s^{-1}, while Atkinson et al. (1981) found an upper limit of 6×10^{-21} cm^3 molecule^{-1} s^{-1}.

The reaction of peroxyacetyl nitrate with O_3 has also been studied. A rate coefficient of 5.4×10^{-20} cm^3 molecule^{-1} s^{-1} was reported by Pate et al. (1976). In light of the slow reaction of PAN with OH (see later in this chapter), this reaction is potentially important in the atmosphere, since the ratio of concentrations of OH and ozone is typically of the order 10^6. However, no information is available regarding the temperature dependence of this reaction, so an evaluation of its influence in the free troposphere (where the thermal decay of PAN becomes negligible) is not possible.

The reactions of alcohols with ozone also appear to be slow. No gas-phase rate coefficients are available, but the reactions have been studied in solution (Rakovski and Cherneva, 1990, and references therein). For alcohols with α-hydrogen atoms the rate coefficients are on the order of 10^{-21} cm^3 molecule^{-1} s^{-1}, while for *tert*-butyl alcohol (2-methyl-2-propanol) the rate coefficient is considerably slower, suggesting that the reaction occurs at the α-site. Reported activation energies are in the range 38–44 kJ mol^{-1}.

V-C. Reactions of Hydroxy Radicals with Products of Alkane Reactions

The reactions of OH radicals with carbonyls have generally been less widely studied than those of the parent hydrocarbons. Fewer measurements exist at high temperatures and, as in the case of the alkanes, measurements at sub-ambient temperatures have only recently become available. The reported temperature coefficients are usually small, and sometimes negative. Where data are available, no evidence has been found for pressure dependences of the reactions of OH radicals with aliphatic aldehydes and ketones.

V-C-1. Reactions of Hydroxy Radicals with Aldehydes

V-C-1.1. OH + Formaldehyde (Methanal, CH_2O)

This reaction has been the subject of a number of studies which cover a wide range of temperature (202–576 K). The measurements are given in table V-C-1. The direct studies of Atkinson and Pitts (1978), Stief et al. (1980), Temps and Wagner (1984), Yetter et al. (1989), and Sivakumaran et al. (2003) all agree to within ±10%. The relative rate measurement by Niki et al. (1984) also agrees with the above measurements. They actually studied the reaction of OH with the ^{13}C isotope, since normal formaldehyde was produced in their reaction system. However, since the secondary isotope effect is expected to be small (see below), that study can be combined with the other, direct studies. At 298 K a rate coefficient of 8.5×10^{-12} cm^3 molecule^{-1} s^{-1} is recommended (Sivakumaran et al. 2003).

The careful measurements made by Sivakumaran et al. (2003) show that the Arrhenius plot is curved, with a minimum rate coefficient at around 315 K. Figure V-C-1 shows the available measurements, along with a nonlinear fit to the data of Atkinson and Pitts (1978), Stief et al. (1980), Temps and Wagner (1984), Niki et al. (1984), Yetter et al. (1989), and Sivakumaran et al. (2003). We recommend a rate coefficient of $k = 1.25 \times 10^{-17}\ T^2 \exp(615/T)$ over the range 200–400 K, identical to that given by Atkinson et al. (2005b). The time-resolved studies of Temps and Wagner (1984) and Sivakumaran et al. (2003) indicate that the major products are formyl radical and water, rather than formic acid and hydrogen atoms which account for less than 4% of the reaction. Furthermore, chamber studies by Morrison and Heicklen (1980),

Table V-C-1. Rate coefficients (k, cm^3 $molecule^{-1}$ s^{-1}) for reaction of OH with formaldehyde (CH_2O)

$10^{12} \times A$	B (K)	$10^{12} \times k$	T (K)	Technique[a]	Reference	Temperature Range (K)
		14.0 ± 3.5	298	DF–MS	Morris and Niki (1971a)	
		15 ± 3	298	DF–RR (C_3H_6)[b]	Morris and Niki (1971b)	
		15 ± 2	298	RR (C_2D_4/C_2H_4)[c]	Niki et al. (1978)	
12.5	88 ± 151	9.4 ± 1.0	299	FP–RF	Atkinson and Pitts (1978)	299–426
		9.4 ± 1.0	356			
		10.3 ± 1.1	426			
70	700	6.0 ± 1.0	268	DF–MS (relative to OH+OH)	Smith (1978)	268–334
		6.5 ± 0.8	298			
		9.6 ± 1.5	334			
10.5 ± 1.1	0 ± 100	11.2 ± 1.0	228	FP–RF	Stief et al. (1980)	228–362
		10.3 ± 0.9	257			
		9.9 ± 1.1	298			
		10.5 ± 1.5	362			
		8.1 ± 1.7	296	DF–LMR	Temps and Wagner (1984)	
		8.4 ± 1.1	299 ± 2	RR (C_2H_4)[c]	Niki et al. (1984)	
16.6	86 ± 40	12.5 ± 0.11	298 ± 3	PLP–LIF	Zabarnick et al. (1988)	296–576
		13.9 ± 0.4	378			
		13.3 ± 0.3	473			
		14.5 ± 2.0	572 ± 5			
		8.0 ± 1.2	298	DF–RF	Yetter et al. (1989)	
		10.8 ± 0.05	202	PLP–LIF	Sivakumaran et al. (2003)	202–399
		9.49 ± 0.07	212			
		9.89 ± 0.06	222			
		9.69 ± 0.05	233			
		9.28 ± 0.06	243			
		9.31 ± 0.06	253			
		8.72 ± 0.07	263			
		8.81 ± 0.09	273			
		8.37 ± 0.18	276			
		8.46 ± 0.07	284			
		8.06 ± 0.17	288			
		8.46 ± 0.40	298 ± 2			
		8.56 ± 0.09	308			
		8.60 ± 0.09	312			
		8.20 ± 0.11	323			
		8.88 ± 0.07	334			
		8.51 ± 0.14	336			
		8.67 ± 0.08	366			
		9.28 ± 0.14	399			

[a] Abbreviations used for the techniques are defined in table II-A-1.

[b] Placed on an absolute basis using $k(C_3H_6) = 1.7 \times 10^{-11}$ cm^3 $molecule^{-1}$ s^{-1} in 1 Torr He.

[c] Placed on an absolute basis using $k(C_2H_4) = 8.52 \times 10^{-12}$ cm^3 $molecule^{-1}$ s^{-1} at 1 bar and 298 K.

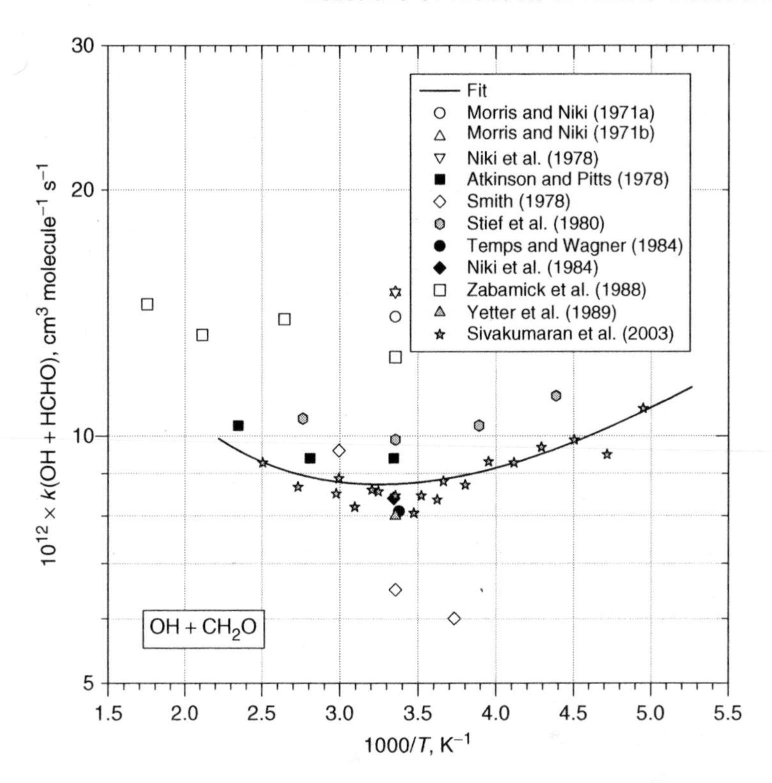

Figure V-C-1. Arrhenius plot of the rate coefficient for the reaction of OH with formaldehyde.

Niki et al. (1984), and D'Anna et al. (2003) all indicate that the yield of formic acid is no more than a few percent.

$$OH + CH_2O \rightarrow H_2O + HCO$$

$$OH + CH_2O \rightarrow [HOCH_2O\bullet]^{\dagger} \rightarrow H + HCOOH$$

The latter pathway could be envisaged as proceeding by addition of the OH to the carbonyl group, and the observed negative temperature dependence does suggest that an attractive potential is involved. However, calculations show that the barrier to addition of OH to the carbonyl is too high for such a rapid reaction to occur (Soto and Page, 1990; Alvárez-Idaboy et al., 2001; D'Anna et al., 2003), and it is more likely that the negative temperature dependence is simply associated with the formation of a strong hydrogen bond between the O-atom in the carbonyl and the hydrogen atom in the hydroxy radical.

D'Anna et al. (2003) and Feilberg et al. (2004) have measured the isotope effects for a number of isotopic variants of formaldehyde with OH (table V-C-2). Expressed relative to normal CH_2O, the two studies can be combined to give the following ratios: $CH_2O/CD_2O = 1.65 \pm 0.03$; $CH_2O/CDHO = 1.28 \pm 0.02$; $CH_2O/^{13}CH_2O = 0.96 \pm 0.02$; $CH_2O/CH_2^{18}O = 0.97 \pm 0.02$, with our estimate of the three standard deviations uncertainty.

Table V-C-2. Reactions of OH with isotopically substituted formaldehyde and acetaldehyde at 298 ± 2 K

Reactants	Ratio of Rate Coefficients	Reference
$CH_2O/CDHO$	1.28 ± 0.02	Feilberg et al. (2004)
CH_2O/CD_2O	1.62 ± 0.08	D'Anna et al. (2003)
	1.66 ± 0.02	Feilberg et al. (2004)
$CH_2O/^{13}CH_2O$	0.97 ± 0.11	D'Anna et al. (2003)
	0.952 ± 0.008	Feilberg et al. (2004)
$^{13}CH_2O/CD_2O$	1.64 ± 0.12	D'Anna et al. (2003)
$CH_2O/CH_2^{18}O$	0.967 ± 0.009	Feilberg et al. (2004)
CH_3CHO/CH_3CDO	1.31 ± 0.10	Taylor et al. (1996)
	1.42 ± 0.10	D'Anna et al. (2003)
	1.73 ± 0.10	Taylor et al. (2006)
CH_3CHO/CD_3CHO	1.13 ± 0.04	D'Anna et al. (2003)
	0.88 ± 0.10	Taylor et al. (1996)
CH_3CHO/CD_3CDO	1.21 ± 0.10	Taylor et al. (1996)
	1.65 ± 0.08	D'Anna et al. (2003)

V-C-1.2. OH + Acetaldehyde (Ethanal, CH_3CHO)

The reaction of OH with acetaldehyde has been well characterized over the temperature range 200–400 K. The measurements are given in table V-C-3. Earlier studies defined the room temperature rate coefficient as being $(1.5 \pm 0.2) \times 10^{-11}$ cm^3 molecule^{-1} s^{-1}, and evidence for a weak, negative temperature dependence was found. However, the studies were not of sufficient precision to fully characterize the temperature dependence. Sivakumaran and Crowley (2003) have performed a thorough study of the temperature dependence, which unifies all the previous data and shows that the reaction indeed has a negative temperature dependence with no obvious signs of curvature at atmospheric temperatures, in contrast to formaldehyde. Figure V-C-2 shows the available measurements, along with an Arrhenius fit to the data of Morris and Niki (1971b), Morris et al. (1971), Niki et al. (1978), Michael et al. (1985), Scollard et al. (1993), Tyndall et al. (1995), D'Anna et al. (2001a), and Sivakumaran and Crowley (2003). We recommend a rate coefficient $k = 5.15 \times 10^{-12} \exp[(324 \pm 40)/T]$ cm^3 molecule^{-1} s^{-1} over the temperature range 200–500 K, with an uncertainty of $\pm 15\%$. The measurements of Taylor et al. (1996, 2006), which extend to higher temperatures, indicate that the rate coefficient passes though a minimum and increases rapidly above 600 K.

It has been known for many years that the major products of the OH-acetaldehyde reaction are acetyl radical and water:

$$OH + CH_3CHO \rightarrow H_2O + CH_3CO \quad \text{(a)}$$

$$OH + CH_3CHO \rightarrow H_2O + CH_2CHO \quad \text{(b)}$$

Table V-C-3. Rate coefficients (k, cm^3 $molecule^{-1}$ s^{-1}) for the reaction of OH with acetaldehyde (CH_3CHO)

$10^{12} \times A$	B (K)	$10^{12} \times k$	T (K)	Technique[a]	Reference	Temperature Range (K)
		15 ± 4	300	DF–MS	Morris et al. (1971)	
		15 ± 3	298	DF–RR (C_3H_6)[b]	Morris and Niki (1971b)	
		16 ± 2	298 ± 2	RR (C_2H_4)[c]	Niki et al. (1978)	
6.87	−257 ± 51	16.0 ± 1.6	299	FP–RF	Atkinson and Pitts (1978)	299–426
		14.4 ± 1.5	355			
		12.4 ± 1.3	426			
		12.8 ± 4.3	298 ± 4	RR (C_2H_4)[c]	Kerr and Sheppard (1981)	
7.1 ± 0.2	−165 ± 91	14.0 ± 3.1	253	FP–RF	Semmes et al. (1985)	253–424
		12.2 ± 2.7	298			
		10.7 ± 2.3	356			
		11.0 ± 2.3	424			
5.52 ± 0.80	−307 ± 52	19.8 ± 1.2	244	DF–RF	Michael et al. (1985)	244–528
		18.4 ± 1.9	259			
		17.3 ± 1.6	273			
		14.7 ± 2.8	298			
		13.0 ± 1.2	333			
		14.0 ± 0.8	355			
		13.0 ± 0.4	367			
		15.0 ± 1.0	373			
		11.6 ± 0.6	393			
		11.7 ± 0.8	420			
		10.6 ± 0.6	424			
		11.0 ± 0.6	433			
		11.5 ± 0.8	466			
		10.4 ± 0.4	468			
		10.4 ± 0.4	492			
		9.2 ± 1.4	499			
		9.9 ± 0.4	528			
8.6 ± 1.8	−200 ± 60	16.9 ± 3.3	298	DF–RF/LIF	Dóbé et al. (1989)	297–517
		15.1 ± 3.2	390			
		13.8 ± 2.7	423			
		13.4 ± 2.7	467			
		17 ± 3	298	PLP–RF	Balestra-Garcia et al. (1992)	
		16.2 ± 1.0	298 ± 2	RR (C_2H_4)[c]	Scollard et al. (1993)	
		14.4 ± 2.5	298	DF–LIF	Tyndall et al. (1995)	
4.31 ± 0.22	−309 ± 19			PLP–LIF	Taylor et al. (1996)	295–550
		14.4 ± 0.7	298	RR (1-C_4H_8)[d]	D'Anna et al. (2001a)	

(continued)

Table V-C-3. *(continued)*

$10^{12} \times A$	B (K)	$10^{12} \times k$	T (K)	Technique[a]	Reference	Temperature Range (K)
4.4	-366 ± 30	26.9 ± 2.0	202(He)	PLP–LIF	Sivakumaran and Crowley (2003)	202–348
		24.9 ± 0.6	211			
		22.7 ± 0.2	224			
		21.4 ± 0.2	234			
		19.0 ± 0.2	244			
		18.2 ± 0.2	253			
		17.3 ± 0.3	263			
		17.3 ± 0.2	273			
		15.9 ± 0.3	283			
		15.5 ± 0.5	296 ± 2			
		16.7 ± 0.3	273(Ar)			
		16.6 ± 0.2	278			
		16.1 ± 0.2	284			
		15.3 ± 0.3	289			
		15.1 ± 0.2	292			
		15.2 ± 0.3	296			
		14.4 ± 0.2	304			
		14.1 ± 0.2	314			
		13.8 ± 0.3	319			
		13.6 ± 0.3	324			
		13.0 ± 0.2	333			
		12.5 ± 0.2	342			
		12.6 ± 0.2	348			
		16.2 ± 1.6	297	PLP–LIF	Taylor et al. (2006)	297–860
		10.8 ± 1.3	383			

[a] Abbreviations used for the techniques are defined in table II-A-1.
[b] Placed on an absolute basis using $k(C_3H_6) = 1.7 \times 10^{-11}$ cm^3 molecule^{-1} s^{-1} in 1 Torr He.
[c] Placed on an absolute basis using $k(C_2H_4) = 8.52 \times 10^{-12}$ cm^3 molecule^{-1} s^{-1} at 1 bar and 298 K.
[d] Placed on an absolute basis using k(1-butene) $= 3.14 \times 10^{-11}$ cm^3 molecule^{-1} s^{-1} at 1 bar and 298 K.

$$OH + CH_3CHO \rightarrow [HOCH(CH_3)O\bullet]^{\dagger} \rightarrow H + CH_3C(O)OH \quad (c)$$

$$OH + CH_3CHO \rightarrow [HOCH(CH_3)O\bullet]^{\dagger} \rightarrow CH_3 + HCOOH \quad (d)$$

This has been confirmed in recent studies through experiments designed to quantify both the major and potential minor products. Butkovskaya et al. (2004) determined that the total yield of H_2O is $(97.7 \pm 4.7)\%$, with abstraction from the methyl group accounting for 5.5% of the reaction. Vandenberk and Peeters (2003) measured an H_2O yield of 0.89 ± 0.06, and showed that less than 3% formic acid is produced. Cameron et al. (2002) found a yield of CH_3CO of 0.93 ± 0.18, and showed that the yields of CH_3 and H are less than 0.03 and 0.02, respectively. Wang et al. (2003) found no strong evidence for the production of CH_3 or H in time-resolved experiments. In chamber studies, Tyndall et al. (2002) were unable to detect either HCOOH or $CH_3C(O)OH$, and placed upper limits of 10% on both these species. Hence, we recommend that channel (a) be assigned a branching fraction of $(95 \pm 5)\%$ and channel (b) $(5 \pm 5)\%$ for atmospheric purposes.

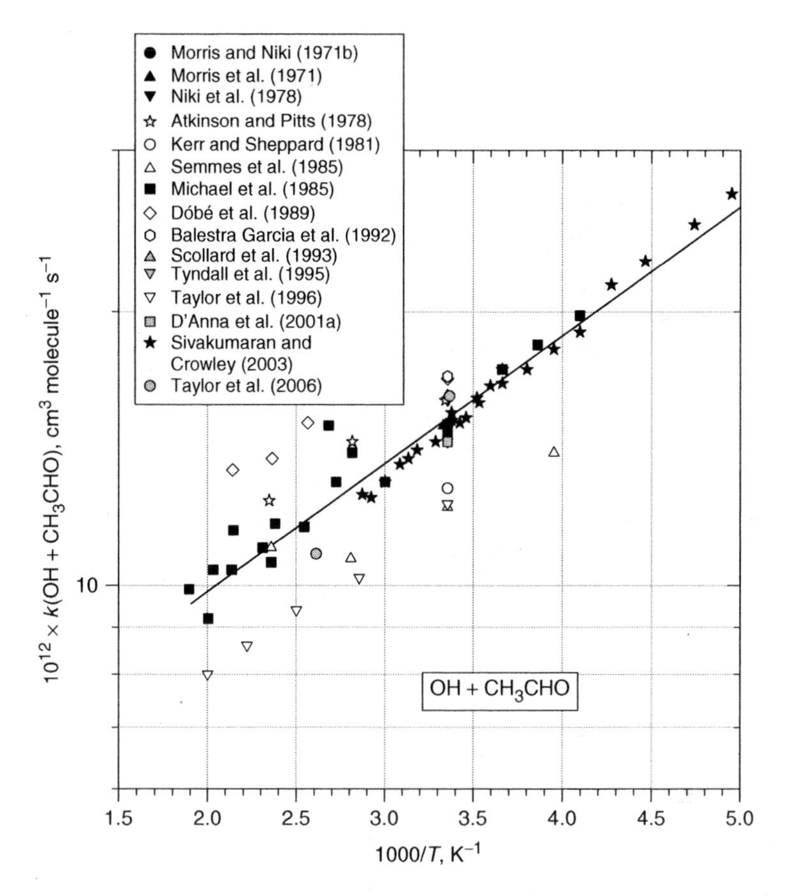

Figure V-C-2. Arrhenius plot of the rate coefficient for the reaction of OH with acetaldehyde.

The absence of addition–displacement products such as H and CH_3 suggests strongly that a stable intermediate radical $HOCH(CH_3)(O\bullet)$ is not formed. As in the case of formaldehyde, the reaction probably proceeds through formation of a hydrogen-bonded complex (Alvárez-Idaboy et al., 2001; D'Anna et al., 2003), which facilitates reaction to give water and the acetyl radical. The existence of the hydrogen bond gives rise to the observed negative temperature dependence.

Taylor et al. (1996, 2006) made direct measurements of the rates for OH radical reaction with the isotopically substituted acetaldehydes CH_3CDO, CD_3CHO, and CD_3CDO. D'Anna et al. (2003) measured rate coefficients for OH with the same compounds, using the relative rate method with FTIR detection. The results of these studies, which are summarized in table V-C-2, are in reasonable agreement. The relative rate experiments of D'Anna et al. give a more direct measurement of the kinetic isotope effect and are thus probably more reliable. Introduction of a D-atom in the aldehyde group caused a reduction in the rate coefficient of 30–40%, while deuteration of the methyl group led to less than 15% change in both studies. Since the reaction proceeds via a complex, it is not obvious whether the deuterium kinetic isotope effect should be large or small (Smith and Ravishankara, 2002). The quantum chemical calculations of D'Anna et al. (2003) agree quite well with their measurements, and it should also be noted that the isotope effect found by D'Anna et al. for CD_3CDO is identical to that for DCDO.

V-C-1.3. OH + Propanal (Propionaldehyde, C_2H_5CHO)

The rate coefficient for the reaction of OH with propanal has been measured by Morris and Niki (1971b), Niki et al. (1978), Audley et al. (1981), Kerr and Sheppard (1981), Semmes et al. (1985), Papagni et al. (2000), Thévenet et al. (2000), D'Anna et al. (2001a), and Le Crâne et al. (2005), using both direct and relative methods. The measurements are given in table V-C-4. The agreement is very good, with most values at 298 ± 3 K ranging from 1.7×10^{-11} cm^3 molecule^{-1} s^{-1} to 2.1×10^{-11} cm^3 molecule^{-1} s^{-1}. The smallest of these values comes from Semmes et al., who noted some difficulties in handling both acetaldehyde and propanal. The weighted average of

Table V-C-4. Rate coefficients (k, cm^3 molecule^{-1} s^{-1}) for the reaction of OH with propanal (C_2H_5CHO)

$10^{12} \times A$	B (K)	$10^{12} \times k$	T (K)	Technique[a]	Reference	Temperature Range (K)
		27 ± 5	298	DF–RR (C_3H_6)[b]	Morris and Niki (1971b)	
		22 ± 2	298 ± 2	RR (C_2H_4)[c]	Niki et al. (1978)	
		17.9 ± 2.3	298	RR (CH_3CHO)[d]	Audley et al. (1981)	
		18 ± 1	298 ± 4	RR (C_2H_4)[b]	Kerr and Sheppard (1981)	
		17.1 ± 2.4	298	FP–RF	Semmes et al. (1985)	
		20.2 ± 1.4	298	RR (MVK)[e]	Papagni et al. (2000)	
5.3 ± 0.5	(405 ± 30)	28.5 ± 2.4	240	PLP–LIF	Thévenet et al. (2000)	240–372
		25.7 ± 1.0	253			
		23.7 ± 0.5	273			
		20 ± 3	298			
		18.6 ± 0.8	323			
		17.3 ± 0.8	348			
		15.5 ± 1.5	372			
		19.0 ± 1.5	298 ± 2	RR (C_3H_6)[f]	D'Anna et al. (2001a)	
		18.6 ± 2.1	295 ± 2	RR (C_2H_4)[b]	Le Crâne et al. (2005)	
		17.8 ± 1.9		RR (C_3H_6)[f]		

[a] Abbreviations used for the techniques are defined in table II-A-1.
[b] Placed on an absolute basis using $k(C_3H_6) = 1.7 \times 10^{-11}$ cm^3 molecule^{-1} s^{-1} in 1 Torr He.
[c] Placed on an absolute basis using $k(C_2H_4) = 8.52 \times 10^{-12}$ cm^3 molecule^{-1} s^{-1} at 1 bar and 298 K.
[d] Placed on an absolute basis using $k(CH_3CHO) = 1.5 \times 10^{-11}$ cm^3 molecule^{-1} s^{-1}.
[e] Placed on an absolute basis using k(methyl vinyl ketone) $= 2.04 \times 10^{-11}$ cm^3 molecule^{-1} s^{-1}.
[f] Placed on an absolute basis using $k(C_3H_6) = 2.63 \times 10^{-11}$ cm^3 molecule^{-1} s^{-1} at 1 bar and 298 K.

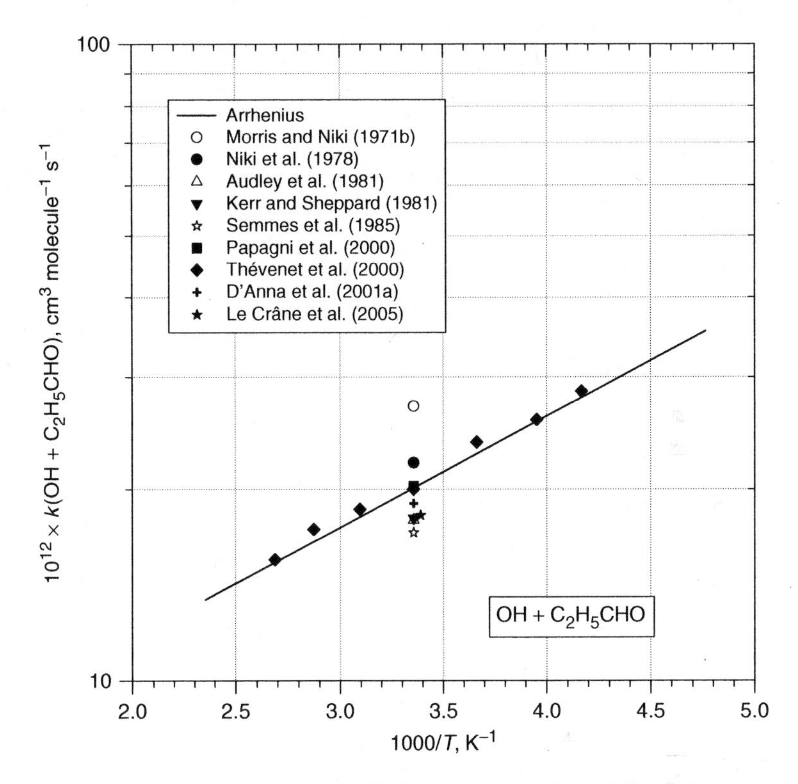

Figure V-C-3. Arrhenius plot of the rate coefficient for the reaction of OH with propanal.

the other values is $(1.95 \pm 0.15) \times 10^{-11}$ cm^3 molecule^{-1} s^{-1}. The only study of the temperature dependence is that of Thévenet et al., who found a temperature coefficient of -405 K between 240 and 372 K.

Figure V-C-3 shows the available measurements, along with a fit to the data of Niki et al. (1978), Kerr and Sheppard (1981), Papagni et al. (2000), Thévenet et al. (2000), D'Anna et al. (2001a), and Le Crâne et al. (2005). We recommend $k = (5.15 \pm 0.75) \times 10^{-12} \exp(400/T)$ cm^3 molecule^{-1} s^{-1} for atmospheric purposes.

On the basis of the observed trends in aldehyde rate coefficients (see later in the chapter) we expect that abstraction from the aldehyde group accounts for a large fraction of the overall reaction. Vandenberk and Peeters (2003) found that the overall yield of H_2O from the reaction is 100 ± 10%, and that the formation of formic acid in an addition-elimination mechanism accounts for less than 3%.

A number of rate coefficients for OH with aldehydes have been measured by Audley et al. (1981). They used a dark, heterogeneous source of OH, flowing H_2O_2 and NO_2 over boric acid-coated glass walls. The characteristics of this source are uncertain, and while some of their rate coefficients agree with the consensus, others differ considerably. Their rate coefficients are included in the tables, but have not been considered further in the derivation of preferred values, because of uncertainties in the methodology.

Table V-C-5. Rate coefficients (k, cm^3 molecule^{-1} s^{-1}) for the reaction of OH with butanal (n-C_3H_7CHO)

$10^{12} \times A$	B (K)	$10^{12} \times k$	T (K)	Technique[a]	Reference	Temperature Range (K)
		25.2 ± 3.2	298	RR (CH_3CHO)[b]	Audley et al. (1981)	
		24 ± 1	298 ± 4	RR (C_2H_4)[c]	Kerr and Sheppard (1981)	
5.7 ± 0.3	-411 ± 164	30.8 ± 4.2	258	FP–RF	Semmes et al. (1985)	258–422
		20.6 ± 3.0	298			
		18.2 ± 2.6	361			
		15.4 ± 2.3	422			
		24.7 ± 1.5	298	RR (MVK)[d]	Papagni et al. (2000)	
		23.8 ± 1.5	298 ± 2	RR (1-C_4H_8)[e]	D'Anna et al. (2001a)	
		28.8 ± 2.6	298	PLP–LIF	Albaladejo et al. (2002)	

[a] Abbreviations used for the techniques are defined in table II-A-1.
[b] Placed on an absolute basis using $k(CH_3CHO) = 1.5 \times 10^{-11}$ cm^3 molecule^{-1} s^{-1}.
[c] Placed on an absolute basis using $k(C_2H_4) = 8.52 \times 10^{-12}$ cm^3 molecule^{-1} s^{-1} at 1 bar and 298 K.
[d] Placed on an absolute basis using k(methyl vinyl ketone) $= 2.04 \times 10^{-11}$ cm^3 molecule^{-1} s^{-1}.
[e] Placed on an absolute basis using k(1-butene) $= 3.14 \times 10^{-11}$ cm^3 molecule^{-1} s^{-1} at 1 bar and 298 K.

V-C-1.4. OH + Butanal (n-Butyraldehyde, $CH_3CH_2CH_2CHO$)

The rate coefficient for the reaction of OH with butanal has been the subject of six studies (Audley et al., 1981; Kerr and Sheppard, 1981; Semmes et al., 1985; Papagni et al., 2000; D'Anna et al., 2001a; Albaladejo et al., 2002). The measurements are given in table V-C-5. The studies are in reasonable agreement, with measurements ranging from 2.1×10^{-11} to 2.9×10^{-11} with a mean value of $(2.44 \pm 0.3) \times 10^{-11}$ cm^3 molecule^{-1} s^{-1} (1-σ). The only study of the temperature dependence was that of Semmes et al., who found an activation temperature of -411 K, almost identical to that for propanal. The rate coefficients are shown in figure V-C-4. A fit to the data (other than those of Audley et al.) gives $k = 5.15 \times 10^{-12} \exp(458/T)$ cm^3 molecule^{-1} s^{-1}.

V-C-1.5. OH + 2-Methylpropanal [Isobutyraldehyde, $(CH_3)_2CHCHO$]

The rate coefficient for this reaction has been studied seven times (Audley et al., 1981; Kerr and Sheppard, 1981; Semmes et al., 1985; Dóbé et al., 1989; Stemmler et al., 1997; Thévenet et al., 2000; D'Anna et al., 2001a). The measurements are given in table V-C-6 and figure V-C-5. The value reported by Dóbé et al. is roughly 70% larger than the others, which have an average of $(2.59 \pm 0.22) \times 10^{-11}$ cm^3 molecule^{-1} s^{-1}. The temperature dependence was measured by Semmes et al. (255–423 K) and Thévenet et al. (243–372 K), who found activation temperatures of -393 K and -390 K, respectively. A fit to the data other than those of Audley et al. (1981) and Dóbé et al. (1980) leads to an expression $k = (6.7 \pm 1.0) \times 10^{-12} \exp(408/T)$ cm^3 molecule^{-1} s^{-1} for atmospheric purposes, where the uncertainty has been set at $\pm 15\%$.

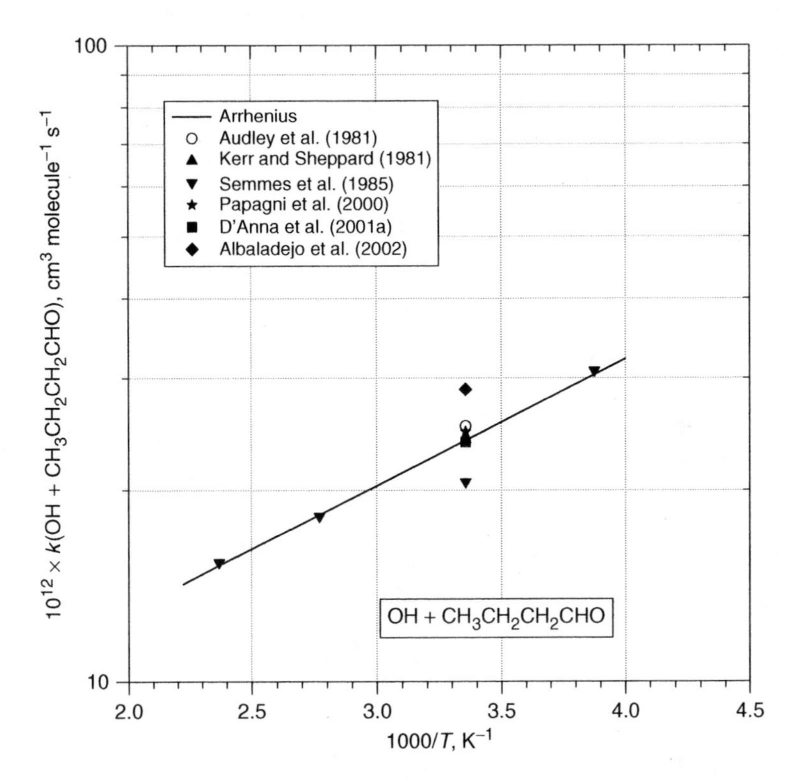

Figure V-C-4. Arrhenius plot of the rate coefficient for the reaction of OH with butanal.

V-C-1.6. OH + Pentanal (Valeraldehyde, n-C_4H_9CHO)

The reaction of OH with pentanal has been measured at 298 K by Kerr and Sheppard (1981), Semmes et al. (1985), Papagni et al. (2000), Thévenet et al. (2000), D'Anna et al. (2001a), and Albaladejo et al. (2002). The results are given in table V-C-7. These are in good agreement, with a mean value of $(2.70 \pm 0.18) \times 10^{-11}$ cm^3 molecule^{-1} s^{-1}. The temperature dependence was measured by Semmes et al. (253–410 K), who found a temperature coefficient of −451 K, and by Thevenet et al. (243–372 K), who found a somewhat weaker dependence of −306 K. A fit to the combined data set (figure V-C-6) leads to a recommended expression $k = (8.4 \pm 1.3) \times 10^{-12} \exp(350/T)$ cm^3 molecule^{-1} s^{-1} for use in atmospheric models.

V-C-1.7. OH + 2-Methylbutanal [$CH_3CH_2CH(CH_3)CHO$]

This reaction has been studied only by D'Anna et al. (2001a). They measured a rate coefficient of $(3.28 \pm 0.09) \times 10^{-11}$ cm^3 molecule^{-1} s^{-1} at 298 ± 2 K relative to 1-butene as the reference. Additional studies are clearly required on this reaction.

Table V-C-6. Rate coefficients (k, cm^3 $molecule^{-1}$ s^{-1}) for the reaction of OH with 2-methylpropanal [$(CH_3)_2CHCHO$]

$10^{12} \times A$	B (K)	$10^{12} \times k$	T (K)	Technique[a]	Reference	Temperature Range (K)
		17.4 ± 2.0	298	RR (CH_3CHO)[b]	Audley et al. (1981)	
		27 ± 5	298 ± 4	RR (C_2H_4)[c]	Kerr and Sheppard (1981)	
6.8 ± 0.3	−(393 ± 125)	33.4 ± 4.5	255	FP–RF	Semmes et al. (1985)	255–423
		24.2 ± 3.3	298			
		19.7 ± 2.7	354			
		18.2 ± 2.7	423			
15.8 ± 5.0	−(313 ± 145)	46.3 ± 7.3	298	DF–RF	Dóbé et al. (1989)	298–519
		39.7 ± 6.3	349			
		32.7 ± 7.8	420			
		32.6 ± 6.1	452			
		40.5 ± 6.1	471			
		43.0 ± 6.0	519			
		25.8 ± 0.7	298	RR (Dipropyl ether)[d]	Stemmler et al. (1997)	
7.3 ± 1.9	−(390 ± 78)	38.5 ± 0.9	243	PLP–LIF	Thévenet et al. (2000)	243–372
		34.2 ± 1.3	253			
		29.9 ± 1.1	273			
		26 ± 4	298			
		25.8 ± 0.8	323			
		21.9 ± 1.0	348			
		22.0 ± 1.1	372			
		26.5 ± 2.1	298 ± 2	RR (1-C_4H_8)[e]	D'Anna et al. (2001a)	

[a] Abbreviations used for the techniques are defined in table II-A-1.
[b] Placed on an absolute basis using $k(CH_3CHO) = 1.5 \times 10^{-11}$ cm^3 $molecule^{-1}$ s^{-1}.
[c] Placed on an absolute basis using $k(C_2H_4) = 8.52 \times 10^{-12}$ cm^3 $molecule^{-1}$ s^{-1} at 1 bar and 298 K.
[d] Placed on an absolute basis using k(dipropyl ether) $= 1.85 \times 10^{-11}$ cm^3 $molecule^{-1}$ s^{-1}.
[e] Placed on an absolute basis using k(1-butene) $= 3.14 \times 10^{-11}$ cm^3 $molecule^{-1}$ s^{-1} at 1 bar and 298 K.

V-C-1.8. OH + 3-Methylbutanal [Isovaleraldehyde, $(CH_3)_2CHCH_2CHO$]

This reaction has been studied by Audley et al. (1981), Kerr and Sheppard (1981), Semmes et al. (1985), Glasius et al. (1997), and D'Anna et al. (2001a). All the measurements were made at ambient temperatures (~295 K). The results are given in table V-C-8. The measurement of Glasius et al. was made relative to isoprene, and the large difference in rate coefficients may have led to an overestimation. The other values are in quite good agreement, and the mean value of $(2.7 \pm 0.1) \times 10^{-11}$ cm^3 $molecule^{-1}$ s^{-1} for temperatures near 298 K is recommended.

V-C-1.9. OH + 2,2-Dimethylpropanal [Pivaldehyde, $(CH_3)_3CCHO$]

The reaction of OH with 2,2-dimethylpropanal has been studied six times. The results are given in table V-C-9. The reported rate coefficients range from 8.5×10^{-12} to

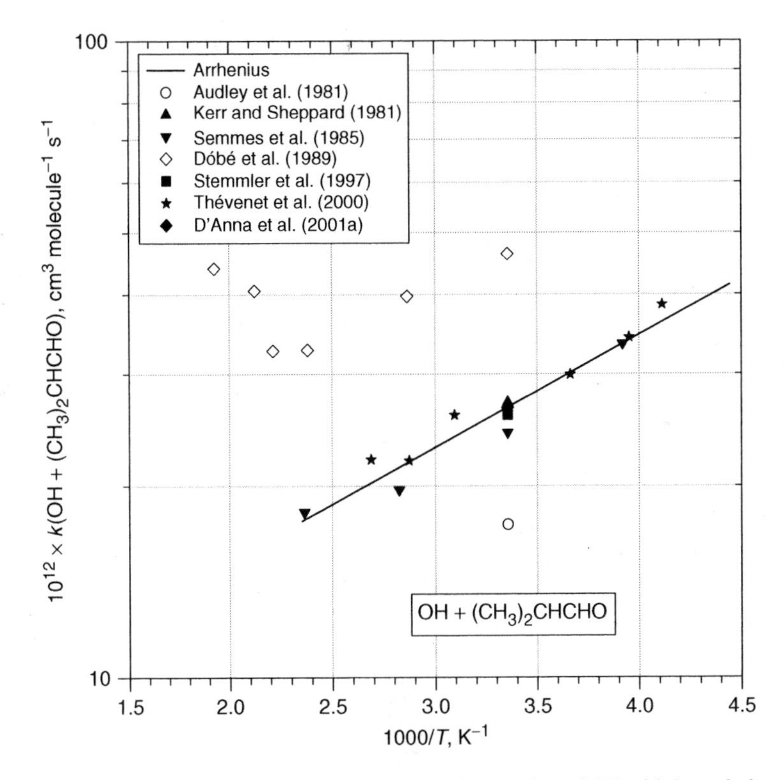

Figure V-C-5. Arrhenius plot of the rate coefficient for the reaction of OH with 2-methylpropanal.

5.2×10^{-11} cm^3 molecule^{-1} s^{-1}. The temperature-dependent data of Semmes et al. (1985) and Thévenet et al. (2000) overlap quite well, and a weighted least-squares fit to the data of Semmes et al. (1985), Thévenet et al. (2000), and D'Anna et al. (2001a) leads to a recommendation of $k = 5.3 \times 10^{-12}$ exp(500/T) cm^3 molecule^{-1} s^{-1}. The corresponding value at 298 K is 2.8×10^{-11} cm^3 molecule^{-1} s^{-1}. The data and fit are shown in figure V-C-7.

V-C-1.10. OH + Hexanal (n-C_5H_{11}CHO)

The reaction of OH with hexanal has been measured three times (Papagni et al., 2000, D'Anna et al., 2001a; Albaladejo et al., 2002). The measurements are given in table V-C-10. The first two measurements were made using relative-rate techniques whereas the last was made using flash photolysis with LIF detection of OH. The measurements are in quite good agreement, and each individual study shows a 5–10% increase for the rate coefficient of hexanal vs. that of pentanal. We thus recommend a rate coefficient at 298 K of $(2.9 \pm 0.3) \times 10^{-11}$ cm^3 molecule^{-1} s^{-1}, on the basis of the recommendation for pentanal and the internally consistent data set of D'Anna et al. (2001a).

Table V-C-7. Rate coefficients (k, cm^3 molecule^{-1} s^{-1}) for the reaction of OH with pentanal (n-C_4H_9CHO)

$10^{12} \times A$	B (K)	$10^{12} \times k$	T (K)	Technique[a]	Reference	Temperature Range (K)
		13.8 ± 1.7	298	RR (CH_3CHO)[b]	Audley et al. (1981)	
		26 ± 4	298 ± 4	RR (C_2H_4)[c]	Kerr and Sheppard (1981)	
6.3 ± 0.2	−451 ± 108	38.9 ± 5.7	253	FP–RF	Semmes et al. (1985)	253–410
		26.9 ± 3.9	298			
		23.3 ± 3.4	355			
		19.0 ± 2.8	410			
		29.9 ± 1.9	298	RR (MVK)[d]	Papagni et al. (2000)	
9.9 ± 1.9	−306 ± 56	33.4 ± 0.5	243	PLP–LIF	Thévenet et al. (2000)	243–372
		33.3 ± 0.6	253			
		31.8 ± 0.6	263			
		30.3 ± 1.0	273			
		28.2 ± 1.3	298			
		25.1 ± 0.4	313			
		23.9 ± 0.5	333			
		23.5 ± 0.7	353			
		22.3 ± 0.7	372			
		26.1 ± 1.4	298 ± 2	RR (1-C_4H_8)[e]	D'Anna et al. (2001a)	
		24.8 ± 2.4	298	PLP–LIF	Albaladejo et al. (2002)	

[a] Abbreviations used for the techniques are defined in table II-A-1.
[b] Placed on an absolute basis using $k(CH_3CHO) = 1.5 \times 10^{-11}$ cm^3 molecule^{-1} s^{-1}.
[c] Placed on an absolute basis using $k(C_2H_4) = 8.52 \times 10^{-12}$ cm^3 molecule^{-1} s^{-1} at 1 bar and 298 K.
[d] Placed on an absolute basis using k(methyl vinyl ketone) $= 2.04 \times 10^{-11}$ cm^3 molecule^{-1} s^{-1}.
[e] Placed on an absolute basis using k(1-butene) $= 3.14 \times 10^{-11}$ cm^3 molecule^{-1} s^{-1} at 1 bar and 298 K.

V-C-1.11. OH + Isomers of Hexanal ($C_5H_{11}CHO$)

D'Anna et al. (2001a) measured rate coefficients of OH with a series of branched C_6-aldehydes at ambient temperatures (~298 K) using the relative-rate technique (table V-C-11). These are the only reported values for these rate coefficients. The values obtained at 298 ± 2 K (in units of 10^{-11} cm^3 molecule^{-1} s^{-1}) are: 2-methylpentanal, 3.32; 3-methylpentanal, 2.91; 4-methylpentanal, 2.63; 3,3-dimethylbutanal, 2.14; and 2-ethylbutanal, 4.02. Recommended uncertainties on these values are ±25%.

V-C-1.12. OH + Heptanal (n-$C_6H_{13}CHO$)

The rate coefficient for the reaction of OH with heptanal has been measured by Albaladejo et al. (2002). They obtained a rate coefficient at 298 K of $(2.96 \pm 0.23) \times 10^{-11}$ cm^3 molecule^{-1} s^{-1} (table V-C-11). This value is consistent with those of the other straight chain aldehydes discussed in the preceding sections.

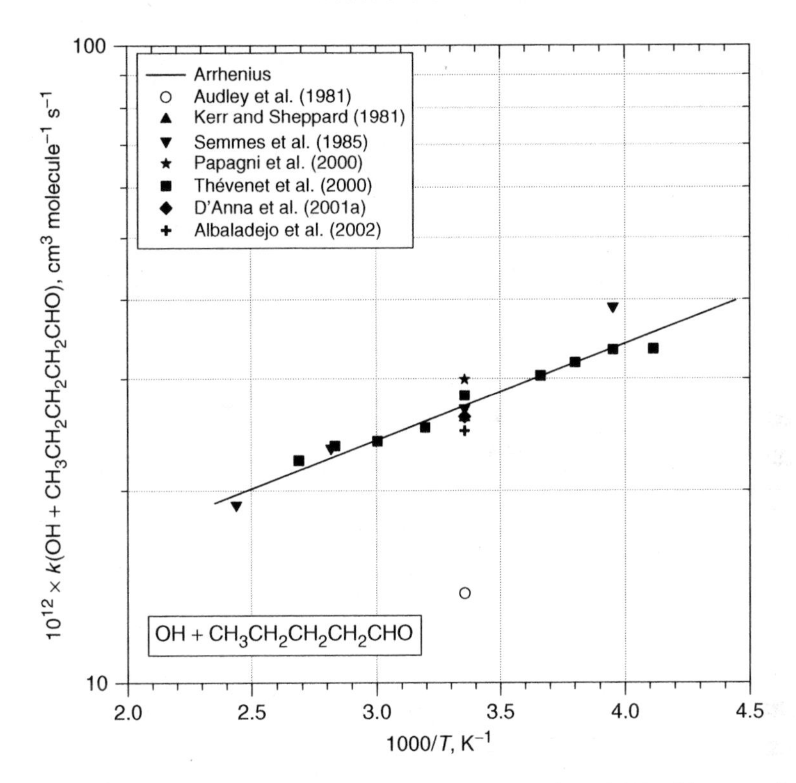

Figure V-C-6. Arrhenius plot of the rate coefficient for the reaction of OH with pentanal.

Table V-C-8. Rate coefficients (k, cm^3 $molecule^{-1}$ s^{-1}) for the reaction of OH with 3-methylbutanal [$(CH_3)_2CHCH_2CHO$]

$10^{12} \times k$	T (K)	Technique[a]	Reference
18.4 ± 2.0	298	RR (CH_3CHO)[b]	Audley et al. (1981)
27 ± 1	298 ± 4	RR (C_2H_4)[c]	Kerr and Sheppard (1981)
25.8 ± 4.0	298	FP–RF	Semmes et al. (1985)
40 ± 0.7	298	RR (isoprene)[d]	Glasius et al. (1997)
27.9 ± 0.7	298 ± 2	RR (1-C_4H_8)[e]	D'Anna et al. (2001a)

[a] Abbreviations used for the techniques are defined in table II-A-1.
[b] Placed on an absolute basis using $k(CH_3CHO) = 1.5 \times 10^{-11}$ cm^3 $molecule^{-1}$ s^{-1}.
[c] Placed on an absolute basis using $k(C_2H_4) = 8.52 \times 10^{-12}$ cm^3 $molecule^{-1}$ s^{-1} at 1 bar and 298 K.
[d] Placed on an absolute basis using $k(\text{isoprene}) = 1.0 \times 10^{-10}$ cm^3 $molecule^{-1}$ s^{-1}.
[e] Placed on an absolute basis using $k(\text{1-butene}) = 3.14 \times 10^{-11}$ cm^3 $molecule^{-1}$ s^{-1} at 1 bar and 298 K.

V-C-1.13. OH + 2,3-Dimethylpentanal [$CH_3CH_2CH(CH_3)CH(CH_3)CHO$]

Tuazon et al. (2003) studied the mechanism of the reaction of OH radicals with 2,3-dimethylpentanal; the value is included in table V-C-11. They determined the

Table V-C-9. Rate coefficients (k, cm^3 $molecule^{-1}$ s^{-1}) for the reaction of OH with 2,2-dimethylpropanal [$(CH_3)_3CCHO$]

$10^{12} \times A$	B (K)	$10^{12} \times k$	T (K)	Technique[a]	Reference	Temperature Range (K)
		8.5 ± 0.8	298	RR (CH_3CHO)[b]	Audley et al. (1981)	
		21 ± 6	298 ± 4	RR (C_2H_4)[c]	Kerr and Sheppard (1981)	
6.7 ± 0.3	-423 ± 154	33.9 ± 6.4	254	FP–RF	Semmes et al. (1985)	254–425
		30.6 ± 4.4	298			
		21.8 ± 3.2	354			
		17.6 ± 2.9	425			
13.3 ± 6.7	-409 ± 205	51.7 ± 13.6	300	DF–RF	Dóbé et al. (1989)	300–405
		40.2 ± 15.1	386			
		33.2	405			
9.9 ± 1.9	-306 ± 56	43.9 ± 1.4	243	PLP–LIF	Thévenet et al. (2000)	243–372
		40.9 ± 1.2	253			
		37.8 ± 0.7	263			
		33.5 ± 1.8	273			
		27.4 ± 2.3	298			
		23.8 ± 0.3	323			
		21.5 ± 0.6	348			
		22.9 ± 0.8	358			
		20.4 ± 0.4	372			
		26.8 ± 1.2	298 ± 2	RR (1-C_4H_8)[d]	D'Anna et al. (2001a)	

[a] Abbreviations used for the techniques are defined in table II-A-1.
[b] Placed on an absolute basis using $k(CH_3CHO) = 1.5 \times 10^{-11}$ cm^3 $molecule^{-1}$ s^{-1}.
[c] Placed on an absolute basis using $k(C_2H_4) = 8.52 \times 10^{-12}$ cm^3 $molecule^{-1}$ s^{-1} at 1 bar and 298 K.
[d] Placed on an absolute basis using k(1-butene) $= 3.14 \times 10^{-11}$ cm^3 $molecule^{-1}$ s^{-1} at 1 bar and 298 K.

rate coefficient relative to that of 2-methyl propene. Using the current value for the reference reaction leads to a recommended rate coefficient of $(4.2 \pm 1.0) \times 10^{-11}$ cm^3 $molecule^{-1}$ s^{-1} at 298 K. The authors quoted a fairly large uncertainty as a result of difficulties in the analytical procedure, and we have added an extra 10% to account for uncertainties in the reference reaction. Tuazon also studied the product distribution, and found that acetaldehyde and 2-butanone were the main products along with a small yield (5.4%) of 3-methyl-2-pentanone. The latter product arises exclusively from abstraction at the 2-position, adjacent to the carbonyl group:

$$OH + CH_3CH_2CH(CH_3)CH(CH_3)CHO\ (+O_2) \rightarrow H_2O + CH_3CH_2CH(CH_3)C(O_2)(CH_3)CHO$$

$$CH_3CH_2CH(CH_3)C(O_2)(CH_3)CHO + NO \rightarrow CH_3CH_2CH(CH_3)C(O\bullet)(CH_3)CHO + NO_2$$

$$CH_3CH_2CH(CH_3)C(O\bullet)(CH_3)CHO \rightarrow HCO + CH_3CH_2CH(CH_3)C(O)(CH_3)$$

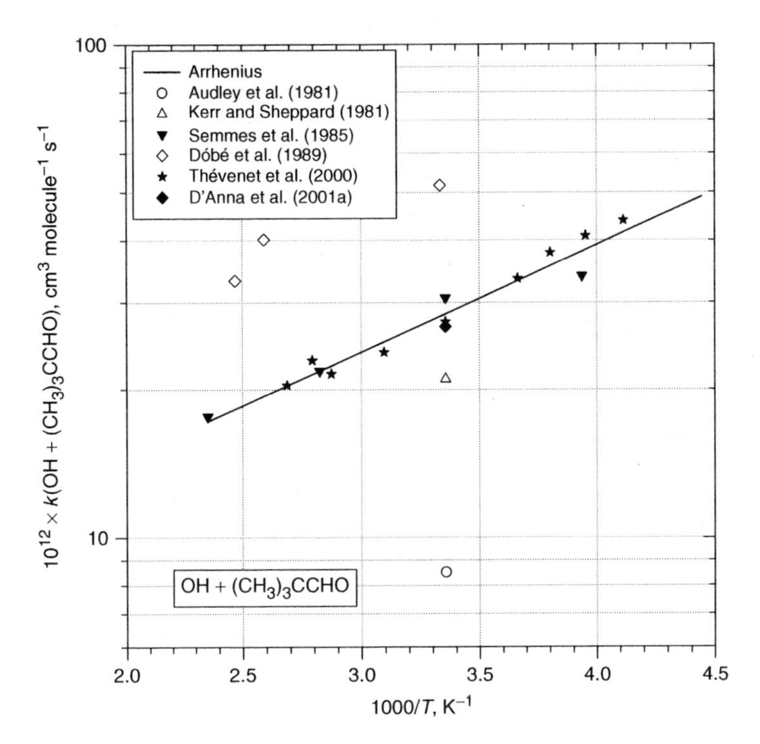

Figure V-C-7. Arrhenius plot of the rate coefficient for the reaction of OH with 2,2-dimethylpropanal.

Table V-C-10. Rate coefficients (k, cm^3 $molecule^{-1}$ s^{-1}) for the reaction of OH with hexanal (n-$C_5H_{11}CHO$)

$10^{12} \times k$	T (K)	Technique[a]	Reference
31.7 ± 1.5	298	RR (MVK)[b]	Papagni et al. (2000)
28.6 ± 1.3	298 ± 2	RR (1-C_4H_8)[c]	D'Anna et al. (2001a)
26.0 ± 2.1	298	PLP–LIF	Albaladejo et al. (2002)

[a] Abbreviations used for the techniques are defined in table II-A-1.
[b] Placed on an absolute basis using k(methyl vinyl ketone) $= 2.04 \times 10^{-11}$ cm^3 $molecule^{-1}$ s^{-1}.
[c] Placed on an absolute basis using k(1-butene) $= 3.14 \times 10^{-11}$ cm^3 $molecule^{-1}$ s^{-1} at 1 bar and 298 K.

OH-attack at the aldehyde group should lead to either three acetaldehyde molecules, or one molecule each of acetaldehyde and 2-butanone (via the 2-butoxy radical, see chapter IV):

$$OH + CH_3CH_2CH(CH_3)CH(CH_3)CHO\ (+O_2) \rightarrow H_2O + CH_3CH_2CH(CH_3)CH(CH_3)C(O)O_2$$

$$CH_3CH_2CH(CH_3)CH(CH_3)C(O)O_2 + NO \rightarrow NO_2 + CO_2 + CH_3CH_2CH(CH_3)C{\bullet}H(CH_3)$$

Table V-C-11. Rate coefficients (k, cm^3 $molecule^{-1}$ s^{-1}) for the reaction of OH with higher aldehydes

Molecule	$10^{12} \times k$	T (K)	Technique[a]	Reference
2-Methylpentanal	33 ± 2.1	298 ± 2	RR ($1-C_4H_8$)[b]	D'Anna et al. (2001a)
3-Methylpentanal	29.1 ± 1.5	298 ± 2	RR ($1-C_4H_8$)[b]	D'Anna et al. (2001a)
4-Methylpentanal	26.3 ± 1.1	298 ± 2	RR ($1-C_4H_8$)[b]	D'Anna et al. (2001a)
3,3-Dimethylbutanal	21.4 ± 1.4	298 ± 2	RR ($1-C_4H_8$)[b]	D'Anna et al. (2001a)
2-Ethylbutanal	40.2 ± 4.0	298 ± 2	RR ($1-C_4H_8$)[b]	D'Anna et al. (2001a)
Heptanal	29.6 ± 2.3	298	PLP–LIF	Albaladejo et al. (2002)
2,3-Dimethylpentanal	42 ± 10	298	RR ($CH_2{=}C(CH_3)_2$)[c]	Tuazon et al. (2003)
Nonanal	32 ± 5	298	RR (octane)[d]	Bowman et al. (2003)

[a] Abbreviations used for the techniques are defined in table II-A-1.
[b] Placed on an absolute basis using k(1-butene) = 3.14×10^{-11} cm^3 $molecule^{-1}$ s^{-1} at 1 bar and 298 K.
[c] Placed on an absolute basis using k(2-methylpropene) = 5.14×10^{-11} cm^3 $molecule^{-1}$ s^{-1} at 1 bar and 298 K.
[d] Placed on an absolute basis using k(octane) = 7.81×10^{-12} cm^3 $molecule^{-1}$ s^{-1} at 1 bar and 298 K.

$$CH_3CH_2CH(CH_3)C{\bullet}H(CH_3) + O_2 \rightarrow CH_3CH_2CH(CH_3)CH(O_2)(CH_3)$$

$$CH_3CH_2CH(CH_3)CH(O_2)(CH_3) + NO \rightarrow CH_3CH_2CH(CH_3)CH(O{\bullet})(CH_3) + NO_2$$

$$CH_3CH_2CH(CH_3)CH(O{\bullet})(CH_3) \rightarrow CH_3CH_2C{\bullet}H(CH_3) + CH_3CHO$$

$$CH_3CH_2C{\bullet}H(CH_3) + O_2 + NO \rightarrow CH_3CH_2CH(O{\bullet})(CH_3) + NO_2$$

$$CH_3CH_2CH(O{\bullet})(CH_3) + O_2 \rightarrow C_2H_5C(O)CH_3 + HO_2$$

$$CH_3CH_2CH(O{\bullet})(CH_3) \rightarrow CH_3CHO + C_2H_5 \rightarrow\rightarrow 2CH_3CHO$$

Acetaldehyde and 2-butanone will be produced together from abstraction at the 3-position, while attack at the 4-position should lead to three acetaldehyde molecules. The results are broadly consistent with (45–60)% abstraction at the aldehyde group, along with a considerable fraction at the 3- and 4-sites (approx. 20–30%), although a unique mechanism cannot be derived from the observed product distribution.

V-C-1.14. OH + Nonanal (n-$C_8H_{17}CHO$)

The atmospheric chemistry of nonanal was studied by Bowman et al. (2003). Using the relative-rate method, they measured a rate coefficient ratio of (4.1 ± 0.3) relative to octane. Using the more recent value for the octane rate coefficient recommended here (7.81×10^{-12} cm^3 $molecule^{-1}$ s^{-1}), we calculate a rate coefficient of (3.2 ± 0.5) $\times 10^{-11}$ cm^3 $molecule^{-1}$ s^{-1} for nonanal at 298 K, where a 15% uncertainty has been assigned (table V-C-11). Bowman et al. also investigated the product distribution from nonanal oxidation. By conducting the oxidation in the presence of excess NO_2

and detecting the PAN-analogue formed, they showed that only 50% of the reaction proceeds by abstraction at the aldehydic hydrogen atom:

$$OH + C_8H_{17}CHO \rightarrow H_2O + C_8H_{17}CO$$

$$C_8H_{17}CO + O_2 \rightarrow C_8H_{17}C(O)O_2$$

$$C_8H_{17}C(O)O_2 + NO_2 \rightarrow C_8H_{17}C(O)O_2NO_2$$

Using the total rate coefficient above, this corresponds to a partial rate coefficient for abstraction at the —CHO group of $(1.6–1.7) \times 10^{-11}$ cm^3 $molecule^{-1}$ s^{-1}, which is very similar to the overall rate coefficient for OH + acetaldehyde. This result suggests that a partial rate coefficient of $(1.5–1.7) \times 10^{-11}$ cm^3 $molecule^{-1}$ s^{-1} for reaction at the —CHO group might be common to all aldehydes.

V-C-1.15. General Observations and Trends in OH-Aldehyde Reactions

A negative temperature dependence seems to be common to all the reactions of OH with aldehydes. The temperature coefficient ($-E/R$) is in range 350–450 K except for pivaldehyde (2,2-dimethylpropanal), which is somewhat more negative. Formaldehyde shows a change in behavior around room temperature, switching from a negative temperature coefficient below room temperature to a positive one above. This may be indicative of a change of reaction mechanism, but not necessarily of products. Where mechanistic studies have been performed, abstraction from the —CHO group is dominant, with a partial rate coefficient of roughly 1.5×10^{-11} cm^3 $molecule^{-1}$ s^{-1}. This value also applies to unsaturated aldehydes such as acrolein and methacrolein (Orlando et al., 1999b; Orlando and Tyndall, 2002b). Presumably, as the chain length increases the contribution from the aldehyde group remains approximately constant, and hence the fractional contribution decreases (Bowman et al., 2003). More product studies of intermediate-size aldehydes are required to refine the site-specific information. For larger aldehydes, the work of D'Anna et al. (2001a) provides a self-consistent data base, all measured relative to 1-butene. The values of Papagni et al. (2000) tend to be somehat higher, but within 10% of the results of D'Anna et al.

V-C-2. Reactions of Hydroxy Radicals with Ketones

V-C-2.1. OH + Acetone [2-Propanone, $CH_3C(O)CH_3$]

The reaction of OH with acetone has been studied over a wide range of temperature, and there is now a broad consensus on the kinetics and mechanism of this reaction (see table V-C-12). Early studies showed that the reaction at ambient temperatures is fairly slow, with a small temperature dependence above room temperature (Wallington and Kurylo, 1987a; Le Calvé et al., 1998). Later studies indicated curvature in the Arrhenius plot; in fact, the rate coefficient showed a minimum value near 240 K (Wollenhaupt et al., 2000; Gierczak et al., 2003). The studies of Wallington and Kurylo (1987a),

Table V-C-12. Rate coefficients (k, cm^3 molecule^{-1} s^{-1}) for the reaction of OH with acetone [$CH_3C(O)CH_3$]

$10^{12} \times A$	B (K)	$10^{12} \times k$	T (K)	Technique[a]	Reference	Temperature Range (K)
		<0.5	300	RR (C_2H_4)[b]	Cox et al. (1980b)	
		0.61 ± 0.09	298	RR (n-C_6H_{14})[c]	Chiorboli et al. (1983)	
		0.27 ± 0.08	303	RR (C_2H_4)[d]	Kerr and Stocker (1986)	
1.7 ± 0.4	600 ± 75	0.145 ± 0.015	240	FP–RF	Wallington and Kurylo (1987a)	240–440
		0.216 ± 0.016	296			
		0.292 ± 0.023	350			
		0.407 ± 0.030	400			
		0.436 ± 0.050	440			
		0.132 ± 0.010	243	PLP–LIF	Le Calvé et al. (1998)	243–372
		0.140 ± 0.007	253			
		0.153 ± 0.010	273			
		0.184 ± 0.015	298			
		0.203 ± 0.017	323			
		0.249 ± 0.023	348			
		0.283 ± 0.010	372			
		0.148 ± 0.005	202	PLP–LIF	Wollenhaupt et al. (2000)	202–355
		0.141 ± 0.004	209			
		0.144 ± 0.003	220 ± 2			
		0.138 ± 0.002	225 ± 1			
		0.138 ± 0.004	228			
		0.129 ± 0.004	236 ± 2			
		0.142 ± 0.002	247			
		0.137 ± 0.003	255 ± 1			
		0.147 ± 0.001	267			
		0.150 ± 0.002	274			
		0.156 ± 0.002	281			
		0.167 ± 0.002	289			
		0.169 ± 0.004	295			
		0.196 ± 0.002	311			
		0.208 ± 0.002	321			
		0.226 ± 0.003	332			
		0.250 ± 0.003	344			
		0.267 ± 0.003	355			
		0.125 ± 0.002	221	PLP–RF		
		0.135 ± 0.015	233			
		0.141 ± 0.008	236			
		0.142 ± 0.015	245 ± 2			
		0.145 ± 0.010	260 ± 2			
		0.177 ± 0.003	298			
		0.209 ± 0.004	318			
		0.274 ± 0.015	361 ± 1			
		0.354 ± 0.003	395			
		0.173 ± 0.005	298	DF–RF	Vasvári et al. (2001)	

Table V-C-12. *(continued)*

$10^{12} \times A$	B (K)	$10^{12} \times k$	T (K)	Technique[a]	Reference	Temperature Range (K)
		0.136 ± 0.010	199	PLP–LIF	Gierczak et al. (2003)	199–383
		0.146 ± 0.010	208			
		0.151 ± 0.020	213			
		0.139 ± 0.010	224 ± 1			
		0.137 ± 0.008	232			
		0.149 ± 0.008	238			
		0.144 ± 0.004	244			
		0.148 ± 0.006	250			
		0.142 ± 0.002	259			
		0.155 ± 0.020	271			
		0.177 ± 0.010	296			
		0.207 ± 0.005	323			
		0.246 ± 0.026	333			
		0.277 ± 0.010	355 ± 2			
		0.341 ± 0.014	383			
		0.156 ± 0.008	298	PLP–LIF	Yamada et al. (2003)	298–832
		0.158 ± 0.007	310			
		0.182 ± 0.018	324			
		0.197 ± 0.006	339			
		0.213 ± 0.012	355			
		0.221 ± 0.026	373			
		0.285 ± 0.018	390			
		0.369 ± 0.013	414			
		0.455 ± 0.027	439			
		0.476 ± 0.034	467			
		0.605 ± 0.062	498			
		0.822 ± 0.017	534			
		0.949 ± 0.017	575			
		1.19 ± 0.12	628			
		1.56 ± 0.11	681			
		2.11 ± 0.13	685			
		1.93 ± 0.13	710			
		3.25 ± 0.16	832			
		0.149 ± 0.029	253	RR (CH_3CHF_2)[e]	Raff et al. (2005)	253–373
		0.160 ± 0.018	263			
		0.177 ± 0.017	273			
		0.172 ± 0.022	283			
		0.186 ± 0.013	293			
		0.208 ± 0.022	303			
		0.198 ± 0.016	308			
		0.192 ± 0.011	313			
		0.220 ± 0.018	323			
		0.217 ± 0.015	333			
		0.246 ± 0.012	343			
		0.294 ± 0.023	373			

(continued)

Table V-C-12. *(continued)*

$10^{12} \times A$	B (K)	$10^{12} \times k$	T (K)	Technique[a]	Reference	Temperature Range (K)
3.92 ± 0.81	938 ± 70	0.152 ± 0.012	271	DF–LIF	Davis et al. (2005)	271–402
		0.164 ± 0.014	287			
		0.173 ± 0.006	300			
		0.218 ± 0.010	325			
		0.246 ± 0.006	341			
		0.254 ± 0.014	342			
		0.311 ± 0.014	371			
		0.373 ± 0.014	402			

[a] Abbreviations used for the techniques are defined in table II-A-1.
[b] Placed on an absolute basis using $k(C_2H_4) = 8.52 \times 10^{-12}$ cm^3 molecule^{-1} s^{-1} at 1 bar and 298 K.
[c] Placed on an absolute basis using $k(n\text{-}C_6H_{14}) = 5.3 \times 10^{-12}$ cm^3 molecule^{-1} s^{-1}.
[d] Placed on an absolute basis using $k(C_2H_4) = 8.32 \times 10^{-12}$ cm^3 molecule^{-1} s^{-1} at 1 bar and 303 K.
[e] Placed on an absolute basis using $k(CH_3CHF_2) = 2.10 \times 10^{-18} T^2 \exp(-503/T)$ cm^3 molecule^{-1} s^{-1}.

Le Calvé et al. (1998), Wollenhaupt et al. (2000), Vasvári et al. (2001), Gierczak et al. (2003), Yamada et al. (2003), Raff et al. (2005), and Davis et al. (2005) give a consistent picture for the rate coefficient between 200 K and 800 K. The reported measurements of Raff et al. lie somewhat higher than the others; however, if the rate coefficient for the reference compound (CH_3CHF_2) recommended in chapter VI is used, the agreement is improved. The room temperature rate coefficient recommended by the IUPAC panel is 1.8×10^{-13} cm^3 molecule^{-1} s^{-1}, and for the temperature range 195–440 K, a sum of two exponentials is recommended: $k = 8.8 \times 10^{-12} \exp(-1320/T) + 1.7 \times 10^{-14} \times \exp(423/T)$ cm^3 molecule^{-1} s^{-1}. This expression is a fit to the data of Wollenhaupt et al. (2000), but it represents most of the above data to within 15% (as shown in figure V-C-8). There is no evidence for any pressure dependence for reaction of OH with this or any of the ketones.

The reaction mechanism has been the subject of some controversy. Earlier studies assumed that the reaction proceeds via abstraction of a hydrogen atom (Wallington and Kurylo, 1987a; Le Calvé et al., 1998). However, Wollenhaupt and Crowley (2000) and Vasvári et al. (2001) both found evidence for up to 50% yield of methyl radicals, presumably occurring via an addition–elimination pathway:

$$OH + CH_3C(O)CH_3 \rightarrow H_2O + CH_3C(O)CH_2$$

$$OH + CH_3C(O)CH_3 \rightarrow [HOC(CH_3)_2O\bullet]^\dagger \rightarrow CH_3 + CH_3C(O)OH$$

The strong curvature in the Arrhenius plot was also taken as evidence for the occurrence of two parallel reaction pathways. However, in the last five years several product studies have shown that abstraction of H to give the 2-oxo-1-propyl (acetonyl) radical is the major, if not only, pathway under atmospheric conditions. Vandenberk et al. (2002), Tyndall et al. (2002), Talukdar et al. (2003), and Raff et al. (2005) all found yields of acetic acid, $CH_3C(O)OH$, of less than 10% at room temperature. Talukdar et al. used the discharge flow technique with detection, using chemical ionization mass spectrometry to show that the $CH_3C(O)OH$ yield is less than 1% between 237

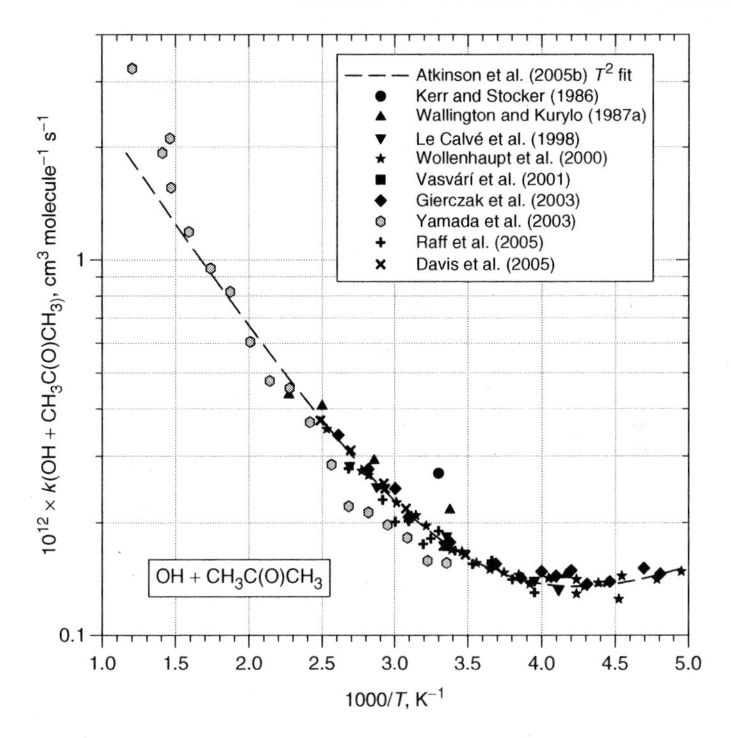

Figure V-C-8. Arrhenius plot of the rate coefficient for the reaction of OH with acetone.

and 353 K. Tyndall et al. also showed that the acetic acid yield is less than 10% at 251 K. The results of Raff et al. indicate a slight disagreement with the other studies: they found a 10% yield of acetic acid at 298 K and a slight negative temperature dependence, with the yield varying between (12 ± 5)% at 273 K and (4.7 ± 0.4)% at 353 K. The yield for the major channel was measured by Talukdar et al. (2003) and by Turpin et al. (2003), both of whom found yields of $CH_3C(O)CH_2$ greater than 85%.

The reaction is thought to proceed by a loose hydrogen-bonded complex, similar to the reactions of OH with the aldehydes, leading to the negative temperature dependence below room temperature. However, Talukdar et al. (2003), Yamada et al. (2003), and Raff et al. (2005) found a considerable deuterium isotope effect (factors between 5 and 7), indicating that the rate-limiting step involves the breaking of a C—H (C—D) bond at all temperatures. Theoretical calculations on the system indicate that the barrier to addition of OH to the carbonyl group is too high for an addition–abstraction mechanism to be operative (Vandenberk et al., 2002; Yamada et al., 2003). Hence, it appears likely that the mechanism switches from abstraction via a hydrogen-bonded complex at low temperature to direct abstraction at high temperature.

V-C-2.2. OH + 2-Butanone [Methyl Ethyl Ketone, $CH_3C(O)C_2H_5$]

The reaction of OH with 2-butanone has been studied seven times. The measurements are given in table V-C-13. Other than the early measurements of Winer et al. (1976)

Table V-C-13. Rate coefficient (k, cm^3 molecule^{-1} s^{-1}) for reaction of OH with 2-butanone [$CH_3C(O)CH_2CH_3$]

$10^{12} \times A$	B (K)	$10^{12} \times k$	T (K)	Technique[a]	Reference	Temperature Range (K)
		3.4 ± 1.0	305	RR	Winer et al. (1976)	
		<2.7	300	RR (C_2H_4)[b]	Cox et al. (1980b)	
		0.95 ± 0.09	298	RR (C_2H_4)[b]	Cox et al. (1981)	
		0.93 ± 0.17	297	RR (C_3H_8)[c]	Edney et al. (1986)	
2.3 ± 1.1	170 ± 120	1.23 ± 0.10	240	FP–RF	Wallington and Kurylo (1987a)	240–440
		1.15 ± 0.10	296			
		1.41 ± 0.09	350			
		1.55 ± 0.07	400			
		1.65 ± 0.09	440			
1.19 ± 0.18	60 ± 61	1.24 ± 0.08	243	PLP–LIF	Le Calvé et al. (1998)	243–372
		1.20 ± 0.05	253			
		1.19 ± 0.05	263			
		1.21 ± 0.09	273			
		1.19 ± 0.18	298			
		1.12 ± 0.11	313			
		1.24 ± 0.09	333			
		1.19 ± 0.09	353			
		1.38 ± 0.10	372			
1.35 ± 0.35	78 ± 52	0.99 ± 0.04	228	PLP–LIF	Jiménez et al. (2005a)	228–388
		0.80 ± 0.06	238			
		1.00 ± 0.06	248			
		0.91 ± 0.09	253			
		0.98 ± 0.04	273			
		1.04 ± 0.07	298			
		0.93 ± 0.11	313			
		1.10 ± 0.08	318			
		0.98 ± 0.11	343			
		1.20 ± 0.12	363			
		1.09 ± 0.40	388			

[a] Abbreviations used for the techniques are defined in table II-A-1.
[b] Placed on an absolute basis using $k(C_2H_4) = 8.52 \times 10^{-12}$ cm^3 molecule^{-1} s^{-1} at 1 bar and 298 K.
[c] Placed on an absolute basis using $k(C_3H_8) = 1.08 \times 10^{-12}$ cm^3 molecule^{-1} s^{-1}.

and Cox et al. (1980b), the determinations are in quite good agreement. The mean of the later values is $(1.1 \pm 0.2) \times 10^{-12}$ cm^3 molecule^{-1} s^1 near 298 K. The temperature dependence has been measured by Wallington and Kurylo (1987a), Le Calvé et al. (1998), and Jiménez et al. (2005a); all reported weak, positive temperature dependences, with no real evidence for curvature since the variation with temperature was slight. A nonlinear fit to the data shown in figure V-C-9 gives $k = 2.70 \times 10^{-18}\ T^2 \exp(444/T)$ cm^3 molecule^{-1} s^1, similar to the expression given by Atkinson et al. (2005b). The simple Arrhenius fit shown gives $k = 1.83 \times 10^{-12} \times \exp(-155/T)$ cm^3 molecule^{-1} s^1, which is preferred for atmospheric purposes. The uncertainty is estimated to be ±20%.

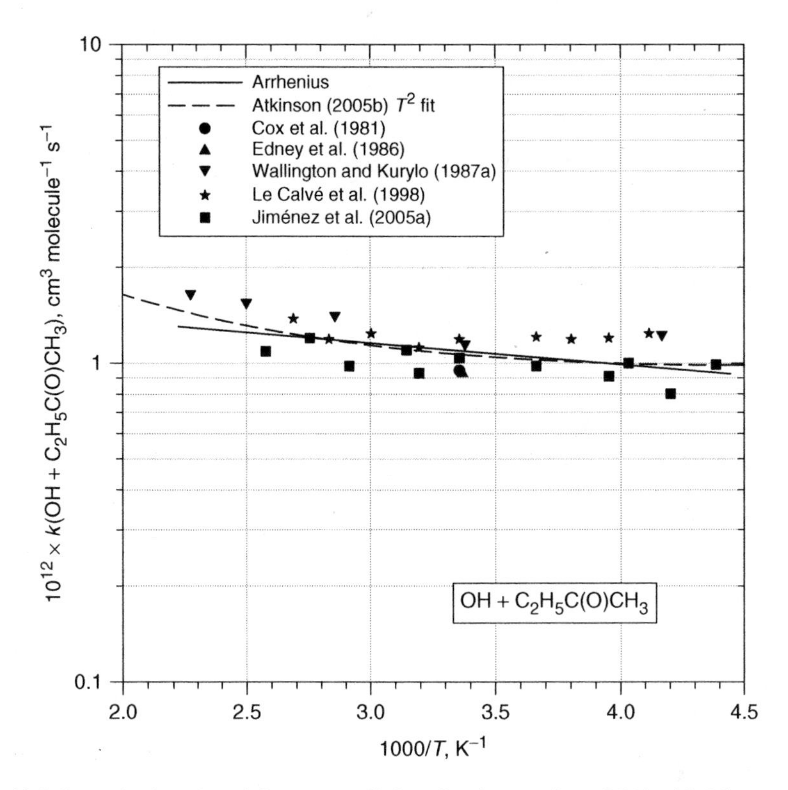

Figure V-C-9. Arrhenius plot of the rate coefficient for the reaction of OH with 2-butanone.

Cox et al. (1981) measured the yield of acetaldehyde from the OH-initiated oxidation of 2-butanone to be 62%. This should represent the yield of abstraction from the $-CH_2-$ group, and corresponds to a partial rate coefficient of 6.6×10^{-13} cm^3 $molecule^{-1}$ s^1, which is some 30% slower than the equivalent rate coefficient in an unsubstituted alkane. This deactivation is reflected in the group value $F(=C=O) = 0.75$ in table II-E-2.

V-C-2.3. OH + 2-Pentanone [Methyl n-Propyl Ketone, n-$C_3H_7C(O)CH_3$]

The reaction of OH with 2-pentanone has been studied by Atkinson et al. (1982a), Wallington and Kurylo (1987a), Atkinson and Aschmann (1988b), Atkinson et al. (2000), and Jiménez et al. (2005a). The measurements are given in table V-C-14. The agreement among the three measurements of Atkinson's group, which were all made relative to cyclohexane, is excellent. When normalized using the same rate coefficient for the reference reaction (6.85×10^{-12} cm^3 $molecule^{-1}$ s^1), they are (in units of 10^{-12} cm^3 $molecule^{-1}$ s^1): 4.29 ± 0.13, 4.66 ± 0.24, and 4.34 ± 0.29. Wallington and Kurylo's absolute measurement (made only at 296 K) is $(4.0 \pm 0.3) \times 10^{-12}$ cm^3 $molecule^{-1}$ s^1. The more recent measurements of Jiménez et al. are somewhat smaller, with a value of $(3.14 \pm 0.40) \times 10^{-12}$ cm^3 $molecule^{-1}$ s^1 at 298 K. Jiménez et al.

Table V-C-14. Rate coefficients (k, cm^3 $molecule^{-1}$ s^{-1}) for the reaction of OH with 2-pentanone [$CH_3C(O)CH_2CH_2CH_3$]

$10^{12} \times k$	T (K)	Technique[a]	Reference	Temperature Range (K)
4.29 ± 0.13	299	RR (c-C_6H_{12})[b]	Atkinson et al. (1982a)	
4.00 ± 0.29	296	FP–RF	Wallington and Kurylo (1987a)	
4.66 ± 0.24	296	RR (c-C_6H_{12})[b]	Atkinson and Aschmann (1988b)	
4.34 ± 0.29	298	RR (c-C_6H_{12})[b]	Atkinson et al. (2000)	
6.14 ± 0.16	248	PLP–LIF	Jiménez et al. (2005a)[c]	248–388
5.94 ± 0.56	253			
4.95 ± 0.18	257			
4.65 ± 0.36	263			
4.64 ± 0.14	268			
4.26 ± 0.32	273			
3.99 ± 0.66	283			
3.59 ± 0.10	288			
3.14 ± 0.40	298			
2.98 ± 0.38	308			
2.90 ± 0.02	323			
2.72 ± 0.01	334			
2.71 ± 0.04	343			
2.49 ± 0.20	354			
2.56 ± 0.08	373			
2.64 ± 0.06	388			

[a] Abbreviations used for the techniques are defined in table II-A-1.
[b] Placed on an absolute basis using k(cyclohexane) = 6.85×10^{-12} cm^3 $molecule^{-1}$ s^{-1}.
[c] $k(T) = 2.2 \times 10^{-12} + 1.73 \times 10^{-15} \exp(1917/T)$.

also measured the rate coefficient between 248 and 388 K. Their measurements indicate some curvature in the Arrhenius plot, superimposed on a negative temperature dependence (similar to acetone). Jiménez et al. reported a 3-parameter fit to the data: $k = 6.74 \times 10^{-19}\ T^2 \exp(1215/T)\ cm^3\ molecule^{-1}\ s^{-1}$; this is shown in figure V-C-10. However, over the range of atmospheric temperatures the curvature is small. We recommend an Arrhenius fit to all the data measured at 333 K and below. This leads to the expression: $k = 3.2 \times 10^{-13} \exp(719/T)\ cm^3\ molecule^{-1}\ s^{1}$, with a rate coefficient of $3.6 \times 10^{-12}\ cm^3\ molecule^{-1}\ s^{-1}$ at 298 K. The uncertainty is estimated to be 25%, as a result of the discrepancy between the data of Jiménez et al. and the earlier measurements.

Atkinson et al. (2000) studied the mechanism of the reaction, and showed that greater than 19% of the reaction occurs at the α-CH_2 group, and greater than 63% at the β-CH_2 group. Further discussion of the activation at the β-carbon will be made subsequently.

V-C-2.4. OH + 3-Pentanone [Diethyl Ketone, $C_2H_5C(O)C_2H_5$]

The reaction of OH with 3-pentanone has been studied three times. The measurements are given in table V-C-15. The flash photolysis study of Wallington and

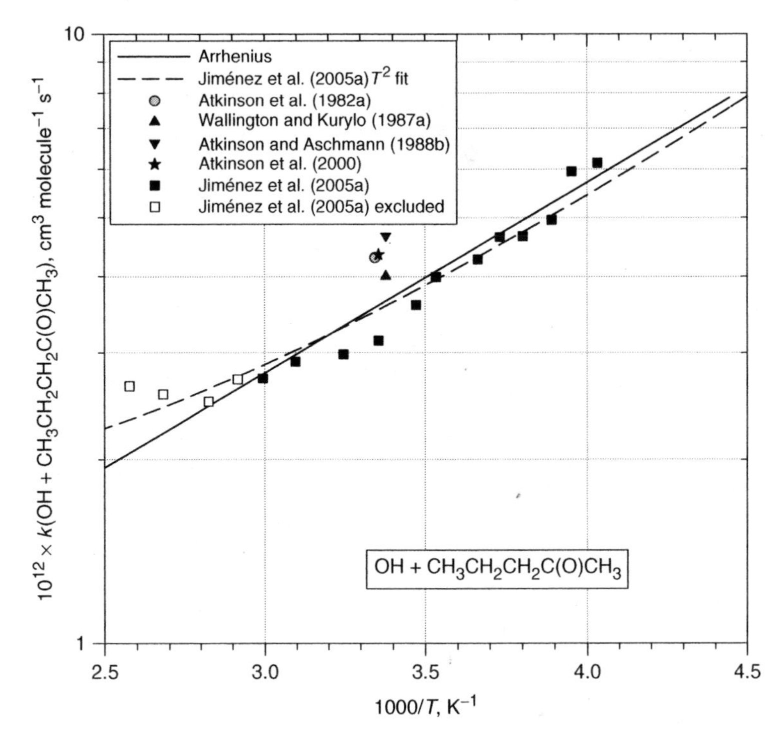

Figure V-C-10. Arrhenius plot of the rate coefficient for the reaction of OH with 2-pentanone.

Table V-C-15. Rate coefficients (k, cm^3 $molecule^{-1}$ s^{-1}) for the reaction of OH with 3-pentanone [$CH_3CH_2C(O)CH_2CH_3$]

$10^{12} \times A$	B (K)	$10^{12} \times k$	T (K)	Technique[a]	Reference	Temperature Range (K)
		1.69 ± 0.31	299	RR (c-C_6H_{12})[b]	Atkinson et al. (1982a)	
2.8 ± 0.3	10 ± 35	2.85 ± 0.17	240	FP–RF	Wallington and Kurylo (1987a)	240–440
		2.74 ± 0.13	296			
		2.91 ± 0.17	350			
		2.79 ± 0.32	400			
		2.78 ± 0.40	440			
		1.92 ± 0.13	296	RR (c-C_6H_{12})[b]	Atkinson and Aschmann (1988b)	

[a] Abbreviations used for the techniques are defined in table II-A-1.
[b] Placed on an absolute basis using k(cyclohexane) = 6.85×10^{-12} cm^3 $molecule^{-1}$ s^{-1}.

Kurylo (1987a) lies approximately 30% higher than the two relative-rate studies of Atkinson et al. (1982a) and Atkinson and Aschmann (1988b) (figure V-C-11). The reasons for the discrepancy are unknown. It seems unlikely that the rate coefficient for 3-pentanone would be more than twice that for 2-butanone, which contains a single —$C(O)C_2H_5$ group. Hence, we tentatively recommend a rate coefficient at 298 K of

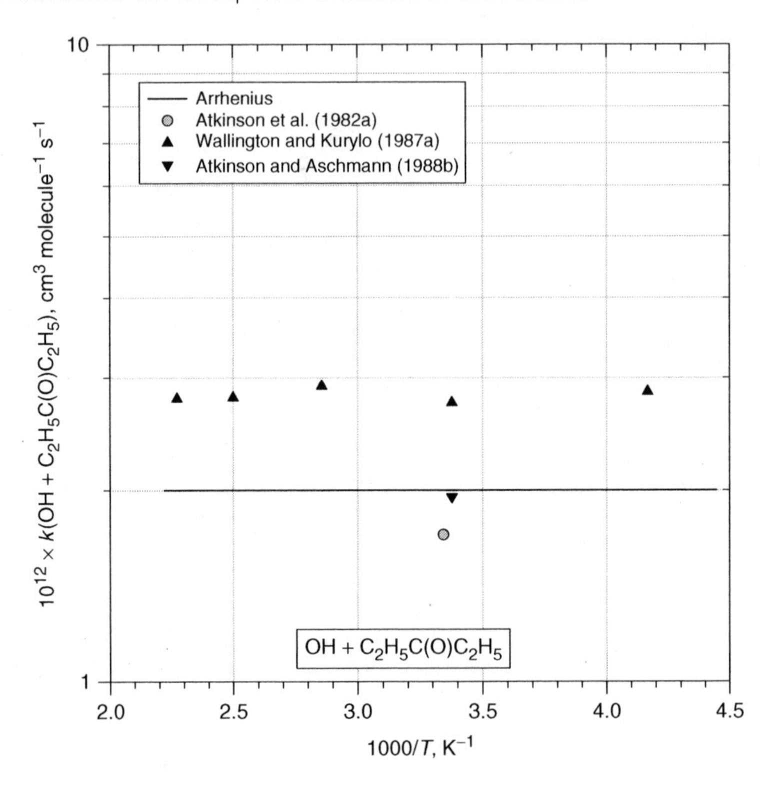

Figure V-C-11. Arrhenius plot of the rate coefficient for the reaction of OH with 3-pentanone.

$(2.0 \pm 0.5) \times 10^{-12}$ cm^3 molecule^{-1} s^{-1} on the basis of the later study from Atkinson's group, but with an increased uncertainty. Wallington and Kurylo measured the rate coefficient as a function of temperature, finding a rate coefficient that was essentially independent of temperature over the range 240–440 K. Hence, we recommend that the room-temperature value of Atkinson and Aschmann (1988b) be used for atmospheric conditions.

V-C-2.5. OH + 3-Methyl-2-butanone [Methyl Isopropyl Ketone, $(CH_3)_2CHC(O)CH_3$]

The reaction of OH with 3-methyl-2-butanone has been studied by Le Calvé et al. (1998). The measurements are given in table V-C-16. They determined a room-temperature rate coefficient of $(2.87 \pm 0.29) \times 10^{-12}$ cm^3 molecule^{-1} s^{-1}, using flash photolysis with LIF detection of OH radicals. They also measured the rate coefficient as a function of temperature, finding a small negative temperature dependence, with some evidence for curvature; the rate coefficient was essentially independent of temperature above room temperature. As shown in figure V-C-12, we have fitted their data to an Arrhenius expression, $k = 1.45 \times 10^{-12} \exp(219/T)$ cm^3 molecule^{-1} s^{-1}, for use in atmospheric models, with an uncertainty of $\pm 20\%$.

Table V-C-16. Rate coefficients (k, cm^3 molecule^{-1} s^{-1}) for the reaction of OH with 3-methyl-2-butanone [$CH_3C(O)CH(CH_3)_2$]

$10^{12} \times A$	B (K)	$10^{12} \times k$	T (K)	Technique[a]	Reference	Temperature Range (K)
1.58 ± 0.35	−193 ± 65	3.68 ± 0.15	253	PLP–LIF	Le Calvé et al. (1998)	253–372
		3.26 ± 0.08	263			
		3.26 ± 0.08	273			
		2.84 ± 0.12	298			
		2.78 ± 0.11	323			
		2.74 ± 0.11	348			
		2.75 ± 0.10	372			

[a] Abbreviations used for the techniques are defined in table II-A-1.

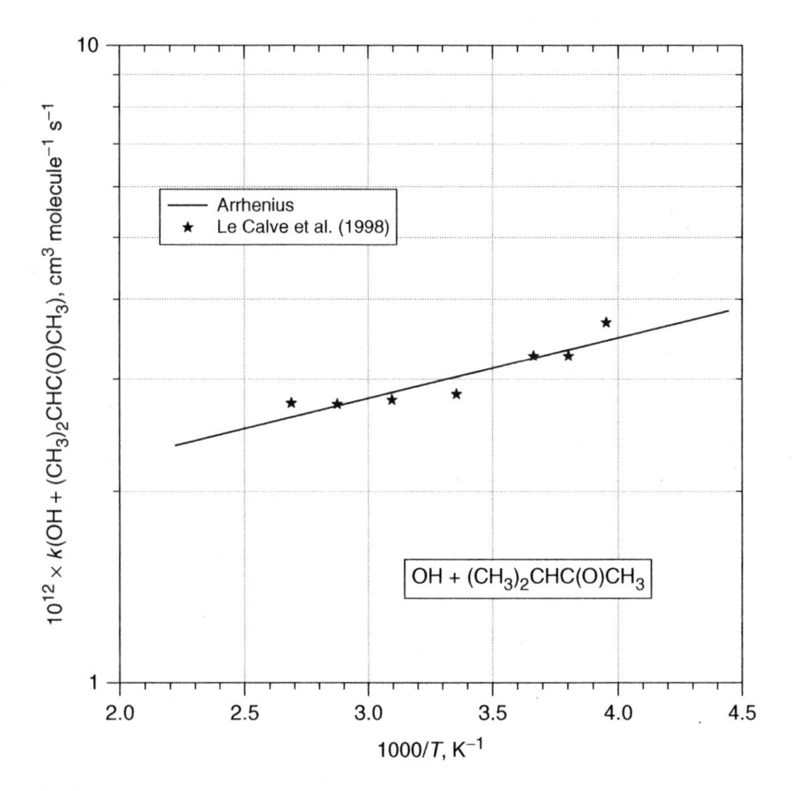

Figure V-C-12. Arrhenius plot of the rate coefficient for the reaction of OH with 3-methyl-2-butanone.

V-C-2.6. OH + 2-Hexanone [Methyl n-Butyl Ketone, $CH_3C(O)CH_2CH_2CH_2CH_3$]

This reaction has been the subject of four studies. The measurements are given in table V-C-17. The absolute studies of Wallington and Kurylo (1987a) and Jiménez et al. (2005a) are in excellent agreement, reporting rate coefficients at 298 ± 3 K of

Table V-C-17. Rate coefficients (k, cm^3 $molecule^{-1}$ s^{-1}) for the reaction of OH with 2-hexanone [n-$C_4H_9C(O)CH_3$]

$10^{12} \times k$	T (K)	Technique[a]	Reference	Temperature Range (K)
8.28 ± 0.56	299	RR (c-C_6H_{12})[b]	Atkinson et al. (1982a)	
6.64 ± 0.56	296	FP–RF	Wallington and Kurylo (1987a)	
8.36 ± 0.42	296	RR (c-C_6H_{12})[b]	Atkinson and Aschmann (1988b)	
10.2 ± 1.20	263	PLP–LIF	Jiménez et al. (2005a)[c]	263–405
7.84 ± 0.84	273			
7.13 ± 0.87	279			
7.34 ± 0.94	284			
6.41 ± 0.20	286			
6.30 ± 0.42	288			
6.37 ± 1.40	298			
5.97 ± 0.16	313			
5.63 ± 0.10	316			
5.32 ± 0.76	328			
4.62 ± 0.46	349			
4.84 ± 0.52	353			
4.70 ± 0.20	370			
4.61 ± 1.40	380			
5.05 ± 0.26	390			
4.50 ± 0.64	405			

[a] Abbreviations used for the techniques are defined in table II-A-1.
[b] Placed on an absolute basis using k(cyclohexane) $= 6.85 \times 10^{-12}$ cm^3 $molecule^{-1}$ s^{-1}.
[c] $k(T) = 2.13 \times 10^{-18} T^2 \exp(1038/T)$.

$(6.64 \pm 0.56) \times 10^{-12}$ and $(6.37 \pm 1.40) \times 10^{-12}$ cm^3 $molecule^{-1}$ s^{-1}, respectively. However, the relative rate determinations of Atkinson et al. (1982a), $(8.28 \pm 0.56) \times 10^{-12}$, and Atkinson and Aschmann (1988b), $(8.36 \pm 0.42) \times 10^{-12}$ cm^3 $molecule^{-1}$ s^{-1}, are about 30% higher. The reasons for this discrepancy are unknown. Jiménez et al. (2005a) noted a distinct negative temperature dependence to the rate coefficient over the entire range of their measurements (263–405 K), as shown by the dashed curve in figure V-C-13, $k = 2.13 \times 10^{-18}$ $T^2 \exp(1038/T)$ cm^3 $molecule^{-1}$ s^{-1}. However, their lowest temperature point appears to have introduced a large curvature into the data. An Arrhenius fit to all the remaining data below 380 K, shown in figure V-C-13, gives a temperature-dependent expression $k = 1.06 \times 10^{-12} \exp(540/T)$ cm^3 $molecule^{-1}$ s^{-1} which provides a good representation of their data from 263 K to 353 K. The corresponding value at 298 K is 6.5×10^{-12} cm^3 $molecule^{-1}$ s^{-1}, with an uncertainty of ±20%.

V-C-2.7. OH + 3-Hexanone [Ethyl n-Propyl Ketone, $CH_3CH_2C(O)CH_2CH_2CH_3$]

This reaction has been studied only by Atkinson et al. (1982a), using the relative rate method. They found a rate coefficient ratio, relative to cyclohexane, of 0.919 ± 0.038.

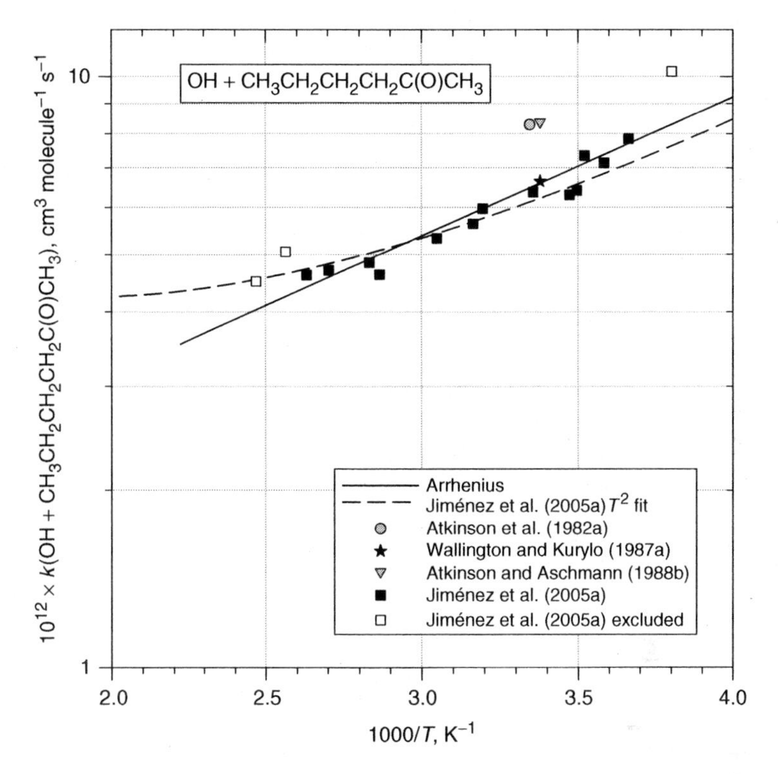

Figure V-C-13. Arrhenius plot of the rate coefficient for the reaction of OH with 2-hexanone.

Using the value of the rate coefficient for OH + cyclohexane recommended in chapter II leads to a rate coefficient for 3-hexanone of $(6.3 \pm 1.5) \times 10^{-12}$ cm^3 molecule^{-1} s^{-1} at 298 ± 3 K, where an uncertainty of ±25% has been assigned. No recommendation for the temperature dependence can be made at present, although it is expected to be slightly negative.

V-C-2.8. OH + 3-Methyl-2-pentanone [Methyl sec-Butyl Ketone, $CH_3C(O)CH(CH_3)CH_2CH_3$]

Tuazon et al. (2003) measured the rate coefficient for the reaction of OH with 3-methyl-2-pentanone relative to that of octane. Using the current recommendation for the reference reaction, and the measured ratio 0.789 ± 0.009, we derive $k = (6.2 \pm 1.5) \times 10^{-12}$ cm^3 molecule^{-1} s^{-1} for a temperature of 298 K, where the uncertainty has been set at ±25%.

V-C-2.9. OH + 4-Methyl-2-pentanone [Methyl Isobutyl Ketone, $CH_3C(O)CH_2CH(CH_3)_2$]

The reaction of OH with 4-methyl-2-pentanone has been measured in six studies: Winer et al. (1976), Cox et al. (1980b, 1981), Atkinson et al. (1982a), O'rji and Stone (1992),

Table V-C-18. Rate coefficients (k, cm^3 $molecule^{-1}$ s^{-1}) for the reaction of OH with 4-methyl-2-pentanone [$CH_3C(O)CH_2CH(CH_3)_2$]

$10^{12} \times A$	B (K)	$10^{12} \times k$	T (K)	Technique[a]	Reference	Temperature Range (K)
		15.0 ± 5.0	305	RR ($CH_2{=}C(CH_3)_2$)[b]	Winer et al. (1976)	
		13.1	300	RR (C_2H_4)[c]	Cox et al. (1980b)	
		13.0 ± 0.3	296	RR (C_2H_4)[c]	Cox et al. (1981)	
		13.1 ± 0.7	299	RR (c-C_6H_{12})[d]	Atkinson et al. (1982a)	
		14.0 ± 0.7	297	RR (C_3H_6)[e]	O'rji and Stone (1992)	
0.76 ± 0.13	−834 ± 46	21.3 ± 1.0	253	PLP–LIF	Le Calvé et al. (1998)	253–372
		18.2 ± 0.2	263			
		16.2 ± 0.3	273			
		12.1 ± 0.5	298			
		10.6 ± 0.2	313			
		9.25 ± 0.5	333			
		8.28 ± 0.14	353			
		7.57 ± 0.3	372			

[a] Abbreviations used for the techniques are defined in table II-A-1.
[b] Placed on an absolute basis using k(2-methylpropene) = 5.14×10^{-11} cm^3 $molecule^{-1}$ s^{-1} at 1 bar and 298 K.
[c] Placed on an absolute basis using $k(C_2H_4) = 8.52 \times 10^{-12}$ cm^3 $molecule^{-1}$ s^{-1} at 1 bar and 298 K.
[d] Placed on an absolute basis using k(cyclohexane) $= 6.85 \times 10^{-12}$ cm^3 $molecule^{-1}$ s^{-1}.
[e] Placed on an absolute basis using $k(C_3H_6) = 2.63 \times 10^{-11}$ cm^3 $molecule^{-1}$ s^{-1} at 1 bar and 298 K.

and Le Calvé et al. (1998); the measurements are given in table V-C-18. The results are in good agreement, ranging between 1.2×10^{-11} and 1.5×10^{-11} cm^3 $molecule^{-1}$ s^{-1} at 298 K. Le Calvé et al. also measured the temperature dependence of the reaction between 253 and 372 K, finding a smooth decrease of the rate coefficient with increasing temperature (figure V-C-14). An Arrhenius fit to all the data other than the point of Winer et al. gives $k = 8.0 \times 10^{-13} \exp(828/T)$ cm^3 $molecule^{-1}$ s^{-1}, leading to a rate coefficient at 298 K of 1.28×10^{-11} cm^3 $molecule^{-1}$ s^{-1}, with an estimated uncertainty of ±15%. Cox et al. (1981) measured an acetone yield of 90%, suggesting that most of the OH attack occurs at the tertiary 4-position.

V-C-2.10. OH + 3,3-Dimethyl-2-butanone [Methyl tert-Butyl Ketone, $(CH_3)_3CC(O)CH_3$]

The reaction of OH with 3,3-dimethyl-2-butanone has been studied only by Wallington and Kurylo (1987a), who used flash photolysis with resonance fluorescence detection. They determined a rate coefficient of $(1.21 \pm 0.05) \times 10^{-12}$ cm^3 $molecule^{-1}$ s^{-1} at 298 ± 3 K. In the absence of other measurements, we recommend this rate coefficient, but with an increased uncertainty $k = (1.2 \pm 0.3) \times 10^{-12}$ cm^3 $molecule^{-1}$ s^{-1}. The reaction probably has a negative temperature dependence similar to that of 4-methyl-2-pentanone. Most of the reaction is expected to occur at the *tert*-butyl group.

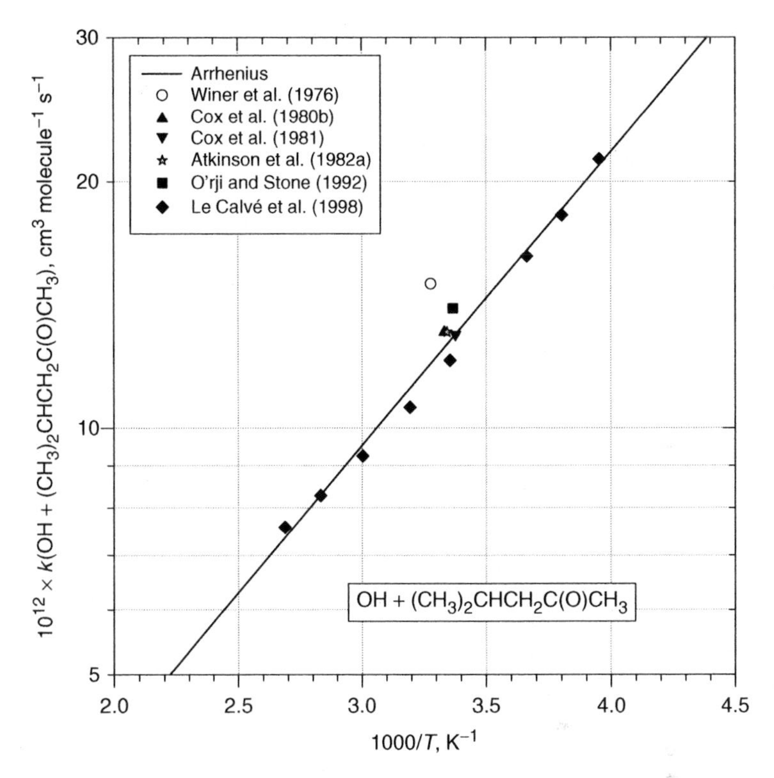

Figure V-C-14. Arrhenius plot of the rate coefficient for the reaction of OH with 4-methyl-2-pentanone.

V-C-2.11. OH + 2-Heptanone [Methyl n-Pentyl Ketone, $CH_3C(O)CH_2CH_2CH_2CH_2CH_3$]

The reaction of OH with 2-heptanone has been the subject of three studies. The measurements are given in table V-C-19. The relative rate study of Atkinson et al. (2000), normalized to the rate coefficient for cyclohexane given in chapter II, leads to a value of $(1.11 \pm 0.11) \times 10^{-11}$ cm^3 molecule^{-1} s^{-1}. The two absolute studies of Wallington and Kurylo (1987a) and Jiménez et al. (2005a) both give considerably smaller values, $(8.67 \pm 0.84) \times 10^{-12}$ and $(8.22 \pm 1.10) \times 10^{-12}$ cm^3 molecule^{-1} s^{-1}, respectively, at ambient temperatures. The reason for the discrepancy between absolute and relative methods is unknown. On the basis of the good agreement between the two direct methods, they form the basis of our recommendation. Jiménez et al. also measured the rate coefficient between 260 and 405 K, finding a monotonic decrease in k with increasing temperature up to about 330 K, and a flat minimum value above 330 K. The dashed curve in figure V-C-15 is our $1/k^2$-weighted nonlinear fit to the data of Jiménez et al., which gives $k = 4.75 \times 10^{-18} T^2 \exp(880/T)$ cm^3 molecule^{-1} s^{-1}. The data below 330 K, along with the point of Wallington and Kurylo, have been fitted to an Arrhenius expression to give a recommended rate coefficient $k = 2.8 \times 10^{-12} \exp(317/T)$ cm^3

Table V-C-19. Rate coefficients (k, cm^3 $molecule^{-1}$ s^{-1}) for the reaction of OH with 2-heptanone [n-$C_5H_{11}C(O)CH_3$]

$10^{12} \times k$	T (K)	Technique[a]	Reference	Temperature Range (K)
8.67 ± 0.84	296	FP–RF	Wallington and Kurylo (1987a)	
11.1 ± 1.10	298	RR (c-C_6H_{12})[b]	Atkinson et al. (2000)	
10.2 ± 0.60	260	PLP–LIF	Jiménez et al. (2005a)[c]	260–405
9.46 ± 0.36	263			
9.25 ± 0.28	267			
8.53 ± 0.76	273			
8.62 ± 0.46	283			
8.20 ± 0.08	287			
8.22 ± 1.10	298			
7.54 ± 0.42	313			
7.55 ± 0.40	323			
7.55 ± 0.26	333			
7.05 ± 0.12	343			
6.97 ± 0.50	354			
7.19 ± 0.44	374			
7.17 ± 0.02	405			

[a] Abbreviations used for the techniques are defined in table II-A-1.
[b] Placed on an absolute basis using k(cyclohexane) = 6.85×10^{-12} cm^3 $molecule^{-1}$ s^{-1}.
[c] $k(T) = 1.54 \times 10^{-6} T^2 \exp(1144/T)$.

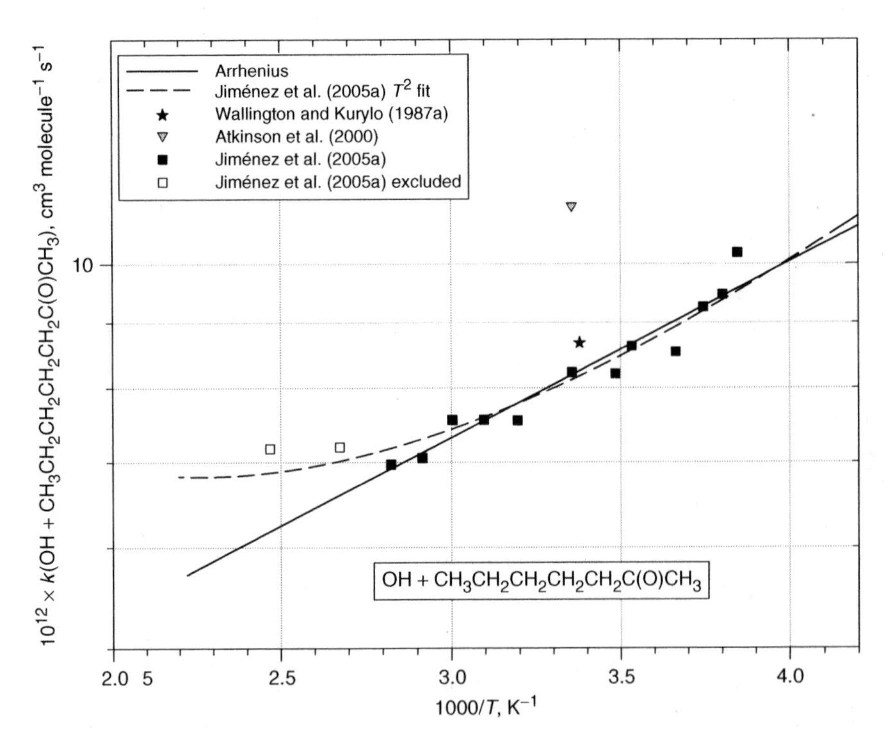

Figure V-C-15. Arrhenius plot of the rate coefficient for the reaction of OH with 2-heptanone.

molecule^{-1} s^{-1} over the range 260–350 K. The value at 298 K is $k = 8.2 \times 10^{-12}$ cm^3 molecule^{-1} s^{-1}, with an uncertainty of ±15%.

Atkinson et al. (2000) studied the product distribution from the reaction of OH with 2-heptanone, using a combination of detection techniques. In addition to the simple aldehydes formaldehyde (38% yield), acetaldehyde (15%), propanal (10%), butanal (7%), and pentanal (9%), evidence was found for a number of dicarbonyls and hydroxydicarbonyls, presumably formed following isomerization of the initially formed alkoxy radicals. The yields of the identified products were consistent with the calculated fractions of attack at the various sites as calculated from the SAR of Kwok and Atkinson (1995), but it was not possible to derive a unique mechanism from the observed products.

V-C-2.12. OH + 5-Methyl-2-hexanone [Methyl Isopentyl Ketone, $(CH_3)_2CHCH_2CH_2C(O)CH_3$]

This reaction has been studied only by Le Calvé et al. (1998) (table V-C-20), who derived a rate coefficient of $(1.03 \pm 0.10) \times 10^{-11}$ cm^3 molecule^{-1} s^{-1} at ambient temperatures. They also measured the temperature dependence of the reaction. The rate coefficient decreased monotonically with increasing temperature. We have fitted an Arrhenius expression 1.24×10^{-12} exp$(675/T)$ cm^3 molecule^{-1} s^{-1} to the data (figure V-C-16), which leads to a rate coefficient at 298 K of $k = (1.20 \pm 0.25) \times 10^{-12}$ cm^3 molecule^{-1} s^{-1}. No product information is available for this reaction.

V-C-2.13. OH + 2,4-Dimethyl-3-pentanone [Di-isopropyl Ketone, $(CH_3)_2CHC(O)CH(CH_3)_2$]

The reaction of OH with 2,4-dimethyl-3-pentanone has been studied only by Atkinson et al. (1982a) at 299 ± 2 K. They used the relative rate technique, with cyclohexane as the reference compound. Using our recommendation for the rate coefficient of

Table V-C-20. Rate coefficients (k, cm^3 molecule^{-1} s^{-1}) for the reaction of OH with 5-methyl-2-hexanone [$CH_3C(O)CH_2CH_2CH(CH_3)_2$]

$10^{12} \times A$	B (K)	$10^{12} \times k$	T (K)	Technique[a]	Reference	Temperature Range (K)
1.33 ± 0.63	−649 ± 140	17.0 ± 0.4	263	PLP–LIF	Le Calvé et al. (1998)	263–372
		15.4 ± 0.9	273			
		13.8 ± 0.5	283			
		12.8 ± 0.3	286			
		13.3 ±0.3	291			
		12.2 ± 0.3	293			
		10.3 ± 0.5	298			
		9.77 ± 0.98	323			
		8.90 ± 0.26	348			
		8.00 ± 0.50	372			

[a] Abbreviations used for the techniques are defined in table II-A-1.

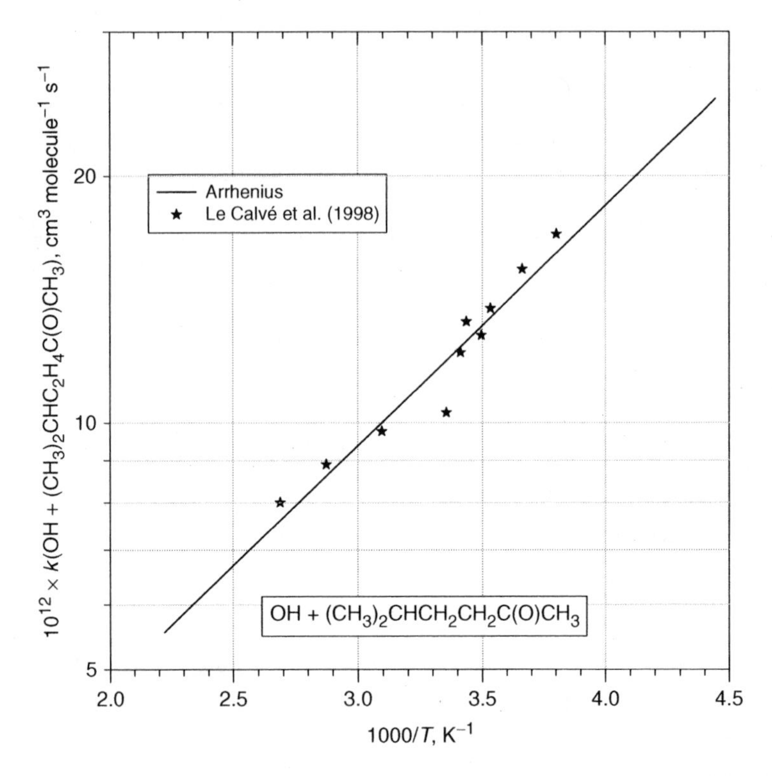

Figure V-C-16. Arrhenius plot of the rate coefficient for the reaction of OH with 5-methyl-2-hexanone.

OH + cyclohexane, 6.85×10^{-12} cm^3 molecule^{-1} s^{-1}, we recommend a rate coefficient of $(4.9 \pm 1.0) \times 10^{-12}$ cm^3 molecule^{-1} s^{-1} at 298 K, where the uncertainties have been increased to 20% to reflect the limited amount of data available. No product information is available for this reaction.

V-C-2.14. OH + 2-Octanone [$CH_3C(O)CH_2CH_2CH_2CH_2CH_2CH_3$]

This reaction was studied at 296 K by Wallington and Kurylo (1987a), using flash photolysis with resonance fluorescence detection of OH. They found a rate coefficient of $(1.10 \pm 0.09) \times 10^{-11}$ cm^3 molecule^{-1} s^{-1}. In the absence of other data, we recommend this value, with the uncertainty increased to ±20%.

V-C-2.15. OH + 2-Nonanone [$CH_3C(O)CH_2CH_2CH_2CH_2CH_2CH_2CH_3$]

This reaction was studied at 296 K by Wallington and Kurylo (1987a), using flash photolysis with resonance fluorescence detection of OH. They found a rate coefficient of $(1.22 \pm 0.13) \times 10^{-11}$ cm^3 molecule^{-1} s^{-1}. In the absence of other data, we recommend this value, with an uncertainty of ±20%.

Table V-C-21. Rate coefficients (k, cm^3 $molecule^{-1}$ s^{-1}) for the reaction of OH with 2,6-dimethyl-4-heptanone [$(CH_3)_2CHCH_2C(O)CH_2CH(CH_3)_2$]

$10^{12} \times k$	T (K)	Technique[a]	Reference
25 ± 8	305	RR ($CH_2{=}C(CH_3)_2$)[b]	Winer et al. (1976)
25.1 ± 1.4	299	RR (c-C_6H_{12})[c]	Atkinson et al. (1982a)

[a] Abbreviations used for the techniques are defined in table II-A-1.
[b] Placed on an absolute basis using k(2-methylpropene) = 5.14×10^{-11} cm^3 $molecule^{-1}$ s^{-1} at 1 bar and 298 K.
[c] Placed on an absolute basis using k(cyclohexane) = 6.85×10^{-12} cm^3 $molecule^{-1}$ s^{-1}.

V-C-2.16. OH + 2,6-Dimethyl-4-heptanone [Di-isobutyl Ketone, $(CH_3)_2CHCH_2C(O)CH_2CH(CH_3)_2$]

The reaction of OH with 2,6-dimethyl-4-heptanone has been measured by Winer et al. (1976) and Atkinson et al. (1982a), using the relative rate technique. The measurements are given in table V-C-21. Winer et al. measured $(2.5 \pm 0.8) \times 10^{-11}$ cm^3 $molecule^{-1}$ s^{-1}. Atkinson et al. found a rate coefficient ratio of 3.66 ± 0.19 relative to cyclohexane. Using the rate coefficient for cyclohexane recommended in this study leads to a value $k = (2.5 \pm 0.5) \times 10^{-11}$ cm^3 $molecule^{-1}$ s^{-1} at 298 K, with a ±20% uncertainty. The rate coefficient is roughly twice that of 4-methyl 2-pentanone, which just contains one isobutyl group and a relatively unreactive methyl group. Most of the reaction is expected to occur at the tertiary C—H bonds.

V-C-2.17. OH + 2-Decanone [n-$C_8H_{17}C(O)CH_3$]

This reaction was studied at 296 K by Wallington and Kurylo (1987a), using flash photolysis with resonance fluorescence detection of OH. They found a rate coefficient of $(1.32 \pm 0.12) \times 10^{-11}$ cm^3 $molecule^{-1}$ s^{-1} at 296 K. In the absence of other data, we recommend this value, with an increased uncertainty of ±20%.

V-C-2.18. General Observations on the Reactions of OH with Ketones

It has been known for many years (Atkinson et al., 1982a; Le Calvé et al., 1998) that the presence of a carbonyl group deactivates the hydrogen atoms at the adjacent (α-) carbon atom, while activating the hydrogen atoms at the β-carbon atom. The magnitude of the β-activation appears to be roughly a factor of 4 (see table II-E-2 for the structure–reactivity factors). The effect of structure beyond the β-carbon is not clear. Figure V-C-17 shows a plot of the rate coefficients for the reactions of the 2-ketones with OH. As noted in the text, the relative rate measurements of Atkinson and co-workers are roughly 30% higher than the absolute determinations of Wallington and Kurylo (1987a) and Jiménez et al. (2005a), for unknown reasons. If the rate coefficients within a given set of measurements are assumed to be internally consistent, it can be seen that the rate coefficients increase rapidly up to about C_8 (which contains five methylene groups in the longer alkyl chain). This linear increase implies

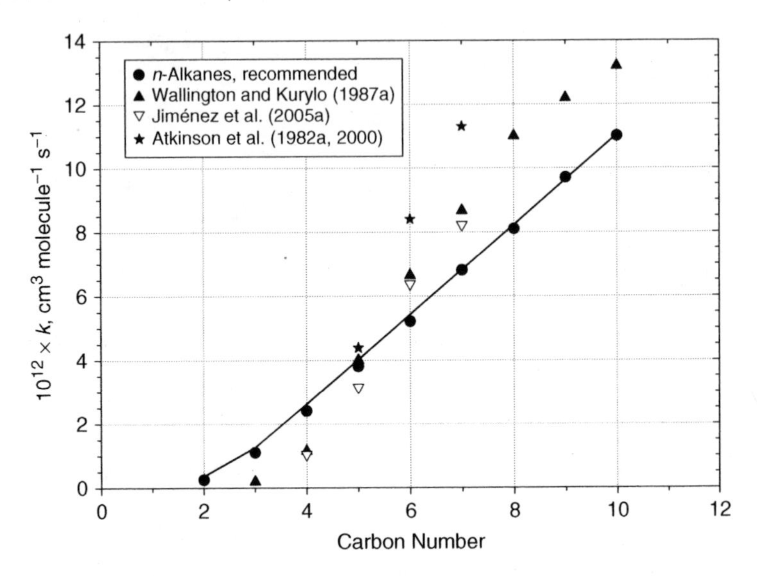

Figure V-C-17. Plot of the rate coefficients for the reaction of OH with 2-ketones as a function of the number of carbon atoms in the molecule. Also shown for slope comparison are the rate coefficients for *n*-alkanes. The ketones show activation up to approximately C_8.

either that the activation due to the carbonyl group extends up to five carbon atoms away, or that the magnitude of the activation at the β- (and possibly γ-) carbon atoms increases with the size of the attached group. Beyond 2-octanone, the slope of the data of Wallington and Kurylo is similar to that for the *n*-alkanes, which are also shown in the figure, indicating that the activation has ceased. Currently, the reason for the activation is not fully understood. The α-hydrogen atoms are probably weakened, as a result of the resonance stabilization of the vinoxy-type radical formed. However, this effect is not expected to extend beyond two carbon atoms. More recent interpretations of the negative temperature dependence suggest that the reaction proceeds through a weakly bound, hydrogen-bonded intermediate, which facilitates attack at the β-hydrogen atoms through a 7-centered intermediate. The observation of rate coefficients which possess a minimum near room temperature (and some of which clearly increase at high temperature) suggests that two parallel channels may be operating, a complex mechanism and a direct abstraction, both of which lead to the same result, abstraction of a hydrogen atom. Because of the complex nature of the reaction mechanism, it should not be expected that a simple SAR exists. Likewise, it is difficult to predict the magnitude of the (negative) temperature dependence relevant for atmospheric conditions, since it is the result of two reactions with opposing temperature dependences.

Overall, it is surprising that so few studies of the kinetics and mechanisms of the reactions of OH with simple ketones have been made. Further experiments on both the temperature dependence and reaction products are clearly required before predictive rules can be formulated and used with any confidence.

V-C-3. Reactions of Hydroxy Radicals with Alcohols

The reactions of simple alcohols with the hydroxy radical have been studied over a wide range of temperature, using both absolute and relative methods. The results generally show an increasing rate coefficient with the length of the alkyl chain, and also indicate an activating effect by the hydroxy group, which extends several carbon atoms away. In recent years, some rate coefficients have been obtained for multifunctional alcohols, which can be formed in alkane oxidation.

V-C-3.1. OH + Methanol (Methyl Alcohol, CH_3OH)

The reaction of OH with methanol has been the subject of some 20 studies, spanning temperatures from 210 to 1000 K (in addition, some high-temperature studies are available in the combustion literature). The measurements are given in table V-C-22. The rate coefficients near 298 K are in very good agreement ($\pm18\%$ of the mean), considering the potential difficulties in handling methanol. On the basis of the studies of Overend and Paraskevopoulos (1978), Ravishankara and Davis (1978), Tuazon et al. (1983), Wallington and Kurylo (1987b), Hess and Tully (1989), Nelson et al. (1990a), Pïcquet et al. (1998), Jiménez et al. (2003), and Dillon et al. (2005), we recommend a rate coefficient of $(9.5 \pm 1.2) \times 10^{-13}$ cm^3 molecule^{-1} s^{-1} for ambient temperatures. The temperature dependence of the reaction has been measured by Hägele et al. (1983), Meier et al. (1984, 1985a), Greenhill and O'Grady (1986), Wallington and Kurylo (1987b), Hess and Tully (1989), Jiménez et al. (2003), and Dillon et al. (2005). The studies of Hess and Tully and Dillon et al. indicate that the Arrhenius plot is curved, and Dillon et al. recommend a rate coefficient of $3.82 \times 10^{-19} T^{2.4} \exp[(300 \pm 14)/T]$ cm^3 molecule^{-1} s^{-1} between 210 and 860 K. Atkinson et al. (2005b) preferred a T^2 fit, $k = 6.38 \times 10^{-18} T^2 \exp(144/T)$ cm^3 molecule^{-1} s^{-1}, which is shown in figure V-C-18. For atmospheric purposes the simple Arrhenius expression $4.0 \times 10^{-12} \exp(-430/T)$ cm^3 molecule^{-1} s^{-1} represents the data of Wallington and Kurylo (1987b), Hess and Tully (1989), Jiménez et al. (2003), and Dillon et al. (2005) to better than $\pm10\%$.

The reaction of OH with CD_3OH was studied by Greenhill and O'Grady (1986) and McCaulley et al. (1989) at 298 K, and by Hess and Tully (1989) between 293 and 862 K. The studies of Greenhill and O'Grady and Hess and Tully showed an isotope effect of about a factor of 2, while that of McCaulley et al. was greater than 5. These results, while not in very good agreement, suggest that the abstraction mostly occurs at the CH_3-group. This was confirmed by the studies of Hägele et al. (1983), who found CH_3O yields of 11% at 298 K and 22% at 393 K, Meier et al. (1984, 1985a), who found yields of CH_3O of 17% (mass spectrometry) and 25% (laser-induced fluorescence), and by Pagsberg et al. (1988), who found a CH_2OH yield of 83% (UV absorption). Hence, it appears that >75% of the reaction occurs at the methyl group at room temperature and below.

V-C-3.2. OH + Ethanol (Ethyl Alcohol, C_2H_5OH)

The reaction of hydroxy radicals with ethanol has been studied between 216 K and 1020 K. The measurements are given in table V-C-23. The results are somewhat more

Table V-C-22. Rate coefficients (k, cm^3 $molecule^{-1}$ s^{-1}) for the reaction of OH with methanol (CH_3OH)

$10^{12} \times A$	B (K)	$10^{12} \times k$	T (K)	Technique[a]	Reference	Temperature Range (K)
		0.95 ± 0.10	292	RR (n-C_4H_{10})[b]	Campbell et al. (1976)	
		1.06 ± 0.10	296 ± 2	FP–RA	Overend and Paraskevopoulos (1978)	
		1.00 ± 0.10	298	FP–RF	Ravishankara and Davis (1978)	
		1.1	300	RR (C_2H_4)[c]	Barnes et al. (1982)	
		0.87 ± 0.07	300 ± 3	RR (CH_3OCH_3)[d]	Tuazon et al. (1983)	
12 ± 3	810 ± 50	0.79 ± 0.20	298	PLP–RF	Hägele et al. (1983)	295–420
11 ± 3	798 ± 45	0.77	300	DF–LIF	Meier et al. (1984, 1985a)	300–1020
		1.1	300	RR (toluene)[e]	Klöpffer et al. (1986)	
8.0 ± 1.9	664 ± 88	0.54 ± 0.04	260	FP–RA	Greenhill and O'Grady (1986)	260–803
		0.76 ± 0.04	292			
		0.75 ± 0.08	300			
		1.13 ± 0.05	331			
		1.44 ± 0.06	362			
		1.44 ± 0.09	453			
		1.35 ± 0.08	465			
		2.06 ± 0.17	570			
		2.67 ± 0.24	597			
		2.79 ± 0.25	669			
		5.76 ± 0.59	803			
4.8 ± 1.2	480 ± 70	0.657 ± 0.046	240	FP–RF	Wallington and Kurylo (1987b)	240–440
		0.861 ± 0.047	296			
		1.25 ± 0.08	350			
		1.41 ± 0.12	400			
		1.62 ± 0.14	440			
		0.88 ± 0.18	298	PR–UVabs	Pagsberg et al. (1988)	
		1.01 ± 0.10	298	DF–LIF	McCaulley et al. (1989)	
$k(T)$[f]		0.934 ± 0.041	294	PLP–LIF	Hess and Tully (1989)	294–866
		1.09 ± 0.05	332			
		1.33 ± 0.06	380			
		1.69 ± 0.07	441			
		2.10 ± 0.09	505			
		2.31 ± 0.10	528			
		3.01 ± 0.13	626			
		3.96 ± 0.20	709			
		5.05 ± 0.24	787			
		6.18 ± 0.32	866			
		0.90 ± 0.09	298 ± 2	PR–UVabs	Nelson et al. (1990a)	
		0.91 ± 0.21		RR (c-C_6H_{12})[g]		
		0.90 ± 0.08	298 ± 4	RR (n-C_5H_{12})[h]	Picquet et al. (1998)	
		0.82 ± 0.07	298 ± 2	RR (c-C_6H_{12})[g]	Oh and Andino (2001)	
		1.0 ± 0.1		RR (C_2H_5OH)[i]		
		0.812 ± 0.054	296	RR (C_2H_4, C_2H_2)[c,j]	Sørensen et al. (2002)	

Table V-C-22. (*continued*)

$10^{12} \times A$	B (K)	$10^{12} \times k$	T (K)	Technique[a]	Reference	Temperature Range (K)
3.6 ± 0.8	415 ± 70	0.63 ± 0.15	235	PLP–LIF	Jiménez et al. (2003)	235–360
		0.673 ± 0.005	244			
		0.729 ± 0.033	251			
		0.689 ± 0.075	260 ± 1			
		0.759 ± 0.025	273			
		0.866 ± 0.036	285			
		0.93 ± 0.11	298			
		0.953 ± 0.014	315			
		1.03 ± 0.027	330			
		1.13 ± 0.05	345			
		1.29 ± 0.12	360			
$k(T)$[k]		0.599 ± 0.020	210	PLP–LIF	Dillon et al. (2005)	210–351
		0.523 ± 0.015	214			
		0.638 ± 0.024	218 ± 1			
		0.620 ± 0.009	223			
		0.64 ± 0.20	227 ± 1			
		0.647 ± 0.025	231 ± 1			
		0.69 ± 0.20	240 ± 1			
		0.766 ± 0.021	253			
		0.789 ± 0.011	263			
		0.808 ± 0.008	273			
		0.93 ± 0.07	298			
		1.10 ± 0.01	322			
		1.12 ± 0.02	328			
		1.22 ± 0.02	351			

[a] Abbreviations used for the techniques are defined in table II-A-1.
[b] Placed on an absolute basis using $k(n\text{-}C_4H_{10}) = 2.36 \times 10^{-12}$ cm^3 molecule^{-1} s^{-1}.
[c] Placed on an absolute basis using $k(C_2H_4) = 8.52 \times 10^{-12}$ cm^3 molecule^{-1} s^{-1} at 1 bar and 298 K.
[d] Placed on an absolute basis using $k(CH_3OCH_3) = 2.75 \times 10^{-12}$ cm^3 molecule^{-1} s^{-1}.
[e] Placed on an absolute basis using k(toluene) $= 5.6 \times 10^{-12}$ cm^3 molecule^{-1} s^{-1}.
[f] $k(T) = 5.89 \times 10^{-20}\, T^{2.65} \exp(833/T)$.
[g] Placed on an absolute basis using k(cyclohexane) $= 6.85 \times 10^{-12}$ cm^3 molecule^{-1} s^{-1}.
[h] Placed on an absolute basis using $k(n\text{-}C_5H_{12}) = 3.92 \times 10^{-12}$ cm^3 molecule^{-1} s^{-1}.
[i] Placed on an absolute basis using $k(C_2H_5OH) = 3.3 \times 10^{-12}$ cm^3 molecule^{-1} s^{-1}.
[j] Placed on an absolute basis using $k(C_2H_2) = 8.1 \times 10^{-13}$ cm^3 molecule^{-1} s^{-1} at 1 bar and 298 K.
[k] $k(T) = 6.67 \times 10^{-18} T^2 \exp(140/T)$.

scattered than those for methanol, with room-temperature rate coefficients covering more than a factor of 2. Excluding the very low values of Meier et al. (1985a, b, 1987), the mean of the values in the table is $(3.26 \pm 0.39) \times 10^{-12}$ cm^3 molecule^{-1} s^{-1} at 298 ± 5 K (1-σ scatter, unweighted). This value is very close to the mean of the two recent measurements by Jiménez et al. (2003) and Dillon et al. (2005). The rate coefficient has a very weak dependence on temperature, being almost independent of temperature from 216 K to 380 K (figure V-C-19). However, when the high-temperature data of Hess and Tully (1988) are included, strong upward curvature is indicated. Atkinson et al. (2005b) recommend a nonlinear Arrhenius expression $k = 6.7 \times 10^{-18} T^2 \exp(511/T)$ cm^3 molecule^{-1} s^{-1}. However, this expression exhibits a minimum near 250 K, which is not seen in any of the individual data sets. For the atmospheric temperature regime, we fit the data of Wallington and Kurylo (1987b),

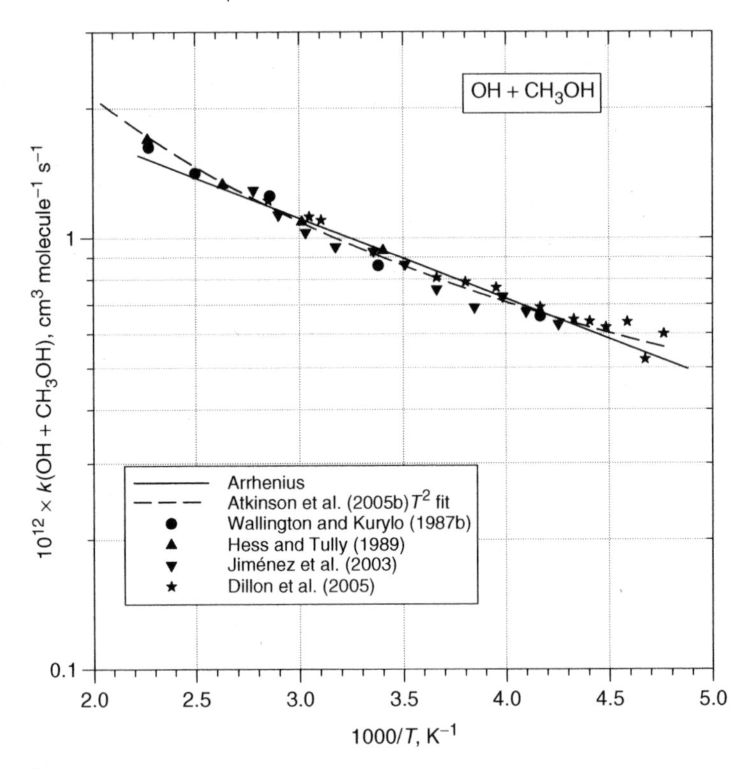

Figure V-C-18. Arrhenius plot of the rate coefficient for the reaction of OH with methanol (selected data).

Hess and Tully (1988) below 500 K, Jiménez et al. (2003), and Dillon et al. (2005): $k = 4.7 \times 10^{-12} \exp(-97)/T)\,\text{cm}^3\ \text{molecule}^{-1}\ \text{s}^{-1}$, with an uncertainty of 20%.

The reaction mechanism has been investigated by Meier et al. (1985a, b). By detection of $CH_3CH(OH)$ radicals, using mass spectrometry, they determined that the branching ratio for abstraction at the CH_2-group at 298 K is 0.75 ± 0.15. The weak temperature dependence of the overall rate coefficient is also consistent with the majority of the reaction occurring at the more weakly bonded methylene hydrogen atoms.

V-C-3.3. OH + 1-Propanol (n-Propyl Alcohol, $CH_3CH_2CH_2OH$)

There have been eight studies of this reaction. The measurements are given in table V-C-24. Other than the result of Campbell et al. (1976), which appears to be anomalously low, the ambient temperature data are in good agreement, with an unweighted mean of the data at 298 K of $(5.45 \pm 0.50) \times 10^{-12}\,\text{cm}^3\ \text{molecule}^{-1}\ \text{s}^{-1}$. The temperature dependence has been measured by Yujing and Mellouki (2001) from 263 to 362 K, and by Cheema et al. (2002) from 273 to 343 K. The data of Cheema et al. indicate a positive temperature dependence with $E/R = 260$ K, while those of Yujing and Mellouki show a weak negative dependence. A fit to all the data shown in figure V-C-20 gives $k = 6.35 \times 10^{-12} \exp(-30/T)\,\text{cm}^3\ \text{molecule}^{-1}\ \text{s}^{-1}$, which returns

Table V-C-23. Rate coefficients (k, cm^3 $molecule^{-1}$ s^{-1}) for the reaction of OH with ethanol (C_2H_5OH)

$10^{12} \times A$	B (K)	$10^{12} \times k$	T (K)	Technique[a]	Reference	Temperature Range (K)
		3.0 ± 0.3	292	RR (n-C_4H_{10})[b]	Campbell et al. (1976)	
		3.74 ± 0.36	296 ± 2	FP–RA	Overend and Paraskevopoulos (1978)	
		2.62 ± 0.36	298	FP–RF	Ravishankara and Davis (1978)	
5.2 ± 1.0	274 ± 90	2.1	300	DF–LIF	Meier et al. (1985a,b, 1987)	300–1000
12.5 ± 2.4	360 ± 52	2.84 ± 0.15	255	FP–RA	Greenhill and O'Grady (1986)	255–459
		3.40 ± 0.14	273			
		3.80 ± 0.24	289			
		3.40 ± 0.17	293			
		4.26 ± 0.19	331			
		4.26 ± 0.18	360			
		5.21 ± 0.36	369			
		5.63 ± 0.48	459			
		3.6 ± 0.4	303	RR (C_2H_4)[c]	Kerr and Stocker (1986)	
7.4 ± 3.2	240 ± 110	2.75 ± 0.14	240	FP–RF	Wallington and Kurylo (1987b)	240–440
		3.33 ± 0.23	296			
		3.25 ± 0.39	350			
		4.07 ± 0.40	400			
		4.58 ± 0.29	440			
$k(T)$[d]		3.26 ± 0.14	293	PLP–LIF	Hess and Tully (1988)	293–750
		3.33 ± 0.14	326.5			
		3.63 ± 0.15	380			
		3.94 ± 0.6	441			
		4.65 ± 0.19	520.5			
		4.78 ± 0.23	544			
		5.47 ± 0.34	599			
		3.04 ± 0.25	298 ± 2	PR–UVabs	Nelson et al. (1990a)	
		3.16 ± 0.48		RR (c-C_6H_{12})[e]		
		3.68 ± 0.11	298 ± 4	RR (n-C_6H_{14})[f]	Picquet et al. (1998)	
		2.7 ± 0.3	298 ± 2	RR (c-C_6H_{12})[e]	Oh and Andino (2001)	
		2.7 ± 0.2		RR (p-xylene)[g]		
		3.47 ± 0.32	296	RR (C_2H_4, C_2H_2)[c,h]	Sørensen et al. (2002)	
		3.40 ± 0.25	295 ± 2	RR (C_3H_8)[i]	Wu et al. (2003)	
4.3 ± 0.7	85 ± 35	2.80 ± 0.23	227	PLP–LIF	Jiménez et al. (2003)	227–360
		3.11 ± 0.10	243			
		3.20 ± 0.11	260			
		3.24 ± 0.08	273			
		3.20 ± 0.10	285			
		3.09 ± 0.15	298			
		3.27 ± 0.10	315			
		3.20 ± 0.25	330			
		3.37 ± 0.22	345			
		3.60 ± 0.11	360			

(continued)

Table V-C-23. *(continued)*

$10^{12} \times A$	B (K)	$10^{12} \times k$	T (K)	Technique[a]	Reference	Temperature Range (K)
4.0	42 ± 14	3.38 ± 0.14	216	PLP–LIF	Dillon et al. (2005)	216–368
		3.40 ± 0.15	224 ± 2			
		3.38 ± 0.10	238 ± 1			
		3.48 ± 0.09	243			
		3.25 ± 0.13	247			
		3.40 ± 0.06	252			
		3.38 ± 0.11	258			
		3.46 ± 0.07	263			
		3.33 ± 0.10	270			
		3.35 ± 0.07	273			
		3.23 ± 0.12	279			
		3.46 ± 0.09	284			
		3.38 ± 0.15	298 ± 2			
		3.61 ± 0.09	319			
		3.72 ± 0.09	339			
		3.57 ± 0.07	353			
		3.47 ± 0.13	358			
		3.67 ± 0.24	363			
		3.83 ± 0.08	368			

[a] Abbreviations used for the techniques are defined in table II-A-1.
[b] Placed on an absolute basis using $k(n\text{-}C_4H_{10}) = 2.36 \times 10^{-12}$ cm^3 molecule^{-1} s^{-1}.
[c] Placed on an absolute basis using $k(C_2H_4) = 8.52 \times 10^{-12}$ cm^3 molecule^{-1} s^{-1} at 1 bar and 298 K.
[d] $k(T) = 6.25 \times 10^{-18} T^2 \exp(530/T)$.
[e] Placed on an absolute basis using k(cyclohexane) $= 6.85 \times 10^{-12}$ cm^3 molecule^{-1} s^{-1}.
[f] Placed on an absolute basis using $k(n\text{-}C_6H_{14}) = 5.27 \times 10^{-12}$ cm^3 molecule^{-1} s^{-1}.
[g] Placed on an absolute basis using $k(p$-xylene$) = 1.43 \times 10^{-11}$ cm^3 molecule^{-1} s^{-1}.
[h] Placed on an absolute basis using $k(C_2H_2) = 8.1 \times 10^{-13}$ cm^3 molecule^{-1} s^{-1} at 1 bar and 298 K.
[i] Placed on an absolute basis using $k(C_3H_8) = 1.08 \times 10^{-12}$ cm^3 molecule^{-1} s^{-1}.

the recommended 298 K value $k = 5.72 \times 10^{-12}$ cm^3 molecule^{-1} s^{-1} with an uncertainty of 20%. Azad and Andino (1999) measured product yields for propanal (72 ± 6)% and acetaldehyde (18 ± 3)%, confirming that abstraction of a hydrogen atom primarily occurs at the α-CH_2 site containing the hydroxy group.

V-C-3.4. OH + 2-Propanol [Isopropyl Alcohol, $(CH_3)_2CHOH$]

This reaction has been studied directly from 240 K to 502 K. The measurements are given in table V-C-25. Above 502 K, Dunlop and Tully (1993b) found that regeneration of OH complicated the analysis, but additional data could be obtained up to 750 K by using isotopically substituted OH radicals, which were not subject to regeneration. The data of Dunlop and Tully (1993b) and Yujing and Mellouki (2001) are in excellent agreement over their region of overlap, whereas the data of Dunlop and Tully show a clear upturn above 400 K. The data of Wallington and Kurylo (1987b), while not showing such a clear trend with temperature, agree to better than 15% with the other two temperature-dependent studies. Other than the high value of Lloyd et al. (1976b), the values near 298 K are all within 15% of one another. For atmospheric purposes, we recommend the expression $k = 4.30 \times 10^{-12} \exp(60/T)$ cm^3 molecule^{-1} s^{-1} from

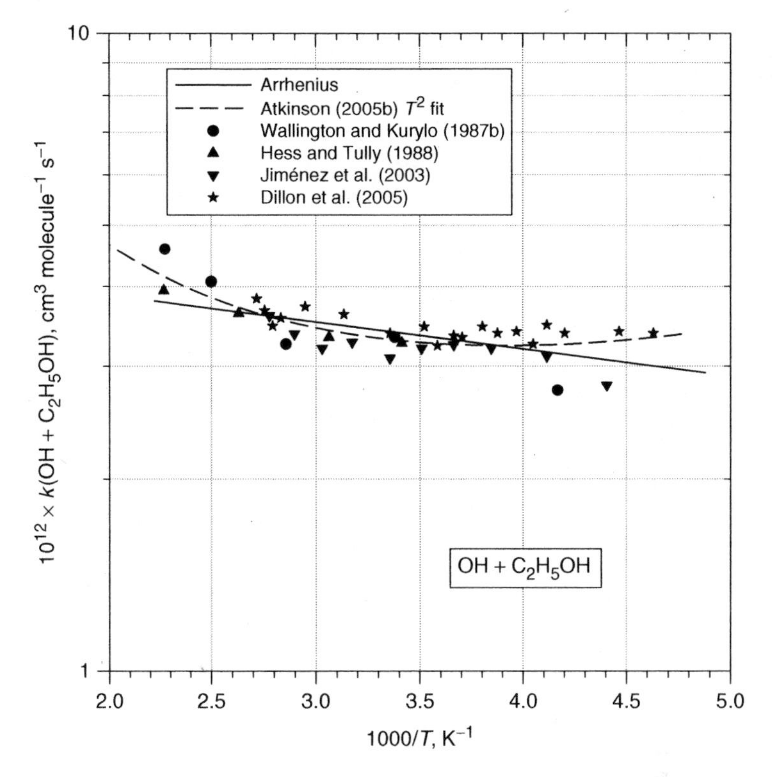

Figure V-C-19. Arrhenius plot of the rate coefficient for the reaction of OH with ethanol.

a fit to the data shown in figure V-C-21. The value of the rate coefficient at 298 K is 5.27×10^{-12} cm^3 $molecule^{-1}$ s^{-1}, with an uncertainty of $\pm15\%$.

Tully and Dunlop also investigated the reaction of OH radicals with $(CD_3)_2CHOH$, $(CH_3)_2CDOH$, $(CD_3)_2CDOD$, and $(CH_3)_2CHOD$. By combining the results for the various isotopologs, they derived branching ratios for attack at the various sites as a function of temperature. They found that abstraction from the β-methyl groups accounts for 12% of the reaction at 293 K (and probably less at lower temperatures), while reaction at the α-CH group is expected to dominate at all temperatures.

V-C-3.5. OH + 1-Butanol (n-Butyl Alcohol, $CH_3CH_2CH_2CH_2OH$)

This reaction has been studied between 263 and 362 K, using both relative and absolute techniques. The measurements are given in table V-C-26. The mean of the measurements of Wallington and Kurylo (1987b), Nelson et al. (1990a), Oh and Andino (2001, relative to cyclohexane), Yujing and Mellouki (2001), Cavalli et al. (2002), and Wu et al. (2003) is $(8.3 \pm 0.4) \times 10^{-12}$ cm^3 $molecule^{-1}$ s^{-1}. The data of Yujing and Mellouki show a weak negative temperature dependence. A fit to the data in figure V-C-22 yields $k = 5.43 \times 10^{-12}$ $\exp(133/T)$ cm^3 $molecule^{-1}$ s^{-1} over the atmospheric temperature range, with an uncertainty of 15%. The products of the reaction were studied by Cavalli et al. (2002), who found 52% butanal, 23% propanal,

Table V-C-24. Rate coefficients (k, cm^3 $molecule^{-1}$ s^{-1}) for the reaction of OH with 1-propanol ($CH_3CH_2CH_2OH$)

$10^{12} \times A$	B (K)	$10^{12} \times k$	T (K)	Technique[a]	Reference	Temperature Range (K)
		3.8 ± 0.3	292	RR (n-C_4H_{10})[b]	Campbell et al. (1976)	
		5.33 ± 0.53	296 ± 2	FP–RA	Overend and Paraskevopoulos (1978)	
		5.34 ± 0.29	296	FP–RF	Wallington and Kurylo (1987b)	
		5.64 ± 0.48	298 ± 2	PR–UVabs	Nelson et al. (1990a)	
		5.03 ± 0.41		RR (c-C_6H_{12})[c]		
		5.3 ± 0.3	298 ± 2	RR (c-C_6H_{12})[c]	Oh and Andino (2000, 2001, 2002)	
		6.4 ± 0.5		RR (p-xylene)[d]		
4.68 ± 0.64	−68 ± 41	6.13 ± 0.34	263	PLP–LIF	Yujing and Mellouki (2001)	263–372
		6.12 ± 0.17	273			
		5.83 ± 0.18	298			
		5.66 ± 0.18	323			
		5.64 ± 0.13	348			
		5.79 ± 0.26	372			
		5.35 ± 0.19	273	RR (2,3-dimethylbutane)[e]	Cheema et al. (2002)	273–343
		5.84 ± 0.17	298			
		6.11 ± 0.33	313			
		6.53 ± 0.17	343			
		5.47 ± 0.44	295 ± 2	RR (C_3H_8)[f]	Wu et al. (2003)	

[a] Abbreviations used for the techniques are defined in table II-A-1.
[b] Placed on an absolute basis using $k(n\text{-}C_4H_{10}) = 2.36 \times 10^{-12}$ cm^3 $molecule^{-1}$ s^{-1}.
[c] Placed on an absolute basis using $k(\text{cyclohexane}) = 6.85 \times 10^{-12}$ cm^3 $molecule^{-1}$ s^{-1}.
[d] Placed on an absolute basis using $k(p\text{-xylene}) = 1.43 \times 10^{-11}$ cm^3 $molecule^{-1}$ s^{-1}.
[e] Placed on an absolute basis using $k(2,3\text{-dimethylbutane}) = 1.25 \times 10^{-11} \exp(-212/T)$ cm^3 $molecule^{-1}$ s^{-1}.
[f] Placed on an absolute basis using $k(C_3H_8) = 1.08 \times 10^{-12}$ cm^3 $molecule^{-1}$ s^{-1}.

13% acetaldehyde, and 44% formaldehyde. The first three products correspond roughly to the fractional attack at the 1-, 2-, and 3-sites. The oxy radical formed at the 3- site can also react with O_2 to form 4-hydroxy-2-butanone, but this is probably less important than decomposition to acetaldehyde and two molecules of formaldehyde:

$$CH_3CH(O\bullet)CH_2CH_2OH \rightarrow CH_3CHO + \bullet CH_2CH_2OH$$

$$\bullet CH_2CH_2OH + O_2 + NO \rightarrow\rightarrow 2CH_2O$$

$$CH_3CH(O\bullet)CH_2CH_2OH + O_2 \rightarrow CH_3C(O)CH_2CH_2OH$$

V-C-3.6. OH + 2-Butanol [sec-Butyl Alcohol, $CH_3CH(OH)CH_2CH_3$]

This reaction has been studied at room temperature, using the relative rate technique, by Chew and Atkinson (1996) and Baxley and Wells (1998), and as a function of

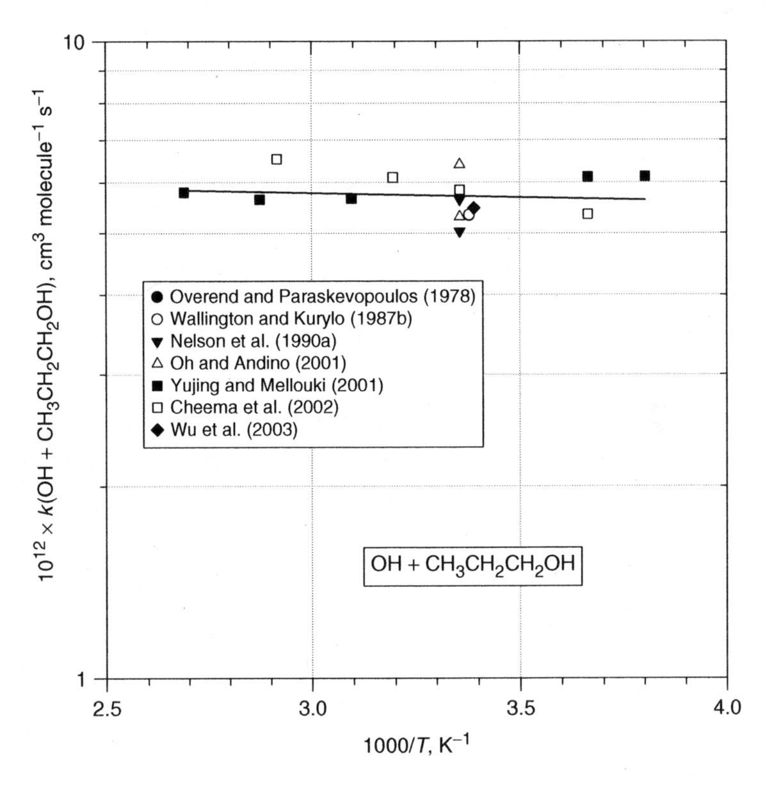

Figure V-C-20. Arrhenius plot of the rate coefficient for the reaction of OH with 1-propanol.

temperature by Jiménez et al. (2005b), using flash photolysis with laser-induced fluorescence. The measurements are given in table V-C-27. The results are in excellent agreement, and a fit to the combined data set (shown in figure V-C-23) yields the expression $k = 2.74 \times 10^{-12}$ exp(330/T) cm^3 molecule^{-1} s^{-1}, with a value of $(8.5 \pm 1.3) \times 10^{-12}$ cm^3 molecule^{-1} s^{-1} at 298 K.

Chew and Atkinson (1996) also studied the reaction mechanism, finding a yield of 2-butanone of (70 ± 7)%. This yield should correspond to attack at the 2-site, which will lead to formation of 2-butanone in essentially 100% yield. Baxley and Wells (1998) found a yield of (60 ± 2)% for this product. Attack at the 3-site (leading to 2 molecules of acetaldehyde) is predicted to account for the majority of the rest of the reaction, and Baxley and Wells found an acetaldehyde yield of (29 ± 4)%.

V-C-3.7. OH + 2-Methyl-1-propanol [Isobutyl Alcohol, $(CH_3)_2CHCH_2OH$]

This reaction has been studied twice, at room temperature by Wu et al. (2003), and between 241 and 370 K by Mellouki et al. (2004). The measurements are given in table V-C-28. The determinations using different techniques are in very good agreement, and we recommend a value $k = (9.6 \pm 1.5) \times 10^{-12}$ cm^3 molecule^{-1} s^{-1} at 298 K.

Table V-C-25. Rate coefficients (k, cm^3 $molecule^{-1}$ s^{-1}) for the reaction of OH with 2-propanol [$CH_3CH(OH)CH_3$]

$10^{12} \times A$	B (K)	$10^{12} \times k$	T (K)	Technique[a]	Reference	Temperature Range (K)
		7.2 ± 2.2	305 ± 2	RR ($CH_2{=}C(CH_3)_2$)[b]	Lloyd et al. (1976b)	
		5.48 ± 0.55	296 ± 2	FP–RA	Overend and Paraskevopoulos (1978)	
		4.7	300	RR (toluene)[c]	Klöpffer et al. (1986)	
5.8 ± 1.9	30 ± 90	5.12 ± 0.31	240	FP–RF	Wallington and Kurylo (1987b)	240–440
		5.81 ± 0.34	296			
		5.27 ± 0.38	350			
		5.16 ± 0.44	400			
		5.75 ± 0.55	440			
		5.69 ± 1.09	298 ± 2	PR–UVabs	Nelson et al. (1990a)	
		5.29 ± 0.70		RR (c-C_6H_{12})[d]		
		5.10 ± 0.21	293	PLP–LIF	Dunlop and Tully (1993b)	293–502
		4.89 ± 0.20	326			
		4.68 ± 0.20	378			
		4.73 ± 0.22	438			
		4.86 ± 0.24	502			
2.80 ± 0.39	−184 ± 40	5.58 ± 0.20	253	PLP–LIF	Yujing and Mellouki (2001)	253–372
		5.74 ± 0.25	263			
		5.51 ± 0.14	273			
		5.30 ± 0.17	283			
		5.17 ± 0.23	298			
		4.83 ± 0.12	323			
		4.66 ± 0.16	348			
		4.71 ± 0.12	372			
		5.31 ± 0.39	295 ± 2	RR (C_3H_8)[e]	Wu et al. (2003)	

[a] Abbreviations used for the techniques are defined in table II-A-1.
[b] Placed on an absolute basis using k(2-methylpropene) = 5.14×10^{-11} cm^3 $molecule^{-1}$ s^{-1}.
[c] Placed on an absolute basis using k(toluene) = 5.6×10^{-12} cm^3 $molecule^{-1}$ s^{-1}.
[d] Placed on an absolute basis using k(cyclohexane) = 6.85×10^{-12} cm^3 $molecule^{-1}$ s^{-1}.
[e] Placed on an absolute basis using $k(C_3H_8) = 1.08 \times 10^{-12}$ cm^3 $molecule^{-1}$ s^{-1}.

A fit to the combined data set (figure V-C-24) gives $k = 2.8 \times 10^{-12} \exp(365/T)\, cm^3$ $molecule^{-1}$ s^{-1}.

V-C-3.8. OH + 2-Methyl-2-propanol [tert-Butyl Alcohol, $(CH_3)_3COH$]

The reaction of OH with 2-methyl-2-propanol has been studied by Wallington et al. (1988a), Saunders et al. (1994), Téton et al. (1996b), and Wu et al. (2003). The measurements are given in table V-C-29. The temperature dependence of the reaction was measured by Wallington et al. from 240 to 400 K and by Téton et al. from 253 to 372 K.

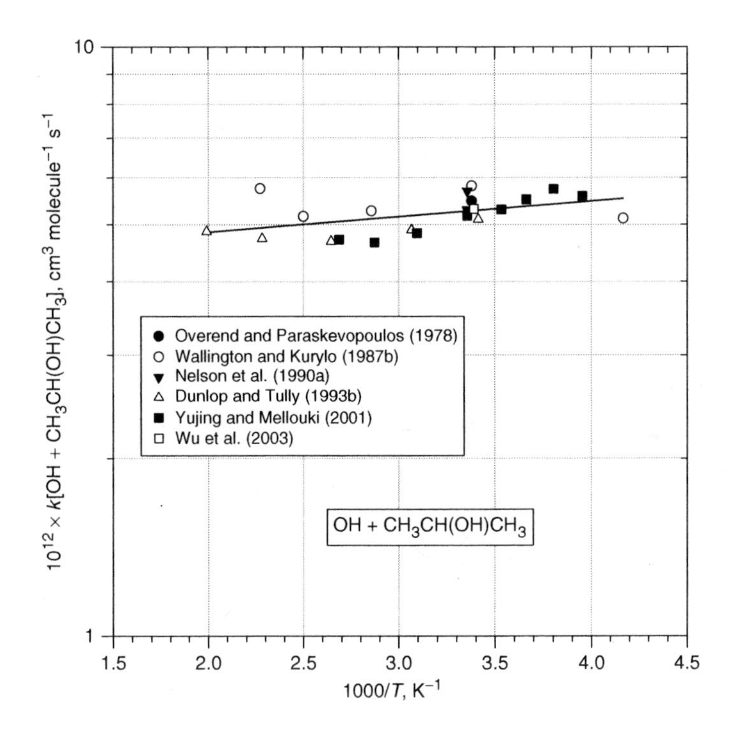

Figure V-C-21. Arrhenius plot of the rate coefficient for the reaction of OH with 2-propanol.

Table V-C-26. Rate coefficients (k, cm^3 molecule^{-1} s^{-1}) for the reaction of OH with 1-butanol (n-C_4H_9OH)

$10^{12} \times A$	B (K)	$10^{12} \times k$	T (K)	Technique[a]	Reference	Temperature Range (K)
		6.7 ± 1.3	292	RR (n-C_4H_{10})[b]	Campbell et al. (1976)	
		8.31 ± 0.63	296	FP–RF	Wallington and Kurylo (1987b)	
		7.80 ± 0.20	298 ± 2	PR–UVabs	Nelson et al. (1990a)	
		7.83 ± 0.66		RR (c-C_6H_{12})[c]		
		8.8 ± 0.3	298 ± 2	RR (c-C_6H_{12})	Oh and Andino (2001)	
		10.3 ± 0.4		RR (p-xylene)[d]		
5.30 ± 1.62	-146 ± 92	9.51 ± 0.44	263	PLP–LIF	Yujing and Mellouki (2001)	263–372
		9.36 ± 0.27	273			
		8.47 ± 0.34	298			
		8.11 ± 0.20	323			
		8.24 ± 0.24	348			
		8.09 ± 0.14	372			
		7.53 ± 0.78	298	RR (c-C_6H_{12})[c]	Cavalli et al. (2002)	
		8.66 ± 0.66	295 ± 2	RR (C_3H_8)[e]	Wu et al. (2003)	

[a] Abbreviations used for the techniques are defined in table II-A-1.
[b] Placed on an absolute basis using $k(n\text{-}C_4H_{10}) = 2.36 \times 10^{-12}$ cm^3 molecule^{-1} s^{-1}.
[c] Placed on an absolute basis using k(cyclohexane) $= 6.85 \times 10^{-12}$ cm^3 molecule^{-1} s^{-1}.
[d] Placed on an absolute basis using k(p-xylene) $= 1.43 \times 10^{-11}$ cm^3 molecule^{-1} s^{-1}.
[e] Placed on an absolute basis using $k(C_3H_8) = 1.08 \times 10^{-12}$ cm^3 molecule^{-1} s^{-1}.

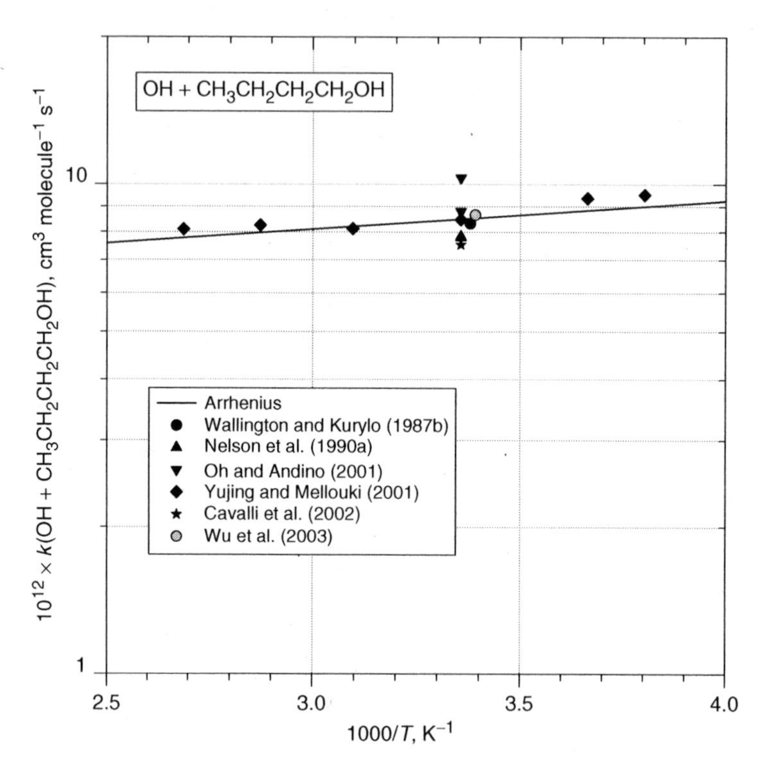

Figure V-C-22. Arrhenius plot of the rate coefficient for the reaction of OH with 1-butanol.

Table V-C-27. Rate coefficients (k, cm^3 $molecule^{-1}$ s^{-1}) for the reaction of OH with 2-butanol [$CH_3CH(OH)CH_2CH_3$]

$10^{12} \times A$	B (K)	$10^{12} \times k$	T (K)	Technique[a]	Reference	Temperature Range (K)
		8.49 ± 0.48	296 ± 2	RR (c-C_6H_{12})[b]	Chew and Atkinson (1996)	
		8.85 ± 0.13	297 ± 3	RR (n-C_9H_{20})[c]	Baxley and Wells (1998)	
		7.80 ± 0.45		RR (n-$C_{12}H_{26}$)[d]		
2.76 ± 1.20	−328 ± 124	9.98 ± 0.30	263	PLP–LIF	Jiménez et al. (2005b)	263–354
		9.77 ± 1.46	268			
		9.34 ± 0.40	273			
		8.40 ± 0.40	278			
		8.07 ± 0.40	283			
		7.90 ± 0.50	288			
		8.77 ± 1.46	298			
		8.51 ± 2.80	303			
		8.32 ± 1.10	321			
		7.93 ± 0.40	327			
		6.99 ± 0.48	338			
		6.64 ± 0.46	354			

[a] Abbreviations used for the techniques are defined in table II-A-1.
[b] Placed on an absolute basis using k(cyclohexane) = 6.85×10^{-12} cm^3 $molecule^{-1}$ s^{-1}.
[c] Placed on an absolute basis using $k(n\text{-}C_9H_{20}) = 9.75 \times 10^{-12}$ cm^3 $molecule^{-1}$ s^{-1}.
[d] Placed on an absolute basis using $k(n\text{-}C_{12}H_{26}) = 1.36 \times 10^{-11}$ cm^3 $molecule^{-1}$ s^{-1}.

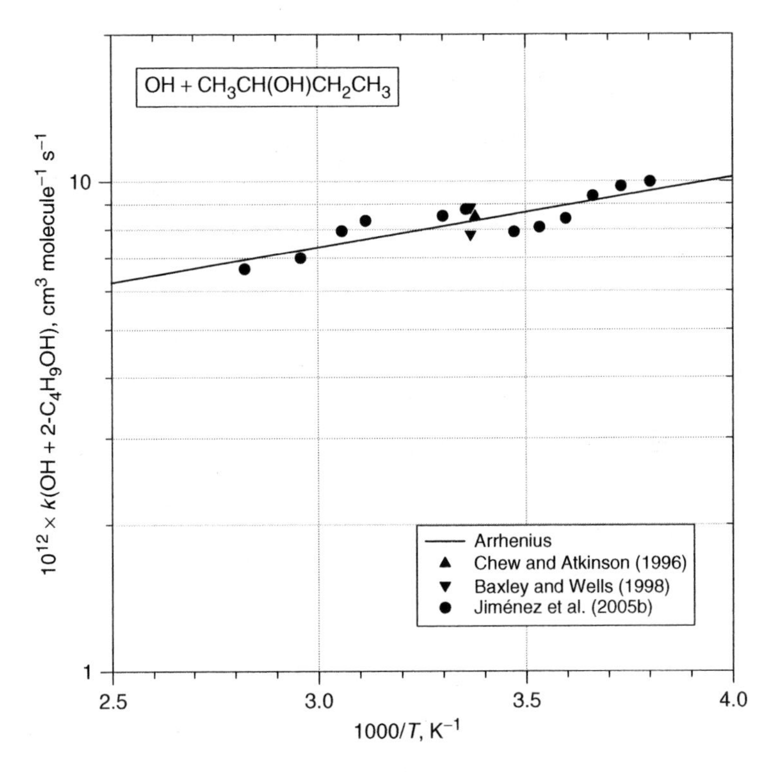

Figure V-C-23. Arrhenius plot of the rate coefficient for the reaction of OH with 2-butanol.

Table V-C-28. Rate coefficients (k, cm^3 $molecule^{-1}$ s^{-1}) for the reaction of OH with 2-methyl-1-propanol [$(CH_3)_2CHCH_2OH$]

$10^{12} \times A$	B (K)	$10^{12} \times k$	T (K)	Technique[a]	Reference	Temperature Range (K)
		9.08 ± 0.35	295 ± 2	RR (C_3H_8)[b]	Wu et al. (2003)	
		9.43 ± 0.45		RR (c-C_6H_{12})[c]		
		8.5 ± 0.1	298 ± 2	RR (1-butanol)[d]	Mellouki et al. (2004)	
3.1 ± 0.9	−352 ± 82	13.8 ± 0.2	241	PLP–LIF	Mellouki et al. (2004)	241–370
		12.3 ± 0.5	253			
		12.4 ± 0.4	260			
		10.6 ± 0.1	271			
		10.0 ± 1.0	297 ± 1			
		8.7 ± 0.2	323			
		8.7 ± 0.2	348			
		8.3 ± 0.2	370			

[a] Abbreviations used for the techniques are defined in table II-A-1.
[b] Placed on an absolute basis using $k(C_3H_8) = 1.08 \times 10^{-12}$ cm^3 $molecule^{-1}$ s^{-1}.
[c] Placed on an absolute basis using k(cyclohexane) $= 6.85 \times 10^{-12}$ cm^3 $molecule^{-1}$ s^{-1}.
[d] Placed on an absolute basis using k(1-butanol) $= 8.5 \times 10^{-12}$ cm^3 $molecule^{-1}$ s^{-1}.

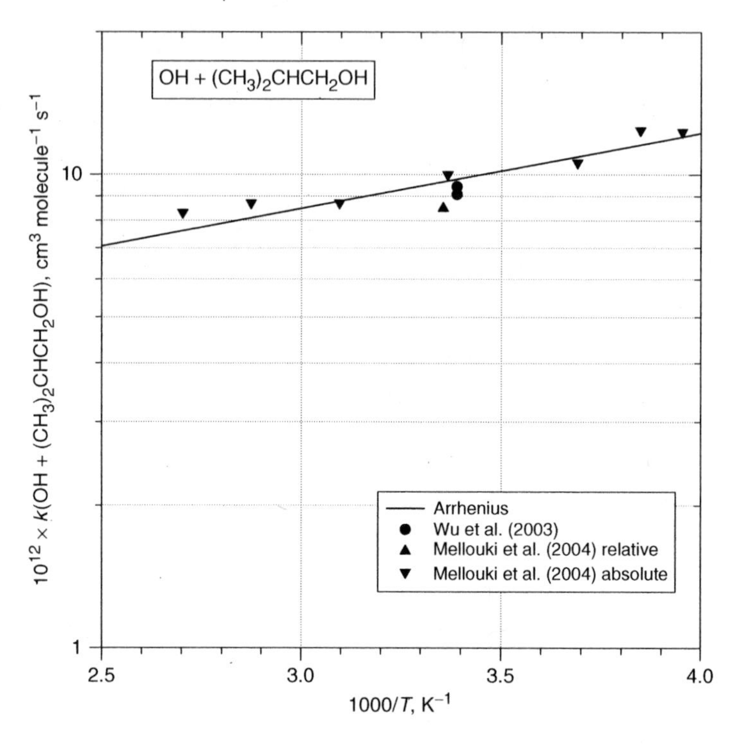

Figure V-C-24. Arrhenius plot of the rate coefficient for the reaction of OH with 2-methyl-1-propanol.

Table V-C-29. Rate coefficients (k, cm^3 $molecule^{-1}$ s^{-1}) for the reaction of OH with 2-methyl-2-propanol [$(CH_3)_3COH$]

$10^{12} \times A$	B (K)	$10^{12} \times k$	T (K)	Technique[a]	Reference	Temperature Range (K)
3.3 ± 1.6	310 ± 150	1.00 ± 0.06	240	FP–RF	Wallington et al. (1988a)	240–440
		1.07 ± 0.08	298			
		1.23 ± 0.08	350			
		1.63 ± 0.07	400			
		1.77 ± 0.7	440			
		0.81 ± 0.17	298	DF–LIF	Saunders et al. (1994)	
2.66 ± 0.48	270 ± 130	1.08 ± 0.10	298	PLP–LIF	Téton et al. (1996b)	253–372
		1.11 ± 0.07	295 ± 2	RR (C_3H_8)[b]	Wu et al. (2003)	

[a] Abbreviations used for the techniques are defined in table II-A-1.
[b] Placed on an absolute basis using $k(C_3H_8) = 1.08 \times 10^{-12}$ cm^3 $molecule^{-1}$ s^{-1}.

There is some evidence for curvature in both data sets. The data of Saunders et al. are somewhat lower than the other measurements, which are otherwise in excellent agreement (figure V-C-25). The values from Téton et al. have been estimated from a figure in their paper. A fit to the data of Wallington et al., Téton et al., and Wu et al. yields $k = 3.9 \times 10^{-12}$ $\exp(-387/T)$ cm^3 $molecule^{-1}$ s^{-1} for atmospheric purposes, with a

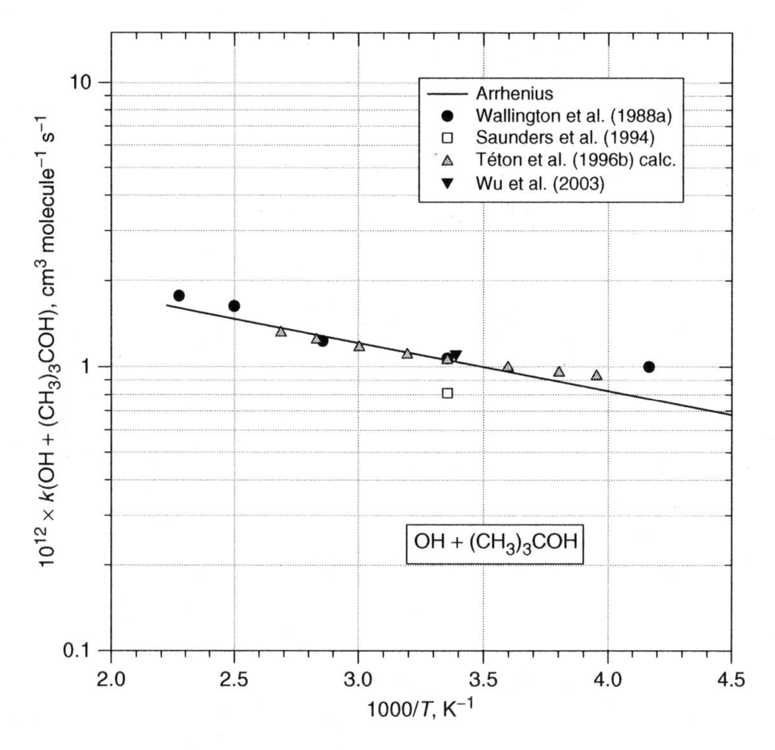

Figure V-C-25. Arrhenius plot of the rate coefficient for the reaction of OH with 2-methyl-2-propanol.

value at 298 K of $(1.05 \pm 0.15) \times 10^{-12}$ cm^3 molecule^{-1} s^{-1}. The reaction likely proceeds mainly by abstraction at the methyl groups, and Japar et al. (1990) observed equimolar amounts of formaldehyde and acetone in the oxidation.

V-C-3.9. OH + 1-Pentanol (n-Pentyl Alcohol, $CH_3CH_2CH_2CH_2CH_2OH$)

The reaction of OH with 1-pentanol has been studied at room temperature, using both absolute and relative techniques. The measurements are given in table V-C-30, and are in good agreement: absolute values of Wallington and Kurylo (1987b) and Nelson et al. (1990a), and relative rate determinations of Cavalli et al. (2000), Oh and Andino (2001), and Wu et al. (2003). An unweighted mean of these data leads to the value $k = (1.1 \pm 0.2) \times 10^{-11}$ cm^3 molecule^{-1} s^{-1} at 298 K.

Cavalli et al. (2000) studied the product distribution from the oxidation of 1-pentanol, finding pentanal ($40 \pm 8\%$), butanal ($16 \pm 4\%$), propanal ($8 \pm 2\%$), acetaldehyde ($18 \pm 4\%$), and formaldehyde ($25 \pm 1\%$), although large corrections were made for secondary consumption of the products. The pentanal, butanal, and propanal are markers for OH attack at the 1-, 2-, and 3-sites. Attack at the 4-site should lead to the formation of 4-hydroxypentanal, which was not identified in this study. Small amounts of 1-hydroxy-3-pentanone and 3-hydroxypropanal could also come from the 3-site.

Table V-C-30. Rate coefficients (k, cm^3 $molecule^{-1}$ s^{-1}) for the reaction of OH with 1-pentanol (n-$C_5H_{11}OH$)

$10^{12} \times k$	T (K)	Technique[a]	Reference
10.8 ± 1.1	296	FP–RF	Wallington and Kurylo (1987b)
12.0 ± 1.6	298 ± 2	PR–UVabs	Nelson et al. (1990a)
9.60 ± 1.21		RR (c-C_6H_{12})[b]	
10.1 ± 1.1	298	RR (c-C_6H_{12})[b]	Cavalli et al. (2000)
11.4 ± 0.6	298 ± 2	RR (c-C_6H_{12})	Oh and Andino (2001)
14.0 ± 0.7		RR (p-xylene)[c]	
12.2 ± 1.0	295 ± 2	RR (C_3H_8)[d]	Wu et al. (2003)
12.2 ± 0.5		RR (c-C_6H_{12})[b]	

[a] Abbreviations used for the techniques are defined in table II-A-1.
[b] Placed on an absolute basis using k(cyclohexane) $= 6.85 \times 10^{-12}$ cm^3 $molecule^{-1}$ s^{-1}.
[c] Placed on an absolute basis using k(p-xylene) $= 1.43 \times 10^{-11}$ cm^3 $molecule^{-1}$ s^{-1}.
[d] Placed on an absolute basis using $k(C_3H_8) = 1.08 \times 10^{-12}$ cm^3 $molecule^{-1}$ s^{-1}.

V-C-3.10. OH + 2-Pentanol [$CH_3CH(OH)CH_2CH_2CH_3$]

The reaction has been studied twice, by Wallington et al. (1988b) using flash photolysis, and by Baxley and Wells (1998) using a relative rate technique against two reference molecules. The data are given in table V-C-31. The measurement of Baxley and Wells that was made relative to dodecane agrees very well with the absolute measurement, whereas that made relative to heptane is some 20% lower. We thus recommend $k = (1.2 \pm 0.2) \times 10^{-11}$ cm^3 $molecule^{-1}$ s^{-1} at 298 K.

Baxley and Wells measured the product distribution from the reaction of OH with 2-pentanol. They found yields of 2-pentanone (41 ± 4%), propanal (14 ± 2%), and acetaldehyde (40 ± 4%). The 2-pentanone comes from attack at the 2-site, while propanal follows from attack at the 3-site and should be accompanied by acetaldehyde coproduct.

$$OH + CH_3CH_2CH_2CH(OH)CH_3 \rightarrow H_2O + CH_3CH_2CH_2C\bullet(OH)CH_3$$

$$CH_3CH_2CH_2C\bullet(OH)CH_3 + O_2 \rightarrow HO_2 + CH_3CH_2CH_2C(O)CH_3$$

Table V-C-31. Rate coefficients (k, cm^3 $molecule^{-1}$ s^{-1}) for the reaction of OH with 2-pentanol [$CH_3CH(OH)CH_2CH_2CH_3$]

$10^{12} \times k$	T (K)	Technique[a]	Reference
11.8 ± 0.8	298	FP–RF	Wallington et al. (1988b)
9.4 ± 1.1	297 ± 3	RR (n-C_7H_{16})[b]	Baxley and Wells (1998)
11.9 ± 0.8		RR (n-$C_{12}H_{26}$)[c]	

[a] Abbreviations used for the techniques are defined in table II-A-1.
[b] Placed on an absolute basis using $k(n\text{-}C_7H_{16}) = 6.50 \times 10^{-12}$ cm^3 $molecule^{-1}$ s^{-1}.
[c] Placed on an absolute basis using $k(n\text{-}C_{12}H_{26}) = 1.36 \times 10^{-11}$ cm^3 $molecule^{-1}$ s^{-1}.

$$OH + CH_3CH_2CH_2CH(OH)CH_3\ (+O_2) \rightarrow H_2O + CH_3CH_2CH(O_2)CH(OH)CH_3$$

$$CH_3CH_2CH(O_2)CH(OH)CH_3 + NO \rightarrow CH_3CH_2CH(O\bullet)CH(OH)CH_3 + NO_2$$

$$CH_3CH_2CH(O\bullet)CH(OH)CH_3 \rightarrow CH_3CH_2CHO + CH_3C\bullet H(OH)$$

$$CH_3C\bullet H(OH) + O_2 \rightarrow CH_3CHO + HO_2$$

Attack at the 4-site, followed by decomposition of the alkoxy radical, should lead to the formation of two molecules of acetaldehyde. A potential unidentified product from this reaction is 4-hydroxy-2-pentanone, $CH_3C(O)CH_2CH(OH)CH_3$, also from the 4-site.

V-C-3.11. OH + 3-Pentanol [$CH_3CH_2CH(OH)CH_2CH_3$]

The reaction of OH with 3-pentanol has been studied only at 298 K by Wallington et al. (1988b), using flash photolysis with resonance fluorescence detection of OH. On the basis of this study we recommend $k = (1.2 \pm 0.3) \times 10^{-11}$ cm^3 molecule^{-1} s^{-1} at 298 K. No product studies have been reported.

V-C-3.12. OH + 3-Methyl-1-butanol [Isopentyl Alcohol, $CH_3CH(CH_3)CH_2CH_2OH$]

The reaction of OH with 3-methyl-1-butanol has been studied at 296 ± 3 K by Saunders et al. (1994) using the discharge flow technique, Wu et al. (2003), using the relative rate method, and by Mellouki et al. (2004), who used both relative and absolute methods. The measurements are given in table V-C-32 and shown in figure V-C-26.

Table V-C-32. Rate coefficients (cm^3 molecule^{-1} s^{-1}) for the reaction of OH with 3-methyl-1-butanol [$(CH_3)_2CHCH_2CH_2OH$]

$10^{12} \times A$	B (K)	$10^{12} \times k$	T (K)	Technique[a]	Reference	Temperature Range (K)
		13.1 ± 2.6	298	DF–LIF	Saunders et al. (1994)	
		13.8 ± 0.5	295 ± 2	RR (C_3H_8)[b]	Wu et al. (2003)	
		13.0 ± 1.1		RR (c-C_6H_{12})[c]		
		13.0 ± 1.3	298 ± 2	RR (1-butanol)[d]	Mellouki et al. (2004)	
2.8 ± 0.9	−503 ± 98	19.1 ± 0.9	263	PLP–LIF	Mellouki et al. (2004)	263–368
		17.7 ± 0.2	272			
		15.0 ± 1.0	297 ± 1			
		12.8 ± 0.3	322			
		11.6 ± 0.3	353			
		11.3 ± 0.4	368			

[a] Abbreviations used for the techniques are defined in table II-A-1.
[b] Placed on an absolute basis using $k(C_3H_8) = 1.08 \times 10^{-12}$ cm^3 molecule^{-1} s^{-1}.
[c] Placed on an absolute basis using k(cyclohexane) $= 6.85 \times 10^{-12}$ cm^3 molecule^{-1} s^{-1}.
[d] Placed on an absolute basis using k(1-butanol) $= 8.5 \times 10^{-12}$ cm^3 molecule^{-1} s^{-1}.

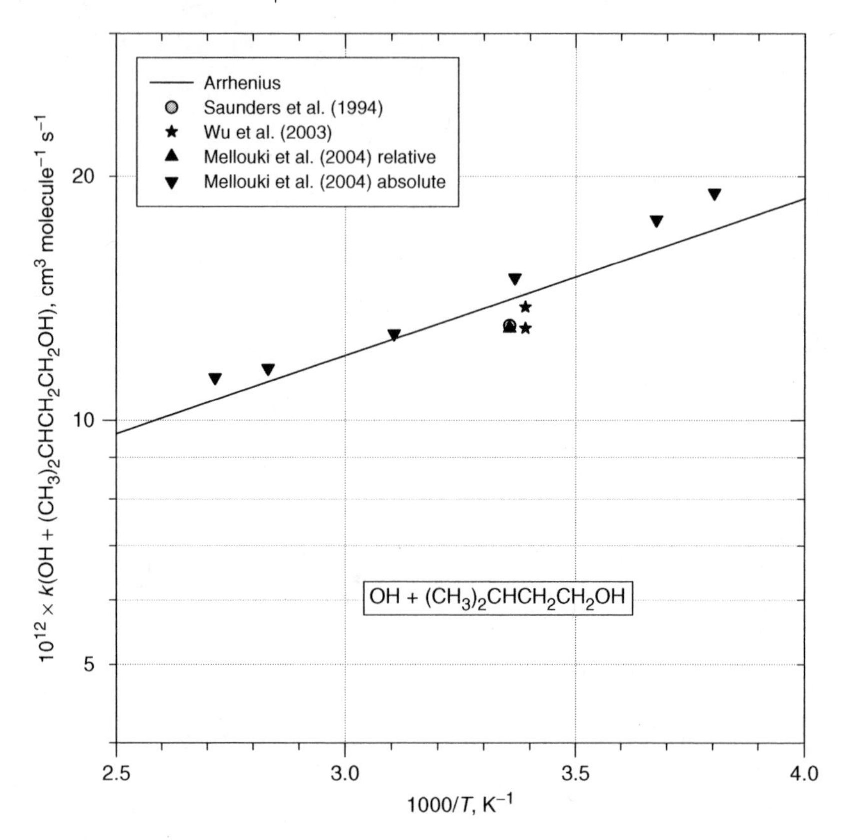

Figure V-C-26. Arrhenius plot of the rate coefficient for the reaction of OH with 3-methyl-1-butanol.

The measurements agree quite well, and we recommend $k = (1.41 \pm 0.20) \times 10^{-11}$ cm^3 $molecule^{-1}$ s^{-1} at 298 K. A fit to all the data gives $k = 3.1 \times 10^{-12}$ $\exp(445/T)$ cm^3 $molecule^{-1}$ s^{-1}. No product studies have been reported.

V-C-3.13. OH + 2-Methyl-2-butanol [tert-Amyl Alcohol, $(CH_3)_2C(OH)CH_2CH_3$]

This reaction has been studied by Jiménez at al. (2005b) who measured the rate coefficient as a function of temperature (263–354 K), using laser flash photolysis with pulsed laser-induced fluorescence. The measurements are given in table V-C-33, and are quite scattered (figure V-C-27). An Arrhenius fit to their data gives $k = 1.98 \times 10^{-12}$ $\exp(160/T)$ cm^3 $molecule^{-1}$ s^{-1}, leading to a k value at 298 K of 3.40×10^{-12} cm^3 $molecule^{-1}$ s^{-1}, with an uncertainty of ±35%.

V-C-3.14. OH + 3-Methyl-2-butanol [$CH_3CH(OH)CH(CH_3)_2$]

The reaction of OH with 3-methyl-2-butanol has been studied at 298 K by Wallington et al. (1988b) and by Mellouki et al. (2004), who used a relative technique at 298 K and

Table V-C-33. Rate coefficients (k, cm^3 $molecule^{-1}$ s^{-1}) for the reaction of OH with 2-methyl-2-butanol [$(CH_3)_2C(OH)CH_2CH_3$]

$10^{12} \times A$	B (K)	$10^{12} \times k$	T (K)	Technique[a]	Reference	Temperature Range (K)
2.37 ± 70	115 ± 210	3.91 ± 0.98	263	PLP–LIF	Jiménez et al. (2005b)	263–354
		3.58 ± 0.34	266			
		3.39 ± 0.80	268			
		3.89 ± 1.10	278			
		3.85 ± 0.80	283			
		3.15 ± 0.80	288			
		3.64 ± 0.60	298			
		3.00 ± 0.60	303			
		3.94 ± 0.32	313			
		2.84 ± 0.98	321			
		3.67 ± 0.24	334			
		2.78 ± 0.06	343			
		3.78 ± 0.52	354			

[a] Abbreviations used for the techniques are defined in table II-A-1.

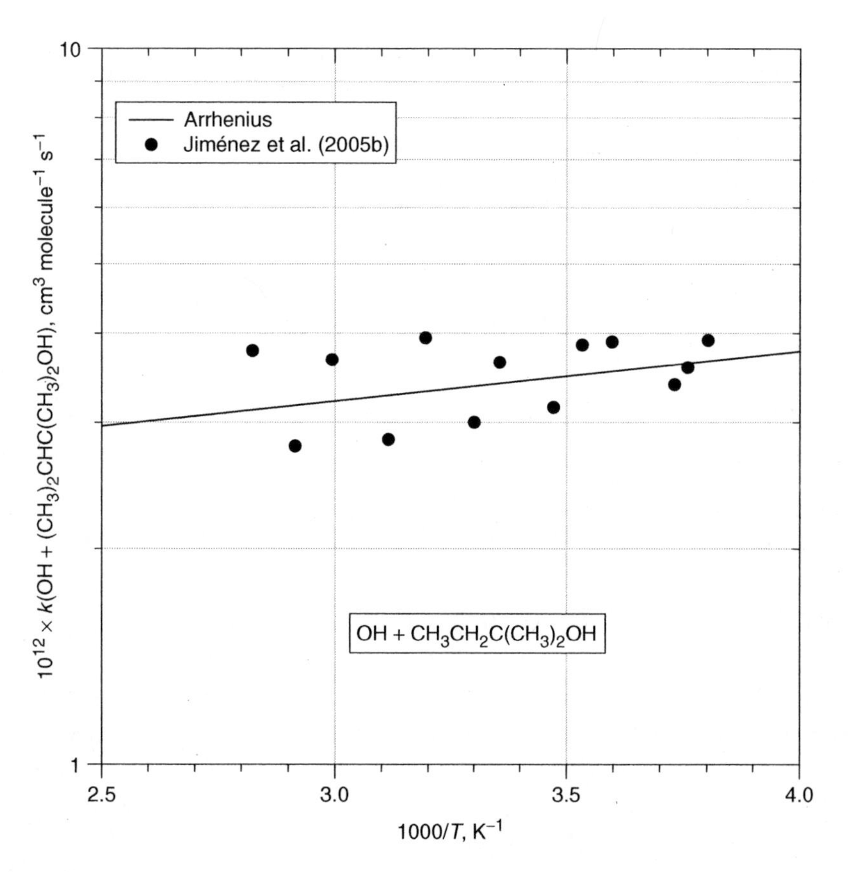

Figure V-C-27. Arrhenius plot of the rate coefficient for the reaction of OH with 2-methyl-2-butanol.

Table V-C-34. Rate coefficients (k, cm^3 $molecule^{-1}$ s^{-1}) for the reaction of OH with 3-methyl-2-butanol [$CH_3CH(OH)CH(CH_3)_2$]

$10^{12} \times A$	B (K)	$10^{12} \times k$	T (K)	Technique[a]	Reference	Temperature Range (K)
		12.4 ± 0.7	298	FP–RF	Wallington et al. (1988b)	
		12.2 ± 0.3	298	RR (1-butanol)[b]	Mellouki et al. (2004)	
2.6 ± 0.6	456 ± 65	15.2 ± 0.8	254	PLP–LIF	Mellouki et al. (2004)	254–373
		14.2 ± 1.2	263			
		14.1 ± 0.6	273			
		12.6 ± 0.5	283			
		11.9 ± 0.6	298			
		10.6 ± 0.4	326			
		9.5 ± 0.4	352			
		9.5 ± 0.5	366			
		7.9 ± 0.8	373			

[a] Abbreviations used for the techniques are defined in table II-A-1.
[b] Placed on an absolute basis using k(1-butanol) = 8.5×10^{-12} cm^3 $molecule^{-1}$ s^{-1}.

an absolute technique as a function of temperature. The measurements, which are in excellent agreement, are given in table V-C-34 and shown in figure V-C-28. On the basis of these studies we recommend $k = 2.5 \times 10^{-12} \exp(464/T)$ cm^3 $molecule^{-1}$ s^{-1}, with a rate coefficient $k = 1.19 \times 10^{-11}$ cm^3 $molecule^{-1}$ s^{-1} at 298 K (uncertainty ±15%). No product studies have been reported.

V-C-3.15. OH + 1-Hexanol [n-Hexyl Alcohol, $CH_3CH_2CH_2CH_2CH_2CH_2OH$]

There are six measurements of this reaction reported in the literature, and the values are somewhat scattered. The measurements are given in table V-C-35. The earlier measurements of Wallington et al. (1988b) and Nelson et al. (1990a) agree quite well, but the later values of Aschmann and Atkinson (1998) and Oh and Andino (2001) are higher. As in the case of 1-butanol and 1-pentanol, Oh and Andino's measurement, made relative to *p*-xylene, appears to be high. The unweighted mean of the other values is $(1.30 \pm 0.25) \times 10^{-11}$ cm^3 $molecule^{-1}$ s^{-1} at 298 K. No product studies have been published.

V-C-3.16. OH + 2-Hexanol [$CH_3CH(OH)CH_2CH_2CH_2CH_3$]

The reaction of OH with 2-hexanol has been studied once at 298 K by Wallington et al. (1988b), using flash photolysis with resonance fluorescence detection of OH. On the basis of this study we recommend $k = (1.2 \pm 0.3) \times 10^{-11}$ cm^3 $molecule^{-1}$ s^{-1} at 298 K. No product studies have been reported.

V-C-3.17. OH + 2-Methyl-2-pentanol [$CH_3C(CH_3)(OH)CH_2CH_2CH_3$]

The study of this reaction by Bethel et al. (2003) is the only one in the literature. The rate coefficient determined was (7.11 ± 0.53) at 298 ± 2 K, using a relative rate

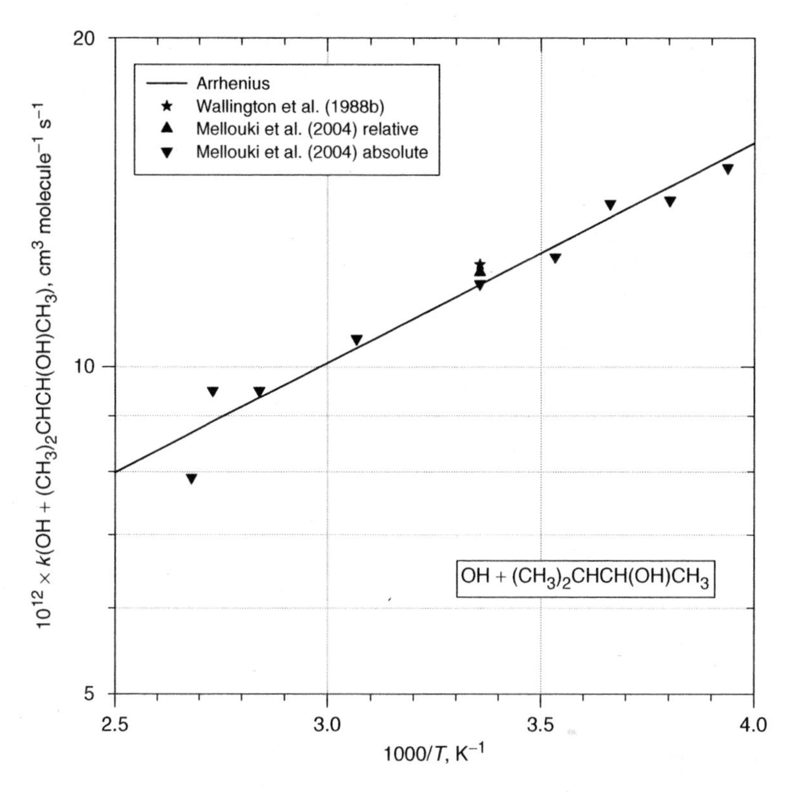

Figure V-C-28. Arrhenius plot of the rate coefficient for the reaction of OH with 3-methyl-2-butanol.

Table V-C-35. Rate coefficients (k, cm^3 $molecule^{-1}$ s^{-1}) for the reaction of OH with 1-hexanol (n-$C_6H_{13}OH$)

$10^{12} \times k$	T (K)	Technique[a]	Reference
12.4 ± 0.7	296	FP–RF	Wallington et al. (1988b)
12.2 ± 2.4	298 ± 2	PR–UVabs	Nelson et al. (1990a)
11.8 ± 1.1		RR (c-C_6H_{12})[b]	
15.1 ± 1.5	298	RR (c-C_6H_{12})[b]	Aschmann and Atkinson (1998)
13.6 ± 0.7	298 ± 2	RR (c-C_6H_{12})[b]	Oh and Andino (2001)
17.6 ± 1.1		RR (p-xylene)[c]	

[a] Abbreviations used for the techniques are defined in table II-A-1.
[b] Placed on an absolute basis using k(cyclohexane) = 6.85×10^{-12} cm^3 $molecule^{-1}$ s^{-1}.
[c] Placed on an absolute basis using k(p-xylene) = 1.43×10^{-11} cm^3 $molecule^{-1}$ s^{-1}.

method with octane as the reference gas. On the basis of that study we recommend $k = (7.1 \pm 1.4) \times 10^{-12}$ cm^3 $molecule^{-1}$ s^{-1} at 298 K, where the rate coefficient has been normalized to the value given for octane in chapter II, and the uncertainty is ±20%. The identified products of the reaction are acetaldehyde (22 ± 3%), propanal (35 ± 4%), acetone (56 ± 6%), and 2-pentanone (5 ± 1%), along with ~8% hydroxynitrates.

Acetone appears as a coproduct of both acetaldehyde and propanal, and its yield equals the sum of those molecules. The results appear to be consistent with greater than 35% attack at the 3-position, and greater than 22% at the 4-position.

V-C-3.18. OH + 4-Methyl-2-pentanol [$CH_3CH(OH)CH_2CH(CH_3)_2$]

The study of this reaction by Bethel et al. (2003) is the only one in the literature. The rate coefficient, measured relative to octane, was $(1.71 \pm 0.9) \times 10^{-11}$ cm^3 molecule^{-1} s^{-1}. On the basis of that study we recommend $(1.7 \pm 0.3) \times 10^{-11}$ cm^3 molecule^{-1} s^{-1} at 298 K, where the rate coefficient has been normalized to the value given for octane in chapter II, and the uncertainty is ±20%. The identified products of the reaction are acetaldehyde (37 ± 6%), acetone (26 ± 3%), 2-methylpropanal (11 ± 1%), and 4-methyl-2-pentanone (25 ± 2%), along with ~7% of hydroxynitrates. The 4-methyl-2-pentanone should arise exclusively from attack at the 2-position. Acetaldehyde appears as a coproduct of both acetone and 2-methylpropanal, and its yield equals the sum of those molecules. The results appear to be consistent with 25% attack at the 2-position, 11% attack at the 3-position, and greater than 25% at the 4-position. The alkoxy radical at the 4-position is expected to isomerize, in addition to decomposing to give acetone, acetaldehyde and formaldehyde. The observed products indicate more than twice as much abstraction at the 4-position than is predicted by the SAR relationship, indicating an activation effect extending at least three carbon atoms down the chain (Bethel et al., 2001, 2003).

V-C-3.19. OH + 2,3-Dimethyl-2-butanol [$(CH_3)_2C(OH)CH(CH_3)_2$]

This reaction has been studied by Jiménez at al. (2005b), who measured the rate coefficient as a function of temperature (268–354 K), using laser flash photolysis with pulsed laser-induced fluorescence. The measurements are given in table V-C-36. A fit to the data is shown in figure V-C-29, yielding $1.45 \times 10^{-12} \exp(554/T)$ cm^3 molecule^{-1} s^{-1}.

Table V-C-36. Rate coefficients (k, cm^3 molecule^{-1} s^{-1}) for the reaction of OH with 2,3-dimethyl-2-butanol [$(CH_3)_2C(OH)CH(CH_3)_2$]

$10^{12} \times k$	T (K)	Technique[a]	Reference	Temperature Range (K)
12.3 ± 1.42	268	PLP–LIF	Jiménez et al. (2005b)[b]	268–354
11.4 ± 0.22	273			
10.9 ± 0.66	278			
9.65 ± 0.76	288			
9.01 ± 0.90	292			
9.08 ± 1.00	298			
8.39 ± 1.74	303			
8.30 ± 0.70	321			
7.34 ± 1.20	338			
7.51 ± 0.90	354			

[a] Abbreviations used for the techniques are defined in table II-A-1.
[b] $k = 5.60 \times 10^{-12} + 1.52 \times 10^{-14} e^{(1620/T)}$.

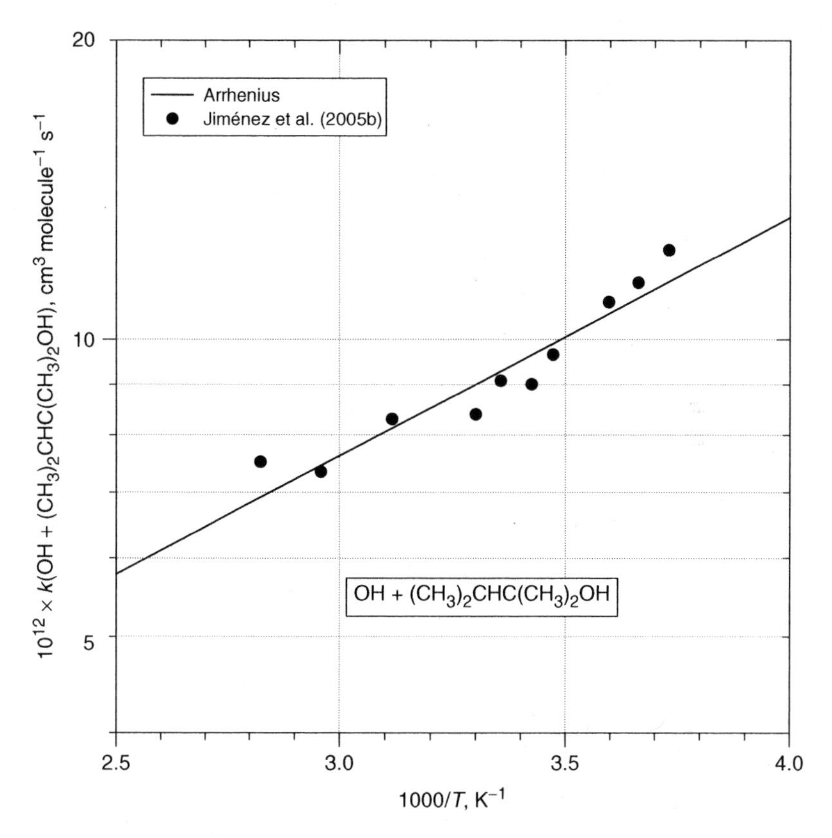

Figure V-C-29. Arrhenius plot of the rate coefficient for the reaction of OH with 2,3-dimethyl-2-butanol.

At 298 K the rate coefficient $k = 9.3 \times 10^{-12}$ cm^3 $moleclule^{-1}$ s^{-1}, with an uncertainty of $\pm 20\%$.

V-C-3.20. OH + 1-Heptanol (n-Heptyl Alcohol, n-$C_7H_{15}OH$)

The room temperature measurements of Wallington et al. (1988b) and Nelson et al. (1990a) are in good agreement, within $\pm 4\%$. The measurements are given in table V-C-37. We recommend a rate coefficient at 298 K of $(1.3 \pm 0.2) \times 10^{-11}$ cm^3 $molecule^{-1}$ s^{-1}, the unweighted mean of the two determinations. No product studies are available.

V-C-3.21. OH + 2,4-Dimethyl-2-pentanol [$(CH_3)_2C(OH)CH_2CH(CH_3)_2$]

The single study of this reaction is by Atkinson and Aschmann (1995), who measured $k = (1.1 \pm 0.3) \times 10^{-11}$ cm^3 $molecule^{-1}$ s^{-1} at 296 ± 2 K, relative to the value for hexane derived in chapter II. Products observed were acetone ($92 \pm 15\%$), 2-methylpropanal ($21 \pm 22\%$), 4-methyl-2-pentanone ($5 \pm 1\%$) and 4-hydroxy-4-methyl-2-pentanone ($12 \pm 2\%$).

Table V-C-37. Rate coefficients (k, cm^3 $molecule^{-1}$ s^{-1}) for the reaction of OH with 1-heptanol (n-$C_7H_{15}OH$)

$10^{12} \times k$	T (K)	Technique[a]	Reference
13.6 ± 1.3	298	FP–RF	Wallington et al. (1988b)
12.5 ± 1.4	298 ± 2	RR (c-C_6H_{12})[b]	Nelson et al. (1990a)

[a] Abbreviations used for the techniques are defined in table II-A-1.
[b] Placed on an absolute basis using k(cyclohexane) = 6.85×10^{-12} cm^3 $molecule^{-1}$ s^{-1}.

V-C-3.22. OH + 1-Octanol (n-$C_8H_{17}OH$)

The only determination of this rate coefficient is that of Nelson et al. (1990a), who measured $k = (1.32 \pm 0.15) \times 10^{-11}$ cm^3 $molecule^{-1}$ s^{-1} at 298 ± 2 K, using the relative rate technique with cyclohexane as the reference compound.

V-C-3.23. OH + 3,5-Dimethyl-3-hexanol [$CH_3CH_2C(CH_3)(OH)CH_2CH(CH_3)_2$]

The single study of this reaction is by Atkinson and Aschmann (1995), who measured $k = (1.3 \pm 0.4) \times 10^{-11}$ cm^3 $molecule^{-1}$ s^{-1} at 296 ± 2 K relative to the value for hexane derived in chapter II. Products observed were acetone ($12 \pm 3\%$), 2-butanone ($28 \pm 2\%$), 2-methylpropanal ($17 \pm 2\%$), 4-methyl-2-pentanone ($16 \pm 1\%$), and 4-hydroxy-4-methyl-2-pentanone ($25 \pm 2\%$).

V-C-4. Reactions of Hydroxy Radicals with Methyl Hydroperoxide

Methyl hydroperoxide is the only organic peroxide for which kinetics with the OH radical have been measured (Niki et al., 1983; Vaghjiani and Ravishankara, 1989a; Blitz et al., 2005). The measurements are summarized in table V-C-38. Niki et al. measured the overall rate coefficient, and carried out product studies to determine the branching ratio for the reaction:

$$OH + CH_3OOH \rightarrow H_2O + CH_3O_2 \tag{a}$$

$$OH + CH_3OOH \rightarrow H_2O + (CH_2OOH) \rightarrow H_2O + CH_2O + OH \tag{b}$$

Niki et al. (1983) found an overall rate coefficient of 1.0×10^{-11} cm^3 $molecule^{-1}$ s^{-1}, using the relative rate technique, and a branching ratio $k_a/k_b = 1.30 \pm 0.25$. Vaghjiani and Ravishankara (1989a) followed the loss of OH directly. By the use of isotopically substituted OH, they determined both the overall rate coefficient ($k_a + k_b$) and that for channel (a), which does not regenerate OH. Their expression for the overall rate coefficient, $k = (2.93 \pm 0.30) \times 10^{-12} \exp((190 \pm 14)/T)$ leads to a value at 298 K of $k = 5.54 \times 10^{-12}$ cm^3 $molecule^{-1}$ s^{-1}. Their branching ratio at room temperature, calculated from the individual rate coefficients, is $k_a/k_b = 2.2 \pm 0.4$. Thus, both the

Table V-C-38. Rate coefficients (k, cm^3 $molecule^{-1}$ s^{-1}) for the reaction of OH with methyl hydroperoxide (CH_3OOH)

$10^{12} \times A$	B (K)	$10^{12} \times k$	T (K)	Technique[a]	Reference	Temperature Range (K)
Overall reaction[c]						
		10	298	RR (C_2H_4)[b]	Niki et al. (1983)	
2.93 ± 0.30	−190 ± 14	6.93 ± 0.26	223	PLP–LIF	Vaghjiani and Ravishankara (1989a)	223–348
		6.45 ± 0.20	244			
		6.29 ± 0.23	249			
		5.38 ± 0.20	298			
		4.97 ± 0.48	348			
		5.06 ± 0.14	373			
		4.61 ± 0.41	423			
Reaction (a)[c]						
		5.6 ± 1.5	298	FTIR	Niki et al. (1983)	
1.78 ± 0.25	−220 ± 21	5.13 ± 0.19	203	PLP–LIF	Vaghjiani and Ravishankara (1989a)	203–348
		5.00 ± 0.29	223			
		4.33 ± 0.54	244			
		3.85 ± 0.23	298			
		3.29 ± 0.32	348			
		9.0 ± 0.2	295	PLP–LIF	Blitz et al. (2005)	

[a] Abbreviations used for the techniques are defined in table II-A-1.
[b] Placed on an absolute basis using $k(C_2H_4) = 8.52 \times 10^{-12}$ cm^3 $molecule^{-1}$ s^{-1} at 1 bar and 298 K.
[c] $OH + CH_3OOH \rightarrow H_2O + CH_3O_2$ (a)
$OH + CH_3OOH \rightarrow H_2O + (CH_2OOH) \rightarrow H_2O + CH_2O + OH$ (b)

overall rate coefficient and the individual channels are in some disagreement with the work of Niki et al. Recently, Blitz et al. (2005) have reported the rate coefficient for channel (a) as $(9.0 \pm 0.2) \times 10^{-12}$ cm^3 $molecule^{-1}$ s^{-1} at 295 K, which is more than a factor of 2 larger than that of Vaghjiani and Ravishankara. Clearly, the rate coefficient for this reaction has not yet been fully defined, and further work is required.

V-C-5. Reactions of OH with Alkyl Nitrates

Rate coefficients have been reported in the literature for the reactions of OH radicals with a number of simple aliphatic alkyl nitrates. More recently, data have been obtained for difunctional hydroxynitrates. The simple alkyl nitrates will be discussed in this section, and hydroxynitrates in section V-C-7. The nitrate group is strongly electron withdrawing, and the rate coefficients for reaction with alkyl nitrates are much slower than those for the corresponding alcohols. The deactivation at the carbon atom bearing the nitrate group is the largest, being a factor of ~25 compared to abstraction at an unsubstituted carbon atom (Becker and Wirtz, 1989, Kwok and Atkinson, 1995). The deactivation at the β-carbon atom is roughly a factor of 5. Temperature-dependent data are available for only the smallest nitrates, which have been studied by a combination of both direct and relative techniques. The slowness of the reactions makes both absolute

Table V-C-39. Rate coefficients (k, cm^3 $molecule^{-1}$ s^{-1}) for the reaction of OH with methyl nitrate (CH_3ONO_2)

$10^{14} \times A$	B (K)	$10^{14} \times k$	T (K)	Technique[a]	Reference	Temperature Range
		3.4 ± 0.4	298	DF–RF	Gaffney et al. (1986)	
		37 ± 9	303	RR (C_2H_4)[b]	Kerr and Stocker (1986)	
		8.5 ± 1.6	298 ± 2	RR ($CH(CH_3)_3$)[c]	Nielsen et al. (1991)	298–393
0.88	-1050 ± 180	32 ± 5	298	PR–UVabs		
		22 ± 6	323			
		15 ± 3	358			
		14 ± 3	393			
82 ± 16	1020 ± 60	0.89 ± 0.02	221	PLP–LIF	Talukdar et al. (1997b)	221–414
		1.06 ± 0.05	233			
		1.43 ± 0.06	254			
		1.91 ± 0.10	273			
		2.33 ± 0.14	298			
		3.13 ± 0.08	324			
		3.91 ± 0.16	348			
		5.11 ± 0.34	373			
		6.56 ± 0.16	394			
		7.69 ± 0.20	414			
41 ± 8	604 ± 121	4.8 ± 0.8	298	DF–RF	Shallcross et al. (1997)	298–423
		7.7 ± 1.5	333			
		8.8 ± 2.1	373			
		8.0 ± 1.8	398			
		10.1 ± 2.0	423			
		3.0 ± 0.32	307	RR (several refs.)	Kakesu et al. (1997)	

[a] Abbreviations used for the techniques are defined in table II-A-1.
[b] Placed on an absolute basis using $k(C_2H_4) = 8.52 \times 10^{-12}$ cm^3 $molecule^{-1}$ s^{-1} at 1 bar and 298 K.
[c] Placed on an absolute basis using $k(CH(CH_3)_3) = 2.12 \times 10^{-12}$ cm^3 $molecule^{-1}$ s^{-1}.

and relative studies difficult, and large uncertainties remain for the rate coefficients of most of the reactions. For the reactions of OH radicals with alkyl nitrates, we have largely followed the recommendations of the IUPAC panel (Atkinson et al., 2005b).

V-C-5.1. OH + Methyl Nitrate (Nitrooxymethane, CH_3ONO_2)

Several measurements of this rate coefficient exist, but there is quite a large spread in the data (see table V-C-39 and figure V-C-30). Absolute measurements were made by Gaffney et al. (1986), Nielsen et al. (1991), Talukdar et al. (1997b), and Shallcross et al. (1997). The last three groups also measured temperature dependences. Relative-rate measurements were made by Kerr and Stocker (1986), Nielsen et al. (1991), and Kakesu et al. (1997). The lowest measured values, those of Gaffney et al., Talukdar et al. and Kakesu et al., are in reasonable agreement, and the recommendation of Atkinson et al. (2005b) is based on a unit-weighted fit of the data of Talukdar et al. between 221 and 298 K: $k = 4.0 \times 10^{-13} \exp(-845/T)$ cm^3 $molecule^{-1}$ s^{-1}, with an

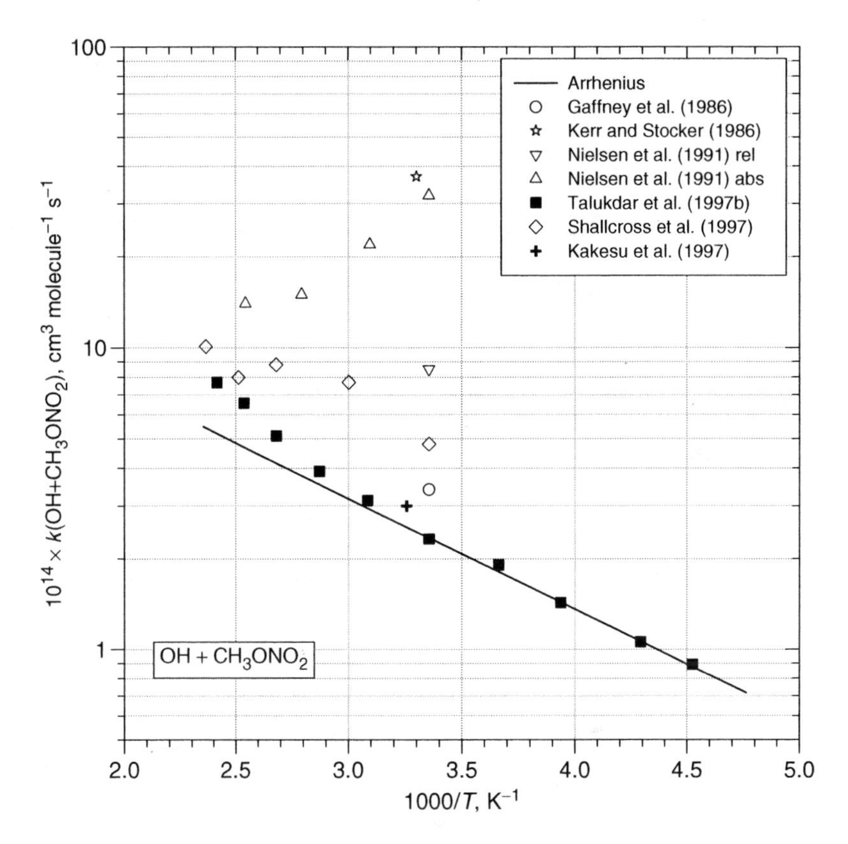

Figure V-C-30. Arrhenius plot of the rate coefficient for the reaction of OH with methyl nitrate.

uncertainty of roughly a factor of 2. The measurements of Nielsen et al. (1991) and Kerr and Stocker (1986) lie considerably above those of Talukdar et al. (1997b) while the absolute measurements of Shallcross et al. (1997) are roughly a factor of 2 higher. It should be noted that, while Nielsen et al. reported a rate coefficient of 3.4×10^{-13} cm^3 molecule^{-1} s^{-1} using the relative-rate technique, the data in their figure 2a indicate a value of 8.5×10^{-14} cm^3 molecule^{-1} s^{-1}. It is likely that the early relative rate techniques had difficulties in measuring a rate coefficient that small, and/or that reference compounds that were too reactive were used. Talukdar also examined the kinetics of various isotopically labeled species, and concluded that the overall reaction proceeds by hydrogen abstraction, with no pressure dependence.

V-C-5.2. OH + Ethyl Nitrate (Nitrooxyethane, $C_2H_5ONO_2$)

The rate coefficient for the reaction of OH with ethyl nitrate has been measured by Kerr and Stocker (1986), Nielsen et al. (1991), who made both an absolute and a relative rate study, by Talukdar et al. (1997b), Shallcross et al. (1997), and Kakesu et al. (1997). The data are summarized in table V-C-40. Temperature-dependent rate coefficients have been reported by Nielsen et al., Talukdar et al., and Shallcross et al. The results of

Table V-C-40. Rate coefficients (k, cm^3 $molecule^{-1}$ s^{-1}) for the reaction of OH with ethyl nitrate ($C_2H_5ONO_2$)

$10^{13} \times A$	B (K)	$10^{13} \times k$	T (K)	Technique[a]	Reference	Temperature Range
		4.9 ± 2.1	303	RR (C_2H_4)[b]	Kerr and Stocker (1986)	
0.47	−716 ± 138	4.2 ± 0.3	298 ± 2	RR ($CH(CH_3)_3$)[c]	Nielsen et al. (1991)	298–393
		5.3 ± 0.6	298	PR–UVabs		
		3.7 ± 0.5	338			
		3.3 ± 0.5	373			
$k(T)$[d]		1.26 ± 0.04	223	PLP–LIF	Talukdar et al. (1997b)	223–394
		1.25 ± 0.04	233			
		1.37 ± 0.06	252			
		1.58 ± 0.02	272			
		1.77 ± 0.08	297 ± 2			
		2.10 ± 0.05	323			
		2.44 ± 0.10	348			
		2.81 ± 0.06	374			
		3.13 ± 0.04	394			
0.33 ± 0.07	699 ± 140	3.2 ± 0.5	298	DF–RF	Shallcross et al. (1997)	298–373
		3.2 ± 0.6	313			
		5.1 ± 0.9	373			
		2.2 ± 0.8	304 ± 6	RR (several refs.)	Kakesu et al. (1997)	

[a] Abbreviations used for the techniques are defined in table II-A-1.
[b] Placed on an absolute basis using $k(C_2H_4) = 8.52 \times 10^{-12}$ cm^3 $molecule^{-1}$ s^{-1} at 1 bar and 298 K.
[c] Placed on an absolute basis using $k(CH(CH_3)_3) = 2.12 \times 10^{-12}$ cm^3 $molecule^{-1}$ s^{-1}.
[d] $k(T) = 3.68 \times 10^{-12}\exp(-1077/T) + 5.32 \times 10^{-14}\exp(126/T)$.

Talukdar et al. and Kakesu et al. are in very good agreement at room temperature, and again are the lowest measured values (figure V-C-31). Over the entire range studied, 223–394 K, Talukdar et al. observed a curved Arrhenius plot. Atkinson et al. (1982b) recommend an Arrhenius fit to the data of Talukdar below room temperature, leading to $k = 6.7 \times 10^{-13} \exp(-395/T)$ cm^3 $molecule^{-1}$ s^{-1}, with an uncertainty of roughly a factor of 2, as shown in the figure. This expression corresponds to a room-temperature rate coefficient of 1.8×10^{-13} cm^3 $molecule^{-1}$ s^{-1}. Talukdar also examined the kinetics of various isotopically labeled species, and concluded that the overall reaction proceeds by hydrogen abstraction, with no pressure dependence.

V-C-5.3. OH + 1-Propyl Nitrate (1-Nitrooxypropane, n-$C_3H_7ONO_2$)

The reaction of OH with 1-propyl nitrate has been studied by Kerr and Stocker (1986), Atkinson and Aschmann (1989), and Nielsen et al. (1991). Nielsen et al. reported both relative and absolute rate coefficients, the latter of which were measured as a function of temperature (table V-C-41). The results are in fair agreement, and Atkinson et al. (2005b) recommend the use of the value measured by Atkinson and Aschmann (1989), $k = (5.8 \pm 1.0) \times 10^{-13}$ cm^3 $molecule^{-1}$ s^{-1}, although an unweighted mean of

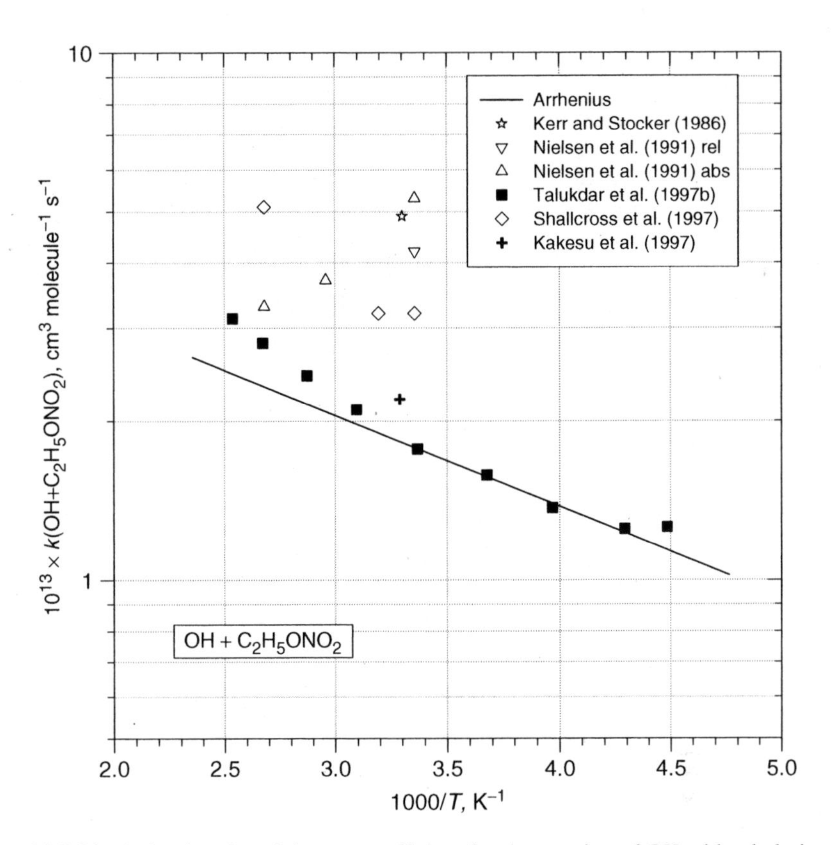

Figure V-C-31. Arrhenius plot of the rate coefficient for the reaction of OH with ethyl nitrate.

Table V-C-41. Rate coefficients (k, cm^3 $molecule^{-1}$ s^{-1}) for the reaction of OH with 1-propyl nitrate (n-$C_3H_7ONO_2$)

$10^{13} \times A$	B (K)	$10^{13} \times k$	T (K)	Technique[a]	Reference	Temperature Range
		7.2 ± 2.3	303	RR (C_2H_4)[b]	Kerr and Stocker (1986)	
		5.8 ± 1.0	298	RR (c-C_6H_{12})[c]	Atkinson and Aschmann (1989)	
5.0	−(140 ± 144)	7.0 ± 0.8	298 ± 2	RR ($CH(CH_3)_3$)[d]	Nielsen et al. (1991)	298–368
		8.2 ± 0.8	298	PR–UV abs		
		7.2 ± 1.5	338			
		7.6 ± 0.5	368			

[a] Abbreviations used for the techniques are defined in table II-A-1.
[b] Placed on an absolute basis using $k(C_2H_4) = 8.52 \times 10^{-12}$ cm^3 $molecule^{-1}$ s^{-1} at 1 bar and 298 K.
[c] Placed on an absolute basis using k(cyclohexane) $= 6.85 \times 10^{-12}$ cm^3 $molecule^{-1}$ s^{-1}.
[d] Placed on an absolute basis using $k(CH(CH_3)_3) = 2.12 \times 10^{-12}$ cm^3 $molecule^{-1}$ s^{-1}.

Table V-C-42. Rate coefficients (k, cm^3 $molecule^{-1}$ s^{-1}) for the reaction of OH with 2-propyl nitrate [$(CH_3)_2CHONO_2$]

$10^{13} \times k$	T (K)	Technique[a]	Reference	Temperature Range
1.68 ± 0.42	299 ± 2	RR (c-C_6H_{12})[b]	Atkinson et al. (1982b)	
5.4 ± 2.1	295 ± 2	RR (n-C_4H_{10})[c]	Becker and Wirtz (1989)	
3.83 ± 0.49	298 ± 2	RR (c-C_6H_{12})[b]	Atkinson and Aschmann (1989)	
2.33 ± 0.10	233	PLP–LIF	Talukdar et al. (1997b)[d]	233–395
2.38 ± 0.08	253			
2.66 ± 0.16	272			
2.88 ± 0.19	298 ± 2			
3.13 ± 0.12	325			
3.53 ± 0.06	350			
3.82 ± 0.14	375			
4.13 ± 0.14	395			

[a] Abbreviations used for the techniques are defined in table II-A-1.
[b] Placed on an absolute basis using k(cyclohexane) $= 6.85 \times 10^{-12}$ cm^3 $molecule^{-1}$ s^{-1}.
[c] Placed on an absolute basis using $k(n\text{-}C_4H_{10}) = 2.36 \times 10^{-12}$ cm^3 $molecule^{-1}$ s^{-1}.
[d] $k = 4.3 \times 10^{-12}\exp(-1250/T) + 2.5 \times 10^{-13}\exp(-32/T)$.

the four determinations gives $(7.0 \pm 1.0) \times 10^{-13}$ cm^3 $molecule^{-1}$ s^{-1}, which agrees with the recommendation of Atkinson et al. within the combined uncertainties.

Nielsen et al. (1991) measured a weak negative temperature dependence for the rate coefficient (indepedent of temperature within experimental error). By analogy with the other nitrates studied by Talukdar et al. (1997b) and Shallcross et al. (1997), a weak positive temperature dependence is expected; consequently, no recommendation is made for this reaction.

V-C-5.4. OH + 2-Propyl Nitrate [2-Nitrooxypropane, $(CH_3)_2CHONO_2$]

The reaction of OH with 2-propyl nitrate has been studied by Atkinson et al. (1982b), Becker and Wirtz (1989), Atkinson and Aschmann (1989), and Talukdar et al. (1997b). The results are given in table V-C-42. The measurements of Talukdar et al. are the only direct study and the only measurements made as a function of temperature. There is a spread of about a factor of 3 in the 298 K values. The Arrhenius plot of Talukdar shows slight upward curvature between 233 and 395 K (figure V-C-32). Atkinson et al. (2005b) recommend an Arrhenius fit to the data of Talukdar et al. below 300 K, $k = 6.2 \times 10^{-13}$ $\exp(-230/T)$ cm^3 $molecule^{-1}$ s^{-1}, with an uncertainty of roughly a factor of 1.5 in the rate coefficient at 298 K and an uncertainty in E/R of ± 300 K. This expression corresponds to a room temperature rate coefficient of 2.9×10^{-13} cm^3 $molecule^{-1}$ s^{-1}.

V-C-5.5. OH + 1-Butyl Nitrate (1-Nitrooxybutane, n-$C_4H_9ONO_2$)

The reaction of OH with 1-butyl nitrate has been studied at ambient temperatures only by Atkinson et al. (1982b), Atkinson and Aschmann (1989), and Nielsen et al. (1991),

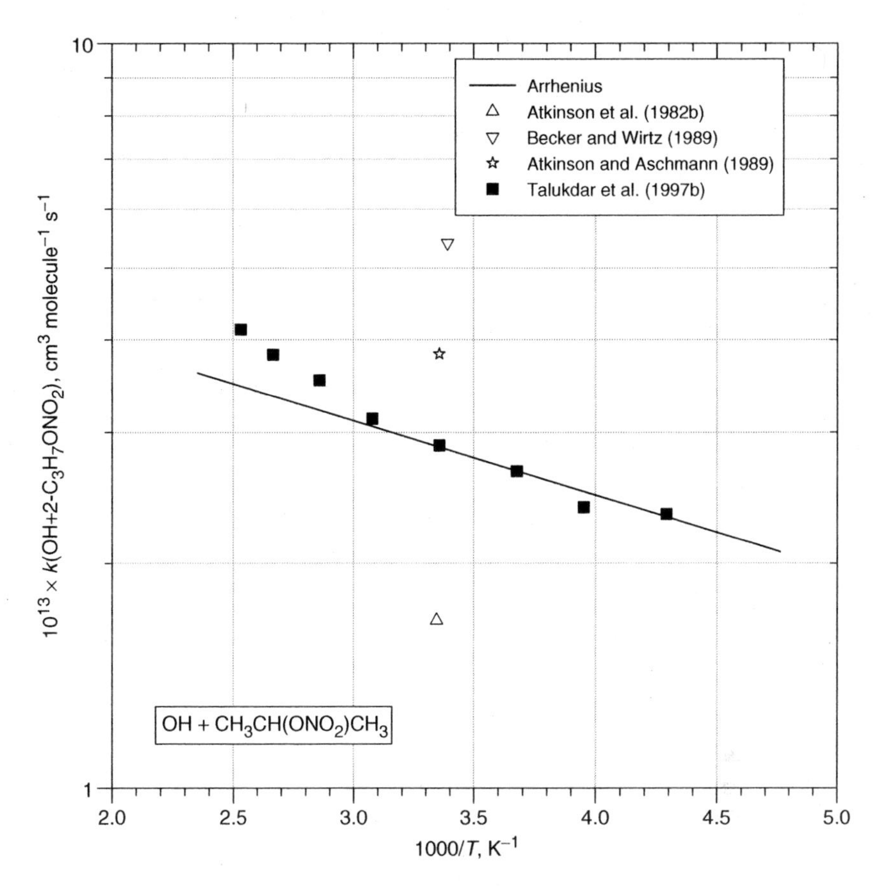

Figure V-C-32. Arrhenius plot of the rate coefficient for the reaction of OH with 2-propyl nitrate.

who used both an absolute and a relative technique (table V-C-43). The results are in good agreement. For ambient temperatures Atkinson et al. (2005b) recommend a rate coefficient of 1.6×10^{-12} cm^3 molecule^{-1} s^{-1}, with an uncertainty of ~50%. A weak temperature dependence is expected for this reaction.

V-C-5.6. OH + 2-Butyl Nitrate [2-Nitrooxybutane, $C_2H_5CH(ONO_2)CH_3$]

The reaction of OH with 2-butyl nitrate has been studied by Atkinson et al. (1982b) and Atkinson and Aschmann (1989) (table V-C-44). The determination of Atkinson and Aschmann supersedes the earlier measurement from that group. A rate coefficient of 8.6×10^{-13} cm^3 molecule^{-1} s^{-1} is recommended, with an uncertainty of ~50%. By analogy with the rate coefficient for the reaction of OH with 2-propyl nitrate, a small temperature coefficient below 298 K is expected.

V-C-5.7. Reaction of OH with Higher Alkyl Nitrates

Single determinations have been made at 298 ± 2 K for rate coefficients of reactions of OH with a series of alkyl nitrates up to 3-octyl nitrate. All measurements were made

Table V-C-43. Rate coefficients (k, cm^3 molecule^{-1} s^{-1}) for the reaction of OH with 1-butyl nitrate (n-$C_4H_9ONO_2$)

$10^{13} \times k$	T (K)	Technique[a]	Reference
13.1 ± 1.0	299 ± 2	RR (c-C_6H_{12})[b]	Atkinson et al. (1982b)
16.5 ± 1.8	298 ± 2	RR (c-C_6H_{12})[b]	Atkinson and Aschmann (1989)
14.7 ± 0.8	298 ± 2	RR ($CH(CH_3)_3$)[c]	Nielsen et al. (1991)
17.4 ± 1.9	298 ± 2	PR–UV abs	

[a] Abbreviations used for the techniques are defined in table II-A-1.
[b] Placed on an absolute basis using k(cyclohexane) = 6.85×10^{-12} cm^3 molecule^{-1} s^{-1}.
[c] Placed on an absolute basis using $k(CH(CH_3)_3) = 2.12 \times 10^{-12}$ cm^3 molecule^{-1} s^{-1}.

Table V-C-44. Rate coefficients (k, cm^3 molecule^{-1} s^{-1}) for the reaction of OH with 2-butyl nitrate [$C_2H_5CH(CH_3)ONO_2$]

$10^{13} \times k$	T (K)	Technique[a]	Reference
6.4 ± 1.0	299 ± 2	RR (c-C_6H_{12})[b]	Atkinson et al. (1982b)
8.6 ± 1.5	298 ± 2	RR (c-C_6H_{12})[b]	Atkinson and Aschmann (1989)

[a] Abbreviations used for the techniques are defined in table II-A-1.
[b] Placed on an absolute basis using k(cyclohexane) = 6.85×10^{-12} cm^3 molecule^{-1} s^{-1}.

using relative techniques. The measurements are summarized in table V-C-45. The rate coefficients increase with the size of the alkyl group, and the 2-substituted isomers have higher rate coefficients than the 3-substituted isomers. These trends are in accord with the deactivating effect of the nitrate group, which becomes less strong with increasing separation down the alkyl chain.

V-C-5.8. OH + Peroxyacetyl Nitrate [1-Nitroperoxyethanal, $CH_3C(O)O_2NO_2$, PAN]

A number of measurements of the rate coefficient for OH with $CH_3C(O)O_2NO_2$ have been made by Winer et al. (1977), Wallington et al. (1984), Tsalkani et al. (1988) and Talukdar et al. (1995). Through the years, the reported rate coefficients have gradually fallen, and the current recommendation is for an upper limit of 3×10^{-14} cm^3 molecule^{-1} s^{-1} at 298 K (Talukdar et al., 1995). The earlier measurements were presumably too high as a result of the presence of reactive impurities.

V-C-6. Reactions of Hydroxy Radicals with Hydroxycarbonyls

A number of hydroxycarbonyls are formed in the oxidation of alkanes. The species formed are predominantly 1,4-hydroxycarbonyls as a result of isomerization reactions. Some 1,3-hydroxycarbonyls are formed if the initial attack is at a tertiary site, in which case the oxy radical formed after the first isomerization cannot itself isomerize.

Table V-C-45. Rate coefficients (k, cm^3 $molecule^{-1}$ s^{-1}) for the reaction of OH with higher alkyl nitrates

Molecule	$10^{13} \times k$	T (K)	Technique[a]	Reference
2-Methyl-1-propyl nitrate	15.3 ± 1.8	295 ± 2	RR (n-C_4H_{10})[b]	Becker and Wirtz (1989)
1-Pentyl nitrate	26.8 ± 0.8	298 ± 2	RR ($CH(CH_3)_3$)[c]	Nielsen et al. (1991)
2-Pentyl nitrate	17.3 ± 1.1	299 ± 2	RR (c-C_6H_{12})[d]	Atkinson et al. (1982b)
3-Pentyl nitrate	11.3 ± 1.8	299 ± 2	RR (c-C_6H_{12})[d]	Atkinson et al. (1982b)
2-Methyl-1-butyl nitrate	23.4 ± 1.4	295 ± 2	RR (n-C_4H_{10})[b]	Becker and Wirtz (1989)
3-Methyl-1-butyl nitrate	23.8 ± 3.3	295 ± 2	RR (n-C_4H_{10})[b]	Becker and Wirtz (1989)
3-Methyl-2-butyl nitrate	18.2 ± 1.4	295 ± 2	RR (n-C_4H_{10})[b]	Becker and Wirtz (1989)
2-Hexyl nitrate	29.5 ± 1.5	298 ± 2	RR (c-C_6H_{12})[d]	Atkinson et al. (1982b)
3-Hexyl nitrate	25.1 ± 2.0	298 ± 2	RR (c-C_6H_{12})[d]	Atkinson et al. (1982b)
3-Heptyl nitrate	34.4 ± 4.0	298 ± 2	RR (c-C_6H_{12})[d]	Atkinson et al. (1982b)
3-Octyl nitrate	36.2 ± 7.4	298 ± 2	RR (c-C_6H_{12})[d]	Atkinson et al. (1982b)

[a] Abbreviations used for the techniques are defined in table II-A-1.
[b] Placed on an absolute basis using $k(n\text{-}C_4H_{10}) = 2.36 \times 10^{-12}$ cm^3 $molecule^{-1}$ s^{-1}.
[c] Placed on an absolute basis using $k(CH(CH_3)_3) = 2.12 \times 10^{-12}$ cm^3 $molecule^{-1}$ s^{-1}.
[d] Placed on an absolute basis using $k(\text{cyclohexane}) = 6.85 \times 10^{-12}$ cm^3 $molecule^{-1}$ s^{-1}.

Following a decomposition reaction, a 1,3-hydroxycarbonyl may be formed. The published data on both 1,3- and 1,4-hydoxycarbonyls are collected in table V-C-46. A number of these rate coefficients were obtained by following the production and secondary loss of the hydroxycarbonyl as a function of the loss of the parent hydrocarbon; e.g., see Baker et al. (2005).

1,4-Hydoxycarbonyls can also undergo a cyclization to form a hemiacetal, with subsequent elimination of water to form a dihydrofuran (Cavalli et al., 2000, Martin et al., 2002, Holt et al., 2005). The steps are reversible and depend on the concentration of water vapor. In the dihydrofuran state these molecules are very reactive with OH, NO_3, and O_3 (Martin et al., 2002). It is also possible that the dihydrofurans can lead to the formation or growth of aerosol particles in the atmosphere (Lim and Ziemann, 2005).

V-C-7. Reactions of Hydroxy Radicals with Hydroxynitrates

The oxidation of alkanes leads to alkyl nitrates in the reaction of alkylperoxy radicals with NO (see discussion of the mechanism in chapter IV-A). Following the isomerization of alkoxy radicals, 1,4-hydroxynitrates can be formed from the δ-hydroxy-alkylperoxy radicals (Arey et al., 2001, Reisen et al., 2005):

$$RO_2 + NO \rightarrow RO + NO_2$$

$$RO_2 + NO + M \rightarrow RONO_2 + M$$

Table V-C-46. Rate coefficients (k, cm^3 $molecule^{-1}$ s^{-1}) for the reaction of OH with 1,3- and 1,4-hydroxycarbonyls at 297 ± 3 K

Molecule	$10^{12} \times k$	Technique[a,b]	Reference
3-Hydroxypropanal	19.9 ± 2.9	RR (*cis*-3-hexen-1-ol)	Baker et al. (2004)
3-Hydroxybutanal	29.5 ± 2.4	RR (1,3-butanediol)	Baker et al. (2004)
4-Hydroxy-2-butanone	7.4 ± 0.7	RR (octane)	Aschmann et al. (2000)
	13.9 ± 2.8	RR (1,3-butanediol)	Baker et al. (2004)
5-Hydroxy-2-pentanone	16 ± 4	RR (4-methyl-2-pentanone)	Aschmann et al. (2003)
	15 ± 8	RR (pentane)	Baker et al. (2005)
4-Hydroxy-3-methyl, 2-pentanone	14.6 ± 0.7	RR (octane)	Aschmann et al. (2000)
5-Hydroxy-2-hexanone	21 ± 6	RR (hexane)	Baker et al. (2005)
6-Hydroxy-3-hexanone	16 ± 8	RR (hexane)	Baker et al. (2005)
4-Hydroxy-4-methyl-2-pentanone	3.8 ± 1.1	RR (butane)	Atkinson and Aschmann (1995)
	3.6 ± 0.6	RR (3 reference compounds)	Magneron et al. (2003)
	<4.2	RR (2-methyl-2,4-pentanediol)	Baker et al. (2004)
5-Hydroxy-2-heptanone	20 ± 7	RR (heptane)	Baker et al. (2005)
6-Hydroxy-3-heptanone	17 ± 7	RR (heptane)	Baker et al. (2005)
1-Hydroxy-4-heptanone	10 ± 7	RR (heptane)	Baker et al. (2005)
5-Hydroxy-2-octanone	24 ± 9	RR (octane)	Baker et al. (2005)
6-Hydroxy-3-octanone	16 ± 9	RR (octane)	Baker et al. (2005)
7-Hydroxy-4-octanone	20 ± 9	RR (octane)	Baker et al. (2005)

[a] Abbreviations used for the techniques are defined in table II-A-1.
[b] Analysis involves production and secondary removal of product; rate coefficient does not scale simply with that of the parent compound.

Rate coefficients for reactions of OH radicals with a series of 1,3- and 1,4-hydroxynitrates have been measured by Treves and Rudich (2003). All these data are collected in table V-C-47. In general, the agreement between the rate coefficients predicted using the parameters of Kwok and Atkinson (1995) for monofunctional alcohols and nitrates does not predict the values for the difunctional compounds well. An enhancement in the rate coefficient by the OH group is observed, as well as a decrease due to the nitrate group, but in cases where these effects are in opposition the effect is difficult to predict.

V-D. Reactions of Nitrate Radicals (NO_3) with Products of Alkane Oxidation

Nitrate radicals, NO_3, can be significant oxidants in the nighttime atmosphere. As a result of its large visible absorption cross section, NO_3 is subject to rapid photolysis and is not usually present in significant amounts during daylight hours. However, under

Table V-C-47. Rate coefficients (k, cm^3 $molecule^{-1}$ s^{-1}) for the reaction of OH with hydroxy nitrates

Molecule	$10^{12} \times k$	T (K)	Technique[a]	Reference
2-Nitrooxy-1-propanol	6.7 ± 1.3	296 ± 2	RR (1-pentanol, 1-octanol)[b]	Treves and Rudich (2003)
1-Nitrooxy-2-propanol	5.1 ± 1.0			
2-Nitrooxy-1-butanol	7.4 ± 1.5			
1-Nitrooxy-2-butanol	7.0 ± 1.4			
4-Nitrooxy-2-butanol	10.4 ± 4.0			
3-Nitrooxy-1-butanol	11.9 ± 4.6			
4-Nitrooxy-1-butanol	12.6 ± 2.5			
1-Nitrooxy-2-pentanol	9.8 ± 1.9			
5-Nitrooxy-2-pentanol	32.1 ± 6.4			
4-Nitrooxy-1-pentanol	28.6 ± 5.7			
6-Nitrooxy-1-hexanol	30.9 ± 6.1			

[a] Abbreviations used for the techniques are defined in table II-A-1.

[b] Two different reference compounds were used in these studies, but it is not stated which was used for which compound.

urban conditions of high NO_x, non-negligible daytime NO_3 concentrations are possible. NO_3 normally exists in equilibrium with NO_2 and N_2O_5:

$$N_2O_5 \leftrightarrow NO_2 + NO_3$$

$$K_{eq} = [NO_2][NO_3]/[N_2O_5]$$

$$K_{eq} = 3.3 \times 10^{26} \exp(-10,990/T) \text{ molecule cm}^{-3}$$

$$= 3.2 \times 10^{10} \text{ molecule cm}^{-3} \text{ at } 298\text{ K}.$$

Reactions of NO_3 with aldehydes are relatively slow, with rate coefficients in the range 5×10^{-16} cm^3 $molecule^{-1}$ s^{-1} for formaldehyde, to 4.5×10^{-14} cm^3 $molecule^{-1}$ s^{-1} for 2-ethylbutanal. Temperature dependences are positive, with activation temperatures of 1500–2500 K. It is thought that the reaction proceeds predominantly by abstraction of the aldehydic hydrogen atom, although the rapid increase of the rate coefficient with the size of the alkyl group indicated above suggests that these are not simple reactions.

Absolute rate coefficients for NO_3 with simple aldehydes have been measured using the discharge flow technique, with detection of NO_3 either by resonance fluorescence (Dlugokencky and Howard, 1989; Cabañas et al., 2001, 2003) or by resonance absorption (Ullerstam et al., 2000; D'Anna et al., 2001a,b). The other studies have used variations on the relative rate technique, or measurements of N_2O_5 decay coupled with observations of reaction products. The relative rate studies used alkenes as reference compounds, and the rate coefficients were placed on an absolute basis using determinations of the reference reaction by Dlugokencky and Howard (1989) and by Canosa-Mas et al. (1999). The measurements of N_2O_5 decay take advantage of the equilibrium relation above, which gives:

$$d[N_2O_5]/dt = -k_{org}[NO_3][\text{organic}] = -k_{org}[\text{organic}]K_{eq}[N_2O_5]/[NO_2]$$

Thus, the first-order loss of N_2O_5 can be related to the rate coefficient provided the equilibrium constant and NO_2 concentration are known. One drawback of this approach is that the temperature must be determined with a high accuracy. K_{eq} changes by 15% per degree Kelvin around room temperature, and this sensitivity translates directly into uncertainties in the k_{org} determinations. Since many such experiments are carried out using large environmental chambers without active temperature control, the results are subject to an uncertainty associated with measurement of the temperature in the chamber (often ±2 K). In some cases (Cantrell et al., 1985, 1986, Doussin et al., 2003), NO_3 has been measured in situ by visible absorption spectrometry, relaxing the need for accurate measurement of the temperature. A comprehensive review of some of the earlier measurements of NO_3 with organics is given by Wayne et al. (1991).

V-D-1. Reactions of Nitrate Radicals with Aldehydes

V-D-1.1. NO_3 + Formaldehyde (Methanal, CH_2O)

The reaction of NO_3 with CH_2O has been studied four times, all at room temperature using environmental chambers. The studies of Cantrell et al. (1985) and Doussin et al. (2003) included in situ optical detection of NO_3 in addition to FTIR detection of stable species. As can be seen from table V-D-1, even after normalization to a common expression for the equilibrium constant the data show a fair amount of scatter, probably more so than the originally reported data. The studies of Atkinson et al. (1984d), Cantrell et al. (1985) using direct detection of NO_3, and Doussin et al. (2003) are in good agreement, and an average value of $(5.6 \pm 1.8) \times 10^{-16}$ cm^3 $molecule^{-1}$ s^{-1} can be derived. Cantrell et al. point out the possibility of secondary loss of NO_3 through its reaction with HO_2, which led to a large correction to their measured values. D'Anna et al. (2003) showed that the sole products of this reaction are HNO_3 + HCO.

Measurements of rate coefficients for NO_3 with isotopically substituted formaldehydes have been made by D'Anna et al. (2003) and Feilberg et al. (2004). The measurements were made relative to unsubstituted CH_2O using FTIR analysis. The obtained ratios are: $k(CH_2O)/k(CDHO) = 1.78 \pm 0.09$ (Feilberg et al.); $k(CH_2O)/k(CD_2O) = 3.00 \pm 0.15$ (average from both studies); $k(CH_2O)/k(^{13}CH_2O) = 0.96 \pm 0.05$ (average from both studies); $k(CH_2O)/k(CH_2^{18}O) = 0.98 \pm 0.05$ (Feilberg et al.) where the

Table V-D-1. Rate coefficients (k, cm^3 $molecule^{-1}$ s^{-1}) for the reaction of NO_3 with formaldehyde

$10^{16} \times k$	T (K)	Technique[a,b]	Reference
5.4	298 ± 1	N_2O_5 decay	Atkinson et al. (1984d)
6.4	298 ± 2	NO_3 abs	Cantrell et al. (1985)
9.0		N_2O_5 decay/FTIR	
12.1	295 ± 2	N_2O_5 decay/FTIR	Hjorth et al. (1988)
5.2 ± 0.9	298 ± 2	NO_3 abs/FTIR	Doussin et al. (2003)

[a] The rate coefficients have been normalized to a common K_{eq} where possible.
[b] Abbreviations used for the techniques are defined in table II-A-1.

uncertainties in the ratios have been widened to 5%. The results show a strong primary kinetic isotope effect that is derived from the abstraction of H by NO_3, but also non-negligible secondary isotope effects.

V-D-1.2. NO_3 + Acetaldehyde (Ethanal, CH_3CHO)

The reaction of NO_3 with acetaldehyde has been studied in chambers at 298 ± 2 K by Morris and Niki (1974), Atkinson et al. (1984d), Cantrell et al. (1986), D'Anna et al. (2001b), and Doussin et al. (2003), and as a function of temperature, using absolute methods, by Dlugokencky and Howard (1989), D'Anna et al. (2001b), and Cabañas et al. (2001). The rate coefficients measured at ambient temperatures agree quite well, with a mean of $(2.6 \pm 0.7) \times 10^{-15}$ cm^3 molecule^{-1} s^{-1} (table V-D-2). While the ambient temperature measurement of Cabañas et al. agrees with the others, the data from experiments above ambient temperatures diverge. It is likely that this study suffered from the adverse effects of secondary chemistry, caused by the removal of more than one NO_3 by the rapid reactions RC(O) + NO_3 and R + NO_3.

Table V-D-2. Rate coefficients (k, cm^3 molecule^{-1} s^{-1}) for the reaction of NO_3 with acetaldehyde

$10^{12} \times A$	B (K)	$10^{15} \times k$	T (K)	Technique[a]	Reference[a]	Temperature Range
		3.0 ± 1.0	300	N_2O_5 decay	Morris and Niki (1974)	
		2.23	298 ± 1	N_2O_5 decay	Atkinson et al. (1984d)	
		2.7	299 ± 1	N_2O_5 decay/ FTIR	Cantrell et al. (1986)	
1.44 ± 0.18	1860 ± 300	1.26 ± 0.04	264	DF–LIF	Dlugokencky and Howard (1989)	264–374
		2.74 ± 0.07	298			
		5.27 ± 0.17	332			
		10.0 ± 0.20	374			
62 ± 75	2830 ± 870	3.2 ± 0.8	298	DF–LIF	Cabañas et al. (2001)	298–433
		21.1 ± 3.5	345			
		42.0 ± 4.3	389			
		74.4 ± 14.3	433			
		2.62 ± 0.29	298 ± 2	RR (1-butene)[b]	D'Anna et al. (2001b)	
	2020 ± 160	0.95 ± 0.13	263	DF–LPA	D'Anna et al. (2001b)	263–363
		1.50 ± 0.13	275			
		2.49 ± 0.15	296			
		3.66 ± 0.12	322			
		5.11 ± 0.34	348			
		10.0 ± 0.5	363			
		2.1 ± 0.7	298	NO_3 abs/FTIR	Doussin et al. (2003)	

[a] Abbreviations used for the techniques are defined in table II-A-1.
[b] Placed on an absolute basis using k(1-butene) = 1.35×10^{-14} cm^3 molecule^{-1} s^{-1}.

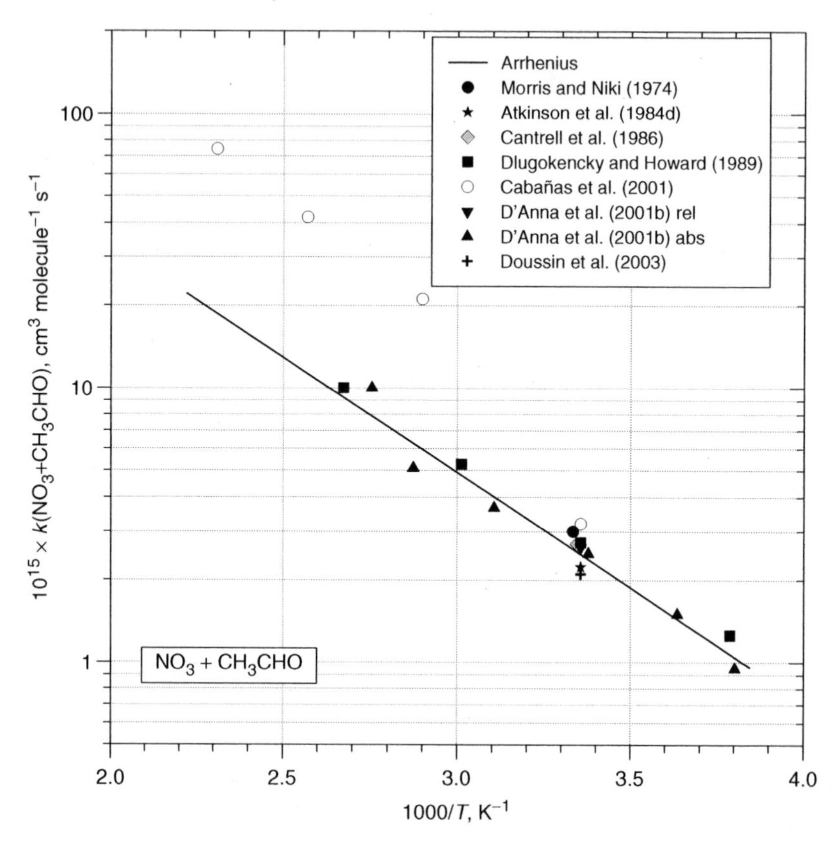

Figure V-D-1. Arrhenius plot of the rate coefficient for the reaction of NO_3 with acetaldehyde.

The preferred Arrhenius expression (figure V-D-1) is derived from a fit to the data of Dlugokencky and Howard and D'Anna et al., along with the room temperature values, $k = 1.6 \times 10^{-12} \exp(-1930/T)$ cm^3 $molecule^{-1}$ s^{-1}, leading to a value at 298 K of $k = 2.5 \times 10^{-15}$ cm^3 $molecule^{-1}$ s^{-1}, with an estimated uncertainty of 20%.

Rate coefficients have been measured at 298 K by D'Anna et al. (2001b, 2003) for various isotopologs of acetaldehyde, using the relative rate technique. The ratios obtained were $k(CH_3CHO)/k(CH_3CDO) = 2.39 \pm 0.07$, $k(CH_3CHO)/k(CD_3CHO) = 1.19 \pm 0.11$, and $k(CH_3CHO)/k(CD_3CDO) = 2.51 \pm 0.09$. A large primary isotope effect on the carbonyl hydrogen is again seen, consistent with this being the main site of attack.

V-D-1.3. NO_3 + Propanal (Propionaldehyde, C_2H_5CHO)

The reaction of NO_3 with propanal has been measured by D'Anna and Nielsen (1997), D'Anna et al. (2001a), Papagni et al. (2000), all using relative methods, and by Cabañas et al. (2001) using an absolute technique (table V-D-3). The results are in reasonably good agreement, with a room temperature value of $(6.5 \pm 1.0) \times 10^{-15}$ cm^3 $molecule^{-1}$ s^{-1}. As in the case of acetaldehyde, the Arrhenius plot of the results of

Table V-D-3. Rate coefficients (k, cm^3 $molecule^{-1}$ s^{-1}) for the reaction of NO_3 with propanal

$10^{12} \times A$	B (K)	$10^{15} \times k$	T (K)	Technique[a]	Reference	Temperature Range
		5.80 ± 0.48	298	RR (C_3H_6)[b]	D'Anna and Nielsen (1997)	
		6.18 ± 0.57	298	RR (C_3H_6)[b]	D'Anna et al. (2001a)	
17 ± 10	2250 ± 190	6.0 ± 0.6	298	DF–LIF	Cabañas et al. (2001)	298–433
		24 ± 6	345			
		62 ± 8	389			
		92 ± 12	433			
		7.49 ± 0.42	296	RR (MACR)[c]	Papagni et al. (2000)	

[a] Abbreviations used for the techniques are defined in table II-A-1.
[b] Placed on an absolute basis using $k(C_3H_6) = 9.5 \times 10^{-15}$ cm^3 $molecule^{-1}$ s^{-1}.
[c] Placed on an absolute basis using k(methacrolein) $= 3.5 \times 10^{-15}$ cm^3 $molecule^{-1}$ s^{-1}.

Cabañas et al. seems to curve above room temperature, which leads to very large A-factors. At present, no temperature dependence can be recommended.

V-D-1.4. NO_3 + Butanal (n-Butyraldehyde, $CH_3CH_2CH_2CHO$)

The rate coefficient for the reaction of NO_3 with butanal has been studied by D'Anna and Nielsen (1997), Ullerstam et al. (2000), D'Anna et al. (2001a), Papagni et al. (2000), and Cabañas et al. (2001). Absolute measurements as a function of temperature were made by Ullerstam et al. and Cabañas et al. (table V-D-4). At 298 K the recommended rate coefficient is $(1.15 \pm 0.15) \times 10^{-14}$ cm^3 $molecule^{-1}$ s^{-1}. Ullerstam et al. included O_2 in their flow-tube experiments to suppress secondary reactions which remove NO_3, and showed that the presence of O_2 led to lower rate coefficients. The temperature dependence reported by Cabañas et al. is much larger than that of Ullerstam et al., probably as a result of secondary chemistry. Figure V-D-2 illustrates the effect of the addition of O_2 on the measurements. It can be seen that the results of Cabañas et al. are similar to those measured by Ullerstam et al. in the absence of O_2. Ullerstam et al. showed that the effect of temperature is independent of the presence or absence of O_2 (see dashed line in figure V-D-2). On the basis of the temperature effect reported by Ullerstam et al., and consistent with other rate coefficients for NO_3 reactions with aldehydes, we recommend $k = 1.5 \times 10^{-12} \exp(-1460/T)$ cm^3 $molecule^{-1}$ s^{-1}.

V-D-1.5. NO_3 + 2-Methylpropanal [Isobutyraldehyde, $(CH_3)_2CHCHO$]

The rate coefficient for the reaction of NO_3 with 2-methylpropanal has been studied by D'Anna and Nielsen (1997), Ullerstam et al. (2000), D'Anna et al. (2001a), and Cabañas et al. (2001). A rate coefficient of $(1.2 \pm 0.2) \times 10^{-14}$ cm^3 $molecule^{-1}$ s^{-1} is preferred at 298 K (table V-D-5). The temperature dependence has been measured by Ullerstam et al. and by Cabañas et al. (figure V-D-3). Again, the temperature

Table V-D-4. Rate coefficients (k, cm^3 molecule^{-1} s^{-1}) for the reaction of NO_3 with butanal

$10^{12} \times A$	B (K)	$10^{15} \times k$	T (K)	Technique[a]	Reference	Temperature Range
		10.9 ± 0.9	298	RR (1-butene)[b]	D'Anna and Nielsen (1997)	
		10.4 ± 1.1	297	RR (C_3H_6)[c]	Ullerstam et al. (2000)	
1.2 ± 0.1	1460 ± 270	4.4 ± 0.1	267	DF (vis abs)	Ullerstam et al. (2000)	267–332
		11 ± 1	296			
		13 ± 2	332			
		12.3 ± 1.0	298	RR (1-butene)[b]	D'Anna et al. (2001a)	
		11.8 ± 0.6	296	RR (MACR)[d]	Papagni et al. (2000)	
76 ± 98	2470 ± 500	14.6 ± 1.6	298	DF–LIF	Cabañas et al. (2001)	298–433
		72 ± 12	345			
		142 ± 6	389			
		166 ± 2	433			

[a] Abbreviations used for the techniques are defined in table II-A-1.
[b] Placed on an absolute basis using k(1-butene) = 1.35×10^{-14} cm^3 molecule^{-1} s^{-1}.
[c] Placed on an absolute basis using $k(C_3H_6) = 9.5 \times 10^{-15}$ cm^3 molecule^{-1} s^{-1}.
[d] Placed on an absolute basis using k(methacrolein) = 3.5×10^{-15} cm^3 molecule^{-1} s^{-1}.

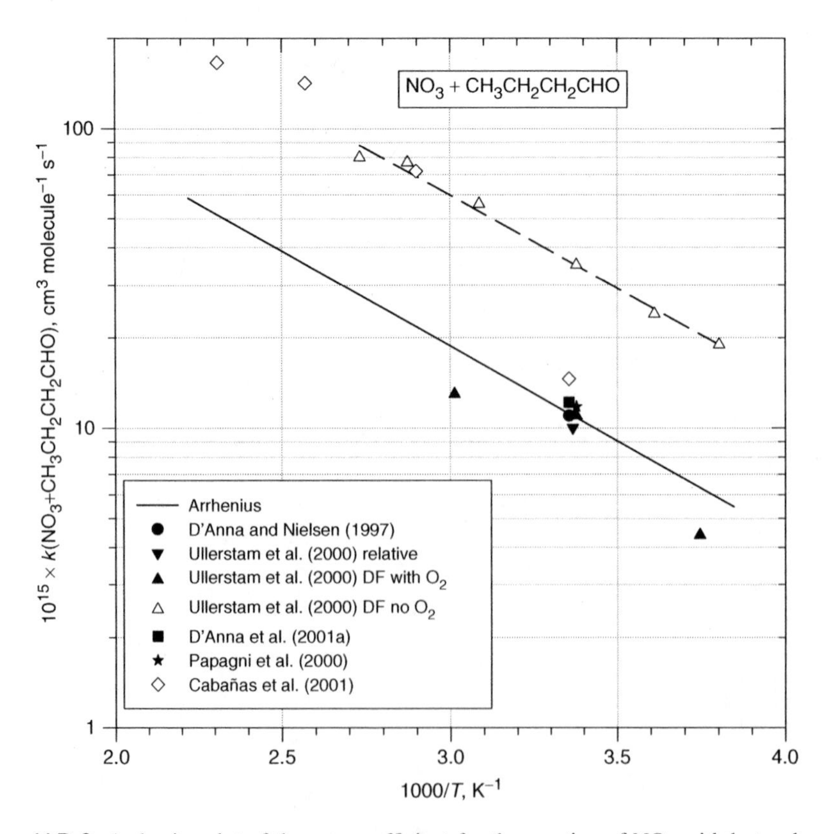

Figure V-D-2. Arrhenius plot of the rate coefficient for the reaction of NO_3 with butanal.

Table V-D-5. Rate coefficients (k, cm^3 $molecule^{-1}$ s^{-1}) for the reaction of NO_3 with 2-methylpropanal

$10^{12} \times A$	B (K)	$10^{15} \times k$	T (K)	Technique[a]	Reference	Temperature Range
		12.1 ± 0.6	298	RR (1-butene)[b]	D'Anna and Nielsen (1997)	
		12 ± 1	297	RR (C_3H_6)[c]	Ullerstam et al. (2000)	
2.9 ± 1.5	1720 ± 140	4.6 ± 1.0	267	DF (visible abs)	Ullerstam et al. (2000)	267–332
		9 ± 1	296			
		16 ± 3	332			
		12.6 ± 1.4	298	RR (1-butene)[b]	D'Anna et al. (2001a)	
100 ± 50	2600 ± 380	16.1 ± 1.3	298	DF–LIF	Cabañas et al. (2003)	298–433
		88 ± 12	345			
		140 ± 21	389			
		260 ± 12	433			

[a] Abbreviations used for the techniques are defined in table II-A-1.
[b] Placed on an absolute basis using k(1-butene) = 1.35×10^{-14} cm^3 $molecule^{-1}$ s^{-1}.
[c] Placed on an absolute basis using $k(C_3H_6) = 9.5 \times 10^{-15}$ cm^3 $molecule^{-1}$ s^{-1}.

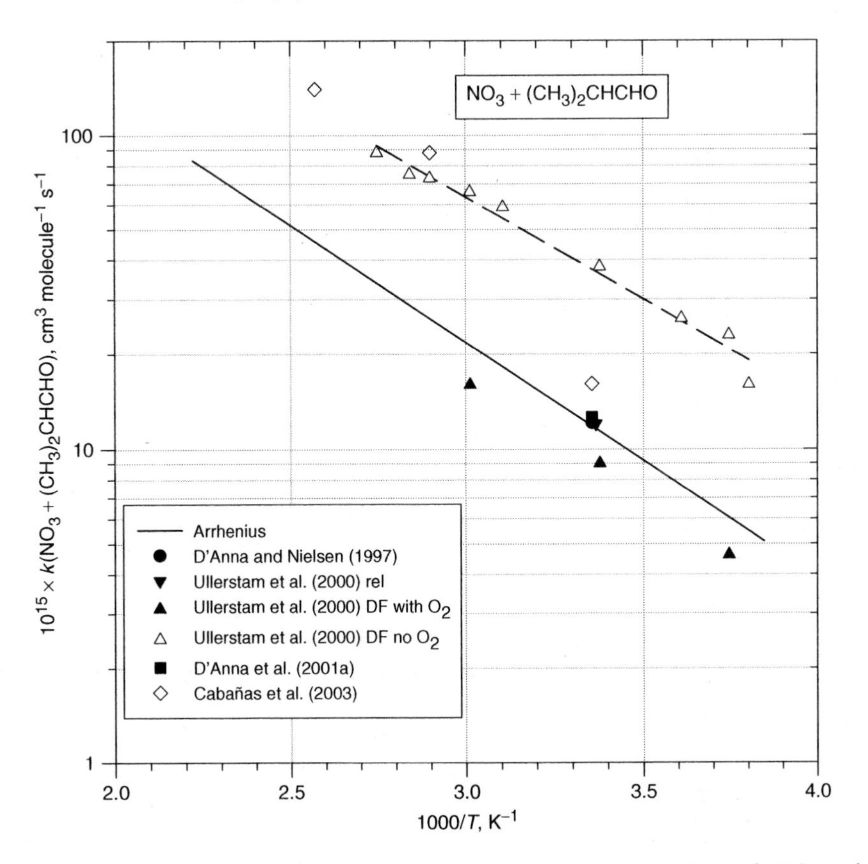

Figure V-D-3. Arrhenius plot of the rate coefficient for the reaction of NO_3 with 2-methylpropanal.

Table V-D-6. Rate coefficients (k, cm^3 $molecule^{-1}$ s^{-1}) for the reaction of NO_3 with pentanal

$10^{12} \times A$	B (K)	$10^{15} \times k$	T (K)	Technique[a]	Reference	Temperature Range
		14.6 ± 0.9	298	RR (1-butene)[b]	D'Anna and Nielsen (1997)	
		17.1 ± 2.1	298	RR (1-butene)[b]	D'Anna et al. (2001a)	
28 ± 14	2190 ± 160	17.5 ± 0.6	298	DF–LIF	Cabañas et al. (2001)	298–433
		54.8 ± 5.2	345			
		98.1 ± 12.6	389			
		133 ± 24	433			
		14.9 ± 1.7	298	RR (MACR)[c]	Papagni et al. (2000)	

[a] Abbreviations used for the techniques are defined in table II-A-1.
[b] Placed on an absolute basis using k(1-butene) = 1.35×10^{-14} cm^3 $molecule^{-1}$ s^{-1}.
[c] Placed on an absolute basis using k(methacrolein) = 3.5×10^{-15} cm^3 $molecule^{-1}$ s^{-1}.

dependence inferred by Ullerstam et al. is preferred, leading to $k = 3.8 \times 10^{-12} \exp(-1720/T)$ cm^3 $molecule^{-1}$ s^{-1}.

V-D-1.6. NO_3 + Pentanal (Valeraldehyde, n-C_4H_9CHO)

The rate coefficient for the reaction of NO_3 with pentanal has been measured by D'Anna and Nielsen (1997), D'Anna et al. (2001a), Papagni et al. (2000) and Cabañas et al. (2001). The latter group performed the only direct study and the only measurements as a function of temperature (see table V-D-6). The room temperature values are in fairly good agreement, and a rate coefficient $k = 1.6 \times 10^{-14}$ cm^3 $molecule^{-1}$ s^{-1} is recommended. In light of the strong downward curvature seen in the Arrhenius plot of Cabañas et al., no recommendation is made for the temperature dependence.

V-D-1.7. NO_3 + 2-Methylbutanal [$CH_3CH_2CH(CH_3)CHO$]

This reaction has been measured by D'Anna et al. (2001a) and Cabañas et al. (2003), who also studied the reaction as a function of temperature (see table V-D-7). The room-temperature values agree quite well. A fit to all the data gives $k = 2.8 \times 10^{-11} \exp(-2100/T)$ cm^3 $molecule^{-1}$ s^{-1}, as shown in figure V-D-4. However, considering the non-linear Arrhenius behavior shown by the other molecules measured by Cabañas et al. (2001, 2003), an uncertainty of a factor of two should be applied.

V-D-1.8. NO_3 + 3-Methylbutanal [$CH_3CH(CH_3)CH_2CHO$]

The reaction of NO_3 with 3-methylbutanal has been measured by Glasius et al. (1997), D'Anna et al. (2001a), and Cabañas et al. (2003). The latter group measured the rate coefficient as a function of temperature. The results are given in table V-D-8; they vary from 1.3×10^{-14} to 2.5×10^{-14} cm^3 $molecule^{-1}$ s^{-1}, with the measurement of Glasius et al. being the lowest. A fit to the data of D'Anna et al. and Cabañas et al., shown in

Table V-D-7. Rate coefficients (k, cm^3 $molecule^{-1}$ s^{-1}) for the reaction of NO_3 with 2-methylbutanal

$10^{12} \times A$	B (K)	$10^{15} \times k$	T (K)	Technique[a]	Reference	Temperature Range
		26.8 ± 1.1	298	RR (1-butene)[b]	D'Anna et al. (2001a)	
55 ± 31	2300 ± 550	25.6 ± 4.9	298	DF–LIF	Cabañas et al. (2003)	298–433
		59.2 ± 5.0	345			
		145 ± 2.4	389			
		224 ± 20	433			

[a] Abbreviations used for the techniques are defined in table II-A-1.
[b] Placed on an absolute basis using k(1-butene) = 1.35×10^{-14} cm^3 $molecule^{-1}$ s^{-1}.

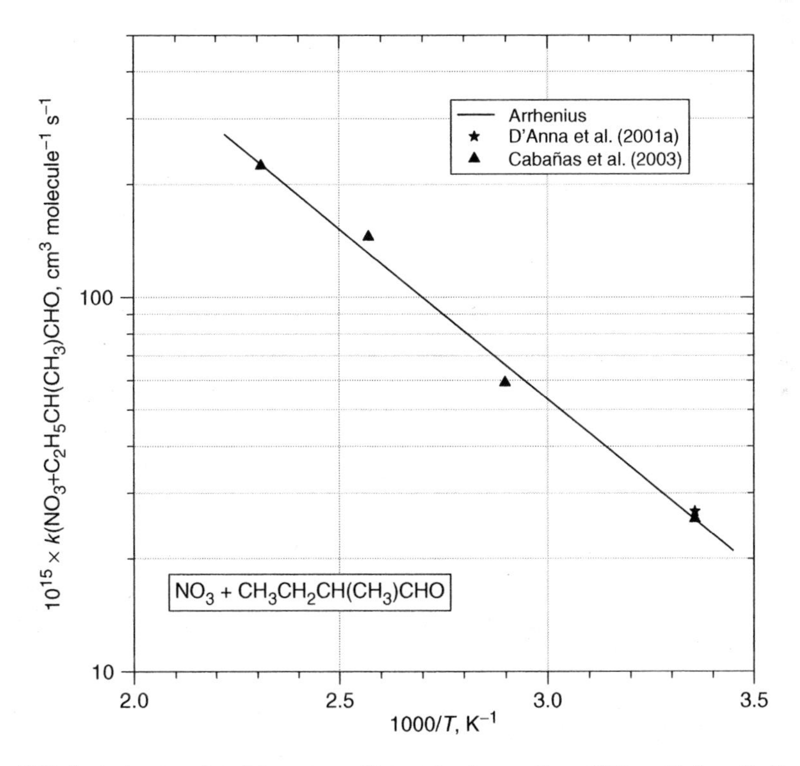

Figure V-D-4. Arrhenius plot of the rate coefficient for the reaction of NO_3 with 2-methylbutanal.

figure V-D-5, gives $k = 4.7 \times 10^{-11} \exp(-2300/T)$ cm^3 $molecule^{-1}$ s^{-1}, with an uncertainty of a factor of two at all temperatures. Over the range of temperatures studied, the rate coefficients for 3-methylbutanal and 2-methylbutanal are indistinguishable.

V-D-1.9. Reaction of NO_3 with 2,2-Dimethylpropanal [Pivaldehyde, $(CH_3)_3CCHO$]

The reaction of NO_3 with 2,2-dimethylpropanal has been studied by D'Anna and Nielsen (1997), D'Anna et al. (2001a), and Cabañas et al. (2003). Values at ambient

Table V-D-8. Rate coefficients (k, cm^3 $molecule^{-1}$ s^{-1}) for the reaction of NO_3 with 3-methylbutanal

$10^{12} \times A$	B (K)	$10^{15} \times k$	T (K)	Technique[a]	Reference	Temperature Range
		13 ± 3	300	RR (1-butene)[b]	Glasius et al. (1997)	
		18.6 ± 0.8	298	RR (1-butene)[b]	D'Anna et al. (2001a)	
37 ± 14	2180 ± 230	24.5 ± 2.0	298	DF–LIF	Cabañas et al. (2003)	298–433
		71.7 ± 3.8	345			
		110 ± 9.3	389			
		257 ± 19	433			

[a] Abbreviations used for the techniques are defined in table II-A-1.
[b] Placed on an absolute basis using k(1-butene) = 1.35×10^{-14} cm^3 $molecule^{-1}$ s^{-1}.

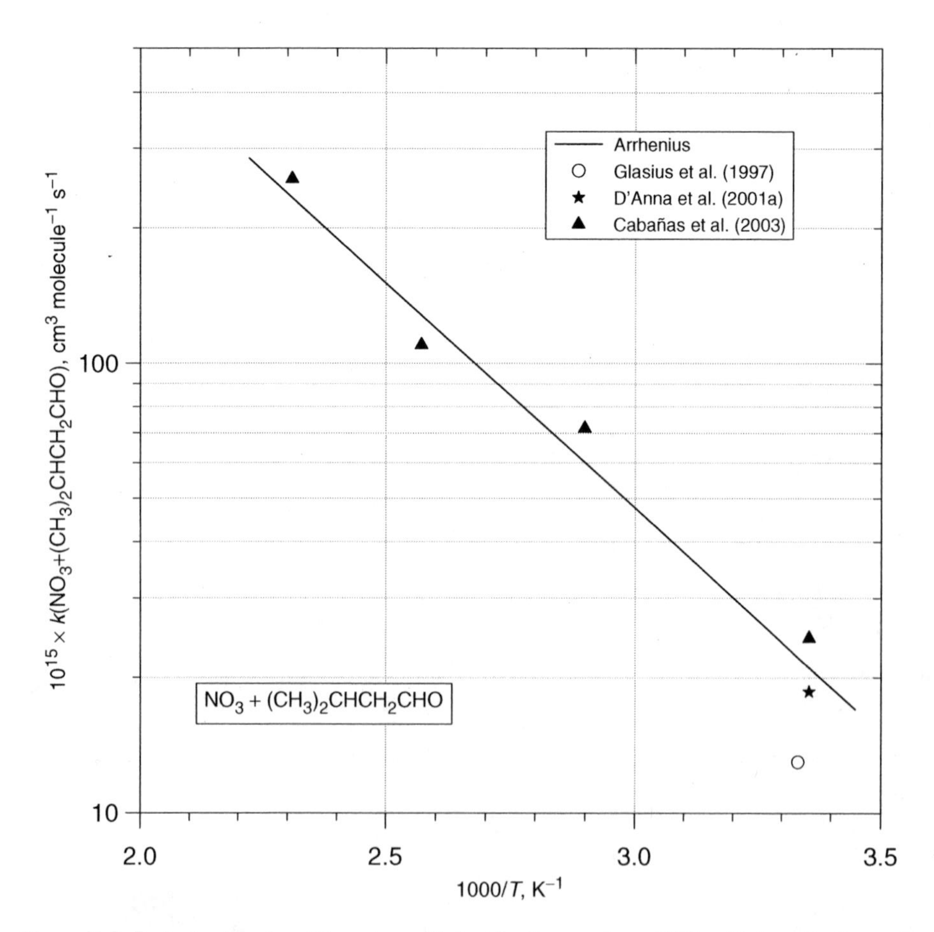

Figure V-D-5. Arrhenius plot of the rate coefficient for the reaction of NO_3 with 3-methylbutanal.

Table V-D-9. Rate coefficients (k, cm^3 $molecule^{-1}$ s^{-1}) for the reaction of NO_3 with 2,2-dimethylpropanal

$10^{12} \times A$	B (K)	$10^{15} \times k$	T (K)	Technique[a]	Reference	Temperature Range
		22.9 ± 0.9	298	RR (1-butene)[b]	D'Anna and Nielsen (1997)	
		25.9 ± 2.1	298	RR (1-butene)[b]	D'Anna et al. (2001a)	
27 ± 14	2060 ± 360	28.0 ± 3.5	298	DF–LIF	Cabañas et al. (2003)	298–433
		75.0 ± 9.7	345			
		134 ± 14	389			
		273 ± 28	433			

[a] Abbreviations used for the techniques are defined in table II-A-1.
[b] Placed on an absolute basis using k(1-butene) = 1.35×10^{-14} cm^3 $molecule^{-1}$ s^{-1}.

temperatures (table V-D-9) are in the range $(2.3–2.8) \times 10^{-14}$ cm^3 $molecule^{-1}$ s^{-1}. A fit to all the data, shown in figure V-D-6, yields $k = 4.5 \times 10^{-11} \exp(-2225/T)$ cm^3 $molecule^{-1}$ s^{-1}, which again is indistinguishable from the rate coefficients for 2-methylbutanal and 3-methylbutanal with NO_3. An uncertainty of a factor of two is assigned to this reaction.

V-D-1.10. NO_3 + Hexanal ($CH_3CH_2CH_2CH_2CH_2CHO$)

The rate coefficient for reaction of NO_3 with hexanal has been measured by D'Anna and Nielsen (1997), Papagni et al. (2000), D'Anna et al. (2001a), and Cabañas et al. (2001). Results are given in table V-D-10. The values for ambient temperatures are in excellent agreement, with an unweighted mean of $(1.6 \pm 0.3) \times 10^{-14}$ cm^3 $molecule^{-1}$ s^{-1} (2-σ uncertainty). As observed with other NO_3–aldehyde reactions, the temperature-dependent data of Cabañas et al. increase rapidly, then roll off at higher temperatures; consequently, no recommendation is made for other temperatures.

V-D-1.11. Reactions of NO_3 with Isomers of Hexanal

Rate coefficients for the reactions of NO_3 with 2-methylpentanal, 3-methylpentanal, 4-methylpentanal, 3,3-dimethylbutanal, and 2-ethylbutanal have been measured at 298 ± 2 K by D'Anna and Nielsen (1997) and D'Anna et al. (2001a). 3,3-Dimethylbutanal is the only compound for which duplicate measurements are available. The results are included in table V-D-10. The presence of a methyl group in the 2- and 3-positions appears to accelerate the reactions in comparison with the straight-chain species, while the rate coefficient for 2-ethylbutanal is a further 50% faster. No recommendation is made for the temperature dependence of these reactions.

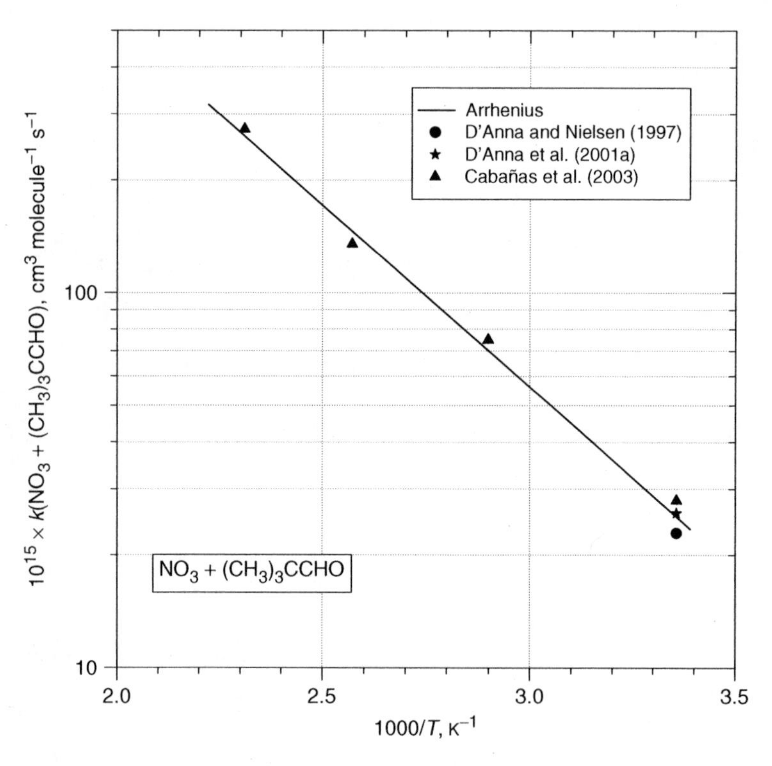

Figure V-D-6. Arrhenius plot of the rate coefficient for the reaction of NO_3 with 2,2-dimethylpropanal.

Table V-D-10. Rate coefficients (k, cm^3 $molecule^{-1}$ s^{-1}) for the reaction of NO_3 with isomers of hexanal

Molecule	$10^{15} \times k$	T (K)	Technique[a]	Reference
Hexanal	17.3 ± 1.8	298	RR (1-butene)[b]	D'Anna and Nielsen (1997)
	14.9 ± 1.3	296	RR (1-butene, MACR)[b,c]	Papagni et al. (2000)
	15.4 ± 1.9	298	RR (1-butene)[b]	D'Anna et al. (2001a)
	18 ± 4	298	DF–LIF	Cabañas et al. (2001)
	59 ± 10	345		
	164 ± 6	389		
	220 ± 2	433		
2-Methylpentanal	26.9 ± 1.1	298	RR (1-butene)[b]	D'Anna et al. (2001a)
3-Methylpentanal	24.1 ± 0.8	298	RR (1-butene)[b]	D'Anna et al. (2001a)
4-Methylpentanal	16.8 ± 0.9	298	RR (1-butene)[b]	D'Anna et al. (2001a)
3,3-Dimethylbutanal	20.0 ± 1.4	298	RR (1-butene)[b]	D'Anna and Nielsen (1997)
	17.1 ± 1.4	298	RR (1-butene)[b]	D'Anna et al. (2001a)
2-Ethylbutanal	44.6 ± 2.3	298	RR (1-butene)[b]	D'Anna et al. (2001a)

[a] Abbreviations used for the techniques are defined in table II-A-1.
[b] Placed on an absolute basis using k(1-butene) $= 1.35 \times 10^{-14}$ cm^3 $molecule^{-1}$ s^{-1}.
[c] Placed on an absolute basis using k(methacrolein) $= 3.5 \times 10^{-15}$ cm^3 $molecule^{-1}$ s^{-1}.

Table V-D-11. Rate coefficients (k, cm^3 $molecule^{-1}$ s^{-1}) for the reaction of NO_3 with heptanal

$10^{12} \times A$	B (K)	$10^{15} \times k$	T (K)	Technique[a]	Reference	Temperature Range
78 ± 10	2400 ± 500	23.7 ± 0.12	298	DF–LIF	Cabañas et al. (2003)	298–433
		79 ± 19	345			
		161 ± 36	389			
		301 ± 40	433			

[a] Abbreviations used for the techniques are defined in table II-A-1.

V-D-1.12. NO_3 + Heptanal (n-$C_6H_{13}CHO$)

The reaction of NO_3 with heptanal has been measured by Cabañas et al. (2003) as a function of temperature, using the discharge flow technique (see table V-D-11). In the absence of other data, their expression is recommended: $k = (7.8 \pm 1.9) \times 10^{-11}$ $\exp(-2400/T)$ cm^3 $molecule^{-1}$ s^{-1}, with an uncertainty of a factor of two at all temperatures.

V-D-1.13. General Comments on the Reactions of NO_3 with Aldehydes

On the basis of the careful measurements of Dlugokencky and Howard (1989) and Ullerstam et al. (2000), an activation temperature of around 1500 K is expected for these reactions. Arrhenius plots of the data of Cabañas et al. (2001) for the larger aldehydes often show curvature and lead to high apparent temperature dependences, resulting in very high A-factors. The later measurements of Cabañas et al. (2003) show less curvature than their earlier measurements, but the overall activation temperature is still higher (2200–2400 K) than those of Dlugokencky and Howard and Ullerstam et al. However, the measurements of Cabañas et al. (2003) have been taken at face value and used to derive a recommendation. It should be noted, though, that a smaller temperature coefficient and A-factor are expected.

Interestingly, the reaction of NO_3 with formaldehyde has never been studied as a function of temperature. New, careful measurements of all these reactions are required at room temperature and below. Secondary chemistry appears to be an issue in both relative and direct measurements since the formation of reactive free radicals occurs in both systems. Coupled with the slowness of the reactions, this makes the measurements very difficult. The reactions are thought to proceed by abstraction of the aldehydic hydrogen atom, leading to formation of nitric acid in the atmosphere, along with loss of NO_x.

V-D-2. Reactions of Nitrate Radicals with Other Oxygenates

Rate coefficients for NO_3 radicals with acetone and with a series of alcohols have also been reported. These reactions are all fairly slow, and the determinations are thus subject to complications from secondary chemistry. Hence, the reported rate coefficients may be upper limits to the true rate coefficient, even in cases where the authors attempted to account for secondary removal. The data base for these reactions has been evaluated by the IUPAC panel (Atkinson et al., 2005b), and we follow their recommendations.

V-D-2.1. NO_3 + Acetone [2-Propanone, $CH_3C(O)CH_3$]

An upper limit to the rate coefficient for this reaction was reported by Boyd et al. (1991), who used the stopped-flow technique with optical detection of NO_3. An observed rate coefficient of $(1.7 \pm 0.5) \times 10^{-17}$ cm^3 molecule^{-1} s^{-1} was reported, which the authors thought to be too high as a result of secondary removal of NO_3. Atkinson et al. (2005b) give a more conservative upper limit of $k < 3 \times 10^{-17}$ cm^3 molecule^{-1} s^{-1}.

V-D-2.2. NO_3 + Methanol (Methyl Alcohol, CH_3OH)

Measurements of this rate coefficient have been made by Wallington et al. (1987a), Canosa-Mas et al. (1989), and Langer and Ljungström (1995). All were direct determinations, with direct detection of NO_3 radicals using visible absorption. Atkinson et al. (2005b) reevaluated the data of Langer and Ljungström to give the preferred expression $k = 9.4 \times 10^{-13} \exp(-2650/T)$ cm^3 molecule^{-1} s^{-1}, with a value of 1.3×10^{-16} cm^3 molecule^{-1} s^{-1} for temperatures near 298 K.

V-D-2.3. NO_3 + Ethanol (Ethyl Alcohol, C_2H_5OH)

The reaction of NO_3 radicals with ethanol was studied by Wallington et al. (1987a) and Langer and Ljungström (1995). Wallington et al. reported an upper limit of 9×10^{-16} cm^3 molecule^{-1} s^{-1} at 298 K, whereas the value of Langer and Ljungström at 298 K is approximately a factor of 2 higher. Atkinson et al. (2005b) recommend a conservative upper limit of $k < 2 \times 10^{-15}$ cm^3 molecule^{-1} s^{-1} at 298 K. No recommendation is made for the temperature dependence.

V-D-2.4. NO_3 + 1-Propanol (n-Propyl Alcohol, $CH_3CH_2CH_2OH$)

Chew et al. (1998) measured the rate coefficient for NO_3 with 1-propanol to be 2.1×10^{-15} cm^3 molecule^{-1} s^{-1} at 298 K. The formation of propanal was observed, with a yield >0.55, and possibly approaching unity after correction for its reaction with NO_3. Chew et al. chose to quote their rate coefficient as an upper limit to account for the fact that loss of the alcohol via a possibly heterogeneous reaction with N_2O_5 may have occurred.

V-D-2.5. NO_3 + 2-Propanol [Isopropyl Alcohol, $CH_3CH(OH)CH_3$]

The reaction of NO_3 radicals with 2-propanol has been studied by Wallington et al. (1987a) and Langer and Ljungström (1995) using direct techniques, and by Chew et al. (1998) using a relative technique. The latter experiments should be less prone to interference by secondary chemistry, since they only involve measurement of the loss of starting material. An upper limit of 2×10^{-15} cm^3 molecule^{-1} s^{-1} was reported by both Wallington et al. and by Chew et al., which is lower than the measurement of Langer and Ljungström. Chew et al. reported the formation of acetone as a reaction product with a yield of 0.76 ± 0.09:

$$NO_3 + CH_3CH(OH)CH_3 \rightarrow HNO_3 + CH_3C(OH)CH_3$$

$$CH_3C(OH)CH_3 + O_2 \rightarrow HO_2 + CH_3C(O)CH_3$$

It is likely that abstraction at the central carbon atom accounts for 100% of the reaction, and that the acetone formation thus represents a true estimate of the total loss of 2-propanol in the gas phase. On the basis of these assumptions, Atkinson et al. (2005b) recommend a rate coefficient of $k = 1.4 \times 10^{-15}$ cm^3 molecule^{-1} s^{-1}, with an uncertainty of a factor of 2; no recommendation is given for the temperature dependence.

V-D-2.6. NO_3 + 1-Butanol (n-Butyl Alcohol, $CH_3CH_2CH_2CH_2OH$)

Chew et al. (1998) measured the rate coefficient for NO_3 with 1-butanol to be 2.1×10^{-15} cm^3 molecule^{-1} s^{-1} at 298 K. Butanal was observed as a product, with a yield of 22%, although its loss due to reaction with NO_3 could lead to the actual yield being higher. Chew et al. chose to quote their rate coefficient as an upper limit to account for the fact that loss of the alcohol via a possibly heterogeneous reaction with N_2O_5 may be occurring.

V-D-2.7. NO_3 + 2-Butanol [sec-Butyl Alcohol, $CH_3CH_2CH(OH)CH_3$]

The reaction of NO_3 radicals with 2-butanol was studied at 298 K by Chew et al. (1998), who reported an upper limit of 2.9×10^{-15} cm^3 molecule^{-1} s^{-1}. 2-Butanone was observed as a product, with a yield of 0.79 ± 0.09, presumably from the reaction:

$$NO_3 + CH_3CH_2CH(OH)CH_3 \rightarrow HNO_3 + CH_3CH_2C(OH)CH_3$$

$$CH_3CH_2C(OH)CH_3 + O_2 \rightarrow HO_2 + CH_3CH_2C(O)CH_3$$

Assuming that abstraction at the 2-position dominates the reactivity, and that formation of 2-butanone represents the gas-phase loss of 2-butanol, Atkinson et al. (2005b) recommend a rate coefficient of 2.1×10^{-15} cm^3 molecule^{-1} s^{-1}, with an uncertainty of a factor of 2; no recommendation is given for the temperature dependence.

V-D-2.8. NO_3 + 4-Heptanol [n-$C_3H_7CH(OH)CH_2CH_2CH_3$]

Chew et al. (1998) measured the rate coefficient for NO_3 with 4-heptanol to be 6.0×10^{-15} cm^3 molecule^{-1} s^{-1} at 298 K. 4-Heptanone was observed as a product, with a yield of 75%, corresponding to abstraction of a hydrogen atom at the 4-position. Chew et al. chose to quote their rate coefficient as an upper limit to account for the fact that loss of the alcohol via a possibly heterogeneous reaction with N_2O_5 may have been occurring.

VI

Atmospheric Chemistry of the Haloalkanes

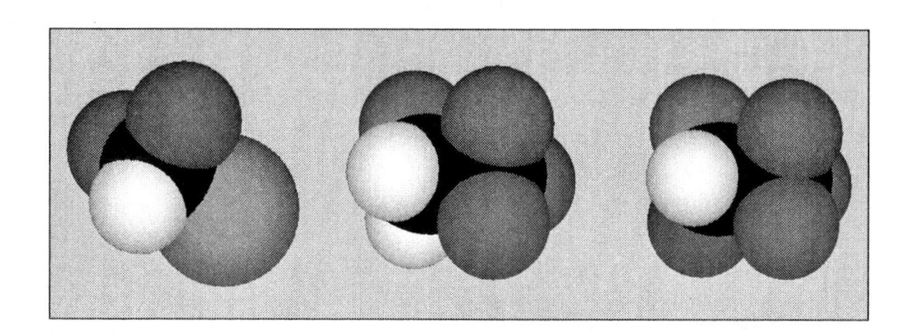

As with alkanes, reaction with OH is the dominant loss mechanism for haloalkanes. However, unlike the case for alkanes, photolysis can be an important loss mechanism for brominated and iodinated alkanes in the troposphere, chlorinated alkanes in the stratosphere, and fluorinated alkanes in the mesosphere. UV spectra and photolysis lifetimes for haloalkanes are discussed in chapter VII. In this chapter we first review the available kinetic data for reactions of OH radicals with haloalkanes (section VI-A). Reactions with Cl, $O(^3P)$, NO_3, and O_3 are then considered in section VI-B. Rules for the empirical estimation of rate coefficients of OH radicals with the haloalkanes are presented in section VI-C. The atmospheric chemistry of haloalkylperoxy and haloalkoxy radicals is described in sections VI-D and VI-E, respectively. Finally, the major products of the OH-initiated oxidation of haloalkanes are discussed in section VI-F.

VI-A. Rate Coefficients for OH-Radical Reactions with the Haloalkanes

VI-A-1. Introduction and Nomenclature

Abstraction of fluorine-, chlorine-, and bromine-containing atoms by OH radicals from haloalkanes is endothermic by approximately 249, 116, and 87 kJ mol^{-1} (calculated for $OH + CH_3X \rightarrow HOX + CH_3$ using data from Sander et al., 2003), respectively, and this is not a significant reaction pathway. Reaction of OH radicals with fluorine-, chlorine-, and bromine-containing haloalkanes proceeds solely via an H-atom abstraction mechanism. Abstraction of iodine atoms by OH radicals has a modest endothermicity, 26 kJ mol^{-1} for CH_3I and 14 kJ mol^{-1} for CF_3I (Sander et al., 2003), and for compounds with few, or no, available C—H reaction sites and relatively weak C—I bonds (e.g., CF_3I), iodine abstraction can be a significant reaction mechanism (Gilles et al., 2000).

There have been numerous absolute and relative rate studies of the gas-phase reactions of OH radicals with haloalkanes. These studies have been driven largely by the need to understand the atmospheric chemistry and hence the environmental impact of haloalkanes under consideration as replacements for chlorofluorocarbons. Many haloalkanes, particularly the highly fluorinated ones, have a low reactivity towards OH radicals; they react with rate coefficients at ambient temperature of the order of 10^{-15}–10^{-14} cm^3 molecule^{-1} s^{-1}. Absolute rate studies of such haloalkanes are particularly challenging because they require a very high level of reactant purity. Impurities in haloalkane samples can react with OH radicals with rate coefficients of the order of 10^{-11}cm^3 molecule^{-1} s^{-1}. In such cases, the impurity must be present at a level of less than 1 part in 10^4–10^5 in the sample to avoid more that a 10% overestimation of k(OH). It is challenging to ensure such sample purity. Relative rate studies are not sensitive to the presence of reactive impurities, and these provide a useful and complementary method to study the reactivity of OH radicals with haloalkanes.

In the following sections we review the available literature for the kinetics of reactions of OH radicals with haloalkanes. With the exception of some early studies which significantly overestimated the reactivity of certain haloalkanes (probably because of the presence of reactive impurities), there is generally good agreement between the results of the various studies. In the following sections we present the available literature data for halogenated alkanes on a compound-by-compound basis. The compounds are grouped by class in the following order: hydrofluorocarbons, hydrochlorocarbons, hydrobromocarbons, hydroiodocarbons, hydrochlorofluorocarbons, hydrobromofluorocarbons, hydrofluoroiodocarbons, hydrochlorobromocarbons, hydrochloroiodocarbons, hydrobromoiodocarbons, hydrochlorobromofluorocarbons, and finally perhalocarbons. For internal consistency, the data from relative rate studies have been reevaluated using reference rate coefficients recommended in this book and reference rate coefficients for OH + alkane reactions recommended by Atkinson (2003). Arrhenius parameters for relative rate studies given in the tables below were obtained by fitting the Arrhenius expression to the re-evaluated kinetic data. Uncertainties quoted for relative rate studies reflect uncertainties (where given) in the published rate constant ratios and do not include uncertainties in the reference rate coefficients. In all other cases, quoted uncertainties in the rate coefficient tables are those published in the individual studies.

In most cases, within the experimental uncertainties, the simple Arrhenius expression $k(\mathrm{OH}) = A \exp(-B/T)$ cm^3 molecule^{-1} s^{-1} provides a reasonable description of the kinetic data reported over the temperature range of approximately 200–400 K. There are no available kinetic data for reactions of OH radicals with haloalkanes at temperatures below 200 K. Where data are available at elevated temperatures, significant curvatures of Arrhenius plots are often observed. In such cases, a modified version of the Arrhenius expression, $k(\mathrm{OH}) = AT^2 \exp(-B/T)$ cm^3 molecule^{-1} s^{-1} has been fitted to the data. In the fitting procedures, non-linear least-squares regression was used, with each data point given a weight of $1/k_{OH}^2$. This procedure is mathematically equivalent to a least-squares fit in which the sum of the squares of the ratio $(k_{measured} - k_{fit})/k_{fit}$ is minimized, thereby providing a fit that minimizes relative error. Values at 298 K derived from the Arrhenius expressions are given with an estimate of the uncertainty associated with k(OH) at 298 K. Modern kinetic techniques are capable of measuring rate coefficients for reactions of hydroxyl radicals with haloalkanes with an accuracy of approximately 10–20%. Where only one measurement has been reported, we typically assign a conservative uncertainty of $\pm 25\%$. When several studies have been reported and the results are in good agreement, we typically assign an uncertainty of $\pm 15\%$. For reactions where different studies give significantly different results and there is no obvious reason to prefer a subset of the reported results, we typically assign uncertainties $>15\%$ (the exact value reflecting our judgment of the likely uncertainty). In the tables associated with each reaction, abbreviations are used to describe experimental techniques employed; these are defined in table II-A-1.

The International Union of Pure and Applied Chemistry (IUPAC) has rules for naming all organic compounds, including the haloalkanes. In the IUPAC convention the substituents in a molecule are assigned numbers (unless the compound can be defined without the need for a number). Thus, CH_3CH_2F is fluoroethane, while $CH_3CH_2CH_2F$ and CH_3CHFCH_3 are 1- and 2-fluoropropane, respectively. The prefixes di-, tri-, tetra-, penta-, hexa-, etc are used to name compounds containing more than one given substituent atom. Numbers are assigned to give the lowest sum; thus CH_3CCl_3 is 1,1,1-trichloroethane, and $CH_2BrCH_2CHBr_2$ is 1,1,3-tribromoethane. If a compound can be described using the same set of numbers, the substituent first in the alphabet is placed first; thus $CH_2ICH_2CH_2Cl$ is 1-chloro-3-iodopropane. Prefixes (di-, tri-, tetra-, etc.) are ignored in the alphabetization, so $CF_3CH_2CCl_3$ is 1,1,1-trichloro-3,3,3-trifluoropropane. If there is a conflict in priority between the lowest sum in the numbering system and alphabetization of the substituent names, the numbering takes precedence; thus CF_3CCl_2H is 2,2-dichloro-1,1,1-trifluoroethane. Finally, if a given substituent occupies all possible sites in the molecule the prefix "per-" is used; hence $CF_3CF_2CF_3$ is perfluoropropane (rather than octafluoropropane).

The IUPAC naming system for the haloalkanes, while logical and clear to the chemical community, is considered by some non-chemists to be rather cumbersome and not particularly useful. Hence, other systems of naming halocarbons have been developed by industrial users that appear more complicated at first sight but have gained in popular use. The American Society of Heating Refrigeration and Air Conditioning Engineers (ASHRAE) and the American National Standards Institute (ANSI) have developed an alternative haloalkane numbering system (ANSI/ASHRAE Standard 34-1992) which is in widespread use. In the system a compound is denoted by several uppercase letters ("Composition Denoting Prefix") which define the elements present in the molecule, followed by several numbers which describe how many of each substituents are present, followed by one, or more, lower case letters to differentiate different isomers (if necessary). The Composition Denoting Prefix always ends in "C" to indicate the presence of carbon. The presence of hydrogen, fluorine, chlorine, bromine, and iodine is indicated by H, F, C, B, and I. Hydrogen is always placed first, carbon always last, and the other elements are placed alphabetically. Hence, CF_3CCl_2H is a hydrochlorofluorocarbon and is given the prefix HCFC, CF_3CHF_2 and CF_3H are hydrofluorocarbons and have the prefix HFC, and CF_2Cl_2 is a chlorofluorocarbon with the prefix CFC.

In principle there are about 30 different prefixes which can be used to describe different classes of halogenated alkanes, but most are not in common use since most classes are not used in the refrigeration industry. Consistent with their prevalence in the literature, we use the ANSI/ASHRAE haloalkane numbering system only for CFCs, FCs, HCFCs, and HFCs. In the haloalkane numbering system, the numbers following the prefix have the following meanings: first number is number of carbon atoms minus 1, second number is the number of hydrogen atoms plus 1, and the third number is the number of fluorine atoms. For substituted methanes, the first number is zero and is omitted; CF_3H is HFC-23 while CF_2Cl_2 is CFC-12. For substituted ethanes, a lower-case letter suffix is sometimes added to distinguish between different isomers. The isomer with the smallest difference in the sum of the masses of the substituents on each carbon has no letter, the compound with next smallest difference is labeled "a", the compound with the next smallest difference is labeled "b", etc. Examples include CF_2HCH_2Cl (HCFC-142), $CFClHCH_2F$ (HCFC-142a), and CF_2ClCH_3 (HCFC-142b). For chlorine- and fluorine-substituted propanes two lower-case letters are added as a suffix to distinguish different isomers. The first letter defines the substituents on the central carbon atom, "a" for —CCl_2—, "b" for —CClF—, "c" for —CF_2—, "d" for —CHCl—, "e" for —CHF—, and "f" for —CH_2— (ordered by decreasing mass). The second letter reflects the difference in mass of the substituents on the two terminal carbon atoms with "a" used for the isomer with the smallest difference, "b" for the next smallest, and so on. For example, $CHF_2CHFCHF_2$ is HFC-245ea, CF_3CHFCH_2F is HFC-245eb, $CF_3CF_2CCl_2H$ is HCFC-225ca, and CF_2ClCF_2CHFCl is HCFC-225cb. For linear compounds with four or more carbons, letters are added using the "a" to "f" system noted above for the methylene carbons and the following for the methyl groups: "j" for —CCl_3, "k" for —CCl_2F, "l" for —CCF_2, "m" for —CF_3, "n" for —$CHCl_2$, "o" for —CH_2Cl, "p" for —CHF_2, "q" for —CH_2F, "r" for —CHClF, and "s" for —CH_3. Letters begin at one end of the molecule and address each carbon in turn in a fashion that first minimizes the number of letters and second uses a combination of letters which appear as early as possible in the alphabet, e.g., $CF_3CF_2CH_2CH_2F$ is HFC-356mcf. Hyphens are added to avoid confusion when double digits are required,e.g., $CF_3CHFCHFCF_2CF_3$ has ten fluorine atoms and is HFC-43-10mee.

Bromofluorocarbons and bromochlorofluorocarbons are effective fire-suppression agents that are useful in military and other applications. The US Army Corps of Engineers has developed a separate naming system for such compounds. The class of compounds is known as Halons; individual compounds are designated "Halon-xyza" where "x" is the number of carbon atoms, "y" is the number of fluorine atoms, "z" is the number of chlorine atoms, and "a" is the number of bromine atoms in the molecule. Examples include Halon-1211 ($CBrClF_2$) and Halon-1301 ($CBrF_3$). The Halon naming system has the advantage of simplicity but the disadvantage that it does not distinguish between isomers. Thus, Halon-2402 could, in principle, refer to either $CBrF_2CBrF_2$ or CBr_2FCF_3. For historic reasons associated with its commercial use under the name "Halon-2402", only the $CBrF_2CBrF_2$ isomer is named Halon-2402. We adopt the conventional uses of the names Halon 2402 ($CBrF_2CBrF_2$) and Halon 2311 ($CF_3CHClBr$) here.

VI-A-2. Hydrofluorocarbons

VI-A-2.1. OH + CH_3F (HFC-41) → H_2O + CH_2F

The rate coefficients of Howard and Evenson (1976), Nip et al. (1979), Jeong et al. (1984), Bera and Hanrahan (1988), Schmoltner et al., (1993), Wallington and Hurley (1993), Møgelberg et al. (1994a), Hsu and DeMore (1995), and DeMore (1996) are given in table VI-A-1. Minor corrections to the work of Jeong and Kaufman (1982) are described by Jeong et al. (1984). As is seen from figure VI-A-1, there is generally good agreement between the results reported from absolute and relative rate studies using a wide variety of techniques. The relative rate measurements of Wallington and Hurley (1993) and Møgelberg et al. (1994a) lie approximately 20–30% below those of the more comprehensive relative rate studies by Hsu and DeMore (1995) and DeMore (1996). The absolute rate measurements by Nip et al. (1979) and Schmoltner et al. (1993) are somewhat above the other measurements. We choose to exclude the results from Wallington and Hurley (1993) and Møgelberg et al. (1994a) and fit an Arrhenius expression to the remaining data at $T < 400$ K in figure VI-A-1 to give $k(OH + CH_3F) = 1.94 \times 10^{-12} \exp(-1361/T)\, cm^3\, molecule^{-1}\, s^{-1}$. This expression gives $k(OH + CH_3F) = 2.02 \times 10^{-14}\, cm^3\, molecule^{-1}\, s^{-1}$ at 298 K. Significant curvature in the Arrhenius plot is evident over the temperature range 210–480 K; a fit of the modified Arrhenius expression gives $k(OH + CH_3F) = 3.14 \times 10^{-18} T^2 \exp(-788/T)\, cm^3\, molecule^{-1}\, s^{-1}$. This expression yields $k(OH + CH_3F) = 1.98 \times 10^{-14}\, cm^3\, molecule^{-1}\, s^{-1}$ at 298 K. The uncertainty in $k(OH + CH_3F)$ at 298 K is estimated to be ±15%.

VI-A-2.2. OH + CH_2F_2 (HFC-32) → H_2O + CHF_2

The rate coefficients of Howard and Evenson (1976), Clyne and Holt (1979b), Nip et al. (1979), Jeong et al. (1984), Bera and Hanrahan (1988), Talukdar et al. (1991a), Schmoltner et al. (1993), Hsu and DeMore (1995), and Szilágyi et al. (2000) are given in table VI-A-2. Minor corrections to the work of Jeong and Kaufman (1982) are described by Jeong et al. (1984). As is seen from figure VI-A-2, there is generally good agreement between the results reported from absolute and relative rate studies using a wide variety of techniques. For reasons that are unclear, the room-temperature rate coefficients reported by Howard and Evenson (1976) and Clyne and Holt (1979b) lie significantly below the other studies. We choose to exclude the room-temperature results of Howard and Evenson (1976) and Clyne and Holt (1979b) and fit an Arrhenius expression to the data at $T < 400$ K in figure VI-A-2 to give $k(OH + CH_2F_2) = 1.33 \times 10^{-12} \exp(-1432/T)\, cm^3\, molecule^{-1}\, s^{-1}$. This expression gives $k(OH + CH_2F_2) = 1.09 \times 10^{-14}\, cm^3\, molecule^{-1}\, s^{-1}$ at 298 K. A fit of the

Table VI-A-1. Rate coefficients $k = A\ e^{-B/T}$ (cm^3 $molecule^{-1}$ s^{-1}) for the reaction: $OH + CH_3F$ (HFC-41) $\rightarrow H_2O + CH_2F$

$10^{12} \times A$	B (K)	$10^{14} \times k$	T (K)	Technique[e]	Reference	Temperature Range (K)
		1.60 ± 0.35	296	DF–LMR	Howard and Evenson (1976)	296
		2.18 ± 0.18	297	FP–RA	Nip et al. (1979)	297
8.11 ± 1.35	1887 ± 61	1.40 ± 0.09	292	DF–RF	Jeong and Kaufman (1982) Jeong et al. (1984)	292–480
		2.50 ± 0.18	330			
		3.86 ± 0.33	356			
		4.76 ± 0.31	368			
		5.48 ± 0.66	385			
		8.56 ± 0.66	416			
		13.1 ± 1.10	455			
		17.1 ± 1.10	480			
		1.71 ± 0.24	298	PR–UVabs	Bera and Hanrahan (1988)	298
1.75	1300 ± 100	0.91 ± 0.05	243	PLP/FP–LIF	Schmoltner et al. (1993)	243–373
		0.62 ± 0.08	243			
		0.76 ± 0.06	243			
		1.06 ± 0.09	251			
		1.18 ± 0.04	258			
		0.94 ± 0.06	258			
		1.53 ± 0.06	273			
		1.21 ± 0.16	273			
		1.48 ± 0.18	273			
		2.18 ± 0.10	295			
		2.32 ± 0.32	297			
		2.09 ± 0.08	298			
		2.18 ± 0.12	299			
		3.66 ± 0.20	323			
		3.12 ± 0.15	323			
		3.01 ± 0.40	323			
		4.22 ± 0.36	347			
		4.37 ± 0.32	348			
		3.53 ± 0.26	348			
		3.90 ± 0.22	348			
		5.63 ± 0.44	373			
		5.56 ± 0.26	373			
		1.71 ± 0.08[a]	296 ± 2	RR (C_2H_2)	Wallington and Hurley (1993)	296
		1.41 ± 0.07[b]	296 ± 2	RR (CH_4)		
		1.35 ± 0.13[b]	296	RR (CH_4)	Møgelberg et al. (1994a)	296
1.19	1228	1.91[c]	298	RR (CH_3CHF_2)	Hsu and DeMore (1995)	298–363
		2.25	308			
		2.59	321			
		2.98	333			
		3.46	345			
		3.98	363			

Table VI-A-1. *(continued)*

$10^{12} \times A$	B (K)	$10^{14} \times k$	T (K)	Technique[e]	Reference	Temperature Range (K)
3.19	1492	2.47[d]	308	RR (CH_3Cl)	DeMore (1996)	308–393
		3.66	333			
		4.16	343			
		5.27	363			
		6.17	378			
		7.05	393			

[a] Placed on an absolute basis using $k(OH + C_2H_2) = 8.49 \times 10^{-13}$ cm^3 molecule^{-1} s^{-1}, (Sørensen et al., 2003).
[b] Placed on an absolute basis using $k(OH + CH_4) = 1.85 \times 10^{-20}\ T^{2.82} \exp(-987/T)$ cm^3 molecule^{-1} s^{-1}.
[c] Placed on an absolute basis using $k(OH + CH_3CHF_2) = 1.21 \times 10^{-12} \exp(-1055/T)$ cm^3 molecule^{-1} s^{-1}.
[d] Placed on an absolute basis using $k(OH + CH_3Cl) = 2.71 \times 10^{-12} \exp(-1269/T)$ cm^3 molecule^{-1} s^{-1}.
[e] Abbreviations of the experimental techniques employed are defined in table II-A-1.

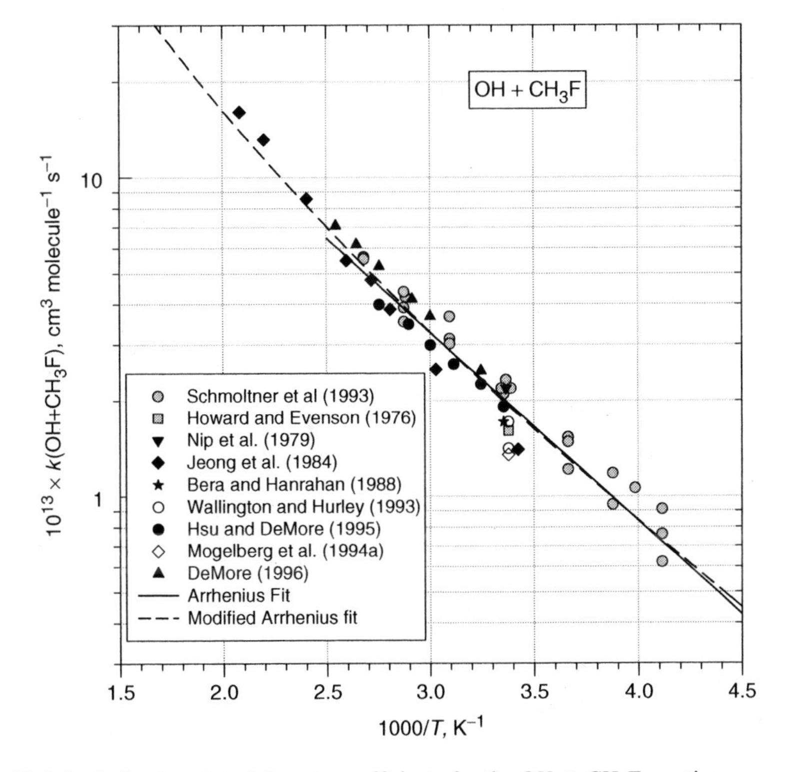

Figure VI-A-1. Arrhenius plot of the rate coefficients for the OH + CH_3F reaction.

modified Arrhenius expression to all the data (except the room-temperature data noted above) gives $k(OH + CH_2F_2) = 2.77 \times 10^{-18} T^2 \exp(-933/T)$ cm^3 molecule^{-1} s^{-1}. This expression gives $k(OH + CH_2F_2) = 1.07 \times 10^{-14}$ cm^3 molecule^{-1} s^{-1} at 298 K. The uncertainty in $k(OH + CH_2F_2)$ at 298 K is estimated to be $\pm 15\%$.

Table VI-A-2. Rate coefficients $k = A\ e^{-B/T}$ (cm^3 $molecule^{-1}$ s^{-1}) for the reaction: $OH + CH_2F_2$ (HFC-32) $\rightarrow H_2O + CHF_2$

$10^{12} \times A$	B (K)	$10^{14} \times k$	T (K)	Technique[b]	Reference	Temperature Range (K)
		0.78 ± 0.12	296	DF–LMR	Howard and Evenson (1976)	296
7.41	2100 ± 200	0.58 ± 0.03	293	DF–RF	Clyne and Holt (1979b)	293–429
		1.61 ± 0.50	327			
		2.41 ± 0.35	368			
		6.03 ± 0.40	429			
		1.17 ± 0.14	297	FP–RA	Nip et al. (1979)	297
4.37 ± 0.58	1766 ± 50	0.429 ± 0.038	250	DF–RF	Jeong and Kaufman (1982) Jeong et al. (1984)	250–492
		1.12 ± 0.075	298			
		2.10 ± 0.14	336			
		4.34 ± 0.27	384			
		7.27 ± 0.46	432			
		9.51 ± 0.66	464			
		14.1 ± 1.2	492			
		0.88 ± 0.14	298	PR–UVabs	Bera and Hanrahan (1988)	298
1.57 ± 0.21	1470 ± 100	0.196 ± 0.022	222	FP–LIF	Talukdar et al. (1991a)	222–381
		0.227 ± 0.013	223			
		0.398 ± 0.034	245			
		0.672 ± 0.020	273			
		1.09 ± 0.03	296			
		1.92 ± 0.04	336			
		2.66 ± 0.16	358			
		3.43 ± 0.18	381			
		0.252 ± 0.025	223	PLP/FP–LIF	Schmoltner et al. (1993)	223–298
		1.09 ± 0.03	298			
0.98	1337	1.10[a]	297	RR (CH_3CHF_2).	Hsu and DeMore (1995)	297–383
		1.27	309			
		1.45	317			
		1.52	323			
		1.89	338			
		2.32	357			
		2.97	383			
		1.00 ± 0.03	298	DF–RF	Szilágyi et al. (2000)	298

[a] Placed on an absolute basis using $k(OH + CH_3CHF_2) = 1.21 \times 10^{-12} \exp(-1055/T)$ cm^3 $molecule^{-1}$ s^{-1}.
[b] Abbreviations of the experimental techniques employed are defined in table II-A-1.

VI-A-2.3. $OH + CF_3H$(HFC-23) $\rightarrow H_2O + CF_3$

The rate coefficients of Howard and Evenson (1976), Clyne and Holt (1979b), Nip et al. (1979), Jeong et al. (1984), Bera and Hanrahan (1988), Schmoltner et al. (1993), Hsu and DeMore (1995), Medhurst et al. (1997), and Chen et al. (2003) are given in table VI-A-3 and plotted in figure VI-A-3. The Arrhenius fit parameters

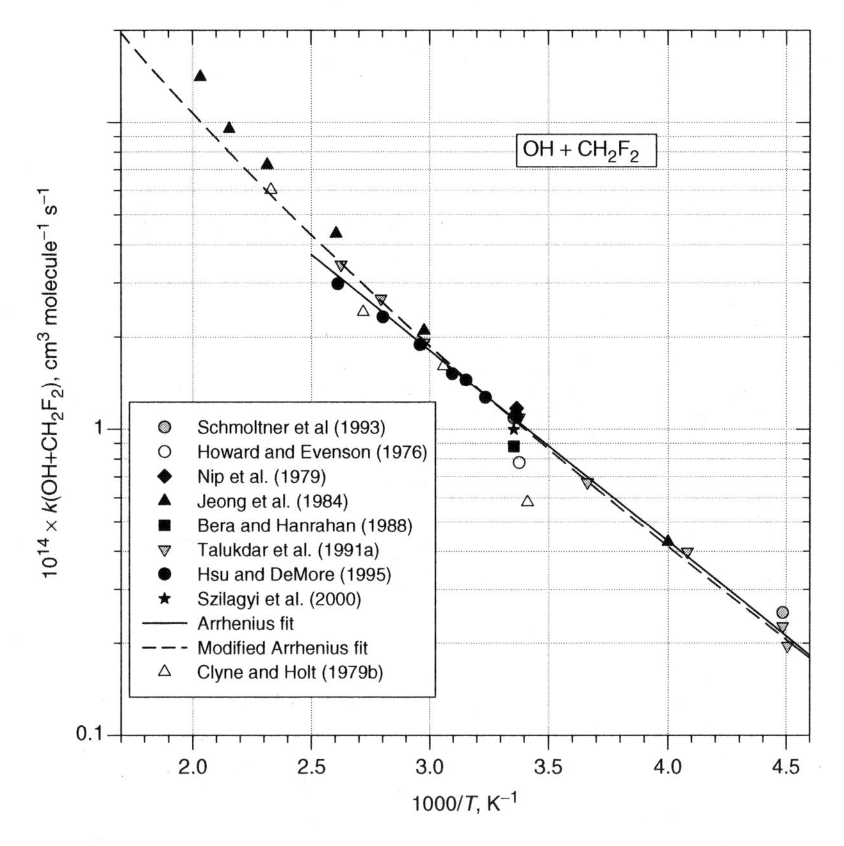

Figure VI-A-2. Arrhenius plot of the rate coefficients for the OH + CH_2F_2 reaction.

Table VI-A-3. Rate coefficients $k = A\ e^{-B/T}$ (cm^3 $molecule^{-1}$ s^{-1}) for the reaction: OH + CF_3H (HFC-23) → H_2O + CF_3

$10^{12} \times A$	B (K)	$10^{16} \times k$	T (K)	Technique[d]	Reference	Temperature Range (K)
		2.0 ± 0.2	296	DF–LMR	Howard and Evenson (1976)	
		13 ± 4	296	DF–RF	Clyne and Holt (1979b)	296–430
		14 ± 6	430			
		3.49 ± 1.66	297	FP–RA	Nip et al. (1979)	297
2.98 ± 1.07	2908 ± 156	16.9 ± 1.1	387	DF–RF	Jeong and Kaufman (1982)	387–480
		23.7 ± 1.7	410		Jeong et al. (1984)	
		33.1 ± 2.7	428			
		44.8 ± 2.9	447			
		56.4 ± 3.6	465			
		71.9 ± 4.5	480			
		23 ± 4	298	PR–UVabs	Bera and Hanrahan (1988)	298

(continued)

Table VI-A-3. *(continued)*

$10^{12} \times A$	B (K)	$10^{16} \times k$	T (K)	Technique[d]	Reference	Temperature Range (K)
0.693	2300 ± 100	0.77 ± 0.07	252	PLP/FP–LIF	Schmoltner et al. (1993)	252–374
		1.07 ± 0.10	262			
		1.60 ± 0.12	271			
		1.67 ± 0.14	272			
		1.70 ± 0.19	273			
		2.17 ± 0.22	284			
		2.93 ± 0.26	296			
		2.96 ± 0.12	298			
		2.96 ± 0.23	298			
		3.03 ± 0.18	298			
		4.01 ± 0.10	308			
		6.21 ± 0.30	324			
		11.0 ± 0.3	353			
		14.9 ± 0.4	374			
0.58	2321	2.51[a]	298	RR (CH_3CHF_2)	Hsu and DeMore (1995)	298–383
		2.30	298			
		3.85	317			
		5.30	330			
		7.25	347			
		9.82	365			
		13.7	383			
1.1	2300 ± 207	1.5 ± 0.2	298	PLP–LIF	Medhurst et al. (1997)	298–753
		11 ± 4	351			
		24 ± 2	393			
		41 ± 4	454			
		44 ± 6	454			
		50 ± 3	454			
		60 ± 2	454			
		72 ± 14	485			
		90 ± 20	529			
		130 ± 130	534			
		250 ± 110	585			
		240 ± 20	594			
		300 ± 40	625			
		400 ± 30	673			
		400 ± 30	673			
		370 ± 30	673			
		280 ± 40	673			
		390 ± 30	722			
		490 ± 40	753			
0.34	2071	0.99[b]	253	RR (CF_3CHF_2)	Chen et al. (2003)	253–343
		1.47	268			
		2.21	283			
		3.37	298			
		4.50	313			

Table VI-A-3. *(continued)*

$10^{12} \times A$	B (K)	$10^{16} \times k$	T (K)	Technique[d]	Reference	Temperature Range (K)
		6.35	328			
		8.21	343			
1.11	2420	0.74[c]	253			
		1.40	268			
		2.14	283	RR (CHF_2Cl)		
		3.43	298			
		4.83	313			
		6.93	328			
		9.31	343			

[a] Placed on an absolute basis using $k(OH + CH_3CHF_2) = 1.21 \times 10^{-12} \exp(-1055/T)$ cm^3 molecule^{-1} s^{-1}.
[b] Placed on an absolute basis using $k(OH + CF_3CHF_2) = 5.10 \times 10^{-13} \exp(-1666/T)$ cm^3 molecule^{-1} s^{-1}.
[c] Placed on an absolute basis using $k(OH + CHF_2Cl) = 1.48 \times 10^{-12} \exp(-1710/T)$ cm^3 molecule^{-1} s^{-1}.
[d] Abbreviations of the experimental techniques employed are defined in table II-A-1.

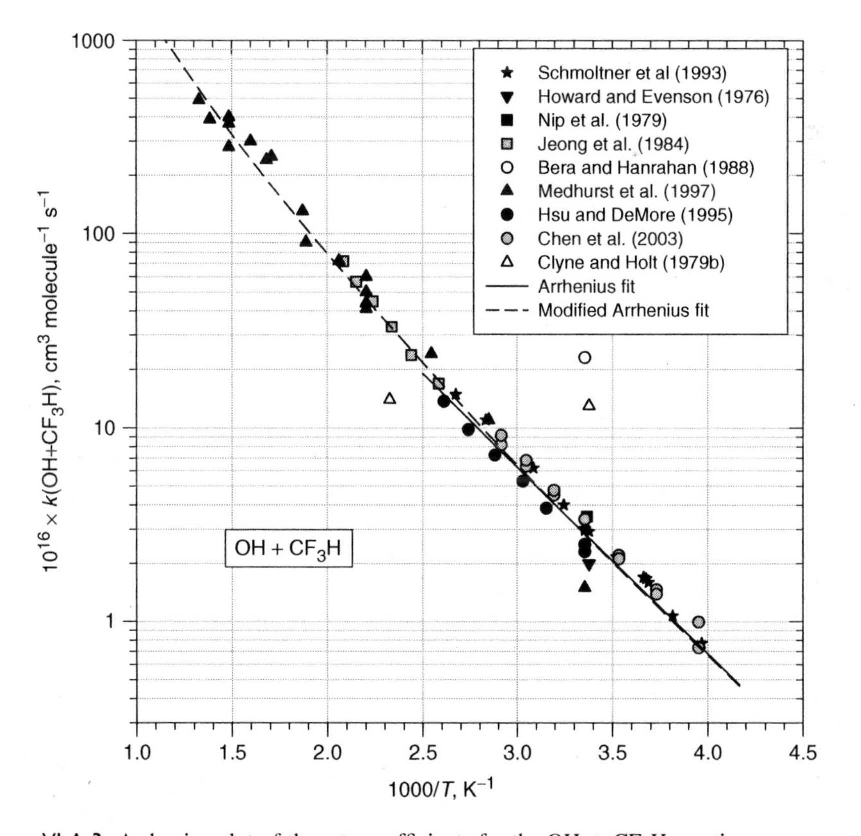

Figure VI-A-3. Arrhenius plot of the rate coefficients for the OH + CF_3H reaction.

and a typographical error in the 465 K data reported by Jeong and Kaufman (1982) are corrected in Jeong et al. (1984). Rate coefficients measured at room temperature by Clyne and Holt (1979b) and Bera and Hanrahan (1988) are approximately an order of magnitude greater than those reported in the other studies. Reactive impurities in the CF_3H sample probably explain the overestimation of $k(OH+CF_3H)$ by Bera and Hanrahan (1988) but would not explain the 430 K data point of Clyne and Holt (1979b). The rate data reported by Clyne and Holt (1979b) and Bera and Hanrahan (1988) are not considered further. A least-squares fit to the remaining data at $T < 400$ K gives $k(OH+CF_3H) = 4.89 \times 10^{-13} \exp(-2220/T)$ cm^3 molecule^{-1} s^{-1}. This expression gives $k(OH+CF_3H) = 2.84 \times 10^{-16}$ cm^3 molecule^{-1} s^{-1} at 298 K. A fit of the modified Arrhenius expression gives $k(OH+CHF_3) = 9.27 \times 10^{-19}T^2 \exp(-1691/T)$ cm^3 molecule^{-1} s^{-1}. This expression gives $k(OH+CHF_3) = 2.83 \times 10^{-16}$ cm^3 molecule^{-1} s^{-1} at 298 K. The uncertainty in $k(OH+CHF_3)$ is estimated to be $\pm$ 20%.

VI-A-2.4. OH + CH_3CH_2F(HFC-161) → H_2O + CH_3CHF (a); OH + CH_3CH_2F(HFC-161) → H_2O + CH_2CH_2F (b)

The rate coefficients of Nip et al. (1979), Schmoltner et al. (1993), Hsu and DeMore (1995), and Kozlov et al. (2003b) are given in table VI-A-4 and plotted in figure VI-A-4. The absolute rate data from Nip et al. (1979), Schmoltner et al. (1993), and Kozlov et al. (2003b) are in good agreement over the entire temperature range over which they can be compared. The relative rate results from Hsu and DeMore (1995) are in good agreement with the other studies at $T > 330$ K but for temperatures < 300 K are significantly lower (approximately 25% at 298 K). A fit of the Arrhenius expression to the $T < 400$ K data from Nip et al. (1979), Schmoltner et al. (1993), and Kozlov et al. (2003b) gives $k(OH+CH_3CH_2F) = 2.62 \times 10^{-12} \exp(-738/T)$ cm^3 molecule^{-1} s^{-1}. This expression gives $k(OH+CH_3CH_2F) = 2.20 \times 10^{-13}$ cm^3 molecule^{-1} s^{-1} at 298 K. Significant curvature in the Arrhenius plot is evident over the temperature range 210–480 K; a fit of the modified Arrhenius expression gives $k(OH+CH_3CH_2F) = 4.97 \times 10^{-18}T^2 \exp(-210/T)$ cm^3 molecule^{-1} s^{-1}. This expression gives $k(OH+CH_3CH_2F) = 2.18 \times 10^{-13}$ cm^3 molecule^{-1} s^{-1} at 298 K. The uncertainty in $k(OH+CH_3CH_2F)$ at 298 K is estimated to be $\pm$15%.

Singleton et al. (1980) studied the products following the 184.9 nm (low-pressure Hg lamp) photolysis of H_2O in the presence of CH_3CH_2F at 297 K. The recombination products of α- and β-fluoroethyl radicals (mainly 2,3 and 1,3 difluorobutane) accounted for 80–90% of the CH_3CH_2F loss. After correction for chemistry associated with H atoms (formed in H_2O photolysis) in the system, Singleton et al. (1980) concluded that 85 $\pm$ 3% of the reaction of OH radicals with CH_3CH_2F proceeds via attack on the —CH_2F group. In light of the complexity of the chemical system employed by Singleton et al. (1980) and the need for corrections to account for H-atom chemistry, further studies of the relative importance of abstraction from the —CH_2F and —CH_3 groups in CH_3CH_2F (preferably as a function of temperature) are needed.

Table VI-A-4. Rate coefficients $k = A\ e^{-B/T}$ (cm^3 molecule^{-1} s^{-1}) for the reaction: $OH + CH_3CH_2F$ (HFC-161) → products

$10^{12} \times A$	B (K)	$10^{13} \times k$	T (K)	Technique[b]	Reference	Temperature Range (K)
		2.32 ± 0.37	297	FP–RA	Nip et al. (1979)	297
2.32	750 ± 100	1.30 ± 0.04	243	PLP/FP–LIF	Schmoltner et al. (1993)	243–373
		1.44 ± 0.10	258			
		1.68 ± 0.10	273			
		1.77 ± 0.05	273			
		2.17 ± 0.18	294			
		2.13 ± 0.18	296			
		2.05 ± 0.06	296.5			
		2.02 ± 0.06	297			
		2.09 ± 0.10	297			
		2.53 ± 0.28	324			
		2.90 ± 0.12	324			
		2.59 ± 0.10	325			
		3.07 ± 0.20	344			
		3.35 ± 0.08	344			
		3.45 ± 0.14	373			
12.6	1285	1.41[a]	285	RR (C_2H_6)	Hsu and DeMore (1995)	285–358
		1.77	298			
		1.63	298			
		1.63	298			
		1.98	308			
		2.22	318			
		2.46	329			
		2.85	338			
		3.33	351			
		3.42	358			
1.1	2300 ± 207	0.794 ± 0.017	210	FP–RF	Kozlov et al. (2003b)	210–480
		0.962 ± 0.0087	220			
		1.058 ± 0.020	230			
		1.348 ± 0.032	250			
		1.747 ± 0.0237	272			
		2.204 ± 0.0334	298			
		2.879 ± 0.0515	330			
		4.042 ± 0.0948	370			
		5.767 ± 0.11	420			
		8.104 ± 0.135	480			

[a] Placed on an absolute basis using $k(OH + C_2H_6) = 1.49 \times 10^{-17}\ T^2\ \exp(-499/T)\ cm^3$ molecule^{-1} s^{-1}.
[b] Abbreviations of the experimental techniques employed are defined in table II-A-1.

VI-A-2.5. $OH + CH_3CHF_2$(HFC-152a) → $H_2O + CH_3CF_2$ (a); $OH + CH_3CHF_2$(HFC-152a) → $H_2O + CH_2CHF_2$ (b)

The rate coefficients of Howard and Evenson (1976), Handwerk and Zellner (1978), Clyne and Holt (1979b), Nip et al. (1979), Brown et al. (1990b), Liu et al (1990), Gierczak et al. (1991), Nielsen (1991), DeMore (1992), Hsu and DeMore (1995), Kozlov et al. (2003b), Wilson et al. (2003), and Taketani et al. (2005) are given in

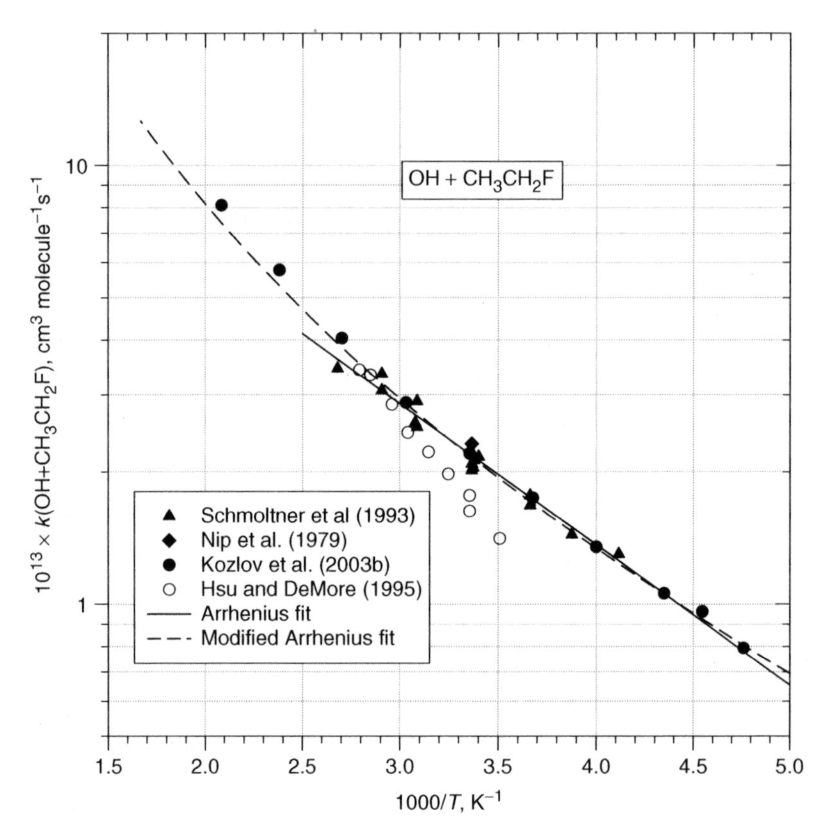

Figure VI-A-4. Arrhenius plot of the rate coefficients for the OH + CH_3CH_2F reaction.

table VI-A-5. As is seen from figure VI-A-5, the absolute rate data from Howard and Evenson (1976), Handwerk and Zellner (1978), Nip et al. (1979), Gierczak et al. (1991), and Kozlov et al. (2003b) and the relative rate data of Hsu and DeMore (1995), Wilson et al. (2003), and Taketani et al. (2005) are in good agreement over the temperature range over which they can be compared. The absolute rate data of Clyne and Holt (1979b) and Nielsen (1991) are systematically about 50–100% higher than the body of data shown in figure VI-A-5 and may have been affected by the presence of reactive impurities. The data from Brown et al. (1990b) are consistent with those from other studies but have unusually large uncertainties (particularly the data points at 220 and 303 K). A fit of the Arrhenius expression to the entire data set except that of Clyne and Holt (1979b), Brown et al. (1990b), and Nielsen (1991) for $T < 400$ K gives $k(OH + CH_3CHF_2) = 1.21 \times 10^{-12} \exp(-1055/T)$ cm^3 molecule^{-1} s^{-1}. This expression gives $k(OH + CH_3CHF_2) = 3.51 \times 10^{-14}$ cm^3 molecule^{-1} s^{-1} at 298 K. Significant curvature in the Arrhenius plot is evident over the temperature range 210–480 K; a fit of the modified Arrhenius expression gives $k(OH + CH_3CHF_2) = 2.10 \times 10^{-18} T^2 \exp(-503/T)$ cm^3 molecule^{-1} s^{-1}. This expression gives $k(OH + CH_3CHF_2) = 3.45 \times 10^{-14}$ cm^3 molecule^{-1} s^{-1} at 298 K. The uncertainty in $k(OH + CH_3CHF_2)$ at 298 K is estimated to be $\pm 15\%$.

Table VI-A-5. Rate coefficients $k = A\,e^{-B/T}$ (cm^3 $molecule^{-1}$ s^{-1}) for the reaction: $OH + CH_3CHF_2$ (HFC-152a) → products

$10^{12} \times A$	B (K)	$10^{14} \times k$	T (K)	Technique[f]	Reference	Temperature Range (K)
		3.12 ± 0.70	296	DF–LMR	Howard and Evenson (1976)	296
		3.50 ± 0.60	293	FP–RA	Handwerk and Zellner (1978)	293
2.92	1200 ± 100	4.66 ± 0.16	293	DF–RF	Clyne and Holt (1979b)	293–417
		7.16 ± 0.16	323			
		10.1 ± 0.08	363			
		16.4 ± 0.05	417			
		3.7 ± 0.40	297	FP–RA	Nip et al. (1979)	
1.42	1050 ± 250	1.20 ± 1.67	220	DF–RF	Brown et al. (1990b)	220–423
		3.10 ± 0.31	279			
		5.60 ± 2.30	303			
		5.56 ± 0.81	343			
		8.30 ± 1.14	375			
		11.18 ± 1.54	423			
1.50	1100 ± 200	2.99 ± 0.27	270	FP–RF	Liu et al. (1990)	270–400
		4.22 ± 0.45	298			
		5.32 ± 0.67	330			
		6.81 ± 0.83	350			
		7.30 ± 0.66	375			
		8.84 ± 0.74	400			
1.0	980 ± 50	1.05 ± 0.09	212	DF–LMR, FP–LIF	Gierczak et al. (1991)	212–422.5
		1.27 ± 0.11	224			
		1.73 ± 0.17	242			
		1.98 ± 0.16	258			
		2.95 ± 0.24	277			
		3.27 ± 0.26	293			
		4.02 ± 0.33	300			
		6.40 ± 0.52	349			
		9.12 ± 0.77	388			
		11.4 ± 0.10	402			
		11.1 ± 0.10	409			
		12.7 ± 0.11	422.5			
3.9	1370 ± 260	4.7 ± 1.1	295	PR–RA	Nielsen (1991)	295–388
		5.3 ± 0.4	323			
		8.0 ± 1.0	358			
		11.3 ± 0.3	388			
		3.86 ± 0.67[a]	298	RR (C_2H_6)	DeMore (1992)	298
		3.45 ± 0.34[b]	298	RR (CH_4)		
2.53	1317	3.05[b]	298	RR (CH_4)	Hsu and DeMore (1995)	298–358
		3.16	303			
		3.39	308			

(continued)

Table VI-A-5. *(continued)*

$10^{12} \times A$	B (K)	$10^{14} \times k$	T (K)	Technique[f]	Reference	Temperature Range (K)
		3.75	308			
		3.76	308			
		3.87	318			
		4.77	333			
		4.83	333			
		6.56	358			
		6.33	358			
0.74	1319	3.35[c]	298	RR (CCl_3CH_3)	Hsu and DeMore (1995)	
		3.29	298			
		3.31	298			
		3.26	298			
		3.71	308			
		3.90	313			
		4.11	313			
		4.84	333			
		5.61	333			
		7.00	358			
0.936	998 ± 56	0.84 ± 0.031	210	FP–RF	Kozlov et al. (2003b)	210–480
(210–298K)	(210–298 K)	0.99 ± 0.017	220			
		1.19 ± 0.019	230			
		1.66 ± 0.021	250			
3.24	1372 ± 89	2.38 ± 0.036	272			
(298–480 K)	(298–480 K)	3.38 ± 0.047	298			
		4.92 ± 0.074	330			
		7.71 ± 0.099	370			
		12.05 ± 0.19	420			
		19.36 ± 0.26	480			
2.17	1703	3.00 ± 0.04[a]	286	RR (C_2H_6)	Wilson et al. (2003)	286–403
		3.46 ± 0.12	299			
		4.90 ± 0.34	324			
		5.89 ± 0.53	343			
		7.89 ± 0.16	373			
		10.45 ± 0.35	403			
1.04	1000	3.39 ± 0.18[d]	290	RR (Cyclopropane)		
		4.24 ± 0.12	313			
		4.80 ± 0.20	332			
		6.20 ± 0.14	353			
		6.95 ± 0.46	372			
		8.26 ± 0.06	391			
		3.06[e]	295	RR (C_2H_2)	Taketani et al. (2005)	295

[a] Placed on an absolute basis using $k(OH + C_2H_6) = 1.49 \times 10^{-17}\ T^2 \exp(-499/T)\ cm^3\ molecule^{-1}\ s^{-1}$.
[b] Placed on an absolute basis using $k(OH + CH_4) = 1.85 \times 10^{-20}\ T^{2.82} \exp(-987/T)\ cm^3\ molecule^{-1}\ s^{-1}$.
[c] Placed on an absolute basis using $k(OH + CH_3CCl_3) = 1.81 \times 10^{-12} \exp(-1543/T)\ cm^3\ molecule^{-1}\ s^{-1}$.
[d] Placed on an absolute basis using $k(OH + Cyclopropane) = 4.21 \times 10^{-18}\ T^2 \exp(-454/T)\ cm^3\ molecule^{-1}\ s^{-1}$.
[e] Placed on an absolute basis using $k(OH + C_2H_2) = 8.49 \times 10^{-13}\ cm^3\ molecule^{-1}\ s^{-1}$ (Sørensen et al. 2003).
[f] Abbreviations of the experimental techniques employed are defined in table II-A-1.

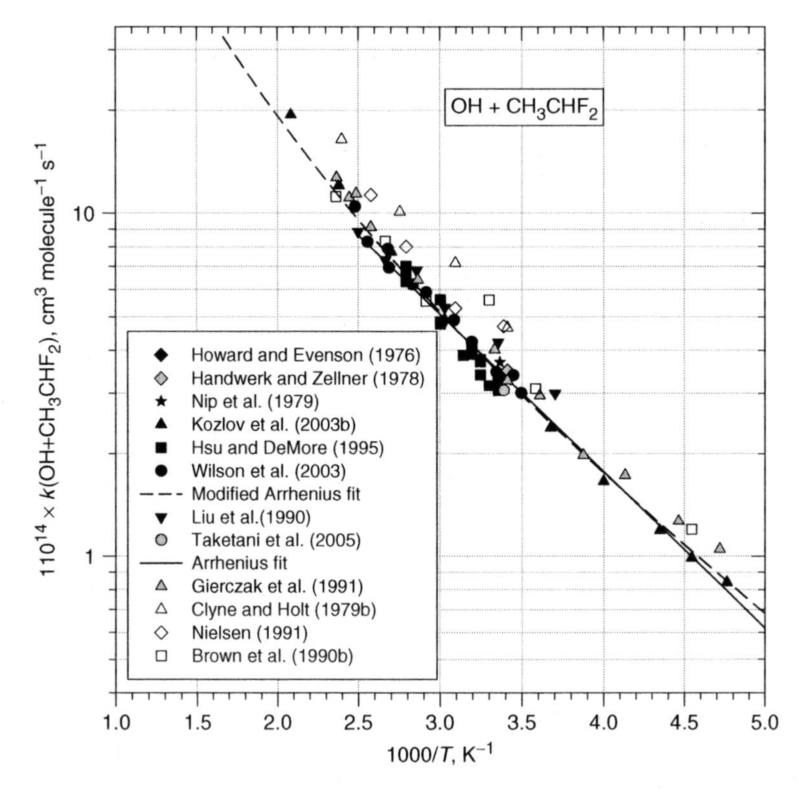

Figure VI-A-5. Arrhenius plot of the rate coefficients for the OH + CH_3CHF_2 reaction.

On the basis of the absence of observable $CF_2HC(O)H$ product following OH-initiated oxidation of CH_3CHF_2 in air, Taketani et al. (2005) report $k_a/(k_a + k_b) > 0.75$ and $k_b/(k_a + k_b) < 0.25$ at 295 K. This result is consistent with the upper limit of $k_b/(k_a + k_b) < 0.10$ inferred by kinetic arguments from Wilson et al. (2003) but slightly lower than that of $k_b/(k_a + k_b) = 0.26$ deduced by Kozlov et al. (2003b) from kinetic data at ambient temperature.

VI-A-2.6. OH + CH_2FCH_2F(HFC-152) → H_2O + $CHFCH_2F$

The rate coefficients of Martin and Paraskevopoulos (1983), Kozlov et al. (2003b), and Wilson et al. (2003) are given in table VI-A-6. As is seen in figure VI-A-6, the absolute rate data from Martin and Paraskevopoulos (1983), Kozlov et al. (2003b), and relative rate data of Wilson et al. (2003) are in good agreement over the temperature range over which they can be compared. Significant curvature in the Arrhenius plot is evident over the temperature range 210–480 K; a fit of the modified Arrhenius expression gives $k(OH + CH_2FCH_2F) = 2.84 \times 10^{-18} T^2 \exp(-277/T)$ cm^3 molecule^{-1} s^{-1}.

Table VI-A-6. Rate coefficients $k = A\ e^{-B/T}$ (cm^3 $molecule^{-1}$ s^{-1}) for the reaction: $OH + CH_2FCH_2F$ (HFC-152) → products

$10^{12} \times A$	B (K)	$10^{13} \times k$	T (K)	Technique[d]	Reference	Temperature Range (K)
		1.12 ± 0.12	298	FP–RA	Martin and Paraskevopoulos (1983)	298
1.02	706 ± 70	0.368 ± 0.011	210	FP–RF	Kozlov et al. (2003b)	210–480
(210–298K)	(210–298 K)	0.419 ± 0.011	220			
		0.461 ± 0.006	230			
		0.577 ± 0.009	250			
5.11	1190 ± 106	0.766 ± 0.023	272			
(298–480 K)	(298–480 K)	0.992 ± 0.018	298			
		1.358 ± 0.022	330			
		1.956 ± 0.083	370			
		2.961 ± 0.051	420			
		4.503 ± 0.060	480			
4.05	1135	0.87 ± 0.07[a]	293	RR (C_2H_6)	Wilson et al. (2003)	287–409
		1.12 ± 0.03	319			
		1.19 ± 0.04	322			
		1.41 ± 0.02	339			
		1.65 ± 0.01	355			
		1.84 ± 0.02	368			
		2.39 ± 0.07	397			
1.84	865	0.90 ± 0.02[b]	287	RR		
		1.05 ± 0.02	300	(Cyclopropane)		
		1.27 ± 0.01	324			
		1.46 ± 0.01	349			
		1.81 ± 0.01	373			
		2.27 ± 0.01	409			
2.98	1038	0.86 ± 0.01[c]	292	RR		
		1.05 ± 0.01	311	(CH_3CHF_2)		
		1.67 ± 0.02	360			
		2.12 ± 0.06	393			

[a] Placed on an absolute basis using $k(OH + C_2H_6) = 1.49 \times 10^{-17}\ T^2 \exp(-499/T)$ cm^3 $molecule^{-1}$ s^{-1}.
[b] Placed on an absolute basis using k(OH + Cyclopropane) $= 4.21 \times 10^{-18}\ T^2 \exp(-454/T)$ cm^3 $molecule^{-1}$ s^{-1}.
[c] Placed on an absolute basis using $k(OH + CH_3CHF_2) = 2.10 \times 10^{-18}\ T^2 \exp(-503/T)$ cm^3 $molecule^{-1}$ s^{-1}.
[d] Abbreviations of the experimental techniques employed are defined in table II-A-1.

This expression gives $k(OH + CH_2FCH_2F) = 9.96 \times 10^{-14}$ cm^3 $molecule^{-1}$ s^{-1} at 298 K, with an estimated uncertainty of ±15%.

VI-A-2.7. $OH + CH_3CF_3$(HFC-143a) → $H_2O + CH_2CF_3$

The rate coefficients of Clyne and Holt (1979b), Martin and Paraskevopoulos (1983), Talukdar et al. (1991a), Hsu and DeMore (1995), and Orkin et al. (1996) are given in table VI-A-7. The data from Clyne and Holt lie substantially above those from the other studies (see figure VI-A-7). It appears that the measurements by Clyne and Holt (1979b) were affected by reactive impurities. A fit of the Arrhenius expression to the data of Martin and Paraskevopoulos (1983), Talukdar et al. (1991a),

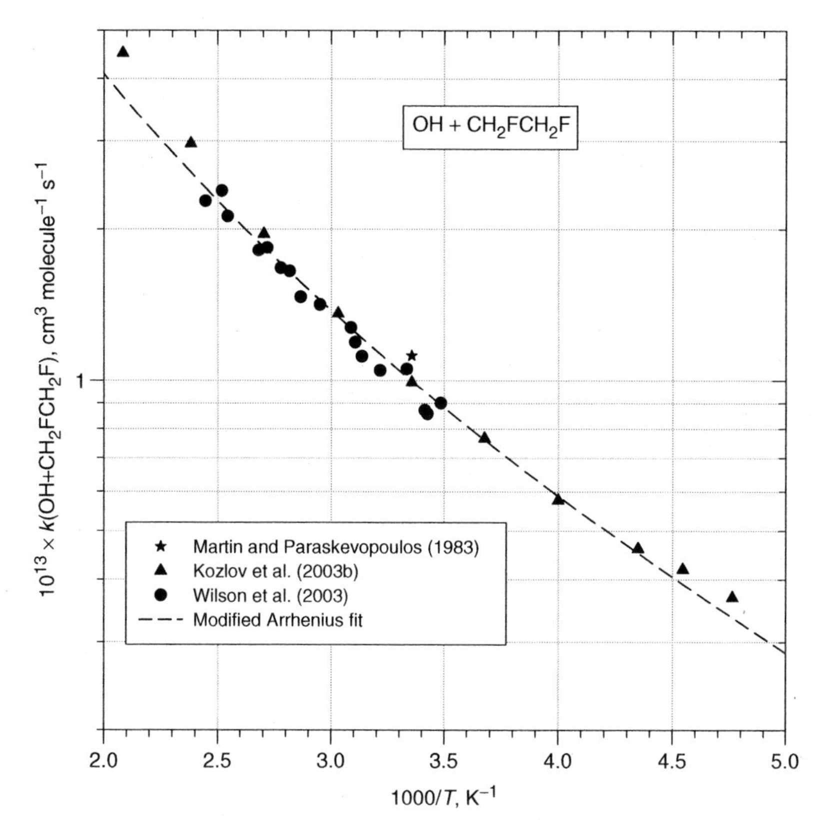

Figure VI-A-6. Arrhenius plot of the rate coefficients for the $OH + CH_2FCH_2F$ reaction.

Hsu and DeMore (1995), and Orkin et al. (1996) shown in figure VI-A-7 gives $k(OH + CH_3CF_3) = 1.06 \times 10^{-12} \exp(-1998/T)$ cm^3 molecule^{-1} s^{-1}. This expression gives $k(OH + CH_3CF_3) = 1.30 \times 10^{-15}$ cm^3 molecule^{-1} s^{-1} at 298 K. A fit of the modified Arrhenius expression gives $k(OH + CH_3CF_3) = 1.57 \times 10^{-18} T^2 \exp(-1400/T)$ cm^3 molecule^{-1} s^{-1}. This expression gives $k(OH + CH_3CF_3) = 1.27 \times 10^{-15}$ cm^3 molecule^{-1} s^{-1} at 298 K, with an estimated uncertainty of $\pm 15\%$.

VI-A-2.8. $OH + CH_2FCHF_2$(HFC-143) $\rightarrow H_2O + CH_2FCF_2$ (a); $OH + CH_2FCHF_2$(HFC-143) $\rightarrow H_2O + CHFCHF_2$ (b)

The rate coefficients of Clyne and Holt (1979b), Martin and Paraskevopoulos (1983), and Barry et al. (1995) are given in table VI-A-8. The data from Clyne and Holt (1979b) lie substantially above those from the other studies, probably reflecting the presence of reactive impurities (see figure VI-A-8). A fit of the Arrhenius expression to the data of Martin and Paraskevopoulos (1983) and Barry et al. (1995), shown in figure VI-A-8 gives $k(OH + CH_2FCHF_2) = 4.11 \times 10^{-12} \exp(-1648/T)$ cm^3 molecule^{-1} s^{-1}. This expression gives $k(OH + CH_2FCHF_2) = 1.63 \times 10^{-14}$ cm^3 molecule^{-1} s^{-1} at 298 K, with an estimated uncertainty of $\pm 25\%$.

Table VI-A-7. Rate coefficients $k = A\ e^{-B/T}$ (cm^3 $molecule^{-1}$ s^{-1}) for the reaction: $OH + CH_3CF_3$(HFC-143a) $\rightarrow H_2O + CH_2CF_3$

$10^{12} \times A$	B (K)	$10^{15} \times k$	T (K)	Technique[c]	Reference	Temperature Range (K)
69.2	3200 ± 500	< 1.0	293	DF–RF	Clyne and Holt (1979b)	293–425
		4.7 ± 1.5	333			
		12.9 ± 3.5	378			
		38.4 ± 12.3	425			
		1.73 ± 0.43	298	FP–RA	Martin and Paraskevopoulos (1983)	298
2.12 (261–374 K)	2200 ± 200 (261–374K)	0.16	223	DF–LMR	Talukdar et al. (1991a)	223–374
		0.23	236	FP–LIF		
		0.33	248			
		0.45	261			
		0.66	266			
		0.61	273			
		1.08	294			
		1.12	296			
		1.29	296			
		1.37	297			
		1.45	298			
		1.70	301			
		2.40	334			
		3.82	346			
		4.36	357			
		5.67	374			
1.90	2179	1.36[a]	298	RR (CH_4)	Hsu and DeMore (1995)	298–403
		1.27	298			
		1.84	314			
		2.15	328			
		3.72	346			
		5.76	374			
		8.53	403			
1.11	2040	1.18[b]	298	RR (CHF_2CF_3)		
		1.60	313			
		2.31	330			
		3.31	351			
		5.36	383			
0.91	1979 ± 65	1.24	298	FP–RF	Orkin et al. (1996)	298–370
		1.88	318			
		2.49	330			
		3.44	353			
		4.51	370			

[a] Placed on an absolute basis using $k(OH + CH_4) = 1.85 \times 10^{-20}\ T^{2.82} \exp(-987/T)$ cm^3 $molecule^{-1}$ s^{-1}.
[b] Placed on an absolute basis using $k(OH + CHF_2CF_3) = 5.10 \times 10^{-13} \exp(-1670/T)$ cm^3 $molecule^{-1}$ s^{-1}.
[c] Abbreviations of the experimental techniques employed are defined in table II-A-1.

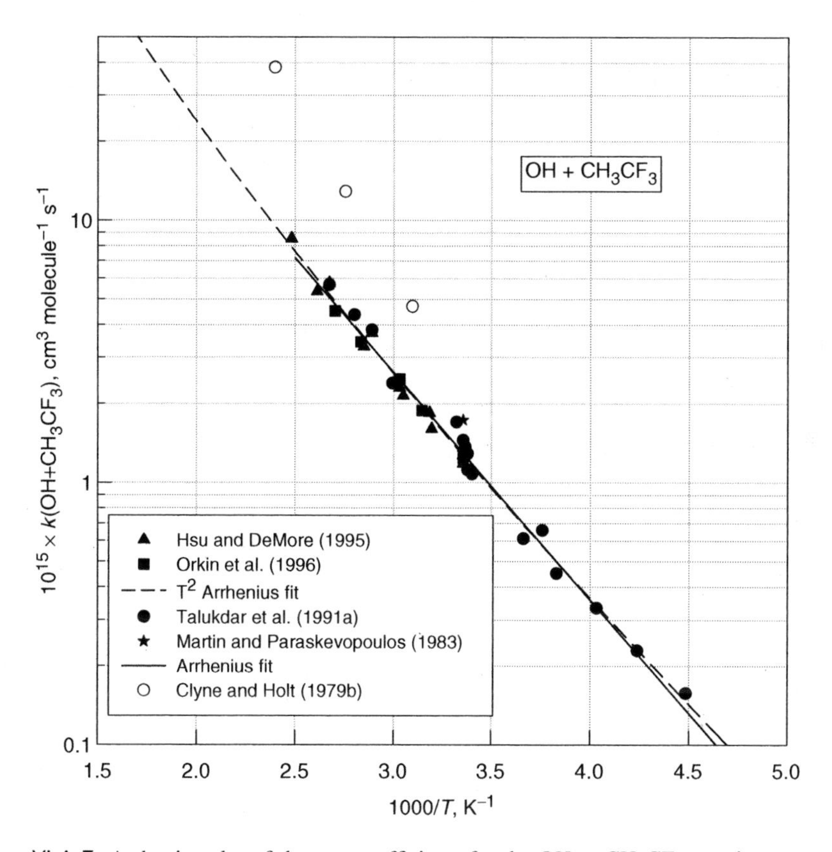

Figure VI-A-7. Arrhenius plot of the rate coefficients for the OH + CH_3CF_3 reaction.

Table VI-A-8. Rate coefficients $k = A\ e^{-B/T}$ (cm^3 $molecule^{-1}$ s^{-1}) for the reaction: OH + CH_2FCHF_2 (HFC-143) → products

$10^{12} \times A$	B (K)	$10^{14} \times k$	T (K)	Technique[b]	Reference	Temperature Range (K)
1.48 ± 1.2	1000 ± 100	4.98 ± 0.82	293	DF–RF	Clyne and Holt (1979b)	293–441
		4.68 ± 0.40	294			
		6.74 ± 0.43	335			
		9.09 ± 0.42	383			
		18.9 ± 0.6	441			
		1.83 ± 0.18	298	FP–RA	Martin and Paraskevopoulos (1983)	298
4.12	1655	1.07[a]	278	RR (CH_3CCl_3)	Barry et al. (1995)	278–323
		1.31	288			
		1.61 ± 0.05	298			
		2.14	313			
		2.41	323			

[a] Placed on an absolute basis using $k(OH + CH_3CCl_3) = 1.81 \times 10^{-12}\ \exp(-1543/T)\ cm^3\ molecule^{-1}\ s^{-1}$.
[b] Abbreviations of the experimental techniques employed are defined in table II-A-1.

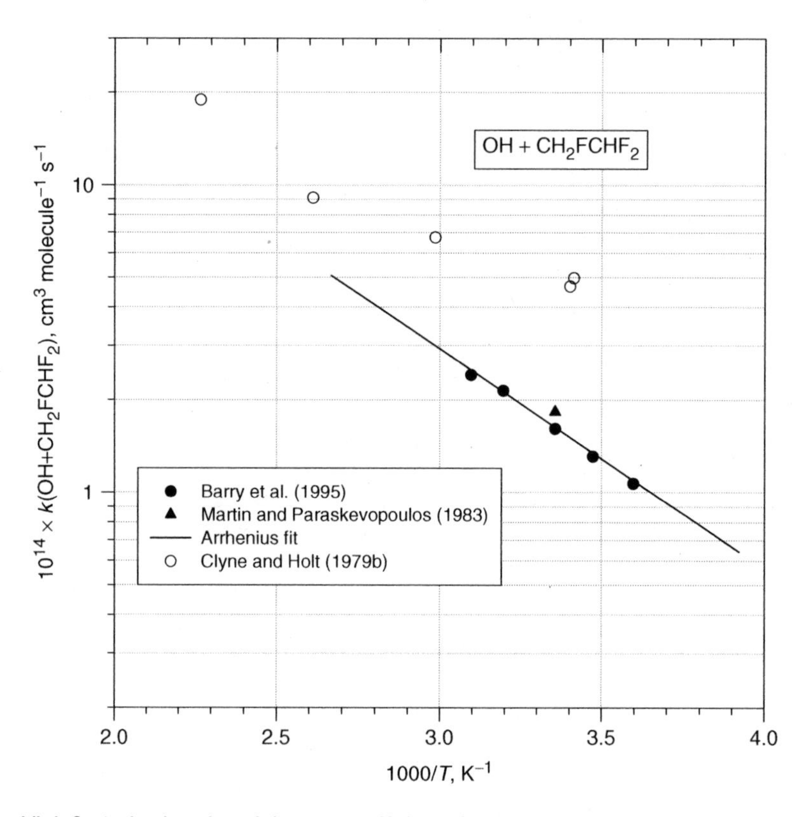

Figure VI-A-8. Arrhenius plot of the rate coefficients for the OH + CH_2FCHF_2 reaction.

VI-A-2.9. OH + CH_2FCF_3(HFC-134a) → H_2O + $CHFCF_3$

The rate coefficients of Clyne and Holt (1979b), Martin and Paraskevopoulos (1983), Jeong et al. (1984), Brown et al. (1990b), Liu et al. (1990), Gierczak et al. (1991), Zhang et al. (1992a), Orkin and Khamaganov (1993a), Leu and Lee (1994), Bednarek et al. (1996), DeMore (1993a), and Barry et al. (1995) are given in table VI-A-9 and plotted in figure VI-A-9. As seen from this figure, the data set from Clyne and Holt (1979b) and Jeong et al. (1984), the low-temperature data from Brown et al. (1990b), and the high-temperature data from Liu et al. (1990) lie systematically above the results from the other studies. An Arrhenius fit to the data, excluding those from Clyne and Holt, Jeong et al., Brown et al., and Liu et al., gives $k(OH+CH_2FCF_3) = 1.01 \times 10^{-12} \exp(-1640/T)\,cm^3\,molecule^{-1}\,s^{-1}$. This expression gives $k(OH+CH_2FCF_3) = 4.11 \times 10^{-15}\,cm^3\,molecule^{-1}\,s^{-1}$ at 298 K. A fit of the modified Arrhenius expression gives $k(OH+CH_2FCF_3) = 1.29 \times 10^{-18} T^2 \exp(-989/T)\,cm^3\,molecule^{-1}\,s^{-1}$. This expression gives $k(OH+CH_2FCF_3) = 4.15 \times 10^{-15}\,cm^3\,molecule^{-1}\,s^{-1}$ at 298 K, with an estimated uncertainty of $\pm 15\%$.

Table VI-A-9. Rate coefficients $k = A\ e^{-B/T}$ (cm^3 $molecule^{-1}$ s^{-1}) for the reaction: $OH + CH_2FCF_3$ (HFC-134a) → products

$10^{12} \times A$	B (K)	$10^{15} \times k$	T (K)	Technique[e]	Reference	Temperature Range (K)
3.2	1800 ± 200	5.5 ± 0.7	294	DF–RF	Clyne and Holt (1979b)	294–429
		13.2 ± 1.0	327			
		16.4 ± 3.1	344			
		19.2 ± 0.8	358			
		38.3 ± 4.9	393			
		42.0 ± 4.7	424			
		36.4 ± 3.8	429			
		5.2 ± 0.6	298	FP–RF	Martin and Paraskevopoulos (1983)	298
1.1	1424 ± 35	3.93 ± 0.24	249	DF–RF	Jeong et al. (1984)	249–473
		4.41 ± 0.04	250			
		5.52 ± 0.35	268			
		7.73 ± 0.71	291			
		8.23 ± 0.55	295			
		8.44 ± 0.73	298			
		15.4 ± 1.2	342			
		25.4 ± 1.7	380			
		39.4 ± 2.6	430			
		45.6 ± 2.9	447			
		64.4 ± 4.0	473			
5.8	1350	1.7 ± 0.1	231	DF–RF	Brown et al. (1990b)	231–423
		3.2 ± 1.0	253			
		4.6 ± 0.2	280			
		6.9 ± 1.5	301			
		8.5 ± 1.4	338			
		19 ± 3	373			
		29 ± 18	423			
3.7	1990 ± 280	2.63 ± 0.8	270	FP–RF	Liu et al. (1990)	270–400
		5.18 ± 0.7	298			
		8.08 ± 0.6	330			
		13.1 ± 1.4	350			
		18.1 ± 2.3	375			
		27.2 ± 3.2	400			
0.57	1430 ± 60	0.99 ± 0.085	223	DF–LMR FP–LIF	Gierczak et al. (1991)	223–450
		1.55 ± 0.13	243			
		2.76 ± 0.27	273			
		4.57 ± 0.33	294			
		4.26 ± 0.29	296			
		4.35 ± 0.38	297			
		4.50 ± 0.37	298			
		7.70 ± 0.69	324			
		14.1 ± 1.1	376			
		20.7 ± 1.8	404			
		25.0 ± 2.0	425			
		31.8 ± 2.6	450			
		2.38 ± 0.22	270	DF–RF	Zhang et al. (1992a)	270

(*continued*)

Table VI-A-9. *(continued)*

$10^{12} \times A$	B (K)	$10^{15} \times k$	T (K)	Technique[e]	Reference	Temperature Range (K)
$1.03^{+0.18}_{-0.15}$	1588 ± 52	5.0 ± 0.44	298	DF–EPR	Orkin and Khamaganov (1993a)	298–460
		8.3 ± 0.66	330			
		11.4 ± 0.8	360			
		14.2 ± 1.1	370			
		33.1 ± 2.4	460			
0.989 ± 0.138	1640 ± 112	1.62 ± 0.08	255	DF–RF	Leu and Lee (1994)	255–424
		2.53 ± 0.13	273			
		3.88 ± 0.11	298			
		8.60 ± 0.27	347			
		11.3 ± 0.4	375			
		22.8 ± 0.4	424			
		4.6 ± 0.5	295	PLP–UVabs	Bednarek et al. (1996)	295
1.03	1667	3.66[a]	298	RR (CHF_2CF_3)	DeMore (1993a)	297–357
		4.02	300			
		4.43	304			
		5.24	313			
		6.18	329			
		7.29	337			
		7.24	341			
		8.93	350			
		9.37	351			
1.69	1821	4.22[b]	297	RR (CH_4)		
		3.70	298			
		3.44	298			
		3.27	298			
		4.11	298			
		3.89	298			
		3.79	298			
		6.07	323			
		9.13	346			
		9.30	353			
1.49	1763	4.09[c]	298	RR (CH_3CCl_3)		
		4.10	298			
		4.10	298			
		3.96	298			
		4.58	308			
		6.47	323			
		6.69	328			
		8.57	340			
		9.56	349			
		9.48	351			
		10.94	357			
		4.19 ± 0.51[c]	298	RR (CH_3CCl_3)	Barry et al. (1995)	298
		4.55 ± 0.20[d]	298	RR (CH_2FCHF_2)		

[a] Placed on an absolute basis using $k(OH + CHF_2CF_3) = 5.1 \times 10^{-13} \exp(-1666/T)$ cm^3 molecule^{-1} s^{-1}.
[b] Placed on an absolute basis using $k(OH + CH_4) = 1.85 \times 10^{-20}\, T^{2.82} \exp(-987/T)$ cm^3 molecule^{-1} s^{-1}.
[c] Placed on an absolute basis using $k(OH + CH_3CCl_3) = 1.81 \times 10^{-12} \exp(-1543/T)$ cm^3 molecule^{-1} s^{-1}.
[d] Placed on an absolute basis using $k(OH + CH_2FCHF_2) = 4.11 \times 10^{-12} \exp(-1648/T)$ cm^3 molecule^{-1} s^{-1}.
[e] Abbreviations of the experimental techniques employed are defined in table II-A-1.

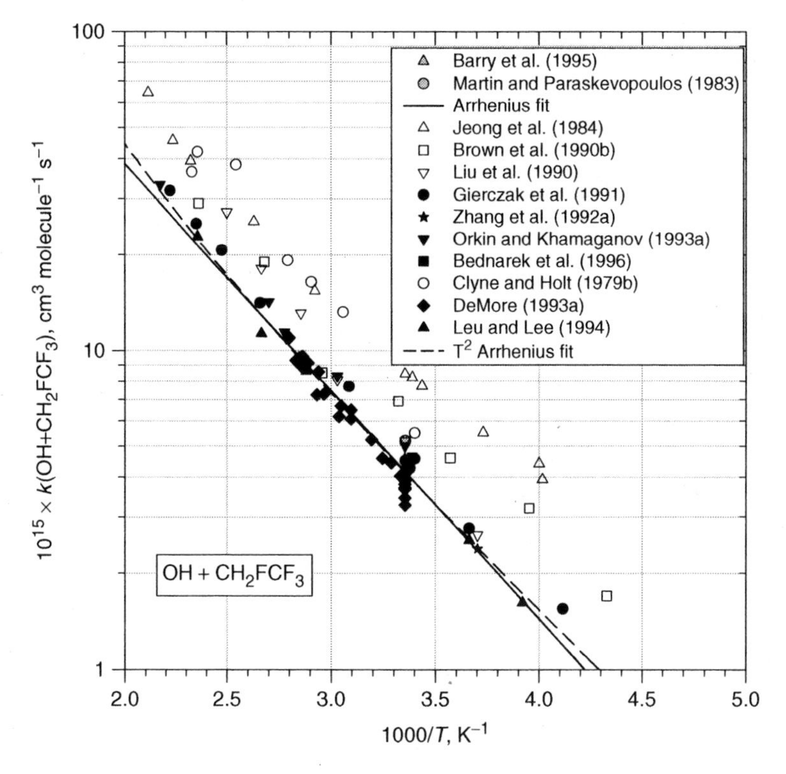

Figure VI-A-9. Arrhenius plot of the rate coefficients for the $OH + CH_2FCF_3$ reaction.

VI-A-2.10. $OH + CHF_2CHF_2(HFC\text{-}134) \rightarrow H_2O + CHF_2CF_2$

The rate coefficients of Clyne and Holt (1979b) and DeMore (1993a) are given in table VI-A-10 and plotted in figure VI-A-10. As seen from this figure, the data from Clyne and Holt are consistent with, but more scattered than, those from DeMore. An Arrhenius fit to the data of DeMore gives $k(OH + CHF_2CHF_2) = 1.27 \times 10^{-12} \times \exp(-1598/T)\,cm^3\ molecule^{-1}\ s^{-1}$. This expression gives $k(OH + CHF_2CHF_2) = 5.96 \times 10^{-15} cm^3\ molecule^{-1}\ s^{-1}$ at 298 K, with an estimated uncertainty of $\pm 25\%$.

VI-A-2.11. $OH + CHF_2CF_3(HFC\text{-}125) \rightarrow H_2O + CF_2CF_3$

The rate coefficients of Clyne and Holt (1979b), Martin and Paraskevopoulos (1983), Brown et al. (1990b), Talukdar et al. (1991a), and DeMore (1993c) are given in table VI-A-11. As is seen from figure VI-A-11, the absolute rate data from Martin and Paraskevopoulos (1983) and Talukdar et al. (1991a) and relative rate data of DeMore (1993c) are in good agreement over the temperature range over which they can be compared. However, the absolute rate data of Clyne and Holt (1979b) and Brown et al. (1990b) are systematically higher than the other

Table VI-A-10. Rate coefficients $k = A\ e^{-B/T}$ (cm^3 $molecule^{-1}$ s^{-1}) for the reaction: $OH + CHF_2CHF_2$ (HFC-134) → products

$10^{12} \times A$	B (K)	$10^{15} \times k$	T (K)	Technique[d]	Reference	Temperature Range (K)
2.76	1800 ± 400	5.3 ± 1.5	294	DF–RF	Clyne and Holt (1979b)	294–434
		18.8 ± 2.7	333			
		21.2 ± 4.1	389			
		48.2 ± 3.6	434			
1.88	1712	5.91[a]	298	RR (CH_3CCl_3)	DeMore (1993a)	298–358
		6.08	298			
		5.93	298			
		6.41	298			
		5.99	298			
		5.72	298			
		7.50	308			
		8.51	318			
		10.65	330			
		12.54	340			
		13.07	346			
		15.22	356			
1.07	1560	6.08[b]	298	RR (CHF_2CF_3)		
		5.31	298			
		5.45	298			
		6.70	305			
		6.77	305			
		7.71	313			
		8.64	320			
		9.41	335			
		10.97	345			
		13.43	355			
		13.42	355			
		14.55	358			
1.44	1627	5.92[c]	298	RR(CH_2FCF_3)		
		6.09	298			
		7.06	303			
		7.65	310			
		9.12	323			
		11.17	330			
		13.31	349			
		14.94	356			

[a] Placed on an absolute basis using $k(OH + CH_3CCl_3) = 1.81 \times 10^{-12} \exp(-1543/T)$ cm^3 $molecule^{-1}$ s^{-1}.
[b] Placed on an absolute basis using $k(OH + CHF_2CF_3) = 5.10 \times 10^{-13} \exp(-1666/T)$ cm^3 $molecule^{-1}$ s^{-1}.
[c] Placed on an absolute basis using $k(OH + CH_2FCF_3) = 1.01 \times 10^{-12} \exp(-1640/T)$ cm^3 $molecule^{-1}$ s^{-1}.
[d] Abbreviations of the experimental techniques employed are defined in table II-A-1.

studies, with the difference decreasing with increasing temperature. This trend suggests that the studies by Clyne and Holt (1979b) and Brown et al. (1990b) were affected by the presence of reactive impurities. A fit of the Arrhenius expression to the data from Martin and Paraskevopoulos (1983), Talukdar et al. (1991a), and DeMore (1993c) shown in figure VI-A-11 gives $k(OH + CHF_2CF_3) = 5.10 \times 10^{-13}$

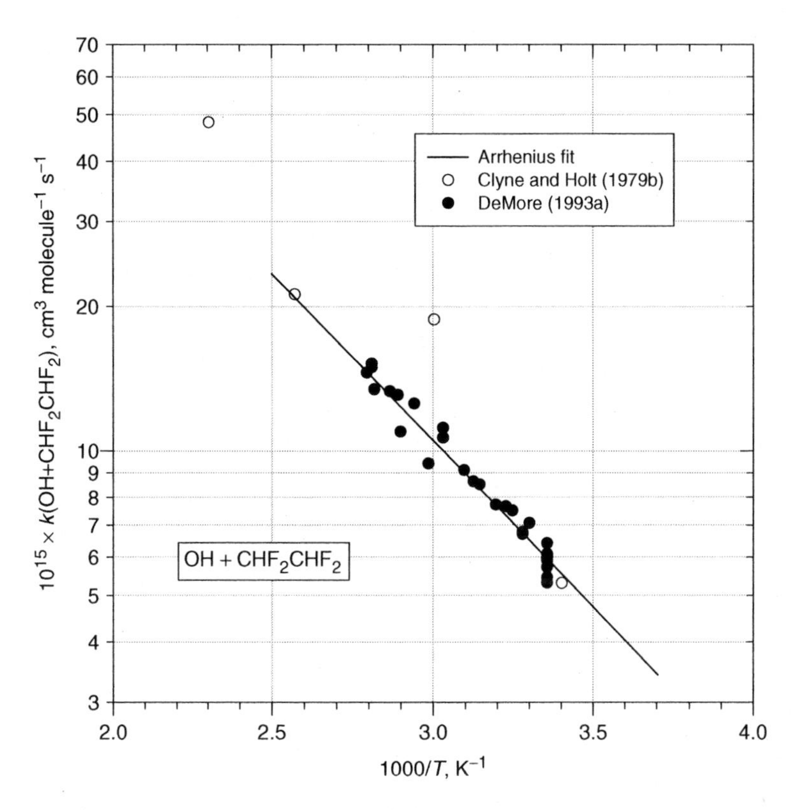

Figure VI-A-10. Arrhenius plot of the rate coefficients for the OH + CHF_2CHF_2 reaction.

Table VI-A-11. Rate coefficients $k = A\ e^{-B/T}$ (cm^3 $molecule^{-1}$ s^{-1}) for the reaction: OH + CHF_2CF_3 → products

$10^{13} \times A$	B (K)	$10^{15} \times k$	T (K)	Technique[b]	Reference	Temperature Range (K)
1.70	1100 ± 100	5.0 ± 2.2	294	DF–RF	Clyne and Holt (1979b)	294–441
		4.9 ± 1.4	294			
		6.2 ± 1.8	336			
		11.3 ± 3.3	378			
		15.8 ± 2.9	441			
		2.49 ± 0.27	298	FP–RA	Martin and Paraskevopoulos (1983)	298
2.8	1350	1.6 ± 2.2	226	DF–RF	Brown et al. (1990b)	226–423
		1.6 ± 1.4	257			
		2.9 ± 1.0	303			
		5.2 ± 0.8	341			

(continued)

Table VI-A-11. *(continued)*

$10^{13} \times A$	B (K)	$10^{15} \times k$	T (K)	Technique[b]	Reference	Temperature Range (K)
		8.2 ± 4.8	373			
		9.9 ± 14.5	423			
5.41	1700 ± 100	0.272 ± 0.028	220	DF–LMR	Talukdar et al.	220–364
		0.360 ± 0.022	228	PLP–LIF	(1991a)	
		0.376 ± 0.014	229			
		0.532 ± 0.038	245			
		0.751 ± 0.025	255			
		1.03 ± 0.03	273			
		1.70 ± 0.10	297			
		2.06 ± 0.26	299			
		2.89 ± 0.16	325			
		4.42 ± 0.03	343			
		4.51 ± 0.20	355			
		5.87 ± 0.20	364			
		1.64 ± 0.21[a]	298	RR (CH_4)	DeMore (1993c)	298

[a] Placed on an absolute basis using $k(OH + CH_4) = 1.85 \times 10^{-20}\, T^{2.82} \exp(-987/T)\, cm^3\, molecule^{-1}\, s^{-1}$.
[b] Abbreviations of the experimental techniques employed are defined in table II-A-1.

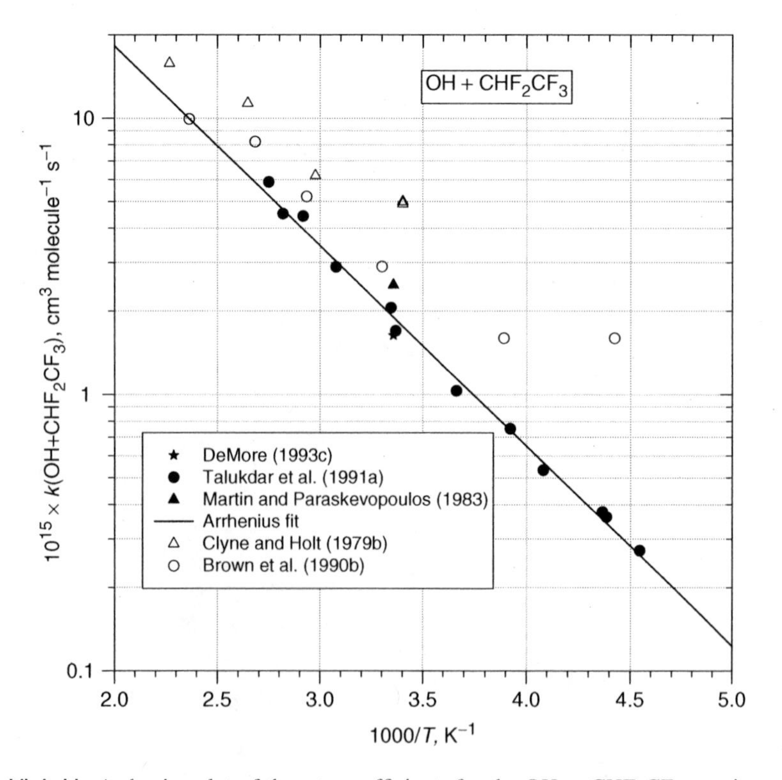

Figure VI-A-11. Arrhenius plot of the rate coefficients for the OH + CHF_2CF_3 reaction.

Table VI-A-12. Rate coefficients $k = A\ e^{-B/T}$ ($cm^3\ molecule^{-1}\ s^{-1}$) for the reaction: $OH + CF_3CH_2CH_3$ (HFC-263fb) → products

$10^{13} \times A$	B (K)	$10^{14} \times k$	T (K)	Technique[a]	Reference	Temperature Range (K)
		4.00 ± 0.13	295	DF–LIF	Nelson et al. (1995)	295
4.36 ± 0.72	1290 ± 40	1.96 ± 0.04	238	PLP–LIF	Rajakumar et al. (2006)	238–373
		1.97 ± 0.02	238			
		2.71 ± 0.06	256			
		3.96 ± 0.08	274			
		5.42 ± 0.04	297			
		5.55 ± 0.14	297			
		5.54 ± 0.04	297			
		5.48 ± 0.08	297			
		8.03 ± 0.12	323			
		10.7 ± 0.10	348			
		14.3 ± 0.10	373			

[a] Abbreviations of the experimental techniques employed are defined in table II-A-1.

$\exp(-1666/T)\ cm^3\ molecule^{-1}\ s^{-1}$. This expression gives $k(OH+CHF_2CF_3) = 1.90\times 10^{-15}\ cm^3\ molecule^{-1}\ s^{-1}$ at 298 K, with an estimated uncertainty of ±20%.

VI-A-2.12. $OH + CF_3CH_2CH_3$ (HFC-263fb) → Products

The rate coefficients of Nelson et al. (1995) and Rajakumar et al. (2006) are given in table VI-A-12 and plotted in figure VI-A-12. The rate coefficient of Nelson et al. (1995) lies approximately 25% lower than the values reported by Rajakumar et al. (2006). Such a difference is just within the combined likely uncertainties associated with the two studies. An Arrhenius fit to the combined data set gives $k(OH+CF_3CH_2CH_3) = 3.94\times 10^{-12} \times \exp(-1274/T)\ cm^3\ molecule^{-1}\ s^{-1}$. This expression gives $k(OH+CF_3CH_2CH_3) = 5.48 \times 10^{-14}\ cm^3\ molecule^{-1}\ s^{-1}$ at 298 K with an estimated uncertainty of ± 25%.

VI-A-2.13. $OH + CF_3CHFCH_2F$ (HFC-245eb) → Products

The rate coefficients of Nelson et al. (1995) and Rajakumar et al. (2006) are given in table VI-A-13 and plotted in figure VI-A-13. The rate coefficient of Nelson et al. (1995) lies approximately 20% lower than the values reported by Rajakumar et al. (2006). Such a difference is within the combined likely uncertainties associated with the two studies. An Arrhenius fit to the combined data set gives $k(OH+CF_3CHFCH_2F) = 1.23\times 10^{-12} \exp(-1257/T)\ cm^3\ molecule^{-1}\ s^{-1}$. This expression gives $k(OH+CF_3CHFCH_2F) = 1.81 \times 10^{-14}\ cm^3\ molecule^{-1}\ s^{-1}$ at 298 K, with an estimated uncertainty of ±20%.

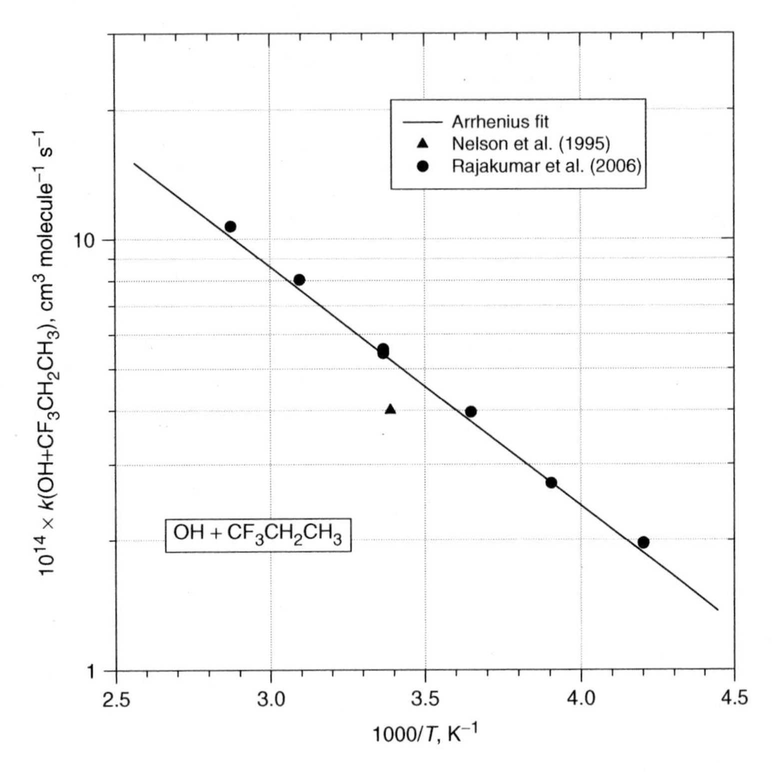

Figure VI-A-12. Arrhenius plot of the rate coefficients for the OH + $CF_3CH_2CH_3$ reaction.

Table VI-A-13. Rate coefficients $k = A\ e^{-B/T}$ (cm^3 $molecule^{-1}$ s^{-1}) for the reaction: OH + CF_3CHFCH_2F (HFC-245eb) → products

$10^{13} \times A$	B (K)	$10^{14} \times k$	T (K)	Technique[a]	Reference	Temperature Range (K)
		1.37 ± 0.03	294	DF–LIF	Nelson et al. (1995)	294
1.23 ± 0.18	1250 ± 40	0.63 ± 0.01	238	PLP–LIF	Rajakumar et al. (2006)	238–374
		0.63 ± 0.01	238			
		0.94 ± 0.01	254			
		1.22 ± 0.02	272			
		1.81 ± 0.04	297			
		1.79 ± 0.08	297			
		1.79 ± 0.06	297			
		1.80 ± 0.08	297			
		2.50 ± 0.06	324			
		3.43 ± 0.08	348			
		3.39 ± 0.12	349			
		4.58 ± 0.08	374			

[a] Abbreviations of the experimental techniques employed are defined in table II-A-1.

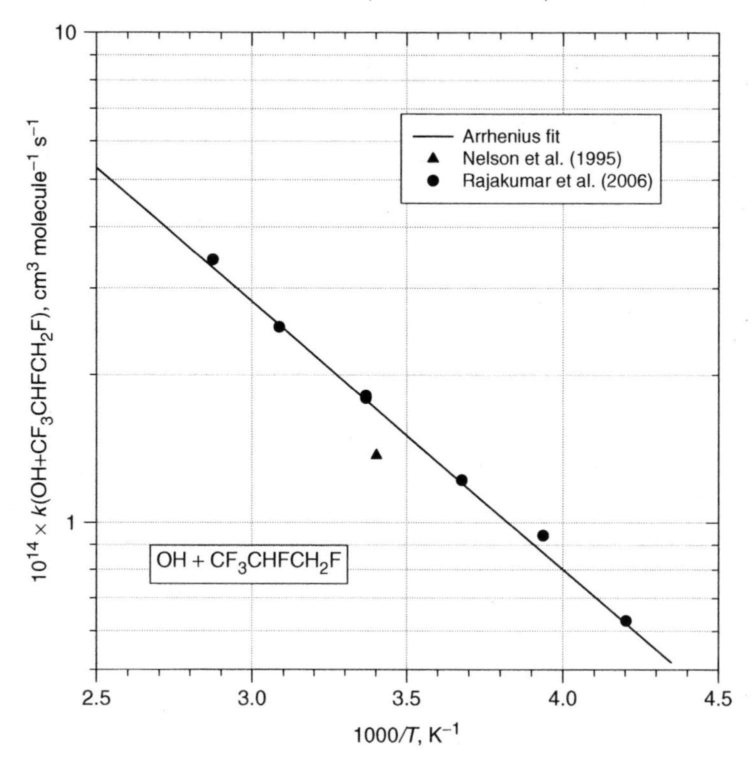

Figure VI-A-13. Arrhenius plot of the rate coefficients for the OH + CF_3CHFCH_2F reaction.

VI-A-2.14. OH + $CHF_2CHFCHF_2$ (HFC-245ea) → Products

The rate coefficients of Nelson et al. (1995) and Rajakumar et al. (2006) are given in table VI-A-14 and plotted in figure VI-A-14. The rate coefficient of Nelson et al. (1995) lies approximately 20% lower than the values reported by Rajakumar et al. (2006). Such a difference is probably within the combined uncertainties associated with the two studies. An Arrhenius fit to the combined data set gives $k(OH + CHF_2CHFCHF_2) = 1.93 \times 10^{-12} \exp(-1387/T)$ cm^3 molecule^{-1} s^{-1}. This expression gives $k(OH + CHF_2CHFCHF_2) = 1.84 \times 10^{-14}$ cm^3 molecule^{-1} s^{-1} at 298 K, with an estimated uncertainty of ±25%.

VI-A-2.15. OH + $CF_3CH_2CHF_2$ (HFC-245fa) → H_2O + CF_3CHCHF_2 (a); H_2O + $CF_3CH_2CF_2$ (b)

The rate coefficients of Nelson et al. (1995) and Orkin et al. (1996) are given in table VI-A-15 and plotted in figure VI-A-15. Results from the two studies are in excellent agreement. An Arrhenius fit to the combined data set gives $k(OH + CF_3CH_2CHF_2) = 6.71 \times 10^{-13} \exp(-1354/T)$ cm^3 molecule^{-1} s^{-1}. This expression gives $(OH + CF_3CH_2CHF_2) = 7.14 \times 10^{-15}$ cm^3 molecule^{-1} s^{-1} at 298 K, with estimated uncertainties of ±15%.

Table VI-A-14. Rate coefficients $k = A\ e^{-B/T}$ (cm^3 $molecule^{-1}$ s^{-1}) for the reaction: OH + $CHF_2CHFCHF_2$ (HFC-245ea) → products

$10^{13} \times A$	B (K)	$10^{14} \times k$	T (K)	Technique[a]	Reference	Temperature Range (K)
		1.52 ± 0.15	295	DF–LIF	Nelson et al. (1995)	295
1.91 ± 0.42	1375 ± 100	0.60 ± 0.01	238	PLP–LIF	Rajakumar et al. (2006)	238–374
		0.79 ± 0.01	254			
		1.17 ± 0.06	272			
		2.02 ± 0.04	297			
		1.90 ± 0.04	297			
		1.93 ± 0.02	297			
		1.87 ± 0.06	297			
		2.68 ± 0.08	324			
		3.70 ± 0.08	348			
		4.75 ± 0.06	374			

[a] Abbreviations of the experimental techniques employed are defined in table II-A-1.

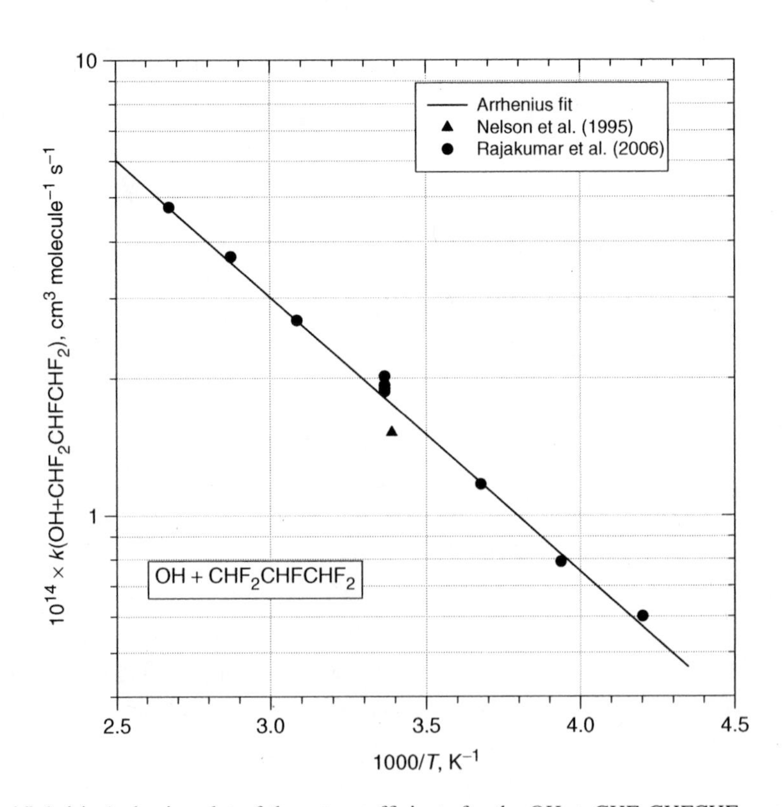

Figure VI-A-14. Arrhenius plot of the rate coefficients for the OH + $CHF_2CHFCHF_2$ reaction.

Table VI-A-15. Rate coefficients $k = A\ e^{-B/T}$ (cm^3 $molecule^{-1}$ s^{-1}) for the reaction: $OH + CF_3CH_2CHF_2$ (HFC-245fa) → products

$10^{13} \times A$	B (K)	$10^{15} \times k$	T (K)	Technique[a]	Reference	Temperature Range (K)
		6.12 ± 0.22	294	DF–LIF	Nelson et al. (1995)	294
$6.32^{+0.89}_{-0.078}$	1331 ± 43	4.73 ± 0.17	273	FP–RF	Orkin et al. (1996)	273–370
		5.86 ± 0.55	284			
		7.24 ± 0.02	298			
		9.26 ± 0.65	313			
		11.4 ± 0.49	330			
		14.0 ± 0.5	350			
		15.5 ± 0.5	360			
		17.2 ± 0.9	370			

[a] Abbreviations of the experimental techniques employed are defined in table II-A-1.

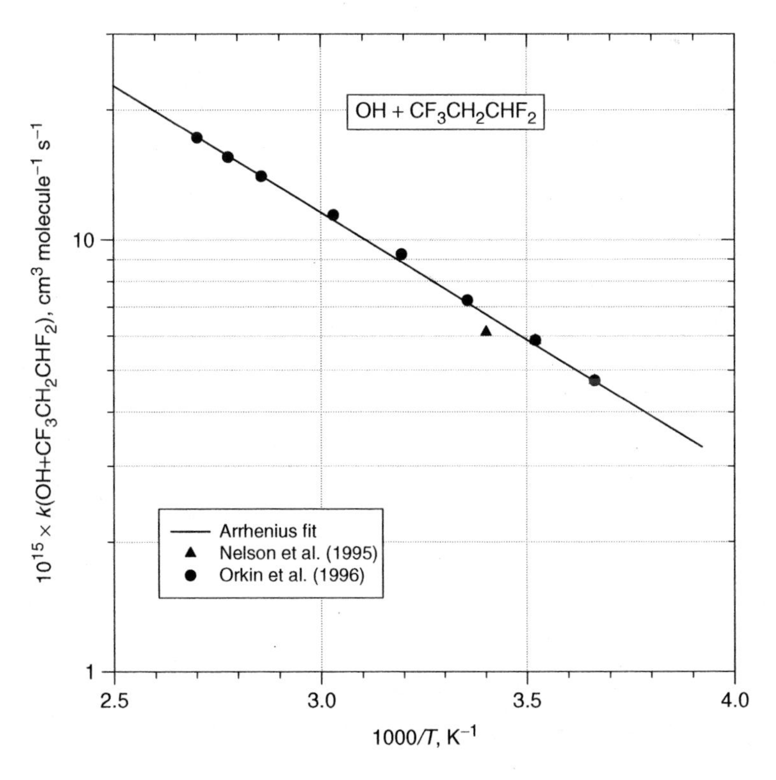

Figure VI-A-15. Arrhenius plot of the rate coefficients for the $OH + CF_3CH_2CHF_2$ reaction.

VI-A-2.16. $OH + CHF_2CF_2CH_2F$ (HFC-245ca) → $H_2O + CHF_2CF_2CHF$ (a); $OH + CHF_2CF_2CH_2F$ (HFC-245ca) → $H_2O + CF_2CF_2CH_2F$ (b)

The rate coefficients of Zhang et al. (1994) and Hsu and DeMore (1995) are given in table VI-A-16 and plotted in figure VI-A-16. The results from the absolute rate study of Zhang et al. are systematically larger and have a less pronounced temperature dependence than those from the relative rate study by Hsu and DeMore. It seems likely that the study by Zhang et al. (1994) is complicated by the presence of reactive impurities. An Arrhenius fit to the data from Hsu and DeMore (1995) gives $k(OH+CHF_2CF_2CH_2F) = 2.48 \times 10^{-12} \exp(-1731/T)$ cm^3 molecule^{-1} s^{-1}. This expression leads to $k(OH+CHF_2CF_2CH_2F) = 7.44 \times 10^{-15}$ cm^3 molecule^{-1} s^{-1} at 298 K, with an estimated uncertainty of ±25%.

VI-A-2.17. $OH + CF_3CF_2CH_2F$ (HFC-236cb) → $H_2O + CF_3CF_2CHF$

The rate coefficients of Garland et al. (1993) are given in table VI-A-17 and plotted in figure VI-A-17. An Arrhenius fit to the data of Garland et al. (1993) gives $k(OH+CF_3CF_2CH_2F) = 2.61 \times 10^{-13} \exp(-1087/T)$ cm^3 molecule^{-1} s^{-1}. This expression yields $k(OH+CF_3CF_2CH_2F) = 6.80 \times 10^{-15}$ cm^3 molecule^{-1} s^{-1} at 298 K, with an estimated uncertainty of ±25%.

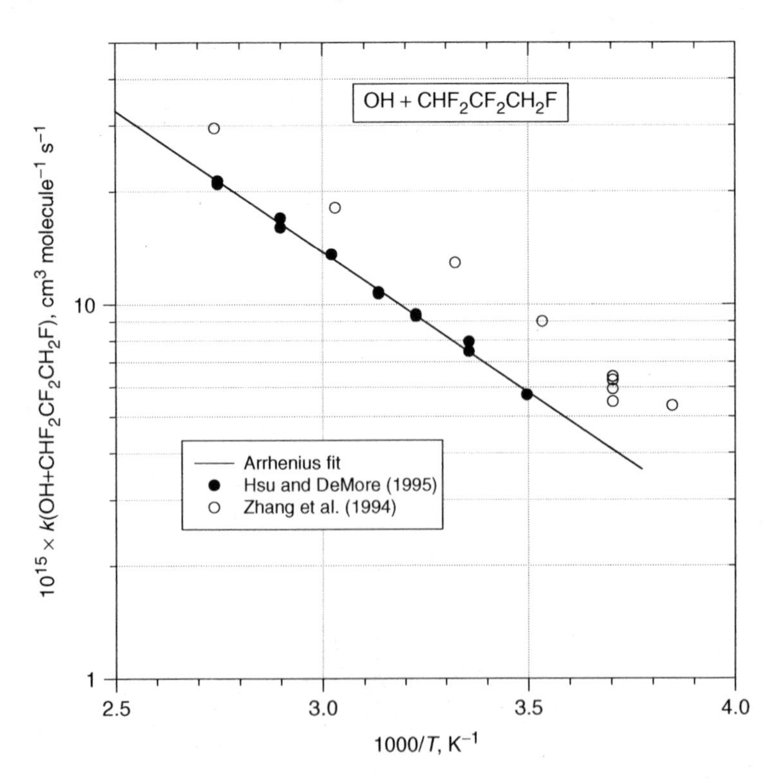

Figure VI-A-16. Arrhenius plot of the rate coefficients for the $OH + CHF_2CF_2CH_2F$ reaction.

Table VI-A-16. Rate coefficients $k = A\ e^{-B/T}$ (cm^3 $molecule^{-1}$ s^{-1}) for the reaction: $OH + CHF_2CF_2CH_2F$ (HFC-245ca) → products

$10^{12} \times A$	B (K)	$10^{15} \times k$	T (K)	Technique[b]	Reference	Temperature Range (K)
0.36	1613 ± 135	5.35 ± 0.90	260	FP–RF	Zhang et al. (1994)	260–365
		6.41 ± 1.10	270			
		5.48 ± 0.48	270			
		6.27 ± 0.80	270			
		5.93 ± 0.65	270			
		9.00 ± 1.14	283			
		12.9 ± 1.3	301			
		18.1 ± 1.7	330			
		29.5 ± 3.5	365			
2.48	1731	5.72[a]	286	RR (CH_4)	Hsu and DeMore (1995)	286–364
		5.71	286			
		7.94	298			
		7.49	298			
		9.41	310			
		9.28	310			
		10.8	319			
		10.7	319			
		13.6	331			
		13.6	331			
		16.0	345			
		17.0	345			
		21.4	364			
		21.0	364			

[a] Placed on an absolute basis using $k(OH + CH_4) = 1.85 \times 10^{-20}\ T^{2.82}\ \exp(-987/T)$ cm^3 $molecule^{-1}$ s^{-1}.
[b] Abbreviations of the experimental techniques employed are defined in table II-A-1.

Table VI-A-17. Rate coefficients $k = A\ e^{-B/T}$ (cm^3 $molecule^{-1}$ s^{-1}) for the reaction: $OH + CF_3CF_2CH_2F$ (HFC-236cb) → products

$10^{13} \times A$	B (K)	$10^{15} \times k$	T (K)	Technique[a]	Reference	Temperature Range (K)
2.6	1107 ± 202	4.3 ± 0.5	256	PLP–LIF	Garland et al. (1993)	256–314
		3.7 ± 0.5	257			
		4.6 ± 0.5	277			
		6.4 ± 0.6	295			
		8.9 ± 0.9	314			

[a] Abbreviations of the experimental techniques employed are defined in table II-A-1.

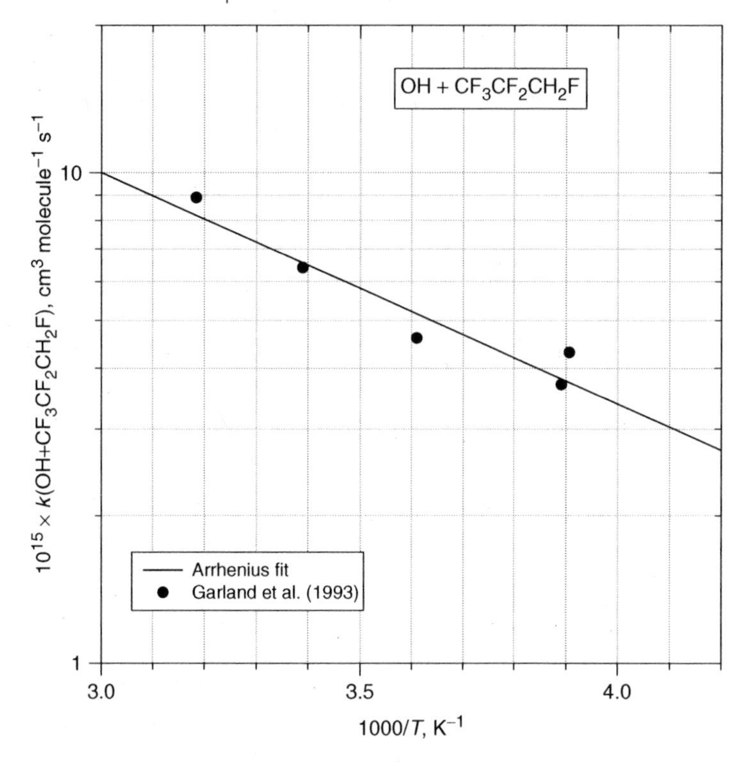

Figure VI-A-17. Arrhenius plot of the rate coefficients for the OH + $CF_3CF_2CH_2F$ reaction.

VI-A-2.18. OH + $CF_3CH_2CF_3$ (HFC-236fa) → H_2O + CF_3CHCF_3

The rate coefficients of Garland et al. (1993), Nelson et al. (1995), Hsu and DeMore (1995), Garland and Nelson (1996), Gierczak et al. (1996), and Barry et al. (1997) are given in table VI-A-18 and plotted in figure VI-A-18. Data from Gierczak et al. (1996) are not considered here, since they were acquired using $CF_3CH_2CF_3$ samples that were not treated with H_2SO_4 (to remove olefins) and Gierczak et al. did not include them in the Arrhenius fit. The data of Garland et al. (1993), Nelson et al. (1995), and the lowest temperature data point from Garland and Nelson (1996) are substantially larger than those of the other studies. As discussed by Garland and Nelson (1996), the data from Garland et al. (1993) and Nelson et al. (1995) are believed to have been affected by reactive impurities. The lowest temperature data point from Garland and Nelson (1996) is questionable because of possible incomplete thermalization of the reaction cell. The measurement by Barry et al. (1997) was relative to $k(OH + CF_3CH_2CF_2CH_3)$, which is itself rather uncertain (see section VI-A-2.21). An Arrhenius fit to the remaining data of Garland and Nelson (1996), Gierczak et al. (1996), and Hsu and DeMore (1995) gives $k(OH + CF_3CH_2CF_3) = 1.28 \times 10^{-12} \exp(-2464/T)$ cm^3 molecule^{-1} s^{-1}. This expression gives $(OH + CF_3CH_2CF_3) = 3.28 \times 10^{-16}$ cm^3 molecule^{-1} s^{-1} at 298 K, with an estimated uncertainty of ±20%.

Table VI-A-18. Rate coefficients $k = A\ e^{-B/T}$ (cm^3 $molecule^{-1}$ s^{-1}) for the reaction: $OH + CF_3CH_2CF_3$ (HFC-236fa) → products

$10^{12} \times A$	B (K)	$10^{16} \times k$	T (K)	Technique[c]	Reference	Temperature Range (K)
0.02 ± 0.01	906 ± 151	6.0 ± 0.6	257	PLP–LIF	Garland et al. (1993)	257–311
		6.6 ± 0.8	270			
		10.9 ± 0.9	311			
		5.1 ± 0.7	294	DF–LIF	Nelson et al. (1995)	294
0.56	2202	3.77[a]	298	RR (CHF_2CF_3)	Hsu and DeMore (1995)	298–367
		3.66	298			
		4.10	306			
		4.53	312			
		6.10	323			
		7.19	333			
		9.45	344			
		11.06	354			
		11.49	355			
		14.87	367			
		3.60 ± 0.27	281	PLP–LIF	Garland and Nelson (1996)	281–337
		4.15 ± 0.33	298			
		8.07 ± 0.90	325			
		8.82 ± 0.75	337			
1.6 ± 0.4	2540 ± 150	1.35 ± 0.24	273	FP–LIF	Gierczak et al. (1996)	273–398
		1.93 ± 0.18	285			
		3.31 ± 0.26	296			
		3.09 ± 0.50	297			
		3.43 ± 0.16	299			
		3.90 ± 0.32	301			
		5.09 ± 0.52	312			
		5.09 ± 0.16	312			
		6.14 ± 0.32	323			
		6.62 ± 0.49	328			
		7.51 ± 0.44	333			
		9.07 ± 0.43	344			
		11.8 ± 0.8	353			
		14.7 ± 1.6	363			
		18.4 ± 0.7	373			
		21.7 ± 1.1	383			
		27.9 ± 1.8	398			
		4.4 ± 1.5[b]	298	RR ($CF_3CH_2CF_2CH_3$)	Barry et al. (1997)	298

[a] Placed on an absolute basis using $k(OH + CHF_2CF_3) = 5.10 \times 10^{-13} \exp(-1666/T)$ cm^3 $molecule^{-1}$ s^{-1}.
[b] Placed on an absolute basis using $k(OH + CF_3CH_2CF_2CH_3) = 7.34 \times 10^{-15}$ cm^3 $molecule^{-1}$ s^{-1}.
[c] Abbreviations of the experimental techniques employed are defined in table II-A-1.

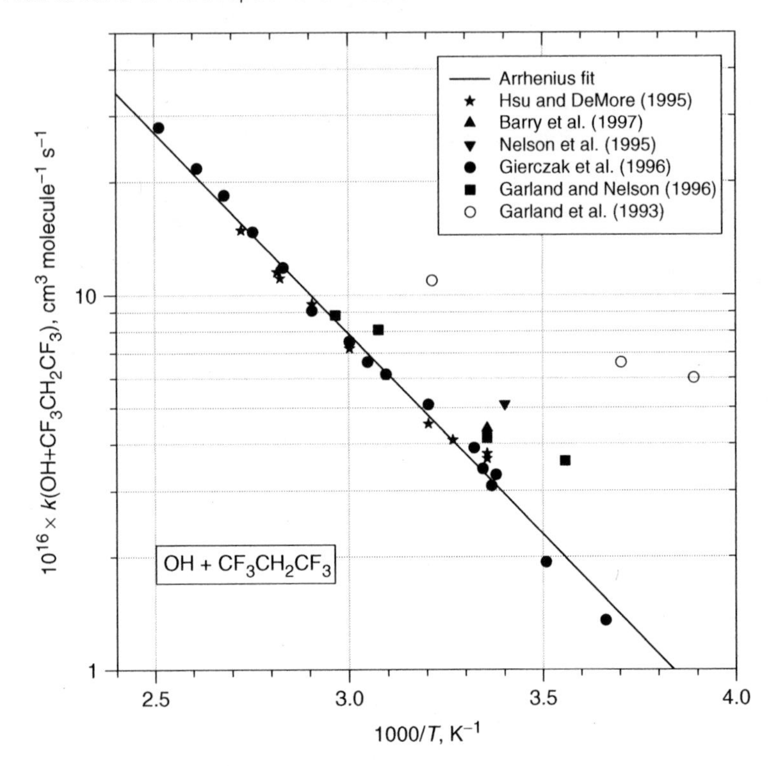

Figure VI-A-18. Arrhenius plot of the rate coefficients for the OH + $CF_3CH_2CF_3$ reaction.

VI-A-2.19. OH + $CF_3CHFCHF_2$ (HFC-236ea) → H_2O + CF_3CHFCF_2; OH + $CF_3CHFCHF_2$ (HFC-236ea) → H_2O + CF_3CFCHF_2

The rate coefficients of Garland et al. (1993), Zhang et al. (1994), Nelson et al. (1995), and Hsu and DeMore (1995) are given in table VI-A-19 and plotted in figure VI-A-19. The data of Garland et al., and Zhang et al., lie systematically higher than those of Nelson et al. and Hsu and DeMore and may have been affected by the presence of reactive impurities. An Arrhenius fit to the data of Nelson et al. (1995) and Hsu and DeMore (1995) gives $k(OH+CF_3CHFCHF_2) = 1.34 \times 10^{-12} \exp(-1658/T)\,cm^3\,molecule^{-1}\,s^{-1}$. This expression gives $k(OH+CF_3CHFCHF_2) = 5.14\times 10^{-15}\,cm^3\,molecule^{-1}\,s^{-1}$ at 298 K, with an estimated uncertainty of ±25%.

VI-A-2.20. OH + CF_3CHFCF_3 (HFC-227ea) → H_2O + CF_3CFCF_3

The rate coefficients of Nelson et al. (1993), Zellner et al. (1994), Zhang et al. (1994), Hsu and DeMore (1995), and Tokuhashi et al. (2004) are given in table VI-A-20 and plotted in figure VI-A-20. The results from Nelson et al., Zellner et al., Zhang et al., and Hsu and DeMore are in excellent agreement. The results from Tokuhashi et al. lie approximately 20% below the other studies but are in agreement within the likely combined experimental uncertainties. An Arrhenius fit to the combined data

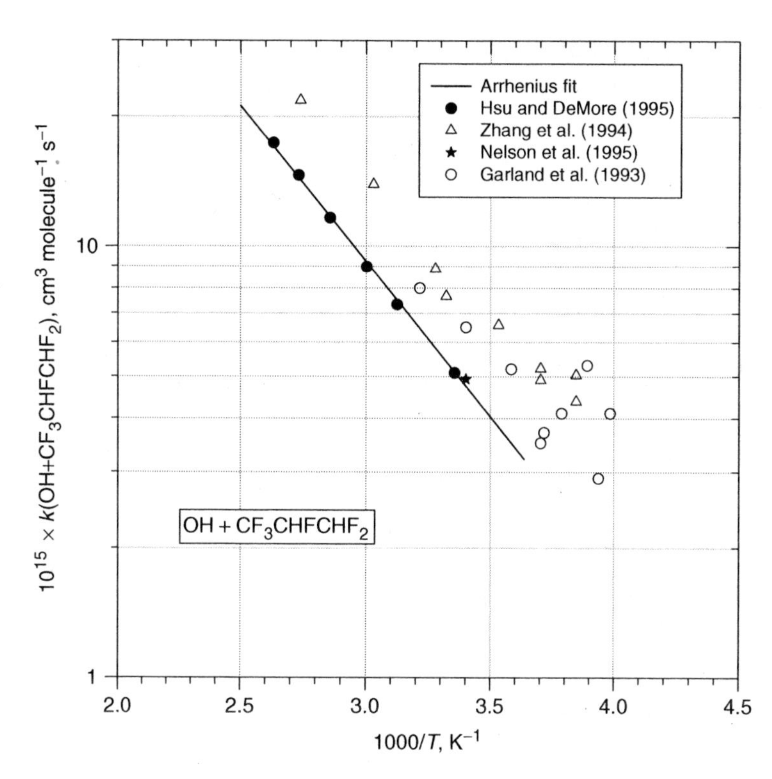

Figure VI-A-19. Arrhenius plot of the rate coefficients for the OH + $CF_3CHFCHF_2$ reaction.

Table VI-A-19. Rate coefficients $k = A\ e^{-B/T}$ (cm^3 $molecule^{-1}$ s^{-1}) for the reaction: OH + $CF_3CHFCHF_2$ (HFC-236ea) → products

$10^{12} \times A$	B (K)	$10^{15} \times k$	T (K)	Technique[b]	Reference	Temperature Range (K)
0.2	1007 ± 151	4.1 ± 0.4	251	PLP–LIF	Garland et al. (1993)	251–311
		2.9 ± 0.3	254			
		5.3 ± 0.5	257			
		4.1 ± 0.4	264			
		3.7 ± 0.3	269			
		3.5 ± 0.3	270			
		5.2 ± 0.4	279			
		6.5 ± 0.5	294			
		8.0 ± 0.7	311			
1.05	1434 ± 161	5.04 ± 0.76	260	FP–RF	Zhang et al. (1994)	260–365
		4.38 ± 0.83	260			
		5.22 ± 0.59	270			
		4.91 ± 0.54	270			
		6.58 ± 0.87	283			

(continued)

Table VI-A-19. *(continued)*

$10^{12} \times A$	B (K)	$10^{15} \times k$	T (K)	Technique[b]	Reference	Temperature Range (K)
		7.67 ± 0.99	301			
		8.87 ± 1.26	305			
		13.9 ± 1.6	330			
		21.8 ± 2.3	365			
		4.93 ± 0.25	294	DF–LIF	Nelson et al. (1995)	294
1.52	1703	5.09[a]	298	RR (CH_4)	Hsu and DeMore (1995)	298–380
		7.33	320			
		8.96	333			
		11.6	350			
		14.6	366			
		17.3	380			

[a] Placed on an absolute basis using $k(OH + CH_4) = 1.85 \times 10^{-20}\, T^{2.82} \exp(-987/T)\, cm^3\, molecule^{-1}\, s^{-1}$.
[b] Abbreviations of the experimental techniques employed are defined in table II-A-1.

Table VI-A-20. Rate coefficients $k = A\, e^{-B/T}$ ($cm^3\, molecule^{-1}\, s^{-1}$) for the reaction: $OH + CF_3CHFCF_3$ (HFC-227ea) → products

$10^{12} \times A$	B (K)	$10^{15} \times k$	T (K)	Technique[c]	Reference	Temperature Range (K)
0.37	1615 ± 190	1.44 ± 0.12	294	DF–LIF	Nelson et al. (1993)	294–369
		1.65 ± 0.28	295			
		1.66 ± 0.60	297			
		2.41 ± 0.13	321			
		2.53 ± 0.11	323			
		3.23 ± 0.24	343			
		3.32 ± 0.20	343			
		3.25 ± 0.63	346			
		4.93 ± 0.61	368			
		4.72 ± 0.33	369			
0.38 ± 0.08	1596 ± 77	1.8 ± 0.2	298	PLP–LIF	Zellner et al. (1994)	298–463
		3.1 ± 0.3	334			
		5.6 ± 1.1	378			
		8.0 ± 0.4	409			
		12.0 ± 0.2	463			
0.36	1613 ± 135	0.93 ± 0.07	270	FP–RF	Zhang et al. (1994)	270–365
		1.56 ± 0.11	296			
		1.60 ± 0.11	296			
		1.82 ± 0.27	313			
		2.70 ± 0.15	330			
		4.64 ± 0.55	365			
0.79	1853	1.49[a]	296	RR (CH_4)	Hsu and DeMore (1995)	296–398
		2.50	320			
		3.27	338			
		4.46	355			
		4.39	361			
		7.65	398			
0.43	1634	1.89[b]	298	RR (CHF_2CF_3)		

Table VI-A-20. *(continued)*

$10^{12} \times A$	B (K)	$10^{15} \times k$	T (K)	Technique[c]	Reference	Temperature Range (K)
		2.04	310			
		2.75	323			
		3.92	347			
		4.52	358			
		4.72	358			
		4.77	367			
0.619 ± 0.207	1830 ± 100	0.424 ± 0.037	250	FP–LIF	Tokuhashi et al.	250–430
		0.473 ± 0.041	250	PLP–LIF	(2004)	
		0.737 ± 0.065	273			
		1.23 ± 0.08	298			
		1.26 ± 0.09	298			
		2.33 ± 0.12	331			
		4.70 ± 0.22	375			
		9.02 ± 0.29	430			
		9.58 ± 0.36	430			

[a] Placed on an absolute basis using $k(OH + CH_4) = 1.85 \times 10^{-20}\ T^{2.82} \exp(-987/T)\,cm^3\ molecule^{-1}\ s^{-1}$.
[b] Placed on an absolute basis using $k(OH + CHF_2CF_3) = 5.10 \times 10^{-13} \exp(-1666/T)\,cm^3\ molecule^{-1}\ s^{-1}$.
[c] Abbreviations of the experimental techniques employed are defined in table II-A-1.

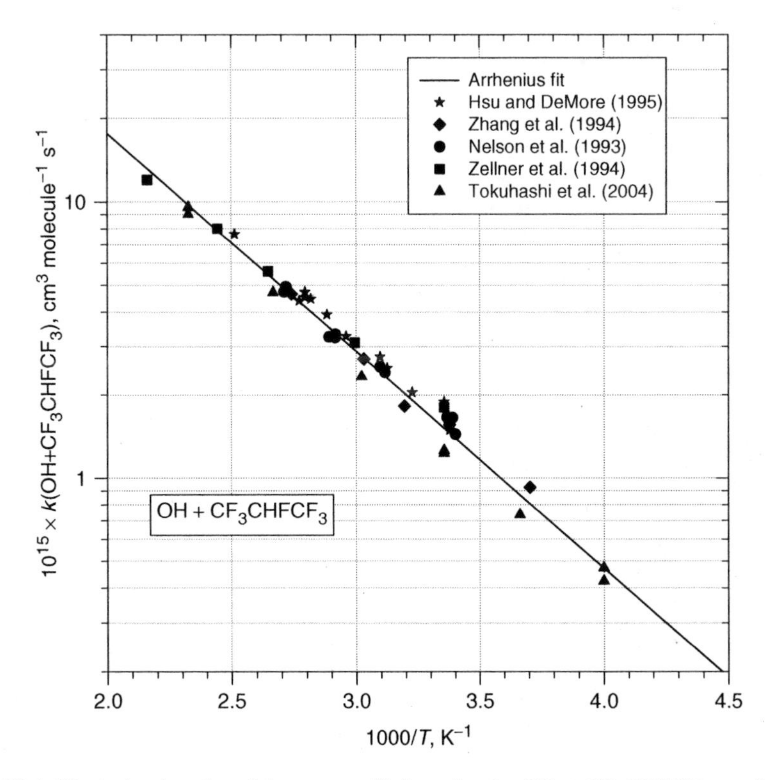

Figure VI-A-20. Arrhenius plot of the rate coefficients for the $OH + CF_3CHFCF_3$ reaction.

set gives $k(\mathrm{OH}+\mathrm{CF_3CHFCF_3}) = 6.55 \times 10^{-13} \exp(-1809/T)\,\mathrm{cm^3\ molecule^{-1}\ s^{-1}}$. This expression gives $k(\mathrm{OH}+\mathrm{CF_3CHFCF_3}) = 1.51 \times 10^{-15}\mathrm{cm^3\ molecule^{-1}\ s^{-1}}$ at 298 K, with an estimated uncertainty of ±20%. CF_3CFHCF_3 contains only half the number of C–H bonds and is more highly fluorinated than $CF_3CH_2CF_3$; both factors typically lead to reduced reactivity. However, CF_3CFHCF_3 is approximately five times more reactive than $CF_3CH_2CF_3$ at 298 K. Computational studies to investigate this unexpected effect would be of interest.

VI-A-2.21. OH + $CF_3CH_2CF_2CH_3$ (HFC-365mfc) → H_2O + $CF_3CHCF_2CH_3$ (a); OH + $CF_3CH_2CF_2CH_3$ (HFC-365mfc) → H_2O + $CF_3CH_2CF_2CH_2$ (b)

The rate coefficients of Mellouki et al. (1995) and Barry et al. (1997) are given in table VI-A-21 and plotted in figure VI-A-21. As is seen from the figure, the results from the absolute rate study by Mellouki et al. lie systematically above those from the relative rate study of Barry et al. A possible explanation of this discrepancy is the presence of a small amount of reactive impurity in the HFC-365mfc sample used by Mellouki et al. This explanation should lead to a trend of increasing discrepancy with decreasing temperature, but no such trend is evident. There is no obvious reason to prefer either study. An Arrhenius fit to the combined data set gives $k(\mathrm{OH}+\mathrm{CF_3CH_2CF_2CH_3}) = 2.04 \times 10^{-12} \exp(-1677/T)\,\mathrm{cm^3\ molecule^{-1}\ s^{-1}}$. This expression yields $k(\mathrm{OH}+\mathrm{CF_3CH_2CF_2CH_3}) = 7.34 \times 10^{-15}\mathrm{cm^3\ molecule^{-1}\ s^{-1}}$ at 298 K, with an uncertainty estimated to be ±30%.

Table VI-A-21. Rate coefficients $k = A\ e^{-B/T}$ ($\mathrm{cm^3\ molecule^{-1}\ s^{-1}}$) for the reaction: OH + $CF_3CH_2CF_2CH_3$ (HFC-365mfc)

$10^{12} \times A$	B (K)	$10^{15} \times k$	T (K)	Technique[b]	Reference	Temperature Range (K)
1.68 ± 0.21	1585 ± 80	4.65 ± 0.43	269	PLP–LIF	Mellouki et al. (1995)	269–370
		5.11 ± 0.68	273			
		5.05 ± 0.51	273			
		5.12 ± 0.62	273			
		6.37 ± 0.58	283			
		7.75 ± 0.71	293			
		7.50 ± 0.74	295			
		7.55 ± 0.78	295			
		7.72 ± 0.70	296			
		8.26 ± 0.71	298			
		9.11 ± 0.74	298			
		9.82 ± 0.83	308			
		13.14 ± 1.26	322			
		11.72 ± 1.09	323			
		16.60 ± 1.53	348			
		19.52 ± 1.82	358			

Table VI-A-21. *(continued)*

$10^{12} \times A$	B (K)	$10^{15} \times k$	T (K)	Technique[b]	Reference	Temperature Range (K)
		24.49 ± 2.23	368			
		23.40 ± 1.97	370			
2.12	1756	3.73[a]	278	RR (CH_3CCl_3)	Barry et al. (1997)	278–323
		4.86	287			
		5.92	298			
		7.27	312			
		9.45	323			

[a] Placed on an absolute basis using $k(OH + CH_3CCl_3) = 1.81 \times 10^{-12} \exp(-1543/T)$ cm^3 molecule^{-1} s^{-1}.
[b] Abbreviations of the experimental techniques employed are defined in table II-A-1.

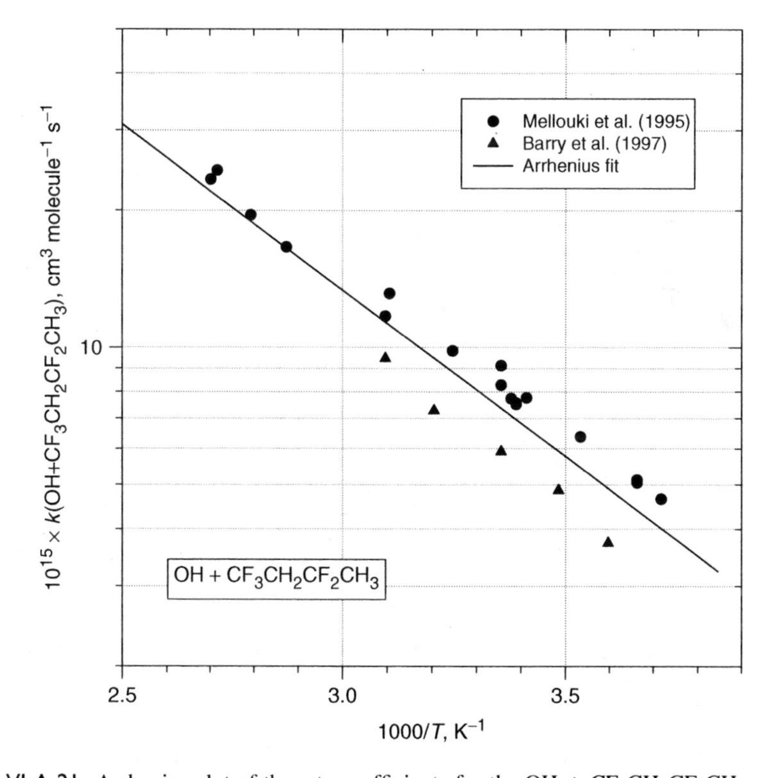

Figure VI-A-21. Arrhenius plot of the rate coefficients for the OH + $CF_3CH_2CF_2CH_3$ reaction.

VI-A-2.22. OH + $CF_3CH_2CH_2CF_3$ (HFC-356mff) → H_2O + $CF_3CHCH_2CF_3$

The rate coefficients of Zhang et al. (1994) and Nelson et al. (1995) are given in table VI-A-22 and plotted in figure VI-A-22. As is seen from the figure, the results from the two studies are in good agreement. An Arrhenius fit gives

$k(OH + CF_3CH_2CH_2CF_3) = 2.53 \times 10^{-12} \exp(-1701/T)$ cm^3 molecule^{-1} s^{-1}. This expression gives $k(OH + CF_3CH_2CH_2CF_3) = 8.40 \times 10^{-15}$ cm^3 molecule^{-1} s^{-1} at 298 K, with an estimated uncertainty of ±15%. Finally, note that Zhang et al. (1994) incorrectly named $CF_3CH_2CH_2CF_3$ as HFC-356ffa; the correct name is HFC-356mff; see section VI-A-1 for a description of the ANSI/ASHRAE haloalkane naming rules.

VI-A-2.23. OH + $CF_3CF_2CH_2CH_2F$ (HFC-356mcf) → Products

The rate coefficients of Nelson et al. (1995) are given in table VI-A-23 and plotted in figure VI-A-23. An Arrhenius fit gives $k(OH + CF_3CF_2CH_2CH_2F) = 1.66 \times 10^{-12} \exp(-1097/T)$ cm^3 molecule^{-1} s^{-1}. This expression gives $k(OH + CF_3CF_2CH_2CH_2F) = 4.18 \times 10^{-14}$ cm^3 molecule^{-1} s^{-1} at 298 K, with an uncertainty estimated to be ±25%.

VI-A-2.24. OH + $CHF_2CF_2CF_2CHF_2$ (HFC-338pcc) → H_2O + $CHF_2CF_2CF_2CF_2$

The rate coefficients of Zhang et al. (1992b) and Schmoltner et al. (1993) are given in table VI-A-24 and plotted in figure VI-A-24. The results of the two studies are in

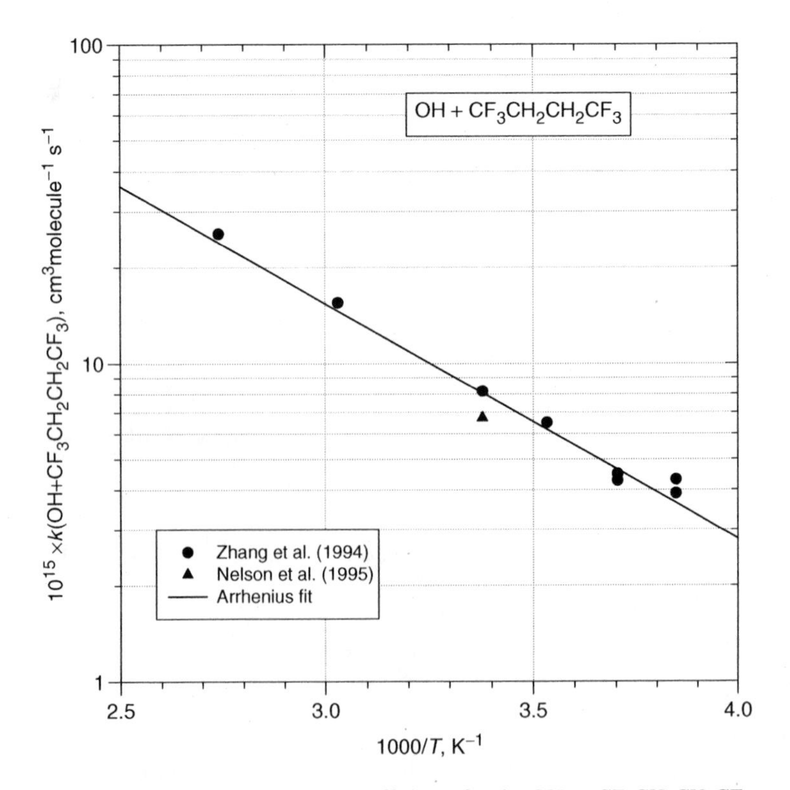

Figure VI-A-22. Arrhenius plot of the rate coefficients for the OH + $CF_3CH_2CH_2CF_3$ reaction.

good agreement. An Arrhenius fit gives $k(OH + CHF_2CF_2CF_2CHF_2) = 7.82 \times 10^{-13} \exp(-1548/T)$ cm^3 molecule^{-1} s^{-1}. This expression gives $k(OH + CHF_2CF_2CF_2CHF_2) = 4.34 \times 10^{-15}$ cm^3 molecule^{-1} s^{-1} at 298 K with an estimated uncertainty of ±20%.

Table VI-A-22. Rate coefficients $k = A\ e^{-B/T}$ (cm^3 molecule^{-1} s^{-1}) for the reaction: $OH + CF_3CH_2CH_2CF_3$ (HFC-356mff)

$10^{12} \times A$	B (K)	$10^{15} \times k$	T (K)	Technique[a]	Reference	Temperature Range (K)
2.94	1734 ± 87	4.31 ± 0.62	260	FP–RF	Zhang et al. (1994)	260–365
		3.90 ± 0.50	260			
		4.49 ± 0.51	270			
		4.28 ± 0.58	270			
		6.51 ± 0.69	283			
		8.17 ± 1.04	296			
		15.5 ± 1.1	330			
		25.5 ± 1.2	365			
		6.73 ± 0.14	296	DF–LIF	Nelson et al. (1995)	296

[a] Abbreviations of the experimental techniques employed are defined in table II-A-1.

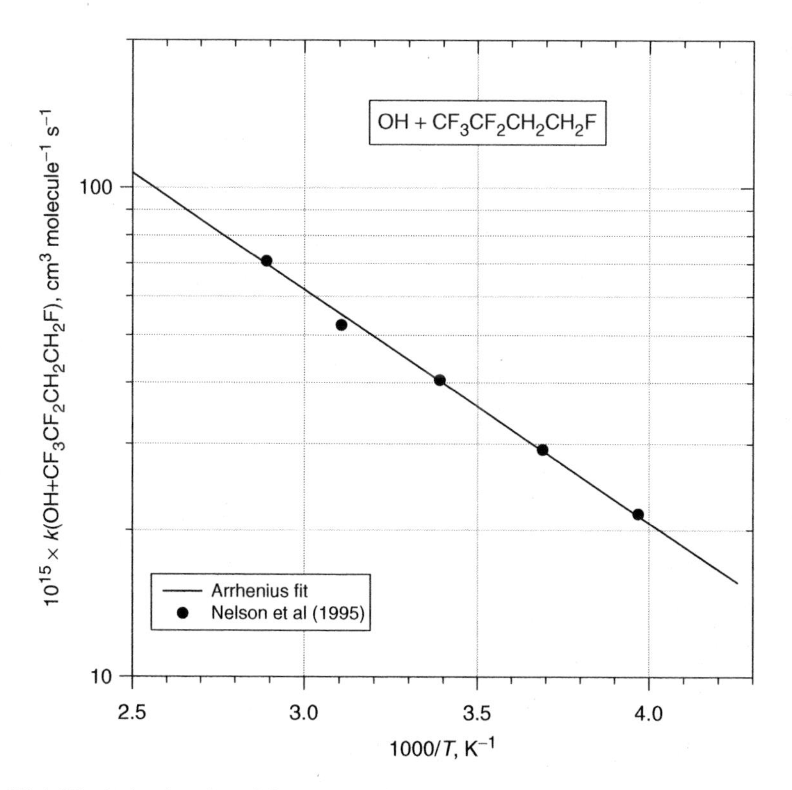

Figure VI-A-23. Arrhenius plot of the rate coefficients for the $OH + CF_3CF_2CH_2CH_2F$ reaction.

Table VI-A-23. Rate coefficients $k = A\ e^{-B/T}$ (cm^3 $molecule^{-1}$ s^{-1}) for the reaction: OH + $CF_3CF_2CH_2CH_2F$ (HFC-356mcf)

$10^{12} \times A$	B (K)	$10^{15} \times k$	T (K)	Technique[a]	Reference	Temperature Range (K)
1.74	1108	21.6 ± 1.1	252	DF–LIF	Nelson et al. (1995)	252–366
		29.2 ± 1.7	271			
		40.5 ± 2.3	295			
		52.4 ± 3.4	322			
		70.7 ± 4.7	346			
		85.2 ± 3.8	366			

[a] Abbreviations of the experimental techniques employed are defined in table II-A-1.

Table VI-A-24. Rate coefficients $k = A\ e^{-B/T}$ (cm^3 $molecule^{-1}$ s^{-1}) for the reaction: OH + $CHF_2CF_2CF_2CHF_2$ (HFC-338pcc) → H_2O + $CHF_2CF_2CF_2CF_2$

$10^{12} \times A$	B (K)	$10^{15} \times k$	T (K)	Technique[a]	Reference	Temperature Range (K)
0.78 ± 0.34	1510 ± 260	1.26 ± 0.50	245	FP–RF	Zhang et al. (1992b)	245–419
		3.22 ± 0.33	266			
		2.79 ± 0.24	266			
		4.18 ± 0.30	296			
		7.80 ± 1.20	335			
		11.2 ± 1.7	380			
		25.3 ± 2.6	419			
0.771 ± 0.34	1550 ± 60	1.04 ± 0.03	232	PLP–LIF	Schmoltner et al. (1993)	232–378
		1.41 ± 0.10	243			
		1.33 ± 0.05	243			
		1.82 ± 0.09	251			
		1.63 ± 0.04	252			
		1.78 ± 0.07	253			
		1.47 ± 0.12	253			
		2.11 ± 0.08	262			
		2.37 ± 0.23	263			
		2.47 ± 0.05	272			
		2.71 ± 0.11	273			
		3.61 ± 0.16	294			
		4.18 ± 0.17	294			
		4.07 ± 0.18	295			
		4.27 ± 0.12	296			
		3.86 ± 0.27	297			
		6.21 ± 0.39	323			
		6.44 ± 0.34	324			
		6.74 ± 0.22	328			
		9.93 ± 0.39	350			
		9.63 ± 0.62	351			
		13.55 ± 0.60	378			
		13.35 ± 0.56	378			

[a] Abbreviations of the experimental techniques employed are defined in table II-A-1.

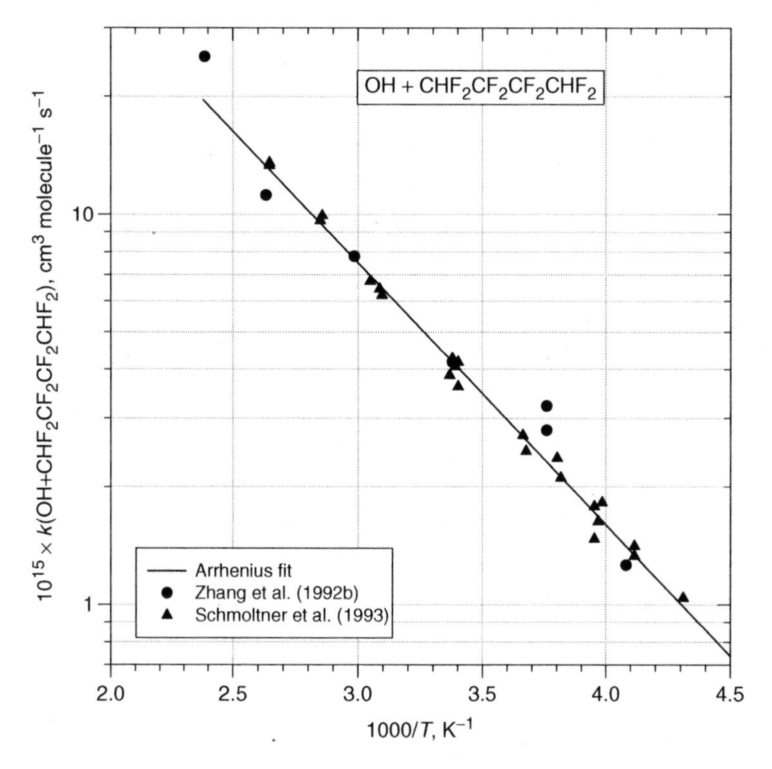

Figure VI-A-24. Arrhenius plot of the rate coefficients for the OH + $CHF_2CF_2CF_2CHF_2$ reaction.

VI-A-2.25. OH + $CF_3CH_2CF_2CH_2CF_3$ (HFC-458mfcf) → H_2O + $CF_3CH_2CF_2CHCF_3$

The rate coefficients of Nelson et al. (1995) are given in table VI-A-25 and plotted in figure VI-A-25. An Arrhenius fit gives $k(OH+CF_3CH_2CF_2CH_2CF_3) = 1.12 \times 10^{-12} \exp(-1804/T)\,cm^3\,molecule^{-1}\,s^{-1}$. This expression gives $k(OH+CF_3CH_2CF_2CH_2CF_3) = 2.63 \times 10^{-15} cm^3\,molecule^{-1}\,s^{-1}$ at 298 K, with an uncertainty estimated to be ±25%.

VI-A-2.26. OH + $CF_3CHFCHFCF_2CF_3$ (HFC-43-10mee) → Products

The rate coefficients of Zhang et al. (1992b) and Schmoltner et al. (1993) are given in table VI-A-26 and plotted in figure VI-A-26. The data from Schmoltner et al. (1993) lie below, but agree with the data from Zhang et al. (1992b) within the experimental uncertainties. An Arrhenius fit gives $k(OH+CF_3CHFCHFCF_2CF_3) = 5.68 \times 10^{-13} \exp(-1534/T)\,cm^3\,molecule^{-1}\,s^{-1}$. This expression gives $k(OH+CF_3CHFCHFCF_2CF_3) = 3.30 \times 10^{-15} cm^3\,molecule^{-1}$ s^{-1} at 298 K, with an estimated uncertainty of ±25%.

Table VI-A-25. Rate coefficients $k = A\ e^{-B/T}$ (cm^3 $molecule^{-1}$ s^{-1}) for the reaction: $OH + CF_3CH_2CF_2CH_2CF_3$ (HFC-458mfcf)

$10^{12} \times A$	B (K)	$10^{15} \times k$	T (K)	Technique[a]	Reference	Temperature Range (K)
1.23	1833	1.75 ± 0.11	278	DF–LIF	Nelson et al. (1995)	278–354
		2.55 ± 0.15	298			
		3.50 ± 0.20	313			
		4.86 ± 0.22	333			
		7.05 ± 0.33	354			

[a] Abbreviations of the experimental techniques employed are defined in table II-A-1.

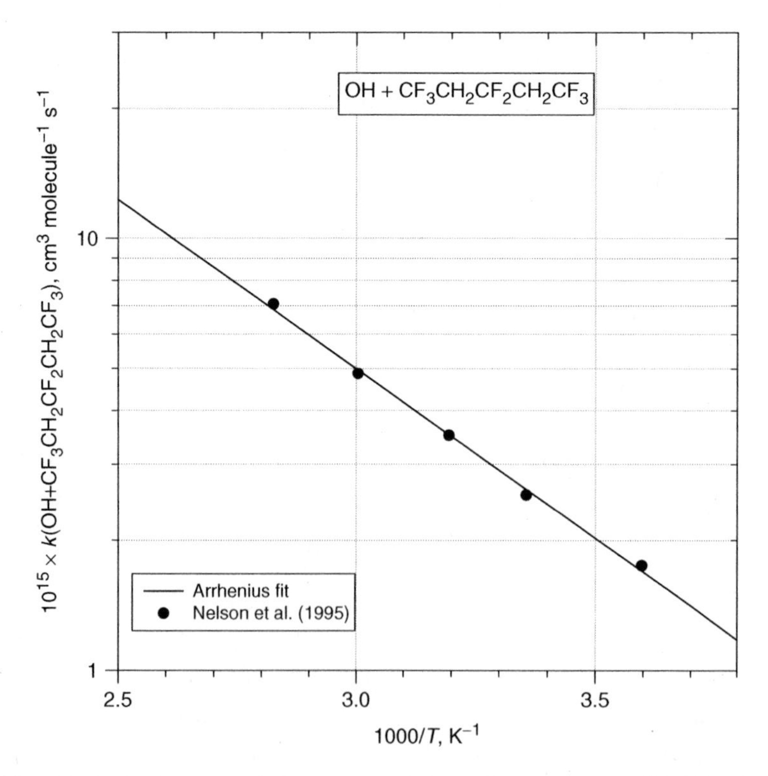

Figure VI-A-25. Arrhenius plot of the rate coefficients for the $OH + CF_3CH_2CF_2CH_2CF_3$ reaction.

Table VI-A-26. Rate coefficients $k = A\ e^{-B/T}$ ($cm^3\ molecule^{-1}\ s^{-1}$) for the reaction: $OH + CF_3CHFCHFCF_2CF_3$ (HFC-43-10mee)

$10^{12} \times A$	B (K)	$10^{15} \times k$	T (K)	Technique[a]	Reference	Temperature Range (K)
0.421 ± 0.128	1400 ± 180	1.56 ± 0.39	250	FP–RF	Zhang et al. (1992b)	250–400
		1.94 ± 0.29	270			
		2.67 ± 0.36	270			
		3.87 ± 0.38	295			
		6.30 ± 0.97	330			
		8.87 ± 1.03	365			
		12.8 ± 2.3	400			
0.646	1600 ± 50	1.16 ± 0.05	251.5	PLP–LIF	Schmoltner et al. (1993)	252–375
		1.45 ± 0.07	263			
		2.10 ± 0.16	278			
		2.88 ± 0.20	296			
		4.56 ± 0.42	324			
		6.54 ± 0.70	349			
		9.27 ± 0.54	375			

[a] Abbreviations of the experimental techniques employed are defined in table II-A-1.

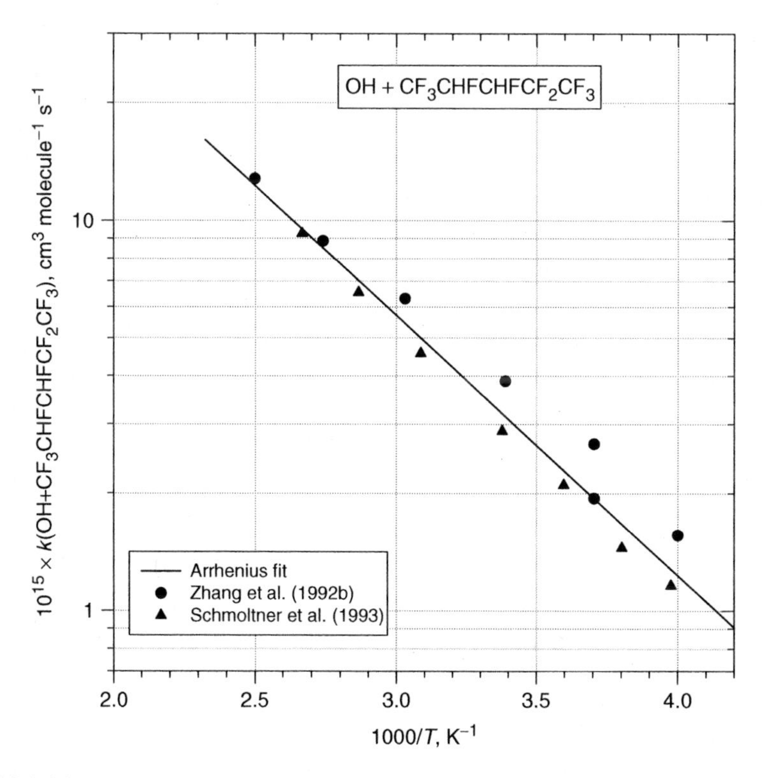

Figure VI-A-26. Arrhenius plot of the rate coefficients for the $OH + CF_3CHFCHFCF_2CF_3$ reaction.

VI-A-2.27. $OH + CF_3CF_2CH_2CH_2CF_2CF_3$ (HFC-55-10mcff) $\rightarrow H_2O + CF_3CF_2CH_2CHCF_2CF_3$

A single study of the rate coefficient of this compound has been made by Nelson et al. (1995) who used DF-LIF techniques. Their data lead to $k(OH + CF_3CF_2CH_2CF_2CF_3) = (7.87 \pm 0.38) \times 10^{-15}$ cm^3 molecule^{-1} s^{-1} at 295 K. Nelson et al. estimated an activation energy of 3.2 kcal (13.4 kJ) mol^{-1} and derived a value of $k(OH + CF_3CF_2CH_2CF_2CF_3) = (8.3 \pm 0.9) \times 10^{-15}$ cm^3 molecule^{-1} s^{-1} at 298 K, this value is recommended, with an estimated uncertainty of ±25%.

VI-A-2.28. $OH + CHF_2(CF_2)_4CF_3$ (HFC-52-13p) $\rightarrow$ Products

Chen et al. (2004) conducted absolute and relative rate studies of the $OH + CHF_2(CF_2)_4CF_3$ reaction. The rate coefficients of Chen et al. are given in table VI-A-27 and plotted in figure VI-A-27. While the absolute rate data are in general slightly (10%) lower than the relative rate results, the data obtained using the two experimental methods are consistent within the combined uncertainties. An Arrhenius fit gives $k(OH + CHF_2(CF_2)_4CF_3) = 5.67 \times 10^{-13} \times \exp(-1706/T)$ cm^3 molecule^{-1} s^{-1}. This expression gives $k(OH + CHF_2(CF_2)_4CF_3) = 1.85 \times 10^{-15}$ cm^3 molecule^{-1} s^{-1} at 298 K, with an estimated uncertainty of ±20%.

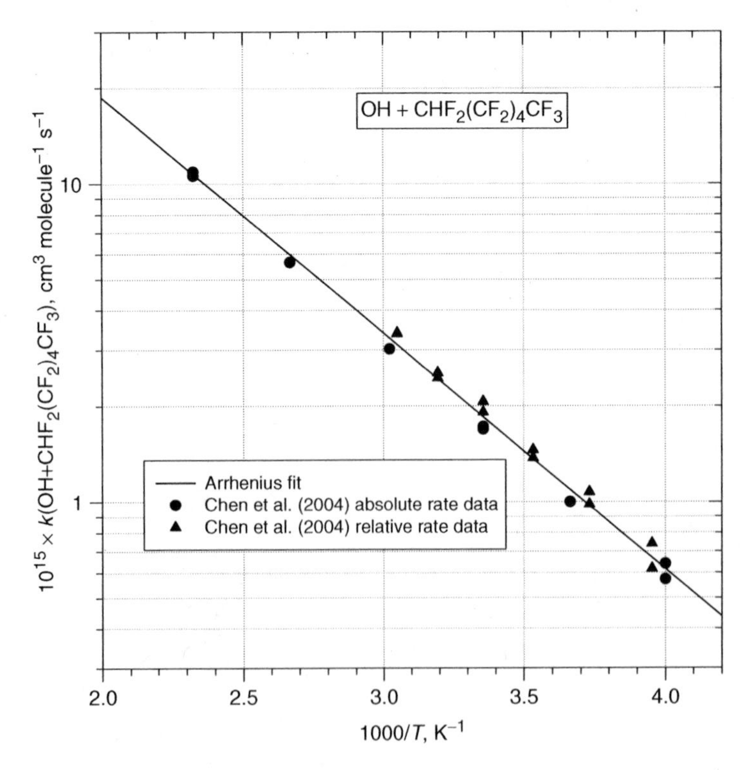

Figure VI-A-27. Arrhenius plot of the rate coefficients for the $OH + CHF_2(CF_2)_4CF_3$ reaction.

Table VI-A-27. Rate coefficients $k = A\ e^{-B/T}$ (cm^3 molecule^{-1} s^{-1}) for the reaction: $OH + CHF_2(CF_2)_4CF_3$ (HFC-52-13p)

$10^{12} \times A$	B (K)	$10^{15} \times k$	T (K)	Technique[c]	Reference	Temperature Range (K)
		0.571 ± 0.043	250	FP–LIF	Chen et al. (2004)	250–430
		1.00 ± 0.05	273			
		1.69 ± 0.07	298			
		3.02 ± 0.12	331			
		5.65 ± 0.23	375			
		10.9 ± 0.4	430			
		0.639 ± 0.043	250	PLP–LIF		
		1.72 ± 0.07	298			
		10.6 ± 0.3	430			
		0.62[a]	253	RR (CHF_2Cl)		
		0.98	268			
		1.37	283			
		1.91	298			
		2.45	313			
		3.40	328			
		0.74[b]	253	RR (CH_2FCF_3)		
		1.08	268			
		1.46	283			
		2.07	298			
		2.54	313			
		3.38	328			

[a] Placed on an absolute basis using $k(OH + CHF_2Cl) = 1.48 \times 10^{-12} \exp(-1710/T)$ cm^3 molecule^{-1} s^{-1}.
[b] Placed on an absolute basis using $k(OH + CH_2FCF_3) = 1.01 \times 10^{-12} \exp(-1640/T)$ cm^3 molecule^{-1} s^{-1}.
[c] Abbreviations of the experimental techniques employed are defined in table II-A-1.

VI-A-3. Hydrochlorocarbons

VI-A-3.1. $OH + CH_3Cl \rightarrow H_2O + CH_2Cl$

The rate coefficients of Cox et al. (1976a), Howard and Evenson (1976), Perry et al. (1976), Davis et al. (1976), Paraskevopoulos et al. (1981), Jeong and Kaufman (1982), Jeong et al. (1984), Brown et al. (1990c), Taylor et al. (1993), Herndon et al. (2001), and Hsu and DeMore (1994) are given in table VI-A-28 and plotted in figure VI-A-28. As is seen from the figure, the result of Cox et al. lies somewhat above the results from the other studies, but it is consistent with them within the experimental uncertainties. Significant curvature is evident in the Arrhenius plot in figure VI-A-28, and this is particularly pronounced for the data at $T > 400$ K. A fit of the Arrhenius expression to the $T < 400$ K data (except Cox et al., 1976a) gives $k(OH + CH_3Cl) = 2.71 \times 10^{-12} \exp(-1269/T)$ cm^3 molecule^{-1} s^{-1}. At 298 K this expression gives $k(OH + CH_3Cl) = 3.83 \times 10^{-14}$ cm^3 molecule^{-1} s^{-1}. A fit of the modified Arrhenius expression gives $k(OH + CH_3Cl) = 7.28 \times 10^{-18}\ T^2 \exp(-842/T)$ cm^3 molecule^{-1} s^{-1}. At 298 K this expression gives $k(OH + CH_3Cl) = 3.83 \times 10^{-14}$ cm^3 molecule^{-1} s^{-1}, with an uncertainty estimated to be ± 20%.

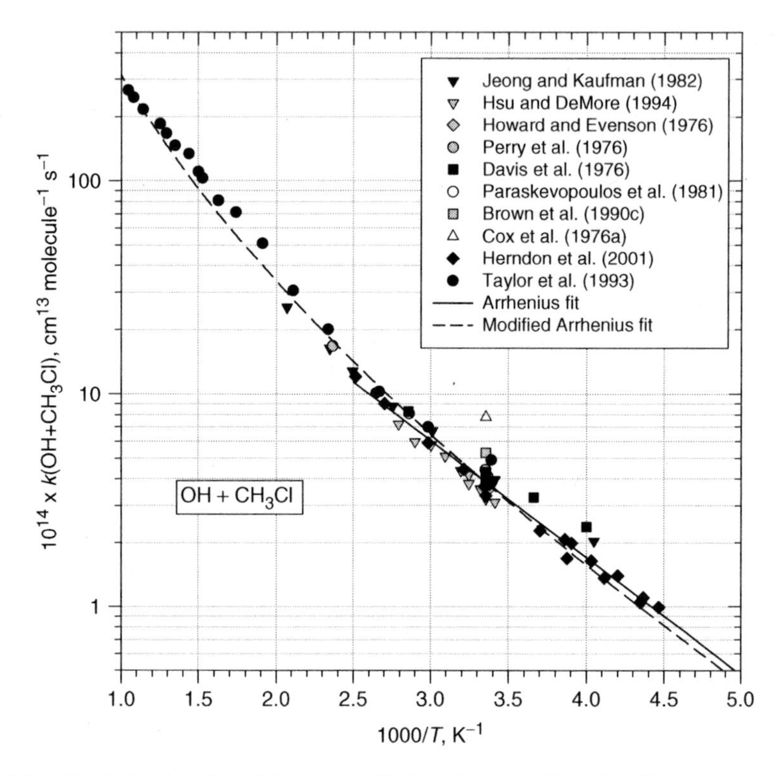

Figure VI-A-28. Arrhenius plot of the rate coefficients for the OH + CH_3Cl reaction.

Table VI-A-28. Rate coefficients $k = A\ e^{-B/T}$ (cm^3 $molecule^{-1}$ s^{-1}) for the reaction: OH + CH_3Cl → products

$10^{12} \times A$	B (K)	$10^{14} \times k$	T (K)	Technique[c]	Reference	Temperature Range (K)
		7.8 ± 3.9[a]	298	RR (CH_4)	Cox et al. (1976a)	298
		3.6 ± 0.8	296	DF–LMR	Howard and Evenson (1976)	296
4.1	1359 ± 151	4.4 ± 0.5	298.4	FP–RF	Perry et al. (1976)	298–423
		8.1 ± 0.8	349.3			
		16.8 ± 1.7	422.6			
1.8	1098 ± 36	2.38 ± 0.14	250	FP–RF	Davis et al. (1976)	250–350
		3.26 ± 0.06	273			
		3.98 ± 0.04	298			
		4.29 ± 0.21	298			
		8.28 ± 0.28	350			
		4.10 ± 0.69	297	FP–RA	Paraskevopoulos et al. (1981)	297
3.04 ± 0.43	1263 ± 45	2.03 ± 0.15	247	DF–RF	Jeong and Kaufman (1982)	247–483
		3.95 ± 0.26	293		Jeong et al. (1984)	
		6.68 ± 0.46	332			

Table VI-A-28. *(continued)*

$10^{12} \times A$	B (K)	$10^{14} \times k$	T (K)	Technique[c]	Reference	Temperature Range (K)
		8.74 ± 0.58	363			
		12.8 ± 0.90	401			
		16.3 ± 1.3	426			
		25.4 ± 2.0	483			
		5.3 ± 0.8	298	DF–RF	Brown et al. (1990c)	298
$k(T)$[d]		4.9 ± 0.6	295	PLP–LIF	Taylor et al. (1993)	295–955
		7.0 ± 0.6	335			
		10.3 ± 1.9	375			
		10.1 ± 1.1	378			
		20.1 ± 1.9	428			
		30.5 ± 3.7	474			
		50.8 ± 2.5	524			
		71.0 ± 3.7	575			
		80.8 ± 3.4	615			
		103.0 ± 3.5	655			
		110.0 ± 6.8	667			
		133.5 ± 8.1	695			
		146.3 ± 4.8	742			
		167.0 ± 4.6	775			
		185.0 ± 6.8	800			
		217.0 ± 10.3	875			
		247.1 ± 11.9	925			
		267.0 ± 31.7	955			
3.08	1342	3.10[b]	293	RR (CH_3CHF_2)	Hsu and DeMore (1994)	293–358
		3.22	298			
		3.54	298			
		3.57	301			
		4.17	308			
		3.79	308			
		4.37	313			
		5.10	323			
		5.75	333			
		5.96	345			
		7.20	358			
$k(T)$[e]		0.99 ± 0.15	224	PLP–LIF	Herndon et al. (2001)	224–398
		1.10 ± 0.20	229			
		1.04 ± 0.08	230			
		1.05 ± 0.08	230			
		1.39 ± 0.10	238			
		1.36 ± 0.09	243			
		1.64 ± 0.12	248			
		1.99 ± 0.05	256			
		1.68 ± 0.09	258			
		2.07 ± 0.17	259			
		2.28 ± 0.12	270			
		3.76 ± 0.14	296			
		3.79 ± 0.10	297			

(continued)

Table VI-A-28. *(continued)*

$10^{12} \times A$	B (K)	$10^{14} \times k$	T (K)	Technique[c]	Reference	Temperature Range (K)
		3.32 ± 0.10	298			
		3.69 ± 0.12	299			
		4.43 ± 0.16	311			
		5.9 ± 0.1	335			
		9.01 ± 0.36	370			
		12.1 ± 0.3	398			

[a] Placed on an absolute basis using $k(OH+CH_4) = 6.4 \times 10^{-15}$ cm^3 molecule^{-1} s^{-1}.
[b] Placed on an absolute basis using $k(OH + CH_3CHF_2) = 2.10 \times 10^{-18}\, T^2 \exp(-503/T)$ cm^3 molecule^{-1} s^{-1}.
[c] Abbreviations of the experimental techniques employed are defined in table II-A-1.
[d] $k = (4.64 \pm 0.58) \times 10^{-12} \times (T/300)^{0.89} \times \exp[(-1447 \pm 75)/T]$.
[e] $k = 1.24 \times 10^{-13}\, T^{0.5} \exp[(-1210 \pm 52)/T]$.

VI-A-3.2. $OH + CH_2Cl_2 \rightarrow H_2O + CHCl_2$

The rate coefficients of Cox et al. (1976a), Howard and Evenson (1976), Perry et al. (1976), Davis et al. (1976), Jeong and Kaufman (1982), Jeong et al. (1984), Taylor et al. (1993), Hsu and DeMore (1994), Villenave et al. (1997), and Herndon et al. (2001) are given in table VI-A-29 and plotted in figure VI-A-29. As is seen from the figure, there is substantial scatter in the reported data. CH_2Cl_2 as commonly supplied is mixed with small amounts of stabilizers (typically epoxides or amylenes) that are much more reactive towards OH radicals than CH_2Cl_2. The presence of such stabilizers is problematic for absolute rate studies but should not affect relative rate studies. This may explain why results from the absolute rate studies of Howard and Evenson (1976), Perry et al. (1976), Jeong and Kaufman (1982), and Taylor et al. (1993) lie higher than those from the other studies. For reasons that are unclear, there is a significant difference between the relative rate results of Hsu and DeMore (1994) obtained using the two different reference compounds. An Arrhenius fit to the data of Cox et al. (1976a), Davis et al. (1976), Villenave et al. (1997), and Herndon et al. (2001) gives $k(OH+CH_2Cl_2) = 1.97 \times 10^{-12} \exp(-877/T)$ cm^3 molecule^{-1} s^{-1}. At 298 K this expression gives $k(OH+CH_2Cl_2) = 1.03 \times 10^{-13}$ cm^3 molecule^{-1} s^{-1}, with an uncertainty estimated to be $\pm 25\%$.

VI-A-3.3. $OH + CHCl_3 \rightarrow H_2O + CCl_3$

The rate coefficients of Cox et al. (1976a), Howard and Evenson (1976), Davis et al. (1976), Jeong and Kaufman (1982), Jeong et al. (1984), Taylor et al. (1993), and Hsu and DeMore (1994) are given in table VI-A-30 and plotted in figure VI-A-30. The result from Cox et al. is consistent with but less precise than the other studies. As is seen from figure VI-A-30, there is reasonable agreement between the studies. An Arrhenius fit to all the data with the exception of that from Cox et al. (1976a) gives $k(OH+CHCl_3) = 2.86 \times 10^{-12} \exp(-1011/T)$ cm^3 molecule^{-1} s^{-1}. At 298 K this expression gives $k(OH+CHCl_3) = 9.62 \times 10^{-14}$ cm^3 molecule^{-1} s^{-1}, with an estimated uncertainty of $\pm 20\%$.

Table VI-A-29. Rate coefficients $k = A\ e^{-B/T}$ ($cm^3\ molecule^{-1}\ s^{-1}$) for the reaction: $OH + CH_2Cl_2 \rightarrow$ products

$10^{12} \times A$	B (K)	$10^{13} \times k$	T (K)	Technique[d]	Reference	Temperature Range (K)
		0.95 ± 0.48[a]	298	RR (CH_4)	Cox et al. (1976a)	298
		1.55 ± 0.34	296	DF–LMR	Howard and Evenson (1976)	296
		1.45 ± 0.20	298.5	FP–RF	Perry et al. (1976)	298.5
4.27	1094 ± 82	0.475 ± 0.057	245	FP–RF	Davis et al. (1976)	245–375
		1.04 ± 1.2	298			
		1.16 ± 0.5	298			
		2.23 ± 0.05	375			
5.57 ± 0.77	1041 ± 45	0.959 ± 0.069	251	DF–RF	Jeong and Kaufman (1982) Jeong et al. (1984)	251–455
		1.53 ± 0.095	292			
		2.08 ± 0.14	323			
		2.76 ± 0.19	342			
		3.52 ± 0.24	384			
		4.50 ± 0.29	415			
		6.09 ± 0.38	455			
$k(T)$[e]		1.47 ± 0.18	295	PLP–LIF	Taylor et al. (1993)	295–955
		1.61 ± 0.30	309			
		2.48 ± 0.23	335			
		2.42 ± 0.46	340			
		2.94 ± 0.56	376			
		4.39 ± 0.58	415			
		4.33 ± 0.42	425			
		6.15 ± 1.38	455			
		6.51 ± 1.06	474			
		8.58 ± 1.18	495			
		8.18 ± 1.00	535			
		10.14 ± 0.76	575			
		13.53 ± 1.30	615			
		14.81 ± 0.83	655			
		16.56 ± 1.16	695			
		18.45 ± 1.33	735			
		21.30 ± 1.28	775			
		25.90 ± 1.17	896			
		31.60 ± 3.92	955			
1.58	856	0.83[b]	293	RR (CH_3CHF_2)	Hsu and DeMore (1994)	293–368
		0.86	293			
		0.89	298			
		0.99	308			
		1.08	318			
		1.23	331			
		1.28	343			
		1.35	349			
		1.37	354			

(*continued*)

Table VI-A-29. (*continued*)

$10^{12} \times A$	B (K)	$10^{13} \times k$	T (K)	Technique[d]	Reference	Temperature Range (K)
		1.51	360			
		1.46	360			
1.24	690	1.23[c]	298	RR(CH_3CH_2F)		
		1.23	298			
		1.23	298			
		1.35	310			
		1.41	319			
		1.46	328			
		1.65	341			
		1.74	353			
		1.94	368			
2.61	944 ± 29	0.86 ± 0.07	277	FP–RF	Villenave et al. (1997)	277–370
		1.10 ± 0.02	298			
		1.51 ± 0.04	330			
		2.02 ± 0.05	370			
$k(T)$[f]		0.38 ± 0.03	219	PLP–LIF	Herndon et al. (2001)	219–394
		0.40 ± 0.02	220			
		0.40 ± 0.02	220			
		0.39 ± 0.02	220			
		0.41 ± 0.03	227			
		0.45 ± 0.04	238			
		0.58 ± 0.01	249			
		0.62 ± 0.05	254			
		0.70 ± 0.10	267			
		0.89 ± 0.04	284			
		0.98 ± 0.03	294			
		0.97 ± 0.03	296			
		1.00 ± 0.09	297			
		1.07 ± 0.05	303			
		1.08 ± 0.04	307			
		1.18 ± 0.09	320			
		1.35 ± 0.04	327			
		1.45 ± 0.10	340			
		1.79 ± 0.05	363			
		1.63 ± 0.10	364			
		1.91 ± 0.07	374			
		2.10 ± 0.20	394			

[a] Placed on an absolute basis using $k(OH + CH_4) = 6.4 \times 10^{-15}$ cm^3 molecule^{-1} s^{-1}.
[b] Placed on an absolute basis using $k(OH + CH_3CHF_2) = 2.10 \times 10^{-18}\, T^2 \exp(-503/T)$ cm^3 molecule^{-1} s^{-1}.
[c] Placed on an absolute basis using $k(OH + CH_3CH_2F) = 4.97 \times 10^{-18}\, T^2 \exp(-210/T)$ cm^3 molecule^{-1} s^{-1}.
[d] Abbreviations of the experimental techniques employed are defined in table II-A-1.
[e] $k = (2.01 \pm 0.17) \times 10^{-12} \times (T/300)^{1.09} \exp[(-771 \pm 48)/T]$.
[f] $k = 1.24 \times 10^{-13} \times T^{0.5} \times \exp[(-1210 \pm 52)/T]$.

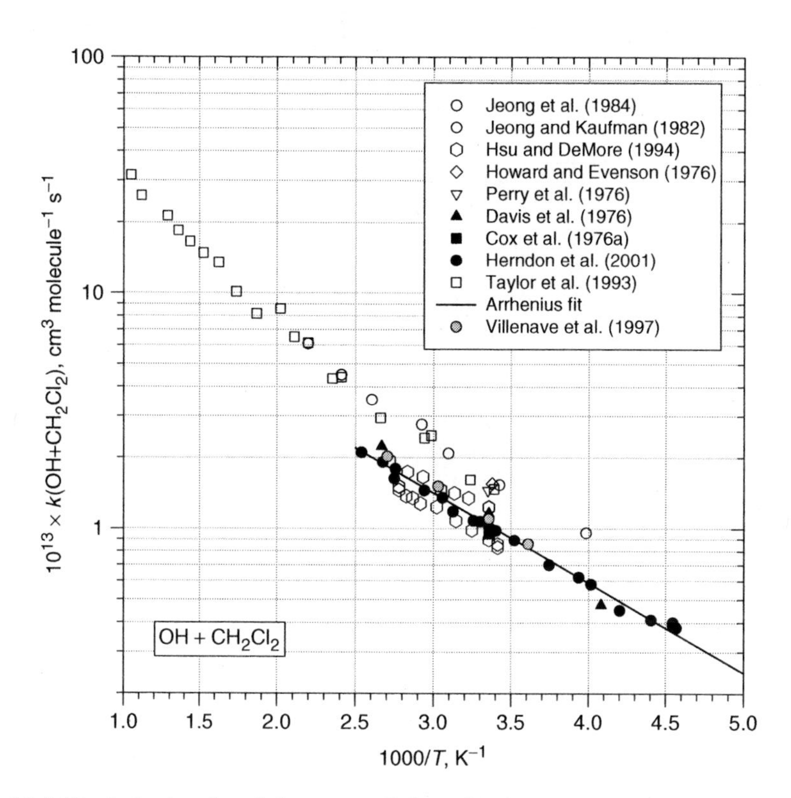

Figure VI-A-29. Arrhenius plot of the rate coefficients for the OH + CH_2Cl_2 reaction.

Table VI-A-30. Rate coefficients $k = A\ e^{-B/T}$ (cm^3 molecule^{-1} s^{-1}) for the reaction: OH + $CHCl_3$ → products

$10^{12} \times A$	B (K)	$10^{13} \times k$	T (K)	Technique[c]	Reference	Temperature Range (K)
		1.53 ± 0.77[a]	298	RR (CH_4)	Cox et al. (1976a)	298
		1.01 ± 0.15	296	DF–LMR	Howard and Evenson (1976)	296
4.69	1134 ± 108	0.44 ± 0.03	245	FP–RF	Davis et al. (1976)	245–375
		1.10 ± 0.05	298			
		1.14 ± 0.07	298			
		2.18 ± 0.14	375			
5.63 ± 0.68	1183 ± 45	0.551 ± 0.041	249	DF–RF	Jeong and Kaufman (1982) Jeong et al. (1984)	249–487
		1.01 ± 0.065	298			
		1.60 ± 0.10	339			
		2.32 ± 0.16	370			
		3.08 ± 0.20	411			
		4.48 ± 0.27	466			
		5.50 ± 0.39	487			

(continued)

Table VI-A-30. (*continued*)

$10^{12} \times A$	B (K)	$10^{13} \times k$	T (K)	Technique[c]	Reference	Temperature Range (K)
$k(T)$[d]		1.12 ± 0.10	295	PLP–LIF	Taylor et al. (1993)	295–775
		1.42 ± 0.27	327			
		1.71 ± 0.20	360			
		2.04 ± 0.12	390			
		2.42 ± 0.13	429			
		2.89 ± 0.45	452			
		2.98 ± 0.18	456			
		3.26 ± 0.18	480			
		4.40 ± 0.31	540			
		4.79 ± 0.37	600			
		6.56 ± 0.61	670			
		6.81 ± 0.50	700			
		7.02 ± 0.43	730			
		8.47 ± 0.50	775			
0.86	657	0.88[b]	288	RR (CH_3CHF_2)	Hsu and DeMore (1994)	288–357
		0.97	293			
		1.01	293			
		0.91	298			
		0.88	298			
		0.90	298			
		1.03	308			
		1.12	318			
		1.12	328			
		1.25	338			
		1.33	351			
		1.37	357			

[a] Placed on an absolute basis using $k(OH + CH_4) = 6.4 \times 10^{-15}$ cm^3 molecule^{-1} s^{-1}.
[b] Placed on an absolute basis using $k(OH + CH_3CHF_2) = 2.10 \times 10^{-18}\ T^2 \exp(-503/T)$ cm^3 molecule^{-1} s^{-1}.
[c] Abbreviations of the experimental techniques employed are defined in table II-A-1.
[d] $k = (2.71 \pm 0.23) \times 10^{-12} \times (T/300)^{1.52} \times \exp[(-261 \pm 42)/T]$.

VI-A-3.4. $OH + CH_3CH_2Cl \rightarrow H_2O + CH_3CHCl$ (a); $OH + CH_3CH_2Cl \rightarrow H_2O + CH_2CH_2Cl$ (b)

The rate coefficients of Howard and Evenson (1976), Paraskevopoulos et al. (1981), Kasner et al. (1990), Markert and Nielsen (1992), and Herndon et al. (2001) are given in table VI-A-31 and plotted in figure VI-A-31. As is seen from figure VI-A-31, there is significant scatter in the reported data. While there is excellent agreement in the room-temperature rate coefficients reported by Howard and Evenson, Paraskevopoulos et al., Kasner et al., and Markert and Nielsen, there is disagreement between the temperature dependences reported by Kasner et al. and Markert and Nielsen. The data reported by Herndon et al. lie substantially (25% at room temperature) below those from the other studies. The data of Markert and Nielsen are more scattered than those from the other studies and are excluded from further consideration. A fit of the modified Arrhenius expression to the data of Howard and Evenson (1976), Paraskevopoulos et al. (1981), Kasner et al. (1990), and Herndon et al. (2001) gives

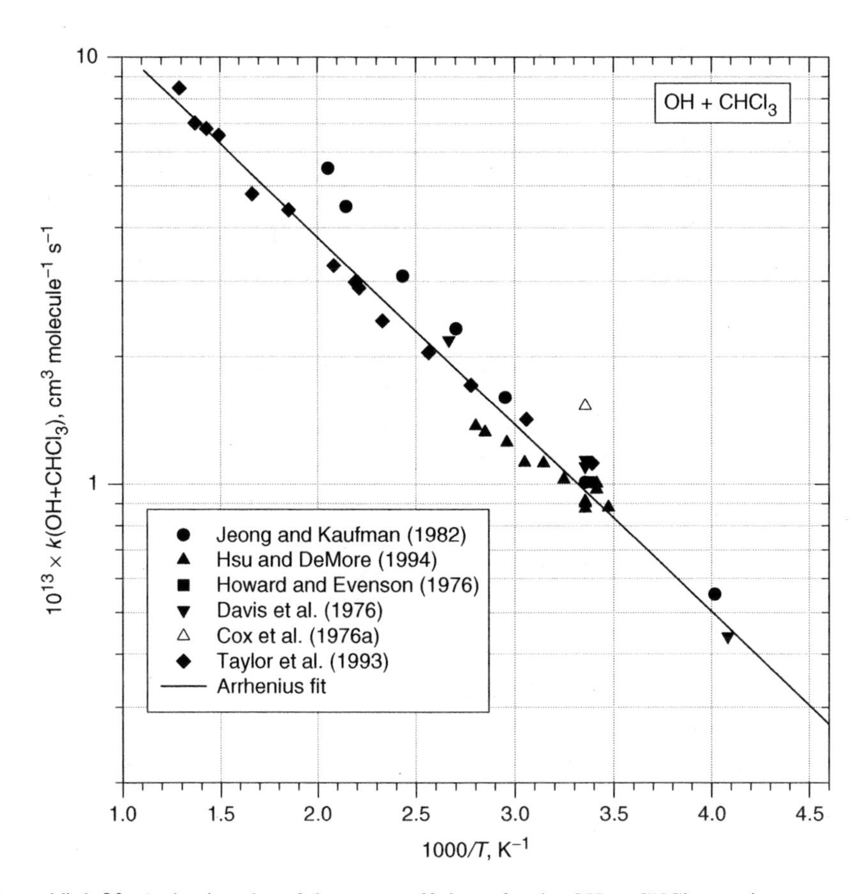

Figure VI-A-30. Arrhenius plot of the rate coefficients for the OH + $CHCl_3$ reaction.

Table VI-A-31. Rate coefficients $k = A\ e^{-B/T}$ (cm^3 $molecule^{-1}$ s^{-1}) for the reaction: OH + CH_3CH_2Cl → products

$10^{12} \times A$	B (K)	$10^{13} \times k$	T (K)	Technique[a]	Reference	Temperature Range (K)
		3.9 ± 0.7	296	DF–LMR	Howard and Evenson (1976)	296
		3.94 ± 0.53	297	FP–RA	Paraskevopoulos et al. (1981)	297
$k(T)$[b]		4.08 ± 0.31	295	FP–LIF	Kasner et al. (1990)	295–789
		5.62 ± 0.51	331			
		7.98 ± 0.70	382			
		9.46 ± 0.43	421			
		10.1 ± 0.69	447			
		12.4 ± 1.61	490			
		16.0 ± 3.93	548			
		19.3 ± 2.51	595			
		23.1 ± 1.10	600			

(*continued*)

Table VI-A-31. (*continued*)

$10^{12} \times A$	B (K)	$10^{13} \times k$	T (K)	Technique[a]	Reference	Temperature Range (K)
		24.3 ± 2.88	641			
		27.5 ± 3.95	687			
		32.7 ± 3.27	710			
		32.4 ± 2.22	714			
		35.1 ± 6.81	733			
		39.7 ± 4.05	769			
		42.9 ± 5.96	789			
24	1082 ± 361	4.3 ± 0.5	295	PR–RA	Markert and Nielsen (1992)	295–360
		6.6 ± 0.4	307			
		8.3 ± 0.8	315			
		8.2 ± 1.6	333			
		9.5 ± 1.3	360			
$k(T)$[c]		1.45 ± 0.04	223	PLP–LIF	Herndon et al. (2001)	223–426
		1.45 ± 0.03	223			
		1.78 ± 0.04	241			
		2.24 ± 0.16	266			
		2.50 ± 0.10	283			
		2.92 ± 0.12	296			
		3.65 ± 0.06	323			
		4.99 ± 0.20	373			
		5.90 ± 0.16	399			
		7.24 ± 0.20	426			

[a] Abbreviations of the experimental techniques employed are defined in table II-A-1.
[b] $k = (2.96 \pm 2.94) \times 10^{-13}\ (T/300)^{2.59} \times \exp[(-28 \pm 75)/T]$.
[c] $k = 1.5 \times 10^{-13}\ T^{0.5} \times \exp(-637/T)$.

$k(OH + C_2H_5Cl) = 8.80 \times 10^{-18}\ T^2 \exp(-259/T)\ cm^3\ molecule^{-1}\ s^{-1}$. At 298 K this expression gives $k(OH + C_2H_5Cl) = 3.28 \times 10^{-13}\ cm^3\ molecule^{-1}\ s^{-1}$, with an estimated uncertainty of ±25%.

VI-A-3.5. $OH + CH_2ClCH_2Cl \rightarrow H_2O + CH_2ClCHCl$

The rate coefficients of Howard and Evenson (1976), Arnts et al. (1989), Taylor et al. (1991), and Qiu et al. (1992) are given in table VI-A-32 and plotted in figure VI-A-32. As is seen from this figure, the results from all studies are in good agreement. A fit of the modified Arrhenius expression gives $k(OH + CH_2ClCH_2Cl) = 1.08 \times 10^{-17}\ T^2 \exp(-410/T)\ cm^3\ molecule^{-1}\ s^{-1}$. At 298 K this expression gives $k(OH + CH_2ClCH_2Cl) = 2.42 \times 10^{-13}\ cm^3\ molecule^{-1}\ s^{-1}$, with an uncertainty estimated to be ±15%.

VI-A-3.6. $OH + CH_3CHCl_2 \rightarrow H_2O + CH_3CCl_2$ (a); $H_2O + CH_2CHCl_2$ (b)

The rate coefficients of Howard and Evenson (1976) and Jiang et al. (1992) are given in table VI-A-33 and plotted in figure VI-A-33. As is seen from the figure,

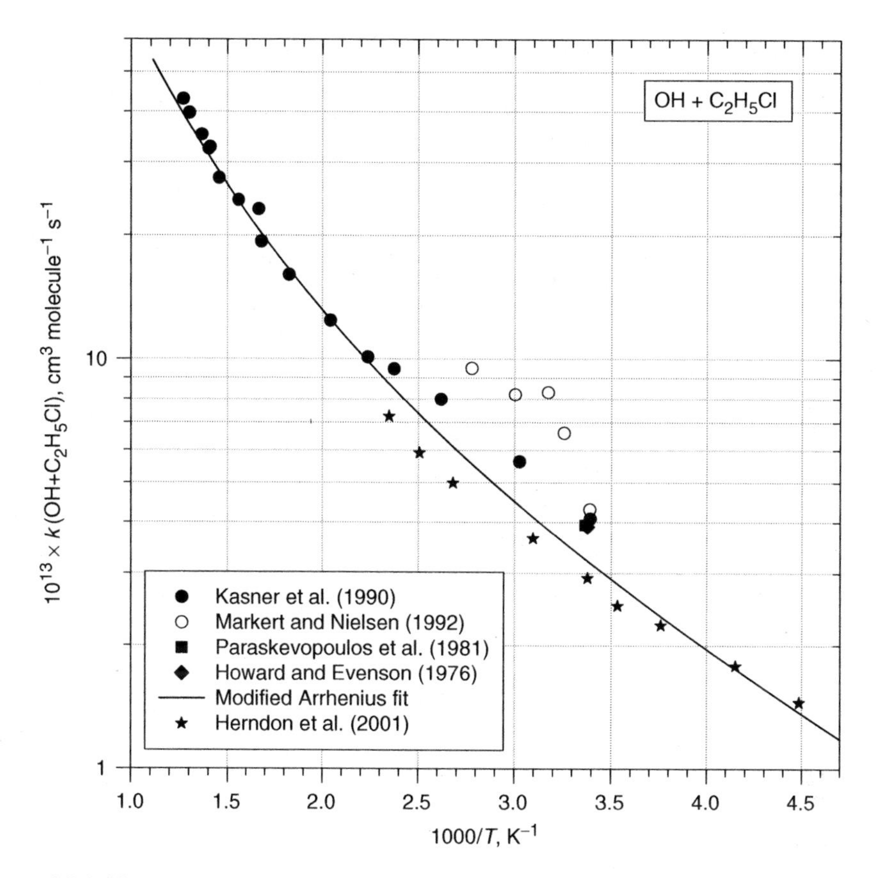

Figure VI-A-31. Arrhenius plot of the rate coefficients for the OH + C_2H_5Cl reaction.

Table VI-A-32. Rate coefficients $k = A\ e^{-B/T}$ (cm^3 $molecule^{-1}$ s^{-1}) for the reaction: OH + CH_2ClCH_2Cl → products

$10^{12} \times A$	B (K)	$10^{13} \times k$	T (K)	Technique[b]	Reference	Temperature Range (K)
		2.2 ± 0.5	296	DF–LMR	Howard and Evenson (1976)	296
		2.23 ± 0.22[a]	297	RR (C_2H_6)	Arnts et al. (1989)	297
$k(T)$[c]		2.48 ± 0.38	292	PLP–LIF	Taylor et al. (1991)	292–775
		3.02 ± 0.40	311			
		3.41 ± 0.32	321			
		4.35 ± 0.30	340			
		5.01 ± 0.30	359			
		6.01 ± 0.57	377			
		6.83 ± 0.39	407			
		7.52 ± 0.63	431			
		9.08 ± 0.74	458			
		10.9 ± 1.08	493			

(*continued*)

Table VI-A-32. (*continued*)

$10^{12} \times A$	B (K)	$10^{13} \times k$	T (K)	Technique[b]	Reference	Temperature Range (K)
		15.1 ± 0.56	529			
		20.6 ± 0.71	570			
		19.5 ± 1.48	574			
		24.4 ± 2.12	627			
		29.3 ± 3.64	696			
		34.7 ± 3.50	771			
		32.5 ± 2.23	775			
10.5 ± 1.6	1141 ± 107	2.09	292	FP–LIF	Qiu et al. (1992)	292–363
		2.14	295			
		3.20	321			
		3.45	333			
		3.59	343			
		4.14	358			
		4.44	363			

[a] Placed on an absolute basis using $k(OH + C_2H_6) = 1.49 \times 10^{-17}\ T^2\ \exp(-499/T)$.
[b] Abbreviations of the experimental techniques employed are defined in table II-A-1.
[c] $k = (4.08 \pm 0.78) \times 10^{-12}\ (T/300)^{1.0} \times\ \exp[(-825 \pm 88)/T]$.

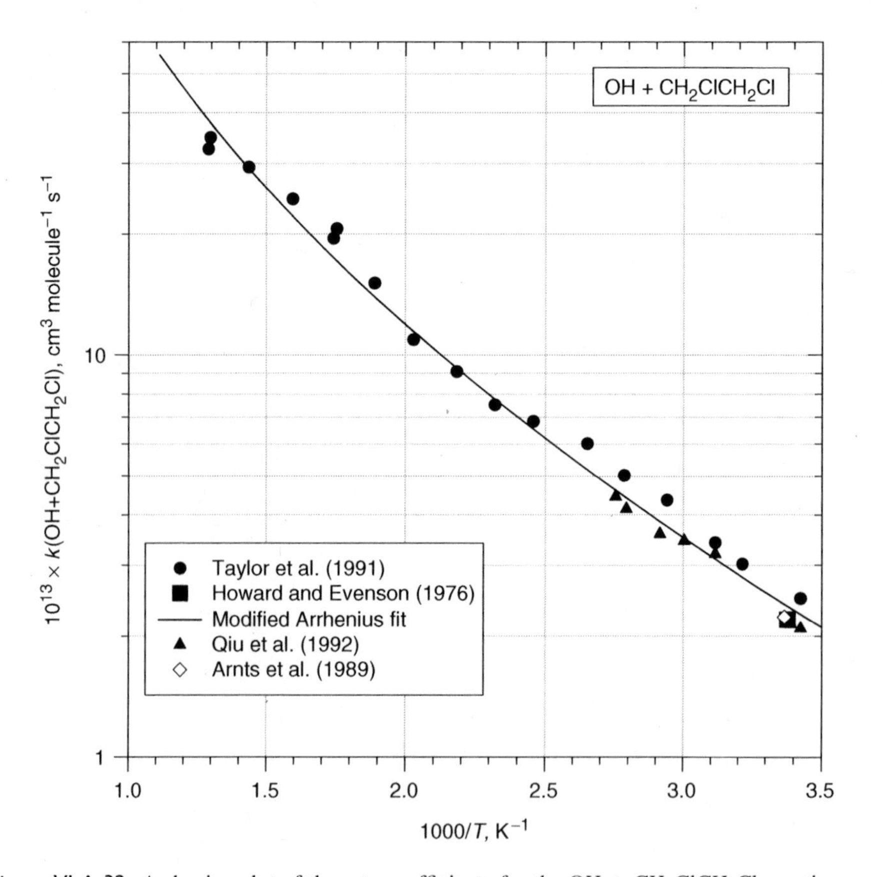

Figure VI-A-32. Arrhenius plot of the rate coefficients for the $OH + CH_2ClCH_2Cl$ reaction.

there is good agreement between the studies. A fit of the modified Arrhenius expression to the data of Howard and Evenson (1976) and Jiang et al. (1992) gives $k(OH + CH_3CHCl_2) = 2.49 \times 10^{-18}\ T^2 \exp(64/T)\ cm^3\ molecule^{-1}\ s^{-1}$. At 298 K this expression gives $k(OH + CH_3CHCl_2) = 2.74 \times 10^{-13}\ cm^3\ molecule^{-1}\ s^{-1}$, with an uncertainty of ±15%.

VI-A-3.7. $OH + CH_2ClCHCl_2 \rightarrow H_2O + CH_2ClCCl_2$ (a); $OH + CH_2ClCHCl_2 \rightarrow H_2O + CHClCHCl_2$ (b)

The rate coefficients of Jeong et al. (1984) and Taylor et al. (1992) are given in table VI-A-34 and plotted in figure VI-A-34. As is seen from the figure, there is significant disagreement between the results of the two studies. There is no obvious reason to prefer either study. A fit of the modified Arrhenius expression gives $k(OH + CH_2ClCHCl_2) = 3.70 \times 10^{-18}\ T^2 \exp(-108/T)\ cm^3\ molecule^{-1}\ s^{-1}$. At 298 K this expression gives $k(OH + CH_2ClCHCl_2) = 2.29 \times 10^{-13}\ cm^3\ molecule^{-1}\ s^{-1}$, with an uncertainty of ±30%.

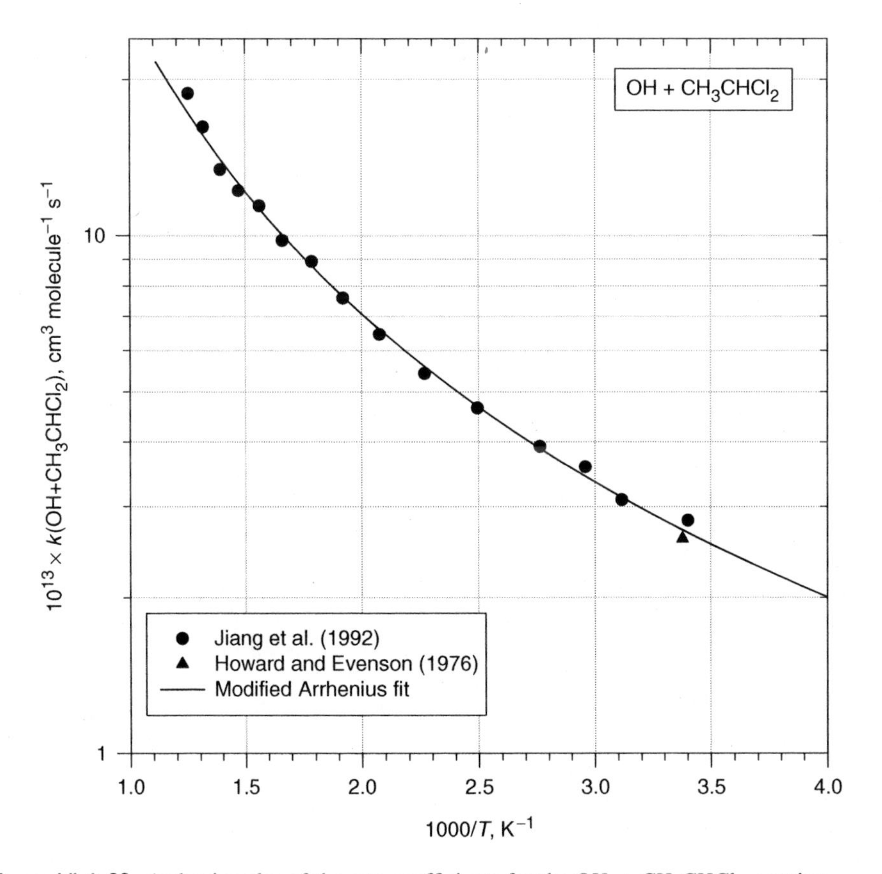

Figure VI-A-33. Arrhenius plot of the rate coefficients for the $OH + CH_3CHCl_2$ reaction.

Table VI-A-33. Rate coefficients $k = A\ e^{-B/T}$ (cm^3 $molecule^{-1}$ s^{-1}) for the reaction: $OH + CH_3CHCl_2 \rightarrow$ products

$10^{12} \times A$	B (K)	$10^{13} \times k$	T (K)	Technique[a]	Reference	Temperature Range (K)
		2.6 ± 0.6	296	DF–LMR	Howard and Evenson (1976)	296
$k(T)$[b]		2.82 ± 0.14	294	PLP–LIF	Jiang et al. (1992)	294–800
		3.09 ± 0.20	321			
		3.58 ± 0.19	338			
		3.92 ± 0.06	362			
		4.65 ± 0.19	401			
		5.42 ± 0.12	441			
		6.45 ± 0.21	482			
		7.58 ± 0.22	521			
		8.91 ± 0.38	560			
		9.78 ± 0.27	602			
		11.4 ± 0.44	640			
		12.2 ± 0.94	680			
		13.4 ± 1.21	719			
		16.2 ± 1.80	760			
		18.8 ± 0.76	800			

[a] Abbreviations of the experimental techniques employed are defined in table II-A-1.
[b] $k = (8.29 \pm 0.36) \times 10^{-14}\ (T/300)^{2.67} \times \exp[387 \pm 18)/T]$.

Table VI-A-34. Rate coefficients $k = A\ e^{-B/T}$ (cm^3 $molecule^{-1}$ s^{-1}) for the reaction: $OH + CH_2ClCHCl_2 \rightarrow$ products

$10^{12} \times A$	B (K)	$10^{13} \times k$	T (K)	Technique[a]	Reference	Temperature Range (K)
1.65 ± 0.27	483 ± 110	2.84 ± 0.21	277	DF–RF	Jeong et al. (1984)	277–461
		3.18 ± 0.20	295			
		3.76 ± 0.23	322			
		4.36 ± 0.28	346			
		4.68 ± 0.29	386			
		4.92 ± 0.31	400			
		5.27 ± 0.35	424			
		5.76 ± 0.37	461			
$k(T)$[b]		1.84 ± 0.07	295	PLP–LIF	Taylor et al. (1992)	295–850
		1.96 ± 0.06	299			
		2.70 ± 0.15	325			
		3.20 ± 0.09	339			
		3.12 ± 0.21	360			
		4.28 ± 0.23	400			
		4.30 ± 0.25	404			
		4.67 ± 0.35	447			

Table VI-A-34. *(continued)*

$10^{12} \times A$	B (K)	$10^{13} \times k$	T (K)	Technique[a]	Reference	Temperature Range (K)
		6.30 ± 0.50	480			
		8.60 ± 0.70	540			
		10.80 ± 0.44	600			
		12.90 ± 0.47	651			
		14.20 ± 0.55	670			
		19.70 ± 0.42	722			
		22.10 ± 1.62	775			
		25.80 ± 3.12	850			

[a] Abbreviations of the experimental techniques employed are defined in table II-A-1.
[b] $k = (1.63 \pm 0.22) \times 10^{-13}\ (T/300)^{2.64} \times \exp[(70 \pm 55)/T]$.

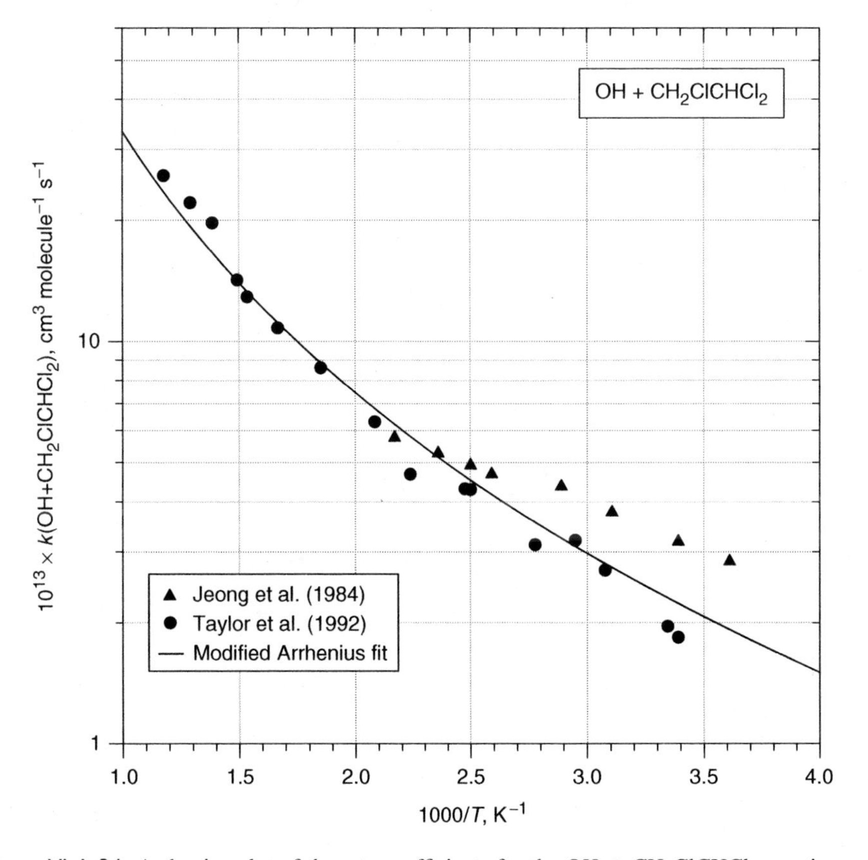

Figure VI-A-34. Arrhenius plot of the rate coefficients for the OH + $CH_2ClCHCl_2$ reaction.

VI-A-3.8. $OH + CH_3CCl_3 \rightarrow H_2O + CH_2CCl_3$

The rate coefficients reported by Cox et al. (1976a), Howard and Evenson (1976), Chang and Kaufman (1977), Watson et al. (1977), Clyne and Holt (1979a), Kurylo et al. (1979), Jeong et al. (1984), Nelson et al. (1984, 1990b), DeMore (1992), Finlayson-Pitts et al. (1992), Jiang et al. (1992). Talukdar et al. (1992), Lancar et al. (1993), and Cavalli et al. (1998) are given in table VI-A-35 and plotted in figure VI-A-35. Data from Jeong and Kaufman (1979) are superseded by those of Jeong et al. (1984) and are not considered further here. As is seen from figure VI-A-35, the data from Cox et al., Watson et al., Chang and Kaufman, and Clyne and Holt appear to overestimate the reactivity of CH_3CCl_3, probably reflecting the effect of reactive impurities in the CH_3CCl_3 samples used. Excluding these studies and fitting the Arrhenius expression to the data at $T < 400$ K gives $k(OH+CH_3CCl_3) = 1.81 \times 10^{-12} \exp(-1543/T)$ cm^3 molecule^{-1} s^{-1}. At 298 K this expression gives $k(OH+CH_3CCl_3) = 1.02 \times 10^{-14}$ cm^3 molecule^{-1} s^{-1}. A fit of the modified Arrhenius expression gives $k(OH+CH_3CCl_3) = 4.07 \times 10^{-18}\ T^2 \exp(-1064/T)$ cm^3 molecule^{-1} s^{-1}. At 298 K this expression gives $k(OH+CH_3CCl_3) = 1.02 \times 10^{-14}$ cm^3 molecule^{-1} s^{-1}, with an estimated uncertainty of $\pm 15\%$.

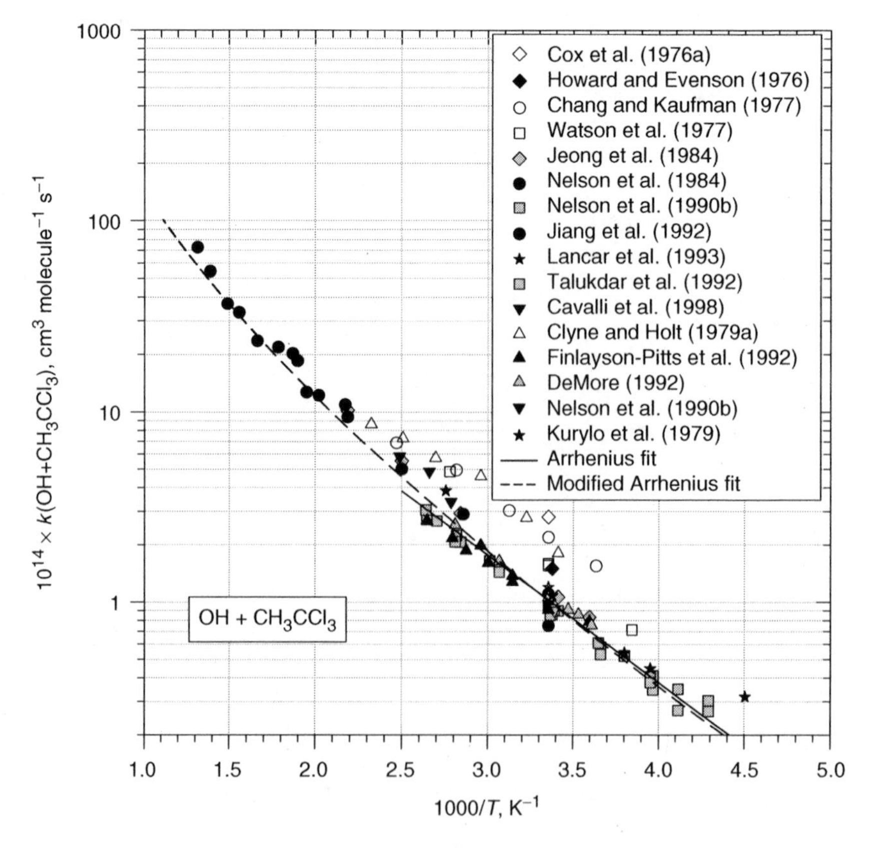

Figure VI-A-35. Arrhenius plot of the rate coefficients for the $OH + CH_3CCl_3$ reaction.

Table VI-A-35. Rate coefficients $k = A\ e^{-B/T}$ (cm^3 $molecule^{-1}$ s^{-1}) for the reaction: $OH + CH_3CCl_3 \rightarrow$ products

$10^{12} \times A$	B (K)	$10^{14} \times k$	T (K)	Technique[c]	Reference	Temperature Range (K)
		2.6 ± 1.3	298	RR (CH_4)[b]	Cox et al. (1976a)	298
		1.50 ± 0.3	296	DF–LMR	Howard and Evenson (1976)	296
1.95 ± 0.24	1331 ± 37	1.55 ± 0.22	275	DF–RF	Chang and Kaufman (1977)	275–405
		2.19 ± 0.26	298			
		3.03 ± 0.30	320			
		4.94 ± 0.48	355			
		6.87 ± 0.40	405			
3.72	1627	0.712 ± 0.094	260	FP–RF	Watson et al. (1977)	260–360
		1.59 ± 0.16	298			
		1.56 ± 0.04	298			
		4.85 ± 0.58	360			
$2.40^{+0.91}_{-0.66}$	1394 ± 113	1.81 ± 0.16	293	DF–RF	Clyne and Holt (1979a)	293–430
		2.78 ± 0.74	310			
		4.59 ± 0.56	338			
		5.73 ± 0.51	371			
		7.29 ± 0.44	399			
		8.63 ± 0.40	430			
5.4 ± 1.8	1810 ± 100	0.318 ± 0.095	222	FP–RF	Kurylo et al. (1979)	222–363
		0.447 ± 0.135	253			
		0.540 ± 0.145	263			
		1.08 ± 0.20	296			
		3.85 ± 0.75	363			
5.04 ± 0.96	1797 ± 65	0.83 ± 0.07	278	DF–RF	Jeong et al. (1984)	278–457
		1.06 ± 0.11	293			
		2.93 ± 0.19	352			
		5.52 ± 0.41	400			
		10.20 ± 0.65	457			
		0.77[a]	298	RR (CH_3Cl)	Nelson et al. (1984)	298
		0.958 ± 0.305[a]	298	RR (CH_3Cl)	Nelson et al. (1990b)	298
5.4 ± 3.0	1797 ± 448	3.36 ± 0.30	359	PR–UVabs	Nelson et al. (1990b)	359–402
		4.85 ± 0.50	376			
		5.83 ± 0.20	402			
		0.75 ± 0.04[b]	277	RR (CH_4)	DeMore (1992)	277–356
		0.86 ± 0.06	283			
		0.91 ± 0.07	288			
		1.06 ± 0.04	298			
		1.64 ± 0.10	326			
		2.56 ± 0.05	356			
0.91 ± 0.46	1337 ± 147	0.789 ± 0.074	278	DF–RF	Finlayson-Pitts et al. (1992)	278–378
		1.12 ± 0.16	298			
		0.911 ± 0.098	298			
		1.38 ± 0.42	318			
		1.28 ± 0.09	318			
		1.61 ± 0.16	333			
		1.99 ± 0.24	338			
		1.87 ± 0.25	348			

(continued)

Table VI-A-35. (*continued*)

$10^{12} \times A$	B (K)	$10^{14} \times k$	T (K)	Technique[c]	Reference	Temperature Range (K)
		2.17 ± 0.11	358			
		2.68 ± 0.28	378			
$k(T)$[d]		1.1 ± 0.1	298	FP–LIF	Jiang et al. (1992)	298–761
		2.9 ± 0.6	350			
		5.0 ± 1.0	400			
		9.4 ± 0.6	457			
		10.9 ± 2.9	460			
		12.2 ± 1.2	495			
		12.7 ± 0.9	513			
		18.5 ± 1.3	527			
		20.2 ± 1.0	535			
		21.8 ± 2.8	560			
		23.5 ± 3.8	601			
		33.1 ± 8.1	642			
		36.8 ± 3.6	671			
		54.5 ± 2.0	720			
		72.6 ± 3.9	761			
1.75 ± 0.34	1550 ± 60	0.304 ± 0.019	233	FP–LIF	Talukdar et al. (1992)	233–379
		0.267 ± 0.018	233			
		0.270 ± 0.019	243			
		0.349 ± 0.016	243			
		0.355 ± 0.026	252			
		0.346 ± 0.030	252			
		0.410 ± 0.013	252			
		0.405 ± 0.017	253			
		0.377 ± 0.018	253			
		0.522 ± 0.030	263			
		0.520 ± 0.022	263			
		0.579 ± 0.022	273			
		0.530 ± 0.026	273			
		0.609 ± 0.030	274			
		0.898 ± 0.064	293			
		0.899 ± 0.100	295			
		0.910 ± 0.030	296			
		0.859 ± 0.027	296			
		0.851 ± 0.023	297			
		1.49 ± 0.05	326			
		1.44 ± 0.10	326			
		1.65 ± 0.05	332			
		2.07 ± 0.06	352			
		2.24 ± 0.11	355			
		2.08 ± 0.08	356			
		2.67 ± 0.05	370			
		2.71 ± 0.05	378			
		3.05 ± 0.11	379			
		1.2 ± 0.2	298	DF–EPR	Lancar et al. (1993)	298
		0.90 ± 0.18	295	RR (CH_4)[b]	Cavalli et al. (1998)	295

[a] Placed on an absolute basis using $k(OH + CH_3Cl) = 3.83 \times 10^{-14}$ cm^3 molecule^{-1} s^{-1}.

[b] Placed on an absolute basis using $k(OH + CH_4) = 1.85 \times 10^{-20}\ T^{2.82}\ \exp(-987/T)$ cm^3 molecule^{-1} s^{-1}.

[c] Abbreviations of the experimental techniques employed are defined in table II-A-1.

[d] $k = 3.95 \times 10^{-13} \times (T/298)^{2.08} \times \exp(-1068/T)$.

The 222 K datum point by Kurylo et al. (1979) lies above that expected based from an extrapolation of the higher temperature data and may have been impacted by the presence of reactive impurities. Excluding this datum point from the fits gives $k(OH+CH_3CCl_3) = 1.95 \times 10^{-12} \exp(-1567/T)$ and $k(OH+CH_3CCl_3) = 4.07 \times 10^{-18}\ T^2 \times \exp(-1064/T)\,cm^3\,molecule^{-1}\,s^{-1}$, which are not materially different from those given above.

VI-A-3.9. $OH + CHCl_2CHCl_2 \rightarrow H_2O + CCl_2CHCl_2$

The rate coefficients of Qiu et al. (1992) and Jiang et al. (1993) are given in table VI-A-36 and plotted in figure VI-A-36. Data from Xing et al. (1992) are duplicated in Qiu et al. (1992) and are not considered here. As is seen in the figure, the data of Qiu et al. (1992) lie 50–100% above those from Jiang et al. (1993). There is no obvious reason to prefer either study and no recommendation of $k(OH+CHCl_2CHCl_2)$ is possible at this time.

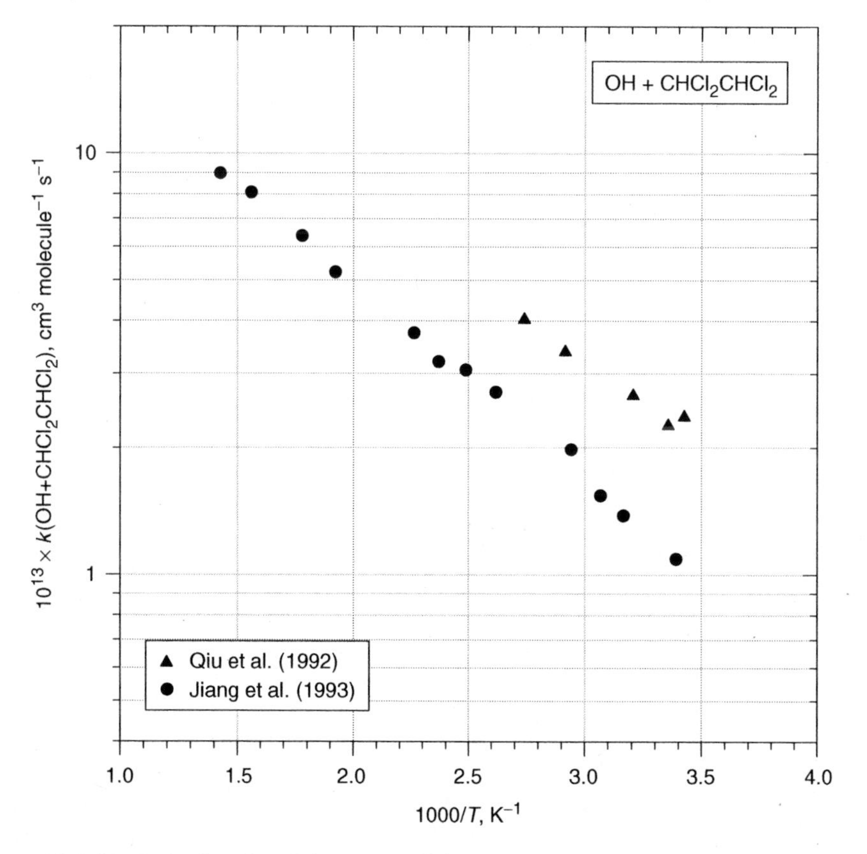

Figure VI-A-36. Arrhenius plot of the rate coefficients for the $OH + CHCl_2CHCl_2$ reaction.

Table VI-A-36. Rate coefficients $k = A\ e^{-B/T}$ (cm^3 $molecule^{-1}$ s^{-1}) for the reaction: $OH + CHCl_2CHCl_2 \rightarrow H_2O + CCl_2CHCl_2$

$10^{12} \times A$	B (K)	$10^{13} \times k$	T (K)	Technique[a]	Reference	Temperature Range (K)
3.70 ± 0.46	816 ± 90	2.37	292	DF–RF	Qiu et al. (1992)	292–365
		2.26	298			
		2.66	312			
		3.38	343			
		4.03	365			
$k(T)$[b]		1.09	295	FP–LIF	Jiang et al. (1993)	295–701
		1.38	316			
		1.54	326			
		1.98	340			
		2.71	382			
		3.06	402			
		3.20	422			
		3.74	442			
		5.22	520			
		6.36	562			
		8.06	640			
		8.96	701			

[a] Abbreviations of the experimental techniques employed are defined in table II-A-1.
[b] $k = (2.72 \pm 0.42) \times 10^{-12} \times (T/300)^{0.22} \exp[(-915 \pm 62)/T]$.

Table VI-A-37. Rate coefficients $k = A\ e^{-B/T}$ (cm^3 $molecule^{-1}$ s^{-1}) for the reactions of OH with various chloroalkanes for which only a single study is available

Chloroalkane	$10^{13} \times k$	T (K)	Technique[b]	Reference
$CHCl_2CCl_3$	2.33	292	DF–RF	Qiu et al. (1992)
$CH_2ClCH_2CH_2Cl$	7.30 ± 0.29[b]	305	RR (Cyclohexane)	Donaghy et al. (1993)
tert-C_4H_9Cl	3.84 ± 0.42[b]	298	RR (Cyclohexane)	Donaghy et al. (1993)

[a] Abbreviations of the experimental techniques employed are defined in table II-A-1.
[b] Placed on an absolute basis using $k(OH + Cyclohexane) = 3.26 \times 10^{-17} T^2 \exp(262/T)\ cm^3\ molecule^{-1}\ s^{-1}$.

VI-A-3.10. $OH + CHCl_2CCl_3 \rightarrow H_2O + CCl_2CCl_3$

The only rate coefficient determination is that of Qiu et al. (1992), given in table VI-A-37, and this is recommended with an uncertainty of ±25%.

VI-A-3.11. $OH + CH_3CH_2CH_2Cl \rightarrow$ Products

The rate coefficients of Markert and Nielsen (1992) and Donaghy et al. (1993) are given in table VI-A-38 and plotted in figure VI-A-37. Results from these two studies are in agreement within the experimental uncertainties. A fit of the Arrhenius expression to the combined data set gives $k(OH + CH_3CH_2CH_2Cl) = 9.51 \times 10^{-11} \exp(-1402/T)\ cm^3\ molecule^{-1}\ s^{-1}$. At 298 K this expression gives $k(OH + CH_3CH_2CH_2Cl) = 8.61 \times 10^{-13}\ cm^3\ molecule^{-1}\ s^{-1}$, with an estimated uncertainty of ±25%.

Table VI-A-38. Rate coefficients $k = A\ e^{-B/T}$ (cm^3 molecule^{-1} s^{-1}) for the reaction: $OH + CH_3CH_2CH_2Cl \rightarrow$ products

$10^{12} \times A$	B (K)	$10^{12} \times k$	T (K)	Technique[d]	Reference	Temperature Range (K)
100	1443 ± 481	0.96 ± 0.04	295	PR–UVabs	Markert and Nielsen (1992)	295–353
		0.68 ± 0.07	295			
		1.47 ± 0.06	317			
		1.76 ± 0.02	328			
		1.16 ± 0.07	331			
		1.80 ± 0.20	333			
		1.60 ± 0.03	353			
		1.05 ± 0.09[a]	303	RR (cyclohexane)	Donaghy et al. (1993)	303–308
		1.02 ± 0.03[b]	305	RR (2-methylpropane)		
		0.85 ± 0.03[c]	308	RR (pentane)		

[a] Placed on an absolute basis using $k(OH + \text{cyclohexane}) = 3.26 \times 10^{-17} T^2 \exp(262/T)$ cm^3 molecule^{-1} s^{-1}.
[b] Placed on an absolute basis using $k(OH + \text{2-methylpropane}) = 1.17 \times 10^{-17} T^2 \exp(213/T)$ cm^3 molecule^{-1} s^{-1}.
[c] Placed on an absolute basis using $k(OH + \text{pentane}) = 2.52 \times 10^{-17} T^2 \exp(158/T)$ cm^3 molecule^{-1} s^{-1}.
[d] Abbreviations of the experimental techniques employed are defined in table II-A-1.

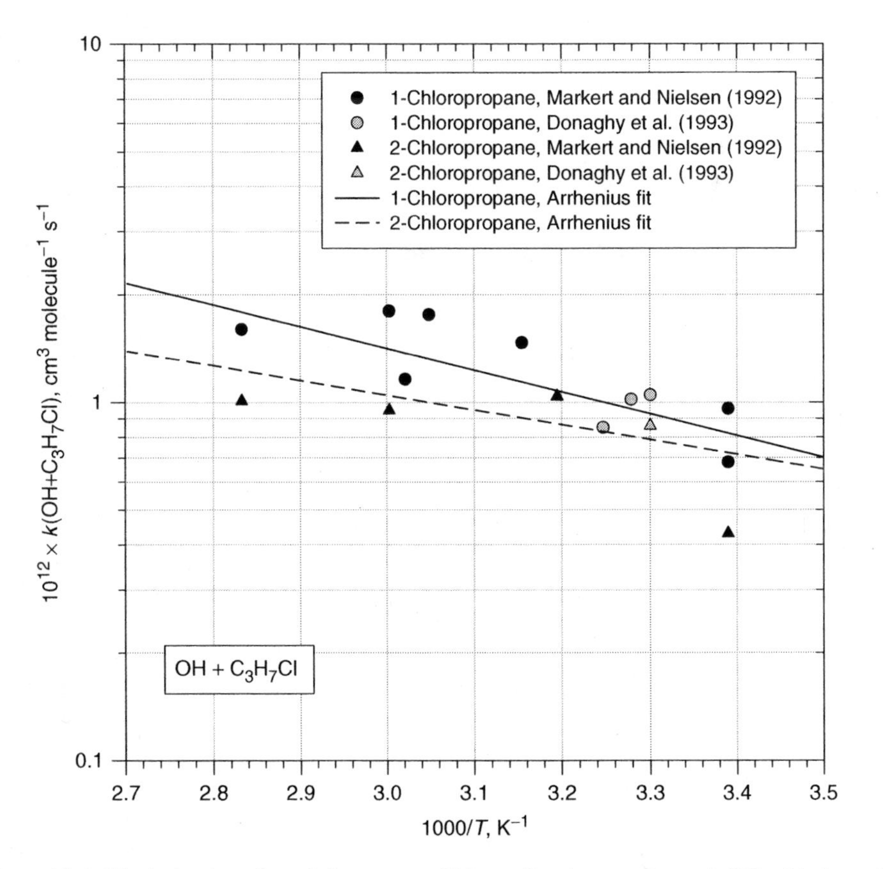

Figure VI-A-37. Arrhenius plot of the rate coefficients for the reactions of OH with 1- and 2-chloropropane.

VI-A-3.12. OH + $CH_3CHClCH_3$ → Products

The rate coefficients of 2-C_3H_7Cl as measured by Markert and Nielsen (1992) and Donaghy et al. (1993) are given in table VI-A-39 and plotted in figure VI-A-37 with those for 1-C_3H_7Cl. Results from the two $CH_3CHClCH_3$ studies are in agreement within the likely combined experimental uncertainties. A fit of the Arrhenius expression to the combined data set gives $k(OH + CH_3CHClCH_3) = 1.83 \times 10^{-11} \exp(-953/T)$ cm^3 molecule^{-1} s^{-1}. At 298 K this expression gives $k(OH + CH_3CHClCH_3) = 7.47 \times 10^{-13}$ cm^3 molecule^{-1} s^{-1} with an estimated uncertainty of ±35%.

VI-A-3.13. OH + $CH_2ClCH_2CH_2Cl$ → Products

The rate coefficient of Donaghy et al. (1993), given in table VI-A-37, is the only study available for this reaction. It is recommended with an estimated uncertainty of ±25%.

VI-A-3.14. OH + tert-C_4H_9Cl → H_2O + $CH_2C(CH_3)_2Cl$

The rate coefficient of Donaghy et al. (1993), given in table VI-A-37, is the only study available for this reaction. It is recommended with an estimated uncertainty of ±25%.

VI-A-3.15. OH + $CH_3CH_2CH_2CH_2Cl$ → Products

The rate coefficients of Markert and Nielsen (1992) are given in table VI-A-40 and plotted in figure VI-A-38, where they can be compared with those for 1-$C_5H_{11}Cl$ and 1-$C_6H_{13}Cl$. A fit of the Arrhenius expression to the 1-C_4H_9Cl data of Markert and Nielsen (1992) gives $k(OH + CH_3CH_2CH_2CH_2Cl) = 3.49 \times 10^{-11} \exp(-941/T)$ cm^3 molecule^{-1} s^{-1}. At 298 K this expression gives $k(OH + CH_3CH_2CH_2CH_2Cl) = 1.48 \times 10^{-12}$ cm^3 molecule^{-1} s^{-1} with an estimated uncertainty of ±35%.

Table VI-A-39. Rate coefficients $k = A\,e^{-B/T}$ (cm^3 molecule^{-1} s^{-1}) for the reaction: OH + $CH_3CHClCH_3$ → products

$10^{12} \times A$	B (K)	$10^{12} \times k$	T (K)	Technique[b]	Reference	Temperature Range (K)
50	1323 ± 842	0.43 ± 0.04	295	PR–UVabs	Markert and Nielsen (1992)	295–353
		1.04 ± 0.07	313			
		0.95 ± 0.03	333			
		1.01 ± 0.04	353			
		0.86 ± 0.05[a]	303	RR (Cyclohexane)	Donaghy et al. (1993)	303

[a] Placed on an absolute basis using $k(OH + Cyclohexane) = 3.26 \times 10^{-17} T^2 \exp(262/T)$ cm^3 molecule^{-1} s^{-1}.
[b] Abbreviations of the experimental techniques employed are defined in table II-A-1.

Table VI-A-40. Rate coefficients $k = A\ e^{-B/T}$ (cm^3 molecule^{-1} s^{-1}) for the reaction: $OH + CH_3CH_2CH_2CH_2Cl \rightarrow$ products

$10^{12} \times A$	B (K)	$10^{13} \times k$	T (K)	Technique[a]	Reference	Temperature Range (K)
28	842 ± 481	1.69 ± 0.09	295	PR–UVabs	Markert and Nielsen (1992)	295–353
		2.10 ± 0.10	295			
		1.14 ± 0.06	295			
		1.28 ± 0.05	295			
		2.31 ± 0.07	320			
		2.34 ± 0.09	331			
		2.14 ± 0.06	353			

[a] Abbreviations of the experimental techniques employed are defined in table II-A-1.

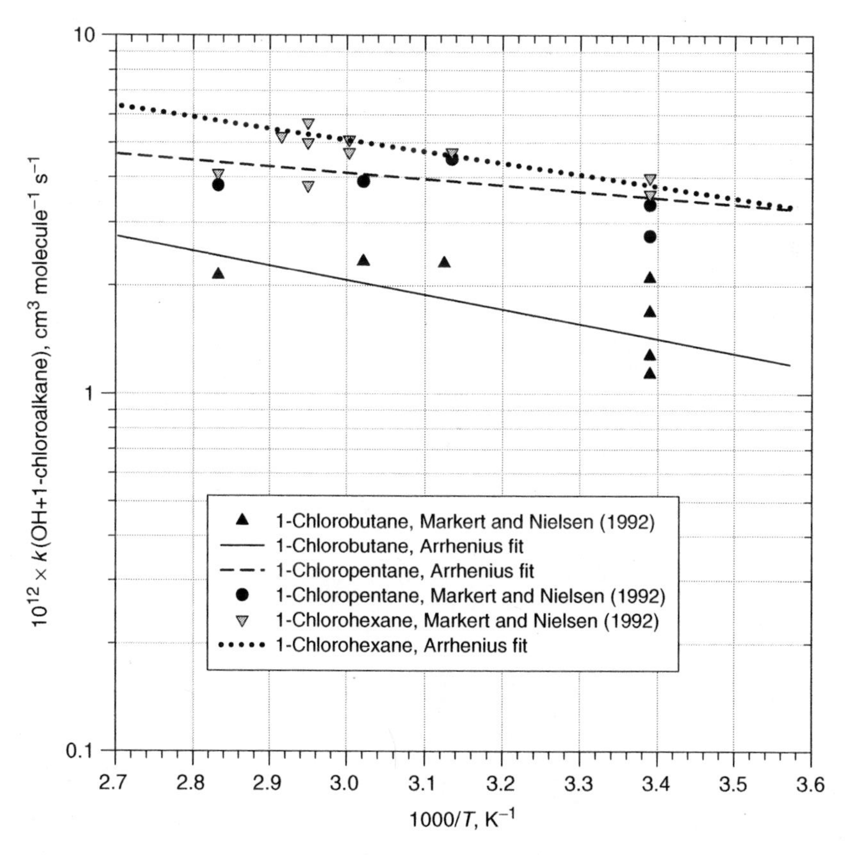

Figure VI-A-38. Arrhenius plot of the rate coefficients for the reaction of OH with n-$C_4H_{10}Cl$, n-$C_5H_{11}Cl$ and n-$C_6H_{13}Cl$.

VI-A-3.16. OH + $CH_3(CH_2)_3CH_2Cl$ → Products

The rate coefficients of Markert and Nielsen (1992) are given in table VI-A-41 and plotted in figure VI-A-38, where they can be compared with those of 1-C_4H_9Cl and 1-$C_6H_{13}Cl$. A fit of the Arrhenius expression to the 1-$C_5H_{11}Cl$ data of Markert and Nielsen (1992) gives $k(OH+CH_3(CH_2)_3CH_2Cl) = 1.40 \times 10^{-11} \exp(-407/T)$ cm^3 molecule^{-1} s^{-1}. At 298 K this expression gives $k(OH+CH_3(CH_2)_3CH_2Cl) = 3.57 \times 10^{-12}$ cm^3 molecule^{-1} s^{-1}, with an estimated uncertainty of ±35%.

VI-A-3.17. OH + $CH_3(CH_2)_4CH_2Cl$ → Products

The rate coefficients of Markert and Nielsen (1992) are given in table VI-A-42 and plotted in figure VI-A-38, where they can be compared with those for 1-C_4H_9Cl and 1-$C_5H_{11}Cl$. A fit of the Arrhenius expression to the 1-$C_6H_{13}Cl$ data of Markert and Nielsen (1992) gives $k(OH+CH_3(CH_2)_4CH_2Cl) = 4.79 \times 10^{-11} \exp(-747/T)$ cm^3 molecule^{-1} s^{-1}. At 298 K this expression gives $k(OH+CH_3(CH_2)_4CH_2Cl) = 3.91 \times 10^{-12}$ cm^3 molecule^{-1} s^{-1}, with an estimated uncertainty of ±20%.

VI-A-4. Hydrobromoalkanes

VI-A-4.1. OH + CH_3Br → H_2O + CH_2Br

The rate coefficients of Howard and Evenson (1976), Davis et al. (1976), Mellouki et al. (1992), Zhang et al. (1992c), Chichinin et al. (1994), and Hsu and DeMore (1994) are given in table VI-A-43 and plotted in figure VI-A-39. As is seen from the figure, the results of Davis et al. lie somewhat above the results from the other studies, with the difference increasing at low temperatures. The simplest explanation for this discrepancy is the presence of reactive impurities in the CH_3Br sample used by Davis et al. A fit of the Arrhenius expression to the data (except those from Davis et al. (1976)) gives $k(OH+CH_3Br) = 2.39 \times 10^{-12} \exp(-1301/T)$ cm^3 molecule^{-1} s^{-1}. At 298 K this expression gives $k(OH+CH_3Br) = 3.04 \times 10^{-14}$ cm^3 molecule^{-1} s^{-1}, with uncertainties estimated at ±15%.

Table VI-A-41. Rate coefficients $k = A\,e^{-B/T}$ (cm^3 molecule^{-1} s^{-1}) for the reaction: OH + $CH_3(CH_2)_3CH_2Cl$ → products

$10^{12} \times A$	B (K)	$10^{13} \times k$	T (K)	Technique[a]	Reference	Temperature Range (K)
17	481 ± 361	3.36 ± 0.09	295	PR–UVabs	Markert and Nielsen (1992)	295–353
		2.75 ± 0.08	295			
		4.5 ± 0.2	319			
		3.9 ± 0.3	331			
		3.8 ± 0.2	353			

[a] Abbreviations of the experimental techniques employed are defined in table II-A-1.

Table VI-A-42. Rate coefficients $k = A\ e^{-B/T}$ (cm^3 molecule^{-1} s^{-1}) for the reaction: $OH + CH_3(CH_2)_4CH_2Cl \rightarrow$ products

$10^{12} \times A$	B (K)	$10^{13} \times k$	T (K)	Technique[a]	Reference	Temperature Range (K)
16	361 ± 241	3.6 ± 0.3	295	PR–UV abs	Markert and Nielsen (1992)	295–353
		4.0 ± 0.3	295			
		4.7 ± 0.1	319			
		5.1 ± 0.5	333			
		4.7 ± 0.4	333			
		5.7 ± 0.4	339			
		5.0 ± 0.4	339			
		3.8 ± 0.3	339			
		5.2 ± 0.1	343			
		4.1 ± 0.2	353			

[a] Abbreviations of the experimental techniques employed are defined in table II-A-1.

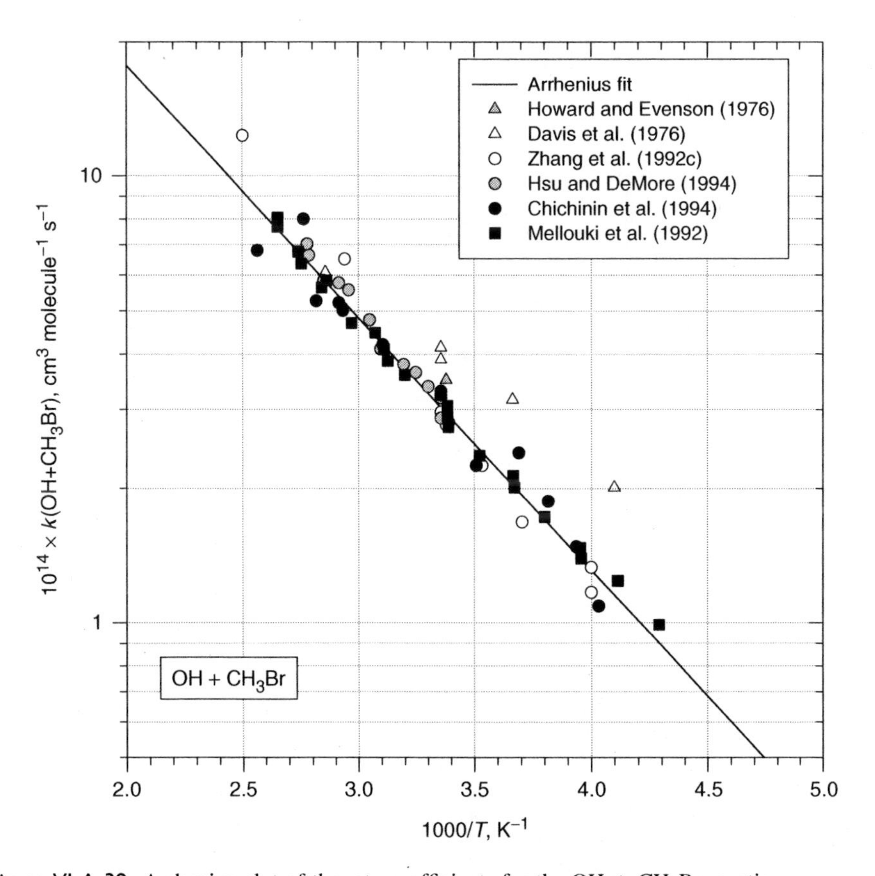

Figure VI-A-39. Arrhenius plot of the rate coefficients for the $OH + CH_3Br$ reaction.

Table VI-A-43. Rate coefficients $k = A\ e^{-B/T}$ ($cm^3\ molecule^{-1}\ s^{-1}$) for the reaction: $OH + CH_3Br \rightarrow$ products

$10^{12} \times A$	B (K)	$10^{14} \times k$	T (K)	Technique[b]	Reference	Temperature Range (K)
		3.5 ± 0.8	296	DF–LMR	Howard and Evenson (1976)	296
0.793 ± 0.079	889 ± 59	2.01 ± 0.12	244	FP–RF	Davis et al. (1976)	244–350
		3.16 ± 0.15	273			
		3.89 ± 0.03	298			
		4.14 ± 0.43	298			
		6.08 ± 0.40	350			
2.35	1300 ± 150	0.99	233	PLP–LIF	Mellouki et al. (1992)	233–377
		1.24	243			
		1.39	253			
		1.47	253			
		1.72	263			
		2.01	272			
		2.14	273			
		2.37	284			
		2.74	295			
		2.90	296			
		3.06	296			
		3.58	313			
		3.86	320			
		4.10	321			
		4.46	325			
		4.69	336			
		5.84	349			
		5.63	352			
		6.36	363			
		6.76	365			
		7.68	377			
		8.06	377			
5.79	1560 ± 150	1.17 ± 0.29	250	FP–RF	Zhang et al. (1992c)	250–400
		1.33 ± 0.19	250			
		1.68 ± 0.30	270			
		2.25 ± 0.56	283			
		2.81 ± 0.31	295			
		2.77 ± 0.74	296			
		2.96 ± 0.83	298			
		6.51 ± 0.97	340			
		12.3 ± 1.3	400			
1.86 ± 0.48	1230 ± 150	1.09 ± 0.16	248	DF–EPR	Chichinin et al. (1994)	248–390
		1.48 ± 0.22	254			
		1.87 ± 0.28	262			
		2.40 ± 0.36	271			
		2.25 ± 0.34	285			
		3.24 ± 0.49	298			
		3.30 ± 0.50	298			
		4.20 ± 0.63	322			
		5.00 ± 0.75	341			

Table VI-A-43. *(continued)*

$10^{12} \times A$	B (K)	$10^{14} \times k$	T (K)	Technique[b]	Reference	Temperature Range (K)
		5.20 ± 0.78	343			
		5.25 ± 0.79	355			
		8.00 ± 1.20	362			
		6.80 ± 1.02	390			
3.22	1389	2.87[a]	298	RR (CH_3CHF_2)	Hsu and DeMore (1994)	298–360
		3.19	298			
		3.38	303			
		3.63	308			
		3.79	313			
		4.11	323			
		4.77	328			
		5.55	338			
		5.76	343			
		5.83	351			
		6.63	359			
		7.03	360			

[a] Placed on an absolute basis using $k(OH + CH_3CHF_2) = 2.10 \times 10^{-18}\, T^2 \exp(-503/T)\, cm^3\, molecule^{-1}\, s^{-1}$.
[b] Abbreviations of the experimental techniques employed are defined in table II-A-1.

VI-A-4.2. $OH + CH_2Br_2 \rightarrow H_2O + CHBr_2$

The rate coefficients of Mellouki et al. (1992), Orlando et al. (1996b), DeMore (1996), and Zhang et al. (1997) are given in table VI-A-44 and plotted in figure VI-A-40. As is seen in the figure, there is substantial scatter in the reported data. The value of Orlando et al. lies significantly below those from the other studies. A fit of the Arrhenius expression to the data of Mellouki et al. (1992), Zhang et al. (1997), and DeMore (1996) gives $k(OH + CH_2Br_2) = 2.52 \times 10^{-12} \exp(-900/T)\, cm^3\, molecule^{-1}\, s^{-1}$. At 298 K this expression gives $k(OH + CH_2Br_2) = 1.23 \times 10^{-13}\, cm^3\, molecule^{-1}\, s^{-1}$, with an estimated uncertainty of ± 20%.

VI-A-4.3. $OH + CHBr_3 \rightarrow H_2O + CBr_3$

The rate coefficients of DeMore (1996) are given in table VI-A-45 and plotted in figure VI-A-41. An Arrhenius fit gives $k(OH + CHBr_3) = 1.48 \times 10^{-12} \exp(-612/T)\, cm^3\, molecule^{-1}\, s^{-1}$. At 298 K this expression gives $k(OH + CHBr_3) = 1.90 \times 10^{-13}\, cm^3\, molecule^{-1}\, s^{-1}$, with an estimated uncertainty of ±25%.

VI-A-4.4. $OH + CH_3CH_2Br \rightarrow H_2O + CH_3CHBr$ (a); $OH + CH_3CH_2Br \rightarrow H_2O + CH_2CH_2Br$ (b)

The rate coefficients of Qiu et al. (1992), Donaghy et al. (1993), and Herndon et al. (2001) are given in table VI-A-46 and plotted in figure VI-A-42. As is seen from figure VI-A-42, there is substantial scatter in the reported room-temperature

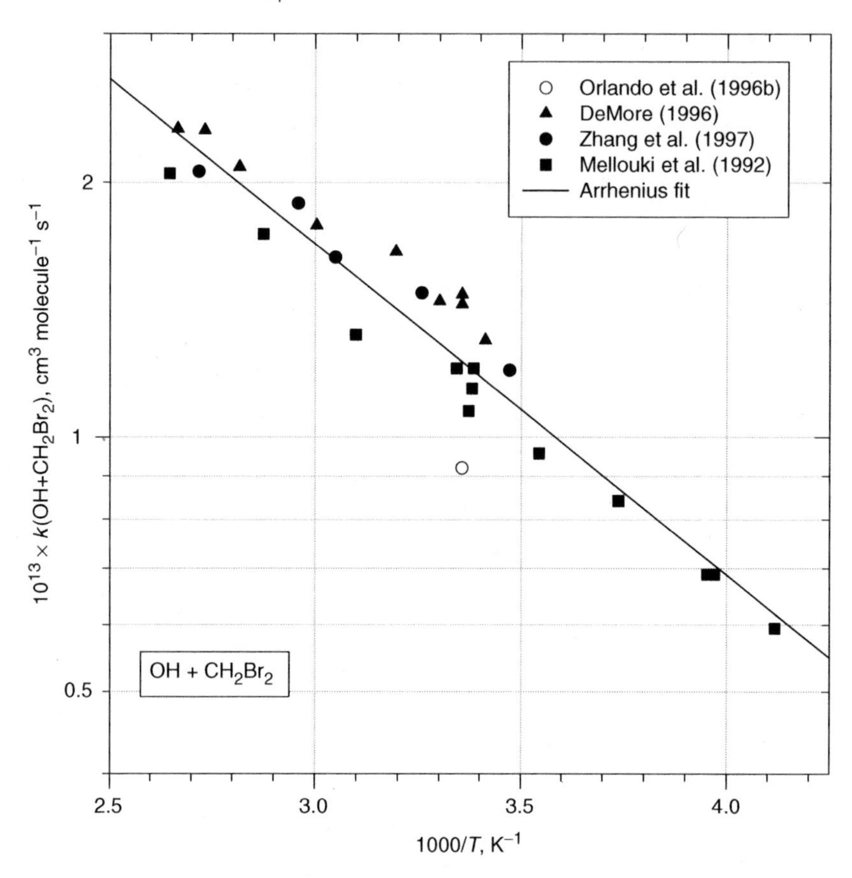

Figure VI-A-40. Arrhenius plot of the rate coefficients for the $OH + CH_2Br_2$ reaction.

rate coefficients. The temperature dependence reported by Qiu et al. is substantially greater than that reported by Herndon et al. There is no obvious reason to exclude the results from any of the studies. A fit of the Arrhenius expression to the data from Qiu et al. (1992), Donaghy et al. (1993), and Herndon et al. (2001) gives $k(OH + C_2H_5Br) = 4.76 \times 10^{-12} \exp(-795/T)\,cm^3\,molecule^{-1}\,s^{-1}$. At 298 K this expression gives $k(OH + C_2H_5Br) = 3.30 \times 10^{-13}\,cm^3\,molecule^{-1}\,s^{-1}$, with an estimated uncertainty of $\pm 25\%$.

VI-A-4.5. $OH + CH_2BrCH_2Br \rightarrow H_2O + CH_2BrCHBr$

The rate coefficients of Howard and Evenson (1976), Arnts et al. (1989), and Qiu et al. (1992) are given in table VI-A-47 and plotted in figure VI-A-43. There is reasonable agreement between the results of the three studies. A fit of the Arrhenius expression to the combined data set gives $k(OH + CH_2BrCH_2Br) = 7.69 \times 10^{-12} \exp(-1056/T)\,cm^3\,molecule^{-1}\,s^{-1}$. At 298 K this expression gives $k(OH + CH_2BrCH_2Br) = 2.22 \times 10^{-13}\,cm^3\,molecule^{-1}\,s^{-1}$, with an estimated uncertainty of $\pm 20\%$.

Table VI-A-44. Rate coefficients $k = A\ e^{-B/T}$ (cm^3 $molecule^{-1}$ s^{-1}) for the reaction: $OH + CH_2Br_2 \rightarrow$ products

$10^{12} \times A$	B (K)	$10^{13} \times k$	T (K)	Technique[c]	Reference	Temperature Range (K)
1.91	840 ± 100	0.59	243	PLP–LIF	Mellouki et al. (1992)	243–378
		0.69	252			
		0.69	253			
		0.84	268			
		0.96	282			
		1.21	296			
		1.14	296			
		1.07	297			
		1.21	299			
		1.32	323			
		1.74	348			
		2.05	378			
		0.92 ± 0.06[a]	298	RR (CH_3COCH_3)	Orlando et al. (1996b)	298
1.66	734	1.30[b]	293	RR (CH_2Cl_2)	DeMore (1996)	293–375
		1.44	298			
		1.47	298			
		1.45	303			
		1.66	313			
		1.78	333			
		2.09	355			
		2.31	366			
		2.31	375			
1.51 ± 0.37	720 ± 60	1.20 ± 0.18	288	DF–RF	Zhang et al. (1997)	298–368
		1.48 ± 0.24	307			
		1.63 ± 0.26	328			
		1.89 ± 0.30	338			
		2.06 ± 0.33	368			

[a] Placed on an absolute basis using $k(OH + CH_3C(O)CH_3) = 1.7 \times 10^{-13}$ cm^3 $molecule^{-1}$ s^{-1}.
[b] Placed on an absolute basis using $k(OH + CH_2Cl_2) = 1.97 \times 10^{-12} \exp(-877/T)$ cm^3 $molecule^{-1}$ s^{-1}.
[c] Abbreviations of the experimental techniques employed are defined in table II-A-1.

Table VI-A-45. Rate coefficients $k = A\ e^{-B/T}$ (cm^3 $molecule^{-1}$ s^{-1}) for the reaction: $OH + CHBr_3 \rightarrow$ products

$10^{12} \times A$	B (K)	$10^{13} \times k$	T (K)	Technique[b]	Reference	Temperature Range (K)
1.48	612	2.05[a]	298	RR (CH_2Cl_2)	DeMore (1996)	298–366
		1.79	298			
		2.11	312			
		2.13	323			
		2.36	333			
		2.50	342			
		2.69	357			
		2.76	366			

[a] Placed on an absolute basis using $k(OH + CH_2Cl_2) = 1.97 \times 10^{-12} \exp(-877/T)$ cm^3 $molecule^{-1}$ s^{-1}.
[b] Abbreviations of the experimental techniques employed are defined in table II-A-1.

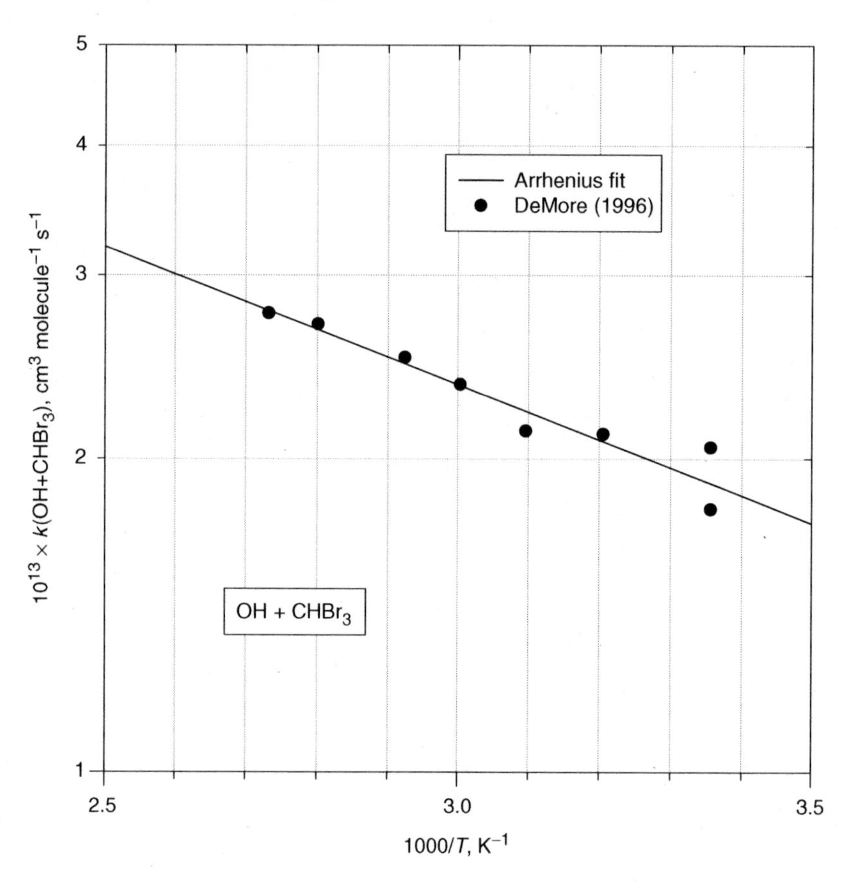

Figure VI-A-41. Arrhenius plot of the rate coefficients for the $OH + CHBr_3$ reaction.

Table VI-A-46. Rate coefficients $k = A\ e^{-B/T}$ (cm^3 $molecule^{-1}$ s^{-1}) for the reaction: $OH + CH_3CH_2Br \rightarrow$ products

$10^{12} \times A$	B (K)	$10^{13} \times k$	T (K)	Technique[b]	Reference	Temperature Range (K)
2.77 ± 0.34	1344 ± 86	2.37	292	DF–RF	Qiu et al. (1992)	292–366
		3.05	298			
		4.53	325			
		5.22	340			
		6.99	366			
		2.25 ± 0.08[a]	300	RR (C_2H_6)	Donaghy et al. (1993)	300
$k(T)$[c]		1.82 ± 0.07	233	PLP–LIF	Herndon et al. (2001)	233-422
		2.09 ± 0.07	243			
		2.21 ± 0.07	243			
		2.48 ± 0.05	258			
		2.55 ± 0.10	268			
		3.56 ± 0.19	296			

Table VI-A-46. (*continued*)

$10^{12} \times A$	B (K)	$10^{13} \times k$	T (K)	Technique[b]	Reference	Temperature Range (K)
		3.47 ± 0.06	296			
		3.42 ± 0.06	297			
		3.29 ± 0.18	297			
		3.17 ± 0.12	297			
		4.15 ± 0.10	321			
		4.58 ± 0.27	338			
		4.91 ± 0.14	346			
		4.65 ± 0.23	351			
		5.11 ± 0.14	358			
		5.74 ± 0.13	375			
		6.88 ± 0.26	399			
		8.09 ± 0.34	421			
		7.83 ± 0.15	422			

[a] Placed on an absolute basis using $k(OH + C_2H_6) = 1.49 \times 10^{-17}\, T^2 \exp(-499/T)\, cm^3\, molecule^{-1}\, s^{-1}$.
[b] Abbreviations of the experimental techniques employed are defined in table II-A-1.
[c] $k = 1.7 \times 10^{-13}\, T^{0.5} \times \exp(-641/T)$.

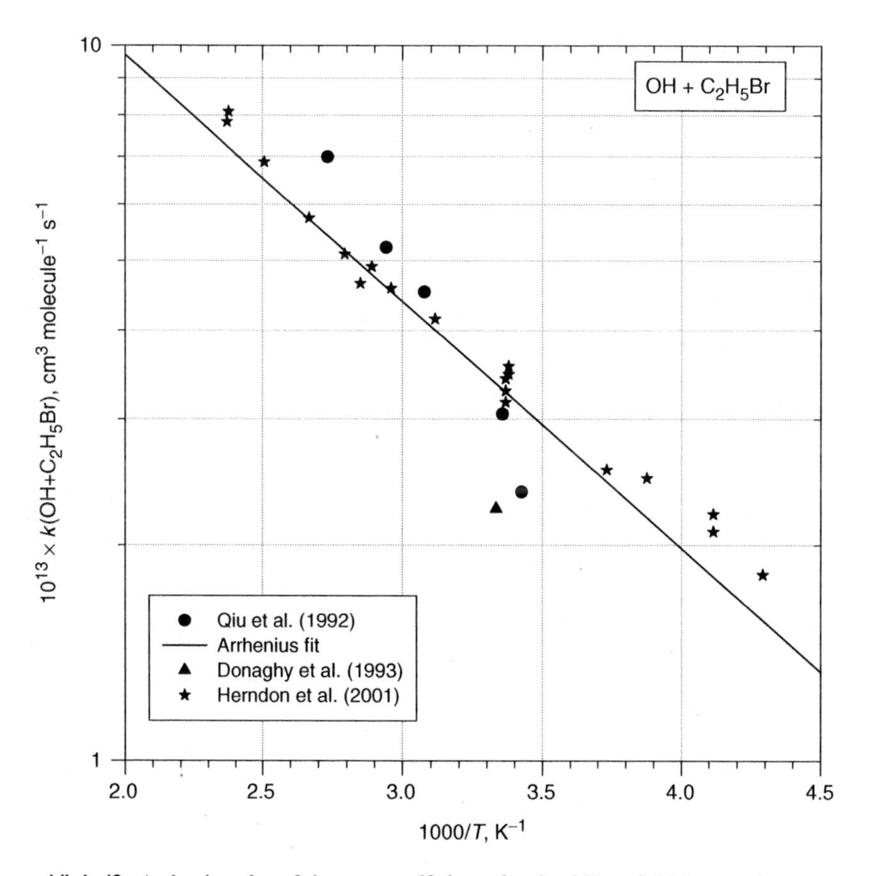

Figure VI-A-42. Arrhenius plot of the rate coefficients for the OH + C_2H_5Br reaction.

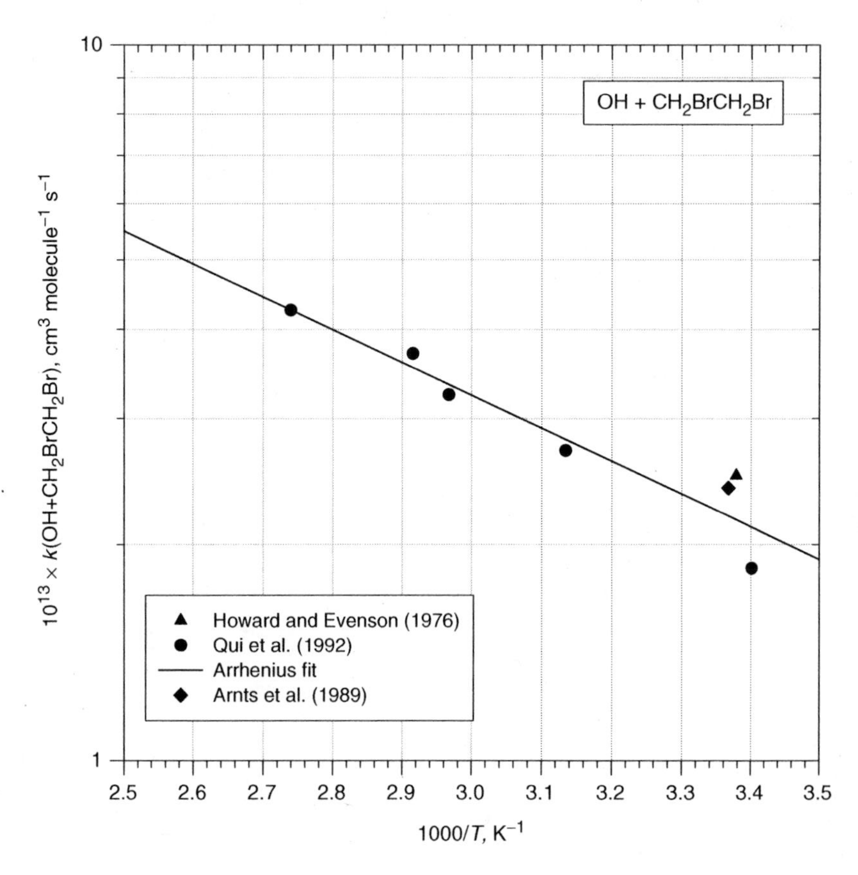

Figure VI-A-43. Arrhenius plot of the rate coefficients for the OH + CH_2BrCH_2Br reaction.

Table VI-A-47. Rate coefficients $k = A\ e^{-B/T}$ (cm^3 $molecule^{-1}$ s^{-1}) for the reaction: OH + CH_2BrCH_2Br → products

$10^{12} \times A$	B (K)	$10^{13} \times k$	T (K)	Technique[b]	Reference	Temperature Range (K)
		2.50 ± 0.55	296	DF–LMR	Howard and Evenson (1976)	296
		2.40 ± 0.17[a]	297	RR (C_2H_6)	Arnts et al. (1989)	297
14.6 ± 2.6	1283 ± 136	1.86	294	DF–RF	Qiu et al. (1992)	294–365
		2.71	319			
		3.24	337			
		3.70	343			
		4.26	365			

[a] Placed on an absolute basis using $k((OH + C_2H_6) = 1.49 \times 10^{-17}\ T^2\ \exp(-499/T)$.
[b] Abbreviations of the experimental techniques employed are defined in table II-A-1.

VI-A-4.6. OH + $CH_3CH_2CH_2Br$ → Products

The rate coefficients of Donaghy et al. (1993), Téton et al. (1996a), Nelson et al. (1997), Herndon et al. (2001), Gilles et al. (2002), and Kozlov et al. (2003a) are given in table VI-A-48 and plotted in figure VI-A-44. As is seen in the figure, there is good agreement between the studies of Donaghy et al., Téton et al., Nelson et al., and Kozlov et al. The results from Herndon et al. (2001) for $T > 270$ K lie systematically lower than the other studies. Herndon et al. argue that the 248 nm laser photolysis of H_2O_2 used to produce OH radicals by Téton et al. could have photolysed $CH_3CH_2CH_2Br$, giving radicals that would react with OH and leading to an overestimation of the rate coefficient. However, the data from Téton et al. at $T < 270$ K are in good agreement with the results of Herndon et al. Also, the relative rate study of Donaghy et al. (1993), discharge flow study of Nelson et al. (1997), and the flash photolysis study by Kozlov et al. (2003a) are in good agreement with the work of Téton et al. (1996a). As is seen from figure VI-A-44, there is evidence for significant curvature in the Arrhenius plot. A fit of the modified Arrhenius expression to the combined data set of Donaghy et al. (1993), Téton et al. (1996a), Nelson et al. (1997), Gilles et al. (2002), and Kozlov et al. (2003a) gives $k(\mathrm{OH} + \mathrm{CH_3CH_2CH_2Br}) = 7.57 \times 10^{-18}\ T^2\ \exp(134/T)\,\mathrm{cm^3\ molecule^{-1}\ s^{-1}}$. At 298 K this expression gives $k(\mathrm{OH} + \mathrm{CH_3CH_2CH_2Br}) = 1.05 \times 10^{-12}\,\mathrm{cm^3\ molecule^{-1}\ s^{-1}}$, with an estimated uncertainty of ±20%.

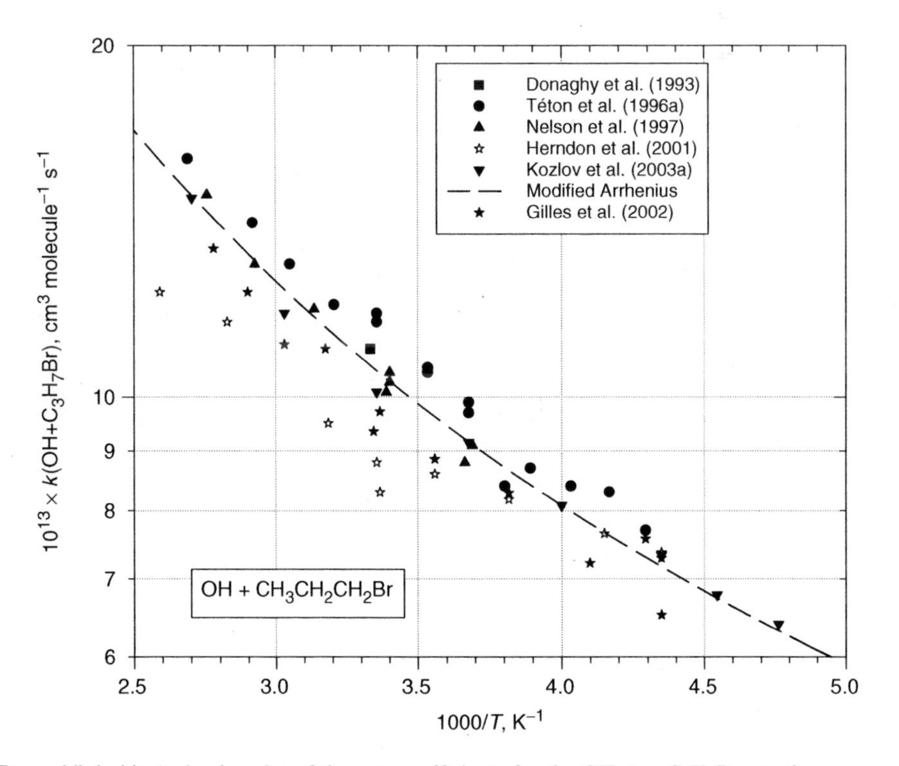

Figure VI-A-44. Arrhenius plot of the rate coefficients for the OH + n-C_3H_7Br reaction.

Table VI-A-48. Rate coefficients $k = A\ e^{-B/T}$ ($cm^3\ molecule^{-1}\ s^{-1}$) for the reaction: $OH + CH_3CH_2CH_2Br \rightarrow$ products

$10^{12} \times A$	B (K)	$10^{13} \times k$	T (K)	Technique[b]	Reference	Temperature Range (K)
		11.0 ± 0.6[a]	300	RR (cyclohexane)	Donaghy et al. (1993)	300
5.29 ± 0.29	456 ± 31	7.7 ± 0.3	233	PLP–LIF	Téton et al. (1996a)	233–372
		8.3 ± 0.6	240			
		8.4 ± 0.3	248			
		8.7 ± 0.4	257			
		8.4 ± 0.3	263			
		9.7 ± 0.4	272			
		9.9 ± 0.5	272			
		10.5 ± 0.5	283			
		10.6 ± 0.7	283			
		11.6 ± 0.5	298			
		11.8 ± 0.6	298			
		12.0 ± 0.4	312			
		13.0 ± 0.7	328			
		14.1 ± 0.7	343			
		16.0 ± 0.7	372			
5.75 ± 0.9	504 ± 50	9.1 ± 0.5	271	DF–LIF	Nelson et al. (1997)	271–363
		8.8 ± 0.5	273			
		10.5 ± 0.3	294			
		10.3 ± 0.5	294			
		10.1 ± 1.0	295			
		11.9 ± 0.5	319			
		13.0 ± 0.3	342			
		14.9 ± 0.8	363			
$k(T)$[c]		7.37 ± 0.25	230	PLP–LIF	Herndon et al. (2001)	230–386
		7.29 ± 0.60	230			
		7.65 ± 0.15	241			
		8.18 ± 0.40	262			
		8.6 ± 0.4	281			
		8.3 ± 0.4	297			
		8.8 ± 0.4	298			
		9.5 ± 0.4	314			
		11.6 ± 0.5	354			
		12.3 ± 0.9	386			
		6.52 ± 0.38	230	PLP–LIF	Gilles et al. (2002)	230–360
		7.57 ± 0.20	233			
		7.22 ± 0.18	244			
		8.29 ± 0.18	262			
		8.86 ± 0.25	281			
		9.72 ± 0.32	297			
		9.35 ± 0.34	299			
		11.0 ± 0.46	315			
		11.1 ± 0.71	330			
		12.3 ± 0.64	345			
		13.4 ± 0.19	360			

Table VI-A-48. (*continued*)

$10^{12} \times A$	B (K)	$10^{13} \times k$	T (K)	Technique[b]	Reference	Temperature Range (K)
$k(T)$[d]		6.40 ± 0.24	210	FP–RF	Kozlov et al. (2003a)	210–480
		6.78 ± 0.40	220			
		7.33 ± 0.18	230			
		8.08 ± 0.08	250			
		9.15 ± 0.09	272			
		10.1 ± 0.15	298			
		11.8 ± 0.2	330			
		14.8 ± 0.5	370			
		19.0 ± 0.3	420			
		24.2 ± 0.5	480			

[a] Relative to $k(\mathrm{OH+Cyclohexane}) = 3.26 \times 10^{-17}\ T^2 \exp(262/T)\,\mathrm{cm^3\ molecule^{-1}\ s^{-1}}$.
[b] Abbreviations of the experimental techniques employed are defined in table II-A-1.
[c] $k = 9.1 \times 10^{-14}\ T^{0.5} \times \exp(-157/T)$.
[d] $k = 2.99 \times 10^{-13} \times \exp(369/T) \times (T/298)^{2.79}$.

From an analysis of kinetic data for the reaction of OH radicals with $CH_3CH_2CH_2Br$, $CD_3CH_2CH_2Br$, $CD_3CH_2CD_2Br$, $CH_3CD_2CD_2Br$, and $CD_3CD_2CD_2Br$, Gilles et al. (2002) deduced branching ratios at 298 K for abstraction at the 1-, 2-, and 3-positions in $CH_3CH_2CH_2Br$ of 0.32 ± 0.08, 0.56 ± 0.04, and 0.12 ± 0.08, respectively.

VI-A-4.7. OH + $CH_3CHBrCH_3$ → Products

The rate coefficients of Donaghy et al. (1993), Téton et al. (1996a), Herndon et al. (2001), and Kozlov et al. (2003a) are given in table VI-A-49 and plotted in figure VI-A-45. As is seen in this figure, there is substantial scatter in the reported rate coefficients. The absolute rate studies by Téton et al. and Herndon et al. report similar temperature dependencies, but the absolute values differ by approximately 20–30%. The relative rate study of Donaghy et al. at room temperature and the absolute rate study of Kozlov et al. at 210–480 K give results which lie between those of Téton et al. and Herndon et al. There is no obvious reason to prefer any of the studies. A fit of the Arrhenius expression to the combined data set gives $k(\mathrm{OH+CH_3CHBrCH_3}) = 2.90 \times 10^{-12}\ \exp(-367/T)\,\mathrm{cm^3\ molecule^{-1}\ s^{-1}}$. At 298 K this expression gives $k(\mathrm{OH+CH_3CHBrCH_3}) = 8.46 \times 10^{-13}\ \mathrm{cm^3\ molecule^{-1}\ s^{-1}}$, with an estimated uncertainty of ±20%.

VI-A-4.8. OH + $CH_3CH_2CH_2CH_2Br$ → Products

The single rate coefficient measurement of Donaghy et al. (1993) is given in table VI-A-50, and this is recommended with an uncertainty estimated to be ±25%.

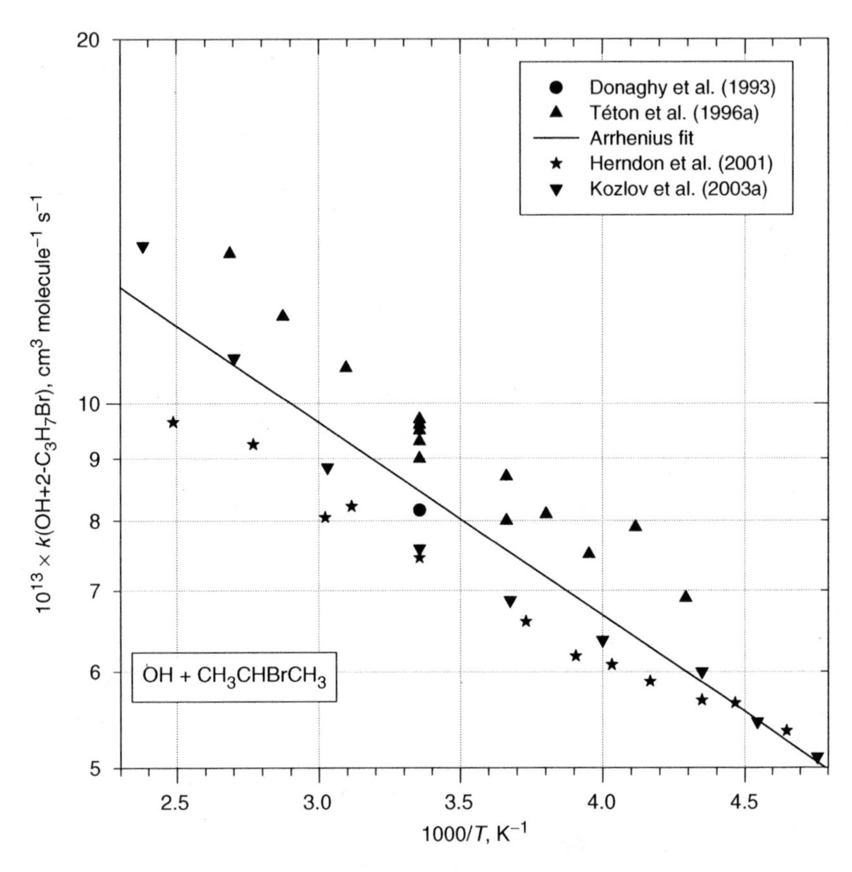

Figure VI-A-45. Arrhenius plot of the rate coefficients for the $OH+2\text{-}C_3H_7Br$ reaction.

Table VI-A-49. Rate coefficients $k = A\ e^{-B/T}$ (cm^3 $molecule^{-1}$ s^{-1}) for the reaction: $OH + CH_3CHBrCH_3 \rightarrow$ products

$10^{12} \times A$	B (K)	$10^{13} \times k$	T (K)	Technique[b]	Reference	Temperature Range (K)
		8.16 ± 0.35[a]	298	RR (Cyclohexane)	Donaghy et al. (1993)	298
3.58 ± 0.47	392 ± 75	6.9 ± 0.3	233	FP–LIF	Téton et al. (1996a)	233–372
		7.9 ± 0.5	243			
		7.5 ± 0.5	253			
		8.1 ± 0.5	263			
		8.0 ± 0.7	273			
		8.7 ± 0.2	273			
		9.7 ± 0.3	298			
		9.0 ± 0.4	298			
		9.3 ± 0.5	298			
		9.5 ± 0.6	298			
		9.6 ± 1.0	298			
		10.7 ± 0.3	323			

Table VI-A-49. *(continued)*

$10^{12} \times A$	B (K)	$10^{13} \times k$	T (K)	Technique[b]	Reference	Temperature Range (K)
		11.8 ± 0.5	348			
		13.3 ± 0.5	372			
$k(T)$[c]		5.36 ± 0.30	215	PLP–LIF	Herndon et al. (2001)	215–402
		5.65 ± 0.20	224			
		5.68 ± 0.17	230			
		5.89 ± 0.11	240			
		6.08 ± 0.16	248			
		6.18 ± 0.24	256			
		6.6 ± 0.3	268			
		7.45 ± 0.15	298			
		8.22 ± 0.40	321			
		8.05 ± 0.50	331			
		9.25 ± 0.39	361			
		9.65 ± 0.40	402			
		5.1 ± 0.3	210	FP–RF	Kozlov et al. (2003a)	210–480
		5.45 ± 0.10	220			
		6.00 ± 0.10	230			
		6.37 ± 0.10	250			
		6.87 ± 0.10	272			
		7.58 ± 0.12	298			
		8.85 ± 0.20	330			
		10.9 ± 0.2	370			
		13.5 ± 0.3	420			
		17.7 ± 0.3	480			

[a] Relative to k(OH + Cyclohexane) = $3.26 \times 10^{-17}\, T^2 \exp(262/T)$ cm^3 molecule^{-1} s^{-1}.
[b] Abbreviations of the experimental techniques employed are defined in table II-A-1.
[c] $k = 7.0 \times 10^{-14}\, T^{0.5} \times \exp(-145/T)$.
[d] $k = 1.66 \times 10^{-13} \times \exp(461/T) \times (T/298)^{2.95}$.

Table VI-A-50. Rate coefficients $k = A\, e^{-B/T}$ (cm^3 molecule^{-1} s^{-1}) for the reaction: OH + $CH_3(CH_2)_nBr$ → products ($n = 3$–5)

Alkyl bromide	$10^{12} \times k$	T (K)	Technique[a]	Reference
1-C_4H_9Br	2.29 ± 0.07[b]	299	RR (Cyclohexane)	Donaghy et al. (1993)
1-$C_5H_{11}Br$	3.71 ± 0.13[b]	304	RR (Cyclohexane)	Donaghy et al. (1993)
1-$C_6H_{13}Br$	5.48 ± 0.19[b]	306	RR (Cyclohexane)	Donaghy et al. (1993)

[a] Abbreviations of the experimental techniques employed are defined in table II-A-1.
[b] Relative to k(OH + Cyclohexane) = $3.26 \times 10^{-17}\, T^2 \exp(262/T)$ cm^3 molecule^{-1} s^{-1}.

VI-A-4.9. OH + $CH_3(CH_2)_3CH_2Br$ → Products

The single measurement of the rate coefficient by Donaghy et al. (1993) is given in table VI-A-50 and is recommended with an uncertainty estimated to be ±25%.

VI-A-4.10. $OH + CH_3(CH_2)_4CH_2Br \rightarrow$ Products

The single measurement of the rate coefficient by Donaghy et al. (1993) is given in table VI-A-50 and is recommended with an uncertainty estimated to be ±25%.

VI-A-5. Hydroiodoalkanes

VI-A-5.1. $OH + CH_3I \rightarrow H_2O + CH_2I$

The rate coefficients of Brown et al. (1990c), Gilles et al. (1996), and Cotter et al. (2003) are given in table VI-A-51 and plotted in figure VI-A-46. The data from Brown et al., Gilles et al., and Cotter et al. are in agreement within the experimental uncertainties. A fit of the Arrhenius expression to the combined data gives $k(OH+CH_3I) = 2.26 \times 10^{-12} \times \exp(-1001/T)\,cm^3\,molecule^{-1}\,s^{-1}$. At 298 K this expression gives $k(OH+CH_3I) = 7.86 \times 10^{-14}\,cm^3\,molecule^{-1}\,s^{-1}$, with an estimated uncertainty of ±30%.

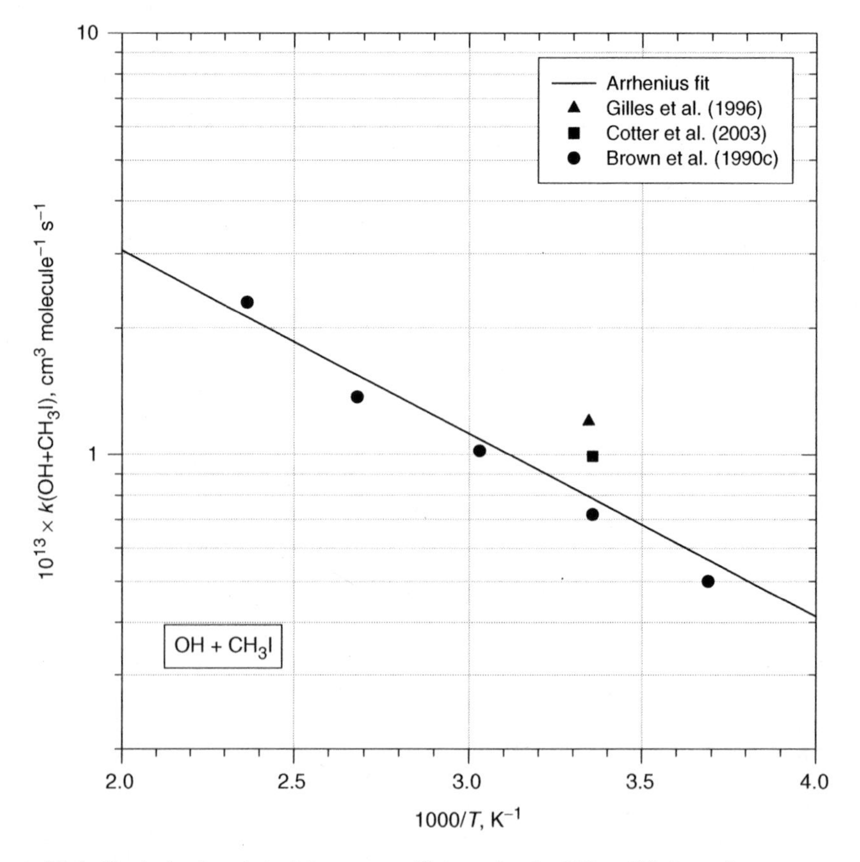

Figure VI-A-46. Arrhenius plot of the rate coefficients for the $OH + CH_3I$ reaction.

VI-A-5.2. OH + CH_3CH_2I → Products

The one determination of the rate coefficient for this reaction is that of Cotter et al. (2003), who employed DR–RF techniques in experiments at 298 K: $k = (7.7 \pm 1.0) \times 10^{-13}$ cm^3 molecule^{-1} s^{-1}. This value is recommended, with an estimated uncertainty of ±25%.

VI-A-5.3. OH + $CH_3CH_2CH_2I$ → Products

The rate coefficients of Carl and Crowley (2001) and Cotter et al. (2003) are given in table VI-A-52. In light of possible secondary loss of OH, Cotter et al. consider their value as an upper limit, and hence it is consistent with the work of Carl and Crowley. On the basis of the study by Carl and Crowley (2001), we recommend $k(OH + CH_3CH_2CH_2I) = 1.47 \times 10^{-12}$ cm^3 molecule^{-1} s^{-1}, with an estimated uncertainty of ±25%.

VI-A-5.4. OH + CH_3CHICH_3 → Products

The rate coefficients of Carl and Crowley (2001) and Cotter et al. (2003) are given in table VI-A-53. In light of possible secondary loss of OH, Cotter et al. consider their value as an upper limit, and hence it is consistent with the work of Carl and

Table VI-A-51. Rate coefficients $k = A\,e^{-B/T}$ (cm^3 molecule^{-1} s^{-1}) for the reaction: OH + CH_3I → products

$10^{12} \times A$	B (K)	$10^{13} \times k$	T (K)	Technique[a]	Reference	Temperature Range (K)
3.1	1119 ± 205	0.5 ± 0.26	271	DF–RF	Brown et al. (1990c)	271–423
		0.72 ± 0.07	298			
		1.02 ± 0.44	330			
		1.37 ± 0.48	373			
		2.30 ± 0.72	423			
		1.2 ± 0.4	299	PLP–LIF	Gilles et al. (1996)	299
		0.99 ± 0.20	298	DF–RF	Cotter et al. (2003)	298

[a] Abbreviations of the experimental techniques employed are defined in table II-A-1.

Table VI-A-52. Rate coefficients $k = A\,e^{-B/T}$ (cm^3 molecule^{-1} s^{-1}) for the reaction: OH + $CH_3CH_2CH_2I$ → products

$10^{12} \times k$	T (K)	Technique[a]	Reference	Temperature Range (K)
1.47 ± 0.08	298	PLP–RF	Carl and Crowley (2001)	298
2.5 ± 0.3	298	DF–RF	Cotter et al. (2003)	298

[a] Abbreviations of the experimental techniques employed are defined in table II-A-1.

Crowley (2001). On the basis of the study by Carl and Crowley (2001), we recommend $k(OH + CH_3CHICH_3) = 1.22 \times 10^{-12}$ cm^3 molecule^{-1} s^{-1}, with an estimated uncertainty of ±25%.

VI-A-6. Hydrochlorofluorocarbons

VI-A-6.1. OH + CH_2FCl (HCFC-31) → H_2O + CHFCl

The rate coefficients of Howard and Evenson (1976), Watson et al. (1977), Handwerk and Zellner (1978), Paraskevopoulos et al. (1981), Jeong and Kaufman (1982), Jeong et al. (1984), and DeMore (1996) are given in table VI-A-54 and plotted in figure VI-A-47. As is seen from the figure, there is reasonable agreement between the results from all of the studies. A fit of the Arrhenius expression to the combined data set gives $k(OH + CH_2FCl) = 2.80 \times 10^{-12} \exp(-1254/T)$ cm^3 molecule^{-1} s^{-1}. At 298 K this expression gives $k(OH + CH_2FCl) = 4.17 \times 10^{-14}$ cm^3 molecule^{-1} s^{-1}, with uncertainties estimated at ±15%.

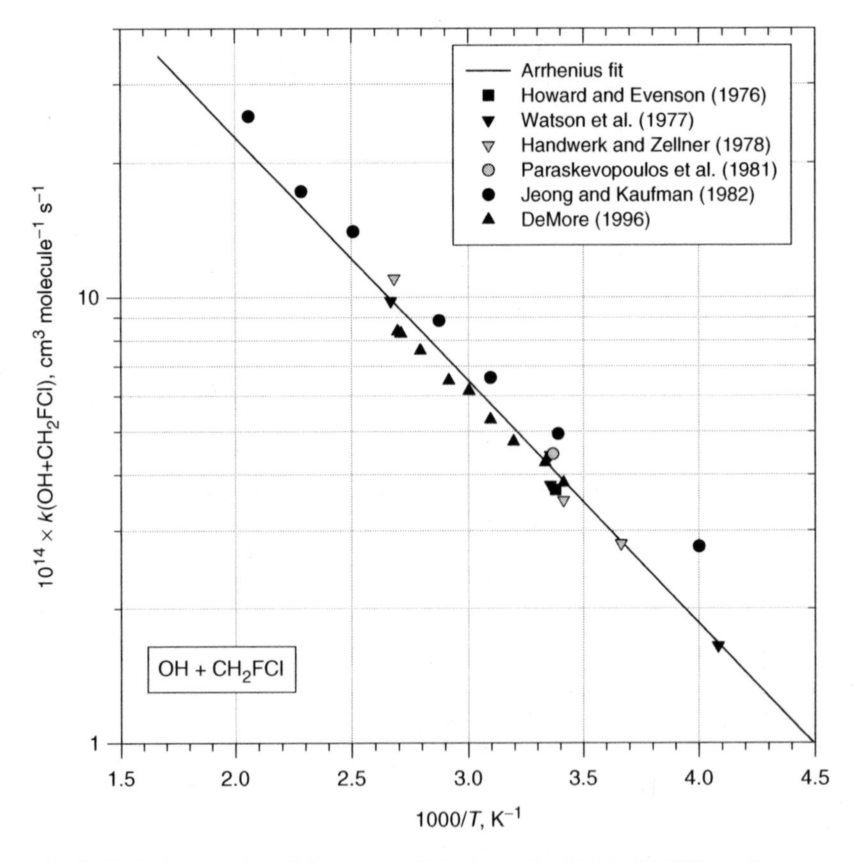

Figure VI-A-47. Arrhenius plot of the rate coefficients for the OH + CH_2FCl reaction.

Table VI-A-53. OH + CH_3CHICH_3 → products

$10^{12} \times k$	T (K)	Technique[a]	Reference
1.22 ± 0.06	298	PLP–RF	Carl and Crowley (2001)
1.60 ± 0.2	298	DF–RF	Cotter et al. (2003)

[a] Abbreviations of the experimental techniques employed are defined in table II-A-1.

Table VI-A-54. Rate coefficients $k = A\ e^{-B/T}$ (cm^3 $molecule^{-1}$ s^{-1}) for the reaction: OH + CH_2FCl (HCFC-31) → products

$10^{12} \times A$	B (K)	$10^{14} \times k$	T (K)	Technique[b]	Reference	Temperature Range (K)
		3.7 ± 0.6	296	DF–LMR	Howard and Evenson (1976)	296
2.84	1259 ± 50	1.65 ± 0.36	245	FP–RF	Watson et al. (1977)	245–375
		4.42 ± 0.24	298			
		3.79 ± 0.30	298			
		9.80 ± 0.34	375			
3.1 ± 0.9	1320 ± 100	2.8 ± 0.5	273	FP–RA	Handwerk and Zellner (1978)	273–373
		3.5 ± 0.7	293			
		11.0 ± 2.0	373			
		4.45 ± 0.67	297	FP–RA	Paraskevopoulos et al. (1981)	297
2.37 ± 0.29	1137 ± 40	2.76 ± 0.18	250	DF–RF	Jeong and Kaufman (1982) Jeong et al. (1984)	250–486
		4.94 ± 0.30	295			
		6.60 ± 0.40	323			
		8.85 ± 0.55	348			
		14.0 ± 0.9	399			
		17.2 ± 1.1	438			
		25.4 ± 1.7	486			
1.60	1094	3.83[a]	293	RR (CH_2Cl_2)	DeMore (1996)	293–371
		4.26	300			
		4.75	313			
		5.32	323			
		6.17	333			
		6.51	343			
		7.58	358			
		8.29	369			
		8.37	371			

[a] Placed on an absolute basis using $k(OH + CH_2Cl_2) = 1.97 \times 10^{-12} \exp(-877/T)$ cm^3 $molecule^{-1}$ s^{-1}.
[b] Abbreviations of the experimental techniques employed are defined in table II-A-1.

VI-A-6.2. OH + CHF_2Cl (HCFC-22) → H_2O + CF_2Cl

The rate coefficients of Atkinson et al. (1975), Howard and Evenson (1976), Watson et al. (1977), Chang and Kaufman (1977), Handwerk and Zellner (1978), Clyne and Holt (1979b), Paraskevopoulos et al. (1981), Jeong and Kaufman (1982),

Jeong et al. (1984), Orkin and Khamaganov (1993a), Yujing et al. (1993), Hsu and DeMore (1995), Fang et al. (1996), Cavalli et al. (1998), and Kowalczyk et al. (1998) are given in table VI-A-55 and plotted in figure VI-A-48. For reasons that are unclear, the results of Howard and Evenson and Cavalli et al. lie below the body of other data, the results from Clyne and Holt show a greater temperature dependence than the other studies, and the result from Kowalczyk et al. lies approximately a factor of two above the other studies. There is significant curvature in the Arrhenius plot for temperatures > 500 K. Excluding the data from Howard and Evenson (1976), Clyne and Holt (1979b), Cavalli et al., and Kowalczyk et al. (1998) from further consideration, a fit of the Arrhenius expression to the combined data set at $T < 500$ K gives $k(OH + CHF_2Cl) = 1.48 \times 10^{-12} \exp(-1710/T)$ cm^3 molecule^{-1} s^{-1}. At 298 K this expression gives $k(OH + CHF_2Cl) = 4.77 \times 10^{-15}$ cm^3 molecule^{-1} s^{-1}. A fit of the modified Arrhenius expression to the data over the entire temperature range gives $k(OH + CHF_2Cl) = 1.51 \times 10^{-18}\ T^2 \exp(-997/T)$ cm^3 molecule^{-1} s^{-1}. At 298 K this expression gives $k(OH + CHF_2Cl) = 4.73 \times 10^{-15}$ cm^3 molecule^{-1} s^{-1}, with an estimated uncertainty at 298 K of ±15%.

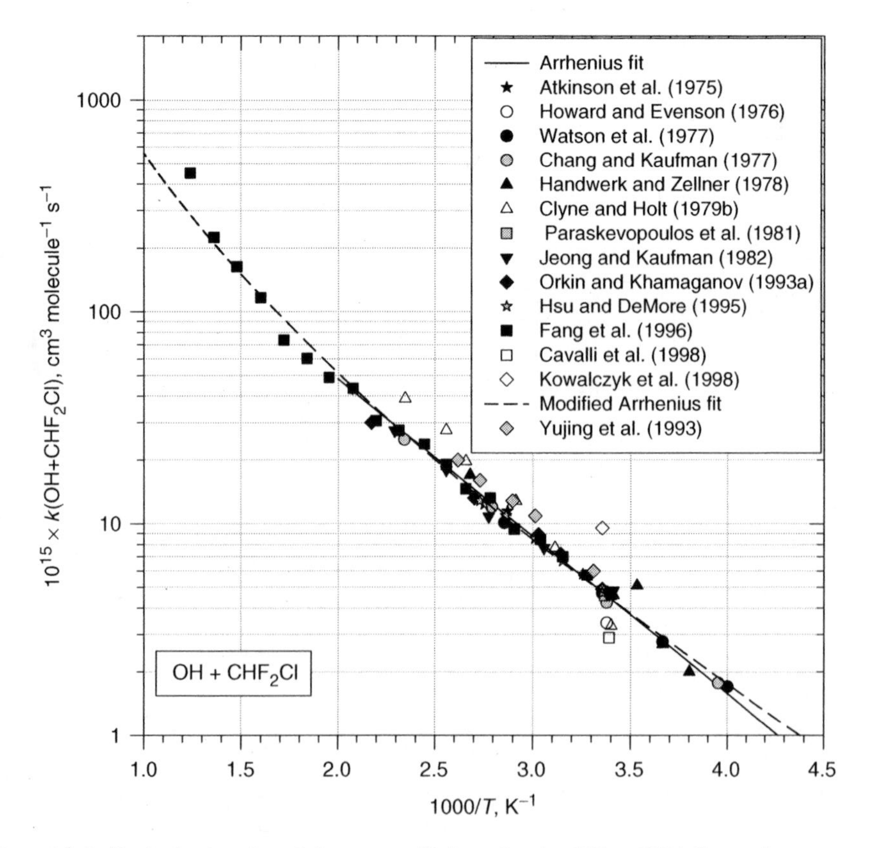

Figure VI-A-48. Arrhenius plot of the rate coefficients for the OH + CHF_2Cl reaction.

Table VI-A-55. Rate coefficients $k = A\ e^{-B/T}$ (cm^3 $molecule^{-1}$ s^{-1}) for the reaction: $OH + CHF_2Cl$ (HCFC-22) → products

$10^{12} \times A$	B (K)	$10^{15} \times k$	T (K)	Technique[b]	Reference	Temperature Range (K)
1.21	1636 ± 151	4.8	296.9	FP–RF	Atkinson et al. (1975)	297–434
		11.5	348.0			
		27.1	433.7			
		3.4 ± 0.7	296	DF–LMR	Howard and Evenson (1976)	296
0.925 ± 0.100	1575 ± 71	1.70 ± 0.4	250	FP–RF	Watson et al. (1977)	245–350
		2.77 ± 0.38	273			
		4.84 ± 0.46	298			
		4.69 ± 1.06	298			
		10.1 ± 0.8	350			
1.20	1657	1.77 ± 0.02	253	DF–RF	Chang and Kaufman (1977)	253–427
		4.25 ± 0.28	296			
		12.0 ± 0.3	358			
		24.9 ± 1.0	427			
2.1 ± 0.6	1780 ± 150	2.0	263	FP–RA	Handwerk and Zellner (1978)	263–373
		2.7	273			
		5.1	283			
		4.6 ± 0.8	293			
		17.0 ± 3.0	373			
9.5	2300 ± 200	3.3 ± 0.7	294	DF–RF	Clyne and Holt (1979b)	294–426
		7.7 ± 1.2	321			
		12.8 ± 1.1	343			
		19.7 ± 0.7	376			
		27.7 ± 1.7	391			
		39.0 ± 0.7	426			
		4.58 ± 0.59	297	FP–RA	Paraskevopoulos et al. (1981)	297
1.27 ± 0.21	1661 ± 60	4.83 ± 0.32	293	DF–RF	Jeong and Kaufman (1982) Jeong et al. (1984)	293–482
		7.68 ± 0.48	327			
		10.8 ± 0.75	360			
		17.9 ± 1.4	391			
		27.5 ± 1.8	436			
		43.9 ± 2.7	482			
1.74 ± 0.2	1700 ± 39	5.99	302	DF–RF	Yujing et al. (1993)	302–382
		10.90	331.8			
		12.80	345			
		16.00	366			
		19.92	382			
$0.81^{+1.5}_{-1.2}$	1516 ± 53	4.9 ± 0.45	298	DF–EPR	Orkin and Khamaganov (1993a)	298–460
		7.2 ± 0.51	318			
		8.9 ± 0.64	330			
		13.2 ± 0.96	370			
		30.0 ± 2.2	460			

(continued)

Table VI-A-55. (*continued*)

$10^{12} \times A$	B (K)	$10^{15} \times k$	T (K)	Technique[b]	Reference	Temperature Range (K)
0.96	1573	4.97[a]	298	RR (CH_4)	Hsu and DeMore (1995)	298–366
		4.71	298			
		5.79	307			
		6.66	317			
		8.53	332			
		10.96	349			
		12.34	363			
		12.81	366			
$k(t)$[c]		4.65 ± 0.66	294	PLP–LIF	Fang et al. (1996)	294–807
		5.73 ± 0.08	305			
		6.99 ± 0.88	317			
		8.43 ± 1.24	329			
		9.43 ± 0.58	344			
		13.23 ± 1.00	359			
		14.60 ± 1.40	376			
		18.99 ± 1.00	391			
		23.75 ± 2.20	409			
		27.60 ± 2.20	432			
		30.66 ± 1.80	455			
		43.58 ± 3.20	481			
		49.02 ± 3.60	512			
		60.23 ± 4.80	544			
		73.59 ± 3.60	582			
		116.7 ± 6.0	625			
		163.7 ± 14.0	676			
		224.8 ± 26.0	735			
		450.7 ± 30.0	807			
		2.9 ± 0.7[a]	295	RR (CH_4)	Cavalli et al. (1998)	295
		9.55	298	US–UV abs	Kowalczyk et al. (1998)	298

[a] Placed on an absolute basis using $k(OH + CH_4) = 1.85 \times 10^{-20}\ T^{2.82} \exp(-987/T)\ cm^3\ molecule^{-1}\ s^{-1}$.
[b] Abbreviations of the experimental techniques employed are defined in table II-A-1.
[c] $k = (3.10 \pm 0.2) \times 10^{-18}\ T^{1.94\pm0.01} \times \exp[(-1112 \pm 26)/T)]$.

VI-A-6.3. OH + $CHFCl_2$ (HCFC-21) → H_2O + $CFCl_2$

The rate coefficients of Howard and Evenson (1976), Perry et al. (1976), Watson et al. (1977), Chang and Kaufman (1977), Clyne and Holt (1979b), Paraskevopoulos et al. (1981), Jeong and Kaufman (1982), and Fang et al. (1996) are given in table VI-A-56 and plotted in figure VI-A-49. The data from Clyne and Holt lie systematically higher than those from other studies and are not considered further. An Arrhenius fit to the remaining data at $T < 400$ K gives $k(OH + CHFCl_2) = 1.17 \times 10^{-12} \exp(-1100/T)\ cm^3\ molecule^{-1}\ s^{-1}$. At 298 K this expression gives $k(OH + CHFCl_2) = 2.92 \times 10^{-14}\ cm^3\ molecule^{-1}\ s^{-1}$. A fit of the modified Arrhenius expression to the entire temperature range gives $k(OH + CHFCl_2) = 2.05 \times 10^{-18}\ T^2 \exp(-551/T)\ cm^3\ molecule^{-1}\ s^{-1}$. At 298 K this expression gives $k(OH + CHFCl_2) = 2.87 \times 10^{-14}\ cm^3\ molecule^{-1}\ s^{-1}$, with an uncertainty at 298 K estimated to be ±15%.

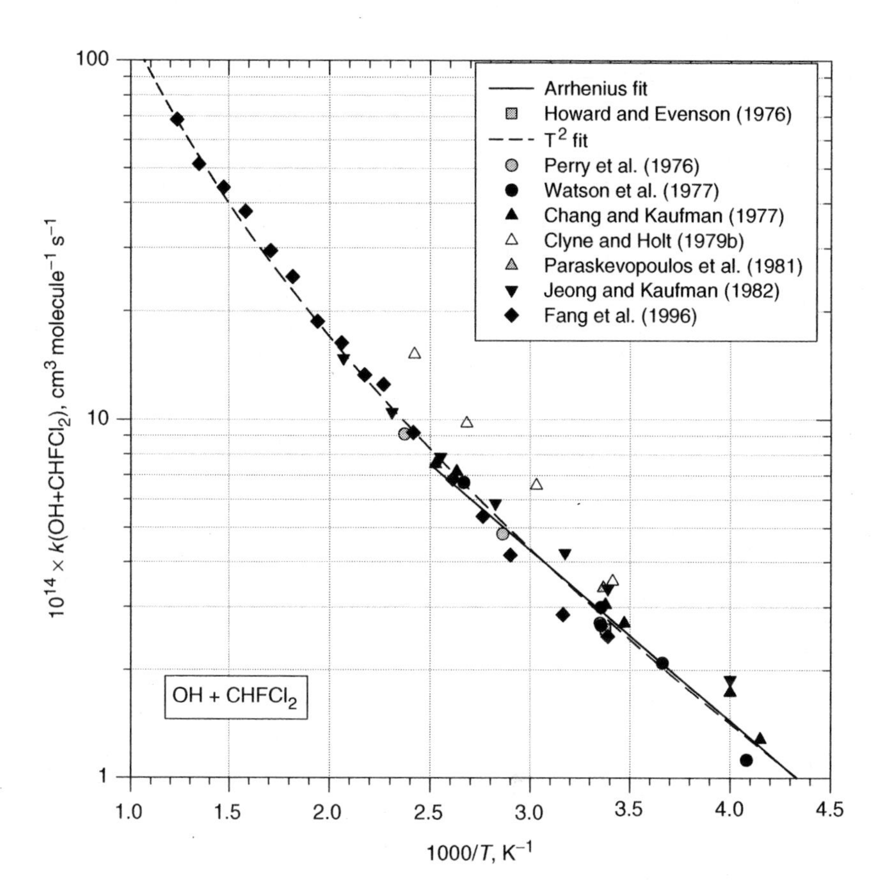

Figure VI-A-49. Arrhenius plot of the rate coefficients for the OH + $CHFCl_2$ reaction.

Table VI-A-56. Rate coefficients $k = A\ e^{-B/T}$ (cm^3 $molecule^{-1}$ s^{-1}) for the reaction: OH + $CHFCl_2$ → products

$10^{12} \times A$	B (K)	$10^{14} \times k$	T (K)	Technique[a]	Reference	Temperature Range (K)
		2.6 ± 0.4	296	DF–LMR	Howard and Evenson (1976)	296
1.75	1253 ± 151	2.7 ± 0.3	298.4	FP–RF	Perry et al. (1976)	298–422
		4.8 ± 0.5	349.5			
		9.1 ± 0.9	421.7			
1.87 ± 0.20	1245 ± 26	1.12 ± 0.12	245	FP–RF	Watson et al. (1977)	245–375
		2.09 ± 0.18	273			
		2.99 ± 0.24	298			
		2.66 ± 0.58	298			
		6.68 ± 0.82	375			
1.16 ± 0.17	1073 ± 40	1.28 ± 0.25	241	DF–RF	Chang and Kaufman (1977)	241–396
		1.73 ± 0.23	250			

(continued)

Table VI-A-56. (*continued*)

$10^{12} \times A$	B (K)	$10^{14} \times k$	T (K)	Technique[a]	Reference	Temperature Range (K)
		2.70 ± 0.20	288			
		3.04 ± 0.11	296			
		7.17 ± 0.16	380			
		7.52 ± 0.29	396			
4.8	1400 ± 100	3.54 ± 0.26	293	DF–RF	Clyne and Holt (1979b)	293–413
		6.57 ± 0.22	330			
		9.77 ± 0.38	373			
		15.2 ± 1.0	413			
		3.39 ± 0.87	297	FP–RA	Paraskevopoulos et al. (1981)	297
1.19 ± 0.15	1052 ± 45	1.88 ± 0.14	250	DF–RF	Jeong and Kaufman (1982) Jeong et al. (1984)	250–483
		3.37 ± 0.22	295			
		4.25 ± 0.27	315			
		5.85 ± 0.36	354			
		7.86 ± 0.48	392			
		10.5 ± 0.65	433			
		14.8 ± 1.0	483			
$k(T)$[b]		4.65 ± 0.66	294	PLP–LIF	Fang et al. (1996)	294–807
		5.73 ± 0.08	305			
		6.99 ± 0.88	317			
		8.43 ± 1.24	329			
		9.43 ± 0.58	344			
		13.23 ± 1.00	359			
		14.60 ± 1.40	376			
		18.99 ± 1.00	391			
		23.75 ± 2.20	409			
		27.60 ± 2.20	432			
		30.66 ± 1.80	455			
		43.58 ± 3.20	481			
		49.02 ± 3.60	512			
		60.23 ± 4.80	544			
		73.59 ± 3.60	582			
		116.7 ± 6.0	625			
		163.7 ± 14.0	676			
		224.8 ± 26.0	735			
		450.7 ± 30.0	807			

[a] Abbreviations of the experimental techniques employed are defined in table II-A-1.
[b] $k = (3.13 \pm 0.2) \times 10^{-18}\ T^{1.93 \pm 0.01} \times \exp[(-552 \pm 18)/T]$.

VI-A-6.4. $OH + CH_3CF_2Cl(HCFC\text{-}142b) \rightarrow H_2O + CH_2CF_2Cl$

The rate coefficients of Cox et al. (1976a), Howard and Evenson (1976), Watson et al. (1977), Handwerk and Zellner (1978), Clyne and Holt (1979b), Paraskevopoulos et al. (1981), Brown et al. (1990b), Liu et al. (1990), Gierczak et al. (1991), Zhang et al. (1992a), Mörs et al. (1996) and Fang et al. (1997) are given in table VI-A-57 and plotted in figure VI-A-50. The data from Clyne and Holt are systematically higher than those from the other studies. The temperature dependence reported by Brown et al. is

Table VI-A-57. Rate coefficients $k = A\ e^{-B/T}$ (cm^3 $molecule^{-1}$ s^{-1}) for the reaction: CH_3CF_2Cl (HCFC-142b) → products

$10^{12} \times A$	B (K)	$10^{15} \times k$	T (K)	Technique[b]	Reference	Temperature Range (K)
		3.7 ± 1.9[a]	298	RR (CH_4)	Cox et al. (1976a)	298
		2.83 ± 0.42	296	DF–LMR	Howard and Evenson (1976)	296
1.15 ± 0.15	1748 ± 30	1.92	273	FP–RF	Watson et al. (1977)	273–375
		3.12	298			
		3.42	298			
		10.90	375			
1.8 ± 0.5	1790 ± 150	3.7 ± 0.7	293	FP–RA	Handwerk and Zellner (1978)	293–373
		14 ± 3	373			
3.3	1800 ± 300	8.4 ± 1.8	293	DF–RF	Clyne and Holt (1979b)	293–417
		6.0 ± 0.7	293			
		12.0 ± 1.1	323			
		14.4 ± 3.7	363			
		30.9 ± 1.5	380			
		40.6 ± 2.7	417			
		4.63 ± 1.73	297	FP–RA	Paraskevopoulos et al. (1981)	297
0.96	1650 ± 250	2.14 ± 0.7	270	FP–RF	Liu et al. (1990)	270–400
		4.02 ± 1.0	298			
		6.60 ± 0.8	330			
		7.97 ± 1.1	350			
		11.5 ± 1.4	375			
		17.0 ± 2.7	400			
0.58	1100	1.5 ± 0.9	231	DF–RF	Brown et al. (1990b)	231–423
		2.6 ± 0.5	266			
		3.7 ± 1.4	303			
		6.9 ± 2.0	339			
		10.3 ± 4.0	373			
		15.8 ± 7.1	423			
2.95 ± 0.25	1750 ± 75	0.455 ± 0.047	223	DF–LMR	Gierczak et al. (1991)	223–427
		0.845 ± 0.071	243	FP–LIF		
		1.71 ± 0.14	273			
		2.99 ± 0.24	297			
		2.96 ± 0.30	299			
		10.9 ± 0.9	374			
		20.7 ± 1.8	427			
		2.45 ± 0.31	270	FP–RF	Zhang et al. (1992a)	270
		2.6 ± 0.4	293	PLP–UV abs	Mörs et al. (1996)	293
$k(T)$[c]		3.77 ± 0.4	295	PLP–LIF	Fang et al. (1997)	295–808
		3.96 ± 0.4	305			
		4.93 ± 0.4	318			
		6.50 ± 0.4	329			
		9.65 ± 0.4	344			
		7.97 ± 0.4	350			
		12.11 ± 0.4	359			

(*continued*)

Table VI-A-57. *(continued)*

$10^{12} \times A$	B (K)	$10^{15} \times k$	T (K)	Technique[b]	Reference	Temperature Range (K)
		14.52 ± 0.4	376			
		19.16 ± 0.4	393			
		17.00 ± 0.4	400			
		23.47 ± 0.4	411			
		31.39 ± 0.4	434			
		39.93 ± 0.4	459			
		54.99 ± 0.4	484			
		70.41 ± 0.4	511			
		99.50 ± 0.4	545			
		144.30 ± 0.4	585			
		198.64 ± 0.4	625			
		311.95 ± 0.4	676			
		666.80 ± 0.4	737			
		922.43 ± 0.4	808			

[a] Placed on an absolute basis using $k(OH + CH_4) = 1.85 \times 10^{-20}\ T^{2.82}\ \exp(-987/T)\ cm^3\ molecule^{-1}\ s^{-1}$.
[b] Abbreviations of the experimental techniques employed are defined in table II-A-1.
[c] $k = 2.05 \times 10^{-30}\ T^{6.01} \times \exp[(-308 \pm 522)/T]$.

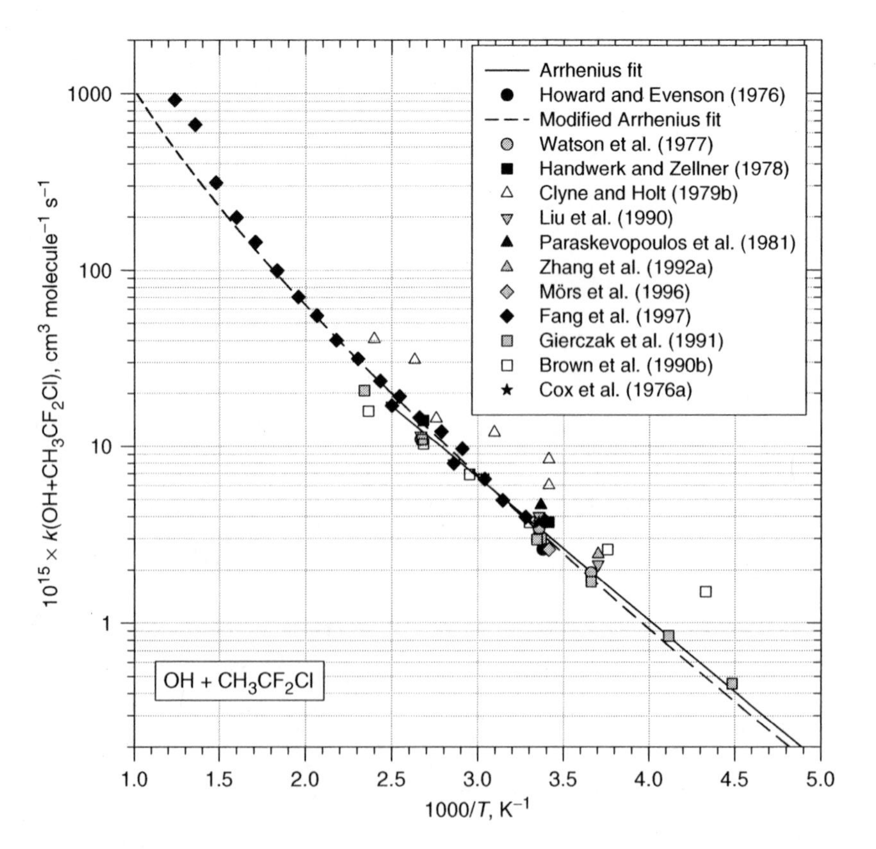

Figure VI-A-50. Arrhenius plot of the rate coefficients for the OH + CH_3CF_2Cl reaction.

substantially lower than that observed in other studies. The data from Clyne and Holt (1979b) and Brown et al. (1990b) are not considered further. A fit of the Arrhenius expression to the remaining data at $T < 400$ K gives $k(OH + CH_3CF_2Cl) = 1.75 \times 10^{-12} \exp(-1857/T)$ cm^3 molecule^{-1} s^{-1}. At 298 K this expression gives $k(OH + CH_3CF_2Cl) = 3.44 \times 10^{-15}$ cm^3 molecule^{-1} s^{-1}. A fit of the modified Arrhenius expression to data over the entire temperature range gives $k(OH + CH_3CF_2Cl) = 4.29 \times 10^{-18} T^2 \exp(-1417/T)$ cm^3 molecule^{-1} s^{-1}. This fit does not describe the data above 700 K well and is only recommended for $T < 700$ K. At 298 K this expression gives $k(OH + CH_3CF_2Cl) = 3.28 \times 10^{-15}$ cm^3 molecule^{-1} s^{-1}. The uncertainty in $k(OH + CH_3CF_2Cl)$ at 298 K is estimated to be $\pm$ 20%.

VI-A-6.5. OH + CH_3CFCl_2 (HCFC-141b) → H_2O + CH_2CFCl_2

The rate coefficients of Liu et al. (1990), Brown et al. (1990b), Talukdar et al. (1991a), Zhang et al. (1992a), Lancar et al. (1993), Huder and DeMore (1993), and Mörs et al. (1996) are given in table VI-A-58 and plotted in figure VI-A-51. As noted by

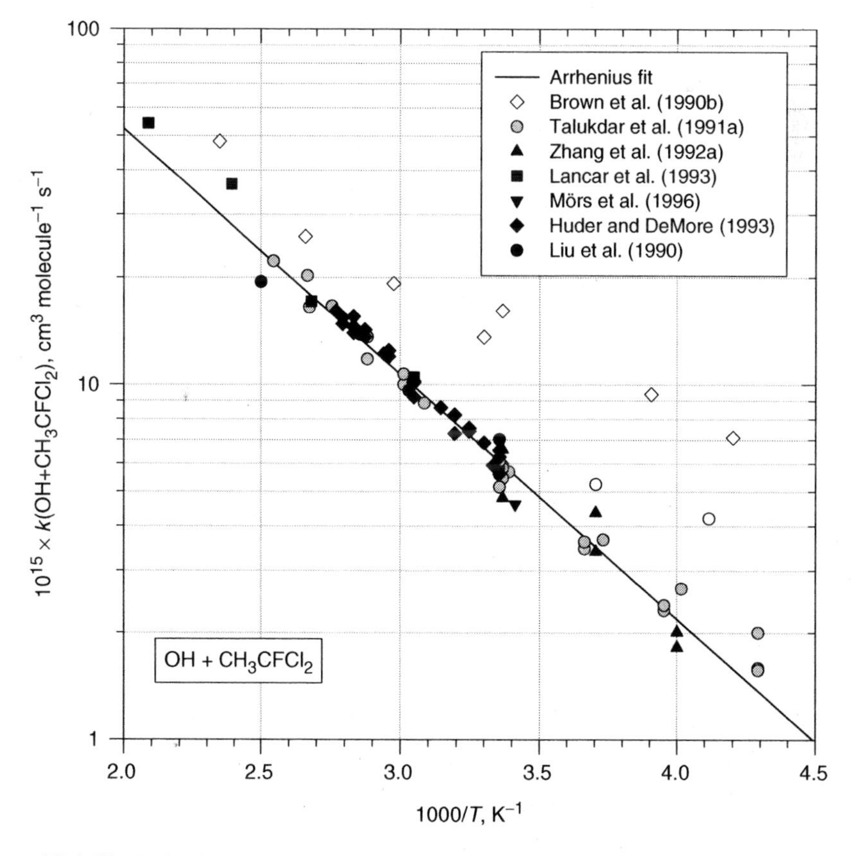

Figure VI-A-51. Arrhenius plot of the rate coefficients for the OH + CH_3CFCl_2 reaction.

Table VI-A-58. Rate coefficients $k = A\ e^{-B/T}$ (cm^3 $molecule^{-1}$ s^{-1}) for the reaction: CH_3CFCl_2 (HCFC-141b) → products

$10^{12} \times A$	B (K)	$10^{15} \times k$	T (K)	Technique[c]	Reference	Temperature Range (K)
0.42	1200 ± 300	4.20 ± 1.2	243	FP–RF	Liu et al. (1990)	243–400
		5.24 ± 0.8	270			
		7.01 ± 1.2	298			
		9.62 ± 1.3	330			
		13.80 ± 0.2	350			
		19.40 ± 2.7	400			
0.58	1100	7.1 ± 4.7	238	DF–RF	Brown et al. (1990b)	238–426
		9.4 ± 4.4	256			
		16.1 ± 5.5	297			
		13.6 ± 2.7	303			
		19.2 ± 5.0	336			
		26.0 ± 7.8	376			
		48.2 ± 6.0	426			
1.47 ± 0.32	1640 ± 100	2.00 ± 0.06	233	DF–LMR	Talukdar	233–393
		1.59 ± 0.10	233	PLP–LIF	et al. (1991a)	
		2.66 ± 0.23	249			
		2.31 ± 0.30	253			
		3.66 ± 0.85	268			
		3.45 ± 0.32	273			
		5.68 ± 0.13	295			
		5.91 ± 0.15	297			
		5.81 ± 0.44	297			
		5.58 ± 0.22	298			
		10.00 ± 0.64	332			
		13.60 ± 0.20	347			
		20.20 ± 0.30	375			
		1.57 ± 0.07	233			
		2.39 ± 0.06	253			
		3.61 ± 0.12	273			
		5.45 ± 0.14	297			
		5.15 ± 0.24	298			
		6.09 ± 0.12	298			
		8.87 ± 0.43	324			
		10.70 ± 0.24	332			
		11.79 ± 0.24	347			
		16.60 ± 0.22	363			
		16.47 ± 0.72	374			
		22.20 ± 0.86	393			
1.42 ± 0.60	1623 ± 293	1.82 ± 0.65	250	FP–RF	Zhang et al. (1992a)	250–297
		2.01 ± 0.90	250			
		4.36 ± 0.81	270			
		3.39 ± 0.82	270			
		6.57 ± 0.70	297			
		4.80 ± 1.46	297			
2.4 ± 0.6	1790 ± 100	5.7 ± 1.5	298	DF–EPR	Lancar et al. (1993)	298–479
		10.5 ± 0.7	328			

(continued)

Table VI-A-58. (*continued*)

$10^{12} \times A$	B (K)	$10^{15} \times k$	T (K)	Technique[c]	Reference	Temperature Range (K)
		17.1 ± 0.8	373			
		36.5 ± 2.0	418			
		54.2 ± 3.0	479			
1.97	1738	5.59[a]	298	RR (CH_4)	Huder and DeMore (1993)	298–361
		5.91	300			
		7.39	308			
		8.20	313			
		7.29	313			
		9.23	328			
		10.11	328			
		11.98	338			
		14.24	348			
		13.95	353			
		14.79	358			
1.43	1615	6.53[b]	298	RR (CH_3CCl_3)		
		6.25	298			
		6.26	298			
		6.86	303			
		7.55	308			
		7.52	308			
		8.23	313			
		8.61	318			
		10.23	328			
		12.45	338			
		12.19	340			
		13.62	348			
		14.52	353			
		15.51	353			
		15.41	358			
		16.05	361			
		4.6 ± 0.8	293	PLP–UV abs	Mörs et al. (1996)	293

[a] Placed on an absolute basis using $k(OH + CH_4) = 1.85 \times 10^{-20}\ T^{2.82}\ \exp(-987/T)\ cm^3\ molecule^{-1}\ s^{-1}$.
[b] Placed on an absolute basis using $k(OH + CH_3CCl_3) = 1.81 \times 10^{-12}\ \exp(-1543/T)\ cm^3\ molecule^{-1}\ s^{-1}$.
[c] Abbreviations of the experimental techniques employed are defined in table II-A-1

Talukdar et al., the presence of $CH_2{=}CCl_2$ impurity can be problematic in studies involving CH_3CFCl_2. The data from Brown et al. (1990b) are systematically higher than those from the other studies, suggesting problems associated with the presence of reactive impurities. The sub-ambient temperature data from Liu et al. (1990) are superseded by data from Zhang et al. (1992a). A fit of the Arrhenius expression to the combined data set, excluding data from Brown et al. (1990b) and the sub-ambient temperature data from Liu et al. (1990), gives $k(OH + CH_3CFCl_2) = 1.25 \times 10^{-12} \times \exp(-1588/T)\ cm^3\ molecule^{-1}\ s^{-1}$. At 298 K this expression gives $k(OH + CH_3CFCl_2) = 6.06 \times 10^{-15}\ cm^3\ molecule^{-1}\ s^{-1}$, with an uncertainty estimated to be ±15%.

VI-A-6.6. $OH + CH_2ClCF_3$ *(HCFC-133a)* $\rightarrow H_2O + CHClCF_3$

The rate coefficients of Howard and Evenson (1976), Handwerk and Zellner (1978), Clyne and Holt (1979b), Fang et al. (1999), and DeMore (2005) are given in table VI-A-59 and plotted in figure VI-A-52. The data from Clyne and Holt are less precise and more scattered than the other results. The data from Howard and Evenson and from DeMore lie significantly below the results from Handwerk and Zellner and Fang et al. DeMore (2005) argues that the lower values for $k(OH+CH_2ClCF_3)$ measured in his work and that of Howard and Evenson (1976) should be preferred as they are more consistent with expectations based upon correlations of $k(298\,K)$ and the Arrhenius exponential factor B in the expression $k=A\ \exp(-B/T)\,cm^3\ molecule^{-1}\ s^{-1}$. On the other hand, it is difficult to dismiss the absolute rate data from Handwerk and Zellner (1978) and Fang et al. (1999) which are in such good agreement. An Arrhenius fit of the $T < 500\,K$ data from the entire data set except the results from Clyne and Holt (1979b) gives $k(OH+CH_2ClCF_3) = 1.63 \times 10^{-12}\ \exp(-1426/T)\,cm^3\ molecule^{-1}\ s^{-1}$. At 298 K this expression gives $k(OH+CH_2ClCF_3) = 1.36 \times 10^{-14}\ cm^3\ molecule^{-1}\ s^{-1}$. A fit of the modified Arrhenius expression over the entire temperature range gives

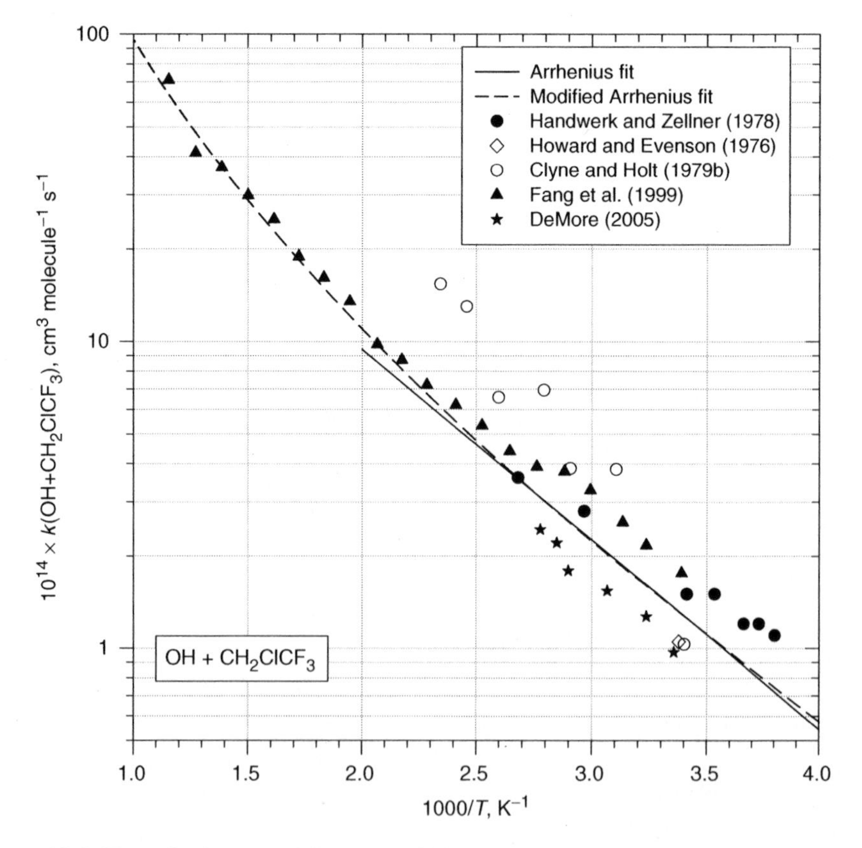

Figure VI-A-52. Arrhenius plot of the rate coefficients for the $OH + CH_2ClCCF_3$ reaction.

Table VI-A-59. Rate coefficients $k = A\ e^{-B/T}$ (cm^3 molecule^{-1} s^{-1}) for the reaction: $OH + CH_2ClCF_3$ (HCFC-133a) → $H_2O + CHClCF_3$

$10^{12} \times A$	B (K)	$10^{14} \times k$	T (K)	Technique[c]	Reference	Temperature Range (K)
		1.05 ± 0.23	296	DF–LMR	Howard and Evenson (1976)	296
1.1	1281 ± 85	1.1 ± 0.2	263	DF–EPR	Handwerk and Zellner (1978)	263–373
		1.2 ± 0.2	268			
		1.2 ± 0.2	273			
		1.5 ± 0.3	283			
		1.5 ± 0.3	293			
		2.8[a]	337			
		3.6 ± 0.8	373			
33 ± 15	2300 ± 300	1.03 ± 0.30	294	DF–EPR	Clyne and Holt (1979b)	294–427
		3.83 ± 0.57	322			
		3.86 ± 0.31	344			
		6.94 ± 0.33	358			
		6.58 ± 0.25	385			
		13.0 ± 1.2	407			
		15.4 ± 1.3	427			
$k(T)$[d]		1.76 ± 0.25	295	DF–EPR	Fang et al. (1999)	295–866
		2.17 ± 0.20	309			
		2.58 ± 0.27	319			
		3.28 ± 0.27	334			
		3.77 ± 0.31	347			
		3.91 ± 0.24	362			
		4.39 ± 0.14	378			
		5.33 ± 0.19	396			
		6.22 ± 0.45	415			
		7.23 ± 0.38	438			
		8.71 ± 0.27	460			
		9.79 ± 0.36	484			
		13.50 ± 0.32	514			
		16.13 ± 0.38	546			
		18.90 ± 0.37	581			
		24.96 ± 0.88	620			
		29.86 ± 1.25	666			
		36.82 ± 2.60	721			
		41.21 ± 3.28	786			
		71.08 ± 2.49	866			
1.45	1483	0.97[b]	298	RR (CH_3CCl_3)	DeMore (2005)	298–360
		1.27	309			
		1.54	326			
		1.79	345			
		2.21	351			
		2.44	360			

[a] Single experiment, no uncertainties quoted.
[b] Placed on an absolute basis using $k(OH + CH_3CCl_3) = 1.81 \times 10^{-12} \exp(-1543/T)$ cm^3 molecule^{-1} s^{-1}.
[c] Abbreviations of the experimental techniques employed are defined in table II-A-1.
[d] $k = (3.06 \pm 4.02) \times 10^{-18}\ T^{1.91 \pm 0.03} \times \exp[(-644 \pm 313)/T]$.

$k(OH + CH_2ClCF_3) = 2.09 \times 10^{-18}\ T^2\ \exp(-780/T)\ cm^3\ molecule^{-1}\ s^{-1}$. At 298 K this expression gives $k(OH + CH_2ClCF_3) = 1.35 \times 10^{-14}\ cm^3\ molecule^{-1}\ s^{-1}$, with an estimated uncertainty of ±35%.

VI-A-6.7. OH + CH_2ClCF_2Cl (HCFC-132b) → H_2O + $CHClCF_2Cl$

The rate coefficients of Watson et al. (1979), Jeong et al. (1984), and Fang et al. (1999) are given in table VI-A-60 and plotted in figure VI-A-53. Watson et al. conducted a GC analysis of their sample and detected the presence of haloalkene impurity at a level of approximately 450 ppm. We adopt the values of $k(OH + CH_2ClCF_2Cl)$ reported by Watson et al. (1979), assuming that the haloalkane impurities react with OH radicals with a rate coefficient of $5 \times 10^{-12}\ cm^3\ molecule^{-1}\ s^{-1}$. As is seen from figure VI-A-53, the data from Jeong et al. lie systematically above those of Watson et al. and Fang et al., with the difference increasing at lower temperatures. The simplest explanation of this behavior is the presence of reactive impurities in the work of Jeong et al. A fit of the Arrhenius expression to the combined data set of Watson et al. (1979) and Fang et al. (1999) at $T < 400$ K gives $k(OH + CH_2ClCF_2Cl) = 4.04 \times 10^{-12} \times \exp(-1611/T)\ cm^3\ molecule^{-1}\ s^{-1}$. At 298 K this expression gives $k(OH + CH_2ClCF_2Cl) = 1.81 \times 10^{-14}\ cm^3\ molecule^{-1}\ s^{-1}$. A fit of the Arrhenius expression to the combined data set of Watson et al. (1979) and Fang et al. (1999) over the entire temperature range gives $k(OH + CH_2ClCF_2Cl) = 5.31 \times 10^{-18}\ T^2\ \exp(-985/T)\ cm^3\ molecule^{-1}\ s^{-1}$. At 298 K this expression gives $k(OH + CH_2ClCF_2Cl) = 1.73 \times 10^{-14}\ cm^3\ molecule^{-1}\ s^{-1}$, with an uncertainty estimated to be ±25%.

Table VI-A-60. Rate coefficients $k = A\ e^{-B/T}$ ($cm^3\ molecule^{-1}\ s^{-1}$) for the reaction: OH + CH_2ClCF_2Cl (HCFC-132b) → products

$10^{12} \times A$	B (K)	$10^{14} \times k$	T (K)	Technique[a]	Reference	Temperature Range (K)
3.37	1578	0.612	250	FP–RF	Watson et al. (1979)	250–350
		1.67	298			
		3.72	350			
2.02 ± 0.24	1263 ± 35	1.42 ± 0.11	249	DF–RF	Jeong et al. (1984)	249–473
		1.60 ± 0.10	253			
		1.91 ± 0.16	267			
		2.72 ± 0.18	295			
		2.42 ± 0.16	297			
		4.31 ± 0.28	333			
		5.95 ± 0.37	365			
		8.06 ± 0.51	383			
		10.40 ± 0.65	418			
		16.00 ± 1.15	473			
$k(T)$[b]		1.84 ± 0.07	295	PLP–LIF	Fang et al. (1999)	295–788
		2.27 ± 0.25	309			
		2.86 ± 0.26	320			
		3.83 ± 0.41	335			

Table VI-A-60. *(continued)*

$10^{12} \times A$	B (K)	$10^{14} \times k$	T (K)	Technique[a]	Reference	Temperature Range (K)
		4.07 ± 0.12	347			
		4.62 ± 0.16	363			
		5.95 ± 0.28	378			
		6.52 ± 0.23	398			
		7.93 ± 0.41	418			
		10.07 ± 0.28	439			
		11.70 ± 0.37	461			
		14.29 ± 0.57	486			
		19.11 ± 0.78	517			
		26.28 ± 1.55	550			
		30.67 ± 0.64	585			
		41.69 ± 2.12	625			
		58.81 ± 3.54	668			
		101.98 ± 3.53	723			
		125.29 ± 12.37	788			

[a] Abbreviations of the experimental techniques employed are defined in table II-A-1.
[b] $k = (8.53 \pm 4.06) \times 10^{-19} \times T^{2.28 \pm 0.18} \exp[(-937 \pm 296)/T]$.

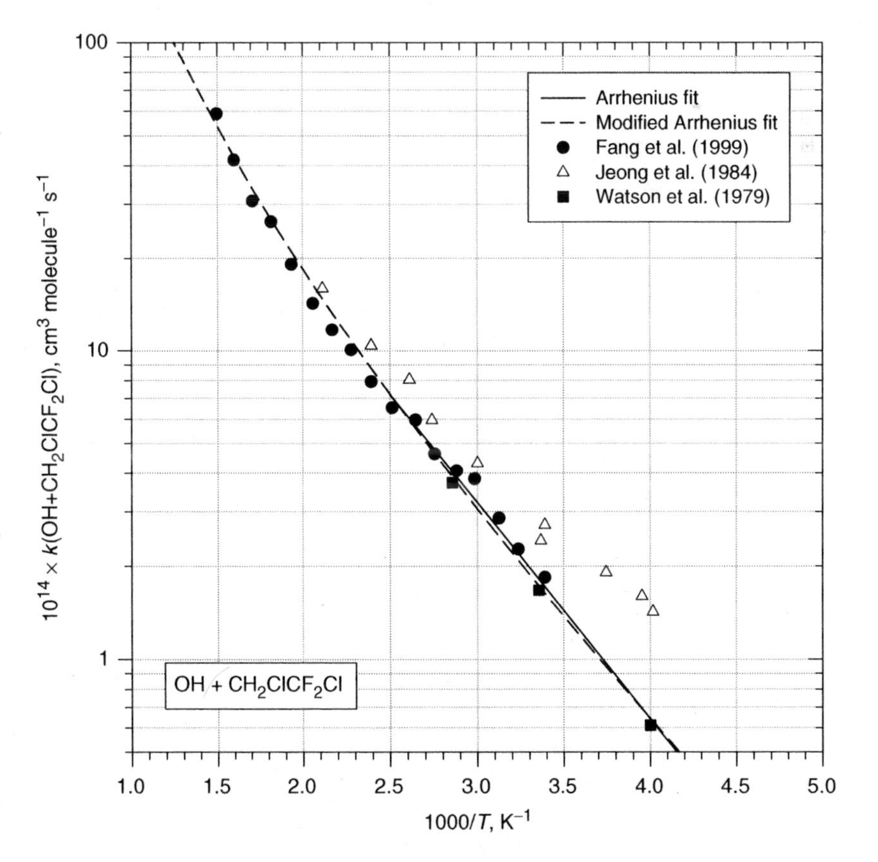

Figure VI-A-53. Arrhenius plot of the rate coefficients for the OH + CH_2ClCF_2Cl reaction.

VI-A-6.8. $OH + CHFClCF_3$ *(HCFC-124)* $\rightarrow H_2O + CFClCF_3$

The rate coefficients of Howard and Evenson (1976), Watson et al. (1979), Gierczak et al. (1991), Hsu and DeMore (1995), and Yamada et al. (2000) are given in table VI-A-61 and plotted in figure VI-A-54. With the possible exception of the data from Hsu and DeMore, there is good agreement between the results from the studies. An Arrhenius fit of the $T < 500$ K data gives $k(OH+CHFClCF_3) = 7.27 \times 10^{-13} \exp(-1303/T)$ cm^3 molecule^{-1} s^{-1}. At 298 K this expression gives $k(OH+CHFClCF_2Cl) = 9.17 \times 10^{-15}$ cm^3 molecule^{-1} s^{-1}. A fit of the modified Arrhenius expression to the entire data set gives $k(OH+CHFClCF_3) = 1.09 \times 10^{-18}\ T^2 \exp(-710/T)$ cm^3 molecule^{-1} s^{-1}. At 298 K this expression gives $k(OH+CHFClCF_3) = 8.94 \times 10^{-15}$ cm^3 molecule^{-1} s^{-1}, with an estimated uncertainty of $\pm 15\%$.

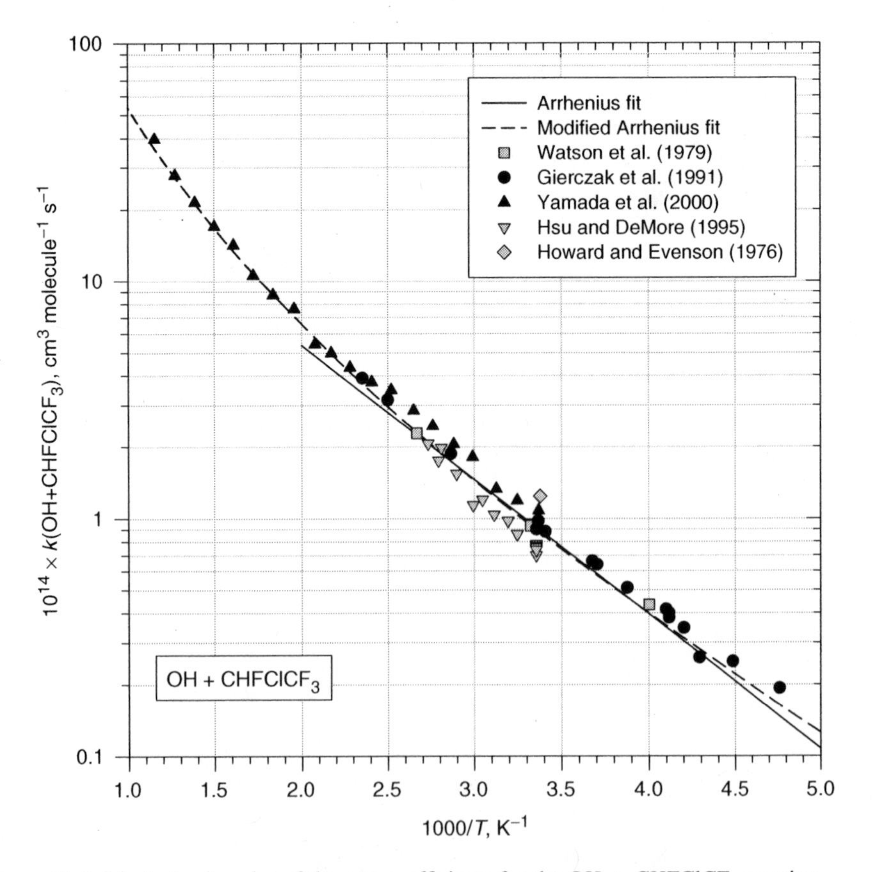

Figure VI-A-54. Arrhenius plot of the rate coefficients for the $OH + CHFClCF_3$ reaction.

Table VI-A-61. Rate coefficients $k = A\ e^{-B/T}$ (cm^3 $molecule^{-1}$ s^{-1}) for the reaction: $OH + CHClFCF_3$ (HCFC-124) → products

$10^{12} \times A$	B (K)	$10^{14} \times k$	T (K)	Technique[c]	Reference	Temperature Range (K)
		1.24 ± 0.19	296	DF–LMR	Howard and Evenson (1976)	296
0.613 ± 0.040	1244 ± 90	0.433 ± 0.019	250	FP–RF	Watson et al. (1979)	250–375
		0.944 ± 0.031	301			
		0.930 ± 0.110	301			
		2.28 ± 0.16	375			
0.445	1150 ± 60	0.193 ± 0.016	210	DF–LMR	Gierczak et al. (1991)	210–425
		0.250 ± 0.02	223	FP–RF		
		0.260 ± 0.022	233			
		0.346 ± 0.028	238			
		0.382 ± 0.038	243			
		0.399 ± 0.041	243			
		0.414 ± 0.034	244			
		0.510 ± 0.042	258			
		0.640 ± 0.052	270			
		0.661 ± 0.072	272			
		0.880 ± 0.074	294			
		0.975 ± 0.080	297			
		0.900 ± 0.075	298			
		1.87 ± 0.15	349			
		3.16 ± 0.26	400			
		3.91 ± 0.31	425			
1.58	1593	0.74[a]	298	RR (CH_4)	Hsu and DeMore (1995)	298–366
		0.78	298			
		0.98	313			
		1.20	328			
		1.54	345			
		1.98	356			
		1.75	358			
		2.06	366			
0.49	1252	0.77[b]	298	RR (CHF_2CHF_2)		
		0.70	298			
		0.75	298			
		0.74	298			
		0.86	308			
		1.04	321			
		1.14	334			
$k(T)$[d]		1.08 ± 0.16	297	PLP–LIF	Yamada et al. (2000)	297–867
		1.19 ± 0.11	308			
		1.33 ± 0.05	320			
		1.81 ± 0.10	334			
		2.05 ± 0.05	347			
		2.45 ± 0.11	362			
		2.85 ± 0.11	378			
		3.48 ± 0.16	397			
		3.76 ± 0.17	415			
		4.33 ± 0.16	438			

(continued)

Table VI-A-61. (*continued*)

$10^{12} \times A$	B (K)	$10^{14} \times k$	T (K)	Technique[c]	Reference	Temperature Range (K)
		4.98 ± 0.20	460			
		5.43 ± 0.24	481			
		7.64 ± 0.27	511			
		8.74 ± 0.49	545			
		10.57 ± 0.57	580			
		14.18 ± 0.40	621			
		17.00 ± 0.69	667			
		21.45 ± 0.61	721			
		27.79 ± 1.14	786			
		39.59 ± 1.29	867			

[a] Placed on an absolute basis using $k(OH + CH_4) = 1.85 \times 10^{-20}\, T^{2.82} \exp(-987/T)\, cm^3\, molecule^{-1}\, s^{-1}$.
[b] Placed on an absolute basis using $k(OH + CHF_2CHF_2) = 1.27 \times 10^{-12} \exp(-1598/T)\, cm^3\, molecule^{-1}\, s^{-1}$.
[c] Abbreviations of the experimental techniques employed are defined in table II-A-1.
[d] $k = (7.72 \pm 0.60) \times 10^{-20}\, T^{2.35 \pm 0.06} \times \exp[(-458 \pm 30)/T]$.

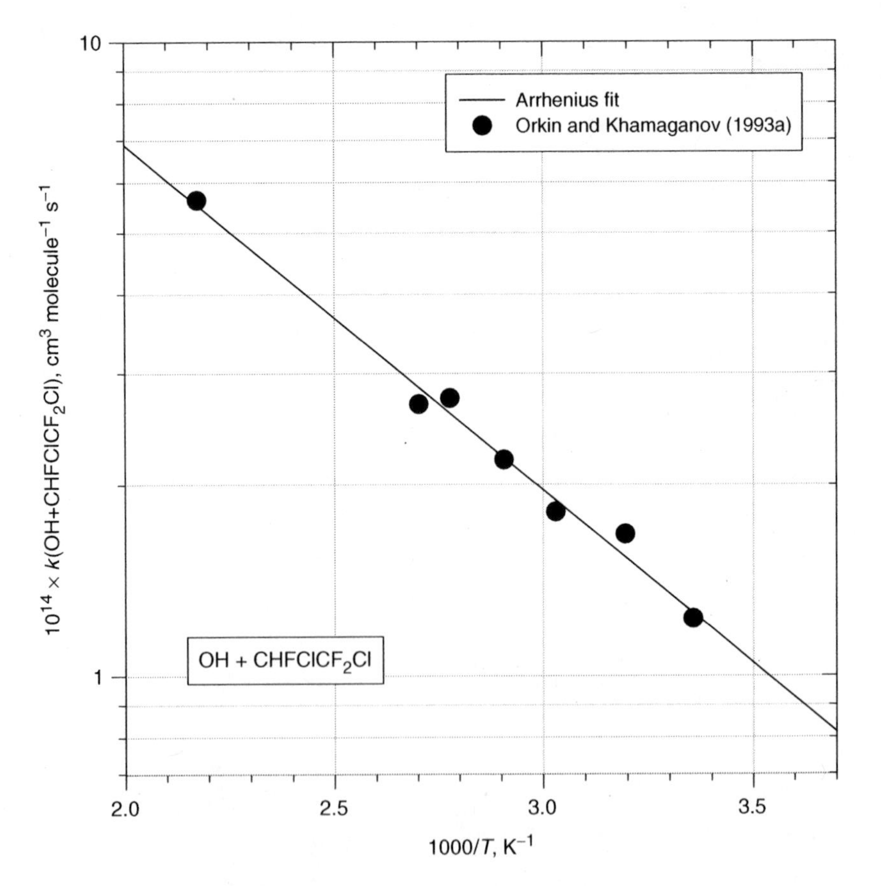

Figure VI-A-55. Arrhenius plot of the rate coefficients for the $OH + CHFClCF_2Cl$ reaction.

VI-A-6.9. $OH + CHFClCF_2Cl$ (HCFC-123a) → $H_2O + CFClCF_2Cl$

The rate coefficients of Orkin and Khamaganov (1993a) are given in table VI-A-62 and plotted in figure VI-A-55. An Arrhenius fit gives $k(OH+CHFClCF_2Cl) = 8.44 \times 10^{-13} \times \exp(-1254/T)$ cm^3 molecule^{-1} s^{-1}. At 298 K this expression gives $k(OH+CHFClCF_2Cl) = 1.26 \times 10^{-14}$ cm^3 molecule^{-1} s^{-1} with an estimated uncertainty of ±25%.

VI-A-6.10. $OH + CF_3CCl_2H$ (HCFC-123) → Products

The rate coefficients of Howard and Evenson (1976), Watson et al. (1979), Clyne and Holt (1979b), Liu et al. (1990), Brown et al. (1990b), Gierczak et al. (1991), Nielsen (1991), Hsu and DeMore (1995), and Yamada et al. (2000) are given in table VI-A-63 and plotted in figure VI-A-56. As is seen in the figure, with the exception of the data

Table VI-A-62. Rate coefficients $k = A\,e^{-B/T}$ (cm^3 molecule^{-1} s^{-1}) for the reaction: $OH + CHFClCF_2Cl$ (HCFC-123a) → $H_2O + CFClCF_2Cl$

$10^{12} \times A$	B (K)	$10^{15} \times k$	T (K)	Technique[a]	Reference	Temperature Range (K)
$0.92^{+2.5}_{-2.0}$	1281 ± 85	1.23 ± 0.10	298	DF–EPR	Orkin and Khamaganov (1993a)	298–460
		1.67 ± 0.12	313			
		1.81 ± 0.13	330			
		2.19 ± 0.16	344			
		2.74 ± 0.22	360			
		2.68 ± 0.19	370			
		5.63 ± 0.41	460			

[a] Abbreviations of the experimental techniques employed are defined in table II-A-1.

Table VI-A-63. Rate coefficients $k = A\,e^{-B/T}$ (cm^3 molecule^{-1} s^{-1}) for the reaction: $OH + CF_3CCl_2H$ → products

$10^{12} \times A$	B (K)	$10^{14} \times k$	T (K)	Technique[b]	Reference	Temperature Range (K)
		2.84 ± 0.43	296	DF–LMR	Howard and Evenson (1976)	296
1.4 ± 0.4	1102^{+157}_{-106}	1.62 ± 0.05	245	FP–RF	Watson et al. (1979)	245–375
		3.78 ± 0.15	298			
		3.30 ± 0.20	298			
		7.20 ± 0.35	375			
1.12 ± 0.05	1000 ± 100	3.86 ± 0.19	293	DF–RF	Clyne and Holt (1979b)	293–429
		5.86 ± 0.15	329			
		8.01 ± 0.33	366			
		11.1 ± 0.4	429			

(*continued*)

Table VI-A-63. *(continued)*

$10^{12} \times A$	B (K)	$10^{14} \times k$	T (K)	Technique[b]	Reference	Temperature Range (K)
0.63	850 ± 250	2.50 ± 0.26	270	FP–RF	Liu et al. (1990)	270–400
		3.52 ± 0.28	298			
		4.70 ± 0.66	330			
		4.99 ± 0.62	350			
		7.36 ± 0.92	375			
		8.84 ± 0.74	400			
1.18	900	4.5 ± 0.9	232	DF–RF	Brown et al. (1990b)	232–426
		7.0 ± 3.0	269			
		5.9 ± 0.6	303			
		8.6 ± 4.0	339			
		10.5 ± 1.1	376			
		14.5 ± 1.7	426			
0.65	840 ± 40	1.13 ± 0.11	213	DF–LMR	Gierczak et al. (1991)	213–380
		1.58 ± 0.29	223	FP–RF		
		1.48 ± 0.13	223			
		1.89 ± 0.16	229			
		2.20 ± 0.24	243			
		2.20 ± 0.18	251			
		3.05 ± 0.25	273			
		4.18 ± 0.36	295			
		3.66 ± 0.30	296			
		3.72 ± 0.34	297			
		3.64 ± 0.34	298			
		4.80 ± 0.39	322			
		7.20 ± 0.58	380			
3.9	1370 ± 260	4.3 ± 1.0	295	PR–UVabs	Nielsen (1991)	295–385
		6.3 ± 1.0	323			
		6.5 ± 1.3	358			
		9.7 ± 1.3	385			
0.34	690	3.32[a]	298	RR (CH_3CHF_2)	Hsu and DeMore (1995)	298–359
		3.41	298			
		3.77	313			
		4.00	320			
		3.88	324			
		4.35	332			
		4.77	345			
		4.86	358			
		4.91	359			
		5.06	359			
$k(T)$[c]		3.67 ± 0.24	296	PLP–LIF	Yamada et al. (2000)	296–866
		4.14 ± 0.22	309			
		4.78 ± 0.11	319			
		5.93 ± 0.21	335			
		6.13 ± 0.45	348			
		6.86 ± 0.18	363			
		7.92 ± 0.32	378			
		9.92 ± 0.16	397			
		10.63 ± 0.21	415			
		13.32 ± 0.46	439			

Table VI-A-63. *(continued)*

$10^{12} \times A$	B (K)	$10^{14} \times k$	T (K)	Technique[b]	Reference	Temperature Range (K)
		14.77 ± 0.21	460			
		15.29 ± 0.36	480			
		19.83 ± 0.42	511			
		22.61 ± 0.86	545			
		25.17 ± 0.93	581			
		29.21 ± 0.42	620			
		36.50 ± 1.09	667			
		45.45 ± 1.11	722			
		55.52 ± 1.90	786			
		86.38 ± 2.01	866			

[a] Placed on an absolute basis using $k(OH + CH_3CHF_2) = 1.21 \times 10^{-12} \exp(-1055/T)$ cm^3 molecule^{-1} s^{-1}.
[b] Abbreviations of the experimental techniques employed are defined in table II-A-1.
[c] $k = (2.20 \pm 0.25) \times 10^{-19} \times T^{2.26 \pm 0.10} \times \exp[(-226 \pm 51)/T]$.

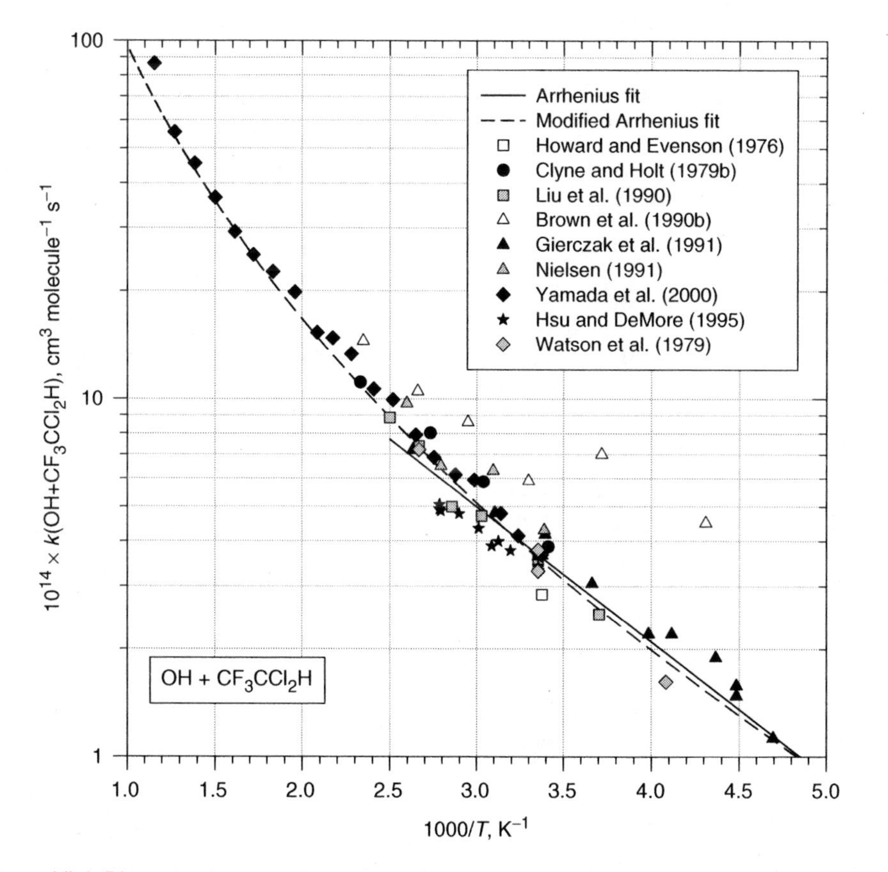

Figure VI-A-56. Arrhenius plot of the rate coefficients for the $OH + CF_3CCl_2H$ reaction.

point from Howard and Evenson and the low-temperature results from Brown et al., there is good agreement between the studies. Excluding the results from Howard and Evenson (1976) and Brown et al., a fit of the Arrhenius expression to the combined data set for $T < 400$ K gives $k(OH + CF_3CCl_2H) = 6.74 \times 10^{-13} \exp(-867/T)$ cm^3 molecule^{-1} s^{-1}. At 298 K this expression gives $k(OH + CF_3CCl_2H) = 3.67 \times 10^{-14}$ cm^3 molecule^{-1} s^{-1}. A fit of the modified Arrhenius expression to all data except those of Howard and Evenson (1976), Brown et al. (1990b), and Hsu and DeMore (1995) gives $k(OH + CF_3CCl_2H) = 1.40 \times 10^{-18}\ T^2 \exp(-370/T)$ cm^3 molecule^{-1} s^{-1}. At 298 K this expression gives $k(OH + CF_3CCl_2H) = 3.59 \times 10^{-14}$ cm^3 molecule^{-1} s^{-1} with an uncertainty estimated to be ±20%.

VI-A-6.11. OH + $CHCl_2CF_2Cl$ (HCFC-122) → H_2O + CCl_2CF_2Cl

The rate coefficients of Orkin and Khamaganov (1993a) and DeMore (1996) are given in table VI-A-64 and plotted in figure VI-A-57. The results from the two studies are in excellent agreement. A fit of the Arrhenius expression to the combined data set gives $k(OH + CHCl_2CF_2Cl) = 1.08 \times 10^{-12} \exp(-916/T)$ cm^3 molecule^{-1} s^{-1}. At 298 K this expression gives $k(OH + CHCl_2CF_2Cl) = 4.99 \times 10^{-14}$ cm^3 molecule^{-1} s^{-1} with an estimated uncertainty of ±15%.

Table VI-A-64. Rate coefficients $k = A\,e^{-B/T}$ (cm^3 molecule^{-1} s^{-1}) for the reaction: OH + $CHCl_2CF_2Cl$ (HCFC-122) → products

$10^{12} \times A$	B (K)	$10^{14} \times k$	T (K)	Technique[c]	Reference	Temperature Range (K)
$1.13^{+2.1}_{-1.6}$	918 ± 52	5.30 ± 0.41	298	DF–EPR	Orkin and Khamaganov (1993a)	298–460
		6.75 ± 0.48	330			
		9.38 ± 0.72	370			
		15.9 ± 1.1	460			
0.65	755	5.73[a]	313	RR (CF_3CCl_2H)	DeMore (1996)	303–371
		7.05	333			
		7.73	358			
		8.56	371			
0.99	885	5.32[b]	303	RR (CH_2Cl_2)		
		7.53	343			
		7.38	343			
		7.61	343			
		8.60	363			

[a] Relative to $k(OH + CF_3CCl_2H) = 6.74 \times 10^{-13} \exp(-867/T)$ cm^3 molecule^{-1} s^{-1}.
[b] Relative to $k(OH + CH_2Cl_2) = 1.97 \times 10^{-12} \exp(-877/T)$ cm^3 molecule^{-1} s^{-1}.
[c] Abbreviations of the experimental techniques employed are defined in table II-A-1.

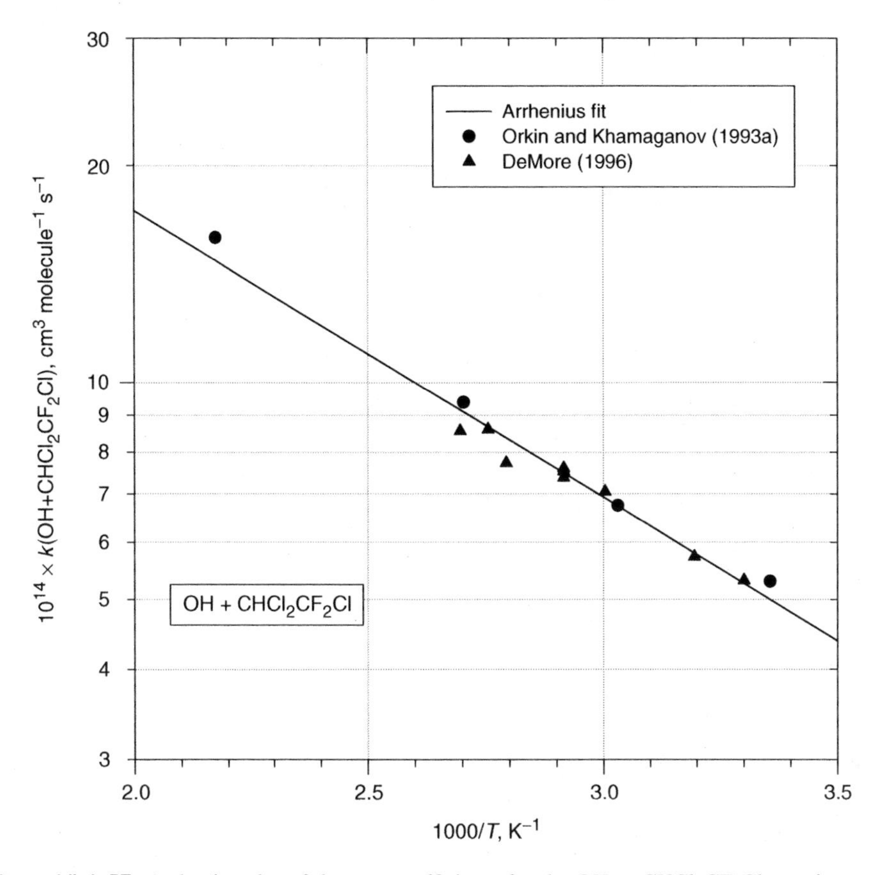

Figure VI-A-57. Arrhenius plot of the rate coefficients for the OH + $CHCl_2CF_2Cl$ reaction.

VI-A-6.12. OH + $CHFClCFCl_2$ (HCFC-122a) → H_2O + $CFClCFCl_2$

The rate coefficients of Hsu and DeMore (1995) are given in table VI-A-65 and plotted in figure VI-A-58. An Arrhenius fit gives $k(OH+CHFClCFCl_2) = 3.78 \times 10^{-13}\ \exp(-938/T)\,cm^3\,molecule^{-1}\,s^{-1}$. At 298 K this expression gives $k(OH+CHFClCFCl_2) = 1.62 \times 10^{-14}\,cm^3\,molecule^{-1}\,s^{-1}$ with an estimated uncertainty of ±25%.

VI-A-6.13. OH + $CH_3CF_2CFCl_2$ (HCFC-243cc) → H_2O + $CH_2CF_2CFCl_2$

The rate coefficients of Nelson et al. (1992) are given in table VI-A-66 and plotted in figure VI-A-59. An Arrhenius fit gives $k(OH+CH_3CF_2CFCl_2) = 7.92 \times 10^{-13}\ \exp(-1732/T)\,cm^3\,molecule^{-1}\,s^{-1}$. At 298 K this expression gives $k(OH+CH_3CF_2CFCl_2) = 2.37 \times 10^{-15}\,cm^3\,molecule^{-1}\,s^{-1}$ with an estimated uncertainty of ±25%.

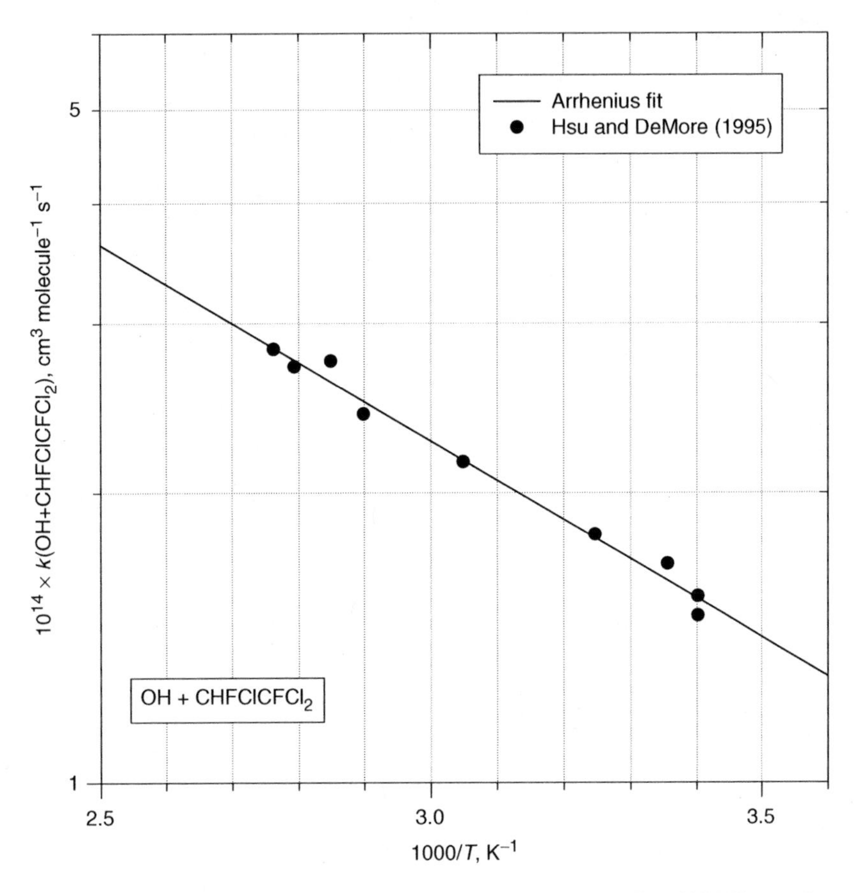

Figure VI-A-58. Arrhenius plot of the rate coefficients for the OH + $CHFClCFCl_2$ reaction.

Table VI-A-65. Rate coefficients $k = A\ e^{-B/T}$ (cm^3 $molecule^{-1}$ s^{-1}) for the reaction: OH + $CHFClCFCl_2$ (HCFC-122a) → products

$10^{12} \times A$	B (K)	$10^{14} \times k$	T (K)	Technique[b]	Reference	Temperature Range (K)
0.38	938	1.49[a]	294	RR (CH_3CHF_2)	Hsu and DeMore (1995)	294–362
		1.56	294			
		1.69	298			
		1.81	308			
		2.15	328			
		2.42	345			
		2.74	351			
		2.71	358			
		2.82	362			

[a] Placed on an absolute basis using $k(OH+CH_3CHF_2) = 1.21 \times 10^{-12}\ \exp(-1055/T)\,cm^3\ molecule^{-1}\ s^{-1}$.
[b] Abbreviations of the experimental techniques employed are defined in table II-A-1.

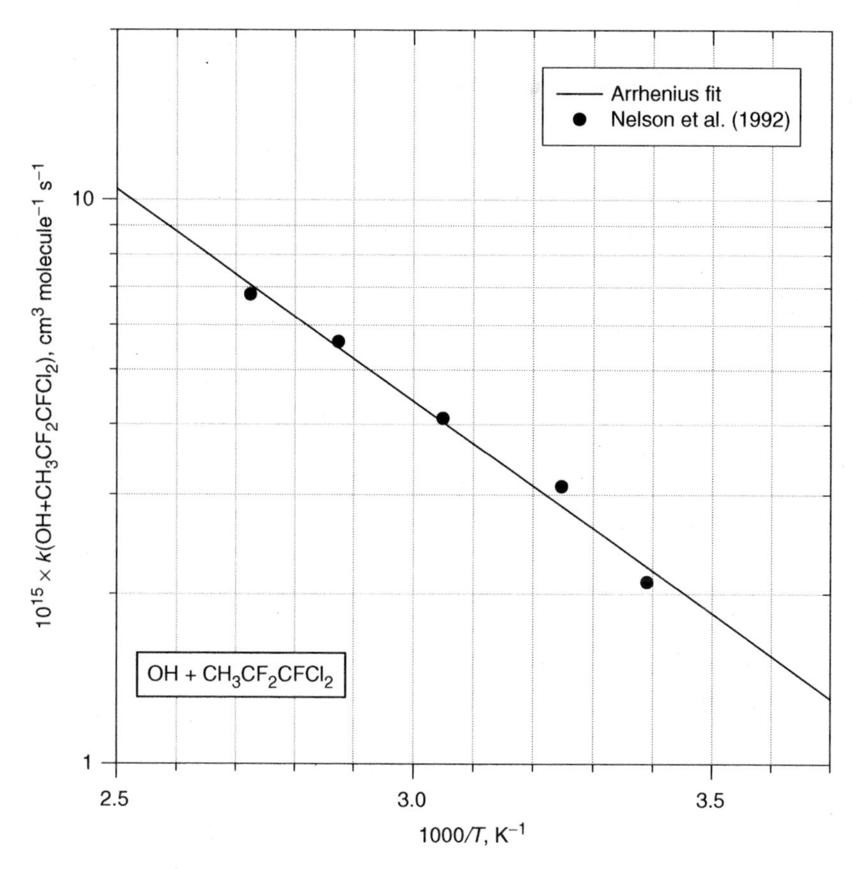

Figure VI-A-59. Arrhenius plot of the rate coefficients for the OH + $CH_3CF_2CFCl_2$ reaction.

Table VI-A-66. Rate coefficients $k = A\ e^{-B/T}$ (cm^3 $molecule^{-1}$ s^{-1}) for the reaction: OH + $CH_3CF_2CFCl_2$ (HCFC-243cc) → H_2O + $CH_2CF_2CFCl_2$

$10^{12} \times A$	B (K)	$10^{15} \times k$	T (K)	Technique[a]	Reference	Temperature Range (K)
0.71 ± 0.28	1690 ± 230	2.1 ± 0.2	295	DF–LIF	Nelson et al. (1992)	295–367
		3.1 ± 0.3	308			
		4.1 ± 0.4	328			
		5.6 ± 0.7	348			
		6.8 ± 0.7	367			

[a] Abbreviations of the experimental techniques employed are defined in table II-A-1.

VI-A-6.14. OH + $CF_3CH_2CFCl_2$ (HCFC-234fb) → H_2O + $CF_3CHCFCl_2$

The only rate coefficient measurement for this reaction is that of Barry et al. (1997), who used the relative rate technique with $CF_3CH_2CF_2CH_3$ as the reference compound. Using our recommended $k(OH+CF_3CH_2CF_2CH_3) = 7.34 \times 10^{-15}$ cm^3 molecule^{-1} s^{-1}, we derive a recommended $k(OH+CF_3CH_2CFCl_2) = (9.5 \pm 3.0) \times 10^{-16}$ cm^3 molecule^{-1} s^{-1} at 298 K.

VI-A-6.15. OH + $CF_3CF_2CCl_2H$ (HCFC-225ca) → H_2O + $CF_3CF_2CCl_2$

The rate coefficients of Brown et al. (1990a), Zhang et al. (1991), and Nelson et al. (1992) are given in table VI-A-67 and plotted in figure VI-A-60. The temperature dependence reported by Brown et al. is significantly lower than measured by Zhang et al. and Nelson et al. The presence of reactive impurities in the work of Brown et al. is a likely explanation of the discrepancy. A fit of the Arrhenius expression to the data from Zhang et al. (1991) and Nelson et al. (1992) gives $k(OH+CF_3CF_2CCl_2H) = 1.47 \times 10^{-12} \exp(-1223/T)$ cm^3 molecule^{-1} s^{-1}. At 298 K this expression gives $k(OH+CF_3CF_2CCl_2H) = 2.43 \times 10^{-14}$ cm^3 molecule^{-1} s^{-1} with an estimated uncertainty of ±15%.

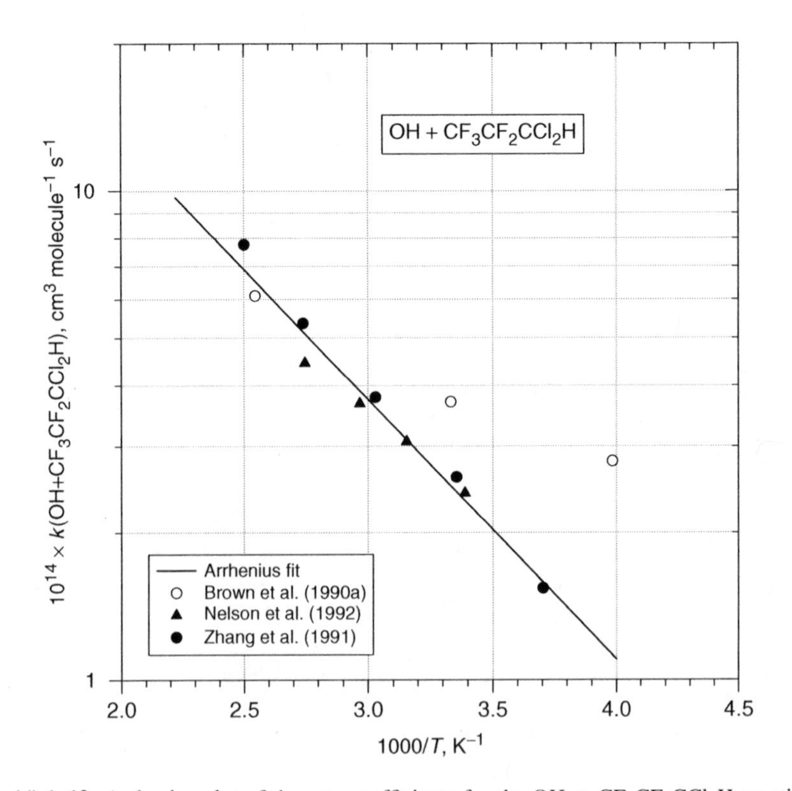

Figure VI-A-60. Arrhenius plot of the rate coefficients for the OH + $CF_3CF_2CCl_2H$ reaction.

Table VI-A-67. Rate coefficients $k = A\ e^{-B/T}$ (cm^3 $molecule^{-1}$ s^{-1}) for the reaction: $OH + CF_3CF_2CCl_2H \rightarrow$ products

$10^{12} \times A$	B (K)	$10^{13} \times k$	T (K)	Technique[a]	Reference	Temperature Range (K)
0.23	550 ± 750	2.8 ± 1.3	251	DF–RF	Brown et al. (1990a)	251–393
		3.7 ± 0.8	300			
		6.1 ± 1.1	393			
1.92 ± 0.52	1290 ± 90	1.54 ± 0.29	270	FP–RF	Zhang et al. (1991)	270–400
		2.60 ± 0.29	298			
		3.78 ± 0.31	330			
		5.36 ± 0.65	365			
		7.76 ± 0.90	400			
0.65 ± 0.13	970 ± 115	2.41 ± 0.24	295	DF–LIF	Nelson et al. (1992)	295–364
		3.07 ± 0.33	317			
		3.67 ± 0.36	337			
		4.45 ± 0.46	364			

[a] Abbreviations of the experimental techniques employed are defined in table II-A-1.

VI-A-6.16. $OH + CF_2ClCF_2CHFCl$ (HCFC-225cb) $\rightarrow$ $H_2O + CF_2ClCF_2CFCl$

The rate coefficients of Zhang et al. (1991) and Nelson et al. (1992) are given in table VI-A-68 and plotted in figure VI-A-61. There is excellent agreement between the results from the two studies. An Arrhenius fit gives $k(OH + CF_2ClCF_2CHFCl) = 5.43 \times 10^{-13} \exp(-1227/T)\,cm^3\,molecule^{-1}\,s^{-1}$. At 298 K this expression gives $k(OH + CF_2ClCF_2CHFCl) = 8.84 \times 10^{-15}\,cm^3\,molecule^{-1}\,s^{-1}$ with an estimated uncertainty of ±15%.

Table VI-A-68. Rate coefficients $k = A\ e^{-B/T}$ (cm^3 $molecule^{-1}$ s^{-1}) for the reaction: $OH + CF_2ClCF_2CHFCl \rightarrow$ products

$10^{12} \times A$	B (K)	$10^{14} \times k$	T (K)	Technique[a]	Reference	Temperature Range (K)
0.675 ± 0.37	1300 ± 180	0.86 ± 0.11	298	FP–RF	Zhang et al. (1991)	298–400
		1.10 ± 0.14	312			
		1.27 ± 0.10	331			
		1.53 ± 0.22	350			
		2.20 ± 0.25	375			
		2.69 ± 0.40	400			
0.39 ± 0.07	1120 ± 125	0.90 ± 0.11	295	DF–LIF	Nelson et al. (1992)	295–374
		1.13 ± 0.12	316			
		1.40 ± 0.14	336			
		1.82 ± 0.19	364			
		1.96 ± 0.20	374			

[a] Abbreviations of the experimental techniques employed are defined in table II-A-1.

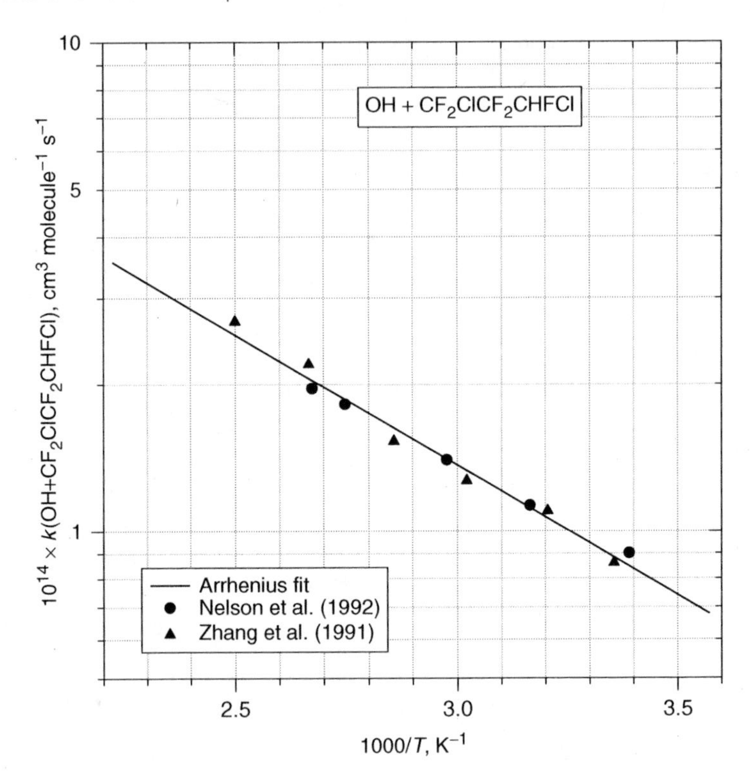

Figure VI-A-61. Arrhenius plot of the rate coefficients for the OH + CF_2ClCF_2CHFCl reaction.

VI-A-7. Hydrobromofluorocarbons

VI-A-7.1. OH + CHF_2Br (Halon-1201) → H_2O + CF_2Br

The rate coefficients reported by Brown et al. (1990a), Talukdar et al. (1991b), Orkin and Khamaganov (1993b), and Hsu and DeMore (1995) are given in table VI-A-69 and plotted in figure VI-A-62. Within the stated uncertainties, the results from the studies are in good agreement. The results from Brown et al. lie slightly above and have larger uncertainties than the other data. Fitting an Arrhenius expression to the data of Talukdar et al. (1991b), Orkin and Khamaganov (1993b), and Hsu and DeMore (1995) gives $k(OH+CHF_2Br) = 1.24 \times 10^{-12} \exp(-1435/T)\,cm^3\,molecule^{-1}\,s^{-1}$. At 298 K this expression gives $k(OH+CHF_2Br) = 1.00 \times 10^{-14}\,cm^3\,molecule^{-1}\,s^{-1}$ with an uncertainty estimated to be ±15%.

VI-A-7.2. OH + CF_3CH_2Br (Halon-2301) → H_2O + CF_3CHBr

The rate coefficients of Nelson et al. (1993) and Orkin and Khamaganov (1993b) are given in table VI-A-70 and plotted in figure VI-A-63. As is seen in the figure, there is reasonable agreement between the studies. A fit of the Arrhenius expression to the combined data of Nelson et al. (1993) and Orkin and Khamaganov (1993b) gives $k(OH+CF_3CH_2Br) = 1.51 \times 10^{-12} \exp(-1359/T)\,cm^3\,molecule^{-1}\,s^{-1}$. At 298 K this

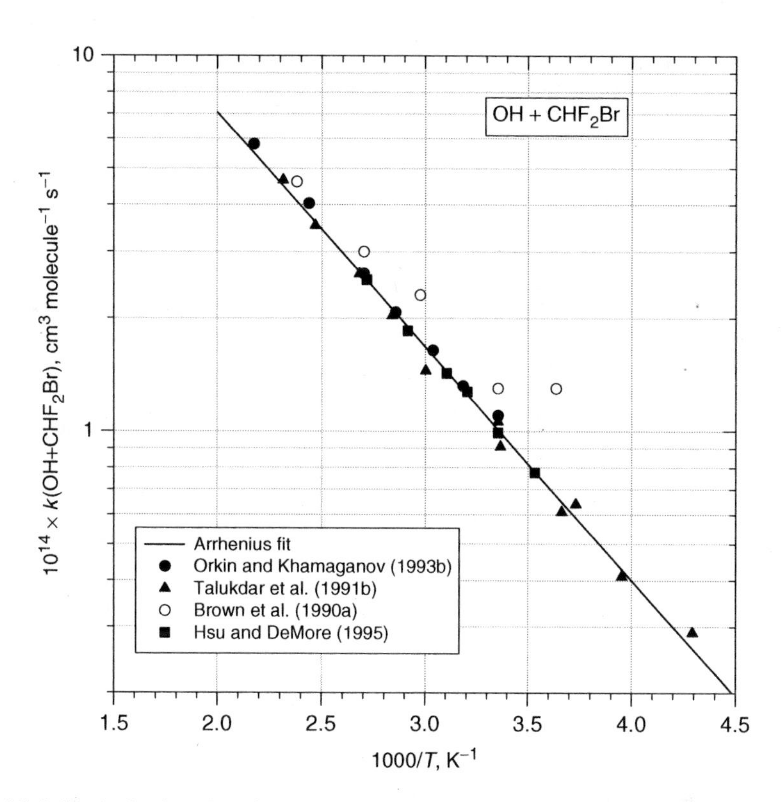

Figure VI-A-62. Arrhenius plot of the rate coefficients for the OH + CHF_2Br reaction.

Table VI-A-69. Rate coefficients $k = A\ e^{-B/T}$ (cm^3 $molecule^{-1}$ s^{-1}) for the reaction: OH + CHF_2Br (Halon-1201) → H_2O + CF_2Br

$10^{12} \times A$	B (K)	$10^{14} \times k$	T (K)	Technique[b]	Reference	Temperature Range (K)
0.44	1050 ± 400	1.3 ± 0.9	275	DF–RF	Brown et al. (1990a)	275–420
		1.3 ± 0.3	298			
		2.3 ± 0.3	336			
		3.0 ± 0.3	370			
		4.6 ± 0.9	420			
0.76 ± 0.16	1300 ± 100	0.29 ± 0.01	233	PLP–LIF	Talukdar et al. (1991b)	233–432
		0.41 ± 0.01	253	DF–LMR		
		0.64 ± 0.05	268			
		0.61 ± 0.01	273			
		0.91 ± 0.01	297			
		1.06 ± 0.08	298			
		1.45 ± 0.04	333			
		2.03 ± 0.12	352			
		2.62 ± 0.08	373			
		3.52 ± 0.23	405			
		4.65 ± 0.17	432			

(continued)

Table VI-A-69. *(continued)*

$10^{12} \times A$	B (K)	$10^{14} \times k$	T (K)	Technique[b]	Reference	Temperature Range (K)
$0.93^{+0.10}_{-0.09}$	1326 ± 33	1.10 ± 0.09	298	DF–EPR	Orkin and Khamaganov (1993b)	298–460
		1.32 ± 0.09	314			
		1.64 ± 0.12	329			
		2.07 ± 0.15	350			
		2.62 ± 0.19	370			
		4.03 ± 0.28	410			
		5.80 ± 0.5	460			
1.24	1434	0.78[a]	283	RR (CH_4)	Hsu and DeMore (1995)	283–368
		0.99	298			
		1.27	312			
		1.43	322			
		1.85	343			
		2.52	368			

[a] Placed on an absolute basis using $k(OH + CH_4) = 1.85 \times 10^{-20}\, T^{2.82} \exp(-987/T)$ cm^3 molecule^{-1} s^{-1}.
[b] Abbreviations of the experimental techniques employed are defined in table II-A-1.

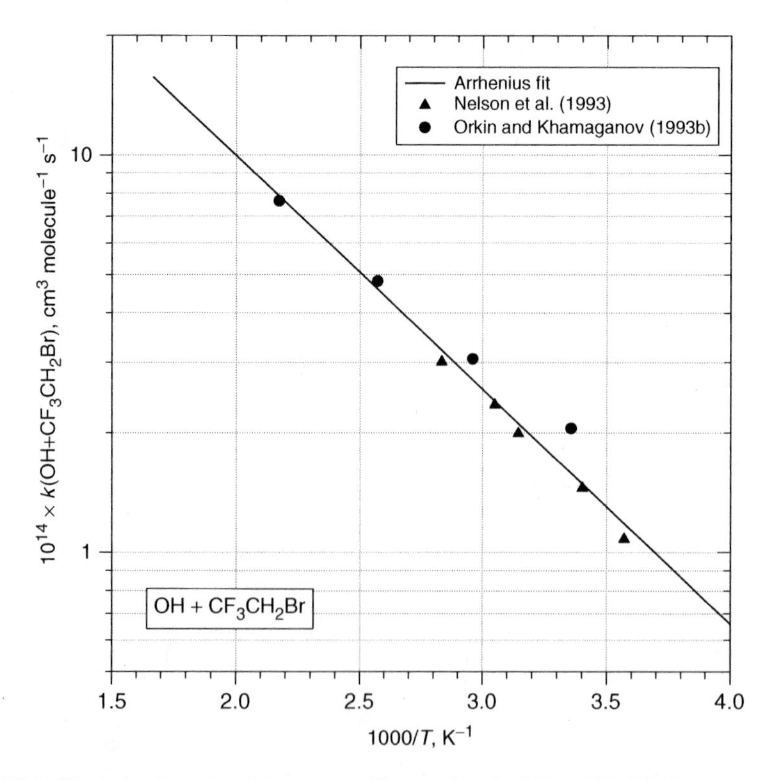

Figure VI-A-63. Arrhenius plot of the rate coefficients for the OH + CF_3CH_2Br reaction.

Table VI-A-70. Rate coefficients $k = A\,e^{-B/T}$ (cm^3 $molecule^{-1}$ s^{-1}) for the reaction: $OH + CF_3CH_2Br$ (Halon-2301) $\rightarrow$ $H_2O + CF_3CHBr$

$10^{12} \times A$	B (K)	$10^{14} \times k$	T (K)	Technique[a]	Reference	Temperature Range (K)
1.39 ± 0.60	1350 ± 189	1.08 ± 0.08	280	DF–LIF	Nelson et al. (1993)	280–353
		1.45 ± 0.13	294			
		1.99 ± 0.07	318			
		2.35 ± 0.15	328			
		3.01 ± 0.10	353			
$0.85^{+0.09}_{-0.08}$	1113 ± 35	2.05 ± 0.16	298	DF–RF	Orkin and Khamaganov (1993b)	298–460
		3.06 ± 0.21	338			
		4.81 ± 0.34	389			
		7.65 ± 0.55	460			

[a] Abbreviations of the experimental techniques employed are defined in table II-A-1.

expression gives $k(OH + CF_3CH_2Br) = 1.58 \times 10^{-14}$ cm^3 $molecule^{-1}$ s^{-1} with an estimated uncertainty of $\pm 25\%$.

VI-A-7.3. $OH + CF_3CHFBr$ (Halon-2401) $\rightarrow$ $H_2O + CF_3CFBr$

The rate coefficients of Brown et al. (1990a) and Orkin and Khamaganov (1993b) are given in table VI-A-71 and plotted in figure VI-A-64. As is seen from this figure, there is excellent agreement between the two studies. A fit of the Arrhenius expression to the combined data gives $k(OH + CF_3CHFBr) = 7.84 \times 10^{-13} \exp(-1149/T)$ cm^3 $molecule^{-1}$ s^{-1}. At 298 K this expression gives $k(OH + CF_3CHFBr) = 1.66 \times 10^{-14}$ cm^3 $molecule^{-1}$ s^{-1} with an estimated uncertainty of $\pm 15\%$.

VI-A-8. Fluoroiodocarbons

VI-A-8.1. $OH + CF_3I$ $\rightarrow$ Products

The rate coefficients of Garraway and Donovan (1979), Brown et al. (1990c), Berry et al. (1998), and Gilles et al. (2000) are given in table VI-A-72 and plotted in figure VI-A-65. As is seen in the figure, the lower temperature data from Berry et al. and the data of Garraway and Donovan lie significantly above those from the other studies. A significant dependence of the measured rate constant on the laser flash energy was reported by Berry et al. (1998) and was attributed to loss of OH via reaction with CF_3I photolysis products. Berry et al. extrapolated the measured rate coefficients to zero flash energy to obtain the quoted values. The uncertainties in such an analysis decrease with increasing temperature. The higher temperature data from Berry et al. are expected to be more reliable than the low temperature data and are in good agreement with an extrapolation of the data from Gilles et al. A fit of the Arrhenius expression to the data from Gilles et al. (2000), Brown et al.(1990c), and the $T > 390$ K data of Berry et al. (1998) gives $k(OH + CF_3I) = 4.05 \times 10^{-11} \exp(-2209/T)$ cm^3 $molecule^{-1}$ s^{-1}.

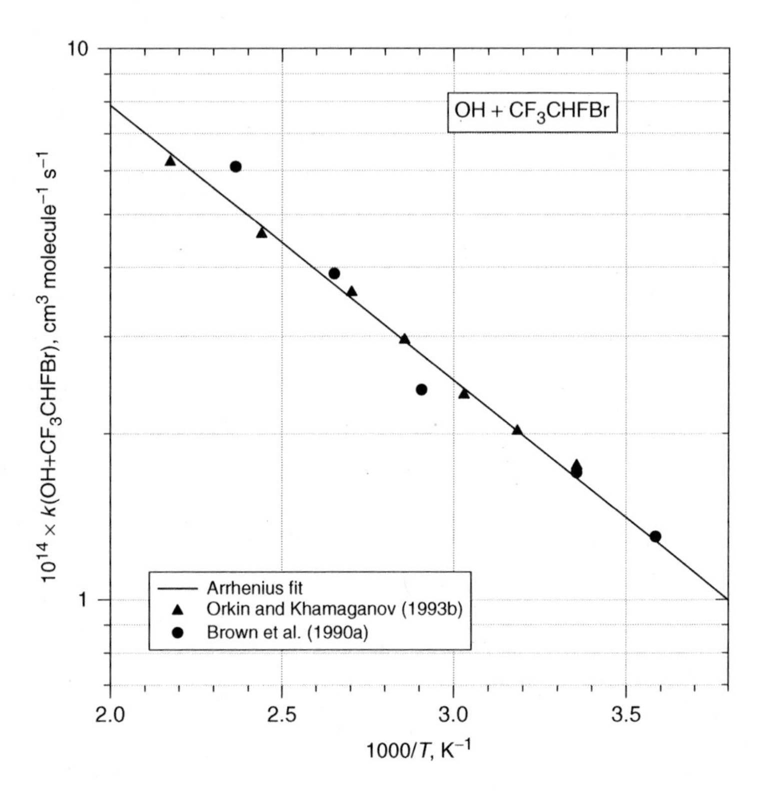

Figure VI-A-64. Arrhenius plot of the rate coefficients for the OH + CF_3CHFBr reaction.

Table VI-A-71. Rate coefficients $k = A\ e^{-B/T}$ (cm^3 $molecule^{-1}$ s^{-1}) for the reaction: OH + CF_3CHFBr (Halon-2401) → H_2O + CF_3CFBr

$10^{12} \times A$	B (K)	$10^{14} \times k$	T (K)	Technique[a]	Reference	Temperature Range (K)
1.13	1250 ± 350	1.3 ± 0.4	279	DF–RF	Brown et al. (1990a)	279–423
		1.7 ± 0.3	298			
		2.4 ± 0.5	344			
		3.9 ± 0.5	377			
		6.1 ± 1.3	423			
$0.72^{+0.07}_{-0.06}$	1113 ± 35	1.75 ± 0.17	298	DF–RF	Orkin and Khamaganov (1993b)	298–460
		2.02 ± 0.14	314			
		2.35 ± 0.16	330			
		2.96 ± 0.27	350			
		3.61 ± 0.43	370			
		4.60 ± 0.32	410			
		6.22 ± 0.48	460			

[a] Abbreviations of the experimental techniques employed are defined in table II-A-1.

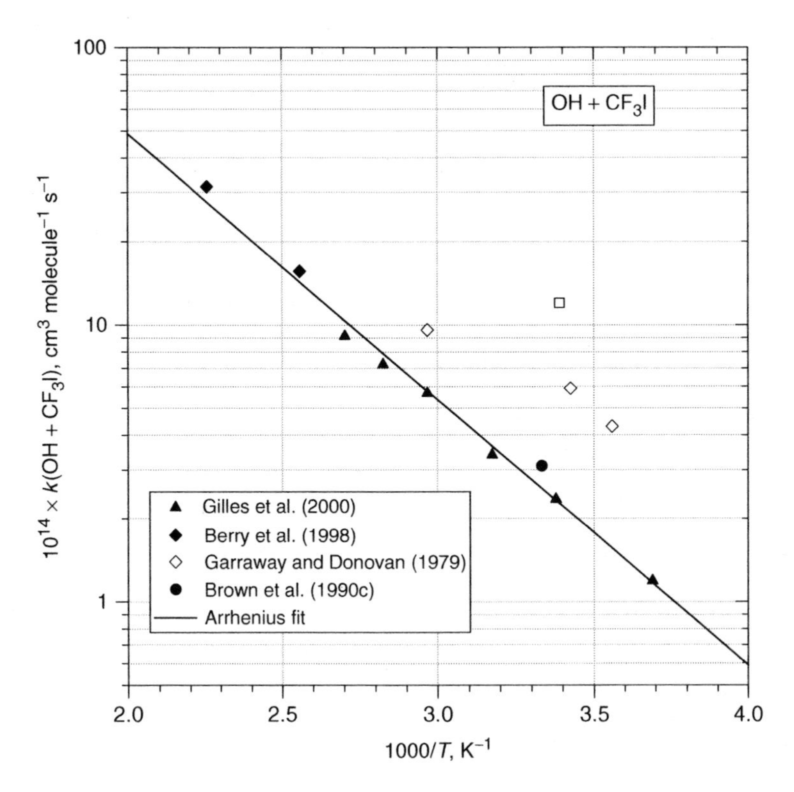

Figure VI-A-65. Arrhenius plot of the rate coefficients for the OH + CF_3I reaction. The $T > 390$ K data from Berry et al. (1998) were included in the fit; open symbols were not included in the fit (see text for details).

Table VI-A-72. Rate coefficients $k = A\ e^{-B/T}$ (cm^3 $molecule^{-1}$ s^{-1}) for the reaction: OH + CF_3I → products

$10^{12} \times A$	B (K)	$10^{14} \times k$	T (K)	Technique[a]	Reference	Temperature Range (K)
		12 ± 0.2	295	FP–RA	Garraway and Donovan (1979)	295
		3.1 ± 0.5	300	DF–RA	Brown et al. (1990c)	300
5.8	1359	4.3 ± 1.4	281	FP–RF	Berry et al. (1998)	281–443
		5.9 ± 1.4	292			
		9.6 ± 2.7	337			
		15.6 ± 3.4	391			
		31.5 ± 9.8	443			
21 ± 8	2000 ± 140	1.19 ± 0.19	271	PLP–LIF	Gilles et al. (2000)	271–370
		2.35 ± 0.54	296			
		3.40 ± 0.27	315			
		5.67 ± 0.31	337			
		7.21 ± 0.48	354			
		9.12 ± 0.22	370			

[a] Abbreviations of the experimental techniques employed are defined in table II-A-1.

At 298 K this expression gives $k(OH + CF_3I) = 2.44 \times 10^{-14}$ cm^3 molecule^{-1} s^{-1}, with an uncertainty estimated to be ±20%.

VI-A-9. Hydrobromochlorocarbons

VI-A-9.1. OH + $CH_2ClBr \rightarrow H_2O$ + CHClBr

The rate coefficients of DeMore (1996), Orkin et al. (1997), and Bilde et al. (1998) are given in table VI-A-73 and plotted in figure VI-A-66. As is seen in the figure, there is good agreement between the studies. A fit of the Arrhenius expression to the combined data set gives $k(OH + CH_2ClBr) = 1.94 \times 10^{-12} \exp(-844/T)$ cm^3 molecule^{-1} s^{-1}. At 298 K this expression gives $k(OH + CH_2ClBr) = 1.14 \times 10^{-13}$ cm^3 molecule^{-1} s^{-1}, with an estimated uncertainty of ±15%.

VI-A-9.2. OH + $CHCl_2Br \rightarrow H_2O + CCl_2Br$

The rate coefficient of Bilde et al. (1998) is the only available measurement. They used a relative rate technique with CH_2Br_2 as the reference compound. Using $k(OH + CH_2Br_2) = 2.52 \times 10^{-12} \exp(-900/T)$ cm^3 molecule^{-1} s^{-1}, we derive the recommended value of 1.24×10^{-13} cm^3 molecule^{-1} s^{-1} with an estimated uncertainty of ±25%.

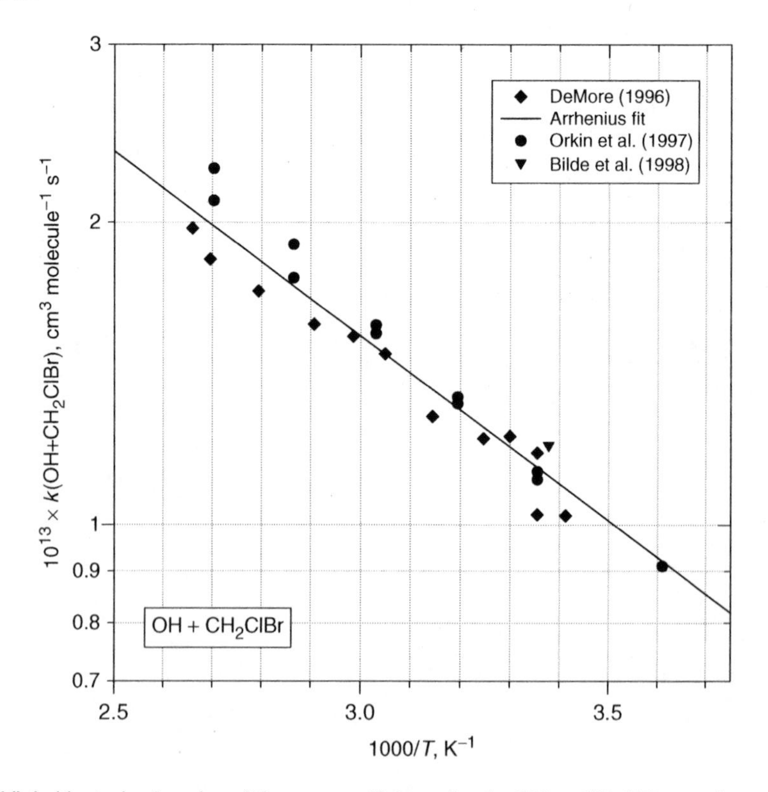

Figure VI-A-66. Arrhenius plot of the rate coefficients for the OH + CH_2ClBr reaction.

Table VI-A-73. Rate coefficients $k = A\ e^{-B/T}$ (cm^3 $molecule^{-1}$ s^{-1}) for the reaction: $OH + CH_2ClBr \rightarrow H_2O + CHClBr$

$10^{12} \times A$	B (K)	$10^{13} \times k$	T (K)	Technique[c]	Reference	Temperature Range (K)
1.67	809	1.02[a]	293	RR (CH_2Cl_2)	DeMore (1996)	293–376
		1.18	298			
		1.02	298			
		1.23	303			
		1.22	308			
		1.28	318			
		1.48	328			
		1.54	335			
		1.58	344			
		1.71	358			
		1.84	371			
		1.97	376			
$3.04^{+0.8}_{-0.6}$	978 ± 72	0.91	277	FP–RF	Orkin et al. (1997)	277–370
		1.11	298			
		1.34	313			
		1.58	330			
		1.76	349			
		2.10	370			
		1.13	298	DF–EPR		
		1.32	313			
		1.55	330			
		1.90	349			
		2.26	370			
		1.2 ± 0.1[b]	296	RR (CH_2Br_2)	Bilde et al. (1998)	296

[a] Relative to $k(OH + CH_2Cl_2) = 1.97 \times 10^{-12} \exp(-877/T)$ cm^3 $molecule^{-1}$ s^{-1}.
[b] Relative to $k(OH + CH_2Br_2) = 2.52 \times 10^{-12} \exp(-900/T)$ cm^3 $molecule^{-1}$ s^{-1}.
[c] Abbreviations of the experimental techniques employed are defined in table II-A-1.

VI-A-9.3. $OH + ClCH_2CHBrCH_2Br \rightarrow$ Products

The rate coefficient of Tuazon et al. (1986) is the only value available. They used a relative rate technique with CH_3OCH_3 as the reference compound. Taking $k(OH + CH_3OCH_3) = 5.7 \times 10^{-12} \exp(-215/T)$ cm^3 $molecule^{-1}$ s^{-1} (Atkinson et al., 2005b), we derive the recommended value of $k(OH + ClCH_2CHBrCH_2Br) = 4.05 \times 10^{-13}$ cm^3 $molecule^{-1}$ s^{-1} with an estimated uncertainty of ±25%.

VI-A-10. Hydrobromochlorofluoroalkanes

VI-A-10.1. $OH + CF_3CHClBr$ (Halon-2311) $\rightarrow H_2O + CF_3CClBr$

The rate coefficients of Brown et al. (1989; 1990b), and Orkin and Khamaganov (1993b) are given in table VI-A-74 and plotted in figure VI-A-67. As is seen in the figure, there is reasonable agreement between the studies. A fit of the Arrhenius expression to the combined data set gives $k(OH + CF_3CHClBr) = 9.01 \times 10^{-13} \exp(-857/T)$ cm^3 $molecule^{-1}$ s^{-1}. At 298 K this expression gives $k(OH + CF_3CHClBr) = 5.04 \times 10^{-14}$ cm^3 $molecule^{-1}$ s^{-1} with an estimated uncertainty of ±25%.

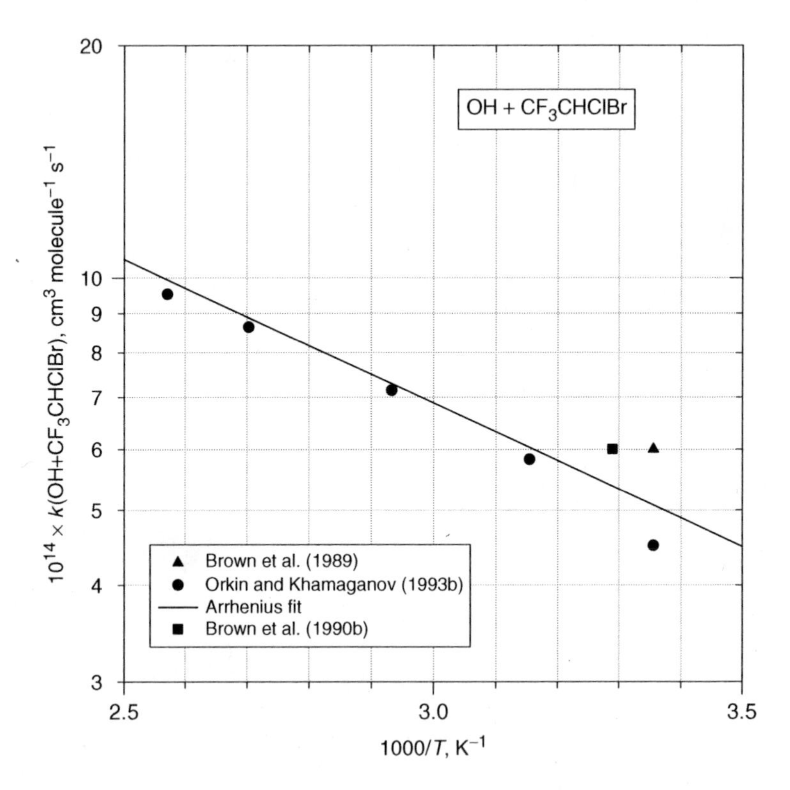

Figure VI-A-67. Arrhenius plot of the rate coefficients for the OH + $CF_3CHClBr$ reaction.

Table VI-A-74. Rate coefficients $k = A\ e^{-B/T}$ (cm^3 $molecule^{-1}$ s^{-1}) for the reaction: OH + $CF_3CHClBr$ (Halon-2311) → H_2O + CF_3CClBr

$10^{12} \times A$	B (K)	$10^{14} \times k$	T (K)	Technique[a]	Reference	Temperature Range (K)
		6.0 ± 0.4	298	DF–RF	Brown et al. (1989)	298
		6.0 ± 0.4	303	DF–RF	Brown et al. (1990b)	303
$1.285^{+1.5}_{-1.2}$	995 ± 38	4.50 ± 0.40	298	DF–RF	Orkin and Khamaganov (1993b)	298–460
		5.82 ± 0.41	317			
		7.15 ± 0.5	341			
		8.64 ± 0.6	370			
		9.53 ± 0.67	389			
		11.1 ± 0.78	410			
		14.9 ± 1.4	460			

[a] Abbreviations of the experimental techniques employed are defined in table II-A-1.

VI-A-10.2. OH + $CF_2BrCHFCl \rightarrow H_2O + CF_2BrCFCl$

The rate coefficients of DeMore (1996) are given in table VI-A-75 and plotted in figure VI-A-68. A fit of the Arrhenius expression to the combined data set gives $k(OH + CF_2BrCHFCl) = 9.80 \times 10^{-13} \exp(-1263/T)\,cm^3\,molecule^{-1}\,s^{-1}$. At 298 K this expression gives $k(OH + CF_2BrCHFCl) = 1.41 \times 10^{-14}\,cm^3\,molecule^{-1}\,s^{-1}$ with an estimated uncertainty of ±25%.

VI-A-11. PFCs, Perchlorocarbons, CFCs, Bromofluorocarbons, and Bromochlorofluorocarbons

Upper limits for the rate coefficients of reactions of OH radicals with perfluorocarbons (PFCs), perchlorocarbons, chlorofluorocarbons (CFCs), bromochlorofluorocarbons, and bromofluorocarbons are given in table VI-A-76. These compounds do not contain any abstractable hydrogen atoms, and the strength of the C—F (547 kJ mol^{-1}), C—Cl (288 kJ mol^{-1}), and C—Br (263 kJ mol^{-1}) bonds (calculated for $CX_4 = CX_3 + X$, using data from Sander et al., 2003) renders halogen abstraction thermodynamically unfavorable. Hence perfluorocarbons, perchloroalkanes, bromofluorocarbons, and chlorofluorocarbons are unreactive towards OH radicals. Equating the endothermicity with the activation energy of the reaction $OH + CX_4 \rightarrow CX_3 + HOX$, and assuming an *A*-factor of 5×10^{-12}, gives upper limits to the room-temperature rate coefficients for

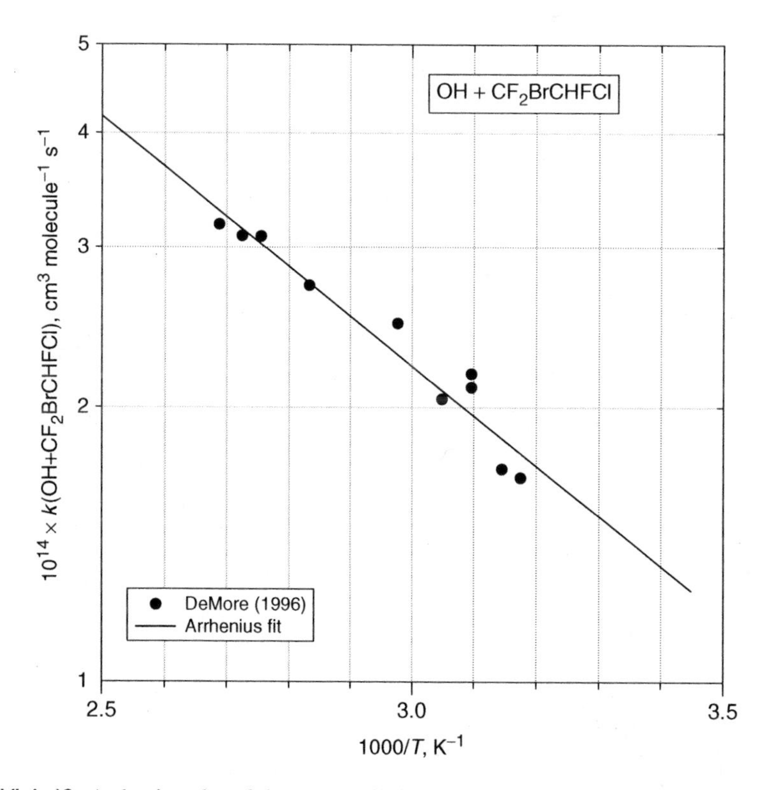

Figure VI-A-68. Arrhenius plot of the rate coefficients for the $OH + CF_2BrCHFCl$ reaction.

Table VI-A-75. Rate coefficients $k = A\ e^{-B/T}$ (cm^3 $molecule^{-1}$ s^{-1}) for the reaction: $OH + CF_2BrCHFCl \rightarrow H_2O + CF_2BrCFCl$

$10^{12} \times A$	B (K)	$10^{14} \times k$	T (K)	Technique[b]	Reference	Temperature Range (K)
0.98	1263	1.67[a]	315	RR (CH_3CCl_3)	DeMore (1996)	315–372
		1.71	318			
		2.10	323			
		2.18	323			
		2.04	328			
		2.47	336			
		2.72	353			
		3.08	363			
		3.09	367			
		3.18	372			

[a] Relative to $k(OH + CH_3CCl_3) = 1.81 \times 10^{-12} \exp(-1543/T)\ cm^3\ molecule^{-1}\ s^{-1}$.
[b] Abbreviations of the experimental techniques employed are defined in table II-A-1.

Table VI-A-76. Rate coefficients $k = A\ e^{-B/T}$ (cm^3 $molecule^{-1}$ s^{-1}) for the reaction: $OH + C_xF_yCl_zBr_a \rightarrow$ products

Compound	k	T (K)	Technique[a]	Reference
CF_4 (FC-14)	$< 4 \times 10^{-16}$	296	DF–LMR	Howard and Evenson (1976)
CF_3Cl (CFC-13)	$< 7 \times 10^{-16}$	296	DF–LMR	Howard and Evenson (1976)
CF_2Cl_2 (CFC-12)	$< 1 \times 10^{-15}$	296–424	FP–RF	Atkinson et al. (1975)
	$< 9 \times 10^{-17}$	298	RR	Cox et al. (1976a)
	$< 4 \times 10^{-16}$	296	DF–LMR	Howard and Evenson (1976)
	$< 6 \times 10^{-16}$	478	DF–RF	Chang and Kaufman (1977)
	$< 1 \times 10^{-15}$	293	DF–RF	Clyne and Holt (1979b)
$CFCl_3$ (CFC-11)	$< 1 \times 10^{-15}$	296–424	FP–RF	Atkinson et al. (1975)
	$< 4 \times 10^{-17}$	298	RR	Cox et al. (1976a)
	$< 5 \times 10^{-16}$	296	DF–LMR	Howard and Evenson (1976)
	$< 5 \times 10^{-16}$	480	DF–RF	Chang and Kaufman (1977)
	$< 1 \times 10^{-15}$	293	DF–RF	Clyne and Holt (1979b)
CCl_4	$< 4 \times 10^{-15}$	296	DF–LMR	Howard and Evenson (1976)
	$< 9 \times 10^{-17}$	298	RR	Cox et al. (1976a)
	$< 1 \times 10^{-15}$	293	DF–RF	Clyne and Holt (1979b)
CF_3Br (Halon-1301)	$< 1.2 \times 10^{-16}$	293, 298	DF–LMR/PLP–LIF	Burkholder et al. (1991)
	$< 2 \times 10^{-17}$	298	DF–EPR	Orkin and Khamaganov (1993b)
	$< 4 \times 10^{-16}$	460		
CF_2Br_2 (Halon-1202)	$< 5 \times 10^{-16}$	298	DF-LMR/PLP-LIF	Burkholder et al. (1991)
CF_2ClBr (Halon-1211)	$< 1 \times 10^{-15}$	293	DF–RF	Clyne and Holt (1979b)
	$< 2 \times 10^{-16}$	293	DF–LMR/PLP–LIF	Burkholder et al. (1991)
	$< 9 \times 10^{-17}$	297		
	$< 7 \times 10^{-17}$	373		
	$< 2 \times 10^{-16}$	424		

Table VI-A-76. (*continued*)

Compound	k	T (K)	Technique[a]	Reference
CF_2BrCF_2Br (Halon-2402)	$< 1.5 \times 10^{-16}$	293, 298	DF–LMR/PLP–LIF	Burkholder et al. (1991)
$CFCl_2CFCl_2$ (CFC-112)	$< 3 \times 10^{-16}$	296	DF–LMR	Howard and Evenson (1976)
	$< 3 \times 10^{-16}$	298	FP–RF	Watson et al. (1977)
$CFCl_2CF_2Cl$ (CFC-113)	$< 3 \times 10^{-16}$	296	DF–LMR	Howard and Evenson (1976)
	$< 3 \times 10^{-16}$	298	FP–RF	Watson et al. (1977)

[a] Abbreviations of the experimental techniques employed are defined in table II-A-1.

abstraction of fluorine, chlorine, and bromine atoms of 3.2×10^{-70}, 1.3×10^{-21}, and 2.2×10^{-21} cm^3 molecule^{-1} s^{-1}, respectively. For molecules with two or more carbon atoms, an exothermic reaction channel exists; substitution of a C_xX_{2x+1} group by OH (e.g., $C_2F_6 + OH \rightarrow CF_3OH + CF_3$, $\Delta H = -70$ kJ mol^{-1}; $C_2Cl_6 + OH \rightarrow CCl_3OH + CCl_3$, $\Delta H = -117$ kJ mol^{-1}). However, steric factors render substitution reactions unfavorable, and there is no evidence that such substitution reactions occur (for any alkane, substituted or not). Hence, based upon these arguments, we recommend an upper limit for the rate constant of reaction of OH radicals with perfluorocarbons, perchloroalkanes, perbromoalkanes, bromofluorocarbons, and chlorofluorocarbons of $k(OH + C_xF_yCl_zBr_a) < 10^{-20}$ cm^3 molecule^{-1} s^{-1} at 298 K.

VI-A-12. Summary of Recommended Kinetic Data

Recommended kinetic data for reaction of OH radicals with haloalkanes are given in table VI-A-77. The majority of compounds listed in this table react with OH radicals with pre-exponential A-factors in the range $(1–2) \times 10^{-12}$ cm^3 molecule^{-1} s^{-1} and exponential B-factors of 1000–2000 (corresponding to activation energies of approximately 8–17 kJ mol^{-1}). For compounds in which a large fraction of the available sites is blocked by unreactive halogen atoms (e.g., CHF_2CF_3, CF_3CHFCF_3, CF_3CCl_2H, CF_3CHFBr), the A-factors are particularly small ($< 1 \times 10^{-12}$ cm^3 molecule^{-1} s^{-1}), reflecting the low probability that the incoming OH radical will encounter an abstractable H-atom during its collision with the haloalkane. Increased fluorine substitution tends to increase the strength of the remaining C—H bonds, leading to larger activation barriers and lower reactivities (e.g., CHF_3, $CF_3CH_2CF_3$). Large alkanes with modest halogenation (e.g., $CH_3CH_2CH_2CH_2Cl$) have pre-exponential A-factors (and activation barriers) which are more typical of unsubstituted alkanes; halogenation has a significant local influence but modest overall influence on the reactivity of such molecules. Finally, we note that uncertainties given in table VI-A-77 are the estimated uncertainties of the rate coefficient at 298 K. Since room temperature (approximately 298 K) is the temperature at which most kinetic studies have been performed, the kinetic data base is most certain at 298 K. The Arrhenius parameters recommended in table VI-A-77 can be used to calculate rate coefficients for temperatures other than 298 K (associated uncertainties will be greater than those estimated for 298 K).

Table VI-A-77. Summary of Arrhenius parameters, $k = A\ \exp(-B/T)$ and $k = CT^2\ \exp(-D/T)\ cm^3\ molecule^{-1} s^{-1}$, and rate coefficients at 298 K for reactions of OH with haloalkanes. Where available the k(298 K) value was evaluated from values of A and B

Haloalkane	$10^{12} \times A$ ($cm^3 molecule^{-1} s^{-1}$)	B (K)	$10^{18} \times C$ ($cm^3 molecule^{-1} s^{-1}$)	D (K)	k(298K) ($cm^3 molecule^{-1} s^{-1}$)
CH_3F (HFC-41)	1.94	1361	3.14	788	$(2.02 \pm 0.30) \times 10^{-14}$
CH_2F_2 (HFC-32)	1.33	1432	2.77	933	$(1.09 \pm 0.16) \times 10^{-14}$
CHF_3 (HFC-23)	0.49	2200	0.927	1691	$(2.84 \pm 0.57) \times 10^{-16}$
CH_3CH_2F (HFC-161)	2.62	738	4.97	210	$(2.20 \pm 0.33) \times 10^{-13}$
CH_2FCH_2F (HFC-152)			2.84	277	$(9.96 \pm 1.50) \times 10^{-14}$
CH_3CHF_2 (HFC-152a)	1.21	1055	2.10	503	$(3.51 \pm 0.53) \times 10^{-14}$
CH_2FCHF_2 (HFC-143)	4.11	1648			$(1.63 \pm 0.41) \times 10^{-14}$
CH_3CF_3 (HFC-143a)	1.06	1998	1.57	1400	$(1.30 \pm 0.20) \times 10^{-15}$
CHF_2CHF_2 (HFC-134)	1.27	1598			$(5.96 \pm 1.49) \times 10^{-15}$
CF_3CH_2F (HFC-134a)	1.01	1640	1.29	989	$(4.11 \pm 0.62) \times 10^{-15}$
CHF_2CF_3 (HFC-125)	0.51	1666			$(1.90 \pm 0.38) \times 10^{-15}$
$CF_3CH_2CH_3$ (HFC-263fb)	3.94	1274			$(5.48 \pm 1.37) \times 10^{-14}$
CF_3CHFCH_2F (HFC-245eb)	1.23	1257			$(1.81 \pm 0.36) \times 10^{-14}$
$CHF_2CHFCHF_2$ (HFC-245ea)	1.93	1387			$(1.84 \pm 0.46) \times 10^{-14}$
$CF_3CH_2CHF_2$ (HFC-245fa)	0.671	1354			$(7.14 \pm 1.07) \times 10^{-15}$
$CF_3CH_2CF_3$ (HFC-236fa)	1.28	2464			$(3.28 \pm 0.66) \times 10^{-16}$
$CF_3CHFCHF_2$ (HFC-236ea)	1.34	1658			$(5.14 \pm 1.29) \times 10^{-15}$
$CF_3CF_2CH_2F$ (HFC-236cb)	0.261	1087			$(6.80 \pm 1.70) \times 10^{-15}$
CF_3CHFCF_3 (HFC-227ea)	0.655	1809			$(1.51 \pm 0.30) \times 10^{-15}$
$CHF_2CF_2CH_2F$ (HFC-245ca)	2.48	1731			$(7.44 \pm 1.86) \times 10^{-15}$
$CF_3CH_2CF_2CH_3$ (HFC-365mfc)	2.04	1677			$(7.34 \pm 2.20) \times 10^{-15}$
$CF_3CH_2CH_2CF_3$ (HFC-356mff)	2.53	1701			$(8.40 \pm 1.26) \times 10^{-15}$
$CF_3CF_2CH_2CH_2F$ (HFC-356mcf)	1.66	1097			$(4.18 \pm 1.05) \times 10^{-14}$
$CHF_2CF_2CF_2CHF_2$ (HFC-338pcc)	0.782	1548			$(4.34 \pm 0.87) \times 10^{-15}$
$CF_3CH_2CF_2CH_2CF_3$ (HFC-458mfcf)	1.12	1804			$(2.63 \pm 0.66) \times 10^{-15}$
$CF_3CHFCHFCF_2CF_3$ (HFC-43-10mee)	0.568	1534			$(3.30 \pm 0.66) \times 10^{-15}$
$CF_3CF_2CH_2CH_2CF_2CF_3$ (HFC-55-10mcff)					$(8.30 \pm 0.21) \times 10^{-15}$
$CHF_2CF_2CF_2CF_2CF_2CF_3$ (HFC-52-13p)	0.567	1706			$(1.85 \pm 0.36) \times 10^{-15}$

Table VI-A-77. *(continued)*

Haloalkane	$10^{12} \times A$ (cm^3 molecule^{-1} s^{-1})	B (K)	$10^{18} \times C$ (cm^3 molecule^{-1} s^{-1})	D (K)	k(298K) (cm^3 molecule^{-1} s^{-1})
CH_3Cl	2.71	1269	7.28	842	$(3.83 \pm 0.77) \times 10^{-14}$
CH_2Cl_2	1.97	877			$(1.03 \pm 0.26) \times 10^{-13}$
$CHCl_3$	2.86	1011			$(9.62 \pm 1.92) \times 10^{-14}$
CH_3CH_2Cl			8.80	259	$(3.28 \pm 0.82) \times 10^{-13}$
CH_2ClCH_2Cl			10.8	410	$(2.42 \pm 0.36) \times 10^{-13}$
CH_3CHCl_2			2.49	−64	$(2.74 \pm 0.41) \times 10^{-13}$
$CH_2ClCHCl_2$			3.70	108	$(2.29 \pm 0.69) \times 10^{-13}$
CH_3CCl_3	1.81	1543	4.07	1064	$(1.02 \pm 0.15) \times 10^{-14}$
CCl_3CCl_2H					$(2.33 \pm 0.58) \times 10^{-13}$
$CH_3CH_2CH_2Cl$	95.1	1402			$(8.61 \pm 2.15) \times 10^{-13}$
$CH_3CHClCH_3$	18.3	953			$(7.47 \pm 2.61) \times 10^{-13}$
$CH_2ClCH_2CH_2Cl$					$(7.30 \pm 1.83) \times 10^{-13}$
$CH_3CH_2CH_2CH_2Cl$	34.9	941			$(1.48 \pm 0.52) \times 10^{-12}$
$(CH_3)_3CCl$					$(3.84 \pm 0.96) \times 10^{-13}$
$CH_3(CH_2)_3CH_2Cl$	14	407			$(3.57 \pm 1.25) \times 10^{-12}$
$CH_3(CH_2)_4CH_2Cl$	47.9	747			$(3.91 \pm 0.78) \times 10^{-12}$
CH_3Br	2.39	1301			$(3.04 \pm 0.46) \times 10^{-14}$
CH_2Br_2	2.52	900			$(1.23 \pm 0.25) \times 10^{-13}$
$CHBr_3$	1.48	612			$(1.90 \pm 0.48) \times 10^{-13}$
CH_3CH_2Br	4.76	795			$(3.30 \pm 0.83) \times 10^{-13}$
CH_2BrCH_2Br	7.69	1056			$(2.22 \pm 0.44) \times 10^{-13}$
$CH_3CH_2CH_2Br$			7.57	−134	$(1.05 \pm 0.21) \times 10^{-12}$
$CH_3CHBrCH_3$	2.90	367			$(8.46 \pm 1.69) \times 10^{-13}$
$CH_3CH_2CH_2CH_2Br$					$(2.29 \pm 0.57) \times 10^{-12}$
$CH_3(CH_2)_3CH_2Br$					$(3.71 \pm 0.93) \times 10^{-12}$
$CH_3(CH_2)_4CH_2Br$					$(5.48 \pm 1.37) \times 10^{-12}$
CH_3I	2.26	1001			$(7.86 \pm 2.36) \times 10^{-14}$
CH_3CH_2I					$(7.7 \pm 1.9) \times 10^{-13}$
$CH_3CH_2CH_2I$					$(1.47 \pm 0.37) \times 10^{-12}$
CH_3CHICH_3					$(1.22 \pm 0.31) \times 10^{-12}$
CH_2FCl (HCFC-31)	2.80	1254			$(4.17 \pm 0.63) \times 10^{-14}$
CHF_2Cl (HCFC-22)	1.48	1710	1.51	997	$(4.77 \pm 0.72) \times 10^{-15}$
$CHFCl_2$ (HCFC-21)	1.17	1100	2.05	551	$(2.92 \pm 0.44) \times 10^{-14}$
CH_3CF_2Cl (HCFC-142b)	1.75	1857	4.29	1417	$(3.44 \pm 0.69) \times 10^{-15}$
CH_3CFCl_2 (HCFC-141b)	1.25	1588			$(6.06 \pm 0.91) \times 10^{-15}$
CH_2ClCF_3 (HCFC-133a)	1.63	1426	2.09	780	$(1.36 \pm 0.48) \times 10^{-14}$
CH_2ClCF_2Cl (HCFC-132b)	4.04	1611	5.31	985	$(1.81 \pm 0.45) \times 10^{-14}$
$CHFClCF_3$ (HCFC-124)	0.727	1303	1.09	710	$(9.17 \pm 1.38) \times 10^{-15}$

(continued)

Table VI-A-77. *(continued)*

Haloalkane	$10^{12} \times A$ (cm^{3}molecule^{-1} s^{-1})	B (K)	$10^{18} \times C$ (cm^{3}molecule^{-1} s^{-1})	D (K)	k(298K) (cm^{3}molecule^{-1} s^{-1})
$CHFClCF_2Cl$ (HCFC-123a)	0.844	1254			$(1.26 \pm 0.31) \times 10^{-14}$
CF_3CCl_2H (HCFC-123)	0.674	867	1.40	370	$(3.67 \pm 0.73) \times 10^{-14}$
$CHCl_2CF_2Cl$ (HCFC-122)	1.08	916			$(4.99 \pm 0.75) \times 10^{-14}$
$CHFClCFCl_2$ (HCFC-122a)	0.378	938			$(1.62 \pm 0.41) \times 10^{-14}$
$CH_3CF_2CFCl_2$ (HCFC-243cc)	0.792	1732			$(2.37 \pm 0.59) \times 10^{-15}$
$CF_3CH_2CFCl_2$ (HCFC-234fb)					$(9.5 \pm 3.0) \times 10^{-16}$
$CF_3CF_2CCl_2H$ (HCFC-225ca)	1.47	1223			$(2.43 \pm 0.36) \times 10^{-14}$
CF_2ClCF_2CHFCl (HCFC-225cb)	0.543	1227			$(8.84 \pm 1.33) \times 10^{-15}$
CHF_2Br (Halon-1201)	1.24	1435			$(1.00 \pm 0.15) \times 10^{-14}$
CF_3CH_2Br (Halon-2301)	1.51	1359			$(1.58 \pm 0.40) \times 10^{-14}$
CF_3CHFBr (Halon-2401)	0.784	1149			$(1.66 \pm 0.25) \times 10^{-14}$
CF_3I	40.5	2209			$(2.44 \pm 0.49) \times 10^{-14}$
CH_2ClBr	1.94	844			$(1.14 \pm 0.17) \times 10^{-13}$
$CHCl_2Br$					$(1.24 \pm 0.31) \times 10^{-13}$
$ClCH_2CHBrCH_2Br$					$(4.05 \pm 1.01) \times 10^{-13}$
$CF_3CHClBr$ (Halon-2311)	0.901	857			$(5.04 \pm 1.26) \times 10^{-14}$
$CF_2BrCHFCl$	0.98	1263			$(1.41 \pm 0.35) \times 10^{-14}$

VI-B. Cl, O(^3P), NO_3, and O_3 Reactions with the Haloalkanes

With a few exceptions, the reactions of Cl, O(^3P), NO_3 radicals with haloalkanes proceed via hydrogen abstraction mechanisms. The H—X bond strength in HCl, OH, and HNO_3 (432, 430, and 426 kJ mol^{-1}, respectively) are substantially weaker than that in H_2O (497 kJ mol^{-1}) and consequently hydrogen abstraction is 65, 67, and 71 kJ mol^{-1} less exothermic in reactions involving Cl, O(^3P), and NO_3 radicals than those involving OH radicals. In contrast to reactions involving OH radicals, abstraction of hydrogen from haloalkanes by Cl, O(^3P), and NO_3 radicals is typically thermoneutral or perhaps endothermic. This difference in thermochemistry has important ramifications. As illustrated by the examples in table VI-B-1, the reactions of Cl, O(^3P), and NO_3 radicals with haloalkanes have substantial activation barriers, proceed slowly at ambient temperature, and are not a significant atmospheric loss mechanism for haloalkanes. Many haloalkanes react faster with Cl atoms than with OH radicals (compare data in tables VI-A-77 and VI-B-1). However, the average Cl-atom concentration in the troposphere is several orders of magnitude lower than that of OH, and hence reaction with Cl-atoms is not a significant atmospheric loss mechanism for haloalkanes.

Table VI-B-1. Rate coefficients (k, cm^3 $molecule^{-1}$ s^{-1}) for selected reactions of Cl, $O(^3P)$, and NO_3 radicals with haloalkanes

Reaction	A	E_a/R (K)	k(298K)[a]	Reference
$Cl + CH_3F \rightarrow CH_2F + HCl$	2.0×10^{-11}	1200	3.5×10^{-13}	Sander et al. (2003)
$Cl + CH_3Cl \rightarrow CH_2Cl + HCl$	3.2×10^{-11}	1250	4.8×10^{-13}	Sander et al. (2003)
$Cl + CH_3Br \rightarrow CH_2Br + HCl$	1.7×10^{-11}	1080	4.5×10^{-13}	Sander et al. (2003)
$Cl + CH_2F_2 \rightarrow CHF_2 + HCl$	1.2×10^{-11}	1630	5.0×10^{-14}	Sander et al. (2003)
$Cl + CHF_3 \rightarrow CF_3 + HCl$			3.0×10^{-18}	Sander et al. (2003)
$Cl + CHFCl_2 \rightarrow CFCl_2 + HCl$	5.9×10^{-12}	2430	1.7×10^{-15}	Sander et al. (2003)
$Cl + CH_3CCl_3 \rightarrow CH_2CCl_3 + HCl$	2.8×10^{-12}	1790	7.0×10^{-15}	Sander et al. (2003)
$O(^3P) + CHCl_3 \rightarrow CCl_3 + OH$	4.98×10^{-12}	2500	1.1×10^{-15}	Herron (1988)
$O(^3P) + CF_3CH_2Cl \rightarrow CF_3CHCl + OH$	3.32×10^{-12}	2600	5.4×10^{-16}	Herron (1988)
$NO_3 + CH_3Cl \rightarrow CH_2Cl + HNO_3$			$< 1 \times 10^{-18}$	Boyd et al. (1991)
$NO_3 + CH_2Cl_2 \rightarrow CHCl_2 + HNO_3$			$< 4.8 \times 10^{-18}$	Boyd et al. (1991)
$NO_3 + CHCl_3 \rightarrow CCl_3 + HNO_3$	8.52×10^{-13}	2814	6.75×10^{-17}	Canosa-Mas et al. (1989)

The exceptions suggested in the previous discussions are the reactions of Cl-atoms with iodine- and, to a lesser extent, bromine-containing haloalkanes, and reaction of $O(^3P)$ atoms with iodine-containing haloalkanes (Cotter et al., 2003). The reactions of Cl-atoms with iodine- and bromine-containing compounds (e.g., CH_3Br, CH_3I, C_2H_5I, CH_2ClI, CH_2BrI, and CH_2I_2) proceed via two channels: hydrogen atom abstraction and reversible adduct formation (Ayhens et al., 1997; Piety et al.,1998; Bilde and Wallington, 1998; Enami et al., 2005; Orlando et al., 2005). Hydrogen abstraction proceeds with a positive activation energy whereas adduct formation proceeds with a negative activation energy (Ayhens et al., 1997; Piety et al., 1998). Adduct formation increases in importance at low temperatures and is competitive for atmospherically relevant temperatures and pressures.

The main atmospheric fate of CH_xI–Cl and CH_xBr–Cl adducts is decomposition to regenerate reactants (Ayhens et al., 1997; Bilde and Wallington, 1998).

The reactions of Cl-atoms with CBr_2Cl_2 and $CBrCl_3$ have been shown by Bilde et al. (1998) to proceed via bromine atom abstraction at 296 K with rate coefficients of $(1.8 \pm 0.4) \times 10^{-13}$ and $(7.1 \pm 0.4) \times 10^{-14}$ cm^3 $molecule^{-1}$ s^{-1}, respectively. The abstraction of a Br-atom from a haloalkane by Cl-atoms is not commonly encountered. For example, abstraction of a Br-atom by Cl from CH_3Br, CF_3Br, and CH_2Br_2 is endothermic by 71–75 kJ mol^{-1}. This type of reaction appears to be unique to the species CBr_xCl_{4-x}, for which Br abstraction by Cl is calculated to be only 4–8 kJ mol^{-1} endothermic. Clyne and Walker (1973) studied the reaction of Cl with $CBrCl_3$ at elevated temperatures and reported a rate coefficient of 2.3×10^{-13} at 652 K and an approximate value of 2×10^{-14} cm^3 $molecule^{-1}$ s^{-1} at 300 K, in fair agreement with the finding of Bilde et al. (1998). Combining the results from Clyne and

Walker (1973) and Bilde et al. (1998) gives an activation energy of 5.4 kJ mol^{-1} for bromine abstraction in the Cl + $CBrCl_3$ reaction. Low concentrations of Cl-atoms in the troposphere and slow rates of reaction render loss of CBr_xCl_{4-x} via reaction with Cl-atoms an insignificant atmospheric process. As discussed in chapter VII, photolysis is often the dominant atmospheric fate of CBr_xCl_{4-x}.

VI-C. Rules for Empirical Estimation of Rate Coefficients for OH–Haloalkane Reactions

Recognition of the difficulties and expense associated with the experimental measurement of rate coefficients for a large number of potential CFC replacement compounds has spurred the development of several approaches to the estimation of rate coefficients for OH–haloalkane reactions: empirical methods (e.g., Atkinson, 1987; Kwok and Atkinson, 1995; DeMore, 1996), semi-empirical methods (e.g., Cooper et al.,1990, 1992; Bartolotti and Edney, 1994; Percival et al.,1995; Klamt,1993; Dhanya and Saini,1997), and ab initio theoretical methods (e.g., Rayez et al.,1993: Martell and Boyd,1995). Hydroxyl radicals are electrophilic reactants and correlations have been demonstrated between rate coefficients of OH–haloalkane reactions and HOMO energy (Dhanya and Saini, 1997), molecular orbital energies (Klamt, 1993), Mulliken charges (Cooper et al., 1990, 1992), ionization potential (Percival et al., 1995), and the reactivity of other electrophilic reactants such as $O(^3P)$ (Gaffney and Levine,1979) and Cl-atoms (Andersen et al., 2005). Steady advancements in available computational power are facilitating increasing sophisticated ab initio theoretical treatments which promise to provide a valuable tool to estimate OH–haloalkane rate coefficients in the future.

As a result of its simplicity, wide applicability, and ease of implementation in computer code, the most widely used method of estimating rate coefficients for reactions of OH radicals with haloalkanes is the empirical structure–additivity relationship (SAR) approach developed by Atkinson (1986a) and Kwok and Atkinson (1995). As discussed in chapter II, this method treats the rate of abstraction from a group within a molecule (rather than a singular hydrogen atom). From a simultaneous fit of all the available data for the alkanes as a function of temperature, Kwok and Atkinson obtained the following group rate coefficients at 298 K: $k(\text{—CH}_3) = 1.36 \times 10^{-13}$; $k(\text{—CH}_2\text{—}) = 9.34 \times 10^{-13}$; $k(\text{>CH—}) = 1.94 \times 10^{-12}$ cm^3 molecule^{-1} s^{-1}. The rate coefficients for hydrogen abstraction from —CH_3, —CH_2—, and >CH— groups depend on the identity of the substituents attached to these groups. The quantitative effect of a given substituent is represented by its substituent factor, F(X). The rate coefficient for hydrogen abstraction from a methyl group with the substituent X, $k(\text{CH}_3\text{—X})$, is $k(\text{—CH}_3) \times F(X)$. The rate coefficient for hydrogen abstraction from a methylene group with substituents X and Y, $k(\text{X—CH}_2\text{—Y})$, is $k(\text{—CH}_2\text{—}) \times F(X) \times F(Y)$. Finally, the rate coefficient for hydrogen abstraction from a tertiary C—H bond with substituents X, Y, and Z, $k({}^{\text{X}}_{\text{Y}}\text{—CH—Z})$, is $k(\text{>CH—}) \times F(X) \times F(Y) \times F(Z)$. Substituent factors for haloalkanes given by Kwok and Atkinson (1995) are listed in table VI-C-1. As an example using this approach, the reactivity of CF_3CHFCH_2F (HFC-245eb) is estimated as $k(\text{>CH—}) \times F(\text{—CF}_3) \times F(\text{—CH}_2\text{F}) \times F(\text{—F}) + k(\text{—CH}_2\text{—}) \times$

Table VI-C-1. Substituent factors $F(X)$ at 298 K from Kwok and Atkinson (1995)

Group X	F(X)
$—CH_3$	1.00
$—CH_2—$, $>CH—$, $>C<$	1.23
—F	0.094
—Cl	0.38
—Br	0.28
—I	0.53
$—CH_2F$	0.61
$—CHF_2$	0.13
$—CF_3$	0.071
—CHF—	0.21
$—CF_2—$	0.018
$—CF_2Cl$	0.031
$—CFCl_2$	0.044
$—CH_2Cl$, $—CHCl_2$, —CHCl—, $>CCl—$	0.36
$—CCl_3$	0.069
—CHBr—, $—CH_2Br$	0.46

F(—CHF—) × F(—F) $= 1.94 \times 10^{-12} \times 0.071 \times 0.61 \times 0.094 + 9.34 \times 10^{-13} \times 0.21 \times 0.094 = 2.63 \times 10^{-14}$; this can be compared to the value of $k(OH + CF_3CHFCH_2F) = 1.81 \times 10^{-14}$ cm^3 $molecule^{-1}$ s^{-1} that is recommended in section VI-A-2.13.

Figure VI-C-1 shows a plot of the k(OH) values predicted using the SAR method with the substituent values given in table VI-C-1 vs. those recommended from the experimental data presented in section VI-A. As is seen from figure VI-C-1, the SAR method describes the reactivity of the majority of haloalkanes to within a factor of two. However, for many of the haloalkanes shown in the figure the reactivity predicted by the SAR method differs by more than a factor of two from the measured reactivity; for $CF_3CH_2CH_2CF_3$ the values differ by a factor of 20. Most of the outliers are highly fluorinated alkanes. As illustrated in figure VI-C-1, the SAR method using the substituent values given in table VI-C-1 tends to overestimate the reactivity of compounds containing CF_3— groups and underestimate the reactivity of molecules containing $—CF_2—$ groups.

The SAR method developed by Atkinson and co-workers was last refined by Kwok and Atkinson (1995). In the last decade the data base for haloalkanes has improved substantially, and we are now in a position to offer suggestions for improvement to the method. In the SAR method the reactivities of $—CH_3$,$—CH_2—$, and $>CH—$ groups are assumed to be affected only by α-substituents. Thus, comparing the reactivity of the molecule $CF_3CH_2CH_3$ to that of the corresponding alkane (propane), the reactivity of the $—CH_2—$ group is modified by the CF_3— substituent while the reactivity of the $—CH_3$ group is unmodified from that in propane. As discussed in section VI-A-2.12, the rate coefficient for reaction of OH radicals with $CF_3CH_2CH_3$ at 298 K is $k(OH + CF_3CH_2CH_3) = (5.48 \pm 1.37) \times 10^{-14}$ cm^3 $molecule^{-1}$ s^{-1}. This value is substantially lower than the reactivity of either of the methyl groups in propane;

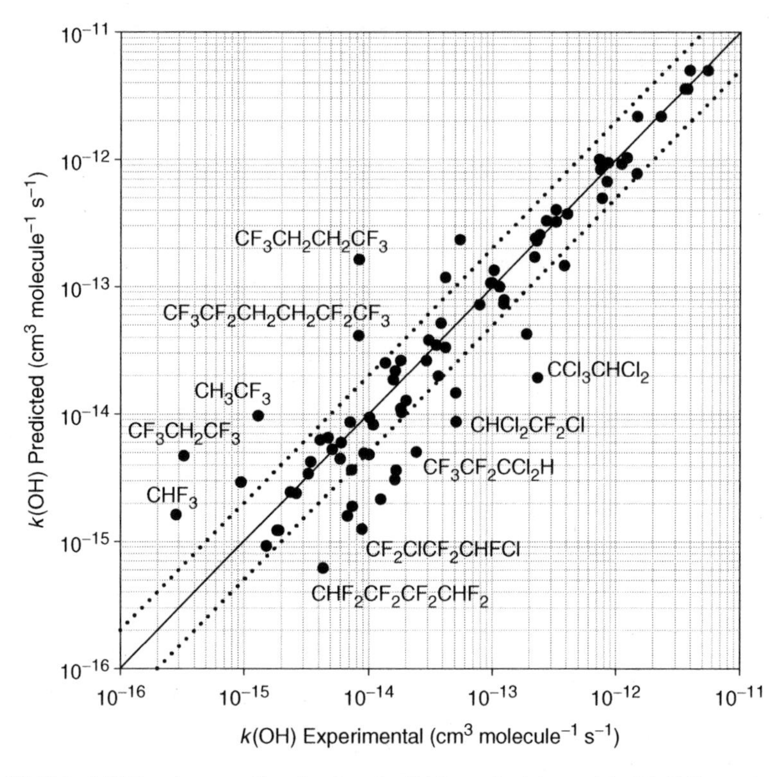

Figure VI-C-1. k(OH) values predicted using the SAR method (see table VI-C-1) vs. those recommended from the experimental data presented in section VI-A. The solid line represents perfect agreement, the dotted lines show disagreement by a factor of two.

$F(\text{—CH}_2\text{—}) \times k(\text{—CH}_3) = 1.23 \times 1.36 \times 10^{-13} = 1.67 \times 10^{-13}$ cm^3 molecule^{-1} s^{-1}. Even if we were to make the extreme assumption that the CF_3— substituent totally deactivates the —CH_2— group, we would still not be able to explain the reactivity of $CF_3CH_2CH_3$ without invoking a significant deactivating influence of the β-CF_3— substituent on the —CH_3 group. A simple way to account for the substantial, long-range deactivating effect of the CF_3— substituent but avoid adding more substituent factors into the scheme is to treat it as an F— substituent when in a position β to the hydrogen which is being abstracted. Given their similar inductive effects, it seems reasonable to acknowledge the likely long-range deactivating effects of —CF_2— and CF_2Cl— substituents by also treating them as F— substituents. Thus, in this modified approach, estimation of the rate coefficient for the $CF_3CH_2CH_2CF_3$ (HFC-356mff) molecule becomes $2 \times k(\text{—CH}_2\text{—}) \times F(\text{—CF}_3) \times F(\text{—CH}_2\text{F}) = 2 \times 9.34 \times 10^{-13} \times 0.071 \times 0.61 = 8.09 \times 10^{-14}$; compare this with the result using the unmodified approach: $2 \times k(\text{—CH}_2\text{—}) \times F(\text{—CF}_3) \times F(\text{—CH}_2\text{—}) = 2 \times 9.34 \times 10^{-13} \times 0.071 \times 1.23 = 1.63 \times 10^{-13}$ cm^3 molecule^{-1} s^{-1}. figure VI-C-2 shows the effect of treating β—CF_3—, —CF_2—, and CF_2Cl— substituents as F— substituents. As is seen from the figure, this rather modest refinement of the SAR approach leads to a significant improvement in its predictive power.

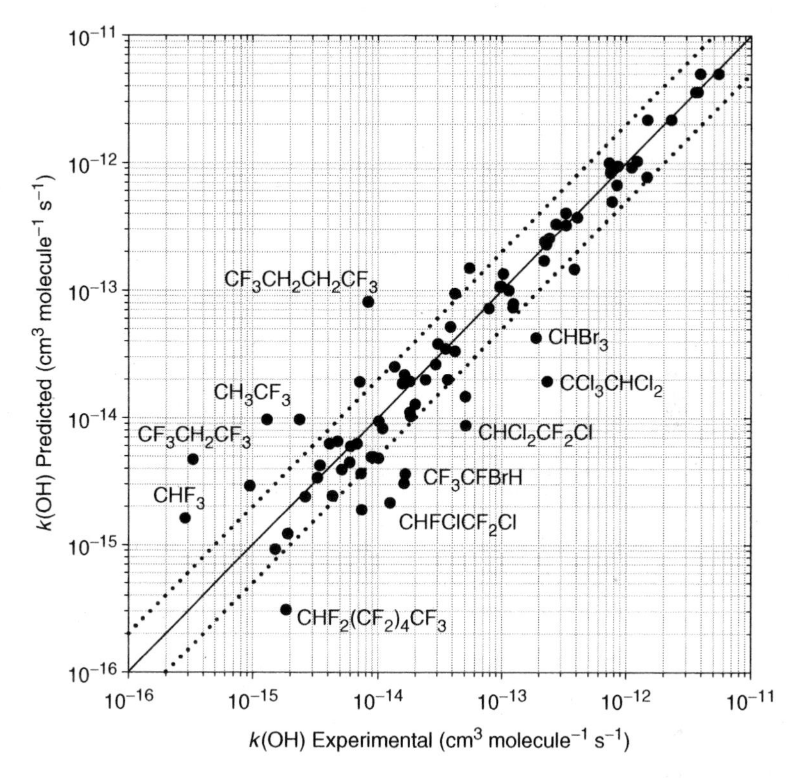

Figure VI-C-2. k(OH) values predicted using the SAR method, using substituent values from Kwok and Atkinson (1995) given in table VI-C-1 with a modification to account for long-range effects of CF_3—, —CF_2—, and CF_2Cl— substituents noted in the text, vs. those recommended from the experimental data presented in section VI-A. The solid line represents perfect agreement, the dotted lines show disagreement by a factor of two.

While the substituent factors given in table VI-C-1 are purely empirical (based upon achieving the best fit to the available data base), chemical intuition would suggest that there might be a progression along the series —CH_2—, —CHF—, —CF_2—, and —CF_3, reflecting an increasing inductive effect with increasing fluorine substitution. The values given in table VI-C-1 follow the order 1.23, 0.21, 0.018, and 0.071. The substituent factor F(—CF_2—) = 0.018 appears to be anomalous and leads to the underestimation noted above for the reactivity of compounds containing the —CF_2— group. This anomaly probably reflects the rather limited kinetic data base available in 1995 upon which Kwok and Atkinson based their F(—CF_2—) substituent factor. The data base in table VI-A-77 offers the opportunity to suggest improvements in the substituent factors used in SAR calculation of haloalkane reactivity. We have derived a new set of substituent factors to describe the reactivity of OH radicals with haloalkanes. The method uses the following steps: (i) adopt the group rate coefficients and substituent factors recommended in chapter II [k(—CH_3) = 1.35×10^{-13}; k(—CH_2—) = 7.8×10^{-13}; k(>CH—) = 1.37×10^{-12} cm^3 molecule^{-1} s^{-1} at 298 K, F(—CH_3) = 1.00, F(—CH_2—) = 1.35]; (ii) constrain the substituent factors to decrease in the order

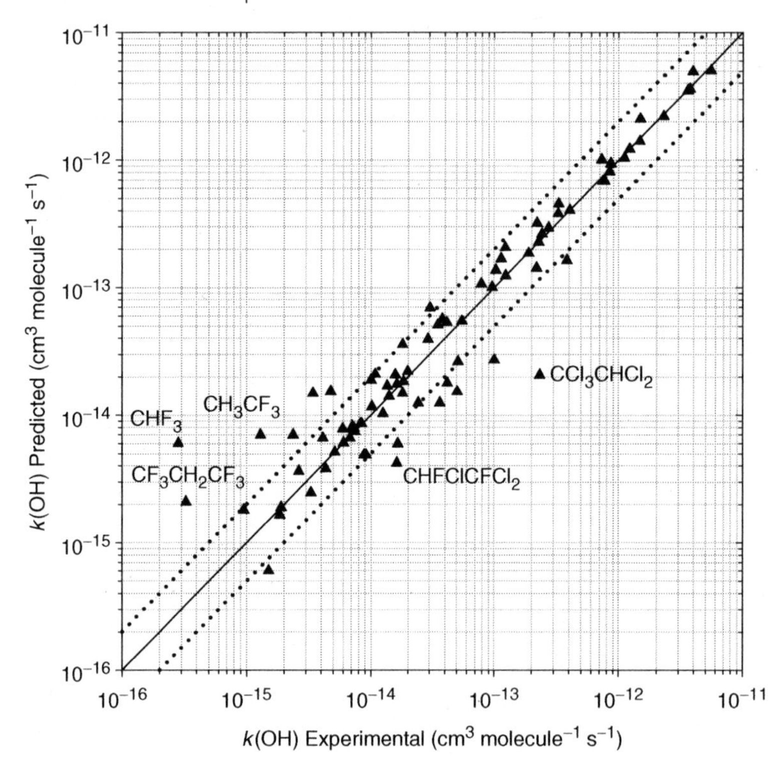

Figure VI-C-3. k(OH) values predicted using the SAR method, using substituent values recommended in this work (see table VI-C-2) with long-range effects of CF_3—, —CF_2—, and CF_2Cl— substituents (see text), vs. those recommended from the experimental data presented in section VI-A. The solid line represents perfect agreement, the dotted lines show disagreement by a factor of two.

$F(—CH_3—) \geq F(—CH_2F) \geq F(—CHF_2) \geq F(—CF_3)$ and $F(—CH_2—) \geq F(—CHF—) \geq F(—CF_2—)$; (iii) introduce new substituent factors $F(CF_2Br—)$, $F(CH_2I—)$, and $F(—CHI—)$ to describe Halons and iodoalkanes; there are very few compounds with these groups so they were not included in the fit but were instead adjusted manually to fit the data; and (iv) incorporate long-range effects of the β-substituents CF_3—, —CF_2—, and CF_2Cl— by treating them as F— substituents. Variation of the substituent factors to minimize the sum of differences between predicted (k_{SAR}) and recommended (k_{rec}) rate coefficients $[\Sigma|\ln(k_{rec}/k_{SAR})|]$ at 298 K for the compounds given in table VI-A-77 gives the substituent factors in table VI-C-2. Figure VI-C-3 shows a plot of the k(OH) values calculated using the SAR method with the substituent values given in table VI-C-2, including the longer range effects of β-substituents: CF_3—, —CF_2—, and CF_2Cl—. As is seen from comparison of figures VI-C-1 and VI-C-3, the substituent factors given in table VI-C-2 provide a better description of the reactivity of OH radicals towards haloalkanes than those derived by Kwok and Atkinson from the more limited kinetic data base available in 1995. Using the substituent factors in table VI-C-2, the reactivities of OH radicals with most haloalkanes are reproduced within a factor of two.

Table VI-C-2. Substituent factors $F(X)$ at 298 K derived in the present work

Group X	F(X)
$—CH_3$	1.000
$—CH_2—$, $—CH—$, $>C<$	1.350
$—F$	0.164
$—Cl$	0.419
$—Br$	0.514
$—I$	0.792
$—CH_2F$	0.106
$—CHF_2$	0.106
$—CF_3$	0.052
$—CHF—$	0.216
$—CF_2—$	0.045
$—CF_2Cl$	0.110
$—CFCl_2$	0.045
$—CH_2Cl$, $—CHCl_2$, $—CHCl—$, $>CCl—$	0.403
$—CCl_3$	0.086
$—CHBr—$, $—CH_2Br$	0.400
$—CF_2Br$	0.150
$—CH_2I$, $—CHI—$	0.500

VI-D. Atmospheric Chemistry of Haloalkylperoxy Radicals

VI-D-1. Introduction

The atmospheric oxidation of haloalkanes containing one or more C—H bonds is initiated largely by reaction with OH radicals in the troposphere (and to a lesser extent in the stratosphere). Reaction with Cl-atoms makes a minor contribution in the troposphere and Cl- and $O(^1D)$-atoms make minor contributions in the stratosphere. Direct photolysis can initiate the atmospheric oxidation of haloalkanes possessing chromophores (C—Br or C—I bonds) which absorb at wavelengths available in the troposphere (see chapter VII). All of these processes lead to the formation of haloalkyl radicals. As with alkyl radicals (see chapter IV), addition of O_2 to haloalkyl radicals proceeds with rate coefficients in the range 10^{-12}–10^{-11} cm^3 $molecule^{-1}$ s^{-1} under atmospheric conditions. In 1 atmosphere of air at 298 K haloalkyl radicals have a lifetime of 20–200 ns with respect to addition of O_2 to give the corresponding haloperoxy radicals. Addition of O_2 is the sole atmospheric fate of haloalkyl radicals. This process is illustrated for the simplest haloalkane (CH_3F) in the following reactions:

$$OH + CH_3F \rightarrow CH_2F + H_2O$$

$$CH_2F + O_2 + M \rightarrow CH_2FO_2 + M$$

Figure VI-D-1 shows the mechanism of the atmospheric oxidation of CH_3F. As illustrated in this figure, the atmospheric fate of peroxy radicals is reaction with NO, NO_2, HO_2 radicals, or other peroxy radicals (Wallington et al., 1992b; Lightfoot et al., 1992; Wallington et al., 1997; Tyndall et al., 2001).

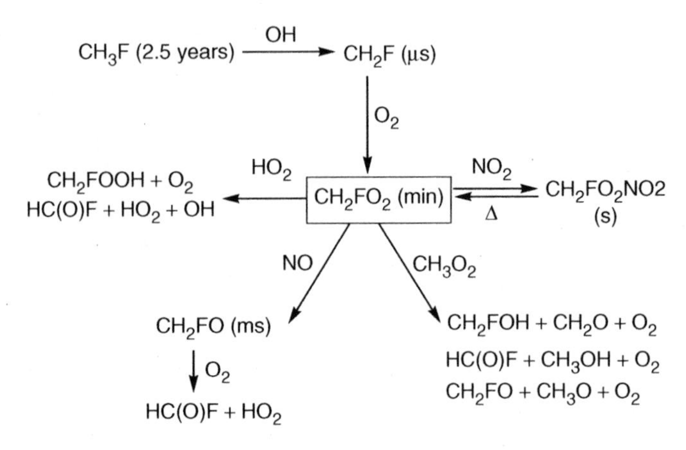

Figure VI-D-1. Atmospheric oxidation mechanism for CH_3F (HFC-41), illustrating the importance of peroxy radical chemistry. Values in parentheses are approximate lifetimes.

The reactions of haloalkylperoxy radicals with NO are rapid (proceeding with rate coefficients in the range $1–2 \times 10^{-11}$ cm^3 molecule^{-1} s^{-1} at 298 K), and their rates generally increase with decreasing temperature. The reactivities of halomethylperoxy and haloethylperoxy radicals given in table VI-D-1 can be compared to values of $k(CH_3O_2{+}NO) = (7.6 \pm 1.1) \times 10^{-12}$ and $k(C_2H_5O_2{+}NO) = (9.2 \pm 1.4) \times 10^{-12}$ cm^3 molecule^{-1} s^{-1} at 298 K recommended in section IV-A. In reactions with NO, haloalkylperoxy radicals are approximately 50–100% more reactive than the corresponding unsubstituted alkylperoxy radicals. The reactions of halomethylperoxy radicals with NO give NO_2 and the corresponding alkoxy radical as the major products. The formation of organic nitrates in the reaction of haloalkylperoxy radicals with NO appears to be less important than in the corresponding reactions of alkylperoxy radicals of similar size. Typically, organic nitrates are not reported during laboratory studies in which haloalkylperoxy radicals are formed in the presence of NO_x (e.g., Wallington et al., 1997; Hasson and Smith, 1999; Vésine et al., 2000; Orlando and Tyndall, 2002a; Andersen et al., 2004a).

The addition of NO_2 to haloalkylperoxy radicals gives thermally unstable haloalkylperoxy nitrates (e.g., $CF_xCl_{3-x}O_2NO_2$). These molecules have lifetimes, with respect to thermal decomposition back to reactants, which range from approximately 10 s at 298 K to 10^4 s (3 hr) at 253 K (Lightfoot et al., 1992). The net result of the reaction of haloalkylperoxy radicals with NO_2 is to sequester NO_x for a short time. Unlike their more stable and more abundant acylperoxy nitrate (e.g., $CH_3C(O)O_2NO_2$) counterparts, the haloalkylperoxy nitrates do not play any significant role in transport of NO_x. Haloacylperoxy nitrates (e.g., $CF_xCl_{3-x}C(O)O_2NO_2$ have stabilities that are comparable to those of acylperoxy nitrates, but the haloacylperoxy nitrates are far less abundant and do not play any significant role in transport of NO_x. While the reactions of haloalkylperoxy radicals with NO_2 are of no practical importance in atmospheric chemistry, they can be important in laboratory experiments and hence are considered here.

As with alkylperoxy radicals, the reaction of haloalkylperoxy radicals with HO_2 radicals leads to the formation of haloalkyl hydroperoxides (CH_2FOOH in the case of

Table VI-D-1. Rate coefficients (k, cm^3 $molecule^{-1}$ s^{-1}) for reactions of haloalkylperoxy radicals with NO

RO_2	$10^{12} \times k$	T (K)	Pressure Range (Torr)	Technique[a]	Reference
CH_2FO_2	12.5 ± 1.3	298	750 (SF_6)	PR–UVabs	Sehested et al. (1993b)
CHF_2O_2	12.6 ± 1.6	298	750 (SF_6)	PR–UVabs	Sehested et al. (1993b)
CF_3O_2	17.8 ± 3.6	295	2–5 (He)	DF–MS	Plumb and Ryan (1982)
	21.4	228	1–10 (N_2)	PLP–MS	Dognon et al. (1985)
	18.9	243			
	14.2	298			
	12.0	373			
	10.0	433			
	16.9 ± 2.6	298	750 (SF_6)	PR–UVabs	Sehested and Nielsen (1993)
	15.3 ± 2.0	297	760 (He)	FT–MS	Bevilacqua et al. (1993)
	15.0	298	100 (He)	FP–LIF	Ravishankara et al. (1994)
	15.7 ± 3.1	293	2–30 (N_2)	FP–MS	Bhatnagar and Carr (1994)
	15.7 ± 3.8	298	5 (O_2)	PLP–LIF	Turnipseed et al. (1994)
	17.6 ± 3.5	298	1–3 (He)	DF–LIF	Bourbon et al. (1996)
	16 ± 3	298	1 (He)	DF–MS	Louis et al. (1999)
CH_2FCFHO_2	> 8.7	298	750 (SF_6)	PR–UVabs	Wallington et al. (1994b)
$CF_3CH_2O_2$	12 ± 3	298	750 (SF_6)	PR–UVabs	Nielsen et al. (1994)
	12.3 ± 3.2	290	2 (He)	DF–MS	Peeters and Pultau (1993)
	18.4	198	30 (O_2)	PLP–TDL	Olkhov and Smith (2003)
	14.8	218			
	13.5	238			
	12.9	258			
	11.8	278			
	9.8	298			
$CHF_2CF_2O_2$	> 9.7 ± 1.3	298	750 (SF_6)	PR–UVabs	Sehested et al. (1993b)
CF_3CFHO_2	> 12.8 ± 3.6	298	750 (SF_6)	PR–UVabs	Wallington and Nielsen (1991)
	11 ± 3	290	2 (He)	DF–MS	Peeters and Pultau (1992)
	15.6 ± 4.1	290	2 (He)	DF–MS	Peeters and Pultau (1993)
	13.1 ± 3.0	324	12–25 (N_2)	FP–MS	Bhatnagar and Carr (1995a)
	9.4	200	30 (O_2)	PLP–TDL	Olkhov and Smith (2003)
	12.2	220			
	12.8	235			
	13.2	257			
	13.5	278			
	13.3	298			

(continued)

Table VI-D-1. *(continued)*

RO_2	$10^{12} \times k$	T (K)	Pressure Range (Torr)	Technique[a]	Reference
$CF_3CF_2O_2$	$> 10.7 \pm 1.5$	298	750 (SF_6)	PR–UVabs	Sehested et al. (1993b)
	14.8	198	30 (O_2)	PLP–TDL	Olkhov and Smith (2003)
	14.7	207			
	14.3	216			
	13.9	225			
	14.5	236			
	14.8	245			
	14.4	256			
	13.6	264			
	12.7	275			
	13.1	286			
	13.6	295			
	13.7	298			
$CH_3CF_2CH_2O_2$	8.5 ± 1.9	296	750 (SF_6)	PR–UVabs	Møgelberg et al. (1995c)
$CF_3CHO_2CF_3$	11 ± 3	296	750 (SF_6)	PR–UVabs	Møgelberg et al. (1995d)
$CF_3CF_2CFHO_2$	> 8	296	750 (SF_6)	Pr–UV abs	Møgelberg et al. (1995a)
$CF_3CFO_2CF_3$	21 ± 9	296	750 (SF_6)	PR–UVabs	Møgelberg et al. (1996b)
CH_2ClO_2	18.7 ± 2.0	298	750 (SF_6)	PR–UVabs	Sehested et al. (1993b)
CCl_3O_2	18.6 ± 2.8	295	2–8 (He/O_2)	DF–MS	Ryan and Plumb (1984)
	22.5	228	1–10 (N_2)	PLP–MS	Dognon et al. (1985)
	19.0	243			
	17.2	298			
	13.5	373			
	10.0	413			
$CCl_3CH_2O_2$	> 6.1	295	750 (SF_6)	PR–UVabs	Platz et al. (1995)
CH_2BrO_2	10.7 ± 1.1	298	750 (SF_6)	PR–UVabs	Sehested et al. (1993b)
CF_2ClO_2	16 ± 3	298	1–10 (N_2)	PLP–MS	Dognon et al. (1985)
	13.1 ± 1.2	298	750 (SF_6)	PR–UVabs	Sehested et al. (1993b)
$CFCl_2O_2$	16 ± 2	298	1–6 (O_2)	PLP–MS	Lesclaux and Caralp (1984)
	23.1	228	1–10 (N_2)	PLP–MS	Dognon et al. (1985)
	18.4	248			
	14.5	298			
	11.4	373			
	9.3	413			
CH_3CFClO_2	27 ± 5	263	8–20 (N_2)	FP–MS	Wu and Carr (1996a)
	21 ± 6	263			
	21 ± 4	298			
	17 ± 5	298			
	18 ± 3	321			
	14 ± 4	321			

Table VI-D-1. *(continued)*

RO_2	$10^{12} \times k$	T (K)	Pressure Range (Torr)	Technique[a]	Reference
$CF_2ClCH_2O_2$	12.8 ± 1.1	298	750 (SF_6)	PR–UVabs	Sehested et al. (1993b)
	11.8 ± 3.0	290	2 (He)	DF–MS	Peeters and Pultau (1993)
$CFCl_2CH_2O_2$	11.8 ± 1.0	298	750 (SF_6)	PR–UVabs	Sehested et al. (1993b)
	16 ± 4	263	8–20 (N_2)	FP–MS	Wu and Carr (1996a)
	13 ± 4	263			
	13 ± 4	298			
	12 ± 4	298			
	12 ± 2	321			
	9 ± 3	321			
CF_3CClHO_2	10 ± 3	296	750 (SF_6)	PR–UVabs	Møgelberg et al. (1995b)
$CF_3CCl_2O_2$	14.5 ± 3.9	290	2 (He)	DF–MS	Peeters and Pultau (1993)
$FC(O)O_2$	25 ± 8	295	750 (SF_6)	PR–UVabs	Wallington et al. (1994a)
$CF_3C(O)O_2$	> 9.9	295	750 (SF_6)	PR–UVabs	Wallington et al. (1994c)
	55 ± 12	220	104 (N_2)	FP–UVabs	Maricq et al. (1996)
	34 ± 7	254	100 (N_2)		
	28 ± 5	295	102 (N_2)		
	24 ± 5	324	101 (N_2)		

[a] Abbreviations of the experimental techniques employed are defined in table II-A-1.

CH_2FO_2). However, in marked contrast to alkylperoxy radicals, which give hydroperoxides as the dominant product, the reaction of haloalkylperoxy radicals, with HO_2 radicals has been observed to give carbonyl compounds as significant (sometimes dominant) products (Wallington et al., 1994d; Catoire et al., 1996; Wallington et al., 1996c). Furthermore, as indicated in figure VI-D-1, there is evidence (but not proof) that the channel leading to the carbonyl product also generates an OH radical in some systems (Wallington et al., 1994c; Catoire et al., 1996; Wu and Carr, 1999; McGivern et al., 2004). The kinetics and mechanism of the reactions of haloalkyl peroxy radicals with HO_2 radicals remain rather poorly understood.

Finally, haloalkylperoxy radicals can undergo cross reactions with alkylperoxy radicals. CH_3O_2 is the most abundant alkylperoxy radical in the atmosphere.The products of reaction of CH_2FO_2 radicals with CH_3O_2 are shown in figure VI-D-1. Reactions between peroxy radicals proceed via both a disproportionation and a radical propagating mechanism. In the disproportionation mechanism a hydrogen atom is transferred between the radicals via the six-membered intermediate first proposed by Russell (1957) and illustrated in figure VI-D-2. For α-hydrogen-containing haloalkylperoxy radicals such as CH_2FO_2, two disproportionation channels are possible, with the CH_2FO_2 radical either donating or receiving a hydrogen atom (see figure VI-D-2). There are few available data concerning such peroxy radical cross reactions (Villenave and Lesclaux, 1996). Peroxy cross reactions can have a significant practical relevance. It has been suggested that the cross reaction between perfluoroalkylperoxy radicals and CH_3O_2

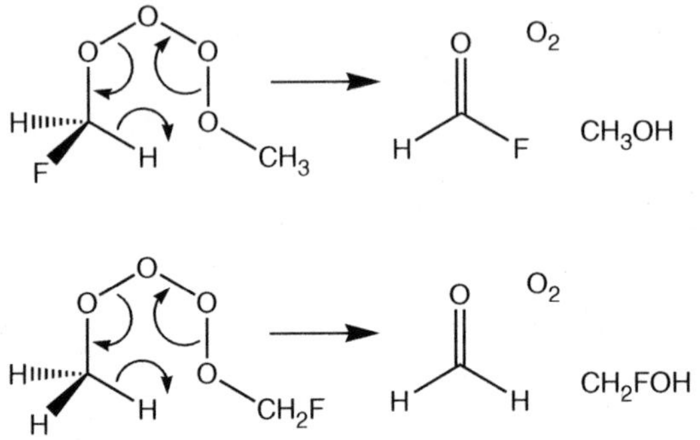

Figure VI-D-2. Mechanism for molecular product formation in the $CH_2FO_2 + CH_3O_2$ reaction.

leads to a significant formation of bioaccumulative and potentially toxic perfluorocarboxylic acids in the atmosphere (Ellis et al., 2004; Wallington et al., 2006). Further work is needed to confirm or refute this suggestion.

In the following sections the reactions of haloalkylperoxy radicals with NO, NO_2, HO_2, and CH_3O_2 are discussed. We begin with a general discussion and then proceed to consider individual reactions.

VI-D-2. Reactions of the Haloalkylperoxy Radicals with NO

As discussed in section IV-A, the reaction of alkylperoxy radicals with NO occurs via the formation of a short-lived peroxynitrite intermediate, which can decompose back to reactants, decompose to give RO and NO_2 products, or rearrange to form an organic nitrate:

$$RO_2 + NO \rightarrow [ROONO]^* \leftrightarrow RO + NO_2$$

$$\rightarrow RONO_2$$

The published data, although limited, show that the yields of organic nitrates in the reactions of haloalkylperoxy radicals with NO are substantially less than those observed for alkylperoxy radicals of the same mass (Nishida et al., 2004).

The available kinetic data for the reactions of haloperoxy radicals with NO are listed in table IV-D-1. As is seen from this table, the reactions of haloalkylperoxy radicals with NO proceed with rate coefficients in the range $1\text{–}2 \times 10^{-11}$ cm^3 molecule^{-1} s^{-1} at ambient temperature with no discernable effect of total pressure over the range 1–760 Torr.

The reaction of RO_2 with NO proceeds via the formation of a short-lived ($10^{-10}\text{–}10^{-9}$ s) vibrationally excited peroxy nitrite, ROONO*, which decomposes to give $RO + NO_2$. Although the lifetime of this vibrationally excited peroxy nitrite is short, it is nevertheless long enough for efficient intramolecular energy transfer (occurring on a time scale of the order of 10^{-12} s^{-1}). Thus, energy distributions among the reaction products are essentially statistical and nascent alkoxy radicals formed following decomposition of ROONO* carry a significant fraction of the reaction exothermicity. There are three factors that render chemical activation more important in the case of haloalkylperoxy than with alkylperoxy radicals. First, the reactions of haloalkylperoxy radicals with NO tend to be more exothermic, e.g., the reaction of NO with CF_3O_2 is 19.7 kJ mol^{-1} more exothermic than with CH_3O_2 (Sander et al., 2003). Second, the mass of the halogen atoms increases the heat capacity of the RO— group and increases the fraction of the reaction exothermicity contained in the RO radical. Third, haloalkoxy radicals often (but not always) have decomposition pathways with relatively low activation barriers that are comparable to the initial excitation of the RO radical.

As is seen from table VI-D-1, a significant fraction of the available kinetic data for reactions of haloalkylperoxy radicals with NO comes from pulsed radiolysis studies by Nielsen and coworkers (Møgelberg et al., 1995b,c,d, 1996b; Nielsen et al., 1994; Platz et al., 1995; Sehested et al., 1993b; Wallington and Nielsen, 1991; Wallington et al., 1994a,b,c). In these studies kinetic data were extracted from the observed rate of formation of NO_2 following pulsed radiolysis of SF_6/haloalkane/NO/O_2 mixtures. The pulsed electron beam results in formation of F-atoms which abstract hydrogen from the haloalkane, giving haloalkyl radicals that add O_2 to give haloalkylperoxy radicals. Reaction with NO gives NO_2 and an alkoxy radical. If the alkoxy radical decomposes (or reacts with O_2) on a time-scale comparable to that of the experimental observations, secondary peroxy (or HO_2) radicals will be formed, and these will react with NO, leading to secondary formation of NO_2. For example, in a study of the $CF_3CF_2O_2$ + NO reaction, an NO_2 yield of 215 ± 16 % was reported by Møgelberg et al. (1997b). Secondary NO_2 formation stretches out the time-scale for NO_2 formation and the rise time provides a lower limit for $k(RO_2+NO)$. This effect can be accounted for by modeling the NO_2 formation as described by Peeters and Pultau (1992, 1993), Bhatnagar and Carr (1995a), and Olkhov and Smith (2003). No such modeling was performed by Nielsen and coworkers, and their studies of CH_2FCHFO_2, $CHF_2CF_2O_2$, CF_3CFHO_2, $CF_3CF_2O_2$, $CCl_3CH_2O_2$, and $CF_3C(O)O_2$ radicals provide lower limits for $k(RO_2+NO)$.

VI-D-2.1. CH_2FO_2 and CHF_2O_2 + NO

Sehested et al. (1993b) used a pulsed radiolysis technique to monitor the formation of NO_2 following the formation of CH_2FO_2 and CHF_2O_2 in the presence of NO in 1000 mbar (750 Torr) of SF_6 at 295 K. Rate coefficients of $k(CH_2FO_2+NO) = (1.25 \pm 0.13) \times 10^{-11}$ and $k(CHF_2O_2+NO) = (1.26 \pm 0.16) \times 10^{-11}$ cm^3 molecule^{-1} s^{-1} were reported.

VI-D-2.2. CF_3O_2 + NO

The reaction of CF_3O_2 radicals with NO has been the subject of nine kinetic studies (Plumb and Ryan, 1982; Dognon et al, 1985; Sehested and Nielsen, 1993; Bevilacqua et al., 1993; Ravishankara et al., 1994; Bhatnagar and Carr, 1994; Turnipseed et al., 1994; Bourbon et al., 1996; Louis et al., 1999). The results from all studies are in agreement within the experimental uncertainties. The average of the published values gives a value of $k(CF_3O_2+NO) = 1.61 \times 10^{-11}$ cm^3 molecule^{-1} s^{-1} at 298 K, independent of total pressure over the range 1–760 Torr. Dognon et al. (1985) studied the effect of temperature over the range 228–433 K and reported $k(CF_3O_2+NO) = (1.45 \pm 0.20) \times 10^{-11} (T/298)^{-(1.2 \pm 0.2)}$ cm^3 molecule^{-1} s^{-1}. Using the temperature effect reported by Dognon et al. (1985) and fixing the room temperature value to be the average from all studies, we derive the recommendation of $k(CF_3O_2+NO) = 1.61 \times 10^{-11} (T/298)^{-1.2}$ cm^3 molecule^{-1} s^{-1}.

At 4.2 K, in a matrix isolation study, Clemitshaw and Sodeau (1987) used FTIR spectroscopy to identify CF_3OONO as an intermediate in the reaction of CF_3O_2 radicals with NO. Nishida et al. (2004) have shown that in 20–700 Torr of N_2/O_2 diluent at 296 K the reaction proceeds via two channels:

$$CF_3O_2 + NO \rightarrow CF_3O + NO_2 \quad \text{(a)}$$

$$CF_3O_2 + NO + M \rightarrow CF_3ONO_2 + M \quad \text{(b)}$$

The CF_3ONO_2 forming channel is in the fall-off regime at pressures of 50 Torr and below, while for pressures above 100 Torr there was no discernable effect of pressure on the CF_3ONO_2 yield. Nishida et al. (2004) report a nitrate yield of $k_b/(k_a + k_b) = 0.0167 \pm 0.0027$ at atmospheric pressure. It is of interest to compare this finding with the nitrate yield in the reaction of *tert*-butylperoxy radicals with NO discussed in chapter IV. The reaction of $(CH_3)_3CO_2$ radicals with NO gives a nitrate yield of approximately 18%. It is well established that the nitrate yield in reactions of unsubstituted peroxy radicals with NO increases with size of the peroxy radical. The mass of CF_3O_2 is similar to that of $(CH_3)_3CO_2$, whereas its nitrate yield is approximately an order of magnitude lower. The lower nitrate yield in the CF_3O_2 + NO reaction presumably reflects the effect of the electronic withdrawing influence of F atoms and/or the lower density of states in the CF_3O_2 radical. Computational studies are needed to provide an improved theoretical understanding of the factors impacting the nitrate yield in this (and other) reactions.

VI-D-2.3. CF_2ClO_2 + NO

The kinetics and mechanism of this reaction have been studied by Dognon et al. (1985) using laser photolysis mass spectroscopy and by Sehested et al. (1993b) using a pulse radiolysis UV absorption method. At 298 K rate coefficients of $(1.6 \pm 0.3) \times 10^{-11}$ and $(1.31 \pm 0.12) \times 10^{-11}$ cm^3 molecule^{-1} s^{-1} were derived. The agreement between these studies demonstrates the absence of any effect of total pressure over the range 1 Torr of N_2 to 750 Torr of SF_6. Dognon et al. (1985) studied the effect of temperature over the range 230–430 K and reported $k(CF_2ClO_2+NO) = (1.6 \pm 0.3) \times 10^{-11} (T/298)^{-(1.5 \pm 0.4)}$ cm^3 molecule^{-1} s^{-1}. Using the temperature effect reported by Dognon et al. (1985) and constraining the 298 K value to be the average of those reported by Dognon et al. (1985) and Sehested et al. (1993b), we recommend $k(CF_2ClO_2+NO) = 1.46 \times 10^{-11} (T/298)^{-1.5}$ cm^3 molecule^{-1} s^{-1}. No evidence for reaction channels other than that leading to the alkoxy radical and NO_2 has been reported:

$$CF_2ClO_2 + NO \rightarrow CF_2ClO + NO_2$$

VI-D-2.4. $CF_3CH_2O_2 + NO$

The kinetics of the reaction of $CF_3CH_2O_2$ radicals with NO have been studied by Peeters and Pultau (1993), Nielsen et al. (1994), and Olkhov and Smith (2003). As is seen from table VI-D-1 and figure VI-D-3, consistent results were obtained in the three studies. A linear least-squares fit to the combined data set gives $k(CF_3CH_2O_2{+}NO) = 4.37 \times 10^{-12} \exp(275/T)\,cm^3\,molecule^{-1}\,s^{-1}$. At 298 K this expression gives $k(CF_3CH_2O_2{+}NO) = 1.10 \times 10^{-11}\,cm^3\,molecule^{-1}\,s^{-1}$ with an uncertainty estimated to be ± 30%. No evidence for reaction channels other than that leading to the alkoxy radical and NO_2 has been reported.

$$CF_3CH_2O_2 + NO \rightarrow CF_3CH_2O + NO_2$$

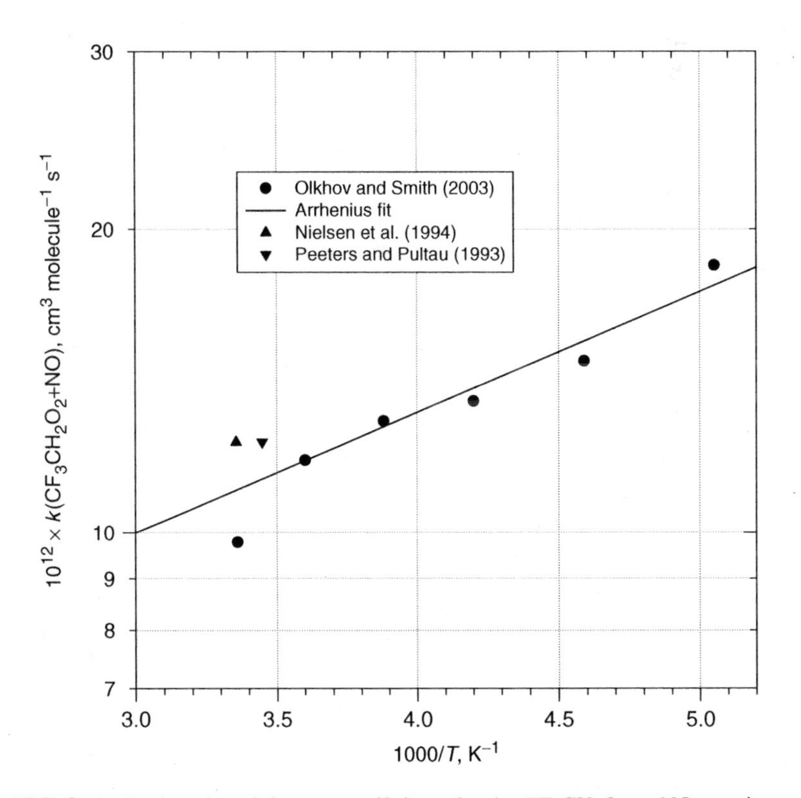

Figure VI-D-3. Arrhenius plot of the rate coefficients for the $CF_3CH_2O_2$ + NO reaction.

VI-D-2.5. $CF_3CFHO_2 + NO$

The reaction of CF_3CFHO_2 radicals with NO has been the subject of five kinetic studies. As is seen from table VI-D-1 and figure VI-D-4, the results from all the studies are in agreement within the experimental uncertainties. Wallington and Nielsen (1991) studied the reaction using a pulse radiolysis technique with UV absorption at 400 nm used to monitor the rate of NO_2 formation at 298 K. As discussed in section VI-D-2, the value of $k(CF_3CFHO_2+NO)$ reported by Wallington and Nielsen (1991) should be treated as a lower limit. Peeters and Pultau (1992, 1993) used a discharge flow technique with CF_3CFHO_2 radicals formed via reaction of F-atoms with HFC-134a in the presence of O_2, and the resulting NO_2 formation was observed using mass spectroscopy. Bhatnagar and Carr (1995a) used flash photolysis to create CF_3CFHO_2 radicals in the presence of NO and then followed the resulting NO_2 formation using mass spectroscopy. Olkhov and Smith (2003) monitored formation of HC(O)F and CF_2O and loss of NO following the pulsed photolysis of CF_3CFHI in 30 Torr of O_2 diluent at 200–298 K. The temporal behaviors of HC(O)F, CF_2O, and NO were fitted using a simple model with three fit parameters: (i) the rate constant $k(CF_3CFHO_2+NO)$, (ii) the fraction of CF_3CFHO radicals undergoing prompt decomposition, and (iii) the rate of thermal decomposition of CF_3CFHO radicals.

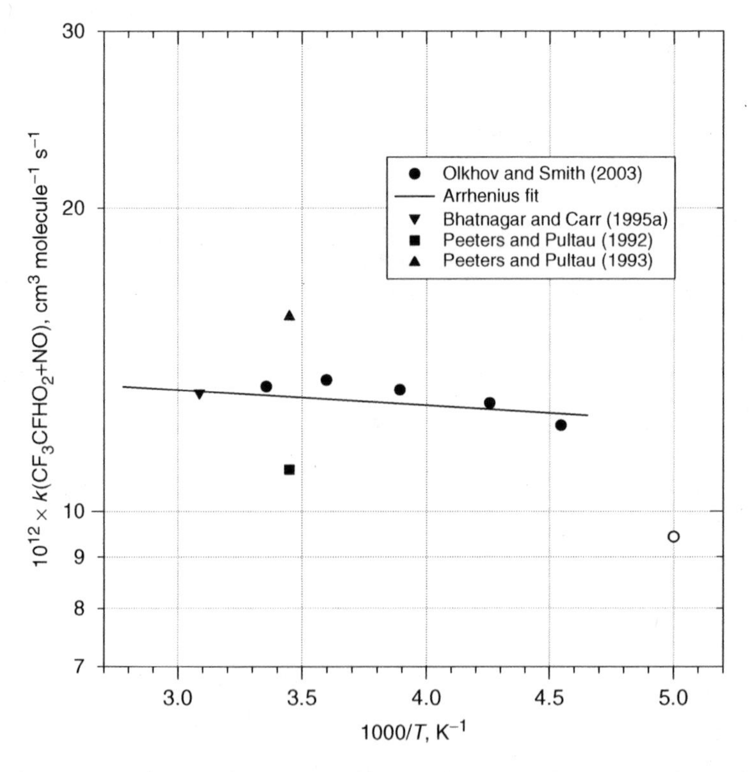

Figure VI-D-4. Arrhenius plot of the rate coefficients for the $CF_3CFHO_2 + NO$ reaction. The open circle is the 200 K data point from Olkhov and Smith and is not included in the fit.

Olkhov and Smith (2003) studied the kinetics of the reaction over the temperature range 200–298 K and reported a very small increase in the rate of reaction with increased temperature. The value of $k(CF_3CFHO_2+NO)$ obtained at 200 K was significantly below that obtained by extrapolating the high-temperature data. The kinetic data acquired near ambient temperature by Peeters and Pultau (1992, 1993), Bhatnagar and Carr (1995a), and Olkhov and Smith (2003) are in agreement within the experimental uncertainties. Excluding the 200 K data point from Olkhov and Smith (2003), the open circle in Figure IV-D-4, a linear least-squares fit to the combined data set from Peeters and Pultau (1992, 1993), Bhatnagar and Carr (1995a), and Olkhov and Smith (2003) gives $k(CF_3CFHO_2+NO) = 1.47 \times 10^{-11} \exp(-35/T)$ cm^3 molecule^{-1} s^{-1}. At 298 K this expression gives $k(CF_3CFHO_2+NO) = 1.31 \times 10^{-11}$ cm^3 molecule^{-1} s^{-1}, with an uncertainty estimated to be $\pm$ 25%.

Wallington et al. (1996b) studied the products following the reaction of CF_3CFHO_2 with NO in 120–1100 Torr of N_2/O_2 diluent at 238–295 K. A substantial fraction of the CF_3CFHO radicals produced in the reaction of CF_3CFHO_2 radicals with NO were found to be chemically activated, and these decompose promptly (on a time-scale sufficiently short that bimolecular reaction with O_2 cannot compete). It was determined that at 295, 263, and 243 K in 800 Torr total pressure prompt decomposition accounted for 64^{+12}_{-23}%, 59 $\pm$ 5%, and 57 $\pm$ 5%, respectively, of the CF_3CFHO radicals produced in the reaction of CF_3CFHO_2 radicals with NO. Olkhov and Smith (2003) monitored the formation of products in real-time following the pulse formation of CF_3CFHO_2 radicals in the presence of NO and concluded that at 298 and 198 K the fraction of CF_3CFHO radicals that undergoes prompt decomposition was 48 $\pm$ 4% and 42 $\pm$ 12%, respectively. Theoretical studies by Schneider et al. (1998), Somnitz and Zellner (2001), and Wu and Carr (2003) have shown that "prompt" decomposition occurs on a time-scale of 10^{-10}–10^{-9}s. There have been no reports (experimental or theoretical) indicating that formation of the nitrate $CF_3CFHONO_2$ is of any significance. The evidence suggests that reaction of CF_3CFHO_2 with NO proceeds via a single channel, giving NO_2 and CF_3CFHO radicals. A substantial fraction (approximately 40–70%) of the nascent CF_3CFHO radicals possesses sufficient internal excitation to decompose promptly:

$$CF_3CFHO_2 + NO \rightarrow CF_3CFHO + NO_2$$

VI-D-2.6. $CF_3CF_2O_2 + NO$

The kinetics of the reaction of $CF_3CF_2O_2$ radicals with NO have been studied by Sehested et al. (1993b) and Olkhov and Smith (2003). The rate constant reported by Olkhov and Smith at ambient temperatures is consistent with the lower limit reported by Sehested et al. Olkhov and Smith (2003) studied the reaction over the temperature range 198–298 K and found a slight increase in rate as the temperature was decreased (see figure VI-D-5). A least-squares linear analysis of the Olkhov and Smith (2003) data gives $k(CF_3CF_2O_2+NO) = 1.08 \times 10^{-11} \exp[(64 \pm 31)/T]$ cm^3 molecule^{-1} s^{-1}. At 298 K this expression gives

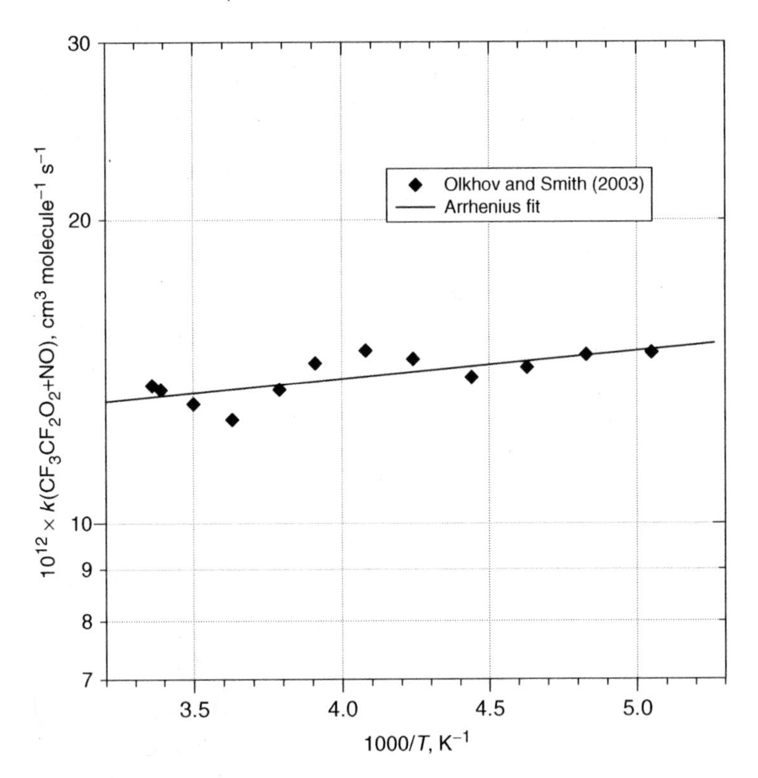

Figure VI-D-5. Arrhenius plot of the rate coefficients for the $CF_3CF_2O_2$ + NO reaction.

$k(CF_3CF_2O_2+NO) = 1.34 \times 10^{-11}$ cm^3 molecule^{-1} s^{-1} with an uncertainty estimated to be ± 25%.

$$CF_3CF_2O_2 + NO \rightarrow CF_3CF_2O + NO_2$$

VI-D-2.7. CCl_3O_2 + NO

The kinetics of this reaction have been studied by Ryan and Plumb (1984) using discharge-flow mass spectroscopy at 298K and by Dognon at al. (1985) using laser photolysis mass spectroscopy over the temperature range 228–413 K. The results from the two studies are in good agreement (see table VI-D-1). Using the temperature dependence reported by Dognon et al. (1985) of $k(CCl_3O_2+NO) = (1.7 \pm 0.7) \times 10^{-11} \times (T/298)^{-(1.0 \pm 0.2)}$ and constraining the 298 K value to be the average of those reported by Ryan and Plumb (1984) and Dognon et al. (1985), we derive the recommendation of $k(CCl_3O_2+NO) = 1.78 \times 10^{-11}$ $(T/298)^{-1.0}$ cm^3 molecule^{-1} s^{-1}. No evidence for reaction channels other than that leading to the alkoxy radical and NO_2 has been reported:

$$CCl_3O_2 + NO \rightarrow CCl_3O + NO_2$$

VI-D-2.8. $CFCl_2O_2$ + NO

The kinetics of the reaction of $CFCl_2O_2$ radicals with NO have been studied by Lesclaux and Caralp (1984) and Dognon et al. (1985), using pulsed laser photolysis of $CFCl_3$ in the presence of O_2 and NO combined with time-resolved mass spectroscopy. Dognon et al. attribute the approximately 10% difference in room-temperature rate coefficients reported in the two studies to an underestimation of the NO partial pressure in their first investigation. Dognon et al. (1985) studied the effect of temperature over the range 228–413 K and reported $k(CFCl_2O_2+NO) = (1.45 \pm 0.20) \times 10^{-11} (T/298)^{-(1.3 \pm 0.2)}$ cm^3 $molecule^{-1}$ s^{-1}, which we adopt as our recommendation. No evidence for reaction channels other than that leading to the alkoxy radical and NO_2 has been reported:

$$CFCl_2O_2 + NO \rightarrow CFCl_2O + NO_2$$

VI-D-2.9. $FC(O)O_2$ and $CF_3C(O)O_2$ + NO

These reactions have been studied by Wallington et al. (1994a, 1995b) and Maricq et al. (1996), using two different techniques. At 298 K the lower limit of $k(CF_3C(O)O_2+NO) > 9.9 \times 10^{-12}$ from Wallington et al. (1995b) is consistent with the measurement of $k(CF_3C(O)O_2+NO) = (2.8 \pm 0.5) \times 10^{-11}$ cm^3 $molecule^{-1}$ s^{-1} by Maricq et al. (1996). The Arrhenius expression $k(CF_3C(O)O_2+NO) = 4.0 \times 10^{-12} \exp(563/T)$ cm^3 $molecule^{-1}$ s^{-1} of Maricq et al. (1996) is recommended. At 298 K this expression gives $k = 2.6 \times 10^{-11}$ cm^3 $molecule^{-1}$ s^{-1}. For the $FC(O)O_2$ + NO reaction the only available data are from Wallington et al. (1994a), who reported $k = (2.5 \pm 0.8) \times 10^{-11}$ cm^3 $molecule^{-1}$ s^{-1} at 295 K. The similarity in the reactivities of $FC(O)O_2$ and $CF_3C(O)O_2$ radicals is not surprising in light of the structural similarity of the two molecules. The reactivities of both fluorinated acylperoxy radicals are greater than that of $CH_3C(O)O_2$, which reacts with NO with a rate constant of 2.0×10^{-11} cm^3 $molecule^{-1}$ s^{-1} (Atkinson et al., 2005a).

VI-D-2.10. Other RO_2 + NO Reactions

Kinetic data for the reactions of CH_2FCFHO_2, $CHF_2CF_2O_2$, $CF_2ClCH_2O_2$, $CFCl_2CH_2O_2$, $CF_3CHO_2(\bullet)CF_3$, $CH_3CF_2CH_2O_2$, $CCl_3CH_2O_2$, and CF_3CClHO_2 radicals with NO in 1000 mbar of SF_6 at 295–298 K have been obtained using pulse radiolysis techniques (Møgelberg et al., 1995b,c,d; Platz et al., 1995; Sehested et al., 1993b; Wallington et al., 1994b). Discharge-flow mass spectroscopy techniques have been used to study the reactions of $CF_2ClCH_2O_2$ and $CF_3CCl_2O_2$ radicals with NO in 2 Torr of helium at 290 K (Peeters and Pultau, 1993). Flash photolysis and mass spectrometry techniques were used to study the reactions of CH_3CFClO_2 and $CFCl_2CH_2O_2$ radicals with NO in 8–20 Torr of nitrogen diluent at 263–321 K (Wu and Carr, 1996a). The results are listed in table VI-D-1. The reactions proceed rapidly to form NO_2 and alkoxy radicals as the major products.

VI-D-3. The Reactions of Haloalkylperoxy Radicals with NO_2

Alkylperoxy radicals react with NO_2 to give peroxynitrates,

$$RO_2 + NO_2 + M \rightarrow ROONO_2 + M$$

As combination reactions they are expected to have a positive pressure dependence and negative temperature dependence. While the overall reaction can be written as a simple addition, in reality the situation is more complex, with the initial formation of an excited adduct which can either decompose back to reactants or be deactivated by collision with a third body capable of removing some of the vibrational excitation:

$$RO_2 + NO_2 \leftrightarrow RO_2NO_2^* \quad \text{(a, –a)}$$

$$RO_2NO_2^* + M \rightarrow RO_2NO_2 + M^* \quad \text{(b)}$$

In the high-pressure limit with sufficient diluent gas, M, to deactivate all of the excited adducts, the overall rate of reaction is independent of pressure and the bimolecular rate constant is k_a. In the low-pressure limit where quenching of the excited adduct does not significantly impact the equilibrium established by reactions (a) and (−a), the overall reaction rate will be linearly dependent on total pressure with a termolecular rate constant given by $(k_a k_b)/k_{-a}$. Under atmospheric conditions most addition reactions are not at their high or low-pressure limits but instead are somewhere in between. Troe and co-workers (Troe, 1983; Gilbert et al., 1983) have carried out extensive theoretical studies of addition reactions and have shown that their "effective" bimolecular rate coefficients in the "fall-off" region between high-and low-pressure limiting behavior can be described by the three-parameter expression (1) given in section IV-A.

As is shown in figure VI-D-6, the kinetics of the reactions of haloalkylperoxy radicals with NO_2 are in the fall-off region under atmospheric conditions. Haloalkylperoxy radicals are more massive than alkylperoxy radicals, have a higher density of states over which to distribute the energy associated with formation of the RO_2–NO_2 bond, form a longer lived initial RO_2–NO_2 adduct, and hence for a given third-body pressure have an effective bimolecular rate constant which is closer to the high-pressure limit (compare $CF_xCl_{3-x}O_2$ with CH_3O_2 reactivity in figure VI-D-6). Destriau and Troe (1990) conducted a theoretical study and concluded that the high-pressure limit for the reactions of all peroxy radicals with NO_2 is approximately 7.5×10^{-12} cm^3 molecule^{-1} s^{-1}, independent of temperature. The experimental data discussed in the following sections are consistent with this conclusion within the likely experimental uncertainties. Finally, it should be noted that for many haloalkylperoxy radicals (e.g., CF_3O_2, CF_2ClO_2, $CFCl_2O_2$, and CCl_3O_2) the measured kinetic data span a range of total pressure which is far from the high-pressure limit. Further measurements at higher total pressures are required to better define the high-pressure limiting rate coefficients for these reactions.

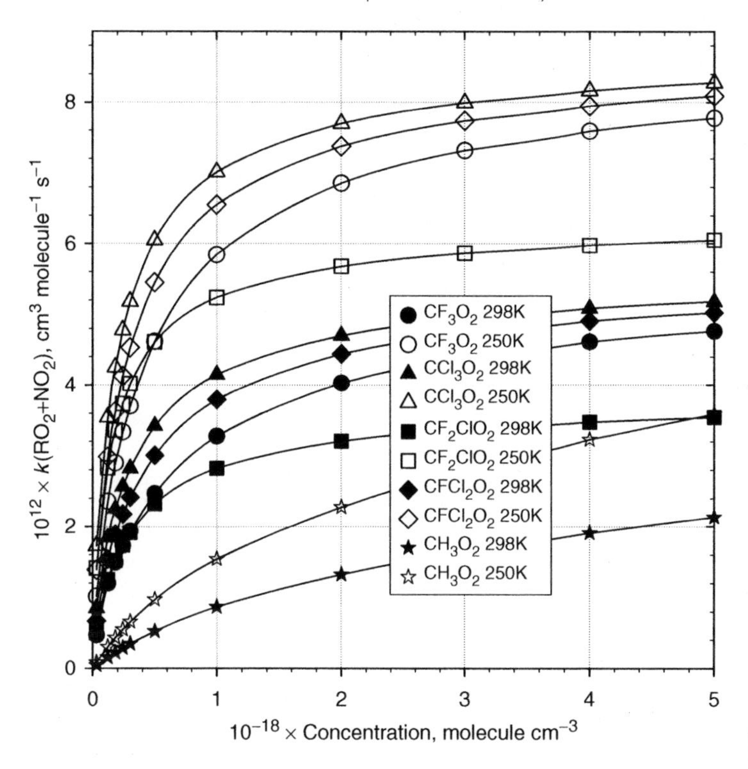

Figure VI-D-6. Kinetic data for reactions of haloalkylperoxy radicals with NO_2 at 298 K (filled symbols) and 250 K (open symbols), calculated using the recommended rate data as a function of total pressure (5×10^{18} molecule cm^{-3} = 154 Torr at 298 K and 130 Torr at 250K).

VI-D-3.1. $CF_3O_2 + NO_2$

The kinetics of the reaction $CF_3O_2 + NO_2$ have been studied by Caralp et al. (1988) using pulsed laser photolysis and time-resolved mass spectroscopy in the pressure range 1–10 Torr over the temperature range 233–373 K. These workers fitted their kinetic data to the Troe formalism with a temperature-dependent F_c value given by $F_c = \exp(-T/416)$. DeMore et al. (1994) have shown that the data of Caralp et al. (1988) can equally well be fitted by the Troe formalization with F_c held constant at $F_c = 0.6$. For the sake of simplicity we recommend use of the expression (1) given in the introduction to section VI-A, with $k_o = (2.2 \pm 0.5) \times 10^{-29} (T/300)^{\{-5 \pm 1\}}$ cm^6 molecule^{-2} s^{-1}, $k_\infty = (6.0 \pm 1.0) \times 10^{-12} (T/300)^{\{-2.5 \pm 1\}}$ cm^3 molecule^{-1} s^{-1}, and $F_c = 0.6$. Use of this expression gives the rate data at 298 and 250 K shown in figure VI-D-6. It can be seen from this figure that: (i) the reaction approaches the high-pressure limit at relatively modest pressures, (ii) because of its much larger k_o value the reaction of CF_3O_2 with NO approaches the high-pressure limit more rapidly than the analogous reaction of CH_3O_2 radicals, and (iii) the rate of reaction increases substantially (by a factor of 1.5–2.0 over the pressure range considered) as the temperature is reduced from 298 to 250 K.

VI-D-3.2. $CCl_3O_2 + NO_2$

The kinetics of the reaction $CCl_3O_2 + NO_2$ have been studied by Caralp et al. (1988) using pulsed laser photolysis and time-resolved mass spectroscopy in the pressure range 1–10 Torr over the temperature range 233–373 K. These workers fitted their kinetic data to the Troe formalism with a temperature-dependent F_c value given by $F_c = \exp(-T/260)$. DeMore et al. (1994) have shown that the data of Caralp et al. (1988) can equally well be fitted by the Troe formalization with F_c held constant at $F_c = 0.6$. For the sake of simplicity and consistency with the bulk of the data reported in the present review, we recommend use of the expression given by DeMore et al. (1994), relation (1) given in the introduction to section IV-A, with $k_o = (5.0 \pm 1.0) \times 10^{-29}\ (T/300)^{\{-5\pm 1\}}$ cm^6 molecule^{-2} s^{-1}, $k_\infty = (6.0 \pm 1.0) \times 10^{-12}\ (T/300)^{\{-2.5\pm 1\}}$ cm^3 molecule^{-1} s^{-1}, and $F_c = 0.6$. As is shown in figure VI-D-6, the kinetics of the reaction of CCl_3O_2 radicals with NO_2 are similar to those of the reaction of CF_3O_2 with NO_2.

VI-D-3.3. $CF_2ClO_2 + NO_2$

The kinetics of the reaction $CF_2ClO_2 + NO_2$ have been studied by Moore and Carr (1990) and Wu and Carr (1991) using flash photolysis coupled with time-resolved mass spectrometry over the pressure range 1–10 Torr in CF_2ClBr bath gas at 248–324 K. Reflecting improvements to the experimental apparatus, Wu and Carr (1991) prefer their high pressure (9 and 10 Torr) data to those from Moore and Carr (1990). At pressures of 8 Torr and below there is good agreement between the two studies.

Sander et al. (2003) have shown that the data from Wu and Carr (1991) can be represented well using the Troe expression (1) given in the introduction to section IV-A, with $k_o = (3.3 \pm 0.7) \times 10^{-29}\ (T/300)^{\{-6.7\pm 1.3\}}$ cm^6 molecule^{-2} s^{-1}, $k_\infty = (4.1 \pm 1.9) \times 10^{-12}\ (T/300)^{\{-2.8\pm 0.7\}}$ cm^3 molecule^{-1} s^{-1}, and $F_c = 0.6$. While the recommended value of the high-pressure limiting rate constant for the $CF_2ClO_2 + NO_2$ reaction is about 30% lower than those for other $CF_xCl_{3-x}O_2 + NO_2$ reactions, it is consistent within the quoted uncertainties. We recommend use of equation (1) given in the introduction to section IV-A (DeMore et al., 1994) with $k_o = (3.5 \pm 0.5) \times 10^{-29}\ (T/300)^{\{-5\pm 1\}}$ cm^6 molecule^{-2} s^{-1}, $k_\infty = (6.0 \pm 1.0) \times 10^{-12}$ $(T/300)^{\{-2.5\pm 1\}}$ cm^3 molecule^{-1} s^{-1}, and $F_c = 0.6$.

VI-D-3.4. $CFCl_2O_2 + NO_2$

The kinetics of the reaction of $CFCl_2O_2$ with NO_2 have been studied by Lesclaux and Caralp (1984) and Caralp et al. (1988) using laser flash photolysis of $CFCl_3$ in the presence of O_2 and NO_2. Rate constant data were measured over the pressure range 1–12 Torr of oxygen at 233, 298, and 373 K. At 298 K the results from both studies by Caralp and coworkers are in agreement. At the pressures investigated, this reaction is in the fall-off region between second- and third-order kinetic behavior. Caralp et al. (1988) fitted their kinetic data to the Troe formalism with a temperature dependent F_c value given by $F_c = \exp(-T/342)$. DeMore et al. (1994) have shown that the data of Caralp et al. (1988) can equally well be fitted by the Troe formalization with F_c held constant at $F_c = 0.6$. For the sake of simplicity we recommend use of the expression given by DeMore et al. (1994); use equation (1) given in the introduction to section IV-A, with $k_o = (3.5 \pm 0.5) \times 10^{-29} (T/300)^{\{-5\pm1\}}$ cm^6 molecule^{-2} s^{-1}, $k_\infty = (6.0 \pm 1.0) \times 10^{-12} (T/300)^{\{-2.5\pm1\}}$ cm^3 molecule^{-1} s^{-1}, and $F_c = 0.6$. At 760 Torr and 300 K this expression gives $k(CFCl_2O_2 + NO_2) = 5.4 \times 10^{-12}$ cm^3 molecule^{-1} s^{-1}. As might be expected from the molecular structures, the reactivity of $CFCl_2O_2$ radicals lies between those for CF_3O_2 and CCl_3O_2 radicals (see figure VI-D-6).

VI-D-3.5. Other RO_2 + NO_2 Reactions

The kinetics of the reactions of CF_3CFHO_2, $CF_3CH_2O_2$, $CH_3CF_2CH_2O_2$, $CCl_3CH_2O_2$, CF_3CClHO_2, $CF_3C(O)O_2$, and $FC(O)O_2$ radicals with NO_2 have been studied using pulse radiolysis coupled with time-resolved UV absorption spectroscopy at 750 Torr of SF_6 and at ambient temperature. The results are listed in table VI-D-2. The kinetics of the reaction of $FC(O)O_2$ radicals with NO_2 were studied in 375–750 Torr of SF_6; there was no discernable effect of pressure over the range studied. The reactions of alkylperoxy and haloalkylperoxy radicals with NO_2 are near their high-pressure limits in 750 Torr of N_2, and the third-body efficiency of SF_6 is greater than that of N_2. Hence, rate coefficients measured in 750 Torr of SF_6 are expected to be close to the high-pressure limits. The rate coefficients measured in 750 Torr of SF_6 for reactions of CF_3CFHO_2, $CF_3CH_2O_2$, $CF_3C(O)O_2$, $FC(O)O_2$, $CH_3CF_2CH_2O_2$, $CCl_3CH_2O_2$, and CF_3CClHO_2 are indistinguishable and have an average value of 6.1×10^{-12} cm^3 molecule^{-1} s^{-1}. These results are consistent with the conclusion of Destriau and Troe (1990) that reactions of all peroxy radicals with NO_2 have a high-pressure limit of approximately 7.5×10^{-12} cm^3 molecule^{-1} s^{-1}.

Table VI-D-2. Rate coefficients (k, cm^3 molecule^{-1} s^{-1}) for reactions of haloalkylperoxy radicals with NO_2

RO_2	$10^{12} \times k$	T (K)	Pressure Range (Torr)	Technique[a]	Reference
CF_3O_2	1.34	233	0.8 (O_2)	PLP–MS	Caralp et al. (1988)
	2.03		1.6		
	3.95		4.7		
	4.55		6.3		
	5.15		7.9		
	0.52	298	1		
	0.92		2		
	1.1		3		
	1.47		6		
	1.76		10		
	0.37	373	2.5		
	0.62		7.5		
	0.69		12.5		

(*continued*)

Table VI-D-2. *(continued)*

RO_2	$10^{12} \times k$	T (K)	Pressure Range (Torr)	Technique[a]	Reference
$CF_3CH_2O_2$	5.8 ± 1.1	298	750 (SF_6)	PR–UVabs	Nielsen et al. (1994)
CF_3CFHO_2	5.0 ± 0.5	298	750 (SF_6)	PR–UVabs	Møgelberg et al. (1994b)
$CH_3CF_2CH_2O_2$	6.8 ± 0.5	295	750 (SF_6)	PR–UVabs	Møgelberg et al. (1995c)
CCl_3O_2	3.25	233	1.2 (O_2)	PLP–MS	Caralp et al. (1988)
	3.85		1.6		
	4.87		2.3		
	5.87		3.3		
	1.22	298	2		
	1.58		3		
	2.45		6		
	3.17		10		
	0.91	373	5		
	1.04		2.6		
	1.32		4.3		
$CCl_3CH_2O_2$	6.5 ± 0.4	295	750 (SF_6)	PR–UVabs	Platz et al. (1995)
CF_2ClO_2	0.75 ± 0.13	298	1 (CF_2ClBr)	FP–MS	Moore and Carr (1990)
	0.82 ± 0.33		1.5		
	1.27 ± 0.04		2		
	1.37 ± 0.07		3		
	1.51 ± 0.13		4		
	1.41 ± 0.18		5		
	1.60 ± 0.23		6		
	1.72 ± 0.48		7		
	1.52 ± 0.40		8		
	2.18 ± 0.71		9		
	2.83 ± 0.20		10		
	1.63 ± 0.17	248	1 (CF_2ClBr)	FP–MS	Wu and Carr (1991)
	2.41 ± 0.46		2		
	3.15 ± 0.14		3		
	3.59 ± 0.39		4		
	3.61 ± 0.32		5		
	3.70 ± 0.41		6		
	3.73 ± 0.33		7		
	3.81 ± 0.33		8		
	4.01 ± 0.26		9		
	4.13 ± 0.18		10		
	0.65 ± 0.12	286	1 (CF_2ClBr)	FP–MS	
	1.17 ± 0.23		2		
	1.42 ± 0.10		3		
	1.64 ± 0.03		4		
	1.64 ± 0.03		5		
	1.73 ± 0.07		6		
	1.71 ± 0.14		7		
	1.93 ± 0.19		8		

Table VI-D-2. *(continued)*

RO_2	$10^{12} \times k$	T (K)	Pressure Range (Torr)	Technique[a]	Reference
	2.05 ± 0.21		9		
	2.13 ± 0.12		10		
	0.61 ± 0.05	298	1 (CF_2ClBr)	FP–MS	
	1.09 ± 0.09		2		
	1.26 ± 0.10		3		
	1.31 ± 0.47		4		
	1.39 ± 0.53		5		
	1.46 ± 0.53		6		
	1.55 ± 0.50		7		
	1.57 ± 0.41		8		
	1.74 ± 0.42		9		
	1.86 ± 0.33		10		
	0.38 ± 0.03	324	1 (CF_2ClBr)	FP–MS	
	0.67 ± 0.07		2		
	0.78 ± 0.07		3		
	0.83 ± 0.08		4		
	0.88 ± 0.05		5		
	0.93 ± 0.18		6		
	1.02 ± 0.16		7		
	1.06 ± 0.17		8		
	1.11 ± 0.17		9		
	1.23 ± 0.07		10		
$CFCl_2O_2$	1.45	298	3.5 (O_2)	PLP–MS	Lesclaux and Caralp (1984)
	1.75	233	0.8 (O_2)	PLP–MS	Caralp et al. (1988)
	2.27		1.4		
	2.74		2.0		
	4.31		3.9		
	5.30		5.9		
	5.90		7.8		
	0.72	298	1 (O_2)	PLP–MS	
	0.92		1.5		
	0.93		2.1		
	1.17		2.5		
	1.52		3.1		
	1.42		3.5		
	1.85		5.0		
	2.1		6.5		
	2.41		10.0		
	0.43	373	3.1 (O_2)		
	0.72		6.3		
	0.84		12.5		
CF_3CClHO_2	6.4 ± 1.5	296	750 (SF_6)	PR–UVabs	Møgelberg et al. (1995b)
$CF_3C(O)O_2$	6.6 ± 1.3	295	750 (SF_6)	PR–UVabs	Wallington et al. (1994f)
$FC(O)O_2$	5.5 ± 0.6	295	750 (SF_6)	PR–UVabs	Wallington et al. (1995b)

[a] Abbreviations of the experimental techniques employed are defined in table II-A-1.

VI-D-4. The Reactions of Haloalkylperoxy with HO_2 Radical

The available kinetic data for the reactions of haloalkylperoxy radicals with HO_2 are listed in table VI-D-3. As is seen from inspection of this table, the reactions of haloalkylperoxy with HO_2 radicals proceed with rate coefficients that are typically in the range $5–10 \times 10^{-12}$ cm^3 $molecule^{-1}$ s^{-1} at 298 K and have rates that increase with decreasing temperature. The negative temperature dependence is indicative of a reaction characterized by a long-range attraction between the reactants, suggesting that the reaction proceeds, at least in part, via the formation of an intermediate adduct which rearranges and decomposes to give the observed products.

Table VI-D-3. Rate coefficients (k, cm^3 $molecule^{-1}$ s^{-1}) for reactions of haloalkylperoxy radicals with HO_2

RO_2	$10^{12} \times k$	T (K)	Technique[a]	Reference
CF_3O_2	< 2	296	FP–UV abs	Hayman and Battin-Leclerc (1995)
	4 ± 2	295	PR–UVabs	Sehested et al. (1997)
CF_3CFHO_2	11 ± 4	210	FP–UVabs	Maricq et al. (1994)
	7.8 ± 4	233		
	7.5 ± 2	243		
	8.9 ± 3.5	253		
	6.8 ± 2.5	273		
	4.7 ± 1.7	295		
	3.0 ± 1.5	313		
	2.8 ± 1.5	323		
	2.0 ± 0.6	363		
	4.0 ± 2.0	296	FP–UVabs	Hayman and Battin-Leclerc (1995)
	5.5 ± 1.5	295	PR–UVabs	Sehested et al. (1997)
$CF_3CF_2O_2$	1.2 ± 0.4	296	FP–UVabs	Hayman and Battin-Leclerc (1995)
CH_2ClO_2	9.3 ± 1.4	255	FP–UVabs	Catoire et al. (1994)
	7.3 ± 0.8	273		
	4.9 ± 0.6	298		
	5.0 ± 0.6	307		
	3.7 ± 0.4	323		
	2.6 ± 0.2	390		
	1.9 ± 0.2	460		
	1.6 ± 0.2	588		
$CHCl_2O_2$	6.54 ± 0.60	286	FP–UVabs	Catoire et al. (1996)
	5.46 ± 0.77	300		
	4.36 ± 0.55	347		
	3.22 ± 0.19	375		
	2.82 ± 0.60	440		
CCl_3O_2	4.9 ± 0.6	300	FP–UVabs	Catoire et al. (1996)
	3.9 ± 0.4	336		
	3.2 ± 0.4	374		
CH_2BrO_2	6.7 ± 3.8	255	FP–UVabs	Villenave and Lesclaux (1995)
CF_2ClO_2	3.4 ± 1.0	296	FP–UVabs	Hayman and Battin-Leclerc (1995)
$CF_2ClCH_2O_2$	6.8 ± 2.0	296	FP–UVabs	Hayman and Battin-Leclerc (1995)
$CFCl_2CH_2O_2$	9.2 ± 2.6	296	FP–UVabs	Hayman and Battin-Leclerc (1995)

[a] Abbreviations of the experimental techniques employed are defined in table II-A-1.

While formation of the hydroperoxide, ROOH, is the dominant reaction channel for alkylperoxy radicals, it is often a minor channel for reaction of haloalkylperoxy radicals with HO_2. Carbonyl compounds have been identified as significant (sometimes dominant) products of the reactions of haloalkylperoxy radicals with HO_2 (Wallington et al., 1994d; Catoire et al., 1996; Wallington et al., 1996c). Wang and coworkers (Hou and Wang, 2005; Hou et al., 2005) have offered a theoretical framework which rationalizes the broad features of the experimentally observed product yield trends. On the basis of their computation studies Wang and coworkers have shown that $RO_2 + HO_2$ reactions proceed on both singlet and triplet surfaces. Figure VI-D-7 shows the mechanism of the $CH_2FO_2 + HO_2$ reaction proposed by Hou and Wang (2005). On the triplet surface the adduct shown on the left-hand side of figure VI-D-7 lies 37.2 kJ mol^{-1} below the reactants and can pass over a barrier that lies slightly (8.3 kJ mol^{-1}) *below* the energy of the reactants to give CH_2FOOH and O_2 products. On the singlet surface the tetroxide adduct shown on the right-hand side of this figure lies 59.1 kJ mol^{-1} below the reactants and can proceed via a 5-membered ring transition state (shown crudely in figure VI-D-7) which lies slightly (9.7 kJ mol^{-1}) *above* the energy of the reactants to give HC(O)F, HO_2, and OH products. Wang and coworkers studied the CH_2FO_2, CH_2ClO_2, $CHCl_2O_2$, and $CCl_3O_2 + HO_2$ reactions and showed that both the depth of the well in which the tetroxide adduct sits, and the barrier for its rearrangement into products via the 5-membered cyclic transition state, are sensitive to F— and Cl— substitutents. The presence of an F— substitutent deepens the well, lowers the activation barrier for decomposition of the tetroxide into products, and opens up a reaction channel for CH_2FO_2 which is essentially unavailable for CH_3O_2 radicals. The experimental evidence for the formation of haloalkoxy radicals as products of the

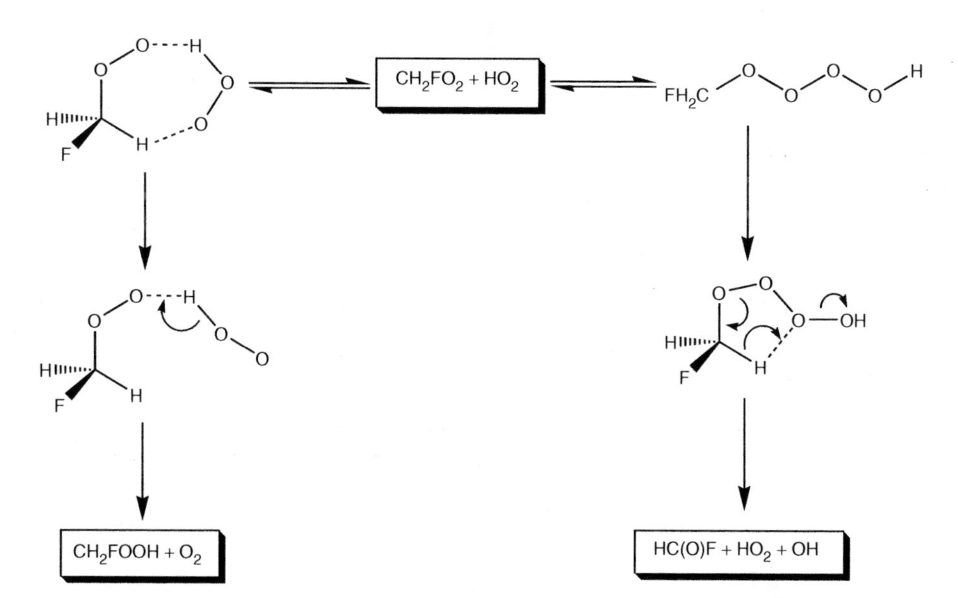

Figure VI-D-7. Mechanism of the $CH_2FO_2 + HO_2$ reaction.

reactions of haloalkylperoxy radicals with HO_2 is rationalized by Wang and coworkers in terms of the formation of chemically activated haloalkylhydroperoxides (ROOH) which, unlike their unsubstituted alkylhydroperoxide counterparts, possess sufficient internal excitation to decompose into RO + OH.

McGivern et al. (2004) have calculated that the OH radical-forming pathways for reactions of $CHBr_2O_2$ and CBr_3O_2 radicals with HO_2 are exothermic by 11.3 and 27.2 kJ mol^{-1}, respectively, and account for 36 and 66% of the overall reaction at 300 K and 760 Torr total pressure. Further work is needed to improve our understanding of the kinetics and mechanisms of the reaction of haloalkylperoxy radicals with HO_2 radicals.

VI-D-4.1. $CH_2FO_2 + HO_2$

Wallington et al. (1994d) used Fourier-transform infrared spectroscopy to identify CH_2FOOH and HC(O)F as products of this reaction. In 700 Torr of air at 295 ± 2 K, branching ratios of 0.38 ± 0.07 for the CH_2FOOH-forming channel and 0.71 ± 0.11 for the HC(O)F channel were reported. As indicated in figure VI-D-7, from their computational study Hou and Wang (2005) ascribed the products reported by Wallington et al. (1994d) to two reaction channels:

$$CH_2FO_2 + HO_2 \rightarrow CH_2FOOH + O_2$$

$$CH_2FO_2 + HO_2 \rightarrow HC(O)F + HO_2 + OH$$

VI-D-4.2. $CF_3O_2 + HO_2$

As part of a study of the reaction of CF_3CFHO_2 with HO_2 radicals, Hayman and Battin-Leclerc (1995) found it necessary to include the $CF_3O_2 + HO_2$ reaction with a rate constant of $< 2 \times 10^{-12}$ cm^3 molecule^{-1} s^{-1} at 296 K. In contrast, in a direct study of the $CF_3O_2 + HO_2$ reaction using pulse radiolysis techniques, Sehested et al. (1997) report a rate coefficient of $(4 \pm 2) \times 10^{-12}$ cm^3 molecule^{-1} s^{-1} at 295 K. The study by Sehested et al. (1997) is more direct, less prone to systematic error, and is preferred.

VI-D-4.3. $CF_3CHFO_2 + HO_2$

The kinetics and mechanism of the $CF_3CHFO_2 + HO_2$ reaction have been studied by Maricq et al. (1994), Hayman and Battin-Leclerc (1995), and Sehested et al. (1997). Maricq et al. used laser flash photolysis techniques and measured rate coefficients over the temperature range 210–363 K at 200–220 Torr total pressure. Hayman and Battin-Leclerc used a flash photolysis method and studied the reaction at 296 K. Sehested et al. used pulse radiolysis techniques and studied the kinetics at 295 K. As is seen from table VI-D-3 and figure VI-D-8, the results from all studies are in agreement within the experimental uncertainties. A fit to all the data gives our recommended value of $k(CF_3CHFO_2{+}HO_2) = 2.24 \times 10^{-13} \exp(-850/T)$ cm^3 molecule^{-1} s^{-1}. At 298 K this gives $k(CF_3CHFO_2{+}HO_2) = 3.9 \times 10^{-12}$ cm^3 molecule^{-1} s^{-1}, with an estimated uncertainty of ± 50%.

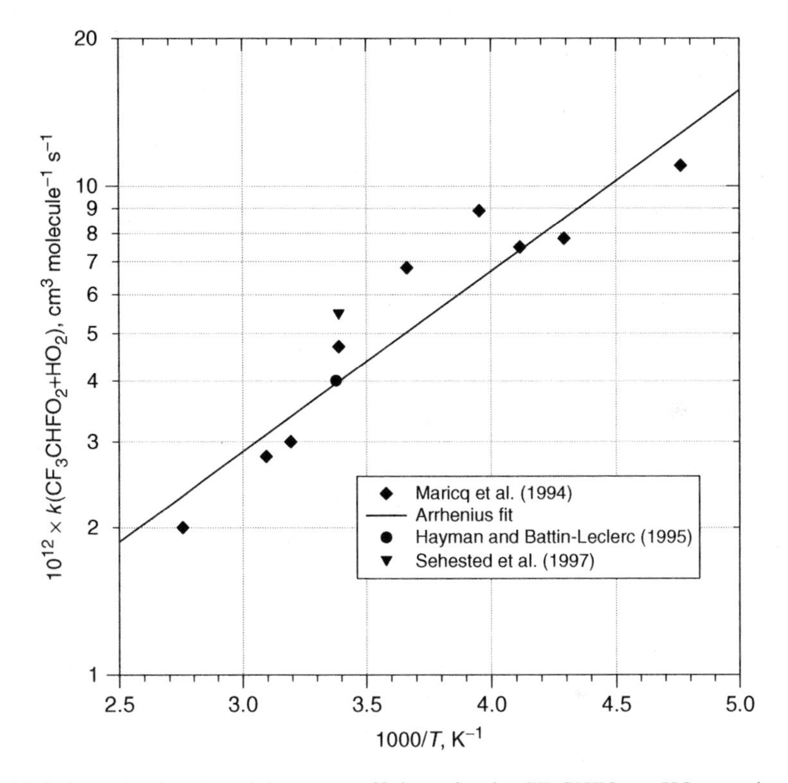

Figure VI-D-8. Arrhenius plot of the rate coefficients for the $CF_3CHFO_2 + HO_2$ reaction.

Maricq et al. (1994) reported that the yield of $CF_3C(O)F$ following the Cl-atom-initiated oxidation of CF_3CH_2F in 700 Torr of air at 295 K was unaffected by the presence of H_2, and hence HO_2 radicals. It was concluded that $< 5\%$ of the reaction of CF_3CHFO_2 with HO_2 radicals gives $CF_3C(O)F$. By implication, $> 95\%$ of the reaction gives the hydroperoxide, $CF_3CFHOOH$. Given the behavior of other haloalkylperoxy radicals in which reaction with HO_2 leads to the formation of substantial yields of carbonyl compounds, it is surprising that analogous behavior is not observed for CF_3CHFO_2 radicals. Further work is needed to better understand this system.

$$CF_3CHFO_2 + HO_2 \rightarrow CF_3CHFOOH + O_2$$

VI-D-4.4. $CH_2ClO_2 + HO_2$

Wallington et al. (1996a) used Fourier-transform infrared spectroscopy to identify CH_2ClOOH and $HC(O)Cl$ as products of this reaction. In 700 Torr of air at 295 ± 2 K, branching ratios of 0.27 ± 0.05 for the CH_2ClOOH-forming channel and 0.73 ± 0.12 for the $HC(O)Cl$ channel were reported.

Catoire et al. (1994) used a flash photolysis technique to determine rate coefficients over the range 255–588 K at atmospheric pressure. As seen from figure VI-D-9, the highest temperature data point from the study by Catoire et al. lies significantly above the value expected from an extrapolation of the lower temperature data. A linear least-squares fit through the lower temperature data gives $k(CH_2ClO_2+HO_2) = 2.49 \times 10^{-13} \exp(908/T)\,cm^3\,molecule^{-1}\,s^{-1}$. At 298 K this expression gives $k(CH_2ClO_2+HO_2) = 5.24 \times 10^{-12}\,cm^3\,molecule^{-1}\,s^{-1}$ with an uncertainty estimated to be ± 30%.

As part of a flash photolysis mass spectrometry kinetic study of reactions of chloromethoxy radicals, Wu and Carr (1999) modeled the formation of CH_2ClO radicals following the formation of CH_2ClO_2 radicals in N_2/O_2 diluent. Wu and Carr used the kinetic data of Catoire et al. (1994). It was found that satisfactory fits to the observed CH_2ClO profiles could be achieved only if it was assumed that the $CH_2ClO_2 + HO_2$ reaction gave approximately a 50% yield of CH_2ClOOH and a 50% yield of CH_2ClO radicals. Wu and Carr (1999) quote a CH_2ClOOH yield of 50 ± 23%, which is consistent with the result of Wallington et al. (1996c) within the large uncertainties.

Hou et al. (2005) conducted a computational study and determined that under atmospheric conditions there are two major product channels: formation of CH_2ClOOH

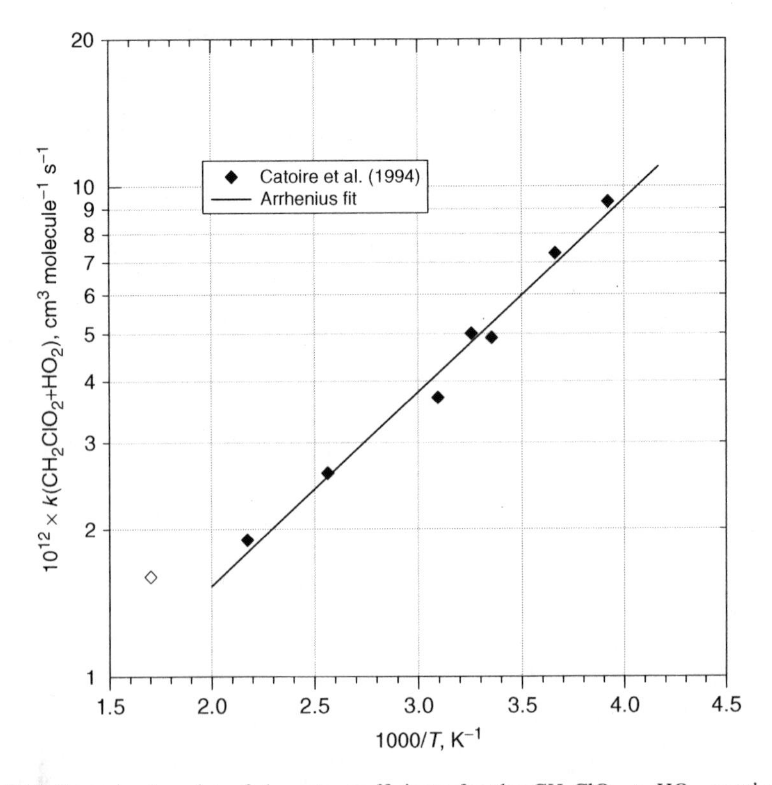

Figure VI-D-9. Arrhenius plot of the rate coefficients for the $CH_2ClO_2 + HO_2$ reaction. The data point represented by the open symbol was not used in determining the least-squares fit of the data.

and O_2 on the triplet surface and formation of HC(O)Cl, HO_2, and OH on the singlet surface. Hou et al. rationalized the experimental observations of CH_2ClO radical formation in terms of the formation of a chemically activated CH_2ClOOH molecule which decomposes to give CH_2ClO and OH.

$$CH_2ClO_2 + HO_2 \rightarrow CH_2ClOOH^* + O_2$$

$$CH_2ClO_2 + HO_2 \rightarrow HC(O)Cl + HO_2 + OH$$

$$CH_2ClOOH^* + M \rightarrow CH_2ClOOH + M^*$$

$$CH_2ClOOH^* \rightarrow CH_2ClO + OH$$

VI-D-4.5. $CHCl_2O_2 + HO_2$

Catoire et al. (1996) have studied the kinetics and products of the $CHCl_2O_2 + HO_2$ reaction. Kinetic data measured over the temperature range 286–440 K at atmospheric pressure are shown in figure VI-D-10. A linear least-squares fit gives $k(CHCl_2O_2 + HO_2) = 5.45 \times 10^{-13} \exp(700/T)$ cm^3 molecule^{-1}. At 298 K this expression gives $k(CHCl_2O_2 + HO_2) = 5.71 \times 10^{-12}$ cm^3 molecule^{-1} s^{-1} with an uncertainty estimated to be ± 30%. Two products were identified in experiments performed in 700 Torr of air at 296 K: HC(O)Cl and CCl_2O in yields of 63 ± 11% and 26 ± 6%, respectively.

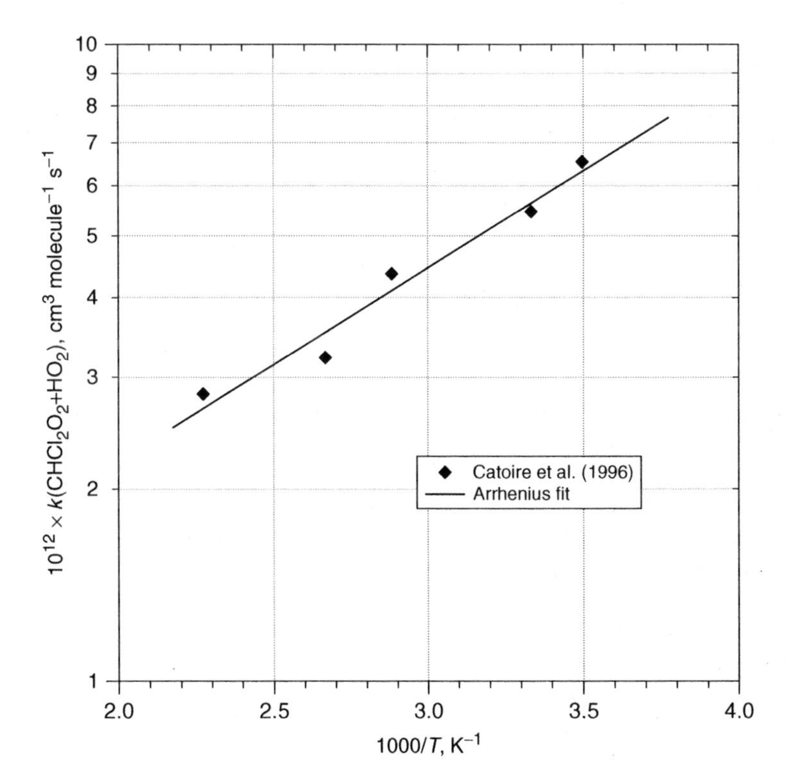

Figure VI-D-10. Arrhenius plot of the rate coefficients for the $CHCl_2O_2 + HO_2$ reaction.

There was no observable (< 10% yield) formation of CH_2ClOOH. Catoire et al. (1994) did not observe any kinetic evidence for Cl-atom or OH-radical formation in the system and concluded that reaction occurred via two pathways giving molecular (as opposed to radical) products.

Hou et al. (2005) conducted a computational study and determined that under atmospheric conditions there are two major product channels: formation of $CHCl_2OOH$ and O_2 on the triplet surface and formation of CCl_2O, HO_2, and OH on the singlet surface. Hou et al. rationalized the experimental observation of HC(O)Cl in terms of the formation of chemically activated $CHCl_2OOH$ which decomposes to give OH radicals and $CHCl_2O$, which eliminates a Cl-atom to give HC(O)Cl:

$$CHCl_2O_2 + HO_2 \rightarrow CHCl_2OOH^* + O_2$$

$$CHCl_2O_2 + HO_2 \rightarrow CCl_2O + HO_2 + OH$$

$$CHCl_2OOH^* \rightarrow CHCl_2O + OH$$

$$CHCl_2O \rightarrow HC(O)Cl + Cl$$

The conclusion by Hou et al (2005) that there is a significant yield of OH radicals is not consistent with the conclusion by Catoire et al. (1996) that there is no significant (<10%) yield of OH radicals. Further work on this reaction is needed.

VI-D-4.6. $CCl_3O_2 + HO_2$

Catoire et al. (1996) have studied the kinetics and products of the $CCl_3O_2 + HO_2$ reaction. Kinetic data measured over the temperature range 300–374 K at atmospheric pressure are shown in figure VI-D-11. A linear least-squares fit gives $k(CCl_3O_2 + HO_2) = 5.70 \times 10^{-13} \exp(646/T)\, cm^3\, molecule^{-1}\, s^{-1}$. At 298 K this expression gives $k(CCl_3O_2 + HO_2) = 4.98 \times 10^{-12}\, cm^3\, molecule^{-1}\, s^{-1}$ with an uncertainty estimated to be ±30%. CCl_2O was the sole product observed following the $CCl_3O_2 + HO_2$ reaction in 700 Torr of air at 296 K. The yield of CCl_2O was indistinguishable from 100%.

Catoire et al. (1994) did not observe any kinetic evidence for Cl-atom or OH-radical formation in the system and concluded that reaction occurred to give molecular (as opposed to radical) products. In marked contrast, Hou et al. (2005) concluded from a computational study that the reaction proceeds essentially exclusively to give chemically activated CCl_3OOH which then decomposes to give CCl_3O and OH radicals. The conclusion by Hou et al. (2005) that there is a significant yield of OH radicals is not consistent with the conclusion by Catoire et al. (1996) that there is no significant (<10%) yield of OH radicals. Further work on this reaction is needed.

VI-D-4.7. $CH_2BrO_2 + HO_2$

Villenave and Lesclaux (1995) used a flash photolysis technique and reported $k(CH_2BrO_2 + HO_2) = (6.7 \pm 3.8) \times 10^{-12}\, cm^3\, molecule^{-1}\, s^{-1}$ at ambient temperature.

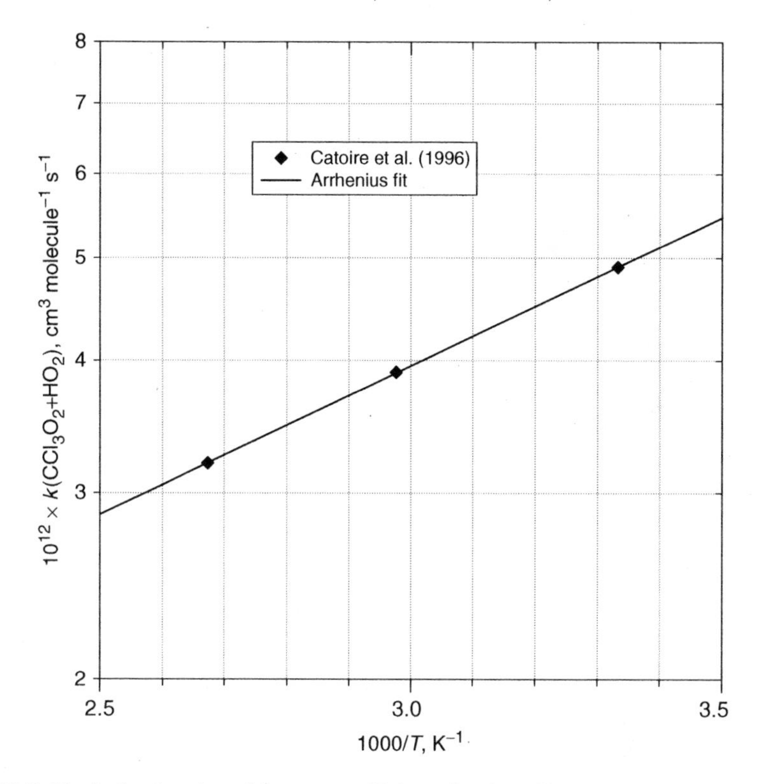

Figure VI-D-11. Arrhenius plot of the rate coefficients for the $CCl_3O_2 + HO_2$ reaction.

The formation of an unidentified product which absorbs strongly at 250–280 nm was reported. None of the expected products (CH_2BrOOH, HC(O)Br, or HOBr) are believed to absorb sufficiently in this wavelength region to account for the residual absorption, and no recommendation is made for this reaction. Chen et al. (1995) report the formation of CH_2BrOOH in a yield $> 85\%$ in 700 Torr of air at 297 K. Mc Givern et al. (2004) conducted a computational study and report that the channel giving $CH_2BrO + O_2 + OH$ is endothermic by 5.0 kJ mol^{-1} (1.2 kcal mol^{-1}) and that this channel accounts for 6.2% of the reaction:

$$CH_2BrO_2 + HO_2 \rightarrow CH_2BrOOH + O_2$$

$$CH_2BrO_2 + HO_2 \rightarrow CH_2BrO + O_2 + OH$$

VI-D-4.8. CF_2ClO_2, $CF_2ClCH_2O_2$, $CFCl_2CH_2O_2$, and $C_2F_5O_2 + HO_2$

Hayman and Battin-Leclerc (1995) used a laser flash photolysis technique to study the rates of these reactions at 298 K. It was found that $CF_2ClCH_2O_2$ and $CFCl_2CH_2O_2$ have reactivities towards HO_2 radicals which are indistinguishable from that of $C_2H_5O_2$. In contrast, the reactivity of CF_2ClO_2 and $C_2F_5O_2$ radicals is substantially less than those of the non-halogenated analogues, CH_3O_2 and $C_2H_5O_2$. The rate coefficient $k(C_2F_5O_2+HO_2) = (1.2 \pm 0.4) \times 10^{-12}$ cm^3 molecule^{-1} s^{-1} is the lowest measured for this class of reactions. Electron-withdrawing groups on the carbon atom which bears the O—O(•) group decrease the reactivity of the RO_2 radical towards HO_2 radicals.

VI-D-5. The Reactions of Haloalkylperoxy Radicals with CH_3O_2

Information is available for the reactions of two haloalkylperoxy radicals, CH_2ClO_2 and CCl_3O_2, with CH_3O_2 radicals. The kinetic data are given in table VI-D-4. The $CH_2ClO_2 + CH_3O_2$ reaction can proceed via three channels; unfortunately no information is available concerning their relative importance:

$$CH_2ClO_2 + CH_3O_2 \rightarrow CH_2ClO + CH_3O + O_2$$

$$CH_2ClO_2 + CH_3O_2 \rightarrow CH_2ClOH + CH_2O + O_2$$

$$CH_2ClO_2 + CH_3O_2 \rightarrow HC(O)Cl + CH_3OH + O_2$$

The $CCl_3O_2 + CH_3O_2$ reaction can proceed via two channels:

$$CCl_3O_2 + CH_3O_2 \rightarrow CCl_3O + CH_3O + O_2 \quad \text{(a)}$$

$$CCl_3O_2 + CH_3O_2 \rightarrow CCl_3OH + CH_2O + O_2 \quad \text{(b)}$$

Catoire et al. (1996) determined $k_a/(k_a + k_b) = 0.5 \pm 0.2$ at room temperature in one atmosphere of N_2/O_2 diluent.

Madronich and Calvert (1990) have proposed, and Villenave and Lesclaux (1996) have presented experimental evidence supporting the notion that the rates of peroxy radical cross reactions can be estimated from the geometric average of the like RO_2—RO_2 reaction rates. Hence, the rate of the cross reaction between CF_3O_2 and CH_3O_2 radicals would be estimated as:

$$k(CF_3O_2 + CH_3O_2) = 2\sqrt{k(CF_3O_2 + CF_3O_2) \times k(CH_3O_2 + CH_3O_2)}$$

Adopting this approach to estimate rate coefficients for reactions of CH_2ClO_2 and CCl_3O_2 with CH_3O_2, Villenave and Lesclaux (1996) calculate $k(CH_2ClO_2+CH_3O_2) = 2.3 \times 10^{-12}$ and $k(CCl_3O_2+CH_3O_2) = 2.4 \times 10^{-12}$ cm^3 molecule^{-1} s^{-1}. Comparison with the experimental measurements given in table VI-D-3 show that the estimated value for $k(CH_2ClO_2+ CH_3O_2)$ is very close to the measured value, while for $k(CCl_3O_2 + CH_3O_2)$ the estimated value is approximately a factor of three lower than that measured. Clearly, further work is needed to refine our understanding of the kinetics and mechanism of the reactions of haloalkylperoxy radicals with CH_3O_2.

Table VI-D-4. Rate Coefficients (k, cm^3 $molecule^{-1}$ s^{-1}) for reactions of haloalkylperoxy radicals with CH_3O_2

RO_2	$10^{12} \times k$	T (K)	Pressure Range (Torr)	Technique[a]	Reference
CH_2ClO_2	2.5 ± 0.5	298	760 (N_2)	PLP–UV abs	Villenave and Lesclaux (1996)
CCl_3O_2	6.6 ± 1.7	298	760 (N_2)	FP–UV abs	Catoire et al. (1996)

[a] Abbreviations of the experimental techniques employed are defined in table II-A-1.

Finally, it is worth noting that Wallington et al. (2006) have argued that the reaction of $C_8F_{17}O_2$ with CH_3O_2 radicals initiates a sequence of reactions (given below) to give $C_7F_{15}C(O)OH$ (perfluorooctanoic acid, PFOA):

$$C_8F_{17}O_2 + CH_3O_2 \rightarrow C_8F_{17}OH + CH_2O + O_2$$

$$C_8F_{17}OH \rightarrow C_7F_{15}C(O)F + HF$$

$$C_7F_{15}C(O)F + H_2O \rightarrow C_7F_{15}C(O)OH + HF$$

Wallington et al. (2006) argue that analogous reactions may explain the perfluorocarboxylic acid pollution observed in remote locations. Hence, reactions of haloalkylperoxy radicals with CH_3O_2 could be of significant practical importance. There are very few kinetic or mechanistic data to assess this possibility. Further work is needed in this area.

VI-E. Atmospheric Chemistry of the Haloalkoxy Radicals

VI-E-1. Introduction

Haloalkoxy radicals are generated during the OH-initiated atmospheric oxidation of all haloalkanes. The fate of haloalkoxy radicals largely determines the first-generation atmospheric oxidation products of the haloalkanes. In the present section, we build upon the discussion presented in chapter IV of the chemistry of unsubstituted alkoxy radicals and explore the atmospheric chemistry of haloalkoxy radicals. As with unsubstituted alkoxy radicals, reaction with O_2, unimolecular decomposition via C—C bond cleavage, and isomerization are possible atmospheric fates of haloalkoxy radicals. In addition, haloalkoxy radicals can undergo unimolecular decomposition via elimination of Cl, Br, or HCl. In several important cases (e.g., CF_3CFHO, CH_2ClO), the barriers for decomposition of haloalkoxy radicals are sufficiently small that chemical activation resulting from their formation from reaction of peroxy radicals with NO is sufficient to have a significant influence. Table VI-E-1 lists the dominant atmospheric fates for the haloalkoxy radicals studied to date and related thermodynamic and reaction dynamic data. Loss of haloalkoxy radicals via reaction with O_2 and decomposition are discussed in sections VI-E-2 and VI-E-3, respectively. The effect of chemical activation is addressed in section VI-E-4. The fate of individual haloalkoxy radicals is considered in section VI-E-5.

Table VI-E-1. Dominant reaction pathways for haloalkoxy radicals in 760 Torr of air at 298 K (from Orlando et al., 2003b)

Radical	Dominant Reaction Pathway(s)	Products	Reaction Enthalpy (kJ mol^{-1})	Activation Energy (kJ mol^{-1})	Approximate Rate (s^{-1})	References
CH_2FO	Reaction with O_2	$HC(O)F + HO_2$				Wallington et al. (1992b) Tuazon and Atkinson (1993a)
CHF_2O	Reaction with O_2	$CF_2O + HO_2$				Nielsen et al. (1992a)
CH_2ClO	Reaction with O_2 Decomposition	$HC(O)Cl + HO_2$ $HCO + HCl$	 −43.9			Sanhueza and Heicklen (1975a) Niki et al. (1980) Wallington et al. (1995a) Bilde et al. (1999)
$CHCl_2O$	Decomposition	$HC(O)Cl + Cl$	−43.9	9.2		Sanhueza and Heicklen (1975a) Niki et al. (1980) Catoire et al, (1996) Hou et al. (1999)
CCl_3O	Decomposition	$CCl_2O + Cl$	−70.3	5.4	$> 1 \times 10^5$	Lesclaux et al. (1987) Li and Francisco (1989)
CH_2BrO	Decomposition	$CH_2O + Br$	−25	< 29	$> 1 \times 10^7$	Chen et al. (1995) Orlando et al. (1996a)
$CHBr_2O$	Decomposition	$HC(O)Br + Br$	−75	< 29	$> 4 \times 10^6$	Orlando et al. (1996b)
CBr_3O	Decomposition	$CBr_2O + Br$	−90.0	0		Kamboures et al. (2002) McGivern et al. (2002)
$CHFClO$	Decomposition	$HC(O)F + Cl$	−48.1	6.3	$> 3 \times 10^3$	Tuazon and Atkinson (1993a) Bhatnagar and Carr (1996) Wang et al. (1999)
CF_2ClO	Decomposition	$CF_2O + Cl$			$> 3 \times 10^4$	Carr et al. (1986)
$CFCl_2O$	Decomposition	$FC(O)Cl + Cl$		< 46		Lesclaux et al. (1987)
$CHBrClO$	Decomposition	$HC(O)Cl + Br$			$> 2 \times 10^6$	Bilde et al. (1997)

CBr_2ClO	Decomposition	$BrC(O)Cl + Br$				Bilde et al. (1998)
CCl_2BrO	Decomposition	$CCl_2O + Br$ $BrC(O)Cl + Cl$				Bilde et al. (1998)
CH_3CHClO	Decomposition	$CH_3CO + HCl$	–51.5	31		Shi et al. (1993) Maricq et al. (1993) Kaiser and Wallington (1995) Orlando and Tyndall (2002a)
CH_2ClCH_2O	Reaction with O_2	$ClCH_2CHO + HO_2$				Wallington et al. (1990) Yarwood et al. (1992) Orlando et al. (1998)
$CH_2ClCHClO$	Decomposition Reaction with O_2	$ClCH_2CO + HCl$ $ClCH_2COCl + HO_2$			$> 4 \times 10^5$	Wallington et al. (1996a)
CCl_3CCl_2O	Decomposition	$CCl_3 + CCl_2O$ $CCl_3C(O)Cl + Cl$				Møgelberg et al. (1995e) Thüner et al. (1999)
CCl_3CH_2O	Reaction with O_2	$CCl_3CHO + HO_2$	25	59	2×10^3	Platz et al. (1995)
	Decomposition	$CCl_3 + CH_2O$ (minor)				
CH_3CHBrO	Decomposition	$CH_3CHO + Br$				Orlando and Tyndall (2002a)
CF_3CHClO	Decomposition	$CF_3CO + HCl$ $CF_3 + HC(O)Cl$ $CF_3CHO + Cl$				Møgelberg et al. (1995b)
	Reaction with O_2	$CF_3C(O)Cl + HO_2$				
$CF_3CH(O)CF_3$	Reaction with O_2	$CF_3C(O)CF_3 + HO_2$				Møgelberg et al. (1995d)
$CH_3CF_2CH_2O$	Reaction with O_2	$CH_3CF_2CHO + HO_2$				Møgelberg et al. (1995c)
$CF_3CH_2CF_2O$	Decomposition	$CF_3CH_2 + COF_2$				Chen et al. (1997)
CH_2FCHFO	Decomposition	$CH_2F + HCOF$	–7.5	34.7	$> 6 \times 10^4$	Wallington et al. (1994b) Hou et al. (2000a)
$CFCl_2CH_2O$	Reaction with O_2	$CFCl_2CHO + HO_2$				Tuazon and Atkinson (1994)

(*continued*)

Table VI-E-1. *(continued)*

Radical	Dominant Reaction Pathway(s)	Products	Reaction Enthalpy (kJ mol^{-1})	Activation Energy (kJ mol^{-1})	Approximate Rate (s^{-1})	References
CF_3CH_2O	Reaction with O_2	$CF_3CHO + HO_2$				Nielsen et al. (1994) Somnitz and Zellner (2001)
CF_3CF_2O	Decomposition	$CF_3 + COF_2$	–38	33	5.3×10^6	Sehested et al. (1993a) Somnitz and Zellner (2001)
CF_3CHFO	Decomposition Reaction with O_2	$CF_3 + HCOF$ $CF_3C(O)F + HO_2$			1.0×10^5	Edney and Driscoll (1992) Tuazon and Atkinson (1993a) Wallington et al. (1992b, 1996a) Somnitz and Zellner (2001)
CF_3CCl_2O	Decomposition	$CF_3C(O)Cl + Cl$				Tuazon and Atkinson (1993a) Hayman et al. (1994)
CF_3CFClO	Decomposition	$CF_3C(O)F + Cl$		< 63	$> 1 \times 10^3$	Tuazon and Atkinson (1993a) Edney and Driscoll (1992) Bhatnagar and Carr (1995b)
$CF_3CBrClO$	Decomposition	$CF_3C(O)Cl + Br$				Bilde et al. (1998)
CCl_3CHClO	Decomposition	$CCl_3 + HCOCl$ $CCl_3CO + HCl$				Møgelberg et al. (1996a)
CH_3CHFO	Decomposition Reaction with O_2	$CH_3 + HC(O)F$ $CH_3C(O)F + HO_2$	0	50		Hou et al. (2000a)
CH_3CF_2O	Decomposition	$CH_3 + CF_2O$			2×10^3	Tuazon and Atkinson (1993a) Taketani et al. (2005)
$CF_3CF_2CCl_2O$	Decomposition	$CF_3CF_2C(O)Cl + Cl$				Tuazon and Atkinson (1994)
CF_2ClCF_2CFClO	Decomposition	$CF_2ClCF_2C(O)F + Cl$				Tuazon and Atkinson (1994)

VI-E-2. Reaction with O_2

The majority of the kinetic data concerning haloalkoxy radicals comes from indirect relative rate experimental studies. As discussed in section VI-E-5, the rate of reaction with O_2 has been measured relative to the rate of decomposition for CH_2ClO, CH_3CHClO, $CH_2ClCHClO$, CCl_3CH_2O, CF_3CHClO, CF_3CFHO, $CH_3CHClCH(O)CH_3$, and $CH_3CH_2CH(O)CH_2Cl$ radicals. Rate coefficients have been measured directly for the reactions of O_2 with the radicals, CH_2ClO (Wu and Carr, 1999, 2001); CF_2ClCH_2O (Mörs et al., 1996); and $CFCl_2CH_2O$ (Wu and Carr, 1996b; Mörs et al., 1996). Figure VI-E-1 shows an Arrhenius plot of these data. Wu and Carr (1996b, 1999, 2001) undertook experiments in 5–35 Torr N_2 diluent for CH_2ClO and 10–35 Torr for $CFCl_2CH_2O$. No discernable effect of total pressure for these reactions was observed. The experiments of Mörs et al. (1996) were conducted at 35 Torr total pressure. The lines through the data in figure VI-E-1 are linear least-squares fits to the data from Wu and Carr which give $k(CH_2ClO+O_2) = 2.06 \times 10^{-12} \exp(-940/T)$ and $k(CFCl_2CH_2O+O_2) = 2.53 \times 10^{-15} \exp(-960/T)$ cm^3 molecule^{-1} s^{-1} over the temperature ranges 265–306 and 251–341 K, respectively. These expressions give $k(CH_2ClO+O_2) = 8.6 \times 10^{-14}$ cm^3 molecule^{-1} s^{-1} and $k(CFCl_2CH_2O+O_2) = 9.9 \times 10^{-17}$ cm^3 molecule^{-1} s^{-1} at 296 K.

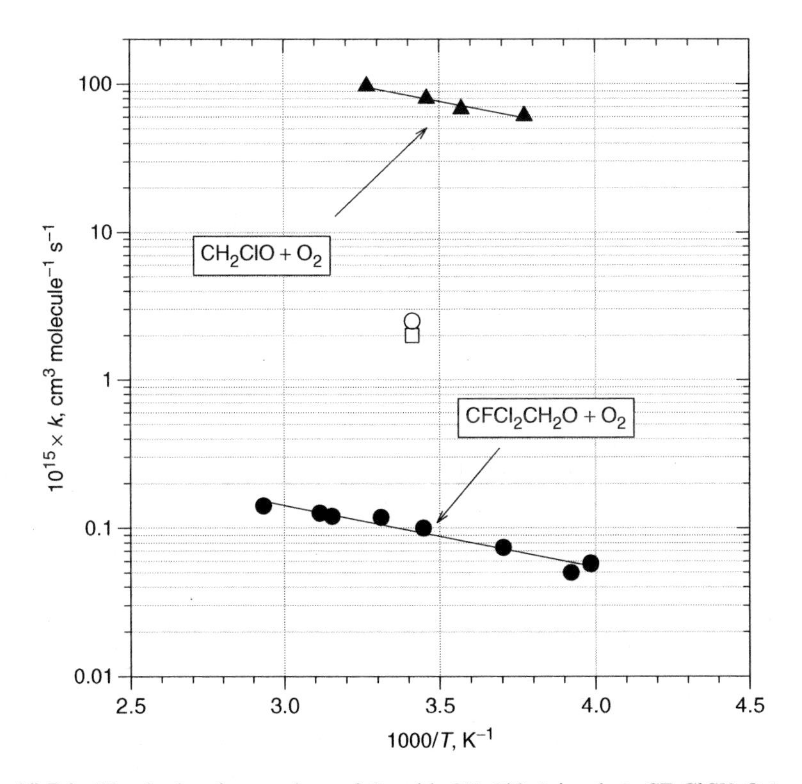

Figure VI-E-1. Kinetic data for reactions of O_2 with CH_2ClO (triangles), CF_2ClCH_2O (square), and $CFCl_2CH_2O$ (circles) radicals. Filled symbols are data from Wu and Carr (1996b, 2001), open symbols are from Mörs et al. (1996).

As part of their study of the reaction with O_2, Wu and Carr (1999) also measured the kinetics of HCl elimination from CH_2ClO radicals. The rate of HCl elimination increased linearly with total pressure over the range 5–35 Torr of N_2 diluent at 289 K, and a rate constant of $(1.4 \pm 0.4) \times 10^{-15}$ cm^3 molecule^{-1} s^{-1} was derived at the low-pressure limit. This value can be combined with the value of $k(CH_2ClO{+}O_2) = (8.0 \pm 0.8) \times 10^{-14}$ cm^3 molecule^{-1} s^{-1} at 289 K, measured in the same study, to give the rate constant ratio $k_{elim}/k_{CH2ClO+O2} = 0.0175 \pm 0.006$. This result is consistent with the rate constant ratio derived at the low-pressure limit by Kaiser and Wallington (1994) of $k_{elim}/k_{CH2ClO+O2} = 0.023$ at 296 K. Given this agreement, we recommend $k_{CH2ClO+O2} = 2.06 \times 10^{-12} \exp(-940/T)$ cm^3 molecule^{-1} s^{-1} from Wu and Carr (2001). As is seen from figure VI-E-1, there is a large discrepancy between the results for the $CFCl_2CH_2O + O_2$ reaction reported by Wu and Carr (1996b) and by Mörs et al. (1996), and no recommendation is made.

The 1000-fold difference in reactivity between CH_2ClO and $CFCl_2CH_2O$ radicals and the low *A*-factor for the reaction of $CFCl_2CH_2O$ radicals with O_2 reported by Wu and Carr (2001) are very striking. As is shown in figure VI-E-1, the available absolute rate kinetic data base for reaction of haloalkoxy radicals with O_2 is very sparse. Further experimental and theoretical studies of the reaction of haloalkoxy radicals with O_2 are required to improve our understanding of these reactions.

VI-E-3. Decomposition Reactions

Haloalkoxy radicals can decompose via three possible pathways: halogen atom elimination, HCl elimination, or C—C bond scission. We first consider the effect of halogen substitution in the α–position. The presence of a halogen in a position alpha to the alkoxy radical site opens up the possibility of cleavage of the carbon–halogen bond; for example, in the case of bromomethoxy,

$$CH_2BrO \rightarrow Br + CH_2O$$

In the case of fluorine, rupture of a C—F bond is thermodynamically unfavorable and does not occur under atmospheric conditions. For example, decomposition of the CH_2FO radical to F and CH_2O is endothermic by 142 kJ mol^{-1}. The presence of F-atoms in the α-position does, however, have a positive effect on the rate of the beta carbon–carbon bond scission process. For example, Somnitz and Zellner (2001) have calculated that the endothermicity and concomitant barriers to carbon–carbon bond cleavage decrease quite dramatically with increased F-atom substitution for the following series of reactions:

$$CF_3CH_2O \rightarrow CF_3 + CH_2O$$

$$CF_3CHFO \rightarrow CF_3 + HC(O)F$$

$$CF_3CF_2O \rightarrow CF_3 + CF_2O$$

Similar trends are evident by comparison of the decomposition of ethoxy ($\Delta H_{rxn} = 54\,kJ\,mol^{-1}$, $E_a = 75\,kJ\,mol^{-1}$, Orlando et al., 2003b) and 1-fluoro-ethoxy radicals ($\Delta H_{rxn} = 0\,kJ\,mol^{-1}$, $E_a = 50\,kJ\,mol^{-1}$, Hou et al., 2000a).

Cleavage of C—X bonds is much more favored in the cases where X = Cl, Br, or I. This can be understood from a simple consideration of the thermodynamics involved, since the enthalpies for rupture of the C—X bond in $X—CH_2O$ radicals are +142, +25, −33, and $-79\,kJ\,mol^{-1}$ for X = F, Cl, Br, and I, respectively. In fact, in all cases where Br atoms are present in the alpha-position, rupture of the C—Br is the dominant atmospheric loss process. Orlando et al. (1996a) and Chen et al. (1995) have shown that CH_2O is formed in $> 90\%$ yield in the Cl-atom-initiated oxidation of methyl bromide:

$$Cl + CH_3Br \rightarrow CH_2Br + HCl$$

$$CH_2Br + O_2 \rightarrow CH_2BrO_2$$

$$CH_2BrO_2 + NO \rightarrow CH_2BrO + NO_2$$

$$CH_2BrO \rightarrow Br + CH_2O$$

Orlando et al. (1996b), Kamboures et al. (2002), and McGivern et al. (2002, 2004) have shown that HC(O)Br and CBr_2O are the end-products of the Cl-atom-initiated oxidation of dibromomethane and bromoform, respectively. The low dissociation barriers to these dissociation processes suggests that chemical activation may play a role when brominated alkoxy radicals are generated from the exothermic reaction of the corresponding peroxy radical with NO, though studies of this possibility have not been conducted to date.

Carbon–chlorine bond cleavages are not as facile as is the case for C—Br, but these processes are still sufficiently rapid to occur in a number of cases, for example in the chemistry of $CFCl_2O$, CF_2ClO, CHFClO, CF_3CCl_2O, and CF_3CFClO radicals. Alkoxy radicals containing a single α-chlorine atom and at least one α-hydrogen atom, such as CH_2ClO and CH_3CHClO, provide something of a special case, as three-centered elimination of HCl has been demonstrated to be a favorable process for these species (Shi et al., 1993; Maricq et al., 1993; Wallington et al., 1995a; Bilde et al., 1999; Orlando and Tyndall, 2002a):

$$CH_3CHClO \rightarrow CH_3CO + HCl$$

$$CH_2ClO \rightarrow HCO + HCl$$

These processes can actually be thought of as C—Cl bond cleavages in their initial stages, with a late occurrence of HCl formation. Elimination of HCl from thermalized CH_3CHClO and CH_2ClO radicals occurs at rates of approximately 10^6 and $10^4\,s^{-1}$, respectively. Chemical activation is important in the atmospheric chemistry of these species.

For alkoxy radicals possessing multiple chlorine atoms in the α-position, C—Cl cleavage dominates the chemistry, as in the case of $CHCl_2O$, $CFCl_2O$, and CF_3CCl_2O

radicals where elimination of a Cl-atom is exothermic (approximately 42 kJ mol^{-1}) and occurs with a rather low barrier (approximately 8–21 kJ mol^{-1}):

$$CHCl_2O \rightarrow HC(O)Cl + Cl$$

The presence of an α-chlorine atom does not guarantee that dissociation via C—Cl cleavage (or via HCl elimination) will be the exclusive atmospheric loss process. For example, CF_3CHClO and CCl_3CCl_2O radicals also undergo decomposition via C—C bond cleavage, and CF_3CHClO and CH_2ClO radicals also undergo reaction with O_2:

$$CCl_3CCl_2O \rightarrow CCl_3 + CCl_2O$$

$$CCl_3CCl_2O \rightarrow CCl_3C(O)Cl + Cl$$

$$CF_3CHClO \rightarrow CF_3 + HC(O)Cl$$

$$CF_3CHClO \rightarrow CF_3CO + HCl$$

$$CF_3CHClO + O_2 \rightarrow CF_3COCl + HO_2$$

$$CH_2ClO \rightarrow HCO + HCl$$

$$CH_2ClO + O_2 \rightarrow HC(O)Cl + HO_2$$

Dissociation is not a very favorable process for the β-chlorinated or β-brominated species. For example, the observation of large yields of chloroacetaldehyde from the Cl-initiated oxidation of ethene (Kleindienst et al., 1989; Wallington et al., 1990; Yarwood et al., 1992; Orlando et al., 1998) indicates that reaction with O_2 is the major atmospheric fate of $ClCH_2CH_2O$ radicals:

$$Cl + CH_2CH_2 \rightarrow CH_2ClCH_2$$

$$CH_2ClCH_2 + O_2 \rightarrow CH_2ClCH_2O_2$$

$$CH_2ClCH_2O_2 + NO \rightarrow CH_2ClCH_2O + NO_2$$

$$CH_2ClCH_2O + O_2 \rightarrow CH_2ClCHO + HO_2$$

Yarwood et al. (1992) have observed similar chemistry following Br-atom addition to ethene. Kukui and Le Bras (2001) have estimated that C—C bond scission occurs with a rate coefficient on the order of about 50 s^{-1} and proceeds through a barrier of about 67 kJ mol^{-1}. Even for larger β-chlorinated or brominated radicals, decomposition via C—C bond cleavage is not very favorable. As discussed by Kukui and Le Bras (2001) and Orlando et al. (2003b), the presence of a chlorine or bromine atom in the beta-position has little effect on the behavior of a given alkoxy radical relative to its non-halogenated counterpart. For example, in one atmosphere of air at 298 K, the rate coefficient ratio for decomposition to reaction with O_2, k_{diss}/k_{O2}, for the 2-butoxy radical (approximately 4×10^{18} molecule cm^{-3}) is similar to the analogous rate coefficient ratio $k_{diss}/k_{O2} = (1.6 \pm 0.4) \times 10^{19}$ molecule cm^{-3}obtained by Orlando et al. (2003a) for the $CH_3CH(Cl)CH(O)CH_3$ radical.

From inspection of table VI-E-1, it is apparent that the substitution of multiple Cl-atoms beta to the alkoxy radical site leads to a substantial decrease in the endothermicity of the C—C bond rupture process, with a more modest lowering of the barrier to decomposition; compare ethoxy radical decomposition, $\Delta H_r = 54\,kJ\,mol^{-1}$, $E_a = 75\,kJ\,mol^{-1}$, with 2, 2, 2-trichloroethoxy, $\Delta H_{rxn} = 25\,kJ\,mol^{-1}$, $E_a = 59$ kJ mol^{-1}. As discussed in section IV-B-3, the activation energy barrier for decomposition of unsubstituted alkoxy radicals via C—C bond scission, E_a, is related to the reaction endothermicity, ΔH_r, and the ionization energy of the leaving group, *IP*, by the formula $E_a = (A*IP-B) + C\,(\Delta H_r)$. For a given leaving group the activation barrier decreases with decreasing reaction endothermicity and decreasing ionization potential of the leaving group (see section IV.B-3). From the data base for unsubstituted alkoxy radicals, one might expect that chlorinated radicals would be very good leaving groups, given the progressive decrease in ionization potential along the series $CH_3 \rightarrow CH_2Cl \rightarrow CHCl_2 \rightarrow CCl_3$. However, as discussed by Orlando et al. (2003b), this is not the case; dissociation reactions involving chloromethyl and 1-chloroethyl radicals as leaving groups follow much the same trend as reactions involving methyl radical as the leaving group, despite the lower ionization potential of the chlorinated radicals.

Less extensive data are available for unimolecular dissociation reactions involving fluorinated methyl radicals as the leaving group. A theoretical study by Somnitz and Zellner (2001) of CF_3CH_2O radical decomposition suggests a very high barrier to this process, 105 kJ mol^{-1}, despite the fact that the endothermicity is not substantially different for ethoxy radical decomposition. In contrast, the available data for decomposition of the series of radicals $CF_{3-n}H_nCHFO$ (see table VI-E-1) indicates a slight decrease in activation energy with increasing fluorine substitution, despite a fairly constant endothermicity. Clearly, a quantitative description of structure–reactivity relationships for these fluorinated radicals will not be possible without further data on the endothermicity and energy barriers to these processes. As a last point, we note that some C—C bond scission reactions involving fluorinated alkoxy radicals are quite exothermic, such as in the case of CF_3CF_2O decomposition (Somnitz and Zellner, 2001), yet it appears that there is a minimum barrier in the neighborhood of 33 kJ mol^{-1} for these processes, regardless of the exothermicity.

VI-E-4. Chemical Activation of Haloalkoxy Radicals

Experimental (Wallington et al., 1996b; Møgelberg et al., 1997b; Bilde et al., 1998; Orlando et al., 1998; Bilde et al., 1999; Thüner et al., 1999; Voicu et al., 2001) and computational studies (Orlando et al., 1998; Schneider et al., 1998; Somnitz and Zellner, 2001) have shown that chemical activation plays an important role in the atmospheric chemistry of certain haloalkoxy radicals. Haloalkoxy radicals are formed in the atmosphere via reaction of peroxy radicals (RO_2) with either NO or other peroxy radicals (e.g., CH_3O_2):

$$RO_2 + NO \rightarrow RO + NO_2$$

$$RO_2 + CH_3O_2 \rightarrow RO + CH_3O + O_2$$

$$\rightarrow \text{non-radical products (carbonyls, alcohols)}$$

The reaction of RO_2 with CH_3O_2 radicals to give RO and CH_3O radicals is close to thermoneutral, and consequently the alkoxy radicals produced have little or no internal excitation. In contrast, formation of RO radicals in the reaction of RO_2 with NO is exothermic, with reaction enthalpies that range from $-46\,\text{kJ mol}^{-1}$ for simple unsubstituted alkyl peroxy radicals (e.g., CH_3O_2, $C_2H_5O_2$) to approximately $-71\,\text{kJ mol}^{-1}$ for certain halogenated alkyl peroxy radicals (e.g., CH_2FO_2, CF_3O_2, CF_3CFHO_2). The reaction of RO_2 with NO proceeds via the formation of a short-lived (10^{-10} s) vibrationally excited peroxy nitrite, ROONO*, which decomposes to give RO + NO_2. Although the lifetime of the vibrationally excited peroxy nitrite, ROONO*, is short (10^{-10}–10^{-9} s), it is nevertheless long enough for efficient intramolecular energy transfer (occurring on a time-scale of the order of 10^{-12} s^{-1}). Thus, energy distributions among the reaction products are essentially statistical and nascent alkoxy radicals formed following decomposition of ROONO* carry a significant fraction of the reaction exothermicity. For many haloalkoxy radicals the barrier to decomposition, or isomerization, is comparable to, or less than, the exothermicity of the RO_2 + NO reaction. Under such conditions it is likely that a significant fraction of the nascent RO radicals will decompose "promptly" (on a time-scale $<10^{-9}$ s). The remaining RO radicals will be thermalized by collisions with the diluent gas and will undergo thermal decomposition, thermal isomerization or bimolecular reaction with O_2. The overall mechanism can be represented as:

$$RO_2 + NO \rightarrow ROONO^*$$

$$ROONO^* \rightarrow RO^* + NO_2$$

$$ROONO^* + M \rightarrow ROONO + M$$

$$ROONO \rightarrow RO + NO_2$$

$$RO^* \rightarrow \text{decomposition or isomerization}$$

$$RO^* + M \rightarrow RO + M$$

$$RO \rightarrow \text{decomposition, isomerization, or reaction with } O_2$$

For chemical activation to play a significant role in the atmospheric chemistry of a given RO radical three conditions must be met: (i) the lifetime of ROONO* must be long enough ($>10^{-12}$ s) to enable efficient intramolecular energy flow but short enough ($<10^{-9}$ s) that collisions with the diluent gas do not remove a significant fraction of the ROONO* excitation; (ii) the RO radical must possess a decomposition, or isomerization, pathway with an activation barrier which is comparable to, or lower than, the excitation of the nascent RO* radical (determined by exothermicity of the reaction, $RO_2 + NO \rightarrow RO + NO_2$, and the relative heat capacities of the RO and NO_2 fragments); and (iii) the overall rate of prompt decomposition of RO* must be sufficiently fast (i.e., the *A*-factor for decomposition must be sufficiently large) that collisions with the diluent gas do not remove a significant fraction of the RO* excitation prior to "prompt" decomposition.

Table VI-E-2 provides a list of alkoxy radicals for which chemical activation effects have been reported. Consistent with the discussion above, in all cases the activation barriers for decomposition are modest ($\leq 42\,kJ\,mol^{-1}$). For some alkoxy radicals (e.g., $CBrCl_2O$ and CCl_3CCl_2O) chemical activation alters the relative importance of competing decomposition pathways. In such cases the experimentally observed change in product distribution provides a lower limit for the importance of chemical activation. For example, while Thüner et al. (1999) have reported that chemical activation of CCl_3CCl_2O radicals increases the relative importance of C—C bond scission at

Table VI-E-2. Haloalkoxy radicals showing chemical activation effects in one atmosphere of air at 296 K, adapted from Orlando et al. (2003b)

RO	Decomposition Pathway	Barrier Height ($kJ\,mol^{-1}$)	Fraction undergoing prompt decomposition[a]	Reference
CH_3CF_2O	$CH_3 + COF_2$		0.30	Taketani et al. (2005)
CF_3CFHO	$CF_3 + HC(O)F$	31–38	$0.64^{+0.12}_{-0.23}$[b]	Wallington et al. (1996b) Schneider et al. (1998) Wu and Carr (2003)
CF_3CF_2CFHO	$CF_3CF_2 + HC(O)F$	ND[i]	$0.67^{+0.19}_{-0.22}$	Møgelberg et al. (1997a)
CH_2ClO	$HCO + HCl$ $Cl + CH_2O$	ND	0.44[c]	Bilde et al. (1999)
CH_3CHClO	$CH_3CO + HCl$	25	~ 0.5	Shi et al. (1993) Hou et al. (2000b) Orlando and Tyndall (2002a)
$CH_2ClCHClO$	$CH_2ClC(O) + HCl$	ND	≥ 0.19[d]	Voicu et al. (2001)
CCl_3CCl_2O	$CCl_3 + COCl_2$	ND	≥ 0.19[e]	Hasson and Smith (1999) Thüner et al. (1999)
$CBrCl_2O$	$Cl + BrC(O)Cl$	ND	≥ 0.21[f]	Bilde et al. (1998)
$CHClBrCCl_2O$	$CHClBr + CCl_2O$	ND	≥ 0.17[g]	Ramacher et al. (2001)
CCl_2BrCCl_2O	$CCl_2Br + CCl_2O$	ND	≥ 0.33[h]	Ramacher et al. (2001)

[a] Fraction of alkoxy radicals formed in $RO_2 + NO$ reaction that undergo prompt decomposition.
[b] Based on increased HC(O)F and COF_2 yields (Wallington et al., 1996b).
[c] Based on combined yield of CH_2O and CO (Bilde et al., 1999).
[d] Based on $CH_2ClC(O)Cl$ yield (Voicu et al., 2001).
[e] Based on decreased yield of $CCl_3C(O)Cl$ (Thüner et al., 1999).
[f] Based on increased yield of BrC(O)Cl (Bilde et al., 1998).
[g] Based on decreased yield of CHClBrC(O)Cl (Ramacher et al., 2001).
[h] Based on decreased yield of $CCl_2BrC(O)Cl$ (Ramacher et al., 2001).
[i] Not determined.

the expense of Cl-atom elimination, it is likely that chemical activation increases the rate of both processes. Hence, estimates of the importance of chemical activation in such systems, derived from observed changes in product distributions, provide lower limits for the importance of prompt decomposition of RO* radicals. In at least one case, that of the CH_2ClO radical, chemical activation allows for access to a dissociation channel not open to the thermalized radicals. While thermalized CH_2ClO radicals decompose via three-centered elimination of HCl (in competition with their reaction with O_2), activated CH_2ClO^* radicals can also undergo Cl-atom elimination (Bilde et al., 1999).

VI-E-5. Atmospheric Fate of Individual Haloalkoxy Radicals

The major atmospheric fates of individual haloalkoxy radicals are discussed in this section. As with the discussion of the kinetics of reaction of OH radicals with haloalkanes in section VI-A, we discuss sequentially alkoxy radicals derived from hydrofluorocarbons, hydrochlorocarbons, hydrobromocarbons, hydroiodocarbons, hydrochlorofluorocarbons, and hydrobromofluorocarbons. Unless otherwise stated, product yields are molar yields (moles of product formed per mole of haloalkane consumed).

In the majority of the experimental studies discussed below, the haloalkoxy radicals were prepared by reacting Cl-atoms with the appropriate haloalkane in the presence of O_2. Photolysis of molecular chlorine provides a convenient laboratory source of Cl-atoms. The reaction of Cl-atoms with haloalkanes generally proceeds via hydrogen abstraction to give alkyl radicals. Under ambient conditions the addition of O_2 to alkyl radicals proceeds with a rate constant of the order of 10^{-12}–10^{-11} cm^3 $molecule^{-1}$ s^{-1}. In one atmosphere of air the O_2 partial pressure is 5.2×10^{18} molecule cm^{-3} and the lifetime of alkyl radicals with respect to addition of O_2 is 20–200 ns. Conversion of haloalkyl radicals into haloalkyl peroxy radicals is an efficient and irreversible process. The self-reaction of peroxy radicals proceeds via two channels: a radical channel giving two alkoxy radicals, and, for alkyl peroxy radicals containing an α-hydrogen, a molecular disproportionation channel giving an alcohol- and carbonyl-containing product. For haloalkyl peroxy radicals the radical channel appears to dominate. Thus, as an example, the atmospheric chemistry of CF_3CFHO radicals can be studied using the UV irradiation of $CF_3CFH_2/Cl_2/O_2/N_2$ mixtures. In this system the following reactions occur:

$$Cl + CF_3CFH_2 \rightarrow HCl + CF_3CFH$$

$$CF_3CFH + O_2 + M \rightarrow CF_3CFHO_2 + M$$

$$CF_3CFHO_2 + CF_3CFHO_2 \rightarrow CF_3CFHO + CF_3CFHO + O_2$$

$$CF_3CFHO_2 + CF_3CFHO_2 \rightarrow CF_3C(O)F + CF_3CFHOH + O_2$$

$$CF_3CFHO + M \rightarrow CF_3 + HC(O)F + M$$

$$CF_3CFHO + O_2 \rightarrow CF_3C(O)F + HO_2$$

CF_3CFHO radicals undergo loss via C—C bond scission and reaction with O_2. The HC(O)F yield provides a marker for C—C bond scission, and the $CF_3C(O)F$ yield provides a marker for reaction with O_2. Monitoring the yields of HC(O)F and $CF_3C(O)F$ as functions of temperature, total pressure, and O_2 partial pressure gives information on the relative importance of C—C bond scission and reaction with O_2 as atmospheric loss mechanisms for CF_3CFHO radicals. In the real atmosphere the concentration of CF_3CFHO_2 radicals will be extremely low, and, unlike in smog chamber experiments, their RO_2–RO_2 reactions will not be important. In the atmosphere CF_3CFHO_2 radicals are converted into CF_3CFHO radicals via reaction with other peroxy radicals (e.g., CH_3O_2) and via reaction with NO. The heats of formation of alkoxy radicals are very similar to those of the corresponding peroxy radicals (ΔH_f (CF_3O_2) $= -146 \pm 4$, ΔH_f (CF_3O) $= -149 \pm 2$, ΔH_f (CH_3O_2) $=$ 2.1 ± 1.2, ΔH_f (CH_3O) $= 17 \pm 4\,\text{kJ mol}^{-1}$; Sander et al., 2003). Consequently, the RO_2-RO_2 reactions of peroxy radicals to give alkoxy radicals are close to thermoneutral and the nascent alkoxy radicals have little or no internal excitation. In contrast, there is a large difference between the heats of formation of NO and NO_2 (ΔH_f (NO) $= 91.29 \pm 0.17$, ΔH_f (NO_2) $= 34.18 \pm 0.42\,\text{kJ mol}^{-1}$; Sander et al., 2003). Nascent alkoxy radicals formed in the reaction of RO_2 with NO are chemically activated (see section VI-E-4). When a dissociation channel with a modest barrier is available, chemical activation associated with the RO_2 + NO reaction can be important in determining the atmospheric fate of the alkoxy radical. Such is the case for CF_3CFHO radicals (Wallington et al., 1996b) and the following reactions need to be added to those above to describe the chemistry in smog chamber experiments using $CF_3CFH_2/Cl_2/NO/O_2/N_2$ mixtures:

$$CF_3CFHO_2 + NO \leftrightarrow CF_3CFHOONO^*$$

$$CF_3CFHOONO^* \rightarrow CF_3CFHO^* + NO_2$$

$$CF_3CFHOONO^* + M \rightarrow CF_3CFHOONO + M$$

$$CF_3CFHOONO \rightarrow CF_3CFHO + NO_2$$

$$CF_3CFHO^* \rightarrow CF_3 + HC(O)F$$

$$CF_3CFHO^* + M \rightarrow CF_3CFHO + M$$

For a complete understanding of the relative importance of decomposition vs. reaction with O_2 for an alkoxy radical such as CF_3CFHO, it is desirable to conduct measurements of the product yields following Cl-atom-initiated oxidation of CF_3CFH_2 as functions of temperature, total pressure, and O_2 partial pressure in the presence and absence of NO. In most cases the available data for a given haloalkoxy radical do not cover the atmospherically relevant range of temperature and pressure (290–220 K, 1013–25 mbar for 0–25 km altitude in the U.S. Standard Atmosphere). In many studies the possible effect of chemical activation has not been investigated. The data base concerning the atmospheric chemistry of haloalkoxy radicals is still rather sparse, and further studies under atmospherically relevant conditions are needed to provide a better understanding of the chemistry of haloalkoxy radicals. The published data concerning the atmospheric fates of individual haloalkoxy radicals are summarized in the following sections.

VI-E-5.1. CH_2FO

Wallington et al. (1992b), Edney and Driscoll (1992), and Tuazon and Atkinson (1993a) have studied the products of Cl-atom-initiated oxidation of CH_3F in 700 Torr of air at 296–298 K and found essentially 100% yield of HC(O)F. Elimination of F is endothermic by 142 kJ mol^{-1} and will not occur. Reaction with O_2 is the atmospheric fate of CH_2FO radicals:

$$CH_2FO + O_2 \rightarrow HC(O)F + HO_2$$

VI-E-5.2. CHF_2O

Atkinson et al. (1976) reported the formation of CF_2O in a yield close to unity following the 184.9 nm photolysis of CHF_2Cl in 100–735 Torr of air at 295 K. Photolysis of CHF_2Cl proceeds via Cl-atom elimination, giving CHF_2 which, in the presence of O_2, is converted into a CHF_2O radical. Nielsen et al. (1992a) have shown that the Cl-atom-initiated oxidation of CH_2F_2 in 700 Torr of air at 298 K gives COF_2 in a yield that is indistinguishable from 100%. The atmospheric fate of CHF_2O radicals is reaction with O_2 since decomposition reactions for CHF_2O radicals have high activation barriers (Hou et al, 1999):

$$CHF_2O + O_2 \rightarrow COF_2 + HO_2$$

VI-E-5.3. CF_3O

The usual modes of loss are not available for CF_3O radicals. Elimination of a fluorine atom is endothermic by 96 kJ mol^{-1} and will not occur under atmospheric conditions. There are no hydrogen atoms for O_2 to abstract; abstraction of F giving FO_2 radicals is endothermic by 42 kJ mol^{-1} and is not significant. The atmospheric fate of CF_3O radicals is reaction with NO giving COF_2 and FNO, and reaction with CH_4 giving CF_3OH and CH_3 radicals (Sehested and Wallington, 1993; Wallington et al., 1994c; Good and Francisco, 2003):

$$CF_3O + NO \rightarrow COF_2 + FNO$$

$$CF_3O + CH_4 \rightarrow CF_3OH + CH_3$$

The reaction of CF_3O with NO proceeds with a rate constant of $k(CF_3O{+}NO) = 3.7 \times 10^{-11} \exp(110/T)$ cm^3 molecule^{-1} s^{-1}. At 298 K,$k(CF_3O{+}NO) = 5.4 \times 10^{-11}$ cm^3 molecule^{-1} s^{-1} (Atkinson et al., 2005a). The reaction of CF_3O with CH_4 proceeds with a rate constant of $k(CF_3O{+}CH_4) = 2.6 \times 10^{-12} \exp(-1420/T)$ cm^3 molecule^{-1} s^{-1}. At 298 K, $k(CF_3O{+}CH_4) = 2.2 \times 10^{-14}$ cm^3 molecule^{-1} s^{-1} (Atkinson et al., 2005a).

The global background concentration of CH_4 is approximately 2 ppm; hence, at 298 K, CF_3O radicals have a lifetime of approximately 1 s with respect to reaction with CH_4. The atmospheric concentration of NO varies widely in time and space. In locations influenced by polluted urban air the concentration of NO is of the order of 10–100 ppb, and CF_3O radicals have a lifetime in such environments of approximately 0.01–0.1 s. In remote locations the NO concentrations decrease to ~1–10 ppt, and CF_3O radicals have a lifetime with respect to reaction with NO of 100–1000 s. The relative importance of reaction with NO and CH_4 as loss mechanisms for CF_3O radicals varies with location. The atmospheric fate of CF_3OH is heterogeneous decomposition to give COF_2 and HF (Wallington and Schneider, 1994), which is believed to take place on a time-scale dictated by the time taken for gas to come into contact with surfaces (of the order of 10 days). Hence, irrespective of whether CF_3O radicals react with NO or CH_4, their fate is conversion into COF_2.

VI-E-5.4. CH_3CHFO

Hou et al. (2000a) conducted a theoretical study and concluded that C—C bond scission is the fate of the CH_3CHFO radical and that this process proceeds with an activation barrier of 48.5 kJ mol^{-1}. It was calculated that in 100 Torr total pressure of N_2 diluent decomposition occurs at a rate of approximately 14 s^{-1} at 230 K and 1.4×10^4 s^{-1} at 298 K. It seems likely that decomposition and reaction with O_2 are competing atmospheric fates of CH_3CHFO radicals:

$$CH_3CHFO + M \rightarrow CH_3 + HC(O)F + M$$

$$CH_3CHFO + O_2 \rightarrow CH_3C(O)F + HO_2$$

VI-E-5.5. CH_2FCHFO

Wallington et al. (1994b) studied the Cl-atom-initiated oxidation of CH_2FCH_2F in 700 Torr of air diluent at 296 K and observed essentially complete (182 ± 19% molar yield) conversion into HC(O)F. From the time dependence of NO_2 formation following the generation of CH_2FCHFO_2 radicals in the presence of NO, a lower limit of $> 6 \times 10^4$ s^{-1} was established for the rate of decomposition of CH_2FCHFO radicals. Hou et al. (2000a) conducted a theoretical study and concluded that C—C bond scission is the fate of the CH_2FCHFO radical. Hou et al. calculated an activation barrier of 34.7 kJ mol^{-1} for this process and estimated a rate of C—C bond rupture of 2×10^6 s^{-1} at 296 K in 700 Torr total pressure, which is consistent with the lower limit derived by Wallington et al. (1994b). The atmospheric fate of CH_2FCHFO radicals is decomposition via C—C bond scission with the resulting CH_2F radicals being converted into HC(O)F (see section VI-E-5.1):

$$CH_2FCHFO + M \rightarrow CH_2F + HC(O)F + M$$

VI-E-5.6. CH_3CF_2O

Edney and Driscoll (1992) reported the formation of COF_2 in a yield of $100 \pm 5\%$ following the Cl-atom-initiated oxidation of CH_3CHF_2 in 700 Torr of air at 298 K. Tuazon and Atkinson (1993a) observed COF_2 in a yield of 92 % following Cl-atom-initiated oxidation of CH_3CHF_2 in 740 Torr of air at 298 K. Taketani et al. (2005) reported the formation of COF_2 in a yield of $97 \pm 5\%$ from the Cl-atom-initiated oxidation of CH_3CHF_2 in 700 Torr of air at 295 K in the absence of NO. From the CH_2ClCHF_2 yield following UV irradiation of $CH_3CHF_2/Cl_2/N_2$ mixtures it has been concluded that the reaction of Cl-atoms proceeds 99.2% via hydrogen atom abstraction from the —CHF_2 group and 0.8% via hydrogen atom abstraction from the —CH_3 group. Hence the products following Cl-atom-initiated oxidation of CH_3CHF_2 largely reflect the chemistry associated with the CH_3CF_2O radical. Taketani et al. (2005) observed a decrease in the yield of COF_2 in the presence of NO and a corresponding increase in the formation of $CH_3C(O)F$. This observation was attributed to reaction of CH_3CF_2O radicals with NO, giving $CH_3C(O)F$ and FNO. Experiments performed in the presence of increased NO concentrations gave increased yields of $CH_3C(O)F$; however, a limit was reached beyond which the $CH_3C(O)F$ yields did not increase, and the COF_2 yields did not decrease further. This behavior was ascribed to the formation of chemically activated $CH_3CF_2O^*$ radicals (see section VI-E-4) which undergo prompt decomposition on a time-scale ($< 10^{-9}$ s) that is too short for reaction with NO to compete.

The atmospheric fate of CH_3CF_2O radicals is decomposition via C—C bond scission:

$$CH_3CF_2O + M \rightarrow CH_3 + CF_2O + M$$

VI-E-5.7. CH_2FCF_2O and CHF_2CHFO

Barry et al. (1995) investigated the products following Cl-atom-initiated oxidation of CH_2FCHF_2 (HFC-143) in 1 atmosphere of N_2/O_2 diluent at 255–330 K and reported the formation of HC(O)F and COF_2 products in molar yields indistinguishable from 100%. The reaction of Cl-atoms with CH_2FCHF_2 occurs approximately 50% via hydrogen abstraction from the CH_2F— group and 50% from the CHF_2 (Tschuikow-Roux et al., 1985b) and will lead to the formation of CH_2FCF_2O and CHF_2CHFO radicals. The observation that HC(O)F and COF_2 are the dominant products even in 760 Torr of O_2 at 255 K shows that decomposition via C—C bond scission is the atmospheric fate of CH_2FCF_2O and CHF_2CHFO radicals:

$$CH_2FCF_2O + M \rightarrow CH_2F + CF_2O + M$$

$$M + CHF_2CHFO \rightarrow CHF_2 + HC(O)F + M$$

VI-E-5.8. CF_3CH_2O

Nielsen et al. (1994) studied the F-atom-initiated oxidation of CF_3CH_3 in 700 Torr of N_2/O_2 diluent with $[O_2] = 1$, 5, or 147 Torr at 296 K. CF_3CHO was observed as the major product in a yield of 101 ± 8 %. From the absence of any observable effect of O_2 partial pressure on the CF_3CHO yield, a lower limit of $k_{O2}/k_{diss} > 2.5 \times 10^{-17}$ cm^3 molecule^{-1} was established for the rate constant ratio of reaction with O_2 vs. dissociation via C—C bond scission in 700 Torr, at 295 K. Nielsen et al. (1994) calculated that reaction with O_2 accounts for > 99.3% of the fate of CF_3CH_2O radicals formed in the atmosphere. In a study of the Cl-atom-initiated oxidation of $CF_3CH_2CF_2CH_3$ in 700 Torr of air at 297 K, Chen et al. (1997) reported the formation of CF_3CHO in a yield of 97 ± 5%. As discussed by Chen et al. (1997), the oxidation of $CF_3CH_2CF_2CH_3$ proceeds via the formation of CF_3CH_2O radicals. Somnitz and Zellner (2001) conducted a theoretical study of the decomposition of CF_3CH_2O radicals and calculate a rate for C—C bond scission of 7.4×10^{-5} s^{-1} in 1 atmosphere total pressure at 300 K. The observed formation of CF_3CHO in the studies by Nielsen et al. (1994) and Chen et al. (1997) and the theoretical work by Somnitz and Zellner (2001) show that the atmospheric fate of CF_3CH_2O radicals is reaction with O_2:

$$CF_3CH_2O + O_2 \rightarrow CF_3CHO + HO_2$$

VI-E-5.9. CHF_2CF_2O

Nielsen et al. (1992c) studied the Cl-atom-initiated oxidation of CHF_2CHF_2 in 700 Torr of air at 295 K and reported the formation of CF_2O in a yield of 196 ± 24%. The atmospheric fate of CHF_2CF_2O radicals is via C—C bond scission to give CHF_2 radicals and CF_2O. CHF_2 radicals are converted into additional CF_2O (see section VI.E.5.2), which explains the essentially 200% molar yield of CF_2O:

$$CHF_2CF_2O + M \rightarrow CHF_2 + CF_2O + M$$

VI-E-5.10. CF_3CHFO

Reflecting the commercial importance of its parent haloalkane CF_3CFH_2 (R-134a, HFC-134a), there have been more studies of the atmospheric chemistry of this compound than of any other haloalkoxy radical. Experimental studies have been performed by Wallington et al. (1992b), Edney and Driscoll (1992), Maricq and Szente (1992), Tuazon and Atkinson (1993b), Rattigan et al. (1994), Meller et al. (1994), Bhatnagar and Carr (1995a), Wallington et al. (1996b), Bednarek et al. (1996), Møgelberg et al. (1997b), Hasson et al. (1998), and Wu and Carr (2003). Computational investigations have been reported by Schneider et al. (1998), Somnitz and Zellner (2001), and Wu and Carr (2003). The studies find that decomposition via C—C bond scission and reaction with O_2 are competing loss mechanisms for CF_3CFHO radicals under atmospheric conditions. Decomposition gives HC(O)F and CF_3 radicals, while reaction with O_2 gives $CF_3C(O)F$:

$$CF_3CFHO + M \rightarrow CF_3 + HC(O)F + M$$

$$CF_3CFHO + O_2 \rightarrow CF_3C(O)F + HO_2$$

The relative yields of HC(O)F and $CF_3C(O)F$ following the Cl-atom-initiated oxidation of CF_3CFH_2 have been used to study the relative importance of decomposition and reaction with O_2 as fates for CF_3CFHO radicals. The unimolecular decomposition (via C—C bond scission) of thermalized CF_3CFHO radicals is a pressure-dependent process and has a rate which increases with increased total pressure. The bimolecular reaction of CF_3CFHO radicals with O_2 does not show any dependence on total pressure. Wallington et al. (1992b, 1996b) have used the observed formation of HC(O)F and $CF_3C(O)F$ to study the pressure dependence of the ratio of the rate coefficient for dissociation vs. reaction with O_2, k_{diss}/k_{O2}, for CF_3CFHO radicals. Figure VI-E-2 shows a plot of the ratio of k_{diss}/k_{O2} vs. total pressure at 295 K from Wallington et al. (1992b, 1996b). The solid curve in figure VI-E-2 is a least-squares fit of the three parameter expression proposed by Troe and coworkers (Troe, 1983; Gilbert et al., 1983):

$$k(\mathrm{M}) = k_o[\mathrm{M}]/(1 + (k_o[\mathrm{M}]/k_\infty) \times 0.6^{\{1+[\log_{10}(k_o[\mathrm{M}]/k_\infty)]^2\}^{-1}}$$

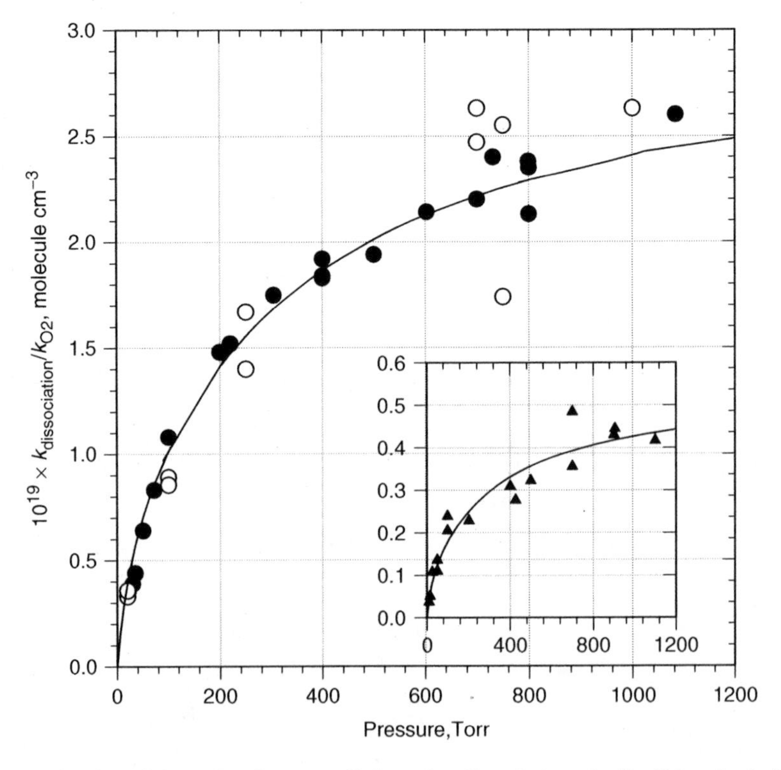

Figure VI-E-2. Plot of the ratio of rate coefficients for dissociation via C—C bond scission vs. reaction with O_2 of CF_3CFHO radicals as a function of total pressure of N_2 diluent at 295 K, from Wallington et al. (1996b), filled symbols, and Wallington et al. (1992b), open symbols. The insert shows data obtained at 269 K (axes have the same units as for the 295 K data).

which gives $k_o = (10.2 \pm 2.5)$ and $k_\infty = (3.45 \pm 0.28) \times 10^{19}$ molecule cm^{-3}, where $k_o = k_{diss,o}/k_{O2}$ and $k_\infty = k_{diss,\infty}/k_{O2}$. These results are in agreement with values of $k_o = 12.3$ and $k_\infty = 3.0 \times 10^{19}$ molecule cm^{-3} evaluated at 296 K from the temperature-dependent expressions given by Meller et al. (1994).

The temperature dependence of k_{O2}/k_{diss} has been measured by Wallington et al. (1992a), Edney and Driscoll (1992), Tuazon and Atkinson (1993b), Rattigan et al. (1994), Meller et al. (1994), Wallington et al. (1996b), Bednarek et al. (1996), Hasson et al. (1998) , and Wu and Carr (2003). As discussed by Wallington and Kaiser (1999), to facilitate comparison of the results it is important to account for different pressures used in the various studies. Using the approach described by Wallington and Kaiser (1999), the results from Wallington et al. (1992a), Edney and Driscoll (1992), Tuazon and Atkinson (1993b), Rattigan et al. (1994), Meller et al. (1994), Wallington et al. (1996b), Bednarek et al. (1996), Hasson et al. (1998), and Wu and Carr (2003) have been normalized to 760 Torr and are plotted in figure VI-E-3. As is seen in this figure, the results from the various studies are generally in very good agreement. The results from Rattigan et al. (1994) are systematically lower than those from the other studies. While the results from Wu and Carr (2003) lie somewhat higher than those from other studies, it should be noted that low total pressures were used in this study (20 and 35 Torr) and some of the discrepancy is probably attributable to uncertainties in normalizing the data to 760 Torr. A linear least-squares fit to the data from Wallington et al. (1992a), Edney and Driscoll (1992), Tuazon and Atkinson (1993b), Meller et al. (1994), Wallington et al. (1996b), Bednarek et al. (1996), and Hasson et al. (1998) gives $k_{O2}/k_{diss} = 2.4 \times 10^{-25} \exp(3590/T)$ cm^3 molecule^{-1} s^{-1}. This expression gives $k_{O2}/k_{diss} = 4.1 \times 10^{-20}$ cm^3 molecule^{-1} s^{-1} at 298 K. In one atmosphere of air $k_{O2}[O_2]/k_{diss} = 0.21$, and 83% of CF_3CFHO radicals undergo decomposition to give CF_3 radicals and HC(O)F with the remaining fraction reacting with O_2 to give $CF_3C(O)F$.

Schneider et al. (1998), Somnitz and Zellner (2001), and Wu and Carr (2003) have conducted theoretical studies of CF_3CFHO radicals. Schneider et al. (1998) calculated a barrier for C—C bond scission of 44.8 kJ mol^{-1} and showed that chemical activation of CF_3CFHO radicals, formed in the reaction of CF_3CFHO_2 radicals with NO, is important. Somnitz and Zellner (2001) conducted a theoretical study of the decomposition of CF_3CFHO radicals and calculated a rate for C—C bond scission of 1.1×10^4 s^{-1} in 1 atmosphere total pressure at 300 K, a barrier of 50.6 kJ mol^{-1}, and concluded that chemical activation effects are important. Wu and Carr (2003) calculated a barrier to C—C bond scission of 40 kJ mol^{-1}. Wu and Carr also conducted RRKM calculations using activation barriers of 44.8 (Schneider et al., 1998) and 39.7 kJ mol^{-1} and reported rates for C—C bond scission of 7.9×10^4 and 4.1×10^5s^{-1}, respectively, at 760 Torr and 295 K. Wu and Carr (2003) reported that they could not reconcile the rapid formation of HC(O)F observed in their lowest temperature (259 K) experiments with the activation barrier of 50.6 kJ mol^{-1} reported by Somnitz and Zellner (2001), but could explain their observations using activation barriers of 44.8 (Schneider et al., 1998) or 39.7 kJ mol^{-1} (Wu and Carr, 2003). Calculated rates for C—C bond scission in 1 atmosphere total pressure at 298 ± 2 K vary by a factor of approximately 40 over the range $(0.11–4.1) \times 10^5$ s^{-1}. The rapid formation of HC(O)F observed by

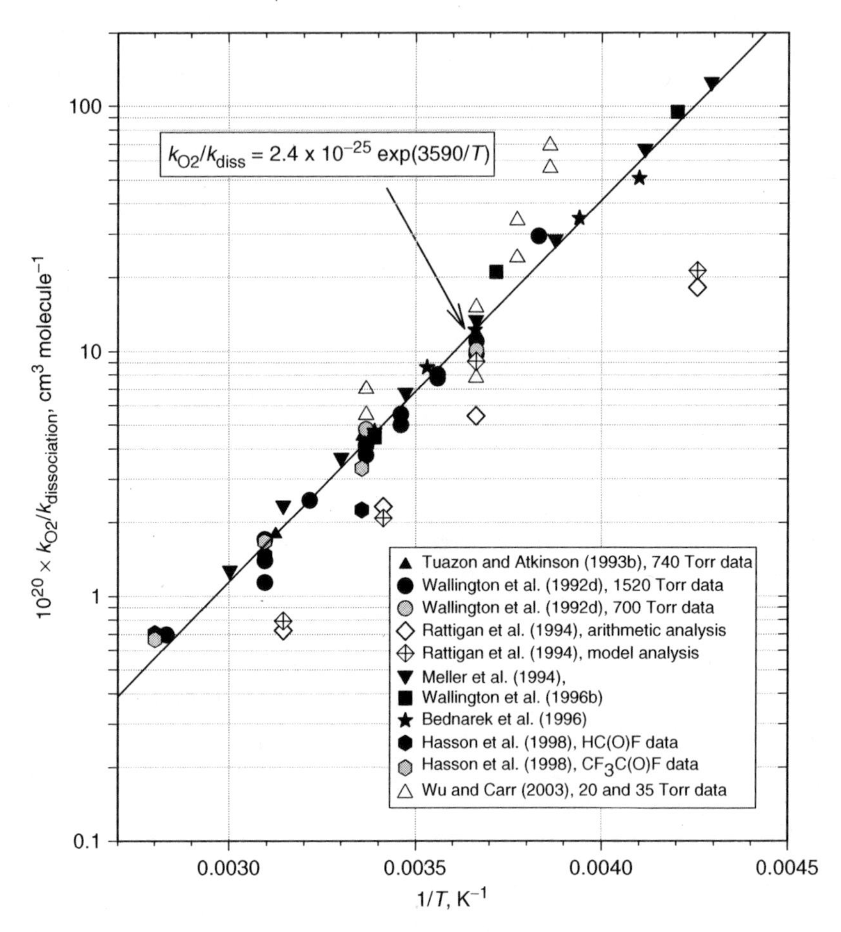

Figure VI-E-3. Arrhenius plot of the ratio of the rate coefficients for the reaction of CF_3CFHO radicals with O_2 vs. decomposition via C—C bond scission. The data have been normalized to an effective total pressure of 760 Torr using the approach of Wallington and Kaiser (1999).

Wu and Carr (2003) suggests that values towards the upper end of this range are more likely.

Maricq and Szente (1992) and Bednarek et al. (1996) have reported experimental measurements of the rate of C—C bond scission. Maricq and Szente conducted experiments in 230 Torr of N_2 diluent at 216–372 K and report $k_{diss} = (3.7 \pm 0.7) \times 10^7 \times \exp[-(2200 \pm 150)/T]$ s^{-1} with a measured value of $k_{diss} = (2 \pm 1) \times 10^4$ s^{-1} at 297 K. Normalizing this value to a pressure of 760 Torr, using the data in figure VI-E-1, gives $k_{diss} = (3.0 \pm 1.5) \times 10^4$ s^{-1} at 297 K. This result falls at the lower end of the $(0.11–4.1) \times 10^5$ s^{-1} range of values derived from theoretical studies. While this agreement is encouraging, the temperature dependence reported by Maricq and Szente (1992) is inconsistent with both the theoretical studies and the experimental measurements shown in figure VI-E-1. The activation barrier measured by Maricq and Szente (1992) of 18.4 ± 1.3 kJ mol^{-1} lies well below the range of 39.7–50.6 kJ mol^{-1}

from the computational studies. The rate constant ratio $k_{O2}/k_{diss} = 2.4 \times 10^{-25}$ $\exp(3590/T)$ cm^3 molecule^{-1} s^{-1}, derived from the work of Wallington et al. (1992a), Edney and Driscoll (1992), Tuazon and Atkinson (1993b), Meller et al. (1994), Wallington et al. (1996b), Bednarek et al. (1996), and Hasson et al. (1998) shows that the activation barrier for C—C bond scission is 29.7 kJ mol^{-1} greater than that for reaction with O_2. From the limited data base for haloalkoxy radicals (see section VI-E-1), it seems reasonable to assume an activation barrier of approximately 8 kJ mol^{-1} for reaction with O_2. Hence, the experimental data from Wallington et al. (1992a), Edney and Driscoll (1992), Tuazon and Atkinson (1993b), Meller et al. (1994), Wallington et al. (1996b), Bednarek et al. (1996), and Hasson et al. (1998) suggest a barrier to C–C bond scission of approximately 38 kJ mol^{-1}, substantially larger than the value of 18.4 kJ mol^{-1} reported by Maricq and Szente (1992).

Bednarek et al. (1996) reported $k_{diss} = 1.8 \times 10^4$ s^{-1} in 50 mbar (37.5 Torr) at 295 K; normalizing this result to 760 Torr gives $k_{diss} = 7 \times 10^4$ s^{-1}. This result falls within the $(0.11–4.1) \times 10^5$ s^{-1} range of values calculated from the theoretical studies. On the basis of the ambient temperature measurement of Maricq and Szente (1992), the measurement of Bednarek et al. (1996), and the computational studies of Schneider et al. (1998), Somnitz and Zellner (2001), and Wu and Carr (2003), we recommend a value of $k_{diss} = 1 \times 10^5$ s^{-1} in one atmosphere of air at 298 K, with an estimated uncertainty of a factor of 3. Combining $k_{O2}/k_{diss} = 4.1 \times 10^{-20}$ cm^3 molecule^{-1} s^{-1} with $k_{diss} = 1 \times 10^5$ s^{-1} gives $k_{O2} = (4.1^{+8.2}_{-2.7}) \times 10^{-15}$ cm^3 molecule^{-1} s^{-1} at 298 K.

The experimental observations of Wallington et al. (1996b) and Olkhov and Smith (2003) and the theoretical calculations of Schneider et al. (1998) and Somnitz and Zellner (2001) indicate that the reaction of CF_3CFHO_2 radicals with NO produces chemically activated CF_3CFHO radicals. Wallington et al. (1996b) observed an increase in the yields of COF_2 and HC(O)F, and a decrease in the yield of $CF_3C(O)F$, when CF_3CFHO radicals were produced in the highly exothermic reaction of CF_3CFHO_2 radicals with NO rather than in the thermoneutral RO_2–RO_2 reaction of CF_3CFHO_2 radicals. It was deduced that in 1 atmosphere of air at 295 K, 64^{+12}_{-23} % of the CF_3CFHO radicals produced in the reaction of CF_3CFHO_2 radicals with NO decompose on a time-scale sufficiently short that bimolecular reaction with O_2 cannot compete. Olkhov and Smith (2003) monitored the formation of HC(O)F and CF_2O and the loss of NO following the pulsed photolysis of CF_3CFHI in 30 Torr of O_2 diluent at 198–298 K. The temporal behaviors of HC(O)F, CF_2O, and NO were fitted using a simple model with three fit parameters: (i) the rate constant $k(CF_3CFHO_2+NO)$, (ii) the fraction of CF_3CFHO radicals undergoing prompt decomposition, and (iii) the rate of thermal decomposition of CF_3CFHO radicals (k_{diss}). It was concluded that at 298 and 198 K the fraction of CF_3CFHO radicals which undergo prompt decomposition is 48 ± 4% and 42 ± 12%, respectively. The computational studies of Schneider et al. (1998) and Somnitz and Zellner (2001) conclude that chemical activation plays an important role in the fate of CF_3CFHO radicals. As with other peroxy radicals, the reaction of CF_3CFHO_2 radicals with NO proceeds via the formation of a short-lived adduct. In the transition state for the CF_3CFHO_2 + NO reaction there are 27 vibrational degrees of freedom, whereas there are only 18 vibrations in the CF_3CFHO radical.

If reaction exothermicity (71 kJ mol^{-1}, Schneider et al., 1998) is distributed according to the ratio of 18/27, then approximately two-thirds of the exothermicity, or about 46 kJ mol^{-1}, will be deposited in the nascent CF_3CFHO radical. The excitation in the nascent CF_3CFHO radical is comparable to the barrier to decomposition and a significant fraction of the CF_3CFHO radicals can decompose on a sufficiently short time-scale ($< 10^{-9}$ s) that collisional deactivation cannot compete.

From the rise time and yield of NO_2 following the flash photolytic generation of CF_3CFHO_2 radicals in the presence of NO, Bhatnagar and Carr (1995a) deduced that the decomposition of CF_3CFHO radicals was complete within the 50 ms experimental time-scale. Assuming > 90% decomposition within 50 ms gives a rate of $> 46\ s^{-1}$ for the decomposition of CF_3CFHO radicals. In a similar fashion, in a pulsed radiolysis study Møgelberg et al. (1997b) deduced that $79 \pm 20\%$ of CF_3CFHO radicals formed in the CF_3CFHO_2 + NO reaction decompose within 3μs. The results from Møgelberg et al. (1997b) show that > 59% of CF_3CFHO radicals decompose within 3 μs, and hence the rate of decomposition of chemically activated CF_3CFHO radicals is $> 3 \times 10^5\ s^{-1}$. The computational studies of Schneider et al. (1998) and Somnitz and Zellner (2001) and the experimental studies of Wallington et al. (1992a, 1996b) and Møgelberg et al. (1997b) provide a consistent picture of the importance of chemical activation in the atmospheric chemistry of CF_3CFHO radicals.

The atmospheric fate of CF_3CFHO radicals is decomposition via C—C bond scission and reaction with O_2:

$$CF_3CFHO + M \rightarrow CF_3 + HC(O)F + M$$

$$CF_3CFHO + O_2 \rightarrow CF_3C(O)F + HO_2$$

VI-E-5.11. CF_3CF_2O

Edney and Driscoll (1992) reported the formation of COF_2 in a yield of $109 \pm 5\%$ following the Cl-atom-initiated oxidation of CF_3CF_2H in 700 Torr of air at 298 K. Tuazon and Atkinson (1993a) observed COF_2 in a yield of approximately 100% following Cl-atom-initiated oxidation of CF_3CF_2H in 740 Torr of air at 298 K. Sehested et al. (1993a) reported the formation of 100% yield of COF_2 following Cl-atom-initiated oxidation of CF_3CF_2H in 700 Torr of air at 295 K. Somnitz and Zellner (2001) have calculated a rate of $5.3 \times 10^6\ s^{-1}$ for the rate of decomposition via C—C bond scission in 1 atmosphere of air at 300 K. The atmospheric fate of CF_3CF_2O radicals is decomposition via C—C bond scission:

$$CF_3CF_2O + M \rightarrow CF_3 + CF_2O + M$$

VI-E-5.12. $CH_3CF_2CH_2O$

Møgelberg et al. (1995c) studied the Cl-atom-initiated oxidation of $CH_3CF_2CH_3$ in 700 Torr of N_2/O_2 at 296 K with O_2 partial pressures of 5 or 147 Torr. CH_3CF_2CHO was observed in a yield of $97 \pm 5\%$ and was independent of $[O_2]$ over the range studied.

The observation that even for an O_2 partial pressure of 5 Torr unimolecular decomposition does not compete with reaction with O_2 serves to define the atmospheric fate of $CH_3CF_2CH_2O$ radicals. Compared to a partial O_2 pressure of 5 Torr, in the atmosphere the O_2 concentration is higher (for altitudes below 25 km) and the temperature is lower. Both factors will suppress the importance of unimolecular decomposition. The sole atmospheric fate of $CH_3CF_2CH_2O$ radicals is reaction with O_2.

$$CH_3CF_2CH_2O + O_2 \rightarrow CH_3CF_2CHO + HO_2$$

VI-E-5.13. $CF_3CH_2CF_2O$

Chen et al. (1997) studied the Cl-atom-initiated oxidation of $CF_3CH_2CHF_2$ in 700 Torr of air at 297 K. Reaction of Cl-atoms was determined to occur mainly ($\geq 95\%$) at the terminal carbon, leading to the formation of $CF_3CH_2CF_2O$ radicals. The observed products included COF_2 and CF_3CHO. The yield of CF_3CHO was $97 \pm 5\%$, leading to the conclusion that the fate of $CF_3CH_2CF_2O$ radicals is decomposition via C—C bond scission with the resulting CF_3CH_2 being converted into CF_3CH_2O radicals that then react with O_2 to give CF_3CHO (see section VI-E-5.8). The atmospheric fate of $CF_3CH_2CF_2O$ radicals is decomposition via C–C bond scission:

$$CF_3CH_2CF_2O + M \rightarrow CF_3CH_2 + CF_2O + M$$

VI-E-5.14. $CF_3CH(O\bullet)CF_3$

Møgelberg et al. (1995d) studied the F-atom-initiated oxidation of $CF_3CH_2CF_3$ in 700 Torr of N_2/O_2 at 296 K with O_2 partial pressures of 5 or 147 Torr. There was no discernable effect of $[O_2]$ on the yield of the major product $CF_3C(O)CF_3$ observed in the system; this leads to the conclusion that reaction with O_2 accounts for $> 99\%$ of the atmospheric fate of $CF_3CH(O\bullet)CF_3$ radicals:

$$CF_3CH(O\bullet)CF_3 + O_2 \rightarrow CF_3C(O)CF_3 + HO_2$$

VI-E-5.15. $CF_3CF(O\bullet)CF_3$

Zellner et al. (1994) monitored the time evolution of NO_2 formation following the laser pulsed photolysis of H_2O_2, generating OH radicals which initiated the oxidation of CF_3CFHCF_3 in O_2/NO mixtures at 37 Torr total pressure and 298 K. It was concluded that $CF_3CF(O\bullet)CF_3$ radicals decompose at a rate of $(2 \pm 1) \times 10^3\ s^{-1}$. Zellner et al. (1994) identified $CF_3C(O)F$ and CF_2O in yields of $100 \pm 5\%$ and $80 \pm 20\%$, respectively, following the Cl-atom-initiated oxidation of CF_3CFHCF_3 in 1 atmosphere of air in the presence of NO. Møgelberg et al. (1996b) studied the F-atom-initiated oxidation of CF_3CFHCF_3 in 700 Torr of N_2/O_2 with $[O_2] = 5$–147 Torr and observed the formation of $CF_3C(O)F$ in a yield of $94 \pm 6\%$. The atmospheric fate of $CF_3CF(O\bullet)CF_3$ radicals is decomposition via C–C bond scission:

$$CF_3CF(O\bullet)CF_3 + M \rightarrow CF_3C(O)F + CF_3 + M$$

VI-E-5.16. $CF_3CH_2CF_2CH_2O$

Barry et al. (1997) studied the Cl-atom-initiated oxidation of $CF_3CH_2CF_2CH_3$ in 1 atmosphere of N_2/O_2 diluent at 273–330 K. The major products were COF_2 (in a yield close to 200%), CO_2, and traces of $CF_3O_3CF_3$(formed from the association of CF_3O and CF_3O_2 radicals; Nielsen et al., 1992b). Kinetic arguments, based on measured rate constant ratios of $k(Cl{+}CF_3CH_2CCl_2F)/k(Cl{+}CF_3CH_2CF_2CH_3) = 0.032 \pm 0.003$ and $k(Cl{+}CF_3CH_2CF_3)/k(Cl{+}CF_3CH_2CF_2CH_3) < 0.01$, were used by Barry et al. (1997) to suggest that the majority of hydrogen atom abstraction by Cl-atoms from $CF_3CH_2CF_2CH_3$ occurs from the —CH_3 group. The observed products are then attributable largely to chemistry associated with the $CF_3CH_2CF_2CH_2O$ radical. There are two possible fates for $CF_3CH_2CF_2CH_2O$ radicals: reaction with O_2 to give $CF_3CH_2CF_2CHO$, or decomposition via C—C bond scission to give $CF_3CH_2CF_2$ radicals and CH_2O. $CF_3CH_2CF_2$ radicals will be converted into the corresponding alkoxy radicals which will eliminate COF_2 to give CF_3CH_2 radicals which will react further to give CF_3CHO. Hence, decomposition of $CF_3CH_2CF_2CH_2O$ radicals via C—C bond scission would give CH_2O, CF_2O, and CF_3CHO as primary products in essentially 100% yield. The 200% yield of CF_2O shows that Barry et al. (1997) were measuring secondary, rather than primary, products. This is reasonable given the >1000-fold difference in the reactivity of Cl-atoms towards the parent $CF_3CH_2CF_2CH_3$ and possible products $CF_3CH_2CF_2CHO$ and CF_3CHO, assuming that $CF_3CH_2CF_2CHO$ and CF_3CHO have similar reactivity (Andersen et al., 2004a).

By analogy with CF_3CH_2O radicals discussed in section VI-E-5.8, and as noted by Barry et al. (1997), it seems likely that the atmospheric fate of $CF_3CH_2CF_2CH_2O$ radicals is reaction with O_2 and that subsequent oxidation of the primary product $CF_3CH_2CF_2CHO$ gives CO_2, CF_2O, and CF_3CHO as secondary products. Oxidation of CF_3CHO then gives additional CO_2 and CF_2O as tertiary products.

$$CF_3CH_2CF_2CH_2O + O_2 \rightarrow CF_3CH_2CF_2CHO + HO_2$$

VI-E-5.17. CH_2ClO

Studies by Sanhueza and Heicklen (1975a) and Niki et al. (1980) showed that the Cl-atom-initiated oxidation of CH_3Cl in 1 atmosphere of air at 298 K in the absence of NO leads to essentially complete conversion of CH_3Cl into HC(O)Cl. It can be concluded that reaction with O_2 is essentially the sole fate of CH_2ClO radicals under these conditions:

$$CH_2ClO + O_2 \rightarrow HC(O)Cl + HO_2$$

Work by Kaiser and Wallington (1994) and Wu and Carr (1999) has established that under conditions of low O_2 partial pressure three-center elimination of HCl becomes a competitive loss process for CH_2ClO radicals:

$$CH_2ClO + M \rightarrow HCO + HCl + M$$

In 700 Torr total pressure of N_2 diluent at 296 K the rate coefficient ratio for reaction with O_2 vs. elimination of HCl is $k_{O2}/k_{HCl} = 5.0 \times 10^{-18}$ cm^3 molecule^{-1}. In 700 Torr of air at 296 K $k_{O2}[O_2]/k_{HCl} = 24$, and elimination of HCl accounts for approximately 4% of the loss of CH_2ClO radicals. The activation barrier for HCl elimination is 30 kJ mol^{-1} larger than that for reaction with O_2. At 700 Torr total pressure $k_{O2}/k_{HCl} = 5.6 \times 10^{-23} \exp(3300/T)$ cm^3 molecule^{-1}; hence in 700 Torr of air at 273 K, $k_{O2}[O_2]/k_{HCl} = 250$, and elimination of HCl accounts for approximately 0.4% of the loss of CH_2ClO radicals.

Wu and Carr (1999, 2001) conducted absolute rate studies of the reaction of CH_2ClO radicals with O_2 in 5–35 Torr of N_2 diluent at 265–306 K. There was no discernable effect of total pressure on the rate of the reaction. Rate coefficients reported by Wu and Carr (2001) are shown in the Arrhenius plot in figure VI-E-1. A linear least-squares fit gives $k_{CH2ClO+O2} = 2.06 \times 10^{-12} \exp(-940/T)$ cm^3 molecule^{-1} s^{-1}. This expression gives $k_{CH2ClO+O2} = 8.8 \times 10^{-14}$ cm^3 molecule^{-1} s^{-1} at 298 K. In one atmosphere of air at 298 K CH_2ClO radicals have a lifetime with respect to reaction with O_2 of 2.2 μs.

Bilde et al. (1999) observed that chemical activation of CH_2ClO radicals formed in the reaction of CH_2ClO_2 radicals with NO plays an important role in determining their fate. It was shown that 12% of CH_2ClO radicals undergo prompt decomposition by eliminating a Cl-atom, approximately 30% undergo prompt decomposition by eliminating HCl, with the remaining approximately 60% of the radicals becoming deactivated. As discussed above, the dominant fate of thermalized CH_2ClO radicals is reaction with O_2. The main atmospheric fate of CH_2ClO radicals is reaction with O_2; chemically excited CH_2ClO^* radicals undergo decomposition via elimination of HCl and Cl-atoms:

$$CH_2ClO + O_2 \rightarrow HC(O)Cl + HO_2$$

$$CH_2ClO^* \rightarrow HCO + HCl$$

$$CH_2ClO^* \rightarrow CH_2O + Cl$$

VI-E-5.18. $CHCl_2O$

Studies by Sanhueza and Heicklen (1975a) and Niki et al. (1980) showed that the Cl-atom-initiated oxidation of CH_2Cl_2 in 1 atmosphere of air at 298 K in the absence of NO leads to essentially complete conversion of CH_2Cl_2 into HC(O)Cl. Catoire et al. (1996) reported that variation of the temperature from 298 to 250 K had no discernable impact on the reaction chain length in the Cl-atom-initiated oxidation of CH_2Cl_2 in the presence of O_2. The studies of Sanhueza and Heicklen (1975a), Niki et al. (1980), and Catoire et al. (1996) show that Cl-atom elimination is the atmospheric fate of $CHCl_2O$ radicals:

$$CHCl_2O \rightarrow HC(O)Cl + Cl$$

VI-E-5.19. CCl_3O

Atkinson et al. (1976) reported the formation of CCl_2O in a yield close to unity following the 184.9 nm photolysis of CCl_4 in 100–735 Torr of air at 295 K. Photolysis of CCl_4 proceeds via Cl-atom elimination, giving CCl_3 which, in the presence of O_2, is converted into a CCl_3O radical. Lesclaux et al. (1987) studied the chlorine-atom-initiated chain photooxidation of CCl_3H at 233 K. NO was added to compete with Cl-atom elimination as a fate for CCl_3O radicals. Cl-atom elimination was determined to proceed with a rate $> 1 \times 10^5$ s^{-1} at 233 K. Li and Francisco (1989) conducted a theoretical study and concluded that elimination of a Cl-atom is exothermic by 70.3 kJ mol^{-1} and proceeds over a small barrier of 5.4 kJ mol^{-1}. Catoire et al. (1996) observed the formation of CCl_2O in 100% yield following either Cl- or F-atom-initiated oxidation of CCl_3H in 700 Torr of air at 296 K. The experimental and theoretical work shows that elimination of a Cl-atom is the sole atmospheric fate of CCl_3O radicals.

$$CCl_3O \rightarrow CCl_2O + Cl$$

VI-E-5.20. CH_3CHClO

Shi et al. (1993), Maricq et al. (1993), Kaiser and Wallington (1995), and Orlando and Tyndall (2002a) have shown that elimination of HCl and reaction with O_2 are competing loss processes for CH_3CHClO under atmospheric conditions. Orlando and Tyndall (2002a) have measured $k_{O2}/k_{HCl} = 3.2 \times 10^{-24}$ $\exp(2240/T)$ cm^3 $molecule^{-1}$ s^{-1} over the temperature range 220–298 K. At 298 K this expression gives $k_{O2}/k_{HCl} = 5.9 \times 10^{-21}$ cm^3 $molecule^{-1}$ s^{-1} and in 1 atmosphere of air $k_{O2}[O_2]/k_{HCl} = 0.03$. For conditions typical of the upper troposphere (e.g., 10 km altitude in the U.S. Standard Atmosphere, 223 K, 265 mbar pressure) $k_{O2}[O_2]/k_{HCl} = 0.51$, and reaction with O_2 assumes increased importance but is not the major loss mechanism. Assuming $k_{O2} = 6 \times 10^{-14} \exp(-550/T)$ cm^3 $molecule^{-1}$ s^{-1} (based upon the analogous reaction of C_2H_5O radicals), Orlando and Tyndall (2002a) derived $k_{HCl} = 1.9 \times 10^{10} \exp(-2790/T)$ cm^3 $molecule^{-1}$ s^{-1} in one atmosphere total pressure. The activation barrier for HCl elimination is approximately 25 kJ mol^{-1}. Orlando and Tyndall (2002a) showed that the CH_3CHClO radical is subject to a chemical activation effect; when produced in the exothermic reaction of CH_3CHClO_2 radicals with NO, about 50% of the nascent CH_3CHClO radicals decompose promptly via HCl elimination before collisional deactivation can occur.

Hou et al. (2000b) conducted a theoretical study of CH_3CHClO radicals and concluded that: (i) the barrier for HCl elimination is 31 kJ mol^{-1}, (ii) elimination of HCl is the dominant atmospheric fate of CH_3CHClO radicals, and (iii) chemical activation is important; vibrationally excited CH_3CHClO radicals have a lifetime of the order of picoseconds, while thermalized CH_3CHClO radicals have a lifetime of the order of microseconds.

The atmospheric fate of CH_3CHClO radicals is HCl elimination and reaction with O_2:

$$CH_3CHClO + M \rightarrow CH_3CO + HCl + M$$

$$CH_3CHClO + O_2 \rightarrow CH_3C(O)Cl + HO_2$$

VI-E-5.21. CH_2ClCH_2O

The observation of large yields of chloroacetaldehyde from the Cl-initiated oxidation of ethene (Kleindienst et al., 1989; Wallington et al., 1990; Yarwood et al., 1992; Orlando et al., 1998) and the computational finding of Kukui and Le Bras (2001) of a large barrier (71 kJ mol^{-1}) for C—C bond scission, show that reaction with O_2 is the major atmospheric fate of $ClCH_2CH_2O$ radicals. Orlando et al. (1998) showed that chemical activation is not an important factor in the fate of these radicals.

$$CH_2ClCH_2O + O_2 \rightarrow CH_2ClCHO + HO_2$$

VI-E-5.22. $CH_2ClCHClO$

Wallington et al. (1996a) have shown that HCl elimination and reaction with O_2 are competing loss processes for $CH_2ClCHClO$ radicals with $k_{O2}/k_{HCl} = 2.3 \times 10^{-20}$ cm^3 $molecule^{-1}$ at 296 K. In 700 Torr of air $k_{O2}[O_2]/k_{HCl} = 0.11$ and HCl elimination dominates the fate of $CH_2ClCHClO$ radicals. By analogy to the temperature dependence of HCl elimination from CH_3CHClO radicals measured by Orlando and Tyndall (2002a), it is anticipated that reaction with O_2 dominates the fate of $CH_2ClCHClO$ radicals for conditions typical of the upper troposphere. Also, by analogy to the behavior of CH_3CHClO radicals observed by Orlando and Tyndall (2002a), it is anticipated that chemical activation will be important for $CH_2ClCHClO$ radicals. Wallington et al. (1996a) deduced a lower limit for the rate of HCl elimination from chemically activated $CH_2ClCHClO$ radicals of $> 4 \times 10^5$ s^{-1}.

The atmospheric fate of $CH_2ClCHClO$ radicals is HCl elimination and reaction with O_2:

$$CH_2ClCHClO + M \rightarrow CH_2ClCO + HCl + M$$

$$CH_2ClCHClO + O_2 \rightarrow CH_2ClC(O)Cl + HO_2$$

VI-E-5.23. $CHCl_2CHClO$

Sanhueza and Heicklen (1975b) studied the products of Cl-initiated oxidation of *cis-* and *trans-*CHCl=CHCl in 6–630 Torr of N_2/O_2 diluent with $[O_2] = 6$–88 Torr at 304 K. A long-chain oxidation was observed, leading to HC(O)Cl, HCl, CO, and CCl_2O products. The reaction of Cl-atoms with CHCl=CHCl occurs via an addition mechanism to give $CHCl_2CHCl$ radicals which will add O_2, and subsequent RO_2–RO_2 reactions give $CHCl_2CHClO$ radicals. Variation of $[O_2]$ over the range 6–88 Torr had no discernable effect on the products or oxidation chain length, which indicates that reaction with O_2 is not an important fate of $CHCl_2CHClO$ radicals. To rationalize the formation of HCl and CO, Sanhueza and Heicklen (1975b) suggested decomposition of $CHCl_2CHClO$ radicals into CO, HCl, and $CHCl_2$ radicals. This decomposition presumably occurs in two steps: HCl elimination to give $CHCl_2CO$ radicals, followed by decomposition of $CHCl_2CO$ radicals to give $CHCl_2$ radicals and CO. A competing fate of $CHCl_2CO$ radicals is reaction with O_2 (Méreau et al., 2001). $CHCl_2$ radicals react further in the system to give HC(O)Cl and a Cl-atom which propagates the chain. The small yield of CCl_2O seen by Sanhueza and Heicklen (1975b) probably reflects secondary reaction of Cl-atoms with both HC(O)Cl and CO, which gives ClCO radicals that can react with the relatively high concentrations of Cl_2 used in the experiments (0.7–6.0 Torr) to give CCl_2O. The atmospheric fate of $CHCl_2CHClO$ radicals under ambient conditions is dominated by HCl elimination. Reaction with O_2 may play a role at the lower temperatures in the upper troposphere.

$$CHCl_2CHClO + M \rightarrow CHCl_2 + HC(O)Cl + M$$

$$CHCl_2CHClO + O_2 \rightarrow CHCl_2C(O)Cl + HO_2$$

VI-E-5.24. CCl_3CH_2O

Nelson et al. (1990b) and Platz et al. (1995) have identified CCl_3CHO as the main product of the Cl-atom-initiated oxidation of CCl_3CH_3 in one atmosphere of air at 296 K. Platz et al. found that decomposition via C—C bond scission competes with reaction with O_2 and measured the rate constant ratio $k_{diss}/k_{O2} = (2.4 \pm 0.8) \times 10^{17}$ molecule cm^{-3} at 296 K. In one atmosphere of air $k_{diss}/k_{O2}[O_2] = 0.05$, and decomposition accounts for approximately 5% of the loss of CCl_3CH_2O radicals. For the lower temperatures in the upper troposphere dissociation will account for much less than 5% of the CCl_3CH_2O radicals. The main atmospheric fate of CCl_3CH_2O radicals is reaction with O_2:

$$CCl_3CH_2O + O_2 \rightarrow CCl_3CHO + HO_2$$

VI-E-5.25. CCl_3CCl_2O

Huybrechts et al. (1967), Mathias et al. (1974), Møgelberg et al. (1995e), Ariya et al. (1997), Hasson and Smith (1999), and Thüner et al. (1999) have shown that decomposition via Cl-atom elimination and decomposition via C—C bond scission are competing loss mechanisms for CCl_3CCl_2O radicals over the temperature range 230–373 K. There is agreement between all studies that, for CCl_3CCl_2O radicals formed with little or no internal excitation, approximately 85% of CCl_3CCl_2O radicals undergo Cl-atom elimination with the remaining 15% decomposing via C—C bond scission. Elimination of a Cl-atom gives $CCl_3C(O)Cl$; decomposition via C—C bond scission gives $COCl_2$. The lack of a discernable effect of temperature on the yields of $CCl_3C(O)Cl$ and $COCl_2$ is interesting as it indicates that the activation barriers for the two decomposition processes are indistinguishable. The $CCl_3C(O)Cl : COCl_2$ product ratio of 85 : 15 can be equated to the ratio of the *A*-factors for the two competing pathways.

Hasson and Smith (1999) and Thüner et al. (1999) have shown that chemical activation effects are important. The reaction of $CCl_3CCl_2O_2$ radicals with NO produces chemically activated CCl_3CCl_2O radicals. Approximately 65% of chemically activated CCl_3CCl_2O radicals decompose via Cl-atom elimination, and the remaining 35% are lost via C—C bond scission. The atmospheric fate of CCl_3CCl_2O radicals is decomposition via Cl-atom elimination and C—C bond scission:

$$CCl_3CCl_2O + M \rightarrow CCl_3C(O)Cl + Cl$$

$$CCl_3CCl_2O + M \rightarrow CCl_3 + CCl_2O$$

VI-E-5.26. $CH_2ClCHClCHClO$

Voicu et al. (2001) studied the products of the Cl-atom-initiated oxidation of $CH_2ClCHClCH_2Cl$ in 700 Torr total pressure of N_2/O_2 diluent at 295 K. The reaction of Cl-atoms with $CH_2ClCHClCH_2Cl$ proceeds 39% via hydrogen abstraction from the terminal positions. From the lack of $CH_2ClCHClC(O)Cl$ and presence of CO_2, Voicu et al. (2001) concluded that the elimination of HCl is an important atmospheric fate of $CH_2ClCHClCHClO$ radicals:

$$CH_2ClCHClCHClO + M \rightarrow CH_2ClCHClCO + HCl + M$$

VI-E-5.27. $CH_2ClCCl(O\bullet)\ CH_2Cl$

Voicu et al. (2001) studied the products of the Cl-atom-initiated oxidation of $CH_2ClCHClCH_2Cl$ in 700 Torr total pressure of N_2/O_2 diluent at 295 K. The reaction of Cl-atoms with $CH_2ClCHClCH_2Cl$ proceeds 61% via hydrogen abstraction from the central position. In experiments conducted in the presence of NO, the combined yield of $CH_2ClC(O)CH_2Cl$ and $CH_2ClC(O)Cl$ was 63 ± 6% and was independent of $[O_2]$ over the range 16–554 Torr. Voicu et al. (2001) concluded that decompositions via Cl elimination and C—C bond scission are competing atmospheric loss mechanisms for $CH_2ClCl(O\bullet)CH_2Cl$ radicals.

$$CH_2ClCCl(O\bullet)CH_2Cl + M \rightarrow CH_2ClC(O)CH_2Cl + Cl + M$$

$$CH_2ClCCl(O\bullet)CH_2Cl + M \rightarrow CH_2ClC(O)Cl + CH_2Cl + M$$

VI-E-5.28. $CH_3CHClCH(O\bullet)CH_3$

Orlando et al. (2003a) conducted a product study of the Cl-atom-initiated oxidation of *trans*-2-butene in one atmosphere of N_2/O_2 diluent (with $[O_2] = 25$–600 Torr) at 298 K. The observed products were 3-chloro-2-butanone, CH_3CHO, CH_2O, $CH_3C(O)Cl$,

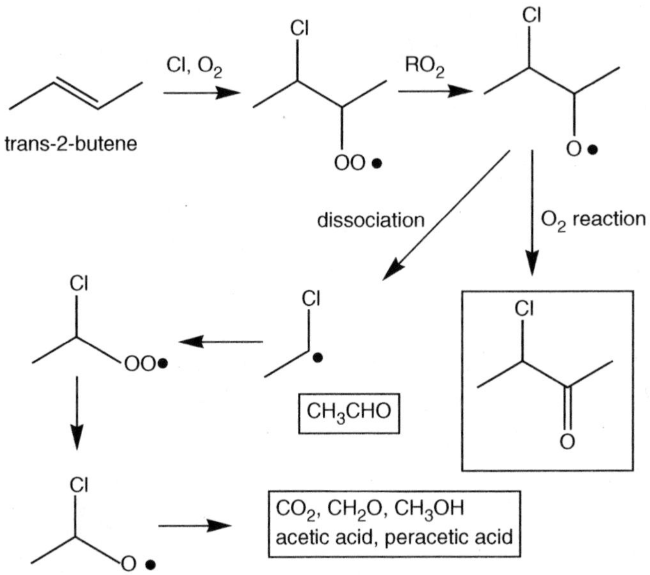

Figure VI-E-4. Mechanism for Cl-atom-initiated oxidation of *trans*-2-butene (only addition channels shown), illustrating competing paths for the $CH_3CHClCH(O\bullet)CH_3$ radical and resulting products (shown in boxes).

CH_3OH, and CO_2. Hydrogen atom abstraction is expected to be of minor importance in the reaction of Cl-atoms with *trans*-2-butene. The products observed by Orlando et al. (2003a) are those expected following addition of Cl to the double bond (see figure VI-E-4). The reaction products were found to be O_2-dependent, with the 3-chloro-2-butanone yield increasing with $[O_2]$ at the expense of the yields of the other products. The observed dependence on $[O_2]$ can be rationalized by the mechanism given in figure VI-E-4. Reaction with O_2 and decomposition via C—C bond scission are competing fates for $CH_3CHClCH(O\bullet)CH_3$ radicals. Reaction with O_2 gives 3-chloro-2-butanone. Decomposition gives CH_3CHO and 1-chloroethyl radicals which react further to give 1-chloroethoxy radicals; these eliminate HCl (see section VI-E-5.20) and, to a much lesser extent, react with O_2, leading to the products observed by Orlando et al. (2003a).

From the observed dependence of the yields of 3-chloro-2-butanone and CH_3CHO on $[O_2]$, a value for the rate constant ratio $k_{diss}/k_{O2} = (1.6 \pm 0.4) \times 10^{19}$ molecule cm^{-3} was derived by Orlando et al. (2003a). In one atmosphere of air at 298 K, $[O_2] = 5.2 \times 10^{18}$ molecule cm^{-3}, $k_{diss}/(k_{O2}[O_2]) = 3.1$, and hence approximately 75% of $CH_3CHClCH(O)CH_3$ radicals decompose via C—C bond scission with the remainder undergoing reaction with O_2:

$$CH_3CHClCH(O\bullet)CH_3 + M \rightarrow CH_3CHCl + CH_3CHO + M$$

$$CH_3CHClCH(O\bullet)CH_3 + O_2 \rightarrow CH_3CHClC(O)CH_3 + HO_2$$

VI-E-5.29. $CH_3CH_2CH(O\bullet)CH_2Cl$ and $CH_3CH_2CHClCH_2O$

Orlando et al. (2003a) conducted a product study of the Cl-atom-initiated oxidation of 1-butene in 700 Torr total pressure of N_2/O_2 diluent (with $[O_2]$ = 25–600 Torr) at 298 K. In all experiments CH_2ClCHO, HC(O)Cl, CH_2O, CO_2, and an unidentified carbonyl-containing product were observed. The yields of CH_2ClCHO, HC(O)Cl, CH_2O, and CO_2 decreased with increasing $[O_2]$, showing that reaction with O_2 competes with at least one other process that forms these products. Decomposition of the $CH_3CH_2CH(O)CH_2Cl$ radical is the only likely pathway to generate CH_2ClCHO and HC(O)Cl in the system. Scission of the terminal C—C bond gives CH_2Cl radicals which react further to give HC(O)Cl (section VI-E-5.17). Scission of the middle C—C bond gives CH_2ClCHO. These C—C bond scission processes compete with reaction with O_2 for the $CH_3CH_2CH(O\bullet)CH_2Cl$ radicals:

$$CH_3CH_2CH(O\bullet)CH_2Cl + M \rightarrow CH_3CH_2CHO + CH_2Cl + M$$

$$CH_3CH_2CH(O\bullet)CH_2Cl + M \rightarrow CH_3CH_2 + CH_2ClCHO + M$$

$$CH_3CH_2CH(O\bullet)CH_2Cl + O_2 \rightarrow CH_3CH_2C(O)CH_2Cl + HO_2$$

From the observed dependence of the CH_2ClCHO and HC(O)Cl yields on $[O_2]$, Orlando et al. (2003a) deduced rate constant ratios of $k_{diss1}/k_{O2} = 1.1 \times 10^{18}$ and $k_{diss2}/k_{O2} = 8.5 \times 10^{18}$ molecule cm^{-3}, where k_{diss1} and k_{diss2} are the rates of scission of the terminal and middle C—C bonds, respectively. In one atmosphere of air at 298 K, $[O_2] = 5.2 \times 10^{18}$ molecule cm^{-3}, $k_{diss1}/(k_{O2}[O_2]) = 0.2$, and $k_{diss2}/(k_{O2}[O_2]) =$ 1.6. Hence, approximately 7% of $CH_3CH_2CH(O)CH_2Cl$ radicals decompose to give CH_3CH_2CHO and CH_2Cl radicals, 57% decompose to give CH_3CH_2 radicals and CH_2ClCHO, and the remainder react with O_2 to give $CH_3CH_2C(O)CH_2Cl$.

VI-E-5.30. CH_2BrO

Chen et al. (1995) and Orlando et al. (1996a) have studied the Cl-atom-initiated oxidation of CH_3Br in 700 Torr of air at 228–298 K in the presence and absence of NO. From the observed yields of CH_2O it was concluded that elimination of Br-atoms is the dominant fate of CH_2BrO under all experimental conditions. Reaction with O_2 and elimination of HBr were determined to contribute less than 5% each to the loss of CH_2BrO radicals. McGivern et al. (2004) performed a computational study of CH_2BrO radicals and found that elimination of Br is exothermic by 40 kJ mol^{-1} and proceeds with essentially no activation barrier. The atmospheric fate of CH_2BrO radicals is Br-atom elimination:

$$CH_2BrO + M \rightarrow CH_2O + Br + M$$

VI-E-5.31. $CHBr_2O$

Orlando et al. (1996b) observed the formation of HC(O)Br from the Cl-atom-initiated oxidation of CH_2Br_2 in 700 Torr of N_2/O_2 diluent at 298 K in the presence and absence of NO. It was concluded that Br-atom elimination occurs at a rate $> 4 \times 10^6\ s^{-1}$ and is the sole atmospheric fate of $CHBr_2O$ radicals. McGivern et al. (2004) performed a computational study of $CHBr_2O$ radicals and found that elimination of Br is exothermic by 83 kJ mol^{-1} and proceeds with essentially no activation barrier. The atmospheric fate of $CHBr_2O$ radicals is Br-atom elimination:

$$CHBr_2O + M \rightarrow HC(O)Br + Br + M$$

VI-E-5.32. CBr_3O

Kamboures et al. (2002) identified CBr_2O in a yield of $104 \pm 5\%$ in the Cl-atom-initiated oxidation of CBr_3H in 11 Torr total pressure of $Cl_2/O_2/CBr_3H$ mixtures with 5 Torr partial pressure of O_2. McGivern et al. (2002) calculated that elimination of Br from CBr_3O is exothermic by 90 kJ mol^{-1} and proceeds without an activation barrier. Elimination of Br is the sole atmospheric fate of CBr_3O radicals:

$$CBr_3O \rightarrow CBr_2O + Br$$

VI-E-5.33. CH_3CHBrO

Orlando and Tyndall (2002a) have shown that Br-atom elimination is the sole fate of CH_3CHBrO radicals at 226–298 K. This conclusion was unaffected by internal excitation in CH_3CHBrO radicals formed in the reaction of CH_3CHBrO_2 with NO.

$$CH_3CHBrO + M \rightarrow CH_3CHO + Br + M$$

VI-E-5.34. CH_2BrCH_2O

Yarwood et al. (1992) conducted a study of the products of the reaction of Br-atoms with ethene in 700 Torr of air at 296 K. The reaction proceeds via an addition mechanism to give CH_2BrCH_2 radicals which then add O_2 to give $CH_2BrCH_2O_2$ radicals. The observed products were CH_2BrCHO, CH_2BrCH_2OOH, and CH_2BrCH_2OH in a combined yield which accounted for all of the ethene loss. As with other peroxy radicals, the RO_2–RO_2 reaction of $CH_2BrCH_2O_2$ will proceed via a radical and a molecular channel:

$$CH_2BrCH_2O_2 + CH_2BrCH_2O_2 \rightarrow CH_2BrCH_2O + CH_2BrCH_2O + O_2$$

$$CH_2BrCH_2O_2 + CH_2BrCH_2O_2 \rightarrow CH_2BrCHO + CH_2BrCH_2OH + O_2$$

From the observations that: (i) the CH_2BrCHO yield was larger (by approximately a factor of 3) than that of CH_2BrCH_2OH; (ii) CH_2BrCH_2OOH was formed (indicating the formation of HO_2 radicals); and (iii) the combined yield of CH_2BrCHO, CH_2BrCH_2OOH, and CH_2BrCH_2OH accounted for the observed loss of ethene, it can be concluded that the atmospheric fate of CH_2BrCH_2O radicals is reaction with O_2:

$$CH_2BrCH_2O + O_2 \rightarrow CH_2BrCHO + HO_2$$

VI-E-5.35. CH_3CH_2CHBrO

Gilles et al. (2002) conducted a kinetic and mechanistic study of the reaction of OH radicals with $CH_3CH_2CH_2Br$. From kinetic data for the reactions of OH with $CH_3CH_2CH_2Br$ isotopomers it was deduced that the branching ratio for hydrogen abstraction from the 1-position is 0.32 ± 0.08 at 298 K. In a product study in 620 Torr of air, it was determined that OH-radical initiation at room temperature in the presence of NO_x leads to the formation of propanal in a molar yield of 0.30 ± 0.15. The experimental evidence suggests that Br elimination is the atmospheric fate of CH_3CH_2CHBrO radicals:

$$CH_3CH_2CHBrO \rightarrow CH_3CH_2CHO + Br$$

VI-E-5.36. $CH_3CH(O\bullet)CH_2Br$

Gilles et al. (2002) conducted a kinetic and mechanistic study of the reaction of OH radicals with $CH_3CH_2CH_2Br$. From kinetic data for the reactions of OH with $CH_3CH_2CH_2Br$ isotopomers it was deduced that the branching ratio for hydrogen abstraction from the 2-position is 0.56 ± 0.04 at 298 K. In a product study in 620 Torr of air, it was determined that OH-radical initiation at room temperature in the presence of NO_x leads to the formation of bromoacetone in a molar yield of 0.50 ± 0.20. The experimental evidence suggests that reaction with O_2 is the atmospheric fate of CH_3CHOCH_2Br radicals:

$$CH_3CH(O\bullet)CH_2Br + O_2 \rightarrow CH_3C(O)CH_2Br + HO_2$$

VI-E-5.37. CH_2IO

Cotter et al. (2001) observed CH_2O as a major product during the Cl-atom-initiated oxidation of CH_3I in 1 atmosphere of air at room temperature and concluded that I-atom elimination is the sole fate of CH_2IO radicals. In their analysis, Cotter et al. assumed that CH_2I radicals (produced in the reaction of Cl-atoms with CH_3I) react with O_2 to give peroxy radicals, CH_2IO_2, which then undergo RO_2–RO_2 reactions to give CH_2IO radicals. However, the assumption that CH_2I radicals react with O_2 to give essentially 100% yield of CH_2IO_2 radicals is not consistent with: (i) the interpretation of product yields observed following photolysis of CH_2I_2 in the presence of O_2 by Schmitt and Comes (1980) and Barnes et al. (1990); (ii) the absence of any pressure dependence in the kinetics of reaction of CH_2I radicals with O_2, noted by Masaki et al. (1995) and Eskola et al. (2006); (iii) the report of IO radicals formed in essentially 100% yield from reaction of CH_2I with O_2 in 5–80 Torr of N_2 diluent at 278–313 K by Enami et al. (2004); and (iv) the report of I-atoms and IO radicals as the primary products of the reaction of CH_2I with O_2 in 6–46 Torr of He at 298 K by Eskola et al. (2006). Furthermore, the observation of CH_2O as a major product by Cotter et al. (2001) in 1 atmosphere of air at room temperature contrasts with the conclusion from Eskola et al. (2006) that CH_2O is, at most, a minor product in 6–46 Torr of He at 298 K. Further work is needed to bridge the gap between the total pressures used in these studies and determine to what degree total pressure can explain the different observed product distributions.

From thermodynamic arguments presented in section VI-E-3 and by analogy to the behavior of CH_2BrO radicals discussed in section VI-E-5.30, it is expected that the fate of CH_2IO radicals is I-atom elimination. However, there is some doubt whether CH_2IO_2 and hence CH_2IO radicals are formed under atmospheric conditions. Further work is needed to clarify the role (if any) of CH_2IO radicals in atmospheric chemistry.

$$CH_2IO + M \rightarrow CH_2O + I + M$$

VI-E-5.38. CH_3CHIO

Cotter et al. (2001) observed CH_3CHO as a major product during the Cl-atom-initiated oxidation of C_2H_5I in 1 atmosphere of air at room temperature. They concluded that CH_3CHO was formed via I-atom elimination from CH_3CHIO radicals. In their analysis, Cotter et al. assumed that CH_3CHI radicals (produced in the reaction of Cl-atoms with C_2H_5I) react with O_2 to give peroxy radicals, CH_3CHIO_2, which then undergo RO_2–RO_2 reactions to give CH_3CHIO radicals. However, as discussed in the previous section, the observation of I-atoms and IO radicals as products of reaction of CH_2I radicals with O_2 by Enami et al. (2004) and Eskola et al. (2006) casts doubt upon the assumptions inherent in the analysis of Cotter et al. (2001). On the basis of the work of Enami et al. (2004) and Eskola et al. (2006), there is a significant probability that CH_3CHI radicals react with O_2 to give CH_3CHO and IO radicals directly. Orlando et al. (2005) studied the Cl-atom-initiated oxidation of C_2H_5I and report the formation of CH_3CHO, $CH_3C(O)OOH$, $CH_3C(O)Cl$, and several other products. $CH_3C(O)OOH$ and $CH_3C(O)Cl$ were secondary products resulting from CH_3CHO oxidation. The CH_3CHO product observed by Cotter et al. (2001) and Orlando et al. (2005) may be attributable to a direct reaction of CH_3CHI radicals with O_2. Further work is needed to clarify our understanding of the mechanism of the reaction of CH_3CHI radicals with O_2 under atmospheric conditions.

From thermodynamic arguments presented in section VI-E-3 and by analogy with the behavior of CH_3CHBrO radicals discussed in section VI-E-5.33, it is expected that the fate of CH_3CHIO radicals is I-atom elimination. However, there is some doubt whether CH_3CHIO_2 and hence CH_3CHIO radicals are formed under atmospheric conditions. Further work is needed to clarify the role (if any) of CH_3CHIO radicals in atmospheric chemistry.

$$CH_3CHIO + M \rightarrow CH_3CHO + I + M$$

VI-E-5.39. $CH_3CI(O\bullet)CH_3$

Cotter et al. (2001) and Orlando et al. (2005) observed $CH_3C(O)CH_3$ as a major product during the Cl-atom-initiated oxidation of 2-iodopropane in 1 atmosphere of air at room temperature. Cotter et al. (2001) concluded that $CH_3C(O)CH_3$ was formed via I-atom elimination from $CH_3CI(O\bullet)CH_3$ radicals. In their analysis, Cotter et al. assumed that CH_3CICH_3 radicals (produced in the reaction of Cl-atoms with CH_3CHICH_3) react with O_2 to give peroxy radicals, $CH_3CI(OO)CH_3$, which then undergo RO_2–RO_2 reactions to give CH_3CIOCH_3 radicals. However, as noted in the two sections above, the work of Enami et al. (2004) and Eskola et al. (2006) casts doubt upon this assumption. The $CH_3C(O)CH_3$ product observed by Cotter et al. (2001) and Orlando et al. (2005) may be attributable to a direct reaction of CH_3CICH_3 radicals with O_2.

From thermodynamic arguments presented in section VI-E-3 and by analogy to the behavior of α-Br-containing alkoxy radicals discussed in that section, it is expected that the fate of $CH_3CI(O)CH_3$ radicals is I-atom elimination. However, there is some doubt whether $CH_3CI(OO)CH_3$ and hence $CH_3CI(O)CH_3$ radicals are formed under atmospheric conditions. Further work is needed to clarify the role (if any) of $CH_3CI(O)CH_3$ radicals in atmospheric chemistry.

$$CH_3CI(O\bullet)CH_3 + M \rightarrow CH_3C(O)CH_3 + I + M$$

VI-E-5.40. CHFClO

Tuazon and Atkinson (1993a) identified HC(O)F as the product (100% yield) of the Cl-atom-initiated oxidation of CH_2FCl in 1 atmosphere of air at 298 K. Bhatnagar and Carr (1996) studied the temporal profile for NO_2 formation and NO loss in the flash photolysis of $CH_2FCl/Cl_2/NO/O_2$. By modeling the NO_2 formation and NO loss, a lower limit of $> 3 \times 10^3$ s^{-1} was established for the elimination of Cl-atoms from CHFClO radicals. Wang et al. (1999) conducted a computational study and found that Cl-atom elimination from CHFClO radicals is exothermic by 48.5 kJ mol^{-1} and proceeds over a small barrier of 5.9 kJ mol^{-1}. The atmospheric fate of CHFClO radicals is elimination of Cl:

$$CHFClO + M \rightarrow HC(O)F + Cl$$

VI-E-5.41. CF_2ClO

Jayanty et al. (1975) and Atkinson et al. (1976) have shown that COF_2 is formed in a quantum yield near unity from the photolysis of CF_2Cl_2 in the presence of O_2. Sanhueza (1977) observed COF_2 production during the Cl-atom-initiated oxidation of CF_2ClH. Suong and Carr (1982) and Carr et al. (1986) produced CF_2Cl radicals by photolysis of $CF_2ClC(O)CF_2Cl$ and identified COF_2 as the main oxidation product. Edney and Driscoll (1992) report the formation of COF_2 in a yield of 111 $\pm$ 6% in the Cl-atom-initiated oxidation of CF_2ClH in 700 Torr of air at 298 K. Tuazon and Atkinson (1993a) observed COF_2 in a yield of 100% following Cl-atom-initiated oxidation of CF_2ClH in 740 Torr of air at 298 K. Wu and Carr (1992) monitored the rate of COF_2 formation in real-time following flash photolysis of $CF_2ClBr/O_2/NO$ mixtures. The rates of Cl-atom elimination from CF_2ClO radicals were determined to be $(6.4 \pm 1.4) \times 10^4$ and $(5.0 \pm 1.1) \times 10^4$ s^{-1} at 298 and 238 K, respectively. The results from all studies indicate that elimination of a Cl-atom is the sole atmospheric fate of the $CFCl_2O$ radical:

$$CF_2ClO \rightarrow COF_2 + Cl$$

VI-E-5.42. $CFCl_2O$

Jayanty et al. (1975) studied the photolysis of $CFCl_3$ (giving $CFCl_2$ radicals) in the presence of O_2 at 298 K and identified FC(O)Cl formation in a yield of $90 \pm 15\%$. Atkinson et al. (1976) reported the formation of FC(O)Cl in a yield close to 100% following the 184.9 nm photolysis of $CHFCl_2$ in 100–735 Torr of air at 295 K. Photolysis of $CFCl_3$ proceeds via Cl-atom elimination, giving $CFCl_2$ which, in the presence of O_2, is converted into a $CFCl_2O$ radical. Lesclaux et al. (1987) studied the chlorine-atom-initiated chain photooxidation of $CFCl_2H$ at 253 K. NO was added to compete with Cl-atom elimination as a fate for $CFCl_2O$ radicals. Cl-atom elimination was determined to proceed with a rate $> 3 \times 10^4$ s^{-1} at 253 K. Wu and Carr (1992) monitored the rate of FC(O)Cl formation in real-time following flash photolysis of $CFCl_3/O_2/NO$ mixtures. The rates of Cl-atom elimination from $CFCl_2O$ radicals were determined to be $(1.2 \pm 0.4) \times 10^5$ and $(1.0 \pm 0.4) \times 10^5$ s^{-1} at 298 and 238 K, respectively. Tuazon and Atkinson (1993a) observed FC(O)Cl in a yield of 100% following Cl-atom-initiated oxidation of $CFCl_2H$ in 740 Torr of air at 298 K. Elimination of a Cl-atom is the sole atmospheric fate of the $CFCl_2O$ radical:

$$CFCl_2O \rightarrow FC(O)Cl + Cl$$

VI-E-5.43. CF_2ClCH_2O

Edney and Driscoll (1992) report the formation of COF_2 in a yield of $98 \pm 3\%$ in the Cl-atom-initiated oxidation of CF_2ClCH_3 in 700 Torr of air at 298 K. Tuazon and Atkinson (1993a) observed COF_2 in a yield of 100% following Cl-atom-initiated oxidation of CF_2ClCH_3 in 740 Torr of air at 298 K. In a subsequent study, Tuazon and Atkinson (1994) showed that CF_2ClCHO is the primary oxidation product, formed in 100% yield, which undergoes further oxidation to give COF_2 as a secondary product. Mörs et al. (1996) monitored the temporal behavior of OH radicals and NO_2 following the pulsed Cl-atom-initiated oxidation of CF_2ClCH_3 in the presence of NO in 50 mbar of O_2 at 293 K. The formation of HO_2 radicals leads (via their reaction with NO) to regeneration of OH. By fitting the OH and, to a lesser degree, NO_2 profiles, values of $k(CF_2ClCH_2O + O_2 \rightarrow CF_2ClCHO + HO_2) = (2.5 \pm 1.5) \times 10^{-15}$ cm^3 $molecule^{-1}$ s^{-1} and $k(CF_2ClCH_2O \rightarrow CF_2Cl + CH_2O) < 1.5 \times 10^3$ s^{-1} were obtained. The atmospheric fate of CF_2ClCH_2O radicals is reaction with O_2:

$$CF_2ClCH_2O + O_2 \rightarrow CF_2ClCHO + HO_2$$

VI-E-5.44. $CFCl_2CH_2O$

Edney et al. (1991) and Tuazon and Atkinson (1993a) observed FC(O)Cl in a yield of 100% following Cl-atom-initiated oxidation of $CFCl_2CH_3$ in one atmosphere of air at ambient temperature. In a subsequent study, Tuazon and Atkinson (1994) showed that $CFCl_2CHO$ is the primary oxidation product, formed in 100% yield, and that this primary product undergoes further oxidation to give the FC(O)Cl observed by Edney et al. (1991) and Tuazon and Atkinson (1993a). Mörs et al. (1996) monitored the temporal behavior of OH radicals and NO_2 following the pulsed Cl-atom-initiated oxidation of $CFCl_2CH_3$ in the presence of NO in 50 mbar of O_2 at 293 K. The formation of HO_2 radicals leads (via their reaction with NO) to regeneration of OH. By fitting the OH and, to a lesser degree, NO_2 profiles values of $k(CFCl_2CH_2O+O_2 \rightarrow CFCl_2CHO+HO_2) = (2.0 \pm 1.0) \times 10^{-15}$ cm^3 molecule^{-1} s^{-1} and $k(CFCl_2CH_2O \rightarrow CFCl_2+CH_2O) < 1.0 \times 10^3$ s^{-1} were obtained.

Wu and Carr (1995) used mass spectrometry to observe the $CFCl_2CH_2O$ radical during a study of the flash photolysis of $CFCl_2CH_3$ in the presence of O_2. Inspection of Figure 2 in their article shows that in the presence of 1 Torr of O_2 at 298 K there was no observable loss of $CFCl_2CH_2O$ over the experimental time-scale of 12 ms. From the data presented by Wu and Carr (1995) it appears that there was $< 5\%$ loss of $CFCl_2CH_2O$ radicals, and hence an upper limit of <5 s^{-1} can be derived for the rate of decomposition via C—C bond scission at 298 K.

Wu and Carr (1996b) studied the kinetics of the reaction of $CFCl_2CH_2O$ radicals with O_2 using flash photolysis–mass spectrometric techniques in experiments conducted in 10–35 Torr of N_2 diluent at 251–341 K. Measured rate coefficients are shown in figure VI-E-1. As seen in this figure and discussed in section VI-E-2, there is a significant discrepancy (a factor of approximately 20) between the results for $k(CFCl_2CH_2O+O_2)$ reported by Mörs et al. (1996) and those of Wu and Carr (1996b). While the origin of this discrepancy is unclear, it is clear that the atmospheric fate of $CFCl_2CH_2O$ radicals is reaction with O_2:

$$CFCl_2CH_2O + O_2 \rightarrow CFCl_2CHO + HO_2$$

VI-E-5.45. CF_3CHClO

Møgelberg et al. (1995b) studied the Cl-atom-initiated oxidation of CF_3CH_2Cl in 700 Torr of air at 296 K and showed that decomposition (via C—C bond scission, HCl elimination, and/or Cl-atom elimination) and reaction with O_2 are competing loss mechanisms for CF_3CHClO radicals. In 700 Torr total pressure at 296 K the rate constant ratio $k_{O2}/k_{decomp} = 2.1 \times 10^{-19}$ cm^3 molecule^{-1} was determined. In one atmosphere of air at 296 K, $k_{O2}[O_2]/k_{decomp} = 1.1$, and decomposition and reaction with O_2 are of comparable importance as loss mechanisms for CF_3CHClO radicals. For the conditions of lower temperature and lower O_2 partial pressure that prevail in the upper troposphere, the rates of both decomposition and reaction with O_2 will decrease. The unimolecular decomposition will be more sensitive to reduced temperature than will reaction with O_2 and the relative importance of reaction with O_2 will increase.

By analogy to the behavior of CH_3CHClO radicals observed by Orlando and Tyndall (2002a), it is anticipated that chemical activation will be of importance for CF_3CHClO radicals.

The atmospheric fate of CF_3CHClO radical is reaction with O_2 and decomposition:

$$CF_3CHClO + O_2 \rightarrow CF_3C(O)Cl + HO_2$$

$$CF_3CHClO + M \rightarrow CF_3C(O) + HCl + M$$

$$CF_3CHClO + M \rightarrow CF_3 + HC(O)Cl + M$$

$$CF_3CHClO + M \rightarrow CF_3C(O)H + Cl + M$$

VI-E-5.46. CF_3CFClO

Edney and Driscoll (1992) report the formation of $CF_3C(O)F$ in a yield of $100 \pm 4\%$ in the Cl-atom-initiated oxidation of CF_3CFClH in 700 Torr of air at 298 K. Tuazon and Atkinson (1993a) observed $CF_3C(O)F$ in a yield of 101% following the Cl-atom-initiated oxidation of CF_3CFClH in 740 Torr of air at 298 K. Bhatnagar and Carr (1995b) monitored the formation of NO_2 following flash photolytic generation of CF_3CFClO_2 radicals in the presence of NO. Bhatnagar and Carr observed a rapid chain reaction leading to complete conversion of NO into NO_2, from which they concluded that Cl-atom elimination occurs at a rate $> 10^3\ s^{-1}$ and that this is the dominant loss for CF_3CFClO radicals. The atmospheric fate of CF_3CFClO radicals is Cl-atom elimination.

$$CF_3CFClO + M \rightarrow CF_3C(O)F + Cl + M$$

VI-E-5.47. CF_3CCl_2O

Edney et al. (1991), Tuazon and Atkinson (1993a), and Hayman et al. (1994) observed $CF_3C(O)Cl$ in a yield indistinguishable from 100% following the Cl-atom-initiated oxidation of CF_3CCl_2H in an atmosphere of air at ambient temperature. Jemi-Alade et al. (1991) conducted a study of the UV absorption spectrum of $CF_3CCl_2O_2$ and the $CF_3CCl_2O_2$–$CF_3CCl_2O_2$ reaction and observed a residual UV absorption which matched that of $CF_3C(O)Cl$. The atmospheric fate of CF_3CCl_2O is Cl-atom elimination:

$$CF_3CCl_2O + M \rightarrow CF_3C(O)Cl + Cl + M$$

VI-E-5.48. $CF_3CF_2CCl_2O$

Tuazon and Atkinson (1994) observed the formation of $CF_3CF_2C(O)Cl$ in 100% yield from the Cl-atom-initiated oxidation of $CF_3CF_2CCl_2H$ (HCFC–225ca) in 740 Torr of air at 298 K. There was no observable $COCl_2$ ($\leq 0.5\%$ yield). The atmospheric fate of $CF_3CF_2CCl_2O$ radicals is elimination of Cl-atoms:

$$CF_3CF_2CCl_2O + M \rightarrow CF_3CF_2C(O)Cl + Cl + M$$

VI-E-5.49. CF_2ClCF_2CFClO

Tuazon and Atkinson (1994) observed the formation of $CF_2ClCF_2C(O)F$ and FC(O)Cl in yields of 100% and 1%, respectively, from the Cl-atom-initiated oxidation of CF_2ClCF_2CHFCl (HCFC–225cb) in 740 Torr of air at 298 K. These results suggest that Cl-atom elimination and decomposition via C—C bond scission are competing fates for CF_2ClCF_2CFClO radicals, with Cl-atom elimination being approximately 100 times more rapid than C–C bond scission. Decomposition via C–C bond scission will lead to the formation of CF_2ClCF_2O which will react further to give two molecules of COF_2. COF_2 absorbs intensely in the infrared at 774 cm^{-1}, and it is surprising that Tuazon and Atkinson (1994) did not report the formation of this product. The dominant atmospheric fate of CF_2ClCF_2CFClO radicals is elimination of Cl-atoms.

$$CF_2ClCF_2CFClO + M \rightarrow CF_2ClCF_2C(O)F + Cl + M$$

VI-E-5.50. $CF_3CF(O\bullet)CF_2Cl$ and $CF_3CFClCF_2O$

Mashino et al. (2000) conducted a product study of the Cl-atom-initiated oxidation of $CF_3CF{=}CF_2$ in 700 Torr of air at 296 K in the presence, and absence, of NO_x. Cl-atoms add to the double bond in $CF_3CF{=}CF_2$, and the resulting C_3F_6Cl radicals then add O_2 to give $CF_3CFClCF_2O_2$ and $CF_3CF(OO)CF_2Cl$ peroxy radicals. It is likely that both peroxy radical isomers will be formed in significant amounts. RO_2–RO_2 reactions between like radicals, and cross reactions, will lead to the formation of the corresponding alkoxy radicals. Mashino et al. (2000) observed the formation of $CF_3C(O)F$ and COF_2 in molar yields that were indistinguishable from 100% and were unaffected by the presence of NO. The atmospheric fate of $CF_3CF(O\bullet)CF_2Cl$ and $CF_3CFClCF_2O$ radicals is decomposition via C—C bond scission. CF_2Cl and CF_3CFCl radicals undergo further reactions leading to CF_2O and $CF_3C(O)F$ (see sections VI-E-5.41 and VI-E-5.46). It is interesting to note that the $CF_3CF(O\bullet)CF_2Cl$ radical eliminates a CF_2Cl radical in preference to a CF_3 radical. Computational work would be of interest to understand this preference.

$$CF_3CF(O\bullet)CF_2Cl + M \rightarrow CF_3C(O)F + CF_2Cl + M$$

$$CF_3CFClCF_2O \rightarrow CF_3CFCl + CF_2O + M$$

VI-E-5.51. $C_4F_9CH(O\bullet)CH_2Cl$, $C_6F_{13}CH(O\bullet)CH_2Cl$, $C_4F_9CHClCH_2O$, $C_6F_{13}CHClCH_2O$

Vésine et al. (2000) conducted a product study of the Cl-atom-initiated oxidation of $C_4F_9CH{=}CH_2$ and $C_6F_{13}CH{=}CH_2$ in one atmosphere of air at ambient temperature in the presence, and absence, of NO. Chlorine atoms react with $C_4F_9CH{=}CH_2$ and $C_6F_{13}CH{=}CH_2$ via an addition mechanism. The inductive effect of the C_xF_{2x+1}— group reduces the electron density on the adjacent carbon atom in the double bond. Electrophilic attack by the incoming Cl-atom is expected to occur preferentially, but not exclusively, on the terminal carbon of the double bond. Addition of Cl-atom to the double bond is followed by addition of O_2 to give the peroxy radicals $C_xF_{2x+1}CH(OO)CH_2Cl$ and $C_xF_{2x+1}CHClCH_2(OO)$. RO_2–RO_2 reactions between like and unlike radicals give the corresponding alkoxy radicals. In the absence of NO_x, Vésine et al. (2000) report the formation of $C_xF_{2x+1}C(O)CH_2Cl$, CF_2O, and CO as major products. The yields of these products were the same for experiments conducted in air or O_2. In the presence of NO_x, the formation of CF_2O, CO, and a "PAN-like" product were observed. The observation of $C_xF_{2x+1}C(O)CH_2Cl$ as a major product with a yield unaffected by $[O_2]$ shows that the fate of $C_xF_{2x+1}CH(O)CH_2Cl$ radicals is reaction with O_2. The upward curving plots of CF_2O and CO formation vs. $C_xF_{2x+1}CH{=}CH_2$ loss indicate that CF_2O and CO are secondary products. The most likely explanation of the observed product yields is that $C_xF_{2x+1}CHClCH_2O$ radicals react with O_2 to give the aldehyde $C_xF_{2x+1}CHClCHO$. In contrast to the ketone, $C_xF_{2x+1}C(O)CH_2Cl$, the aldehyde $C_xF_{2x+1}CHClCHO$ is expected to react rapidly with Cl-atoms, leading to the formation of an acyl radical, $C_xF_{2x+1}CHClCO$. As discussed by Tuazon and Atkinson (1994), haloacyl radicals tend both to react with O_2 to give the corresponding acylperoxy radical and to decompose via CO elimination. The formation of acylperoxy radicals in the presence of NO_x can then explain the "PAN-like" products observed by Vésine et al. (2000). The results from Vésine et al. (2000) indicate that reaction with O_2 is the fate of $C_4F_9CH(O\bullet)CH_2Cl$, $C_6F_{13}CH(O\bullet)CH_2Cl$, $C_4F_9CHClCH_2O$, and $C_6F_{13}CHClCH_2O$ radicals:

$$C_4F_9CH(O\bullet)CH_2Cl + O_2 \rightarrow C_4F_9C(O)CH_2Cl + HO_2$$

$$C_6F_{13}CH(O\bullet)CH_2Cl + O_2 \rightarrow C_6F_{13}C(O)CH_2Cl + HO_2$$

$$C_4F_9CHClCH_2O + O_2 \rightarrow C_4F_9CHClCHO + HO_2$$

$$C_6F_{13}CHClCH_2O + O_2 \rightarrow C_6F_{13}CHClCHO + HO_2$$

VI-E-5.52. CHBrClO

Bilde et al. (1997) observed the formation of HC(O)Cl in essentially 100% yield in the chlorine-atom-initiated oxidation of CH_2BrCl in 700 Torr of N_2/O_2 diluent at 295 K in the presence, and absence, of NO. Variation of the O_2 partial pressure had no discernable effect on the product yields, leading to the conclusion that Br-atom elimination from CHBrClO radicals occurs at a rate $> 2\times10^6\ s^{-1}$. The atmospheric fate of CHBrClO radicals is elimination of Br:

$$CHBrClO + M \rightarrow HC(O)Cl + Br$$

VI-E-5.53. $CBrCl_2O$

Bilde et al. (1998) studied the Cl-atom-initiated oxidation of $CHBrCl_2$ in 700 Torr of air, or O_2, diluent at 253–295 K, with and without added NO. In the absence of NO, the major product ($99\pm5\%$ yield) was $COCl_2$, with a trace amount ($1.8\pm0.7\%$ yield) of COBrCl. In the presence of NO, the yield of $COCl_2$ dropped to $73\pm4\%$ and the

COBrCl yield increased to 21 ± 4%. There was no effect of temperature on the product yields. The results indicate that chemical activation plays an important role in the fate of $CBrCl_2O$ radicals (see section VI-E-4). Without chemical activation, approximately 98% of $CBrCl_2O$ radicals undergo decomposition via elimination of a Br-atom, with the remaining 2% eliminating a Cl-atom. $CBrCl_2O$ radicals formed in the reaction of $CBrCl_2O_2$ with NO are chemically activated; approximately 75% decompose via Br-atom elimination, and 25% decompose via Cl-atom elimination.

$$CBrCl_2O \rightarrow CCl_2O + Br$$

$$CBrCl_2O \rightarrow BrC(O)Cl + Cl$$

VI-E-5.54. CBr_2ClO

Bilde et al. (1998) observed the formation of BrC(O)Cl in essentially 100% yield in the chlorine-atom-initiated oxidation of $CHBr_2Cl$ in 700 Torr of air at 295 K in the presence and absence of NO. Br-atom elimination is the sole atmospheric fate of CBr_2ClO radicals:

$$CBr_2ClO + M \rightarrow BrC(O)Cl + Br$$

VI-E-5.55. $CHClBrCCl_2O$

Catoire et al. (1997) and Ramacher et al. (2001) have studied the products of the Br-atom-initiated oxidation of $CHCl{=}CCl_2$ in an atmosphere of air at 250–298 K. The reaction of Br with $CHCl{=}CCl_2$ takes place via addition, essentially exclusively at the hydrogen-bearing carbon atom, to give $CHClBrCCl_2$ radicals which add O_2 to give the corresponding peroxy radical that undergoes RO_2–RO_2 reactions to give $CHClBrCCl_2O$ radicals. In experiments conducted in the absence of NO, Catoire et al. (1997) and Ramacher et al. (2001) observed CHClBrC(O)Cl in 97% yield and CCl_2O and HC(O)Cl both in 3% yield. There was no discernable effect of temperature on the product yield, showing that the activation barriers for Cl elimination (giving CHClBrC(O)Cl) and C—C bond scission (leading to CCl_2O and HC(O)Cl) are indistinguishable. Ramacher et al. (2001) observed a change in the product yields when NO was present with a decrease in the CHClBrC(O)Cl yield to 80 ± 5% and increases in the CCl_2O and HC(O)Cl yields to 17 ± 4% and ~15%, respectively. Ramacher et al. ascribed this change in product yield distribution to the effect of chemical activation of $CHClBrCCl_2O^*$ radicals formed in the $CHClBrCCl_2O_2$ + NO reaction. The dominant atmospheric fate of $CHClBrCCl_2O$ radicals is Cl-atom elimination; decomposition via C—C bond scission is a minor fate.

$$CHClBrCCl_2O + M \rightarrow CHClBrC(O)Cl + Cl + M$$

$$CHClBrCCl_2O + M \rightarrow CHClBr + CCl_2O + M$$

VI-E-5.56. CCl_2BrCCl_2O

Ariya et al. (1997) and Ramacher et al. (2001) have studied the products of the Br-atom-initiated oxidation of $CCl_2{=}CCl_2$ in an atmosphere of air at 250–298 K. The reaction of Br with $CCl_2{=}CCl_2$ takes place via addition to give CCl_2BrCCl_2 radicals which add O_2 to give the corresponding peroxy radical, and these undergo RO_2–RO_2 reactions to give CCl_2BrCCl_2O radicals. At 298 K in the absence of NO, Ariya et al. reported the formation of $CCl_2BrC(O)Cl$ in $80 \pm 11\%$ yield and CCl_2O in $22 \pm 4\%$ yield. At 250 K in the absence of NO, Ramacher et al. reported the formation of CCl_2O in $30 \pm 7\%$ yield and observed features attributable to $CCl_2BrC(O)Cl$; they concluded that it accounted for the balance of carbon (i.e., 85% molar yield). As with CCl_3CCl_2O and $CHClBrCCl_2O$ radicals, there is no discernable effect of temperature on the products from CCl_2BrCCl_2O radicals, indicating that the activation barriers for Cl-atom elimination and C—C bond scission are indistinguishable. Ramacher et al. (2001) observed a change in the product yields when NO was present with a decrease in the $CCl_2BrC(O)Cl$ yield to $52 \pm 5\%$ and an increase in the CCl_2O yields to $55 \pm 5\%$. Ramacher et al. ascribed this change in product yield distribution to the affect of chemical activation of $CCl_2BrCCl_2O^*$ radicals formed in the $CCl_2BrCCl_2O_2 + NO$ reaction. Given the evidence that the barriers for the two decomposition processes are indistinguishable, the change in product yields observed for chemically activated alkoxy radicals suggests that dynamical factors are important. It appears that excitation in the nascent $CCl_2BrCCl_2O^*$ flows preferentially into the C—C rather than C—Br bond. Dynamical calculations would be of interest to shed more light on this process. The dominant atmospheric fate of CCl_2BrCCl_2O radicals is Cl-atom elimination; decomposition via C—C bond scission is a minor fate.

$$CCl_2BrCCl_2O + M \rightarrow CCl_2BrC(O)Cl + Cl + M$$

$$CCl_2BrCCl_2O + M \rightarrow CCl_2Br + CCl_2O + M$$

VI-E-5.57. $CF_3CBrClO$

Bilde et al. (1998) studied the Cl-atom-initiated oxidation of $CF_3CHBrCl$ in 700 Torr of air at 296 K with and without added NO. $CF_3C(O)Cl$ was observed in a yield of $93 \pm 10\%$; there was no discernable effect of NO. The atmospheric fate of the $CF_3CBrClO$ radical is elimination of a Br-atom:

$$CF_3CBrClO + M \rightarrow CF_3C(O)Cl + Br + M$$

VI-F. Major Products of Atmospheric Oxidation of the Haloalkanes

The atmospheric oxidation of haloalkanes containing one or more C—H bonds is generally initiated by reaction with OH radicals in the troposphere (and to a lesser extent in the stratosphere). Reaction with Cl-atoms can make a minor contribution in the troposphere and Cl- and $O(^1D)$-atoms make minor contributions in the stratosphere. Direct photolysis can initiate the atmospheric oxidation of haloalkanes possessing chromophores (C—Br or C—I bonds) which absorb at wavelengths available in the troposphere (see chapter VII). Haloalkanes that possess neither C—H, C—Br, nor C—I bonds (e.g., CFCs, PFCs) are unreactive in the troposphere and are transported by the atmospheric circulation into the stratosphere. Absorption by the layer of stratospheric ozone at approximately 15–35 km altitude prevents solar radiation with wavelengths shorter than approximately 290 nm from reaching the troposphere. Within the stratosphere, solar radiation with wavelengths shorter than 290 nm is available. Photolysis is an efficient removal mechanism for haloalkanes containing C—Cl bonds in the stratosphere. PFCs do not absorb at the wavelengths of light available in the stratosphere. They survive transport through the stratosphere and into the mesosphere where they are photolyzed by short-wavelength UV radiation. It takes millennia for the atmosphere to circulate through the mesosphere and PFCs have very long atmospheric lifetimes. The atmospheric lifetimes of haloalkanes with respect to reaction with OH radicals are discussed in chapter I.

It is convenient to separate the OH-radical-initiated atmospheric oxidation of haloalkanes into two parts. First, reactions that convert the haloalkanes into halogenated carbonyl species. Second, reactions that remove these carbonyl compounds. A generic scheme for the atmospheric oxidation of a C_2 haloalkane, adapted from Wallington et al. (1994e), is given in figure VI-F-1. Values in parentheses are order-of-magnitude lifetime estimates. Reactions with OH radicals give halogenated alkyl radicals which react with O_2 to give the corresponding peroxy radicals (RO_2) on a time-scale of approximately 1 μs or less. As discussed in section VI-D, peroxy radicals react with four important trace species in the atmosphere: NO, NO_2, HO_2, and CH_3O_2 radicals; the atmospheric lifetime of peroxy radical is of the order of a minute. Reaction with NO_2 gives haloalkoxy peroxynitrates (RO_2NO_2), which are thermally unstable and decompose to regenerate RO_2 radicals and NO_2 within minutes. Reaction with HO_2 radicals gives hydroperoxides (ROOH) and, in some cases, carbonyl products. The relative importance of the hydroperoxide and carbonyl-forming channels is uncertain. The hydroperoxide $CX_3CXYOOH$ is expected to be returned to the CX_3CXYO_x radical pool via reaction with OH and photolysis. Reaction with α-hydrogen-containing peroxy radicals such as CH_3O_2 proceeds via two channels: the first channel gives alkoxy radicals, and the second gives $CX_3CXYOH + CH_2O$ (and $CX_3C(O)X + CH_3OH$, when Y = H). When Y is an F- or Cl-atom, the alcohol CX_3CXYOH will undergo heterogeneous decomposition to give $CX_3C(O)X + HY$. The peroxy radicals derived from haloalkanes react rapidly with NO to give NO_2 and a haloalkoxy radical.

As is indicated in figure VI-F.1, haloalkoxy radicals can decompose via C—C bond scission, react with O_2, eliminate a Cl-, Br-, or I-atom, and/or eliminate HCl. These loss processes largely dictate the first-generation atmospheric oxidation products of the haloalkanes. The primary products of haloalkane oxidation are carbonyl-containing compounds. The carbonyl products represent a break point in the chemistry. The sequence of gas-phase reactions that follow from the initial attack of OH radicals on the parent haloalkane are sufficiently rapid that heterogeneous and aqueous processes

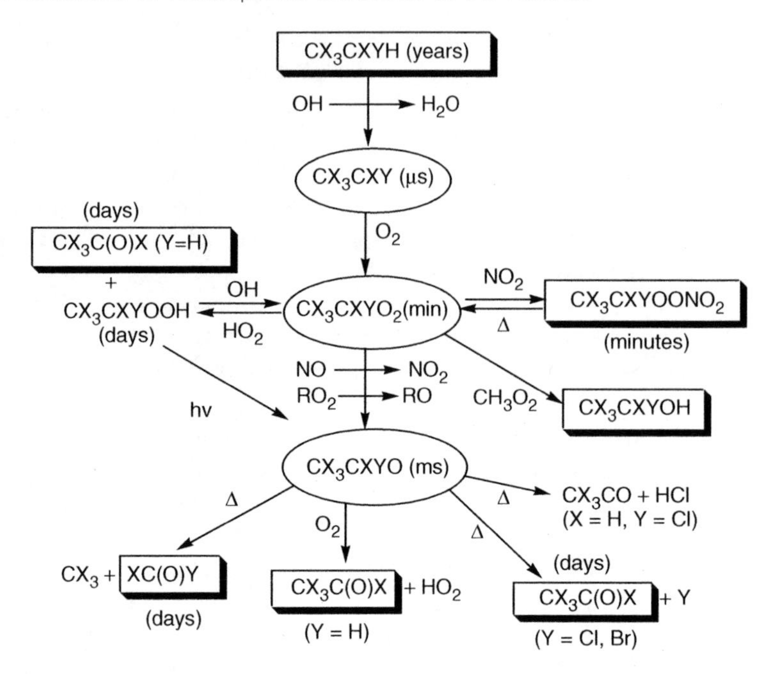

Figure VI-F-1. A generic scheme for the atmospheric oxidation of a C_2 haloalkane, taken from Wallington et al. (1994e).

play no role. In contrast, the lifetimes of the carbonyl products (e.g., HC(O)F, $C(O)F_2$, $CF_3C(O)F$) are, in general, relatively long. Incorporation into water droplets, followed by hydrolysis, plays an important role in the removal of such halogenated carbonyl compounds. In the case of $C(O)F_2$, FC(O)Cl, and $CF_3C(O)F$, reaction with OH radicals and photolysis are too slow to be of any significance in the troposphere. These compounds are removed entirely by incorporation into water droplets, which is believed to occur on a time scale of approximately 1–2 weeks (Wallington et al., 1994e).

The gas-phase oxidation mechanism for $CX_3C(O)H$ and $CF_3C(O)Cl$ is shown in figure VI-F-2. For $CX_3C(O)H$ species, reaction with OH radicals is important. The lifetimes of $CF_3C(O)H$, $CF_2ClC(O)H$, and $CFCl_2C(O)H$ with respect to OH attack have been estimated by Scollard et al. (1993) to be 24, 19, and 11 days, respectively. By analogy to acetaldehyde, it is expected that photolysis is also a sink for $CF_3C(O)H$, $CF_2ClC(O)H$, and $CFCl_2C(O)H$ (see sections VII-F-5 and VII-F-6, Selleväg et al., 2004, 2005; Chiappero et al., 2006). Finally, scavenging by water droplets also probably plays a role in the atmospheric fate of these halogenated aldehydes. Reaction of CF_3CHO with H_2O proceeds via a heterogeneous mechanism to give the hydrate $CF_3CH(OH)_2$. Unimolecular dehydration of $CF_3CH(OH)_2$ has a high barrier (54.6 kcal mol^{-1}; Francisco, 1992) and will be negligibly slow. The hydrate reacts with OH radicals, leading to the formation of $CF_3C(O)OH$ in a yield close to 100% (Andersen et al., 2006). For $CF_3C(O)Cl$, reaction with OH is not feasible. Photolysis of $CF_3C(O)Cl$ is important (photodissociation

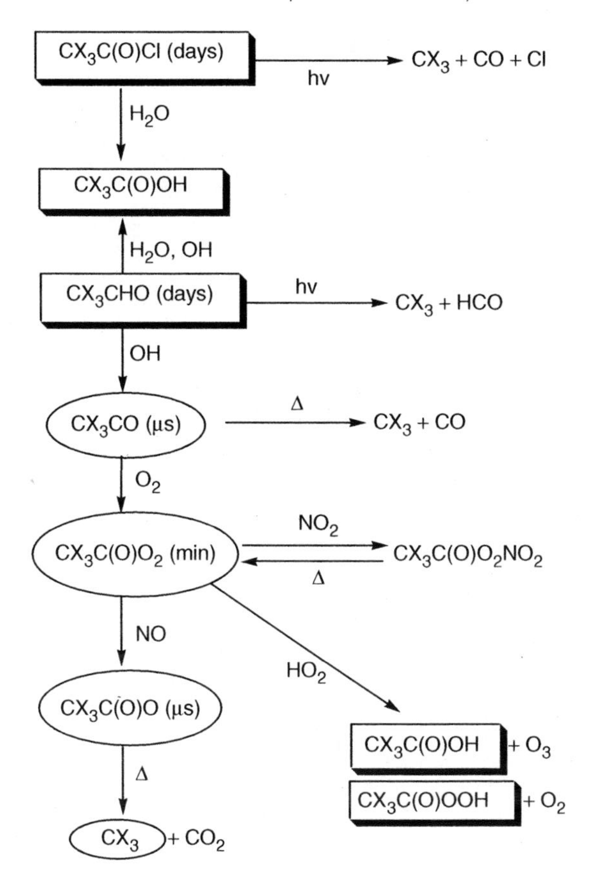

Figure VI-F-2. The gas-phase oxidation mechanism for $CX_3C(O)H$ and $CF_3C(O)Cl$.

lifetime for overhead Sun, ~23 days; see chapter VII), and this competes with incorporation of $CF_3C(O)Cl$ into water droplets followed by hydrolysis to give $CF_3C(O)OH$.

Following reaction with OH radicals, $CF_3C(O)$, $CF_2ClC(O)$, and $CFCl_2C(O)$ radicals can either react with O_2 or decompose to give CO and a halogenated methyl radical (Méreau et al., 2001; Solignac et al., 2006; Hurley et al., 2006). Wallington et al. (1994c) have reported that reaction with O_2 accounts for 99.5% of the atmospheric fate of $CF_3C(O)$ radicals. Tuazon and Atkinson (1994) have determined that in 740 Torr of air at 298 K, reaction with O_2 accounts for 61 ± 5% of CF_2ClCO and 21 ± 5% of $CFCl_2CO$ radicals. Zabel et al. (1994) showed that reaction with O_2 is of minor importance as a loss for CCl_3CO radicals in one atmosphere of air at ambient temperature. The resulting $CX_3C(O)O_2$ radical can react with NO, NO_2, or HO_2 radicals. Reaction with NO gives a $CX_3C(O)O$ radical which rapidly dissociates to give CX_3 radicals and CO_2. Reaction with NO_2 gives a haloacetyl peroxynitrate which undergoes thermal decomposition to regenerate $CX_3C(O)O_2$ and NO_2. As discussed by Andersen et al. (2004b), the reaction of $CX_3C(O)O_2$ radicals with

HO_2 appears to proceed via three reaction channels giving peracid, acid, and radical products:

$$CX_3C(O)O_2 + HO_2 \rightarrow CX_3C(O)OOH + O_2$$

$$CX_3C(O)O_2 + HO_2 \rightarrow CX_3C(O)OH + O_3$$

$$CX_3C(O)O_2 + HO_2 \rightarrow CX_3C(O)O + OH + O_2$$

The peracid is expected to undergo photolysis and reaction with OH radicals returning radicals to the $CX_3C(O)O_x$ pool. In contrast, the major fate of $CX_3C(O)OH$ is rain-out. Perfluorocarboxylic acids such as $CF_3C(O)OH$ and higher analogues (e.g., $C_7F_{15}C(O)OH$, perfluoroactanoic acid, PFOA) are persistent in the environment and are the subject of a significant current research effort.

The atmospheric oxidation of haloalkanes leads to mineralization (conversion to CO_2 and HX) and the formation (generally in minor yields) of halocarboxylic acids. Although halocarboxylic acids are formed only in small yields, the persistence, bioaccumulative nature of longer chain acids, and potential health concerns associated with these acids have led to interest in these compounds. There are four routes by which halocarboxylic acids can be formed in the oxidation of haloalkanes. First, conversion of CX_3CXYO radicals into $CX_3C(O)X$ ($X = F$ or Cl), followed by hydrolysis of $CX_3C(O)X$. Second, reaction of $CX_3C(O)O_2$ with HO_2 radicals. Third, reaction of CX_3CXYO_2 radicals with CH_3O_2 to give CX_3CXYOH (X and Y = Cl or F), followed by elimination of HF or HCl, giving $CX_3C(O)X$ ($X =$ Cl or F) and hydrolysis to $CX_3C(O)OH$ (Ellis et al., 2004; Wallington et al., 2006). Fourth, addition of H_2O to CX_3CHO to give the hydrate $CX_3CH(OH)_2$, reaction with OH to give $CX_3CH(OH)_2$ radicals which then react with O_2 to give $CX_3C(O)OH$ (Andersen et al., 2006). In the following sections the major products of the atmospheric oxidation of haloalkanes are described.

VI-F-1. CH_3F (HFC-41)

Wallington et al. (1992b) and Tuazon and Atkinson (1993a) have studied the products of Cl-atom-initiated oxidation of CH_3F in 700 Torr of air at 296 K and found essentially 100% yield of HC(O)F. The atmospheric oxidation of CH_3F is initiated by reaction with OH radicals, proceeds via reaction of the CHF_2O species with O_2, and gives HC(O)F as the major product. HC(O)F reacts slowly with OH radicals, does not photolyze in the troposphere, and is expected to hydrolyze on contact with rain, cloud, or sea water to give CO_2 and HF (Wallington and Hurley, 1993).

VI-F-2. CH_2F_2 (HFC-32)

Nielsen et al. (1992a) have shown that the Cl-atom-initiated oxidation of CH_2F_2 in 700 Torr of air at 298 K gives COF_2 in a yield that is indistinguishable from 100%. The atmospheric oxidation of CH_2F_2 is initiated by reaction with OH radicals, proceeds via reaction of the CHF_2O species with O_2, and gives COF_2 as the major product. COF_2 does not react with OH radicals, does not photolyze in the troposphere, and is expected to hydrolyze on contact with rain, cloud, or sea water to give CO_2 and HF (Wallington et al., 1994e).

VI-F-3. CF_3H (HFC-23)

Reaction of OH radicals with CF_3H leads to the formation of CF_3O_2 radicals which are converted into CF_3O radicals. Reaction with NO and CH_4 converts CF_3O radicals into COF_2 (see section VI-E-5.3). COF_2 does not react with OH radicals, does not photolyze in the troposphere, and is expected to hydrolyze on contact with rain, cloud, or sea water to give CO_2 and HF (Wallington et al, 1994e).

VI-F-4. CH_3CH_2F (HFC-161)

Using the SAR substituent factors recommended in section VI-C (see table VI-C-2), it is predicted that at 298 K approximately 75% of the reaction of OH radicals with CH_3CH_2F proceeds via hydrogen abstraction from the —CH_2F group, with the balance proceeding via attack on the —CH_3 group. Abstraction from the —CH_2F group will lead to the formation of CH_3CHFO radicals which react with O_2 to give CH_3COF or decompose to give CH_3 radicals and HC(O)F (Hou et al., 2000a). CH_3 radicals will react further to give CH_2O and CH_3OH. Abstraction from the —CH_3 group will lead to the formation of CH_2FCH_2O radicals which, by analogy to the behavior of CH_2ClCH_2O and CH_2BrCH_2O radicals, are expected to react with O_2 to give CH_2FCHO. The atmospheric oxidation of CH_3CH_2F is initiated by reaction with OH and gives HC(O)F, CH_2O, CH_3OH, CH_2FCHO, and CH_3COF as major primary products.

VI-F-5. CH_2FCH_2F (HFC-152) and CH_3CHF_2 (HFC-152a)

Wallington et al. (1994b) investigated the Cl-atom-initiated oxidation of CH_2FCH_2F in 700 Torr of air diluent at 296 K and found essentially complete (182 ± 19%) conversion into HC(O)F. This results from the decomposition of CH_2FCHFO to CH_2F radicals and HC(O)F and the subsequent conversion of CH_2F radicals into another molecule of HC(O)F. The major atmospheric oxidation product of the OH-radical-initiated oxidation of CH_2FCH_2F is HC(O)F.

Using the SAR substituent factors recommended in section VI-C (see table VI-C-2), it is predicted that at 298 K approximately 72% of the reaction of OH radicals with CH_3CHF_2 proceeds via hydrogen abstraction from the —CHF_2, group, with the balance proceeding via attack on the —CH_3 group. Abstraction from the —CHF_2 group will lead to the formation of CH_3CF_2O radicals which decompose to give CH_3 radicals and COF_2 (Edney and Driscoll, 1992; Tuazon and Atkinson, 1993a; Taketani et al., 2005). CH_3 radicals will react further to give CH_2O and CH_3OH. Abstraction from the —CH_3 group will lead to the formation of CHF_2CH_2O radicals which, by analogy to the behavior of CF_3CH_2O radicals, are expected to react with O_2 to give CHF_2CHO.

Photolysis and reaction with OH will convert CHF_2CHO into COF_2. The atmospheric oxidation of CH_3CHF_2 is initiated by reaction with OH and gives COF_2, CHF_2CHO, CH_2O, and CH_3OH as major primary products.

VI-F-6. CH_2FCHF_2 (HFC-143)

Barry et al. (1995) investigated the products following Cl-atom-initiated oxidation of CH_2FCHF_2 in 1 atmosphere of N_2/O_2 diluent at 255–330 K and reported the formation of HC(O)F and COF_2 products in yields indistinguishable from 100%. The reaction of Cl-atoms with CH_2FCHF_2 occurs approximately 50% via hydrogen abstraction from the CH_2F– group and 50% from the –CHF_2 group (Tschuikow-Roux et al., 1985b) and will lead to the formation of CH_2FCF_2O and CHF_2CHFO radicals. The fact that Barry et al. (1995) observed HC(O)F and COF_2 as the dominant products even in 760 Torr of O_2 at 255 K shows that decomposition via C–C bond scission is the sole atmospheric fate of CH_2FCF_2O and CHF_2CHFO radicals. Irrespective of the site of OH-radical-initiated oxidation of CH_2FCHF_2, the same products will be formed: HC(O)F and COF_2. The dominant atmospheric oxidation products are HC(O)F and COF_2.

VI-F-7. CF_3CH_3 (HFC-143a)

Nielsen et al. (1994) studied the F-atom-initiated oxidation of CF_3CH_3 in 700 Torr of N_2/O_2 diluent with $[O_2] = 1$, 5, or 147 Torr at 296 K. CF_3CHO was observed as the major product in a yield of $101 \pm 8\%$. In a study of the Cl-atom-initiated oxidation of $CF_3CH_2CF_2CH_3$ in 700 Torr of air at 297 K, Chen et al. (1997) reported the formation of CF_3CHO in a yield of $97 \pm 5\%$. As discussed by Chen et al. (1997), the oxidation of $CF_3CH_2CF_2CH_3$ proceeds via the formation of CF_3CH_2 radicals, which are the same radicals as those produced during the atmospheric oxidation of CF_3CH_3. Atmospheric oxidation of CF_3CH_3 is initiated by reaction with OH radicals and gives CF_3CHO as the major product.

VI-F-8. CHF_2CHF_2 (HFC-134)

Nielsen et al. (1992c) studied the Cl-atom-initiated oxidation of CHF_2CHF_2 in 700 Torr of air at 295 K and report the formation of CF_2O in a yield of $196 \pm 24\%$. Atmospheric oxidation of CHF_2CHF_2 is initiated by reaction with OH radicals, proceeds to give CHF_2CF_2O radicals which decompose to give CF_2O and CHF_2 radicals which react further (section VI-F-2) to give more CF_2O. Atmospheric oxidation of CHF_2CHF_2 gives CF_2O.

VI-F-9. CF_3CH_2F (HFC-134a)

Reflecting its commercial importance, there have been numerous experimental (Wallington et al., 1992d; Edney and Driscoll, 1992; Maricq and Szente, 1992; Tuazon and Atkinson, 1993b; Rattigan et al., 1994; Meller et al., 1994; Wallington et al., 1996b; Bednarek et al., 1996; Møgelberg et al., 1997b; Hasson et al., 1998; Wu and Carr, 2003) and computational investigations (Schneider et al., 1998; Somnitz and Zellner, 2001; Wu and Carr, 2003) of the atmospheric degradation mechanism of CF_3CH_2F. Studies of the Cl-atom-initiated oxidation of CF_3CH_2F have been performed in 20–5650 Torr of N_2/O_2 diluent at 233–357 K in the presence and absence of NO (Wallington et al., 1992d, 1996b; Edney and Driscoll, 1992; Tuazon and Atkinson, 1993b; Meller et al., 1994; Rattigan et al, 1994; Hasson et al., 1998; Wu and Carr, 2003). Three major carbon-containing products are observed: HC(O)F, $CF_3C(O)F$, and CF_2O. The atmospheric oxidation of CF_3CH_2F is initiated by reaction with OH radicals, leading to the formation of CF_3CFHO_2 which then reacts with other peroxy radicals; CH_3O_2 is the most abundant organic peroxy radical in the atmosphere and is the choice for the example given in the following reactions. CF_3CFHO_2 can also reacts with NO to give CF_3CFHO radicals. In one atmosphere of air at 296 K, approximately 64% of the CF_3CFHO radicals formed in the reaction of CF_3CFHO_2 with NO are produced with sufficient internal excitation to decompose on a time-scale that is sufficiently short ($< 10^{-9}$ s) so that reaction with O_2 cannot compete. The remaining approximately 36% of the CF_3CFHO radicals become thermalized. Bimolecular reaction with O_2 to give $CF_3C(O)F$ and unimolecular decomposition via C—C bond scission are competing fates for thermalized CF_3CFHO radicals.

$$CF_3CH_2F + OH \rightarrow CF_3CHF + H_2O$$

$$CF_3CHF + O_2 + M \rightarrow CF_3CHFO_2 + M$$

$$CF_3CHFO_2 + CH_3O_2 \rightarrow CF_3CHFO + CH_3O + O_2$$

$$CF_3CHFO + NO \rightarrow CF_3CHFO^* + NO_2$$

$$CF_3CHFO + NO \rightarrow CF_3CHFO + NO_2$$

$$CF_3CHFO^* \rightarrow CF_3 + HC(O)F$$

$$CF_3CHFO + O_2 \rightarrow CF_3C(O)F + HO_2$$

$$CF_3CHFO + M \rightarrow CF_3 + HC(O)F + M$$

The rate constant ratio for reaction with O_2 vs. decomposition is $k_{O2}/k_{diss} = 4.1 \times 10^{-20}$ cm^3 molecule^{-1} at 298 K. In one atmosphere of air $k_{O2}[O_2]/k_{diss} = 0.21$ and 83% of the thermalized CF_3CFHO radicals undergo decomposition to give CF_3 radicals and HC(O)F with the remaining fraction reacting with O_2 to give $CF_3C(O)F$. CF_3 radicals react further to give CF_2O (section VI-E-5.3). The major atmospheric degradation products of CF_3CH_2F are HC(O)F, CF_2O, and $CF_3C(O)F$.

VI-F-10. CF_3CF_2H (HFC-125)

Edney and Driscoll (1992), Tuazon and Atkinson (1993a), and Sehested et al. (1993a) have shown that the Cl-atom-initiated oxidation of CF_3CF_2H in air at ambient temperature gives CF_2O and CF_3 radicals via decomposition of CF_3CF_2O radicals.

CF_3 radicals react further to give CF_2O (see section VI-E-5.3). The major atmospheric degradation product of CF_3CF_2H is CF_2O.

VI-F-11. $CH_3CF_2CH_3$ (HFC-272ca)

Møgelberg et al. (1995c) studied the Cl-atom-initiated oxidation of $CH_3CF_2CH_3$ in 700 Torr of N_2/O_2 at 296 K with O_2 partial pressures of 5 or 147 Torr. CH_3CF_2CHO was observed in a yield of 97 ± 5% and was independent of $[O_2]$ over the range studied. Atmospheric oxidation of $CH_3CF_2CH_3$ is initiated by reaction with OH radicals, which leads to the formation of CH_3CF_2CHO as the major product.

VI-F-12. $CF_3CH_2CHF_2$ (HFC-245fa)

Using the SAR substituent factors recommended in section VI-C (see table VI-C-2), it is predicted that approximately 56% of the reaction of OH radicals with $CF_3CH_2CHF_2$ proceeds via hydrogen abstraction from the terminal carbon atom, with the balance proceeding via attack on the —CH_2— group. Chen et al. (1997) have studied the Cl-atom-initiated oxidation of $CF_3CH_2CHF_2$ in 700 Torr of air at 297 K. On the basis of the 95 ± 2% yield of $CF_3CH_2CF_2Cl$ observed in the photochemical chain chlorination following UV irradiation of $CF_3CH_2CHF_2/Cl_2/He$ mixtures, it was concluded that Cl-atoms react mainly at the terminal carbon atom. Chen et al. (1997) observed the formation of CF_3CHO and COF_2 as major products in the Cl-atom-initiated oxidation of $CF_3CH_2CHF_2$ in 700 Torr of air at 297 K. The CF_3CHO yield was 97 ± 5%, indicating that hydrogen abstraction from the terminal carbon gives CF_3CHO and COF_2 products in essentially 100% yield. Hydrogen abstraction from the —CH_2— group will lead to the formation of $CF_3CH(O)CHF_2$ radicals which, by analogy to CF_3CHOCF_3 radicals, are expected to react with O_2 to give $CF_3C(O)CHF_2$. The major atmospheric oxidation products of $CF_3CH_2CHF_2$ are expected to be CF_3CHO, COF_2, and $CF_3C(O)CHF_2$,

VI-F-13. $CF_3CH_2CF_3$ (HFC-236fa)

Møgelberg et al. (1995d) studied the F-atom-initiated oxidation of $CF_3CH_2CF_3$ in 700 Torr of N_2/O_2 at 296 K with O_2 partial pressures of 5 or 147 Torr and observed the formation of $CF_3C(O)CF_3$ in a yield of 81±5%. Atmospheric oxidation of $CF_3CH_2CF_3$ is initiated by reaction with OH radicals, which leads to the formation of $CF_3C(O)CF_3$ as the major product. The limited information on the photochemistry of $CF_3C(O)CF_3$ is described in section VII-G.

VI-F-14. CF_3CHFCF_3 (HFC-227ea) and $CF_3CF_2CF_2H$ (HFC-227ca)

Zellner et al. (1994) identified $CF_3C(O)F$ and CF_2O in yields of 100 ± 5% and 80 ± 20%, respectively, following the Cl-atom-initiated oxidation of CF_3CFHCF_3 in 1 atmosphere of air in the presence of NO. Møgelberg et al. (1996b) studied the

F-atom-initiated oxidation of CF_3CFHCF_3 in 700 Torr of N_2/O_2 with $[O_2] = 5$–147 Torr and observed the formation of $CF_3C(O)F$ in a yield of 94 ± 6%. The atmospheric oxidation of CF_3CFHCF_3 is initiated by reaction with OH radicals, proceeds via decomposition of $CF_3C(O\bullet)FCF_3$ radicals, and leads to the formation of $CF_3C(O)F$ and CF_2O as major products.

Giessing et al. (1996) studied the Cl-atom-initiated oxidation of $CF_3CF_2CF_2H$ in the presence of NO in 5 or 10 Torr of O_2 diluent. COF_2 was the only carbon-containing product observed. The observed formation of COF_2 accounted for 91 ± 5% of the loss of the $CF_3CF_2CF_2H$. The atmospheric oxidation of $CF_3CF_2CF_2H$ is initiated by reaction with OH radicals and proceeds via the formation of $CF_3CF_2CF_2O$ radicals which then undergo "unzipping" reactions which convert $CF_3CF_2CF_2H$ into COF_2. The atmospheric degradation of $CF_3CF_2CF_2H$ gives COF_2 as the major product.

VI-F-15. $CF_3CH_2CF_2CH_3$ (HFC-365mfc)

Using the SAR substituent factors recommended in section VI-C (see table VI-C-2), it is predicted that at 298 K approximately 76% of the reaction of OH radicals with $CF_3CH_2CF_2CH_3$ proceeds via hydrogen abstraction from the CH_3— group, with the balance proceeding via attack on the —CH_2— group. The results of the product study by Barry et al. (1997) suggest that: (i) abstraction from the CH_3— group will lead to the formation of the aldehyde $CF_3CH_2CF_2CHO$ as the dominant primary product; (ii) oxidation of $CF_3CH_2CF_2CHO$ will give CO_2, COF_2, and CF_3CHO as secondary products; and (iii) oxidation of CF_3CHO will give CO_2 and COF_2 as tertiary products. By analogy to the chemistry of $CF_3CH_2CF_3$, it is expected that abstraction from the —CH_2— group in $CF_3CH_2CF_2CH_3$ will lead to the formation of the ketone $CF_3C(O)CF_2CH_3$. Further OH-radical-initiated oxidation of $CF_3C(O)CF_2CH_3$ is expected to produce $CF_3C(O)CF_2CHO$ which, in turn, will undergo oxidation to CO_2 and COF_2. The primary oxidation products of the OH-radical-initiated atmospheric oxidation of $CF_3CH_2CF_2CH_3$ are $CF_3CH_2CF_2CHO$ and $CF_3C(O)CF_2CH_3$.

VI-F-16. CH_3Cl (Methyl Chloride)

Studies by Sanhueza and Heicklen (1975a) and Niki et al. (1980) showed that the Cl-atom-initiated oxidation of CH_3Cl in 1 atmosphere of air at 298 K in the absence of NO leads to essentially complete conversion of CH_3Cl into HC(O)Cl. The atmospheric oxidation of CH_3Cl will be initiated by reaction with OH radicals and will give the same products as Cl-atom-initiated oxidation. Bilde et al. (1999) have shown that in 700 Torr of air in the presence of NO the Cl-atom-initiated oxidation of CH_3Cl gives approximately 60% HC(O)Cl, 30% CO, and 10% CH_2O. The change in product yield reflects the formation of chemically activated CH_2ClO^* in the presence of NO; see section VI-E-4. The major products of the OH-radical-initiated atmospheric oxidation of CH_3Cl are HC(O)Cl, CH_2O, and CO.

VI-F-17. CH_2Cl_2 (Dichloromethane)

Sanhueza and Heicklen (1975a) and Niki et al. (1980) have shown that HC(O)Cl is the major product of the Cl-atom-initiated oxidation of CH_2Cl_2 in 1 atmosphere of air at 298 K. Atmospheric oxidation of CH_2Cl_2 is initiated by reaction with OH radicals and proceeds via the formation of $CHCl_2O$ radicals which eliminate Cl-atoms to give HC(O)Cl as the major product.

VI-F-18. CCl_3H (Chloroform)

Lesclaux et al. (1987) and Li and Francisco (1989) have shown that CCl_2O is the major product of the chlorine-atom-initiated oxidation of CCl_3H. Atmospheric oxidation of CCl_3H is initiated by reaction with OH radicals, leading to the formation of CCl_3O radicals. The atmospheric fate of the CCl_3O radical is elimination of a Cl-atom to give CCl_2O. The atmospheric oxidation of CCl_3H gives CCl_2O as the major product.

VI-F-19. CH_3CH_2Cl (Ethyl Chloride)

Using the SAR substituent factors recommended in section VI-C (see table VI-C-2), it is predicted that at 298 K approximately 89% of the reaction of OH radicals with CH_3CH_2Cl proceeds via hydrogen abstraction from the —CH_2Cl group, with the balance proceeding via attack on the CH_3— group. Shi et al. (1993), Maricq et al. (1993), Kaiser and Wallington (1995), and Orlando and Tyndall (2002a) have shown that hydrogen abstraction from the —CH_2Cl group will lead to the formation of the acetyl radicals, CH_3CO, and HCl as major products and $CH_3C(O)Cl$ as a minor oxidation product. CH_3CO radicals will react further to give CH_2O, CH_3OH, and CO_2. Attack on the CH_3— group leads to formation of $CH_2ClC(O)H$ as the major product (Wallington et al., 1990; Yarwood et al., 1992; Orlando et al., 1998). Atmospheric oxidation of CH_3CH_2Cl is initiated by reaction with OH radicals and gives CH_2O, CH_3OH, CO_2, $CH_3C(O)Cl$, and CH_2ClCHO as major products.

VI-F-20. CH_2ClCH_2Cl (1,2-Dichloroethane)

Wallington et al. (1996a) have shown that $CH_2ClCHClO$ radicals formed during the OH-radical-initiated oxidation of CH_2ClCH_2Cl undergo both reaction with O_2 to give CH_2ClCHO and elimination of HCl to give CH_2ClCO radicals. CH_2ClCO radicals will react further to give CO_2 and HC(O)Cl. The major primary products of the OH-radical-initiated oxidation of CH_2ClCH_2Cl are CH_2ClCHO, CO_2, and HC(O)Cl.

VI-F-21. CCl_3CH_3 (Methyl Chloroform)

Nelson et al. (1990b) and Platz et al. (1995) have shown that the Cl-atom-initiated oxidation of CCl_3CH_3 in one atmosphere of air at ambient temperatures proceeds to give CCl_3CH_2O radicals which then react with O_2 to give CCl_3CHO (chloral) as the major product (> 90%). Trace amounts of CCl_2O were also detected during the Cl-atom-initiated oxidation of CCl_3CH_3. CCl_3CHO is the major product of the OH-radical-initiated atmospheric oxidation of CCl_3CH_3.

VI-F-22. CCl_3CCl_2H

Huybrechts et al. (1967), Mathias et al. (1974), Sanhueza and Heicklen (1975a), Møgelberg et al. (1995e), Ariya et al. (1997), Hasson and Smith (1999), and Thüner et al. (1999) have shown that the OH-radical-initiated oxidation of CCl_3CCl_2H gives $CCl_3C(O)Cl$ and $COCl_2$ in molar yields of approximately 85% and 30% in low NOx environments, and 65% and 30% in high NO_x environments.

VI-F-23. $CH_2ClCHClCH_2Cl$

Using the SAR substituent factors recommended in section VI-C (see table VI-C-2), it is predicted that at 298 K approximately 77% of the reaction of OH radicals with $CH_2ClCHClCH_2Cl$ proceeds via hydrogen abstraction from the —CH_2Cl groups (leading to formation of $CH_2ClCHClCHClO$ radicals) with the balance proceeding via attack on the —CHCl— group (leading to formation of $CH_2ClCCl(O\bullet)CH_2Cl$ radicals). Voicu et al. (2001) showed that HC(O)Cl and CO_2 are the major products following hydrogen abstraction from the —CH_2Cl groups, while $CH_2ClC(O)CH_2Cl$ and $CH_2ClC(O)Cl$ are products following hydrogen abstraction from the —CHCl— group. The atmospheric oxidation of $CH_2ClCHClCH_2Cl$ is initiated by reaction with OH radicals and gives HC(O)Cl and CO_2 as the major products, with $CH_2ClC(O)CH_2Cl$ and $CH_2ClC(O)Cl$ formed in lesser yields.

VI-F-24. CH_3Br (Methyl Bromide)

Chen et al. (1995) and Orlando et al. (1996a) have shown that CH_2O is the major product of the Cl-atom-initiated oxidation of CH_3Br in 700 Torr of air at 228–298 K in the presence and absence of NO. Atmospheric oxidation of CH_3Br is initiated by reaction with OH radicals, proceeds via the formation of CH_2BrO radicals, and gives CH_2O as the major product.

VI-F-25. CH_2Br_2 (Dibromomethane)

Orlando et al. (1996b) observed the formation of HC(O)Br from the Cl-atom-initiated oxidation of CH_2Br_2 in 700 Torr of N_2/O_2 diluent at 298 K in the presence and absence of NO. Similarly, the OH-initiated oxidation of CH_2Br_2 proceeds via the formation of $CHBr_2O$ radicals which eliminate Br-atoms to give the product HC(O)Br. Photolysis of CH_2Br_2 is a slow process within the troposphere (see chapter VII). The major atmospheric loss mechanism for CH_2Br_2 is reaction with OH, and HC(O)Br is the major product. The photodecomposition of HC(O)Br is relatively rapid, even within the lower troposphere (see sections VII-E-7 and VII-H).

VI-F-26. CBr_3H (Bromoform)

Kamboures et al. (2002) identified CBr_2O in a yield of 104 ± 5% in the Cl-atom-initiated oxidation of CBr_3H in 11 Torr total pressure of $Cl_2/O_2/CBr_3H$ mixtures with 5 Torr partial pressure of O_2. The OH-initiated oxidation of $CHBr_3$ gives CBr_2O. Photolysis of $CHBr_3$ will give $CHBr_2$ radicals which will be converted into $CHBr_2O$ radicals. The atmospheric fate of $CHBr_2O$ radicals is elimination of a Br-atom to give HC(O)Br. Loss of $CHBr_3$ via reaction with OH radicals and photolysis are of comparable importance in the atmosphere, and hence both CBr_2O and HC(O)Br are products of CBr_3H atmospheric oxidation.

VI-F-27. CH_3CH_2Br (Ethyl Bromide)

Using the SAR substituent factors recommended in section VI-C (see table VI-C-2), approximately 83% of the reaction of OH radicals with CH_3CH_2Br is predicted to proceed via hydrogen abstraction from the —CH_2Br group, with the balance proceeding via attack on the CH_3— group. Orlando and Tyndall (2002a) have shown that hydrogen abstraction from the —CH_2Br group will lead to the formation of CH_3CHO as the dominant product. Hydrogen abstraction from the CH_3— group gives CH_2CH_2Br radicals which either add O_2 to give $CH_2BrCH_2O_2$ radicals or decompose via Br-atom elimination to give C_2H_4. Orlando and Tyndall (2002a) have reported $k_{Br}/k_{O2} = 1.4 \times 10^{23} \exp(-2800/T)\,cm^3\,molecule^{-1}\,s^{-1}$ at 226–298 K. In one atmosphere of air, $k_{Br}/k_{O2}[O_2] = 2.3$ at 298 K and Br-atom elimination dominates. For conditions typical of the upper troposphere (e.g., 10 km altitude in the U.S. Standard Atmosphere, 223 K, 265 mbar pressure) $k_{Br}/k_{O2}[O_2] = 0.2$ and reaction with O_2 dominates. Yarwood et al. (1992) have shown that $CH_2BrCH_2O_2$ radicals are converted into CH_2BrCH_2O radicals which then react with O_2 to form CH_2BrCHO. The main atmospheric oxidation products of CH_3CH_2Br are CH_3CHO, $CH_3C(O)Br$, C_2H_4, and CH_2BrCHO.

VI-F-28. $CH_3CH_2CH_2Br$ (*n*-Propyl Bromide)

Gilles et al. (2002) conducted a study of the products of the OH-radical-initiated oxidation of $CH_3CH_2CH_2Br$ in 620 Torr of air in the presence of NO_x at room temperature. Propanal and bromoacetone were identified in molar yields of 30 ± 15 and 50 ± 20%, respectively. The propanal and bromoacetone yields were indistinguishable from the branching ratios for hydrogen abstraction by OH radicals at the 1- and 2-positions. At low $[O_2]$, propene was observed as a product, indicating that reaction with O_2 and dissociation via Br-atom elimination are competing loss mechanisms for CH_3CHCH_2Br radicals. Gilles et al. (2002) derived $k_{O2}/k_{diss} = (4.0 \pm 0.6) \times 10^{-18} \times cm^3\,molecule^{-1}$. In one atmosphere of air $k_{O2}[O_2]/k_{diss} = 21$, and reaction with O_2 dominates the atmospheric fate of CH_3CHCH_2Br radicals. Reaction of OH radicals with $CH_3CH_2CH_2Br$ at the 3-position is expected to be of minor (< 20%) importance. The major products of the OH-radical-initiated oxidation of $CH_3CH_2CH_2Br$ are propanal and bromoacetone.

VI-F-29. CH_3I (Methyl Iodide)

As discussed in chapter VII, the major atmospheric loss of CH_3I is by photolysis, giving CH_3 radicals and I-atoms. For 40°N latitude, clear tropospheric skies, and diurnal variation of the Sun, the lifetimes of CH_3I with respect to photolysis are approximately 4.9 and 27 days for June 22 and December 22, respectively (see section VII-E-2). CH_3 radicals will add O_2 to give CH_3O_2 radicals, which are converted into CH_2O and CH_3OH via reaction with NO or other CH_3O_2 radicals. Reaction with OH radicals and Cl-atoms accounts for a minor (approximately 5%) fraction of the CH_3I loss. The reaction of OH radicals or Cl-atoms with CH_3I will give CH_2I radicals. The atmospheric fate of CH_2I radicals is unclear. Cotter et al. (2001) and Enami et al. (2004), while disagreeing on the exact mechanism, have presented evidence that CH_2I radicals react with O_2, leading to CH_2O formation. In contrast, Eskola et al. (2006) conclude that CH_2O is, at most, a minor product. As discussed in section VI-E-5.37, the different findings may reflect the different experimental conditions in the studies. More work is needed in this area. The major products of the atmospheric oxidation of CH_3I are CH_2O and CH_3OH.

VI-F-30. C_2H_5I (Ethyl Iodide)

As discussed in chapter VII, the major atmospheric loss of C_2H_5I is by photolysis, giving C_2H_5 radicals and I-atoms. For 40°N latitude, clear tropospheric skies, and diurnal variation of the Sun, the lifetimes of C_2H_5I with respect to photolysis are approximately 4.2 and 24 days for June 22 and December 22, respectively (see section VII-E-3). C_2H_5 radicals will add O_2 to give $C_2H_5O_2$ radicals, which are converted into CH_3CHO and C_2H_5OH via reaction with NO or other peroxy radicals (e.g., CH_3O_2). Reaction with OH radicals accounts for a minor (approximately 20%, Cotter et al., 2001) fraction of the C_2H_5I loss. Using the SAR substituent factors recommended in section VI-C (see table VI-C-2), approximately 90% of the reaction of OH radicals with C_2H_5I is predicted to proceed via hydrogen abstraction from the —CH_2I group, with the balance proceeding via attack on the CH_3— group. Cotter et al. (2001) and Orlando et al. (2005) have shown that hydrogen abstraction from the —CH_2I group leads to CH_3CHO as the major product. Orlando et al. (2005) have shown that hydrogen abstraction from the —CH_3 group leads to C_2H_4 as the major product. The major products of the atmospheric oxidation of C_2H_5I are CH_3CHO, C_2H_5OH, and C_2H_4.

VI-F-31. CH_3CHICH_3 (*iso*-Propyl Iodide)

As discussed in chapter VII, the major atmospheric loss of CH_3CHICH_3 is by photolysis, giving 2-C_3H_7 radicals and I-atoms. For 40°N latitude, clear tropospheric skies, and diurnal variation of the sun, the lifetimes of C_2H_5I with respect to photolysis are approximately 1.6 and 9.2 days for June 22 and December 22, respectively (see section VII-E-5). 2-C_3H_7 radicals will add O_2 to give 2-$C_3H_7O_2$ radicals, which are converted into $CH_3C(O)CH_3$ and, to a lesser extent, 2-C_3H_7OH via reaction with NO or other peroxy radicals (e.g., CH_3O_2). Reaction with OH radicals accounts for a minor (approximately 25%, Cotter et al., 2001) fraction of the 2-C_3H_7I loss. Using the

SAR substituent factors recommended in section VI-C (see table VI-C-2), approximately 90% of the reaction of OH radicals with CH_3CHICH_3 is predicted to proceed via hydrogen abstraction from the —CHI— group, with the balance proceeding via attack on the CH_3— groups. Cotter et al. (2001) and Orlando et al. (2005) have shown that hydrogen abstraction from the —CHI— group leads to $CH_3C(O)CH_3$ as the major product. Orlando et al. (2005) have shown that hydrogen abstraction from the —CH_3 group gives the CH_3CHICH_2 radical, whose sole fate is elimination of an I-atom to give $CH_3CH{=}CH_2$ (propene, C_3H_6). The major products of the atmospheric oxidation of 2-C_3H_7I are $CH_3C(O)CH_3$, C_3H_6, and 2-C_3H_7OH.

VI-F-32. CH_2FCl (HCFC-31)

Tuazon and Atkinson (1993a) identified HC(O)F in a yield of 100% following the Cl-atom-initiated oxidation of CH_2FCl in one atmosphere of air at 298 K. The oxidation proceeds via the formation of CHFClO radicals which eliminate Cl-atoms. HC(O)F is the major product of the OH-radical-initiated atmospheric oxidation of CH_2FCl.

VI-F-33. CF_2ClH (HCFC-22)

Lesclaux et al. (1987) showed that Cl-atom elimination is the fate of CF_2ClO radicals. Tuazon and Atkinson (1993a) found that COF_2 is formed in 100% yield following the Cl-atom-initiated oxidation of CF_2ClH. The oxidation proceeds via the formation of CF_2ClO radicals which eliminate Cl-atoms. COF_2 is the major product of the atmospheric oxidation of CF_2ClH.

VI-F-34. $CFCl_2H$ (HCFC-21)

Tuazon and Atkinson (1993a) showed that FC(O)Cl is formed in 100% yield following the Cl-atom-initiated oxidation of $CFCl_2H$. The oxidation proceeds via the formation of $CFCl_2O$ radicals which eliminate Cl-atoms. FC(O)Cl is the major product of the OH-radical-initiated atmospheric oxidation of $CFCl_2H$.

VI-F-35. CF_2ClCH_3 (HCFC-142b)

Tuazon and Atkinson (1994) observed CF_2ClCHO as the primary product, formed in a yield indistinguishable from 100% during the Cl-atom-initiated oxidation of CF_2ClCH_3 in 740 Torr of air at 298 K. The oxidation proceeds via the formation of CF_2ClCH_2O radicals, which react with O_2 to give CF_2ClCHO. Subsequent oxidation of CF_2ClCHO leads to the formation of CO, CO_2, and CF_2O as secondary products (Edney and Driscoll, 1992; Tuazon and Atkinson, 1993a). The atmospheric oxidation of CF_2ClCH_3 is initiated by reaction with OH radicals and gives CF_2ClCHO as the major primary product.

VI-F-36. $CFCl_2CH_3$ (HCFC-141b)

Tuazon and Atkinson (1994) observed $CFCl_2CHO$ as the primary product, formed in a yield indistinguishable from 100% during the Cl-atom-initiated oxidation of $CFCl_2CH_3$ in 740 Torr of air at 298 K. The oxidation proceeds via the formation of $CFCl_2CH_2O$ radicals, which react with O_2 to give $CFCl_2CHO$. Subsequent oxidation of $CFCl_2CHO$ leads to the formation of CO, CO_2, and FC(O)Cl as secondary products (Tuazon and Atkinson, 1993a). The atmospheric oxidation of $CFCl_2CH_3$ is initiated by reaction with OH radicals and gives $CFCl_2CHO$ as the major primary product.

VI-F-37. CF_3CH_2Cl (HCFC-133a)

Møgelberg et al. (1995b) studied the Cl-atom-initiated oxidation of CF_3CH_2Cl in 700 Torr of air at 296 K and showed that decomposition and reaction with O_2 are of comparable importance as fates of CF_3CHClO radicals. Unimolecular decomposition is expected to have a much larger activation barrier than that of the bimolecular reaction with O_2. Decomposition will decrease in importance with increasing altitude (and decreasing temperature) within the troposphere. Decomposition leads to the formation of COF_2, while reaction with O_2 gives $CF_3C(O)Cl$. The atmospheric oxidation of CF_3CH_2Cl is initiated by reaction with OH radicals and gives $CF_3C(O)Cl$ and COF_2 as major products.

VI-F-38. CF_3CFClH (HCFC-124)

Edney and Driscoll (1992) and Tuazon and Atkinson (1993a) reported the formation of $CF_3C(O)F$ in a yield indistinguishable from 100% in the Cl-atom-initiated oxidation of CF_3CFClH in 700–740 Torr of air at 298 K. The oxidation proceeds via the formation of CF_3CFClO radicals, which eliminate Cl to give $CF_3C(O)F$. The atmospheric oxidation of CF_3CFClH is initiated by reaction with OH radicals and gives $CF_3C(O)F$ as the major product.

VI-F-39. CF_3CCl_2H (HCFC-123)

Edney et al. (1991), Tuazon and Atkinson (1993a), and Hayman et al. (1994) have observed $CF_3C(O)Cl$ in a yield indistinguishable from 100% following the Cl-atom-initiated oxidation of CF_3CCl_2H in an atmosphere of air at ambient temperature. The oxidation proceeds via the formation of CF_3CCl_2O radicals, which eliminate Cl to give $CF_3C(O)Cl$. The atmospheric oxidation of CF_3CCl_2H is initiated by reaction with OH radicals and gives $CF_3C(O)Cl$ as the major product.

VI-F-40. $CF_3CF_2CCl_2H$ (HCFC-225ca)

Tuazon and Atkinson (1994) observed the formation of $CF_3CF_2C(O)Cl$ in 100% yield from the Cl-atom-initiated oxidation of $CF_3CF_2CCl_2H$ in 740 Torr of air at 298 K. The oxidation proceeds via the formation of $CF_3CF_2CCl_2O$ radicals, which eliminate Cl to give $CF_3CF_2C(O)Cl$. $CF_3CF_2C(O)Cl$ is the major atmospheric oxidation product of $CF_3CF_2CCl_2H$.

VI-F-41. CF_2ClCF_2CFClH (HCFC-225cb)

Tuazon and Atkinson (1994) observed the formation of $CF_2ClCF_2C(O)F$ and FC(O)Cl in yields of 100% and 1%, respectively, from the Cl-atom-initiated oxidation of CF_2ClCF_2CHFCl in 740 Torr of air at 298 K. The oxidation proceeds via the formation of CF_2ClCF_2CFClO radicals. The major fate of CF_2ClCF_2CFClO radicals is elimination of Cl to give $CF_2ClCF_2C(O)F$. $CF_2ClCF_2C(O)F$ is the major atmospheric oxidation product of CF_2ClCF_2CFClH.

VI-F-42. CH_2BrCl

Bilde et al. (1997) observed the formation of HC(O)Cl in essentially 100% yield in the chlorine-atom-initiated oxidation of CH_2BrCl in 700 Torr of N_2/O_2 diluent at 295 K in the presence, and absence, of NO. The oxidation proceeds via the formation of CHBrClO radicals, which eliminate Br to give HC(O)Cl. HC(O)Cl is the major product of the atmospheric oxidation of CH_2BrCl.

VI-F-43. $CBrCl_2H$

Atmospheric oxidation of $CHBrCl_2$ is initiated by photolysis and reaction with OH radicals. Bilde et al. (1998) estimated lifetimes with respect to photolysis, reaction with OH, and removal by both processes of 60, 120, and 40 days, respectively. Photolysis will probably proceed via Br-atom elimination, leading to the formation of $CHCl_2O$ radicals which give HC(O)Cl. Bilde et al. (1998) studied the Cl-atom-initiated oxidation of $CHBrCl_2$ in 700 Torr of air, or O_2, diluent at 253–295 K with and without added NO. In the absence of NO, the major product ($99 \pm 5\%$ yield) is $COCl_2$, with a trace amount ($1.8 \pm 0.7\%$ yield) of BrC(O)Cl. In the presence of NO, the yield of $COCl_2$ dropped to $73 \pm 4\%$ and the COBrCl yield increased to $21 \pm 4\%$. There was no effect of temperature on the product yields. The oxidation proceeds via the formation of $CBrCl_2O$ radicals, which eliminate Br to give $COCl_2$. The OH-radical-initiated oxidation of $CHBrCl_2$ gives $COCl_2$ as the major product and BrC(O)Cl as a minor product. Atmospheric oxidation of $CHBrCl_2$ is initiated by both photolysis and reaction with OH radicals and gives HC(O)Cl and $COCl_2$ as major products and BrC(O)Cl as a minor product.

VI-F-44. CBr_2ClH

Atmospheric oxidation of $CHBr_2Cl$ is initiated by photolysis and reaction with OH radicals. Bilde et al. (1998) estimated lifetimes with respect to photolysis and reaction with OH of 50 and 120 days, respectively. Photolysis will probably proceed via Br-atom elimination, leading to the formation of CHBrClO radicals which give HC(O)Cl. Bilde et al. (1998) observed the formation of BrC(O)Cl in essentially 100%

yield in the chlorine-atom-initiated oxidation of $CHBr_2Cl$ in 700 Torr of air at 295 K in the presence, and absence, of NO. The OH-radical-initiated oxidation of $CHBr_2Cl$ will give BrC(O)Cl as the major product. Atmospheric oxidation of $CHBr_2Cl$ is initiated by photolysis and reaction with OH radicals and gives HC(O)Cl and BrC(O)Cl as major products.

VI-F-45. $CF_3CHBrCl$

Bilde et al. (1998) studied the Cl-atom-initiated oxidation of $CF_3CHBrCl$ in 700 Torr of air at 296 K with and without added NO. $CF_3C(O)Cl$ was observed in a yield of $93 \pm 10\%$, and there was no discernable effect of NO. The major product of the OH-radical-initiated atmospheric oxidation of $CF_3CHBrCl$ is $CF_3C(O)Cl$.

VII

The Primary Photochemical Processes in the Alkanes, the Haloalkanes, and Some of Their Oxidation Products

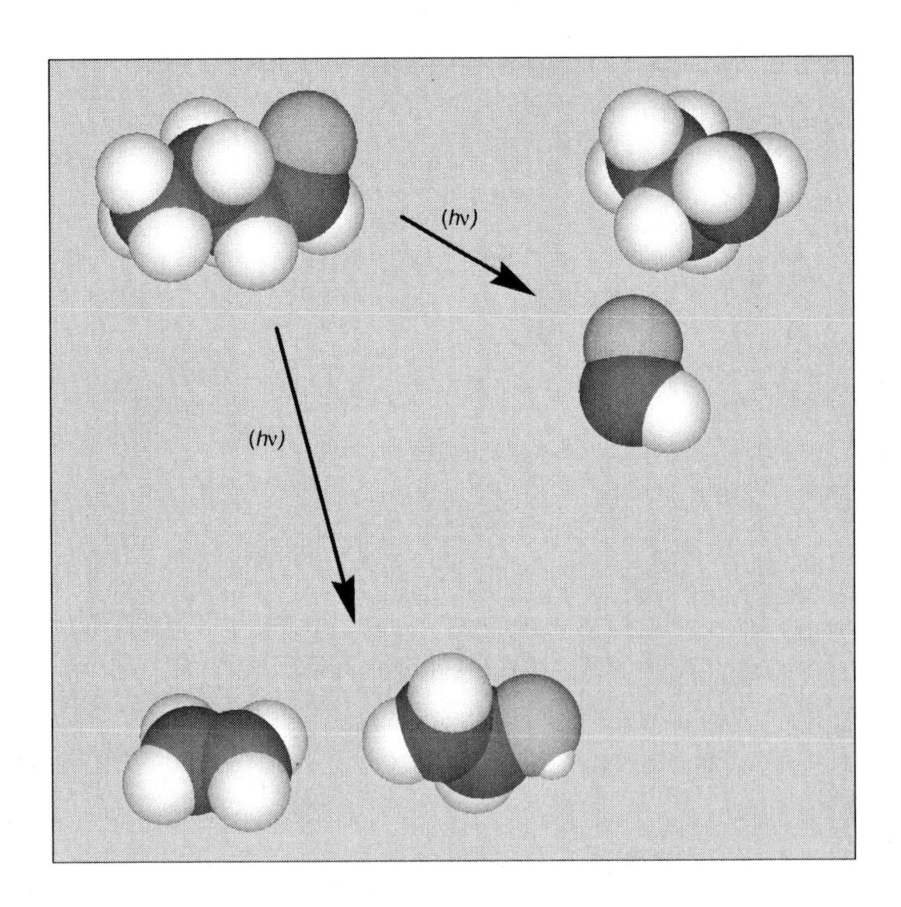

VII-A. Introduction

In this chapter we consider the chemical changes that are induced by absorption of sunlight by the alkanes and some of their oxidation products. The lowest lying electronic transitions of the alkanes ($\sigma \rightarrow \sigma^*$ excitation) occur at wavelengths in the vacuum ultraviolet region ($\lambda < 200\,\text{nm}$) that are not present within the troposphere (see figure VII-A-1). Although absorption is shifted to somewhat longer wavelengths

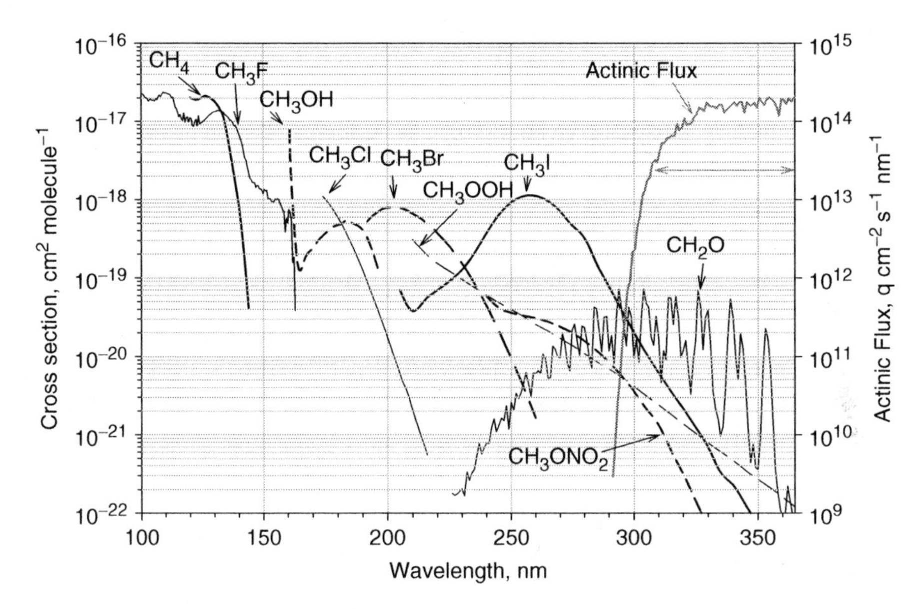

Figure VII-A-1. Cross sections for CH_4 and various substituted methanes and oxidation products of methane; spectra for methane from Okabe and Becker (1963); CH_3OH from Harrison et al. (1959); CH_2O, Calvert et al. (2000); CH_3F from Locht et al. (2000); other compounds, Atkinson et al. (2004). The actinic flux curve is that measured at noon on a cloudless day (June 19, 1998) near Boulder, CO.

as one progresses from methane to ethane to propane, absorption of sunlight by even the higher alkanes, and hence their photodecomposition, are unimportant within the troposphere. However, as discussed in the previous chapters of this book, the alkanes undergo oxidation, initiated largely by OH-radical reactions in the troposphere (see chapter II). The products formed show a shift in absorption to the longer wavelengths. This is illustrated in the absorption data given for methane, its oxidation products, and halogen-substituted methanes in figure VII-A-1. As a CH_3—H bond in methane is replaced with a C—OH, C—Cl, C—Br, C—I, C—ONO_2, or C—O_2H bond, light absorption involving excitation of non-bonding electrons ($n \rightarrow \sigma^*$ transitions) becomes possible. These "forbidden" absorptions are smaller than the allowed $\sigma \rightarrow \sigma^*$ transitions in the alkanes, but the absorption bands shift to the longer wavelengths. Progressing from CH_3OH to CH_3Cl, CH_3Br, and CH_3I, less energy is required to excite the non-bonding electron to the σ^* orbital, and a progressive shift to longer wavelengths in absorption is observed. With CH_3ONO_2, CH_3OOH, and CH_3I, the shift in absorption band is sufficient to catch a small portion of the available distribution of light quanta present within the troposphere. As a carbonyl bond (>C=O) is introduced into the alkanes, a weak $n \rightarrow \pi^*$ absorption appears that has a favorable overlap with the actinic flux present within the troposphere (see Figure VII-A-1).

We review in this chapter the nature of the photodecomposition pathways of several of the major products of alkane atmospheric oxidation, namely, the aliphatic aldehydes (RCHO) in section VII-B; the aliphatic ketones [RC(O)R] in section VII-C, where R = CH_3, C_2H_5, or a larger alkyl group; some cyclic ketones in section VII-D; the halogen-substituted alkanes (alkyl halides) in section VII-E; halogen-substituted aldehydes and ketones in sections VII-F and VII-G, respectively; acyl halides, alkyl nitrates, acylperoxy nitrates, and alkyl hydroperoxides in sections VII-H, VII-I, VII-J, and VII-K, respectively. A summary of the photochemistry results is given in section VII-L. The quantum efficiencies of radical and molecular products formed from these alkane derivatives are important as these can influence significantly ozone generation and other atmospheric chemistry. We restrict detailed discussion to compounds whose tropospheric photolysis is likely sufficiently rapid to play a role in urban and regional air quality issues. Longer lived compounds, such as the CFCs, HCFCs, Halons, etc., which impact significantly the chemistry of the stratosphere, are discussed only briefly.

VII-A-1. Some Problems in the Use of Laboratory Photochemical Data to Derive Information Useful to Atmospheric Scientists

The most useful photochemical studies of the carbonyl compounds for the atmospheric scientist are those in which the quantum yields of compounds that relate directly or indirectly to specific primary processes are determined. Preferably, these studies should be made at several wavelengths of light within the overlapping region of the absorption band of the compound and sunlight within the atmosphere (280–360 nm). It is important as well to simulate atmospheric conditions as closely as possible; thus experiments should be made using dilute mixtures of the aldehyde, ketone, or other compound, in air at various pressures common to the atmosphere (up to 760 Torr of air). However, the analytical techniques and equipment required to do such demanding experiments are not available in many laboratories. The many hundreds of photochemical studies of the aldehydes and ketones that appear in the literature often have been carried out employing sophisticated photochemical techniques, but applied to pure carbonyl compounds. Many of these studies involved quantitative measurements of quantum yields of specific molecules or radicals generated in photolysis at a single wavelength. However, the transformation of these laboratory results into information applicable to atmospheric conditions remains a major problem to the atmospheric scientist. For example, the quenching of the various excited states of the carbonyl compounds by molecules of the compound itself and by air molecules is very different, and this presents a special challenge in relating the usual laboratory measurements to atmospheric conditions.

In this chapter an attempt is made to use the photochemical information from diverse published photochemical experiments and theoretical studies to derive wavelength-dependent information on quantum yields of the primary photodecomposition pathways for atmospheric conditions. Ultimately, this information is coupled to the absorption cross sections of the compound of interest and the solar flux available in the atmosphere to derive the photolysis frequencies (j-values) that define the photodecomposition rates within the atmosphere. The methods used in these calculations are described in the next section.

VII-A-2. Description of the Photolysis Model Used in Photolysis Frequency Estimates

Atmospheric photolysis rate coefficients, photolysis frequencies, or j-values were calculated as in the previous studies of the atmospheric chemistry of the alkenes (Calvert et al., 2000) and the aromatic hydrocarbons (Calvert et al., 2002). Numerical summations were made over wavelength λ,

$$j = \Sigma \langle\sigma\rangle_i \langle\phi\rangle_i \langle F\rangle_i \Delta\lambda_i \tag{1}$$

where $\langle\sigma\rangle_i$, $\langle\phi\rangle_i$, and $\langle F\rangle_i$ are, respectively, the absorption cross section, quantum yield, and spectral actinic flux averaged over small wavelength intervals $\Delta\lambda_i$. Cross sections and quantum yields specific to each photodissociation reaction are discussed later in this chapter.

The spectral actinic flux was calculated with the 4-stream Tropospheric Ultraviolet Visible (TUV) model version 4.1 (Madronich and Flocke, 1998; Lantz et al., 1996). The model code is publicly available and may be downloaded from http://cprm.acd.ucar.edu/models/TUV. Most of the calculations are made for clear sky conditions at an altitude of 0.5 km. Vertical profiles of temperature and number densities for air and ozone, representative of annual means for 45°N, were taken from U.S. Standard Atmosphere (1976), with a total ozone column of 350 Dobson units (1 DU $= 2.69 \times 10^{16}$ molecule cm^{-2} $= 10^{-3}$ atm cm). At 500 m above sea level the temperature was 285 K and air density 2.43×10^{19} molecule cm^{-3}. No attenuation by other atmospheric gases (e.g., NO_2, SO_2) was considered. An aerosol profile typical for continental rural conditions, from Elterman (1968), was specified with a total vertical optical depth of 0.293 at 340 nm. The chosen Lambertian surface albedo of 10% was assumed to be independent of wavelength. Extraterrestrial irradiances are from van Hoosier et al. (1988) for $\lambda < 350$ nm, and from Neckel and Labs (1984) for longer wavelengths, for annual mean Earth–Sun distance. O_3 absorption cross sections over the 185–350 nm range are from Molina and Molina (1986) and include temperature dependence over the wavelength range 240.5–350 nm. O_3 cross sections at other wavelengths and those for Rayleigh scattering are from WMO (1986). O_2 absorption, important at higher altitudes, is taken from WMO (1986) with parameterization of the Schumann–Runge bands according to Kockarts (1994).

The calculation of atmospheric photolysis rate coefficients via equation (1) requires the computation of average values of $\langle\sigma\rangle_i$, $\langle\phi\rangle_i$, and $\langle F\rangle_i$ over sufficiently small wavelength intervals $\Delta\lambda_i$ to minimize numerical integration errors. Because computational cost increases approximately linearly with the number of intervals, excessively small values of $\Delta\lambda_i$ are undesirable. Here, uniform $\Delta\lambda_i = 1$ nm intervals centered on integer (nm) values were used in the 303–314 nm region, with larger intervals (3–5 nm) used in other wavelength regions. Numerical errors induced by the resolution employed have been discussed by Madronich and Weller (1990) and, for CH_2O photodissociation, by

Cantrell et al. (1990a), who showed that acceptable accuracy may be obtained using 5 nm intervals, but this degrades rapidly for wider intervals.

In this book, the calculated photolysis frequencies for each compound are presented graphically as a function of the solar zenith angle, the angle between a vertical line and a line extending to the Sun from the observation point. This angle is a complex function of the latitude and longitude of the observation point, the day of the year, and the time of the day. In figure VII-A-2 some representative plots of solar zenith angle are given as a function of the latitude and the local time of the day for selected times of the year: Spring equinox, March 22, and Fall equinox, September 23, days with equal periods of daylight and darkness (figure VII-A-2a); June 22, Summer solstice, the longest day in the northern hemisphere (figure VII-A-2b); and December 22, Winter solstice, the shortest day in the northern hemisphere (figure VII-A-2c). Solar zenith angles applicable for any given location and point in time obviously vary over a wide range, and choice of the photolysis frequency that is appropriate for a given time and location requires information of the sort presented in figure VII-A-2 as well as the plots of *j*-value vs solar zenith angle presented in this work.

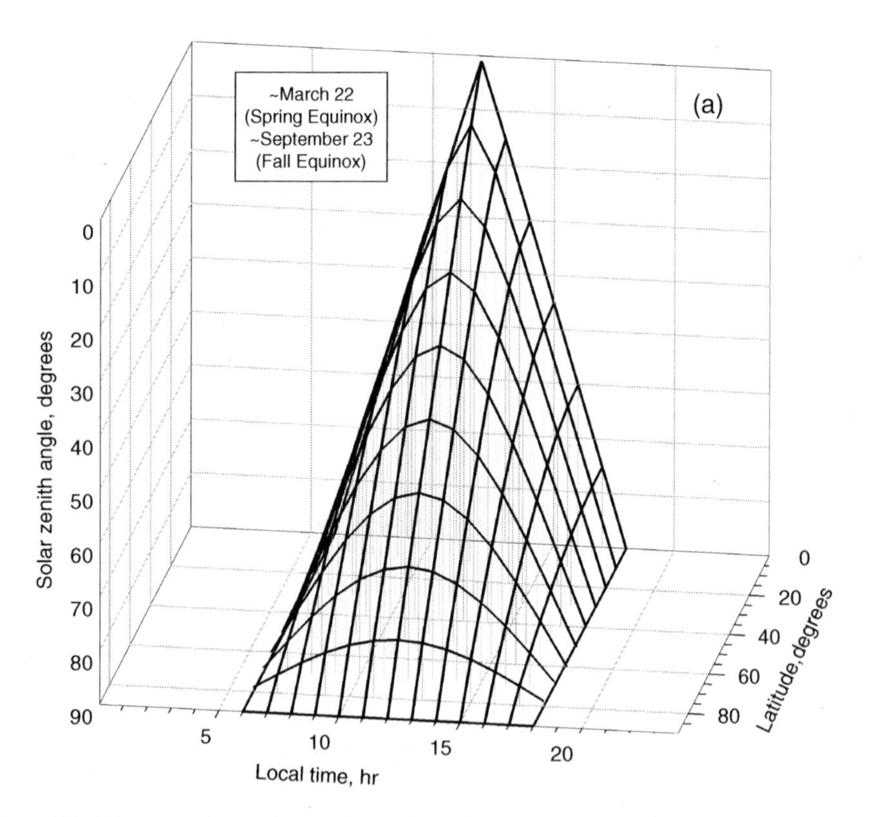

Figure VII-A-2a. Plot of the solar zenith angle as a function of latitude and local time of the day for the periods of Spring and Fall equinox. Note the days and nights are of equal length at all latitudes.

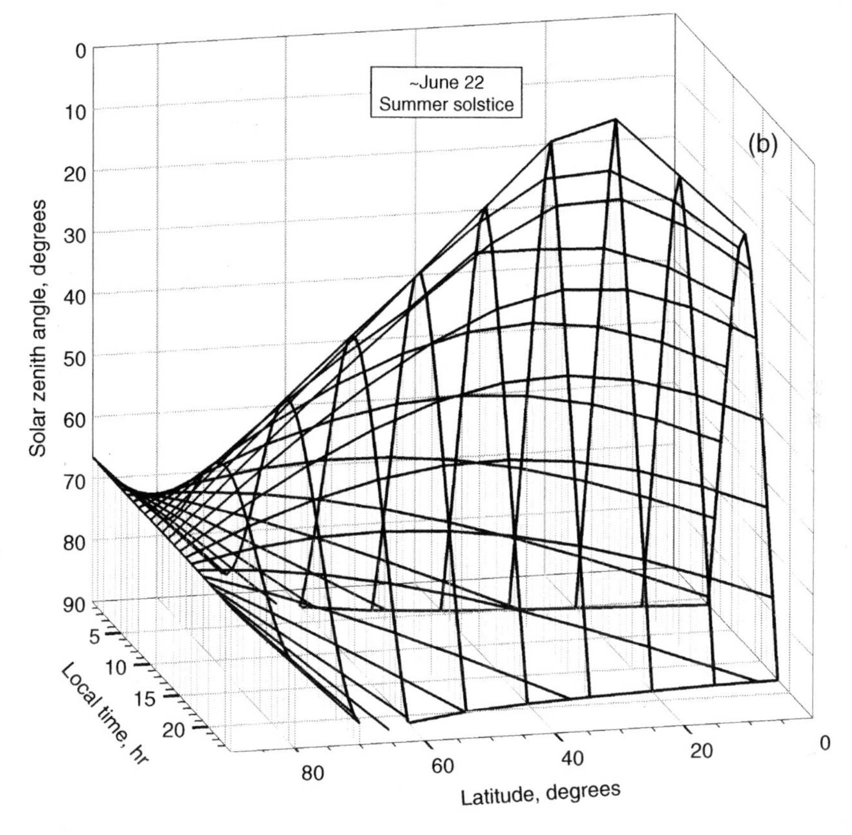

Figure VII-A-2b. Plot of the solar zenith angle as a function of latitude and local time of the day for June 22, summer solstice. Note that the Sun is above the horizon all the day in the high latitudes.

VII-B. Photodecomposition of Aldehydes

VII-B-1. Photodecomposition of Formaldehyde (Methanal, CH_2O)

Formaldehyde (CH_2O) is one of the most important products of alkane atmospheric oxidation. It provides a major source of radicals formed in the atmosphere and is very important in determining the rate of ozone generation. New information on the atmospheric photochemistry of formaldehyde has been reported since the review by Calvert et al. (2000), and a further discussion of the available data is given here. Four different primary processes have been considered:

$$CH_2O + h\nu \rightarrow HCO + H \tag{I}$$

$$\rightarrow H_2 + CO \tag{II}$$

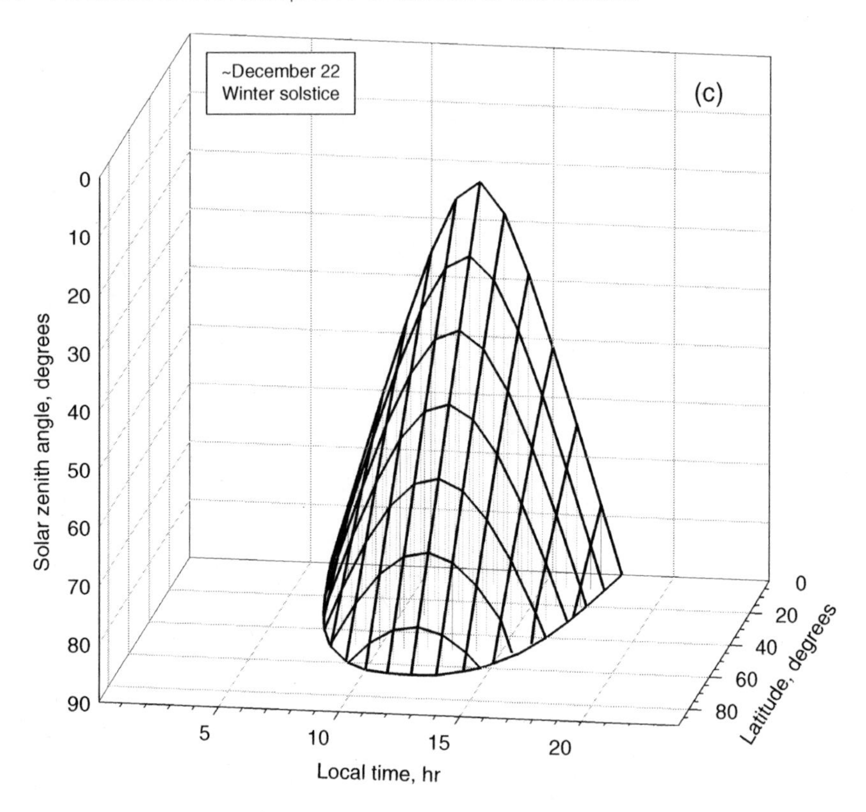

Figure VII-A-2c. Plot of the solar zenith angle as a function of latitude and local time for December 22, Winter solstice in the Northern Hemisphere. Note that the Sun is below the horizon for the entire 24 hr period for locations at high latitudes.

$$\rightarrow H + H + CO \qquad \text{(III)}$$

$$\rightarrow HCOH \qquad \text{(IV)}$$

Photodissociation in reaction (I) can occur from the initially excited singlet state as well as both the vibrationally excited ground state and the triplet state of formaldehyde that are populated from the initially excited singlet state by internal conversion and intersystem crossing, respectively. Primary processes like process (I) in formaldehyde, a photochemically induced split of an aldehyde or ketone molecule into two free radicals, are often called "Norrish Type I" photodissociations. It was R. G. W. Norrish who first suggested the analogous process in 2-hexanone (Norrish and Appleyard, 1934). The molecular channel (II) occurs from the vibrationally excited ground state (e.g., see Chuang et al., 1987; Green et al., 1992; Yamaguchi et al., 1998). Vibrational relaxation, induced by molecular collisions, results in quenching of process (II) in CH_2O when excited at the longer wavelengths ($\lambda > 330$ nm). In contrast, primary process (I) is not sensitive to pressure change from 0 to 1 atm. Only processes (I) and (II) are important in the wavelength region available within the troposphere. The threshold wavelength for process (I) is ~324 nm, whereas process (II) is energetically feasible at all wavelengths absorbed by formaldehyde. Sufficient energy is available in the quantum of radiation at 297 nm for process (III) to occur. Studies of formaldehyde photolysis at 123.6 and 147.0 nm (Glicker and Stief, 1971) suggest that process (III) is important at these short wavelengths; it is estimated that $\phi_{II}+\phi_{III} = 1.0$, with $\phi_{II} = \phi_{III} = 0.5$. The occurrence of either (I) or (III) in the O_2-rich Earth's atmosphere leads to the same products: CH_2O $(+O_2) \rightarrow 2HO_2+CO$, so the differentiation of the extent of processes (I) and (III) is unimportant for atmospheric studies.

The hydroxycarbene (HCOH), a structural isomer of CH_2O, is formed in process (IV); this has been considered as a possible intermediate in formaldehyde photolysis (e.g., see Houston and Moore, 1976; Sodeau and Lee, 1978, Ho and Smith, 1982; Bamford et al., 1985). Process (IV) was originally invoked as a possible pathway to rationalize the delay in the appearance of CO following the occurrence of process (II). Subsequent observations of a vibrationally rich CO product (Bamford et al., 1985) appear to have removed interest in the occurrence of (IV). Theoretical studies confirm that this potential primary process is possible on energetic grounds and that it may play some role in formaldehyde photolysis (e.g., see Kemper et al., 1978). Ab initio calculations of Goddard and Schaefer (1979) suggest that a minimum of 370 kJ mol^{-1} (323 nm) is required for the rearrangement in process (IV). The significance of this channel, if any, in formaldehyde photolysis is unknown. Conceivably, the drop-off of the quantum yield sum, $\phi_I + \phi_{II}$, from a value very near unity in the 290–330 nm range to ~0.8 in the 250–280 nm range may reflect the onset of process (IV). If this scenario occurs, then the observed products of CH_2O photolysis at short wavelengths demand that the hydroxycarbene (HCOH) should not lead to CO or H-atom formation. Conceivably, reaction with O_2 in air could lead to peroxyformic acid formation [HC(O)OOH] and/or other products that have not been detected to date.

VII-B-1.1. Absorption Cross Sections of Formaldehyde

The formaldehyde cross section data reported by Meller and Moortgat (2000) are very similar to those recommended by Calvert et al. (2000). In the latter study, the authors use a combined data set of unweighted averages of the older data of Moortgat and Schneider (given by Atkinson et al., 1997), Cantrell et al. (1990a), and Rogers (1990). The averaged cross sections (Calvert et al., 2000) can be compared with the more recent Meller and Moortgat (2000) data in figure VII-B-1 and, in tabular form, in table VII-B-1. The latter data show somewhat greater band maxima than the averaged σ data recommended by Calvert et al. (2000). Although the average ratio of all the sigma values (excluding zero values) at 298 K over the range from 280 to 365 nm is $\sigma_{M\&M}/\sigma_{C\ et\ al.} = 0.994$, the ratios of the major absorption peaks that dominate the contributions of CH_2O absorption within the atmosphere is 1.17. In both cases the data have been averaged over 1 nm intervals centered at the wavelength given. Meller and Moortgat (2000) determined the σ values over the entire 240–356 nm range, and their data are seemingly the most accurate of the available data. Integrated absoption bands from these data agree well with those of the high-resolution spectra of CH_2O (300–340 nm, 294 and 245 K) reported by Smith et al. (2006).

Table VII-B-1. Comparison of recommended CH_2O cross sections, σ ($cm^2 molecule^{-1}$), 298 K

λ (nm)	$10^{21} \times \sigma$ Meller and Moortgat (2000)	$10^{21} \times \sigma$, Calvert et al. (2000)	λ (nm)	$10^{21} \times \sigma$ Meller and Moortgat (2000)	$10^{21} \times \sigma$, Calvert et al. (2000)
280	23.4	21.8	327	43.7	39.1
281	15.6	14.6	328	12.2	15.0
282	9.72	9.82	329	31.2	25.0
283	7.22	10.8	330	38.7	33.6
284	42.6	36.2	331	14.1	13.0
285	40.5	37.0	332	3.47	3.61
286	21.0	21.8	333	2.14	1.81
287	11.5	14.6	334	1.59	1.36
288	31.7	28.1	335	0.967	1.22
289	32.2	30.1	336	1.26	1.45
290	11.7	14.8	337	3.83	4.05
291	18.4	16.5	338	19.2	17.1
292	7.97	11.4	339	53.8	42.9
293	31.3	29.9	340	31.5	27.9
294	71.5	62.3	341	9.78	10.1
295	40.5	41.9	342	5.09	5.62
296	24.7	25.5	343	19.2	16.2
297	13.7	16.6	344	12.7	11.9
298	42.2	34.5	345	4.37	3.79
299	31.7	30.5	346	1.19	1.10
300	9.64	12.8	347	0.441	0.67
301	16.2	15.4	348	0.754	0.88
302	8.54	13.2	349	0.379	0.50
303	30.2	27.2	350	0.362	0.32
304	72.2	62.6	351	0.893	0.90
305	47.5	47.9	352	7.30	6.31
306	42.9	43.8	353	22.8	18.7
307	17.8	21.9	354	16.4	15.0
308	13.8	15.3	355	6.96	6.86
309	32.5	30.5	356	1.48	1.72
310	17.4	19.5	357	0.345	0.37
311	4.62	7.25	358	0.186	0.06
312	11.9	10.9	359	0.111	0.00
313	9.06	12.3	360	0.087	0.00
314	56.4	42.0	361	0.100	0.39
315	55.7	50.9	362	0.221	0.51
316	25.6	29.5	363	0.141	0.30
317	57.8	47.7	364	0.094	0.21
318	31.5	31.5	365	0.088	0.24
319	9.78	11.5	366	0.085	
320	11.9	10.9	367	0.091	
321	16.0	14.2	368	0.142	
322	7.22	7.65	369	0.297	
323	3.28	3.67	370	0.635	
324	8.58	7.64	371	0.571	
325	15.8	14.8	372	0.198	
326	68.8	52.2	373	0.113	

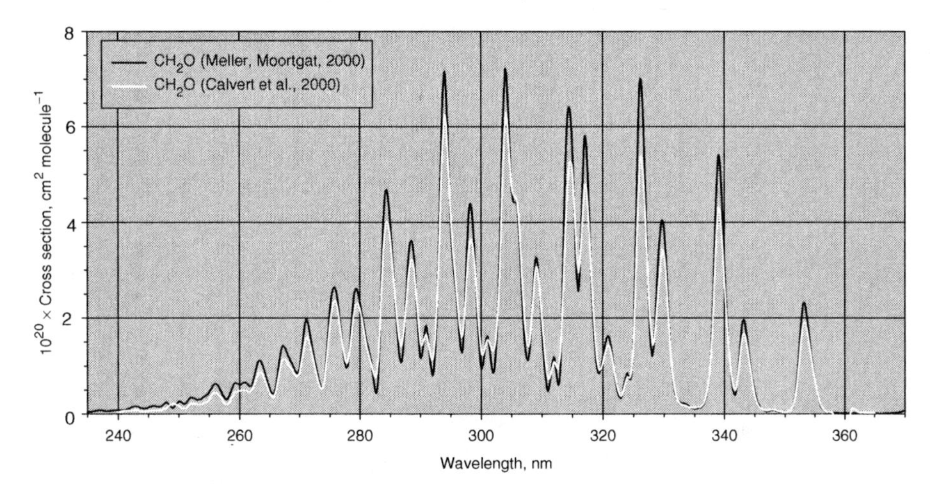

Figure VII-B-1. Comparison of measured cross sections for CH_2O (298 K). White lines are as estimated from the averaged results of Moortgat and Schneider as given by Atkinson et al. (1997), Cantrell et al. (1990a), and Rogers (1990); black lines are as reported by Meller and Moortgat (2000). The measured high resolution data have been lumped into bands of 1 nm widths in this comparison.

VII-B-1.2. Quantum Yields for Formaldehyde Photodecomposition

New information has added to our knowledge of the quantum yields of the primary processes in CH_2O photolysis, and our review of this information should update our previous discussions of formaldehyde photolysis (Calvert et al., 2000). The quantum yields of the photodissociation processes have been estimated in studies using a variety of different techniques. Many studies have been made using various added reagents that react to remove the radical products formed in (I) so that measurements of the H_2 formed can be used to monitor the occurrence of (II). The studies most relevant to atmospheric conditions are those of Moortgat and Warneck (1979) and Moortgat et al. (1983) in which dilute formaldehyde mixtures in air were used. In these studies the quantum yields of both H_2 and CO were measured. The quantum yield of CO was found to be near unity for wavelengths in the 330–285 nm range in both the studies of Moortgat et al. (1983) and Horowitz and Calvert (1978a,b); see figure VII-B-2. Moortgat and Warneck (1979) and Moortgat et al. (1983) measured Φ_{H2} to monitor ϕ_{II}, and $\Phi_{CO} - \Phi_{H2}$ gave estimates of ϕ_I.

Horowitz and Calvert (1978a,b) and Clark et al. (1978) estimated values of ϕ_{II} in formaldehyde photolysis in the presence of various radical scavengers. Clark et al. (1978) studied CH_2O photolysis with added nitric oxide. In the studies of Horowitz and Calvert (1978a,b) the quantum yields of H_2 and CO were measured in dilute mixtures of CH_2O with added radical scavengers (2-methylpropene, trimethylsilane, or NO). In both the Clark et al. and Horowitz and Calvert studies estimates of ϕ_{II} were derived from the measured Φ_{H2} for experiments at high scavenger gas concentrations. The Horowitz and Calvert results with added 2-methylpropene appear to be the most easily interpreted. Photolysis of pure formaldehyde at all but the longest wavelength used (338 nm) gave

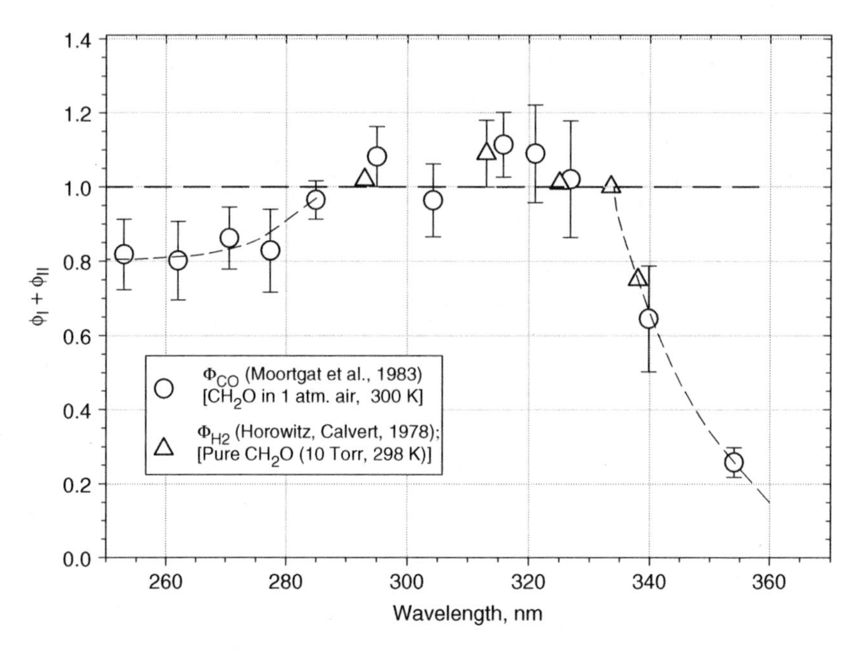

Figure VII-B-2. Experimental estimates of the sum of quantum yields of photodecomposition ($\phi_I + \phi_{II}$) for formaldehyde as a function of wavelength; $CH_2O + h\nu \rightarrow H + HCO$ (I); $CH_2O + h\nu \rightarrow H_2 + CO$ (II).

$\Phi_{H2} \cong 1.0$ (see figure VII-B-2), independent of the pressure. Hence, over most of the range of wavelengths absorbed by formaldehyde, $\phi_I + \phi_{II} = 1.0$, a conclusion supported as well in the studies of Moortgat et al. (1983). Thus Horowitz and Calvert estimated ϕ_I from the difference $1 - \phi_{II}$.

In the study of Smith et al. (2002a) the experiments were designed to measure the radical yields more directly. They estimated the relative quantum yields of process (I) at a pressure of 50 Torr, using a method of chemical amplification (Cantrell et al., 1984, 1993) of the HO_2 radicals that were formed by reaction of the primary products of process (I) with O_2 [$H + O_2 (+ M) \rightarrow HO_2(+ M)$; $CHO + O_2 \rightarrow HO_2 + CO$]. The HO_2 radicals formed in the system reacted with an excess of added NO, and an ensuing chain reaction (involving HO_2, OH, NO, and CH_2O) formed NO_2 and amplified the HO_2 yield by a factor of ~35. The NO_2 formed was proportional to the radical yield from process (I) and was detected by sensitive chemical ionization mass spectroscopy. The relative estimates of ϕ_I at 20 different wavelengths (268.75–338.75 nm) were made in the Smith et al. (2002a) experiments, and these were then normalized to a quantum yield of 0.753 at 303.75 nm, a region of seemingly near-constant value of ϕ_I as estimated by previous workers; see Calvert et al. (2000). The Smith et al. data showed a previously unobserved structure in the wavelength dependence.

In the very high resolution studies, close to the limit for Doppler broadening of CH_2O at 294 K, Pope et al. (2005a) confirmed the presence of resolved fine structure in the absorption bands of CH_2O. The measured maxima were higher than those obtained in previous lower resolution studies, but band-integrated cross sections were in excellent accord with previous measurements. Preliminary studies of relative values of the quantum yields of HCO radical were reported (Pope et al., 2005b), as measured directly by cavity ring-down spectroscopy for a small region (308–320 nm) of the CH_2O absorption.

The various estimates of the quantum yields of primary processes (I) and (II) are compared in figures VII-B-3 and VII-B-4, respectively. The vertical and horizontal bars on the data points from each study give, respectively some measure of the precision of the ϕ measurements and the bandwidth of the radiation used. The data are in reasonable agreement within the uncertainty limits at most wavelengths. In this study we have made "best estimates" of ϕ_I and ϕ_{II} from the combined data sets given in figures VII-B-3 and VII-B-4. We weighted each of the data points equally, drew straight lines between each data point in the order of increasing wavelength, and the intersection of this line with the ordinate at each nm unit of wavelength was taken as the best estimate of ϕ for that wavelength. When measurements at the same wavelength were reported, a simple average of these estimates was used.

Recommendations of the IUPAC panel (Atkinson et al., 2004) are based largely on the normalized Smith et al. (2002a) data for ϕ_I, setting the value of ϕ_I at 0.78 at 302.5 nm (Smith et al., 2002a). To derive their estimates of ϕ_{II}, the panel has assumed that $\phi_{II} = \Phi_T - \phi_I$, where $\Phi_T = \phi_I + \phi_{II}$ as estimated by Moortgat and colleagues for 298 K and 1 atmosphere. For data points given for $\lambda > 340$ nm and $\lambda < 268.75$ nm, the Moortgat and colleagues data alone are used. The tabular listing of recommended

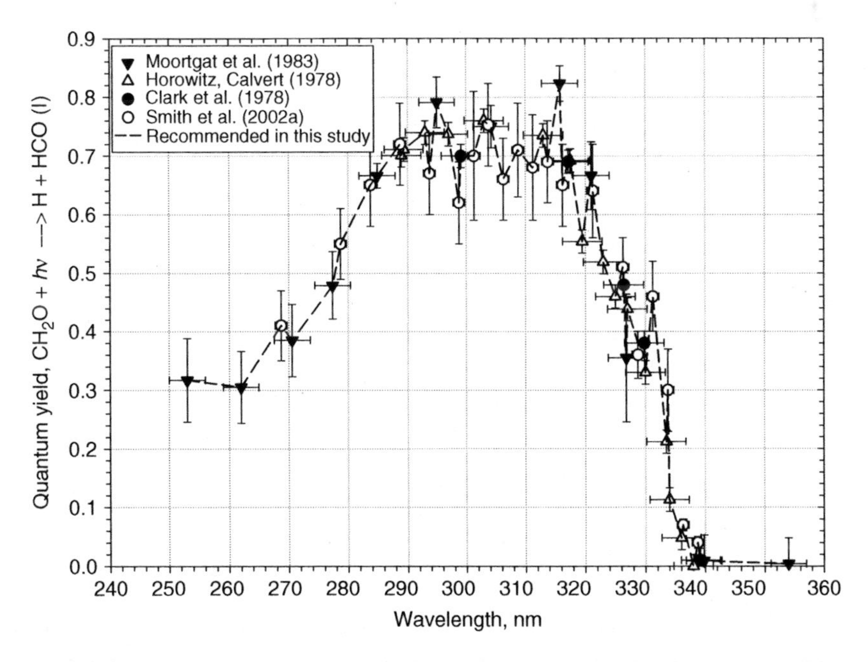

Figure VII-B-3. Comparison of quantum yields (ϕ_I) for process (I), $CH_2O + h\nu \rightarrow HCO + H$, in CH_2O photodecomposition, based on the results of Horowitz and Calvert (1978a,b), Moortgat et al. (1983), Clark et al. (1978), and Smith et al. (2002a).

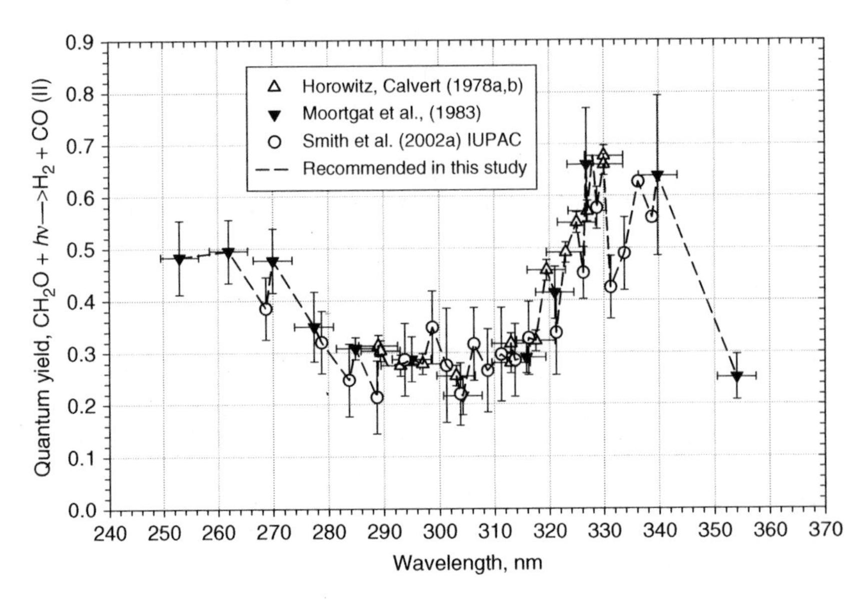

Figure VII-B-4. Comparison of the quantum yields (ϕ_{II}) for primary process (II), $CH_2O + h\nu \rightarrow H_2 + CO$, in CH_2O photodecomposition. Data are from Horowitz and Calvert (1978a,b), Moortgat et al. (1983), and Smith et al. (2002a).

ϕ_I values from the two evaluations is given in table VII-B-2. The data for ϕ_{II} given in figure VII-B-4 are for 298 K in air at 1 atm pressure. At wavelengths greater than 330 nm the values of ϕ_{II} are sensitive to pressure and, to some extent, the temperature. The quenching follows the Stern–Volmer relation, $1/\phi_{II} = 1 + (k_q/k_d)\times$ [M], where [M] is the concentration of air molecules. The quenching coefficients (k_q/k_d, cm^3 molecule^{-1}) for CH_2O mixtures in air, as reported by Moortgat et al. (1983) for 300 K at 329 nm and 353 nm, respectively, are: $(0.26 \pm 0.10) \times 10^{-19}$ and $(1.12 \pm 0.17) \times 10^{-19}$; constants for 220 K at 329 and 353 nm, respectively, are: $(0.39 \pm 0.07) \times 10^{-19}$ and $(2.47 \pm 0.59) \times 10^{-19}$. At a temperature of 300 K, the expected effect of lowering the atmospheric pressure from 760 (sea level) to 600 Torr (typical of Denver, CO) is not dramatic; but it leads to an increase in the quantum yield of process (II) at 329 and 353 nm by about 8 and 16%, respectively. The recommended ϕ_{II} values of Atkinson et al. (2004) and our present study are tabulated in table VII-B-3.

VII-B-1.3. Photolysis Frequencies (j-values) for Formaldehyde

Given in figure VII-B-5a are the recommended j_I and j_{II} values estimated for a clear day as a function of solar zenith angle and several different overhead ozone columns (300–400 Dobson units). We used the preferred Meller and Moortgat (2000) cross sections and the quantum yield data as recommended in this study. In figure VII-B-5b the percentage decrease in the estimated j-values per Dobson unit of the overhead O_3 column is shown as a function of solar zenith angle. The j-values for CH_2O photodissociation are not very sensitive to the ozone column. For solar zenith angles between

Table VII-B-2. Estimates of the wavelength dependence of the quantum yield (ϕ_I) for CH_2O photolysis in primary process (I): $CH_2O + h\nu \rightarrow H + CHO$ at 298 K and 1 atm as evaluated by Atkinson et al. (2004)[a] and in the present study[b]

λ (nm)	ϕ_I (Atkinson et al., 2004)	ϕ_I (This Study)	λ (nm)	ϕ_I (Atkinson et al., 2004)	ϕ_I (This Study)
240	0.270		316 (316.25)[c]	0.673	0.75
250	0.290	0.32	317		0.69
260	0.300	0.31	318		0.66
268.75	0.425	0.41	319		0.58
278.75	0.570	0.56	320		0.60
278		0.52	321 (321.25)[c]	0.663	0.67
279		0.56	322		0.58
280		0.58	323		0.52
281		0.60	324		0.49
282		0.62	325		0.46
283		0.64	326 (326.25)[c]	0.528	0.51
284 (283.75)[c]	0.673	0.65	327		0.50
285		0.67	328		0.40
286		0.68	329 (328.75)[c]	0.373	0.36
287		0.70	330		0.35
288		0.71	331 (321.25)[c]	0.476	0.46
289 (288.75)[c]	0.746	0.71	332		0.37
290		0.72	333		0.24
291		0.73	334 (333.75)[c]	0.311	0.112
292		0.73	335		0.070
293 (293.75)[c]		0.74	336 (336.25)[c]	0.072	0.055
294 (293.75)[c]	0.694	0.70	337		0.035
295		0.79	338		0.010
296		0.76	339 (338.75)[c]	0.041	0.010
297		0.74	340	0.000	0.011
298		0.67	341		0.010
299 (299.75)[c]	0.642	0.65	342		0.008
300		0.70	343		0.008
301 (301.25)[c]	0.725	0.70	344		0.008
302		0.73	345		0.007
303		0.78	346		0.007
304 (303.75)[c]	0.780	0.75	347		0.007
305		0.71	348		0.007
306 (306.25)[c]	0.684	0.66	349		0.006
307		0.68	350	0.000	0.006
308		0.70	351		0.005
309 (308.75)[c]	0.735	0.71	352		0.005
310		0.70	353		0.005
311 (311.25)[c]	0.704	0.68	354		0.005
312		0.72	355		0.005
313		0.73	356		0.000
314 (313.75)[c]	0.715	0.69			
315		0.78			

[a] Based largely on the relative quantum yield data of Smith et al. (2002a) with normalization of the results by fixing $\phi_I = 0.78$ at 302.5 nm.

[b] Based largely on the data of Horowitz and Calvert (1978a,b), Moortgat et al. (1983), and Smith et al. (2002a).

[c] Actual wavelength of excitation used by Smith et al. (2002a).

Table VII-B-3. Comparison of recommendations of the wavelength dependence of the quantum yields (ϕ_{II}) of the primary process $CH_2O + h\nu \rightarrow H_2 + CO$ (II) from IUPAC (Atkinson et al., 2004)[a] and this study[b], temperature 298 K, pressure near 1 atm

λ (nm)	IUPAC (2004)	ϕ_{II} (This Study)	λ (nm)	IUPAC (2004)	ϕ_{II} (This Study)
240	0.49		318		0.35
250	0.49	0.48	319		0.435
260	0.49	0.49	320		0.44
268.75	0.385	0.40	321 (321.25)[c]	0.337	0.338
278.75	0.320	0.32	322		0.415
280		0.30	323		0.49
281		0.27	324		0.52
282		0.27	325		0.55
283		0.25	326 (326.25)[c]	0,452	0.45
284 (283.75)[c]	0.247	0.26	327		0.62
285		0.31	328		0.68
286		0.28	329 (328.75)[c]	0.577	0.60
287		0.25	330		0.67
288		0.23	331 (321.25)[c]	0.424	0.43
289 (288.75)[c]	0.214	0.31	332		0.45
290		0.29	333		0.48
291		0.27	334 (333.75)[c]	0.489	0.51
292		0.28	335		0.56
293 (293.75)[c]	0.286	0.27	336 (336.25)[c]	0.627	0.62
294 (293.75)[c]	0.348	0.29	337		0.61
295		0.29	338		0.58
296		0.29	339 (338.75)[c]	0.559	0.58
297		0.28	340	0.560	0.64[d]
298		0.32	341		0.61[d]
299 (299.75)[c]	0.275	0.35	342		0.58[d]
300		0.26	343		0.55[d]
301 (301.25)[c]	0.220	0.28	344		0.51[d]
302		0.27	345		0.50[d]
303		0.25	346		0.47[d]
304 (303.75)[c]	0.316	0.22	347		0.44[d]
305		0.26	348		0.42[d]
306 (306.25)[c]	0.265	0.32	349		0.38[d]
307		0.30	350		0.36[d]
308		0.28	351		0.33[d]
309 (308.75)[c]	0.265	0.27	352		0.30[d]
310		0.28	353		0.28[d]
311 (311.25)[c]	0.296	0.30	354		0.25[d]
312		0.29	355		0.12[d]
313		0.28	356		0.10[d]
314 (313.75)[c]	0.285	0.29	357		0.07[d]
315		0.29	358		0.04[d]
316 (316.25)[c]	0.327	0.33	359		0.01[d]
317		0.32	360		0.00[d]

[a] Based largely on the relative quantum yield data of Smith et al. (2002a), with normalization of the results by fixing $\phi_I = 0.78$ at 302.5 nm.

[b] Based on the data of Clark et al. (1978), Horowitz and Calvert (1978a,b), Moortgat et al. (1983), and Smith et al. (2002a); see text.

[c] Actual wavelength listed by IUPAC (Athinson et al., 2004).

[d] Pressure and temperature sensitive; estimates are for temperatures near 300 K and 1 atm pressure.

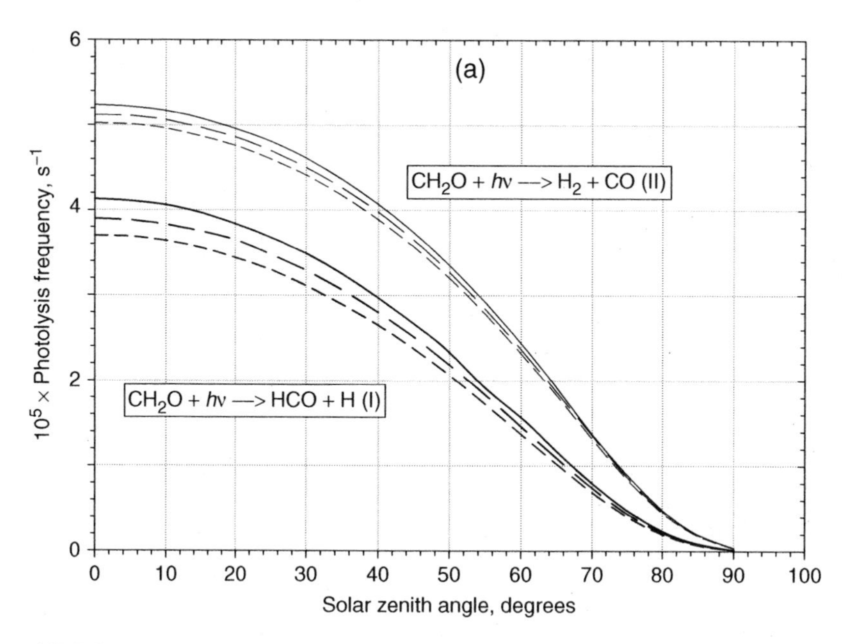

Figure VII-B-5a. Recommended j-values vs. solar zenith angle for formaldehyde for cloud-free conditions in the lower troposphere (~760 Torr); heavy curves are for process (I): $CH_2O + h\nu \rightarrow CHO + H$; light curves are for process (II): $CH_2O + h\nu \rightarrow CO + H_2$; solid curves, long-dashed curves, and short-dashed curves, respectively, are from calculations assuming overhead ozone columns of 300, 350, and 400 Dobson units. The cross sections and quantum yield recommendations from this study were used in the calculations.

0 and 50°, an increase of ozone column by 100 Dobson units leads to approximately 11% and 4% decrease in j_{I}- and j_{II}-values, respectively. The effect of temperature change on the absorption cross section of formaldehyde can be seen in figure VII-B-5c. Note that the effect is relatively small at the wavelengths that contribute the most to the j-values. However, the increase of altitude in the atmosphere affects the j-values for processes (I) and (II) significantly; see figure VII-B-5d. Here the calculations employed the current recommendations and the temperature and pressure effects measured by Moortgat et al. (1983). The ratio k_q/k_d required to derive ϕ_{II} as a function of pressure at long wavelengths was estimated from data measured at wavelengths 329 and 353 nm and temperatures of 300 and 220 K. Estimates required for each wavelength at a desired temperature were obtained by extrapolation using $\ln(k_q/k_d)$ vs. $1/T$ plots, assuming a linear relationship. The temperature profile of the U.S. Standard Atmosphere (1976) was used to define the temperature at a given altitude. The large increase in j-values seen with altitude increase results mainly from two factors: 1) the increased actinic flux at the higher altitudes; and 2) the decreased quenching of process (II) that occurs at the lowered pressure.

In view of the alternative methods of derivation of the preferred values of ϕ_{I} and ϕ_{II} and the different recommended values for cross sections for formaldehyde, we have investigated the sensitivity of the calculated j-values to the use of different combinations

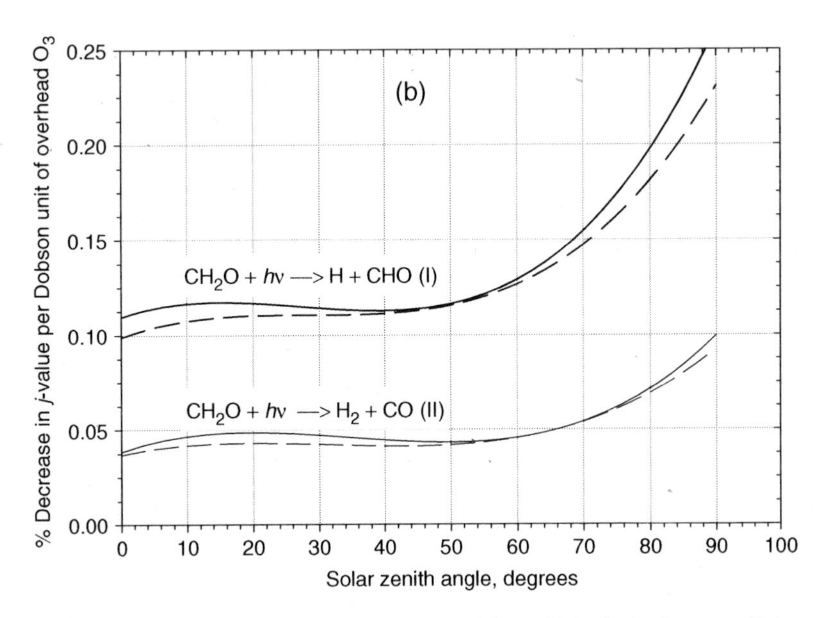

Figure VII-B-5b. Percentage decrease in the calculated formaldehyde j-values per Dobson unit of overhead ozone; the heavy and light curves are for primary processes (I) and (II), respectively; solid curves and dashed curves are for an overhead ozone column of 300–350 DU and 350–400 DU, respectively.

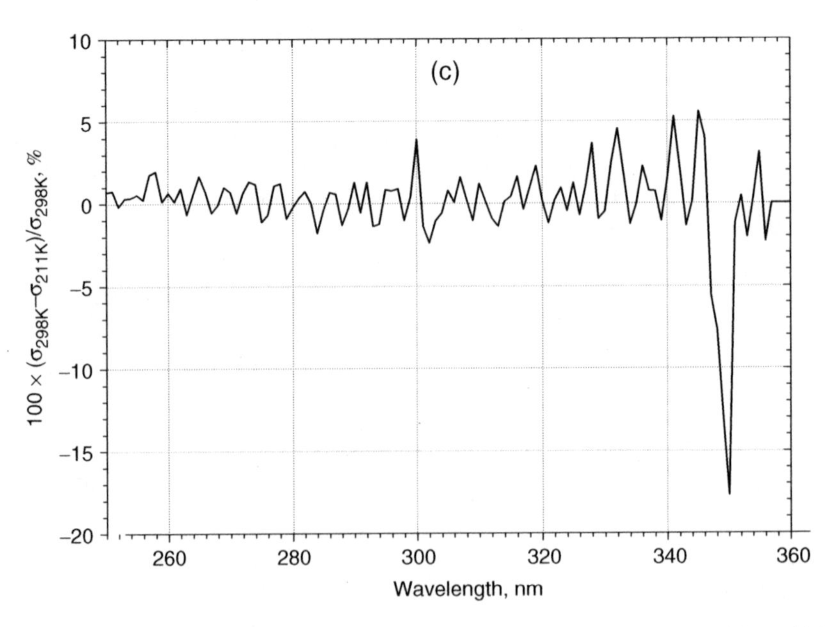

Figure VII-B-5c. Plot of the percentage difference in the absorption cross section of formaldehyde at 298 and 211 K as a function of wavelength, from the data of Meller and Moortgat (2000). Note the differences are small and rather random in sign at the wavelengths that contribute significantly to the j-values.

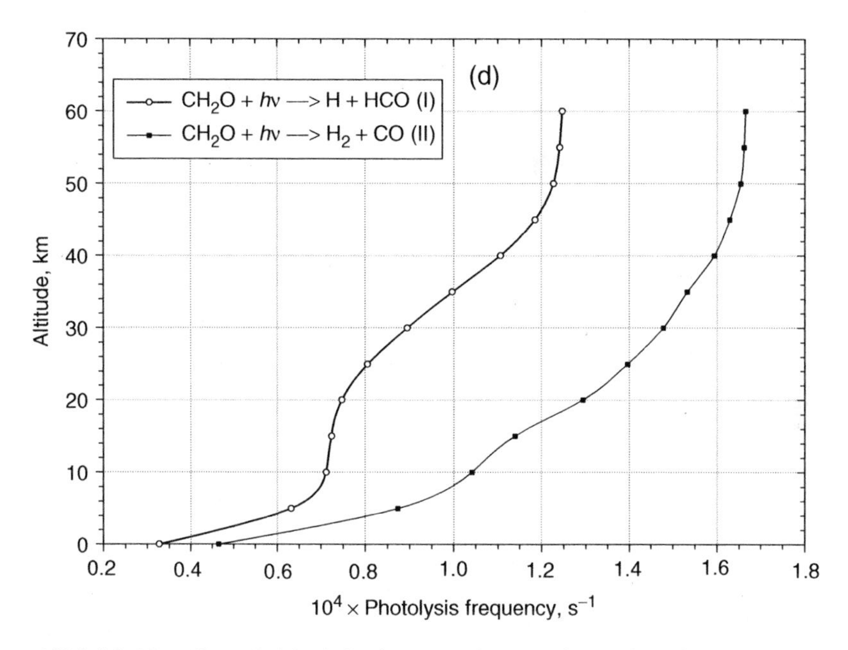

Figure VII-B-5d. The effect of altitude in the atmosphere on the j-values for processes (I) and (II). The calculations used the current recommendations with solar zenith angle fixed at 45° and the temperature and pressure effects characterized by Meller and Moortgat (2000). The large increase seen in j-values at higher altitudes results mainly from the increased actinic flux and the decreased extent of quenching of process (II) at the lowered pressure.

Table VII-B-4. Comparison of photolysis frequencies (j-values) of the formaldehyde photodecomposition modes as estimated using different spectral and quantum yield data, assuming a cloud-free troposphere with overhead Sun ($z = 0$): $CH_2O + h\nu \rightarrow H + HCO$ (I); $CH_2O + h\nu \rightarrow H_2 + CO$ (II)

	Source of data:				
	Cross Section	Quantum Yields	j_I (s^{-1})	j_{II} (s^{-1})	$j_I + j_{II}$ (s^{-1})
1)	Meller & Moortgat (2000)	Recommended in this study	3.90×10^{-5}	5.12×10^{-5}	9.02×10^{-5}
2)	Meller & Moortgat (2000)	IUPAC (2004)	4.09×10^{-5}	4.57×10^{-5}	8.66×10^{-5}
3)	Calvert et al. (2000)	Calvert et al. (2000)	3.47×10^{-5}	5.23×10^{-5}	8.70×10^{-5}
4)	Meller & Moortgat (2000)	Calvert et al. (2000)	3.66×10^{-5}	5.67×10^{-5}	9.33×10^{-5}
5)	Calvert et al. (2000)	IUPAC (2004)	3.88×10^{-5}	4.21×10^{-5}	8.09×10^{-5}

of recommended spectral and quantum yield data. In arriving at our estimates attributed in this study to the IUPAC panel (Atkinson et al., 2004), we used a spline curve fit to the data points the panel listed and determined the ϕ-values at the intersection of these curves with the ordinate at each nm. The estimated j-values for processes (I) and (II) are summarized in table VII-B-4 for the case of overhead Sun with an O_3 column of 350 Dobson units. Some differences in the j-estimates are seen for the different choices of data. For example, compare the data in row 3 with that in row 4, and in row 2 with that in row 5 of table VII-B-4. Here, using the same source of quantum yield data [either IUPAC (2004) or Calvert et al., (2000)], but the newer cross section data of Meller and Moortgat (2000) in place of the averaged cross section data reported by Calvert et al. (2000), leads to about a 5% increase in j_I and a 7% increase in $j_I + j_{II}$ in each case. When Meller and Moortgat cross sections are used and the quantum yield recommendations given in this study are replaced by those based on the IUPAC (2004) recommendations, about a 5% increase in the j_I value and a 4% decrease in $j_I + j_{II}$ results.

Feilberg et al. (2005a) have measured relative photolysis frequencies for various isotopomers of formaldehyde, $H^{12}CH^{16}O$, $H^{13}CH^{16}O$, $H^{12}CH^{18}O$, and $D^{12}CDO$, in photolyses in sunlight in the EUPHORE facility: $j(H^{13}CHO)/j(H^{12}CHO) = 0.894 \pm 0.006$; $j(H^{12}CH^{18}O)/j(H^{12}CH^{16}O) = 0.911 \pm 0.011$; $j(D^{12}CD^{16}O)/j(H^{12}CH^{16}O = 0.597 \pm 0.001$. The inefficient photodecomposition of CD_2O compared to CH_2O observed by Feilberg et al. (2005a) was also observed by McQuigg and Calvert (1969). They used a broad band continuum (278–376 nm) of a xenon flash tube to photolyze pure CH_2O and CD_2O and their mixtures. The average value of $\Phi(H_2 + HD + D_2)/\Phi(CO)$ from 57 experiments was 0.97 ± 0.08. For a given pressure of the pure aldehyde, the volume of product was proportional to the energy of the flash. However, for a given energy of the flash and a given pressure of the pure aldehyde, they found an average ratio $j(CD_2O)/j(CH_2O) \approx 0.5$. Although band positions are different, the integrated magnitude of the cross sections at low resolution as a function of wavelength (278–376) nm are similar for the two formaldehydes, so it is evident that the effective quantum yield of CH_2O is significantly greater than that of CD_2O for the wavelength region available in the lower troposphere.

In a simple molecule such as formaldehyde, spectral lines of different isotopomers often are at distinct, non-overlapping wavelengths that make selective photolyis possible. Yeung and Moore (1972) and Letokhov (1972) recognized this at about the same time. Marling (1977) induced photodecomposition of specific isotopomers of natural formaldehyde using highly resolved laser lines that matched specific isotopomers. He observed effective enrichment of rare isotopes (^{17}O, ^{18}O, ^{13}C, and ^{2}H). Thus ^{18}O was enriched up to ninefold using natural neon lasing at 332.426 nm. Irradiation with a ^{22}Ne-laser at 332.371 nm gave a ninefold enrichment of ^{17}O and a 33-fold enrichment of ^{13}C. Using D_2CO permitted a 44-fold ^{18}O and 27-fold ^{17}O enrichment using radiation at 332.371 nm. Mannik et al. (1979) effected an enrichment of deuterium in natural CH_2O (4 Torr) by a factor of 254 ± 30 using 345.6 nm radiation.

VII-B-2. Photodecomposition of Acetaldehyde (Ethanal, CH_3CHO)

Acetaldehyde is a major product of the oxidation of the alkanes and other hydrocarbons present in the atmosphere. It is important as a source of radicals through its photodecomposition. The atmospheric chemistry reviews by Calvert et al. (2000) and Atkinson et al. (2004) provide critical evaluations and recommendations for cross sections and quantum yields for modeling acetaldehyde atmospheric chemistry, and these should be reviewed by the reader. Here we present a brief summary of the available data and our

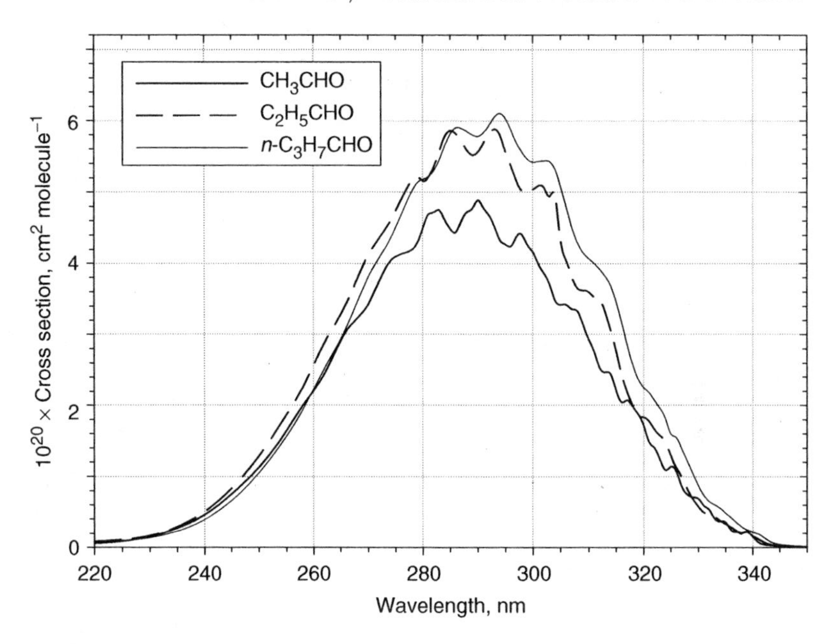

Figure VII-B-6. Absorption cross sections for acetaldehyde, propanal and butanal. Data from Martinez et al. (1992); temperature, 300 K.

current recommendations. Acetaldehyde and the larger aldehydes absorb light available within the troposphere to promote an $n \rightarrow \pi^*$ excitation; see figure VII-B-6.

Three primary processes have been considered in explanation of acetaldehyde photochemistry:

$$CH_3CHO + h\nu \rightarrow CH_3 + CHO \qquad \text{(I)}$$

$$\rightarrow CH_4 + CO \qquad \text{(II)}$$

$$\rightarrow CH_3CO + H \qquad \text{(III)}$$

Process (I) is most important at the long wavelengths of absorbed light, and process (II) becomes increasingly important at the short wavelengths (Blacet and Heldman, 1942; Blacet and Loeffler, 1942). Many experimental studies have helped to elucidate the role of the excited triplet state (T_1) and other states in the photochemistry of acetaldehyde (e.g., see Parmenter and Noyes, 1963; Archer et al., 1973a,b; Gandini and Hackett, 1977b). Internal conversion of the originally formed excited singlet to form a vibrationally excited ground state molecule ($S_1 \rightarrow S_0^*$) and intersystem crossing of the S_1 state to vibrationally excited levels of the first excited triplet state ($S_1 \rightarrow T_1^*$) are of increasing importance for the higher aldehydes. The efficiency of these processes depends on the density of states in the molecule. The density of states increases with the higher number of internal degrees of freedom that accompanies increased molecular complexity. Noble and Lee (1984a,b) concluded that the total rate of non-radiative processes, internal conversion and intersystem crossing, in acetaldehyde and the higher aldehydes is about three orders of magnitude greater that the rate of radiative decay. Theoretical studies (Yadav and Goddard, 1986; Smith et al., 1991; Martell et al., 1997; King et al., 2000; Kurosaki and Yokoyama, 2002) find that the S_1 state formed on excitation to low vibrational levels (longer wavelengths) decays largely to the highly excited ground state (S_0) via internal conversion. For excitation at $\lambda <\sim 317$ nm, population of the first excited triplet state (T_1) and subsequent dissociation by process (I) can occur, although Yadav and Goddard (1986) estimate that a barrier of ~50–63 kJ mol^{-1} exists for this dissociation. If the S_0 state is populated by either internal conversion or intersystem crossing, the vibrationally rich aldehyde molecule may dissociate to radicals, process (I), presumably without a barrier. The transition state for dissociation by (II) is energetically accessible for ~351 kJ mol^{-1} (340 nm), although somewhat tortured geometry may make it dynamically difficult to reach. Kurosaki and Yokoyama (2002) estimate theoretically that the CO formed in process (II) is highly rotationally excited (quantum number, 68), but vibrationally it is almost unexcited (quantum number, 0.15), in qualitative agreement with the experiments of Gherman et al. (2001). The minimum energy for dissociation by primary process (I) lies slightly lower than that for (III); see Yadav and Goddard (1986).

VII-B-2.1. Cross Sections for Acetaldehyde

Different measurements of cross sections for acetaldehyde are in reasonable agreement (Calvert et al., 2000), but those reported by Martinez et al. (1992) are considered to be the most accurate and are recommended here. These are given in figure VII-B-6 and listed in table VII-B-5 for wavelengths present within the troposphere.

VII-B-2.2. Quantum Yields for Primary Processes in Acetaldehyde Photodecomposition

Studies that relate to the quantum efficiency of the photodecomposition of acetaldehyde have continued since the early work of Leighton and Blacet (1933), Blacet and Heldman (1942), and Blacet and Loeffler (1942). They studied photolysis of acetaldehyde and its mixtures with iodine. Later studies confirm the original conclusions of these workers, namely, that radical formation in process (I) is the dominant mode of photodecomposition at the longer wavelengths present within the troposphere. Process (II) increases in importance at the shorter wavelengths. The aldehyde–iodine mixture photolysis, carried out at several wavelengths and temperatures in the range from 57 to 150°C, provide quantitative results related to the occurrence of processes (I) and (II), but the uncertainties in the extent of quenching of the excited states of acetaldehyde by CH_3CHO, I_2, and air molecules restricts the accurate extrapolation of these data to atmospheric conditions.

Quenching of the products of process (I) by added air, N_2, CO_2, or other gases is much less efficient than those of processes (II) and (III), and confirms the view that more than one excited state is involved in the photochemistry of acetaldehyde. Experimental evidence presented by Horowitz et al. (1982) suggests that the vibrationally excited triplet state (T_1) of acetaldehyde is the precursor to process (I), while processes (II)

Table VII-B-5. Preferred cross sections (σ, cm^2 $molecule^{-1}$) for acetaldehyde (ethanal), propanal, and butanal at 298 K (Martinez et al., 1992)

λ (nm)	$10^{20} \times \sigma$, Ethanal	$10^{20} \times \sigma$, Propanal	$10^{20} \times \sigma$, Butanal	λ (nm)	$10^{20} \times \sigma$, Ethanal	$10^{20} \times \sigma$, Propanal	$10^{20} \times \sigma$, Butanal
280	4.50	5.16	5.18	323	1.24	1.58	1.96
281	4.69	5.21	5.22	324	1.09	1.49	1.82
282	4.72	5.35	5.30	325	1.14	1.30	1.69
283	4.75	5.57	5.45	326	1.07	1.13	1.55
284	4.61	5.78	5.65	327	0.868	1.00	1.39
285	4.49	5.86	5.81	328	0.747	0.828	1.22
286	4.44	5.82	5.90	329	0.707	0.685	1.04
287	4.59	5.72	5.89	330	0.688	0.575	0.868
288	4.72	5.59	5.85	331	0.588	0.494	0.734
289	4.77	5.52	5.80	332	0.530	0.466	0.647
290	4.89	5.56	5.78	333	0.398	0.430	0.602
291	4.78	5.68	5.83	334	0.363	0.373	0.554
292	4.68	5.81	5.93	335	0.350	0.325	0.485
293	4.53	5.88	6.05	336	0.238	0.280	0.417
294	4.33	5.80	6.10	337	0.222	0.230	0.346
295	4.27	5.57	6.03	338	0.205	0.185	0.285
296	4.24	5.37	5.97	339	0.219	0.166	0.242
297	4.38	5.16	5.69	340	0.150	0.155	0.215
298	4.41	5.02	5.56	341	0.074	0.110	0.191
299	4.26	5.02	5.46	342	0.042	0.076	0.144
300	4.16	5.04	5.42	343	0.031	0.045	0.093
301	3.99	5.09	5.43	344	0.026	0.031	0.059
302	3.86	5.07	5.44	345	0.021	0.025	0.041
303	3.72	4.94	5.43	346	0.019	0.019	0.031
304	3.48	4.69	5.32	347	0.015	0.016	0.026
305	3.42	4.32	5.08	348	0.016	0.014	0.023
306	3.42	4.04	4.79	349	0.010	0.013	0.018
307	3.36	3.81	4.50	350	0.008	0.010	0.015
308	3.33	3.65	4.29	351	0.007	0.008	0.014
309	3.14	3.62	4.15	352	0.006	0.007	0.012
310	2.93	3.60	4.07	353	0.005	0.005	0.010
311	2.76	3.53	4.00	354	0.005	0.004	0.008
312	2.53	3.50	3.92	355	0.004	0.002	0.006
313	2.47	3.32	3.82	356	0.005	0.001	0.005
314	2.44	3.06	3.69	357	0.003	0.001	0.004
315	2.20	2.77	3.46	358	0.004	0	0.003
316	2.04	2.43	3.17	359	0.002	0	0.002
317	2.07	2.18	2.85	360	0.003	0	0.002
318	1.98	2.00	2.57	361	0.002	0	0.002
319	1.87	1.86	2.37	362	0.001	0	0.001
320	1.72	1.83	2.25	363	0.000	0	0.001
321	1.48	1.78	2.18	364	0.000	0	0.001
322	1.40	1.66	2.07	365	0.000	0	0

and (III) originate from high vibrational levels of the excited singlet state (e.g., see the discussion of Horowitz et al., 1982). From results of their CH_3CHO–O_2 mixture photolysis, Weaver et al. (1976/1977) speculated that the T_1 state of acetaldehyde forms a complex with O_2 that reacts to form CH_3, CO, and HO_2 products. This mechanism has not been adopted in later studies.

The acetaldehyde photodecomposition studies most relevant to atmospheric conditions are those of Meyrahn et al. (1982). They photolyzed dilute mixtures of acetaldehyde in air and measured quantum yields of CO and CH_4. Under atmospheric conditions, process (I), CH_3 and HCO formation, will lead to the initial products CH_3O_2, HO_2, and CO. Methane cannot be formed following (I). However, methane in the products of the CH_3CHO–air mixture photolysis can only arise from the occurrence of process (II). The abstraction of H-atoms from acetaldehyde by CH_3 radicals formed in (I) is unimportant, since its rapid reaction with O_2, $CH_3 + O_2 \rightarrow CH_3O_2$, is its only fate in air. Under these conditions, Φ_{CO} and Φ_{CH4} should provide estimates of $\phi_I + \phi_{II}$ and ϕ_{II}, respectively. Φ_{CO} and Φ_{CH4} measured by Meyrahn et al. (1982) in very dilute acetaldehyde–air mixtures (near 1 atm pressure and 25°C) are shown as large, open circles and open triangles, respectively, in figure VII-B-7.

Horowitz et al. (1982) and Horowitz and Calvert (1982) determined Φ_{CO} (measure of $\phi_I + \phi_{II}$) and Φ_{CH4} (attributed to ϕ_{II}) from photolysis of mixtures of CH_3CHO, O_2, 2-methylpropene, and CO_2. Estimates of the quantum yields of the various processes for acetaldehyde in air at 1 atm were made from experiments with added CO_2 at pressures that were believed to be equivalent to those of air (N_2/O_2) in quenching ability. These measurements of Φ_{CO} and Φ_{CH4} (small black circles and small black triangles) are shown in figure VII-B-7. The values of Φ_{CO} fall somewhat below those of the curve defining the measurements of Meyrahn et al. (1982), although the trend with wavelength is the same. In extrapolating their data to quantum yields for 1 atm of air, Horowitz and Calvert assumed their measured ratio of quenching constants of excited acetaldehyde by CO_2 to that for air (1.29 at 313 nm) was independent of wavelength. Subsequent studies of Horowitz (1986) suggest that this ratio was overestimated, and, as a consequence, the Φ_{CO} values are believed to be somewhat underestimated; see figure VII-B-7.

Simonaitis and Heicklen (1983) measured the quantum yields of CH_3O_2 formation from CH_3CHO (11–25 Torr) laser photolysis in air. The reported estimates of ϕ_I at the highest pressures employed are: 0.33 at 310.5 nm (495 Torr), 0.59 at 302.0 nm (487 Torr), and 0.73 at 294 nm (612 Torr). We have used Stern–Volmer plots of the Meyrahn et al. (1982) data for CH_3CHO–air mixtures measured at 313.0 and 304.4 nm to adjust these Simonaitis and Heicklen (1983) results to 760 Torr of air. We estimate $\phi_I = 0.24$, 0.47, and 0.65 for $\lambda = 310.5$, 302.0, and 294 nm, respectively. These data are plotted in figure VII-B-7 as small, open circles. The accuracy of these results is limited by corrections for HO_2 and $CH_3C(O)O_2$ absorption that are required to derive CH_3O_2 yields, and extrapolation to 1 atm introduces additional error. However, the same trend in the quantum yield with wavelength is seen, and the data are very consistent with the other estimates shown. From studies of CH_3CHO photolysis at 313 nm in O_2 or air mixtures, Weaver et al. (1976/1977) found $\Phi_{co} = 0.15 \pm 0.02$ for 1 atm air (shown as the grey circle in figure VII-B-7). This also is in reasonable agreement with the observations of Meyrahn et al. (1978) and the other data for this wavelength.

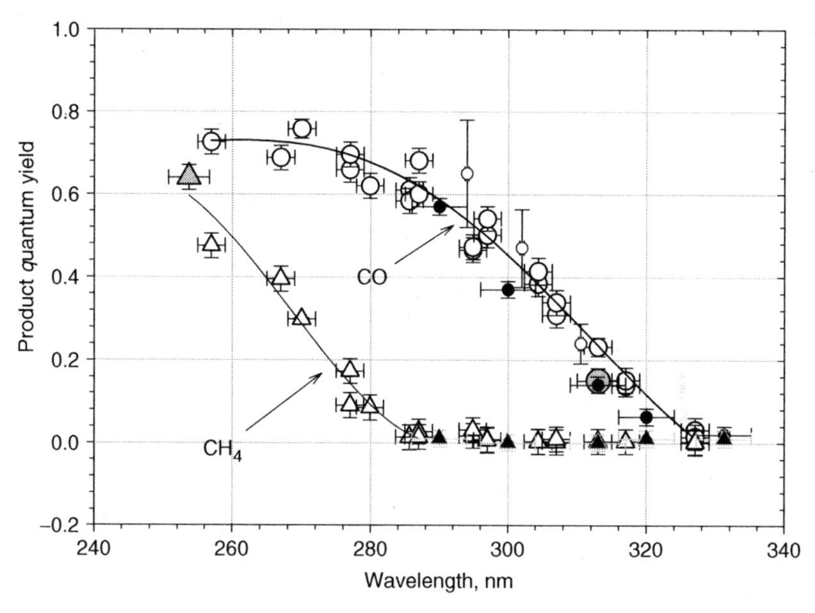

Figure VII-B-7. Comparison of estimates of the quantum yields of products in CH_3CHO photodecomposition. Data for $\Phi_{CO} = \phi_I + \phi_{II}$ (shown as circles) are from photolyses in air (large open circles, Meyrahn et al., 1982); photolyses with added 2-methylpropene, O_2 and CO_2 (small closed circles, Horowitz and Calvert, 1982); photolysis in air at 1 atm (large dark-grey circle, Weaver et al., 1976/1977); Φ_{CH3O2} as measured from laser photolysis in air (487–612 Torr) and adjusted to 1 atm air using Stern–Volmer plots from data of Meyrahn et al., 1982 (small open circles, Simonaitis and Heicklen, 1983). Data for $\Phi_{CH4} = \phi_{II}$ estimates (shown as triangles) are from photolyses in air at 1 atm (open triangles, Meyrahn et al., 1982); photolyses with added O_2, 2-methylpropene, and CO_2 (small black triangles, Horowitz and Calvert, 1982); photolysis with added NO (large dark-grey triangle, Parmenter and Noyes, 1963).The heavy curve and the light curve are spline fits to the data.

Several other measurements of the quantum efficiencies of processes (I) and (II) have been made that are consistent with the more direct data of Meyrahn et al., but are not readily transformed to atmospheric conditions. Thus Gill and Atkinson (1979) observed directly the HCO radical generation in the laser photolysis of acetaldehyde (10 Torr pressure); the wavelength dependence observed is different from that indicated by the data of figure VII-B-7. The onset of the HCO formation occurred at 320 nm, and the relative quantum yields maximized at about 315 nm and then decreased rapidly and continuously with photolysis at the shorter wavelengths (306, 295, 266 nm).

Horowitz (1986) has shown in 300 nm photolysis of HCl, CH_3CHO, and CO_2 mixtures that HCl is an effective trap for CH_3 radicals and H-atoms formed in (I) and (III), respectively ($CH_3 + HCl \rightarrow CH_4 + Cl$; $H + HCl \rightarrow H_2 + Cl$; $Cl + CH_3CHO \rightarrow HCl + CH_3CO$). He indicated that HCl can be used as a convenient and sensitive probe in studies of the yields of radicals in acetaldehyde photodecomposition. The Horowitz estimates of ϕ_I, ϕ_{II}, and ϕ_{III} with $\lambda = 300$ nm (P_{CH3CHO}, 5–76 Torr; P_{HCl}, 0.5–5 Torr) check well with those made previously with added O_2 and CO_2 (Horowitz et al., 1982); however, no estimates under simulated atmospheric conditions were made.

Table VII-B-6. Recommended values for ϕ_I, ϕ_{II}, and ϕ_{III} for acetaldehyde photodecomposition in air at 298 K[a]

λ (nm)	ϕ_I [b]	ϕ_{II} [c]	ϕ_{III} [d]	λ (nm)	ϕ_I [b]	ϕ_{III} [d]
275	0.542	0.182	0.065	303	0.432	0.012
276	0.557	0.162	0.063	304	0.416	0.011
277	0.571	0.144	0.059	305	0.400	0.010
278	0.584	0.125	0.057	306	0.384	0.009
279	0.595	0.108	0.054	307	0.367	0.008
280	0.605	0.092	0.052	308	0.350	0.007
281	0.615	0.076	0.049	309	0.333	0.006
282	0.622	0.062	0.048	310	0.316	0.005
283	0.627	0.049	0.045	311	0.299	0.005
284	0.631	0.037	0.043	312	0.282	0.004
285	0.633	0.027	0.040	313	0.265	0.003
286	0.633	0.018	0.038	314	0.248	0.003
287	0.630	0.011	0.036	315	0.231	0.002
288	0.626	0.005	0.034	316	0.214	0.002
289	0.619	0.002	0.032	317	0.197	0.001
290	0.610	0.00	0.031	318	0.180	0.001
291	0.599	0.00	0.029	319	0.164	0.001
292	0.587	0.00	0.027	320	0.148	0.000
293	0.575	0.00	0.025	321	0.132	
294	0.562	0.00	0.024	322	0.116	
295	0.549	0.00	0.022	323	0.101	
296	0.536	0.00	0.021	324	0.086	
297	0.522	0.00	0.019	325	0.072	
298	0.508	0.00	0.018	326	0.058	
299	0.493	0.00	0.017	327	0.045	
300	0.479	0.00	0.015	328	0.033	
301	0.463	0.00	0.014	329	0.021	
302	0.448	0.00	0.013	330	0.009	

[a] Data are derived from the trend curves shown in figures VII-B-7 and VII-B-8.
[b] $CH_3CHO + h\nu \rightarrow CH_3 + HCO$ (I)
[c] $CH_3CHO + h\nu \rightarrow CH_4 + CO$ (II)
[d] $CH_3CHO + h\nu \rightarrow CH_3CO + H$ (III)

We believe Meyrahn et al. (1982) provide the most relevant data for ϕ_I and ϕ_{II} in acetaldehyde photodecomposition in the atmosphere, since the other data shown in figure VII-B-7 require uncertain corrections to achieve estimates for 1 atm of air. Although results using either N_2 or air gave similar results, we have used only the results of the most relevant data, that from air mixtures, to derive the recommended values given in table VII-B-6.

The occurrence of primary process (III) in acetaldehyde photolysis is supported in theory (e.g., see Yadav and Goddard, 1986) and in the experiments of Horowitz et al. (1982) using CH_3CHO in the absence of oxygen. Evidence excludes possible sources of H_2 other than (III): (1) the absence of ketene as a product of acetaldehyde photolysis (presumably formed in the reaction $CH_3CHO + h\nu \rightarrow CH_2{=}CO + H_2$); (2) the observed kinetics of the suppression of H_2 formation in acetaldehyde–2-methylpropene mixtures (resulting from $H + (CH_3)_2C{=}CH_2 \rightarrow (CH_3)_3C$); (3) the insignificance of H_2 from the photolysis of the CH_2O formed in the CH_3CHO photolysis; and (4) the unimportance of H_2 formation from HCO–HCO encounters that have been shown to give largely CO and CH_2O (Horowitz and Calvert, 1978a,b; Clark et al., 1978). See also Horowitz et al. (1982).

The quantum yields of process (III) for acetaldehyde (with CO_2 at pressures equivalent to 1 atm of air) and excitation at 313, 300, and 290 nm (Horowitz et al., 1982) are given in figure VII-B-8 (black circles). The data shown as triangles in this figure at wavelengths 313, 280.4, 265.4, 253.7, and 238 nm are estimated in the current study using the Φ_{CO} measurements of Leighton and Blacet (1933) for 300 Torr of CH_3CHO, coupled with the ratios Φ_{H2}/Φ_{CO} measured at this pressure by Blacet and Roof (1936) and Blacet and Volman (1938). In using these older data we have assumed that the quenching of process (III) for excitation at short wavelengths and 300 Torr pressures of CH_3CHO is small but similar to that observed for CH_3CHO mixtures in 1 atm of air. This assumption is consistent with expectation from the quenching relations of Horowitz and Calvert (1982) for $\lambda = 290$ nm. For the shorter wavelengths, it is justified by data of Leighton and Blacet (1933); these show Φ_{CO} to be insensitive to acetaldehyde pressure (60–300 Torr) for excitations at 265.4 and 253.7 nm. One expects this to be the case also for excitation at $\lambda = 238$ nm. Although primary process (III) is unimportant for wavelengths of light within the troposphere, it becomes significant at the shorter wavelengths present in the upper atmosphere.

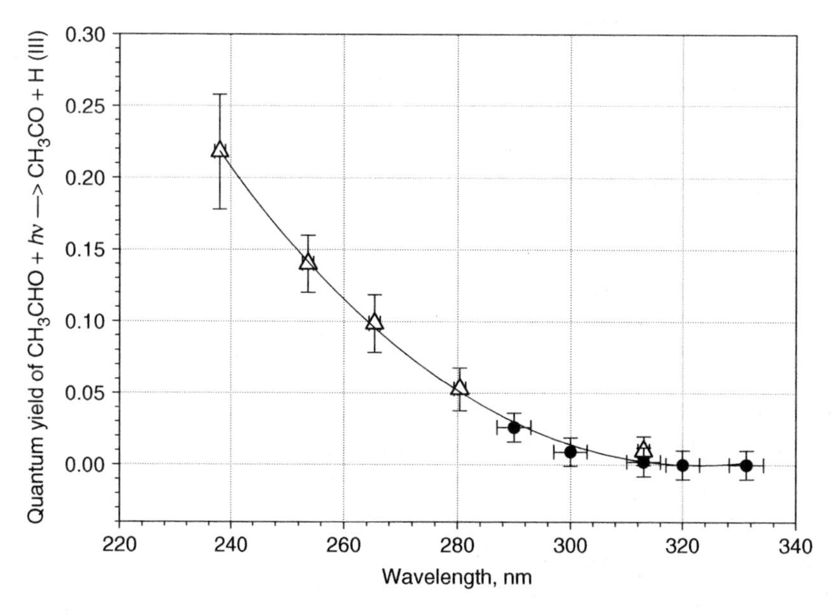

Figure VII-B-8. Estimated quantum yields for primary process (III) in acetaldehyde photodecomposition in air. Data (triangles) are from quantum yields of H_2 measured in pure acetaldehyde (~300 Torr), as estimated from the Φ_{CO} data of Leighton and Blacet (1933) and the Φ_{H2}/Φ_{CO} data of Blacet and Roof (1936) and Blacet and Volman (1938) at 300 Torr. The data given as black circles are the estimates of Horowitz and Calvert (1982) for acetaldehyde–CO_2mixtures with quenching equivalent to that of air at 1 atm pressure. See the text for a description of the data treatment employed.

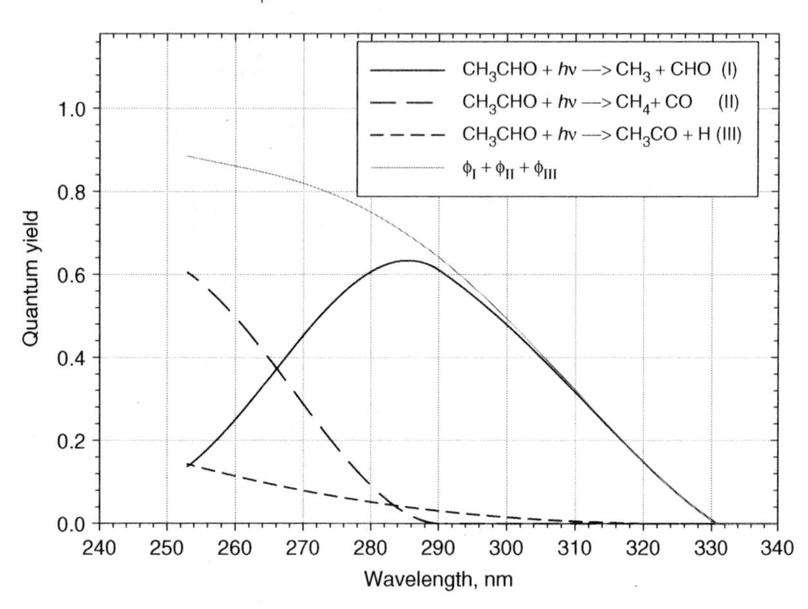

Figure VII-B-9. Plot of the recommended quantum yields for CH_3CHO photodecomposition in air (760 Torr) as a function of wavelength: ϕ_I, ϕ_{II}, ϕ_{III}, ϕ_{total}.

Our recommendations for values of ϕ_I, ϕ_{II}, and ϕ_{III} for use in tropospheric modeling are given in table VII-B-6, and are shown graphically versus wavelength in figure VII-B-9.

VII-B-2.3. Photolysis Frequencies for Acetaldehyde

The current recommended values for the cross sections and quantum yields have been used to calculate the clear-sky values of the photolysis frequencies for acetaldehyde primary processes (I) and (III) as a function of the solar zenith angle (ozone column = 350 DU); see figure VII-B-10. Values of j(II), not given in the figure, are very small for tropospheric conditions.

VII-B-3. Photodecomposition of Propanal (Propionaldehyde, CH_3CH_2CHO)

Propanal (C_2H_5CHO) photodecomposition in the troposphere follows the pattern exhibited by acetaldehyde. Although five decomposition paths have been considered in explanation of its photochemistry, only process (I) is important for wavelengths of excitation within the troposphere:

$$C_2H_5CHO + h\nu \rightarrow C_2H_5 + HCO \qquad (I)$$

$$\rightarrow C_2H_6 + CO \qquad (II)$$

$$\rightarrow C_2H_5CO + H \qquad (III)$$

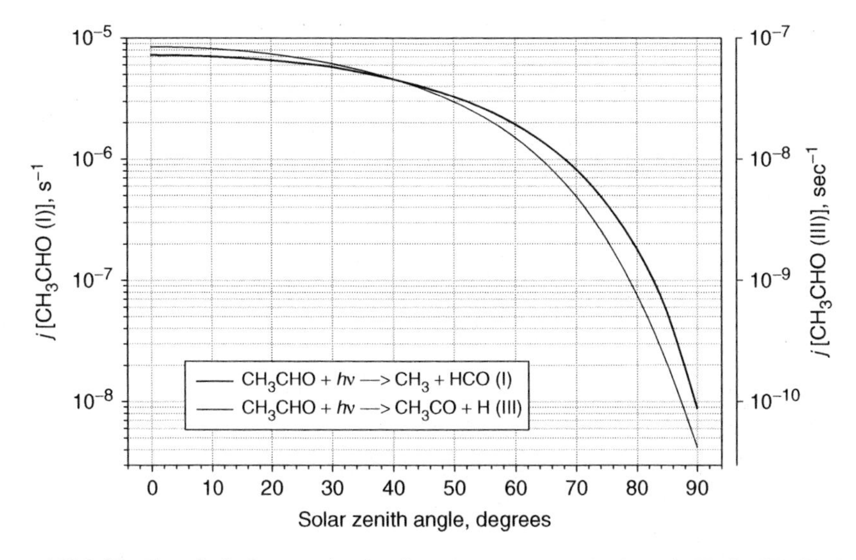

Figure VII-B-10. Photolysis frequencies for the primary processes (I) and (III) in CH_3CHO vs. solar zenith angle; process (II) values (not shown here) are very near zero. Data are representative of a cloudless day in the lower troposphere with an overhead ozone column of 350 DU.

$$\rightarrow CH_3 + CH_2CHO \quad \text{(IV)}$$

$$\rightarrow C_2H_4 + CH_2O \quad \text{(V)}$$

The other primary processes (II)–(V) become significant at wavelengths available in the upper troposphere and the stratosphere. Experimental work has shown that the dissociation into radicals in (I) likely occurs from the excited T_1 state following intersystem crossing, and that the threshold for this process lies at 326.26 nm, $25 \pm 4\,\text{kJ mol}^{-1}$ above the dissociation energy (Terentis et al., 1995; Metha et al., 2002). Process (II) and presumably other molecular processes arise from the vibrationally excited ground state (S_0*) formed by internal conversion. We will consider the experimental evidence concerning the importance of these various processes in this section.

VII-B-3.1. Cross Sections for Propanal

The cross sections for propanal as a function of wavelength are shown with those for acetaldehyde and butanal in figure VII-B-6. The measurements of Martinez et al. (1992) are tabulated in table VII-B-5. These are recommended for use in considerations of the atmospheric photochemistry of propanal.

VII-B-3.2. Quantum Yields of Primary Processes in Propanal

The important primary process (I), $C_2H_5CHO + h\nu \rightarrow C_2H_5 + HCO$, has been studied by a variety of techniques. Quantum yields have been measured for final products of

C_2H_5CHO photolysis in air mixtures (Shepson and Heicklen, 1982a,b), C_2H_5I from C_2H_5CHO–I_2 mixtures (Blacet and Pitts, 1952), $C_2H_5O_2$ from C_2H_5CHO–air mixtures (Heicklen et al., 1986), as well as the initial radical product, HCO (Chen and Zhu, 2001), at low pressures. In spite of these extensive efforts, the experimental estimates of ϕ_I from the data of different research groups that relate to propanal photodissociation under atmospheric conditions are rather scattered, and significant uncertainty in estimates of ϕ_I remains. The available estimates for $\phi_I + \phi_{II}$ are shown in figure VII-B-11. The most direct measurements related to tropospheric conditions (1 atm air, 25°C) are the Φ_{CO} measurements of Shepson and Heicklen (1982b,a), shown as open circles and open square, respectively, in figure VII-B-11. In air mixtures, both processes (I) and (II) lead ultimately to CO formation, since CO is also formed in the rapid reaction CHO + $O_2 \rightarrow HO_2 + CO$ following (I) in air mixtures.

Data from aldehyde–iodine mixtures in studies of Blacet and Pitts (1952) have shown clearly that process (I) is the dominant pathway for photolysis at 313 nm. However, the use of these data to derive ϕ_I and ϕ_{II} estimates for tropospheric conditions (1 atm of air) is problematical. There are uncertainties in both the extent of quenching of the primary photodissociation processes by added iodine, propanal, and air molecules and in extrapolation procedures to convert measurements to representative values for atmospheric conditions. A test of the degree of quenching by iodine can be made by

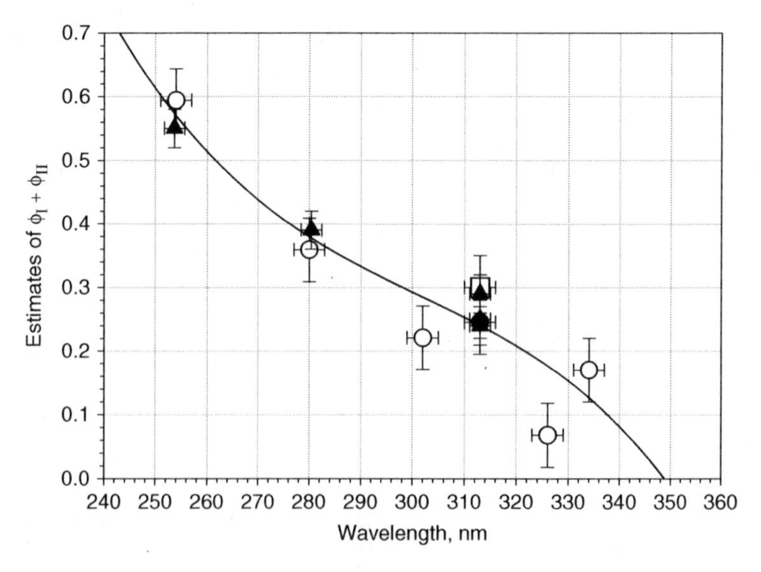

Figure VII-B-11. Estimates of the sum of the quantum efficiencies of primary processes (I) + (II) in the photodissociation of C_2H_5CHO under atmospheric conditions (760 Torr of air). $C_2H_5CHO + h\nu \rightarrow C_2H_5 + HCO$ (I); $C_2H_5CHO + h\nu \rightarrow C_2H_6 + CO$ (II). Data shown as open circles (Shepson and Heicklen, 1982b) and the open square (Shepson and Heicklen, 1982a) are derived from Φ_{CO} measurements of propanal photolysis in air; data plotted as filled triangles are derived from Φ_{CO}measurements in C_2H_5CHO–I_2 mixture photolysis (Blacet and Pitts, 1952), adjusted for air quenching alone, see text. The black curve gives the recommended wavelength dependence of $\phi_I + \phi_{II}$.

coupling the results of Blacet and Pitts with the propanal quenching data presented by Shepson and Heicklen (1982b). For example, the Φ_{CO} data of Shepson and Heicklen show that quenching of 313 nm excited propanal by C_2H_5CHO and by air molecules can be described by the Stern–Volmer relation (1):

$$1/\Phi_{CO} = 1.02 + 0.00573 \times P_{C2H5CHO} + 0.00365 \times P_{air} \qquad (1)$$

where the pressure P is in Torr. Similar functions can be derived from data at other wavelengths. Using equation (1) to test qualitatively the extent of quenching, we find that Φ_{CO} expected for pure propanal in the experiments of Blacet and Pitts (1952) at 110, 201, and 211 Torr are 0.61, 0.46, and 0.45, respectively, while those observed were 0.56, 0.45, and 0.50, respectively. Thus little, if any, influence is seen on propanal photodecomposition at $\lambda \leq 313$ nm with addition of small amounts of iodine (1.7–4.0 Torr). The ability of added gases to quench excited propanal molecules decreases significantly as the exciting wavelength decreases in the order 326 > 313 > 302 > 280 > 254 nm. As in C_2H_5CHO mixtures in air, the relation $\Phi_{CO} = \phi_I + \phi_{II}$ should hold since the HCO formed in iodine mixtures reacts to form CO ultimately ($HCO + I_2 \rightarrow I + HCOI$; $HCOI \rightarrow HI + CO$). The quantum yields of CO, determined in aldehyde–iodine mixtures, can be used to obtain additional estimates of $\phi_I + \phi_{II}$. The efficiency of the photodecomposition as measured in the I_2–C_2H_5CHO mixtures can be adjusted for the propanal quenching to that expected from air alone, by using the quenching relations derived by Shepson and Heicklen [equation (1) and similar relations for 280 and 253.7 nm]. These adjusted data from the Blacet and Pitts (1952) studies at wavelengths 313, 280, and 253.7 nm are shown in figure VII-B-11 by the black triangles. The agreement with the data of Shepson and Heicklen (1982a,b) is reasonably good at each of the three wavelengths common to both studies. The solid line given in the figure is the wavelength dependence of $\phi_I + \phi_{II}$ recommended in this study.

Estimates of the quantum efficiency of primary process (II) are shown in figure VII-B-12. These are derived from Φ_{C2H6} measurements in C_2H_5CHO mixtures in air (shown as black circles, Shepson and Heicklen, 1982b) and in C_2H_5CHO–I_2 mixtures (open triangles, Blacet and Pitts, 1952). The data from the iodine mixture were adjusted, as described previously, to those expected for air at 1 atm using the relations derived from the studies of Shepson and Heicklen. The line is the wavelength dependence of ϕ_{II} recommended in this work.

Estimates of the important primary process (I) for tropospheric conditions (1 atm air) by various methods are shown in figure VII-B-13. The solid curve in this figure (ϕ_I) is derived from the difference between the two recommended trend lines for $\phi_I + \phi_{II}$ (figure VII-B-11) and ϕ_{II} (figure VII-B-12). The black diamonds denote quantum yields of iodoethane from Blacet and Pitts (1952), adjusted for the expected extent of quenching by air as described previously. The open circles are from the measured differences, $\Phi_{CO} - \Phi_{C2H6}$, from the studies of Shepson and Heicklen (1982b). Seemingly, the most direct measures of ϕ_I are from experimental estimates of HCO radical quantum yields determined in C_2H_5CHO–N_2 mixtures (Chen and Zhu, 2001), shown as open triangles, and those based on $C_2H_5O_2$ radical quantum yields from C_2H_5CHO–air mixture

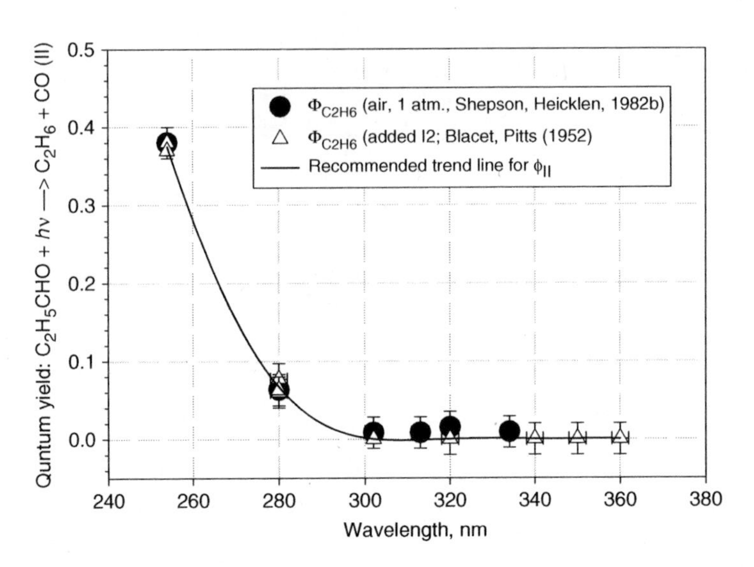

Figure VII-B-12. Experimental estimates of the quantum yields of primary process (II) in C_2H_5CHO photolysis in air (760 Torr): $C_2H_5CHO + h\nu \rightarrow C_2H_6 + CO$. The curve is the wavelength dependence of ϕ_{II} recommended in this work.

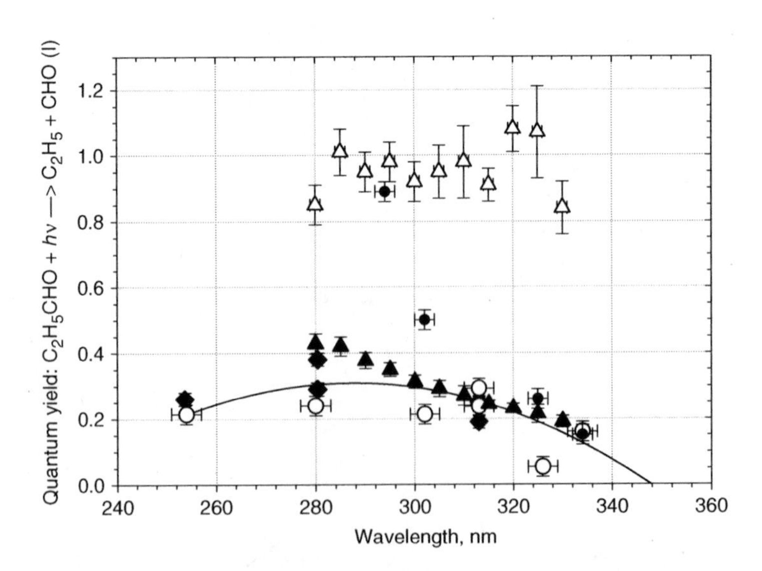

Figure VII-B-13. Experimental estimates of ϕ_I for C_2H_5CHO as a function of wavelength. $\Phi_{CO} - \Phi_{C2H6}$ [C_2H_5CHO in air (760 Torr), Shepson and Heicklen, 1982b, open circles]; Φ_{C2H5O2} [C_2H_5CHO in air (1 atm), Heicklen et al., 1986, small filled circles]; Φ_{C2H5I} [Blacet and Pitts, 1952, I_2–C_2H_5CHO mixtures, adjusted for air (1 atm) quenching, black diamonds]; Φ_{CHO} [C_2H_5CHO zero pressure, Chen and Zhu (2001), open triangles]; Φ_{CHO} adjusted for quenching of process (I) by air (1 atm), assuming Stern–Volmer quenching constants as observed with CH_3CHO and C_2H_5CHO in air (black triangles); see figure VII-B-14b. The solid line is the difference between the two best-fit curves for $\phi_I + \phi_{II}$ and ϕ_{II} given in figures VII-B-11 and VII-B-12, respectively.

photolysis (shown as small black circles in figure VII-B-13). The estimates of ϕ_I by Chen and Zhu at all wavelengths and the data point from the 294 nm photodecomposition by Heicklen et al. (1986) are much larger than those derived from the less direct measurements. However, Heicklen et al. (1986) data obtained for photolyses at wavelengths 302, 325, and 334 nm are in reasonable accord with the less direct estimates of ϕ_I. Both of the two direct radical quantum yield measurements were derived from complex experiments with necessary corrections to the measured data to arrive at the radical yields.

One possible explanation for the differences between the Chen and Zhu (2001) data and the other measurements can be noted. They report that there is no quenching of process (I) by added N_2 (10–400 Torr). As described above, Shepson and Heicklen (1982a,b) report significant quenching of C_2H_5CHO excited at the longer wavelengths by added N_2, O_2, or air. Thus at wavelength 313 nm they report that the Stern–Volmer quenching constant for air is 3.82×10^{-3} $Torr^{-1}$. The sensitivity of the measurement methods of Chen and Zhu may be inadequate to measure the extent of quenching by added N_2. The quenching constants for the small aldehydes (C_2–C_4) for quenching by aldehyde are given in figure VII-B-14a and limited data for quenching by air in figure VII-B-14b. Using the quenching constants for air as defined by the trend line, we have adjusted the measured ϕ_{HCO} values for zero pressure of Chen and Zhu (2001) to those expected for 760 Torr of air. These points are shown as the black triangles in figure VII-B-13. It can be seen that the Chen and Zhu data are brought into reasonable agreement with the other data for ϕ_I by using the estimated quenching by air seen in other studies.

The recommended data for the quantum efficiencies of the primary processes (I) and (II) are also listed in table VII-B-7 as a function of wavelength for each nm as derived from the recommended trend lines in figures VII-B-11 and VII-B-12. The current estimates of ϕ_I and ϕ_{II} for propanal given here and those suggested by the IUPAC panel (Atkinson et al., 2004), derived solely from the ϕ_{HCO} data for zero pressure (Chen and Zhu, 2001), are significantly different. Further studies will be required to clarify these differences.

The primary quantum yields for processes (III) and (V) are small and uncertain. Φ_{H2} measured in the C_2H_5CHO–I_2 mixtures is <0.01 at the longer wavelengths, about the same as is found in the I_2-free photolysis (Blacet and Pitts, 1952). Process (V) becomes significant in the I_2-free photolysis at the shorter wavelengths; e.g., $\Phi_{H2} = 0.05$ and 0.07 at 253.7 and 238 nm, respectively. No recommendations for the wavelength dependences of ϕ_{III} and ϕ_V can be made at this time, but it is clear that both processes are unimportant for tropospheric photolysis of propanal. CH_4 formation presumably occurs following CH_3 radical formation in (IV) and abstraction of an H-atom, since added iodine inhibits its formation. Process (IV) is also unimportant for tropospheric chemistry, but at short wavelengths it becomes significant. See figure VII-B-15.

VII-B-3.3. Photolysis Frequencies (j-values) for Propanal in Air (1 atm)

The *j*-values for primary process (I) for propanal as a function of solar zenith angle are given in figure VII-B-16 as estimated for a cloudless troposphere

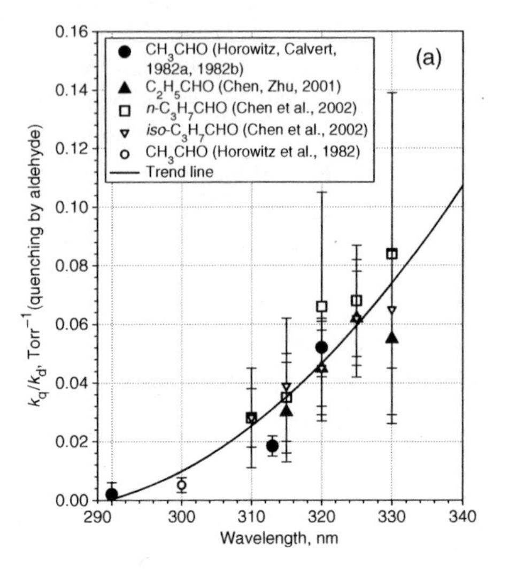

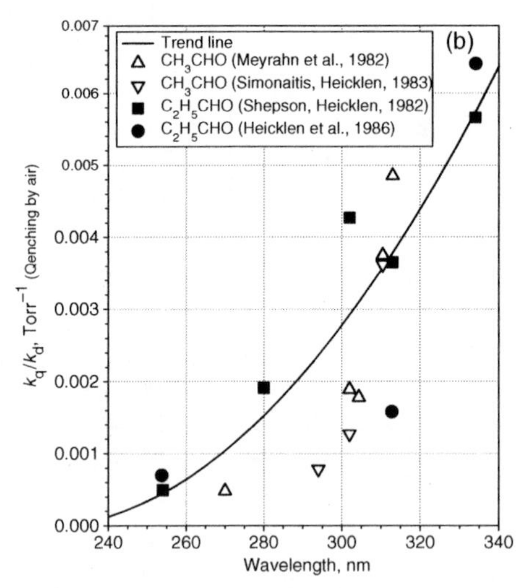

Figure VII-B-14. Wavelength depedence of Stern–Volmer quenching constants (k_q/k_d) from $1/\phi_I$ vs. pressure plots for primary process (I), RCHO + $h\nu \rightarrow$ R + HCO, as determined for the quenching by aldehyde (panel a) and by air (panel b).

(ozone column = 350 DU), using the current recommendations for cross sections and primary quantum yields.

VII-B-4. Photodecomposition of Butanal (n-Butyraldehyde, $CH_3CH_2CH_2CHO$)

As with the other aliphatic aldehydes, butanal photolysis is a source of radicals within the troposphere, but, in addition, a new type of molecular dissociation occurs that is not present in smaller aldehydes. Several primary pathways for photodecomposition have

Table VII-B-7. Recommended values for ϕ_I and ϕ_{II} for propanal photodecomposition in air at 1 atm pressure

λ (nm)	ϕ_I [a]	ϕ_{II} [b]	λ (nm)	ϕ_I [a]
278	0.322	0.081	313	0.269
279	0.324	0.075	314	0.265
280	0.326	0.069	315	0.260
281	0.327	0.063	316	0.255
282	0.329	0.058	317	0.250
283	0.330	0.053	318	0.245
284	0.331	0.048	319	0.239
285	0.331	0.043	320	0.233
286	0.332	0.039	321	0.228
287	0.332	0.035	322	0.222
288	0.332	0.031	323	0.216
289	0.332	0.028	324	0.209
290	0.332	0.024	325	0.203
291	0.331	0.021	326	0.196
292	0.331	0.018	327	0.189
293	0.330	0.015	328	0.182
294	0.329	0.013	329	0.174
295	0.338	0.000	330	0.167
296	0.335	0.000	331	0.159
297	0.332	0.000	332	0.150
298	0.328	0.000	333	0.142
299	0.325	0.000	334	0.133
300	0.321	0.000	335	0.124
301	0.317	0.000	336	0.115
302	0.314	0.000	337	0.105
303	0.310	0.000	338	0.095
304	0.306	0.000	339	0.085
305	0.303	0.000	340	0.075
306	0.299	0.000	341	0.064
307	0.295	0.000	342	0.053
308	0.291	0.000	343	0.041
309	0.287	0.000	344	0.029
310	0.283	0.000	345	0.017
311	0.278	0.000	346	0.005
312	0.274	0.000	347	0.000

[a] $CH_3CH_2CHO + h\nu \rightarrow CH_3CH_2 + CHO$ (I).
[b] $CH_3CH_2CHO + h\nu \rightarrow C_2H_6 + CO$ (II).

been suggested to explain its photochemistry (e.g., see Blacet and Calvert, 1951a,b; Förgeteg et al., 1979):

$$CH_3CH_2CH_2CHO + h\nu \rightarrow CH_3CH_2CH_2 + HCO \quad \text{(I)}$$

$$\rightarrow C_3H_8 + CO \quad \text{(II)}$$

$$\rightarrow CH_2{=}CH_2 + CH_2{=}CHOH \quad \text{(III)}$$

$$\rightarrow CH_3 + CH_2CH_2CHO \quad \text{(IV)}$$

$$\rightarrow C_2H_5 + CH_2CHO \quad \text{(V)}$$

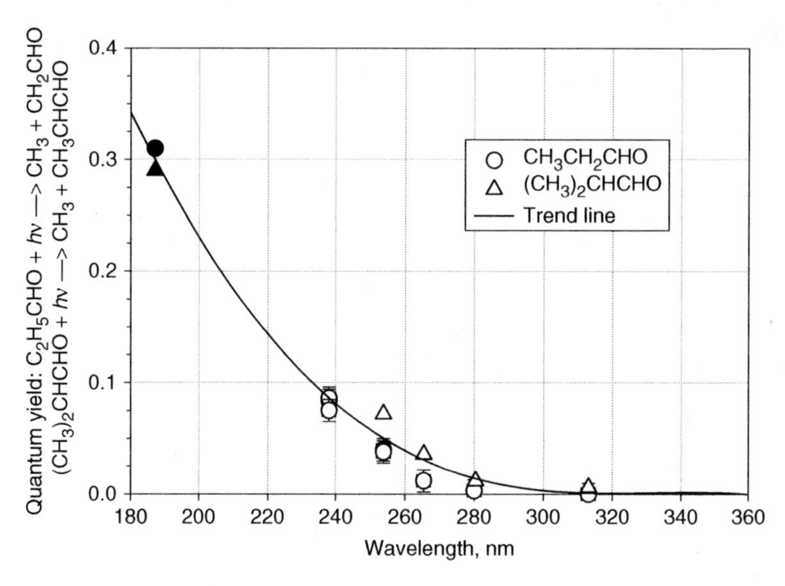

Figure VII-B-15. Quantum yields of primary process (IV) in propanal (open circles, Blacet and Pitts, 1952; black circle, Blacet and Crane, 1954). Quantum yield of primary process (IV) in 2-methylpropanal (open triangles, Blacet and Calvert, 1951b; black triangle, Blacet and Crane, 1954). The trend line (solid curve) is the least-squares fit of all the data.

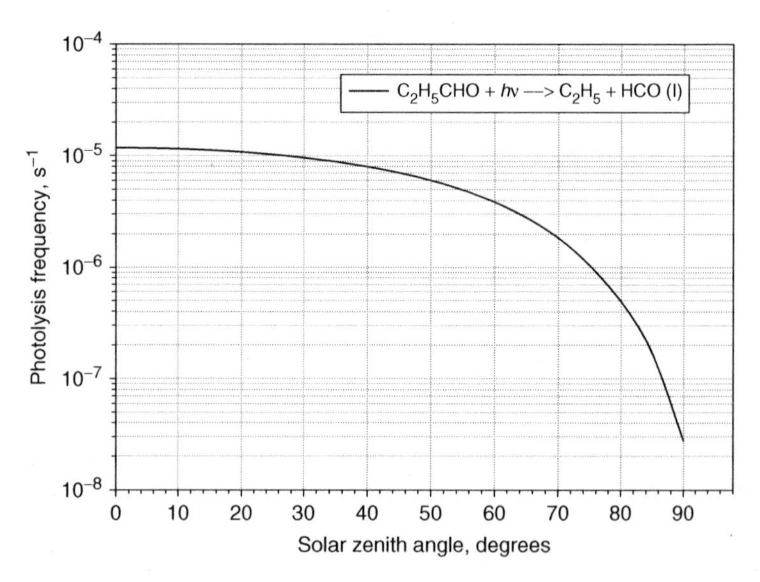

Figure VII-B-16. Estimated photolysis frequencies vs. solar zenith angle for the primary process (I) in propanal photodecomposition in the cloud-free, lower troposphere with an overhead ozone column = 350 DU.

$$\rightarrow CH_3CH_2CH_2CO + H \qquad (VI)$$

$$\rightarrow \textit{cyclo}\text{-butyl alcohol} \qquad (VII)$$

At wavelengths of sunlight within the troposphere, only primary processes (I) and (III) are important. Primary processes (II), (IV), (V), (VI), and (VII) may become significant for photolysis at the shorter wavelengths that are available only in the upper atmosphere. While processes (I), (II), and (VI) are analogous to those observed with acetaldehyde and propanal, process (III) is unique to carbonyl compounds that contain a γ-H-atom, an H-atom attached to the third C-atom down-chain from the carbonyl group. This photodecomposition path is often called a "Norrish type II process" (Bamford and Norrish, 1935; Davis and Noyes, 1947). The initial enol-form of acetaldehyde formed in (III), $CH_2{=}CHOH$, ultimately forms the more stable acetaldehyde molecule (CH_3CHO), a transformation catalyzed by contact with a surface. The commonly accepted mechanism for this type of photodecomposition involves a cyclic, 6-membered ring intermediate:

$CH_3CH_2CH_2CHO + h\nu \longrightarrow$ [6-membered ring intermediate: CH···O···H, CH₂, CH₂, CH₂] $\longrightarrow$ [CH(=CH₂)–O–H + CH₂=CH₂] (III)

The minor process (VII) may occur from the subsequent reaction of an intermediate diradical, $\bullet CH_2CH_2CH_2CH(\bullet)OH$, formed by H-atom abstraction by the excited carbonyl group. The mechanisms of the two processes (III) and (VII) do not appear to involve alternative pathways of decay of the same diradical intermediate. Theoretical calculations of Chen and Fang (2002) on the mechanism of process (I) showed that first excited triplet state is involved, following intersystem crossing ($S_1 \rightarrow T_1$), whereas the 1,5-H shift in primary process (III) proceeds on the S_1 surface. This study also showed that dissociation of the T_1 state may occur by either primary process (I) or (VI), but the barrier for the occurrence of (I), $\sim$57 kJ mol^{-1}, is somewhat lower than that for (VI), $\sim$73 kJ mol^{-1}.

VII-B-4.1. Absorption Cross Sections for Butanal

The cross sections for butanal reflect the characteristic carbonyl excitation and are very similar to those for acetaldehyde and propanal; see figure VII-B-6. Recommended cross section values are those determined by Martinez et al. (1992), given in tabular form in table VII-B-5.

VII-B-4.2. Quantum Yields of the Primary Processes in Butanal

Most of the quantitative experimental studies of the quantum yields of the primary processes in the photolysis of butanal as a function of wavelength have been carried out in the absence of oxygen or air, and it is difficult to convert these to typical atmospheric conditions. Blacet and Calvert (1951a,b) determined the quantum yields of processes

(I), (II), (III), and (IV) at several wavelengths and temperatures (25–150°C) using pure $CH_3CH_2CH_2CHO$ (80–142 Torr) and its mixtures with small amounts of added I_2 (1–5 Torr). The quantum yields of formation of $CH_3CH_2CH_2I$, C_3H_8, and C_2H_4 from the experiments with iodine–butanal mixtures were given as measures of ϕ_I, ϕ_{II}, and ϕ_{III}, respectively. The quantum yield of CO monitored the sum of $\phi_I + \phi_{II}$. Using pure butanal photolysis at 313 nm and a range of temperature (253–529 K), Förgeteg et al. (1978, 1979) measured quantum yields of 20 different reaction products, and derived estimates of the primary yields for all of the suggested primary processes for butanal photolysis at 313 nm and 100 Torr aldehyde pressure. The only study which directly relates to the dilute mixtures of butanal in air is that of Tadić et al. (2001a). This was carried out with broad-band radiation (275–340 nm) at several pressures of air (10–700 Torr). Although the wavelength dependencies of the major processes could not be determined, very significant useful information on the primary processes was obtained.

(i) Primary process (I): $CH_3CH_2CH_2CHO + h\nu \rightarrow CH_3CH_2CH_2 + HCO$. The most direct measurements of the quantum yields of this primary process were made by Chen et al. (2002). Using cavity ring-down spectroscopy, they observed directly the HCO formation and decay following the laser photolysis of butanal at several discrete wavelengths in the range 280–330 nm. A calibration of the formyl radical generation was accomplished using either chlorine photolysis in mixtures with CH_2O or direct CH_2O photolysis. This is a significant yet complex experiment. The values of ϕ_I at zero pressure were determined from experiments with pure butanal (1–10 Torr). These are shown as black circles in figure VII-B-17. Chen et al. (2002) observed quenching of process (I) by butanal molecules (1–10 Torr), but no quenching was observed by added N_2 (8–400 Torr). Hence, the zero-pressure values of ϕ_I were assumed to represent those expected for 1 atm of air. The IUPAC recommendations (Atkinson et al., 2004) are based on this interpretation. However, this view should be questioned. The Chen et al. experimental system was not designed to measure small quenching effects that are expected for nitrogen addition, and their observations may have missed the effect of added nitrogen that has been observed by others.

Experiments with the smaller aldehydes (CH_3CHO and C_2H_5CHO) show clearly that the excited state(s) leading to process (I) are quenched by air, although inefficiently compared to quenching by aldehyde molecules. See figure VII-B-14, where plots of the measured Stern–Volmer quenching constants (k_q/k_d) for process (I) by aldehyde (panel a) and by air molecules (panel b) are shown as a function of wavelength. Although considerable scatter of the data is seen in these plots, the excited aldehyde quenching constants for the four aldehydes, ethanal, propanal, butanal, and 2-methylpropanal, follow the same trend curve within large error limits. The quenching constants for the aldehydes with air that were determined at several wavelengths are available for only two compounds, acetaldehyde and propanal. These data show a trend to smaller values as the excitation of the aldehyde occurs at shorter wavelengths and the lifetime of the excited aldehyde becomes shorter.

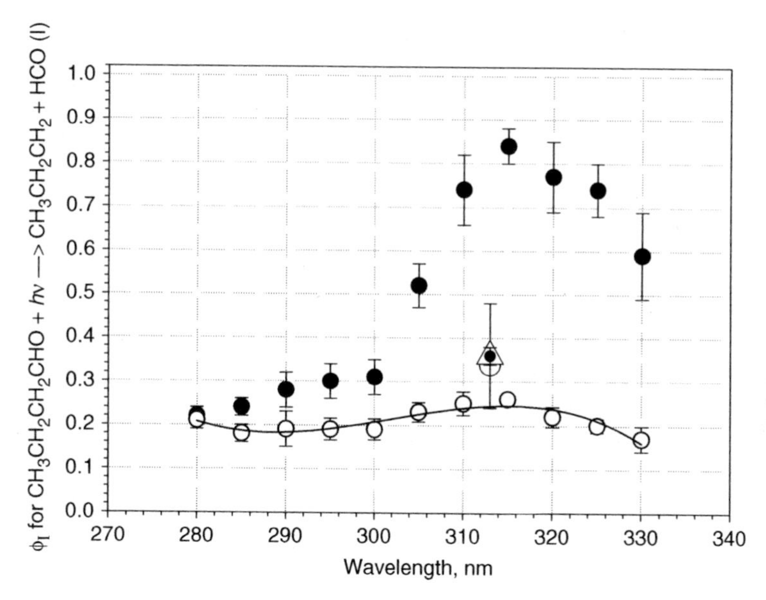

Figure VII-B-17. Plot of the quantum yield of HCO radicals formed in primary process (I) in butanal photolysis at zero pressure as a function of wavelength (large filled circles Chen et al., 2002). Estimates of ϕ_I in air (760 Torr; shown as small open circles) were derived in the present work using the Chen et al. (2002) data and the air quenching rate constants as measured for radical generation in CH_3CHO and C_2H_5CHO. The open triangle and small filled circle at 313 nm are from ϕ_I estimates derived from Φ_{CO} and Φ_{C3H7I}, respectively; these were measured in butanal–I_2 mixtures ($P_{butanal}$ ~100 Torr; Blacet and Calvert, 1951b). The large open circle at 313 nm is from measurements of quantum yields of extensive products of pure butanal photolyses ($P_{butanal} \approx$ 100 Torr; Förgeteg et al., 1979). These data points are uncorrected for differences in quenching by 100 Torr of butanal and 760 Torr of air molecules.

Experiments of Tadić et al. (2001a), using broad-band radiation, show that quenching of process (I) in butanal by air molecules is significant at pressures of the troposphere. Unfortunately, there are no quenching data for butanal as a function of wavelength, so adjustment of the low-pressure quantum yields at specific wavelengths to those at 1 atm of air can only be made indirectly. In view of the similarity in the quenching behavior of the smaller aldehydes seen in figure VII-B-14, we can make a qualitative adjustment to the zero pressure data of butanal using the trend lines of this figure. The ratio of the two trend line curves for quenching by air molecules compared to those for aldehyde molecules increases slowly with decreasing wavelength: 0.059 (340 nm), 0.064 (335 nm), 0.071 (330 nm), 0.079 (325 nm), 0.091 (320 nm), 0.108 (315 nm), 0.132 (310 nm), 0.173 (305 nm), 0.249 (300 nm), 0.429 (295 nm), and 1.28 (290 nm). We have used the trend line values to adjust the Chen et al. (2002) zero-pressure values of ϕ_I to those expected for 1 atm of air. The adjusted values are shown also in figure VII-B-17 as small open circles.

The butanal photodecomposition studies of Tadić et al. (2001a) were carried out in air, and they are the most direct test of the magnitude of ϕ_I for tropospheric conditions. However, the distribution of the wavelengths of light provided by the UV

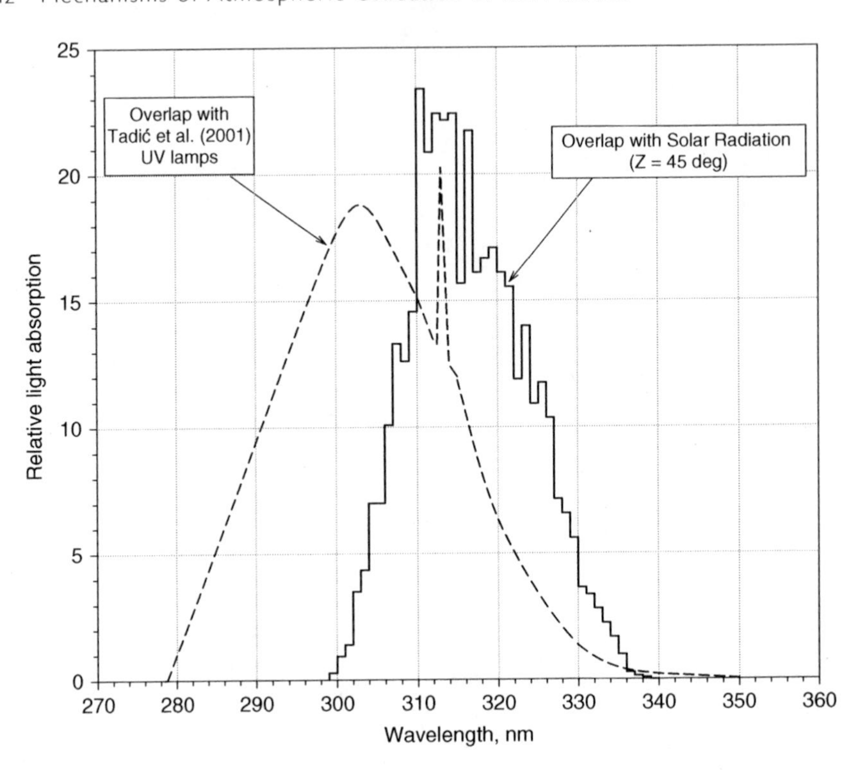

Figure VII-B-18. Comparison of the wavelength dependencies of the overlap of the butanal absorption spectrum with the emission of the TL12 Philips lamps (dashed curve) used by Tadić et al. (2001a), and the overlap with clear-sky actinic flux at solar zenith angle 45°, solid lines.

lamps employed by Tadić et al. is shifted to somewhat shorter wavelengths than that provided by sunlight. Figure VII-B-18 compares the distributions of the overlap of the absorption cross sections of butanal with the radiation from the lamps employed by Tadić et al. and sunlight (solar zenith angle = 45°) on a cloudless day. Our estimates of ϕ_I for 1 atm of air given in figure VII-B-17 are consistent with the measurements of Tadić et al. (2001a). Their results, extrapolated to 760 Torr of air with their observed dependence on air pressure $[1/\phi_{CO} = 1/\phi_I = 3.00 + 1.25 \times 10^{-3}$ $(p_{air}$, Torr)] give an average value of $\phi_I = 0.25$. With the overlap of their UV lamp radiation with butanal absorption and our estimates of the wavelength dependence of ϕ_I in air, given in figure VII-B-17, the value of ϕ_I expected from their broad-band experiments is 0.22, similar to that measured experimentally (0.25) by Tadić et al. (2001a).

In the 313 nm photolysis of butanal–iodine mixtures, Blacet and Calvert (1951b) measured values of Φ_{CO}, a surrogate for ϕ_I at 313 nm where process (II) is unimportant. These values were relatively insensitive to the pressure of iodine (1–6 Torr) and the temperature (63–150°C); the pressure of butanal was ~100 Torr. The average value of all the experiments was $\Phi_{CO} = 0.36 \pm 0.12$. Using a complete product analysis from photolysis of 100 Torr butanal at 313 nm and 298 K, Förgeteg et al. (1979)

estimated ϕ_I to be 0.34 ± 0.02, in good agreement with the Blacet and Calvert (1951a,b) estimates. Förgeteg et al. concluded that there was little, if any, quenching of the butanal photodecomposition from the small amount of iodine added in the Blacet and Calvert study. In four of their thirteen experiments, Blacet and Calvert determined the quantum yields of 1-iodopropane ($CH_3CH_2CH_2I$) as well as CO, another surrogate for ϕ_I. The average ratio of $\Phi_{CH3CH2CH2I}/\Phi_{CO} = 0.98 \pm 0.02$, and the 1-iodopropane quantum yields give $\phi_I = 0.36 \pm 0.02$. The three independent estimates of ϕ_I at 313 nm (uncorrected for quenching by air) are shown in figure VII-B-17 as on open triangle (Φ_{CO}, Blacet and Calvert, 1951b), small closed circle ($\Phi_{CH3CH2CH2I}$, Blacet and Calvert, 1951b) and large open circle (complete product analysis, Förgeteg et al., 1979). These data cannot be adjusted for differences in quenching of 100 Torr of butanal and 760 Torr of air, since the temperature dependence of the required quenching constants is not known. However, for the temperature of 298 K, the constants for the small aldehydes (figure VII-B-14) show that little difference is expected for these two different conditions ($P_{butanal} \sim 100$ Torr; $P_{air} = 760$ Torr). It is seen that these three points at 313 nm lie somewhat above the trend line derived from the adjusted data of Chen et al. (2002), but they are in reasonable accord within the uncertainty of the estimates.

We recommend that the ϕ_I values given by the trend line in figure VII-B-17 be used in atmospheric simulations of butanal photochemistry, but it is clear that there is considerable uncertainty associated with these estimates.

(ii) Primary process (III): $CH_3CH_2CH_2CHO + h\nu \rightarrow CH_2{=}CH_2 + CH_2{=}CHOH$. The most direct observation of the occurrence of this intramolecular rearrangement was made by Tadić et al. (2002); they observed through characteristic infrared bands, the formation of both ethene and vinyl alcohol [$CH_2{=}CH(OH)$, the enol-form of acetaldehyde] in broad-band irradiation of dilute mixtures of butanal in air mixtures. In the Tadić et al. (2001a) laboratory experiments, the enol-form that appears first rearranges to the more stable acetaldehyde structure on a time-frame of an hour or so, presumably at surfaces of the reaction chambers. Most of the studies of primary process (III) have been made using 313 nm radiation, a wavelength near the center of the overlap region between butanal and the actinic flux available within the troposphere; see the distribution shown with solid lines in figure VII-B-18. The quantum yields of process (III) are relatively insensitive to the temperature, pressure of aldehyde, and small amounts of added O_2 or I_2. Blacet and Calvert (1951a,b) reported an average value of $\Phi_{C2H4} = 0.153 \pm 0.015$ in nine experiments over a range of temperatures (25–150°C), pressures of aldehyde (42–104 Torr), and added iodine (0–5 Torr). Extensive product analyses were made in studies of the uninhibited photolysis (pure aldehyde) at 313 nm by Förgeteg et al. (1978, 1979) for a variety of experimental conditions. They found process (III) to be insensitive to light intensity variation over a 160-fold range of light intensities in experiments with 100 Torr of butanal and at 25°C; the average quantum yield reported is: $\Phi_{C2H4} = 0.175 \pm 0.014$ [number of experiments $(n) = 19$]; $\Phi_{CH3CHO} = 0.171 \pm 0.009$ $(n = 13)$. Over the temperature range from 9 to 153°C, the average quantum

yields of the products of process (III) are: $\Phi_{C2H4} = 0.166 \pm 0.005$; $\Phi_{CH3CHO} = 0.159 \pm 0.015$ ($n = 4$).

Tadić et al. (2001a) reported average quantum yields of process (III) from broad-band photolysis (λ distribution given in figure VII-B-18) of dilute mixtures of butanal in air at pressures of 100–700 Torr. The variation of the total quantum yield (ϕ_T) of butanal photodecomposition with pressure of air followed the equation: $1/\phi_T = 1.81 + 1.932 \times 10^{-3} \times P_{air}$ (Torr). The magnitude of the observed Stern–Volmer quenching constant in air is consistent with that expected for excitation at $\lambda \approx 290$ nm for acetaldehyde and propanal quenching by air (figure VII-B-14b). It is in reasonable accord with expectations based on wavelength of maximum overlap in the Tadić et al. experiments, ~303 nm. The fractions of the total decomposition that occurred by process (III) at various pressures of air were reported. Since the wavelength dependence of the quantum yield of ethene in air mixtures was not determined, it is impossible to reconstruct this information from the available average quantum yields in air. However, we can test for the consistency of the existing information to arrive at our best estimates of the ϕ_{III} values in air.

The first attempt at determination of the ϕ_{III} values at various wavelengths in air at 1 atm was made using the function $(\phi_{III})_0 = 1 - (\phi_I)_0 - (\phi_{II})_0$, where the quantities with subscript zero are zero pressure values; this assumes that processes (I) , (II), and (III) alone account for the photodecomposition of butanal. Values for $(\phi_I)_o$ were taken from the Chen et al. (2002) study and those for $(\phi_{II})_0$ from the measurements of Φ_{C3H8} in the I_2-inhibited photolysis of butanal (Blacet and Calvert, 1951b); see figure VII-B-19. The zero-pressure values were then converted to those at 1 atm using

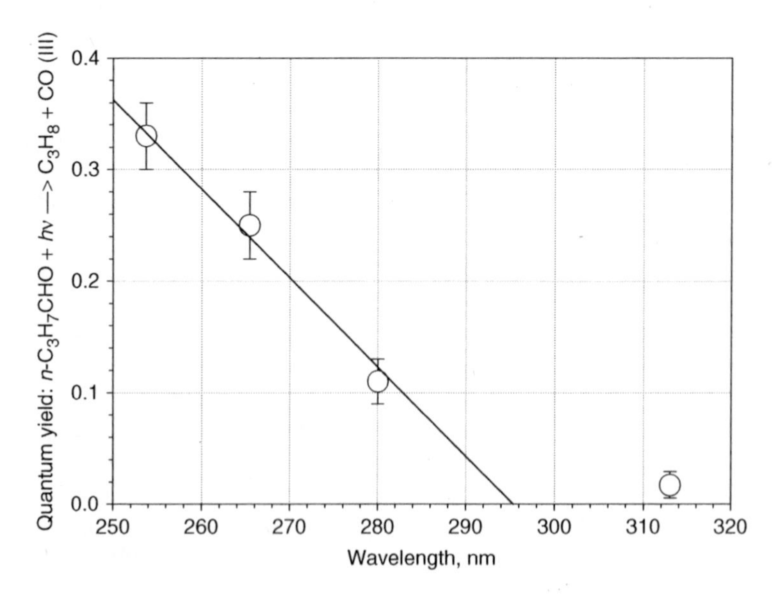

Figure VII-B-19. Estimated values for ϕ_{II} as determined in butanal photolysis with added small amounts of I_2 (Blacet and Calvert, 1951b); the data point at 313 nm was omitted in deriving the best-fit line shown, to be consistent with the experiments of Tadić et al. (2001a) who did not detect C_3H_8 in the broadband photolysis (280–350 nm) of butanal in air.

the quenching rate coefficient for the appropriate λ as taken from the air-quenching data for acetaldehyde and propanal given in figure VII-B-14b, the same procedure that gave promising results for the earlier estimation of ϕ_I data. This led to the prediction of the average $\phi_{III} = 0.21$ for air at 1 atm, inconsistent with the value of 0.12 observed by Tadić et al. in their broad-band studies.

A second attempt to derive the wavelength dependent values of ϕ_{III} at 1 atm of air involved the use of Φ_{C2H4} measurements from iodine-inhibition studies of Blacet and Calvert (1951b). Available data suggest that these are insensitive to the pressure of aldehyde and other experimental variables, and conceivably they may be representative of the zero-pressure values; see figure VII-B-20. We used the approach Described previously to estimate the effect of air quenching, using the data of figure VII-B-14b. The observed Stern–Volmer quenching constant for Φ_{C2H4}, derived from the average Φ_{C2H4} values for 100 and 700 Torr data of Tadić et al. (2001a), 2.83×10^{-3} Torr^{-1}, is consistent with the data of figure VII-B-14b. It corresponds to that expected for a wavelength of 300 nm, near the center of the excitation band used by Tadić et al. Using this second scenario, we estimated an average value of $\Phi_{C2H4} = 0.13$, in accord with the 0.12 estimate by Tadić et al. In view of these considerations and in the absence of more definitive measurements, we recommend estimates of the wavelength dependence of the zero-pressure and 1 atm values of ϕ_{III} shown in figure VII-B-20.

(iii) The quantum yields for the minor primary processes in butanal photolysis. The suggested data for use with process (II) are given in figure VII-B-19. This process is unimportant for tropospheric conditions. Process (VII) has been reported by Förgeteg et al. (1979) in their photolysis studies at 313 nm. Their data suggest a quantum yield of 0.025 at 313 nm. However, the *cyclo*-butyl alcohol product has not been observed in other gas phase studies (Blacet and Calvert, 1951a,b; Tadić et al., 2001a). There is no information related to the extent of occurrence of process (VI), the rupture of the $CH_3CH_2CH_2C(O)$—H bond. This and possibly other primary decomposition modes may account for $\phi_{total} < 1$ at zero-pressure at the short wavelengths, as noted by Tadić et al. (2001a).

The limited data available lead to the approximate values of primary quantum yields for the major primary processes for butanal given in table VII-B-8; these are recommended for current use in tropospheric modeling.

(iv) Photolysis frequencies for butanal in air (1 atm). The *j*-values for the two major primary photodecomposition modes in butanal in 1 atm of cloud-free tropospheric air have been calculated using the recommended cross sections, quantum yields, and solar flux estimates presented in this work (ozone column = 350 DU). These are given for processes (I) and (III) in figure VII-B-21. The *j*-value for primary process (II) for overhead Sun, $j(\text{II}) = 3.2 \times 10^{-11}$ sec^{-1}, shows that process (II) is negligible within the troposphere. Processes (IV) and (V) also have insignificant photolysis frequencies. Those for processes (VI) and (VII) are probably small as well, but the existing data for these is insufficient to allow a definitive estimate.

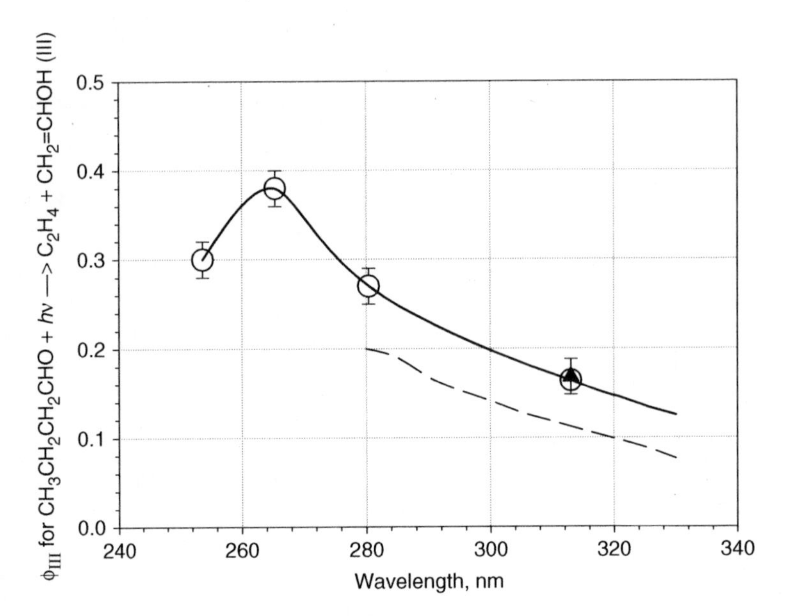

Figure VII-B-20. Estimated values for ϕ_{III} in butanal. The open circles and the solid trend line indicate estimated values of Φ_{C2H4} from butanal–I_2 mixture photolysis (Blacet and Calvert, 1951b); the filled triangle represents the average of 36 experiments of Förgeteg et al. (1978) derived from a complete product analysis of pure butanal photolysis under a range of light intensities and temperatures. The dashed line gives the wavelength dependence of ϕ_{III} estimated in air (760 Torr) using Stern–Volmer quenching constants given in figure VII-B-14b; these quenching constants are consistent with the quenching of process (III) by air molecules observed by Tadić et al. (2001a) in their broadband photolyses of butanal.

VII-B-5. Photodecomposition of 2-Methylpropanal [Isobutyraldehyde, $(CH_3)_2CHCHO$]

The primary photodecomposition processes in 2-methylpropanal follow the pattern of the smaller aldehydes. Both radical generation and fragmentation to form stable molecules occur:

$$(CH_3)_2CHCHO + h\nu \rightarrow (CH_3)_2CH + HCO \quad \text{(I)}$$

$$\rightarrow C_3H_8 + CO \quad \text{(II)}$$

$$\rightarrow (CH_3)_2CHCO + H \quad \text{(III)}$$

$$\rightarrow CH_3 + CH_3CHCHO \quad \text{(IV)}$$

Several studies have demonstrated that process (I) is the dominant mode of photodecomposition for excitation at wavelengths of sunlight within the troposphere (Blacet and Calvert, 1951a,b; Desai et al., 1986; Chen et al., 2002). Processes (II) and (IV) become important at the short wavelengths of absorbed light ($\lambda < 290$ nm).

Table VII-B-8. Recommended (approximate) values for the primary quantum yields for butanal photolysis in air at 1 atm for selected wavelengths

λ (nm)	ϕ_I [a]	ϕ_{II} [b]	ϕ_{III} [c]	λ (nm)	ϕ_I [a]	ϕ_{III} [c]
275	0.251	0.163	0.205	306	0.228	0.125
276	0.241	0.155	0.204	307	0.232	0.123
277	0.231	0.147	0.203	308	0.234	0.122
278	0.222	0.139	0.202	309	0.237	0.120
279	0.214	0.131	0.201	310	0.240	0.118
280	0.207	0.123	0.200	311	0.242	0.116
281	0.202	0.115	0.198	312	0.243	0.114
282	0.197	0.107	0.197	313	0.245	0.112
283	0.192	0.099	0.195	314	0.245	0.110
284	0.189	0.091	0.193	315	0.246	0.108
285	0.186	0.083	0.190	316	0.245	0.106
286	0.184	0.075	0.185	317	0.244	0.104
287	0.183	0.067	0.181	318	0.243	0.102
288	0.182	0.059	0.177	319	0.241	0.101
289	0.182	0.051	0.172	320	0.238	0.099
290	0.183	0.043	0.168	321	0.234	0.097
291	0.184	0.035	0.165	322	0.230	0.095
292	0.185	0.027	0.161	323	0.225	0.093
293	0.187	0.019	0.158	324	0.219	0.091
294	0.189	0.011	0.155	325	0.211	0.089
295	0.191	0.003	0.153	326	0.203	0.086
296	0.194	0.000	0.150	327	0.194	0.084
297	0.197	0.000	0.148	328	0.184	0.081
298	0.200	0.000	0.145	329	0.173	0.078
299	0.204	0.000	0.143	330	0.161	0.076
300	0.207	0.000	0.141	331	0.147	0.074
301	0.211	0.000	0.139	332	0.133	0.070
302	0.214	0.000	0.137	333	0.117	0.067
303	0.218	0.000	0.133	334	0.099	0.063
304	0.222	0.000	0.131	335	0.081	0.060
305	0.225	0.000	0.128	336	0.061	0.057

[a] $CH_3CH_2CH_2CHO + h\nu \rightarrow CH_3CH_2CH_2 + HCO$ (I).
[b] $CH_3CH_2CH_2CHO + h\nu \rightarrow C_3H_8 + CO$ (II).
[c] $CH_3CH_2CH_2CHO + h\nu \rightarrow CH_2{=}CH_2 + CH_2{=}CHOH$ (III).

Process (III) is energetically feasible for $\lambda < 322$ nm, but its extent of occurrence has not been determined, and it is seldom considered in modeling tropospheric chemistry.

VII-B-5.1. Absorption Cross Sections for 2-Methylpropanal

The typical $n \rightarrow \pi^*$ transition is present in 2-methylpropanal; see figure VII-B-22. The magnitude of the cross sections of 2-methylpropanal is similar to those of acetaldehyde and the higher aldehydes, and the unresolved vibrational structure of the excited aldehydes is evident in the spectrum. The recommended cross sections, determined by Martinez et al. (1992), and shown in figure VII-B-22, are given in tabular form in table VII-B-9.

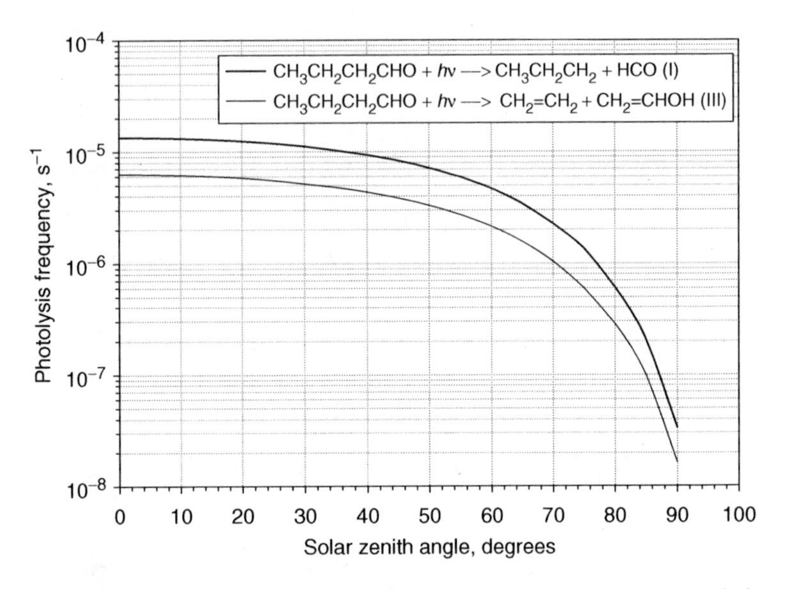

Figure VII-B-21. Plot of the photolysis frequencies (*j*-values) for primary processes in butanal as a function of solar zenith angle for cloudless day in the lower troposphere with an overhead ozone column of 350 DU. Photodecomposition by process (II) is insignificant within the troposphere.

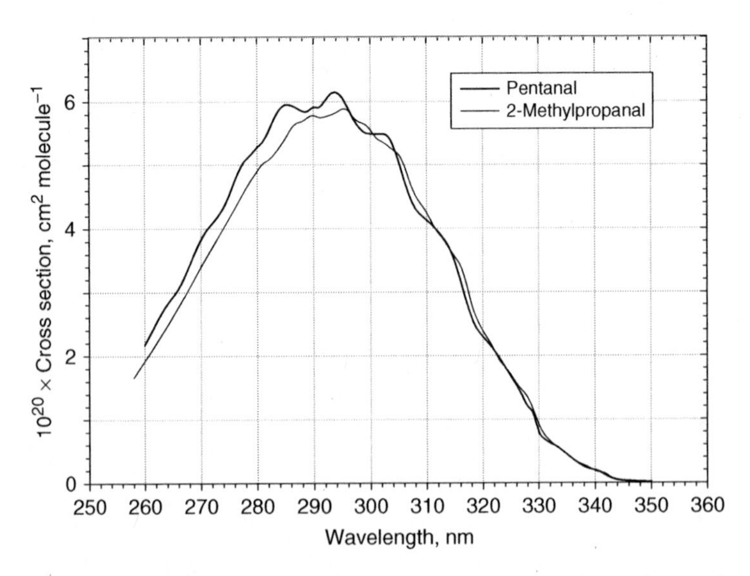

Figure VII-B-22. Plot of absorption cross sections as a function of wavelength for 2-methylpropanal (Martinez et al., 1992) and pentanal (Tadić et al., 2001a).

Table VII-B-9. Recommended absorption cross sections (σ, cm^2 molecule^{-1}) for $(CH_3)_2CHCHO$ (298 K) and quantum yields (ϕ_I) for the primary process $(CH_3)_2CHCHO + h\nu \rightarrow CH_3CHCH_3 + HCO$ (I), in 1 atm of air. ϕ_I(a) and ϕ_I(b) are alternative wavelength dependencies given by the solid line and dashed lines, respectively, in figure VII-B-23

λ (nm)	$10^{20} \times \sigma$	ϕ_I(a)	ϕ_I(b)	λ (nm)	$10^{20} \times \sigma$	ϕ_I(a)	ϕ_I(b)
278	4.63	0.39	0.41	315	3.55	1.00	0.86
279	4.77	0.39	0.40	316	3.41	1.00	0.88
280	4.90	0.38	0.40	317	3.14	1.00	0.91
281	5.02	0.38	0.40	318	2.79	0.99	0.93
282	5.08	0.38	0.40	319	2.55	0.99	0.95
283	5.17	0.38	0.40	320	2.38	0.99	0.96
284	5.29	0.40	0.40	321	2.24	0.98	0.97
285	5.42	0.43	0.40	322	2.09	0.98	0.97
286	5.57	0.49	0.41	323	1.91	0.97	0.97
287	5.66	0.53	0.41	324	1.81	0.97	0.96
288	5.69	0.60	0.41	325	1.69	0.96	0.96
289	5.75	0.67	0.42	326	1.56	0.95	0.94
290	5.78	0.70	0.42	327	1.46	0.93	0.92
291	5.75	0.73	0.43	328	1.34	0.91	0.90
292	5.76	0.77	0.43	329	1.15	0.88	0.88
293	5.79	0.79	0.44	330	0.918	0.84	0.84
294	5.83	0.82	0.45	331	0.760	0.80	0.80
295	5.88	0.84	0.46	332	0.661	0.76	0.76
296	5.86	0.86	0.47	333	0.579	0.72	0.72
297	5.75	0.88	0.48	334	0.510	0.68	0.68
298	5.68	0.90	0.49	335	0.445	0.63	0.63
299	5.64	0.92	0.51	336	0.373	0.59	0.59
300	5.55	0.93	0.52	337	0.317	0.56	0.56
301	5.43	0.95	0.53	338	0.279	0.50	0.50
302	5.37	0.96	0.55	339	0.238	0.46	0.46
303	5.31	0.97	0.56	340	0.187	0.42	0.42
304	5.24	0.98	0.58	341	0.165	0.38	0.38
305	5.17	0.98	0.60	342	0.138	0.34	0.34
306	5.00	0.98	0.62	343	0.077	0.29	0.29
307	4.73	0.99	0.65	344	0.047	0.25	0.25
308	4.52	0.99	0.67	345	0.034	0.21	0.21
309	4.38	1.00	0.70	346	0.025	0.16	0.16
310	4.25	1.00	0.72	347	0.019	0.12	0.12
311	4.08	1.00	0.75	348	0.015	0.08	0.08
312	3.93	1.00	0.78	349	0.013	0.04	0.04
313	3.82	1.00	0.80	350	0.011	0.00	0.00
314	3.68	1.00	0.83				

VII-B-5.2. Quantum Yields of the Primary Processes in 2-Methylpropanal

(i) Primary process (1): $(CH_3)_2CHCHO + h\nu \rightarrow CH_3CHCH_3 + HCO$. The results of several determinations of the quantum yields of the important primary process (I), generating CH_3CHCH_3 and HCO radicals, are shown in figure VII-B-23. Results from photolysis of 2-methylpropanal–iodine mixtures by Blacet and Calvert (1951b) gave values of ϕ_I both from ($\Phi_{CO} - \Phi_{C3H8}$) and $\Phi_{CH3CHICH3}$ determinations at 313 and 265.4 nm (shown as large black circles and open triangles,

respectively, in figure VII-B-23). The quantum yields of process (I) were obtained in 2-methylpropanal–air mixture photolysis by Desai et al. (1986). A complex kinetic mechanism was necessary to treat the data, which ultimately resulted in estimates of $\phi_I = \Phi_{CO} - \Phi_{C3H8}$ at wavelengths 253.7, 280.3, 302.2, 312.8, 326.1, and 334.1 nm. These results are shown as small black circles. Chen et al. (2002) determined the quantum efficiencies of HCO radical formation in process (I) at 11 wavelengths in the 280–330 nm range; their values for zero-pressure are shown as circles in figure VII-B-23. They detected no effect of added nitrogen gas (up to 400 Torr) on the quantum yield and concluded that the zero-pressure quantum yields applied for air at 1 atm as well. Although quenching of excited 2-methylpropanal molecules by air or nitrogen molecules is less efficient than that observed for acetaldehyde and propanal, Stern–Volmer plots of the Φ_{CO} data of Desai et al. (1986) clearly show significant quenching of process (I) in 2-methylpropanal by oxygen; values of k_q/k_d ($Torr^{-1}$) are: 7.3×10^{-4} (253.7 nm); 1.5×10^{-3} (280.3 nm);

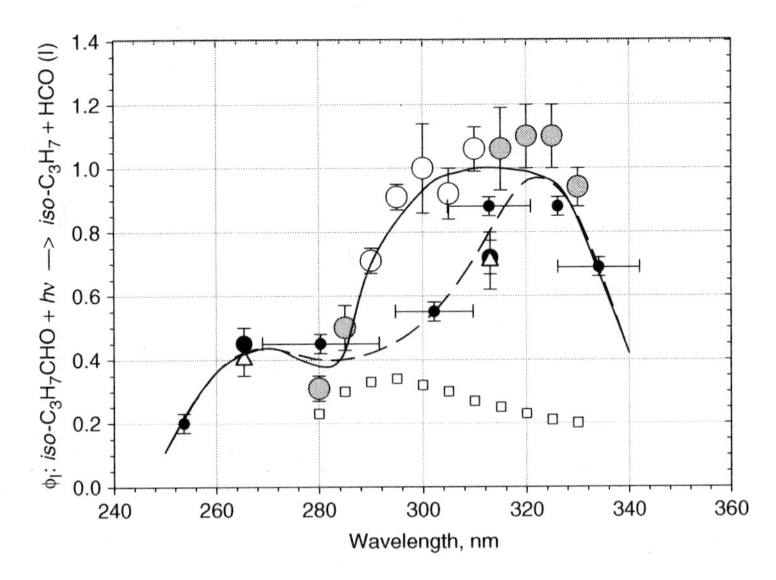

Figure VII-B-23. Estimates of the quantum yield of process (I) in 2-methylpropanal photolysis in air (760 Torr). Data of Chen et al. (2002) are ϕ_I at zero pressure and are indicated by the large open circles and grey circles; the latter highlight the Chen et al. data points that seem compatible with the dashed trend line. Estimates of Desai et al. (1986) are derived from $\Phi_{CO} - \Phi_{C3H8}$from photolysis in air at 1 atm (small black circles). Data from photolysis of $(CH_3)_2CHCHO$-I_2mixtures are derived from measurements of Blacet and Calvert (1951b): $\Phi_{CO} - \Phi_{C3H8}$ (large black circles) and $\Phi_{CH3CHICH3}$ (open triangles).The small squares give the Chen et al. (2002) estimates that have been adjusted for quenching by air (760 Torr), assuming the Stern–Volmer constants of figure VII-B-14b apply. Obviously, the quenching by air is very much less than that observed for CH_3CHO and C_2H_5CHO. The trend line given by the solid curve neglects the other experimental estimates at 313 and 302.2 nm. The wavelength range, shown by the bars on the small circles, represents the full-width at half-maximum used in the Desai et al. (1986) experiments. The two trend lines represent alternative choices of the wavelength dependence of process (I) that are used in deriving photolysis frequencies (figure VII-B-25).

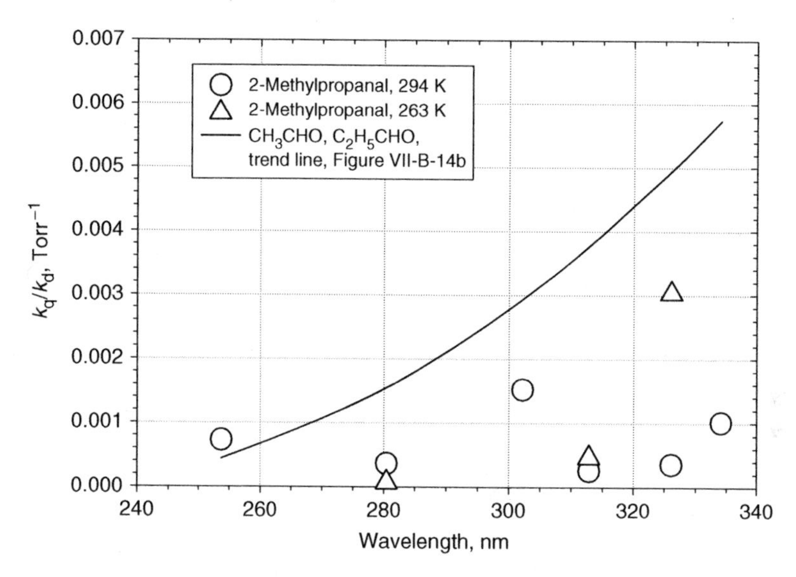

Figure VII-B-24. The data points are Stern–Volmer quenching constants (k_q/k_d) for air quenching of the precursor state(s) responsible for process (I) in 2-methylpropanal (Desai et al., 1986); contrast these points with the trend line of k_q/k_d values from acetaldehyde and propanal shown here and in figure VII-B-14b. Excited 2-methylpropanal molecules are quenched much less efficiently than those of acetaldehyde and propanal, probably as a consequence of the more rapid photodissociation of excited 2-methylpropanal molecules.

2.5×10^{-4} (302.2 nm); 3.6×10^{-4} (312.8 nm); 1.0×10^{-3} (326.1 nm). These values, although imprecise, are significantly lower than those observed for acetaldehyde and propanal at a given wavelength; see figure VII-B-24. The difference is evident also in figure VII-B-23. The data points marked by small squares in the latter figure were derived using the zero-pressure values of ϕ_I determined by Chen et al. (2002) that were adjusted for quenching by air using the same Stern–Volmer constants (figure VII-B-14b) that rationalized the quenching observed with propanal and butanal data reasonably well. The difference between the quenching observed with $(CH_3)_2CHCHO$ and with the smaller aldehydes probably arises from a shorter lifetime of the excited 2-methylpropanal molecule, perhaps reflecting the weaker R—CHO bond.

The choice of 2-methylpropanal quantum yield data to establish the recommended trend in ϕ_I with wavelength is not clear. Certainly the experiments carried out in air mixtures at 1 atm pressure match the conditions of the troposphere better than the zero-pressure study of Chen et al. If we favor the Desai data in the 300 nm region, then a trend curve shown as the dashed line in figure VII-B-23 is derived. In this case, the data of Chen et al. are in reasonable accord with the other two data sets at six of the 11 wavelengths used in their study (indicated by grey-filled circles). If we neglect the data point of Desai et al. at 302.2 nm, then a trend curve shown by the solid line represents better the several estimates shown. In this case the 313 nm estimates of Blacet and Calvert fall significantly below the curve. In the absence of a clear choice, the quantum yield estimates from both trend lines are tabulated in table VII-B-9, and both are used as alternatives to derive *j*-values, as described in the following section. The fall-off in ϕ_I is in part a consequence of the onset of process (II), but this alone cannot explain the data. See the discussion of the next section.

(ii) Primary process (II): $(CH_3)_2CHCHO + h\nu \rightarrow C_3H_8 + CO$. The 2-methylpropanal photodecomposition that leads directly to propane and carbon monoxide is unimportant at the wavelengths of light available within the troposphere. Estimates can be derived from Φ_{C3H8} measured in iodine–2-methylpropanal mixture photolysis (Blacet and Calvert, 1951b): ~0.03 at 313 nm, rising to 0.40 at 265.4 nm.

(iii) Primary process (IV): $(CH_3)_2CHCHO + h\nu \rightarrow CH_3 + CH_3CHCHO$. This process becomes important only at wavelengths of light available in the upper atmosphere. Methane formation in uninhibited $(CH_3)_2CHCHO$ photolysis is believed to arise through H-atom abstraction by methyl radicals formed in (IV). CH_3I formation, observed in the photolysis of $(CH_3)_2CHCHO–I_2$ mixtures at 265.4 nm, supports this conclusion (Blacet and Calvert 1951b). Values of ϕ_{IV} are estimated to be: 0.006 (313 nm); 0.012 (280 nm); 0.036 (265.4 nm); and 0.072 (253.7 nm). The down-chain split of 2-methylpropanal is similar to that seen in propanal; see figure VII-B-15.

(iv) Primary process (III): $(CH_3)_2CHCHO + h\nu \rightarrow (CH_3)_2CHCO + H$. The fall-off of the quantum yield of photodecomposition at short wavelengths seen for processes (I) in figure VII-B-23 probably results from the onset of processes (II) and (IV), but the sum of the efficiencies of processes (I), (II), and (IV) at 265.4 nm is well below unity (~0.8) and may result from the occurrence of process (III). In photolysis of pure $(CH_3)_2CHCHO$, H-atoms formed in (III) would react readily with the aldehyde to form H_2. However, hydrogen formation remains unimportant even at 253.7 nm, where $\Phi_{H2} = 0.03$ (Blacet and Calvert,1951a). The apparent inefficiency of photodecomposition at the shorter wavelengths of absorbed light remains unexplained.

VII-B-5.3. Photolysis Frequencies for 2-Methylpropanal in Air (1 atm)

Process (I) appears to be the only important mode of photodecomposition of $(CH_3)_2CHCHO$ molecules within the troposphere, although the other processes are expected to become important in the upper troposphere and the stratosphere. Only *j*-values for process (I) are considered here. As described in the previous section, two different scenarios related to the wavelength dependence of process (1) are assumed, since the choice between these is unclear. The results of the *j*-value calculations as a function of solar zenith angle are given in figure VII-B-25 for a cloudless day within the lower troposphere (ozone column = 350 DU); scenarios (1) and (2) were used, with ϕ_I estimates given by the solid and the dashed trend lines, respectively, in figure VII-B-23.

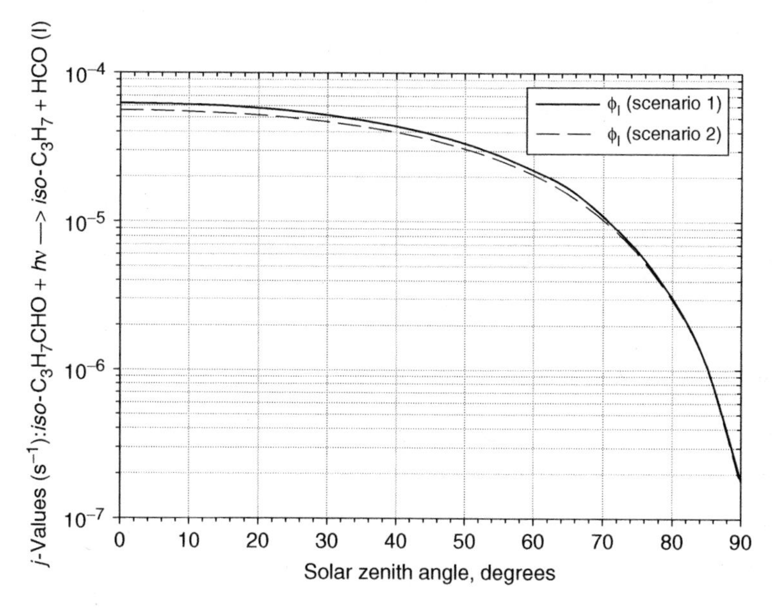

Figure VII-B-25. Estimated photolysis frequencies for $(CH_3)_2CHCHO$ as a function of solar zenith angle. Values are calculated for a cloudless day in the lower troposphere with an overhead ozone column of 350 DU. Two alternative scenarios for ϕ_I wavelength dependence were used in the calculations; see text and figure VII-B-23. The best present estimates of j-values probably lie between the two curves.

VII-B-6. Photodecomposition of Pentanal (Valeraldehyde, $CH_3CH_2CH_2CH_2CHO$)

Several different photodecomposition modes have been considered in explanation of the photolysis of pentanal:

$$CH_3CH_2CH_2CH_2CHO + h\nu \rightarrow CH_3CH_2CH_2CH_2 + HCO \quad \text{(I)}$$

$$\rightarrow CH_3CH_2CH_2CH_3 + CO \quad \text{(II)}$$

$$\rightarrow CH_2{=}CHCH_3 + CH_2{=}CHOH \quad \text{(III)}$$

$$\rightarrow C_2H_5 + CH_2CH_2CHO \quad \text{(IV)}$$

$$\rightarrow CH_3 + CH_2CH_2CH_2CHO \quad \text{(V)}$$

$$\rightarrow CH_3CH_2CH_2CH_2CO + H \quad \text{(VI)}$$

The important primary processes in the troposphere are (I) and (III). As with the lighter aldehydes, it is likely that process (I) occurs from the first excited triplet state of pentanal whereas process (III) arises from the excited singlet. The down-chain fragmentation of pentanal, which becomes important only at short wavelengths, may reflect reactions of highly vibrationally excited ground-state molecules formed by internal conversion.

Table VII-B-10. Cross sections (σ, cm^2 $molecule^{-1}$) for pentanal from Tadić et al. (2001a) and estimated quantum yields for primary process: $CH_3CH_2CH_2CH_2CHO + h\nu$ $CH_3CH_2CH_2CH_2 + HCO$ (I) for pentanal at 298 K

λ (nm)	$10^{21} \times \sigma$	ϕ_I	λ (nm)	$10^{21} \times \sigma$	ϕ_I	λ (nm)	$10^{21} \times \sigma$	ϕ_I
278	50.9	0.048	303	54.6	0.108	328	12.0	0.069
279	51.8	0.053	304	53.1	0.109	329	10.9	0.066
280	52.8	0.058	305	50.6	0.109	330	8.01	0.062
281	53.7	0.060	306	47.6	0.109	331	6.89	0.058
282	55.3	0.064	307	45.0	0.109	332	6.28	0.054
283	57.3	0.068	308	43.1	0.108	333	5.82	0.050
284	58.8	0.072	309	42.1	0.108	334	5.16	0.045
285	59.5	0.075	310	41.3	0.107	335	4.44	0.041
286	59.4	0.078	311	40.5	0.106	336	3.74	0.036
287	59.0	0.081	312	39.5	0.105	337	3.08	0.031
288	58.5	0.084	313	38.4	0.104	338	2.55	0.026
289	58.5	0.087	314	36.9	0.103	339	2.20	0.021
290	59.1	0.089	315	34.6	0.101	340	1.97	0.015
291	59.1	0.092	316	31.7	0.100	341	1.67	0.010
292	60.1	0.094	317	28.5	0.098	342	1.14	0.003
293	61.2	0.096	318	25.8	0.097	343	0.73	0.000
294	61.4	0.098	319	24.0	0.095	344	0.48	
295	60.6	0.100	320	22.9	0.092	345	0.35	
296	59.0	0.101	321	21.9	0.090	346	0.28	
297	57.2	0.103	322	20.8	0.087	347	0.23	
298	55.9	0.120	323	19.5	0.085	348	0.20	
299	55.1	0.105	324	18.1	0.082	349	0.17	
300	54.9	0.106	325	16.7	0.079	350	0.15	
301	54.9	0.107	326	15.3	0.076			
302	54.9	0.108	327	13.7	0.073			

VII-B-6.1. Absorption Cross Sections for Pentanal

Independent determinations of the cross sections by Martinez et al. (1992) and Tadić et al. (2001a) agree well. Those reported by Tadić et al. are plotted in figure VII-B-22 and tabulated in table VII-B-10.

VII-B-6.2. Quantum Yields of the Primary Processes in Pentanal

(i) Primary process (I): $CH_3CH_2CH_2CH_2CHO + h\nu \rightarrow CH_3CH_2CH_2CH_2 + HCO$. Two studies give important information on the extent of the primary processes in pentanal. Cronin and Zhu (1998) measured the quantum yield of formyl radical formation and provided estimates of ϕ_I at zero-pressure. No quenching of the excited states of pentanal by pentanal molecules (2–18 Torr) or added nitrogen molecules (8–480 Torr) was detected in their experiments, and they suggested that the zero-pressure ϕ_I values could be used for atmospheric conditions of 1 atm. These measurements are shown in figure VII-B-26 (open circles) for the 11 different wavelengths of laser excitation that they employed.

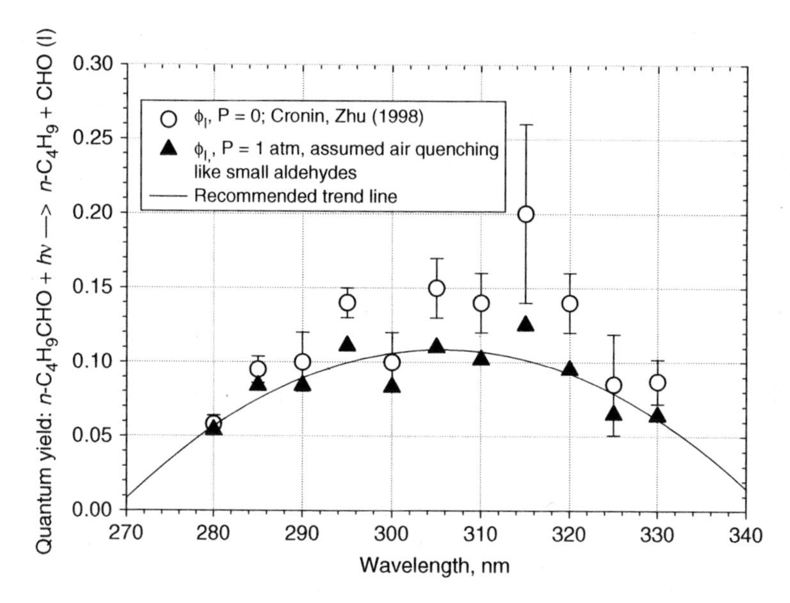

Figure VII-B-26. Plot of the wavelength dependence of ϕ_I for pentanal at zero pressure (Cronin and Zhu, 1998), open circles, and of ϕ_I for air (760 Torr) as estimated using average Stern–Volmer quenching constants for small aldehydes (figure VII-B-17b), black triangles.

Tadić et al. (2001a) studied the broadband photolysis of pentanal under atmospheric conditions, dilute mixtures in air (100–700 Torr). The overlap of the wavelength distribution of the lamps employed with the absorption of pentanal is similar to that shown in figure VII-B-18 for butanal and maximizes at about 304 nm. The proportion of short-wavelength radiation in the lamps is somewhat greater than that in sunlight within the troposphere. Quenching of process (I) by air molecules was observed; using the Tadić et al. data for 100–700 Torr air pressure, we derive a Stern–Volmer quenching constant $k_q/k_d = 3.29 \times 10^{-3}$ Torr^{-1}. This value is in reasonable agreement with that expected from the wavelength distribution of k_q/k_d defined by acetaldehyde and propanal data in figure VII-B-14b. In view of this, we have adjusted the measurements of Cronin and Zhu (1998) using the trend line for k_q/k_d vs. λ given in figure VII-B-14b and the ϕ_I^o estimates of Cronin and Zhu with the relation $1/\phi = 1/\phi_I^o + k_q/k_d \times P$ (Torr); these adjusted values are shown in figure VII-B-26 as closed triangles. The value of the average ϕ_I expected using the lamp distribution of Tadić at al. and the Cronin and Zhu data adjusted for air quenching, $\phi_I = 0.097$, is somewhat larger than the quantum yield estimate of Tadić et al. (2001a) for 760 Torr, $\phi_I = 0.052$. This may suggest that quenching by air is somewhat more effective than we assumed here. The recommended values for the wavelength dependence of ϕ_I for pentanal are those defined by the trend line in figure VII-B-26 and tabulated in table VII-B-10.

(ii) Primary process (III): $CH_3CH_2CH_2CH_2CHO + h\nu \rightarrow CH_2{=}CHCH_3 + CH_2{=}CHOH$. The results of Tadić et al. (2001a) indicate that process (III) is favored

over process (I) for photolysis of pentanal in air. They determined the quantum yields of propene and acetaldehyde formed in photolysis of pentanal in mixtures of air at pressure up to 700 Torr. Initially the enol of acetaldehyde ($CH_2{=}CHOH$) was formed, and this was observed to form the ultimate stable product, acetaldehyde. The 100–700 Torr data for propene yield an estimated average quantum yield, $\phi_{III} = 0.19$, for 760 Torr of air. Cronin and Zhu (1998) also identified qualitatively the products related to the occurrence of process (III) in steady-state photolysis of pentanal in oxygen and nitrogen atmospheres. The dominance of process (III) over (II) was confirmed. Since no data related to the wavelength dependence of ϕ_{III} exists at this writing, we suggest the average value (0.19) obtained in pentanal–air mixtures by Tadić et al. be used for ϕ_{III} at all the absorbed wavelengths within the troposphere.

It is interesting to compare the quantum yield of the analogous primary processes (III) in butanal and pentanal photolyzed in air. The data show that the ratio ϕ_{III} (pentanal)/ϕ_{III} (butanal) ≈ 1.6. This difference may reflect in part the lower barrier for H-atom abstraction from the secondary hydrogen atom in pentanal than from the primary hydrogen atom in butanal as the cyclic transition state progresses to the process (III) products.

(iii) Minor primary processes in pentanal photolysis. There is no evidence that defines the quantum yields for the other suggested primary processes in pentanal photolysis, processes (II), (IV), (V), and (VI), but these are expected to be relatively small; it is recommended that they be assumed to be zero in modeling of tropospheric chemistry.

VII-B-6.3. Photolysis Frequency for Pentanal Photolysis in air (1 atm)

The recommended cross sections and quantum yields for photodecomposition of pentanal were used to calculate the photolysis frequencies as a function of solar zenith angle for a cloudless day in the lower troposphere (ozone column = 350 DU). These results are given in figure VII-B-27.

VII-B-7. Photodecomposition of 3-Methylbutanal [Isovaleraldehyde, $(CH_3)_2CHCH_2CHO$]

By analogy with the information on the smaller aldehydes, one expects the following primary processes to occur in 3-methylbutanal:

$$(CH_3)_2CHCH_2CHO + h\nu \rightarrow (CH_3)_2CHCH_2 + HCO \quad \text{(I)}$$

$$\rightarrow (CH_3)_2CHCH_3 + CO \quad \text{(II)}$$

$$\rightarrow CH_3CH{=}CH_2 + CH_2{=}CHOH \quad \text{(III)}$$

$$\rightarrow (CH_3)_2CHCH_2CO + H \quad \text{(IV)}$$

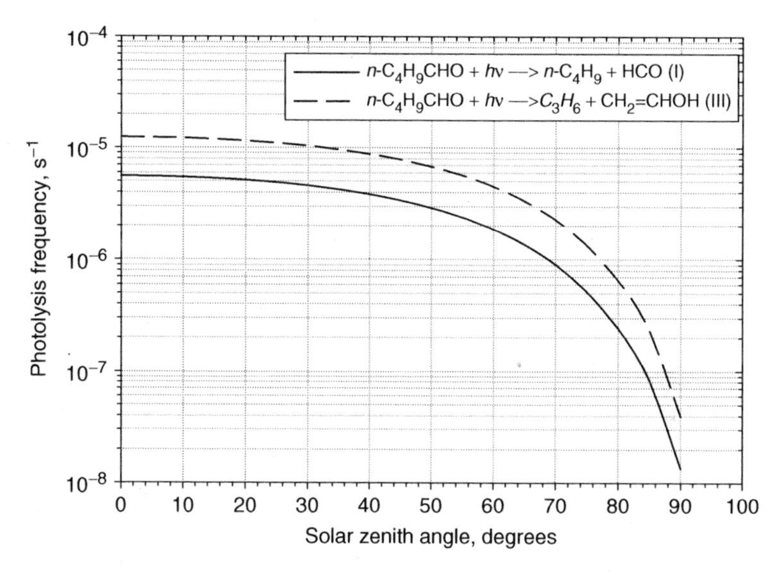

Figure VII-B-27. Plot of the estimated j-values for processes (I) and (III) in pentanal vs. solar zenith angle, as calculated for a cloudless day within the lower troposphere with an overhead ozone column = 350 DU.

Bamford and Norrish (1935) first published information on the primary processes in 3-methylbutanal. They estimated from full-arc photolysis that 47% of the decomposition occurred by process (II), presumably including process (I), while 53% occurred by process (III). More recently, Zhu et al. (1999) have reported quantitative information on the quantum yields of process (I).

VII-B-7.1. Absorption Cross Sections for 3-Methylbutanal

Estimates of the absorption cross sections for 3-methylbutanal were made at eleven wavelengths within the absorption band by Zhu et al. (1999). These data are shown in figure VII-B-28 (black circles). The smoothed spline line that connects the points was drawn to facilitate estimates at other wavelengths needed in modeling the j-determinations. The data match reasonably well those shown for pentanal (light, solid line) from measurements of Tadić et al. (2001a).

VII-B-7.2. Quantum Yields of Primary Processes in 3-Methylbutanal

(i) Primary process (I): $(CH_3)_2CHCH_2CHO + h\nu \rightarrow (CH_3)_2CHCH_2 + HCO$. Quantitative information on this primary process was determined by Zhu et al. (1999), who reported $\Phi_{HCO} = \phi_I$ at zero-pressure at eleven wavelengths. Figure VII-B-29 shows these data as black circles. Zhu et al. report that added nitrogen gas (20–400 Torr) did not quench HCO formation detectably at 280, 295, and 310 nm, but at 325 nm, a pressure of 400 Torr decreased the HCO yield by 25%. This corresponds to a

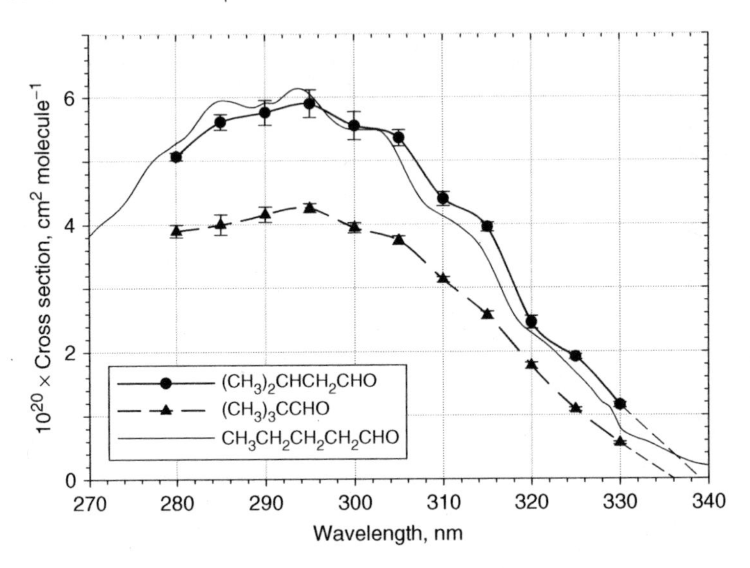

Figure VII-B-28. Approximate absorption cross sections for 3-methylbutanal (line with black circles) and 2,2-dimethylpropanal (dashed line with black triangles); data of Zhu et al (1999). A comparison spectrum of pentanal (light line) is from Tadić et al. (2001). The short dashed lines from 330–340 nm are the extrapolated wavelength dependencies of the cross sections used in j-value calculations.

Stern–Volmer quenching constant, $k_q/k_d \approx 3.8 \times 10^{-3}$ $Torr^{-1}$, similar to those observed for the smaller aldehydes; see figure VII-B-14b. In view of this observation, we have adjusted the measured zero-pressure values of ϕ_I to that expected for 1 atm of air, using the trend line data of figure VII-B-14b to select k_q/k_d values for the various wavelengths. These adjusted values, shown as open circles in figure VII-B-29, together with those derived from the spline curve connecting zero-pressure data of Zhu et al. (filled circles), are tabulated in table VII-B-11, and both were used in estimating the j-values for 3-methylbutanal to illustrate the range of estimated j-values expected from current information.

(ii) Primary process (III): $(CH_3)_2CHCH_2CHO + h\nu \rightarrow CH_3CH{=}CH_2 + CH_2{=}CHOH$. Zhu et al. (1999) suggest that the relatively low values measured for ϕ_I are in accord with the occurrence of (III), and they identified acetaldehyde among the products of steady-state photolysis experiments. No quantitative information on the magnitude of ϕ_{III} is available at this writing, but the full-arc photolysis reported by Bamford and Norrish (1935) suggest its dominance.

(iii) Other primary processes. There is no information related to the significance of primary processes (II) and (IV), but, as with the other aldehydes, it is likely that these occur, especially at the shorter wavelengths.

Table VII-B-11. Approximate cross sections (cm^2 $molecule^{-1}$) at 298 K and quantum yields of primary process (I), $(CH_3)_2CHCH_2CHO + h\nu \rightarrow (CH_3)_2CHCH_2 + HCO$, for 3-methylbutanal. $(\phi_I)^0$ and $(\phi_I)^{760}$ are the quantum yields at zero pressure (Zhu et al., 1999) and the estimated values in 760 Torr of air, assuming quenching occurs by air molecules, similar to that observed in the smaller aldehydes (figure VII-B-14b)

λ (nm)	$10^{20} \times \sigma$	$(\phi_I)^0$	$(\phi_I)^{760}$	λ (nm)	$10^{20} \times \sigma$	$(\phi_I)^0$	$(\phi_I)^{760}$
280	5.07	0.12	0.11	311	4.32	0.35	0.17
281	5.20	0.14	0.12	312	4.26	0.37	0.18
282	5.32	0.15	0.13	313	4.19	0.38	0.18
283	5.43	0.16	0.14	314	4.11	0.39	0.18
284	5.52	0.17	0.14	315	3.96	0.40	0.18
285	5.61	0.17	0.14	316	3.72	0.39	0.17
286	5.66	0.17	0.13	317	3.43	0.36	0.17
287	5.69	0.16	0.13	318	3.10	0.32	0.16
288	5.72	0.15	0.12	319	2.76	0.28	0.15
289	5.73	0.14	0.11	320	2.46	0.26	0.14
290	5.76	0.14	0.11	321	2.30	0.24	0.13
291	5.80	0.15	0.12	322	2.16	0.23	0.13
292	5.85	0.17	0.13	323	2.07	0.22	0.13
293	5.88	0.20	0.14	324	2.00	0.22	0.12
294	5.89	0.21	0.15	325	1.92	0.22	0.12
295	5.90	0.22	0.16	326	1.80	0.22	0.12
296	5.84	0.23	0.16	327	1.67	0.22	0.12
297	5.77	0.23	0.16	328	1.50	0.22	0.11
298	5.69	0.23	0.16	329	1.32	0.21	0.11
299	5.62	0.23	0.16	330	1.16	0.21	0.11
300	5.55	0.23	0.15	331	1.02	0.19	0.11
301	5.53	0.24	0.15	332	0.90	0.17	0.09
302	5.51	0.24	0.15	333	0.75	0.15	0.08
303	5.48	0.25	0.15	334	0.65	0.13	0.07
304	5.45	0.26	0.16	335	0.50	0.10	0.05
305	5.36	0.27	0.16	336	0.35	0.08	0.04
306	5.20	0.28	0.16	337	0.24	0.06	0.03
307	5.00	0.29	0.16	338	0.11	0.04	0.02
308	4.78	0.30	0.16	339	0	0.02	0.01
309	4.53	0.31	0.17	340	0	0	0
310	4.40	0.32	0.17				

VII-B-7.3. Photolysis Frequencies for 3-Methylbutanal within the Troposphere

Both the zero-pressure quantum yields for process (I) reported by Zhu et al. (1999) and those adjusted assuming quenching by air have been used to calculate approximate clear-sky j-values for primary process (I) within the lower troposphere (ozone column = 350 DU); see figure VII-B-30. The appropriate j-values for 1 atm of air remain uncertain, although those given by the line of small dashes is recommended here; the present best estimates may lie between the two curves in the figure.

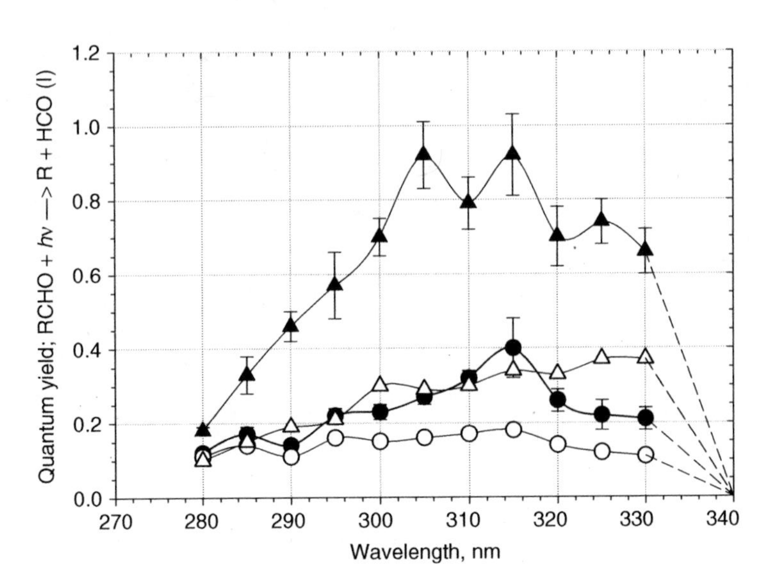

Figure VII-B-29. Quantum yields of HCO formation for 3-methylbutanal (circles) and 2,2-dimethylpropanal (triangles). Filled symbols are data reported at zero pressure. Open symbols are calculated for 760 Torr of air assuming the same quenching efficiency as for the smaller aldehydes. Smoothed spline lines have been drawn to connect the measured points. The short dashed lines are the assumed wavelength dependencies of the quantum yields for the wavelength range 330–340 nm used in *j*-value calculations.

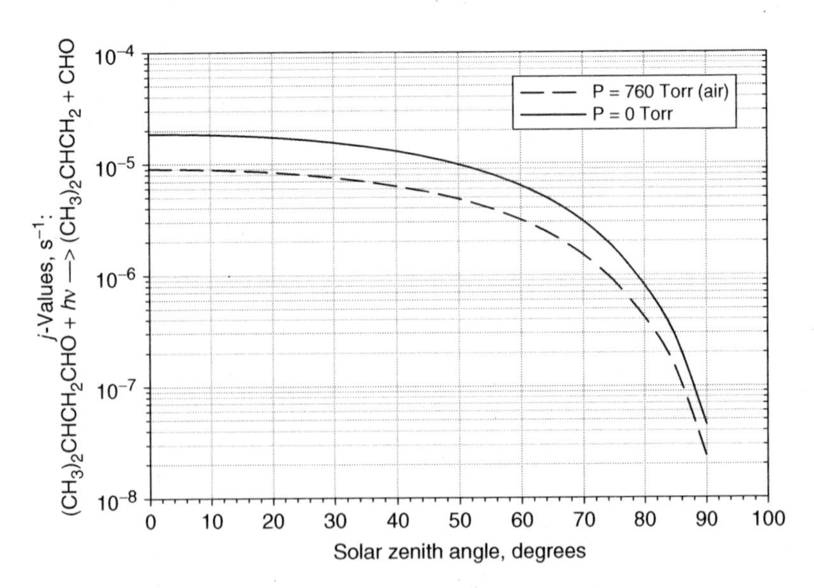

Figure VII-B-30. Approximate photolysis frequencies vs. solar zenith angle calculated for 3-methylbutanal for a cloudless day in the lower troposphere with an overhead ozone column = 350 DU. The solid curve is calculated using the Zhu et al. (1999) quantum yields for zero pressure; the dashed curve is that estimated assuming air quenching (760 Torr) equivalent to that shown by the smaller aldehydes. Present best estimates of *j*-values probably lie between the two curves.

VII-B-8. Photodecomposition of 2,2-Dimethylpropanal [Pivaldehyde, $(CH_3)_3CCHO$]

By analogy with the photochemical behavior of the smaller aldehydes, one expects the following primary photodecomposition processes to occur:

$$(CH_3)_3CCHO + h\nu \rightarrow (CH_3)_3C + HCO \qquad (I)$$

$$\rightarrow HC(CH_3)_3 + CO \qquad (II)$$

$$\rightarrow (CH_3)_3CCO + H \qquad (III)$$

$$\rightarrow CH_3 + (CH_3)_2CCHO \qquad (IV)$$

Primary process (I) is believed to be the only important mode of photodecomposition in the troposphere. The other modes of decay are likely to become important at $\lambda <$ 280 nm in view of the photochemistry of the other similar aldehydes.

VII-B-8.1. Absorption Cross Sections for 2,2,-Dimethylpropanal

Measurements of the absorption spectrum of 2,2-dimethylpropanal were reported for eleven wavelengths in the ultraviolet by Zhu et al. (1999), shown in figure VII-B-28. It is surprising that the absorption is about 40% smaller than those observed with the other aldehydes of similar size. Zhu et al. (1999) rationalized this difference in terms of a possible effect on the π^* energy induced by the bulky $(CH_3)_3C$ group present in 2,2-dimethylpropanal. If indeed the aldehyde sample used by Zhu et al. was 97% pure, as stated, then the observed effect is most interesting, and reasons for diminution of its cross sections from that of aldehydes of similar size should be explored theoretically.

VII-B-8.2. Quantum Yields of the Primary Processes in 2,2-Dimethylpropanal

(i) Primary process (I). Process (I) has been studied by Zhu et al. (1999), who measured the quantum yields of the HCO radical formed in photolysis at eleven wavelengths (280–330 nm). The zero-pressure ϕ_I estimates are summarized in the figure VII-B-29 (black triangles). Variation of the pressure of nitrogen from 20 to 400 Torr gave no detectable quenching of HCO formation, so the zero-pressure values for ϕ_I were suggested to be applicable to air at 1 atm. Indeed, significant quenching may not occur in the experiments of Zhu et al., but the large scatter in the measured ϕ_{HCO} values, indicated by the 1 σ error bars, suggests possible difficulties in observing the small changes in the HCO signal expected as a result of N_2 addition in the Zhu et al. studies. To test the effect of possible quenching by added nitrogen at 1 atm pressure, we have applied the Stern–Volmer quenching constants given by the trend line in figure VII-B-14b, and these data points are shown as open triangles in figure VII-B-29. Both the zero-pressure measurements of ϕ_I, as suggested by the spline curve joining the experimental points, and the pressure-adjusted values with assumed quenching are tabulated in table VII-B-12. The values of ϕ_I are significantly greater for

Table VII-B-12. Approximate cross sections (σ, cm^2 molecule^{-1}) at 298 K and quantum yields of primary process (I), $(CH_3)_3CCHO + h\nu \rightarrow (CH_3)_3C + HCO$, for 2,2-dimethylpropanal. $(\phi_I)^0$ and $(\phi_I)^{760}$ are the quantum yields at zero pressure (Zhu et al., 1999) and the estimated values in 760 Torr of air, assuming quenching occurs by air molecules, similar to that observed in the smaller aldehydes (figure VII-B-14b)

λ (nm)	$10^{20} \times \sigma$	$(\phi_I)^0$	$(\phi_I)^{760}$	λ (nm)	$10^{20} \times \sigma$	$(\phi_I)^0$	$(\phi_I)^{760}$
280	3.90	0.18	0.10	310	3.13	0.79	0.30
281	3.92	0.21	0.11	311	3.01	0.81	0.31
282	3.94	0.24	0.12	312	2.90	0.84	0.32
283	3.95	0.27	0.13	313	2.80	0.88	0.33
284	3.98	0.30	0.14	314	2.68	0.91	0.34
285	4.00	0.33	0.15	315	2.57	0.92	0.34
286	4.02	0.36	0.16	316	2.43	0.90	0.34
287	4.06	0.38	0.17	317	2.27	0.84	0.34
288	4.05	0.41	0.18	318	2.10	0.79	0.32
289	4.13	0.44	0.19	319	1.95	0.74	0.33
290	4.16	0.46	0.19	320	1.78	0.70	0.33
291	4.21	0.48	0.19	321	1.64	0.69	0.34
292	4.24	0.51	0.19	322	1.48	0.70	0.34
293	4.26	0.53	0.19	323	1.35	0.71	0.35
294	4.27	0.55	0.20	324	1.23	0.73	0.36
295	4.26	0.57	0.21	325	1.09	0.74	0.37
296	4.22	0.59	0.23	326	0.97	0.74	0.37
297	4.16	0.61	0.25	327	0.87	0.72	0.37
298	4.08	0.63	0.27	328	0.76	0.70	0.37
299	4.01	0.67	0.29	329	0.66	0.68	0.37
300	3.95	0.70	0.30	330	0.56	0.66	0.37
301	3.91	0.75	0.31	331	0.45	0.60	0.32
302	3.87	0.81	0.31	332	0.37	0.52	0.29
303	3.85	0.87	0.30	333	0.28	0.45	0.25
304	3.81	0.90	0.30	334	0.19	0.40	0.22
305	3.75	0.92	0.29	335	0.08	0.33	0.18
306	3.66	0.91	0.29	336	0	0.27	0.15
307	3.54	0.87	0.29	337		0.20	0.11
308	3.40	0.83	0.29	338		0.14	0.08
309	3.27	0.81	0.29	339		0.07	0.03
				340		0	0

2,2-dimethylpropanal than for 3-methylbutanal, shown in the same figure. This may result, at least in part, from the occurrence of the major process (III) in 3-methylbutanal, a pathway not available in 2,2-dimethylpropanal.

(ii) Other primary processes. Processes (II) and (IV) have not been demonstrated to occur, but they are expected to be significant, especially at wavelengths of absorbed light below 280 nm. By analogy with the increased importance of process (IV) at short wavelengths forming CH_3 radicals by rupture of the CH_3—CH_2CHO bond in propanal and the CH_3—$CH(CH_3)CHO$ bond in 2-methylpropanal (see figure VII-B-15), it seems likely that process (IV) will play an important role in the photolysis of

2,2-dimethylpropanal. An alternative pathway competing with process (I) for photodecomposition, possibly process (III), may account for decrease in the decomposition by process (I) that is observed with 2,2-dimethylbutanal at the shorter wavelengths present within the troposphere (280–300 nm).

VII-B-8.3. Photolysis Frequencies for 2,2-Dimethylpropanal within the Troposphere

The j-values for process (I) have been estimated for clear sky conditions in the lower troposphere (ozone column = 350 DU). Two different scenarios have been used: one assuming no quenching of process (I) in one atmosphere of air; and the other assuming quenching similar to that observed in other small aldehydes. The results of the two calculations are summarized in figure VII-B-31. Assuming that the seemingly low absorption cross sections for 2,2-dimethylpropanal given by Zhu et al. (1999) are correct, then it is probable that the appropriate j-values lie within the area between the two curves in the figure.

VII-B-9. Photodecomposition of 2-Methylbutanal [$CH_3CH_2CH(CH_3)CHO$]

Studies of the photochemistry of 2-methylbutanal and its mixtures with iodine have been reported for excitation at one wavelength, 313 nm (Gruver and Calvert, 1956, 1958). By analogy with the other aldehydes, one expects the following primary processes:

$$CH_3CH_2CH(CH_3)CHO + h\nu \rightarrow CH_3CHCH_2CH_3 + HCO \quad \text{(I)}$$

$$\rightarrow CH_3CH_2CH_2CH_3 + CO \quad \text{(II)}$$

$$\rightarrow CH_2{=}CH_2 + CH_3CH{=}CH(OH) \quad \text{(III)}$$

$$\rightarrow CH_3CH_2CH(CH_3)CO + H \quad \text{(IV)}$$

The results of Gruver and Calvert (1956, 1958) show clearly that processes (I) and (III) are important at 313 nm. Product quantum yields were determined in experiments with the pure aldehyde over a range of temperatures (24.5–348.7°C), aldehyde pressures (23–33 Torr) and iodine pressures (0–25 Torr). The strong temperature dependence of the quantum yields of products in the uninhibited experiments was suppressed in the presence of small amounts of iodine. The average quantum yields of products at 313 nm (8 experiments) were: $\Phi_{C4H10} = 0.004 \pm 0.004$; $\Phi_{CO} = 0.58 \pm 0.05$; $\Phi_{C2H4} = 0.15 \pm 0.02$; $\Phi_{C4H9I} = 0.51 \pm 0.05$. From these results quantum yields of $\phi_{I} \approx 0.55$; $\phi_{II} \approx 0.004$; $\phi_{III} \approx 0.15$ were established. CH_3I was undetectable in the products, so down-chain fragmentation is unimportant at 313 nm. Optical activity was introduced into the aldehyde synthesized for use in the 1958 studies of Gruver and Calvert. Optical activity was not retained in the product $CH_3CHICH_2CH_3$, which indicates either a planar or an easily inverted pyramidal structure of the 2-butyl radical

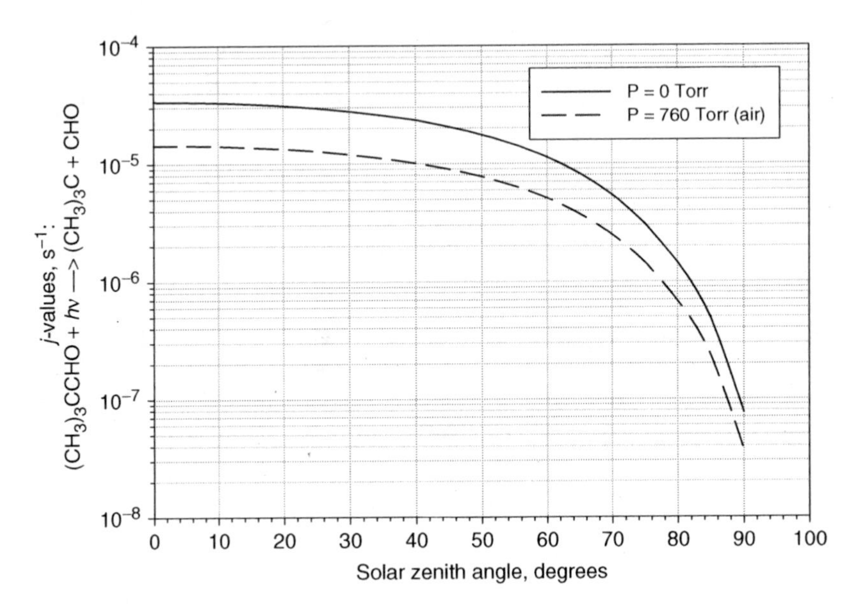

Figure VII-B-31. Approximate photolysis frequencies vs. solar zenith angle calculated for 2,2-dimethylpropanal for a cloudless day in the lower troposphere with an overhead ozone column of 350 DU. The solid curve is calculated using the Zhu et al. (1999) quantum yields for zero pressure; the dashed curve is that estimated assuming air quenching (760 Torr) equivalent to that shown by the smaller aldehydes. Present best estimates of *j*-values may lie between the two curves.

formed in process (I). Note that the quantum efficiency of process (III) is very similar to that observed in the structurally related butanal photolyzed at 313 nm. Since the wavelength dependence of the primary processes has not been determined, *j*-values cannot be calculated.

VII-B-10. Photodecomposition of Hexanal ($CH_3CH_2CH_2CH_2CH_2CHO$)

The primary processes in hexanal photodecomposition are suggested to be:

$$CH_3CH_2CH_2CH_2CH_2CHO + h\nu \rightarrow CH_3CH_2CH_2CH_2CH_2 + HCO \qquad (I)$$

$$\rightarrow CH_3CH_2CH_2CH_2CH_3 + CO \qquad (II)$$

$$\rightarrow CH_3CH_2CH{=}CH_2 + CH_2{=}CHOH \qquad (III)$$

$$\rightarrow CH_3CH_2CH_2CH_2CH_2CO + H \qquad (IV)$$

The occurrence of processes (I) and (III) at wavelengths of sunlight available in the troposphere is well documented. Processes (II) and (IV) may be important at the shorter wavelengths, as observed with the smaller aldehydes.

VII-B-10.1. Absorption Cross Sections for Hexanal

Tadić et al. (2001b) point out that the spectrum of hexanal as determined by Plagens et al. (1998) is very similar to those of butanal and pentanal. In the absence of published data, we recommend the use of the surrogate pentanal spectrum of Tadić et al. (2001b) for atmospheric considerations; see figure VII-B-22 and table VII-B-10.

VII-B-10.2. Quantum Yields of Primary Processes in Hexanal Photodecomposition

(i) Process (I). Tang and Zhu (2004) measured the zero-pressure quantum yields of HCO radical for hexanal at eleven different wavelengths (280–330 nm). These data are shown as open circles in figure VII-B-32. In this study the authors report rather efficient quenching of process (I) by hexanal molecules, yet no quenching by nitrogen was observed for pressures up to 387 Torr. However, the data of Tadić et al. (2001b), from their study of the broad-band photolysis of hexanal, showed quenching of process (I) by air with a Stern–Volmer constant $k_q/k_d = 5.0 \times 10^{-3}$ Torr^{-1}, similar to that expected ($\sim 3 \times 10^{-3}$ Torr^{-1}) for the smaller aldehydes at the wavelength of maximum overlap (303 nm) used by Tadić et al. (2001b). Quenching by air at 1 atm has been observed in most of the aldehydes that have been studied, as discussed in the previous sections of this chapter. If significant air quenching of the zero-pressure ϕ_I values of Tang and Zhu is assumed to occur, as with the smaller aldehydes, then the data points shown as open

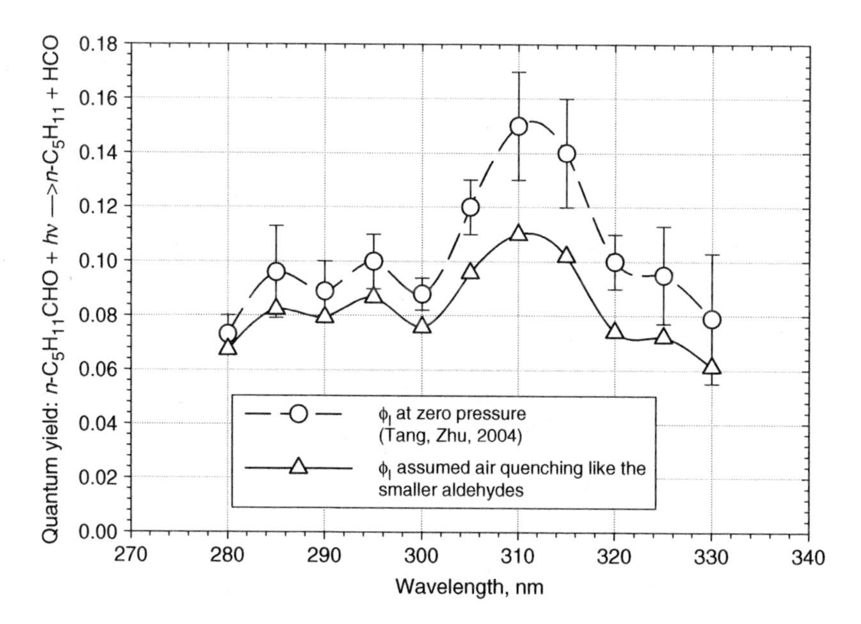

Figure VII-B-32. Quantum yields of primary process (I) in hexanal vs. wavelength; zero pressure values of Tang and Zhu (2004), open circles. The open triangles represent adjusted data assuming hexanal is quenched by air (760 Torr) as observed with the smaller aldehydes. Recommended trend line, solid curve.

Table VII-B-13. Recommended quantum yields for process (I) in hexanal photolysis in air (1 atm): $CH_3CH_2CH_2CH_2CH_2CHO + h\nu \rightarrow CH_3CH_2CH_2CH_2CH_2 + HCO$. In view of the limited data, we recommend that $\phi_{III} = 0.29$ be used for process (III), $CH_3CH_2CH_2CH_2CH_2CHO + h\nu \rightarrow CH_2{=}CHCH_2CH_3 + CH_2{=}CHOH$, independent of wavelength

λ (nm)	Estimated ϕ_I	λ (nm)	Estimated ϕ_I	λ (nm)	Estimated ϕ_I
278	0.060	300	0.076	322	0.072
279	0.064	301	0.078	323	0.072
280	0.067	302	0.081	324	0.072
281	0.073	303	0.086	325	0.072
282	0.075	304	0.091	326	0.071
283	0.079	305	0.096	327	0.069
284	0.081	306	0.100	328	0.067
285	0.083	307	0.104	329	0.064
286	0.082	308	0.107	330	0.061
287	0.081	309	0.109	331	0.057
288	0.080	310	0.110	332	0.053
289	0.079	311	0.112	333	0.048
290	0.079	312	0.110	334	0.044
291	0.081	313	0.108	335	0.038
292	0.083	314	0.106	336	0.034
293	0.085	315	0.102	337	0.027
294	0.087	316	0.097	338	0.020
295	0.087	317	0.089	339	0.014
296	0.085	318	0.083	340	0
297	0.082	319	0.078	341	0
298	0.080	320	0.074	342	0
299	0.077	321	0.072	343	0

triangles in figure VII-B-32 are obtained. If the solid spline curve joining these points is used to derive an average ϕ_I value for the wavelength distribution used by Tadić et al., we find $\phi_I \approx 0.089$. This agrees reasonably well with the estimate of $\phi_I = 0.096$ for 760 Torr of air in the broad-band photolysis of Tadić et al. (2001b). The recommended values for ϕ_I are those shown in figure VII-B-32, and tabulated recommendations are given in table VII-B-13.

(ii) Primary process (III). No dependence on pressure of nitrogen (100–700 Torr) was observed for process (III) by Tadić et al. (2001b), and the average measured $\phi_{III} = 0.29 \pm 0.02$. Processes (I) and (III) probably occur from different electronic states. Presumably, (I) occurs largely from the triplet T_I state, whereas (III) may arise from the first excited singlet. We suggest that modelers of tropospheric chemistry use the estimate $\phi_{III} = 0.29$, independent of wavelength.

VII-B-10.3. Photolysis Frequencies for Hexanal

The recommended data from this study were used to derive the *j*-values for hexanal given in figure VII-B-33.

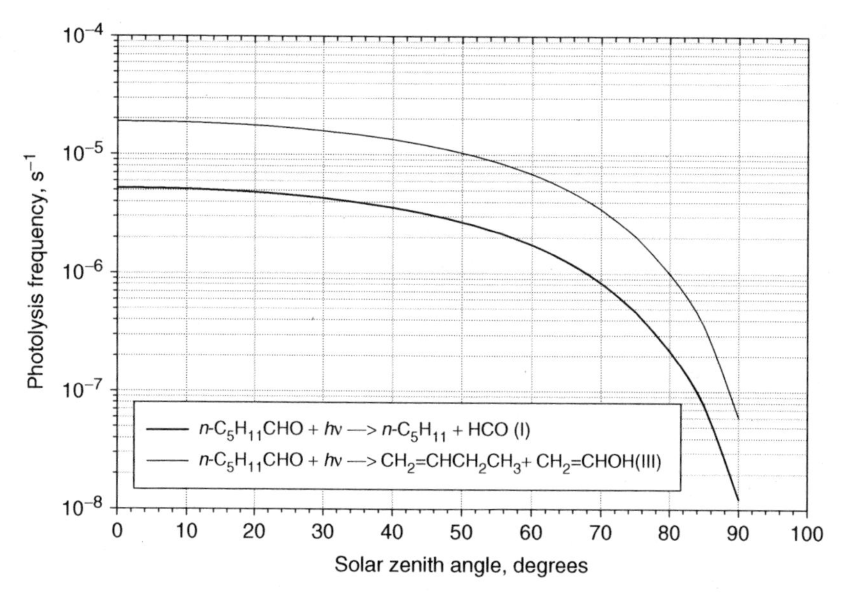

Figure VII-B-33. Approximate photolysis frequencies for primary processes (I) and (III) in hexanal vs. solar zenith angle; curves are calculated for a cloudless day within the lower troposphere with an overhead ozone column = 350 DU.

VII-B-11. Photodecomposition of 3-Methylpentanal [$CH_3CH_2CH(CH_3)CH_2CHO$]

The primary processes in the photodecomposition of 3-methylpentanal are expected to follow the pattern of the lighter aldehydes:

$$CH_3CH_2CH(CH_3)CH_2CHO + h\nu \rightarrow CH_3CH_2CH(CH_3)CH_2 + HCO \quad (I)$$

$$\rightarrow CH_3CH_2CH(CH_3)CH_3 + CO \quad (II)$$

$$\rightarrow cis\text{-or } trans\text{-}CH_3CH{=}CHCH_3 + CH_2{=}CHOH \quad (IIIa)$$

$$\rightarrow CH_2{=}CHCH_2CH_3 + CH_2{=}CHOH \quad (IIIb)$$

$$\rightarrow CH_3CH_2CH(CH_3)CH_2CO + H \quad (IV)$$

This aldehyde provides an interesting test of the mechanism of the Norrish type II process. In this case there are two C-atoms, down-chain from the carbonyl group, that have γ-H atoms; one group is composed of secondary H-atoms and the other of primary H-atoms. To the extent that the former is abstracted by the carbonyl oxygen, path (IIIa) occurs; when the latter is abstracted, path (IIIb) occurs. Since the secondary H-atoms have a weaker C—H bond than the primary H-atoms, one expects process (IIIa) to be favored over (IIIb). Such is the experimental result. The quantum yields of the various butene products were determined by Rebbert and Ausloos (1967) in experiments at 313 and 253.7 nm. The $\phi_{IIIa} + \phi_{IIIb}$ values were relatively insensitive to a 10-fold change in pressure of the aldehyde (1.8–21 Torr), the temperature (27–151°C), and the addition of 83 Torr of cyclohexane. At 313 nm, $\phi_{IIIa} + \phi_{IIIb} = 0.33 \pm 0.02$; $\phi_{IIIa}/\phi_{IIIb} = 9.0(27°C), 9.3 (50°C), 6.8 (151°C)$; at 253.7 nm, $\phi_{IIIa}+\phi_{IIIb} = 0.65\pm0.06$; $\phi_{IIIa}/\phi_{IIIb} = 4.5(28°C), 4.2 (53°C), 3.9 (152°C)$. As one might expect, the change in product ratio with increase in temperature reflects an apparent activation energy difference between abstraction of a primary H-atom [process (IIIb)] and a secondary H-atom [process (IIIa)].

Rebbert and Ausloos (1967) used energy transfer from acetone and benzene triplets to show that the primary process (III) can occur from the triplet state of 3-methylpentanal. The extent of processes (I), (II), and (IV) is not known, but these are expected to occur, as seen in the photodecomposition of aldehydes of analogous structure. Sufficient data for estimation of *j*-values for this aldehyde are not available at this writing.

VII-B-12. Photodecompostion of Heptanal ($CH_3CH_2CH_2CH_2CH_2CH_2CHO$)

Heptanal photodecomposition is expected to follow the pattern of the lighter aldehydes:

$$CH_3CH_2CH_2CH_2CH_2CH_2CHO + h\nu \rightarrow CH_3CH_2CH_2CH_2CH_2CH_2 + HCO \quad (I)$$

$$\rightarrow CH_3CH_2CH_2CH_2CH_2CH_3 + CO \quad (II)$$

$$\rightarrow CH_2{=}CHCH_2CH_2CH_3 + CH_2{=}CHOH \quad (III)$$

$$\rightarrow CH_3CH_2CH_2CH_2CH_2CH_2CO + H \quad (IV)$$

Evidence for the occurrence of primary process (I) has been obtained by Tang and Zhu (2004) and processes (I) and (III) by Tadić et al. (2002). Primary processes (II) and (IV) have not been observed but are expected to occur, especially for absorbed wavelengths < 280 nm.

VII-B-12.1. Absorption Cross Sections for Heptanal

Approximate cross sections for heptanal at eleven different wavelengths (280–350 nm) were reported by Tang and Zhu (2004). These values are somewhat greater than those measured for hexanal, but most values overlapped within the scatter of the measurements. Tadić et al. (2002) reported that the cross sections for the straight-chain aldehydes, as measured by Zabel (1999), are almost identical from butanal to nonanal. Thus the spectra of one of the straight-chain C_4- to C_9-aldehydes could be used as a surrogate for that of heptanal for atmospheric calculations until an accurate absorption spectrum of heptanal becomes generally available.

VII-B-12.2. Quantum Yields of Primary Processes in Heptanal Photodecomposition

(i) Process (I): $CH_3CH_2CH_2CH_2CH_2CH_2CHO + h\nu \rightarrow CH_3CH_2CH_2CH_2CH_2CH_2 + HCO$. The most representative of heptanal photodecomposition in air at one atmosphere are the studies of Tadić et al. (2002), who photolyzed dilute heptanal mixtures in air (100–700 Torr) using broad-band radiation (see figure VII-B-18). Their data, extrapolated to 760 Torr, gives estimates of the weighted-average quantum yields over the 279–350 nm region: $\phi_I = 0.030$. Direct evidence for the occurrence of process (I) in heptanal photolysis was obtained by Tang and Zhu (2004), who measured the zero-pressure values of ϕ_I at 11 wavelengths. These data are given by the black circles in figure VII-B-34. In this case, quenching of process (I) was observed to occur by the aldehyde molecules, and the Stern–Volmer quenching constants were very much larger than those seen for the smaller aldehydes; for example, at 280 nm, 1.1 ± 0.2 and at 330 nm, 2.9 ± 1.2 Torr^{-1}. However, as in their previous studies of aldehyde photolysis, no observable quenching of process (1) occurred in experiments with addition of nitrogen (6–387 Torr). This is surprising, especially in view of the very large quenching of process (I) by heptanal molecules that was observed. Physical quenching of aldehyde excited states by nitrogen is expected to be much less efficient than that of the aldehyde molecule; see figure VII-B-14. However, in the Tadić et al. (2002) studies with

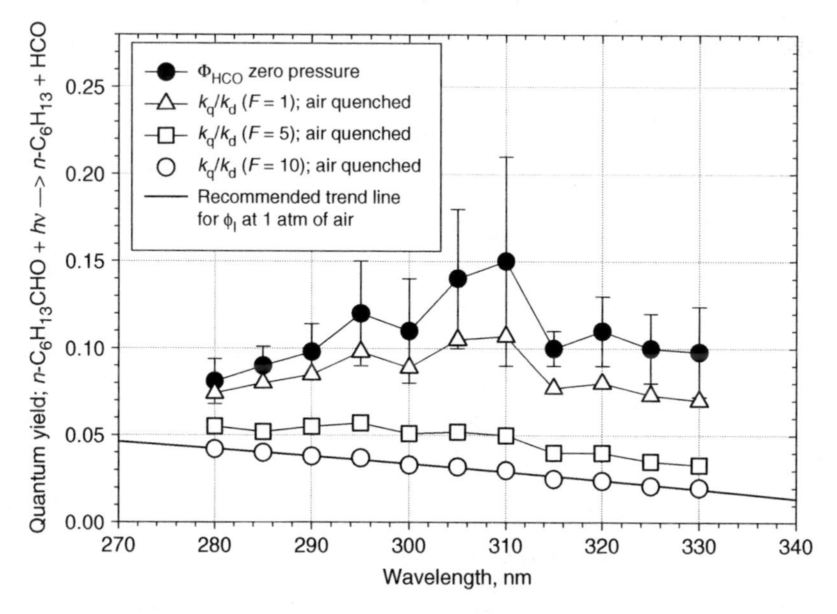

Figure VII-B-34. Plot of the quantum yield of process (I) in heptanal vs. wavelength. Values measured for zero pressure by Tang and Zhu (2004) are shown as black circles. Estimates of ϕ_I adjusted for air quenching (760 Torr) are shown, using different extents of quenching estimated from $k_q/k_d = F \times (k_q/k_d)^0$ where $(k_q/k_d)^0$ is the value estimated for the light aldehydes at a given wavelength (figure VII-B-14b). F is varied from 1 (triangles), 5 (squares), to 10 (open circles). The latter values fit best the broad-band (average) estimate of $\phi_I = 0.030$ at 760 Torr (Tadić et al., 2002), The solid heavy line is the recommended trend curve for ϕ_I in air (760 Torr).

broad-band irradiation, quenching of process (I) by air molecules showed a Stern–Volmer quenching constant of 1.3×10^{-2} Torr^{-1}, which is significantly larger that those observed for the smaller aldehydes. The observations of Tang and Zhu (2004) on effects of pressure on the HCO signal are complicated by the changes in kinetics of the HCO removal mechanism and the vibrational/rotational relaxation of the energy-rich HCO molecule formed in process (I). The inherent scatter in the HCO signal observed by Tang and Zhu (2004) may indicate difficulties in establishing the zero-time signals and make very difficult the measurements of the relatively small quenching constants for nitrogen. In our treatment of the data to arrive at the recommended trend line for ϕ_I with wavelength for 1 atm of air, we have increased the size of the assumed quenching rate constant for excitation at each wavelength from that observed for the simple aldehydes (k_q/k_d), given by the open triangles, by factors (F) of 5 (open squares) and 10 (open circles) in figure VII-B-34 [$k_q/k_d = F \times (k_q/k_d)^0$, where $(k_q/k_d)^0$ is the Stern–Volmer quenching constant as estimated for the small aldehydes in figure VII-B-14b]. The latter case provides the best correlation with the Tadić et al. (2002) data from the broad-band irradiation. A large quenching constant (k_q/k_d) for aldehyde–air collisions is consistent with the unusually large efficiency seen by Tadić et al.

For the wavelength distribution employed by Tadić et al. (2002)(figure VII-B-18), the estimates, from the recommended trend line, give $\phi_{III} = 0.031$, in good agreement with the value measured by Tadić et al., $\phi_I = 0.030$ at 760 Torr of air. Note that the Stern–Volmer quenching constant that describes the effect of air pressure on ϕ_I measured by Tadić et al., $k_q/k_d = 1.3 \times 10^{-2}$ Torr^{-1}, is similar to that used in our calculations (with $F = 10$) for light absorption at 305 nm, near the wavelength of maximum overlap in the Tadić et al. (2002) experiments. The recommended ϕ_I values for heptanal in 1 atm of air are given in table VII-B-14.

(ii) Process (III): $CH_3CH_2CH_2CH_2CH_2CH_2CHO + h\nu \rightarrow CH_2{=}CHCH_2CH_2CH_3 + CH_2{=}CHOH$. The data of Tadić et al. (2002) give $\phi_{III} = 0.12$ at 760 Torr of air. The wavelength dependence of ϕ_{III} is unknown at this writing. For the purposes of tropospheric modeling, we assume that $\phi_{III} = 0.12$, independent of wavelength of light absorbed within the troposphere. The occurrence of processes (II) and (IV) may become important at the short wavelengths of light ($\lambda < 280$ nm); these or other unrecognized pathways of excited molecule decay must exist to rationalize the relatively small fraction of heptanal decay that is accounted for by processes (I) and (III).

VII-B-12.3. Photolysis Frequencies for Heptanal in the Troposphere

Photolysis frequencies for processes (I) and (III) have been calculated for a cloudless day in the lower troposphere as a function of solar zenith angle (ozone column = 350 DU); see figure VII-B-35. The curve fitted to the open circles in figure VII-B-34 is the recommended wavelength dependence of ϕ_I. The effect of different choices of quenching efficiency of air on the j-value for process (I) is seen in figure VII-B-35. Quenching by air is assumed to occur as observed with the small aldehyde (long-dashed curve) or with an efficiency that is 10 times that of the small aldehydes

Table VII-B-14. Recommended quantum yields for process (I) in heptanal photolysis in air (1 atm): $CH_3CH_2CH_2CH_2CH_2CH_2CHO + h\nu \rightarrow CH_3CH_2CH_2CH_2CH_2CH_2 + HCO$. In view of the limited data, we recommend that $\phi_{III} = 0.12$ be used, independent of wavelength, for primary process (III) forming $CH_2{=}CHCH_2CH_2CH_3$ and $CH_2{=}CHOH$ (Tadić et al., 2002)

λ (nm)	Estimated ϕ_I	λ (nm)	Estimated ϕ_I	λ (nm)	Estimated ϕ_I
278	0.043	300	0.034	325	0.022
279	0.042	301	0.033	326	0.021
280	0.042	302	0.033	327	0.021
281	0.041	303	0.032	328	0.020
282	0.041	304	0.032	329	0.020
283	0.041	305	0.031	330	0.019
284	0.041	306	0.031	331	0.019
285	0.040	307	0.030	332	0.018
286	0.040	308	0.030	333	0.017
287	0.039	309	0.029	334	0.017
288	0.039	310	0.029	335	0.016
289	0.038	311	0.029	336	0.016
290	0.038	312	0.028	337	0.015
291	0.038	313	0.028	338	0.015
292	0.037	314	0.027	339	0.014
293	0.037	315	0.027	340	0.014
294	0.036	316	0.026	341	0.013
295	0.036	317	0.026	342	0.013
296	0.035	318	0.025	343	0.012
297	0.035	319	0.025	344	0.011
298	0.035	320	0.024	345	0.011
299	0.034	321	0.024	346	0.010
		322	0.023	347	0.010
		323	0.022	348	0.009
		324	0.022		

(short-dashed curve). In the absence of a measured wavelength dependence of ϕ_{III}, we have assumed that the value of $\phi_{III} = 0.12$, as measured for the average in the broad-band experiments of Tadić et al. (2002), and that is independent of λ.

VII-B-13. The Effect of Structure on the Quantum Yield of Primary Processes in the Aldehydes

Evidence of the influence of aldehyde structure on the quantum yields of the various photodecomposition modes has been cited in this section, and further examples of this influence can be seen in figure VII-B-36 for photolysis at 313 nm. Here, data related to the straight-chain aldehydes are shown for the two major photodissociation processes that occur in tropospheric sunlight: $RCHO + h\nu \rightarrow R + CHO$ (I); and $RCHO + h\nu \rightarrow R'CH{=}CH_2 + CH_2{=}CH(OH)$ (II), where R varies from H to C_6H_{13} and R′ from H to C_3H_7 as one progresses from formaldehyde to heptanal. As the chain length increases there is significant drop in the quantum yield of the radical dissociation pathway (I) that can be seen in figure VII-B-36. While formaldehyde photolysis is a major source of radicals that promote ozone generation in the troposphere, a decline in process (I)

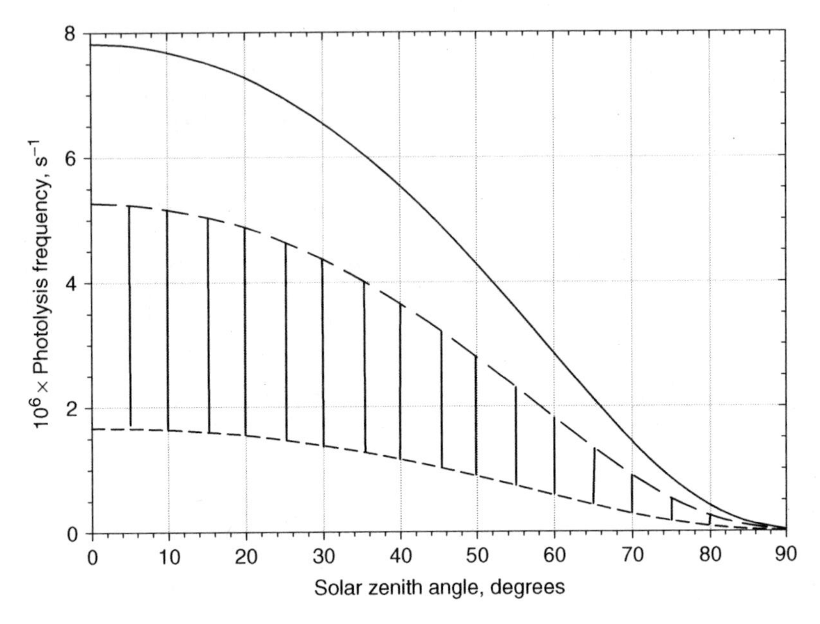

Figure VII-B-35. Plot of the recommended photolysis frequencies vs. solar zenith angle for heptanal primary processes: $CH_3CH_2CH_2CH_2CH_2CH_2CHO + h\nu \rightarrow CH_3CH_2CH_2CH_2CH_2CH_2 +$ HCO (I) and $CH_3CH_2CH_2CH_2CH_2CH_2CHO + h\nu \rightarrow CH_2{=}CHCH_2CH_2CH_3 + CH_2{=}CHOH$ (III); curves are calculated for a cloudless day in the lower troposphere with an overhead ozone column of 350 DU. The long-dashed line gives j-values for process (I) calculated assuming that quenching of process (I) by air is like that in the small aldehydes (see figure VII-B-14b); the short-dashed line (the dependence recommended in this study) gives j-values for process (I) calculated assuming air quenching at 10 times the efficiency shown by small aldehydes. The probable values for process (I) lie within the striped area. The solid curve gives the recommended j-values for primary process (III).

occurs with increasing chain length of the aldehyde until, at heptanal, ϕ_I has fallen so that estimated j-values for radical generation are only about 1/20th of those for CH_2O. However, the higher aldehydes exhibit a new influence on ozone generation with the formation of alkenes and enol-acetaldehyde [$CH_2{=}CH(OH)$] that begins with butanal and increases in quantum efficiency up to hexanal. All of the double-bond-containing products formed in (II) are expected to be reactive with ozone and OH radicals to generate a variety of highly reactive, ozone-generating products (e.g., see Calvert et al., 2000).

VII-C. Photodecomposition of Acylic Ketones

VII-C-1. Acetone [2-Propanone, $CH_3C(O)CH_3$]

The photochemistry and photodecomposition of acetone have been studied extensively over the past three-quarters of a century. Several reviews of the pioneering work have been published (e.g., Noyes et al., 1956; Lee and Lewis, 1980). Absorption of solar radiation in the troposphere ($\lambda > 290$ nm) by acetone and other carbonyl compounds

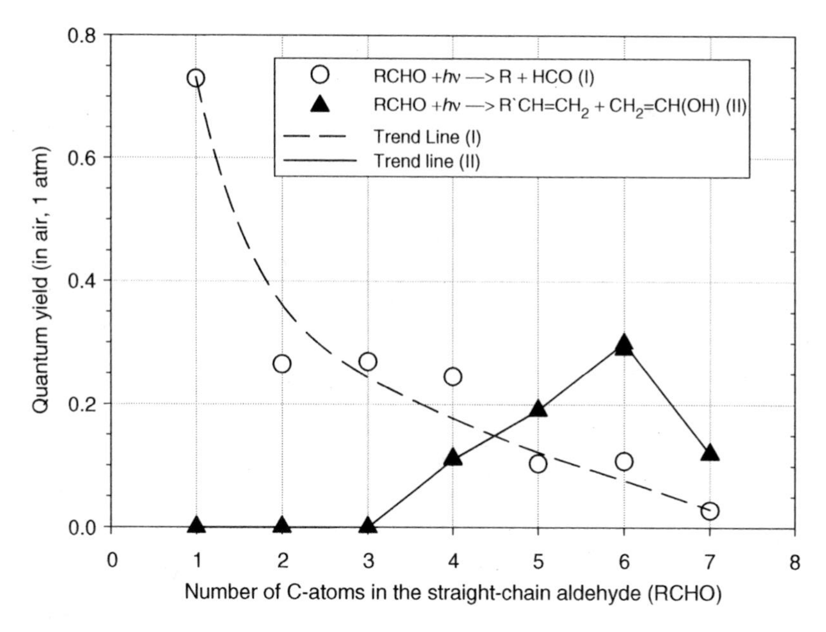

Figure VII-B-36. Plot of the quantum yields of the primary processes for photolysis in air (1 atm) at 313 nm for a series of straight-chain aldehydes (RCHO), where the number of carbon atoms in R is varied from 1 to 7.

forms an excited singlet state (S_1) via an $n \rightarrow \pi^*$ electronic transition. The origin of this transition in acetone is near 328.6 nm. Features of the photodissociation have been elucidated in theoretical studies by Sakurai and Kato (1999) and Diau et al. (2001). The S_1 state is short lived. It rapidly undergoes intersystem crossing to form the triplet state ($S_1 \rightarrow T_1$) or internal conversion to a highly vibrationally excited ground state (S_0*). Many excellent studies of acetone photochemistry have established an important role for the first excited triple in the gas phase photolysis of acetone, particularly at longer wavelengths of absorbed light; [e.g., see Cundall and Davies (1967); Gandini and Hackett (1977a)]. In experiments using pure acetone, photodecomposition can occur from S_1, T_1, or S_0* by several primary photodecomposition pathways:

$$CH_3C(O)CH_3 + h\nu \rightarrow CH_3CO + CH_3 \quad \text{(I)}$$

$$\rightarrow CH_3 + (CH_3CO*) \rightarrow 2CH_3 + CO \quad \text{(II)}$$

$$\rightarrow H + CH_2C(O)CH_3 \quad \text{(III)}$$

$$\rightarrow CH_4 + CH_2{=}CO \quad \text{(IV)}$$

Primary process (I), a "Norrish Type I" photodissociation, results in an α-C—C bond cleavage. Diau et al. (2001) gave a detailed picture of the mechanism of this process in acetone and other simple ketones. Upon excitation to the S_1 state, the $S_1 \rightarrow T_1$ channel of reaction is expected to dominate for excitation below the S_1 bond dissociation threshold (~260 nm). There is an energy barrier along the α-C—C bond-breaking reaction coordinate of the T_1 surface (~75 kJ mol^{-1}) of the excited molecule that prolongs the lifetime of the molecule excited at longer wavelengths. A sudden decrease in lifetime of the fluorescence occurs for excitation at wavelengths in the 313–306 nm range when dissociation becomes energetically feasible. For excitation at shorter wavelengths, the S_1 lifetime decreases smoothly from 2.1 ns at 295 nm to < 1.6 ns at 260 nm. Calculations of Diau et al. (2001) indicate that direct bond dissociation on the S_1 surface must overcome an energy barrier of ~84 kJ mol^{-1}. Direct bond cleavage from the S_1 surface may compete with intersystem crossing at shorter wavelengths ($\lambda < 260$ nm). Processes (I) and (II) are relevant within the troposphere. Processes (III) and (IV) occur only in vacuum UV radiation that is absent in the lower atmosphere.

VII-C-1.1. Absorption Cross Sections of Acetone

The recommended ultraviolet absorption cross sections for acetone are shown in figure VII-C-1, where they can be compared with those for some other simple ketones (Martinez et al., 1992). The agreement of the cross sections measured by Martinez et al. (1992) and Gierczak et al (1998) can be seen by comparison of appropriate columns in table VII-C-1.

VII-C-1.2. Quantum Yields of the Primary Processes in the Photodecomposition of Acetone in Air

Of primary importance in the lower atmosphere are processes (I) and (II) that result in the formation of methyl and acetyl radicals. The CH_3CO^* species has been suggested to

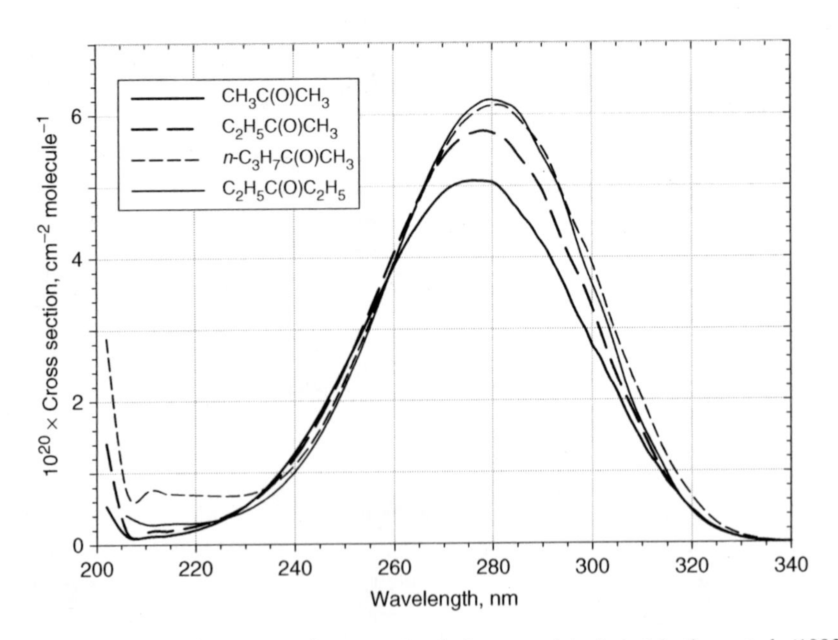

Figure VII-C-1. Absorption spectra for some simple ketones; data from Martinez et al. (1992).

Table VII-C-1. Absorption cross sections (σ, cm^2 $molecule^{-1}$) at 298 K for the simple ketones: acetone [$CH_3C(O)CH_3$] and 2-butanone [$CH_3C(O)C_2H_5$]

	$10^{20} \times \sigma$				$10^{20} \times \sigma$		
λ (nm)	Acetone[a]	Acetone[b]	2-Butanone[c]	λ (nm)	Acetone[a]	Acetone[b]	2-Butanone[c]
278	5.07	4.94	5.77	315	0.858	0.837	0.896
279	5.06	4.92	5.76	316	0.777	0.760	0.794
280	5.05	4.91	5.74	317	0.699	0.684	0.697
281	5.01	4.86	5.72	318	0.608	0.598	0.611
282	4.94	4.79	5.68	319	0.530	0.523	0.531
283	4.86	4.71	5.62	320	0.467	0.455	0.457
284	4.76	4.62	5.54	321	0.407	0.411	0.389
285	4.68	4.54	5.44	322	0.344	0.348	0.328
286	4.58	4.44	5.35	323	0.287	0.294	0.276
287	4.50	4.36	5.26	324	0.243	0.248	0.229
288	4.41	4.28	5.17	325	0.205	0.210	0.189
289	4.29	4.15	5.06	326	0.168	0.174	0.156
290	4.19	4.06	4.94	327	0.135	0.141	0.129
291	4.08	3.95	4.78	328	0.108	0.113	0.105
292	3.94	3.82	4.60	329	0.086	0.091	0.085
293	3.81	3.71	4.42	330	0.067	0.074	0.067
294	3.67	3.57	4.24	331	0.051	0.059	0.054
295	3.52	3.42	4.08	332	0.040	0.047	0.042
296	3.35	3.26	3.93	333	0.031	0.038	0.033
297	3.20	3.31	3.79	334	0.026	0.031	0.025
298	3.07	2.98	3.65	335	0.017	0.025	0.020
299	2.94	2.82	3.48	336	0.014	0.020	0.014
300	2.77	2.67	3.30	337	0.011	0.016	0.011
301	2.66	2.58	3.10	338	0.009	0.014	0.008
302	2.53	2.45	2.89	339	0.006	0.011	0.007
303	2.37	2.30	2.69	340	0.005	0.009	0.005
304	2.24	2.18	2.50	341	0.005	0.007	0.005
305	2.11	2.05	2.33	342	0.003	0.006	0.003
306	1.95	1.89	2.17	343	0.004	0.005	0.003
307	1.80	1.75	2.02	344	0.002	0.004	0.002
308	1.66	1.61	1.88	345	0.002	0.003	0.001
309	1.54	1.49	1.73	346	0.001	0.002	0.001
310	1.41	1.36	1.59	347	0.002	0.002	0.000
311	1.28	1.24	1.42	348	0.001	0.001	0.001
312	1.17	1.14	1.28	349	0.001	0.001	0
313	1.08	1.06	1.14	350	0.001		0
314	0.967	0.94	1.01				

[a] Acetone cross sections as reported by Martinez et al. (1992).
[b] Acetone cross sections as reported by Gierczak et al. (1998).
[c] 2-Butanone cross sections as reported by Martinez et al. (1992).

be an intermediate in process (II). Presumably it is a short-lived, vibrationally rich acetyl radical that dissociates readily to form a second methyl radical and carbon monoxide. The energy thresholds for processes (I) and (II) are 338 and 299 nm, respectively. Processes (III) and (IV) occur to a limited extent in photolysis in the vacuum UV. In the oxygen-free system, acetone photolysis at the longer wavelengths proceeds largely from the T_1 state that lies about 327 kJ mol^{-1} above S_0. However, a significant barrier to the dissociation has been observed experimentally (about 56 kJ mol^{-1}; Zuckermann et al., 1988); this is consistent with the barrier estimated theoretically to be ~75 kJ mol^{-1} by Diau et al. (2001).

In the presence of even small amounts of oxygen, the vibrationally equilibrated triplet state of acetone is quenched effectively; the rate constant for this reaction is large: $k \approx 1.5 \times 10^{-11}$ cm^3 molecule^{-1} s^{-1} (Costela et al., 1986). Presumably quenching by oxygen involves electronic energy transfer: $CH_3C(O)CH_3$ (T_1) + $O_2(^3\Sigma_g^-) \rightarrow$ $CH_3C(O)CH_3(S_0)$+ $O_2(^1\Delta_g)$. In the lower atmosphere, rich in oxygen, vibrationally equilibrated acetone triplets will be removed by oxygen quenching. Presumably dissociation occurs from the vibrationally excited T_1 or S_0 states formed initially. Excellent reviews of the photochemistry of acetone in air have been given by Horowitz (1991) and Emrich and Warneck (2000).

As with the photochemical studies of aldehydes described in section VII-B, most experiments with acetone have been made in the absence of oxygen. A few studies have been made under conditions relevant to acetone photodecomposition within the troposphere, that is, at low acetone concentrations in air at tropospheric pressures (Gardner et al., 1984; Meyrahn et al., 1986; Emrich and Warneck, 2000; Gierczak et al., 1998; Blitz et al., 2004). From the results of broad-band studies of acetone–air mixture photolysis, Gardner et al. (1984) reported that the quantum yields of CO_2 formation and acetone loss (~0.075 at 1 atm) were indistinguishable within the experimental uncertainties, and the ratio $-\Phi_{CH3C(O)CH3}/\Phi_{CO2} = 1.02 \pm 0.04$ (band near 313 nm). Quenching of acetone decomposition by O_2, N_2, and air was observed by Gardner et al. (1984), but the complex form of the quenching reported by Gardner et al. did not follow the Stern–Volmer mechanism, and the quantum yields of products formed at a given pressure of reactants were relatively invariant with wavelength. Both these latter observations differ from results from subsequent studies and are now considered to be in error.

The quenching of electronically excited acetone molecules by collisional energy transfer with air molecules is less efficient than electronic energy transfer to acetone, 2,3-butanedione, and other molecules with triplet-state energies lower than that of acetone. Emrich and Warneck (2000) and Gierczak et al. (1998) demonstrated clearly that collisional deactivation of electronically excited acetone by air follows a Stern–Volmer mechanism. A semilogarithmic plot of the quenching constants (k_q/k_d) vs. $1/\lambda$ is given in figure VII-C-2. The constants follow the trend observed for the light aldehydes (figure VII-B-14b); they decrease regularly as the excitation energy is increased and the lifetime of the excited states decreases.

(i) Primary process (I): $CH_3C(O)CH_3$+ $h\nu$ $\rightarrow$ CH_3+ CH_3CO. Several experimental approaches have been used to define the quantum yields of this process in air. In the studies of Meyrahn et al. (1986) and Emrich and Warneck (2000) dilute mixtures of acetone and NO_2 in air were photolyzed over a wide range of wavelengths (250–330 nm) and different pressures of air. The acetyl radical formed in process (I) reacts rapidly with oxygen in the air-rich mixture, $CH_3CO + O_2 \rightarrow$ $CH_3C(O)O_2$; in the NO_2 containing system used by Emrich and Warneck, the

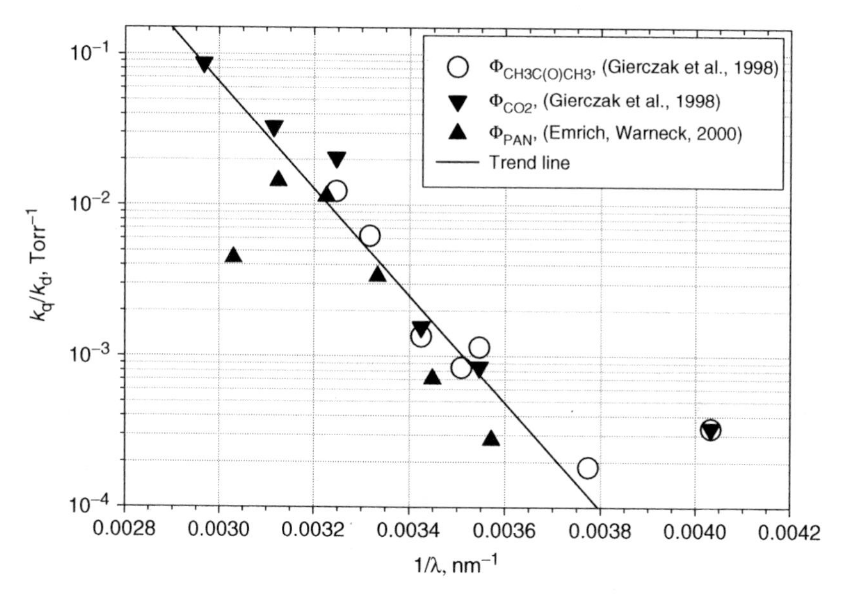

Figure VII-C-2. Semilog plot of Stern–Volmer constants (k_q/k_d) vs. $1/\lambda$ for quenching of excited acetone by collisional energy transfer with air molecules. The line is the least-squares fit to the data excluding the divergent points at $1/\lambda = 0.00403$ and $0.00303\,\text{nm}^{-1}$: $\log_{10}(k_q/k_d) = 9.448 - 3544.3\,(1/\lambda)$.

acetylperoxy radical was captured effectively by NO_2 to form peroxyacetyl nitrate (PAN), $CH_3C(O)O_2 + NO_2 \rightarrow CH_3C(O)OONO_2$. Although PAN is thermally unstable, its decomposition in the gas phase is relatively unimportant at 25°C on the time-scale of the experiments. Since the "hot" acetyl radical is believed to have only a transient lifetime, PAN derived from it is unimportant, and the quantum yields of PAN formation refer only to process (I). However, process (II) is monitored by the formation of CO. In both the studies of Meyrahn et al. (1986) and Emrich and Warneck (2000), the quantum yields of PAN were found to decrease with increasing pressure of air in accordance with the Stern–Volmer quenching mechanism. Shown in figure VII-C-3 are the results of ϕ_I determinations (Φ_{PAN}) by Meyrahn et al. (1986) and Emrich and Warneck (2000), black squares and diamonds, respectively.

Gierczak et al. (1998) also studied the photodecomposition of acetone in dilute mixtures in air in experiments at discrete wavelengths between 248 and 337 nm and at several pressures of air. In these studies the quantum yields of $CH_3C(O)CH_3$ loss and CO_2 formation were monitored. The formation of CO_2 results from CH_3CO radical generation in process (I) and the reaction sequence: $CH_3CO + O_2 \rightarrow CH_3C(O)O_2$; $CH_3C(O)O_2 + CH_3C(O)O_2 \rightarrow 2CH_3CO_2 + O_2$; $CH_3C(O)O_2 + CH_3O_2 \rightarrow CH_3CO_2 + CH_3O + O_2$; $CH_3CO_2 \rightarrow CH_3 + CO_2$. In experiments in which both $-\Phi_{CH3C(O)CH3}$ and Φ_{CO2} could be measured, the authors observed a near equality between these quantities, and both were used as a measure of process (I). However, in view of other evidence of the significant occurrence of process (II), particularly at

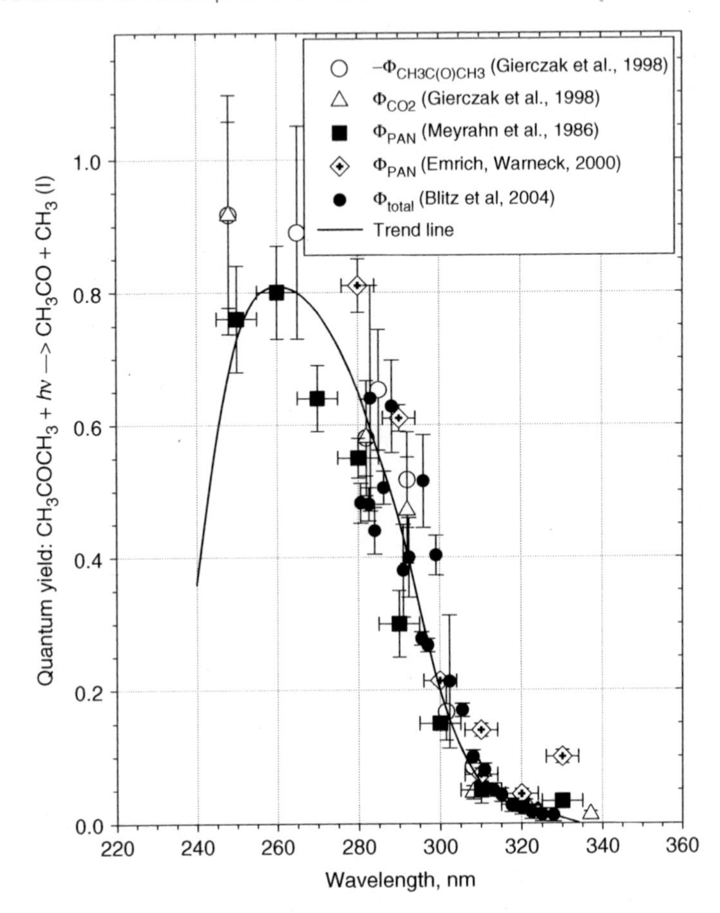

Figure VII-C-3. Experimental determinations of the quantum yield of process (I) in acetone in air (760 Torr) as a function of wavelength.

the shorter wavelengths, it seems more appropriate to use the CO_2 product alone as a monitor of process (I). The acetone loss should reflect the sum of processes (I) and (II) plus any other decomposition pathways. The values of Φ_{CO2} are shown in figure VII-C-3 by open triangles, while $-\Phi_{CH3C(O)CH3}$ values are shown by open circles. For most wavelengths the open circles lie above the trend line in the figure, while the open triangles lie near the line. Only at a wavelength of 248 nm do the two measurements actually coincide. In experiments with very small extent of conversion of a reactant, the measurements of the reactant loss are much less accurate than those of a product formed, since the former depends on the difference between two very similar numbers.

Blitz et al. (2004) have utilized the unique OH radical marker for the CH_3CO radical to determine the quantum yields of photodissociation of acetone as a function of wavelength (279–327.5 nm), temperature (218–295 K), and pressure of added gases (He, 4.5–4050 Torr; N_2, 4.5–998 Torr; and air, 4.5–600 Torr). In all experiments with added N_2 or He, O_2 was added (0.098 Torr) to ensure the production of OH radicals from the reaction of O_2 with CH_3CO (Tyndall et al., 1997b):

$$CH_3CO + O_2 \longrightarrow CH_3C(O)O_2^* \xrightarrow[(a)]{(+M)} CH_3COO_2$$
$$CH_3C(O)O_2^* \xrightarrow{(b)} OH + Product$$

Tyndall et al. (1997b) estimate that $k_b/(k_a + k_b) \approx 0.5$ at 6 Torr pressure and 0.05 at 100 Torr of added gases (N_2, O_2). Although the yield of $CH_3C(O)O_2$ is essentially 100% at the pressures of the troposphere, the small amount of OH radical formed from the $CH_3CO + O_2$ reaction can be detected and was used to monitor the formation of CH_3CO radical in the experiments of Blitz et al. (2004). Their quantum yield estimates for 1 atm and 295 K are shown as the small black circles in figure VII-C-3. Blitz et al. (2004) present a parameterization of their data for process (I) in acetone photolysis as a function of wavelength, pressure, and temperature, but the precision of the measurements can be judged best from the individual measurements. The original data were reproduced from an enlarged view of the data presented in figure 3 of Blitz et al. (2004).

Although there is considerable scatter of the data in figure VII-C-3, the trend of ϕ_I with wavelengths in the 340–260 nm range is clear. The sharp decline in ϕ_I suggested by the trend line at wavelengths below 260 nm may or may not reflect the real trend accurately, but it has been drawn to reflect our acceptance of the significant occurrence of process (II) at short wavelengths and the requirement that $\phi_I + \phi_{II} \leq 1.0$.

The wavelength at which process (I) approaches zero is difficult to determine from the very small product yields. However, an estimate can be made from the data of figure VII-C-4 using the long-wavelength data of Gierczak et al. (1998), Meyrahn et al. (1986), and Blitz et al. (2004) that have been plotted vs. $1/\lambda$. The least-squares lines from each of the data sets give the origin of primary process (I) between 328 and 344 nm. Use of all of the data sets gives an intercept of 334 nm. The wavelength range is consistent with the necessary energy for the occurrence of process (I) that is available in a quantum of radiation with $\lambda \leq 338$ nm. Our recommended quantum yields for process (I) are given in table VII-C-2.

(ii) Primary process (II): $CH_3C(O)CH_3 + h\nu \rightarrow CH_3 + (CH_3CO^*) \rightarrow 2CH_3 + CO$. Carbon monoxide quantum yields from acetone photolysis in the presence of radical scavengers have been interpreted as arising from process (II). However, Φ_{CO} data from photolysis at the longer wavelengths may be an inaccurate measure of this process. A correction should be made for CO generation from the photolysis of formaldehyde, a secondary product of the oxidation of the primary CH_3 radical oxidation. In the experiments of Meyrahn et al. (1986), values of Φ_{CO}were determined. Their estimates of ϕ_{II}, shown in figure VII-C-5 by the open squares, are probably upper limits at the longer wavelengths since, as Meyrahn et al. suggest, CH_2O photolysis could

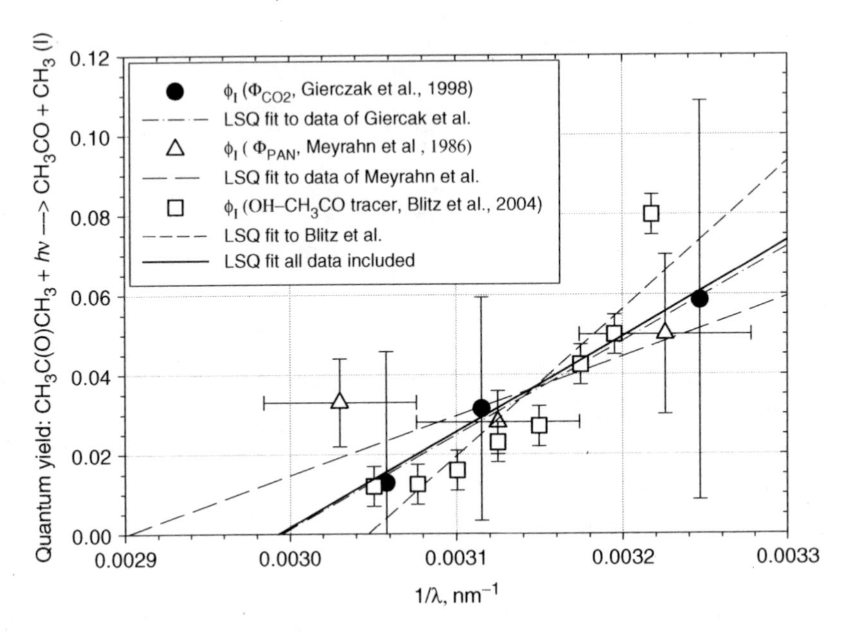

Figure VII-C-4. Plot of the estimates of ϕ_I from photolysis of acetone in air (or N_2) at 760 Torr vs. $1/\lambda$ for data at long wavelengths. The least-squares lines are fits for each individual data set and combined data. The y-axis intercepts are: 344 nm, Meyrahn et al. (1986); 334 nm, Gierczak et al. (1998); 334 nm (all data included); and 328 nm, Blitz et al. (2004). Although the range of intercepts is large, the estimates are consistent with the wavelength of minimum energy for the occurrence of process (I), 338 nm.

contribute to CO formation. Similar limited data from acetone–oxygen mixture photolysis by Pearson (1963) and Kirk and Porter (1962) can be used to estimate ϕ_{II} values which are shown in figure VII-C-5 as inverted open triangles and black circles, respectively.

Horowitz (1991) studied the quantum yields of CH_4 and CO formation in HCl-inhibited photolysis of acetone at 285, 300, and 313 nm and 25°C. HCl is an effective trap for CH_3 radicals, and from the dependence of Φ_{CH4} and Φ_{CO} on the pressures of acetone and added CO_2 an evaluation of the primary photodissociation processes was possible. Values of ϕ_I and ϕ_{II} at zero-pressure were estimated; ϕ_I and ϕ_{II}, respectively, were: 0.90 and 0.10 at 285 nm; 0.94 and 0.06 at 300 nm; 0.28 and 0.03 at 313 nm. Quantum yields of process II were found to be insensitive to the pressure of added acetone (2–80 Torr), while those of process (I) were sensitive to both the pressure of acetone and added CO_2. Horowitz suggested that process (II) originates from high vibrational levels of the S_1 state. His data are shown as open circles in figure VII-C-5.

North et al. (1995) studied photofragmentation of acetone excited at 248 and 193 nm. They estimated that at 248 nm a substantial fraction (30 ± 4%) of the nascent acetyl radicals contains sufficient energy to undergo spontaneous secondary decomposition. The data point is shown as a filled square in figure VII-C-5. They reported that, at 193 nm, the only observed dissociation pathway led to the formation of two methyl radicals and carbon monoxide.

Table VII-C-2. Recommended quantum yields for primary process $CH_3C(O)CH_3 + h\nu \rightarrow CH_3CO + CH_3$ (I) and primary process $CH_3C(O)CH_3 + h\nu \rightarrow (CH_3CO^* + CH_3) \rightarrow 2CH_3 + CO$ (II), in air (760 Torr) as a function of wavelength; see section VII-C-1.3 for the conditions assumed for the alternative cases a, b, and c

λ (nm)	ϕ_I	ϕ_{II} (case a)	ϕ_{II} (case b)	ϕ_{II} (case c)	λ (nm)	ϕ_I	ϕ_{II} (case a)	ϕ_{II} (case b)
278	0.680	0.082	0.068	0.048	310	0.060	0.026	0.004
279	0.665	0.078	0.064	0.046	311	0.056	0.025	0.002
280	0.650	0.075	0.060	0.044	312	0.052	0.024	0.001
281	0.631	0.072	0.057	0.042	313	0.048	0.024	0
282	0.612	0.070	0.054	0.041	314	0.044	0.023	0
283	0.583	0.067	0.048	0.037	315	0.040	0.022	0
284	0.574	0.065	0.045	0.035	316	0.038	0.021	0
285	0.555	0.063	0.042	0.032	317	0.036	0.021	0
286	0.535	0.061	0.041	0.030	318	0.034	0.020	0
287	0.515	0.059	0.040	0.028	319	0.032	0.019	0
288	0.496	0.057	0.039	0.026	320	0.030	0.019	0
289	0.475	0.055	0.037	0.024	321	0.027	0.018	0
290	0.456	0.053	0.036	0.022	322	0.025	0.018	0
291	0.430	0.051	0.034	0.020	323	0.023	0.017	0
292	0.404	0.049	0.033	0.018	324	0.021	0.017	0
293	0.377	0.047	0.031	0.016	325	0.019	0.016	0
294	0.351	0.045	0.030	0.014	326	0.017	0.016	0
295	0.325	0.044	0.028	0.012	327	0.014	0.015	0
296	0.300	0.042	0.027	0.010	328	0.012	0.015	0
297	0.275	0.041	0.025	0.007	329	0.010	0.014	0
298	0.250	0.039	0.024	0.005	330	0.008	0.014	0
299	0.225	0.038	0.022	0.003	331	0.006	0.014	0
300	0.200	0.037	0.021	0	332	0.004	0.013	0
301	0.183	0.035	0.019	0	333	0.002	0.013	0
302	0.166	0.034	0.018	0	334	0	0.012	0
303	0.148	0.033	0.016	0	335	0	0.012	0
304	0.131	0.032	0.015	0	336	0	0.012	0
305	0.119	0.031	0.013	0	337	0	0.011	0
306	0.103	0.030	0.011	0	338	0	0.011	0
307	0.092	0.029	0.010	0	339	0	0.011	0
308	0.082	0.028	0.008	0	340	0	0.011	0
309	0.071	0.027	0.006	0				0

Gandini and Hackett (1977a) studied the quantum yields of CO and 2,3-butanedione products of broad-band acetone photolysis at 15 different wavelength regions (240–305 nm) and an acetone pressure of 13.5 Torr. These data, coupled with the Horowitz observations of the insensitivity of Φ_{CO} to the experimental variables, suggest that the CO formed in the Gandini and Hackett experiments may arise from process (II). On this assumption, the Gandini and Hackett data are plotted in figure VII-C-5 as inverted black triangles. All the data from the variety of experiments appear to form a consistent pattern, with increasing importance of ϕ_{II} with increase in energy of the quantum of light absorbed. The data in figure VII-C-5 form the basis for the values of ϕ_{II} recommended in this study.

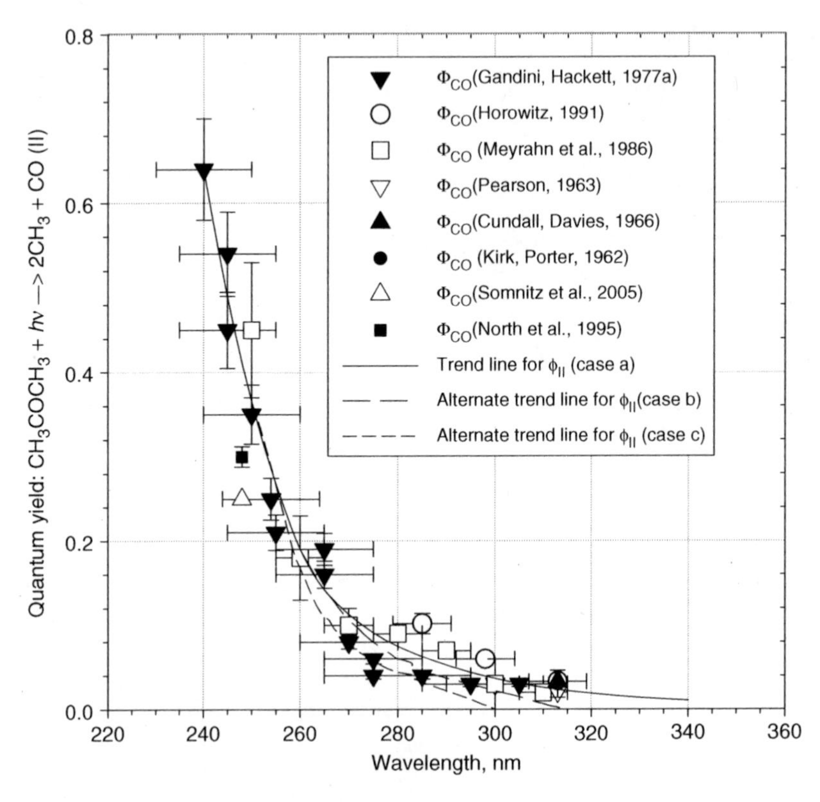

Figure VII-C-5. Plot of experimental estimates of process (II) in acetone–air mixture photolysis as a function of wavelength. The three different trend lines represent different assumptions about the poorly defined cutoff wavelength for process (II). See the text.

Somnitz et al. (2005) reported that the quantum yield of CO from acetone photolysis at 248 nm and 298 K shows a significant dependence on total pressure: $\Phi_{CO} = 0.45$ at 15 Torr of N_2; $\Phi_{CO} = 0.25$ at 675 Torr of N_2. They support the view that formation of CO via a direct and /or concerted channel is unimportant at 248 nm. Their data point for the higher pressure is shown as an open triangle in figure VIII-C-5. These observations may have implications for the interpretation of the data of Blitz at al. (2004) that appears to be based on an assumption of a pressure-independent CO yield at 248 nm.

Consistent with the trend of ϕ_{II} with λ observed in figure VII-C-5, Lightfoot et al. (1988) report that 95% of acetone photodissociation at 193.3 nm occurs by process (II), and they suggest this system is a clean source of CH_3 radicals. Woodbridge et al. (1988) studied the dynamics of process (II) at 193 nm and observed infrared emission from highly vibrationally and rotationally excited CO product. The observed vibrational and rotational distributions fit those expected for temperatures of 2030 K and 3360 K, respectively. The distribution of the rotational and the vibrational excitation can be modeled using a pure impulsive mechanism for the fragmentation of an initially bent acetyl fragment. Woodbridge et al. (1988) point out that these observations support previous arguments that the dissociation occurs by a two-step fragmentation mechanism following single photon excitation.

From the limiting value of Φ_{CO} at high 2-butene concentrations, the results of Cundall and Davies (1966) can be used to derive $\phi_{II} \approx 0.033$ at 313 nm at 20°C (although these authors do not suggest process (II) as the mechanism for CO formation). This datum point is shown as an upright filled triangle in figure VII-C-5.

In the experiments of Meyrahn et al. (1986), carried out in one atmosphere of air, vibrationally equilibrated acetone molecules in the T_1 state must be quenched efficiently by oxygen, and these cannot be the source of CO. The consistency of the Meyrahn Φ_{CO} data with those of other studies carried out in the absence of oxygen suggests that quenching of the parent state of process (II) by air is relatively unimportant, and, to a first approximation, the Φ_{CO} data seen in figure VII-C-5 can be attributed to process (II). One obvious problem with this interpretation is the apparent continuation of the process beyond the expected cut-off near 299 nm. Small quantum yields of CO found in experiments beyond the cut-off may result in part from the photolysis of the formaldehyde product in acetone–air mixtures, as noted previously in this section and by Meyrahn et al. (1986) and Emrich and Warneck (2000). However, Horowitz (1991) and Cundall and Davies (1966) observed CO formation with all CH_3 radicals removed effectively and with no oxygen present, conditions under which formaldehyde could not be formed. If CO formation from acetone does continue well beyond the thermodynamic threshold, then dissociation of the "hot" acetyl radical through the use of internal rovibrational energy as well as collision-assisted dissociation is required. This mechanism accounts for the photodissociation of NO_2, which has been observed to extend well beyond its energy threshold near 398 nm to wavelengths greater than 411 nm (Roehl et al., 1994).

Another apparent problem with the current interpretation of CO origin in process (II) is the magnitude of the $\phi_I + \phi_{II}$ sum at short wavelengths. If the Gierczak et al. (1998) data for ϕ_I at 248 nm are accurate as well as the estimates of Gandini and Hackett (1977a) at 240–260 nm that are attributed to process (II), then the estimated $\phi_I + \phi_{II}$ sum becomes somewhat greater than unity in the 240–258 nm range. The falloff in ϕ_I that we have assumed to occur for $\lambda < 260$ nm is suggested to overcome this problem. Although the single point of Meyrahn et al. (1986) at $\lambda = 250$ nm (see figure VII-C-3) supports this possibility, its reality is unproven experimentally.

Acetone–iodine mixture photolysis at 313 nm and 80–100°C by Pitts and Blacet (1952) gave very small quantum yields of CO (0.0015; 0.003), while at 265.4 nm $\Phi_{CO} = 0.057$ was observed. CH_3I quantum yields varied with temperature: 0.12 (80°C); 0.17 (100°C); 0.28 (177°C). Martin and Sutton (1952) carried out a similar study, using acetone mixtures with radioactive iodine. Photolyses were made at 313 nm, 100 and 120°C, and a range of I_2 and $CH_3C(O)CH_3$ pressures. Again, small quantum yields of CO were found (0.0084–0.02) as well as significant quantum yields of CH_3I. These results appear to support the trend of case (c) in figure VII-C-5 (see discussion in section VII-C-1.3). Recommended values of ϕ_{II} for acetone are given in table VII-C-2. In view of the uncertainties in the long-wavelength tail, we choose the three cases given in figure VII-C-5 as alternative recommendations for tropospheric modelers. In cases (a), (b) and (c) the cut-off occurs at 340, 315, and 300 nm, respectively. Cases (a) and (c) were chosen to represent the extreme cases, whereas case (b) is approximately midway between the extremes and is perhaps the best choice. Values of ϕ_{II} from cases (a), (b), and (c) are shown in figure VII-C-5 and summarized in table VII-C-2.

(iii) Minor processes in acetone photodecomposition. Primary process (III), H-atom formation, has been observed by Takahashi et al. (2004) and Lightfoot et al. (1988) in photolysis of acetone at 193 nm. Lightfoot et al. reported that 3% of acetone photodecomposition at this wavelength proceeds by process (III). Takahashi et al. report a similar finding: $\phi_{III} = 0.039 \pm 0.006$, and the authors report that the nascent kinetic energy of the H-atoms is characterized by a Maxwell–Boltzmann function with a temperature of 5000 K. Primary process (IV) accounts for $\leq 2\%$ of acetone photodecomposition in 193 nm photolysis by Lightfoot et al. (1988).

VII-C-1.3. Photolysis Frequencies for Acetone

Cloudless-sky *j*-values for acetone photodecomposition in the lower troposphere were derived from current recommendations for cross sections and quantum yields (see tables VII-C-1 and VII-C-2). These are given in figure VII-C-6a as a function of solar zenith angle for overhead ozone columns of 300, 350, and 400 DU. Values are shown for primary process (I) (heavy lines), and for the several scenarios given for estimation

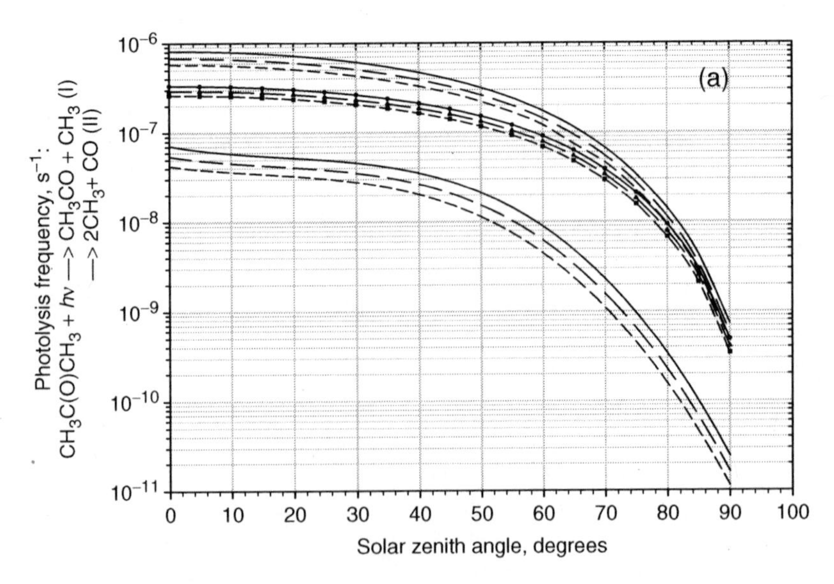

Figure VII-C-6a. Plot of the estimated *j*-values for processes (I) and (II) of acetone for a cloudless day within the lower troposphere at 760 Torr. Process (I), heavy curves; process (II), case (a), light curves with symbols; process (II), case (b), light curves, no symbols. The solid lines, long-dashed lines, and short-dashed lines, respectively, are for overhead ozone columns of 300, 350, and 400 Dobson units. The use of the assumptions given for case (c), curves not shown, gave *j*-values for process (II) that were less than 1/100th of those for case (a).

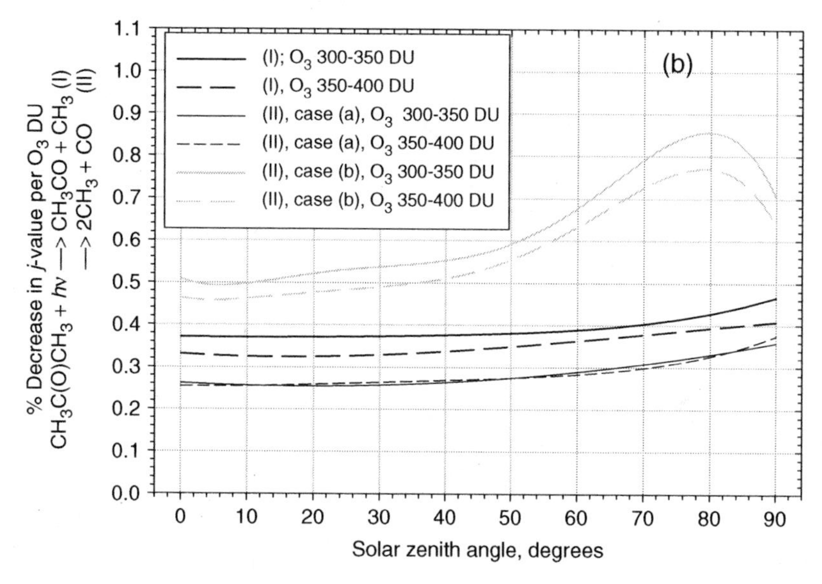

Figure VII-C-6b. Percentage decrease in the j-values for acetone for data presented in figure VII-C-6a for overhead ozone columns of 300–350 and 350–400 Dobson units.

of primary process (II) quantum yields (shown by the light lines). Significant quenching of the dissociation of acetone by process (I) occurs in air at 1 atm, whereas the quantum yields for process (II) are little affected. The estimated j-values for process (II), as calculated using ϕ_{II} estimates based on the solid lines in figure VII-C-5 [case (a), cut-off at 340 nm] are given by light lines marked with symbols in figure VII-C-6a. With this choice of λ-dependence, the predicted fraction of acetone decomposition that occurs by process (II) is significant. However, the j_{II} values are sensitive to the measured magnitude of Φ_{CO} that determines the long-wavelength tail shown for ϕ_{II} in figure VII-C-5. If the alternative descriptions of this long wavelength region shown by the two dashed curves in figure VII-C-5 are used, the estimated j-values are significantly lower. Using the data from case (b), (cut-off wavelength at 315 nm), the j_{II} values given by the light lines near the bottom of figure VII-C-6a are obtained; j-values for process (II) are decreased by about a factor of about 2.4 from those of scenario (a). If one accepts the cut-off to be at the energy limit near 300 nm, case (c), the j-values are decreased by over a factor of 100 from those observed in case (a); these are not shown in the figure. Obviously, additional accurate determinations of Φ_{CO} in acetone–air mixtures at the longer wavelengths are required to evaluate properly the influence of process (II) in tropospheric photodecomposition of acetone.

In figure VII-C-6b the percentage decrease in the j-values is given for increases in the overhead ozone column. Note that the response of the acetone j-values to ozone column changes is much greater than was seen with formaldehyde (figure VII-B-5b). This reflects the greater influence of the longer wavelengths of sunlight absorbed by formaldehyde that are least affected by ozone absorption. Compare

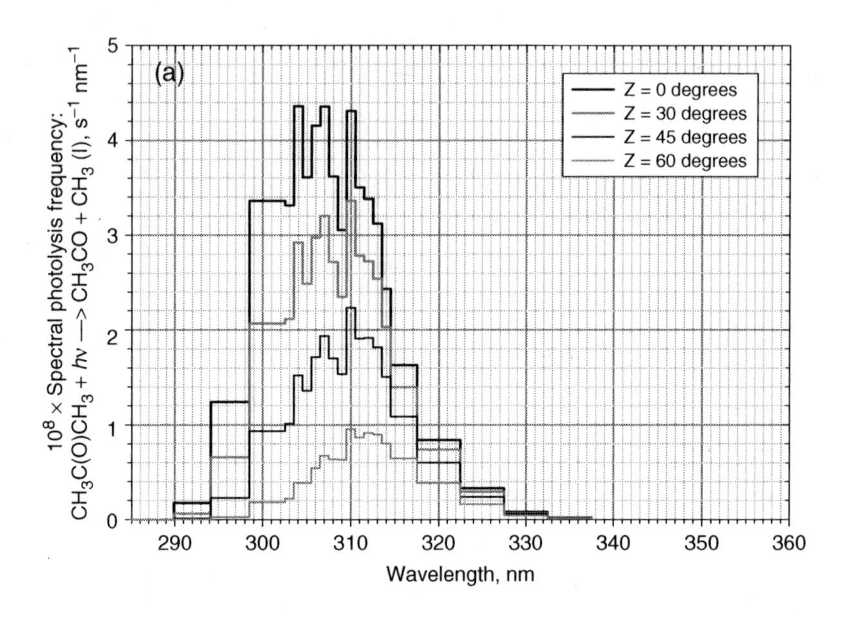

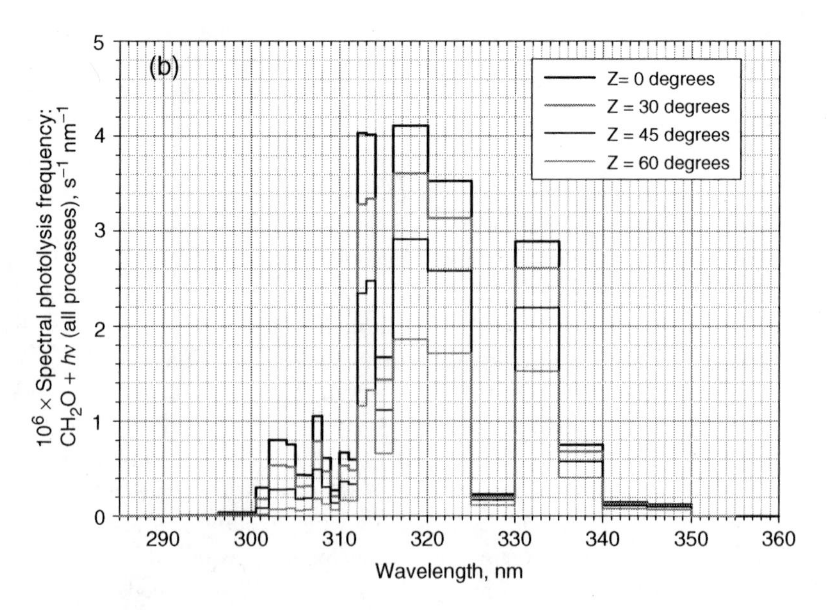

Figure VII-C-7. Comparison of the spectral j-values for acetone (a), formaldehyde (b), and acetaldehyde (c); note the y-scale of the acetone is much smaller than that for formaldehyde and acetaldehyde, and the wavelength response for acetone is shifted to shorter wavelengths than that for the aldehydes. An overhead ozone column of 350 DU was assigned in the calculations.

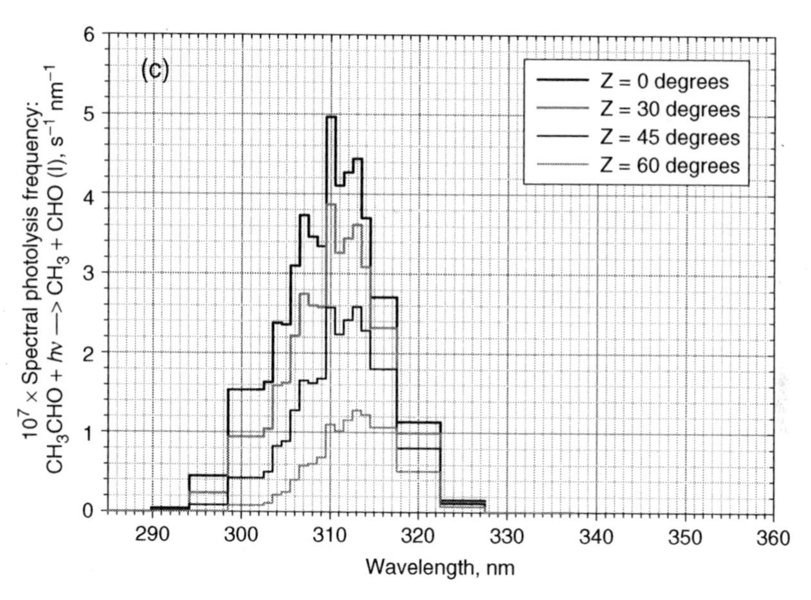

Figure VII-C-7. *(continued)*

the spectral j-values of acetone with those for formaldehyde and acetaldehyde in figure VII-C-7.

Warneck (2001) presented an algorithm for the quantum yield of photodissociation of acetone as a function of air pressure that allows a view of the large effects of lowered pressure at higher altitudes. The data used in the algorithm development included the Φ_{CO2}, $-\Phi_{CH3C(O)CH3}$, and Φ_{PAN} measurements as monitors of (I); process (II) was not considered. The curves calculated from Warneck's algorithm are given in figure VII-C-8. The large increase in quantum yields for acetone photodissociation at higher altitudes results from the lower pressures and large increase in short-wavelength radiation present. The j-values for acetone increase dramatically at the high altitudes from the relatively small values predicted for the lower troposphere. McKeen et al. (1997) have considered these effects on the photodissociation coefficients and have found that acetone photolysis can lead to significant source of odd-hydrogen radicals in the upper troposphere. However, these conclusions may be altered if the effect of temperature on the quantum yield of photodissociation observed by Blitz et al. (2004) is correct. The results of Gierczak et al. (1998) indicate the quantum yield of acetone photodecomposition to be nearly independent of temperature (298–195 K). On the other hand, the results of Blitz et al. (2004) show a significant decrease in the quantum yield with a decrease in temperature, about 80 to 90% at the temperatures of the cold upper troposphere. This difference in results remains unresolved at this writing. If the Blitz et al. (2004) results are correct, then the atmospheric lifetime of acetone increases from 22 to 35 days, with an increase in the global burden from 2.6 to 4.1 Tg (Arnold et al., 2005).

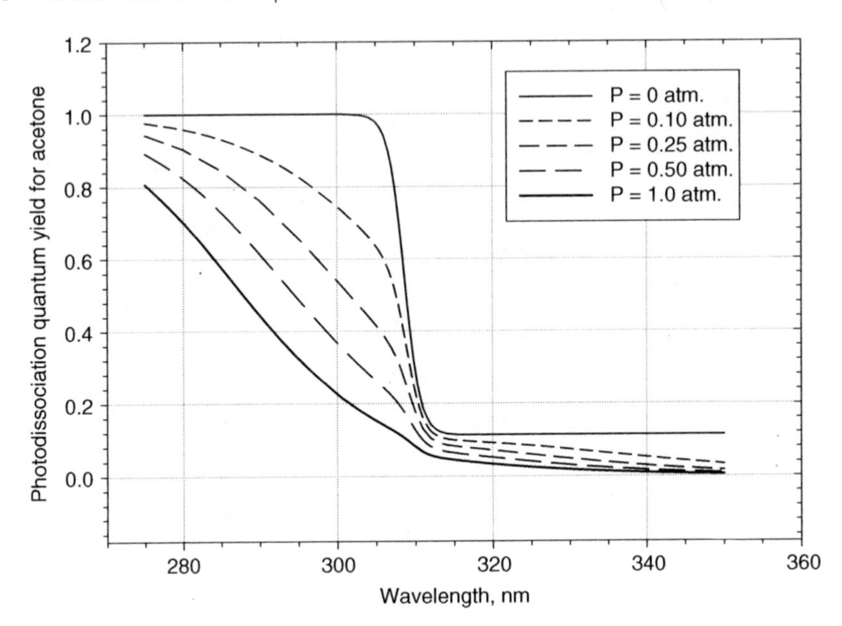

Figure VII-C-8. The effect of pressure of air in the atmosphere (resulting from an altitude change) on the quantum yield of photodecomposition of acetone, as estimated from the algorithm of Warneck (2001).

VII-C-2. 2-Butanone [Methyl Ethyl Ketone, $CH_3C(O)CH_2CH_3$]

Three different photodecomposition pathways have been considered for 2-butanone for excitation in its $n \rightarrow \pi^*$ absorption band:

$$CH_3C(O)C_2H_5 + h\nu \rightarrow CH_3CO + C_2H_5 \quad \text{(I)}$$

$$\rightarrow CH_3 + C_2H_5CO \quad \text{(II)}$$

$$\rightarrow C_2H_5 + (CH_3CO*) \rightarrow C_2H_5 + CH_3 + CO \quad \text{(III)}$$

VII-C-2.1. Absorption Cross Sections for 2-Butanone

The $n - \pi^*$ absorption band for 2-butanone is similar to that of other simple ketones; see figure VII-C-1. The maximum value of the cross section is intermediate between those for acetone and 3-pentanone. The cross sections of 2-butanone as determined by Martinez et al. (1992) are given in table VII-C-1.

VII-C-2.2. Quantum Yields of the Primary Processes in 2-Butanone

There have been very limited studies of the photodissociation modes of 2-butanone, and no studies have been reported of the wavelength dependence of the quantum yields of the individual processes for atmospheric conditions. As with acetone, at temperatures somewhat above 100°C the sum of the primary quantum yields for 313 nm excitation is near unity (Ells and Noyes, 1938; Lissi et al., 1973/1974). Although difference between the dissociation energies for the $CH_3CO—C_2H_5$ and the $CH_3—COC_2H_5$ bonds is relatively small (about 3.2 kJ mol^{-1}), the process (I) is highly favored. Photolysis of 2-butanone–iodine mixtures was carried out by Durham et al. (1949), Pitts and Blacet (1950), and Martin and Sutton (1952). Pitts and Blacet (1950) reported in a preliminary study that $\Phi_{C2H5I}/\Phi_{CH3I} \approx 40$ at 100°C and 21 at 175°C (pressures of ketone and iodine were 90 and 6.3 Torr, respectively). In these experiments $\Phi_{C2H5I} + \Phi_{CH3I} \approx 0.41$ and 0.45 at 100 and 175°C, respectively, and $\Phi_{CO} \approx 0.11$ at 175°C. In similar experiments at 313 nm, Martin and Sutton (1952) used radioactive iodine in mixtures with 2-butanone to attain high sensitivity in analyses of the iodide products at very small conversions (~0.1%). They reported: $\Phi_{C2H5I}/\Phi_{CH3I} \approx 19.3 \pm 0.8$ at 100°C (pressures of ketone and iodine were 88 and 1 Torr, respectively); $\Phi_{C2H5I} + \Phi_{CH3I} \approx 0.88$, and $\Phi_{CO} = 0.004$. Using a lower pressure of ketone (14.8 Torr) but the same I_2 pressure, they found $\Phi_{C2H5I}/\Phi_{CH3I} \approx 10.3$; $\Phi_{C2H5I} + \Phi_{CH3I} \approx 1.07$, and $\Phi_{CO} \approx 0.096$. Some formation of acetyl iodide was evident, but the analytical results for this compound were not reproducible. With excitation at shorter wavelengths, somewhat less discrimination in bond cleavage was found. At 265.4 nm excitation (100°C), Pitts and Blacet reported, $\Phi_{C2H5I}/\Phi_{CH3I} \approx 5.5$ and $\Phi_{C2H5I} + \Phi_{CH3I} \approx 0.52$. At 253.7 nm, Martin and Sutton found $\Phi_{C2H5I}/\Phi_{CH3I} \approx 2.4$ and $\Phi_{CO}/\Phi_{CH3I} \approx 0.78$. The results are consistent with the dominant occurrence of process (I) but with process (II) becoming significant at the shorter wavelengths. They also suggest that fragmentation of the nascent acyl radicals that characterizes process (III) appears to occur, particularly at the short wavelengths.

Energy transfer to 2,3-butanedione from 2-butanone excited at 313 nm was studied by Lissi et al. (1973/1974) through observed phosphorescence of 2,3-butanedione triplets. They found that the quantum yield of triplet 2-butanone formation was near unity at high pressures (400 Torr) of added propane. The triplet lifetime was reported to be 1.3×10^{-5} and 1.8×10^{-5} s at 400 and 14 Torr, respectively. They reported $\Phi_{CO} = 0.9$ at 120°C, 100 Torr pressure, and low light intensity. From 2,3-butanedione photosensitized experiments, the rate coefficient for the photocleavage of 2-butanone triplet (400 Torr total pressure) as estimated by Abuin and Lissi (1975) is represented by the function:

$$\log_{10}(k_{decomp}) = (14.63 \pm 1.49) - (12.710 \pm 2150)/R(\text{kcalmol}^{-1}\text{K}^{-1})T$$

The photochemical study of 2-butanone by Raber and Moortgat (1996) is the most relevant to reactions expected for atmospheric conditions. They carried out broad-band photolysis using TL12 lamps (see figure VII-B-18). Dilute mixtures of the ketone in air were photolyzed at several pressures (52–758 Torr). They determined the nature and the time dependence of the ratio of several products formed to the ketone photodecomposed using FTIR spectroscopy. Products observed were those expected with the dominance of process (I): CO_2, CH_3CHO, CH_2O, and smaller amounts of CH_3OH, $CH_3C(O)OOH$, $HC(O)OH$, and CO. The ratio $\Delta CO/\Delta CH_3C(O)C_2H_5$ showed a positive intercept at zero time that indicates that a fraction of the photodecomposition ($\leq 10\%$) forms CO directly in process (III) even in an atmosphere of air, as observed in acetone.

The quantum yields of 2-butanone that decomposed at the various pressures followed a Stern–Volmer quenching mechanism: $1/(\Phi_{CH3C(O)CH2CH3}) = 0.96 + 2.22 \times 10^{-3} P_{air}$ (Torr). This k_q/k_d value for 2-butanone with broad-band radiation is similar to that estimated for acetone and the simple aldehydes at the wavelength of maximum overlap of the lamp distribution and absorption by the ketone. The intercept of the Stern–Volmer relation is consistent with the observations of previous investigators that the quantum yield of decomposition of excited 2-butanone at zero-pressure is near unity (Lissi et al., 1973/1974).

In separate experiments using a different apparatus, Raber (1992) excited 2-butanone at 310 nm in air containing NO_2 scavenger. Although peroxyacetyl nitrate was the major product, methyl nitrate, ethyl nitrate, and peroxypropionyl nitrate were also detected. The presence of the latter compound confirms the occurrence of process (II) to a minor, but unquantified, extent.

VII-C-2.3. Photolysis Frequencies for 2-Butanone

Since no wavelength dependence of the product quantum yields has been determined, only the *j*-value that describes the quantum yield of photodecomposition of 2-butanone ($\phi_I + \phi_{II} + \phi_{III}$) is calculated here. We have assumed the average quantum yield value for 1 atm of air (0.34 ± 0.15), determined in the broad-band experiments of Raber and Moortgat (1996), applies to all wavelengths absorbed by 2-butanone. Until more detailed information becomes available, we suggest that modelers assume that 10% of the photodecomposition arises from process (III) as suggested from broad-band experiments of Raber and Moortgat. The calculated *j*-values for 2-butanone photodissociation for cloudless conditions in the lower troposphere (ozone column = 350 DU) are shown as a function of solar zenith angle in figure VII-C-9.

VII-C-3. 3-Pentanone [Diethyl Ketone, $CH_3CH_2C(O)CH_2CH_3$]

The primary photodecomposition modes of 3-pentanone follow the pattern of the smaller ketones:

$$C_2H_5C(O)C_2H_5 + h\nu \rightarrow C_2H_5CO + C_2H_5 \quad \text{(I)}$$

$$\rightarrow (C_2H_5CO^*) + C_2H_5 \rightarrow 2C_2H_5 + CO \quad \text{(II)}$$

VII-C-3.1. Absorption Cross Sections for 3-Pentanone

In figure VII-C-1 the absorption spectrum of 3-pentanone can be compared with those of some other simple ketones. The recommended values of the cross sections (Martinez et al., 1992) are given in tabular form in table VII-C-3.

VII-C-3.2. Quantum Yields of Photodecomposition of 3-Pentanone

Early studies employed partially filtered, full arc radiation of pure 3-pentanone. Reported quantum yields of photodecomposition and CO formation were near unity,

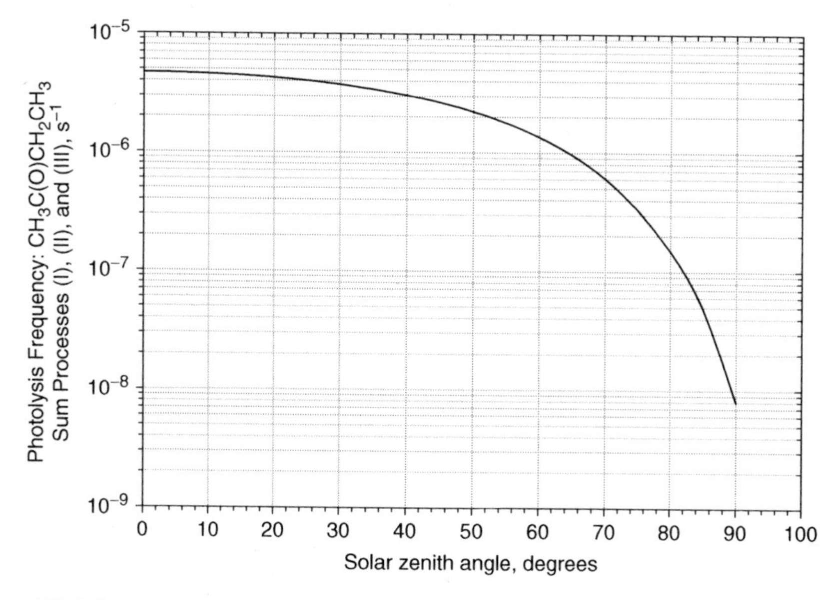

Figure VII-C-9. Photolysis frequencies for the photodecomposition of 2-butanone as a function of solar zenith angle for a cloudless day in the lower troposphere with an overhead ozone column = 350 DU.

even at relatively low temperatures. In experiments at 28 ± 1°C, Davis (1948) reported Φ_{CO} values of 1.03 ± 0.14, 0.99 ± 0.13, and 1.06 ± 0.14, while Dorfman and Sheldon (1949) found Φ_{CO} = 0.97 ± 0.10 at 56.5 ± 2°C and 1.09 ± 0.09 at 120 ± 2°C. It was suggested that the more efficient CO formation seen with 3-pentanone photolysis than with acetone for similar experimental conditions arose from the greater instability of the C_2H_5CO radical toward decomposition compared to CH_3CO. Thermal data for the decomposition reactions, $C_2H_5CO \rightarrow C_2H_5 + CO$ (1) and $CH_3CO \rightarrow CH_3 + CO$ (2), show only a small difference, $\Delta H_2 - \Delta H_1$, in the range 3.2 to 18 kJ mol^{-1} (Vollenweider and Paul, 1986; Kerr and Stocker, 2001/2002). Kutschke et al. (1952) reported Φ_{CO} = 0.62 ± 0.02 at 25°C and values near unity for temperatures ≥ 176°C, very similar to the acetone behavior.

Weir (1961) studied 3-pentanone–2,3-butanedione mixture photolyses at 313 and 253.7 nm. In experiments near 25°C and with excitation at 313 nm, Φ_{CO} was lowered from 0.72 ± 0.02 for pure 3-pentanone at 25 Torr, to 0.45 with 2.4 Torr of 2,3-butanedione addition, as strong emission from 2,3-butanedione triplets appeared. Weir found this same limiting yield of CO is reached in photolysis at high light intensities where C_2H_5CO radical association reactions compete successfully with decomposition. The data suggest strongly that the triplet state of 3-pentanone is involved, at least to some extent, at 313 nm and that a significant portion of the C_2H_5CO radicals formed dissociates readily, presumably in primary process (II). In experiments at 253.7 nm, addition of 1.2 Torr of 2,3-butanedione gave no reduction of the Φ_{CO} = 0.91 seen in pure 3-pentanone photolysis, and no phosphorescence of 2,3-butanedione was observed.

Table VII-C-3. Cross sections (σ, cm^2 $molecule^{-1}$) at 298 K for 2-pentanone and 3-pentanone (Martinez et al., 1992)

	$10^{20} \times \sigma$			$10^{20} \times \sigma$	
λ (nm)	2-Pentanone	3-Pentanone	λ (nm)	2-Pentanone	3-Pentanone
278	6.09	6.18	315	1.20	0.967
279	6.11	6.19	316	1.07	0.836
280	6.12	6.20	317	0.953	0.716
281	6.13	6.19	318	0.845	0.613
282	6.12	6.17	319	0.745	0.522
283	6.09	6.14	320	0.652	0.444
284	6.03	6.11	321	0.560	0.374
285	5.96	6.05	322	0.479	0.317
286	5.88	5.96	323	0.406	0.264
287	5.80	5.84	324	0.343	0.219
288	5.72	5.70	325	0.285	0.181
289	5.63	5.56	326	0.237	0.150
290	5.53	5.43	327	0.194	0.125
291	5.41	5.31	328	0.160	0.101
292	5.26	5.18	329	0.130	0.083
293	5.08	5.04	330	0.103	0.068
294	4.90	4.87	331	0.083	0.054
295	4.72	4.67	332	0.067	0.044
296	4.55	4.44	333	0.051	0.034
297	4.40	4.21	334	0.037	0.025
298	4.25	3.99	335	0.029	0.020
299	4.10	3.80	336	0.022	0.016
300	3.92	3.63	337	0.018	0.013
301	3.72	3.46	338	0.013	0.011
302	3.51	3.28	339	0.009	0.008
303	3.29	3.08	340	0.008	0.002
304	3.08	2.87	341	0.005	0.001
305	2.88	2.63	342	0.002	0.000
306	2.69	2.40	343	0.004	0.000
307	2.52	2.18	344	0.001	0.000
308	2.35	1.98	345	0.001	0.000
309	2.18	1.81	346	0.001	0.000
310	2.02	1.65	347	0.001	0.000
311	1.85	1.51	348	0.000	0.001
312	1.67	1.37	349	0.001	0
313	1.50	1.24	350	0.000	0
314	1.35	1.10	351	0.001	0

Photodecomposition from the excited singlet state formed at 253.7 nm was suggested to rationalize these results.

A few studies of the photolysis of 3-pentanone with added oxygen have been reported, but these have been at low oxygen pressures (Finklestein and Noyes, 1953; Jolley, 1957; Dunn and Kutschke, 1958). Finklestein and Noyes (1953) found that as the O_2 pressure was increased from 0 to 0.2 Torr, Φ_{CO} decreased from near unity to 0.15 and Φ_{CO2} increased from 0 to 1.1 for 2-pentanone pressure of 35 Torr. No further decrease in Φ_{CO} was seen as P_{O2} was increased to 1.25 Torr. These data indicate that a large share of the photodecomposition forms vibrationally excited C_2H_5CO radicals that dissociate quickly. Yet another, larger, share of the C_2H_5CO radicals formed in (I) can be captured by O_2 before dissociation. The $C_2H_5C(O)O_2$ radicals formed lead to CO_2 through various peroxy radical reactions [e.g., $C_2H_5C(O)O_2 + C_2H_5C(O)O_2 \rightarrow (2CH_3CH_2CO_2 + O_2) \rightarrow 2C_2H_5 + 2CO_2 + O_2$]. Several studies involving labeled 3-pentanone [$C_2H_5C^{14}(O)C_2H_5$ (Finklestein and Noyes, 1953), $C_2H_5C(O^{18})C_2H_5$ (Jolley, 1957; Dunn and Kutschke, 1958)] demonstrated that the CO and CO_2 formed in 2-pentanone photolysis with oxygen added arise largely from the carbonyl carbon, and that vibrationally rich propionyl radicals form CO in process (II). Jolley calculated from his results that dissociation of the thermalized C_2H_5CO radical requires 54 ± 17 kJ mol^{-1} more energy than its association reaction with O_2.

Hall et al. (1995) studied the infrared emission from CO and C_2H_5 formed in the photolysis of 3-pentanone at 193 nm. Process (II) occurs with formation of CO with a rotational temperature of ~2100 K and an average vibrational energy in the range of 6.3 to 8.8 kJ mol^{-1}, somewhat lower excitation than that observed in acetone photolysis at 193 nm. However, the ethyl radicals formed appear to have more vibrational energy (~79 kJ mol^{-1} total) than the methyl radicals formed in acetone photolysis with similar excitation. None of the models considered by Hall et al. (1995) for photodissociation were fully satisfactory in explaining the excitation of the CO. The double impulse model, considered in explanation of these results from 3-pentanone and acetone photolysis by Trentelman et al. (1989), is characterized by a pair of closely timed impulses along the two C–C bonds, each releasing half of the available energy to the three-body fragmentation. This leads to 13 kJ mol^{-1} excitation in CO, significantly higher than that observed experimentally yet a much better fit than is found for some alternative models that predict vibrational excitation of CO to be: 0.0004 kJ mol^{-1}(impulsive first step–statistical second step model); 0.17 kJ mol^{-1} (two-step information-theoretic prior); or 0.7 kJ mol^{-1} (soft barrier impulsive model).

VII-C-3.3. Photolysis Frequencies for 3-Pentanone

Unfortunately, there are no data that establish the wavelength dependence of the primary photodissociative processes in 3-pentanone in air at tropospheric pressures. However, it is likely that the quantum efficiency is somewhat greater than that seen for acetone and 2-butanone under similar conditions. Until this detailed experimental information becomes available, it is recommended that *j*-values calculated for 2-pentanone be used as a surrogate for 3-pentanone, with about 10% of the total photodissociation arising from process (II), as is observed for acetone and 2-butanone.

VII-C-4. 2-Pentanone [Methyl *n*-Propyl Ketone, $CH_3C(O)CH_2CH_2CH_3$]

The photochemistry of 2-pentanone follows the pattern of the other small ketones in that photodissociation occurs to form radicals in processes (I), (II), and (III), but an additional photodecomposition pathway appears. This is the smallest ketone that exhibits photodecomposition by the Norrish "Type II" rearrangement in process (IV), which occurs in all aldehydes and ketones that contain a γ-hydrogen atom. Also, process (V) leading to a cyclobutanol derivative has been observed.

$$n\text{-}C_3H_7COCH_3 + h\nu \rightarrow n\text{-}C_3H_7 + CH_3CO \quad \text{(I)}$$
$$\rightarrow CH_3 + CH_3CH_2CH_2CO \quad \text{(II)}$$
$$\rightarrow CH_3 + CO + CH_2CH_2CH_3 \quad \text{(III)}$$
$$\rightarrow [\text{six-membered cyclic transition state: } CH_3\text{–}C(=O\cdots H\cdots CH_2\text{–}CH_2\text{–}CH_2)] \rightarrow CH_3C(OH){=}CH_2 + CH_2{=}CH_2 \quad \text{(IV)}$$
$$\rightarrow HO\text{–}C(CH_3)\text{–}CH_2\text{–}CH_2\text{–}CH_2 \text{ (cyclobutanol ring)} \quad \text{(V)}$$

Excitation of 2-pentanone at long wavelengths ($\lambda \geq 313$ nm) populates the triplet state efficiently through intersystem crossing, $S_1 \rightarrow T_1$, and dissociation apparently occurs from this state. At shorter wavelengths, dissociation is believed to occur either from a vibrationally excited T_1 or the S_1 state.

VII-C-4.1. Absorption Cross Sections for 2-Pentanone

The absorption spectrum of 2-pentanone, as measured by Martinez et al. (1992), is very similar to that of 3-pentanone; see figure VII-C-1. Tabular listing of the cross sections as a function of wavelength is given in table VII-C-3.

VII-C-4.2. Quantum Yields of the Primary Photodissociative Processes in 2-Pentanone

Data directly related to 2-pentanone photolysis in the atmosphere are very limited, but some useful, related information is available for the pure ketone and its mixture with various added compounds.

(i) Processes (I) and (II): $CH_3CH_2CH_2C(O)CH_3 + h\nu \rightarrow CH_3CH_2CH_2 + CH_3CO$ (I); $CH_3CH_2CH_2C(O)CH_3 + h\nu \rightarrow CH_3CH_2CH_2CO + CH_3$(II). The $CH_3CH_2CH_2$—$C(O)CH_3$ bond is somewhat weaker than the $CH_3CH_2CH_2C(O)$—CH_3 bond, and, as in $CH_3C(O)C_2H_5$ photolysis, discrimination in the bond cleavage occurs in 2-pentanone photolysis. A very qualitative measure of the relative importance of processes (I) and (II) can be gained from hydrocarbon product analyses of Nicholson (1954) from 2-pentanone photolysis. His full-arc photolysis at 20°C suggests that $\phi_I/\phi_{II} \approx 8$. Rough estimates of minimum values of $\phi_I + \phi_{II}$ for pure ketone photolysis can be estimated from the summation of the quantum yields of the CH_3- and $CH_3CH_2CH_2$–groups in the

hydrocarbon products for experiments near 20°C where the decomposition reactions of radical products are unimportant. Maximum values for $\phi_{I}+\phi_{II}+\phi_{III}$ are probably well represented by Φ_{CO} as measured at temperatures near 120°C, where nearly complete decomposition of CH_3CO and $CH_3CH_2CH_2CO$ radicals is expected and radical induced decomposition of the ketone that forms CO is unimportant. This reasoning, applied to data of Wettack and Noyes (1968), gives estimates of the approximate range of $\phi_{I}+\phi_{II}+\phi_{III}$ values for pure 2-pentanone photolysis at several wavelengths: 0.11–0.12 (254 nm); 0.19–0.23 (280 nm); 0.27–0.36 (296.5 nm); and 0.16–0.39 (313 nm).

(ii) Process (III): $CH_3CH_2CH_2COCH_3 + h\nu \rightarrow CH_3CH_2CH_2 + CO + CH_3$. At low temperatures (~25°C) and high pressures and high intensities of absorbed light, the decomposition of thermalized acyl radicals should be relatively unimportant and Φ_{CO} values should provide a measure of the maximum values of ϕ_{III}. The data of Wettack and Noyes (1968) give $\phi_{III} \leq 0.06$ (313 nm); 0.08 (296.5 nm); 0.10 (280.4 nm); and 0.05 (254 nm). Φ_{CO} values reported by Michaël and Noyes (1963) for 2-pentanone photolyses at 313 and 253.7 nm are also small: ~0.01 (313 nm) and 0.05 (253.7 nm). These are relatively insensitive to added 2,3-butanedione, an excellent quencher of thermalized, triplet ketone molecules; see figure VII-C-10. It is probable this photodissociation to form CO involves the S_1 state or a vibrationally "hot" acyl radical, as observed with

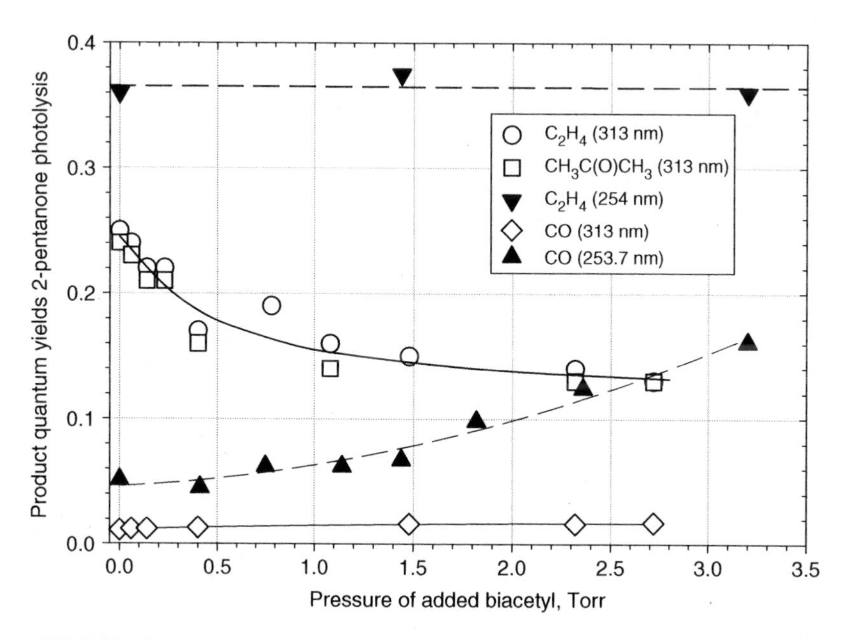

Figure VII-C-10. Quantum yield of C_2H_4, $CH_3C(O)CH_3$, and CO in 2-pentanone as a function of added 2,3-butanedione for photolyses at 313 and 253.7 nm. Data from Michael and Noyes (1963). The rise in the Φ_{CO} values at the higher pressures of 2,3-butanedione probably reflects the increasing importance of CO formation from direct photolysis of the added 2,3-butanedione; the larger increase in CO seen at 253.7 nm reflects the much stronger absorption by 2,3-butanedione at this wavelength.

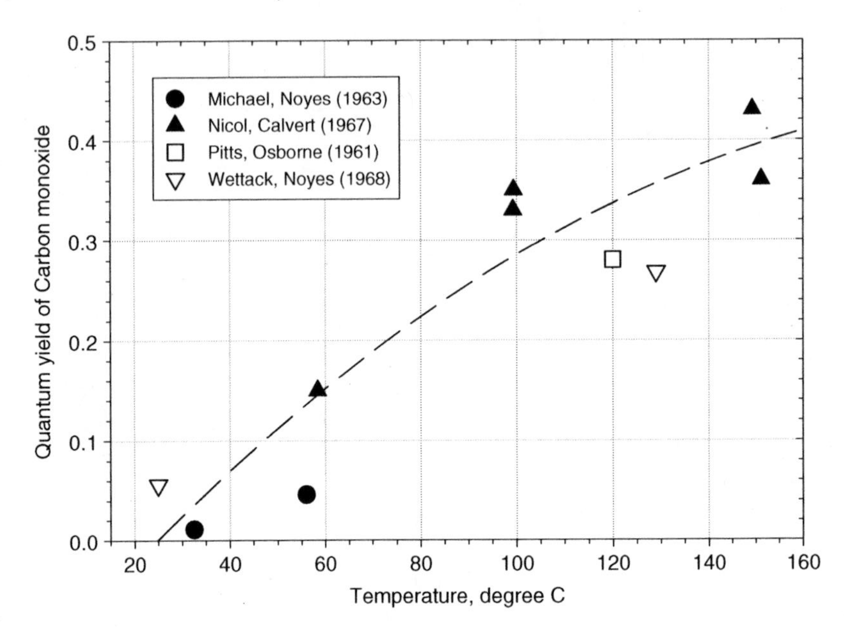

Figure VII-C-11. Plot of the quantum yield of CO in the 313 nm photolysis of 2-pentanone as a function of temperature.

$CH_3C(O)CH_3$ photolysis. We conclude that ϕ_{III} for 2-pentanone is $\sim$ 0.01–0.06 at 313 nm and $\sim$0.05 at 254 nm. As the temperature rises, the decomposition of CH_3CO and $CH_3CH_2CH_2CO$ radicals and CO formation become more important. For photolysis at 313 nm the quantum yields of CO are shown as a function of temperature in figure VII-C-11. For data at temperatures greater than 120°C, $\Phi_{CO} = \phi_I + \phi_{II} + \phi_{III}$, but this result cannot be extrapolated to atmospheric temperatures since ketones often have barriers to photodecomposition.

(iii) Process (IV): $CH_3CH_2CH_2C(O)CH_3 + h\nu \rightarrow CH_2{=}CH_2 + CH_2{=}C(OH)CH_3$. Extensive studies of 2-pentanone photolysis confirm the occurrence of process (IV), forming ethene and the enol-form of acetone. These product quantum yields are insensitive to temperature and pressure changes; see table VII-C-4. Data of Wettack and Noyes (1968) suggest that, within the wavelength region of importance in the troposphere, ϕ_{IV} is also relatively insensitive to wavelength of excitation (313–280 nm); see figure VII-C-12.

The formation of the intermediate enol-form of carbonyl compounds in aldehyde and ketone photolysis was first suggested by Davis and Noyes (1947) and was invoked by Gruver and Calvert (1958) to explain some unexpected results in the deuterium content of acetone formed in the photolysis of 1,1,1,3,3-pentadeutero-2-pentanone (McNesby and Gordon, 1958). The formation of the enol-form of the carbonyls in aldehyde and ketone photolysis has been widely accepted since the report of Srinivasan (1959) that acetone produced in 2-hexanone photolysis could exchange a hydrogen atom at the photochemical reactor walls. Formation of $CH_2{=}C(OH)CH_3$ in process (IV) of

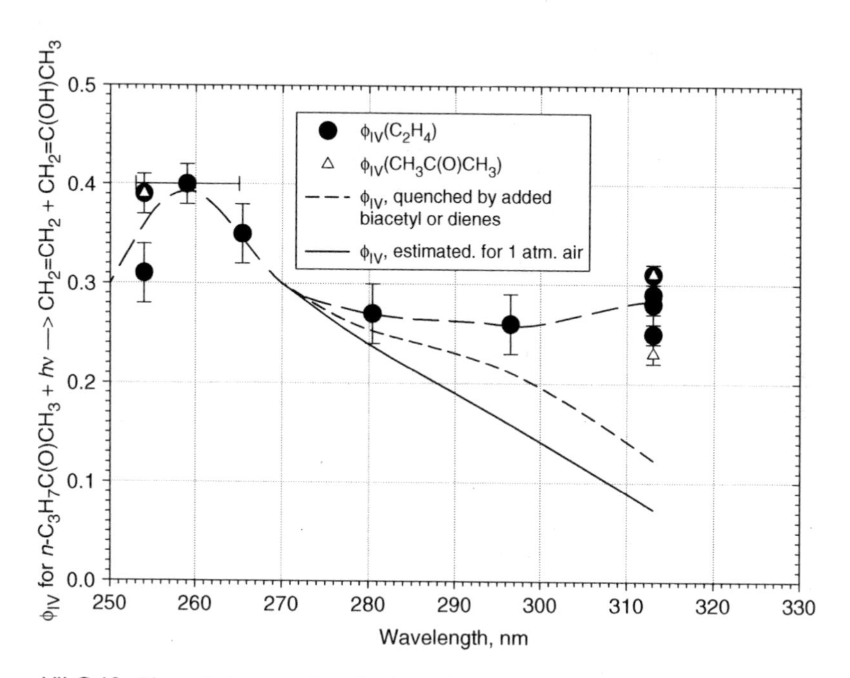

Figure VII-C-12. Plot of the wavelength dependence of the quantum yields of C_2H_4 and $CH_3C(O)CH_3$ in photolysis of 2-pentanone as a function of wavelength, as estimated from the data referenced in table VII-C-4. Data are insensitive to pressures of 2-pentanone and temperature. Experiments with pure ketone, circles and triangles; ϕ_{IV}, unquenched by 2,3-butanedione or dienes, short-dashed line; ϕ_{IV}, adjusted for additional air quenching at 1 atm using k_q/k_d for acetone as interpolated for the selected wavelength, solid line. Both the short-dashed and the solid lines were used in the estimation of j-values for 2-pentanone given in figure VII-C-15.

Table VII-C-4. Quantum yields of process (IV) in 2-pentanone at 313 nm: $CH_3CH_2CH_2C(O)CH_3 + h\nu \rightarrow CH_2{=}CH_2 + CH_2{=}C(OH)CH_3$

λ (nm)	Pressure (ketone, Torr)	Temp. (°C)	Φ_{C2H4}	$\Phi_{CH3C(O)CH3}$	Reference
313	1.3–32	28–152	0.31 ± 0.01	0.31 ± 0.01	Ausloos and Rebbert (1961)
313	5.2–270[a]	32–147	0.28 ± 0.03[b]	—	Borkowski and Ausloos (1961)
313	30.5	33–56	0.25 ± 0.01	0.23 ± 0.01	Michael and Noyes (1963)
313	9–55	25–129	0.29 ± 0.02	—	Wettack and Noyes (1968)
313	Not reported	120	0.28	—	Pitts and Osborne (1961)
313	30–66	58–149	0.25 ± 0.01	—	Nicol and Calvert (1967)
296.5	17–25	25–129	0.26 ± 0.03	—	Wettack and Noyes (1968)
280.4	17–30	25–129	0.27 ± 0.03	—	Wettack and Noyes (1968)
265.4	25	130	0.35	—	Wettack and Noyes (1968)
254–265	26–206	34	0.40 ± 0.02	—	Borkowski and Ausloos (1961)
254	20–32	25–128	0.31 ± 0.03	—	Wettack and Noyes (1968)
254	25	33	0.39 ± 0.02	0.39 ± 0.02	Michael and Noyes (1963)

[a] Deuterated 2-pentanone employed here: $CH_3C(O)CH_2CHDCD_2H$.

[b] Ethene product $D_2C{=}CHD$ and $DHC{=}CHD$ formed here; the ratio of the two isomers was temperature dependent.

2-pentanone has been observed directly through measurement of the enol infrared band at 2.756 μm (McMillan et al., 1964). The average half-life of CH_2=C(OH)CH_3 molecule in a large, dark cell was 3.34 ± 0.11 min., and acetone appeared with a half-time of 3.35 ± 0.13 min. The conversion of the enol of acetone is apparently catalyzed by the cell surface, for when D_2O is adsorbed on this surface $CH_3C(O)CH_2D$ is an observed product. For the conditions employed by McMillan et al., the ketone form of acetone [$CH_3C(O)CH_3$] does not exchange measurably to form $CH_3C(O)CH_2D$ on cell surfaces preconditioned with D_2O.

The ethene and acetone products of process (IV) in 2-pentanone photolysis at long wavelengths (313 nm) can be partially quenched by added 2,3-butanedione, a known efficient quencher of excited carbonyl triplets. The results of Michael and Noyes (1963) are shown in figure VII-C-10. Both ethene and acetone quantum yields decrease significantly on addition of a small amount of 2,3-butanedione. On photolysis, these 2,3-butanedione–2-pentanone mixtures exhibit a strong phosphorescence emission from the 2,3-butanedione triplet. This is compelling evidence that a significant portion, about 50%, of the 2-pentanone molecules excited at 313 nm decompose via the T_1 state of the molecule. Presumably the other half of the molecules that decay by process (IV) come from the excited singlet (S_1) state or, conceivably, from a vibratrionally rich T_1 or S_0 state of short lifetime. Experiments at 253.7 nm show little or no quenching of the ethene and acetone products by added 2,3-butanedione (Michael and Noyes, 1963); see figure VII-C-10. Obviously, at the short wavelengths, the products of (IV) form by dissociation from a very short-lived state of 2-pentanone.

Experiments of Wettack and Noyes (1968) support and extend the studies of Michael and Noyes (1963). Wettack and Noyes report that the addition of 1,3-butadiene or 1,3-pentadiene, both good reactants for carbonyl excited triplets, inhibits the photodecomposition of 2-pentanone; see figure VII-C-13. These authors conclude that about 62% of the photodecomposition by process (IV) in 2-pentanone at 313 nm involves the triplet state. Similar experiments of Wettack and Noyes (1968) show that excitation at 265.4 and 253.7 nm involves photodecomposition by process (IV) of about 18% and <12%, respectively, from non-quenchable, excited 2-pentanone molecules. Both the studies of Michael and Noyes (1963) and Wettack and Noyes (1968) indicate that both the singlet and triplet states of 2-pentanone are probable precursors to process (IV), and that the extent of triplet involvement increases at the long wavelengths.

The effect of oxygen on the primary process (IV) in 2-pentanone has been investigated in a few experiments that are more relevant to atmospheric conditions. O_2 quenches long-lived carbonyl triplet states effectively as $O_2(^1\Delta_g)$ is generated (Nau and Scaiano, 1996). Ausloos and Rebbert (1961) determined the C_2H_4, acetone, and 1-methylcyclobutanol quantum yields in the photolysis of 2-pentanone at 313 nm. The Stern–Volmer plot of the quenching of these products with added pressure of O_2 is given in figure VII-C-14. Shown are data from Ausloos and Rebbert (1961) for C_2H_4 (large black circles) and acetone (open squares). The involvement of at least two excited states with very different quenching efficiencies is evident in the curvature in the plot at low O_2 additions. The extent of quenching by oxygen seen in these studies is consistent with that seen in the experiments with other triplet quenchers (2,3-butanedione and dienes). The slope of the line in the linear region at high O_2 pressures

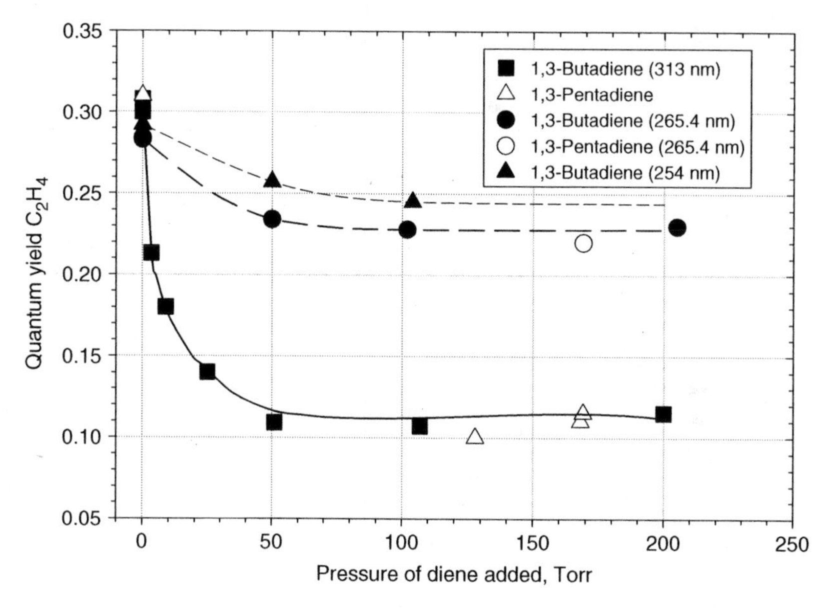

Figure VII-C-13. Quantum yields of C_2H_4 in 2-pentanone photolysis as a function of added 1,3-butadiene or 1,3-pentadiene. Data from Wettack and Noyes (1968); wavelength, 313 nm; temperature, 25°C.

leads to a Stern–Volmer quenching constant, $k_q/k_d \approx 9 \times 10^{-3}$ Torr^{-1}, a value similar to that observed for acetone. Using the relation derived in figure VII-C-2, one finds that this k_q/k_d value corresponds to that expected for acetone excited at 306 nm. Since air at 1 atm pressure should allow the nearly complete quenching of the thermalized T_1 state that is quenchable by added 2,3-butanedione or dienes, the results suggest that rough estimates for quantum yields of products of 2-pentanone photolysis in air may be made using the measured quantum yields of products of process (IV) from the S_1 state, corrected for quenching using the observed k_q/k_d values as estimated for acetone at various wavelengths.

(iv) Process (V): $CH_3CH_2CH_2C(O)CH_3 + h\nu \rightarrow$ 1-methylcyclobutanol. The formation of 1-methylcyclobutanol in the gas phase photodecomposition of 2-pentanone is well established. Ausloos and Rebbert (1961) demonstrated unambiguously that both process (IV) and (V) occur. At 313 nm and 28°C, the ratio ϕ_V/ϕ_{IV} decreased from 0.34 at 32 Torr to 0.088 at 1.3 Torr of 2-pentanone. Conceivably, processes (IV) and (V) could involve the same intermediate diradical, $CH_3C(\bullet)(OH)CH_2CH_2CH_2\bullet$. Such an intermediate could, in theory, decay to ethene and enol-acetone as in (IV) by a single C—C bond fission, or form the 1-methylcyclobutanol by collisional stabilization of the cyclized product at the higher pressures. Although processes (IV) and (V) appear to be related, they apparently do not occur from a common intermediate, since the quenching of the products of the precursor states show very different efficiencies; see figure VII-C-14. Φ_{C2H4} and $\Phi_{CH3C(O)CH3}$ are insensitive

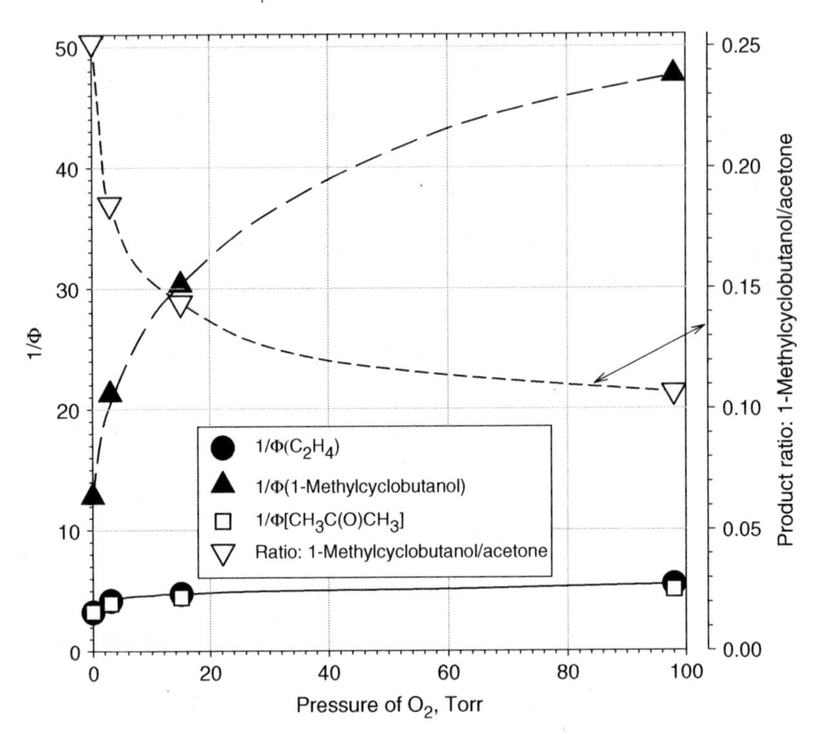

Figure VII-C-14. Effect of added O_2 on the quantum yields of products (left axis) and the ratio of 1-methylcyclobutanol to acetone (right axis) in the 313 nm photolysis of 2-pentanone at 28°C (Ausloos and Rebbert, 1961).

to temperature and pressure and the observed ratio of 1-methylcyclobutanol to ethene varies from near 0.25 in the absence of oxygen to about 0.11 with 100 Torr of O_2 added. The ratio of 1-methylcyclobutanol to ethene decreases with lowered pressure of ketone from 0.335 at 32 Torr to 0.088 at 1.3 Torr. The studies of McMillan et al. (1964) confirm the formation of the cyclobutanol product of (V) but indicate a somewhat smaller quantum yield of this product than that expected from the results of Ausloos and Rebbert (1961). McMillan et al. (1964) suggested that the failure of some workers to detect the cyclobutanol products in ketone and aldehyde photolysis may be due in part to destruction of these alcohols during gas chromatographic analysis.

VII-C-4.3. Estimation of Photolysis Coefficients for 2-Pentanone

A very rough estimate of *j*-values can be made for 2-pentanone with certain unsubstantiated assumptions regarding the effect of 1 atmosphere of air on the quantum yields of products:

(i) Estimates of Wettack and Noyes (1968) and Michael and Noyes (1963) may be used to determine the fraction (E) of the process (IV) that occurs via states that are unquenched by added 2,3-butanedione and diene; these are 56% (average of 52 and 59%) at 313 nm;

18% at 265.4 nm; and 6% (average of 0% and 12%) at 253.7 nm. These E-values fit well a linear relation with λ: $E = -203.6 + 0.8302(\lambda)$, which can be used to estimate E at other wavelengths.

(ii) The weak additional quenching of the "unquenchable" fraction of 2-pentanone by air probably occurs as with other aldehydes and ketones. We assumed this quenching is described by the Stern–Volmer relationship with the k_q/k_d values derived from acetone studies (figure VII-C-2).

(iii) The formation of 1-methylcyclobutanol is assumed to be negligible in air at 1 atm.

(iv) Processes (I) and (II) are assumed to originate from the same state as process (IV), with a ratio of quantum yields, $(\phi_{I} + \phi_{II} + \phi_{III})/\phi_{IV} \approx 0.43$, a very rough estimate from the full arc experiments of Nicholson (1954).

Two scenarios of possible quenching of the primary processes were assumed in estimating the j-values for a cloudless day in the lower troposphere (ozone column = 350 DU); see figure VII-C-15. In case A, we accepted the assumptions (i), (iii), and (iv) above, while in case B all four assumptions were accepted. There is a reasonable probability that actual j-values for processes (I) + (II) + (III) and for process (IV) for 2-pentanone in the atmosphere may lie somewhere between these two estimates. The tabular data for the quantum yields that were derived in the two scenarios are summarized in table VII-C-5. Obviously, more definitive data are necessary to determine these quantities reliably.

VII-C-5. 3-Methyl-2-Butanone [Methyl Isopropyl Ketone, $CH_3C(O)CH(CH_3)_2$]

Limited data on 3-methyl-2-butanone suggest consideration of the following primary photodissociative processes:

$$CH_3C(O)CH(CH_3)_2 + h\nu \rightarrow CH_3CO + CH_3CHCH_3 \quad (I)$$

$$\rightarrow CH_3 + (CH_3)_2CHCO \quad (III)$$

$$\rightarrow CH_3 + CO + CH_3CHCH_3 \quad (III)$$

$$\rightarrow CH_3CHO + CH_3CH = CH_2 \quad (IV)$$

Weir (1962), Zahra and Noyes (1965), and Lissi et al. (1973/1974) have published the only extensive studies of the photolysis of this ketone. The product analyses of Zahra and Noyes provide a rough estimate of the extent of each of these processes at 313 and 253.7 nm. For ketone pressures in the range 3–42 Torr, the product distribution at 313 nm suggests $\phi_{I}/\phi_{II} \approx 6 \pm 327°C$ and 5 ± 2 at 60–62°C; at 253.7 nm $\phi_{I}/\phi_{II} \approx 6$ (27°C), 5 (60°C), and 2 (127°C). The product data of Lissi et al. (1973/1974), from experiments at 313 nm over a 130-fold range of intensities (from the highest to lowest), give the ϕ_{I}/ϕ_{II} estimates: 2.8, 2.3, 2.4, and 1.4. The product distribution reported by Nicholson (1954) from the full arc irradiation of 3-methyl-2-butanone also suggests the dominance of process (I); $\phi_{I}/\phi_{II} \approx 7 \pm 2$. These ratios

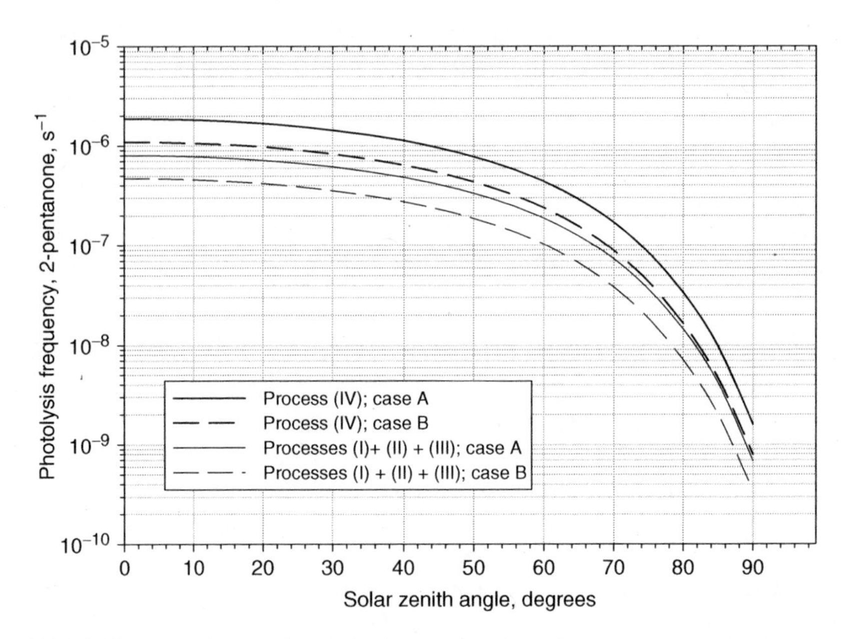

Figure VII-C-15. Approximate photolysis frequencies for primary processes in 2-pentanone vs. solar zenith angle: $CH_3CH_2CH_2C(O)CH_3 + h\nu \rightarrow CH_3CH_2CH_2 + CH_3CO$ (I); $CH_3CH_2CH_2C(O)CH_3 + h\nu \rightarrow CH_3CH_2CH_2CO + CH_3$(II); $CH_3CH_2CH_2C(O)CH_3 + h\nu \rightarrow CH_3CH_2CH_2 + CO + CH_3$ (III); $CH_3CH_2CH_2C(O)CH_3 + h\nu \rightarrow C_2H_4 + CH_2{=}C(OH)CH_3$ (IV). Case A assumes that all processes are quenched in air to the limiting values observed for 2,3-butanedione and diene quenching; case B assumes quenching as in A but additional quenching as observed for acetone–air photolysis. Data were calculated for a cloudless day in the lower troposphere with an overhead ozone column of 350 DU.

are consistent with the somewhat weaker $(CH_3)_2CH{-}C(O)CH_3$ bond compared to $(CH_3)_2CHC(O){-}CH_3$.

The total quantum yield of decomposition of 3-methyl-2-butanone, as estimated by Lissi et al. (1973/1974) from an extrapolation of the product yields to zero absorbed light intensity gives 0.75. This is in good agreement with that derived from $\Phi_{CO} = 0.78$, estimated by Zahra and Noyes (1965) in photolysis at 127°C.

The quantum yields of CO, reported by Zahra and Noyes (1965) at 313 nm and 27°C, are: 0.045 (3 Torr pressure), 0.034 (11 Torr), 0.027 (13 Torr), 0.030 (27.5 Torr), 0.031 (30 Torr), and 0.037 (41.8 Torr). At 253.7 nm, $\Phi_{CO} \approx 0.12$ (27°C, 22 Torr), 0.35 (60°C, 22 Torr), and 0.78 (127°C, 22 Torr). These data probably reflect the occurrence of process (III) to a small extent ($\phi_{III} \approx 0.03 \pm 0.01$ at 313 nm), as seen with the other small ketones.

Process (IV) was suggested by Zahra and Noyes (1965) to explain the presence of CH_3CHO in the products. A 5-membered ring intermediate, similar to the 6-membered ring intermediate suggested for Norrish type II reactions, was proposed to be involved in the reaction. At 313 nm the process (IV) is negligible, $\phi_{IV} < 0.01$ (27–60°C), but at 253.7 nm acetaldehyde yields suggest $\phi_{IV} \approx 0.36 \pm 0.03$ (60–127°C). Nicholson

Table VII-C-5. Approximate estimates of the quantum yields for 2-pentanone photolysis in air at 1 atm pressure as a function of wavelength: $CH_3CH_2CH_2C(O)CH_3 + h\nu \rightarrow CH_3CH_2CH_2 + CH_3CO$ (I); $CH_3CH_2CH_2C(O)CH_3 + h\nu \rightarrow CH_3CH_2CH_2CO + CH_3$ (II); $CH_3CH_2CH_2C(O)CH_3 + h\nu \rightarrow CH_3CH_2CH_2 + CO + CH_3$ (III); $CH_3CH_2CH_2C(O)CH_3 + h\nu \rightarrow CH_2{=}CH_2 + CH_2{=}C(OH)CH_3$ (IV); $CH_3CH_2CH_2C(O)CH_3 + h\nu \rightarrow$ 1-methylcyclobutanol (V); ϕ_V is small but not estimated here

λ (nm)	$\phi_I + \phi_{II} + \phi_{III}$		ϕ_{IV}		λ (nm)	$\phi_I + \phi_{II} + \phi_{III}$		ϕ_{IV}	
	(A)[a]	(B)[b]	(A)[a]	(B)[b]		(A)[a]	(B)[b]	(A)[a]	(B)[b]
278	0.11	0.11	0.26	0.25	304	0.07	0.05	0.17	0.12
279	0.11	0.10	0.26	0.24	305	0.07	0.05	0.16	0.11
280	0.11	0.10	0.26	0.24	306	0.07	0.05	0.16	0.11
281	0.11	0.10	0.25	0.23	307	0.07	0.04	0.16	0.10
282	0.11	0.10	0.25	0.23	308	0.06	0.04	0.15	0.10
283	0.11	0.09	0.25	0.22	309	0.06	0.04	0.15	0.09
284	0.10	0.09	0.24	0.22	310	0.06	0.04	0.14	0.09
285	0.10	0.09	0.24	0.21	311	0.06	0.03	0.14	0.08
286	0.10	0.09	0.24	0.21	312	0.06	0.03	0.13	0.08
287	0.10	0.09	0.24	0.20	313	0.05	0.03	0.12	0.07
288	0.10	0.09	0.24	0.20	314	0.05	0.03	0.12	0.07
289	0.10	0.09	0.23	0.20	315	0.05	0.03	0.11	0.06
290	0.10	0.08	0.23	0.19	316	0.05	0.03	0.11	0.06
291	0.10	0.08	0.23	0.18	317	0.04	0.02	0.10	0.05
292	0.10	0.08	0.23	0.18	318	0.04	0.02	0.09	0.05
293	0.09	0.08	0.22	0.18	319	0.03	0.02	0.08	0.04
294	0.09	0.07	0.22	0.17	320	0.03	0.01	0.07	0.03
295	0.09	0.07	0.22	0.17	321	0.03	0.01	0.06	0.03
296	0.09	0.07	0.21	0.16	322	0.02	0.01	0.05	0.02
297	0.09	0.07	0.21	0.16	323	0.02	0.01	0.05	0.02
298	0.09	0.06	0.20	0.15	324	0.02	0	0.04	0.01
299	0.09	0.06	0.20	0.15	325	0.01	0	0.02	0
300	0.09	0.06	0.20	0.14	326	0.01	0	0.02	0
301	0.08	0.06	0.19	0.14	327	0	0	0.01	0
302	0.08	0.06	0.19	0.13	328	0	0	0	0
303	0.08	0.05	0.18	0.12					

[a] Scenario based on acceptance of assumptions (i), (iii), and (iv) given in section VII-C-4.3.
[b] Scenario based on acceptance of assumptions (i), (ii), (iii) and (iv), section VII-C-4.3.

(1954) also reported CH_3CHO among the products formed in his full-arc photolysis. Lissi et al. (1973/1974) found no evidence for (IV) in their experiments at 313 nm. The evidence for process (IV) is not strong, and we conclude that it is probably unimportant at the wavelengths available within the troposphere.

Zahra and Noyes (1965) found that at 313 nm the addition of small amounts of oxygen (0.03–0.14 Torr) to 26 Torr of ketone (27°C) lowered the emission yield from 3-methyl-2-butanone by 21%, comparable to the effect seen in other light ketones. The emission yields observed by Zahra and Noyes in photolyses of pure ketone and in 2,3-butanedione–ketone mixtures indicate clearly that the triplet state plays an important role in the photolysis of 3-methyl-2-butanone at 313 nm. There was no evidence of the triplet excited state observed in experiments at 253.7 nm. Lissi et al. (1973/1974) carried out experiments at 313 nm both with added 2,3-butanedione and with added *cis*-1,3-pentadiene. These experiments confirm the importance of the triplet precursor to products for photolysis at 313 nm.

The photochemistry of 3-methyl-2-butanone appears to be very similar to that observed in the photolysis of smaller ketones in which an CH_3CO radical is also formed in the primary process. For example with excitation at 313 nm, in 3-methyl-2-butanone $\Phi_{CO} = 0.033 \pm 0.007$ (27°C, pressure 5.5–41.8 Torr; Zahra and Noyes, 1965) and in acetone $\Phi_{CO} = 0.033$(20°C, 50 Torr; Cundall and Davies, 1966). For similar conditions, the two ketones in the same two studies (temperature 60–65°C) give $\Phi_{CO} = 0.23 \pm 0.01$ and 0.29, respectively, and for temperature 127–130°C, $\Phi_{CO} = 0.78$ and 0.98, respectively.

Ausloos and Rebbert (1964) report that the relative efficiency of energy transfer to 2,3-butanedione is strongly dependent on the dissociative lifetime of the excited ketone molecule and varies according to the following sequence: acetone > 2-butanone > 2-pentanone > 3-methyl-2-butanone.

The data available are insufficient to allow a definitive estimate of the quantum efficiencies as a function of wavelength for photolysis of 3-methyl-2-butanone in air that are required to define *j*-values. The photochemistry of 3-methyl-2-butanone may be similar to that observed for 2-butanone and 3-pentanone (sections VII-C-2 and VII-C-3). However, the $CH_3C(O)—CH(CH_3)_2$ bond is somewhat weaker than the $CH_3C(O)^-C_2H_5$ bond, and it seems reasonable that the *j*-values for use with 3-methyl-2-butanone should be somewhat greater than those of acetone and 2-butanone.

VII-C-6. 2-Hexanone [Methyl *n*-Butyl Ketone, $CH_3C(O)CH_2CH_2CH_2CH_3$]

Studies of the photodecomposition of 2-hexanone have been rationalized in terms of the following primary dissociation channels:

$$CH_3C(O)CH_2CH_2CH_2CH_3 + h\nu \rightarrow CH_3CO + CH_2CH_2CH_2CH_3 \quad \text{(I)}$$

$$\rightarrow CH_3CH_2CH_2CH_2CO + CH_3 \quad \text{(II)}$$

$$\rightarrow CH_3 + CO + CH_3CH_2CH_2CH_2 \quad \text{(III)}$$

$$\rightarrow CH_3CH{=}CH_2 + CH_2{=}C(OH)CH_3 \quad \text{(IV)}$$

The photodissociative pathways (I), (II), and (III) are relatively unimportant compared to process (IV). Srinivasan (1959) estimated from experiments at 313 nm and 27°C that $\phi_I + \phi_{II} + \phi_{III} \approx 0.018$, which was unaffected by the addition of up to 4.0 Torr of 2,3-butanedione. A maximum value for this sum at 313 nm is 0.11 (from Φ_{CO} at 103°C), measured by Davis and Noyes (1947), and 0.08 (from Φ_{CO} at 125°C), measured by Michael and Noyes (1963). The product distribution determined in full-arc photolysis at 21°C by Nicholson (1954) suggests $(\phi_I + \phi_{II} + \phi_{III})/\phi_{IV} \approx 0.04$. By analogy with the smaller ketones one expects $\phi_I/\phi_{II} > 1.0$, but Nicholson's product analysis indicates only a small discrimination in bond cleavage. The measurements of Φ_{CO} at 313 nm (20–30°C) reflect a small value for $\phi_{III} \approx 0.02$ (Brunet and Noyes, 1958), but a portion of this value may arise from the photolysis of the primary product, acetone.

The dominance of process (IV) was established in the pioneering study of Norrish and Appleyard (1934); these workers concluded that 2-hexanone decomposes primarily into an equimolecular mixture of propene and acetone. In the several studies that have followed, quantitative data confirmed this interpretation. Measurements of the quantum yields of process (IV) are in relatively good agreement:

(i) *Wavelength 313 nm*: $\phi_{IV} = 0.34$ (27°C, Wilson and Noyes, 1943); 0.45 ± 0.05 (C_3H_6, 103°C, Davis and Noyes, 1947); 0.50 ± 0.07 ($CH_3C(O)CH_3$, 103°C, Davis and Noyes, 1947); 0.46 ± 0.02 (C_3H_6; 28–30°C, Brunet and Noyes, 1958); 0.43 ± 0.06 ($CH_3C(O)CH_3$; 28–30°C, Brunet and Noyes, 1958); 0.41 ± 0.01 (C_3H_6, 35–100°C, Michael and Noyes, 1963); 0.44 ± 0.04 ($CH_3C(O)CH_3$, 35°C, Michael and Noyes, 1963).

(ii) *Wavelength 253.7 nm*: $\phi_{IV} = 0.34$ (C_3H_6, 27°C, Wilson and Noyes, 1943); 0.29 ± 0.01 (C_3H_6, 36–125°C; Michael and Noyes, 1963). The decrease in ϕ_{IV} with decrease in λ appears to be unique among the ketones that have been studied.

Srinivasan (1959) studied the photodecomposition of 2-hexanone-5,5-d_2($CH_3C(O)CH_2CH_2CD_2CH_3$) and found the propene formed was largely C_3H_5D with about 3% $C_3H_4D_2$, in accord with expectations based on the mechanism of process (IV). However, unexpectedly, the acetone was nearly an equimolar mixture of $CH_3C(O)CH_2D$ (45%) and $CH_3C(O)CH_3$ (54%). This result provided the first indication of the involvement of the enol-acetone product ($CH_2{=}C(OH)CH_3$) that could exchange H-atoms for D-atoms on equilibration at the walls of the reaction vessel.

In photolysis at 313 nm, quenching of 2-hexanone by formation of triplet 2,3-butanedione is only 1/20th of that seen for 1-pentanone. The quantum yields of propene and acetone are also little affected by 2,3-butanedione addition: at 35°C, $\Phi_{C3H6} = 0.41 \pm 0.01$ and $\Phi_{CH3C(O)CH3} = 0.44 \pm 0.04$. At 253.7 nm, there is no observable effect of added 2,3-butanedione on the photodecomposition of 2-hexanone, and no evidence for a sensitized emission of 2,3-butanedione was observed (Michael and Noyes, 1963). Obviously, a very short-lived triplet or singlet state is the precursor to all photodecomposition in 2-hexanone.

Brunet and Noyes (1958) found a small effect on the propene and acetone product quantum yields with the addition of 230 and 560 Torr of added oxygen: $\Phi_{C3H6} = 0.39$ and 0.38; $\Phi_{CH3C(O)CH3} = 0.42$ and 0.19, respectively. The decrease observed was rationalized by Brunet and Noyes as probably a consequence of incomplete condensation of the products of the reaction when large quantities of oxygen are removed after irradiation. Ausloos and Rebbert (1961) suggested that the effect could be actual quenching, as they observed in their studies with oxygen added to 2-pentanone. If the decrease is attributed to quenching, then a Stern–Volmer plot using only the propene data gives $k_q/k_d \approx 10^{-3}$ Torr^{-1}, a value similar to those observed for air-quenching of other aldehydes and ketones that have been discussed in earlier sections.

Until the appropriate detailed quantum yield and spectral data become available, we can make only very rough estimates of the expected j-values for 2-hexanone in air, using several as yet untested assumptions:

Scenario (A):

(i) In the pure ketone, the quantum yield of process (IV) decreases in a linear fashion from $\phi_{IV} = 0.43 \pm 0.05$, the average value at 313 nm, to 0.31 ± 0.03 at 253.7 nm.

(ii) The fraction of the ketone that is quenchable by 2,3-butanedione at 313 nm is about 3% and decreases in a linear fashion with wavelength to zero at 253.7 nm.

(iii) The cross sections for light absorption by 2-hexanone are assumed to be equal to those measured for 2-pentanone.

(iv) Air does not quench the ketone further than that expected from assumption (ii).

Scenario (B):

The first three assumptions above apply, and we assume that the reported decrease in ϕ_{IV} with addition of oxygen observed by Brunet and Noyes (1958) is a result of quenching by oxygen. The k_q/k_d values at each wavelength are assumed to be equal to those expected for acetone (figure VII-C-2) quenching by air.

Approximate quantum yields for process (IV) as a function of wavelength are given in table VII-C-6 for the two different scenarios (A) and (B) of wavelength dependence of ϕ_{IV} outlined in this section. The sum of the quantum yields for photodissociation of 2-hexanone by processes (I), (II), and (III) is small, and probably in the range of 4%–18% of ϕ_{IV} at each wavelength.

The formation of a cyclobutanol product ($\Phi \approx 0.075$) has been observed in the solution-phase photodecomposition of 2-hexanone (Coulson and Yang, 1966). An additional process analogous to process (V) in 2-pentanone probably accounts for this. Determination of the quantum efficiency of this process in the gas-phase photochemistry has not been reported, but presumably it is small.

The approximate photolysis frequencies for process (IV) for a cloudless day within the lower troposphere (ozone column = 360 DU) are given as a function of solar zenith angle for two scenarios (A and B) in figure VII-C-16.

VII-C-7. 3-Hexanone (Ethyl *n*-Propyl Ketone, $CH_3CH_2C(O)CH_2CH_2CH_3$)

Photodissociation of 3-hexanone follows the pattern of the smaller ketones:

$$CH_3CH_2CH_2C(O)CH_2CH_3 + h\nu \rightarrow C_2H_5CO + CH_3CH_2CH_2 \quad \text{(I)}$$

$$\rightarrow CH_3CH_2 + CH_3CH_2CH_2CO \quad \text{(II)}$$

$$\rightarrow CH_3CH_2 + CO + CH_3CH_2CH_2 \quad \text{(III)}$$

$$\rightarrow CH_2{=}CH_2 + CH_2{=}C(OH)C_2H_5 \quad \text{(IV)}$$

Data on product quantum yields from photolysis of the pure ketone at 313 nm and at temperatures from 62 to 150°C indicate that there is a small preference in bond breakage favoring process (I): $\phi_I/\phi_{II} \approx 1.6 \pm 0.2$ (Nicol and Calvert, 1967). Φ_{CO} data for

Table VII-C-6. Estimated quantum yields of primary process $CH_3CH_2CH_2CH_2C(O)CH_3 + h\nu \rightarrow CH_3CH{=}CH_2 + CH_2{=}C(OH)CH_3$ (IV); cases A and B represent different scenarios of the dependence of ϕ_{IV} on wavelength of excitation; see text for discussion of quantum yields for this and other processes

λ (nm)	ϕ_{IV} (Case A)[a]	ϕ_{IV} (Case B)[b]	λ (nm)	ϕ_{IV} (Case A)[a]	ϕ_{IV} (Case B)[b]
278	0.36	0.32	314	0.44	0.08
279	0.36	0.32	315	0.44	0.07
280	0.37	0.31	316	0.44	0.07
281	0.37	0.31	317	0.44	0.06
282	0.37	0.31	318	0.45	0.06
283	0.37	0.30	319	0.45	0.05
284	0.38	0.30	320	0.45	0.05
285	0.38	0.29	321	0.45	0.05
286	0.38	0.29	322	0.45	0.04
287	0.38	0.28	323	0.46	0.04
288	0.38	0.27	324	0.46	0.04
289	0.39	0.27	325	0.46	0.03
290	0.39	0.26	326	0.46	0.03
291	0.39	0.25	327	0.47	0.03
292	0.39	0.24	328	0.47	0.03
293	0.39	0.24	329	0.47	0.03
294	0.40	0.23	330	0.47	0.02
295	0.40	0.22	331	0.47	0.02
296	0.40	0.21	332	0.48	0.02
297	0.40	0.20	333	0.48	0.02
298	0.41	0.19	334	0.48	0.02
299	0.41	0.18	335	0.48	0.02
300	0.41	0.18	336	0.48	0.02
301	0.41	0.17	337	0.49	0.01
302	0.41	0.16	338	0.49	0.01
303	0.42	0.15	339	0.49	0.01
304	0.42	0.14	340	0.44	0.01
305	0.42	0.13	341	0.44	0.01
306	0.42	0.13	342	0.44	0.01
307	0.42	0.12	343	0.44	0.01
308	0.43	0.11	344	0.45	0.01
309	0.43	0.10	345	0.45	0.01
310	0.43	0.10	346	0.45	0.01
311	0.43	0.09	347	0.45	0.01
312	0.43	0.09	348	0.45	0.01
313	0.44	0.08	349	0.46	0.01

[a] Scenario A assumes no extra quenching by air (1 atm); only the thermalized triplets are quenched.

[b] Scenario B assumes extra quenching by air (1 atm) with k_q/k_d mimicking the air quenching by acetone; see section VII-C-1.2.

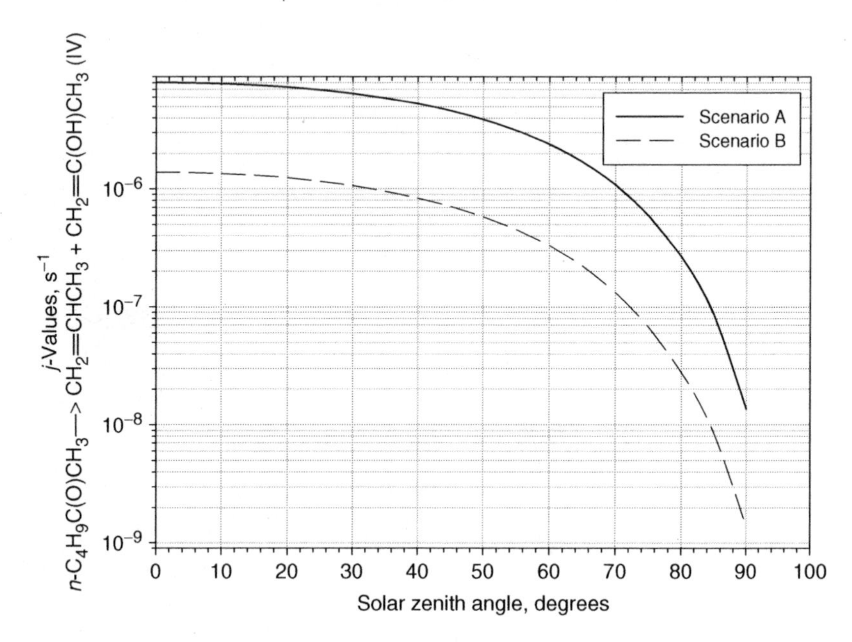

Figure VII-C-16. Approximate j-values for primary process (IV) in 2-hexanone, $CH_3CH_2CH_2CH_2C(O)CH_3 + h\nu \rightarrow CH_2{=}CHCH_3 + CH_2{=}C(OH)CH_3$, as estimated for a cloudless day in the lower troposphere with an overhead ozone column of 350 DU; two different scenarios (A and B) related to different choices for wavelength dependence of ϕ_{IV} were used; see the text.

temperatures greater than 100°C give the estimate: $\phi_I + \phi_{II} + \phi_{III} = 0.61 \pm 0.06$. The quantum yields of C_2H_4 were corrected for a small contribution (1.7–5.8%) from disproportionation reactions involving C_2H_5—C_2H_5 and $C_2H_5 + CH_3CH_2CH_2$ radicals. These values of Φ_{C2H4}, independent of temperature and ketone pressure, suggest that $\phi_{IV} = 0.21 \pm 0.01$ (Nicol and Calvert, 1967). The effect of 1 atm of air on these primary processes is unknown, but in view of the data for the other ketones it is likely that some quenching of the processes will occur. Until such information is available, no reliable estimates of the primary quantum yields or photolysis frequencies in air can be made.

VII-C-8. 4-Methyl-2-pentanone [Methyl Isobutyl Ketone, $CH_3C(O)CH_2CH(CH_3)_2$]

The photodecomposition of 4-methyl-2-pentanone can be rationalized in terms of the following primary processes:

$$CH_3C(O)CH_2CH(CH_3)_2 + h\nu \rightarrow CH_3CO + CH_2CH(CH_3)_2 \quad (I)$$

$$\rightarrow CH_3 + (CH_3)_2CHCH_2CO \quad (II)$$

$$\rightarrow CH_3 + CO + CH_2CH(CH_3)_2 \quad (III)$$

$$\rightarrow CH_2{=}CHCH_3 + CH_2{=}C(OH)CH_3 \quad (IV)$$

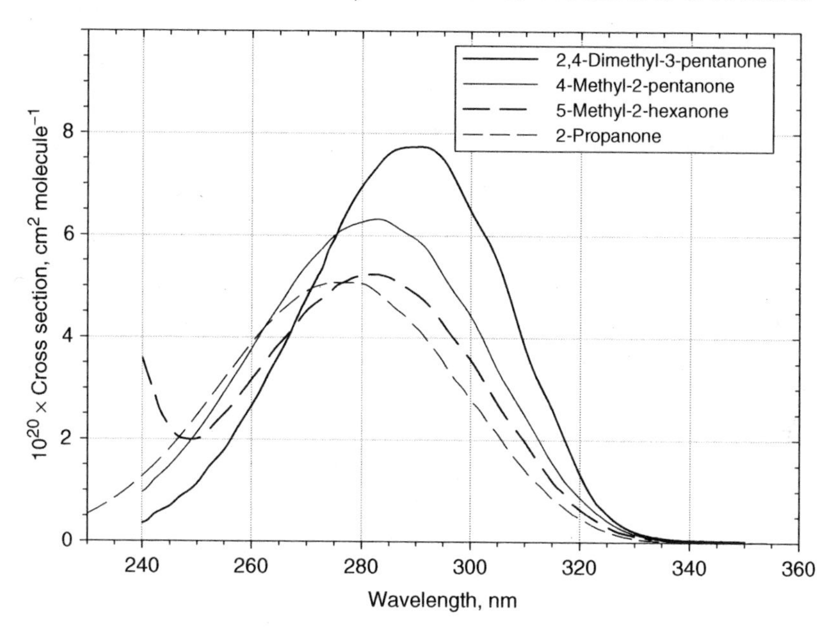

Figure VII-C-17. Comparison of absorption spectra of some branched chain ketones (Yujing and Mellouki, 2000) with that of acetone (Martinez et al., 1992).

The absorption spectrum of 4-methyl-2-pentanone is compared with that of several other ketones in figure VII-C-17, and tabular data are given in table VII-C-7. Data from a single gas-phase, 313 nm photolysis experiment with the pure ketone (120°C, pressure 49 Torr) by Martin and Pitts (1955) suggest that there is little discrimination in bond breakage in (I) and (II). From Φ_{CO}, $\phi_I + \phi_{II} + \phi_{III} \approx 0.15$; from Φ_{C3H6}, $\phi_{IV} \approx 0.35$.

The only extensive study of the photolysis of 4-methyl-2-pentanone was carried out in degassed solutions of hexane confined in Pyrex test tubes that were irradiated with a medium-pressure Hg lamp (Encina et al., 1975). Extrapolation of this information to photolysis in 1 atm of air is impossible. However, the information may shed some light on the gas-phase photolysis. In the hexane solutions, $\Phi_{isobutane} \approx \phi_I \approx 0.034 \pm 0.003$ (76°C); $\Phi_{CH3C(O)CH3} \approx \phi_{IV} \approx 0.21 \pm 0.02$ (−5 to 69°C). Experiments with added 1,3-pentadiene were used to quench excited triplet ketone molecules. The authors estimated from experiments at 20°C that the quantum yield of quenchable triplets excited under their conditions was 0.75 ± 0.03; about 88% of the acetone product of (IV) was formed from the T_1 state, with about 12% from the S_1 state. These results point to the involvement of both the singlet and triplet states in gas-phase photolysis, but the extent of participation of each is unclear. Until the necessary data are available to evaluate properly the wavelength dependence of the quantum yields of the primary processes in air under atmospheric conditions, realistic estimates of j-values for 4-methyl-2-pentanone are not possible.

Table VII-C-7. Absorption cross sections ($10^{20} \times \sigma$, cm^2 $molecule^{-1}$ at 298 K) for (A) 2,4-dimethyl-3-pentanone; (B) 4-methyl-2-pentanone; (C) 5-methyl-2-hexanone; from Yujing and Mellouki (2000)

λ (nm)	$10^{20} \times \sigma$ (A)	$10^{20} \times \sigma$ (B)	$10^{20} \times \sigma$ (C)	λ (nm)	$10^{20} \times \sigma$ (A)	$10^{20} \times \sigma$ (B)	$10^{20} \times \sigma$ (C)
278	6.61	6.19	5.15	315	2.59	1.55	1.19
279	6.79	6.23	5.19	316	2.33	1.39	1.07
280	6.96	6.26	5.21	317	2.07	1.23	0.976
281	7.11	6.29	5.22	318	1.82	1.10	0.823
282	7.24	6.31	5.23	319	1.56	0.981	0.727
283	7.37	6.32	5.23	320	1.32	0.870	0.639
284	7.50	6.30	5.21	321	1.11	0.765	0.560
285	7.60	6.25	5.17	322	0.922	0.667	0.484
286	7.67	6.19	5.12	323	0.760	0.576	0.420
287	7.70	6.12	5.05	324	0.655	0.492	0.357
288	7.72	6.05	4.99	325	0.543	0.413	0.298
289	7.73	5.99	4.92	326	0.446	0.345	0.245
290	7.73	5.92	4.84	327	0.366	0.285	0.201
291	7.73	5.83	4.76	328	0.299	0.235	0.171
292	7.72	5.71	4.66	329	0.245	0.193	0.145
293	7.67	5.56	4.54	330	0.201	0.158	0.120
294	7.59	5.38	4.40	331	0.166	0.129	0.103
295	7.48	5.20	4.25	332	0.137	0.104	0.088
296	7.32	5.04	4.11	333	0.113	0.082	0.074
297	7.12	4.88	3.97	334	0.094	0.064	0.101
298	6.90	4.73	3.83	335	0.079	0.050	0.050
299	6.69	4.59	3.67	336	0.066	0.039	0.043
300	6.48	4.43	3.56	337	0.057	0.035	0.036
301	6.28	4.24	3.40	338	0.049	0.031	0.033
302	6.11	4.03	3.23	339	0.043	0.028	0.030
303	5.93	3.81	3.05	340	0.040	0.026	0.030
304	5.72	3.58	2.87	341	0.037	0.026	0.031
305	5.47	3.36	2.69	342	0.035	0.024	0.043
306	5.18	3.16	2.53	343	0.034	0.025	0.035
307	4.86	2.99	2.37	344	0.033	0.025	0.032
308	4.50	2.82	2.25	345	0.031	0.025	0.033
309	4.14	2.64	2.10	346	0.030	0.023	0.034
310	3.80	2.47	1.96	347	0.029	0.021	0.030
311	3.50	2.29	1.80	348	0.026	0.022	0.029
312	3.24	2.10	1.64	349	0.024	0.022	0.026
313	3.02	1.91	1.48	350	0.022	0.023	0.025
314	2.81	1.72	1.34				

VII-C-9. 4-Methyl-2-hexanone [$CH_3C(O)CH_2CH(CH_3)CH_2CH_3$]

The photolysis of 4-methyl-2-hexanone follows the pattern of photodecompositions observed in other ketones (Ausloos, 1961):

$$CH_3C(O)CH_2CH(CH_3)CH_2CH_3 + h\nu$$

$$\rightarrow CH_3CO + CH_2CH(CH_3)CH_2CH_3 \quad \text{(I)}$$

$$\rightarrow CH_3 + CH_3CH_2CH(CH_3)CH_2CO \quad \text{(II)}$$

$$\rightarrow CH_3 + CO + CH_2CH(CH_3)CH_2CH_3 \quad \text{(III)}$$

$$\rightarrow CH_2{=}C(OH)CH_3 + CH_3CH{=}CHCH_3 \quad \text{(IVa)}$$

$$\rightarrow CH_2{=}C(OH)CH_3 + CH_2{=}CHCH_2CH_3 \quad \text{(IVb)}$$

The gas-phase photodecomposition of 4-methyl-2-hexanone shows an interesting competition between abstraction of secondary and primary H-atoms by the carbonyl O-atom in processes (IVa) and (IVb). The results at 313 nm give $(\phi_{I}+\phi_{II}+\phi_{III})/(\phi_{IVa}+\phi_{IVb}) < 0.02$ at 42°C and <0.10 at 150°. Quantum yields were not determined, but analyses of the *cis*- and *trans*-2-butene and 1-butene in the products were reported for several temperatures (44–258°C) and pressures of the ketone (7–50 Torr). For photolysis at 313 nm, ϕ_{IVb}/ϕ_{IVa} is a function of temperature, increasing from 0.11 at 44°C to 0.17 at 258°C. The ratio *trans*-2-butene/*cis*-2-butene varied from 0.32 to 0.38 as the temperature was increased from 44 to 258°C. An Arrhenius plot of the 1-C_4H_8/2-C_4H_8 ratio is shown in figure VII-C-18. The slope corresponds to a difference in activation energy of 3.3 kJ mol^{-1} for abstraction of a primary H-atom and a secondary H-atom by the carbonyl O-atom in this ketone.

For direct excitation at 253.7 nm the ratio $\phi_{IVb}/\phi_{IVa} = 0.27$, whereas the Hg($^3P_1$)-atom-sensitized decomposition of the ketone gives $\phi_{IVb}/\phi_{IVa} = 0.22$. Direct photolysis near 190 nm gives $\phi_{IVb}/\phi_{IVa} = 0.37 \pm 0.03$. Addition of 2.5 Torr of oxygen in the 313 nm photolysis at 44°C had no effect on the primary process (IV).

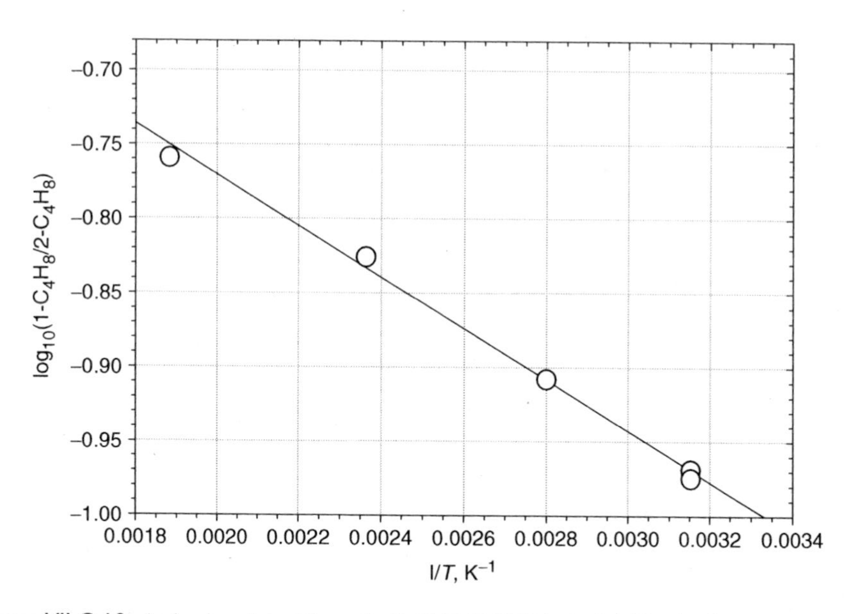

Figure VII-C-18. Arrhenius plot of the ratio (1-C_4H_8/2-C_4H_8) obtained by Ausloos (1961) in the photolysis of 4-methyl-2-hexanone. The slope corresponds to an activation energy difference of 3.3 kJ mol^{-1} for abstraction of a primary H-atom and a secondary H-atom by the carbonyl oxygen in the 313 nm excited ketone.

The appropriate data are not available to calculate realistic j-values for 4-methyl-2-hexanone at this time.

VII-C-10. 4-Heptanone [Di-n-Propyl Ketone, $CH_3CH_2CH_2C(O)CH_2CH_2CH_3$]

The primary photodecomposition pathways for 4-heptanone follow those observed in other ketones:

$$CH_3CH_2CH_2C(O)CH_2CH_2CH_3 + h\nu$$

$$\rightarrow CH_3CH_2CH_2 + CH_3CH_2CH_2CO \qquad \text{(I)}$$

$$\rightarrow CH_3CH_2CH_2 + CO + CH_3CH_2CH_2 \qquad \text{(II)}$$

$$\rightarrow CH_2{=}CH_2 + CH_2{=}C(OH)CH_2CH_2CH_3 \qquad \text{(III)}$$

The first investigators to study the photodecomposition of 4-heptanone were Bamford and Norrish (1935, 1938) who had available only relatively primitive equipment and analytical techniques by today's standards. However, they managed to establish the occurrence of both radical generation in processes (I) and (II) and the Norrish Type II rearrangement in process (III). Using filtered mercury lamp radiation (largely 248–277 nm) and a calibrated thermopile system to determine quantum yields of products of 4-heptanone photodecomposition, Bamford and Norrish (1938) reported: $\Phi_{CO} \approx 0.26$ (15°C), 0.36 (74°C), and 0.37 (100°C); and $\Phi_{C2H4} \approx 0.45$ (15°C), 0.30 (74°C), and 0.29 (100°C).

The photolysis of 4-heptanone at 313 nm was studied later by Masson (1952), who reported: $\Phi_{CO} = 0.31 \pm 0.03$ (at 113°C) and $\Phi_{C2H4} = 0.22$ (8 Torr) and 0.20 (115 Torr), independent of temperature. Nicol and Calvert (1967), in experiments at 313 nm, determined $\Phi_{C2H4} = 0.21 \pm 0.01(61.3 - 151°C)$ and observed that Φ_{CO} increased from 0.26 at 52.3°C to 0.37 at 150°C. Their product mass balance of CO and the observed hydrocarbons (propane, propene, and hexane) supported a mechanism initiated by processes (I), (II), and (III).

From the previous data it is likely that $\phi_{III} = 0.21 \pm 0.01$, independent of temperature and ketone pressure, with $\phi_I + \phi_{II} \leq 0.34 \pm 0.06$. The appropriate data to estimate the j-values for 4-heptanone photodecomposition for tropospheric conditions are not available at this writing.

VII-C-11. 2,4-Dimethyl-3-pentanone [Di-Isopropyl Ketone, $(CH_3)_2CHC(O)CH(CH_3)_2$]

The photodecomposition of 2,4-dimethyl-3-pentanone occurs by the following pathways:

$$(CH_3)_2CHC(O)CH(CH_3)_2 + h\nu \rightarrow (CH_3)_2CH + (CH_3)_2CHCO \qquad \text{(I)}$$

$$\rightarrow (CH_3)_2CH + CO + (CH_3)_2CH \qquad \text{(II)}$$

Whiteway and Masson (1955) studied the photodecomposition of pure 2,4-dimethyl-3-pentanone at 313 nm and 50–150°C; they reported $\Phi_{CO} = \phi_I + \phi_{II} = 0.98 \pm 0.04$. The efficiency of quenching of these processes by air at 1 atm pressure remains uncertain, and j-value estimation is unwarranted at this writing.

VII-C-12. 2-Methyl-3-hexanone [n-Propyl Isopropyl Ketone, $CH_3CH_2CH_2COCH(CH_3)_2$]

The photodecomposition of 2-methyl-3-hexanone probably occurs by the following pathways:

$$CH_3CH_2CH_2COCH(CH_3)_2 + h\nu \rightarrow CH_3CH_2CH_2CO + CH(CH_3)_2 \quad (I)$$

$$\rightarrow CH_3CH_2CH_2 + (CH_3)_2CHCO \quad (II)$$

$$\rightarrow CH_3CH_2CH_2 + CO + CH(CH_3)_2 \quad (III)$$

$$\rightarrow CH_2{=}CH_2 + CH_2{=}C(OH)CH(CH_3)_2 \quad (IV)$$

Quantum yield measurements of Nicol and Calvert (1967) from photolysis of the pure ketone at 313 nm and 59–150°C suggest a preference for bond cleavage in the competition between pathways (I) and (II). The ratio of the quantum yields of 2-methylpentane (derived from 2-propyl + 1-propyl combination) and hexane (formed from 1-propyl + 1-propyl combination), coupled with the assumption of the near equality of the association reactions forming these, suggests [2-propyl]/[1-propyl] $\approx 2.7 \pm 0.5 (59 - 150°C)$; this may reflect $\phi_I > \phi_{II}$. The values of Φ_{CO} are consistent with $\phi_I + \phi_{II} + \phi_{III} = 0.69 \pm 0.06 (100 - 150°C)$. Φ_{C2H4} values show a small increase with temperature, but an average over the temperature range 59–150°C gives the estimate of $\phi_{IV} \approx 0.12 \pm 0.03$. The total $\phi_I + \phi_{II} + \phi_{III} + \phi_{IV} = 0.83 \pm 0.06 (59 - 150°C)$ for photolysis of the pure ketone is near unity, similar to the case of 2,4-dimethyl-3-pentanone.

The necessary data to derive realistic j-values for 2-methyl-3-hexanone in the troposphere are not available at this writing.

VII-C-13. 4,4-Dimethyl-2-pentanone [Methyl Neopentyl Ketone, $CH_3C(O)CH_2C(CH_3)_3$]

4,4-Dimethyl-2-pentanone provides nine H-atoms on its three γ-carbon atoms, and it is interesting to see the effect of this on the efficiency of the Norrish type-II process in the photodecomposition of this ketone. The following primary processes have been suggested to rationalize the data (Martin and Pitts, 1955):

$$CH_3C(O)CH_2C(CH_3)_3 + h\nu \rightarrow CH_3CO + CH_2C(CH_3)_3 \quad (I)$$

$$\rightarrow CH_3 + (CH_3)_3CCH_2CO \quad (II)$$

$$\rightarrow CH_3 + CO + CH_2C(CH_3)_3 \quad (III)$$

$$\rightarrow CH_2{=}C(OH)CH_3 + CH_2{=}C(CH_3)_2 \quad (IV)$$

$$\rightarrow C(CH_3)_3 + CH_2COCH_3 \quad (V)$$

Φ_{CO} data from 313 nm photolysis allow an estimate of $\phi_I + \phi_{II} + \phi_{III} \approx 0.042$–$0.075$ (120–250°C). 2-Methylpropene and acetone quantum yields give $\phi_{IV} = 0.25 \pm 0.02$ (2-methylpropene, 120–250°C) and 0.22 ± 0.04 (acetone, 120–210°C). The presence of 2-methylpropane among the products suggests the occurrence of minor process (V), a bond cleavage down-chain from the energy-absorbing carbonyl group: $\phi_V \leq 0.005$(120°C), 0.02 (250°C).

In photolysis at 265.4 nm, Φ_{CO} data of Martin and Pitts (1955) give $\phi_I + \phi_{II} + \phi_{III} \approx 0.07 \pm 0.01$ (120–250°C). The data for process (IV) appear to show a temperature dependence; the average quantum yields give $\phi_{IV} = 0.31 \pm 0.05$ (2-methylpropene, 120–250°C), and 0.27 ± 0.05 (acetone, 120–250°C). As in photolysis at 313 nm, process (V) is relatively unimportant as judged from the 2-methylpropane yield: $\phi_V \geq 0.007 \pm 0.003$.

No meaningful estimate of *j*-values for 4,4-dimethyl-2-pentanone can be made from the limited data available at this writing.

VII-C-14. 4-Methyl-2-pentanone [Methyl Isobutyl Ketone, $CH_3C(O)CH_2CH(CH_3)_2$]

A single experiment by Martin and Pitts (1955) at 313 nm gave evidence of the following primary processes in 4-methyl-2-pentanone:

$$CH_3C(O)CH_2CH(CH_3)_2 + h\nu \rightarrow CH_3CO + CH_2CH(CH_3)_2 \quad (I)$$

$$\rightarrow CH_3 + (CH_3)_2CHCH_2CO \quad (II)$$

$$\rightarrow CH_3 + CO + CH_2CH(CH_3)_2 \quad (III)$$

$$\rightarrow CH_2{=}C(OH)CH_3 + CH_2{=}CHCH_3 \quad (IV)$$

Estimated quantum yields at 313 nm and 120°C are: $\phi_I + \phi_{II} + \phi_{III} \approx \Phi_{CO} = 0.15$; $\phi_{IV} \approx 0.35$ (propene). Sufficient data are not available to make realisitic estimates of *j*-value for this compound.

VII-C-15. 4-Octanone [*n*-Propyl *n*-Butyl Ketone, $CH_3CH_2CH_2C(O)CH_2CH_2CH_2CH_3$]

4-Octanone photodecomposition can be described in terms of the following processes:

$$CH_3CH_2CH_2C(O)CH_2CH_2CH_2CH_3 + h\nu$$

$$\rightarrow CH_3CH_2CH_2 + CH_3CH_2CH_2CH_2CO \quad (I)$$

$$\rightarrow CH_3CH_2CH_2CO + CH_3CH_2CH_2CH_2 \quad (II)$$

$$\rightarrow CH_3CH_2CH_2 + CO + CH_3CH_2CH_2CH_2 \quad \text{(III)}$$

$$\rightarrow CH_2{=}CH_2 + CH_2{=}C(OH)CH_2CH_2CH_2CH_3 \quad \text{(IV)}$$

$$\rightarrow CH_2{=}CHCH_3 + CH_2{=}C(OH)CH_2CH_2CH_3 \quad \text{(V)}$$

The quantum yields of products from photodecomposition of 4-octanone at 313 nm and temperatures from 76 to 151°C were measured by Nicol and Calvert (1967); the data suggest: $\phi_I + \phi_{II} + \phi_{III} = 0.078 \pm 0.022$ (70–151°C); $\phi_{IV} = 0.030 \pm 0.005$ (temperature dependent); $\phi_V = 0.054 \pm 0.017$ (temperature dependent). An Arrhenius plot of the ratio ϕ_{IV}/ϕ_V gives an activation energy difference of 7.4 kJ mol^{-1} for the abstraction of a primary H-atom and a secondary H-atom by the carbonyl oxygen in 313 nm-excited 4-octanone.

Data are insufficient to estimate realistic *j*-values for this ketone.

VII-C-16. 2-Methyl-4-heptanone [*n*-Propyl Isobutyl Ketone, $(CH_3)_2CHCH_2C(O)CH_2CH_2CH_3$]

Photodissociation studies of 2-methyl-4-heptanone at 313 nm suggest the following primary reaction channels:

$$CH_3CH_2CH_2C(O)CH_2CH(CH_3)_2 + h\nu$$

$$\rightarrow CH_3CH_2CH_2 + (CH_3)_2CHCH_2CO \quad \text{(I)}$$

$$\rightarrow CH_3CH_2CH_2CO + (CH_3)_2CHCH_2 \quad \text{(II)}$$

$$\rightarrow CH_3CH_2CH_2 + CO + (CH_3)_2CHCH_2 \quad \text{(III)}$$

$$\rightarrow CH_2{=}CH_2 + CH_2{=}C(OH)CH_2CH(CH_3)_2 \quad \text{(IV)}$$

$$\rightarrow CH_2{=}CHCH_3 + CH_2{=}C(OH)CH_2CH_2CH_3 \quad \text{(V)}$$

The studies of Nicol and Calvert (1967) provide estimates of the quantum efficiencies of these processes for photolysis at 313 nm: $\phi_I + \phi_{II} + \phi_{III} = 0.21 \pm 0.04$ (75–150°C); $\phi_{IV} = 0.055 \pm 0.003$ (75–150°C); $\phi_V = 0.11 \pm 0.04$ (75–150°C).

There are insufficient data to estimate realistic *j*-values for this ketone.

VII-C-17. 3-Methyl-4-heptanone [*n*-Propyl *sec*-Butyl Ketone, $CH_3CH_2CH_2C(O)CH(CH_3)CH_2CH_3$]

Photodecomposition modes of 3-methyl-4-heptanone follow the pattern of other ketones:

$$CH_3CH_2CH_2C(O)CH(CH_3)CH_2CH_3 + h\nu$$

$$\rightarrow CH_3CH_2CH_2CO + CH_3CHCH_2CH_3 \quad \text{(I)}$$

$$\rightarrow CH_3CH_2CH_2 + CH_3CH_2CH(CH_3)CO \quad \text{(II)}$$

$$\rightarrow CH_3CH_2CH_2 + CO + CH_2CHCH_2CH_3 \quad \text{(III)}$$

$$\rightarrow CH_2{=}CH_2 + CH_2{=}C(OH)CH(CH_3)CH_2CH_3 \quad \text{(IV)}$$

$$\rightarrow CH_2{=}CH_2 + CH_3CH{=}C(OH)CH_2CH_2CH_3 \quad \text{(V)}$$

The studies of Nicol and Calvert (1967) at 313 nm provide the following estimates of the quantum yields of the primary processes: $\phi_I + \phi_{II} + \phi_{III} = 0.39 \pm 0.04$ (77–150°C); $\phi_{IV} + \phi_V = 0.10 \pm 0.02$ (77–150°C).

The data are insufficient to estimate realistic *j*-values for this ketone.

VII-C-18. 2,2-Dimethyl-3-hexanone [*n*-Propyl *tert*-Butyl Ketone, $CH_3CH_2CH_2C(O)C(CH_3)_3$]

Photodecomposition pathways of 2,2-dimethyl-3-hexanone are similar to those of other ketones:

$$(CH_3)_3CC(O)CH_2CH_2CH_3 + h\nu \rightarrow (CH_3)_3C + CH_3CH_2CH_2CO \quad \text{(I)}$$

$$\rightarrow (CH_3)_3CCO + CH_3CH_2CH_2 \quad \text{(II)}$$

$$\rightarrow (CH_3)_3C + CO + CH_3CH_2CH_2 \quad \text{(III)}$$

$$\rightarrow CH_2{=}CH_2 + CH_2{=}C(OH)C(CH_3)_3 \quad \text{(IV)}$$

In the temperature range 76–152°C in the studies of Nicol and Calvert (1967), little discrimination was seen between the bond breakage in processes (I) and (II): $\phi_I/\phi_{II} \approx 0.82 \pm 0.11$. The data provide estimates of the quantum efficiencies of the various reaction channels at 313 nm: $\phi_I + \phi_{II} + \phi_{III} = 0.57 \pm 0.04$ (76–152°C); $\phi_{IV} = 0.097 \pm 0.016$ (76–152°C).

Styrene was the most efficient quencher of triplet excited states of carbonyl compounds among a great variety of olefins employed by Ausloos and Rebbert (1964), yet there was no detectable influence of added styrene (20.1 Torr added to 19.9 Torr of 2,2-dimethyl-3-hexanone) on the ethene and carbon monoxide quantum yields (Nicol and Calvert, 1967). However, an almost quantitative removal of radical products of 1-propyl and 2-methyl-2-propyl radicals resulted. Since Φ_{CO} was unaffected by styrene addition, the radical product suppression must have occurred through the radical addition to styrene; this is consistent also with the induced polymerization of the styrene observed by Nicol and Calvert (1967). Obviously, whatever the precursor states to the photodecomposition might be, they are very short lived.

There are not sufficient data to allow estimates of the *j*-values for this ketone.

VII-C-19. 2,6-Dimethyl-4-heptanone [Di-*iso*-Butyl Ketone, $(CH_3)_2CHCH_2C(O)CH_2CH(CH_3)_2$]

The photodecomposition modes of this ketone follow those of the other ketones:

$$(CH_3)_2CHCH_2C(O)CH_2CH(CH_3)_2 + h\nu$$

$$\rightarrow (CH_3)_2CHCH_2CO + (CH_3)_2CHCH_2 \quad \text{(I)}$$

$$\rightarrow (CH_3)_2CHCH_2 + CO + (CH_3)_2CHCH_2 \quad \text{(II)}$$

$$\rightarrow CH_2{=}CHCH_3 + CH_2{=}C(OH)CH_2CH(CH_3)_2 \quad \text{(III)}$$

The data from the full Hg-arc photolysis of 2,6-dimethyl-4-heptanone (Kraus and Calvert, 1957) allow the following estimates of the average quantum efficiencies (~250–340 nm) of the photodecomposition modes of 2,6-dimethyl-4-heptanone: $\phi_I + \phi_{II} \approx \Phi_{CO} = 0.18 \pm 0.02$ (78–155°C); $\phi_{III} \approx \Phi_{C3H6} = 0.36 \pm 0.02$ (78–155°C).

There are not sufficient data to estimate realistic *j*-values for this ketone.

VII-C-20. 3,5-Dimethyl-4-heptanone [Di-*sec*-Butyl Ketone, $CH_3CH_2CH(CH_3)C(O)CH(CH_3)CH_2CH_3$]

Photodecomposition of 3,5-dimethyl-4-heptanone involves both bond cleavage and Norrish type II reactions:

$$CH_3CH_2CH(CH_3)C(O)CH(CH_3)CH_2CH_3 + h\nu$$

$$\rightarrow CH_3CH_2CHCH_3 + CH_3CH_2CH(CH_3)CO \quad \text{(I)}$$

$$\rightarrow 2CH_3CHCH_2CH_3 + CO \quad \text{(II)}$$

$$\rightarrow CH_2{=}CH_2 + CH_3CH{=}C(OH)CH(CH_3)CH_2CH_3 \quad \text{(III)}$$

Data from full Hg-arc photolysis of 3,5-dimethyl-4-heptanone by Kraus and Calvert (1957) suggest the following estimates of the average quantum yields (250–340 nm, temperature 98–105°C) of the primary photodecomposition processes: $\Sigma(\Phi_{C4H9\text{-groups}})/2 \approx 0.30 \pm 0.03 \approx \phi_I + \phi_{II}$; $\Phi_{C2H4} \approx 0.29 \pm 0.03 \approx \phi_{III}$. There are insufficient data to estimate realistic *j*-values for this ketone.

VII-C-21. 2,2,4,4-Tetramethyl-3-pentanone [Di-*tert*-Butyl Ketone, $(CH_3)_3CC(O)C(CH_3)_3$]

The photodecomposition of 2,2,4,4-tetramethyl-3-pentanone involves only bond cleavage in Norrish type I reactions:

$$(CH_3)_3CC(O)C(CH_3)_3 + h\nu \rightarrow (CH_3)_3C + (CH_3)_3CCO \quad \text{(I)}$$

$$\rightarrow (CH_3)_3C + CO + C(CH_3)_3 \quad \text{(II)}$$

The full Hg-arc photolysis of 2,2,4,4-tetramethyl-3-pentanone was studied by Kraus and Calvert (1957). With the assumption of a unity quantum yield of CO product at the temperatures in the range 114–251°C , the hydrocarbon product analyses give: $\Sigma(\Phi_{C4H9\text{-groups}})/2 \approx \phi_{I} + \phi_{II} = 0.84 \pm 0.05$. It is interesting that the high efficiency of photodissociation of 2,2,4,4-tetramethyl-3-pentanone occurs even in solution. Abuin et al. (1984) studied the photodissociation of this ketone at 313 nm in various homogeneous solvents of different viscosities. A marked effect on the apparent extent of photodissociation resulted from cage reactions of the geminate radicals. Measured values of $\phi_{I} + \phi_{II}$ varied from 0.5 in paraffin oil to 1.0 in heptane solutions. There are not sufficient data to allow realistic estimates of the *j*-values for this ketone.

VII-C-22. Some Effects of Molecular Structure on the Quantum Efficiencies of Photodecomposition Modes in the Ketones

Some interesting relations between molecular structure of the ketone and the quantum yield of primary products can be seen in the summary of quantum yield data for the different ketones of common structure $CH_3CH_2CH_2C(O)R$ given in table VII-C-8. In column (3), observe that the $\phi_{Ia} + \phi_{Ib}$ values show a rather random variation in magnitude over a considerable range (0.10 for R = $CH_3CH_2CH_2CH_2$ to 0.71 for R = CH_3CHCH_3). One might have naively expected Φ_{CO} (taken as $\phi_{Ia} + \phi_{Ib}$) to be highest

Table VII-C-8. Summary of quantum yields for primary processes in the 313 nm photolysis of a series of $CH_3CH_2CH_2COR$ ketones at 150°C (based on the data of Nicol and Calvert, 1967)[a]

$$CH_3CH_2CH_2C(O)R + h\nu \rightarrow CH_3CH_2CH_2 + RCO \quad \text{(Ia)}$$

$$\rightarrow CH_3CH_2CH_2CO + R \quad \text{(Ib)}$$

$$\rightarrow CH_2{=}CH_2 + CH_2{=}C(OH)R \quad \text{(IIa)}$$

$$\rightarrow CH_3CH_2CH_2C(OH){=}CHR' + C_nH_{2n} \quad \text{(IIb)}$$

R	Parent Carbonyl Compound	$\phi_{Ia} + \phi_{Ib}$	ϕ_{IIa}	ϕ_{IIb}	$\phi_{IIa} + \phi_{IIb}$	$\Sigma\phi_i$
H	Butanal	0.35[b]	0.170	—	0.17	0.52
CH_3	2-Pentanone	0.43	0.252	—	0.25	0.68
C_2H_5	3-Hexanone	0.57	0.203	—	0.20	0.77
$CH_3CH_2CH_2$	4-Heptanone	0.37	0.106[c]	0.106[c] (0.099)	0.21	0.58
CH_3CHCH_3	2-Methyl-3-hexanone	0.71	0.155[d]	—	0.16	0.87
$CH_3CH_2CH_2CH_2$	4-Octanone	0.10	0.035	0.248	0.28	0.38
$(CH_3)_2CHCH_2$	2-Methyl-4-heptanone	0.24	0.055	0.151	0.21	0.45
$CH_3CHCH_2CH_3$	3-Methyl-4-heptanone	0.36	0.062[d]	0.061[d]	0.12	0.48
$(CH_3)_3C$	2,2-Dimethyl-3-hexanone	0.54	0.118[d]	—	0.12	0.66

[a] Values are the arithmetic mean of individual values at 150°C except for those ketones for which the type II process is independent of temperature; the tabular values for ϕ_{IIa} and ϕ_{IIb} represent the arithmetic average of all individual values at all temperatures in these cases.
[b] Estimated from butanal photolysis, with added I_2 reported by Blacet and Calvert (1951b).
[c] Values are obtained by dividing the total alkene yield by 2, assuming equal participation by the two equivalent sides of the molecule.
[d] These Norrish type II yields are temperature dependent.

for the 2,2-dimethyl-3-hexanone since this compound has the weakest C—C bond in these molecules: $(CH_3)_3C—C(O)CH_2CH_2CH_3$. However, the Φ_{CO} for the 3-hexanone is even higher. Obviously, other factors in addition to the product radical stability influence ϕ_I values.

Qualitatively, one can conclude from the type II photodecomposition data of table VII-C-8 that the fraction of $CH_3CH_2CH_2C(O)R$ compounds that undergo the type-II process is surprisingly similar for ketones with varied R groups; see column (6) of the table. If the type-I and type-II processes are in competition from the same state, one expects that large variations in ϕ_{II} would occur.

There is no simple correlation evident between the number of γ-H-atoms present in the compound and ϕ_{II} values. For example, although 4-heptanone and the 2-methyl-4-heptanone contain two and three times, respectively, the number of γ-H-atoms as 3-hexanone, the total type-II process quantum yields are nearly the same for all four compounds. However, when there are different competitive type-II processes possible within the same excited molecule, and the different processes involve the transfer of a primary H-atom, there is nearly a statistical distribution of the products that is observed. Thus, in 3-methyl-4-heptanone, one expects equal yields of processes (IIa) and (IIb) from purely statistical reasoning; this is the observed result. In the photolysis of 2-methyl-4-heptanone one expects statistically twice the quantum yield since any one of six γ-H-atoms will lead to propene, while abstraction of any one of three γ-H-atoms may lead to ethene formation. Actually, the propene to ethene ratio is 2.7; apparently some structural preferences beyond that of statistics favor the propene-forming reaction to some degree. In both of these cases considered here, there is no significant activation energy difference between the process, and the quantum yields are temperature independent.

When H-atoms of different type are present in the γ-position, a significant preference is shown for abstraction of the atom of weaker C—H bond dissociation energy, and the quantum efficiencies of the competitive processes are temperature dependent. However, the sum of the two yields, $\phi_{IIa}+\phi_{IIb}$, is substantially independent of temperature. Note that the data for 4-octanone show the propene product is greatly favored over ethene in spite of the adverse 2:3 ratio of the number of secondary to primary H-atoms in the γ-position. An Arrhenius plot of the ratio Φ_{C2H4}/Φ_{C3H6} gives an apparent activation energy difference between the reaction involving primary H-atom abstraction and that involving the secondary H-atom abstraction of $7.4\,kJ\,mol^{-1}$. Ausloos (1961) reported $3.8\,kJ\,mol^{-1}$ from the 313 nm photolysis of 4-methyl-2-hexanone where a similar competition between primary-H- and secondary-H-atoms exists. The discrimination between different types of H-atom is also in accord with the preference found for H-atom over D-atom abstraction in the type-II process in $CH_3C(O)CH_2CDHCD_2H$ photolysis studied by Borkowski and Ausloos (1961).

The $\phi_{IIa}+\phi_{IIb}$ values are independent of temperature for the $CH_3CH_2CH_2C(O)R$ compounds with R=H, CH_3, C_2H_5, $CH_3CH_2CH_2$, $CH_3CH_2CH_2CH_2$, and $(CH_3)_2CH_2CH_2$. However, for the three ketones studied that contain multiple α-alkyl substituents [R=CH_3CHCH_3, $(CH_3)_2CHCH_2$, $(CH_3)_3C$], a temperature dependence of the ϕ_{II} values is seen. This corresponds to an apparent activation energy of ethene formation of about $8\,kJ\,mol^{-1}$ in each case. A somewhat smaller inverse temperature dependence to ϕ_{I} (about $-3\,kJ\,mol^{-1}$) is seen in each of these three ketones, such that the sum $\phi_{I}+\phi_{II}$ is approximately independent of the temperature.

There is an interesting quantitative relation between the extent of the α-alkyl substitution in the $CH_3CH_2CH_2C(O)R$ ketones and the size of ϕ_{II}. For example, the data in table VII-C-8 from experiments at 150° show a regular decrease in the ϕ_{II} values in the order $R = CH_3$ ($\phi_{II}=0.252$), C_2H_5 ($\phi_{II}=0.203$), CH_3CHCH_3 ($\phi_{II}=0.155$), $(CH_3)_3C$ ($\phi_{II}=0.118$). Substitution of one methyl group for a methyl hydrogen atom in 2-pentanone lowers the quantum yield of the type-II process by a factor of 0.80; a second methyl substitution in the α-position lowers ϕ_{II} by a factor of 0.76; a third methyl substitution in the α-position lowers ϕ_{II} again by an identical factor, 0.76. A similar lowering by a factor 0.6 in ϕ_{II} is seen when an α-methyl group is introduced in 3-heptanone. The consistency of the magnitude of the lowering of ϕ_{II} values with α-substitution suggests that the effect of α-methyl groups is real, although its detailed mechanistic explanation is open to question. From the near-constancy of the $\phi_{I}+\phi_{II}$ values for the series of ketones with $R = C_2H_5$, $CH_3CH_2CH_2$, $(CH_3)_2CHCH_2$ (0.20, 0.21, 0.21), it appears that the nature of the α-substituent is not highly important as long as only primary H-atoms are present in the γ-positions.

Nicol and Calvert (1967) suggested the following empirical equations relating ϕ_{II} and molecular structure of the $CH_3CH_2CH_2C(O)R$ ketones listed in table VII-C-8:

$$\phi_{IIa}+\phi_{IIb} = [0.252(0.78)^m + 0.038n][e^{-1.9p/RT}/e^{-2.25p}] \quad (1)$$

$$\phi_{IIa}/\phi_{IIb} = (N_a A_a/N_b A_b)e^{-\Delta E/RT} \quad (2)$$

The first term of equation (1) formulates the effect of α-alkyl substituents on ϕ_{II} and m represents the number of alkyl substituents attached to the α-carbon atom of the R group. The second term increases the calculated ϕ_{II} values for the case of ketones with *sec*-H atoms in the γ-position; n is the number of these atoms. Depending on the interpretation of the reactive state involved, the second term ($0.038n$) may be related either to the additional alkene product from singlet excited state participation in the type-II process or to an enhanced triplet reactivity. Presumably a third term would be required to account for tertiary H-atoms in the γ-position. The exponential term in equation (1) introduces a temperature dependence to the ϕ_{II} values for the highly substituted ketones; $p = 1$ when there are two or more alkyl substituents on the α-carbon atom and $p = 0$ for all other cases. Equation (2) is presumably related to the ratio of the competitive rate constants for abstraction of γ-H-atoms. $\Delta E = 0$ for those cases in which the two sources of process II are of the same type of γ-H-atom. ΔE equal to $5.4\,kJ\,mol^{-1}$ (units of R are $kJ\,mol^{-1}\,K^{-1}$) when both primary and secondary H-atoms are present in the γ-position in the ketone. Presumably a slightly higher value for ΔE will be required when there are tertiary H-atoms involved in the abstraction reaction. N_a and N_b are the numbers of γ-H-atoms of a given type in the $CH_3CH_2CH_2$- and R-group, respectively. A_a and A_b are the pre-exponential factors related to the rate coefficients k_{IIa} and k_{IIb}, expressed as a rate per γ-H-atom. Unless some peculiar geometrical configuration favors the type-II process or the singlet state is involved significantly as well

Table VII-C-9. Comparison of measured ϕ_{IIa} and ϕ_{IIb} values from 313 nm photolysis of the $CH_3CH_2CH_2C(O)R$ ketones with those calculated from equations (1) and (2) in section VII-C-22

R	Parent Carbonyl Compound	Temperature °C	ϕ_{IIa} Measured	ϕ_{IIa} Calculated	ϕ_{IIb} Measured	ϕ_{IIb} Calculated
CH_3	2-Pentanone	58–151	0.252	0.252	—	—
CH_3CH_2	3-Hexanone	62–151	0.203	0.197	—	—
$CH_3CH_2CH_2$	4-Heptanone	61–151	0.106	0.099[a]	0.106	0.099[a]
CH_3CHCH_3	2-Methyl-3-hexanone	150	0.155	0.154	—	—
		100	0.112	0.116	—	—
		60	0.085	0.085	—	—
$CH_3CH_2CH_2CH_2$	4-Octanone	150	0.035	0.034[b]	0.248	0.240[b]
		150	0.035	0.066[a]	0.248	0.206[a]
		110	0.029	0.029[b]	0.243	0.244[b]
		76	0.024	0.025[b]	0..237	0.248[b]
$(CH_3)_2CHCH_2$	2-Methyl-4-heptanone	75–150	0.055	0.066[a]	0.151	0.131[a]
$CH_3CHCH_2CH_3$	3-Methyl-4-heptanone	150	0.062	0.077[a]	0.061	0.077[a]
		110	0.052	0.060[a]	0.052	0.060[a]
		75	0.039	0.047[a]	0.039	0.047[a]
$(CH_3)_3C$	2,2-Dimethyl-3-hexanone	150	0.118	0.120	—	—
		110	0.093	0.093	—	—
		75	0.073	0.074	—	—

[a] Calculated from equations (1) and (2) in section VII-C-22, assuming $A_a = A_b$.
[b] Calculated assuming $A_a = (0.44)A_b$.

as the triplet, one would expect $A_a = A_b$. This assumption appears justified except for 4-octanone, where the choice of $A_a = (0.44)A_b$ was necessary to fit the data.

Reasonable agreement between ϕ_{IIa} and ϕ_{IIb} values calculated from equations (1) and (2) and those measured experimentally can be seen by comparison of the values shown in table VII-C-9. With the large number of variables involved in equations (1) and (2), it is not surprising that a reasonably good match of experimental and calculated ϕ_{II} values can be had, but the exercise allows some insight into the factors that affect ϕ_{II}.

Product quantum yields obtained in photolysis of a given ketone in the liquid phase are not expected to agree quantitatively with those determined in the gas phase. The more efficient vibrational relaxation of the excited states in solutions is a major factor that leads to this difference. However, the studies of Guillet and co-workers provide some useful insight into the effect of molecular structure on primary processes in aliphatic ketones that relate, at least qualitatively, to effects that would be seen in gas-phase photolysis. In this sense, it is important to review their findings here.

Hartley and Guillet (1968b) determined the quantum yields of the type-II reactions for a series of straight-chain, symmetrical ketones of structure RC(O)R, where the total length of the chain of carbon atoms in the ketone varied from 7 (4-heptanone) to

43 (22-tritetracontanone). The results were consistent with the occurrence of the modes of reaction common to the smaller ketones:

$$CH_3(CH_2)_nC(O)(CH_2)_nCH_3 + h\nu$$

$$\rightarrow CH_3(CH_2)_n + CH_3(CH_2)_nCO \quad \text{(I)}$$

$$\rightarrow R'CH{=}CH_2 + CH_2{=}C(OH)(CH_2)_nCH_3 \quad \text{(II)}$$

where R′ is an alkyl group of formula H—$(CH_2)_{n-2}$.

For photolysis at 313 nm in paraffin oil at 120°C, Hartley and Guillet (1968b) reported the following values of Φ_{CO} (= ϕ_I) for symmetrical ketones of total chain length shown in parentheses: 0.35 (7); 0.022 (9); 0.019 (11); 0.018 (13); 0.016 (15); 0.013 (23); 0.012 (43). The very large drop in ϕ_I as the carbon chain length increased from 7 to 9 was rationalized in terms of a competition of the excited molecule between dissociation in (I) or rearrangement in (II). The greatly increased rate of process (II) in 2-nonanone compared to 2-heptanone presumably reflects the abstraction of only secondary H-atoms in this and all the higher ketones, whereas only primary H-atoms are abstracted in 2-heptanone.

Hartley and Guillet (1968b) determined, from the quantum yields of the alkene and ketone products of (II), estimates of ϕ_{II} = 0.15(7); 0.108 (9); 0.092 (13); 0.080 (15); 0.072 (23); 0.072 (43); the total length of the ketone chain is shown in parentheses. As one might expect from process (II), there was a near equality between the 1-hexene and 2-nonanone products of 8-pentadecanone in photolysis in paraffin oil solutions at several temperatures: at 30°C, 0.066 and 0.067, respectively; 60° C, 0.068 and 0.067; 90°C, 0.091 and 0.086; 120°C, 0.087 and 0.091; 145°C, 0.092 and 0.102. The temperature dependence is small, equivalent to about 4 kJ mol^{-1}. The results of these photolyses of very long chain ketones are very similar to those observed by Hartley and Guillet (1968a) in the 313 nm photolysis of ethene–carbon monoxide copolymers that have very long chain ketone structures.

Golemba and Guillet (1972) studied the primary processes in another series of structurally related ketones: linear 2-alkanones [$CH_3C(O)R$] ranging from C_6 (2-hexanone) to C_{19} (2-nonadecanone). These were photolyzed in a variety of hydrocarbon solvents at 313 nm, and quantum yields of photodissociation by both primary processes (I) and (II) were determined. Using 2-undecanone in dodecane solvent, they measured the nonane quantum yields formed by H-atom abstraction from the solvent following the photodissociation: n-$C_9H_{19}C(O)CH_3 + h\nu \rightarrow CH_3CO + n$-$C_9H_{19}$ (I). They found that Φ_{nonane} increased from 0.00036 at 24°C to 0.012 at 117.5°C, where it remained constant at higher temperatures. These data suggested that ϕ_I could be estimated from measurement of the quantum yield of the RH product at temperatures > 117°C in $CH_3C(O)R$ photolysis in hydrocarbon solvents. Measured quantum yields of the alkane product gave estimates of ϕ_I for photolysis at various temperatures: 2-heptanone: 0.0060 (70°C); 0.015 (100°C); 2-octanone: 0.018 (120°C); 2-nonanone: 0.0046 (70°C); 0.011 (100°C); 2-undecanone: 0.0033 (25°C); 0.0019 (70°C); 0.0074 100°C; 0.012 (120°C); 2-nonadecanone: ~0.002 (70°C). It is clear that process (I) is relatively unimportant compared to (II) in all of these ketones, and it appears to decrease somewhat with increase in chain length of the ketone.

Quantum yields of acetone and the corresponding alkene product of the Norrish type-II reaction for the 2-alkanones were found to be near equal in preliminary experiments of Golemba and Guillet (1972). In most experiments, they used the alkene to monitor the extent of the type-II reaction. The results showed a slight decrease in ϕ_{II} with increase in chain length of the ketone, which is shown in parentheses: $\phi_{II} = 0.25$ (6); 0.20 (7); 0.20 (8); 0.20 (9); 0.20 (11); 0.15 (19).

Solution-phase photolyses of the ketones were made with added triplet molecule quencher, *cis*-1,3-cyclooctadiene, at concentrations sufficient to quench virtually all of the triplet reactions. From such experiments, Golemba and Guillet (1972) found the ratios of the quantum yields of the triplet to that of the singlet participation in reaction (II): $\phi_{II}^{T}/\phi_{II}^{S} = 0.65$ with 2-hexanone; 0.61 with 2-octanone; 0.64 with 2-nonadecanone.

The cyclobutanol products of the photolysis of these ketones were analyzed only for 2-nonanone, where the quantum yield $\phi_{cyclobutanol} = 0.02$. Golemba and Guillet (1972) compare this value with that reported from 2-hexanone (0.075; Coulson and Yang, 1966) and surmise that $\phi_{cyclobutanone}$ decreases with increase in chain length.

Our discussion here focuses on the effects of structure on aliphatic ketones. However, it should be noted that Pitts et al. (1968) determined the effect of the length of the alkyl group (R) on the Norrish type-II process seen in 313 nm photolysis of alkyl phenyl ketones $C_6H_5C(O)R$ in benzene or hexane solutions. They found ϕ_{II} decreased from 0.40 for R = 1-C_3H_7 to 0.29 for R = 1-C_8H_{17}.

Additional illustration of the influence of ketone structure on the quantum yields of photodecomposition at 313 nm can be seen in figure VII-C-19. Here the existing quantum yield data for the ketones of structure $CH_3C(O)R$ and RC(O)R are presented vs. the length of the straight-chain alkyl radical. Data from both photolyses in the gas phase and in paraffin solution (Hartley and Guillet, 1968b; Golemba and Guillet, 1972) show similar trends with structural change of the ketone. In figure VII-C-19a it is seen that the radical generation, $CH_3C(O)R + h\nu \rightarrow CH_3CO + R$ (I), in ketones of the structure $CH_3C(O)R$, maximizes for ketones with R = C_2H_5, and a significant decline in quantum efficiency occurs as one progresses to R = $C_{17}H_{35}$. On the other hand, process (II), forming alkene and enol-acetone, $CH_3C(O)R + h\nu \rightarrow R'CH{=}CH_2 + CH_2{=}CH(OH)CH_3$ (II), begins with R = C_3H_7, and the process continues to be the dominant mode of photodecomposition up to the largest ketone in the series studied, $CH_3C(O)C_{17}H_{35}$.

The quantum yield data for the photodecomposition of symmetrical ketones, RC(O)R, at 313 nm show a similar behavior; see figure VII-C-19b. Radical generation, $RC(O)R + h\nu \rightarrow R + RCO$, falls rapidly in progression from the ketone with R = C_3H_7 to R = $C_{17}H_{35}$. However, the quantum yield of the Norrish type-II process in these ketones, $RC(O)R + h\nu \rightarrow R'CH{=}CH_2 + CH_2{=}CH(OH)R''$(II), continues to be significant up to the ketone of structure $C_{17}H_{35}C(O)C_{17}H_{35}$. As with the aldehydes, this process (II) forms highly reactive enols of ketones and alkenes that can enhance ozone generation in the lower troposphere (e.g., see Calvert et al., 2000).

One should gather from our discussion that Norrish type-II photodecomposition in ketones is important in both gas- and liquid-phase systems, but it can also occur in

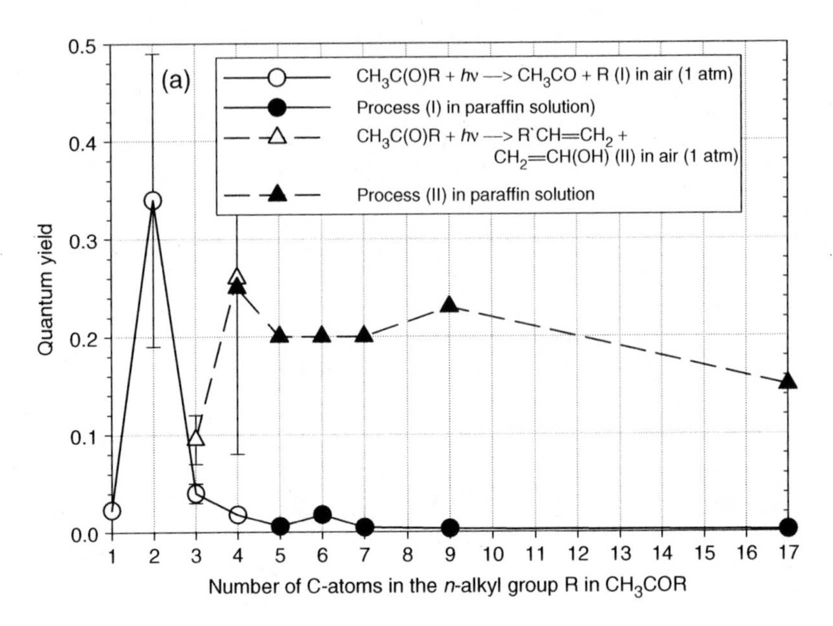

Figure VII-C-19a. Plot of the quantum yields of primary processes in photolysis at 313 nm for ketones of structure $CH_3C(O)R$, where R is a straight-chain alkyl group with the number of C-atoms varied from 1 to 17. Data are shown for photolyses in the gas phase and in paraffin solution (Golemba and Guillet, 1972).

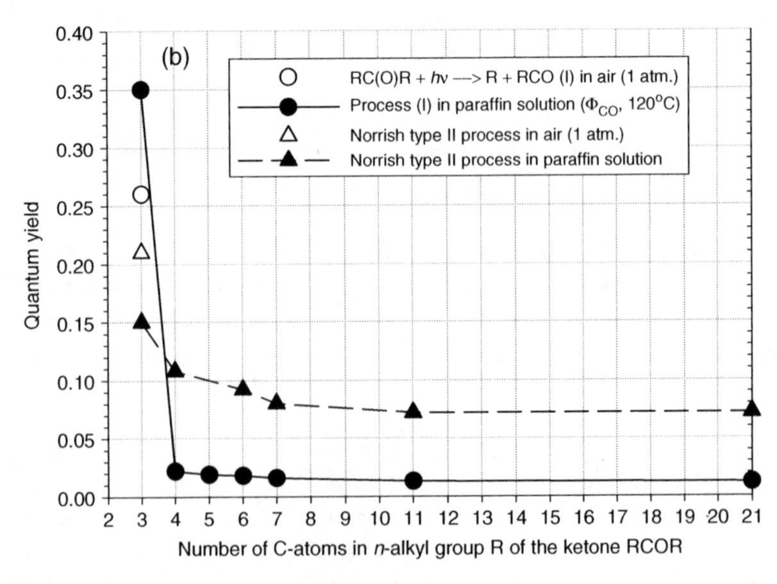

Figure VII-C-19b. Plot of the quantum yields of primary processes in photolysis at 313 nm for ketones of the structure RC(O)R, where R is a straight-chain alkyl group with the number of C-atoms varied from 3 to 21. Data are shown for both photolyses in the gas phase and in paraffin solution (Hartley and Guillet, 1968b). The Norrish type II process can be represented in this case as: $RC(O)R + h\nu \rightarrow R'CH{=}CH_2 + CH_2{=}CH(OH)R''$(II).

solid ethene–CO copolymers with $\phi_{II} = 0.025$ independent of temperature (-25 to $90°C$). However, when temperatures between -50 and $-150°C$ are reached at which free rotation about the C—C bonds is restricted, ϕ_{II} drops from 0.015 to 0.000 (Hartley and Guillet, 1968a).

VII-D. Photodecomposition of Cyclic Ketones

Cyclic alkanes present in the troposphere are oxidized in part to cyclic ketones. The mechanisms of photodecomposition of these ketones are reviewed briefly in this section.

VII-D-1. Cyclopropanone (*cyclo*-C_3H_4O)

Cyclopropanone is the simplest of the cyclic ketones. Studies of the mechanism of its gas phase photochemistry suggest the following photodecomposition pathways:

$$cyclo\text{-}(H_2C)_2C{=}O + h\nu \rightarrow CH_2{=}CH_2 + CO \qquad \text{(I)}$$

$$\rightarrow H_2C{\cdots}C({=}O){\cdots}CH_2 \rightarrow \text{Products} \qquad \text{(II)}$$

The only gaseous products identified are ethene and carbon monoxide, formed in equal amounts (Thomas and Rodriguez, 1971; Rodriguez et al., 1976). Several theoretical studies of the mechanism of process (I) have been given (Martino et al., 1977; Yamabe et al., 1979; Sevin et al., 1981). Oxyallyl, shown as the product in (II), has been suggested as a possible intermediate to polymer formation that appears on the walls of reaction vessels in laboratory studies of cyclopropanone photolysis. This molecule is expected to have a small energy barrier to its recyclization (Olsen et al., 1971; Liberles et al., 1973). If it is formed in (II), its possible fate in the atmosphere is open to conjecture.

VII-D-1.1. Absorption Cross Sections for Cyclopropanone

The absorption spectrum of cyclopropanone was reported by Thomas and Rodriguez (1971), who estimated that the 0–0 transition in this molecule lies near 395 nm. The spectrum given in figure VII-D-1 is approximate, since it has been redrawn from the original publication. Compare this spectrum with that of acetone, the other simple 3-carbon ketone. One can see that the absorption profile of cyclopropanone is shifted to much longer wavelengths than that of acetone. Tabular listing of the approximate cross sections for cyclopropanone is given in table VII-D-1.

VII-D-1.2. Quantum Yields of Photodecomposition of Cyclopropanone

The quantum yields of the products of cyclopropanone are shown as a function of wavelength in figure VII-D-2, as reported in the studies of Thomas and Rodriguez (1971)

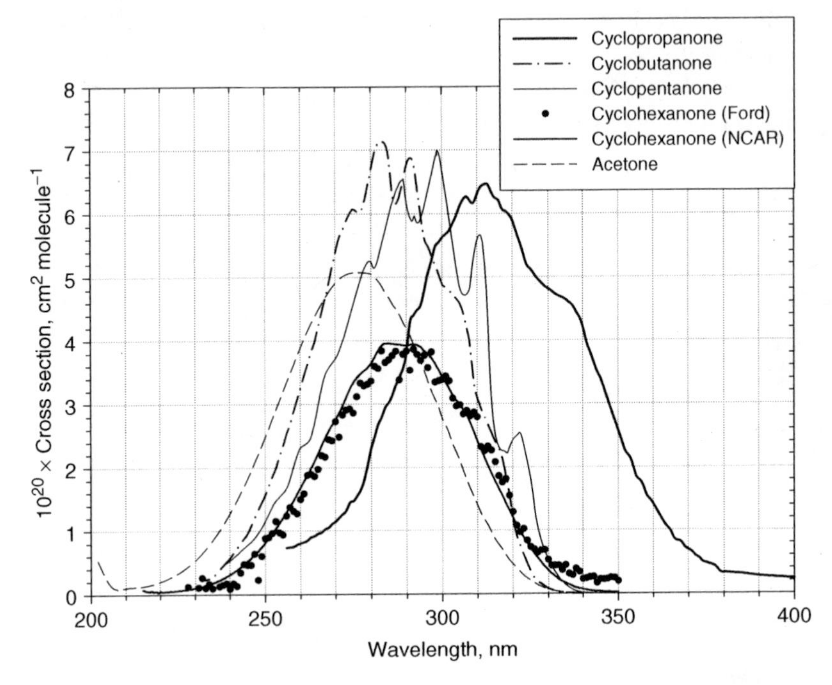

Figure VII-D-1. Approximate absorption cross sections vs. wavelength for cyclopropanone vapor (adapted from Thomas and Rodriguez, 1971); cyclobutanone measured in cyclohexane solution (adapted from Carless and Lee, 1972); cyclopentanone in cyclohexane solution (adapted from Nakashima et al., 1982); cyclohexanone vapor measured independently in two laboratories: National Center for Atmospheric Research (NCAR) and Ford Motor Company (Ford), (Iwasaki et al., 2008); acetone vapor (Atkinson et al., 2004).

and Rodriguez et al. (1976). The major products are ethene and carbon monoxide (circles in figure VII-D-2). The quantum yield of photodestruction of cyclopropanone is near unity at all wavelengths (shown as squares in the figure), so a significant fraction of the products remains unidentified: ~0.41 (365 nm); ~0.36 (334 nm); ~0.29 (313 nm). These unknown products may be formed from oxyallyl reactions or some other intermediate species formed in (II). The tabular listing of the quantum yields of process (I) is given in table VII-D-1.

VII-D-1.3. Approximate Photolysis Frequencies for Cyclopropanone

The cross sections and quantum yields given in table VII-D-1, coupled with the appropriate solar flux data for a cloudless day within the lower troposphere (ozone column = 350 DU), were used to derive the approximate *j*-values for cyclopropanone photodecomposition into C_2H_4 and CO, process (I). These are shown as the solid curve in figure VII-D-3. Given by the dashed curve in this figure are the *j*-values for the total of the photodecomposition processes. These data suggest that the lifetime for photodecomposition in the lower troposphere is about 45 min with an overhead Sun. For 40°N latitude, clear tropospheric skies, and diurnal

Table VII-D-1. Approximate absorption cross sections (cm^2 $molecule^{-1}$) at 293 K and quantum yields as a function of wavelength for the primary process cyclopropanone + $h\nu \rightarrow$ C_2H_4 + CO (I)

λ (nm)	$10^{20} \times \sigma$	ϕ_I	λ (nm)	$10^{20} \times \sigma$	ϕ_I	λ (nm)	$10^{20} \times \sigma$	ϕ_I
278	1.80	1.00	320	5.88	0.68	361	1.32	0.61
279	2.02	1.00	321	5.71	0.67	362	1.22	0.61
280	2.29	1.00	322	5.54	0.67	363	1.13	0.60
281	2.48	1.00	323	5.36	0.67	364	1.05	0.60
282	2.67	1.00	324	5.18	0.66	365	0.97	0.60
283	2.79	1.00	325	5.12	0.66	366	0.86	0.59
284	2.94	1.00	326	5.04	0.66	367	0.83	0.58
285	3.02	1.00	327	4.97	0.65	368	0.80	0.57
286	3.13	1.00	328	4.91	0.65	369	0.73	0.57
287	3.29	1.00	329	4.85	0.65	370	0.69	0.56
289	3.55	1.00	330	4.81	0.65	371	0.65	0.55
290	3.82	1.00	331	4.78	0.65	372	0.59	0.54
291	4.28	0.99	332	4.73	0.64	373	0.55	0.53
292	4.39	0.98	333	4.70	0.64	374	0.50	0.52
293	4.48	0.96	334	4.67	0.64	375	0.48	0.51
294	4.55	0.94	335	4.64	0.64	376	0.46	0.49
295	4.78	0.92	336	4.60	0.64	377	0.44	0.47
296	5.04	0.90	337	4.55	0.64	378	0.40	0.46
297	5.23	0.89	338	4.45	0.64	379	0.33	0.44
298	5.46	0.87	339	4.35	0.64	380	0.32	0.42
299	5.54	0.86	340	4.20	0.64	381	0.32	0.36
300	5.62	0.85	341	3.97	0.63	382	0.32	0.32
301	5.67	0.83	342	3.82	0.63	383	0.31	0.13
302	5.73	0.82	343	3.67	0.63	384	0.31	0
303	5.84	0.81	344	3.51	0.63	385	0.31	0
304	5.96	0.79	345	3.40	0.63	386	0.30	0
305	6.11	0.78	346	3.24	0.63	387	0.29	0
306	6.22	0.78	347	3.07	0.63	388	0.28	0
307	6.26	0.76	348	2.90	0.63	389	0.28	0
308	6.20	0.76	349	2.74	0.63	390	0.27	0
309	6.17	0.75	350	2.58	0.63	391	0.26	0
310	6.30	0.74	351	2.42	0.63	392	0.26	0
311	6.42	0.73	352	2.25	0.63	393	0.26	0
312	6.46	0.72	353	2.18	0.63	394	0.25	0
313	6.46	0.72	354	2.06	0.63	395	0.25	0
314	6.34	0.71	355	1.87	0.62	396	0.24	0
315	6.25	0.71	356	1.80	0.62	397	0.24	0
316	6.21	0.70	357	1.72	0.62	398	0.24	0
317	6.07	0.69	358	1.57	0.62	399	0.23	0
318	6.04	0.69	359	1.49	0.61	400	0.23	0
319	5.98	0.68	360	1.41	0.61			

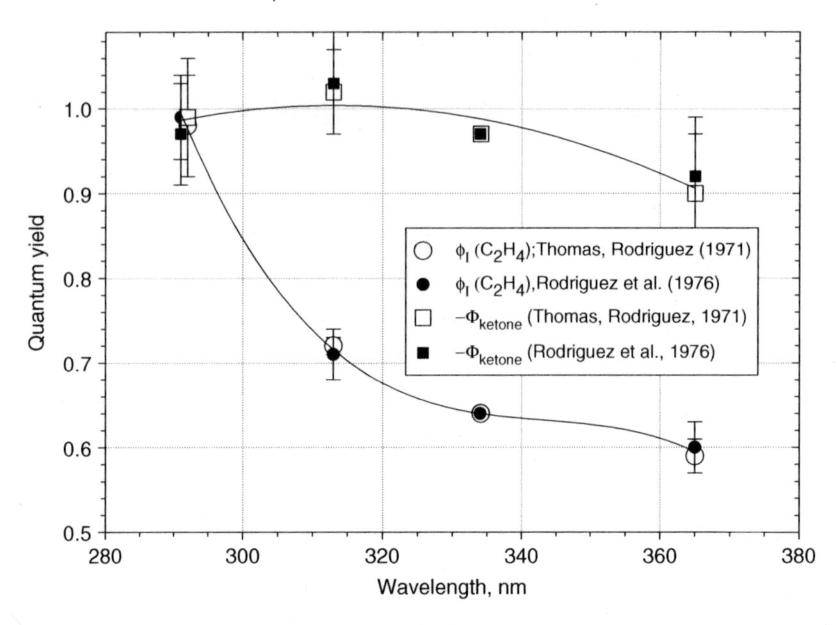

Figure VII-D-2. Quantum yields of photodecomposition of cyclopropanone by process (I), cyclopropanone + $h\nu \rightarrow C_2H_4 + CO$, and those for decomposition by all pathways. Data are from Thomas and Rodriguez (1971) and Rodriguez et al. (1976).

variation of the Sun, the expected average photochemical lifetimes of cyclopropanone are about 2.1 and 8.2 hr for the time periods near June 22 and December 22, respectively.

VII-D-2. Cyclobutanone (*cyclo*-C_4H_6O)

In the gas phase, photodecomposition of cyclobutanone occurs largely through two different product-forming channels:

$$\text{cyclo-}(CH_2CH_2CH_2C{=}O) + h\nu \rightarrow C_2H_4 + CH_2{=}C{=}O \qquad \text{(I)}$$

$$\rightarrow \text{cyclo-}(CH_2CH_2CH_2)^{\ddagger} + CO \qquad \text{(II)}$$

$$\text{cyclo-}(CH_2CH_2CH_2)^{\ddagger} \rightarrow CH_2{=}CHCH_3$$

$$\text{cyclo-}(CH_2CH_2CH_2)^{\ddagger} \xrightarrow{(+M)} \text{cyclo-}(CH_2CH_2CH_2)$$

Studies reported in 1942 gave evidence of these two primary processes (Benson and Kistiakowsky, 1942). Later studies showed clearly that propene is not a primary product

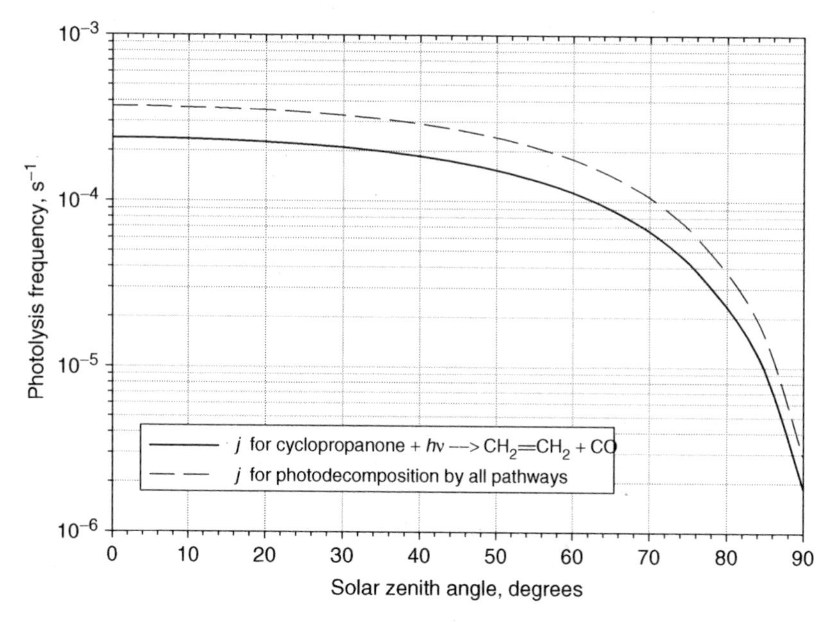

Figure VII-D-3. Approximate j-values as a function of solar zenith angle for the photodecomposition of cyclopropanone on a cloudless day within the troposphere with an overhead ozone column = 350 DU. Quantum yields vs. wavelength data are from Thomas and Rodriguez (1971) and Rodriguez et al. (1976), as shown in figures VII-D-1 and VII-D-2.

but is formed from the initial products of (II), a vibrationally excited cyclopropane molecule (Klemm et al., 1965; Campbell et al., 1967; Campbell and Schlag, 1967). The involvement of both singlet and triplet states has been demonstrated using triplet benzene to photosensitize cyclobutanone (Lee, 1967; Ishakawa and Noyes, 1962a,b; Noyes and Unger, 1966). Theoretical studies have elucidated the nature of the precursor excited states involved in the photodecomposition (e.g., see Baba and Hanazaki, 1984; Hopkinson et al., 1988; Diau et al., 2001; Chen and Ye, 2004). Evidence suggests that photodissociation of cyclobutanone by pathway (I) occurs from a vibrationally excited ground state ($S_0^{\ddagger}$) that is formed by internal conversion from the initially excited S_1 state. Like most ketones, for excitation of cyclobutanone at the longer wavelengths ($\lambda > 320$ nm), the T_1 state is populated by intersystem crossing ($S_1 \rightarrow T_1$), and this state is believed to be the source of process (II), forming ethene and ketene.

VII-D-2.1. Cross Sections for Cyclobutanone

Shown in figure VII-D-1 are the approximate cross sections for cyclobutanone (in cyclohexane solution) vs. wavelength as redrawn from the publication of Carless and Lee (1972). The general structure and magnitude seen in this absorption band are in qualitative agreement with those reported by Benson and Kistiakowsky (1942). Note in the figure that there is much less shift of the absorption band from that of acetone

Table VII-D-2. Approximate cross sections (σ, cm^2 molecule^{-1}) at ~298 K and quantum yields for the primary processes cyclobutanone + $h\nu \rightarrow CH_2{=}CH_2 + CH_2{=}C{=}O$ (I); cyclobutanone + $h\nu \rightarrow$ *cyclo*-$C_3H_6^{\ddagger}$ + CO (II)

λ (nm)	$10^{20} \times \sigma$	ϕ_I	ϕ_{II}	λ (nm)	$10^{20} \times \sigma$	ϕ_I	ϕ_{II}
278	6.15	0.61	0.37	310	3.09	0.69	0.27
279	6.42	0.61	0.37	311	2.93	0.69	0.27
280	6.69	0.62	0.37	312	2.80	0.70	0.26
281	6.95	0.62	0.36	313	2.67	0.70	0.26
282	7.11	0.62	0.36	314	2.54	0.70	0.26
283	7.14	0.63	0.35	315	2.41	0.70	0.26
284	7.11	0.63	0.35	316	2.29	0.70	0.25
285	6.88	0.63	0.35	317	2.14	0.70	0.25
286	6.34	0.63	0.34	318	1.87	0.71	0.25
287	6.15	0.64	0.34	319	1.57	0.71	0.25
288	6.26	0.64	0.33	320	1.18	0.71	0.25
289	6.49	0.64	0.33	321	0.92	0.71	0.25
290	6.76	0.64	0.33	322	0.76	0.71	0.25
291	6.88	0.65	0.32	323	0.62	0.71	0.24
292	6.80	0.65	0.32	324	0.48	0.72	0.24
293	6.42	0.65	0.32	325	0.35	0.72	0.24
294	6.00	0.65	0.31	326	0.25	0.72	0.24
295	5.62	0.66	0.31	327	0.19	0.72	0.24
296	5.49	0.66	0.31	328	0.15	0.72	0.24
297	5.34	0.66	0.30	329	0.11	0.72	0.24
298	5.20	0.66	0.30	330	0.08	0.73	0.24
299	5.04	0.67	0.30	331	0.06	0.73	0.24
300	4.89	0.67	0.29	332	0.04	0.73	0.24
301	4.84	0.67	0.29	333	0.02	0.73	0.23
302	4.78	0.67	0.29	334	0.01	0.73	0.23
303	4.73	0.68	0.28	335	0.01	0.73	0.23
304	4.66	0.68	0.28	336	0.003	0.74	0.23
305	4.58	0.68	0.28	337	0.002	0.74	0.23
306	4.43	0.68	0.28	338	0.002	0.74	0.23
307	4.20	0.68	0.27	339	0.001	0.74	0.23
308	3.90	0.69	0.27	340	0	0.74	0.23
309	3.36	0.69	0.27				

than is observed for cyclopropanone. The tabular listing of the cross sections is given in table VII-D-2.

VII-D-2.2. Quantum Yields of the Primary Processes in Cyclobutanone

Shown as a function of wavelength at two pressures in figure VII-D-4 are the ratios of propene to cyclopropane products that are formed in alternative pathways for reaction of the vibrationally excited cyclopropane molecule. A large increase in this ratio is seen as the excitation energy is increased in photolysis at the shorter wavelengths. A cyclopropane molecule that is vibrationally richer results as the quantum energy is increased. A somewhat more efficient quenching of the vibrationally excited cyclopropane can be seen from the data for 11 Torr of cyclobutanone (dashed curve)

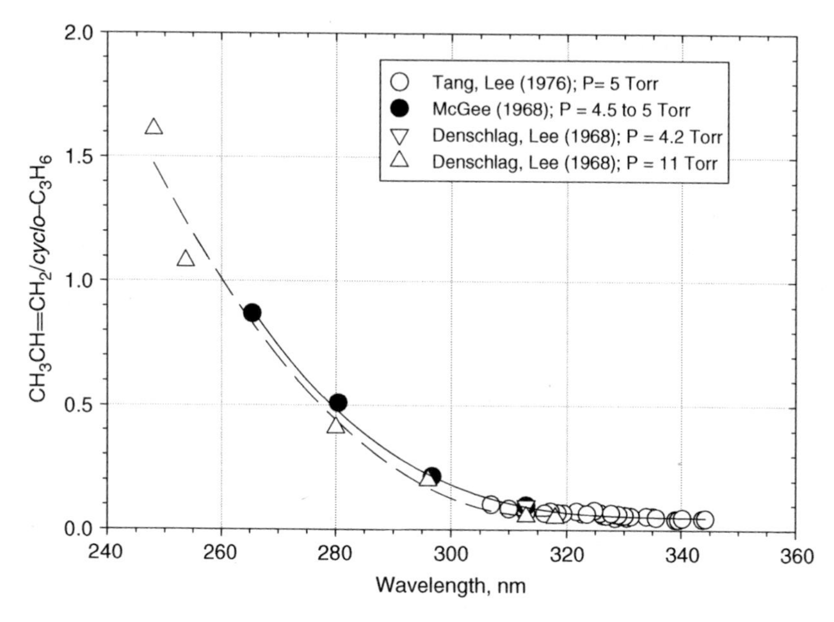

Figure VII-D-4. The ratio of propene to cyclopropane as a function of wavelength in the photolysis of cyclobutanone at two pressures. Propene and cyclopropane can both arise from process (II): cyclopropanone + $h\nu \rightarrow$ excited cyclopropane + CO. Collisional stabilization of the excited cyclopropane becomes less effective (and decomposition more effective) in photolysis at the shorter wavelengths.

compared to those for 5 Torr (solid curve). One expects that an atmosphere of air will result in more significant stabilization of the cyclopropane molecule. For example, with cyclobutanone photolysis (4.5 Torr) at 265.4 nm, the addition of 33.5 Torr of hexane decreased the ratio of propene to cyclopropane from 1.04 to 0.56 (McGee, 1968).

In figure VII-D-5 the ratio of the quantum yields of the two primary processes, ϕ_{II}/ϕ_I, is shown as a function of wavelength. Note the striking change in the slope of the lines that define the ratio as a function of wavelength as 315 nm is approached. The obvious change in mechanism that occurs in this wavelength region has been rationalized theoretically by Diau et al. (2001) and Chen and Ye (2004) through the involvement of an S_0/S_1 conical intersection near 315 nm.

The quantum yields of the two primary processes as estimated by McGee (1968) are given as a function of wavelength in figure VII-D-6. The data give smooth curves over the wavelength range studied (313–265 nm). Process (II) is seen to increase gradually (triangles, heavy solid line) as one progresses to shorter wavelengths, whereas process (I) decreases somewhat (circles, dashed line).The sum of $\phi_I + \phi_{II}$ is unity within the error of the determinations (filled circles, light line) over the wavelength range shown. The values are relatively unaffected by temperature (40–250°C) and pressure. The tabular listing of the quantum yields for processes (I) and (II) is given in table VII-D-2.

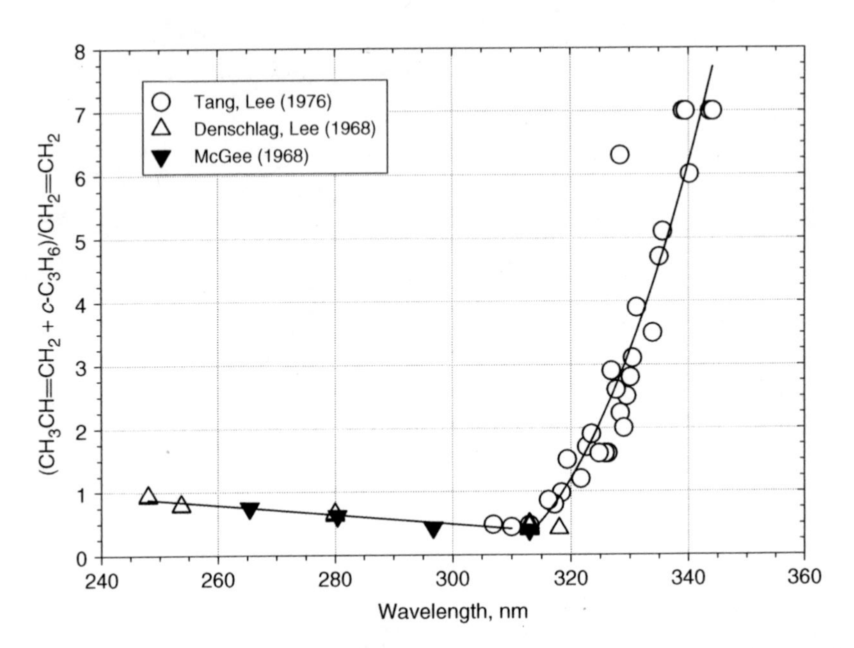

Figure VII-D-5. Plot of the ratio of products (propene + cyclopropane)/ethene (ϕ_{II}/ϕ_{I}) as a function of wavelength in the photolysis of cyclobutanone. Cyclobutanone + $h\nu \rightarrow CH_2{=}CH_2 + CH_2{=}C{=}O$ (I); cyclopropanone + $h\nu \rightarrow$ excited cyclopropane + CO (II).

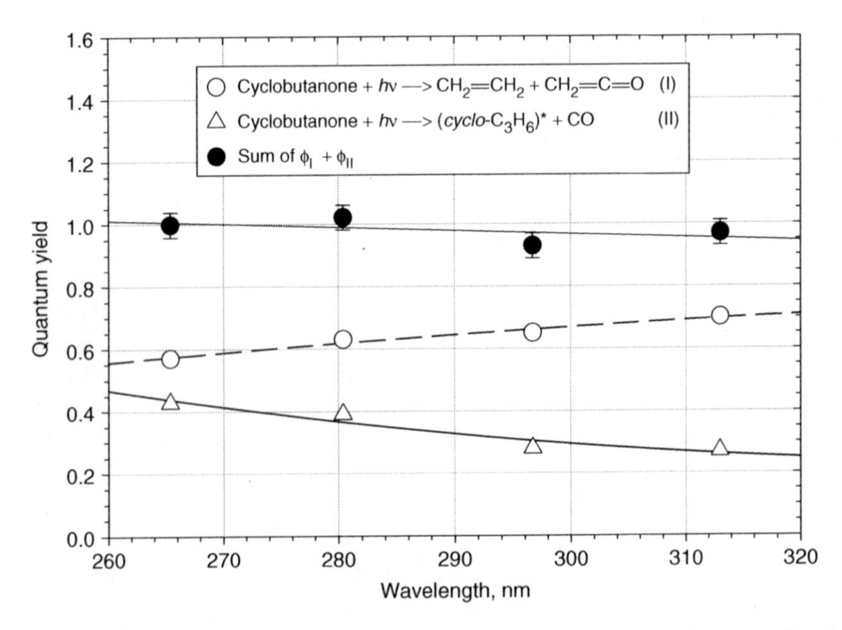

Figure VII-D-6. Plot of the quantum yields of the primary processes in cyclobutanone photolysis as a function of wavelength; data from McGee (1968).

VII-D-2.3. Approximate Photolysis Frequency Estimates for Cyclobutanone

Values of ϕ_I and ϕ_{II} have not been reported for wavelengths greater than 315 nm, so an extrapolation of the trend curve of the data shown in figure VII-D-6 has been used to arrive at estimates for the 315–340 nm region. This procedure will probably overestimate the j-values since the triplet state may be formed in this wavelength region and it may be quenched to some extent by air molecules. We have used the limited cross section and quantum yield data summarized in table VII-D-2, coupled with the appropriate solar flux for a cloudless day within the lower troposphere (ozone column = 350 DU), to estimate approximate j-values for cyclobutanone. The results of these calculations are summarized in figure VII-D-7.

VII-D-3. Alkyl-Substituted Cyclobutanones

There are several studies of the photodecomposition of alkyl-substituted cyclobutanones for which there are insufficient data to estimate j-values, but the results may be helpful in understanding the effects of structure on the photodecomposition modes of the cyclic ketones.

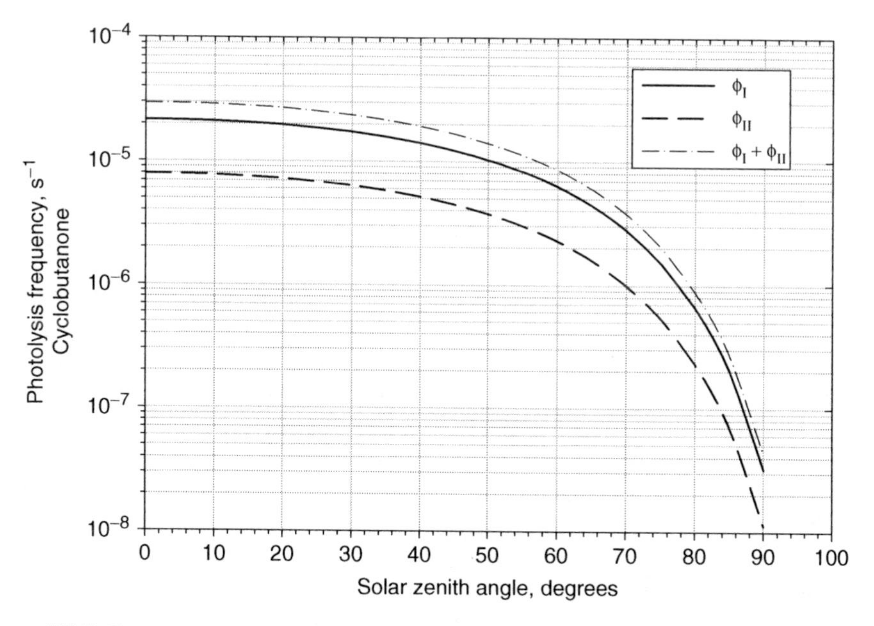

Figure VII-D-7. Approximate j-values for cyclobutanone photodecomposition vs. solar zenith angle for a cloudless day in the lower troposphere with an overhead ozone column = 350 DU. Process (I): cyclobutanone + $h\nu \rightarrow CH_2{=}CH_2 + CH_2{=}C{=}O$; process (II): cyclobutanone + $h\nu \rightarrow$ vibrationally excited cyclopropane + CO.

VII-D-3.1. Photodecomposition of Cis- and Trans-2,3-Dimethylcyclobutanone

$$\text{cis/trans-2,3-dimethylcyclobutanone} + h\nu \rightarrow \dot{C}H_2\text{-}\dot{C}(=O)\text{-}CH(CH_3)\text{-}CH(CH_3) \rightarrow CH_3CH{=}CH_2 + CH_3CH{=}C{=}O \quad \text{(I)}$$

$$\rightarrow CH_2\text{-}\dot{C}(=O)\text{, }CH(CH_3)\text{-}\dot{C}H(CH_3) \rightarrow CH_2{=}C{=}O + CH_3CH{=}CHCH_3 \quad \text{(II)}$$

$$\rightarrow \text{1,2-dimethylcyclopropane} + CO \quad \text{(III)}$$

The biradicals shown may be involved as precursors to the formation of alkenes and ketenes in the photodecomposition processes (I) and (II). Those formed from singlet and triplet excited states are expected to form singlet and triplet biradicals, respectively. The biradical precursor to process (II) appears to be favored over that suggested for process (I), as one might expect from the difference in bond dissociation energies involved. The extent of stereospecificity of the products formed was rationalized by Carless et al. (1972) in terms of the relative dissociative lifetimes of the biradicals of different spin states involved compared to rotation about a C—C bond.

Carless and Lee (1970) and Carless et al. (1972) studied the effects of added ethene on the relative rates of the primary processes in photolysis of *cis*-2,3-dimethylbutanone (1 Torr) at 325 nm. Results are shown in figure VII-D-8. It can be seen that process (III) is relatively independent of the added gas, consistent with its formation from a very short-lived excited singlet state. However, both primary processes (I) and (II) show extensive quenching with added gas, although neither process is quenched completely at the highest pressures used. The efficiency of quenching is greater for process (I) than for (II). The observations show that these processes originate from different vibronic levels of the excited triplet and singlet states of the ketone. A similar effect is seen for the addition of argon in photolysis of *cis*-2,3-dimethylbutanone at wavelength 281 nm; see figure VII-D-9. The order of importance of the primary processes for photolysis at low pressures changes from that observed at 325 nm, $\phi_{I} > \phi_{II} > \phi_{III}$, to $\phi_{II} > \phi_{I} > \phi_{III}$ at 281 nm. At high pressures, however, the unquenchable process (III) increases in importance at each wavelength.

In photolysis at 335 and 313 nm, Carless and Lee (1972) could not detect any of the possible C_5-alkene products that might result from the suggested intermediate

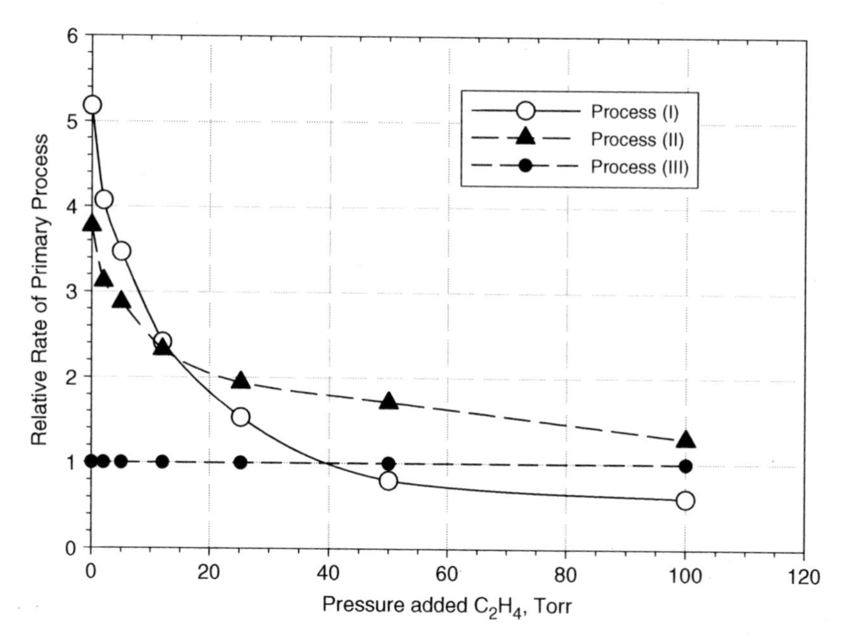

Figure VII-D-8. The effect of added ethene on the relative rates of the primary processes in the 325 nm photolysis of 1 Torr of *cis*-2,3-dimethylcyclobutanone (CB); process (I): CB + $h\nu \rightarrow$ $CH_2{=}CHCH_3 + CH_3CH{=}C{=}O$; process (II): CB + $h\nu \rightarrow CH_3CH{=}CHCH_3 + CH_2{=}C{=}O$; process (III); CB + $h\nu \rightarrow$ dimethylcyclopropane + CO; data are from Carless and Lee (1972).

biradicals with loss of CO. At 281 nm these species amounted to about 3% of the C_5 compounds formed (including dimethylcyclopropane). They also reported that the addition of oxygen (25–210 Torr) in the photolysis at 325 nm quenched the products not as a radical scavenger but as a vibrational deactivator, similar to C_2H_4.

VII-D-3.2. Photodecomposition of 2,4-Dimethylcyclobutanone

The photodecomposition of *cis*- and *trans*-2,4-dimethylcyclobutanone can be described by the following overall reaction scheme:

CH_3–CH–C(=O)–CH(CH_3)–CH_2 (ring) + $h\nu$ → CH_3–$\dot{C}H$ $\dot{C}$(=O) CH_2–CH–CH_3 → $CH_2{=}CHCH_3 + CH_3CH{=}C{=}O$ (I)

→ CH_2 / CH–CH (ring) CH_3 CH_2 + CO (II)

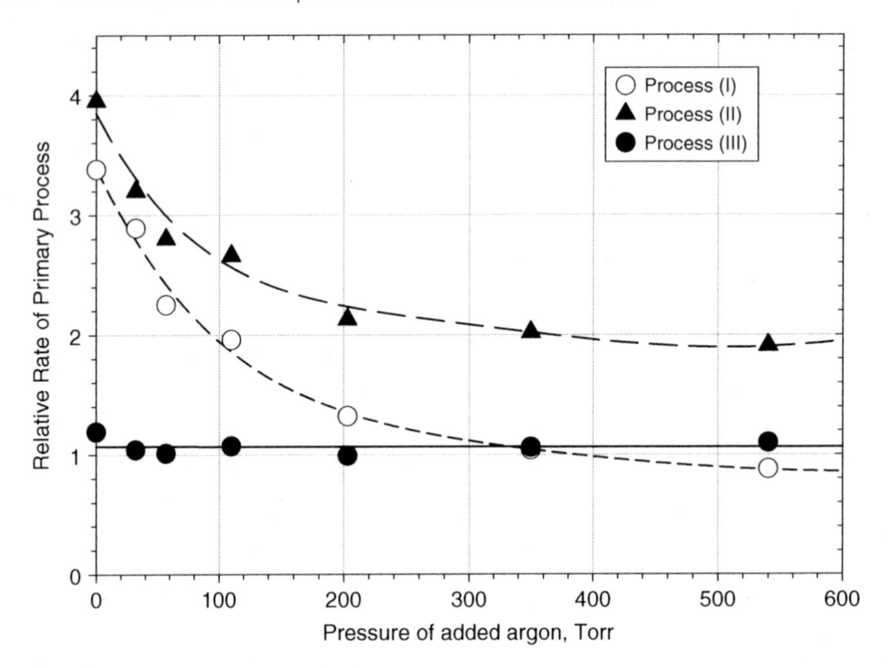

Figure VII-D-9. The effect of added argon on the relative rates of primary processes in the photolysis of 1.0 Torr of *cis*-2,3-dimethylbutanone (CB) at 281 nm; process (I): CB + $h\nu \rightarrow$ $CH_2{=}CHCH_3 + CH_3CH{=}C{=}O$; process (II): CB + $h\nu \rightarrow CH_3CH{=}CHCH_3 + CH_2{=}C{=}O$; process (III): CB + $h\nu \rightarrow$ dimethylcyclopropane + CO; data are from Carless and Lee (1972).

The biradical intermediate shown in process (I) may be formed in either a triplet or a singlet state, reflecting the spin state of the triplet or singlet excited ketone molecule that decomposes. In this case the rupture of either of the C—C bonds next to the carbonyl group leads to the same products. The behaviors of the product yields on variations in wavelength and with added gases were found to be very similar to those observed for 2,3-dimethylcyclobutanone. Added inert quenching gases reduced the propene yield, process (I). At 325 nm the ratio ϕ_I/ϕ_{II} decreases from 8.2 at 0.1 Torr of the ketone to 2.0 with 30 Torr of added C_2H_4. The possible C_5-alkene products that might arise from the biradical intermediate formed with loss of CO were not detected in photolysis at 325 and 313 nm, and they were only ~3% of the total C_5 products (including dimethylcyclopropane) at 281 nm.

VII-D-3.3. 2-n-Propylcyclobutanone

Carless and Lee (1972) have studied the photodecomposition of 2-*n*-propylcyclobutanone at 325 nm. The absorption cross sections of this ketone are significantly larger than those of cyclobutanone ($\sigma_{max} \approx 1.07 \times 10^{-19}$ cm^2 molecule^{-1}), and the maximum in the absorption band is shifted to the longer wavelengths (~295 nm).

The following overall primary processes are consistent with the results of Carless and Lee (1972):

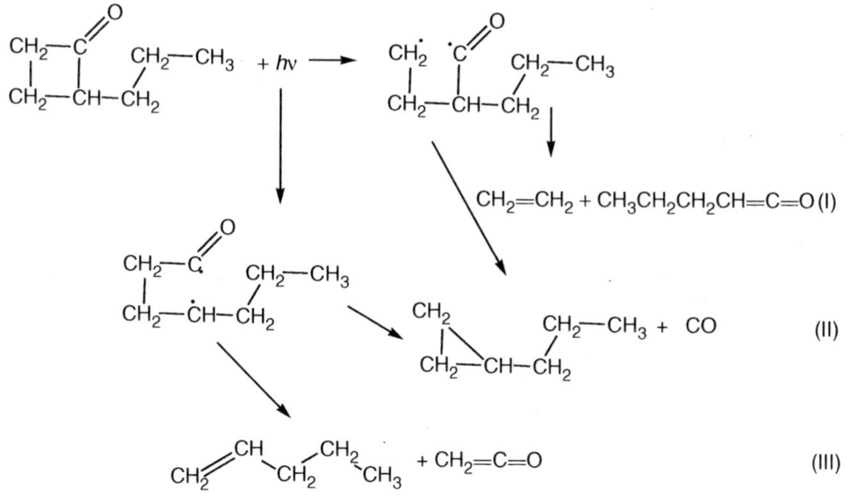

$CH_2{-}C(=O){-}CH(CH_2CH_3){-}CH_2$ (ring) $+ h\nu \rightarrow \dot{C}H_2 \; \dot{C}{=}O \; CH_2{-}CH{-}CH_2{-}CH_2{-}CH_3$

$\rightarrow CH_2{=}CH_2 + CH_3CH_2CH_2CH{=}C{=}O$ (I)

$\rightarrow$ (cyclopropane) $CH_2{-}CH_2{-}CH{-}CH_2{-}CH_3 + CO$ (II)

$\rightarrow CH_2{=}CH{-}CH_2{-}CH_2{-}CH_3 + CH_2{=}C{=}O$ (III)

The various photodecomposition modes are visualized as progressing through the biradical intermediates shown. Carless and Lee (1972) suggest that both the excited ground state and the S_1 state are involved in these processes. The relative yields of the various primary processes at 325 nm are shown as a function of ketone pressure in figure VII-D-10. The extent of quenching of channels (I) and (III) is much greater than that seen for channel (II). As with cyclobutanone and the other alkyl-substituted cyclobutanones, the process forming the cyclopropane derivative is relatively unaffected by pressure of added gases.

The addition of 100 Torr of O_2 in the photolysis of the ketone (0.1 Torr) leads to the suppression of the yield of products of processes (I), (II), and (III) by 96.8, ~0.0, and 23.0%, respectively. Obviously, different vibronic levels and/or excited states are involved in these processes.

It is interesting that although this ketone has two γ-hydrogen atoms in the $CH_3CH_2CH_2$ group attached to the cyclobutanone ring that is adjacent to the carbonyl, there is no evidence that a Norrish type II reaction occurs here. Propene was not reported as a product. This probably reflects the unusual geometry of this ketone. A simple model of the ground-state ketone (CS CHEM3D) suggests that the δ-H-atoms on the methyl group are closer to the carbonyl-oxygen atom (~2.75Å) than the two γ-hydrogen atoms that are necessarily involved in the Norrish type II reaction (3.13 Å). The reaction seems unlikely as well in view of the ring strain introduced by the formation of the expected initial product, the enol-form of cyclobutanone:

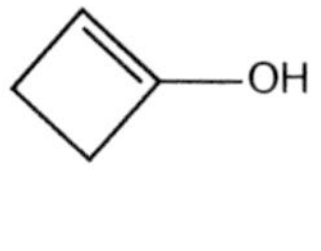

(cyclobutene ring)—OH

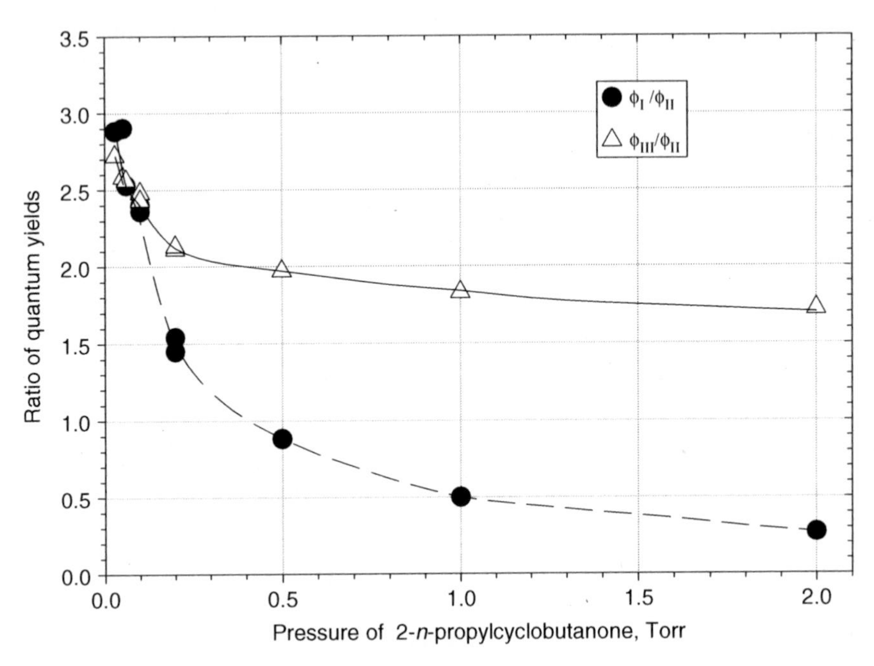

Figure VII-D-10. Plot of the ratio of quantum yields for primary processes in 2-*n*-propylcyclobutanone (PcB) vs. pressure of the ketone; wavelength, 325 nm; temperature, 23°C; process (I): PcB + $h\nu \rightarrow CH_2{=}CH_2 + CH_3CH_2CH_2CH{=}C{=}O$; process (II): PcB + $h\nu \rightarrow$ *n*-propylcyclopropane + CO; process (III): PcB + $h\nu \rightarrow CH_2{=}CHCH_2CH_3 + CH_2{=}C{=}O$; data of Carless and Lee (1972).

VII-D-4. Cyclopentanone (*cyclo*-C_5H_8O)

Shortridge et al. (1971) studied the fluorescence of the cyclic ketones: cyclobutanone, cyclopentanone, cyclohexanone, with excitation at 313 nm. All of these ketones have quantum yields of fluorescence ≈ 0.002, similar to that for acetone. These authors point out that this similarity suggests that the $S_1 \rightarrow T_1$ intersystem crossing rates for these ketones is similar for excitation at the longer wavelengths and that much of the photochemical activity may proceed through the T_1 manifold. The highly resolved fluorescence spectrum of cyclopentanone has been analyzed by Baba and Hanazaki (1984). Müller-Remmers et al. (1984) studied theoretically the stability and reactions of possible biradical intermediates formed from the excited triplet state of cyclopentanone.

The approximate absorption cross sections for cyclopentanone (in cyclohexane solution) are compared with those of the other simple cyclic ketones in figure VII-D-1. The structure and magnitude seen in this absorption are qualitatively similar to those reported for this compound by Benson and Kistiakowsky (1942).

The major products of photodecomposition of cyclopentanone are CO, cyclobutane, ethene, and 4-pentenal. Alternative mechanisms have been considered to rationalize these products. Consider the following reaction schemes:

$$\text{c-}C_5H_8O \text{ (cyclopentanone)} + h\nu \longrightarrow CO + [CH_2\text{-}CH_2\text{-}CH_2\text{-}CH_2]^{\ddagger} \xrightarrow{(+M)} \text{c-}C_4H_8 \quad (\text{I})$$

$$[CH_2\text{-}CH_2\text{-}CH_2\text{-}CH_2]^{\ddagger} \longrightarrow 2\,CH_2{=}CH_2$$

$$\longrightarrow CO + 2\,CH_2{=}CH_2 \quad (\text{II})$$

$$\longrightarrow CH_2{=}CHCH_2CH_2CHO \quad (\text{III})$$

$$\longrightarrow \cdot CH_2CH_2CH_2CH_2\dot{C}O \quad (\text{Ia})$$

Srinivasan (1964) has reviewed the early studies of cyclopentanone photolysis. From the data available at that time, he favored the direct formation of these products by concerted processes (I), (II), and (III). His conclusion was based largely on the rather small effects seen on the rates of product formation in the 313 nm photolysis of cyclopentanone on addition of 35 Torr of oxygen to 10.9 Torr of the ketone (Srinivasan,1961): the product rates changed by factors of 1.05 (C_2H_4), 0.88 (cyclobutane), and 0.26 (pentenal). With 275 Torr of oxygen added, rates relative to those with no gas added changed by factors of 0.34 (C_2H_4), 0.29 (cyclobutane), and 0.87 (pentenal). The change of product rates with methane addition was similar to that of oxygen at high pressures; addition of 195 Torr of CH_4 to 11.6 Torr of ketone changed the rates of formation of products of 313 nm photolysis by the factors 0.36 (ethene), 0.36 (cyclobutane), and 2.33 (pentenal). Srinivasan concluded that the small effect of oxygen was inconsistent with product formation that must pass through the intermediate biradical formed in (Ia). He favored the direct formation of products from the excited singlet state in concerted mechanisms summarized by processes (I), (II), and (III). However, it is not established that biradicals, particularly singlet biradicals, react with oxygen as rapidly as measured for alkyl and acyl radicals; so the step shown as (Ia) above may indeed play a part in product generation in cyclopentanone photodecomposition, as many investigators conclude.

Blades (1970) photolyzed cyclopentanone-2,2,5,5-d_4 at several wavelengths. Photolysis at 254 nm (80°C, 11.2 Torr) gave primarily $C_2H_2D_2$, with some C_2H_3D and C_2H_4. At 0.65 and 1.3 Torr pressure, substantial yields of C_2D_4 and C_2H_4 in addition to $C_2H_2D_2$ were observed, suggesting approximately 14 and 10%, respectively, of decomposition of initially formed cyclobutane. These results confirm the initial formation of a vibrationally rich cyclobutane molecule that can decompose at low pressures or be quenched at the higher pressures:

$$\text{cyclo-}(CD_2C(=O)CD_2CH_2CH_2) + h\nu \xrightarrow{(II)} 2CD_2{=}CH_2 + CO$$

$$\text{cyclo-}(CD_2C(=O)CD_2CH_2CH_2) + h\nu \xrightarrow{(I)} CO + \text{cyclo-}(CD_2CD_2CH_2CH_2)^{\ddagger} \xrightarrow{(+M)} \text{cyclo-}(CD_2CD_2CH_2CH_2)$$

$$\text{cyclo-}(CD_2CD_2CH_2CH_2)^{\ddagger} \rightarrow CD_2{=}CH_2 + CD_2{=}CH_2$$

$$\text{cyclo-}(CD_2CD_2CH_2CH_2)^{\ddagger} \rightarrow CD_2{=}CD_2 + CH_2{=}CH_2$$

If ethene were derived entirely from primary process (II), then CH_2CD_2 would be the only ethene formed. If, however, all the ethene formed in process (I), then, at low pressures, where the vibrationally hot cyclobutane is incompletely quenched, the ethene would have a more random distribution of the deuterium atoms: CD_2CD_2, CH_2CH_2, and CD_2CH_2. The observation of C_2D_4 and C_2H_4 in the photolysis of cyclopentanone-2,2,5,5-d_4 at 254 nm at low pressures, but not at 11.5 Torr pressure, confirms the partial decomposition of excited cyclobutane at low pressures and high photon energy (Blades, 1970).

The variation of the cyclobutane/ethene ratio with pressure for photolysis of cyclopentanone at several wavelengths is shown in figure VII-D-11 (Blades, 1970). Note several interesting features:

(i) Increasing the pressure of the reactant in photolysis at any of the wavelengths increases the ratio of cyclobutane to ethene.
(ii) At a given pressure of ketone, product data at 313, 254, and 239 nm show a progressive trend to greater percentages of cyclobutane decomposing at low pressures as the photon energy increases.
(iii) The extent of quenching is least efficient for the ketone excitation with highest energy quantum (229 nm).
(iv) The ratio appears to approach a similar limiting value at high pressures for experiments at 326, 313, and 254 nm.

The wavelength dependence of the ratio observed by Blades confirms and extends the earlier findings of Klemm et al. (1965), and the trends are consistent with the mechanism described above for ethene formation in part from excited cyclobutane. At 326 nm and 1 Torr, the cyclobutane/ethene ratio is at the high pressure value found at the shorter wavelengths, suggesting no tendency for cyclobutane decomposition. Blades points out that this is in keeping with the fact that the exothermicity of the overall photolysis reaction at this wavelength is nearly equal to the activation energy of the cyclobutane decomposition.

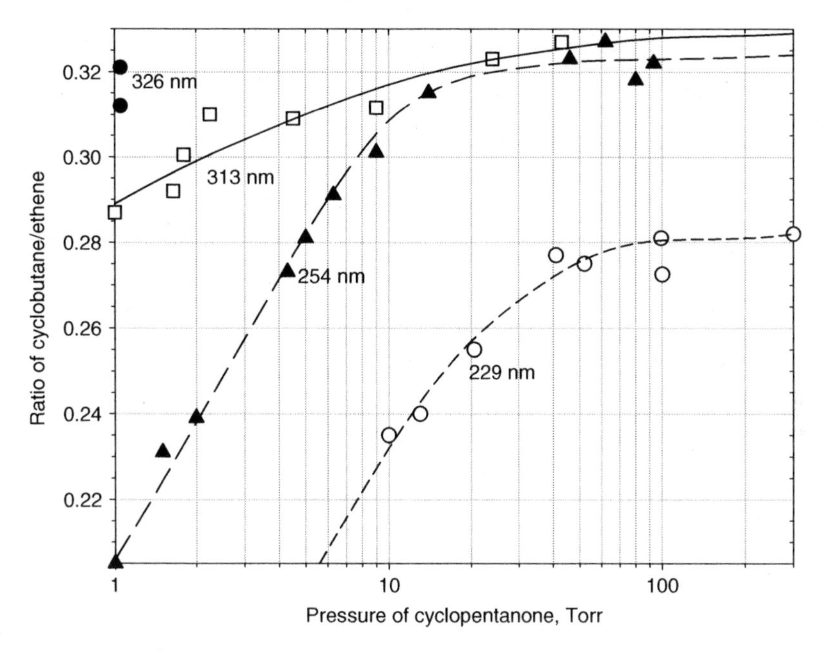

Figure VII-D-11. Ratio of products cyclobutane/ethene vs. ketone pressure for photolysis of cyclopentanone-2,2,5,5-d_4 at several wavelengths (from Blades, 1970).

The wavelength dependence of the ratio of 4-pentenal to CO is also interesting. Blades reported that the product ratio 4-pentenal/CO = 10 at 326 nm, 1.4 at 313 nm, 1.0 at 307 nm, and 0.02 at 254 nm. The trend fits the expected pattern, involving the intermediate biradical (•$CH_2CH_2CH_2CH_2CO$•), where absorption of the higher energy quantum favors decomposition to CO and •$CH_2CH_2CH_2CH_2$• over intramolecular H-atom abstraction to form 4-pentenal. Tsunashima et al. (1973) found that the 326.1 nm direct photolysis and the Cd5($^3P_1$)-sensitized photolysis of cyclopentanone gave similar results, and this favored the triplet state involvement, at least to some extent, in both photolyses.

Vacuum UV photolysis of cyclopentanone at 147 nm shows a dominance of the reaction channels forming CO, ethene, and cyclobutane, with $\phi_I + \phi_{II} = 0.87$. The primary step was suggested to involve CO and •$CH_2CH_2CH_2CH_2$• biradical formation directly (Scala and Ballan, 1972). Becerra and Frey (1988) studied the product distribution in photolysis of cyclopentanone at 208 and 193 nm and concluded that most of the available energy appears as vibrational excitation in the cyclobutane molecule and that CO is not appreciably excited.

The cross section and quantum yield data are too incomplete to provide meaningful estimates of the *j*-values for cyclopentanone at the time of this writing.

VII-D-5. Cyclohexanone (*cyclo*-$C_6H_{10}O$)

The ultraviolet absorption spectrum of cyclohexanone reflects the $n \rightarrow \pi^*$ transition common to all carbonyls; see figure VII-D-1. The maximum in the cyclohexanone cross sections shown here is significantly less than those observed for cyclopropanone, cyclobutanone, and cyclopentanone, and, in fact, all other carbonyls considered in this work. The data derived from gas phase measurements of two different research groups [National Center for Atmospheric Research (NCAR) and Ford Scientific Laboratories (Ford)] are in reasonable agreement (Iwasaki et al., 2008)]. The cyclohexanone cross sections, as measured in cyclohexane solution by Benson and Kistiakowsky (1942) indicated seemingly low values ($\sigma_{max} = 4.2 \times 10^{-20}$cm^2 molecule^{-1}) that were confirmed in the more recent studies of Iwasaki et al. (2008). It is not obvious why the probability for the $n \rightarrow \pi^*$ transition should be significantly less for cyclohexanone than that of the other cyclic ketones and, in fact, most other carbonyl compounds. Theoretical studies will be important in defining the reasons for these differences.

The early studies of the photochemistry of cyclohexanone have been reviewed by Srinivasan (1964), Pitts and Wan (1966), Cundall and Davies (1967), and Wagner and Hammond (1968). The major products are identified as CO, cyclopentane, 1-pentene, and 5-hexenal, but the mechanisms of their formation remain ill defined. Two alternative mechanisms have been discussed through the years. The first invokes intramolecular rearrangements of the original excited ketone (e.g., see Shortridge and Lee, 1970), while the second implies formation of biradicals that are intermediates to the final products (e.g., Benson and Kistiakowski, 1942; Blacet and Miller, 1957).

(i) *Intramolecular rearrangement to form final products:*

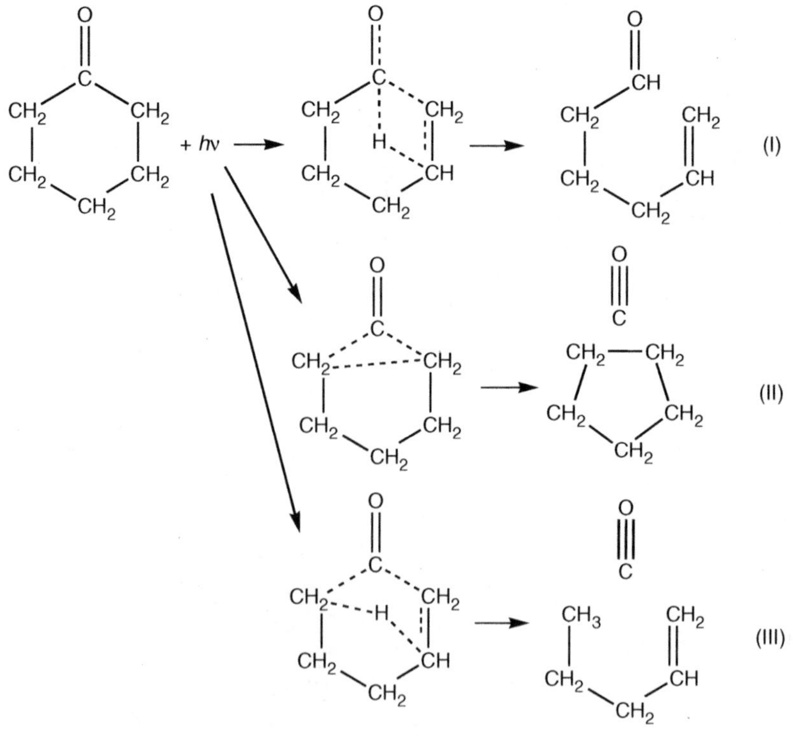

(I)

(II)

(III)

(ii) *Biradical involvement in the formation of final products:*

Cyclohexanone + $h\nu$ → $\cdot C(=O)CH_2CH_2CH_2CH_2\dot{C}H_2$ → [intramolecular H transfer] → $CHO\,CH_2CH_2CH_2CH{=}CH_2$ (Ia)

$\cdot C(=O)CH_2CH_2CH_2CH_2\dot{C}H_2$ → CO + $\dot{C}H_2CH_2CH_2CH_2\dot{C}H_2$ → cyclopentane (IIa)

$\dot{C}H_2CH_2CH_2CH_2\dot{C}H_2$ → [intramolecular H transfer] → $CH_2{=}CHCH_2CH_2CH_3$ (IIIa)

The quantum yields of the products from cyclohexanone photolysis (1 Torr) as a function of wavelength are summarized in figure VII-D-12 (Shortridge and Lee, 1970). Pathway (I), forming 5-hexenal, is seen to be dominant at the longer wavelengths, while channels (II) and (III), forming cyclopentane and 1-pentene, increase in importance at the shorter wavelengths. In figure VII-D-13 the effect of pressure increase (0.1 to 2 Torr) can be seen. Quenching of the precursor state to process (I) is less effective than that for (II) and (III) (Shortridge and Lee, 1970). For photolysis of cyclohexanone (2 Torr) at 297 nm, 95 Torr of added propane or oxygen give very similar quenching of the 1-pentene and cyclopentane product formation. The quantum yields with added propane and oxygen are reduced by factors of 0.16 and 0.17, respectively. However, the 5-hexenal is essentially unaffected by the addition of 95 Torr of oxygen. This behavior is not characteristic of thermalized triplet molecules. Certainly the state that forms the 5-hexenal is short lived. The electronic states involved in the mechanisms are open to question. The studies of Shortridge and Lee (1970) indicate that the generation of triplet cyclohexanone by sensitization with triplet benzene leads to the same products as those formed in direct photolysis. However, cyclohexanone quenches excited singlet benzene with a cross section that is about 50 times that of triplet benzene. Triplet-benzene-sensitized reactions predominate below 1 Torr of ketone, but the singlet-benzene sensitization of excited singlet cyclohexanone becomes very important in experiments above 1 Torr. Thus the results of the benzene sensitization of cyclohexanone decomposition are complex to interpret, and unambiguous conclusions are impossible. Shortridge and Lee (1970) suggest a reaction scheme in

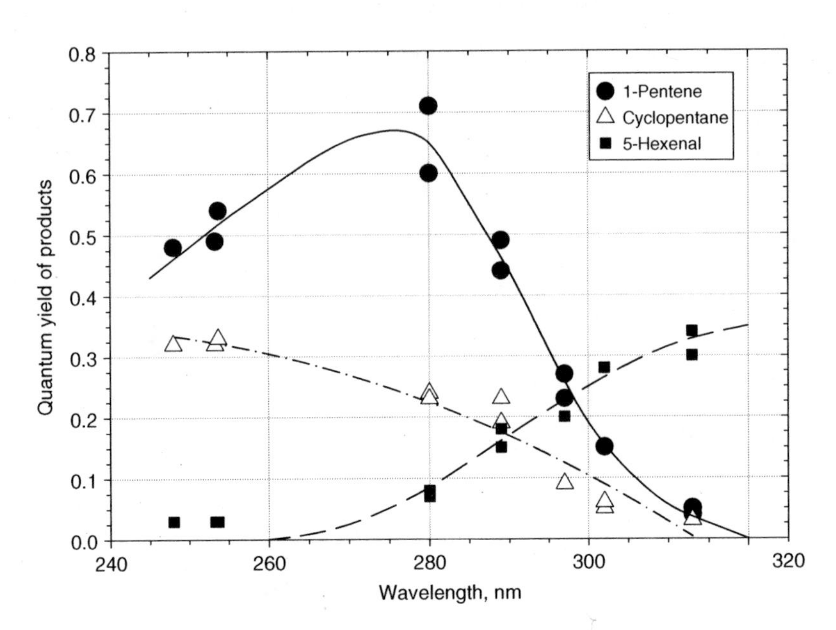

Figure VII-D-12. Quantum yields of products vs. wavelength for photodecomposition of cyclohexanone at 1 Torr pressure (from Shortridge and Lee, 1970).

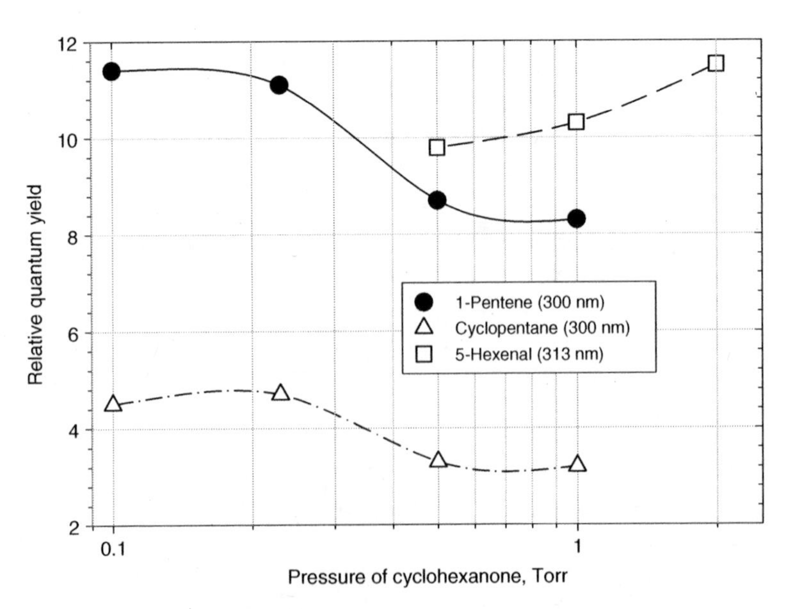

Figure VII-D-13. The effect of pressure on the relative quantum yields of 1-pentene, cyclopentane, and 5-pentenal products from the photolysis of cyclohexanone (Shortridge and Lee, 1970). Note that the data shown for 5-hexenal were determined at 313 nm while those for 1-pentene and cyclopentane were obtained using 300 nm radiation, and the two data sets are not quantitatively intercomparable. The 5-hexenal data were normalized from those given by Shortridge and Lee (1970) so that the relative yields at 1 Torr were more consistent with those given for all products in figure VII-D-12.

which the importance of the $S_1 \rightarrow T_1$ intersystem crossing is stressed. Cundall and Davies (1967) conclude that the remarkable inefficiency of oxygen in affecting the photodecomposition of the cyclic ketones argues most strongly against the participation of triplet states. They state that all the features of the cyclic ketone photochemistry are most consistently explained by a mechanism involving the first excited singlet states with different degrees of vibrational excitation. Srinivasan (1964) concluded from the data available in 1963 that it is probable that only one excited state, an upper singlet, is involved in cyclohexanone photolysis.

The products of the photolysis of the methyl-substituted cyclohexanones offer some insight into the mechanism of cyclohexanone photodecomposition. The lack of stereospecificity in the products of *cis*- and *trans*-methyl-substituted cyclohexanones appears to favor biradical intermediates (Alumbaugh et al., 1965; Badcock et al., 1969). The possible unsaturated ketone products of the photolysis of methylcyclohexanone were of the structures (a), (b), and (c):

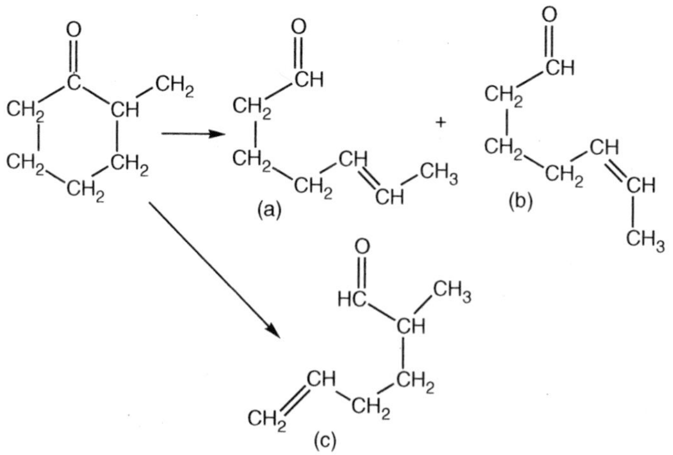

The two major products were identified as structures (a) and (b); in gas phase photolysis at 313 nm and 100°C, $\phi_a/\phi_b = 2.1 \pm 0.2$. The C—C bond rupture on the side of the carbonyl with the CH_3-group was favored heavily, $(\phi_a + \phi_b)/\phi_c \approx 50$ (Badcock et al., 1969). The mechanism of biradical formation best explained these results.

Unfortunately, the product quantum yield and cross section data for cyclohexanone photodecomposition are insufficient to allow reasonable estimates of j-values for its photolysis in the troposphere. However, it can be concluded that the unsaturated aldehydes, affected very little by oxygen addition, probably are important products.

VII-E. Photodecomposition of Alkyl Halides

VII-E-1. Introduction

Alkyl halides, alkanes containing F-, Cl-, Br-, or I-atoms, enter the atmosphere from both natural and man-made sources. Orlando (2003) has reviewed the haloalkane

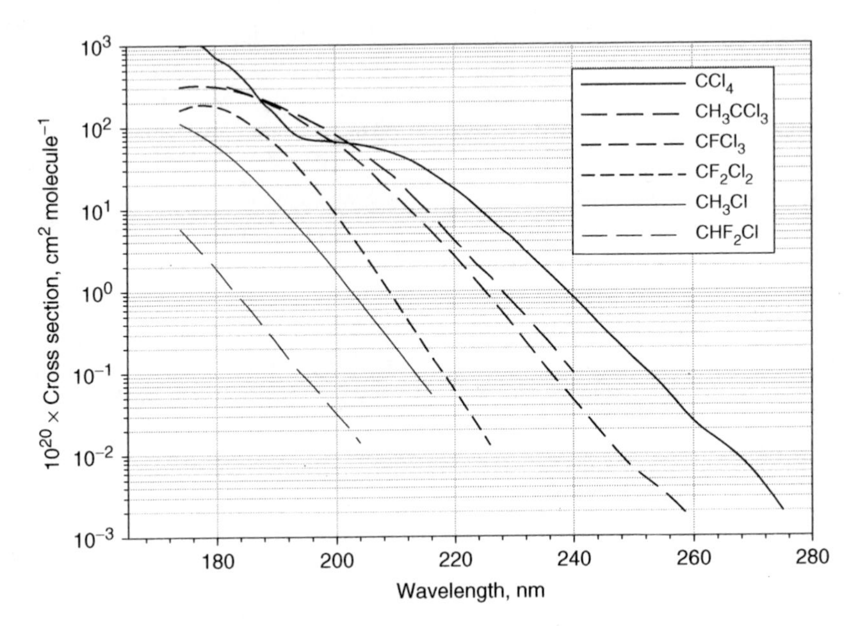

Figure VII-E-1. Absorption cross sections for several Cl- and F-atom-substituted methanes and ethanes; data are from Atkinson et al. (2004). These compounds do not absorb light in the wavelength range available in sunlight within the troposphere.

sources, sinks, distributions, trends, and impacts on atmospheric chemistry; these features are discussed also in chapter VI. In this section, we will review briefly the tropospheric photochemistry of several of the alkyl halides for which tropospheric photolysis represents an important loss process. Absorption cross sections for simple alkanes in which single or multiple H-atoms have been replaced by Cl-atoms ($C_nH_{2n+2-x}Cl_x$), or F-atoms ($C_nH_{2n+2-x-y}Cl_xF_y$) are compared in figure VII-E-1. The light absorption by these and other alkyl halides involves an $n \rightarrow \sigma^*$ transition with the promotion of a non-bonding electron on the halogen to an antibonding sigma orbital of the C—X bond. As is seen in figure VII-A-1, these excitations for alkyl halides (RX) of similar hydrocarbon backbone structure (R) are of lower energy (longer wavelength absorption) as one progresses from X = F, Cl, Br, to I. This reflects in part the trend to lower electron affinity of the halogen atom in this sequence.

As an increasing number of Cl-atoms replace either H-atoms in methane or F-atoms in F-substituted methanes, the cross section increases. The strongest absorber among the compounds shown in figure VII-E-1 is CCl_4. However, even this compound does not absorb sunlight significantly at wavelengths present within the troposphere. The compounds shown in this figure are not removed appreciably by photodissociation within the troposphere. However, many of them are also unreactive toward OH, O_3, and other reactive species present within the lower troposphere, and they live long enough to be transported significantly to the upper troposphere and ultimately to the stratosphere where they are photolyzed and/or react with reactive species

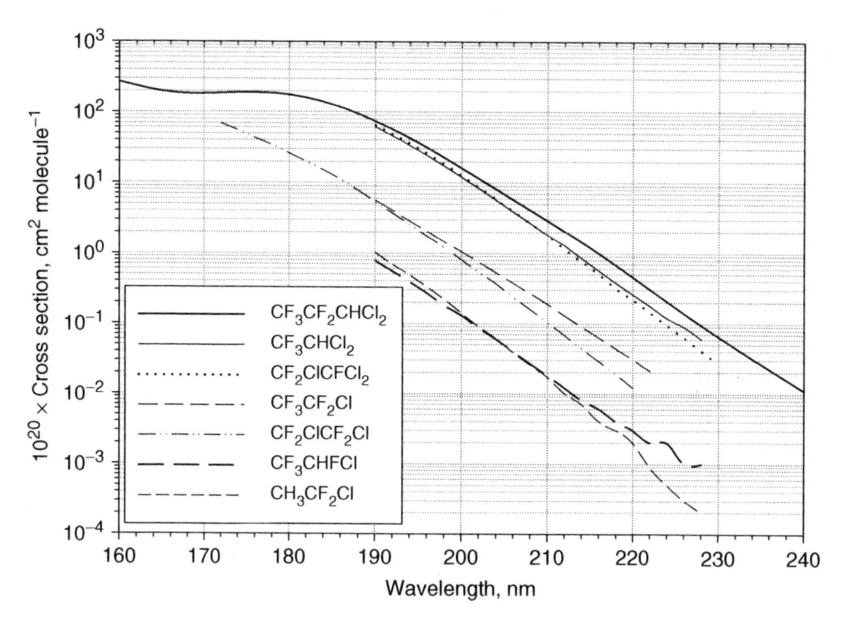

Figure VII-E-2. Absorption cross sections for some F-atom and Cl-atom-substituted ethanes and propane; data are from Atkinson et al. (2004). Absorption within the wavelength range of sunlight available within the troposphere is insignificant for these alkyl halides.

[$O(^1D)$] to generate Cl-atoms that initiate ozone destruction (Molina and Rowland, 1974).

The absorption cross sections for some F-atom and Cl-atom-substituted C_2 and C_3 alkanes are shown in figure VII-E-2. Compounds that contain two Cl-atoms on a given carbon atom, $CF_3CF_2CHCl_2$, CF_3CHCl_2, and $CF_2ClCFCl_2$, have higher cross sections at tropospherically relevant wavelengths than the compounds with a single Cl-atom substitution on a given carbon atom. Again, none of these alkyl halides photolyzes significantly in the troposphere.

In figure VII-E-3 the absorption cross sections of some Br- and F-substituted C_1 and C_2 alkanes are compared. In this case weak absorption of sunlight within the troposphere is expected for some of the species shown ($CHBr_3$, CF_2Br_2, CH_2Br_2). The strongest absorbing species among this group, $CHBr_3$, has a photodissociation lifetime greater than a week within the troposphere, although its photolysis in the troposphere is its major loss mechanism; reaction with OH occurs on a timescale of 100 days or so. Photodecomposition of all of these compounds is important if they are transported to the stratosphere where they will photolyze to provide reactive Br-atoms that efficiently destroy ozone.

The alkanes containing iodine atoms show significantly greater absorption cross sections than structurally similar bromine-, chlorine-, and fluorine-substituted alkanes. Cross sections for some of the naturally occurring iodine-containing methanes are compared in figure VII-E-4. The simple alkyl iodides (RI) are relatively weak absorbers but show a small increase in cross section as the R group grows in size: CH_3I,

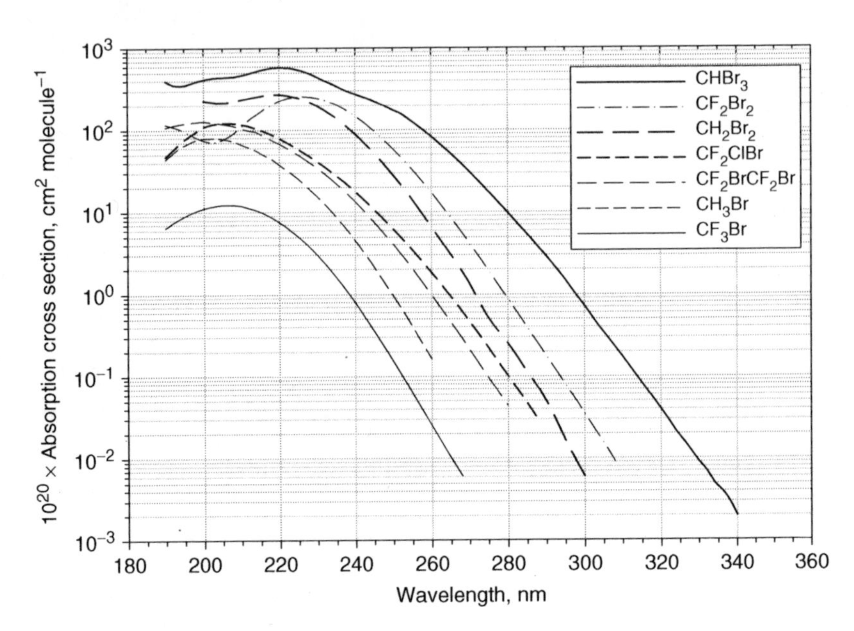

Figure VII-E-3. Absorption cross sections for some Br- and F-atom-substituted methanes and ethane; data are from Atkinson et al. (2004). The absorption regions for CH_2Br_2, $CHBr_3$, and CF_2Br_2 do overlap somewhat the wavelengths available in tropospheric sunlight, and these compounds are expected to have *j*-values of interest to modelers of tropospheric chemistry.

C_2H_5I, $CH_3CH_2CH_2I$, to CH_3CHICH_3. Substitution of a Br-atom or a Cl-atom for H-atoms in CH_3I also enhances the cross sections of CH_2BrI and CH_2ClI. The substitution of a second I-atom in CH_3I leads to CH_2I_2, a species that absorbs tropospheric sunlight strongly. Modeling studies suggest that the photolysis of these species may contribute to the depletion of both ozone and mercury that is observed at the Earth's surface in the springtime polar regions (e.g., see Calvert and Lindberg, 2004a,b).

The structureless absorption bands of the alkyl halides suggest that the excited states of these compounds are very short lived. Limited evidence from alkyl halide studies support the view that photodissociation of C—Cl, C—Br, or C—I bonds occurs with a quantum yield near unity. There appears to be a general acceptance of this view, although this has not been demonstrated specifically for many of the alkyl halides; e.g., see IUPAC (Atkinson et al., 2004) and NASA (DeMore et al., 1997). Majer and Simons (1964), in their review of the photochemistry of halogenated compounds, stated: "The long wavelength absorption is continuous, and results in ready photodissociation, almost certainly with unit quantum efficiency."

When more than one kind of halogen is present in the molecule, the weakest bound halogen is the predominant photodissociation product. Photolysis of an alkyl halide RX, where R is an alkyl group and X is a halogen atom, for wavelengths available within the troposphere, results in dissociation of the R—X bond with formation of the

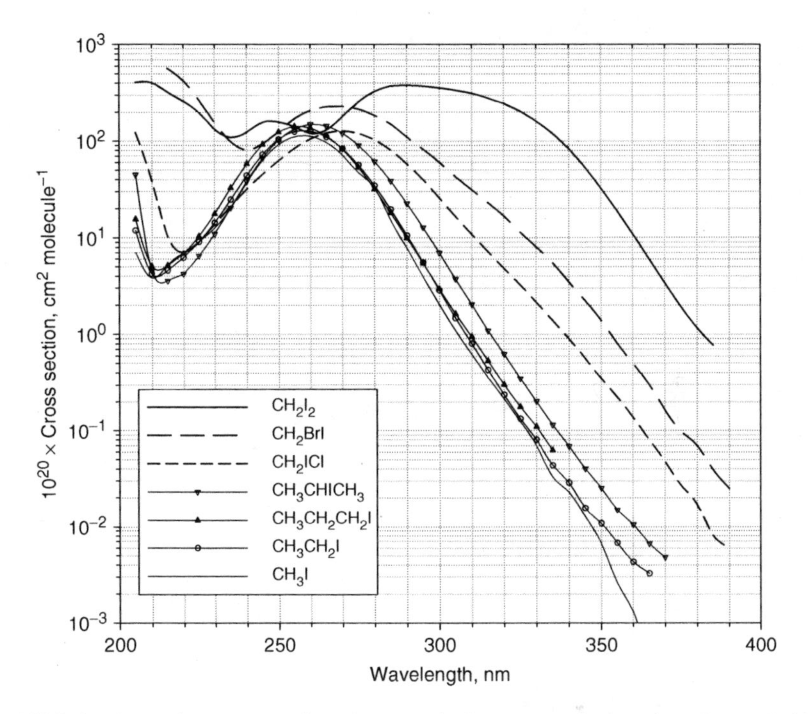

Figure VII-E-4. Absorption cross sections for some iodine-atom-substituted methanes (Atkinson et al., 2004) and alkyl iodides (Roehl et al., 1997) that are formed in natural processes. All of these species absorb tropospheric sunlight to some extent and are of interest to modelers of tropospheric chemistry.

halogen atom in either its ground, $X(^2P_{3/2})$, or low-lying electronically excited state, $X(^2P_{1/2})$:

$$RX + h\nu \rightarrow R + X(^2P_{3/2}) \quad \text{(Ia)}$$

$$\rightarrow R + X(^2P_{1/2}) \quad \text{(Ib)}$$

VII-E-2. Iodomethane (Methyl Iodide, CH_3I)

The absorption spectrum of CH_3I, shown in figure VII-E-4, is given in tabular form in table VII-E-1. The lifetime of the (n, σ^*) excited state of CH_3I is extremely short, estimated to be < 0.1 ps (Dzvonik et al., 1974: Knee et al., 1985). Photodissociation over the range of wavelengths 240–287 nm results in rapid dissociation in primary process (Ia), formation of ground state $I(^2P_{3/2})$, or process (Ib), forming excited $I(^2P_{1/2})$ atoms:

$$CH_3I + h\nu \rightarrow CH_3 + I(^2P_{3/2}) \quad \text{(Ia)}$$

$$\rightarrow CH_3 + I(^2P_{1/2}) \quad \text{(Ib)}$$

Table VII-E-1. Cross sections (σ, cm^2 $molecule^{-1}$) of some simple alkyl iodides: CH_3I, C_2H_5I, $CH_3CH_2CH_2I$, and CH_3CHICH_3; temperature, 298 K; wavelengths 275–350 nm, data of Roehl et al. (1997); 355–365 nm data for CH_3I are from Rattigan et al. (1997)

	CH_3I	C_2H_5I	$CH_3CH_2CH_2I$	CH_3CHICH_3
λ (nm)	$10^{20} \times \sigma$	$10^{20} \times \sigma$	$10^{20} \times \sigma$	$10^{20} \times \sigma$
275	45.9	56.3	53.4	90.2
280	26.9	34.6	32.0	61.4
285	14.6	19.7	18.1	38.6
290	7.54	10.6	9.96	22.6
295	3.79	5.54	5.42	12.8
300	1.98	2.85	2.96	6.94
305	1.07	1.47	1.63	3.73
310	0.603	0.804	0.945	2.04
315	0.352	0.429	0.532	1.09
320	0.212	0.236	0.301	0.627
325	0.122	0.133	0.177	0.348
330	0.0724	0.0805	0.110	0.202
335	0.0415	0.0436	0.0627	0.115
340	0.0225	0.0282		0.0688
345	0.0131	0.0156		0.0402
350	7.38e−3	0.0108		0.0253
355	3.20e−3	6.76e−3		0.0150
360	1.90e−3	4.95e−3		0.0105
365	9.00e−4	3.26e−3		6.66e−3
370				4.79e−3
375				5.35e−3
380				5.30e−3

Progression to shorter wavelengths of excitation increases the importance of process (Ib), forming the higher energy state $I(^2P_{1/2})$; see figure VII-E-5. No significant chemical consequences are expected to result from the formation of the $I(^2P_{1/2})$-atom rather than the $I(^2P_{3/2})$-atom, since the energy difference between the $^2P_{3/2}$ and $^2P_{1/2}$ states of the iodine atom is relatively small, 90.7 kJ mol^{-1}, and collisional equilibration of the distribution of the two I-atom states is expected to occur rapidly. However, it is possible that the reaction $I(^2P_{1/2}) + O_2(^3\Sigma_g^-) \rightarrow I(^2P_{3/2}) + O_2(^1\Delta_g)$ may occur to generate $O_2(^1\Delta_g)$, as the ΔH_{298} for this reaction is only slightly endothermic (+3.4 kJ mol^{-1}), and the recoiling $I(^2P_{1/2})$ atom formed in photolysis has considerable translational energy. However, chemical effects of $O_2(^1\Delta_g)$ in the troposphere are probably unimportant (Calvert et al., 2002).

Most of the energy (~90%) in excess of that required for dissociation and I-atom excitation is channeled into the translational energy of the product fragments CH_3 and I. The rest of the energy is distributed almost exclusively into the ν_2 umbrella mode of the CH_3 radical (see Riley and Wilson, 1972; Guo and Schatz, 1990).

The j-values for CH_3I in a cloudless, lower troposphere (298 K, ozone column = 350 DU) are compared to those of other alkyl halides in figure VII-E-6. These are shown

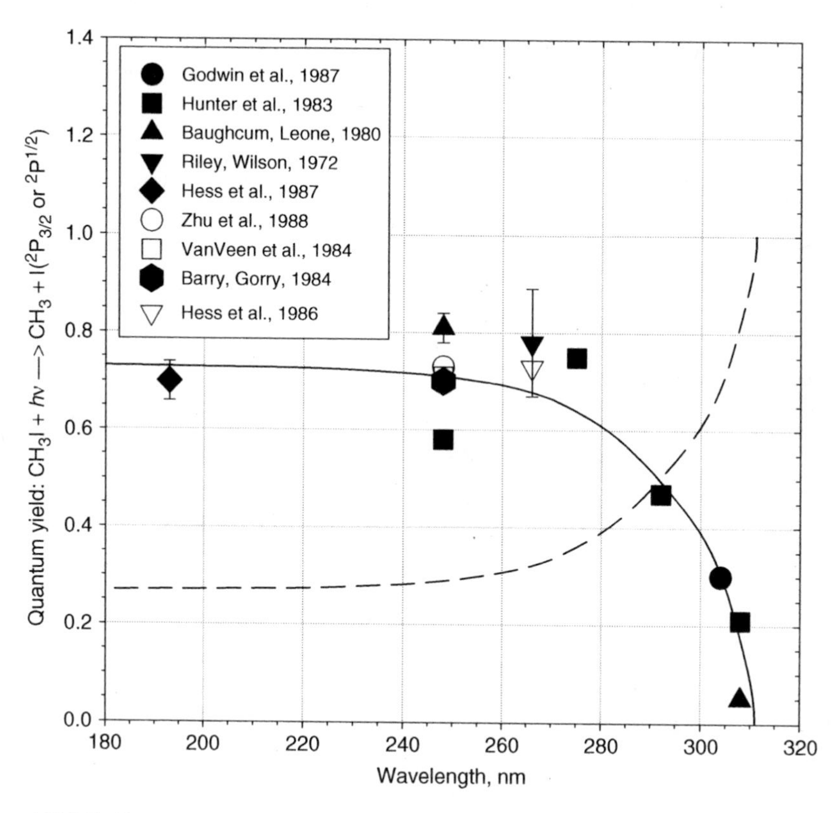

Figure VII-E-5. Quantum yield of I-atoms formed in CH_3I photolysis as a function of wavelength: $I(^2P_{1/2})$, solid curve, and $I(^2P_{3/2})$, dashed curve.

also in figure VII-E-7a for varied values of overhead ozone column (300–400 Dobson units). The sensitivity of the j-values for CH_3I to the magnitude of the overhead ozone column can be seen best in the data of figure VII-E-7b. For the range of solar zenith angles from 0 to 60°, about a 20% decrease in the j-value is seen for an increase of ozone column from 300 to 400 Dobson units. The response is much greater than that seen for formaldehyde (figure VII-B-5). This is expected since the short-wavelength absorption by ozone represents a larger share of the active wavelengths for iodomethane than for formaldehyde. The estimated j-values for CH_3I predict a photochemical lifetime in the lower troposphere of ~ 1.6 days for overhead Sun. For 40°N latitude, clear tropospheric skies, and diurnal variation of the Sun, the expected average photochemical lifetimes of CH_3I are about 4.9 and 27 days for the time periods near June 22 and December 22, respectively.

The cross sections for the alkyl iodides show significant temperature dependence, particularly in the long-wavelength tail. This results in lengthening of the photochemical lifetime in the troposphere where temperatures below 298 K are common. This can be seen in figure VII-E-7c, where the percentage difference between cross sections at 298 and 210 K is shown as a function of wavelength. The effect of altitude on the estimated

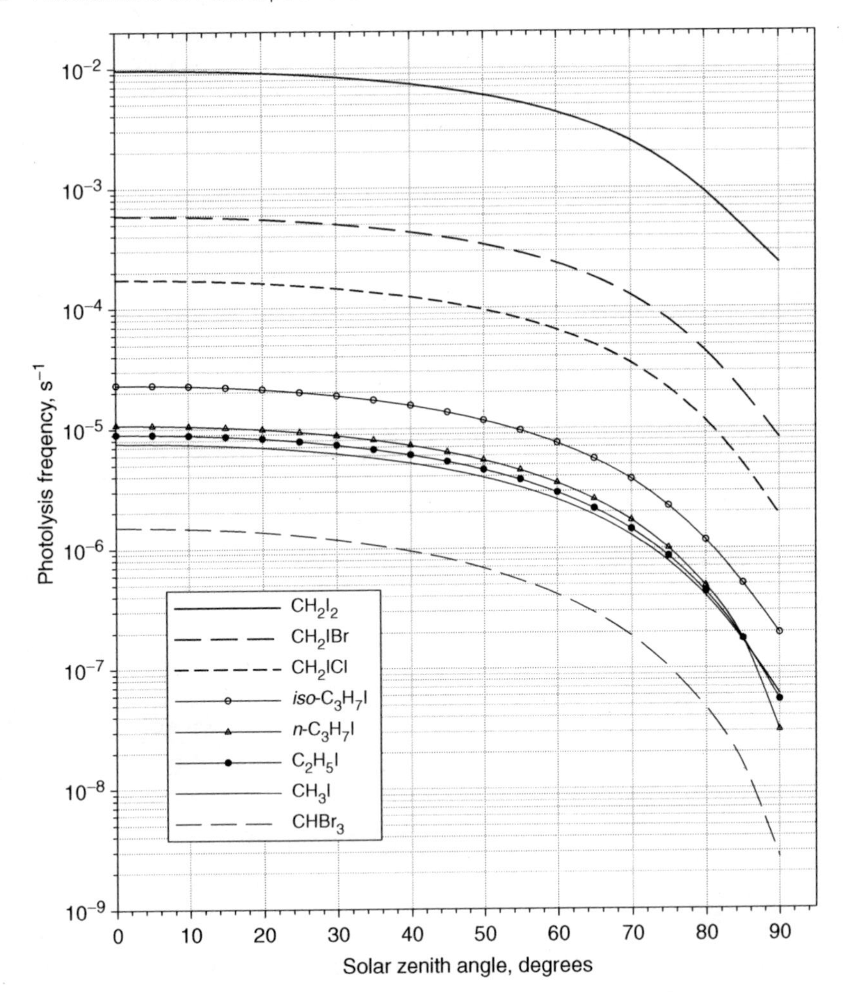

Figure VII-E-6. Photolysis frequencies (j-values) vs. solar zenith angle for the photodissociation of some photochemically active alkyl halides (I-, Br-, and Cl-atom-substituted methanes, ethanes, and propanes). Calculated for a cloudless day within the lower troposphere, assuming the quantum yield of photodissociation to be unity, and the overhead ozone = 350 DU.

j-values for CH_3I is seen in figure VII-E-7d. In these calculations the solar zenith angle (45°) and the overhead O_3 column at ground level (350 DU) have been fixed. A typical temperature profile was used in the calculations (U.S. Standard Atmosphere, 1976). It can be seen that, as altitudes are reached that are in and above the stratospheric ozone reservoir (~25 km), the attenuation of the actinic flux is lowered significantly, and overlap of the the actinic flux distribution with the CH_3I absorption is markedly improved.

The spectral j-values for CH_3I for the lower atmosphere are compared to those of CH_3ONO_2 in figure VII-I-4.

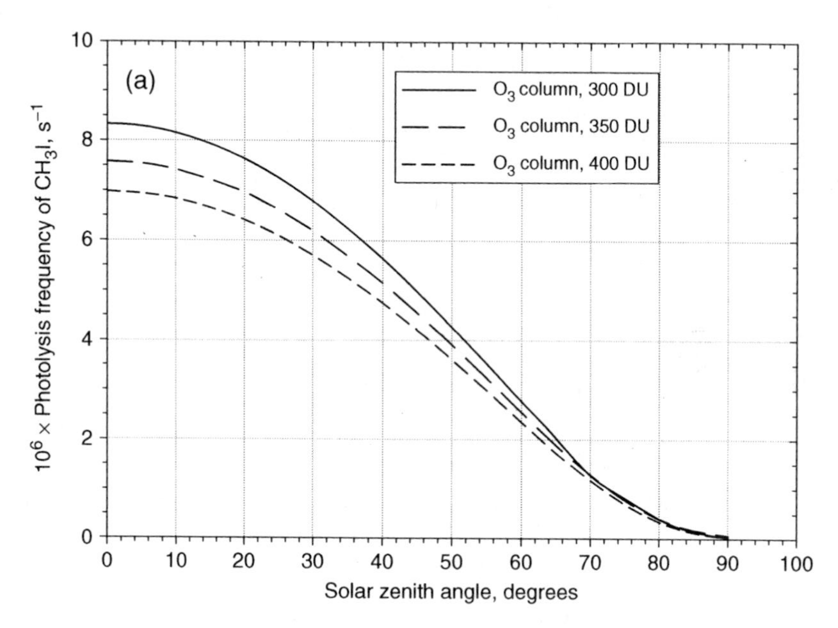

Figure VII-E-7a. Estimated photolysis frequencies for CH_3I as a function of solar zenith angle for a cloudless day in the lower troposphere for several overhead ozone columns.

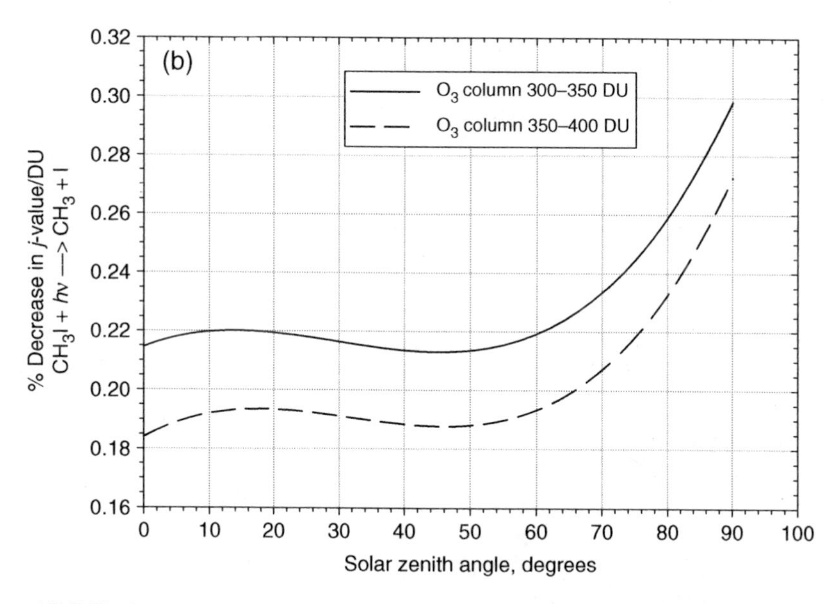

Figure VII-E-7b. Percentage decrease in the j-value per Dobson unit of overhead ozone column for iodomethane photodecomposition on a cloudless day within the troposphere.

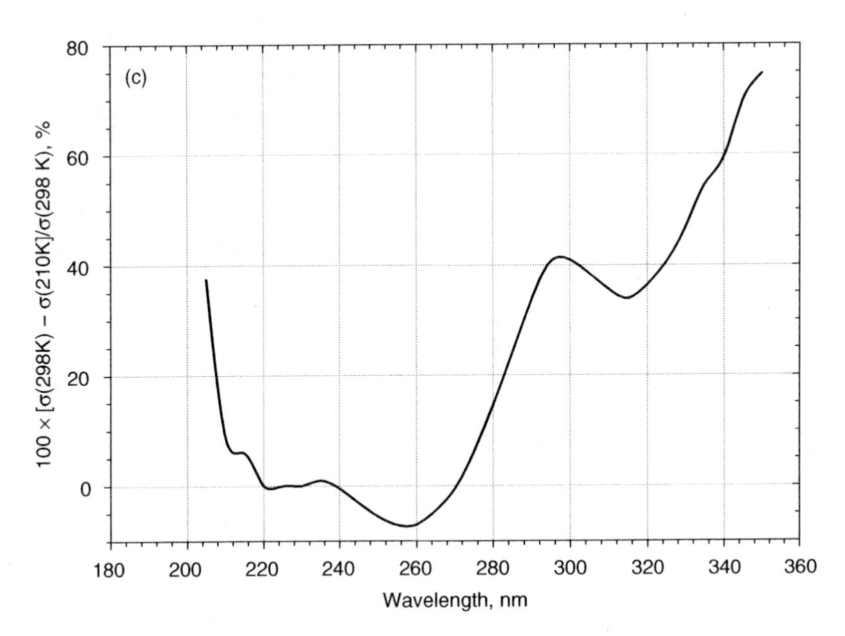

Figure VII-E-7c. Plot of the percentage difference in the absorption cross sections of CH_3I at 298 and 210 K as a function of wavelength, from data of Roehl et al. (1997).

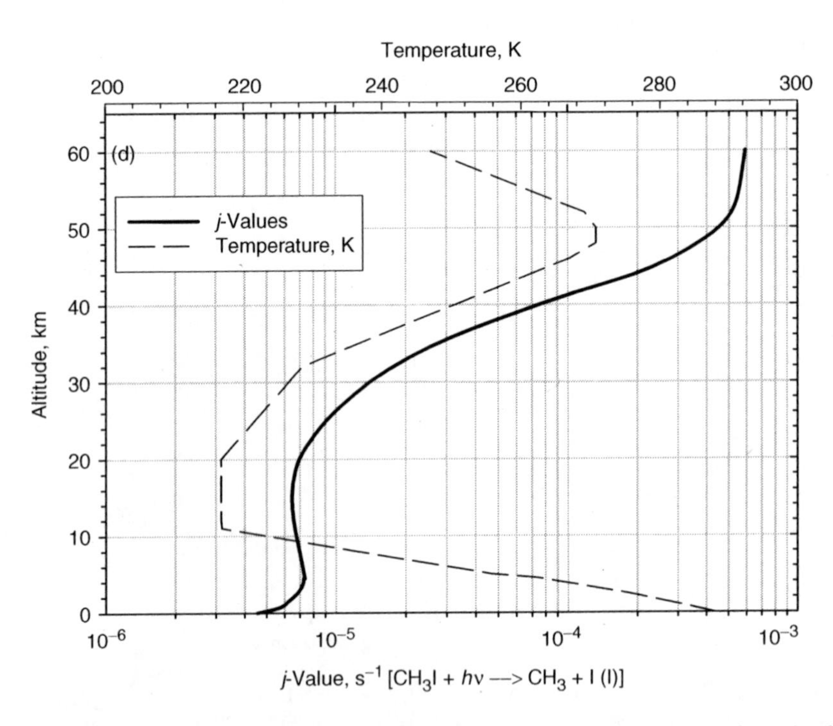

Figure VII-E-7d. Plot of the estimated photolysis frequencies for CH_3I vs. altitude for a cloudless day with solar zenith angle = 45° and vertical ozone column = 350 DU. Also shown is the standard atmosphere temperature variation with altitude.

VII-E-3. Iodoethane (Ethyl Iodide, CH_3CH_2I)

Primary processes (Ia) and (Ib) occur in the photodecomposition of iodoethane and an additional primary process (II) forming C_2H_4 and HI has been suggested (Shepson and Heicklen, 1981):

$$CH_3CH_2I + h\nu \rightarrow C_2H_5 + I(^2P_{3/2}) \quad \text{(Ia)}$$

$$\rightarrow C_2H_5 + I(^2P_{1/2}) \quad \text{(Ib)}$$

$$\rightarrow CH_2{=}CH_2 + HI \quad \text{(II)}$$

The quantum yield of ethene in the photolysis of mixtures of C_2H_5I and I_2 at 313 nm and at 22°C was determined to be 0.0096 ± 0.0002. Shepson and Heicklen (1981) attributed this to process (II). They reasoned that ethene formed in their experiment could not arise from the disproportionation of ethyl radicals, since they are removed rapidly by reaction with $I_2(C_2H_5 + I_2 \rightarrow C_2H_5I + I)$.

A rather surprising result from the study of Shepson and Heicklen (1981) was the conclusion that $\phi_{Ia} + \phi_{Ib} = 0.31 \pm 0.01$, much less than unity. This estimate was based on product quantum yields in the photolysis of C_2H_5I–O_2–He mixtures (acetaldehyde and ethanol measured) and C_2H_5I–$(C_2H_5)_2NOH$ mixtures (ethane measured), together with an assumed mechanism of formation of these products. It is possible that the reformation of C_2H_5I by the rapid reaction $C_2H_5 + I_2 \rightarrow C_2H_5I + I$, or other unanticipated complications, may have led to an underestimation of $\phi_{Ia} + \phi_{Ib}$. These results have not been repeated by other workers, and a quantum yield near unity has been assumed to be more in accord with the continuous, structureless absorption of the C_2H_5I.

No methane was found in Shepson and Heicklen's photolysis of CH_3CH_2I–$(C_2H_5)_2NOH$ mixtures ($\Phi_{CH4} < 0.04$), so the possible down-chain rupture in C_2H_5I photolysis ($C_2H_5I + h\nu \rightarrow CH_3 + CH_2I$) was excluded.

Assuming $\phi_{Ia} + \phi_{Ib} = 1.0$, the value of ϕ_{Ib} has been estimated by Godwin et al. (1987) to be 0.64 (248 nm) and 0.22 (304 nm) with $\phi_{Ia} \approx 0.36$ (248 nm) and 0.78 (304 nm). Riley and Wilson (1972) reported from C_2H_5I photolysis studies at 266.2 nm that the fraction of the energy available after iodine atom excitation which appears in internal excitation of the C_2H_5 radical (~0.37) is significantly higher than that observed in the CH_3 radical from 266.2 nm photolysis of CH_3I (~0.11).

Using the preferred cross sections for C_2H_5I and the appropriate solar flux data for the troposphere, estimates have been made of the *j*-values for C_2H_5I as a function of solar zenith angle for a cloudless day within the troposphere at 298 K, assuming $\phi_{Ia} + \phi_{Ib} = 1.0$. These values can be compared in figure VII-E-6 with those for some other simple iodine- and bromine-containing alkyl halides that are photochemically active in the lower troposphere. The data suggest that the photochemical lifetime for iodoethane will be ~ 1.3 days with overhead Sun (298 K). For 40°N latitude, clear tropospheric skies, and diurnal variation of the Sun, the expected average photochemical lifetimes of iodoethane are about 4.2 and 24 days for the time periods near June 22 and December 22, respectively.

VII-E-4. 1-Iodopropane (*n*-Propyl Iodide, $CH_3CH_2CH_2I$)

As with the simpler alkyl iodides, the photodecomposition of 1-iodopropane is expected to occur largely through processes (Ia) and (Ib):

$$CH_3CH_2CH_2I + h\nu \rightarrow CH_3CH_2CH_2 + I(^2P_{3/2}) \quad \text{(Ia)}$$

$$\rightarrow CH_3CH_2CH_2 + I(^2P_{1/2}) \quad \text{(Ib)}$$

Godwin et al. (1987) estimated $\phi_{Ib} \approx 0.56$ (248 nm) and 0.20 (304 nm) and $\phi_{Ia} \approx 0.44$ (248 nm) and 0.80 (304 nm). For photolysis at 266.2 nm, the fraction of energy available after iodine atom excitation that appears in internal excitation of the propyl fragment (~0.53) is greater than that which appears in the CH_3 (~0.11) and C_2H_5 radicals (~0.37) formed from CH_3I and C_2H_5I photolyses, respectively, at the same wavelength.

The cross section data, coupled with an assumed $\phi_{Ia} + \phi_{Ib} = 1.0$ and the appropriate solar flux data, give the *j*-value estimates for a cloudless troposphere at 298 K as a function of solar zenith angle shown in figure VII-E-6. It is seen that the values for $CH_3CH_2CH_2I$ are somewhat larger than those for C_2H_5I and suggest a photochemical lifetime in the lower troposphere with overhead Sun of about 1.1 days (298 K). For 40°N latitude, clear tropospheric skies, and diurnal variation of the Sun, the expected average photochemical lifetimes of $CH_3CH_2CH_2I$ are about 3.6 and 20 days for the time periods near June 22 and December 22, respectively.

VII-E-5. 2-Iodopropane (Isopropyl Iodide, CH_3CHICH_3)

Photodecomposition of CH_3CHICH_3 is expected to follow the pattern of the smaller alkyl iodides:

$$CH_3CHICH_3 + h\nu \rightarrow CH_3CHCH_3 + I(^2P_{3/2}) \quad \text{(Ia)}$$

$$\rightarrow CH_3CHCH_3 + I(^2P_{1/2}) \quad \text{(Ib)}$$

For photolysis at 266.2 nm, the fraction of energy available after iodine atom excitation that appears in internal excitation of the CH_3CHCH_3 fragment (~0.57) is greater than that which appears in the CH_3 (~0.11), C_2H_5 (~0.37), and $CH_3CH_2CH_2$ radicals (~0.53) formed from CH_3I, C_2H_5I, and $CH_3CH_2CH_2I$ photolyses, respectively, at the same wavelength (Riley and Wilson, 1972).

The cross section data, coupled with an assumed $\phi_{Ia} + \phi_{Ib} = 1.0$ and the appropriate solar flux data, were used to estimate the *j*-value as a function of solar zenith angle for a cloudless day at 298 K; see figure VII-E-6. The *j*-values for CH_3CHICH_3 are somewhat larger than those for the other small alkyl iodides and suggest a photochemical lifetime of about 0.5 day in the lower troposphere with an overhead Sun. For 40°N latitude, clear tropospheric skies, and diurnal variation of the Sun, the expected average photochemical lifetimes of 2-iodopropane are about 1.6 and 9.2 days for the time periods near June 22 and December 22, respectively.

VII-E-6. 2-Methyl-2-iodopropane [*tert*-Butyl Iodide, $(CH_3)_3CI$]

The photodecomposition of $(CH_3)_3CI$ is expected to follow the pattern of the smaller alkyl iodides:

$$(CH_3)_3CI + h\nu \rightarrow (CH_3)_3C + I(^2P_{3/2}) \quad \text{(Ia)}$$

$$\rightarrow (CH_3)_3C + I(^2P_{1/2}) \quad \text{(Ib)}$$

Kim et al. (1997) studied the translation energies and angular distributions of the photodissociation products of 2-methyl-2-iodopropane at 277 and 304 nm. It is interesting, and unexpected, that process (Ia), formation of the ground-state I-atom, was favored, unlike the case of the smaller alkyl iodides. Values of ϕ_{Ia} were estimated to be 0.93 ± 0.03 (277 nm) and 0.92 ± 0.04 (304 nm). The observed rapid dissociation was interpreted in terms of three independent reaction paths on an excited potential energy surface: 1) prompt dissociation along the C—I stretching mode; 2) repulsive mode along the C—I stretching, coupled with some bending motions; and (3) indirect dissociation, probably due to large contributions from the bending motions.

The incomplete spectral and quantum yield data prevent realistic estimates of the *j*-values for 2-methyl-2-iodopropane, but these values are expected to be somewhat greater than those estimated for the simpler iodides.

VII-E-7. Tribromomethane (Bromoform, $HCBr_3$)

The photodissociation of tribromomethane ($CHBr_3$) appears to occur largely by the primary process (I):

$$CHBr_3 + h\nu \rightarrow CHBr_2 + Br(^2P_{3/2} \text{ or } ^2P_{1/2}) \quad \text{(I)}$$

Evidence favoring this process in photolysis at 193 nm was presented by McGivern et al. (2000). The $CHBr_2$ photofragments are produced with sufficient internal energy to undergo secondary dissociation by two alternative pathways: one producing CHBr + Br and the other forming CBr + HBr, with quantum yields of 0.3 and 0.4, respectively. Bayes et al. (2003) measured Br-atom quantum yields from 266 to 344 nm, and concluded that ϕ_I for tribromomethane photodissociation at wavelengths greater than 300 nm is near unity. From their study of tribromomethane photodissociation at 248 nm, Zou et al. (2004) concluded that process (I) is the dominant dissociation channel. Secondary photolysis of $CHBr_2$ leads to a variety of products (HBr + Br, CH + Br_2, and CBr + HBr) that could be interpreted incorrectly as primary products of $CHBr_3$ photodecomposition.

The evidence from several studies points to the generation of Br-atoms in process (I) with a quantum yield near unity (McGivern et al., 2000; Xu et al., 2002a; Bayes et al., 2003; Zou et al., 2004).

The photolysis frequencies for $CHBr_3$ can be estimated from the absorption cross sections (table VII-E-2) and the appropriate solar flux data (ozone column

Table VII-E-2. Cross sections (σ, cm^2 $molecule^{-1}$) of $CHBr_3$ (296 K) and CH_2ICl (298 K). The 1 nm resolution data were determined by interpolation of the 2–5 nm resolution data recommended by Atkinson et al. (2004), using a spline fit

λ (nm)	$CHBr_3$ $10^{20} \times \sigma$	CH_2ICl $10^{20} \times \sigma$	λ (nm)	$CHBr_3$ $10^{20} \times \sigma$	CH_2ICl $10^{20} \times \sigma$	λ (nm)	CH_2ICl $10^{20} \times \sigma$
278	12.6	110	313	0.110	8.6	348	0.35
279	11.3	107	314	0.096	7.8	349	0.30
280	10.0	103	315	0.083	7.2	350	0.30
281	8.90	99	316	0.072	6.7	351	0.30
282	7.80	95	317	0.062	6.2	352	0.29
283	6.90	91	318	0.053	5.6	353	0.28
284	6.13	86	319	0.046	5.2	354	0.27
285	5.45	81	320	0.039	4.8	355	0.26
286	4.82	76	321	0.033	4.5	356	0.25
287	4.26	72	322	0.029	4.2	357	0.24
288	3.75	67	323	0.025	3.8	358	0.23
289	3.30	63	324	0.021	3.5	359	0.21
290	2.86	58	325	0.0185	3.3	360	0.20
291	2.55	54	326	0.0164	3.0	361	0.16
292	2.22	50	327	0.0120	2.8	362	0.14
293	1.95	46	328	0.0090	2.6	363	0.12
294	1.69	42	329	0.0090	2.3	364	0.10
295	1.50	39	330	0.0090	2.2	365	0.08
296	1.26	37	331	0.0083	2.0	366	0.06
297	1.13	34	332	0.0060	1.8	367	0.04
298	0.95	31	333	0.0056	1.6	368	0.02
299	0.82	28	334	0.0050	1.5	369	0.01
300	0.72	25	335	0.0043	1.4	370	0
301	0.63	23.5	336	0.0040	1.3		
302	0.53	21.5	337	0.0035	1.2		
303	0.46	19.5	338	0.0030	1.1		
304	0.395	18.2	339	0.0025	1.0		
305	0.345	16.8	340	0.0020	0.90		
306	0.297	15.5	341	0.0015	0.80		
307	0.260	14.2	342	0.0010	0.75		
308	0.228	13.2	343	0.0005	0.70		
309	0.197	11.9	344	0.0002	0.60		
310	0.170	10.9	345	0	0.55		
311	0.149	10.0	346		0.50		
312	0.127	9.4	347		0.40		

= 350 DU), assuming $\phi = 1.0$; these calculated values for various solar zenith angles on a cloudless day in the lower troposphere are shown in figure VII-E-6 where they can be compared with other halogen-substituted alkanes. It can be seen that the photochemical lifetime of $CHBr_3$ is long, about 7 days for overhead Sun. For 40°N latitude, clear tropospheric skies, and diurnal variation of the Sun, the expected average photochemical lifetimes of $CHBr_3$ are about 27 and 180 days for the time periods near June 22 and December 22, respectively. Nonetheless, this still represents the major atmospheric loss process for this species.

In contrast to the photochemical lifetime of $CHBr_3$, that for CH_2Br_2 has a photochemical lifetime in the troposphere of several years. Its reaction with OH radicals in the troposphere shortens its residence time in the troposphere considerably, so that transport to the stratosphere and initiation of ozone depletion becomes much less serious.

VII-E-8. Chloroiodomethane (CH_2ICl)

The photodissociation of CH_2ICl at 266.2 nm occurs largely by cleavage of the weakest bond in the molecule, CH_2Cl—I (Schmitt and Comes, 1987) in reactions (Ia) and (Ib):

$$CH_2ICl + h\nu \rightarrow I(^2P_{3/2}) + CH_2Cl \quad \text{(Ia)}$$

$$\rightarrow I(^2P_{1/2}) + CH_2Cl \quad \text{(Ib)}$$

$$\rightarrow Cl(^2P_{3/2,1/2}) + CH_2I \quad \text{(II)}$$

This cleavage occurs although the energy in the 266.2 nm quantum is sufficient to break the stronger ICH_2—Cl bond in the molecule in process (II) as well. From extensive product analyses, Schmitt and Comes concluded that secondary photolysis of the primary product, CH_2Cl, leads to Cl-atoms and methylene radical (CH_2). The quantum yield of I_2 was found to be 0.5 ± 0.05, indicating that $\phi_{Ia} + \phi_{Ib} \approx 1.0$. Senapati et al. (2002) determined the quantum yields of $I(^2P_{1/2})$ production in the photodecomposition of CH_2ICl in process (Ib): 0.47 ± 0.02 (222 nm); 0.51 ± 0.01 (236 nm); 0.51 ± 0.02 (266 nm); 0.55 ± 0.03 (280 nm); and 0.38 ± 0.01 (304 nm).

The cross section data for CH_2ICl (table VII-E-2) and the appropriate solar flux data (ozone column = 350 DU) were used to estimate the photolysis frequency of CH_2ICl as a function of solar zenith angle for a cloudless lower troposphere, assuming $\phi_I = 1.0$; see figure VII-E-6. These data indicate that the photochemical lifetime of CH_2ICl is approximately 1.6 hr for an overhead Sun. For 40°N latitude, clear tropospheric skies, and diurnal variation of the Sun, the expected average photochemical lifetimes of CH_2ICl are about 4.9 hr and 1.0 day for the time periods near June 22 and December 22, respectively.

VII-E-9. Bromoiodomethane (CH_2BrI)

Photodecomposition of CH_2BrI provides the opportunity for the alternative pathways that involve rupture of either the CH_2Br—I or the CH_2I—Br bond:

$$CH_2BrI + h\nu \rightarrow CH_2Br + I(^2P_{3/2}) \quad \text{(Ia)}$$

$$\rightarrow CH_2Br + I(^2P_{1/2}) \quad \text{(Ib)}$$

$$\rightarrow CH_2I + Br(^2P_{3/2}, {}^2P_{1/2}) \quad \text{(II)}$$

In studies of Lee and Bersohn (1982) photolysis of CH_2BrI was conducted at 258 nm, and it was estimated that 86% of the dissociation occurred by processes (Ia) and (Ib), with 14% by process (II). Similar results were obtained in the more detailed later studies of Butler et al. (1986); they showed processes (Ia) and (Ib) to be dominant for photolysis at 248 nm, with $\phi_{Ib} + \phi_{Ia} \approx 0.75$. At this wavelength, excitation primarily involves promotion of a non-bonding electron on the iodine atom to an antibonding orbital associated with the C—I bond. Process (II) is a minor channel at 248 nm and presumably so for excitation at wavelengths greater than 248 nm. As one might expect, with excitation at much higher energies available at 193 nm there is less discrimination in the bond breakage, and the occurrence of process (II) becomes more significant: $\phi_{II}/(\phi_{Ia} + \phi_{Ib}) \approx 3.5$. Direct dissociation of CH_2BrI to form CH_2 and IBr at 210 nm may occur, but with a quantum yield of less than 0.06. The results are different for excitation at 210 nm, assigned to an $n(Br) \rightarrow \sigma^*(C—Br)$ transition; photolysis at 210 nm results largely in CH_2I—Br bond fission with little or no primary CH_2Br—I fission (Butler et al., 1986).

There is a large wavelength region in which the cross sections of CH_2BrI overlap the solar spectrum within the troposphere, and this leads to a very short atmospheric lifetime. The cross section data of table VII-E-3 and the appropriate solar flux data (ozone column = 350 DU) were used to calculate the photolysis frequency of CH_2BrI as a function of solar zenith angle for a cloudless day in the lower troposphere, assuming $\phi_I = 1.0$; see figure VII-E-6. These values indicate a photochemical lifetime of CH_2IBr of approximately 30 min for an overhead Sun. For 40°N latitude, clear tropospheric skies, and diurnal variation of the Sun, the expected average photochemical lifetimes of CH_2IBr are about 1.4 and 6.6 hr for the time periods near June 22 and December 22, respectively.

VII-E-10. Diiodomethane (CH_2I_2)

The photodecomposition of CH_2I_2 has been shown to occur by loss of an iodine atom in processes (Ia) and (Ib):

$$CH_2I_2 + h\nu \rightarrow CH_2I + I(^2P_{3/2}) \qquad \text{(Ia)}$$

$$\rightarrow CH_2I + I(^2P_{1/2}) \qquad \text{(Ib)}$$

Kawasaki et al. (1975) used molecular beam photofragmentation to show that the lower lying states of CH_2I_2 dissociate directly in processes (Ia) and (Ib). From extensive product studies of photolysis of CH_2I_2 at 300 nm, Schmitt and Comes (1980) confirmed the occurrence of processes (Ia) and (Ib) and excluded the direct formation of I_2 and CH_2. Koffend and Leone (1981) demonstrated that the fraction of the excited CH_2I_2 molecules decomposed in process (Ib) increases as the excitation energy is increased; see figure VII-E-8.

There is an overlap over a very large wavelength region of the absorption of CH_2I_2 and the solar spectrum available within the troposphere that leads to a very short photochemical lifetime for CH_2I_2. The photolysis frequency of CH_2I_2 has been calculated as a function of solar zenith angle for a cloudless lower troposphere using the cross section data of table VII-E-3 and the appropriate solar flux data (ozone column = 350 DU), assuming $\phi_I = 1.0$; see figure VII-E-6. The j-values for this compound are

Table VII-E-3. Cross sections (σ, cm^2 $molecule^{-1}$) for CH_2BrI and CH_2I_2 at 298 K. The 1 nm resolution data given here were determined by interpolation of the 5 nm resolution data recommended by Atkinson et al. (2004), using a spline fit

λ (nm)	CH_2BrI $10^{20} \times \sigma$	CH_2I_2 $10^{20} \times \sigma$	λ (nm)	CH_2BrI $10^{20} \times \sigma$	CH_2I_2 $10^{20} \times \sigma$	λ (nm)	CH_2BrI $10^{20} \times \sigma$	CH_2I_2 $10^{20} \times \sigma$
278	196	300	318	19.3	257	358	0.60	13.9
279	191	315	319	18.0	252	359	0.55	12.4
280	184	329	320	16.8	243	360	0.50	10.9
281	178	339	321	15.7	237	361	0.45	9.8
282	172	350	322	14.6	228	362	0.40	8.7
283	165	360	323	13.6	221	363	0.35	7.8
284	158	366	324	12.5	212	364	0.30	6.9
285	150	373	325	11.5	204	365	0.26	6.2
286	142	377	326	10.7	196	366	0.24	5.5
287	135	380	327	10.0	186	367	0.23	4.8
288	126	381	328	9.03	177	368	0.22	4.3
289	117	382	329	8.7	169	369	0.21	3.8
290	110	381	330	8.0	162	370	0.20	3.4
291	104	380	331	7.5	153	371	0.10	3.0
292	98	378	332	6.9	144	372	0	2.7
293	92	375	333	6.5	137	373	0	2.5
294	87	374	334	6.0	128	374	0	2.2
295	82	371	335	5.5	120	375	0	1.9
296	78	368	336	5.2	113	376	0	1.7
297	74	365	337	4.6	105	377	0	1.6
298	69	363	338	4.3	99	378	0	1.4
299	65	360	339	3.9	90	379	0	1.3
300	61	356	340	3.5	83	380	0	1.2
301	57	353	341	3.2	78	381	0	1.1
302	53	349	342	2.9	70	382	0	1.0
303	49	345	343	2.6	66	383	0	0.9
304	46	342	344	2.4	60	384	0	0.8
305	43	338	345	2.3	53	385	0	0.7
306	40	334	346	2.1	49	386	0	0.6
307	37	330	347	1.8	45	387	0	0.5
308	35	325	348	1.7	40	388	0	0.4
309	33	319	349	1.6	36	389	0	0.3
310	31.3	315	350	1.5	32.5	390	0	0.2
311	29.6	308	351	1.4	29.5	391	0	0.1
312	28.0	302	352	1.3	26.6	392	0	0
313	26.4	295	353	1.1	24.0			
314	24.6	287	354	1.0	21.5			
315	23.3	280	355	0.9	19.3			
316	21.7	273	356	0.8	17.3			
317	20.5	265	357	0.7	15.5			

the highest of those shown in the figure; they indicate a photochemical lifetime of CH_2I_2 of approximately 1.7 min for an overhead Sun. For 40°N latitude, clear tropospheric skies, and diurnal variation of the Sun, the expected average photochemical lifetimes of CH_2I_2 are about 5.0 and 21 min for the time periods near June 22 and December 22, respectively.

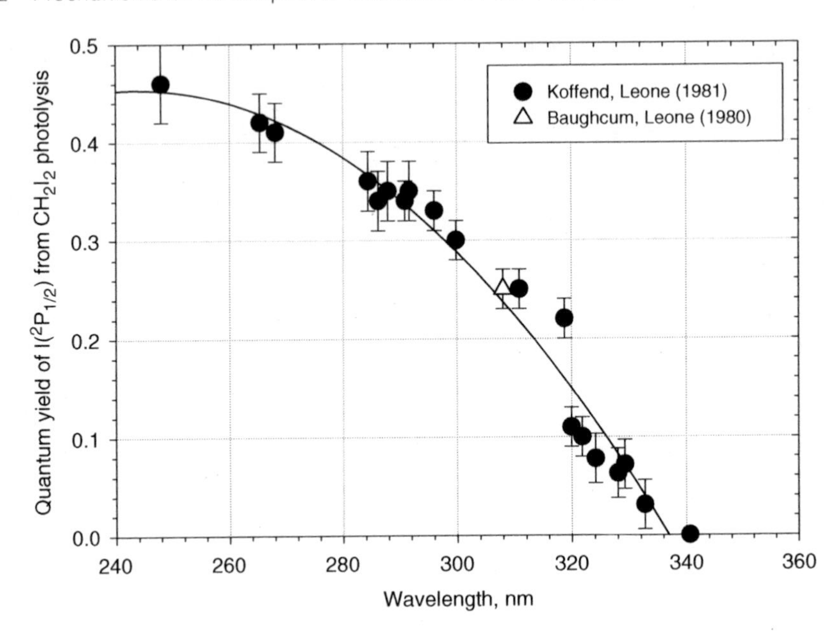

Figure VII-E-8. Quantum yield of $I(^2P_{1/2})$ atom formation as a function of wavelength in the photolysis of CH_2I_2.

VII-F. Photodecomposition of Halogen-Atom- and HO-Substituted Aldehydes

VII-F-1. Formation of Halogen-Atom-Substituted Carbonyls in the Atmosphere

The halogen-containing alkanes that have largely replaced the chlorofluorocarbons always contain H-atoms as well as Cl- and F-atoms, and their oxidation in the troposphere, initiated largely by OH attack on these H-atoms, can lead to aldehydes and ketones that contain chlorine and fluorine. Many of these species are photochemically active in the wavelength region present in sunlight within the troposphere, and thus their photochemistry is of interest. In this section we review the limited data related to the photodecomposition processes in the substituted aldehydes, and in section VII-G the substituted ketones are discussed. In section VII-H the photochemical data related to the acyl halides are discussed.

VII-F-2. Effects of Halogen-Atom- and OH-Substitution on the Absorption Spectra of the Aldehydes

Substitution of an OH-group or Cl-atom for an H-atom in the CH_3-group of acetaldehyde results in significant changes relative to the ultraviolet absorption spectrum of acetaldehyde. This can be seen in figure VII-F-1, where the spectra of 2-hydroxyethanal and 2-chloroethanal are compared with that of the unsubstituted ethanal. The wavelength of maximum absorption in $HOCH_2CHO$ is shifted about 13 nm to *shorter*

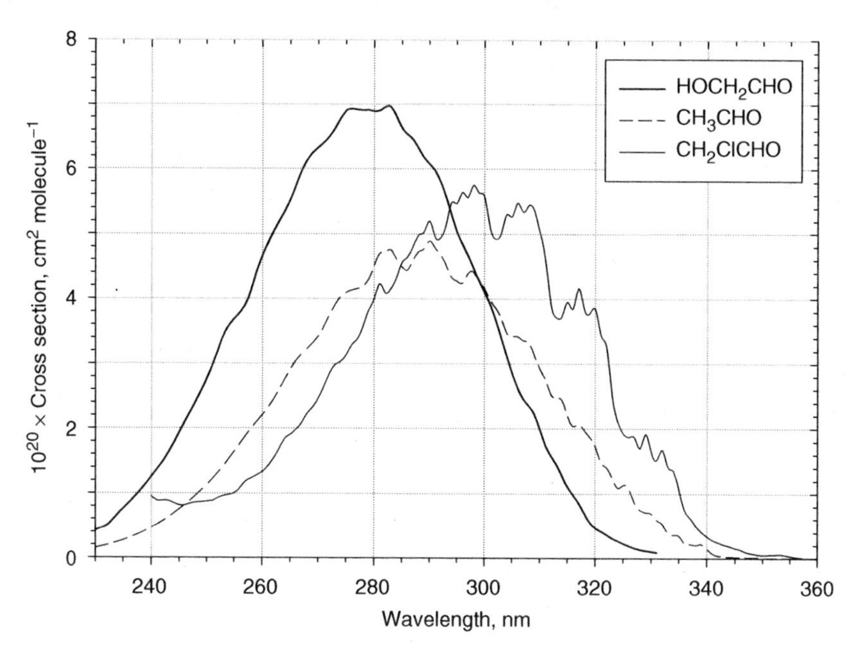

Figure VII-F-1. Comparison of the absorption spectra of some substituted acetaldehydes: $HOCH_2CHO$ at 298 ± 3 K (Magneron et al., 2005), CH_2ClCHO (Libuda, 1992), and CH_3CHO (Martinez et al., 1992); temperature, 298 K.

wavelengths than λ_{max} seen in CH_3CHO. However, λ_{max} in CH_2ClCHO is shifted about 8 nm to *longer* wavelengths, and σ_{max} is increased by about 18%.

The effects of multiple substitution of Cl-atoms for H-atoms in CH_3CHO can be seen in figure VII-F-2. The addition of a second Cl-atom to form $CHCl_2CHO$ does not alter the wavelength of the maximum absorption significantly, but the absorption coefficient is raised about 7% over that present in CH_2ClCHO. A significant and unexpected change in the spectrum occurs with the addition of the third atom of Cl in the $CHCl_2CHO$ molecule. The wavelength of maximum absorption of CCl_3CHO is shifted back to a shorter wavelength, approximately that observed with the unsubstituted acetaldehyde, and the cross section at the maximum is raised to about 2.2 times that in CH_3CHO. With all of the chloroacetaldehydes, CH_2ClCHO, $CHCl_2CHO$, and CCl_3CHO, the absorption of tropospheric sunlight is increased significantly over that of CH_3CHO, and hence their tropospheric photochemistry is potentially important.

In figure VII-F-3 the effect of multiple F- and Cl-atom substitutions for H-atoms on the absorption spectra of CH_3CHO can be seen. Replacement of two H-atoms by F-atoms in CH_3CHO, to form CHF_2CHO, shifts the absorption maximum 19 nm to longer wavelengths while the maximum absorption coefficient is decreased only slightly (~3%). With the introduction of a third F-atom in CHF_2CHO to form CF_3CHO, the maximum in the absorption band is decreased significantly (~35%) from that in CH_3CHO, and it shifts 11 nm to longer wavelengths. Note, however, that the substitution of Cl-atoms for F-atoms in CF_3CHO causes more significant changes. One Cl-atom substitution ($CClF_2CHO$) increases the absorption cross section of CF_3CHO by about a

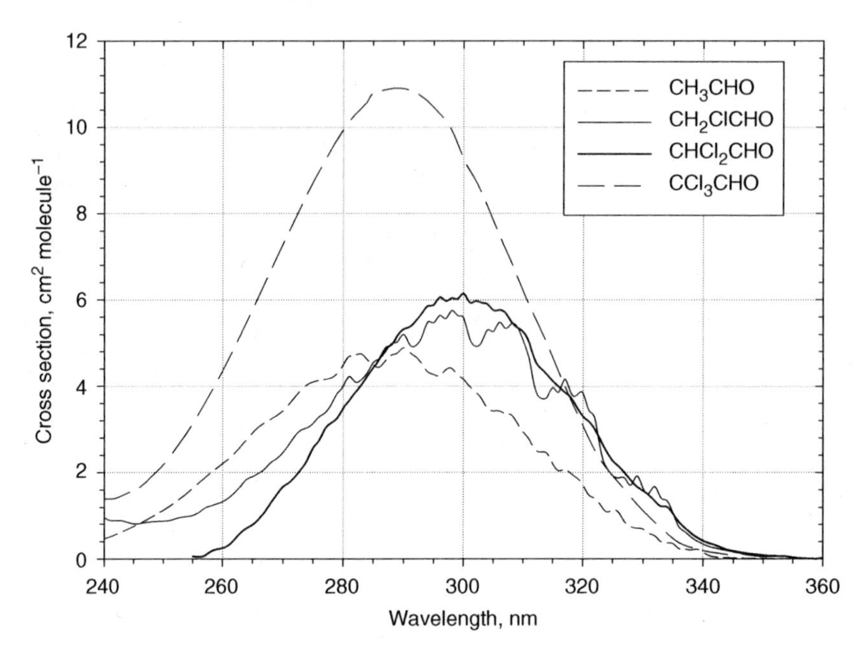

Figure VII-F-2. Comparison of the absorption spectra of the chlorine-atom-substituted acetaldehydes: CH_3CHO (Martinez et al., 1992), CH_2ClCHO (Libuda, 1992), $CHCl_2CHO$ (Libuda, 1992), CCl_3CHO (Talukdar et al., 2001); temperature, 298 K.

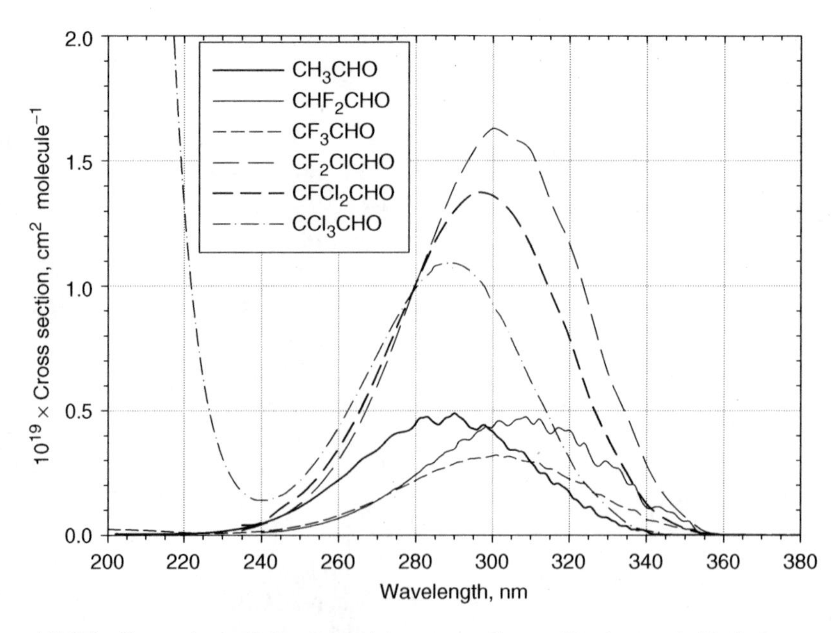

Figure VII-F-3. Comparison of the absorption spectra of some fluorine- and chlorine-atom substituted acetaldehydes: CHF_2CHO (Sellevåg et al., 2005), CF_3CHO (Sellevåg et al., 2004), CF_2ClCHO (Rattigan et al., 1998), $CFCl_2CHO$ (Rattigan et al., 1998), CCl_3CHO (Talukdar et al., 2001), CH_3CHO (Martinez et al., 1992); temperature, 298 K.

factor of 5.1, although the wavelength at maximum absorption is relatively unchanged. The second Cl-atom substitution to form CCl_2FCHO lowers the cross section by about 16% and shifts the absorption by about 5 nm to shorter wavelengths. The third Cl-atom substitution to form CCl_3CHO shifts the absorption maximum by about 6 nm to shorter wavelengths and lowers the cross section by about 20%.

The effect of replacement of all of the H-atoms on the alkyl groups of the simple aldehydes by fluorine atoms can be seen in figure VII-F-4a (Hashikawa et al., 2004). Shown in black curves are the cross section data for the fluoro-aldehydes, $C_xF_{2x+1}CHO$, where $x = 1$–4; compare these spectra with those for the analogous unsubstituted aldehydes shown in the grey curves. As described previously, the cross sections for CF_3CHO are decreased significantly (~35%) from those of CH_3CHO and the maximum is shifted to the longer wavelengths. As the length of the alkyl chain in the fluoro-aldehydes is increased in the sequence CF_3CHO, C_2F_5CHO, $CF_3CF_2CF_2CHO$, to $CF_3CF_2CF_2CF_2CHO$, a significant increase in the absorption maximum is seen, but only a small shift in the wavelength of the maximum occurs. In contrast, the unsubstituted aldehydes (grey curves in figure VII-F-4a) show only a small rise in the absorption maximum as one progresses in the analogous series: ethanal, propanal, butanal, and pentanal, and the cross sections of the larger aldehydes (butanal, pentanal, hexanal, etc.) are very similar (Tadić et al., 2001b; Plagens et al., 1998). In figure VII-F-4b a comparison is given of the spectra of aldehydes with F-atom substitution for H-atoms on the carbon atoms down-chain from the carbonyl group. With CF_3CH_2CHO and n-$C_6F_{13}CH_2CHO$ the wavelengths of maximum absorption are shifted slightly to longer wavelengths than in the analogous unsubstituted aldehyde, CH_3CH_2CHO, and a representative long-chain alkyl aldehyde, hexanal. However, the maximum absorption coefficients are affected greatly by F-atom substitution for H-atoms. The σ_{max} for $C_6F_{13}CH_2CHO$ is much larger than that for the unfluorinated large aldehyde, whereas the σ_{max} for CF_3CH_2CHO is smaller than that for CH_3CH_2CHO. In general, substitution of a Cl- or F-atom in the aldehydes results in a significant increase in the overlap with the wavelength region present in solar radiation within the troposphere. Thus the potential for the importance of photodecomposition of the fluoro-chloro-aldehydes is enhanced over that of the unsubstituted aldehydes.

The photodecomposition modes of the OH- and halogen-substituted acetaldehydes are reviewed in this section.

VII-F-3. 2-Hydroxyethanal (Hydroxyacetaldehyde; Glycolaldehyde)

The absorption spectrum of 2-hydroxyethanal was measured by several research groups (Bacher et al., 2001; Max Planck Institute and LCSR/CNRS-Orléans as reported by Magneron et al., 2005). The spectral data reported by Bacher et al. are about 25% lower than those of the studies reported by Magneron et al. In view of the tendency of 2-hydroxyethanal to adsorb on surfaces, and the good agreement of the two higher cross section measurements, the Magneron et al., (2005) data are recommended. The averaged cross sections reported by the LCSR/CNRS and Mainz groups are shown in figure VII-F-1, where they are compared with those of acetaldehyde. It is seen that a smaller

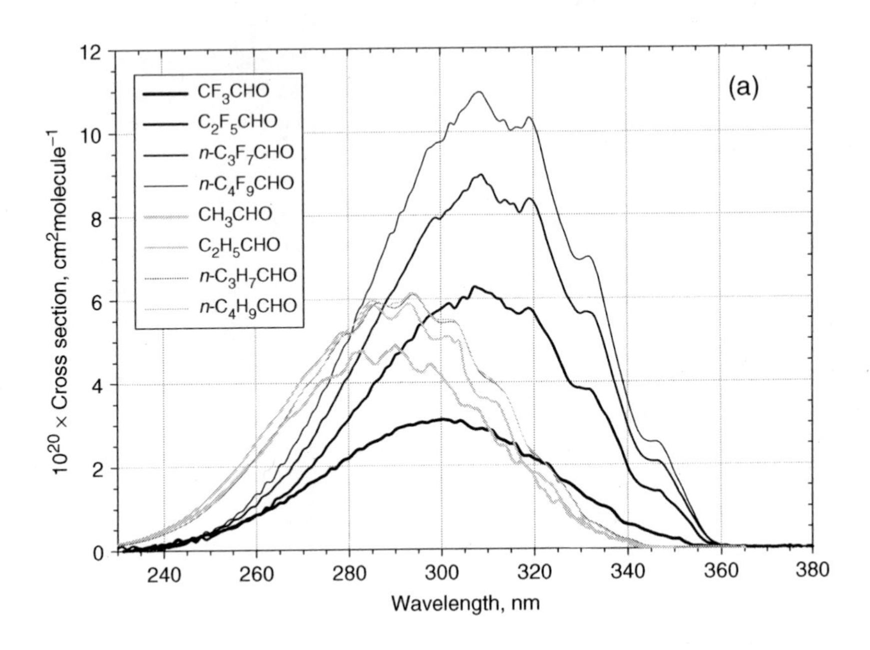

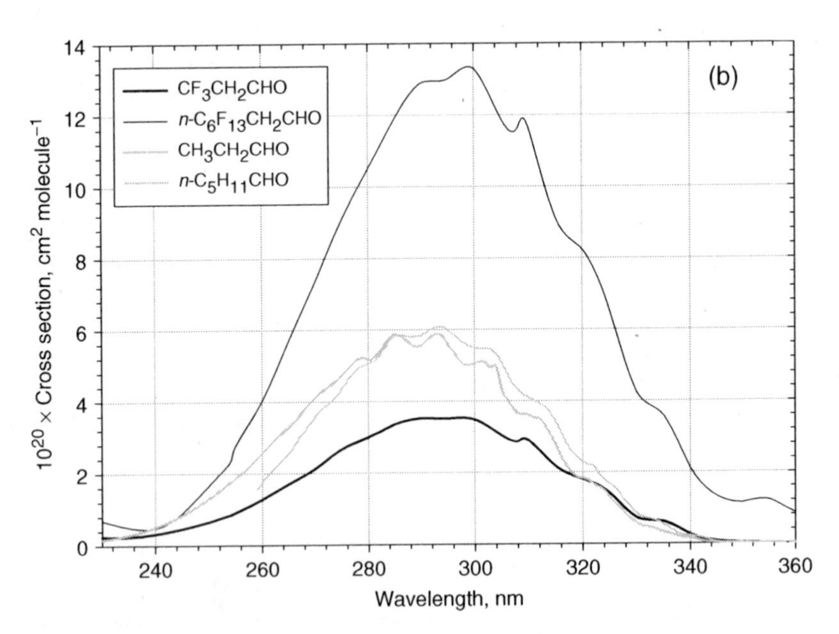

Figure VII-F-4. Comparison of the absorption cross sections as a function of wavelength: Panel (a) compares a series of fluorine-substituted aldehydes [$C_xF_{2x+1}CHO$, x = 1–4 (black lines)] and the corresponding unsubstituted aldehydes (grey lines). Data for the fluoro-aldehydes are from Hashikawa et al. (2004); ethanal, propanal, and butanal data are from Martinez et al. (1992); pentanal data are from Tadić et al. (2001a). In panel (b) the data for CF_3CH_2CHO and n-$C_6F_{13}CH_2CHO$ from Chiappero et al. (2006) (black lines) are compared with those for structurally similar, unfluorinated aldehydes, CH_3CH_2CHO (Martinez et al. (1992) and hexanal (spectrum representative of larger, unsubstituted aldehydes; Plagens et al., 1998) (grey lines).

Table VII-F-1. Absorption cross sections (σ, cm^2 $molecule^{-1}$) for 2-hydroxyethanal at 298 ± 3 K. Data are averages of those determined at the Max Planck Institute and at LCSR/CNRS-Orléans as reported by Magneron et al. (2005)

λ (nm)	$10^{20} \times \sigma$	λ (nm)	$10^{20} \times \sigma$
278	6.91	304	2.86
279	6.91	305	2.66
280	6.90	306	2.46
281	6.93	307	2.35
282	6.96	308	2.23
283	6.79	309	1.99
284	6.64	310	1.75
285	6.55	311	1.60
286	6.46	312	1.45
287	6.32	313	1.28
288	6.18	314	1.11
289	6.08	315	0.985
290	5.98	316	0.86
291	5.78	317	0.705
292	5.58	318	0.55
293	5.31	319	0.485
294	5.04	320	0.420
295	4.86	321	0.365
296	4.67	322	0.31
297	4.48	323	0.27
298	4.28	324	0.23
299	4.10	325	0.195
300	3.92	326	0.16
301	3.67	327	0.14
302	3.41	328	0.12
303	3.14	329	0.11
		330	0.096

fraction of the sunlight within the troposphere is absorbed by 2-hydroxyacetaldehyde than by unsubstituted acetaldehyde. Given in table VII-F-1 is a numerical listing of the average cross sections as measured as a function of wavelength by the groups at Max Planck Institute and LCSR/CNRS-Orléans, as reported by Magneron et al. (2005).

Bacher et al. (2001) and Magneron et al. (2005) have studied the photodecomposition of 2-hydroxyethanal in air. Bacher et al. (2001) employed 240–400 nm radiation from a filtered xenon lamp. Their data suggested that process (I) below is the major photodecomposition mode, analogous to the bond cleavage seen in acetaldehyde photolysis, although some CH_3OH was detected also, presumably by reaction (III). Magneron et al. (2005) also concluded that process (I) is the major mode of photodecomposition, but they considered three additional thermochemically favorable processes as well, (II), (III), and (IV):

$$HOCH_2CHO + h\nu \rightarrow CH_2OH + HCO \quad (I)$$

$$\rightarrow OH + CH_2CHO \quad (II)$$

$$\rightarrow CH_3OH + CO \quad (III)$$

$$\rightarrow H + HOCH_2CO \quad (IV)$$

Several observations of Magneron et al. point to the occurrence of the OH-forming channel (II): 1) the difference between the loss rates of 2-hydroxyethanal measured during photolysis with and without cyclohexane (an OH scavenger); and 2) the high loss rate of di-*n*-butyl ether (OH tracer) observed in the EUPHORE chamber. Methanol observed in the products indicated that process (III) occurs. No evidence for process (IV) has been observed. Magneron et al. (2005) concluded that 90% of the photodecomposition occurs in channels (I) and (II) while about 10% occurs in (III).

Bacher et al. (2001) used acetone and acetaldehyde as reference compounds in photolyses within the reaction chamber to estimate the average quantum yield of process (I) in 1 atm of air: $\phi_I > 0.5$ for $\lambda = 285 \pm 25$ nm. From measured *j*-values in sunlight in the EUPHORE chamber, Magneron et al. (2005) concluded that the effective total quantum yield of photodecomposition was near unity ($\phi = 1.3 \pm 0.3$).

Bacher et al. (2001) pointed out that 2-hydroxyethanal is very soluble in water and that uptake by hydrometeors (liquid or solid water particles, e.g., cloud, rain, fog, snow, etc.) may be an important loss mechanism for this compound in the troposphere. The data of Betterton and Hoffmann (1988) show that 90% of the 2-hydroxyethanal will hydrate in aqueous solution to form $HOCH_2CH(OH)_2$ that does not absorb radiation available in the troposphere.

Estimates of the average photolysis frequency of 2-hydroxyethanal over periods of 5–7 hr exposure to sunlight were made by Magneron et al. (2005) using the EUPHORE photochemical chamber during June 19–21, 1998 and June 15–16, 1999; mixtures of 2-hydroxyethanal with OH radical scavengers were exposed in air for periods of 5–7 hr. The mean *j*-value for photodecomposition by all pathways was reported as $(1.1 \pm 0.3) \times 10^{-5}$ sec^{-1}.

The photolysis frequencies as a function of solar zenith angle for 2-hydroxyethanal photodecomposition in sunlight within the troposphere were calculated using the average of cross sections measured by the Max Planck Institute and by LCSR/CNRS Orléans as reported by Magneron et al. (2005). A total photodecomposition quantum yield of unity and actinic flux appropriate for a cloudless day within the lower troposphere (ozone column, 350 DU) were assumed; see figure VII-F-5. These data give an average *j*-value for the date and time period of the measurements at Valencia of about 1.2×10^{-5} sec^{-1}, consistent with the average value that was measured directly, $(1.1 \pm 0.3) \times 10^{-5}$ sec^{-1}.

VII-F-4. 2,2,2-Trichloroethanal (Trichloroacetaldehyde, Chloral, CCl_3CHO)

Absorption cross sections for 2,2,2-trichloroethanal can be compared with those of other halogen-substituted acetaldehydes in figure VII-F-2. Tabular listing of these cross sections is given in table VII-F-2 for wavelengths present in the troposphere.

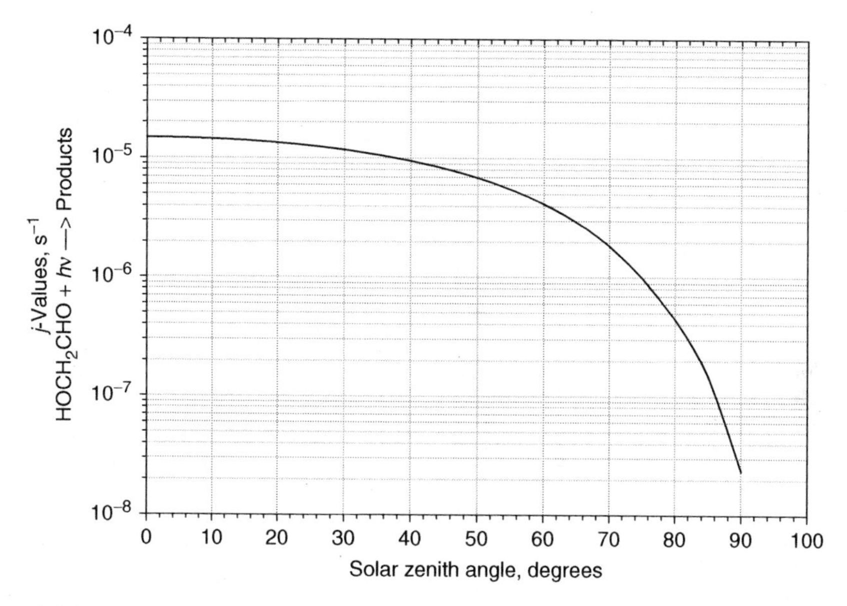

Figure VII-F-5. Photolysis frequencies for $HOCH_2CHO$ as a function of solar zenith angle for a cloudless day within the troposphere with overhead ozone column = 350 DU. The quantum yield of the photodecomposition was assumed to be 1.0, independent of wavelength (Magneron et al., 2005).

The photodecomposition of 2,2,2-trichloroethanal has been studied by Ohta and Mizoguchi (1980) and Talukdar et al. (2001). The following primary photodissociation processes have been considered:

$$CCl_3CHO + h\nu \rightarrow Cl + CCl_2CHO \quad (I)$$

$$\rightarrow CCl_3 + CHO \quad (II)$$

$$\rightarrow CCl_3CO + H \quad (III)$$

$$\rightarrow CHCl_3 + CO \quad (IV)$$

Enthalpy data estimated by Zhu and Bozzelli (2002) suggest the values for ΔH_{298} for processes (I), (II), (III), and (IV) are approximately 250.7, 287.1, 381.3, and $-28.2\,kJ\,mol^{-1}$, respectively. Thus, from consideration of energy alone, process (I) can in theory be competitive with (II), the common photodecomposition mode for aldehydes in the troposphere.

Ohta and Mizoguchi (1980) interpreted their results from photolysis of CCl_3CHO–oxygen mixtures as initiated by process (II), analogous to the photolysis of acetaldehyde. However, the more recent studies of Talukdar et al. (2001) give evidence of the importance of process (I) in forming Cl-atoms. They suggested that the subsequent thermal reactions of the Cl-atoms with CCl_3CHO can lead to the products (CO, ClC(O)Cl, CO_2) observed in O_2–CCl_3CHO mixture photolysis by Ohta and Mizoguchi (1980).

Table VII-F-2. Absorption cross sections (σ, cm^2 $molecule^{-1}$) for CCl_3CHO at 298 K (after Talukdar et al., 2001)

λ (nm)	$10^{20} \times \sigma$	λ (nm)	$10^{20} \times \sigma$
278	9.49	311	5.88
279	9.72	312	5.58
280	9.94	313	5.28
281	10.1	314	4.98
282	10.3	315	4.66
283	10.45	316	4.33
284	10.6	317	4.01
285	10.7	318	3.68
286	10.8	319	3.39
287	10.85	320	3.09
288	10.9	321	2.80
289	10.9	322	2.51
290	10.9	323	2.30
291	10.85	324	2.09
292	10.8	325	1.93
293	10.7	326	1.76
294	10.6	327	1.60
295	10.45	328	1.43
296	10.3	329	1.28
297	10.1	330	1.12
298	9.92	331	0.985
299	9.59	332	0.849
300	9.25	333	0.720
301	9.01	334	0.590
302	8.77	335	0.482
303	8.47	336	0.373
304	8.17	337	0.317
305	7.84	338	0.261
306	7.50	339	0.225
307	7.18	340	0.188
308	6.86	341	0.162
309	6.52	342	0.136
310	6.18	343	0.118
		344	0.100

At wavelength 308 nm Talukdar et al. (2001) observed $\phi_I = 1.3 \pm 0.3$, with $\phi_{III} < 0.002$. At wavelengths 193 and 248 nm, they find $\phi_{III} = 0.04 \pm 0.01$ and < 0.01, respectively. There was no evidence of process (IV) forming $CHCl_3$ even at 248 nm, and it is likely that ϕ_{IV} is unimportant at the wavelengths of sunlight available within the troposphere. The extent to which the excited states of CCl_3CHO are quenched by air molecules remains undetermined.

Approximate photolysis frequencies for CCl_3CHO in tropospheric sunlight have been calculated coupling the cross section data of table VII-F-2 with an assumed value of $\phi_I = 1.0$ at all absorbed wavelengths and the appropriate solar flux (ozone column = 350 DU) for a cloudless troposphere. These results are plotted vs. solar zenith angle in figure VII-F-6.

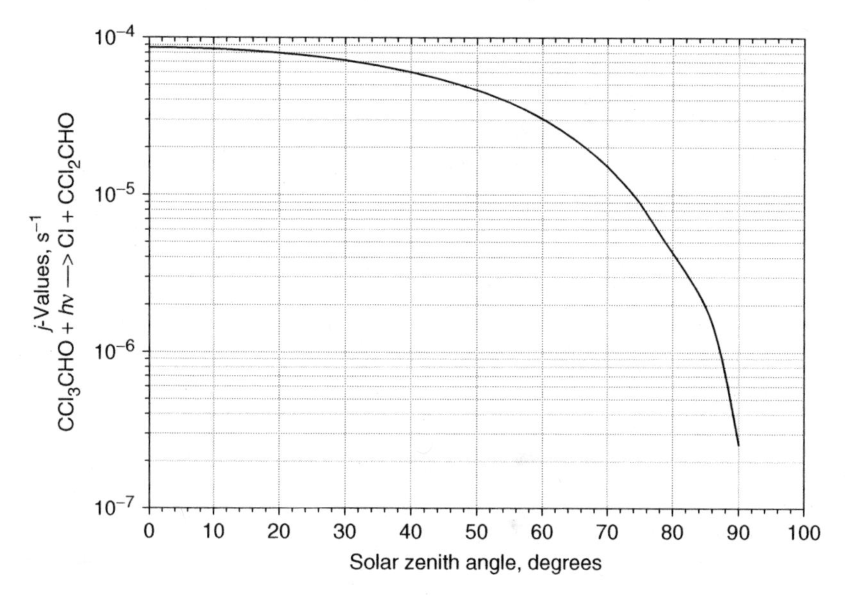

Figure VII-F-6. Approximate photolysis frequencies for CCl_3CHO vs. solar zenith angle for a cloudless day in the lower troposphere with an overhead ozone column = 350 DU; the quantum yield of decomposition to form Cl-atoms was assumed to be unity at all absorbed wavelengths.

The lifetime from OH attack alone would be about 1 day, while its lifetime from photodecomposition, for an overhead Sun, is about 3 hr. For 40°N latitude, clear tropospheric skies, and diurnal variation of the Sun, the expected average photochemical lifetimes of CCl_3CHO are about 10 hr and 2.3 days for the time periods near June 22 and December 22, respectively.

VII-F-5. 2,2,2-Trifluoroethanal (Trifluoroacetaldehyde, CF_3CHO)

Limited quantitative data on the photodecomposition of trifluoroacetaldehyde has been published by several groups (Dodd and Smith, 1957; Morris and Thynne, 1968, Pearce and Whytock, 1971; Sellevåg et al., 2004; Chiappero et al., 2006). The following primary processes can be considered in explanation of the results:

$$CF_3CHO + h\nu \rightarrow CF_3 + CHO \quad \text{(I)}$$

$$\rightarrow CHF_3 + CO \quad \text{(II)}$$

$$\rightarrow CF_3CO + H \quad \text{(III)}$$

VII-F-5.1. Absorption Cross Sections for CF_3CHO

The reported cross sections of CF_3CHO as a function of wavelength that have been given by several groups are in reasonable agreement (Sellevåg et al., 2004;

Hashikawa et al., 2004; Chiappero et al., 2006); the data of Sellevåg et al. (2004) are shown in figures VII-F-3 and VII-F-4, where cross sections for a series of structurally related haloaldehydes are compared. Tabular data for CF_3CHO from Sellevåg et al. (2004) and Chiappero et al. (2006) are compared in table VII-F-3. The cross sections reported by Chiappero are about 10% lower than those of Sellevåg et al. over much of the wavelength range important within the troposphere. An average of the two σ determinations at each wavelength has been used in photolysis frequency calculation given in this work.

VII-F-5.2. Quantum Yields of Primary Processes in CF_3CHO Photolysis

Dodd and Smith (1957) photolyzed pure CF_3CHO at 313 nm (~ 20°C, 30–40 Torr) and concluded that $\phi_I \approx 0.12$ and $\phi_{II} \approx 0.021$. Pearce and Whytock (1971) found no evidence for process (II) in their experiments at 313 nm (30–150°C, pressure 8.6–146 Torr). From photolysis of CF_3CHO at 253.7 nm in air mixtures, Richter et al. (1993) concluded that process (II) occurs to some extent, but very little photodecomposition occurred on irradiation at 366 nm (very little light is absorbed at this wavelength).

Sellevåg et al. (2004) determined the effective quantum yield of photolysis (290–400 nm) to be <0.02 under pseudo-natural conditions (1 atm. air) at the European simulation chamber in Valencia, Spain (EUPHORE). Chiappero et al. (2006) studied the photolysis of a series of structurally related fluoroaldehydes, including CF_3CHO, at two wavelengths: 308 (pulse laser) and 254 nm (CW Hg lamp). They determined the quantum yields of the primary processes in dilute mixtures of CF_3CHO with 700 Torr of nitrogen and 10–50 Torr of NO. The NO was added to scavenge the radical products formed in the photolysis that might otherwise contribute to the unwanted secondary loss of the aldehyde. In the experiments at 308 nm, CH_3CHO and $CH_3C(O)CH_3$ were used as chemical actinometers. The reliability of the method was assured in experiments with equal quanta absorbed by the two actinometer compounds. The measured ratio of the acetone loss to that of acetaldehyde was 0.153 ± 0.017. This checked well with the ratio expected from the 2005 IUPAC recommendations (Atkinson et al., 2005a) for cross sections and quantum yields at 308 nm: 0.179 ± 0.028. The expected ratio calculated using the recommendations for 308 nm in the present study give $\Phi_{-CH3C(O)CH3}/\Phi_{-CH3CHO} = 0.15$ and 0.13, assuming scenarios (a) and (b), respectively, for process (II) in acetone photodecomposition. In photolyses of CF_3CHO at 308 nm the measured amount of CF_3NO product formed was equal to the aldehyde loss within the experimental error: $\Phi_{CF3NO}/\Phi_{-CF3CHO} = 0.98 \pm 0.07$. No CF_3H was detected. Clearly process (I) is the dominant photodecomposition mode at 308 nm. The measured quantum yield of process (I) at 308 nm of 0.17 ± 0.03 is adopted in this study. For unexplained reasons, the experimental measurements of the effective quantum yield of photodecomposition of CF_3CHO (290–400 nm) by Sellevåg et al. ($\phi_I <$ 0.02) appear to be in error.

In their studies of CF_3CHO photodecomposition at 254 nm in its mixture with NO and N_2 (700 Torr), 296 K, Chiappero et al. (2006) used $(CF_3CO)_2O$ as an actinometer. The two observable products at this wavelength, CF_3NO and CF_3H, showed the occurrence of process (II) as well as (I), with $\phi_I = 0.42 \pm 0.07$ and $\phi_{II} = 0.38 \pm 0.07$.

Table VII-F-3. Comparison of absorption cross sections (σ, cm^2 $molecule^{-1}$) for CF_3CHO [Sellevåg et al. (2004) at 298 K (A); Chiappero et al. (2006) at 297 K (B)]

λ (nm)	(A) $10^{20} \times \sigma$	(B) $10^{20} \times \sigma$	λ (nm)	(A) $10^{20} \times \sigma$	(B) $10^{20} \times \sigma$	λ (nm)	(A) $10^{20} \times \sigma$	(B) $10^{20} \times \sigma$
278	2.03	1.83	319	2.33	2.10	360	0.0380	0.01
279	2.11	1.90	320	2.25	2.02	361	0.0360	0
280	2.19	1.97	321	2.19	1.97	362	0.0360	0
281	2.28	2.06	322	2.13	1.91	363	0.0320	0
282	2.35	2.12	323	2.08	1.88	364	0.0330	0
283	2.42	2.18	324	2.06	1.83	365	0.0320	0
284	2.50	2.26	325	1.90	1.67	366	0.0280	0
285	2.57	2.32	326	1.72	1.52	367	0.0250	0
286	2.63	2.37	327	1.64	1.47	368	0.0250	0
287	2.67	2.42	328	1.62	1.44	369	0.0230	0
288	2.73	2.48	329	1.55	1.36	370	0.0220	0
289	2.79	2.54	330	1.44	1.27	371	0.0230	0
290	2.86	2.61	331	1.35	1.18	372	0.0220	0
291	2.92	2.66	332	1.26	1.09	373	0.0210	0
292	2.94	2.69	333	1.18	1.03	374	0.0230	0
293	3.00	2.75	334	1.13	0.98	375	0.0220	0
294	3.05	2.79	335	1.06	0.91	376	0.0220	0
295	3.06	2.81	336	1.01	0.87	377	0.0200	0
296	3.08	2.83	337	0.993	0.86	378	0.0190	0
297	3.08	2.83	338	0.891	0.74	379	0.0190	0
298	3.10	2.86	339	0.730	0.61	380	0.0170	0
299	3.14	2.88	340	0.622	0.51	381	0.0180	0
300	3.17	2.92	341	0.585	0.49	382	0.0180	0
301	3.20	2.93	342	0.569	0.47	383	0.0170	0
302	3.15	2.88	343	0.531	0.43	384	0.0180	0
303	3.12	2.87	344	0.471	0.37	385	0.0170	0
304	3.15	2.89	345	0.425	0.34	386	0.0170	0
305	3.13	2.86	346	0.385	0.30	387	0.0180	0
306	3.07	2.81	347	0.337	0.26	388	0.0180	0
307	3.03	2.76	348	0.310	0.23	389	0.0160	0
308	2.97	2.72	349	0.286	0.21	390	0.0160	0
309	2.95	2.69	350	0.246	0.18	391	0.0160	0
310	2.92	2.67	351	0.235	0.17	392	0.0170	0
311	2.92	2.68	352	0.232	0.15	393	0.0160	0
312	2.91	2.65	353	0.162	0.093	394	0.0140	0
313	2.78	2.51	354	0.0960	0.036	395	0.0140	0
314	2.67	2.43	355	0.0710	0.017	396	0.0150	0
315	2.65	2.41	356	0.0580	0.008	397	0.0140	0
316	2.62	2.37	357	0.0500	0.007	398	0.0130	0
317	2.52	2.28	358	0.0440	0.011	399	0.0110	0
318	2.42	2.18	359	0.0420	0.019	400	0.0130	0

These results are consistent with the occurrence of process (II) observed by Richter et al. (1993) in experiments at 254 nm.

VII-F-5.3. Approximate Photolysis Frequencies for CF_3CHO

The wavelength dependence of the primary processes at wavelengths within the 290–400 nm region of importance in atmospheric photolysis cannot be determined from the limited data at 308 and 254 nm. However, the data from 308 nm, a wavelength near the maximum in the overlap of the actinic flux with the absorption of CF_3CHO, were used to derive qualitative estimates of the effective quantum yield and approximate *j*-values (Chiappero et al., 2006). A similar calculation has been made in this study, and the results are shown in figure VII-F-7 for clear-sky conditions and overhead O_3 column of 350 DU. These data suggest that for an overhead Sun the approximate photochemical lifetime of CF_3CHO in the lower troposphere is about 19 hr.

VII-F-6. 2,2-Difluoroethanal (Difluoroacetaldehyde; CHF_2CHO)

Studies of the photochemistry of CHF_2CHO have been reported by Sellevåg et al. (2005). The absorption spectrum of difluoroacetaldehyde is similar to that for CF_3CHO

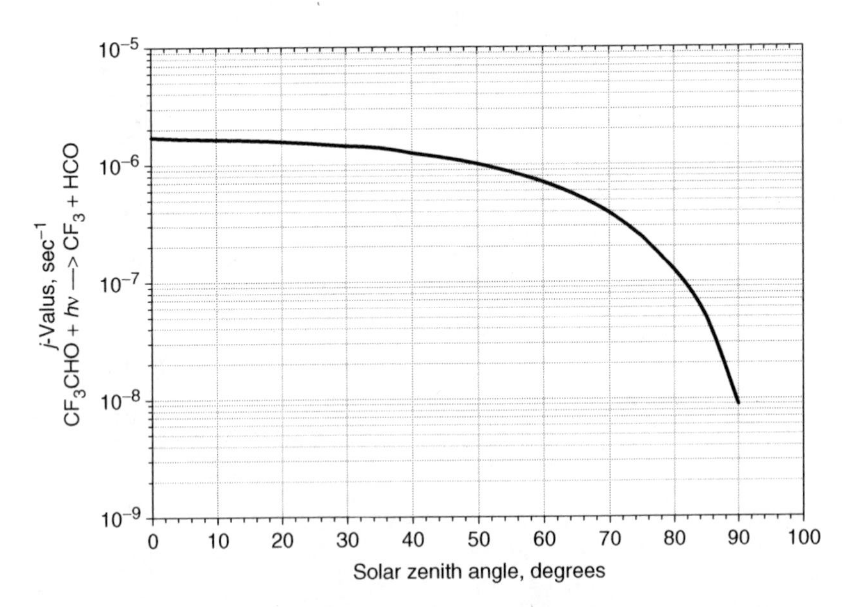

Figure VII-F-7. Approximate photolysis frequencies for CF_3CHO vs. solar zenith angle for a cloudless day in the lower troposphere. The quantum yield of photodecomposition to form CF_3 and HCO radicals is assumed to be 0.17 at all wavelengths of absorbed light; this is equal to the value measured at 308 nm, near the maximum in the overlap of the absorption (Chiappero et al., 2006). We also used in this calculation the actinic flux available within the lower troposphere (overhead ozone column = 350 DU), and an average of the cross section data of Sellevåg et al. (2004) and Chiappero et al. (2006).

(see figure VII-F-3), but the efficiency of its photodecomposition in sunlight appears to be much greater than the low value (<0.02) observed for CF_3CHO by Sellevåg et al. (2004), being closer to that found by Chiappero et al. (2006) for CF_3CHO at 308 nm (0.17). An effective quantum yield of photodecomposition of 0.30 ± 0.05 (290–400 nm) was determined for pseudonatural tropospheric conditions in the European simulation chamber in Valencia, Spain (Sellevåg et al., 2005). The primary photodissociation pathways considered for CHF_2CHO mirror those considered for the other F-substituted acetaldehyde molecules:

$$CHF_2CHO + h\nu \rightarrow CHF_2 + HCO \quad \text{(I)}$$

$$\rightarrow CH_2F_2 + CO \quad \text{(II)}$$

$$\rightarrow CHF_2CO + H \quad \text{(III)}$$

Only CO and CF_2O were detected products of the photolysis of CHF_2CHO–air mixtures at 313 nm (Sellevåg et al., 2005). This suggests that process (II) is unimportant, at least for tropospheric conditions.

Quantum chemical calculations have been made by Sellevåg et al. (2005) to compare theoretical energy barriers for these pathways and the analogous photolytic dissociation pathways of CH_3CHO, CH_2FCHO, CHF_2CHO, and CF_3CHO. Presumably processes (I) and (II) occur from the T_1 and S_0 states, respectively. The theoretical barriers for process (I) relative to the T_1 minimum were estimated to be 53, 54, 56, and 44 kJ mol^{-1} for CH_3CHO, CH_2FCHO, CF_3CHO, and CHF_2CHO, respectively. For process (II) the barrier decreases in the series CH_3CHO, CH_2FCHO, CHF_2CHO, and CF_3CHO, with the calculated CF_3CHO barrier being 36 kJ mol^{-1} lower than the equivalent barrier in CH_3CHO.

We have used the cross section data tabulated in table VII-F-4 and the effective quantum yield of CHF_2CHO photodissociation (0.30) reported by Sellevåg et al. (2005) to calculate the photolysis frequencies as a function of solar zenith angle for a cloudless day within the lower troposphere (ozone column = 350 DU); see figure VII-F-8. These data suggest a photochemical lifetime for overhead Sun of about 6 hr.

VII-F-7. Larger Fluorinated Aldehydes: C_2F_5CHO, $CF_3CF_2CF_2CHO$, $CF_3CF_2CF_2CF_2CHO$, CF_3CH_2CHO, and n-$C_6F_{13}CH_2CHO$

Limited photochemical data for several of the larger fluorinated aldehydes have been reported by Chiappero et al. (2006). The absorption spectra of these molecules are shown in figure VII-F-4, where they are compared with those of the unfluorinated aldehydes. The quantum efficiencies of the primary processes have been determined at a few wavelengths in experiments at 700 Torr pressure of N_2; experimental conditions were similar to those described for CF_3CHO in section VII-F-5 (Chiappero et al., 2006). Photodecomposition proceeds through the

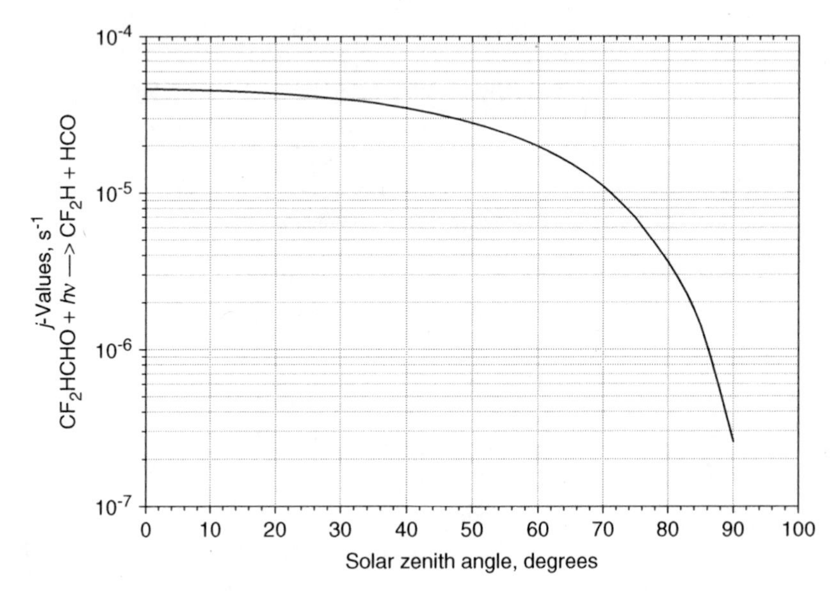

Figure VII-F-8. Approximate photolysis frequencies for CHF_2CHO vs. solar zenith angle for a cloudless day in the lower troposphere with an overhead ozone column = 350 DU; the quantum yield of decomposition to form CF_2H and HCO radicals was assumed to be 0.30 at all wavelengths, in accord with the effective quantum yield estimated by Sellevåg et al. (2005).

following primary processes which are similar to those observed for the unfluorinated aldehydes:

$$RCHO + h\nu \rightarrow R + CHO \qquad (I)$$

$$\rightarrow RH + CO \qquad (II)$$

From experiments at 254 nm, the reported quantum yields for primary process (I) in CF_3CHO, C_2F_5CHO, $CF_3CF_2CF_2CHO$, $CF_3CF_2CF_2CF_2CHO$, and CF_3CH_2CHO are 0.41 ± 0.07, 0.38 ± 0.08, 0.31 ± 0.07, 0.31 ± 0.08, and 0.38 ± 0.09, respectively. Quantum yields for process (II) at 254 nm are 0.38 ± 0.07, 0.43 ± 0.08, 0.32 ± 0.07, 0.29 ± 0.07, and 0.36 ± 0.07, respectively. Chiappero et al. (2006) determined for $C_6F_{13}CH_2CHO$ the total photodecomposition quantum yield: $\phi_I + \phi_{II} = 0.55 \pm 0.09$.

From experiments at 308 nm, Chiappero et al. (2006) report values of ϕ_I for CF_3CHO, $CF_3CF_2CF_2CF_2CHO$, and CF_3CH_2CHO of 0.17 ± 0.03, 0.08 ± 0.02, and 0.04 ± 0.01, respectively. These investigators have assumed the quantum yield data measured at 308 nm to be representative of that over the wavelength region of sunlight absorption, and estimated the photochemical lifetimes of the fluorinated aldehydes (averaged over the entire year) as a function of altitude within the lower atmosphere. For altitudes of 0 and 11.7 km, respectively, the estimated photochemical lifetimes are: CF_3CH_2CHO, ~36 and 13 days; n-$C_6F_{13}CH_2CHO$, ~21 and 8 days; CF_3CHO, ~5.7

Table VII-F-4. Absorption cross sections (σ, cm^2 $molecule^{-1}$) for CHF_2CHO (Sellevåg et al., 2005) at 298 K

λ (nm)	$10^{20} \times \sigma$	λ (nm)	$10^{20} \times \sigma$	λ (nm)	$10^{20} \times \sigma$
278	2.19	319	4.19	360	0.0355
279	2.31	320	4.21	361	0.0283
280	2.43	321	4.04	362	0.0283
281	2.54	322	3.69	363	0.0253
282	2.69	323	3.60	364	0.0241
283	2.81	324	3.69	365	0.0146
284	2.85	325	3.37	366	0.0165
285	2.95	326	3.00	367	0.0165
286	3.11	327	2.78	368	0.0130
287	3.22	328	2.79	369	0.00679
288	3.34	329	2.95	370	0.00700
289	3.45	330	2.92	371	0.00976
290	3.54	331	2.77	372	0.00875
291	3.71	332	2.69	373	0.0133
292	3.82	333	2.64	374	0.00655
293	3.84	334	2.38	375	0.0163
294	3.83	335	2.07	376	0.0215
295	3.97	336	2.02	377	0.0102
296	4.14	337	1.98	378	0.0106
297	4.21	338	1.62	379	0.0116
298	4.33	339	1.29	380	0.0111
299	4.33	340	1.11	381	0.00759
300	4.34	341	1.06	382	0.00408
301	4.54	342	1.12	383	0.00445
302	4.58	343	1.14	384	0.00917
303	4.48	344	1.09	385	0.00634
304	4.39	345	0.974	386	0.00305
305	4.48	346	0.922	387	0.00425
306	4.62	347	0.885	388	0.00384
307	4.66	348	0.735	389	0.00837
308	4.71	349	0.597	390	0.00995
309	4.73	350	0.553	391	0.00904
310	4.53	351	0.557	392	0.00363
311	4.47	352	0.365	393	0.00522
312	4.65	353	0.182	394	0.00722
313	4.53	354	0.111	395	0.00732
314	4.24	355	0.0878	396	0.00457
315	4.07	356	0.0773	397	0.00249
316	4.16	357	0.0626	398	0.00329
317	4.31	358	0.0552	399	0.00152
318	4.27	359	0.0396	400	0.00224

and 2.1 days; C_2F_5CHO ~2.5 and 0.92 days; and *n*-C_4F_9CHO, ~2 and 0.74 days. Obviously, photodecomposition is an important decay pathway for fluoroaldehydes in the lower atmosphere.

VII-G. Photodecomposition of Halogen-Atom- and HO-Substituted Ketones

VII-G-1. Effect of Halogen-Atom and HO Substitution on the Absorption Spectrum of Acetone

As observed with the aldehydes (section VII-F-1), substitution of HO or halogen atoms for H-atoms in ketones has a significant effect on the absorption spectra. It is not surprising that the effects of a given halogen atom or an OH-group substitution for H-atoms in the CH_3 group of $CH_3C(O)CH_3$ are very similar to those seen in CH_3CHO for the same substitution, since the chromophores are very similar. Compare the changes in observed λ_{max} and σ_{max} for similar atom substitutions, summarized in table VII-G-1.

In figure VII-G-1 spectra are compared for acetone molecules where a single halogen atom or OH-group replaces an H-atom in $CH_3C(O)CH_3$. The replacement of an H-atom with an F-atom to form $CH_2FC(O)CH_3$ lowers the cross section significantly (by a factor of 2.5) and a small shift (~4 nm) to longer wavelengths occurs in the absorption maximum. A very different influence is seen with a Cl-atom substitution in acetone that parallels those observed with the aldehydes. The maximum cross section of $CH_2ClC(O)CH_3$ is a factor of about 2.0 larger than that in acetone, and it is shifted about 14 nm to the longer wavelengths. A similar effect is seen with Br-atom substitution in acetone; the maximum absorption cross section is increased by a factor of about 2.3 and it is shifted about 21 nm to the longer wavelengths. The result of Cl- or Br-atom substitution in $CH_3C(O)CH_3$ is a much better overlap of the spectrum

Table VII-G-1. Comparison of the wavelength (λ_{max}) and cross section (cm^2 $molecule^{-1}$) at the maximum in the absorption spectra of substituted acetaldehyde and acetone molecules at temperatures near 298 K

Acetaldehydes	λ_{max} (nm)	$10^{20} \times \sigma_{max}$	Acetones	λ_{max} (nm)	$10^{20} \times \sigma_{max}$
CH_3CHO	290	4.88	$CH_3C(O)CH_3$	278	5.07
$HOCH_2CHO$	277	5.42	$HOCH_2C(O)CH_3$	266	6.74
CH_2ClCHO	298	5.75	$CH_2ClC(O)CH_3$	292	10.2
$CHCl_2CHO$	300	6.14			
CCl_3CHO	289	10.9			
CHF_2CHO	309	4.73			
			$CH_2FC(O)CH_3$	282	2.24
			$CH_2BrC(O)CH_3$	299	11.5
CF_3CHO	301	3.20	$CF_3C(O)CH_3$	289	3.19
			$CF_3C(O)CF_3$	303	3.09
CF_2ClCHO	300	16.3	$CF_2ClC(O)CF_3$	308	18.8
			$CF_2ClC(O)CF_2Cl$	305	18.9
$CFCl_2CHO$	295	13.7	$CFCl_2C(O)CF_2Cl$	305	18.9

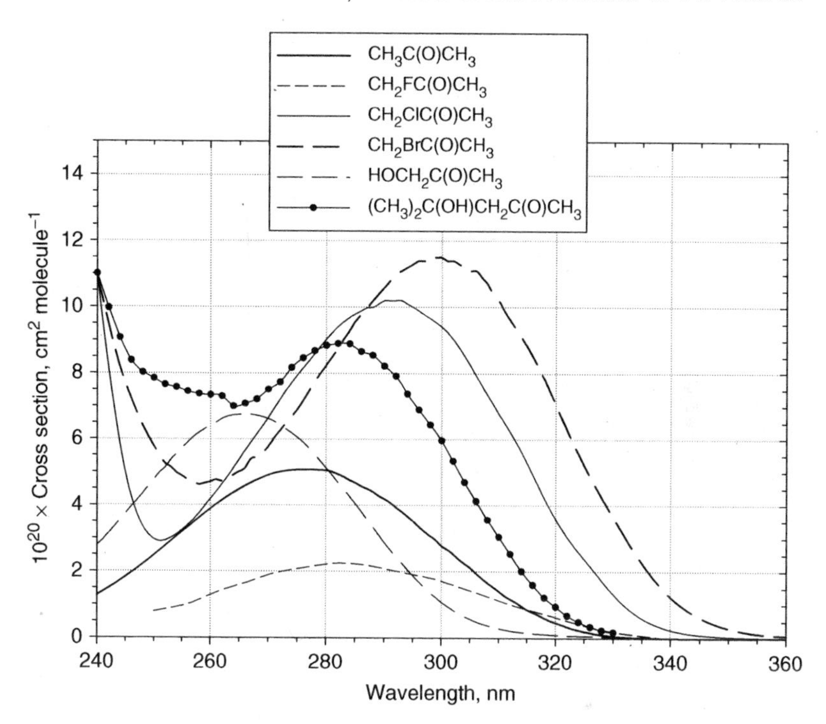

Figure VII-G-1. Comparison of the absorption cross sections for the monosubstituted acetones, $CH_2FC(O)CH_3$ (Hackett and Phillips, 1974a), $CH_2ClC(O)CH_3$ (Burkholder et al., 2002), $CH_2BrC(O)CH_3$ (Burkholder et al., 2002), $HOCH_2C(O)CH_3$(Orlando et al., 1999a), and diacetone alcohol $(CH_3)_2C(OH)CH_2C(O)CH_3$ (Magneron et al., 2003) with those of $CH_3C(O)CH_3$ (Martinez et al., 1992).

with that of the distribution of wavelength of sunlight present within the troposphere, and hence the potential for importance of their tropospheric photodecomposition is enhanced.

Formation of $HOCH_2C(O)CH_3$ by substitution of an OH-group for an H-atom in acetone increases the absorption maximum somewhat (by a factor of about 1.3), as with the aldehydes, and the absorption maximum shifts about 12 nm to the shorter wavelengths. Hence overlap of the absorption of $HOCH_2COCH_3$ with the solar flux present within the troposphere is decreased significantly from that seen for acetone. Note that with introduction of an OH-group two carbon atoms down-chain from the carbonyl in diacetone alcohol [$(CH_3)_2C(OH)CH_2C(O)CH_3$] the maximum cross section is enhanced, but the wavelength of maximum absorption is altered little from that in unsubstituted acetone. One expects a similar behavior for the common 1,4-hydroxycarbonyl compounds; they probably absorb light in about the same wavelength region as the unsubstituted acetone.

The effects on the absorption spectrum of $CH_3C(O)CH_3$ with substitution of multiple F-atoms for H-atoms in $CH_3C(O)CH_3$ can be seen in figure VII-G-2. The data for cross sections of $CF_3C(O)CH_3$ are from Beavan et al. (1978). However, the wavelength distribution of the data from Ausloos and Murad (1961) at specific wavelengths

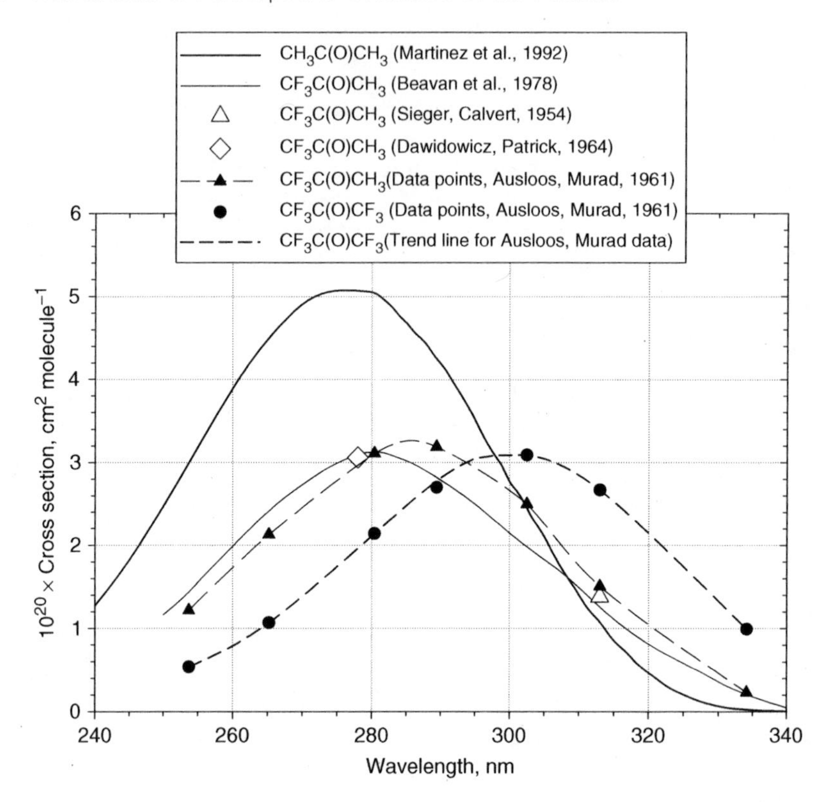

Figure VII-G-2. Comparison of the approximate absorption spectra for F-substituted acetones, $CF_3C(O)CH_3$, $CF_3C(O)CF_3$, with that of $CH_3C(O)CH_3$ near 298 K. The data points for $CF_3C(O)CH_3$ at wavelength 313 nm (triangle, Sieger and Calvert, 1954) and 278 nm (diamond, Dawidowicz and Patrick, 1964) are in reasonable accord with the complete spectral data obtained by Beavan et al. (1978); data points at selected wavelengths for $CF_3C(O)CH_3$, given by Ausloos and Murad (1961), are significantly shifted to longer wavelengths than the Beavan et al. data. Data points shown for $CF_3C(O)CF_3$ are limited to those published for specific wavelengths that are shown by the filled circles. The short-dashed and long-dashed spline lines are drawn to fit the Ausloos and Murad data points and are not experimentally determined regions of data.

(shown as filled triangles) is significantly different, shifted about 5 nm to longer wavelengths. The single data points at 313 nm from Sieger and Calvert (1954) and 278 nm from Dawidowicz and Patrick (1964) are reasonably consistent with both the Beavan et al. and the Ausloos and Murad data. Obviously, further measurements are required to establish a reliable absorption spectrum of CF_3COCH_3. However, it is clear that, with the substitution of three F-atoms on one carbon of $CH_3C(O)CH_3$ to form $CF_3C(O)CH_3$, the maximum cross section of acetone is lowered by a factor of about 1.6, and it is shifted to the longer wavelengths.

The $CF_3C(O)CF_3$ spectral data from Ausloos and Murad (1961) were determined at only selected wavelengths shown by the symbols. The precision of the $CF_3C(O)CF_3$ data points is uncertain, but it is likely that it is reasonably good, since there

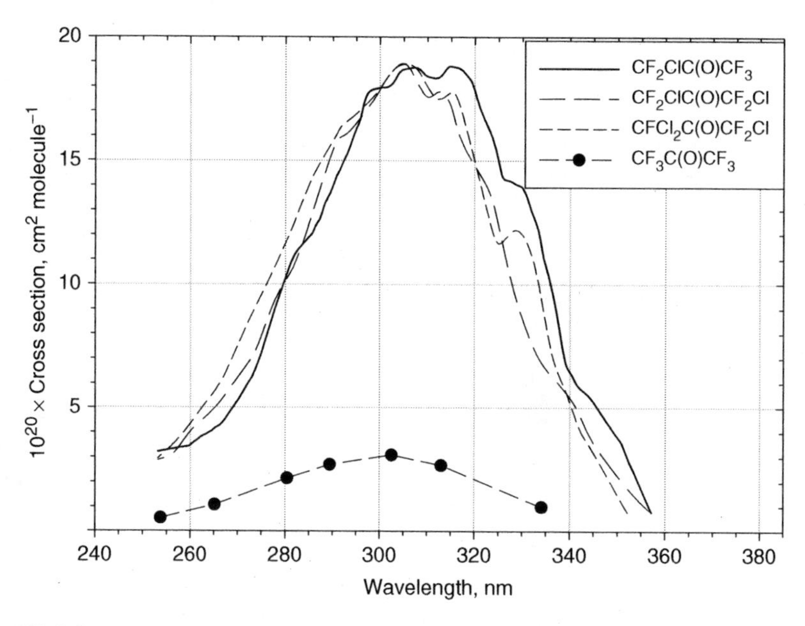

Figure VII-G-3. Comparison of the absorption spectra of some F-, Cl-substituted acetones: $CF_2ClC(O)CF_3$, $CF_2ClC(O)CF_2Cl$, and $CFCl_2C(O)CF_2Cl$ (redrawn from epsilon vs. $1/\lambda$ data of Hackett and Phillips, 1974a); $CF_3C(O)CF_3$ data points from Ausloos and Murad (1961).

is qualitative agreement of their acetone data points with the currently accepted $CH_3C(O)CH_3$ spectra given by Martinez et al. (1992). With all the C—H bonds in acetone replaced with C—F bonds to form $CF_3C(O)CF_3$, the maximum cross section is lowered as in $CF_3C(O)CH_3$, and the wavelength of maximum absorption is shifted to even longer wavelengths. These highly fluorinated acetones absorb a much greater fraction of the sunlight available within the troposphere than unsubstituted acetone.

The effect of Cl-atom substitution for F-atoms in $CF_3C(O)CF_3$ is shown in figure VII-G-3. The cross section data given for $CF_2ClC(O)CF_3$, $CF_2ClC(O)CF_2Cl$, and $CFCl_2C(O)CF_2Cl$ were obtained by transformation of the plots of molar decadic extinction coefficients vs. $1/\lambda$ given in the figures of Hackett and Phillips (1974a). The reconstructed curves of cross section vs. wavelength are likely of lower precision than the data presented graphically by the original investigators. As is seen with the fluoro-chloro-acetaldehydes, a very large increase in the cross section results from the replacement of one F-atom by a Cl-atom in the $CF_3C(O)CF_3$ molecule to form $CF_2ClC(O)CF_3$. However, the wavelength of the absorption maximum shows little change with F-atom replacement by the Cl-atom. The data of figure VII-G-3 show little difference in the spectrum of $CF_2ClC(O)CF_3$ as one and two additional F-atoms are replaced with Cl-atoms. It is clear that substitution of chlorine into the fluoroacetone molecule enhances dramatically the overlap of the absorption of sunlight within the troposphere.

Table VII-G-2. Approximate absorption cross sections (σ, cm^2 $molecule^{-1}$) for hydroxyacetone at 298 K (Orlando et al., 1999a); uncertainties (95% confidence limits) in the cross sections are $\pm$ 10% near the spectral maximum and increase to $\pm$ 35% at the longest wavelengths reported

λ (nm)	$10^{20} \times \sigma$	λ (nm)	$10^{20} \times \sigma$
278	5.54	308	0.379
279	5.33	309	0.331
280	5.12	310	0.287
281	4.91	311	0.249
282	4.69	312	0.219
283	4.48	313	0.192
284	4.27	314	0.177
285	4.05	315	0.157
286	3.82	316	0.142
287	3.58	317	0.133
288	3.34	318	0.117
289	3.11	319	0.104
290	2.87	320	0.095
291	2.66	321	0.087
292	2.45	322	0.078
293	2.26	323	0.072
294	2.06	324	0.067
295	1.87	325	0.063
296	1.69	326	0.065
297	1.52	327	0.057
298	1.36	328	0.051
299	1.22	329	0.051
300	1.08	330	0.046
301	0.961	331	0.041
302	0.843	332	0.037
303	0.743	333	0.036
304	0.652	334	0.037
305	0.569	335	0.035
306	0.493	336	0.031
307	0.431		

In the following sections the limited data on the photodecomposition of the substituted acetone molecules are reviewed.

VII-G-2. 1-Hydroxy-2-propanone [Hydroxyacetone, $HOCH_2C(O)CH_3$]

The absorption spectrum of 1-hydroxy-2-propanone is compared with that of some other substituted ketones in figure VII-G-1, and the approximate cross sections are given in tabular form in table VII-G-2 for wavelengths available within the troposphere. Several possible modes of photodecomposition of hydroxyacetone should be considered in view of the experimental studies:

$$HOCH_2C(O)CH_3 + h\nu \rightarrow HOCH_2 + CH_3CO \qquad (I)$$

$$\rightarrow HOCH_2CO + CH_3 \qquad (II)$$

$$\rightarrow \mathrm{HOCH_2 + CO + CH_3} \quad \text{(III)}$$

$$\rightarrow \mathrm{OH + CH_2C(O)CH_3} \quad \text{(IV)}$$

The minimum energies required to dissociate hydroxyacetone by processes (I), (II), (III), and (IV) are approximately 335, 335, 377, and 377 kJ mol^{-1}, respectively (Chowdhury et al., 2002b). From the product analysis following hydroxyacetone/air photolysis in broad-band experiments (240–420 nm), Orlando et al. (1999a) concluded that at most 50% of the photolysis occurred via channel (I). If the estimated minimum energies for the occurrence of dissociation by (I) and (II) are correct, then it would not be surprising if both (I) and (II) occurred at roughly the same efficiency. Photolysis of hydroxyacetone was made using light from a lamp–filter combination that mimicked the shape of the solar spectrum in the troposphere ($\lambda > 290$ nm); these experiments suggested that $0.3 \pm 0.2 < \phi_{I} + \phi_{II} < 0.6$.

Direct observations of OH formation, presumably from process (IV), were made by Chowdhury et al., (2002a) in the photolysis of $HOCH_2COCH_3$ at 248 nm. The initial rotational state of the OH radical was Boltzmann-like, with a single rotational temperature of 450 ± 40 K, and the average relative translational energy of the photofragments was 36.4 kJ mol^{-1}. Chowdhury et al. concluded that the observation of OH with a modest rotational energy, no vibrational energy, and a large translational energy points to a significant exit energy barrier with the dissociation surface.

Dissociation in the 193 nm photolysis of hydroxyacetone appears to occur by process (II), with the initial $HOCH_2CO$ product dissociating to OH and $CH_2{=}C{=}O$ with a rate coefficient $k = (4.6 \pm 5) \times 10^6$ s^{-1} (Chowdhury et al., 2002b).

The photolysis frequencies of $HOCH_2COCH_3$ photodissociation have been calculated versus solar zenith angle for a cloudless lower troposphere (ozone column = 350 DU), assuming the average quantum yield of dissociation $\phi_{I} + \phi_{II}$ is in the range 0.3 to 0.6. The current best estimates of j-values lie within the area between the two curves shown in figure VII-G-4.

The photodecomposition of another OH-substituted acetone molecule, diacetone alcohol [$(CH_3)_2C(OH)CH_2C(O)CH_3$], has been studied by Magneron et al. (2003) in the EUPHORE facility in Valencia, Spain. In this molecule the OH-group is located 2 carbon atoms down-chain from the carbonyl group. Its estimated photolysis frequency in sunlight was too low to measure ($j < 5 \times 10^{-6}$ s^{-1}).

Although studies of the photochemistry of the 1,4-hydroxycarbonyls, common products of the atmospheric oxidation of the long-chain alkanes, would be interesting, none have been made at the time of this writing. Until such information is available, one might model the atmospheric chemistry of the 1,4-hydroxycarbonyls using photodecomposition modes and j-values that mimic those of the ketone or aldehyde with equivalent lengths of the alkyl chains attached to the carbonyl group. The γ-H atoms in these molecules are bound relatively weakly and should be abstracted easily by the excited carbonyl group on the occurrence of the Norrish type-II process. For example, one might expect the pathways for the photolysis of the molecule $CH_3CH(OH)CH_2CH_2C(O)CH_3$ to be analogous to those seen in 2-hexanone

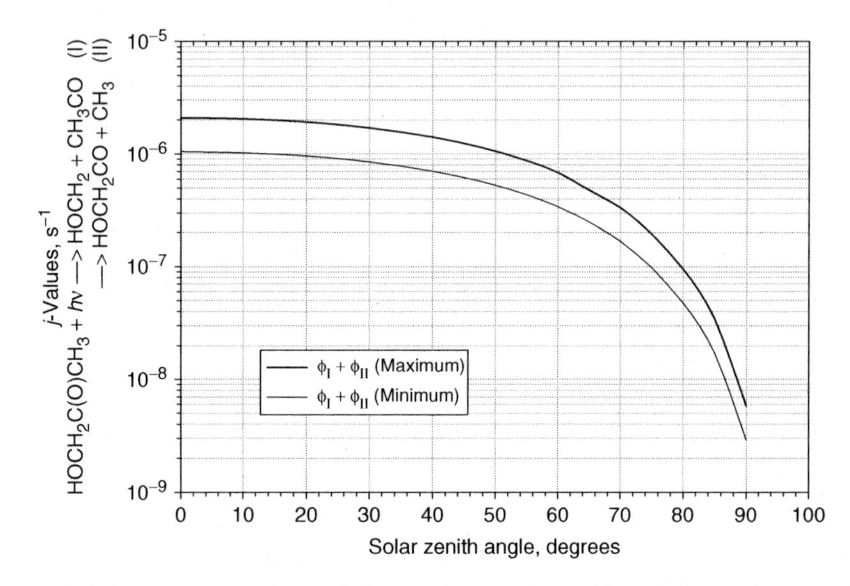

Figure VII-G-4. Approximate photolysis frequencies vs. solar zenith angle for $HOCH_2C(O)CH_3$ photodecomposition on a cloudless day in the lower troposphere with an overhead ozone column = 350 DU. The present best estimate of *j*-values lies between the two curves.

photolysis: a small quantum yield of free radical formation with a more significant pathway forming two molecules of enol-acetone [$CH_3CH(OH){=}CH_2$].

VII-G-3. 1-Chloro-2-propanone [Chloroacetone; $ClCH_2C(O)CH_3$]

The absorption spectrum for 1-chloro-2-propanone can be compared with those of some other substituted acetones in figure VII-G-1, and the cross sections are given in tabular form in table VII-G-3 for wavelengths available within the troposphere.

Photodecomposition of 1-chloro-2-propanone was first studied by Strachan and Blacet (1955), who reported the products and quantum yields from photolysis of the pure ketone at 313 nm and temperatures of 59–335°C. Products identified by mass spectrometry were: HCl, $CH_3C(O)CH_3$, $(CH_3C(O)CH_2)_2$, and $CH_3C(O)CH{=}CHC(O)CH_3$. The following quantum yields of the major product, HCl, were reported from experiments at temperatures of 59, 60, 150, (150), 204, 252, 301 and 335°C, respectively: 0.72, 0.86, 0.78, (0.87), 0.95, 1.02, 1.23, 1.49. Yields of CO and CH_4 were immeasurably small at temperatures $\leq$ 59°C, but these products were formed in increasing yields as the temperature was raised from 60 to 335°C. Photolyses were also carried out in mixtures with NO or I_2. NO suppressed the formation of both acetone and acetonylacetone as the product $CH_3C(O)CH_2NO$ appeared. Photolyses with mixtures of I_2 were less readily interpreted, but I_2 suppressed acetonylacetone formation as $ICH_2C(O)CH_3$ appeared as a product. However, HCl formation was not suppressed completely by I_2 addition (10 Torr). From their studies, Strachan and Blacet concluded that a new type of primary process, process (I), occurred in α-substituted

Table VII-G-3. Absorption cross sections (σ, cm^2 $molecule^{-1}$) for chloroacetone at 298 K for the wavelengths of sunlight available within the troposphere (Burkholder et al., 2002)

λ (nm)	$10^{19} \times \sigma$	λ (nm)	$10^{19} \times \sigma$	λ (nm)	$10^{19} \times \sigma$
278	0.859	306	0.794	334	0.0707
279	0.882	307	0.763	335	0.0598
280	0.903	308	0.734	336	0.0506
281	0.923	309	0.706	337	0.0427
282	0.941	310	0.677	338	0.0361
283	0.956	311	0.650	339	0.0302
284	0.969	312	0.622	340	0.0252
285	0.980	313	0.593	341	0.0212
286	0.989	314	0.561	342	0.0176
287	0.998	315	0.528	343	0.0145
288	1.00	316	0.492	344	0.0120
289	1.01	317	0.457	345	0.0103
290	1.02	318	0.422	346	0.00887
291	1.02	319	0.389	347	0.00757
292	1.02	320	0.358	348	0.00642
293	1.02	321	0.330	349	0.00547
294	1.01	322	0.304	350	0.00458
295	1.00	323	0.280	351	0.00411
296	0.980	324	0.258	352	0.00328
297	0.977	325	0.237	353	0.00319
298	0.966	326	0.216	354	0.00220
299	0.954	327	0.195	355	0.00193
300	0.941	328	0.173	356	0.00138
301	0.925	329	0.152	357	0.00134
302	0.904	330	0.133	358	0.00092
303	0.880	331	0.114	359	0.00155
304	0.853	332	0.0979	360	0.00128
305	0.824	333	0.0832		

carbonyls such as $CH_2ClC(O)CH_3$. Several other primary processes, analogous to those seen in acetone, were considered as minor in experiments at the lower temperatures employed.

$$CH_2ClC(O)CH_3 + h\nu \rightarrow Cl + CH_2C(O)CH_3 \qquad (I)$$

$$\rightarrow CH_2Cl + CH_3CO \qquad (II)$$

$$\rightarrow CH_2ClCO + CH_3 \qquad (III)$$

$$\rightarrow CH_3 + CO + CH_2Cl \qquad (IV)$$

$$\rightarrow CH_3C(O)CH + HCl \qquad (V)$$

The absence of CO and CH_4 in the products from experiments at the lowest temperatures showed that (III) and (IV) were unimportant, and the small quantum yield of CO (0.027, 0.022) observed in experiments at 150°C, where CH_3CO decomposition is rapid, suggests that process (II) is also not important. Process (V) was considered possible in view of tentative identification of $CH_3COCH{=}CHCOCH_3$ in the products and the failure of I_2 to suppress HCl formation completely.

Waschewsky et al. (1994) used crossed laser–molecular beam experiments to monitor the competition between photodissociation pathways in chloroacetone. Their studies proved that C—C bond fission, process (II), competes with C—Cl bond fission, process (I), in $CH_2ClC(O)CH_3$photolysis at 308 nm, although the latter process is favored. Several kJ of translational energy was evident in the translation of the initial products of the two dissociation pathways. The authors point out that this may indicate that, for both channels, dissociation proceeds via a reaction coordinate that has a significant exit barrier (barrier to reverse reaction), so fragments exert a repulsive force on each other as they separate. The pathway to dissociation probably does not involve internal conversion to the ground electronic state, as the barrier to reverse reaction in the ground state is negligible.

The *gauche* (Cl- and O-atoms in closest position) and *anti*-conformers (Cl- and O-atoms in a more distant position) of $CH_2ClC(O)CH_3$ show different efficiencies of C—C and C—Cl bond cleavage; the splitting at the barrier to C—Cl fission is consistently larger (by at least a factor of seven, and as much as a factor of 30) in *gauche* chloroacetone than in the *anti*-conformer for equivalent C=O distances. The authors suggest that the presence of large non-adiabatic effects along the C–C fission reaction coordinate in *anti*-chloroacetone suppresses C–C fission, allowing C–Cl fission to dominate. It is uncertain how these results, obtained at low pressures and somewhat elevated temperatures, would compare with those for chloroacetone photodissociation in air at 1 atm.

Studies by Burkholder et al. (2002) support the conclusion that the photodecomposition of chloroacetone is efficient, although the nature of the primary processes was not determined in their studies. In broad-band photolysis with fluorescent blacklamps (300–400 nm) and in experiments with pulsed excimer lasers (308 and 351 nm), the loss of parent ketone was determined relative to photolysis of NO_2, Cl_2, and CH_2I_2 as reference compounds. Quantum yields of photodissociation (disappearance) of $CH_2ClC(O)CH_3$ at both 351 and 308 nm were determined to be 0.5 ± 0.08. The photolysis rate coefficients were independent of the total cell pressure between 50 and 600 Torr of air. CO and CO_2 produced were about one-half of the chloroacetone lost, consistent with the significant occurrence of processes (II), (III), and/or (IV). Although HCl was not detected in the photolysis with blacklamps, it was detected as a product from excimer laser photolysis by Burkholder et al. (2002) as well as in the studies of Strachan and Blacet (1955). A likely fate of Cl-atoms formed in (I) is the abstraction of H-atoms from the parent ketone $CH_2ClC(O)CH_3$ molecule. Hence the observed HCl may arise following process (I) and/or from process (V). Burkholder et al. (2002) concluded that the effective quantum yields for atmospheric photodissociation are significant, probably greater than 0.5. It seems likely that the photodecomposition of chloroacetone lies between 0.5 and 1.0 over much of the spectral range available in tropospheric sunlight. Assuming this range of quantum yield values and using the cross section data of Burkholder et al. (2002), we have calculated the range of probable photolysis frequencies of $CH_2ClC(O)CH_3$ as a function of solar zenith angle for a cloudless day within the troposphere (ozone column = 350 DU); these data are given in figure VII-G-5.

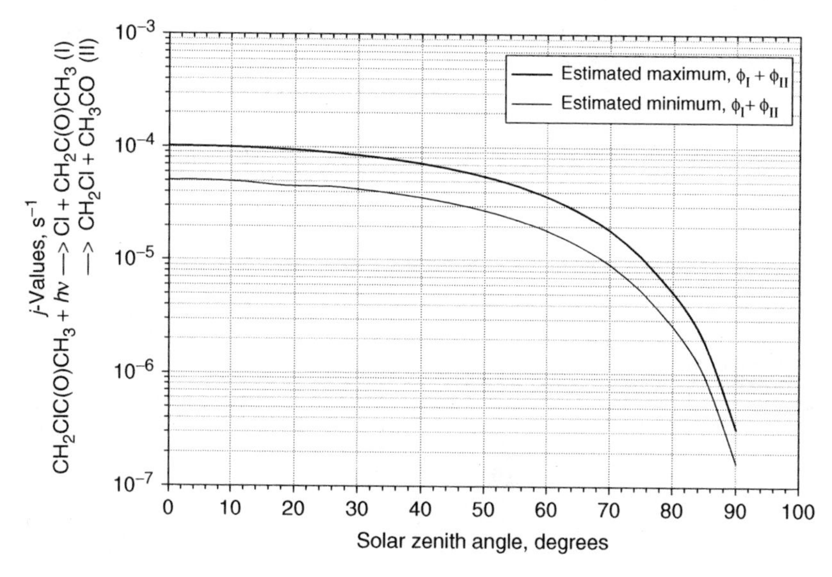

Figure VII-G-5. Approximate photolysis frequencies vs. solar zenith angle for $CH_2ClC(O)CH_3$ photodecomposition on a cloudless day in the lower troposphere with an overhead ozone column = 350 DU. The present best estimate of *j*-values lies between the two curves.

It is interesting to note that in the exploratory study of the photolyses of 3-chloro-2-butanone [$CH_3CHClC(O)CH_3$] and 4-chloro-2-butanone [$CH_2ClCH_2C(O)CH_3$] by Taylor and Blacet (1956), the former compound gave results analogous to those obtained for chloroacetone (C—Cl bond cleavage), while the latter compound appeared to favor C—C bond dissociation, analogous to 2-butanone.

VII-G-4. 1-Bromo-2-propanone [Bromoacetone, $BrCH_2C(O)CH_3$]

The absorption spectrum of 1-bromo-2-propanone is compared with those of other monosubstituted acetones in figure VII-G-1 and the cross sections are listed in tabular form in table VII-G-4 as a function of wavelength. It is apparent that bromoacetone absorbs the wavelengths of sunlight present within the troposphere much more efficiently than fluoroacetone and chloroacetone.

The photodecomposition of 1-bromo-2-propanone has been studied at 308 nm in crossed laser–molecular beam experiments by Waschewsky et al. (1994) and Kash et al. (1994) and in broad-band photolysis (300–400 nm) and laser beam photolysis (308 and 351 nm) by Burkholder et al. (2002). The photodecomposition follows the pattern of 1-chloro-2-propanone. The following primary processes should be considered:

$$CH_2BrC(O)CH_3 + h\nu \rightarrow Br + CH_2C(O)CH_3 \quad \text{(I)}$$

$$\rightarrow CH_2Br + CH_3CO \quad \text{(II)}$$

Table VII-G-4. Absorption cross sections (σ, cm^2 molecule^{-1}) for 1-bromo-2-propanone at 298 K for the wavelengths of sunlight available within the troposphere (Burkholder et al., 2002)

λ (nm)	$10^{19} \times \sigma$	λ (nm)	$10^{19} \times \sigma$	λ (nm)	$10^{19} \times \sigma$
278	0.774	306	1.11	334	0.251
279	0.800	307	1.09	335	0.226
280	0.825	308	1.07	336	0.202
281	0.850	309	1.04	337	0.180
282	0.875	310	1.02	338	0.160
283	0.904	311	0.989	339	0.141
284	0.929	312	0.962	340	0.124
285	0.958	313	0.936	341	0.109
286	0.978	314	0.910	342	0.0955
287	0.998	315	0.882	343	0.0831
288	1.02	316	0.852	344	0.0724
289	1.04	317	0.819	345	0.0627
290	1.06	318	0.783	346	0.0545
291	1.08	319	0.745	347	0.0473
292	1.10	320	0.706	348	0.0408
293	1.11	321	0.666	349	0.0351
294	1.12	322	0.628	350	0.0303
295	1.13	323	0.589	351	0.0259
296	1.14	324	0.555	352	0.0223
297	1.14	325	0.520	353	0.0190
298	1.15	326	0.487	354	0.0164
299	1.15	327	0.455	355	0.0140
300	1.15	328	0.424	356	0.0120
301	1.14	329	0.394	357	0.0108
302	1.14	330	0.364	358	0.00952
303	1.13	331	0.335	359	0.00838
304	1.12	332	0.306	360	0.00784
305	1.11	333	0.277		

$$\rightarrow CH_2BrCO + CH_3 \qquad \text{(III)}$$

$$\rightarrow CH_3 + CO + CH_2Br \qquad \text{(IV)}$$

The results of Waschewski et al. (1994) and Kash et al. (1994) in experiments at 308 nm suggest that, in a thermal population of 1-bromo-2-propanone conformers (*gauche* and *anti*), photodissociation at 308 nm occurs largely from Br—C fission, process (I), and by fission of one of the two C—C bonds, process (II). The angular distribution of Br-atoms shows C—Br fission occurs primarily from the minor *anti* conformer, not the dominant *gauche* structure. Low-level ab initio calculations show C—Br fission is more non-adiabatically suppressed in the *anti* conformer than in the *gauche* form. The conformation dependence of the branching ratio of C–C to C–Br is not determined by the C–Br fission reaction coordinate. In the molecular beam experiments with nozzle temperatures of 100° and 400°C, the experimentally observed ratio was $\phi_I/\phi_{II} \approx 5.7$ and 3.5, respectively. It is uncertain how these results, obtained at low pressures and somewhat elevated temperatures, would compare with those for 1-bromo-2-propanone photodissociation at the lower temperatures of the troposphere with 1 atm of air. No evidence for the occurrence of processes (III) and (IV) was obtained, but, by analogy with the photolysis of acetone, these processes are expected to occur to some extent in photolysis at the shorter wavelengths of absorbed light.

In the experiments of Burkholder et al. (2002), the loss of 1-bromo-2-propanone was determined in broad-band photolysis (308–351 nm) in air. The authors concluded that the quantum yield for the dissociation of bromoacetone is greater than 0.5. The amounts of CO and CO_2 produced were ~0.5 of the bromoacetone lost, so it seems likely that processes (II), (III), and/or (IV) occur to some extent. HBr was not detected. Pulsed laser photolysis of bromoacetone was referenced to photolyses of NO_2, Cl_2, and CH_2I_2 to estimate photodissociation quantum yields of 1.0 ± 0.15 and 1.6 ± 0.25 at 308 and 351 nm, respectively.

In view of the limited results outlined above, only approximate j-values related to total photodissociation can be calculated in the present work. We have used the cross section data of Burkholder et al. (2002) and quantum yield limits of 0.5 and 1.0, together with the appropriate solar flux (ozone column = 350 DU), to derive approximate j-values for 1-bromo-2-propanone photodissociation on a cloudless day in the lower troposphere; see figure VII-G-6.

VII-G-5. 1,1,1-Trifluoro-2-propanone [$CF_3C(O)CH_3$] and 1,1,1,3,3,3-Hexafluoro-2-propanone [$CF_3C(O)CF_3$]

Data necessary to evaluate the photolysis frequencies of these two ketones in air are limited, but it is clear that both ketones follow a pattern of photodissociation related

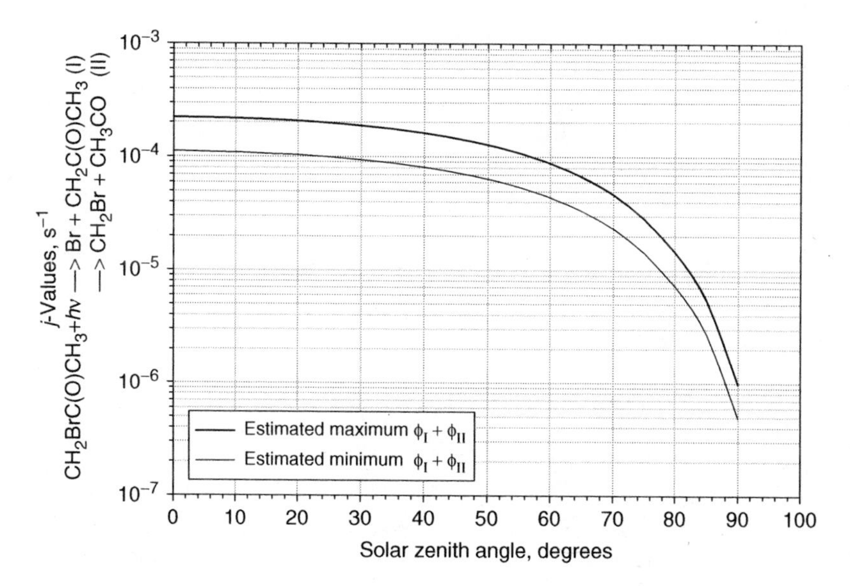

Figure VII-G-6. Approximate photolysis frequencies vs. solar zenith angle for $CH_2BrC(O)CH_3$ photodecomposition on a cloudless day in the lower troposphere with an overhead ozone column = 350 DU. The present best estimate of j-values lies between the two curves.

to that observed for acetone. Two different estimations of the absorption spectrum of $CF_3C(O)CH_3$are given in figure VII-G-2. That shown as the light solid line has been redrawn from the $1/\lambda$ vs. molar extinction coeffient plot of Beavan et al. (1978). The filled triangles given at a few specific wavelengths and connected by the short-dashed trend line are from data determined by Ausloos and Murad (1961). Although the data are of limited accuracy, it is clear that the absorption maximum is shifted somewhat to longer wavelengths than $CH_3C(O)CH_3$ and that the cross sections of $CF_3C(O)CH_3$ are significantly less than those for unsubstituted acetone (factor of about 0.6).

Also shown in figure VII-G-2 are the cross sections for $CF_3C(O)CF_3$ as measured at several discrete wavelengths by Ausloos and Murad (1961). A trend line has been drawn through these data points to provide an approximate envelope for the absorption of $CF_3C(O)CF_3$. Note that the absorption maximum is shifted again to longer wavelengths than that for $CF_3C(O)CH_3$, although the magnitude of the cross section is relatively unchanged from that of $CF_3C(O)CH_3$.

Discussions of the photochemistry and photophysics of 1,1,1-trifluoro-2-propanone and 1,1,1,2,2,2-hexafluoroacetone (efficiency of fluorescence, phosphorescence, intersystem crossing efficiency, lifetimes of excited states, etc.) have been given in the literature; e.g., for $CF_3C(O)CH_3$ see: Ausloos and Murad (1961); Dawidowicz and Patrick (1964); Beavan et al. (1975); Gandini and Hackett (1976/1977); Gandini and Hackett (1977a); Beavan et al. (1978). For $CF_3C(O)CF_3$ see: Ayscough and Steacie (1956); Whytock and Kutschke (1968); Gandini and Kutschke (1968); Gandini et al. (1968a,b,c); Halpern and Ware (1970, 1971); Hackett and Kutschke (1977). The mechanisms considered for light emission and photodecomposition of the fluorine substituted ketones are similar to those described for acetone in section VII-C-1.

The photodecomposition of $CF_3C(O)CH_3$ was first studied by Sieger (1954) and Sieger and Calvert (1954) in photolysis studies at 313 nm, using relatively high pressures of the ketone (250 Torr at 25°C) and temperatures in the range 25–350°C. The measurements were limited to the determination of the quantum yields of CO, C_2H_6, CH_3CF_3, CF_3CF_3, CH_4 and CF_3H. These data suggested that the primary photodissociation occurred with both CF_3—$C(O)CH_3$ and $CF_3C(O)$—CH_3 bond cleavage, with process (II) being somewhat favored.

$$CF_3C(O)CH_3 + h\nu \rightarrow CF_3 + CH_3CO \quad \text{(I)}$$

$$\rightarrow CH_3 + CF_3CO \quad \text{(II)}$$

$$\rightarrow CF_3 + CO + CH_3 \quad \text{(III)}$$

However, in the study of Dawidowicz and Patrick (1964) with full Hg-arc photodecomposition of $CF_3C(O)CH_3$ at 45 ± 5°C, the observed major products were: $CH_3C(O)C(O)CH_3$, C_2H_6, C_2F_6, CH_3CF_3, $CH_3C(O)CH_3$, and $CH_3CF_2CH_3$, with small amounts of CH_4 and CF_3H. From the observed formation of 2,3-butanedione and the absence of any products containing the CF_3CO fragment, these authors concluded that process (I) was the dominant photodecomposition mode. In view of the weak CF_3—CO bond, compared to the CH_3—CO bond, it is possible that the low yield of CF_3CO-containing molecules in the study of Dawidowicz and Patrick (1964) is a result of CF_3CO decomposition at 45°C, and thus these data do not rule out channel (II). Formation of the product $CH_3CF_2CH_3$ in this study remains unexplained. Amphlett and Whittle (1967) determined the activation energy for $CF_3CO \rightarrow CF_3 + CO$ to be ~12–24 kJ mol^{-1}. More recently, Maricq et al. (1995) and Tomas et al. (2000) measured ~52 and ~49 kJ mol^{-1}, respectively, for this activation energy. All these measurements are somewhat lower than those for CH_3CO decomposition ($E_a \approx 72$ kJ mol^{-1}, Baulch et al., 1992). Theoretical studies (Viscolcz and Bérces, 2000; Méreau et al., 2001) support this view. These data suggest that significantly less energy (~22 kJ mol^{-1}) is required for dissociation of $CF_3C(O)CH_3$ by process (I) than by process (II), assuming that the barriers for the two reverse reactions ($R + CO \rightarrow RCO$) are equal. Thus one expects some preference to photodissociation by process (I), particularly at the longer wavelengths of absorbed light. Photodecomposition of $CF_3C(O)CF_3$ follows the pattern outlined for acetone and $CF_3C(O)CH_3$ molecules:

$$CF_3C(O)CF_3 + h\nu \rightarrow CF_3 + CF_3CO \quad \text{(I)}$$

$$\rightarrow CF_3 + CO + CF_3 \quad \text{(II)}$$

Figure VII-G-7 gives an interesting comparison of the quantum yields of CO from $CF_3C(O)CH_3$, $CF_3C(O)CF_3$, and $CH_3C(O)CH_3$ as a function of temperature for photolysis at 313 nm. These data provide some insight into mechanisms of photodissociation of the three ketones. In figure VII-G-7a it can be seen that the Cundall and Davies (1966) data from $CH_3C(O)CH_3$ at 50 Torr pressure lie well above those from the $CF_3C(O)CH_3$ data of Sieger and Calvert (1954), which were determined at much higher ketone pressures (250–477 Torr) but at constant number density. The $CF_3C(O)CF_3$ data of Ayscough and Steacie (1956) for 25 Torr pressure lie above those obtained at about 95 Torr and near those for $CH_3C(O)CH_3$ determined by Cundall and Davies (1966) at 50 Torr pressure. The differences in pressure employed in these studies tend to mask differences in quantum yields that exist since the extent of self-quenching will be different. We have attempted to remove the differences due to quenching by adjusting the $CF_3C(O)CF_3$ data to the same pressures as those used for the $CH_3C(O)CH_3$ and $CF_3C(O)CH_3$ studies. We made use of the Ayscough and Steacie Φ_{CO} measurements as a function of $CF_3C(O)CF_3$ pressure. The k_q/k_d values for the five temperatures that they used (27–219°C) can be summarized in the equation,

$$\log_{10}(k_q/k_d) = 0.175 - 0.0291T + 4.86 \times 10^{-5}T^2$$

This equation was used to derive k_q/k_d values for the temperatures employed in the experiments by Sieger and Calvert ($CF_3C(O)CH_3$) and Cundall and Davies ($CH_3C(O)CH_3$); these were then used to adjust the Ayscough and Steacie reported quantum yield data to the pressures of the ketones employed by Sieger and Calvert and Cundall and Davies. The results are shown in figure VII-G-7b. Note that the originally divergent curves of figure VII-G-7a converge significantly: $CF_3C(O)CF_3$ data, when adjusted to 50 Torr, match reasonably well those of acetone at that

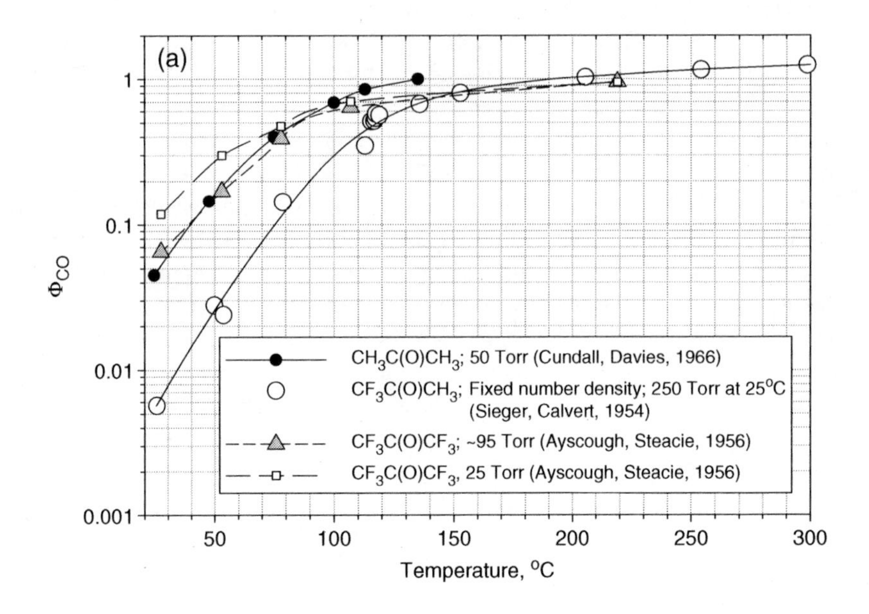

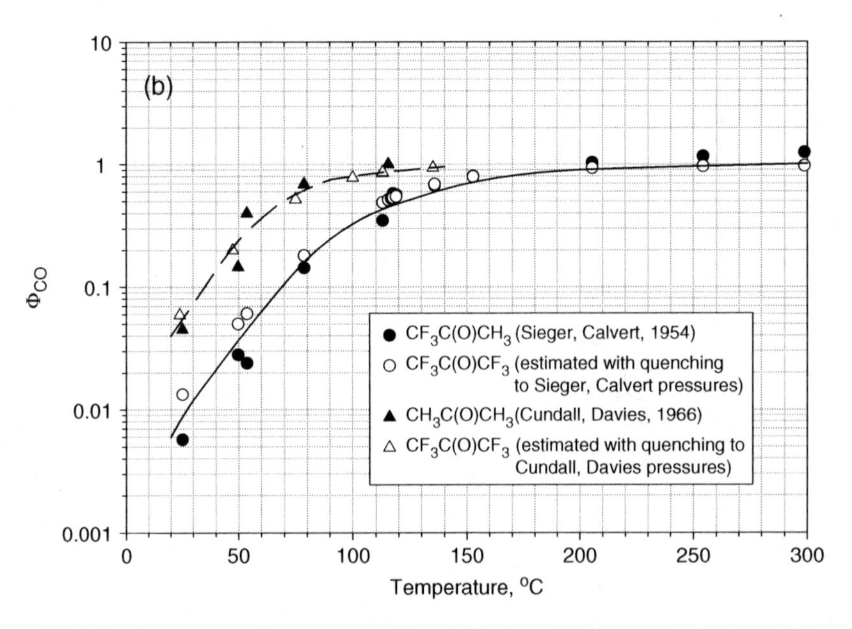

Figure VII-G-7. Comparison of quantum yields of CO from $CF_3C(O)CH_3$, $CH_3C(O)CH_3$, and $CF_3C(O)CF_3$ from photolysis at 313 nm as a function of temperature 23–300°C. (a) Data as determined at several pressures of the ketones. (b) $CF_3C(O)CF_3$ data of Ayscough and Steacie (1956) adjusted for quenching to the pressures of the measurements of Sieger and Calvert (1954) for $CF_3C(O)CH_3$ and Cundall and Davies (1966) for $CH_3C(O)CH_3$. The k_q/k_d data for $CF_3C(O)CF_3$ from Ayscough and Steacie were used in the calculations.

pressure as measured by Cundall and Davies (1966), while adjustment of the $CF_3C(O)CF_3$ data to the higher pressures employed for $CF_3C(O)CH_3$ by Sieger and Calvert leads to significant closure of the $CF_3C(O)CF_3$ and $CF_3C(O)CH_3$ data sets. However, the Φ_{CO} values for $CF_3C(O)CF_3$ still lie somewhat above those for $CH_3C(O)CH_3$ at 50 Torr and $CF_3C(O)CH_3$ at 250 Torr at the lower temperatures (298 K). This could reflect the lower thermal stability of the CF_3CO radical formed in $CF_3C(O)CF_3$ photolysis compared to the somewhat more stable CH_3CO radical formed from $CH_3C(O)CH_3$ and presumably favored in $CF_3C(O)CH_3$ photolysis as well.

Gandini and Hackett (1977a) determined the quantum yields of both CO and 2,3-butanedione in the photolyses of $CF_3C(O)CH_3$ and $CH_3C(O)CH_3$ as a function of wavelength in experiments at 22°C; see figures VII-G-8 and VII-G-9. The Φ_{CO} values determined with both ketones are similar at the longest wavelengths (305–313 nm), and they both approach unity at the shortest wavelengths. Over most of the range of wavelength given in figure VII-G-8, quantum yields from $CF_3C(O)CH_3$ exceed those of $CH_3C(O)CH_3$ by significant factors (~1.3 to 2.0). The differences seen are opposite to those expected from the difference in the extent of self-quenching for the two ketones ($CH_3C(O)CH_3$ was at 13.5 Torr and $CF_3C(O)CH_3$ at 20 Torr). Since most evidence suggests that both ketones favor photodissociation to form CH_3CO radicals in the primary process (I), these data may indicate either that $CF_3C(O)CH_3$ has a somewhat larger ϕ_I than $CH_3C(O)CH_3$, or that a larger fraction of the dissociation in $CF_3C(O)CH_3$ forms CO in the primary process (III).

The 2,3-butanedione quantum yields shown in figure VII-G-9 give a somewhat different picture. $CH_3C(O)CH_3$ has about twice the quantum yield for 2,3-butanedione

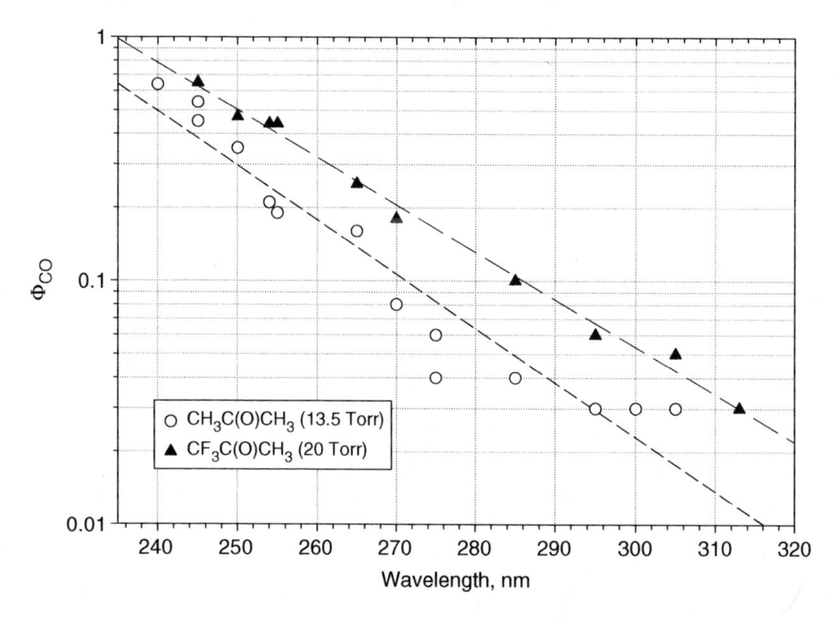

Figure VII-G-8. Quantum yields of CO as a function of wavelength in photolyses of $CF_3C(O)CH_3$ (20 Torr) and $CH_3C(O)CH_3$ (13.5 Torr), as measured by Gandini and Hackett (1977a) at 22°C.

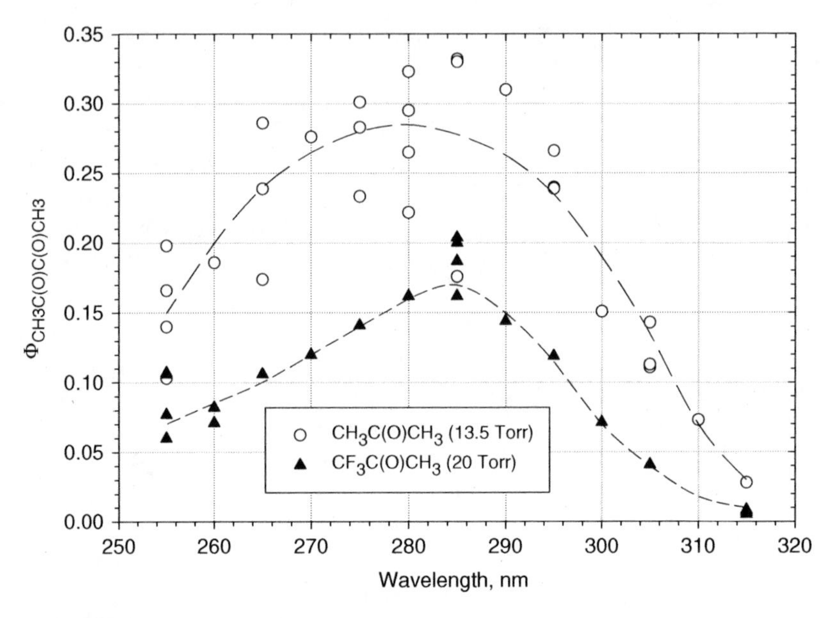

Figure VII-G-9. Quantum yields of 2,3-butanedione from photolyses of $CF_3C(O)CH_3$ (20 Torr) and $CH_3C(O)CH_3$ (13.5 Torr), as measured by Gandini and Hackett (1977a) at 22°C.

formation as has $CF_3C(O)CH_3$, presumably reflecting a greater quantum yield of CH_3CO radicals from $CH_3C(O)CH_3$ at all wavelengths. The presence of neither $CF_3C(O)C(O)CH_3$ nor $CF_3C(O)C(O)CF_3$ was reported in the Gandini and Hackett (1977a) study. The very sensitive analysis method depended on the measurement of the sensitized phosphorescence of 2,3-butanedione. It is not clear whether this technique was unique to the detection of $CH_3C(O)C(O)CH_3$ or some contribution from emission of the fluorinated 2,3-butanediones was possible. Gandini and Hackett comment on "the known instability of the trifluoroacetyl radical at room temperature", and apparently dismiss as unlikely the formation of $CF_3COCOCH_3$ and $CF_3COCOCF_3$. As evidence, they quote the work of Amphlett and Whittle (1967), who demonstrated that CF_3CO is less stable than CH_3CO at 25°C.

In view of the conflicting evidence on the magnitude of the quantum yields of the primary processes in $CF_3C(O)CH_3$, the uncertainties in the absorption cross sections, and the lack of information on the effect of an atmosphere of air on these values, estimates of the photolysis frequencies of these two ketones remain uncertain at the time of this writing. In figure VII-G-10 approximate values for these quantities have been estimated for $CF_3C(O)CH_3$, using the quantum yields as a function of wavelength as estimated in this study for acetone in air ($\phi_I + \phi_{IIb}$) in table VII-C-2, coupled with the cross section data for $CF_3C(O)CH_3$ given both by Beavan et al. (1978) and by Ausloos and Murad (1961) in figure VII-G-2, and the appropriate solar flux (ozone column = 350 DU) for a cloudless day with the lower troposphere. Although the uncertainties and differences in the two sets of sigma values are seemingly significant, the effect on the magnitude of the j-values is seen to be relatively small. However, major uncertainty exists in the quantum yields.

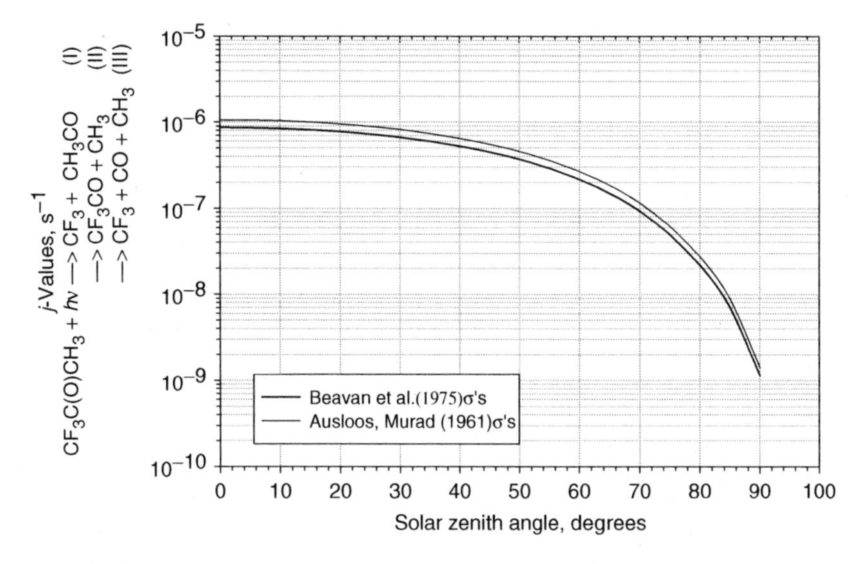

Figure VII-G-10. Approximate photolysis frequencies vs. solar zenith angle for $CF_3C(O)CH_3$ total photodecomposition on a cloudless day in the lower troposphere with an overhead ozone column = 350 DU. Two different cross section sources were used, and $\phi_I + \phi_{II} + \phi_{III}$ values were assumed to be equal to $\phi_I + \phi_{IIb}$ for $CH_3C(O)CH_3$. Very rough estimates of *j*-values lie between the two curves.

Rough estimates of the photolysis frequencies for $CF_3C(O)CF_3$ on a cloudless day within the lower troposphere (ozone column = 350 DU) as a function of solar zenith angle are given in figure VII-G-11. We have used the approximate cross section data of Ausloos and Murad (1961) and quantum yields as a function of wavelength, as estimated for $CH_3C(O)CH_3$ ($\phi_I + \phi_{IIb}$) in table VII-C-2. Major uncertainties exist in both the cross section and the quantum yield data for $CF_3C(O)CF_3$, so the *j*-value data given here are approximate estimates at best.

VII-G-6. Photochemistry of Highly Substituted F- and Cl-Substituted Acetones

The photophysics and limited data on the photochemistry of several F- and Cl-containing, multisubstituted acetone molecules have been reported: e.g., $CF_2Cl(O)CF_3$, $CF_2ClC(O)CF_2Cl$; $CFCl_2C(O)CF_2Cl$, $CFCl_2C(O)CFCl_2$, $CCl_3C(O)CCl_3$ (Majer et al., 1971; Hackett and Phillips, 1974a–d); $CF_3C(O)CH_2Br$ (Majer et al., 1976; Majer and Al-Saigh, 1984). The absorption spectra of several of these molecules, given in figure VII-G-3, show that replacement of F-atoms by one, two, or three Cl-atoms in the $CF_3C(O)CF_3$ molecule leads to a large enhancement of the maximum in the cross section (by about a factor of six). It is clear that these molecules absorb a much larger fraction of the actinic flux available within the troposphere than $CF_3C(O)CH_3$ and $CF_3C(O)CF_3$ molecules, and their photochemistry is probably important within the troposphere. Evidence suggests that photolysis results in a C–C bond rupture, but it is

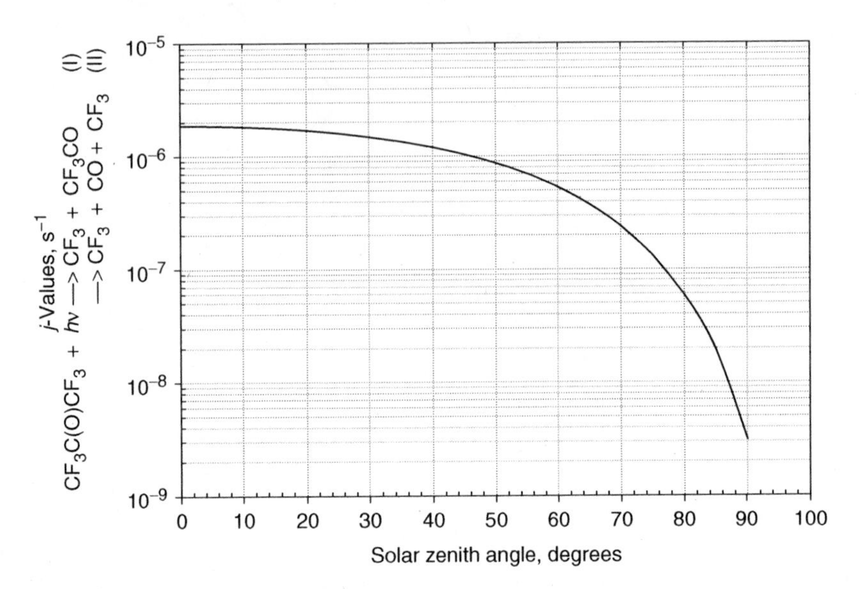

Figure VII-G-11. Approximate photolysis frequencies vs. solar zenith angle for $CF_3C(O)CF_3$ total photodecomposition on a cloudless day in the lower troposphere with an overhead ozone column = 350 DU. Limited cross section data from Ausloos and Murad (1961) were used, and $\phi_I + \phi_{II}$ values were assumed to be equal to $\phi_I + \phi_{IIb}$ for $CH_3C(O)CH_3$.

unclear as to the extent, if any, to which a C–Cl bond rupture competes with this process. The existing information on the modes and efficiencies of photodissociation has not been explored in sufficient detail to allow meaningful estimates of their tropospheric photolysis rates.

VII-G-7. Perfluoro-2-methyl-3-pentanone [$CF_3CF_2C(O)CF(CF_3)_2$]

The absorption spectrum of a highly fluorine-substituted ketone, perfluoro-2-methyl-3-pentanone, is shown in figure VII-G-12. The measurements of Taniguchi et al. (2003), shown as open circles in the figure, and those of D'Anna et al. (2005), given as a heavy line, are seen to be in good agreement. Compare these spectra with that for an unsubsitituted ketone of a somewhat similar structure, 2,4-dimethyl-3-pentanone, $(CH_3)_2CHC(O)CH(CH_3)_2$, given by the light line in figure VII-G-12. As observed with the simpler ketones, the substitution of fluorine for H-atoms shifts the absorption to longer wavelengths and decreases the absorption maximum somewhat.

The photodecomposition of this fluorine-substituted ketone in mixtures with 50–700 Torr of air was first studied by Taniguchi et al. (2003), using a smog chamber with Fourier-transform infrared techniques and sunlamps or blacklamps as the radiation source. These workers followed the decay through the formation of its major products $CF_3C(O)F$ and $FC(O)F$. In similar experiments the rates of photodecomposition of isotopically labeled acetaldehyde ($^{13}CH_3^{13}CHO$) were determined by following the

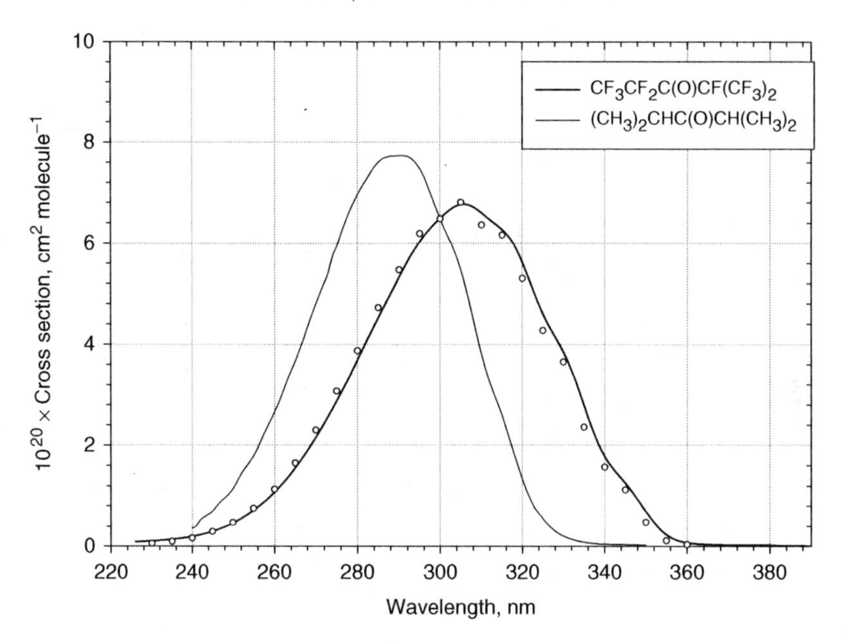

Figure VII-G-12. Comparison of the absorption cross sections for perfluoro-2-methyl-3-pentanone [$CF_3CF_2C(O)CF(CF_3)_2$] and 2,4-dimethyl-3-pentanone; data for perfluoro-2-methyl-3-pentanone, shown as the heavy line, are from D'Anna et al. (2005), and the points shown as open circles are from Taniguchi et al. (2003); the data for 2,3-dimethyl-3-pentanone (light line) are from Yujing and Mellouki (2000).

^{13}CO product. The estimated *j*-values for CH_3CHO in air (Calvert et al., 2000) and the relative rates of photodecomposition were used to estimate that the photochemical lifetime of the perfluoroketone is approximately 1–2 weeks. The estimate is necessarily approximate since the absorption band of the perfluoroketone is shifted to somewhat longer wavelengths than that for acetaldehyde, and this difference in the overlap of absorption with the lamp radiation necessarily complicated the derivation of the *j*-values estimates.

More recently, D'Anna et al. (2005) studied the photodecomposition of perfluoro-2-methyl-3-pentanone in the EUPHORE outdoor chamber in Valencia, Spain. An effective quantum yield of decomposition was estimated to be 0.043 ± 0.011 over the wavelength range 290–400 nm. The primary processes that are likely for this ketone are (I) and (II), with process (I) favored somewhat, particularly at the longer wavelengths:

$$(CF_3)_2CFC(O)CF_2CF_3 + h\nu \rightarrow (CF_3)_2CF + CF_3CF_2CO \quad \text{(I)}$$

$$\rightarrow (CF_3)_2CFCO + CF_2CF_3 \quad \text{(II)}$$

We have used the absorption cross section data reported by D'Anna et al. (2005), shown in table VII-G-5, together with their average estimated quantum yield of dissociation (0.043) and the appropriate solar flux to calculate approximate *j*-values as a function

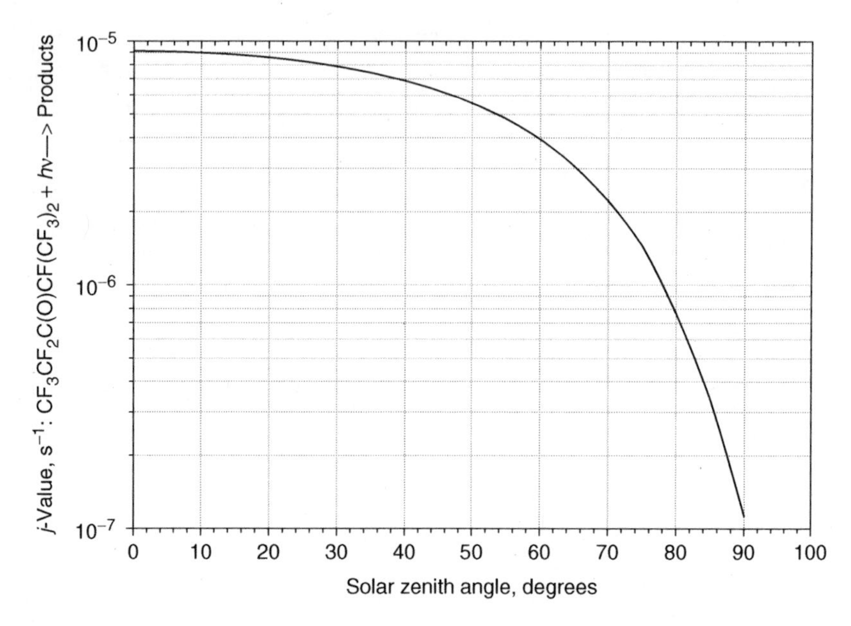

Figure VII-G-13. Estimated photolysis frequencies for $CF_3CF_2C(O)CF(CF_3)_2$ as a function of solar zenith angle. Values are calculated for a cloudless day in the lower troposphere with an overhead ozone column of 350 DU. Cross section data and average quantum yields of photodecomposition of D'Anna et al. (2005) were used in the calculations.

of solar zenith angle for a cloudless day in the lower troposphere; see figure VII-G-13. The experimental *j*-value of $(6.4 \pm 0.3) \times 10^{-6}$ s^{-1}, measured by D'Anna et al. (2005) in Valencia, Spain (39.5°N) on July 14, 2003 from 10.00 to 14.00 GMT, is reasonably consistent with the calculated values given in figure VII-G-13.

VII-H. Photodecomposition of Acyl Halides and Carbonyl Halides

VII-H-1. Spectra of the Acyl Halides and Carbonyl Halides

Acyl halides [HC(O)X, $CX_nH_{3-n}C(O)X$, etc., where X is a halogen atom] are among the products of the tropospheric oxidation of the haloalkanes (see chapter VI). The absorption cross sections for a series of formyl halides, HC(O)F, HC(O)Cl, and HC(O)Br, and several carbonyl halides, FC(O)F, ClC(O)Cl, and BrC(O)Br, are compared with those for unsubstituted CH_2O in figure VII-H-1. Note that all of these halogen-substituted formaldehydes have lower cross sections than the unsubstituted formaldehyde in the long-wavelength region of the tropospheric solar flux. Many of the halogen-atom-substituted formaldehyde derivatives absorb only the short wavelengths of sunlight that are weak in intensity within the troposphere. The long-wavelength absorption bands in these compounds result from an $n \rightarrow \pi^*$ electron promotion as assigned for HC(O)Cl by Judge and Moule (1985) and Ding et al. (1999)

Table VII-G-5. Absorption cross sections (σ, cm^2 $molecule^{-1}$) for $CF_3CF_2C(O)CF(CF_3)_2$ as a function of wavelength (D'Anna et al., 2005)

λ (nm)	$10^{20} \times \sigma$	λ (nm)	$10^{20} \times \sigma$	λ (nm)	$10^{20} \times \sigma$
278	3.35	316	6.19	354	0.296
279	3.51	317	6.09	355	0.229
280	3.68	318	5.96	356	0.176
281	3.85	319	5.80	357	0.136
282	4.02	320	5.62	358	0.106
283	4.19	321	5.42	359	0.085
284	4.36	322	5.20	360	0.069
285	4.52	323	4.99	361	0.058
286	4.68	324	4.79	362	0.049
287	4.84	325	4.60	363	0.042
288	5.01	326	4.44	364	0.037
289	5.17	327	4.29	365	0.033
290	5.33	328	4.15	366	0.030
291	5.48	329	4.01	367	0.029
292	5.63	330	3.85	368	0.027
293	5.78	331	3.68	369	0.025
294	5.90	332	3.48	370	0.024
295	6.03	333	3.27	371	0.024
296	6.14	334	3.03	372	0.023
297	6.24	335	2.79	373	0.023
298	6.33	336	2.55	374	0.023
299	6.41	337	2.33	375	0.023
300	6.49	338	2.12	376	0.023
301	6.57	339	1.94	377	0.022
302	6.64	340	1.78	378	0.020
303	6.70	341	1.65	379	0.019
304	6.74	342	1.54	380	0.017
305	6.77	343	1.44	381	0.015
306	6.77	344	1.35	382	0.013
307	6.76	345	1.25	383	0.011
308	6.72	346	1.15	384	0.009
309	6.66	347	1.04	385	0.008
310	6.60	348	0.928	386	0.007
311	6.53	349	0.809	387	0.007
312	6.47	350	0.690	388	0.006
313	6.40	351	0.576	389	0.005
314	6.34	352	0.471	390	0.005
315	6.27	353	0.376		

and for the $CH_3C(O)Cl$ molecule by Person et al. (1992a,b). It can be seen that a halogen-atom substitution for an H-atom in CH_2O results in a significant change in the absorption spectrum. In figure VII-H-1a note that the maximum in the absorption cross sections shifts to the shorter wavelengths for the formyl halides in the progression: CH_2O, BrC(O)H, ClC(O)H, FC(O)H, while the long-wavelength tail in the absorption shows a decreasing overlap with the available wavelengths of sunlight. HC(O)Cl absorbs sunlight only weakly; among the formyl halides shown in the figure, only HC(O)Br absorbs significantly at the wavelengths available

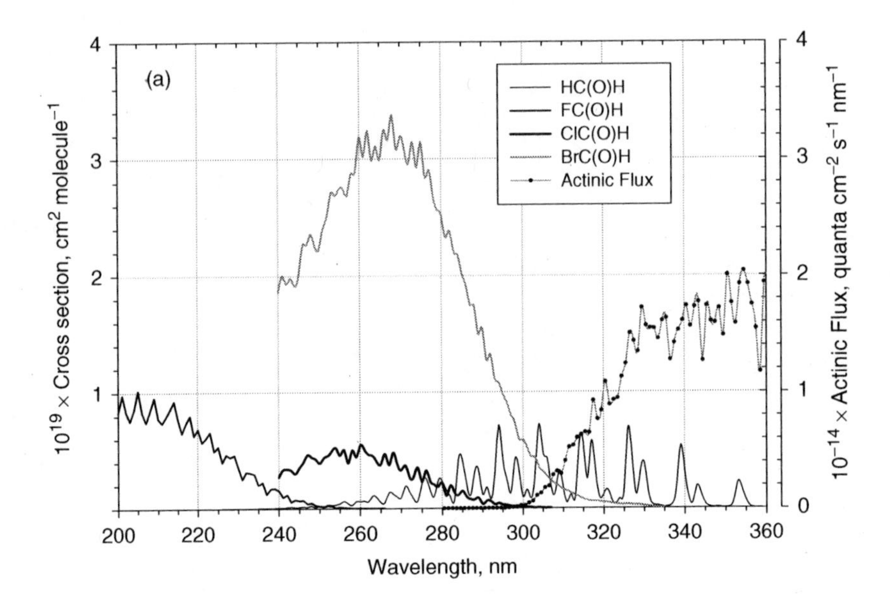

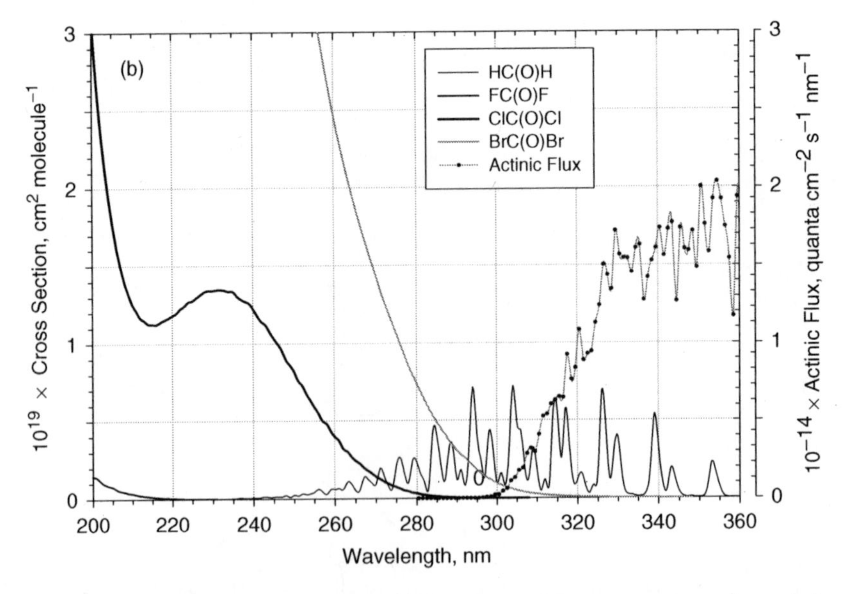

Figure VII-H-1. Comparison of the wavelength dependence of the cross sections of some halogen-atom-substituted formaldehyde molecules and the actinic flux as measured for a cloudless day (June 19, 1998) near Boulder, CO, at noon. Figure VII-H-1a shows the spectra for the formyl halides; figure VII-H-1b gives the spectra for the carbonyl halides. Note that only the HC(O)Br and BrC(O)Br compounds have a significant overlap with the available actinic flux. The long-wavelength tail of the absorption of HC(O)Cl and ClC(O)Cl molecules catches the edge of the flux distribution. Spectral data are from Keller-Rudek and Moortgat (2005).

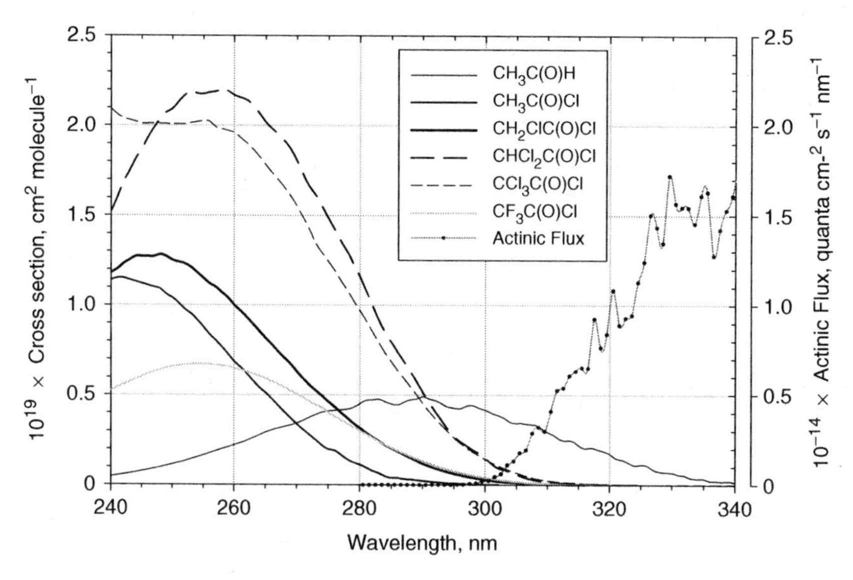

Figure VII-H-2. Comparison of the wavelength dependence of the cross sections of acetaldehyde and some halogen-atom-substituted acetaldehyde molecules, the acetyl chlorides; note the relatively poor overlap of the spectra for most acetyl chlorides with the distribution of the actinic flux as measured for a cloudless day (June 19, 1998) near Boulder, CO, at noon.

within the troposphere. A similar trend is seen for the carbonyl halides in figure VII-H-1b; only the bromine-containing carbonyl [BrC(O)Br] absorbs the wavelengths of tropospheric sunlight significantly. Similar effects are seen in the absorption of the halogen-atom-substituted acetyl chlorides; see figure VII-H-2. Significant overlap of the absorption with the wavelength distribution of the actinic flux occurs only for those acetyl chlorides with the largest numbers of halogen atoms, i.e., $CHCl_2C(O)Cl$, $CCl_3C(O)Cl$, and $CF_3C(O)Cl$. Cross sections are given in table VII-H-1 for some of the species for which photodissociation within the troposphere is somewhat significant.

Although the rates of photodecomposition of many of these compounds are slow in the troposphere, competing removal processes (reaction with OH, hydrolysis in cloud-, rain-, and sea-water) are also slow for many of these compounds. Thus, even slow photolysis in sunlight can be a significant loss mechanism, and consideration of their photodecompositions is of interest to atmospheric chemists.

VII-H-2. Photodecomposition Modes of the Acyl Halides and Carbonyl Halides

The photodecomposition of a few of the acyl halides and carbonyl halides has been studied, and the results give some insight into the mechanisms of these processes. In this section we consider the various alternative photodecomposition modes that have been suggested in rationalizing experimental and theoretical studies.

Table VII-H-1. Absorption cross sections (σ, cm^2 $molecule^{-1}$) as a function of wavelength for a series of formyl halides, carbonyl halides, and acetyl halides at 298 K. Data for HC(O)Br, BrC(O)Br, $CCl_3C(O)Cl$, $CHCl_2C(O)Cl$, and $CH_2ClC(O)Cl$ are from Libuda (1992) and for $CF_3C(O)Cl$ from Meller and Moortgat (1997)

	$10^{20} \times \sigma$					
λ (nm)	HC(O)Br	BrC(O)Br	$CCl_3C(O)Cl$	$CHCl_2C(O)Cl$	$CH_2ClC(O)Cl$	$CF_3C(O)Cl$
278	26.8	8.55	10.83	13.0	3.71	3.44
279	25.6	7.99	10.23	12.3	3.41	3.19
280	25.1	7.44	9.69	11.6	3.15	3.04
281	23.2	6.90	9.12	10.8	2.89	2.80
282	23.8	6.35	8.51	9.95	2.65	2.63
283	22.0	5.81	7.90	9.15	2.43	2.41
284	21.7	5.34	7.33	8.46	2.21	2.25
285	20.1	4.92	6.84	7.92	2.03	2.06
286	19.1	4.50	6.35	7.37	1.85	1.91
287	17.4	4.13	5.89	6.85	1.67	1.74
288	17.3	3.730	5.44	6.30	1.50	1.59
289	14.9	3.38	4.99	5.73	1.33	1.45
290	15.5	3.08	4.52	5.14	1.17	1.32
291	12.7	2.81	4.09	4.61	1.04	1.20
292	13.2	2.58	3.69	4.08	0.918	1.08
293	11.1	2.32	3.32	3.60	0.814	0.969
294	10.8	2.07	2.96	3.10	0.717	0.867
295	9.79	1.82	2.62	2.68	0.628	0.771
296	9.08	1.62	2.32	2.41	0.546	0.681
297	8.02	1.41	2.06	2.17	0.473	0.599
298	7.34	1.26	1.83	1.94	0.409	0.528
299	5.96	1.11	1.62	1.74	0.353	0.458
300	5.86	0.971	1.41	1.48	0.300	0.404
301	5.24	0.858	1.23	1.18	0.248	0.347
302	4.46	0.750	1.06	0.944	0.199	0.304
303	4.37	0.661	0.918	0.843	0.170	0.256
304	3.53	0.577	0.787	0.706	0.157	0.221
305	3.18	0.501	0.672	0.577	0.131	0.186
306	2.77	0.436	0.551	0.458	0.113	0.158
307	2.43	0.385	0.444	0.347	0.0909	0.130
308	2.12	0.330	0.358	0.277	0.0809	0.109
309	1.90	0.301	0.294	0.216	0.0657	0.0879
310	1.65	0.260	0.247	0.177	0.0547	0.0722
311	1.51	0.230	0.201	0.151	0.0522	0.0579
312	1.34	0.204	0.166	0.120	0.0419	0.0472
313	1.13	0.178	0.141	0.095	0.0380	0.0373
314	1.01	0.159	0.121	0.067	0.0311	0.0301
315	0.801	0.143	0.103	0.058	0.0286	0.0234
316	0.687	0.127	0.0849	0.049	0.0214	0.0189
317	0.613	0.114	0.0780	0.052		0.0146
318	0.611	0.0995	0.0657	0.044		0.0123
319	0.576	0.0892	0.0576	0.046		0.0095
320	0.521	0.0789	0.0502	0.035		0.0084
321	0.476	0.0694	0.0466	0.022		0.0063

Table VII-H-1. (*continued*)

	$10^{20} \times \sigma$					
λ (nm)	HC(O)Br	BrC(O)Br	$CCl_3C(O)Cl$	$CHCl_2C(O)Cl$	$CH_2ClC(O)Cl$	$CF_3C(O)Cl$
322	0.424	0.0632	0.0436	0.010		0.0057
323	0.363	0.0578	0.0377	0.0093		0.0041
324	0.307	0.0528	0.0369	0.0099		0.0037
325	0.316	0.0498	0.0342	0.013		0.0025
326	0.325	0.0449	0.0327	0.020		0.0025
327	0.288	0.0397	0.0284	0.025		0.0016
328	0.281	0.0336	0.0269	0.021		0.0017
329	0.271	0.0296	0.0266	0.010		0.0009
330	0.193	0.0264	0.0278	0.0098		
331	0.161	0.0229	0.0259	0.0029		
332	0.108	0.0238	0.0226	0.0029		
333	0.091	0.0204	0.0218	0.0029		
334	0.067	0.0169	0.0166	0.0029		
335	0.073	0.0137	0.0159	0.012		
336	0.080	0.0110	0.0153	0.011		
337	0.086	0.0138	0.0139	0.019		
338	0.118	0.0110	0.0156	0.010		
339	0.166	0.0106	0.0142			
340	0.135	0.0079	0.0171			
341		0.0081	0.0135			
342		0.0082	0.0119			
343		0.0070				
344		0.0070				
345		0.0055				
346		0.0043				
347		0.0016				
348		0.0008				
349		0.0011				
350		0.0006				
351		0.0025				
352		0.0019				
353		0.0002				

VII-H-2.1. The Formyl Halides

The following primary pathways for photodissociation have been considered in theoretical studies of the formyl halides:

$$HC(O)X + h\nu \rightarrow HCO + X \quad \text{(I)}$$

$$\rightarrow H + XCO \quad \text{(II)}$$

$$\rightarrow H + CO + X \quad \text{(III)}$$

$$\rightarrow HX + CO \quad \text{(IV)}$$

$$\rightarrow X\text{—}C\text{—}O\text{—}H \text{ or } H\text{—}C\text{—}O\text{—}X \quad \text{(V)}$$

The dissociation pathways on the ground-state potential energy surface (S_0) for HC(O)F and HC(O)Cl have been reported by Kamiya and Morokuma (1991) and Francisco and Zhao (1992). The results of the latter study suggest that if dissociation occurs from the ground electronic state, process (III) is favored. The products of process (V) are the carbenes, possible isomerization routes for reaction that are predicted in theory. Fang and Liu (2001) reported an ab initio study of the dissociation of HC(O)Cl from the S_0, S_1 and T_1 states. They conclude that the C–Cl bond cleavage, process (I), is the most favored mechanism for HC(O)Cl photodissociation for light absorbed within the 230–320 nm range. The extent of occurrence of processes (IV) and (V) is unknown, but it is expected to be small.

VII-H-2.2. The Carbonyl Halides

It is interesting that FC(O)F has been detected in the stratosphere (Rinsland et al., 1986), and ClC(O)Cl within the lower troposphere of some urban areas (Singh, 1976). Both compounds are believed to have been derived from atmospheric reactions of halogenated hydrocarbons (haloalkanes and/or haloalkenes). Several primary photodissociation processes have been considered for the carbonyl halides:

$$\mathrm{XC(O)X} + h\nu \rightarrow \mathrm{XCO} + \mathrm{X} \quad \text{(I)}$$

$$\rightarrow \mathrm{X} + \mathrm{CO} + \mathrm{X} \quad \text{(II)}$$

$$\rightarrow \mathrm{X_2} + \mathrm{CO} \quad \text{(III)}$$

(i) ClC(O)Cl, Phosgene. The studies of the photodecomposition of ClC(O)Cl by both theoretical (Francisco and Li, 1989) and experimental methods (Wijnen, 1961, 1962; Heicklen, 1965; Okabe, 1977; Maul et al., 1995; Jäger et al., 1996) suggest that process (I) is favored: ClC(O)Cl + $h\nu$ → ClCO + Cl (I). The quantum yield of CO formation at 253.7 nm was found to be unity by Wijnen (1961, 1962), and this was confirmed in experiments reported by Okabe (1977). Jäger et al. (1996) reported from their studies that the quantum yield of photodecomposition of ClC(O)Cl was also unity at 193 nm. Wijnen, Heicklen and others have used the photolysis of ClC(O)Cl as a convenient source of Cl-atoms. Presumably, the processes (I) and/or (II) are important, but the barrier for decomposition of the ClCO molecule (ClCO → CO + Cl) is very small (<27 kJ mol^{-1}; Francisco and Li, 1989), and rapid dissociation of ClCO occurs at 298 K (ClCO → CO + Cl). Thus, the two processes (I) and (II) result ultimately in the same products, 2Cl + CO. If process (III) occurs to any extent, it also will lead ultimately to the same products (CO and Cl-atoms), since Cl_2 undergoes rapid photolysis. The most direct measurement of ClC(O)Cl photolysis was made by Maul et al. (1995). They found from dye-laser photolysis near 238 nm that 15% of the chlorine atoms were formed in the Cl($^2P_{1/2}$) excited state with a mean kinetic energy of 38 kJ mol^{-1}, whereas the Cl($^2P_{3/2}$) ground-state Cl-atoms formed had a mean kinetic energy of 18 kJ mol^{-1}. Analysis of the kinetic energy spectra yielded evidence for a concerted three-body decay; Maul et al. concluded that the formation of the intermediate ClCO

is of minor importance in the dissociation process, at least for the conditions of these experiments.

(ii) BrC(O)Br. Evidence has been presented (Xu et al., 2002b) that the photodissociation of BrC(O)Br at 267 nm also occurs by several channels: (1) process (II), forming both fast and slow Br-atoms and CO via an asynchronously concerted three-body decay process for both ground- and spin-orbit excited bromine atoms; (2) process (I); and (3) process (III), forming molecular Br_2 and CO. No quantum yield studies of BrC(O)Br photolysis have been reported to our knowledge, but in view of the results described for ClC(O)Cl and the weak BrC(O)—Br bond in this molecule, one expects a total quantum yield of decomposition near unity. In view of the instability of the BrCO radical, processes (I) and (II) will both result in formation of two Br-atoms and a CO molecule.

VII-H-2.3. The Acyl Halides

Experimental (Person et al., 1992a,b; Deshmukh and Hess,1994) and theoretical studies (Sumathi and Chandra, 1993) of the photodecomposition of acetyl chloride have been reported. Experimental studies have been reported for bromoacetyl chloride, $BrCH_2C(O)Cl$ (Person et al., 1992a,b) and perfluoroacetyl chloride [$CF_3C(O)Cl$]. The following primary processes have been considered in explanation of the results:

$$CX_nH_{3-n}C(O)X + h\nu \rightarrow CX_nH_{3-n}CO + X \quad \text{(I)}$$

$$\rightarrow CX_nH_{3-n} + XCO \quad \text{(II)}$$

$$\rightarrow CX_nH_{3-n} + CO + X \quad \text{(III)}$$

$$\rightarrow CX_nH_{3-n}X + CO \quad \text{(IV)}$$

(i) Acetyl chloride ($CH_3C(O)Cl$). From results of acetyl chloride photolysis at 248.5 nm in a molecular beam experiment, Person et al. (1992a) reported that the anisotropic angular distributions of the observed Cl product of (I) show that dissociation occurs on a time-scale of less than a rotational period; no significant dissociation by process (II) was observed. The Cl fragment had an average translational energy of 67 kJ mol^{-1}. A possible elimination of HCl in a primary process could be excluded. Further theoretical considerations of the mechanism of the occurrence of process (I) in acetyl chloride photodissociation have been reported by Sumathi and Chandra (1993). The extent of process (IV) is unknown, but is expected to be small.

In their studies of $CH_3C(O)Cl$ photodecomposition at 236 nm, Deshmukh and Hess (1994) investigated the mechanism for preferential cleavage of the C—Cl bond in process (I) over that of C—C bond cleavage in process (II). They estimated that the quantum yield for CH_3 formation in (II) and/or (III) was 0.28 relative to that formed in CH_3I photolysis. Only a small fraction of the available energy is channeled into methyl fragment angular momentum as the rotational state distribution extends only to $N'' = 5$. The fraction of Cl-atoms formed in the $^2P_{1/2}$ state was 0.40 ± 0.02. They concluded that CH_3CO is generated as a primary product in the photodissociation of acetyl chloride with about 126 kJ mol^{-1} of internal energy, well above the ~72 kJ mol^{-1} necessary for dissociation of CH_3CO, and, for photolysis at 236 nm, it subsequently decomposes to CH_3 and CO.

(ii) Bromoacetyl chloride [BrCH$_2$C(O)Cl]. Excitation of bromoacetyl chloride at 248 nm can be attributed to an overlap of $n(O) \rightarrow \pi^*(C{=}O)$ excitation and $n(Br) \rightarrow \sigma^*(C{-}Br)$ transition. In this case both Cl- and Br-atoms were detected by Person et al. (1992b). They reported that fission of the C—Cl and C—Br bonds occurs with a branching ratio of 1.0:1.1.

(iii) Trifluoroacetyl chloride [CF$_3$C(O)Cl]. Studies of the photodecomposition of trifluoroacetyl chloride indicate the dominance of primary process (I) for excitation at wavelengths in the range 254–280 nm. Product studies of Weibel et al. (1995) from the photodecomposition of $CF_3C(O)Cl$ at 254 and 280 nm showed $\Phi_{CF3Cl} + 2\Phi_{C2F6} = 0.98 \pm 0.13$. CF_3H was formed in the photolysis of $CF_3C(O)Cl$ in the presence of added cyclohexane, indicating clearly that CF_3 radicals were present. Their results could not distinguish between the possible primary processes, but they suggest strongly that the total primary photodecomposition yield is near unity.

Maricq and Szente (1995) used time-resolved IR and UV spectroscopy to study the photodissociation of $CF_3C(O)Cl$ in photolysis at 193 and 248 nm. They concluded that the net reaction shown by process (III) described the results best; rapidly dissociating CF_3CO and/or COCl radicals presumably were present as intermediates. At 193 nm the CO is formed with extensive internal excitation ($T_{vib} = 3800 \pm 900$ K), whereas at 248 nm it is formed predominantly in $v = 0$. Quantum yields of the photodecomposition were measured by comparing yields of HCl (formed in experiments with added ethane) and RO_2 (experiments with added O_2) from $CF_3C(O)Cl$ to those from the photodisssociation of CH_3Cl at 193 nm and Cl_2 at 248 nm; they assumed the latter two species have unit photodissociation efficiencies. Values of the total quantum yield of photodecomposition $(\phi_I + \phi_{II} + \phi_{III}) = 1.10 \pm 0.11$ at 193 nm and 0.92 ± 0.08 at 248 nm. They conclude from their results that at 248 nm about 14% of the CF_3CO fragment [product of process (I)] remains intact under the conditions of the experiment.

Meller and Moortgat (1997) measured the quantum yields for the photodissociation of $CF_3C(O)Cl$ as a function of pressure and the quantum yields for Cl_2 and CO in experiments at 254 nm and 296 K. ClC(O)Cl was used as the actinometer. They report a mean value of photodissociation $= 0.95 \pm 0.05$ (760 Torr of N_2 or air or C_2H_6/air mixtures). At varied pressures of air (50 to 760 Torr) the quantum yield showed no obvious change, ranging from 1.00 to 1.06. In photolysis of $CF_3C(O)Cl$ in 760 Torr of air, they measured $\Phi_{Cl2} = 0.502 \pm 0.012$, indicating that all chlorine originating from $CF_3C(O)Cl$ is converted into molecular chlorine. They report a mean quantum yield for dissociation $= 0.96 \pm 0.04$, consistent again with $\phi_I + \phi_{II} + \phi_{III} =$ unity.

At the long wavelengths of absorbed light, there is little excess energy available in the quantum beyond that for dissociation, and one might expect process (I) to be the dominant mode of decay. However, haloacyl radicals ($CX_nH_{3-n}CO$) are unstable with respect to decomposition via CO elimination (Zabel et al., 1994; Tuazon and Atkinson, 1994; Solignac et al., 2006; Hurley et al., 2006; Waterland and Dobbs, 2006). While a small fraction of the $CX_nH_{3-n}CO$ radicals produced in the primary process at the long wavelengths may survive decomposition and react with O_2 in the atmosphere to give haloacylperoxy radicals, the majority will probably decompose, leading to the formation of haloalkylperoxy radicals.

VII-H-2.4. Photolysis frequencies of the acyl halides and halocarbonyls

To our knowledge, there have been no measurements of the quantum yields of photodissociation of acyl halides or carbonyl halides in the long-wavelength tail of the absorption that overlaps the distribution of actinic flux within the troposphere. In this region of absorption, there is little energy available in the quantum of light beyond that for dissociation. For example, the energy available in the quantum of light at the edge of the absorption of CF_3COCl, 328 nm (365 kJ mol^{-1}), is probably close to the $CF_3C(O)—Cl$ bond strength, assuming this bond strength is similar to that for $Cl_3C(O)—Cl$ (339 ± 24 kJ mol^{-1}; Zhu and Bozzelli, 2002). However, no barriers to dissociation of the $n \rightarrow \pi^*$ excited haloacetyl chlorides or carbonyl halides have been reported in theoretical or experimental studies, and dissociation with high efficiency observed at the shorter wavelengths may continue into the long-wavelength absorption region. With this assumption, we can estimate the photolysis frequencies for those acyl halides and carbonyl halides.

Photolysis frequencies for those compounds that show some overlap of the actinic flux with the long-wavelength absorption bands have been estimated using the cross section data of table VII-H-1 (298 K) and an assumed quantum yield of photodissociation of unity over the entire absorption region. These data are shown in figure VII-H-3. The weak absorption of the formyl and acyl halides at wavelengths available within the troposphere leads to much lower j-values than those of the unsubstituted aldehydes. In figure VII-H-3a it is seen that the j-values decrease significantly in the sequence $CH_2O > HC(O)Br > BrC(O)Br > HC(O)Cl$. The estimated tropospheric lifetimes for photodecomposition with an overhead Sun are: CH_2O, 3.1 hr; $HC(O)Br$, 16 hr; $BrC(O)Br$, 4.0 days; $HC(O)Cl$, 72 days; $ClC(O)Cl$ (not shown in figure VII-H-3a), ~16 years. For the haloacetyl chlorides shown in figure VII-H-3b, the j-values are seen to decrease from those of acetaldehyde in the order: $CH_3CHO > CCl_3C(O)Cl > CHCl_2C(O)Cl > CF_3C(O)Cl > CH_2ClC(O)Cl > CH_3C(O)Cl$. Approximate lifetimes for photodecomposition with an overhead Sun are: CH_3CHO, 1.6 days; $CCl_3C(O)Cl$, 4.4 days; $CHCl_2C(O)Cl$, 8.0 days; $CF_3C(O)Cl$, 23 days; $CH_2ClC(O)Cl$, 32 days; $CH_3C(O)Cl$, 8.9 years. The absorption cross sections for these compounds decrease with temperature decrease. As the altitude in the atmosphere increases, the absorption cross sections decrease as a result of the decreasing temperature, while the actinic flux increases. So the net effect of altitude change on the j-values depends on the sizes of the two opposite effects. For $CF_3C(O)Cl$, Meller and Moortgat (1997) have shown

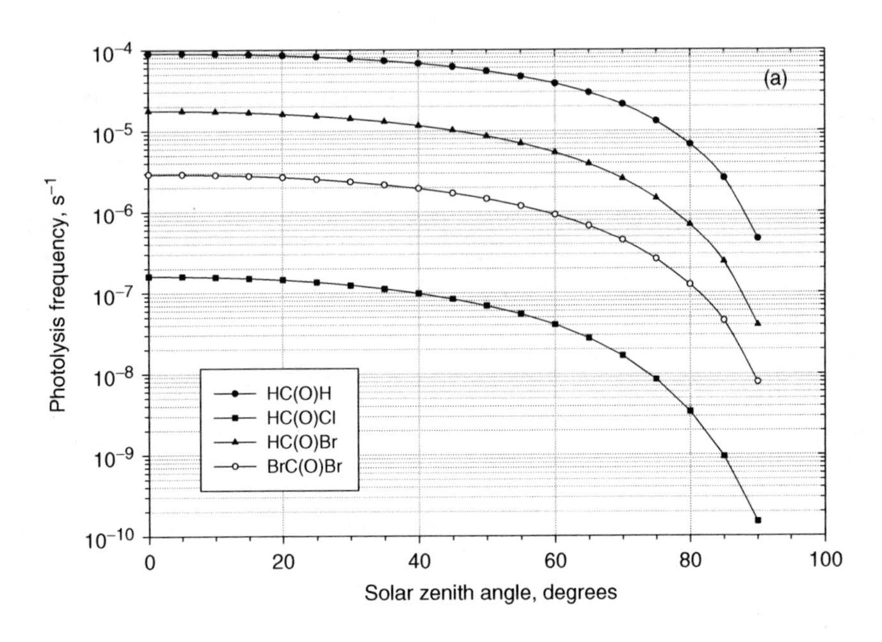

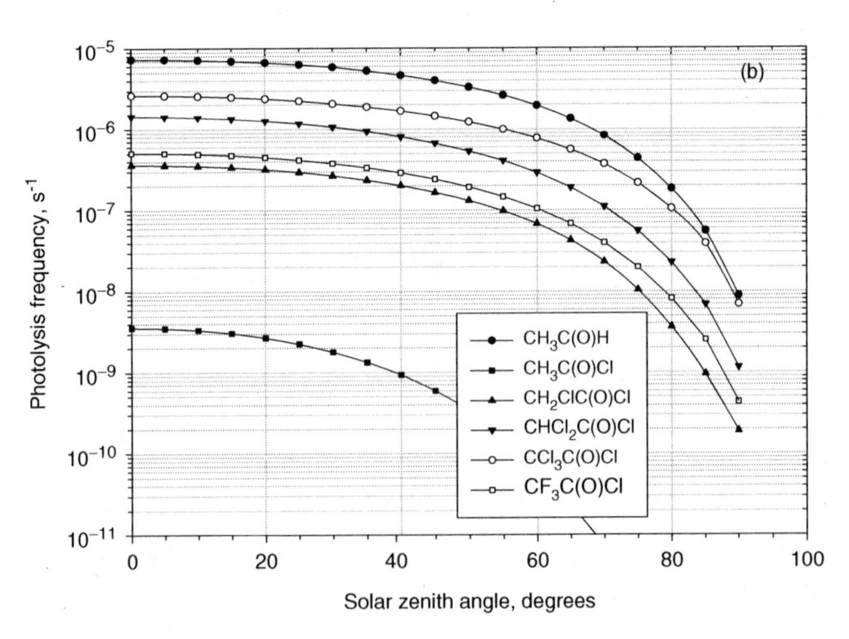

Figure VII-H-3. Comparison of the photolysis frequencies for photodecomposition by all modes vs. solar zenith angle for some acyl halides for which absorption of tropospheric actinic flux is measurable within their long-wavelength tails: (a) formyl halides and carbonyl halides compared to CH_2O; (b) haloacetyl chlorides compared to CH_3CHO. Calculations were made for a cloudless day within the lower troposphere with a vertical ozone column of 350 DU, cross section data of table VII-H-1, and an assumed quantum yield of photodecomposition of unity.

that the calculated j-values remain fairly constant within the troposphere up to 10 km, but rise rapidly at the higher altitudes where the effect of short-wavelength actinic flux increases dramatically.

The acyl halides and carbonyl halides have a relatively high reactivity toward hydrolysis in cloudwater and on moist aerosols and surfaces, and the removal of these species when formed within the troposphere is probably dominated by such reactions. DeBruyn et al. (1992) have shown that the uptake of ClC(O)Cl, FC(O)F, $CF_3C(O)F$, $CCl_3C(O)Cl$, and $CF_3C(O)Cl$ in water is rapid, and it is expected that their hydrolysis in cloudwater, rainwater, or ocean water will be a dominant fate for these species. Wallington and Hurley (1993) suggest that HC(O)F will likely face a similar fate. Wallington et al. (1994e) have estimated that the atmospheric lifetime of FC(O)F, FC(O)Cl, $CF_3C(O)F$, and $CF_3C(O)Cl$ with respect to uptake and hydrolysis in cloud water is 5–30 days. Certainly the heterogeneous removal of most of the acyl and carbonyl halides is competitive with their loss by photodecomposition.

VII-I. Photodecomposition of Alkyl Nitrates ($RONO_2$)

VII-I-1. Mechanism of Photodecomposition of Alkyl Nitrates

The alkyl nitrates, $RONO_2$ (where R is CH_3, C_2H_5, or other alkyl group) are common products of the oxidation of alkanes within the troposphere, particularly in NO_x-rich, polluted air. They show a structureless absorption, with the long-wavelength tail of the absorption overlapping somewhat the solar flux available within the troposphere; see figure VII-I-1 and tabular data in tables VII-I-1 and VII-I-2 (Roberts and Fajer, 1989). There is no significant difference to be seen between the three recent measurements of cross sections for the alkyl nitrates (Roberts and Fajer, 1989; Rattigan et al., 1992; Talukdar et al., 1997a). Note in figure VII-I-1 that the absorption cross sections of the alkyl nitrates increase regularly as the alkyl group (R) grows in size. Methyl nitrate is the weakest absorber while $(CH_3)_3CONO_2$ is the strongest among those shown.

The favored pathway for the photolysis of the alkyl nitrates in the 280–350 nm region leads to the formation of an alkoxy radical and NO_2:

$$RONO_2 + h\nu \rightarrow RO + NO_2 \qquad \text{(I)}$$

Some other photodissociation pathways have been suggested in which the alkyl nitrates ($C_nH_{2n+1}ONO_2$) form either: (1) HONO and an aldehyde or a ketone ($C_nH_{2n}O$); or (2) an alkyl nitrite (RONO) and an O-atom. In photolysis at 193 nm, Yang et al. (1993) found that ~70% of the photodecomposition involves fission of the weak $CH_3O—NO_2$ bond, and the fragment pairs showed a bimodal translational energy distribution. The fast fragment pairs had an average translational energy of 171 kJ mol^{-1}, while the slow component had 79 kJ mol^{-1}. A significant fraction of the NO_2 fragments of the slow component was subject to unimolecular decay to NO and $O(^3P)$. Yang et al. observed an additional primary decay route that led to the formation of methyl nitrite and $O(^1D)$. Photodissociation of methyl nitrate at the longer wavelength of 248 nm follows process (I) exclusively with a translational energy of 71 kJ mol^{-1}(Yang et al., 1993).

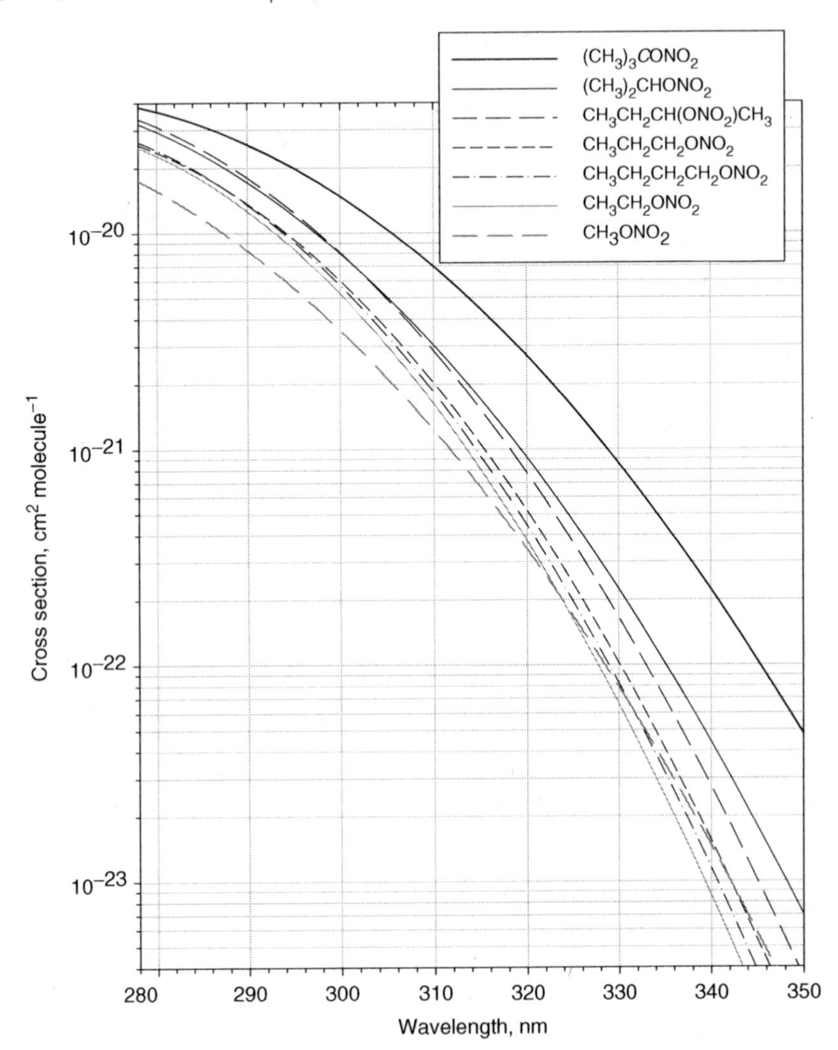

Figure VII-I-1. Cross sections (298 K) for the simple alkyl nitrates as represented by the equations for least-squares fits of the cross section vs. wavelength data of Roberts and Fajer (1989).

Talukdar et al. (1997a) also studied CH_3ONO_2 photolysis at 248 nm and confirmed that the alternative pathways to reaction (I) are relatively unimportant. They found $\phi_I = 0.9 \pm 0.2$(O_2 present as a radical scavenger) and $\phi_I = 1.3 \pm 0.3$ (NO present as the radical scavenger). Direct measurements of NO_2 product formation from $RONO_2$ in sunlight (R = C_2H_5, $CH_3CH_2CH_2$, $CH_3CH_2CH_2CH_2$, and $CH_3CHCH_2CH_3$) were made by Luke and Dickerson (1988) and Luke et al. (1989). Using the well-accepted cross sections of Roberts and Fajer (1989), the calculated photolysis rates were consistent with $\phi_I = 1.0 \pm 0.5$. Zhu and Ding (1997) studied $C_2H_5ONO_2$ photolysis at 308 nm and found $\phi_I = 1.0 \pm 0.1$. The limited available evidence favors a value of unity for process (I) for wavelengths of sunlight available within the troposphere

Table VII-I-1. Cross sections (σ, cm^2 molecule^{-1}) for some simple alkyl nitrates (298–301 K): CH_3ONO_2, $C_2H_5ONO_2$, $CH_3CH_2CH_2ONO_2$, $CH_3CH(ONO_2)CH_3$, and $CH_3CH_2CH_2CH_2ONO_2$ (data from Roberts and Fajer, 1989)

Cross Section Data ($10^{20} \times \sigma$) for $RONO_2$ Where the Appropriate R Heads the Column:

λ (nm)	CH_3	C_2H_5	1-C_3H_7	2-C_3H_7	1-C_4H_9	λ (nm)	CH_3	C_2H_5	1-C_3H_7	2-C_3H_7	1-C_4H_9
278	1.75	2.49	2.56	3.20	2.64	315	0.067	0.081	0.106	0.172	0.094
279	1.66	2.39	2.47	3.08	2.53	316	0.059	0.070	0.093	0.152	0.081
280	1.58	2.29	2.37	2.95	2.43	317	0.052	0.060	0.081	0.135	0.070
281	1.49	2.18	2.27	2.82	2.32	318	0.045	0.052	0.070	0.119	0.061
282	1.41	2.08	2.17	2.69	2.21	319	0.040	0.044	0.060	0.105	0.052
283	1.33	1.97	2.07	2.56	2.10	320	0.035	0.038	0.052	0.092	0.045
284	1.25	1.87	1.96	2.44	1.99	321	0.030	0.032	0.045	0.081	0.038
285	1.17	1.76	1.86	2.31	1.88	322	0.026	0.027	0.038	0.071	0.033
286	1.10	1.66	1.76	2.19	1.77	323	0.023	0.023	0.033	0.062	0.028
287	1.03	1.56	1.66	2.07	1.66	324	0.020	0.019	0.028	0.054	0.023
288	0.958	1.46	1.56	1.95	1.56	325	0.017	0.016	0.024	0.047	0.020
289	0.892	1.36	1.46	1.83	1.45	326	0.015	0.014	0.020	0.040	0.017
290	0.829	1.26	1.37	1.72	1.35	327	0.013	0.011	0.017	0.035	0.014
291	0.768	1.17	1.27	1.61	1.26	328	0.011	0.010	0.015	0.030	0.012
292	0.711	1.08	1.18	1.50	1.17	329	0.009	0.008	0.012	0.026	0.010
293	0.656	1.00	1.10	1.40	1.08	330	0.008	0.007	0.010	0.022	0.008
294	0.605	0.919	1.02	1.30	0.991	331	0.007	0.005	0.009	0.019	0.007
295	0.556	0.843	0.937	1.21	0.911	332	0.006	0.004	0.007	0.016	0.006
296	0.510	0.770	0.862	1.12	0.835	333	0.005	0.004	0.006	0.014	0.005
297	0.467	0.702	0.791	1.03	0.763	334	0.004	0.003	0.005	0.012	0.004
298	0.427	0.638	0.724	0.952	0.695	335	0.003	0.002	0.004	0.010	0.003
299	0.389	0.579	0.661	0.876	0.631	336	0.003	0.002	0.003	0.009	0.003
300	0.354	0.523	0.602	0.804	0.572	337	0.002	0.002	0.003	0.007	0.002
301	0.321	0.471	0.546	0.736	0.517	338	0.002	0.001	0.002	0.006	0.002
302	0.291	0.423	0.494	0.673	0.466	339	0.002	0.001	0.002	0.005	0.001
303	0.263	0.379	0.446	0.614	0.419	340	0.001	0.001	0.002	0.004	0.001
304	0.238	0.339	0.402	0.559	0.375	341	0.001	0.001	0.001	0.004	0.001
305	0.214	0.302	0.361	0.507	0.335	342	0.001	0.001	0.001	0.003	0.001
306	0.192	0.268	0.323	0.460	0.299	343	0.001	0	0.001	0.003	0.001
307	0.172	0.238	0.289	0.415	0.266	344	0.001	0	0.001	0.002	0
308	0.154	0.210	0.257	0.375	0.235	345	0.001	0	0.001	0.002	0
309	0.138	0.185	0.229	0.337	0.208	346	0	0	0	0.001	0
310	0.123	0.162	0.203	0.303	0.183	347	0	0	0	0.001	0
311	0.109	0.142	0.179	0.272	0.161	348	0	0	0	0.001	0
312	0.097	0.124	0.158	0.243	0.141	349	0	0	0	0.001	0
313	0.086	0.108	0.139	0.217	0.124	350	0	0	0	0.001	0
314	0.076	0.094	0.122	0.193	0.108						

(e.g., see Roberts, 1990; Roberts and Fajer, 1989; Atkinson et al., 2004). The structureless continuum shown in the absorption of the alkyl nitrates is consistent with this view, and we recommend that the value of $\phi_I = 1.0$ be assumed for all the alkyl nitrates for the wavelength region of sunlight available within the troposphere.

VII-I-2. Absorption Cross Sections for Simple Alkyl Nitrates

The cross section data of Roberts and Fajer (1989) for a temperature of 298 K for some simple alkyl nitrates are given in figure VII-I-1, and in tabular form in tables VII-I-1 [for CH_3ONO_2, $C_2H_5ONO_2$, $CH_3CH_2CH_2ONO_2$, $CH_3CH(ONO_2)CH_3$, and

Table VII-I-2. Cross sections (σ, cm^2 $molecule^{-1}$) for some alkyl nitrates (298–301 K): $2\text{-}C_4H_9ONO_2$, $(CH_3)_3CONO_2$, $2\text{-}C_5H_{11}ONO_2$, $3\text{-}C_5H_{11}ONO_2$, *cyclo*-$C_5H_9ONO_2$ (data from Roberts and Fajer, 1989)

Cross Section Data ($10^{20} \times \sigma$) for $RONO_2$ Where the Appropriate R Heads the Column:											
λ (nm)	$2\text{-}C_4H_9$	$t\text{-}C_4H_9$	$2\text{-}C_5H_{11}$	$3\text{-}C_5H_{11}$	$c\text{-}C_5H_9$	λ (nm)	$2\text{-}C_4H_9$	$t\text{-}C_4H_9$	$2\text{-}C_5H_{11}$	$3\text{-}C_5H_{11}$	$c\text{-}C_5H_9$
278	3.37	3.83	3.44	3.64	2.66	314	0.175	0.490	0.171	0.172	0.062
279	3.24	3.75	3.31	3.52	2.56	315	0.154	0.446	0.151	0.150	0.052
280	3.11	3.66	3.17	3.39	2.46	316	0.135	0.405	0.132	0.130	0.044
281	2.98	3.56	3.03	3.26	2.35	317	0.118	0.367	0.116	0.112	0.036
282	2.85	3.46	2.89	3.12	2.23	318	0.103	0.332	0.101	0.097	0.030
283	2.71	3.36	2.75	2.98	2.11	319	0.090	0.300	0.088	0.084	0.025
284	2.58	3.25	2.61	2.84	2.00	320	0.078	0.270	0.076	0.072	0.021
285	2.44	3.14	2.47	2.70	1.88	321	0.067	0.243	0.066	0.061	0.017
286	2.31	3.03	2.33	2.56	1.76	322	0.058	0.218	0.057	0.052	0.014
287	2.18	2.92	2.19	2.41	1.64	323	0.050	0.196	0.049	0.045	0.011
288	2.05	2.80	2.06	2.27	1.53	324	0.043	0.175	0.042	0.038	0.009
289	1.92	2.69	1.93	2.13	1.42	325	0.037	0.156	0.036	0.032	0.007
290	1.80	2.57	1.81	2.00	1.31	326	0.032	0.139	0.031	0.027	0.006
291	1.68	2.45	1.68	1.86	1.20	327	0.027	0.124	0.026	0.023	0.005
292	1.57	2.34	1.57	1.73	1.10	328	0.023	0.110	0.023	0.019	0.004
293	1.46	2.22	1.45	1.61	1.01	329	0.020	0.097	0.019	0.016	0.003
294	1.35	2.11	1.35	1.49	0.915	330	0.017	0.086	0.016	0.013	0.002
295	1.25	2.00	1.24	1.37	0.829	331	0.014	0.076	0.014	0.011	0.002
296	1.15	1.89	1.14	1.26	0.748	332	0.012	0.067	0.012	0.009	0.002
297	1.06	1.78	1.05	1.16	0.672	333	0.010	0.059	0.010	0.008	0.001
298	0.972	1.68	0.964	1.06	0.602	334	0.008	0.051	0.008	0.006	0.001
299	0.889	1.58	0.881	0.967	0.538	335	0.007	0.045	0.007	0.005	0.001
300	0.812	1.48	0.804	0.880	0.478	336	0.006	0.039	0.006	0.004	0.001
301	0.740	1.39	0.731	0.798	0.423	337	0.005	0.034	0.005	0.003	
302	0.672	1.29	0.664	0.721	0.374	338	0.004	0.030	0.004	0.003	
303	0.609	1.21	0.601	0.650	0.328	339	0.003	0.026	0.003	0.002	
304	0.550	1.12	0.543	0.584	0.288	340	0.003	0.022	0.003	0.002	
305	0.496	1.04	0.489	0.524	0.251	341	0.002	0.019	0.002	0.001	
306	0.446	0.966	0.440	0.468	0.218	342	0.002	0.017	0.002	0.001	
307	0.401	0.893	0.394	0.417	0.189	343	0.002	0.014	0.002	0.001	
308	0.358	0.825	0.353	0.371	0.163	344	0.001	0.012	0.001	0.001	
309	0.320	0.760	0.314	0.329	0.140	345	0.001	0.011	0.001	0.001	
310	0.285	0.699	0.280	0.290	0.120	346	0.001	0.009	0.001		
311	0.253	0.641	0.248	0.256	0.102	347	0.001	0.008	0.001		
312	0.224	0.587	0.220	0.225	0.087	348	0.001	0.007	0.001		
313	0.198	0.537	0.194	0.197	0.074	349		0.006			

$CH_3CH_2CH_2CH_2ONO_2$] and VII-I-2 [for $CH_3CH(ONO_2)CH_2CH_3$, $(CH_3)_3CONO_2$, $2\text{-}C_5H_{11}ONO_2$, $3\text{-}C_5H_{11}ONO_2$, and *cyclo*-$C_5H_9ONO_2$]. These data are in good agreement with other recent σ measurements of the alkyl nitrates (Rattigan et al., 1992; Talukdar et al.,1997a) and are recommended for use in modeling tropospheric chemistry.

VII-I-3. Photolysis Frequencies for Simple Alkyl Nitrates

We have used the cross section data of Roberts and Fajer (1989) given in tables VII-I-1 and VII-I-2, an assumed value of unity for ϕ_1, and appropriate solar flux data for clear-sky conditions to calculate the j-values for the simple alkyl nitrates (0.5 km elevation,

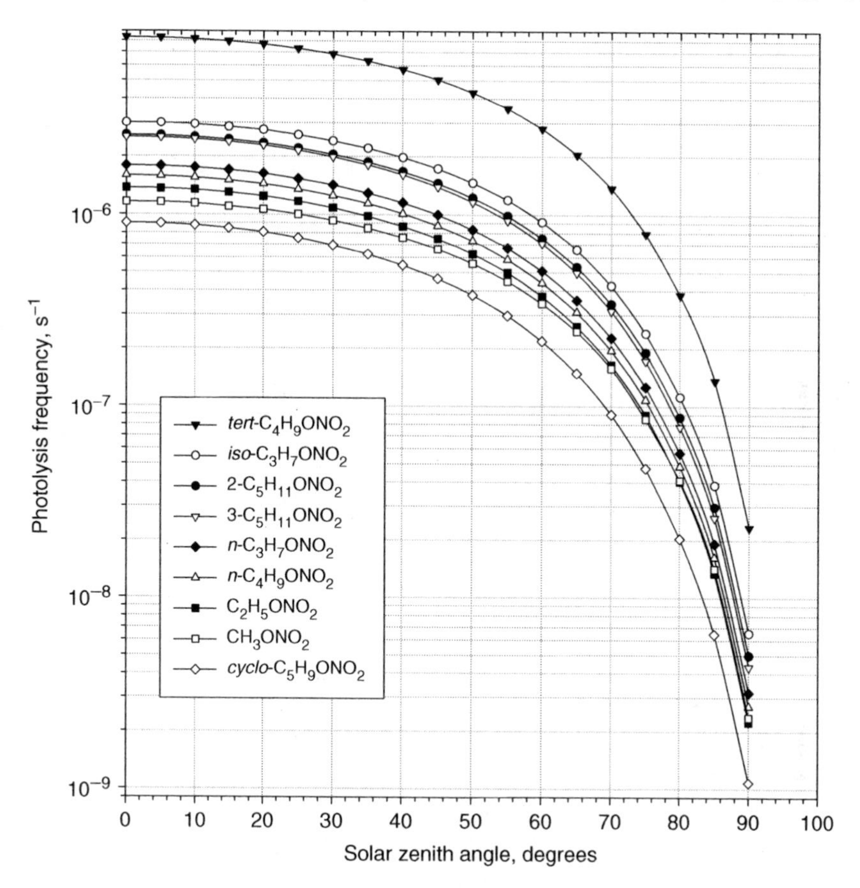

Figure VII-I-2. Comparison of photolysis frequencies for several organic nitrates vs. solar zenith angle, calculated for a cloud-free lower troposphere with an overhead ozone column = 350 DU.

298 K, overhead ozone column = 350 DU) as a function of solar zenith angle; see figure VII-I-2. The photolysis frequencies for CH_3ONO_2, as calculated for a cloudless day in the lower troposphere at several different overhead ozone columns (300–400 Dobson units), are shown in figure VII-I-3a. The sensitivity of the j-values to overhead ozone column is seen best in figure VII-I-3b. It is similar to that observed for CH_3I (figure VII-E-7b), which absorbs light in about the same region. About a 20% decrease in the j-value results from an increase in overhead ozone of 100 Dobson units.

The j-values measured directly can be compared with those calculated in this study (values in parentheses). CH_3ONO_2: surface, solar zenith angle (Z) = 45° (Taylor et al., 1980), $j = 2.2 \times 10^{-6}$ s^{-1}(7.4×10^{-7}). For $C_2H_5ONO_2$: surface, $Z = 0°$ (Luke and Dickerson, 1988), $j = 1.1 \times 10^{-6}$ s^{-1}(1.4×10^{-6}); surface, $Z = 60°$ (Luke and Dickerson, 1988), $j = 0.45 \times 10^{-6}$ s^{-1} (0.38×10^{-6}); surface, O_3 column of 296 Dobson units, temperature 298 K, $Z = 0°$ (Luke et al., 1989), $j = 1.45 \times 10^{-6}$ s^{-1} (1.37×10^{-6}). For $CH_3CH_2CH_2ONO_2$: same conditions, $Z = 0°$ (Luke et al., 1989), $j = 1.33 \times 10^{-6}$ s^{-1} (1.79×10^{-6}). For $CH_3CH(ONO_2)CH_2CH_3$: same

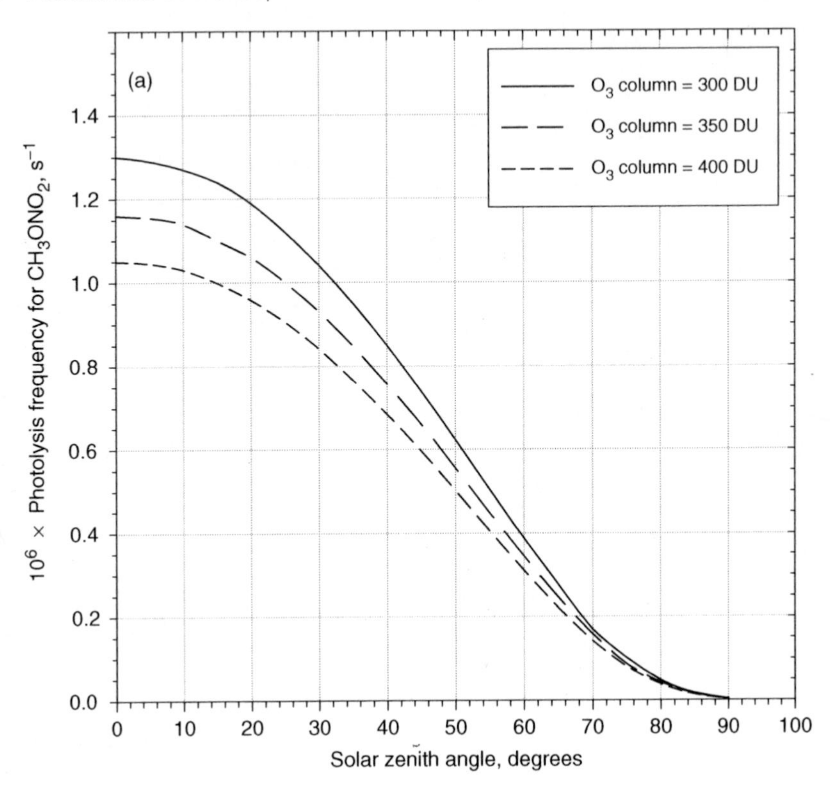

Figure VII-I-3a. Calculated photolysis frequencies for CH_3ONO_2 as a function of solar zenith angle and overhead ozone column. The cross sections of Roberts and Fajer (1989) and assumed quantum yields of photodissociation of unity at all absorbed wavelengths of sunlight were used in the calculation.

conditions, $Z = 0°$ (Luke et al., 1989), $j = 2.54 \times 10^{-6}$ s^{-1} (2.60×10^{-6}). The degree of agreement between the measured j-values and those calculated is reasonable considering the differences in the actual and assumed conditions that must exist (extent of clouds, altitude, and ozone columns), and it is consistent with the assumption of a quantum yield of photodissociation of unity for these alkyl nitrates.

There is significant temperature dependence to the cross sections of CH_3ONO_2 that can be seen in figure VII-I-3c. Of course, this can affect the lifetime of the methyl nitrate present in the various regions of the atmosphere. To illustrate the magnitude of the effects that both changes in temperature (cross section) and actinic flux have in this case, we have calculated j-values from 0 to 20 km altitude. The cross sections for CH_3ONO_2 have been reported only for $\lambda > 240$ nm, so valid j-calculations for altitudes above 20 km could not be made. As one can see from figure VII-I-3d, the changes affect significantly the j-values that are applicable to the conditions present in the upper troposphere.

The spectral j-values for CH_3ONO_2 are compared with those of CH_3I in figure VII-I-4.

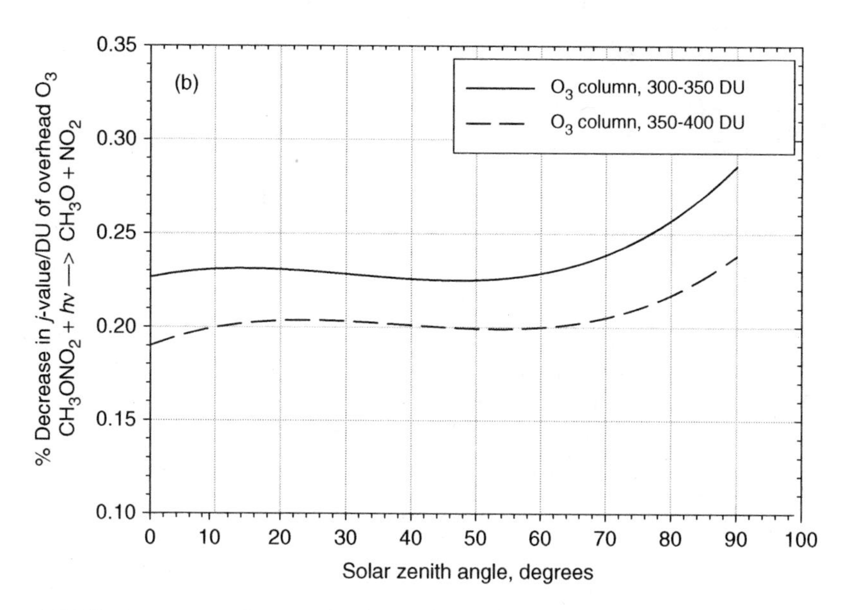

Figure VII-I-3b. Percentage decrease in the j-values of CH_3ONO_2 expected per Dobson unit increase in the overhead ozone column. Data were calculated for a cloudless day within the lower troposphere.

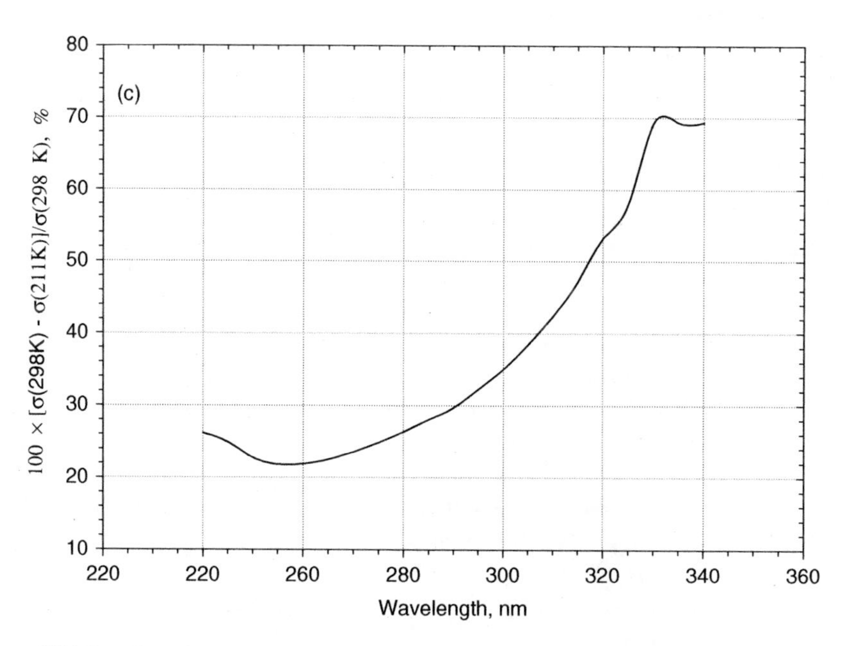

Figure VII-I-3c. Plot of the percentage difference in the absorption cross sections of CH_3ONO_2 at 298 and 211 K as a function of wavelength; from data of Talukdar et al. (1997a).

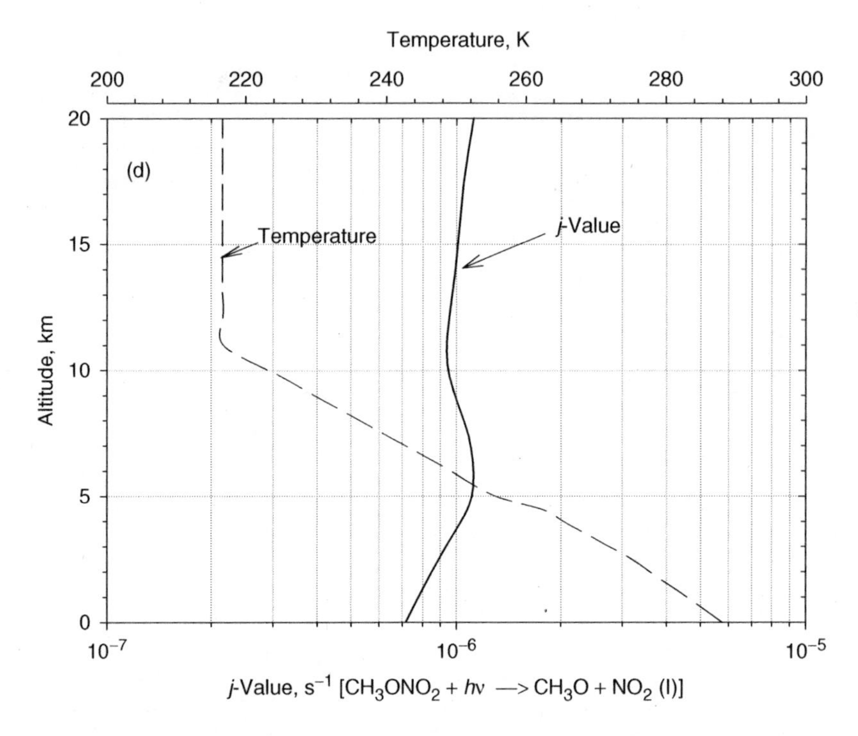

Figure VII-I-3d. Plot of the estimated photolysis frequency for CH_3ONO_2 as a function of altitude for a cloudless day with solar zenith angle = 45° and vertical ozone column = 350 DU. Shown also is the temperature variation with altitude that was employed in the calculations (U.S. Standard Atmosphere, 1976).

VII-J. Photodecomposition of Peroxyacyl Nitrates [$RC(O)OONO_2$]

The compounds with the chemical formula $RC(O)OONO_2$ are commonly named "peroxyacyl nitrates", a name that is not chemically correct and is not favored by many chemists. Roberts (1990) points out that, following the IUPAC nomenclature rules, they should be named as anhydrides of the acid components from which they are formed; thus $CH_3C(O)OONO_2$ would be named ethaneperoxoic nitric anhydride or peroxyacetic nitric anyhydride. Some chemists prefer the name 1-nitroperoxyethanal. In view of its common usage in the extensive literature, here and elsewhere in this book we use the common name peroxyacetyl nitrate and the abbreviation "PAN" for the compound $CH_3C(O)OONO_2$, and peroxypropionyl nitrate and the abbreviation "PPN" for the compound $CH_3CH_2COO_2NO_2$.

The peroxyacyl nitrates [$RC(O)OONO_2$] are common products of the atmospheric oxidation of alkanes and are important for temporary storage and transport of NO_x, particularly within the cooler regions of the troposphere. The cross sections for the structurally simplest first members of this family are shown in figure VII-J-1: peroxyacetyl nitrate [$CH_3C(O)OONO_2$] and peroxypropionyl nitrate [$C_2H_5C(O)OONO_2$)].

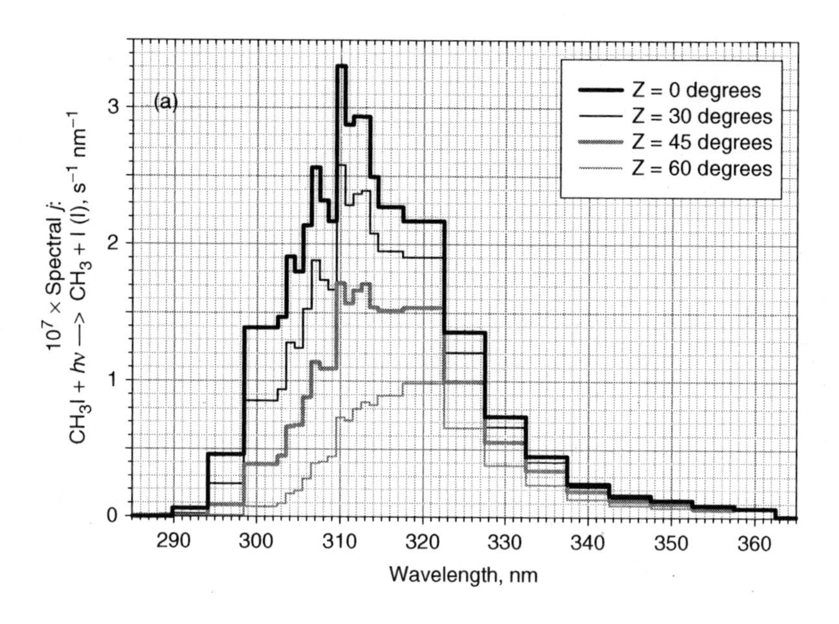

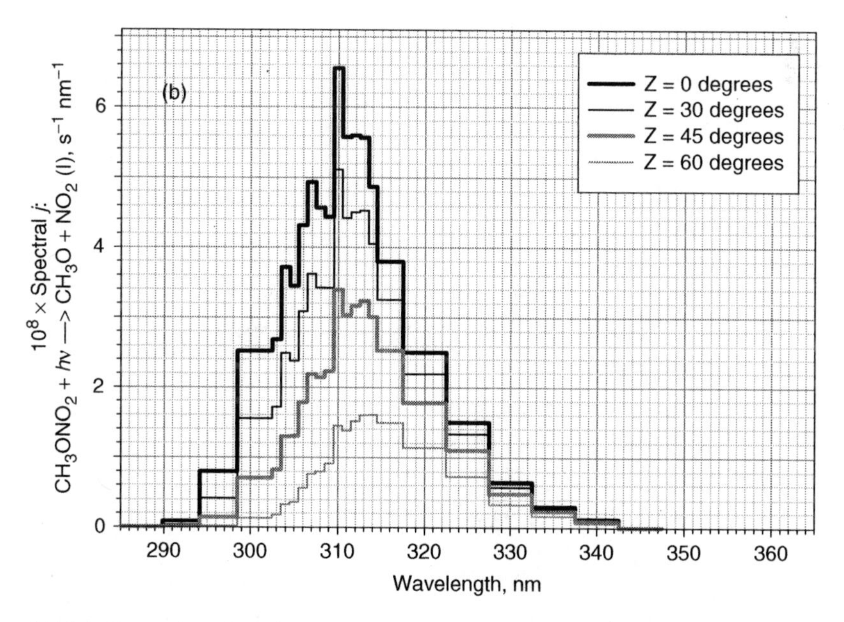

Figure VII-I-4. Spectral j-values for CH_3I (a) and CH_3ONO_2 (b) on a cloudless day in the lower troposphere for overhead ozone column = 350 DU and temperature 298 K. The contribution of the tail in the long-wavelength absorption of CH_3I decreases as the temperature is lowered below 298 K.

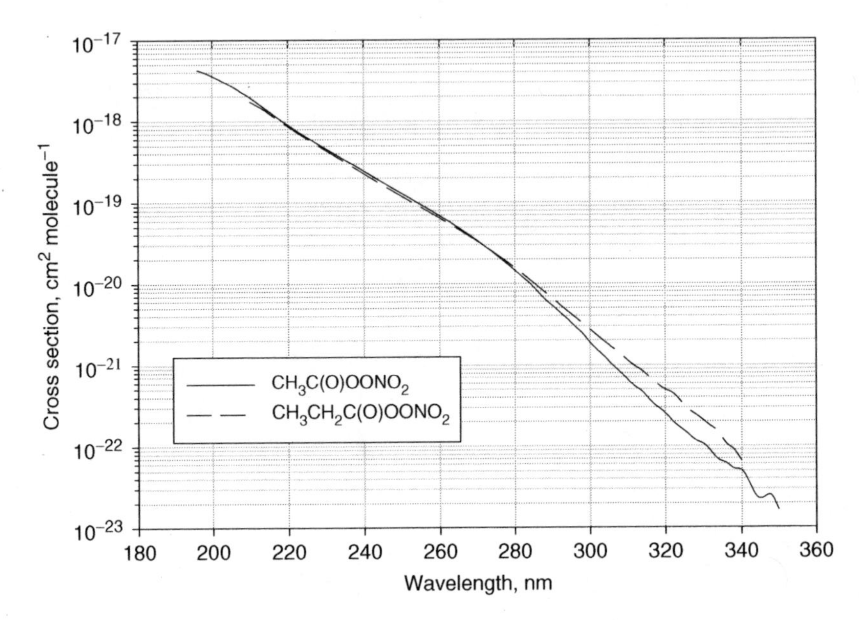

Figure VII-J-1. Absorption cross sections as a function of wavelength: peroxyacetyl nitrate (PAN) at 298 K (Talukdar et al., 1995) and peroxypropionyl nitrate (PPN) at 296 K (Harwood et al., 2003).

The tabular listings of cross sections for these two peroxyacyl nitrates are given in tables VII-J-1 and VII-J-2, respectively.

Peroxyacyl nitrates are highly reactive, unstable at elevated temperatures, and explosions have often been reported in laboratory experiments with these compounds. As a consequence, peroxyacyl nitrates of questionable purity may have been used in many of the early studies of their photochemistry that have proven to be inaccurate. These compounds are not strong absorbers of the wavelengths of sunlight present in the troposphere, but molecules of PAN or PPN that do absorb are efficiently dissociated.

VII-J-1. Peroxyacetyl Nitrate [1-Nitroperoxoethanal, PAN, $CH_3C(O)OONO_2$]

Two pathways of photodecomposition are recognized for $CH_3C(O)OONO_2$:

$$CH_3C(O)OONO_2 + h\nu \rightarrow CH_3C(O)O_2 + NO_2 \quad \text{(I)}$$

$$\rightarrow CH_3CO_2 + NO_3 \quad \text{(II)}$$

Estimates of the enthalpy changes (298 K) involved in the overall chemistry for processes (I) and (II), respectively, are: $\Delta H_I = 118.8 \pm 4.0\,\text{kJ mol}^{-1}$ (1007 nm) and $\Delta H_{II} = 121.2 \pm 8.4\,\text{kJ mol}^{-1}$ (987 nm) (Miller et al., 1999). Thus the quantum energies available even at the long-wavelength limit of the PAN absorption, 360 nm

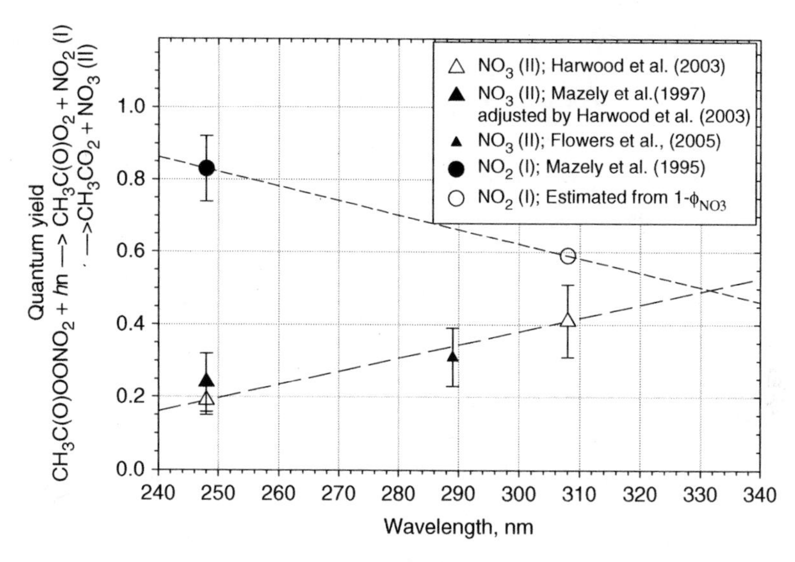

Figure VII-J-2. Estimated quantum yields of the primary processes in $CH_3C(O)OONO_2$ as a function of wavelength. The data of Harwood et al. for NO_3 quantum yields were used to determine the long-dashed line that represents the approximate wavelength dependence of process (II). The open circle shown for ϕ_I at 308 nm was derived assuming $\phi_I = 1 - \phi_{II}$ at 308 nm.

(332 kJ mol^{-1}), are much greater than these dissociation energies. The quantum yields of these two processes have been measured at two wavelengths (248 and 308 nm) by direct observations of the NO_3 and NO_2 products. These data are plotted in figure VII-J-2. Measurements $NO_3(\phi_{II})$ by Harwood et al. (2003) are shown by the open triangles. The $\phi_{II} = 0.24$ data point at 248 nm (black triangle) has been corrected using the currently accepted quantum yield for NO_3 from the reference compound (N_2O_5) used by Mazely et al. (1997). Flowers et al. (2005) measured $\phi_{II} = 0.31 \pm 0.08$ at 289 nm, consistent with the wavelength dependence suggested by measurements of Harwood et al. (2003). The sum of ϕ_I and ϕ_{II} at wavelength 248 nm is near unity (1.02 $\pm$0.10) and, in view of the thermochemistry involved in the bond breaking and the continuous nature of the absorption spectra, it is reasonable to assume that $\phi_I + \phi_{II} \approx 1.0$ for other wavelengths of light absorbed by the peroxyacyl nitrates. The NO_2 quantum yields (ϕ_I) at wavelength 248 nm, as determined by Mazely et al. (1995), represent the only measurements of this quantity. The open circle shown for ϕ_{II} at wavelength 308 nm was derived by assuming $\phi_{II} = 1 - \phi_I$ at 308 nm.

In the calculation of the approximate photolysis frequencies for $CH_3C(O)OONO_2$, the quantum yields for processes (I) and (II) as a function of wavelength were estimated from the trend lines in figure VII-J-2, assuming a simple linear dependence. Approximate photolysis frequencies as a function of solar zenith angle were estimated using these data (table VII-J-1) combined with the measured cross sections (table VII-J-1; Talukdar et al., 1995) and the appropriate actinic flux for a clear day within the lower

Table VII-J-1. Cross sections (σ, cm^2 $molecule^{-1}$; Talukdar et al., 2003) and approximate quantum yields as a function of wavelength for $CH_3C(O)OONO_2$ at 298 K; quantum yield estimates are based on an extrapolation of the results of Mazely et al. (1997) and Harwood et al. (2003)

λ (nm)	$10^{20} \times \sigma$	ϕ_I [a]	ϕ_{II} [b]	λ (nm)	$10^{20} \times \sigma$	ϕ_I [a]	ϕ_{II} [b]	λ (nm)	$10^{20} \times \sigma$	ϕ_I [a]	ϕ_{II} [b]
278	1.74	0.71	0.29	302	0.152	0.615	0.385	326	0.0140	0.518	0.482
279	1.60	0.705	0.295	303	0.139	0.610	0.390	327	0.0129	0.513	0.487
280	1.46	0.701	0.299	304	0.125	0.607	0.393	328	0.0117	0.510	0.490
281	1.34	0.698	0.302	305	0.112	0.602	0.398	329	0.0112	0.506	0.494
282	1.21	0.695	0.305	306	0.0998	0.598	0.402	330	0.0106	0.502	0.498
283	1.11	0.690	0.310	307	0.0907	0.593	0.407	331	0.00959	0.497	0.503
284	1.01	0.686	0.314	308	0.0816	0.590	0.410	332	0.00857	0.495	0.505
285	0.910	0.682	0.318	309	0.0741	0.586	0.414	333	0.00767	0.490	0.510
286	0.810	0.678	0.322	310	0.0666	0.582	0.418	334	0.00676	0.486	0.514
287	0.729	0.673	0.327	311	0.0602	0.578	0.422	335	0.00646	0.483	0.517
288	0.648	0.670	0.330	312	0.0538	0.575	0.425	336	0.00615	0.478	0.522
289	0.593	0.666	0.334	313	0.0500	0.570	0.430	337	0.00571	0.474	0.526
290	0.537	0.662	0.338	314	0.0462	0.566	0.434	338	0.00526	0.470	0.530
291	0.492	0.658	0.342	315	0.0413	0.562	0.438	339	0.00514	0.466	0.534
292	0.447	0.655	0.345	316	0.0363	0.558	0.442	340	0.00502	0.463	0.537
293	0.408	0.651	0.349	317	0.0332	0.553	0.447	341	0.00431	0.458	0.542
294	0.369	0.646	0.354	318	0.0300	0.550	0.450	342	0.00360	0.455	0.545
295	0.333	0.642	0.358	319	0.0276	0.547	0.453	343	0.00301	0.450	0.550
296	0.297	0.638	0.362	320	0.0252	0.542	0.458	344	0.00241	0.446	0.554
297	0.271	0.635	0.365	321	0.0226	0.538	0.462	345	0.00236	0.442	0.558
298	0.245	0.631	0.369	322	0.0199	0.535	0.465	346	0.00231	0.437	0.563
299	0.217	0.626	0.374	323	0.0183	0.530	0.470	347	0.00239	0.435	0.565
300	0.189	0.622	0.378	324	0.0166	0.526	0.474	348	0.00247	0.432	0.568
301	0.171	0.618	0.382	325	0.0153	0.521	0.479	349	0.00206	0.428	0.572
								350	0.00165	0.425	0.575

[a] $CH_3C(O)OONO_2 + h\nu \rightarrow CH_3C(O)O_2 + NO_2$ (I)
[b] $CH_3C(O)OONO_2 + h\nu \rightarrow CH_3CO_2 + NO_3$ (II)

troposphere with an overhead ozone column = 350 DU. The results are plotted in figure VII-J-3a.

There is a significant temperature dependence to the absorption cross sections of PAN that is evident in figure VII-J-3b. The temperature-dependent cross sections and the varied actinic flux alter the j-values for PAN that exist in the upper atmosphere. We have calculated the j-values as a function of altitude, using the temperature-dependent cross sections for PAN measured by Talukdar et al. (1995) and the appropriate actinic flux representative of the upper atmosphere. These calculations are summarized in figure VII-J-3c. Here the solar zenith angle has been fixed at 45° and clear sky conditions have been assumed, with an overhead O_3 column of 350 DU at ground level. Note that a marked increase in the j-values is estimated at the higher altitudes. This results largely from the increase in solar flux that occurs at altitudes near and above the ozone region of the stratosphere. This is seen clearly in figure VII-J-3d, where the spectral j-values at various wavelength regions can be compared for several altitudes. The cutoff of actinic flux for $\lambda < 280$ nm that results from the ozone layer near 20 km restricts the active wavelength regions at the lower altitudes. The wavelength dependence of the absorption cross sections of ozone determines the wavelength range

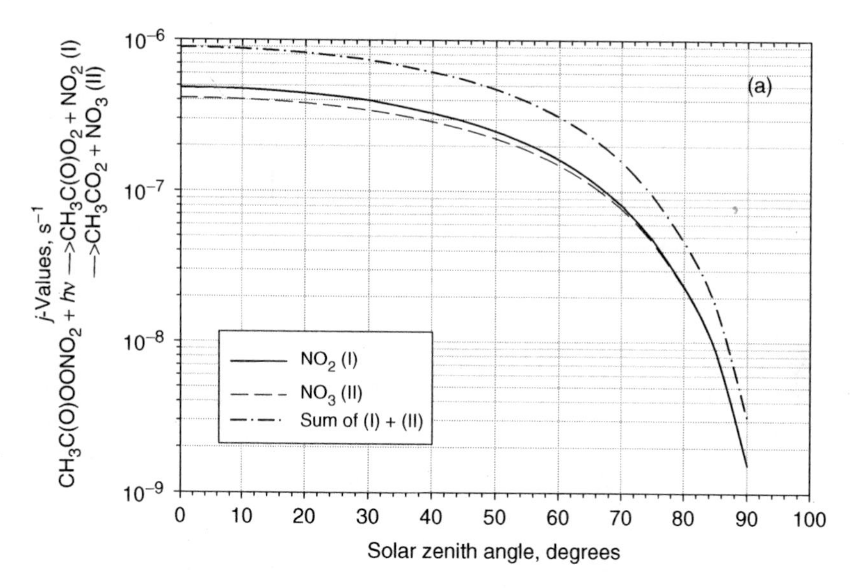

Figure VII-J-3a. Plot of the approximate photolysis frequencies for peroxyacetyl nitrate photodecomposition as a function of solar zenith angle for a cloudless day within the troposphere with an overhead ozone column = 350 DU.

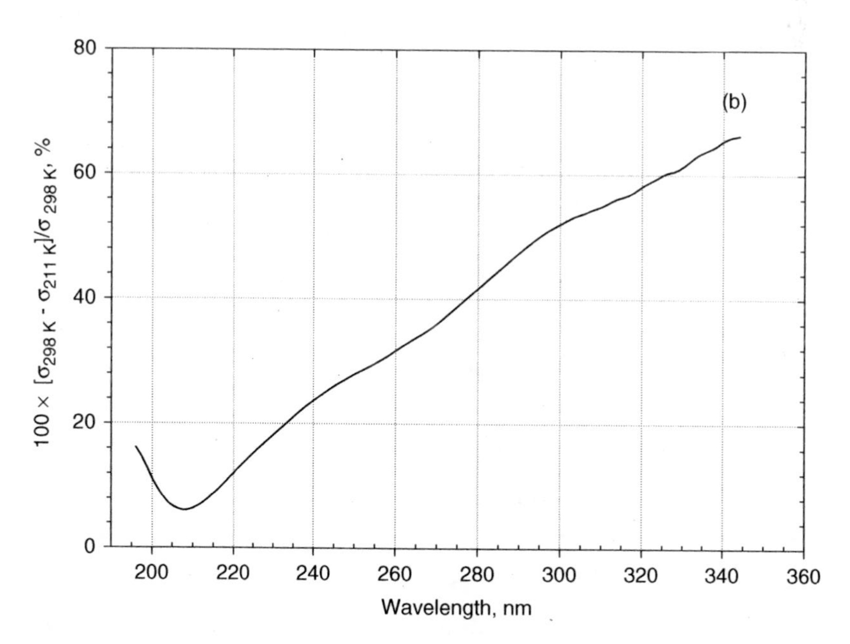

Figure VII-J-3b. Plot of the percentage difference in the absorption cross section of $CH_3C(O)OONO_2$ (PAN) at 298 and 211 K as a function of wavelength; from data of Talukdar et al. (1995).

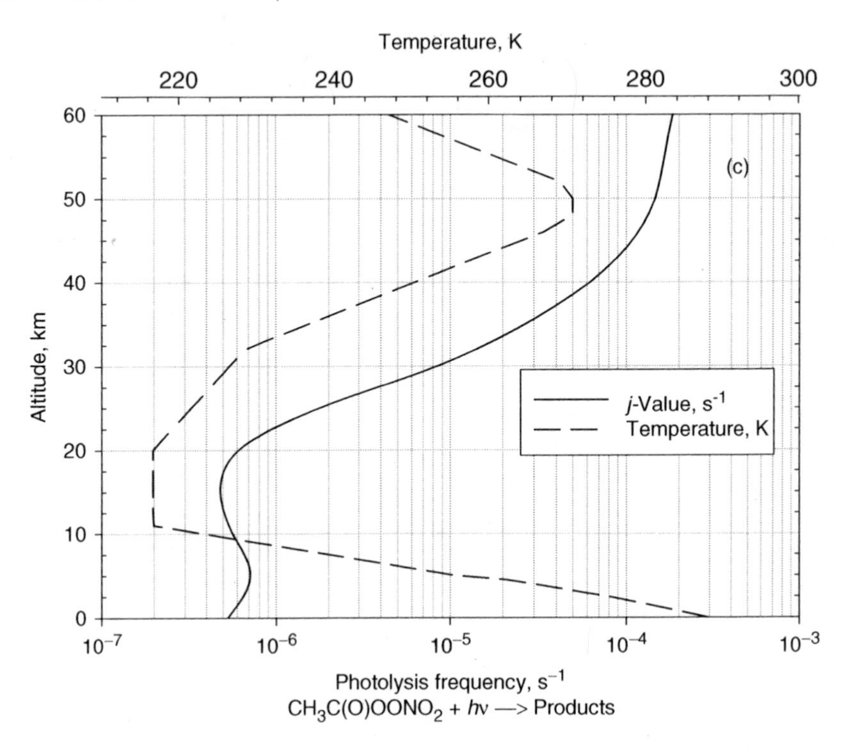

Figure VII-J-3c. Plot of the photolysis frequency for $CH_3C(O)OONO_2$ (PAN) as a function of altitude for a cloudless day with solar zenith angle = 45° and overhead ozone column = 350 DU. Also shown is the approximate atmospheric temperature as a function of altitude (U.S. Standard Atmosphere, 1976).

that is an active input to the *j*-values of PAN. One can see the start of input to the *j*-values from the shorter-wavelength regions even at 15 km. But as one progresses in the sequence: 0, 15, 30, 40, 45, 60 km, the region of wavelengths encompassing those of O_3 absorption becomes less filtered, and these wavelengths give a substantial increase to the expected *j*-values. Note the response region of wavelengths for 40 km altitude mirrors the absorption cross section envelope of ozone. As one approaches the upper regions of the stratospheric ozone band (45–60 km), the net contribution to *j*-values from the short-wavelength region is the dominant source. What began as a photolysis frequency near 5×10^{-7} s^{-1} at ground level has increased to $\sim 2 \times 10^{-4}$ s^{-1} at 60 km. Although the increase seen results largely from the increase in solar flux at high altitudes, the effect of the temperature dependence of the *j*-values should be taken into account for realistic modeling of the *j*-values in the upper atmosphere.

Nizkorodov et al. (2005) measured the C–H overtone transition strengths combined with estimates of the photodissociation cross sections for these transitions and found that near-IR photodissociation of $CH_3C(O)OONO_2$ is less significant ($j \cong 3 \times 10^{-8}$ s^{-1} at noon) in the lower atmosphere than competing sinks of unimolecular decomposition and ultraviolet photolysis.

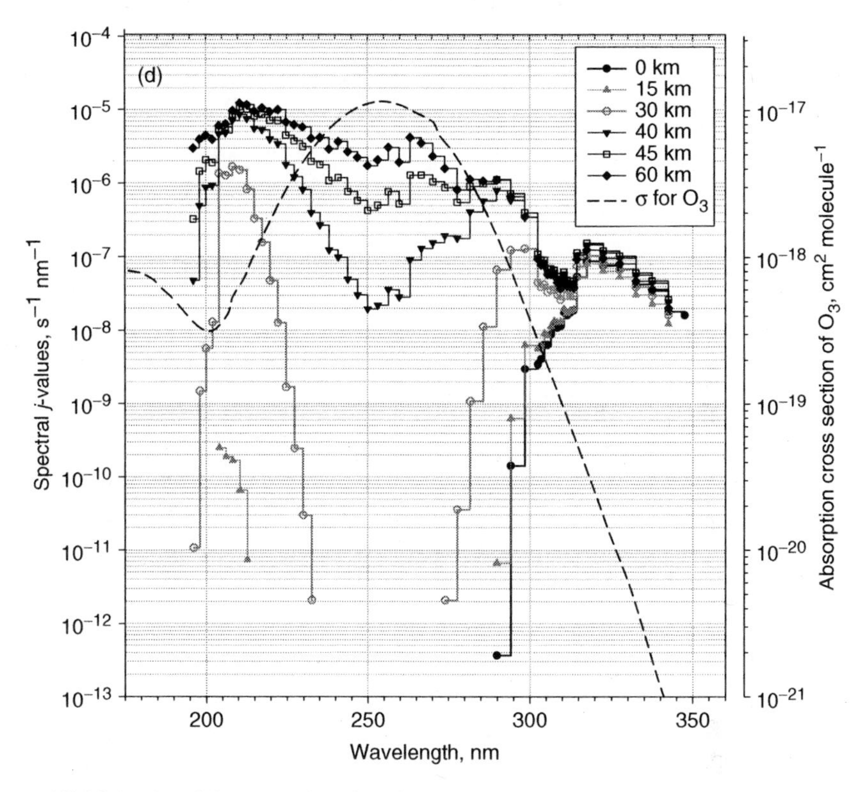

Figure VII-J-3d. Plot of the spectral j-values for $CH_3C(O)OONO_2$ photodecomposition as a function of altitude for a cloudless day with solar zenith angle = 45° and overhead O_3 at ground level = 350 DU. Note the marked change in the contribution of the wavelength ranges as the overhead O_3 column is lowered at high altitudes and the blocked radiation within the O_3 absorption band increases.

VII-J-2. Peroxypropionyl Nitrate [1-Nitroperoxopropanal, PPN, $CH_3CH_2C(O)OONO_2$]

The primary processes in the photodecomposition of peroxypropionyl nitrate are presumed to be like those in the analogous compound, peroxyacetyl nitrate:

$$C_2H_5C(O)OONO_2 + h\nu \rightarrow C_2H_5C(O)O_2 + NO_2 \quad \text{(I)}$$

$$\rightarrow C_2H_5CO_2 + NO_3 \quad \text{(II)}$$

Quantum yields for photodecomposition to form NO_3 have been measured by Harwood et al. (2003): $\phi_{II} = 0.22 \pm 0.04$ at 248 nm and 0.39 ± 0.04 at 308 nm. As with our treatment of the peroxyacetyl nitrate results, we have assumed: (i) $\phi_I = 1 - \phi_{II}$; and (ii) the quantum yields are a linear function of the excitation wavelength. These ϕ_I and ϕ_{II} estimates and σ measurements of Harwood et al. are given in table VII-J-2 as a function of wavelength of sunlight available within the troposphere.

Table VII-J-2. Cross sections (σ, cm^2 molecule^{-1}; Talukdar et al., 2003) and approximate quantum yields as a function of wavelength for peroxypropionyl nitrate at 298 K; quantum yield data are based on an extrapolation of the results of Harwood et al. (2003)

λ (nm)	$10^{20} \times \sigma$	ϕ_I [a]	ϕ_{II} [b]	λ (nm)	$10^{20} \times \sigma$	ϕ_I [a]	ϕ_{II} [b]	λ (nm)	$10^{20} \times \sigma$	ϕ_I [a]	ϕ_{II} [b]
278	1.84	0.307	0.693	299	0.299	0.365	0.635	320	0.0491	0.426	0.574
279	1.71	0.310	0.690	300	0.273	0.368	0.632	321	0.0467	0.428	0.572
280	1.57	0.313	0.687	301	0.251	0.370	0.630	322	0.0443	0.432	0.568
281	1.45	0.315	0.685	302	0.228	0.374	0.626	323	0.0399	0.435	0.565
282	1.33	0.318	0.682	303	0.210	0.377	0.623	324	0.0354	0.438	0.562
283	1.23	0.320	0.680	304	0.192	0.380	0.620	325	0.0318	0.441	0.559
284	1.12	0.323	0.677	305	0.177	0.382	0.618	326	0.0282	0.444	0.556
285	1.03	0.325	0.675	306	0.162	0.385	0.615	327	0.0262	0.446	0.554
286	0.940	0.328	0.672	307	0.149	0.388	0.612	328	0.0242	0.449	0.551
287	0.865	0.332	0.668	308	0.136	0.390	0.610	329	0.0224	0.452	0.548
288	0.790	0.335	0.665	309	0.125	0.394	0.606	330	0.0206	0.455	0.545
289	0.726	0.337	0.663	310	0.114	0.396	0.604	331	0.0190	0.458	0.542
290	0.662	0.340	0.660	311	0.105	0.399	0.601	332	0.0174	0.461	0.539
291	0.607	0.343	0.657	312	0.0962	0.402	0.598	333	0.0160	0.464	0.536
292	0.551	0.346	0.654	313	0.0899	0.405	0.595	334	0.0146	0.467	0.533
293	0.507	0.349	0.651	314	0.0835	0.408	0.592	335	0.0127	0.469	0.531
294	0.462	0.352	0.648	315	0.0762	0.412	0.588	336	0.0107	0.472	0.528
295	0.426	0.355	0.645	316	0.0689	0.415	0.585	337	0.00985	0.475	0.525
296	0.389	0.357	0.643	317	0.0630	0.417	0.583	338	0.00900	0.478	0.522
297	0.357	0.359	0.641	318	0.0571	0.420	0.580	339	0.00780	0.481	0.519
298	0.325	0.362	0.638	319	0.0531	0.423	0.577	340	0.00660	0.484	0.516

[a] $CH_3CH_2C(O)OONO_2 + h\nu \rightarrow CH_3CH_2C(O)O_2 + NO_2$ (I)
[b] $CH_3CH_2C(O)OONO_2 + h\nu \rightarrow CH_3CH_2CO_2 + NO_3$ (II)

Approximate photolysis frequencies were estimated from these data and the appropriate actinic flux at various solar zenith angles for a cloudless day in the troposphere with an overhead ozone column = 350 DU. These are plotted vs. solar zenith angle in figure VII-J-4. The j-values of PPN are greater than those of PAN by about a factor of 1.7, and this leads to a somewhat shorter photochemical lifetime of PPN within the lower troposphere: for overhead Sun, ~7.4 days for PPN compared to ~13 days for PAN.

VII-K. Photodecomposition of Alkyl Hydroperoxides (ROOH)

VII-K-1. Methyl Hydroperoxide (CH_3OOH)

The atmospheric photochemistry of methyl hydroperoxide has been reviewed previously by Calvert et al. (2000) and in earlier reviews (Atkinson et al., 1992b; DeMore et al., 1997). Recommendations all favor the acceptance of the Vaghjiani and Ravishankara (1989b, 1990) and Thelen et al. (1993) data. A single mode of photodissociation, process (I), is suggested for photolysis at wavelengths within the troposphere:

$$CH_3OOH + h\nu \rightarrow CH_3O + OH \quad \text{(I)}$$

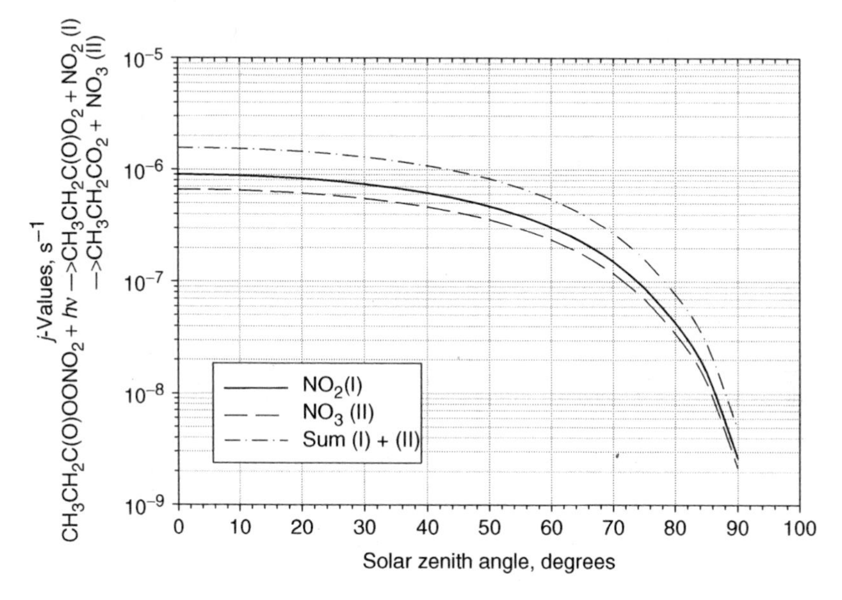

Figure VII-J-4. Plot of the approximate photolysis frequencies for peroxypropionyl nitrate (PPN) photodecomposition as a function of solar zenith angle for a cloudless day within the lower troposphere with an overhead ozone column = 350 DU.

The absorption cross sections of Vaghjiani and Ravishankara (1989b) are recommended; see table VII-K-1 and figure VII-A-1. These authors reported a quantum yield of OH radical production near unity ($\phi_I = 1.00 \pm 0.18$) in photolysis at 248 nm carried out in 300 Torr of N_2; formation of some H-atoms ($\phi = 0.038 \pm 0.007$) was observed. The quantum yield of formation of $O(^3P)$ was less than 0.007. Thelen et al. (1993) confirmed the unit quantum efficiency for process (I) at both 248 and 193 nm. They reported, from photolysis at 193 nm, that the average kinetic energy of the fragments was 280 kJ mol^{-1}; this represents about 65% of the available energy, with an average of about 154 kJ mol^{-1} residing in internal energy of the fragments. Photolysis at 248 nm created fragments with 201 and 96 kJ mol^{-1} of translational and internal energy, respectively. H-atom formation, observed by Vaghjiani and Ravishankara (1990), probably arose from the secondary dissociation of energy-rich CH_3O radicals. According to the data of Thelen et al., the secondary decay of internally excited CH_3O radicals is expected to be efficient (~50%) only for photolysis at wavelength 193 nm. No quantum yield measurements have been reported for $\lambda > 248$ nm. However, there is a large excess of energy available in a quantum of light to rupture the CH_3O—OH bond ($\Delta H \sim 193$ kJ mol^{-1}) over the entire long-wavelength region of CH_3OOH absorption ($\lambda \leq 360$ nm; equivalent to 332 kJ mol^{-1}). In the absence of quantum yield data for irradiation at the longer wavelengths (>248 nm), we have assumed that $\phi_I = 1.0$ for all wavelengths available within the troposphere. The approximate j-values for CH_3OOH as a function of solar zenith angle are given in figure VII-K-1 as calculated for a cloudless day within the lower troposphere with an ozone column of 350 DU. Quantum yields of primary processes analogous to process (I) in CH_3OOH and reliable spectral data

Table VII-K-1. Cross sections (σ, cm^2 $molecule^{-1}$; Vahgjiani and Ravishankara,1989b) for CH_3OOH as a function of wavelength

λ (nm)	$\sigma \times 10^{20}$	λ (nm)	$\sigma \times 10^{20}$
210	31.2	290	0.691
215	20.9	295	0.551
220	15.4	300	0.413
225	12.2	305	0.313
230	9.62	310	0.239
235	7.61	315	0.182
240	6.05	320	0.137
245	4.88	325	0.105
250	3.98	330	0.079
255	3.23	335	0.061
260	2.56	340	0.047
265	2.11	345	0.035
270	1.70	350	0.027
275	1.39	355	0.021
280	1.09	360	0.016
285	0.863	365	0.012

for the higher analogues to methyl hydroperoxide (ROOH, with R = C_2H_5, C_3H_7, etc.) are undetermined at this time. It is common practice in modeling to use the *j*-estimates for CH_3OOH as representative of those of the larger alkyl hydroperoxides. However, somewhat larger *j*-values are probably more realistic since cross sections are expected to grow as one progresses to the larger alkyl hydroperoxides.

VII-L. Summary of Photochemical Processes

The photodecomposition of the various oxidation products of the alkanes and the naturally occurring and anthropogenic alkane derivatives play important roles in the chemistry of the urban, rural, and remote atmospheres. These processes provide radical and other reactive products that help drive the chemistry that leads to ozone generation and other important chemistry in the atmosphere. In this chapter we have reviewed the evidence for the nature of the primary processes that occur in the aldehydes, ketones, alkyl halides, acyl halides, alkyl nitrates, peroxyacyl nitrates, and the alkyl peroxides. Where sufficient data exist, estimates have been made of the rate of the photolytic processes that occur in these molecules by calculation of the photolysis frequencies or *j*-values. These rate coefficients allow estimation of the photochemical lifetimes of the various compounds in the atmosphere as well as the rates at which various reactive products are formed through photolysis.

In table VII-L-1 we have summarized the photolysis frequencies (*j*-values) for each of the molecules for which sufficient data exist to allow reasonable estimations to be made (298 K; latitude 40° N; 500 m altitude; ozone column 350 DU). In the second and third columns the *j*-values for overhead Sun and the photochemical lifetime (1/*j*-value), respectively, are given. Of course, the Sun is rarely at the zenith, and

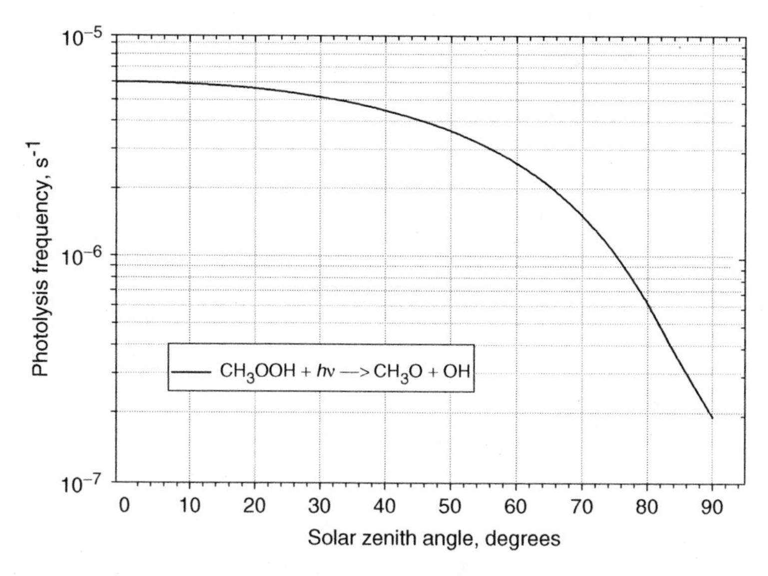

Figure VII-K-1. Estimated photolysis frequencies vs. solar zenith angle for the photodissociation of CH_3OOH in the cloud-free lower troposphere with an overhead ozone column of 350 DU.

a more representative photochemical lifetime is the diurnally averaged number. In the fourth and fifth columns, respectively, the photochemical lifetimes are given as diurnal averages for clear sky conditions in the lower troposphere at 40°N during periods of extended sunshine (near June 22) and those of more limited daylight (near December 22). The calculations of these quantities are made by averaging the j-values over each hour of the day and night for the 24-hour period. As one expects, diurnal-averaged lifetimes are much longer than the lifetimes at solar zenith, on the average about three times for periods near June 22 to about twenty times for winter months (near December 22).

Note the large range in the lifetimes listed in table VII-L-1. The most active molecule photochemically is CH_2I_2, which lives only 103 s with overhead Sun and about 5 min during the June 22 period. The least reactive photochemical species shown is $CH_3C(O)Cl$, which has a photochemical lifetime in the lower troposphere of hundreds of years during periods of low solar flux. Obviously, the loss of this species by other means, such as hydrolysis in water solution, will be the important fate.

The most important class of these molecules in promoting ozone generation within the troposphere is the aldehydes, and formaldehyde leads this list. During the summer months formaldehyde's photochemical lifetime is about 9 hr. Its rate of photodecomposition is competitive with its reaction with OH; for an averaged $[OH] = 10^6$ molecule cm^{-3}, its lifetime for reaction with OH is about 33 hr. Photodecomposition of the other, higher aldehydes is somewhat less competitive with the OH reaction, but still important. This is especially true at the higher altitudes of the troposphere where the photodissociation is often enhanced by the lowered collisional quenching of the excited aldehyde molecules and the increased solar flux.

Table VII-L-1. Summary of the estimated photochemical lifetimes (τ) for alkane oxidation products and derivatives for the cloudless lower troposphere (latitude 40°; 298 K; 500 m altitude; vertical ozone column = 350 DU)

Compound	$10^5 \times j$-Value (Overhead Sun, s^{-1})	τ (Overhead Sun)	τ, Diurnal Cycle (June 22 Time Period)	τ, Diurnal Cycle (Dec. 22, Time Period)
CH_2O	9.01	3.1 hr	9.1 hr	1.66 days
CH_3CHO	0.729	1.6 days	5.3 days	31days
CH_3CH_2CHO	1.18	1.0 day	2.5 days	18 days
$CH_3CH_2CH_2CHO$	1.98	14 hr	1.9 days	10 days
$(CH_3)_2CHCHO$ (max.)[a]	6.23	4.5 hr	14 hr	3.1 days
$(CH_3)_2CHCHO$ (min.)[a]	5.58	5.0 hr	15 hr	3.3 days
$CH_3CH_2CH_2CH_2CHO$	1.79	16 hr	2.0 days	11 days
$(CH_3)_3CCHO$ (max.)[b]	3.36	8.3 hr	1.1 days	6.2 days
$(CH_3)_3CCHO$ (min.)[b]	1.44	19 hr	2.5 days	14 days
$(CH_3)_2CHCH_2CHO$ (max)[c]	1.87	15 hr	4.0 days	22 days
$(CH_3)_2CHCH_2CHO$ (min.)[c]	0.908	1.3 days	2.0 days	11 days
$CH_3CH_2CH_2CH_2CH_2CHO$	2.41	12 hr	1.5 days	8.0 days
$CH_3CH_2CH_2CH_2CH_2CH_2CHO$ (max.)[d]	1.31	21 hr	2.8 days	15 days
$CH_3CH_2CH_2CH_2CH_2CH_2CHO$ (min.)[d]	0.950	1.2 days	3.8 days	20 days
$CH_2(OH)CHO$	1.48	19 hr	2.5 days	14 days
CCl_3CHO	8.64	3.2 hr	10 hr	2.2 days
CF_3CHO	1.34	21 hr	2.6 days	12 days
CHF_2CHO	4.60	6.0 hr	18 hr	3.2 days
$CH_3C(O)CH_3$ (max.)[e]	0.0975	12 days	44 days	1.0 years
$CH_3C(O)CH_3$ (min.)[e]	0.0731	16 days	59 day	1.5 years
$CH_3C(O)CH_2CH_3$	0.468	2.5 days	8.4 days	55 days
$CH_3C(O)CH_2CH_2CH_3$ (max.)[f]	0.267	4.3 days	16 days	4.2 months
$CH_3C(O)CH_2CH_2CH_3$ (min.)[f]	0.157	7.4 days	17 days	7.8 months
$CH_3C(O)CH_2CH_2CH_2CH_3$ (max.)[g]	0.804	1.4 days	7.0 days	31 days
$CH_3C(O)CH_2CH_2CH_2CH_3$ (min.)[g]	0.139	8.3 days	30 days	7.5 months
Cyclopropanone (C_3H_4O)	37.0	45 min	2.0 hr	8.1 hr
Cyclobutanone (C_4H_6O)	2.93	9.5 hr	1.3 days	8.8 days
$CH_2(OH)C(O)CH_3$(max.)[h]	0.209	5.5 days	18 days	3.4 months
$CH_2(OH)C(O)CH_3$ (min.)[h]	0.105	11 days	36 days	6.6 months
$CH_2ClC(O)CH_3$ (max.)[i]	10.1	2.8 hr	8.6 hr	1.9 days
$CH_2ClC(O)CH_3$ (min.)[i]	5.06	5.5 hr	17 hr	3.9 days
$CH_2BrC(O)CH_3$(max.)[j]	22.1	1.3 hr	3.8 hr	18 hr
$CH_2BrC(O)CH_3$ (min.)[j]	11.2	2.5 hr	7.5 hr	1.5 days
$CF_3C(O)CH_3$ (max.)[k]	0.106	11 days	47 days	1.0 year
$CF_3C(O)CH_3$ (min.)[k]	0.0866	13 days	39 days	9.3 months
$CF_3C(O)CF_3$	0.187	6.2 days	21 days	4.5 months
$CF_3CF_2C(O)CF(CF_3)_2$	0.91	1.3 days	3.7 days	16 days
HC(O)Br	1.76	16 hr	2.2 days	14 days
HC(O)Cl	0.0161	2.3 months	8.4 months	5.3 years
BrC(O)Br	0.29	4.0 days	13 days	2.5 months
$CH_3C(O)Cl$	0.000358	8.9 years	56 years	~ 320 years
$Cl_3CC(O)Cl$	0.263	4.4 days	15 days	2.9 months
$CF_3C(O)Cl$	0.0509	23 days	2.8 months	2.2 years
CH_3I	0.758	1.5 days	4.9 days	24 days
CH_3CH_2I	0.90	1.3 days	4.3 days	23 days
$CH_3CH_2CH_2I$	0.108	1.1 days	3.9 days	20 days
CH_3CHICH_3	2.31	12 hr	1.6 days	9.2 days
$CHBr_3$	0.15	7. 7 days	27 days	6.1 months
CH_2ICl	17.4	1.6 hr	4.9 hr	24 hr
CH_2BrI	59.1	28 min	83 min	6.6 hr
CH_2I_2	968	1.7 min	4.9 min	20 min

Table VII-L-2. (*continued*)

Compound	$10^5 \times j$-Value (Overhead Sun, s^{-1})	τ (Overhead Sun)	τ, Diurnal Cycle (June 22 Time Period)	τ, Diurnal Cycle (Dec. 22, Time Period)
CH_3ONO_2	0.116	10 days	34 days	6.7 months
$CH_3CH_2ONO_2$	0.137	8.5 days	29 days	6.5 months
$CH_3CH_2CH_2ONO_2$	0.179	6.5 days	22 days	4.7 months
$CH_3CH(ONO_2)CH_3$	0.300	3.9 days	13 days	2.5 month
$CH_3CH_2CH_2CH_2ONO_2$	0.159	7.3 days	25 days	5.4 months
$(CH_3)_3CONO_2$	0.835	1.4 days	4.5 days	25 days
$CH_3CH(ONO_2)CH_2CH_2CH_3$	0.258	4.5 days	15 days	3.2 months
$CH_3CH_2CH(ONO_2)CH_2CH_3$	0.252	4.6 days	16 days	3.4 months
cyclo-$C_5H_9ONO_2$	0.0900	13 days	47 days	7.1 months
$CH_3C(O)OONO_2$ (PAN)	0.0892	13 days	41 days	6.9 months
$CH_3CH_2C(O)OONO_2$ (PPN)	0.156	7.4 days	24 days	4.1 months
CH_3OOH	0.599	1.9 days	5.6 days	24 days

[a] The two alternative scenarios related to the wavelength dependence of the quantum yields are used here as outlined in section VII-B-5.
[b] Two alternative scenarios related to the wavelength dependence of the quantum yields are used here as outlined in section VII-B-8.
[c] Two alternative scenarios related to the wavelength dependence of the quantum yields are used here as outlined in section VII-B-7.
[d] Two alternative scenarios related to the wavelength dependence of the quantum yields are used here as outlined in section VII-B-12.
[e] Three scenarios related to the wavelength dependence of the quantum yield of process (III) were used as outlined in section VII-C-1.3; only the two most probable cases, (a) and (b), are used here.
[f] Two scenarios related to the wavelength dependence of the quantum yields are used here as outlined in section VII-C-4.
[g] Two scenarios related to the wavelength dependence of the quantum yields are used here as outlined in section VII-C-6.
[h] Two scenarios related to the wavelength dependence of the quantum yields are used here as outlined in section VII-G-2.
[i] Two scenarios related to the wavelength dependence of the quantum yields are used here as outlined in section VII-G-3.
[j] Two scenarios related to the wavelength dependence of the quantum yields are used here as outlined in section VII-G-4.
[k] Two scenarios related to different quantum yields and cross sections are used here as outlined in section VII-G-5.

The ketones also share their decay pathways between OH attack and photodecomposition. With acetone in the lower troposphere during the summer months, photochemical lifetimes between 12 and 16 days are expected, whereas the lifetime for OH attack is over 63 days. The acceleration of the photodecomposition of acetone at the higher altitudes results from lowered collisional deactivation of the excited acetone molecules at the low pressure and higher actinic flux levels; together with the suppression of the reaction with OH as a result of the activation energy barrier, the photodecomposition pathway stays competitive. With the higher ketones the preference for photodecomposition is somewhat less. The lifetime for photodecomposition of $CH_3C(O)C_2H_5$ during summer months is about 8 days, while that for reaction with OH is about 10 days in the lower troposphere. With 2-hexanone the photodecomposition lifetime is about 4–7 days, while that for reaction with OH is about 2 days.

The alkyl nitrates have rather large photochemical lifetimes but these also are competitive with the lifetimes for their reactions with OH in the lower troposphere. CH_3ONO_2, $C_2H_5ONO_2$, $CH_3CH_2CH_2ONO_2$, and $CH_3CH_2CH_2CH_2ONO_2$ have lifetimes of photodissociation during the summer months (40° latitude) in the lower troposphere of about 34, 29, 22, and 25 days, respectively, while the rates of OH attack on these compounds lead to lifetimes of about 30, 24, 17, and 8 days, respectively.

The peroxyacyl nitrates show a similar behavior: PAN has a photochemical lifetime in the lower troposphere during the summer months of about 41 days whereas the lifetime for OH attack is 85–154 days .

Some of the alkyl halides also show pathways of decay that are competitive between OH radical attack and photodissociation. In the lower troposphere during the summer months lifetimes for photodissociation for CH_3I, C_2H_5I, $CH_3CH_2CH_2I$, and $(CH_3)_2CHI$ are about 5, 4, 4, and 1.6 days, while those for OH attack are about 150, 15, 8, and 9 days, respectively.

As we have seen in our discussions in this chapter, photochemistry provides an important part of the atmospheric chemistry that occurs within the troposphere. The direct photochemistry of the alkanes is unimportant within the troposphere, since they do not absorb the available sunlight, However, photochemistry is indirectly responsible for their decay by the dominant pathway of OH attack. It is the photodecomposition of ozone that is the major source of the OH radicals that account for the majority of the observed destruction of the alkanes (RH):

$$O_3 + h\nu \rightarrow O(^1D) + O_2$$

$$O(^1D) + H_2O \rightarrow OH + OH$$

$$OH + RH \rightarrow H_2O + R$$

$$R + O_2 \rightarrow RO_2$$

$$RO_2 + R'O_2, NO, NO_2, HO_2 \rightarrow RO, RONO_2, RO_2NO_2, RO_2H, \text{etc.} \rightarrow \text{other products}$$

VIII

Representation of the Atmospheric Chemistry of Alkanes in Models

VIII-A. Policy Applications of Atmospheric Chemistry Models

VIII-A-1. Development of Ozone Control Strategies

The production of ozone and other oxidants by sunlight-driven atmospheric chemical reactions at levels that may damage human health was first recognized in the Los Angeles basin of the United States of America. Subsequently, episodes of elevated ozone levels, with their attendant human health effects, have been identified in many of the world's population and industrial centers (WHO, 1987). In many of these locations, pollution control strategies have been developed and implemented to reduce population exposure to ozone and thereby reduce public health hazards.

The design of health-based strategies to reduce exposure to ozone involves reducing emissions of either or both of the main ozone precursors: organic compounds and oxides of nitrogen, NO and NO_2 (NO_x). Ozone control strategies have therefore a number of essential elements:

- Emission inventories covering the entire air basin for organic compounds and NO_x for the present day;
- Models that link ozone precursor emissions to ozone and other oxidant formation;

- An inventory of policy measures and control actions that act upon the precursor emissions and are considered feasible at some time in the future;
- An understanding of the target air quality that would be considered acceptable by the policy process.

The models are used to quantify current ozone levels based on present-day emissions and to identify whether target air quality can be reached in the future with the policy measures and control options considered feasible. They therefore play a central role in the development of ozone control policies.

VIII-A-2. Modeling Ozone for Policy Development Purposes

Models that address ozone formation may be complex three-dimensional airshed grid models or relatively simple moving air parcel trajectory models. A review of the available tools is given elsewhere (for example, see Metcalfe and Derwent, 2005). An essential element of any approach to ozone modeling is a chemical mechanism that can relate ozone precursor emissions within an air basin to the subsequent ozone levels formed downwind. The chemical mechanism needs to be accurate enough to direct policy decisions in such a way that when at some time in the future ozone precursor emissions are reduced, target air quality is actually achieved.

Because of the limitations of current understanding of the atmospheric processes involved and the limitations of current computers, compromises have to be made in the currently available ozone modeling tools and the chemical mechanisms incorporated within them. Sophisticated three-dimensional airshed models must necessarily employ highly condensed chemical mechanisms. Simple trajectory models can handle a great amount of chemical complexity with some loss of treatment of some of the important atmospheric transport and diffusion processes. Great care must be taken in the simplification and condensation of the chemical mechanisms used in current policy tools. If the chemical mechanism is incorrect or inadequate in some respects, it may result in the promulgation of inadequate or counter-productive ozone control strategies in a given air basin.

Many chemical mechanisms have been proposed to account for ozone formation in different air basins and over different spatial scales. For example, Derwent (1990) reviews 24 different chemical mechanisms that have been in common use up to that time. The chemical mechanisms chosen here for further study are a sample of the whole range and are meant to illustrate the broad range of treatments that have been applied. Five vastly differing chemical mechanisms are considered here in some detail. These mechanisms span the range from simple to complex and include:

- Carbon Bond Mechanism, CBM (Gery et al., 1989);
- Regional Atmospheric Chemistry Mechanism, RACM (Stockwell et al., 1997);
- State-wide Air Pollution Research Center (SAPRC) mechanism (Carter, 2000);
- Master Chemical Mechanism, MCM (Jenkin et al., 1997);
- Self-generating Explicit Mechanism (Aumont et al., 2005).

The main focus of the discussion in section VIII-B is how the atmospheric chemistry of the alkanes has been specifically treated in the above five mechanisms;

in section VIII-C, we discuss how the differences in these treatments influence policy outcomes for the alkanes. Generally speaking, these mechanisms treat only the reactions of the alkanes with OH radicals, neglecting any alkane reactivity with chlorine atoms and other radical species.

The CBM mechanism (Gery et al., 1989) was chosen because of its highly compact nature, which makes it well suited for application within complex three-dimensional grid airshed models where computer resources are at an absolute premium. The RACM mechanism is aimed at both regional acid deposition and ozone formation and addresses both polluted and remote atmospheres. In contrast to the CBM and RACM mechanisms, the SAPRC and MCM mechanisms are larger in size by orders of magnitude. The SAPRC mechanism (Carter, 2000) addresses the urban scale with its associated high-NO_x conditions whereas the MCM mechanism (Jenkin et al., 1997) addresses the long-range transboundary transport scale, of particular interest within Europe. The Self-generating Explicit Mechanism (Aumont et al., 2005) is highly detailed and sophisticated, representing many millions of chemical processes.

VIII-A-3. Reactivity-Based Ozone Control Strategies

An important aspect of ozone control strategies is the issue of reactivity and of reactivity-based strategies. The term reactivity refers to the ability of an individual organic compound to produce ozone when emitted into a given air basin. Each organic compound exhibits a different propensity to make ozone. Organic compounds may thus be highly reactive if, for a given level of emission, they generate a substantial fraction of the photochemically produced ozone. They may be considered unreactive if, for the same level of emissions, they produce only a negligible fraction of the elevated ozone levels. Alkenes and aromatic compounds are considered to be highly reactive and alkanes are considered to be much less reactive or even unreactive in some cases.

Reactivity-based control strategies or policies would thus seek to reduce ozone levels in a given air basin by substituting the emissions of highly reactive organic compounds with those of much less reactive or unreactive organic compounds. This would imply some quantitative understanding of the relative importance of each organic compound in stimulating photochemical ozone formation. An ordered list of the relative importance of each organic compound is termed a reactivity scale. Generation of reactivity scales is an important first step in the promulgation of reactivity-based policies. It is also an important application for ozone policy models and for the chemical mechanisms that underpin them.

In the United States of America, the policies of the Environmental Protection Agency (EPA) to control the emissions of organic compounds for ambient ozone reduction have been based on reactivity considerations (Dimitriades, 1996, 1999). Ozone policies have not simply been based on the reduction of precursor emissions since it is widely accepted that selective control of organic compound emissions is more cost-effective than an indiscriminate approach. Also, substitution of less reactive organic emissions for more reactive emissions offers additional options when alternative mass-based control options have been exhausted. EPA policy thus allows for negligibly reactive organic emissions to be exempt from all controls, and ethane is taken as the benchmark for distinguishing reactive from negligibly reactive organic compounds (Dimitriades, 1996, 1999).

The EPA guidelines for classifying reactivity use the so-called k_{OH} reactivity method for identifying negligibly reactive organic compounds and the maximum incremental reactivity (MIR) method for ascertaining reactivities higher than that of ethane (Dimitriades, 1999). The k_{OH} method is based on the comparison of the rate coefficient for the hydroxyl OH radical attack on the organic compound with that of ethane. It is based on the assumption that reaction with OH is the rate-determining step in photochemical ozone formation. The strength of this method is that these OH reaction rate coefficients, and their pressure and temperature dependences, are geophysical constants and hence are the same everywhere. In contrast, the MIR method uses a combination of environmental chamber data and computer modeling to assess the relative reactivities of organic compounds and is based on the work of Carter (1990). Relative reactivities are highly dependent on environmental conditions and, in principle, must be determined for each air basin and emission situation. This inherent variability is carefully taken into account in the assessment of the MIR reactivity scale, as discussed in section VIII-C-3 below.

The California Air Resources Board also takes reactivity considerations into account in the formulation of its policies to improve ozone air quality and reduce exceedances of ozone air quality standards. In its Low Emission Vehicle/Clean Fuel Regulations, reactivity adjustment factors are used to place the ozone impacts of the different exhaust emissions of the different alternatively-fueled low-emission vehicles on a similar footing to those using conventional gasoline engines (CARB, 1993). These adjustment factors are calculated using the MIR scale (Carter, 1994).

In contrast to the situation in North America, reactivity considerations have not received the corresponding attention of the policy-makers in Europe. The main policy instrument for air pollution control within Europe is the Convention on Long-Range Transboundary Air Pollution of the United Nations Economic Commission for Europe (UN ECE). Under the 1991 VOC Protocol, parties to the Convention agreed to reduce emissions of volatile organic compounds to reduce population exposure to ozone (UN ECE, 2005). Parties were encouraged to focus their activities on those VOC species that were most active in generating photochemical ozone, and reactivity data were presented within the VOC Protocol from the Maximum Incremental Reactivity (MIR) scale (Carter, 1990) and the Photochemical Ozone Creation Potential (POCP) scale (Derwent and Jenkin, 1991).

However, for the large part, commitments under the UN ECE are achieved through the promulgation and implementation of directives issued by the Commission of the European Communities, at least for the member states of the European Union. The European Commission has based all its control activities for volatile organic compounds on mass-based reductions. So, for example, the Solvents Emissions Directive (Commission of the European Communities, 1999) treats all solvents as equally important despite some having no actual propensity to form ozone photochemically and others being highly reactive.

VIII-A-4. Tropospheric Chemistry Models

It is now well recognized that climate change is driven by increases in the global burdens of a wide range of trace gases such as carbon dioxide, methane, tropospheric ozone, nitrous oxide, and the halocarbons, by changes in the global burden of aerosol particles, and by changes in land use and deforestation (IPCC, 2001). The Intergovernmental Panel on Climate Change (IPCC) has identified methane (CH_4) and tropospheric ozone (O_3) as the second and third most important radiatively active trace gases after carbon dioxide (CO_2). It is therefore of some policy importance to be able to follow accurately trends in both trace gases since pre-industrial times to the present day and from the present day onwards to the end of the 21st century. Measurements on ice cores have allowed accurate records to be constructed of the growth in atmospheric CH_4 levels since pre-industrial times and further back into prehistory (Etheridge et al., 1998). However, because of the inherent reactivity of O_3, no such equivalent methodology is available for tropospheric O_3, and hence there are no reliable indications from observations of pre-industrial O_3 levels. In the absence of observations, therefore, tropospheric chemistry-transport models (CTMs) have been used to quantify the growth in tropospheric ozone levels since pre-industrial times.

In addition, CTMs have been employed to estimate future CH_4 and O_3 levels and hence radiative forcing, so contributing to the development of a firmly based understanding of all the major radiative forcing agents and mechanisms. CTMs have therefore been the main tools for converting trace gas emissions into tropospheric burdens of CH_4 and O_3 in the various global emission scenarios through to the year 2100 and thereby providing policy-makers with best estimates of the radiative forcing from CH_4 and O_3 (IPCC, 2001).

These models are, however, only approximations to reality and are no better than the assumptions and simplifications that have been built into their formulations. There are considerable uncertainties in many of the process descriptions and parameterizations employed, together with the input data used to drive them. The IPCC has established a consensus view of the scope and limitations of these global CTMs through a series of model intercomparisons which have culminated in multi-model studies of the past, present, and future tropospheric distributions of carbon monoxide (CO) (Shindell et al., 2006) and O_3 (Stevenson et al., 2006). In this section, the role of alkanes in the chemical mechanisms underpinning the CTMs currently used for policy-making in the context of global climate change is described.

Table VIII-A-1 presents a summary of the representation of alkane chemistry in a selection of the 25 state-of-the-art global CTMs described by Shindell et al. (2006) and Stevenson et al. (2006). In addition to CH_4, all the CTMs in table VIII-A-1 include ethane (C_2H_6). Some models included propane (C_3H_8) or butane (n-C_4H_{10}) and others included them both. In the models that treated butane, it was generally used to represent all C_4 and higher alkanes. Some models adopted approaches that followed the carbon bond mechanism and lumped the entire family of emitted alkanes into one surrogate species. There is, however, no simple, agreed way to lump together the different alkanes between the different models that use this approach, and this introduces additional uncertainties into the representation of alkane chemistry.

Table VIII-A-1. Representation of the alkanes in a selection of tropospheric CTMs

Model	Institute	Alkane	Notes
A. CHASER_CTM	FRCGC/JAMSTEC	Ethane, propane, butane	Sudo et al. (2002)
B. FRSGC/UCI	FRCGC/JAMSTEC	Ethane, butane	Wild and Prather (2000)
C. GEOS-CHEM	LMCA-EPFL	Ethane, propane, butane	Bey et al. (2003); Horowitz et al. (1998)
D. GISS	NASA/GISS, Columbia University[a]	Lumped family of alkanes	Shindell et al. (2003)
E. GMI/CCM3	NASA GMI		
F. GMI/DAO	NASA GMI		
G. GMI/GISS	NASA GMI		
H. IASB	IASB/Belgium		
I. LLNL/IMPACT	Lawrence Livermore NL	Ethane, propane	Rotman et al. (2004)
J. LMDz/INCA-CTM	LSCE	Ethane, propane, butane	Folberth et al. (2005)
K. MATCH-MPIC/ECMWF	Max Planck IC	Ethane, propane, butane	von Kuhlmann et al. (2003)
L. MATCH-MPIC/NCEP	NCAR		
M. MOZ2-GFDL	GFDL	Ethane, propane, butane	Horowitz et al. (2003)
N. MOZART4	NCAR		
O. MOZECH	Max Planck IM	Ethane, propane, butane	Horowitz et al. (2003)
P. p-TOMCAT	University of Cambridge	Ethane, propane	Law et al. (1998)
Q. STOCHEM-HadAM3	University of Edinburgh	Ethane, propane, butane	Collins et al. (2002),
R. STOCHEM-HadGEM	UK Met Office		Stevenson et al. (2004)
S. TM4	KNMI	Lumped family of alkanes	Houweling et al. (1998)
T. TM5	JRC	Lumped family of alkanes	Houweling et al. (1998)
U. UIO_CTM2	University of Oslo	Ethane, butane	Berntsen and Isaksen (1997)
V. ULAQ	Università L'Aquila	Ethane, propane	Pitari et al. (2002)
X. UM_CAM	University of Cambridge	Ethane, propane	Zeng and Pyle (2003)

The chemical mechanism adopted in the LMDz-INCA model (Folberth et al., 2005) treats 83 chemical species and 264 chemical reactions and appears to be one of the most comprehensive chemical mechanisms employed of all the models in table VIII-A-1. It represents a useful yardstick by which to evaluate the broader range of mechanisms. This mechanism treats hydroxyl (OH) attack only on the secondary CH_2 groups in propane and butane, producing alkyl peroxy radicals whose fates include reactions with nitric oxide (NO) and with nitrate, hydroperoxy, and other organic peroxy radicals. Alkoxy radicals are replaced in the mechanism by their reaction products with O_2 and hence are not specifically represented. First-generation partially oxidized products include hydroperoxides, formaldehyde, acetaldehyde, acetone, and methyl ethyl ketone. These are then subjected to reaction with OH and photolysis. Exactly analogous chemical mechanisms have been adopted in the STOCHEM (Collins et al., 2002) and MOZART (Horowitz et al., 2003) models. These mechanisms should provide a reasonable reflection of the atmospheric chemistry of ethane, propane, and butane.

The main difficulty with the models in table VIII-A-1 lies in the assumption that butane can be used to represent the higher alkanes. Alkoxy radicals derived from the higher alkanes may isomerize, forming difunctional radicals and reaction products, and there are other mechanistic differences that cannot be taken into account by using butane as a surrogate species. At present, it is not straightforward to identify the drawbacks inherent in omitting the higher alkanes in global CTMs.

Alkanes released into the global atmosphere, while only weakly radiatively active themselves (IPCC, 2005), act as indirect greenhouse gases because they perturb the tropospheric distributions of CH_4 and O_3, as opposed to acting as direct greenhouse gases because they themselves are radiatively-active. Collins et al. (2002) used the STOCHEM model (see table VIII-A-1) to quantify the radiative forcing responses of CH_4 and O_3 to emission pulses of ethane, propane, and butane. Emission pulses of the simple alkanes acted to decrease the steady-state concentrations of tropospheric OH radicals and hence increase the global build-up of CH_4. This was associated with a positive radiative forcing. The emission pulses also stimulated tropospheric ozone production and this was also associated with a positive radiative forcing. The global warming potentials (GWPs) of the simple alkanes were found to be 8.4, 6.3, and 7.0, respectively; further details of GWPs are given in section I-G and table I-G-1. Across a range of different classes of organic compounds, Collins et al. (2002) have shown that OH reactivity, the formation of aldehydes rather than ketones, the formation of peroxyacylnitrates, and the number of C—C and C—H bonds in the emitted organic compound each had a marked influence on GWPs. On this basis, it is important in future work to maintain and improve the representation of alkane chemistry in the CTMs and to extend their coverage to the higher alkanes.

VIII-B. Representation of the Atmospheric Chemistry of the Alkanes in Urban and Regional Photochemical Ozone Models

VIII-B-1. Introduction to the Chemical Mechanisms in Urban and Regional Models

Generally speaking, the purpose of a chemical mechanism in any modeling tool is to convert the emissions of the organic compounds into estimates of the exposure levels of pollutants that may cause damage to human health and the environment. The modeling tool may address the urban, regional, or global atmosphere and the damaging pollutants addressed may be ozone and the other atmospheric oxidants on the urban and regional scales. On the global scale, the pollutants may be the radiatively active trace gases: methane and ozone. In stratospheric models, the emissions may be of chlorofluorocarbons and their replacements, and the damage may involve the depletion of the ozone layer. In each case, the chemical mechanism is required to quantify the conversion rates of the emitted organic compounds into ozone, other oxidants, methane, and reactive chlorine compounds and, in turn, to describe quantitatively their atmospheric destruction and removal rates.

Chemical mechanisms, in principle, should contain all the atmospheric chemical processes involved for all the relevant atmospheric constituents. But this ideal is never met in practice because of the limitations of current understanding and the limited capacities of current computers and models. A great deal of simplification is therefore inevitable in the compilation of the chemical mechanisms employed in policy-based models.

In the first level of simplification, a handful of organic compounds is selected from the hundreds or thousands of such compounds believed to be present or observed in the atmosphere. In the second level of simplification, chemical mechanisms for the handful of selected organic compounds are compressed and condensed by omitting particular sets of reactions or by neglecting the subsequent reactions of next-generation reaction products. Both these simplifications have been applied to the chemical mechanisms employed in policy-based models representing the alkanes.

Historically, two methods have been used to condense chemical mechanisms to a size appropriate for application within policy-based models. In the lumped molecule approach, a particular organic compound is used to represent a series of similar compounds. In the lumped structure approach, organic compounds are divided into smaller reaction elements based on the types of carbon bond in each organic compound emitted (Dodge, 2000). Examples of both methods will be described, appropriate to the alkanes.

In the paragraphs below, we describe some of the main chemical mechanisms that have been used to represent alkane chemistry in policy models in urban and regional policy applications. The review is not meant to be fully comprehensive, but it aims to illustrate the range in complexity of the different treatments that have been applied. There is no correct mechanism for the atmospheric chemistry of the alkanes and no clear way of knowing what the impact of the necessary simplifications has been on the policy results derived in a given application.

VIII-B-2. Detailed Treatment of the Alkanes in the Carbon Bond Mechanism (CBM)

The Carbon Bond Mechanism (Gery et al., 1989) is one of the most important mechanisms for modeling photochemical ozone formation because of its widespread application in complex 3-D urban airshed models, particularly in policy applications in North America. The CBM mechanism has been applied widely, because it is the most compact and concise mechanism and because it is firmly and rigorously rooted in environmental chamber studies. Here, the representation of the alkanes within the CBM mechanism is reviewed and the influence this has on the policy importance of the alkanes is assessed.

The CBM uses both condensation methods to great advantage to produce highly compact mechanisms. In a lumped structure mechanism such as CBM, organic compounds are grouped according to their bond types; the alkanes are represented as a one-carbon-atom molecule surrogate called PAR. Most single-bonded carbon atoms, regardless of the compound in which they appear, are represented as PAR. Other surrogates are named ETH, TOL, XYL, representing ethene, toluene, and xylenes. An alkane such as pentane would thus be represented as 5 PAR on a molar basis. An additional non-reactive (NR) surrogate is also included that takes no part in photochemical ozone formation. Each emitted methane and haloalkane molecule is assigned to the NR category, ethane to 0.4 PAR and 1.6 NR and propane to 1.5 each of PAR and NR.

As described in Gery et al. (1989), the CBM mechanism has been evaluated against about 200 different experiments performed in the University of North Carolina and University of California at Riverside smog chambers. Subsequently, CBM has been tested against experiments conducted in the smog chambers of the Tennessee Valley Authority (USA) and the Commonwealth Scientific and Industrial Research Organisation (Australia), as detailed by Dodge (2000). However, the extent to which the representation of alkane chemistry, in particular, in CBM has been evaluated against smog chamber data remains unclear.

The CBM mechanism has undergone a series of updates since the original version of Gery et al. (1989) to take into account new information obtained in kinetic and mechanistic studies. The latest version, recommended for use in policy applications, is that implemented in the Models-3/Community Multiscale Air Quality (CMAQ) modeling system (Gipson and Young, 1999). This version is sometimes called Carbon Bond version 4 (CB4) and contains 93 chemical reactions and 36 chemical species.

The representation of alkanes in CBM (or CB4) can be written as:

$$\mathrm{OH + PAR = RO_2}$$

$$\mathrm{RO_2 + NO = RO + NO_2}$$

$$\mathrm{RO + O_2 = PRODUCT_i + PRODUCT_d + PRODUCT_o + RADICALS}$$

where $\mathrm{PRODUCT_i}$, $\mathrm{PRODUCT_d}$, and $\mathrm{PRODUCT_o}$ represent the organic reaction products formed when alkoxy RO radicals undergo isomerization, decomposition, or reaction with O_2 to form an aldehyde, ketone, or a multifunctional compound (Dodge, 2000). In the condensations and simplifications made to generate the CBM mechanism, the isomerization and decomposition steps were ignored and the assumption made that all alkoxy radicals derived from PAR, and hence alkanes, react with oxygen. By replacing the RO radical with its products, the $RO_2 + NO$ reaction could be written as:

$$\mathrm{RO_2 + NO = PRODUCT_o + NO_2 + RADICALS}$$

Then, to avoid having to treat each separate peroxy radical formed from each surrogate such as PAR, ETH, TOL, XYL and so on, most of the RO_2 radicals in the CBM mechanism were eliminated by substituting the reaction products formed during the NO to NO_2 conversion and adding an operator that converts NO into NO_2 (Dodge, 2000), as follows:

$$\mathrm{PAR + OH = PRODUCT_o + XO_2}$$

$$\mathrm{XO_2 + NO = NO_2}$$

The species XO_2 is a universal peroxy radical operator that acts as a counter-species to keep track of the rate of NO to NO_2 conversion and hence of ozone formation. In addition to XO_2, an additional operator or counter-species XO_2N is used to account for the formation of nitrates during the oxidation of the alkanes.

Overall, then, the representation of the alkanes in CBM (Gipson and Young, 1999) is as follows:

$$\begin{aligned} PAR + OH = {} & 0.870^{*}XO_2 + 0.130^{*}XO_2N + 0.110^{*}HO_2 + 0.110^{*}ALD2 \\ & + 0.760^{*}ROR - 0.110^{*}PAR \end{aligned}$$

with an assigned rate coefficient of 8.1×10^{-13} cm^3 molecule^{-1} s^{-1}. This is a highly concise and compact treatment of the atmospheric chemistry of alkanes and accounts for the popularity of the CBM mechanism and its widespread application in 3-D urban airshed models where computer resources are limited.

VIII-B-3. Detailed Treatment of Alkanes in the Regional Atmospheric Chemistry Mechanism (RACM)

The Regional Atmospheric Chemistry Mechanism (RACM) was created to represent atmospheric chemistry from polluted urban atmosphere through to the remotest and cleanest regions of the lower atmosphere (Stockwell et al., 1997). The mechanism is based on that originally designed for regional acid deposition modeling (RADM-2) by Stockwell et al. (1990), with revisions to take into account laboratory chemical kinetic studies published in the interim. The full RACM mechanism contains 56 organic compounds and 237 reactions and so is somewhat more complex than the CBM but several orders of magnitude less complex than the SAPRC or MCM mechanisms.

The RACM mechanism allows for the description of five alkanes that are termed methane, ethane (ETH), HC3, HC5, and HC8. Methane and ethane describe in detail the atmospheric chemistry of the individual simple alkanes whereas HC3, HC5, and HC8 represent surrogate alkanes. HC3 refers to the members of the alkane class with relatively low OH + alkane rate coefficients in the range below 3.4×10^{-12} cm^3 molecule^{-1} s^{-1}, namely, propane, butane, 2-methylpropane, and 2,2-dimethylbutane. HC5 refers to alkanes with moderately high OH + alkane rate coefficients in the range 3.4×10^{-12} to 6.8×10^{-12} cm^3 molecule^{-1} s^{-1} and includes pentane, 2-methylbutane, methylpentanes, hexane, 2,3-dimethylbutane, 2,2,4-trimethylpentane, and cyclopentane. HC8 refers to alkanes with relatively high OH + alkane rate coefficients above 6.8×10^{-12} cm^3 molecule^{-1} s^{-1} and includes C_7–C_{15} straight and branched chain alkanes and C_6–C_8 cycloalkanes.

Alkanes react only with OH radicals in the RACM mechanism, according to the following non-stoichiometric equations:

$$OH + CH_4 \rightarrow MO_2 + H_2O$$

$$OH + ETH \rightarrow ETHP + H_2O$$

$$\begin{aligned} OH + HC3 \rightarrow {} & 0.583\ HC3P + 0.381\ HO_2 + 0.335\ ALD + 0.036\ ORA1 \\ & + 0.036\ CO + 0.036\ GLY + 0.036\ OH + 0.010\ HCHO + H_2O \end{aligned}$$

$$OH + HC5 \rightarrow 0.75\ HC5P + 0.25\ KET + 0.25\ HO_2 + H_2O$$

$$OH + HC8 \rightarrow 0.951\ HC8P + 0.025\ ALD + 0.024\ HKET + 0.049\ HO_2 + H_2O$$

Peroxy radicals such as MO_2, ETHP, HC3P, HC5P, and HC8P react with NO, NO_2, HO_2, NO_3, and MO_2. Alkoxy radicals may react with O_2, thermally decompose, or undergo isomerization to form new peroxy radicals. The production of an operator radical XO2 accounts for the additional NO to NO_2 conversion step due to isomerization.

Although the reaction products of OH with alkanes are always alkyl peroxy radicals in reality, other species such as HO_2, OH, higher saturated aldehydes (ALD), formic acid (ORA1), carbon monoxide (CO), glyoxal (GLY), formaldehyde (HCHO), acetone and higher saturated ketones (KET), hydroxyketones (HKET), are assumed to be reaction products in the RACM mechanism for alkanes. This is because alcohols, alkynes, esters, ethers, haloalkanes, glycols, and glycol ethers are grouped together with alkanes into the surrogate species HC3, HC5, and HC8. OH radical attack on some of these non-alkane components of HC3, HC5, and HC8 form oxygenated compounds and HO_2 or OH radicals rather than organic peroxy radicals, and so distort the reaction pathways when they are blended with the alkane pathways. The reaction pathways and rate coefficients for the alkane peroxy and alkoxy radical reactions in RACM are also a blend of the behavior of the corresponding peroxy and alkoxy radicals generated by alkanes and oxygenated compounds.

Stockwell et. al. (1997) describe the testing of the RACM mechanism against 20 smog chamber runs with the evacuable chamber of the Statewide Air Pollution Research Center (SAPRC) (Carter et al., 1995). However, it is not clear how well the mechanism performed for the alkanes, specifically. Nevertheless, the mechanism peformed well for the alkenes and aromatics, reproducing the concentrations of O_3, NO_2, and hydrocarbons measured in smog chambers to within about $\pm$ 30%.

The alkane degradation pathways described in the RACM mechanism, however, do not accurately describe what is known about the atmospheric chemistry of alkanes, because they are blended with the pathways of a range of alkynes and oxygenated compounds. While this is of little consequence for the modeling of ozone on the urban and regional scales for the development of broad ozone control strategies, it could distort the estimation of reactivities for the alkanes in a manner that is difficult to assess.

VIII-B-4. Detailed Treatment of the Alkanes in the SAPRC Mechanism

The State-wide Air Pollution Research Center (SAPRC) mechanism is the most extensive and complete chemical mechanism employed anywhere in the world and is the most important for policy proposed in North America because it underpins the maximum incremental reactivity (MIR) scale. Here the representation of the atmospheric chemistry of the alkanes within the SAPRC mechanism is reviewed and the influence this representation has on the reactivity estimates for the alkanes is assessed.

The SAPRC mechanism has been through several cycles of updating: SAPRC-90 (Carter, 1990), SAPRC-93 (Carter, 1995), SAPRC-97 (Carter et al., 1997), and SAPRC-99 (Carter, 2000), upon which latter version the following paragraphs have been based. The mechanism has five major components:

- The base mechanism;
- A set of estimation procedures for the initial reactions of the organic compounds with the main free radicals;
- A mechanism generation system;
- A set of lumping procedures for condensing the mechanism into product species and chemical operators representing NO to NO_2 conversions and organic nitrate formation;
- Lumping procedures that represent complex mixtures of emitted organic compounds.

The base mechanism represents the atmospheric chemical reactions of the inorganic species and radicals, the common organic reaction products, and the intermediate organic radicals that lead to the organic reaction products. Most of the alkanes are not in the base mechanism but are added to the mechanism as explicit reactions. These explicit reactions are generated automatically using a sophisticated mechanism generator and estimation software system. Although many of the estimated rate coefficients and rate coefficient ratios are highly uncertain, this software system provides a consistent set of assumptions for developing best-estimate mechanisms for the alkanes. It allows for the adjustment of rate coefficients and their ratios where laboratory and smog chamber data are available.

The performance of the SAPRC mechanism in simulating ozone formation, rates of nitric oxide (NO) oxidation, and other measures of smog chamber reactivity have been evaluated by conducting model simulations of over 1600 smog chamber experiments. The performance of the mechanisms for the alkanes was considered satisfactory in the experiments that followed individual alkanes and their incremental reactivities.

The SAPRC mechanism has been used to estimate the MIR reactivity values for 156 alkanes, including straight-chain, branched-chain, and cycloalkanes. In addition, MIR values have been estimated for 13 haloalkanes with chlorine and bromine substituents. However, it has proven to be difficult to find mechanisms that result in fully satisfactory simulations of the experimental data for the haloalkanes. It is possible that, although the SAPRC mechanism generates approximate estimates of haloalkane reactivity under high-NO_x conditions, ozone impacts may be overestimated under low-NO_x conditions where they may actually be inhibitors of ozone formation (Carter, 2000). Consequently, this review concentrates on the representation of the atmospheric chemistry of the straight-chain, branched-chain, and cycloalkanes in the SAPRC mechanism.

Reactivity estimates for the alkanes were obtained by using the SAPRC mechanism in a series of single-day box-model scenarios. The mechanism was embedded in a model that described the emission of ozone precursors into a moving box whose vertical dimensions extended from the surface to the top of the atmospheric boundary layer. The model approach is termed the Empirical Kinetic Model Approach (EKMA). The EKMA approach is simple enough in its formulation to allow for chemical mechanisms of huge complexity without the need for any simplification or condensation. The single box is moved over the urban region with its intense precursor emissions during the early morning, then into the downwind region during the afternoon. During the morning, the boundary layer depth increases from a low nighttime value to its maximum value during the middle of the afternoon. As the boundary layer expands, air is entrained from aloft containing ozone and its precursors.

The single-day box-model scenarios were developed for the purpose of assessing how various ozone precursor emission scenarios would affect ozone air quality exceedance in 39 urban areas of the United States of America (Baugues, 1990). The initial concentrations of ozone, organic compound, and NO_X precursors over each air basin and in the aloft air were specified, together with the diurnal cycles in ozone precursor emission rates, atmospheric boundary layer depths, temperatures, and humidities. Photolysis rates were calculated using light intensities and spectral data for the mid-point of the atmospheric boundary layer, using the procedures developed by Jeffries (1991).

The NO_X emission inputs for each of the 39 base-case scenarios were then adjusted so that the final ozone concentrations showed the maximum sensitivity to changes in the emissions of the organic compounds. These conditions are those of maximum incremental reactivity (MIR). They represent relatively high-NO_X conditions where control of the emissions of organic compounds is the most effective means of reducing ozone formation. There are other ranges of NO_X conditions that can be used to assess reactivities; however, only the MIR scale has received widespread acceptance for policy purposes.

For each alkane, the SAPRC mechanism generates the following elements:

- An understanding of the OH attack on the alkane, its total rate coefficient, location of attack on the carbon skeleton of the alkane, and the nature of the carbon radicals produced;
- An understanding of the fate of the above alkyl carbon radicals and the nature of the alkyl peroxy radicals formed;
- The fate of the alkyl peroxy radicals and the formation of nitrate reaction products and alkoxy radicals;
- The reactions of the alkoxy radicals to form first-generation carbonyl products.

In the SAPRC mechanism, few of the organic radicals produced by OH attack on an alkane and its subsequent complete oxidation are explicitly represented in detail. Rapidly reacting organic radicals, which either react along a single reactive channel or whose competing reaction pathways do not depend on the concentrations of other reacting species, are replaced by the set of products they form. Other radicals have specific simplification procedures applied, with the main exceptions being the methyl peroxy and *tert*-butoxy radicals for which detailed representations are employed without simplification.

The representation of the alkyl peroxy radical reactions is complicated because there is a large number of such radicals formed, and they can react with radicals of the same kind and other peroxy radicals under low-NO_X conditions. In the SAPRC mechanism, alkyl peroxy radicals are represented by the set of products that they would ultimately form if they fully reacted in the presence of NO_X and sunlight, and correction terms are calculated to allow for their other effects on the reaction system. These correction terms allow, for example, for the reaction of alkyl peroxy radicals with HO_2 radicals to form alkyl hydroperoxides, the formation of alkyl nitrates through the reactions with NO, and the formation of organic products from $RO_2 + RO_2$ reactions.

In the SAPRC mechanism, few first-generation reaction products of the OH + alkane reactions are represented in detail, with formaldehyde, acetaldehyde, and acetone being the only such carbonyl products represented explicitly. The reactions of the SAPRC model species RCHO represents all C_{3+} aldehydes, with the photolysis, NO_3, and OH reaction behavior following that of acetaldehyde. Methyl ethyl ketone (MEK) is used to represent ketones and other non-aldehyde oxygenated intermediate reaction products that react with OH radicals with OH rate coefficients in the range 5.0×10^{-13} to 5.0×10^{-12} cm^3 molecule^{-1} s^{-1}. Methyl hydroperoxide represents the behavior of all alkyl hydroperoxides (ROOH) formed. A lumped organic nitrate model species, RNO_3, is used to represent the various alkyl nitrate species formed from alkyl peroxy radicals, and this species is consumed primarily by OH radical oxidation and by photolysis.

Total OH radical rate coefficients for the H-atom abstraction reactions of the alkanes are obtained from experimental data or by estimation procedures, using the group additivity method developed by Atkinson (1987) as updated by Kwok and Atkinson (1995) and Kwok et al. (1996). For the 46 alkanes where both experimental and estimated total OH abstraction rate coefficients are available, the average bias was 2% and the average error, independent of sign, was 11% in the estimation methods for the alkanes within the SAPRC mechanism.

An important factor, that needs to be considered in the representation of the atmospheric chemistry of the alkanes, is the branching ratio in the reactions of the alkyl peroxy radicals with NO between the formation of NO_2 and the corresponding alkoxy radical and the addition and rearrangement to form organic nitrate:

$$RO_2 + NO \rightarrow RO + NO_2 \qquad (k_A)$$

$$RO_2 + NO + M \rightarrow RONO_2 + M \qquad (k_N)$$

The rate coefficient ratio $k_N/(k_A + k_N)$ is the nitrate yield and affects the impact of the alkane on ozone formation because nitrate formation is a radical termination process and NO to NO_2 conversion is an ozone formation process. In the SAPRC mechanism, nitrate yields were estimated using the experimental data of Arey et al. (2001) and other smog chamber data when nitrate observations were not available. Temperature and pressure effects were estimated in a relative sense, using the experimental data of Atkinson et al. (1983c).

Alkoxy radicals are critical intermediates in the reactions of alkanes because of the variety of reactions that they undergo in the presence of oxygen (see chapter IV for a complete description). In general, primary and secondary alkoxy radicals can react with oxygen to generate the corresponding carbonyl compounds, C_{2+} alkoxy radicals can react by β-scission forming smaller carbonyl products and radicals, and long-chain alkoxy radicals can undergo 1,5 H-atom shift isomerization reactions, ultimately forming disubstituted reaction products. Long-chain alkoxy radicals react via a cyclic transition state to produce a hydroxyl-substituted carbon-centered radical using the mechanism based on Carter et al. (1976), Baldwin et al. (1977), Carter and

Atkinson (1985) and Atkinson (1994). These estimation procedures indicate that 1,5 H-atom shift isomerization reactions will be relatively rapid and dominate over competing processes where a relatively unstrained 6-member transition state can be formed. Estimation procedures for the β-scission reactions of alkoxy radicals are based on Atkinson (1994, 1997b). Details are given in Carter (2000) for β-scission reactions forming methyl, ethyl, 1-propyl, and 2-propyl radicals and carbonyl compounds.

VIII-B-5. Detailed Treatment of the Alkanes in the Master Chemical Mechanism (MCM)

The Master Chemical Mechanism (MCM) is a large chemical mechanism which represents the chemistry of the radicals and intermediate reaction products formed in the photochemical oxidation of over 100 emitted organic compounds in the presence of NO_x. Few of the commonly made simplification and condensation procedures adopted in many chemical mechanisms are applied, and all the reactions are represented by elementary stoichiometric equations. Nevertheless, some simplifications still have to be made to generate a mechanism that is tractable. These involve, principally, disregarding sites of OH attack on alkanes that have low probability, and simplifying the representation of the myriad of reactions between the many peroxy radicals. Because of the complexity of the chemical mechanism that is produced, its policy applications have necessarily been restricted to implementation within Lagrangian trajectory and empirical kinetic modeling approach (EKMA)-type moving-box models. Because few simplification and condensation procedures have been implemented, the mechanism has been applied mostly in the assessment of the role of individual organic compounds in the formation of photochemical ozone on the regional and transboundary scales in Europe. These multi-day regional-scale episodes tend to occur under low-NO_x conditions that are often far removed from the concentration regimes investigated in environmental chambers. Consequently, the reactivity estimates tend to be different from those appropriate to the intense single-day urban-scale ozone episode conditions found in the cities of the USA. Such differences in reactivity appear strongly in the case of the alkanes.

The master chemical mechanism has been through various cycles of updating and extension and is at version 3.1 (MCM, 2005) at the time of this writing. Currently, MCM v3.1 compiles the atmospheric chemistry of 125 emitted organic compounds, comprising 4351 organic species and 12,691 organic chemical reactions. Of the emitted organic compounds, 22 are alkanes. The explicit reactions of each organic compound are generated using a sophisticated mechanism protocol (Jenkin et al., 1997). Although many of the estimated rate coefficients employed are highly uncertain, the protocol and estimation procedures provide a consistent basis for building up explicit mechanisms for a wide range of organic compounds emitted into the ambient atmosphere in Europe.

The MCM construction protocol defines a series of generic reaction rules that apply to the OH-, O_3-, NO_3-, and photolytically-initiated reactions of organic compounds (Jenkin et al., 1997) and the organic radicals generated by them. The fate of the initially produced organic radicals is to react rapidly with oxygen under atmospheric conditions to form peroxy radicals, and this is assumed in almost all cases in the MCM. There is a small number of exceptions and these are fully treated in the MCM. The MCM goes on to treat the full behavior of the peroxy radical intermediates, the Criegee biradicals, that of the oxy radical intermediates, and the subsequent reactions of the first and subsequent generation degradation products. The initial MCM protocol has been updated for the 107 non-aromatic organic compounds in Saunders et al. (2003) and for the 18 aromatic organic compounds in Jenkin et al. (2003) and Bloss et al. (2005a). In addition, an expert system has been developed for the automatic generation of tropospheric oxidation schemes (Saunders et al., 2003). Figure VIII-B-1 presents a flow chart indicating the major reactions, intermediate classes, and product classes considered in the MCM protocols.

The MCM has been carefully evaluated against laboratory experimental and field campaign data. The butane and isoprene degradation mechanisms in the MCM have been compared against SAPRC environmental chamber data by Pinho et al. (2005); this included an evaluation of the oxidation mechanisms for formaldehyde, acetaldehyde, methyl ethyl ketone, methyl vinyl ketone, and methacrolein. Experimental data from the European Photoreactor (EUPHORE) have been used to evaluate the performance of the MCMv3 for α-pinene (Saunders et al., 2003), benzene, toluene, *p*-xylene and 1,3,5-trimethylbenzene (Bloss et al., 2005b; Wagner et al., 2003), ethylene (Zador et al., 2005), and secondary organic aerosol formation (Jenkin, 2004). Experimental data from the CSIRO indoor environmental chamber for propene and 1-butene

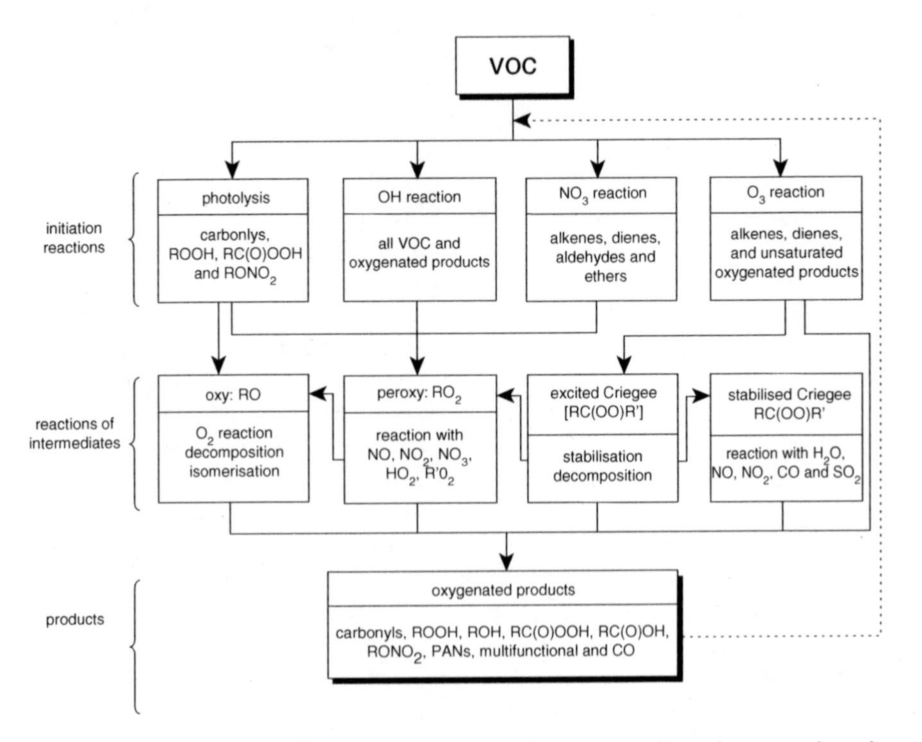

Figure VIII-B-1. Flow chart indicating the major reactions, intermediate classes, and product classes in the MCM protocol (taken from Jenkin et al., 2003).

(Hynes et al., 2005) have been used to test alkene sub-mechanisms. The MCM has been used to estimate OH, HO_2, and RO_2 concentrations during the Berlin Ozone Experiment (BERLIOZ) campaign between July 5 and August 7, 1998, and the results agreed closely with the observations in a high-NO_x regime (Volz-Thomas et al., 2003). The transient radicals OH, HO_2, and RO_2 were estimated using the MCM and compared against observations during the Eastern Atlantic Spring Experiment 1997 at Mace Head, Ireland, and the agreement was found to be best in the cleaner, lower-NO_x air masses (Carslaw et al., 2002).

A detailed representation of the atmospheric chemistry of the alkanes is included in the MCM mechanism. The rate coefficients for OH attack are taken from the literature (Atkinson, 1989, 1994, 1997a; Atkinson et al., 1999) or from estimation methods (Atkinson et al., 1999). The sites of OH attack are identified and the yields of each organic peroxy radical are quantified along with the rate coefficients for each reactive channel for $C_{\leq 6}$ alkanes. However, the number of reactive channels is markedly reduced for $C_{\geq 7}$ alkanes. Each organic peroxy radical (of which there are 902 altogether in MCM v3) is treated separately, and reacts with NO and HO_2 to generate oxy radical and organic hydroperoxide products, respectively. The branches of the $RO_2 + NO$ reaction forming organic nitrates and those of the $RO_2 + HO_2$ reaction forming hydroperoxides are based on experimental data where available and are otherwise treated using estimation techniques. Organic peroxy radicals also react with other radicals to form alcohols, carbonyls, and oxy radicals. Because there are so many possible $RO_2 + R'O_2$ reactions, special techniques have to be employed to handle them accurately but at minimum computational cost. In the MCM, a peroxy radical summation technique is employed, similar to that described by Madronich and Calvert (1990). At every time-step, the total molecular density of organic peroxy radicals is computed and this total peroxy radical reacts with each individual organic peroxy radical with a composite rate coefficient. Reactions of organic peroxy radicals with NO_2 are represented only for peroxyacyl radicals, with the formation of peroxyacyl nitrates.

Oxy radicals may undergo reaction with oxygen, 1,5 H-shift isomerization, or decomposition. Isomerization reactions produce δ-hydroxy alkyl radicals and are treated using estimation procedures outlined by Atkinson (1997b). Hydroxy alkyl radicals react further to generate hydroxyl-substituted carbonyl and nitrate compounds. First-generation reaction products of alkanes are therefore aldehydes, ketones, hydroperoxides, alcohols, and nitrates. Each of these is further processed by the MCM until the final inorganic degradation products, CO, CO_2, and water, are generated.

Table VIII-B-1 presents the numbers of chemical species and chemical reactions required by MCM v3.1 to describe the atmospheric chemistry of a selection of alkanes. The number of species and reactions increases quickly with carbon number and a number of simplifications are essential. The MCM protocol (Jenkin et al., 1997) is designed to incorporate some strategic simplifications in the mechanisms generated. This is achieved by limiting the proliferation of the chemistry related to minor pathways by disregarding sites of OH attack that have low probability and by simplifying the treatment of the degradation of minor reaction products.

Table VIII-B-1. The number of chemical species and reactions required in the MCM v3 to represent the complete degradation of a selected range of alkanes

Alkane	Number of Chemical Species	Number of Chemical Reactions
Methane	17	22
Ethane	43	103
Propane	87	236
Butane	184	538
2-Methylpropane	111	310
Pentane	311	925
2-Methylbutane	299	874
2,2-Dimethylpropane	74	188
Hexane	425	1278
2-Methylpentane	397	1174
3-Methylpentane	289	846
2,2-Dimethylbutane	276	799
2,3-Dimethylbutane	169	478
Heptane	269	785
2-Methylhexane	355	1049
3-Methylhexane	342	984
Octane	313	914
Nonane	367	1083
Decane	369	1081
Undecane	409	1198
Dodecane	411	1204
Cyclohexane	408	1243

For larger molecules, it is usual for a large number of attack positions for OH to exist, and every pathway can reasonably be classified as "minor". In these cases, radical products possessing similarities in their subsequent degradation chemistry are represented by a single radical species in the MCM. The longer chain alkanes ($C_{\geq 7}$) provide a good illustration of this. In the case of decane, for example, attack of OH is collectively dominated by formation of the secondary peroxy radicals 2-, 3-, 4-, and 5-decyl peroxy, with attack at the terminal carbon atom to form the primary peroxy radical, 1-decyl peroxy, accounting for only about 3% of the reaction. The subsequent chemistry of each of the four secondary peroxy radicals proceeds by near-identical pathways. For example, the subsequently formed oxy radicals all react predominantly by 1,5 H-atom shift isomerization, generating four isomeric δ-hydroxyketone products of almost identical reactivity. The rates and mechanisms of the subsequent degradation of these products are therefore also almost identical. In the MCM, therefore, 3-decyl peroxy is used as a single representative for the series of peroxy radicals, and species formed from its further reactions (e.g., the δ-hydroxyketone product, 6-hydroxy-3-decanone) essentially represent the series of isomeric species formed from all the peroxy radicals (e.g., 5-hydroxy-2-decanone, 6-hydroxy-3-decanone, 7-hydroxy-4-decanone, and 8-hydroxy-5-decanone).

To illustrate how the MCM treats the atmospheric degradation of the $C_{\leq 6}$ alkanes, 2-methylpentane [$(CH_3)_2CHCH_2CH_2CH_3$] has been chosen as an example. The complete mechanism for the oxidation of 2-methylpentane to water, CO, and CO_2 contains 397 chemical species and 1174 chemical reactions; see table VIII-B-1. 2-Methylpentane has 6 carbon atoms in a branched configuration and 14 C—H bonds, of which 9 are primary, 4 are secondary, and 1 is tertiary. The bulk of the atmospheric oxidation of 2-methylpentane is initiated by the attack of the OH radical and the MCM recognizes four sites for the OH attack. Of the total flux through the OH + 2-methylpentane reaction, 6% occurs through the attack on the primary C—H bonds in the MCM, 22% on the secondary C—H bonds attached to the less-substituted carbon atom (the 4-position), 27% on the secondary C—H bonds attached to the more-substituted carbon atom (the 3-position), and 45% on the tertiary C-H bond (the 2-position).

The oxidation pathways of 2-methylpentane have been quantified using a photochemical trajectory model (Derwent et al., 1998) to ascertain the identities of the major intermediate reaction products in an air parcel that has undergone a 5-day trajectory across Europe, starting off in Austria and arriving in the United Kingdom. The air parcel continuously picked up ozone precursor emissions and hence reached the trajectory end-point with a complex mixture of unreacted 2-methylpentane parent hydrocarbon and intermediate reaction products. By dividing the mixing ratios of all of the 397 2-methylpentane reaction products by that of 2-methylpentane itself, it was possible to identify the major reaction intermediates at the trajectory end-point. Table VIII-B-2 presents the details of the 21 chemical species, out of the 397, that exhibited mixing ratios that equaled or exceeded that of 2-methylpentane.

There are two main factors that influence the mixing ratio of a given reaction intermediate at the end-point of the 5-day trajectory. The first is the total reaction flux through the reactions that generate that intermediate reaction product. The second is the atmospheric lifetime of that intermediate itself. Of the 21 intermediate reaction products in table VIII-B-2, 7 were ketones, 5 aldehydes, 5 alkyl nitrates, and 4 were peroxyacylnitrates. The common feature of all of these classes of organic compounds is that their atmospheric lifetimes are generally of the order of several hours or longer.

Several of the reaction intermediates identified in table VIII-B-2 are fragments containing 1–3 carbon atoms and hence are difficult to associate with specific degradation pathways of 2-methylpentane. These include acetone (CH_3COCH_3), formaldehyde (HCHO), acetaldehyde (CH_3CHO), propionaldehyde (C_2H_5CHO), peroxyacetyl nitrate (PAN), and peroxypropionyl nitrate PPN. However, the remaining 15 reaction intermediates have unique enough structures so that they can be associated with particular degradation pathways, and these are summarized in table VIII-B-2. Five of these reaction intermediates were formed directly in subsequent reactions of the peroxy radicals formed in the initial attack by OH radicals on 2-methylpentane. These include two ketones, 2-methyl-3-pentanone and 4-methyl-2-pentanone, formed by OH attack at the 3- and 4-positions, respectively. The remaining three reaction intermediates were nitrates: 2-nitrato-2-methyl-pentane, 3-nitrato-2-methyl-pentane, and 4-nitrato-2-methyl-pentane. These were formed from the secondary route of the RO_2 + NO reactions involving the peroxy radicals formed by OH attack at the 2-, 3-, and 4-positions, respectively. Note that the chemistry of 2-methylpentane has not been studied in detail in the laboratory. However, there is a good correspondence between products obtained in the MCM and those outlined in section IV-C-11 which are based on observed and expected behavior of the relevant alkoxy radicals.

Table VIII-B-2. Identification of the major intermediate reaction products found in the atmospheric oxidation of 2-methylpentane in an air parcel traversing Europe in a 5-day trajectory from Austria to the United Kingdom

Intermediate Reaction Product and Structure	Relative Productivity	Notes
$CH_3C(O)CH_3$ Acetone	63.0	Unspecific product
CH_2O Formaldehyde	40.1	Unspecific product
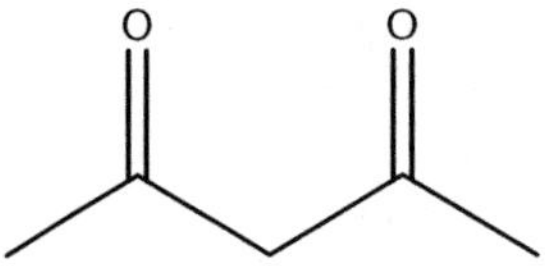 2,4-Pentanedione	14.3	Formed from the product of the isomerization reaction of the alkoxy radical generated following OH attack at the 4-position
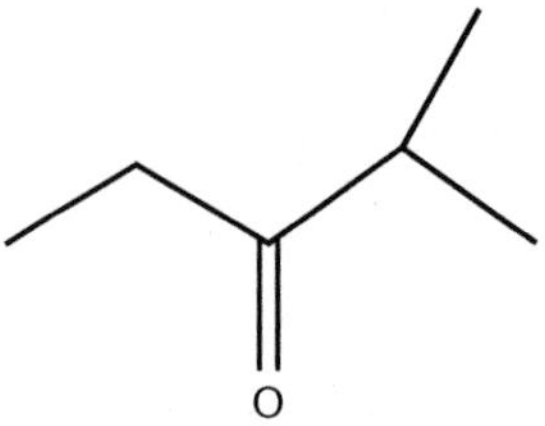 2-Methyl-3-pentanone	11.1	Formed directly from the peroxy radical generated following OH attack at the 3-position
CH_3CHO Acetaldehyde	11.1	Unspecific product
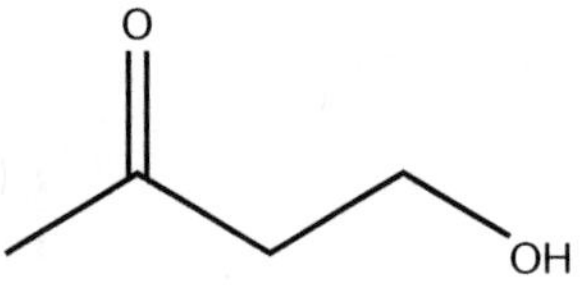 OH 4-Hydroxy-2-butanone	10.8	Formed from the product of the isomerization reaction of the alkoxy radical generated following OH attack at the 2-position
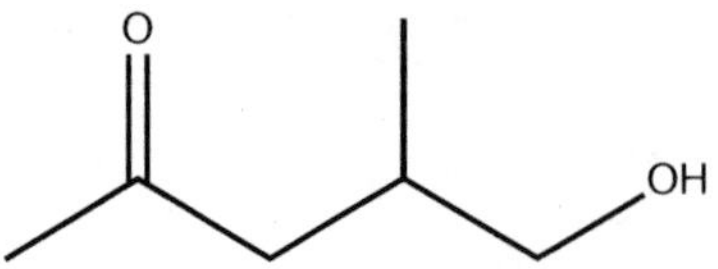 OH 5-Hydroxy-4-methyl-2-pentanone	6.0	Formed from the product of the isomerization reaction of the alkoxy radical generated following OH attack at the 4-position
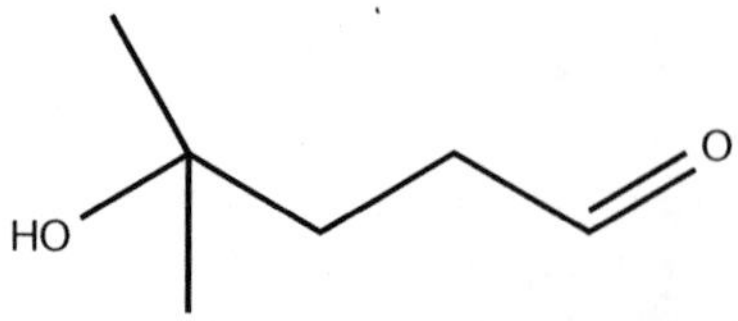 HO 4-Hydroxy-4-methyl-pentanal	6.0	Formed from the product of the isomerization reaction of the alkoxy radical generated following OH attack at the 2-position

Table VIII-B-2. (*continued*)

Intermediate Reaction Product and Structure	Relative Productivity	Notes
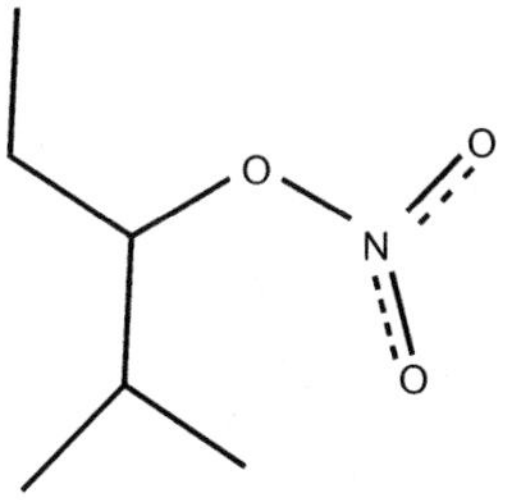 2-Methyl-3-nitrooxy-pentane	5.0	Formed directly from the peroxy radical generated following OH attack at the 3-position
C_2H_5CHO Propionaldehyde	4.8	Unspecific product
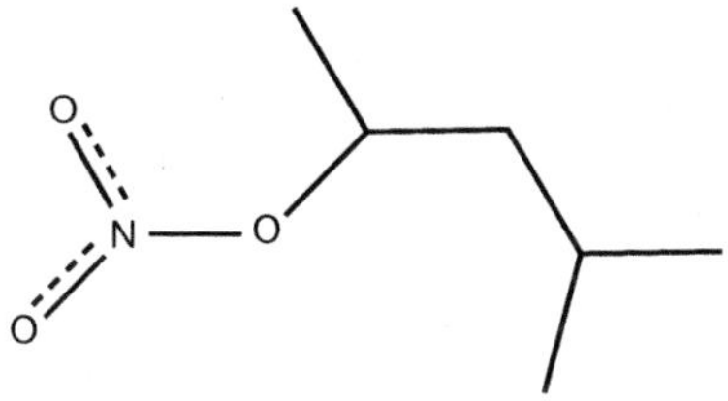 2-Methyl-4-nitrooxy-pentane	4.1	Formed directly from the peroxy radical generated following OH attack at the 4-position
PAN Peroxyacetyl nitrate	3.9	Unspecific product
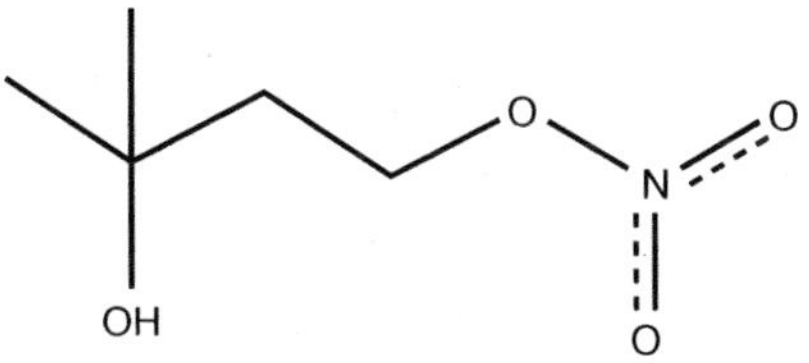 4-Nitrooxy-2-methyl-2-butanol	1.7	Formed from the product of the isomerization reaction of the alkoxy radical generated following OH attack at the 2-position
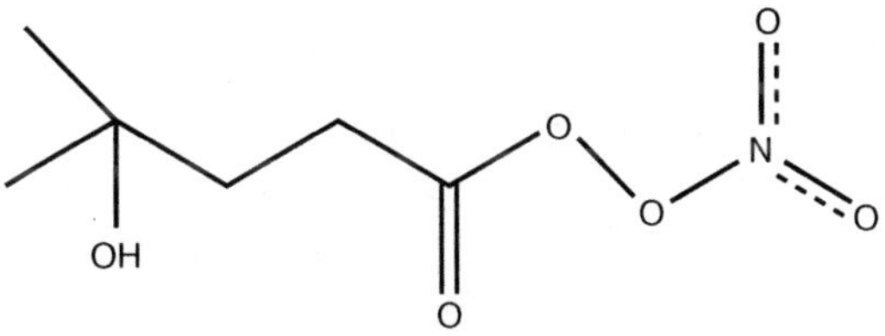 4-Hydroxy-4-methyl-1-nitroperoxypentanal	1.5	Formed from the product of the isomerization reaction of the alkoxy radical generated following OH attack at the 2-position
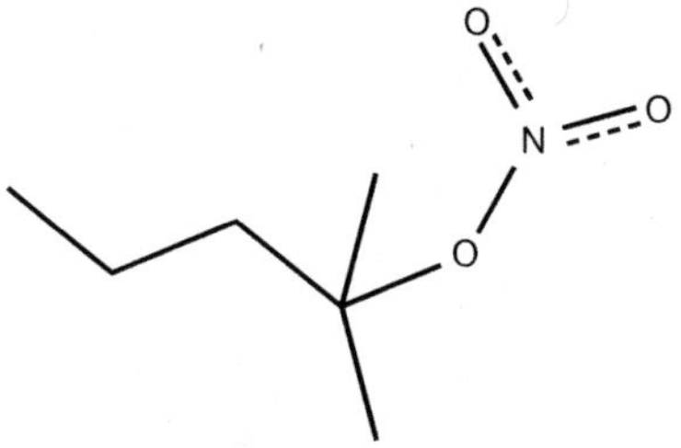 2-Nitrooxy-2-methyl-pentane	1.5	Formed directly from the peroxy radical generated following OH attack at the 2-position

(*continued*)

Table VIII-B-2. (*continued*)

Intermediate Reaction Product and Structure	Relative Productivity	Notes
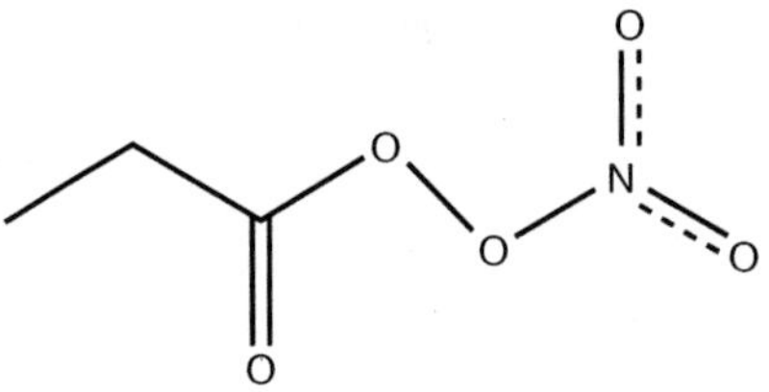 Peroxypropionyl nitrate, PPN	1.4	Unspecific product
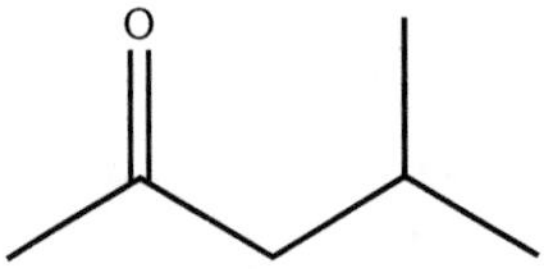 4-Methyl-2-pentanone	1.4	Formed directly from the peroxy radical generated following OH attack at the 4-position
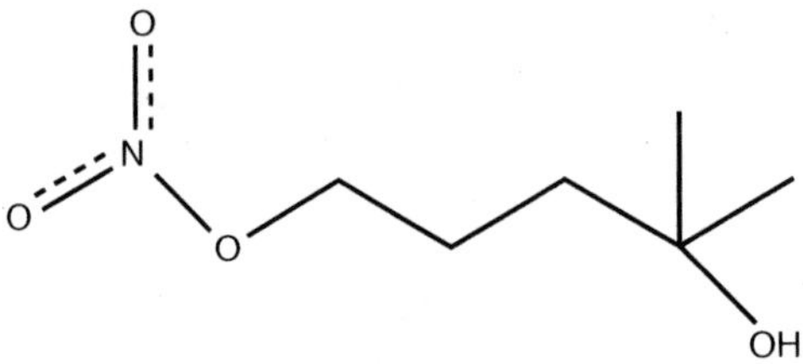 5-Nitrooxy-2-methyl-2-pentanol	1.2	Formed from the product of the isomerization reaction of the alkoxy radical generated following OH attack at the 2-position
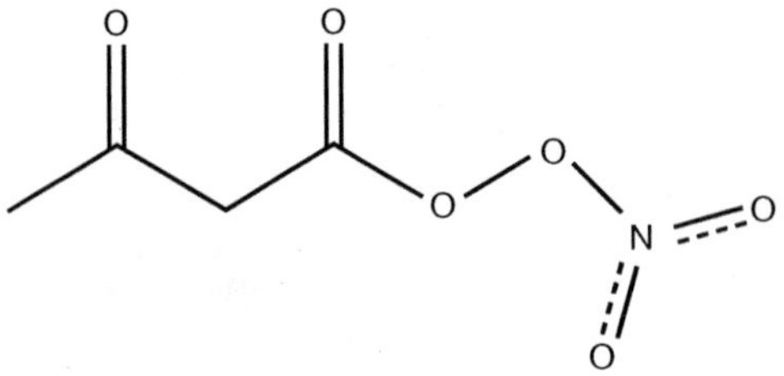 3-Oxo-1-nitroperoxybutanal	1.0	Formed from the product of the isomerization reaction of the alkoxy radical generated following OH attack at the 2-position
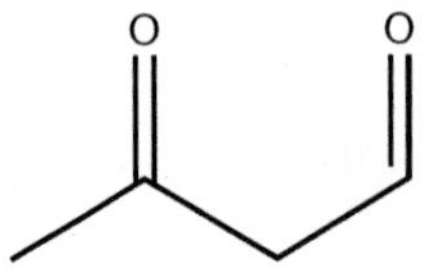 3-Oxo-butanal	1.0	Formed from the product of the isomerization reaction of the alkoxy radical generated following OH attack at the 2-position

[a] The relative productivity column gives the mixing ratio of that species in an air parcel at the arrival point of a five-day trajectory passing over northwestern Europe, expressed as a percentage of that of 2- methylpentane remaining, where the percentage exceeded 1.0%.

The majority of the intermediate reaction products in table VIII-B-2, of which there are 10, are complex di- or tri-functional organic compounds that result from the degradation of hydroxy-substituted peroxy radicals that are formed when alkoxy radicals undergo isomerization reactions. Taking the example of the alkoxy radical formed following OH attack on the tertiary C—H bond (see also figure IV-C-2), this alkoxy radical can isomerize through a 1,5 H-atom shift to form a hydroxyperoxy radical:

$$(CH_3)_2C(O\bullet)CH_2CH_2CH_3 \xrightarrow{(+O_2)} (CH_3)_2C(OH)CH_2CH_2CH_2O_2\bullet$$

Eight of the reaction intermediates in table VIII-B-2 were formed from the degradation of the hydroxyperoxy radical formed in this manner by isomerization of the alkoxy radical formed by the OH attack on the tertiary C—H bond. These include two each of the hydroxyketones, hydroxynitrates, disubstituted PANs, and ketoaldehydes. Finally, two of the intermediate reaction products were formed from the degradation of the hydroxyperoxy radical formed by the isomerization of the alkoxy radical formed by the OH attack at the 4-position. These are 4-oxo-2-pentanone and 5-hydroxy-4-methyl-2-pentanone.

The example of 2-methylpentane shows up the importance of alkoxy radical 1,5 H-atom shift isomerization reactions on the structure and reactivity of the intermediate reaction products generated. With the low molecular mass alkanes such isomerization reactions do not occur, and alkoxy radicals invariably form ketones and aldehydes, often with some fragmentation involved. With the higher molecular mass alkanes, isomerization reactions are favored and intermediate reaction products are often multifunctional compounds without any significant fragmentation. As detailed later in section VIII-C-2, this change in behavior may exert an important impact on reactivity.

VIII-B-6. Self-Generating Explicit Mechanism

Major advances have been made towards the development of more explicit representations of the atmospheric chemistry of hydrocarbons, alkanes included, as replacements for the highly simplified and condensed mechanisms often used in urban and regional models. The National Center for Atmospheric Research Master Mechanism (Madronich and Calvert, 1990; Aumont et al., 2000), with over 5000 reactions to represent the oxidation of 20 emitted hydrocarbons is an important early example of such an explicit mechanism. But even this explicit mechanism neglects many reaction pathways by assuming that they individually may make only a minor contribution. Aumont et al. (2005) have described a "self-generating" approach to develop fully explicit atmospheric chemistry mechanisms for selected organic compounds in a level of detail that would be difficult, if not impossible, to write manually.

The self-generating approach has two main elements (Aumont et al., 2005):

- A generator: this is a computer program that produces the oxidation mechanism for a set of emitted organic compounds based on a predefined protocol;
- A protocol: this is a set of rules that lays out the choice of reaction pathways and estimates the rate coefficients needed in the mechanism on the basis of the molecular structure of the species.

The construction of the chemical scheme generator requires the identification of all of the reactions for each emitted organic compound and for each intermediate reaction product in the complete degradation of each emitted compound into CO and CO_2. These reactions generally include:

- Initiation of the atmospheric degradation by the attack of OH, NO_3, O_3, or photolysis, leading to the formation of peroxy radicals;
- Reactions of peroxy radicals with NO, NO_2, NO_3, HO_2, and with other RO_2 radicals, leading to the formation of stable reaction intermediates or alkoxy radicals (RO);
- Alkoxy radical reactions with O_2, unimolecular decomposition, or isomerization, leading to the formation of stable reaction intermediates or new peroxy radicals.

The generator is therefore a computer program or expert system that mimics the steps by which chemists develop chemical mechanisms manually. First, the generator analyses the chemical structure of the emitted organic compound or stable reaction intermediate to identify reactive sites and determine all reaction pathways. For each identified reaction the generator then searches a database of laboratory measurements to determine whether they are already available. If not, an estimation is made of the reaction pathway and rate coefficient. Details of the reaction are then added to the mechanism and the generator moves on to the next emitted organic compound or reaction intermediate. The complete chemical mechanisms written by the generator are, in principle, no different from those written manually by chemists, given enough time. However, there are several important advantages to the automatic method, namely, speed, accuracy, and ease of updating.

The total number of species generated to describe the full oxidation grows exponentially with increasing carbon number. It is both possible and desirable to set a "de minimis" threshold to cut down the number of trivial reaction pathways. For propane, the number of chemical species in the generated mechanism is about 400, but the number rises to about 460,000 with heptane. In each case, the number of chemical reactions identified is about 10 times the number of species, reaching about 25 million for the oxidation of octane. The exponential growth of the number of generated species as a function of the carbon chain length can be explained by the exponential increase of the total number of species that can be produced by permuting all the possible functional groups produced during the oxidation of the parent compounds. For the alkanes, these groups are non-substituted alkyl groups, ketones, aldehydes, alcohols, hydroperoxides, nitrates, peroxy radicals, alkoxy radicals, carboxylic acids, peracids, peroxyacyl nitrates, and peroxyacyl radicals. The generator does not produce pathways leading to carbon atoms bearing two functionalities.

For the vast majority of the reactions inferred by the generator, there are no direct laboratory measurements, and hence reaction pathways and rate coefficients have to be inferred for them. This is the role of the protocol and the published structure–activity relationships that underpin it. The main feature of the generator described above is to codify the various estimation methods to produce consistent and comprehensive oxidation schemes on a systematic basis. Structure–activity relationships are used to describe VOC + OH, VOC + NO_3, VOC + O_3, VOC photolysis, peroxyacyl nitrate decomposition, peroxy radical, alkoxy radical, and Criegee radical reactions. Details of these relationships are given in Laval-Szopa (2003). In addition, the protocol describes how the steady-state approximation is used to decrease the number of reactive species in the mechanism, if they are highly reactive and possess a unique reaction pathway, by replacing them with their reaction products.

A predefined set of rules can also be provided to the generator as an additional "protocol" devoted to reducing the size of the generated mechanism, namely, rules to lump species and reactions according to their contribution to the budget of some target species. Various commonly used reduction methods were tested and implemented in the generator to simulate the gas-phase chemistry in various tropospheric situations, from polluted to remote conditions (Szopa et al., 2005), ultimately leading to a mechanism small enough to be included in current three-dimensional models.

To evaluate the chemical mechanisms for the alkanes written by the generator, some comparisons were made with the SAPRC (Carter, 2000) and MCM (Saunders et al., 2003) mechanisms for the oxidation of heptane. Concentration–time profiles were compared for a simple box model containing NO_X, O_3, and heptane undergoing atmospheric photolysis at the 10–40 ppb concentration level. The time profiles for certain tested O_3, NO_X, and HO_X species simulated with the self-generating approach, SAPRC, and MCM showed fairly good agreement, with mixing ratios matching to within $\pm$ 20% between all three schemes; see figure VIII-B-2. Secondary reaction products such as formaldehyde showed surprisingly good agreement between the three mechanisms, despite their huge difference in complexity. The representation of heptane in the self-generating approach required 460,000 species and 4,500,000 reactions, compared with 269 species and 789 reactions, (see table VIII-B-1) with the MCM.

VIII-C. Determination of VOC Reactivity Scales

VIII-C-1. Assessment of Reactivities of Alkanes Using the CBM Mechanism

The Reactivity Research Working Group (RRWG, 1999) coordinates policy-relevant research related to VOC reactivity and has overseen three studies undertaken with complex 3-D urban airshed models and performed by Carter et al. (2003), Arunachalam et al. (2003), and Hakami et al. (2003, 2004b). Taken together, these three studies have made a significant contribution to basic understanding of the likely role that reactivity considerations could play in ozone control strategies in the United States of America. These studies have shown that reactivity-based policies should work efficiently on both the urban and regional scales by reducing episodic peak ozone levels and by reducing the exceedances of the ozone air quality standards. The Carter et al. (2003) and Arunachalam et al. (2003) studies both involved the application of the CBM mechanism within 3-D airshed models and so are described in some detail below, with a view to understanding more about the assessment of the reactivities of alkanes with the CBM mechanism.

Carter et al. (2003) describe how they have employed the CAMx model to study reactivity metrics across the eastern United States. The study period chosen was the NARSTO-NE episode which covered four days during July 1995. The model employed

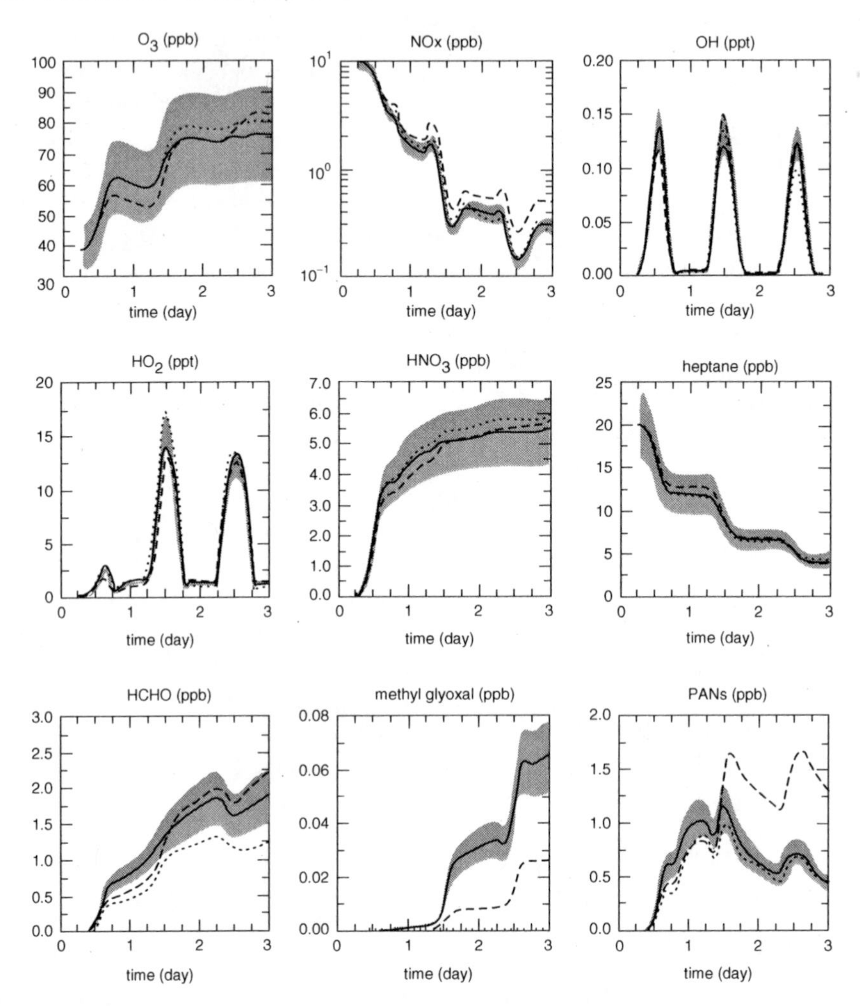

Figure VIII-B-2. Comparison of the time development of a range of reaction products in the photooxidation of *n*-heptane, simulated with three chemical mechanisms: self-generating (solid lines), MCM (dashed lines), and SAPRC (dotted lines); printed with permission from Aumont et al. (2005).

the Carbon Bond Mechanism (CB4) and was run at 4 km × 4 km, 12 km × 12 km, and 36 km × 36 km scales. Reactivity metrics were estimated for four episode days, three grid resolutions, and two ozone air quality standards. In each case, a reactivity scale was produced comprising the following nine VOC species: OLE (olefins), formaldehyde, acetaldehyde, ethene, xylenes, ethanol, PAR (alkanes), ethane, and TOL (toluene).

Carter et al. (2003) comment that the differences between the various reactivity metrics for the same species were less than the differences between the species for the same metric. Hence, in each case, a usable reactivity scale could be constructed from the available results. In general, the use of the 8-hour or 1-hour ozone standards made little difference to the reactivity metrics and the use of cut-offs in the ozone levels of up to 80 ppb also did not significantly change the reactivity metrics. Significant differences were seen between the 4 km × 4 km grid resolutions and the 12 km × 12 km and 36 km × 36 km resolutions because of the better representation at the 4 km × 4 km resolution of urban source regions which are more VOC-sensitive. In summarizing their results, Carter et al. (2003) noted that for most model species the reactivity metrics determined with the SAPRC mechanism were surprisingly close to the relative reactivity metrics derived from the 3-D regional airshed model, using the CBM mechanism for both ethane and the surrogate alkane species PAR.

In addition to the studies of reactivity metrics discussed above, Carter et al. (2003) also performed some VOC substitution experiments with a view to evaluating the impact of the substitution of current VOC emissions with low reactivity or borderline exempt VOCs. Ethane was the chosen surrogate for these low-reactivity VOCs. Varying amounts of ethane were added back to replace the entire man-made VOC emission inventory. This type of substitution led to almost as much ozone reduction as the complete curtailment of man-made VOC emissions. If enough of a VOC of low but positive reactivity such as ethane was added, then ozone levels would eventually equal or exceed current levels. This level of substitution was found to involve replacing all man-made VOC emissions with about five times the mass of ethane.

Arunachalam et al. (2003) have used the Multi-scale Air Quality Simulation Platform (MAQSIP) model (Odman and Ingram, 1996) to investigate the impact of reactivity-based VOC emission control policies on the attainment of ozone air quality standards. The MAQSIP model was run for the August–September 2000 period of the Texas AQS campaign and for a period during June 1996 over the eastern United States. For the Texas study, the model resolutions were 5 km × 5 km, 15 km × 15 km and 45 km × 45 km, whereas for the eastern United States study they were set at 12 km × 12 km and 36 km × 36 km. Two different chemical mechanisms were employed, namely Carbon Bond (CB4) and the Regional Acid Deposition Model (RADM-2) (Stockwell et al., 1990) mechanisms, and seven different ozone exposure metrics were calculated. The RADM-2 mechanism is the precursor to the Regional Atmospheric Chemistry Mechanism (RACM) which was discussed in section VIII-B-3. The reactivity-based substitution scenarios evaluated were: a 15% reduction in man-made VOC emissions, substitution of 15% of all man-made VOC emissions with butanone, and substitution of non-mobile xylene emissions with 2-butoxyethanol. The key finding presented by Arunachalam et al. (2003) was that reactivity-based VOC substitution strategies resulted in reduced ozone levels, particularly in source regions and in areas downwind of major urban and industrial centers. Noticeable changes in ozone were seen with the 15% substitution by butanone in a region dominated by natural biogenic emissions and not usually thought to be VOC-sensitive. The substitution of the highly reactive xylene with 2-butoxyethanol was beneficial in reducing exceedances of the ozone standards, but small increases were noted in some ozone metrics, in some model domains, and on some days. VOC substitution strategies appear to work better on peak ozone levels.

Because of its formulation, the CBM mechanism assumes that the per-carbon reactivities of the alkanes are the same regardless of the alkane. In contrast, the reactivities of the alkanes estimated using the SAPRC or MCM mechanisms, as shown in sections VIII-C-2 and VIII-C-3, increase with increasing carbon number to a maximum at about C_4–C_8 before declining monotonically up to C_{12} in the MCM and C_{22} in SAPRC. Reactivity predictions using the CBM are therefore unlikely to be consistent with either those from the SAPRC and MCM mechanisms or with our current understanding of alkane chemistry.

VIII-C-2. Assessment of the Reactivities of Alkanes Using the SAPRC Mechanism

Estimates of MIR reactivities, averaged over all the 39 urban scenarios with the EKMA model, are tabulated in table VIII-C-1. Clearly, uncertainties are present in these estimates as a consequence of the limits to the adequacy and completeness of the EKMA model scenarios and the experimental data used in the SAPRC base mechanism and the explicit extensions dealing with the alkanes. Carter (2000) assigned subjective uncertainty codes to the MIR values for each alkane, expressing the likelihood that the mechanism estimation procedures and hence MIR reactivities would change significantly as new experimental data became available. For the alkanes, ranges of uncertainty for MIR reactivities could be as large as a factor of two, that is, ±33%. In addition, some MIR reactivities may be particularly sensitive to ambient conditions such as NO_x levels.

MIRs in table VIII-C-1 for the alkanes cover the range from 0.0036 for 1,1,1-trichloroethane (methyl chloroform) to 2.69 for cyclopentane. Only six alkanes exhibit MIRs that are higher than 2.0, and, excepting 3-methylpentane, they are all derivatives of cyclopentane. In the context of the entire MIR reactivity scale, the alkanes as a class appear to show low to medium reactivity. In comparison, the lowest reactivity alkene is ethylene with a MIR of 9.08 and the lowest reactivity aromatic compounds are benzene and toluene with MIRs of 0.81 and 3.97, respectively. On this basis it is not surprising, therefore, that alkanes play only a small role in stimulating photochemical ozone formation on the urban scale in the North American cities, as discussed in chapter I.

Within the range of alkanes detailed in table VIII-C-1, there are some interesting variations with carbon chain length and extent of branching that are illustrated in figure VIII-C-1. For the homologous series of straight-chain alkanes and cycloalkanes, MIRs increase quickly from relatively low values for low carbon numbers, reach maxima with carbon numbers of C_5 and C_6, and then decline steadily with increasing carbon number monotonically to C_{22} and C_{16}, respectively. The MIRs of 2-methyl-substituted alkanes are somewhat higher than the corresponding values of the straight-chain alkanes with the same carbon numbers. Two alkanes appear to be outliers from the main body of points: neopentane and 2,2,3,3-tetramethylbutane. These are both alkanes with only primary C—H bonds and no secondary or tertiary C—H bonds, which illustrates how reactivity is inhibited by the substitution of tertiary C—H bonds with methyl groups.

The main feature of figure VIII-C-1 is that MIRs decline steadily with increasing carbon number from about C_6 to C_8 onwards despite steadily increasing rate coefficients for OH + alkane reactions across the entire plot. Since the fraction of emitted alkane that is reacting within the EKMA scenario follows the reaction rate coefficient for

Table VIII-C-1. Maximum incremental reactivities (MIRs) for alkanes and haloalkanes under the conditions of intense urban-scale photochemical ozone formation in North America (taken from Carter, 2000)

Alkane	MIR
a) Straight-chain alkanes	
Methane (CH_4)	0.01
Ethane (C_2H_6)	0.31
Propane (C_3H_8)	0.56
Butane (n-C_4H_{10})	1.33
Pentane (n-C_5H_{12})	1.54
Hexane (n-C_6H_{14})	1.45
Heptane (n-C_7H_{16})	1.28
Octane (n-C_8H_{18})	1.11
Nonane (n-C_9H_{20})	0.95
Decane (n-$C_{10}H_{22}$)	0.83
Undecane (n-$C_{11}H_{24}$)	0.74
Dodecane (n-$C_{12}H_{26}$)	0.66
Tridecane (n-$C_{13}H_{28}$)	0.62
Tetradecane(n-$C_{14}H_{30}$)	0.58
Pentadecane(n-$C_{15}H_{32}$)	0.56
Hexadecane(n-$C_{16}H_{34}$)	0.52
Heptadecane(n-$C_{17}H_{36}$)	0.49
Octadecane(n-$C_{18}H_{38}$)	0.46
Nonadecane(n-$C_{19}H_{40}$)	0.44
Eicosane(n-$C_{20}H_{42}$)	0.42
Heneicosane(n-$C_{21}H_{44}$)	0.40
Docosane(n-$C_{22}H_{46}$)	0.38
b) Specific branched alkanes	
2-Methylpropane	1.35
2-Methylbutane	1.68
2,2-Dimethylpropane	0.69
2,2-Dimethylbutane	1.33
2,3-Dimethylbutane	1.14
2-Methylpentane	1.80
3-Methylpentane	2.07
2,2,3-Trimethylbutane	1.32
2,2-Dimethylpentane	1.22
2,3-Dimethylpentane	1.55
2,4-Dimethylpentane	1.65
2-Methylhexane	1.37
3,3-Dimethylpentane	1.32
3-Methylhexane	1.86
2,2,3,3-Tetramethylbutane	0.44
2,2,4-Trimethylpentane	1.44
2,2-Dimethylhexane	1.13
2,3,4-Trimethylpentane	1.23
2,3-Dimethylhexane	1.34
2,4-Dimethylhexane	1.80
2,5-Dimethylhexane	1.68
2-Methylheptane	1.20
3-Methylheptane	1.35
4-Methylheptane	1.48
2,2,5-Trimethylhexane	1.33
2,3,5-Trimethylhexane	1.33
2,4-Dimethylheptane	1.48
2-Methyloctane	0.96
3,3-Diethylpentane	1.35
3,5-Dimethylheptane	1.63
4-Ethylheptane	1.44
4-Methyloctane	1.08
2,4-Dimethyloctane	1.09
2,6-Dimethyloctane	1.27
2-Methylnonane	0.86
3,4-Diethylhexane	1.20
3-Methylnonane	0.89
4-Methylnonane	0.99
4-n-Propylheptane	1.24
2,6-Dimethylnonane	0.95
3,5-Diethylheptane	1.21
3-Methyldecane	0.77
4-Methyldecane	0.80
2,6-Diethyloctane	1.09
3,6-Dimethyldecane	0.88
3-Methylundecane	0.70
5-Methylundecane	0.72
3,6-Dimethylundecane	0.82
3,7-Diethylnonane	1.08
3-Methyldodecane	0.64
5-Methyldodecane	0.64
3,7-Dimethyldodecane	0.74
3,8-Diethyldecane	0.68
3-Methyltridecane	0.57
6-Methyltridecane	0.62
3,7-Dimethyltridecane	0.64
3,9-Diethylundecane	0.62
3-Methyltetradecane	0.53
6-Methyltetradecane	0.57
3-Methylpentadecane	0.50
4,8-Dimethyltetradecane	0.58
7-Methylpentadecane	0.51

(*continued*)

Table VIII-C-1. *(continued)*

Alkane	MIR
c) Unspecific branched alkanes	
Branched C_5alkanes	1.68
Branched C_6 alkanes	1.53
Branched C_7alkanes	1.63
Branched C_8 alkanes	1.57
Branched C_9 alkanes	1.25
Branched C_{10} alkanes	1.09
Branched C_{11} alkanes	0.87
Branched C_{12} alkanes	0.80
Branched C_{13} alkanes	0.73
Branched C_{14} alkanes	0.67
Branched C_{15} alkanes	0.60
Branched C_{16} alkanes	0.54
Branched C_{17} alkanes	0.51
Branched C_{18} alkanes	0.48
d) Cycloalkanes	
Cyclopropane	0.103
Cyclobutane	1.05
Cyclopentane	2.69
Cyclohexane	1.46
2-Propylcyclopropane	1.52
Methylcyclopentane	2.42
1,3-Dimethylcyclopentane	2.15
Cycloheptane	2.26
Ethylcyclopentane	2.27
Methylcyclohexane	1.99
1,3-Dimethylcyclohexane	1.72
Cyclooctane	1.73
Ethylcyclohexane	1.75
1-Propylcyclopentane	1.91
1,1,3-Trimethylcyclohexane	1.37
1-Ethyl-4-methylcyclohexane	1.62
1-Propylcyclohexane	1.47
1,3-Diethylcyclohexane	1.34
1,4-Diethylcyclohexane	1.49
1-Methyl-3-*iso*-propylcyclohexane	1.26
1-Butylcyclohexane	1.07
1,3-Diethyl-5-methylcyclohexane	1.11
1-Ethyl-2-*n*-propylcyclohexane	0.95
1-Pentylcyclohexane	0.91
1,3,5-Triethylcyclohexane	1.06
1-Methyl-4-pentylcyclohexane	0.81
1-Hexylcyclohexane	0.75
1,3-Diethyl-5-pentylcyclohexane	0.99
1-Methyl-2-hexylcyclohexane	0.70
1-Heptylcyclohexane	0.66
1,3-Dipropyl-5-ethylcyclohexane	0.94
1-Methyl-4-heptylcyclohexane	0.58
Octylcyclohexane	0.60
1,3,5-Tripropylcyclohexane	0.90
1-Methyl-2-octylcyclohexane	0.60
1-Nonylcyclohexane	0.54
1,3-Dipropyl-5-butylcyclohexane	0.77
1-Methyl-4-nonylcyclohexane	0.55
1-Decylcyclohexane	0.50
e) Unspecific cycloalkanes	
C_6 cycloalkanes	1.46
C_7 cycloalkanes	1.99
C_8 cycloalkanes	1.75
C_9 bicycloalkanes	1.57
C_9 cycloalkanes	1.55
C_{10} bicycloalkanes	1.29
C_{10} cycloalkanes	1.27
C_{11} bicycloalkanes	1.01
C_{11} cycloalkanes	0.99
C_{12} bicycloalkanes	0.88
C_{12} cycloalkanes	0.87
C_{13} bicycloalkanes	0.79
C_{13} cycloalkanes	0.78
C_{14} bicycloalkanes	0.71
C_{14} cycloalkanes	0.71
C_{15} bicycloalkanes	0.69
C_{15} cycloalkanes	0.68
C_{16} cycloalkanes	0.61
f) Haloalkanes	
Chloromethane	0.03
Chloroethane	0.25
Dichloromethane	0.07
Bromomethane	0.02
1,1-Dichloroethane	0.10
1,2-Dichloroethane	0.10
Bromoethane	0.11
Trichloromethane	0.03
1-Bromopropane	0.35
1,1,1-Trichloroethane	0.004
1,1,2-Trichloroethane	0.06
1-Bromobutane	0.60
1,2-Dibromoethane	0.05

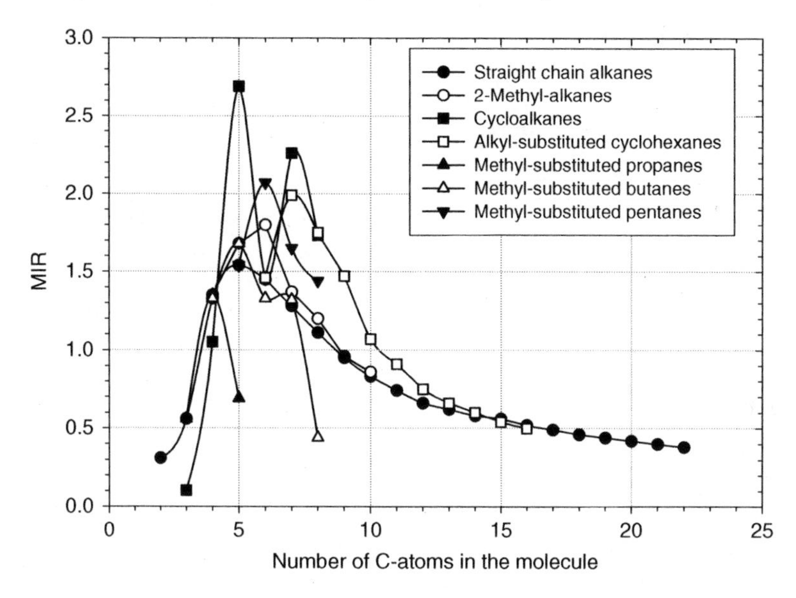

Figure VIII-C-1. Variation in the MIRs for alkanes with increasing carbon number.

OH attack, this also increases steadily across the plot. However, the propensity for the radicals produced in the initial OH attack to produce ozone is declining steadily because of the increasing tendency for the alkyl peroxy radicals to form alkyl nitrates rather than convert NO into NO_2. Alkyl nitrate formation acts as a loss of both HO_x radicals and NO_x. The latter is relatively unimportant in the MIR scenarios because of the relatively high loadings of NO_x employed. Loss of free radicals, however, is important and this leads to the loss of ozone production despite the increased alkane consumption with increasing carbon number. The MIRs for the alkanes show how the detailed understanding of the degradation pathways for the alkanes has been incorporated into the SAPRC mechanism, and this strongly influences reactivity estimates. Without the accurate treatment of alkyl nitrate formation, reactivities for the alkanes would increase steadily with increasing carbon number, following the increasing OH reactivity.

A useful way to judge the overall accuracy of the MIR reactivities, and hence the SAPRC mechanism that underpins them, is to compare them with reactivities taken from a completely independent scale. This has been attempted by Derwent et al. (2001), who compared the MIR reactivities estimated by Carter (1998) with those estimated on the Photochemical Ozone Creation Potential (POCP) scale using the Master Chemical Mechanism. Both studies used the EKMA scenarios developed by Baugues (1990) and addressed the intense single-day urban-scale ozone episode conditions for the cities of the United States of America. Both mechanisms were developed using the same experimental database, though they have radically different philosophies and employ different techniques. Figure VIII-C-2 compares the MIR reactivities from Carter (2000) with the POCP reactivities from Derwent et al. (2001) for the 21 alkanes for which there have been independent assessments on the two scales. The correlation between the two scales is reasonable, and there are few significant differences in the rankings

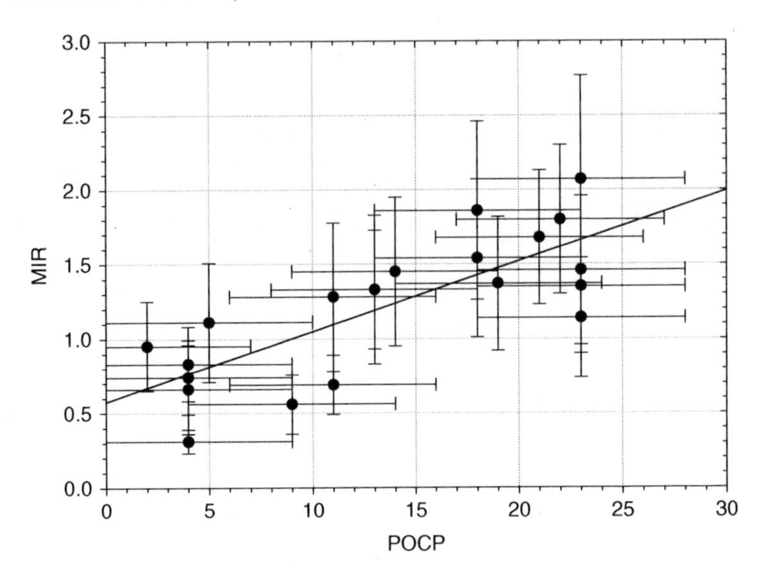

Figure VIII-C-2. Comparison of the MIR and POCP reactivities for selected alkanes, determined for the intense urban-scale photochemical ozone formation conditions of North American cities.

of the different alkanes on the different scales, taking into account the error bars on the respective estimates. On this basis, it is concluded that both mechanisms have been able to represent adequately and consistently the known aspects of the atmospheric chemistry of the alkanes under the high-NO_X conditions appropriate to the intense urban-scale single-day ozone episodes found in North America.

Hakami et al. (2003, 2004a) describe how they have used the urban 3-D airshed models, Urban-to-Regional Multi-scale (URM) model (Boylan et al., 2002) and Multi-scale Air Quality Simulation Platform (MAQSIP) model (Odman and Ingram, 1996), to estimate reactivity metrics using the SAPRC-99 mechanism over the eastern United States of America and in central California. The Maximum Incremental Reactivity MIR-3D, of all of the reactivity metrics tested, gave the best agreement with box-model MIR reactivity metrics for all study areas and episodes. The reactivity metrics for the seven alkanes, ethane, butane, 2-methylpropane, 2-methybutane, methylcyclopentane, 2,2,4-trimethypentane, and dodecane, were estimated using the urban 3-D airshed model and were found to agree closely with comparable box-model reactivities estimated using the same version of the SAPRC mechanism. This level of agreement suggests that MIR reactivities are robust measures of reactivity and that they are likely to be applicable to wider areas than just the Los Angeles basin.

It is instructive to compare the urban 3-D airshed model reactivities reported by Hakami et al. (2003, 2004b) with those by Carter et al. (2003). Despite using different models and chemical mechanisms (SAPRC versus CBM), similar conclusions were reached about the robustness of the reactivity metrics between the different model domains and episodes. Relative metrics for the highly reactive organic compounds were similar between the studies. Necessarily, however, the coverage of the alkanes was limited because of the shortcomings of the CBM mechanism. This comparison does, however, point to the usefulness and robustness of the MIR reactivity metrics in the policy context and their wide applicability in the formulation of ozone control strategies in North America.

VIII-C-3. Assessment of Reactivities of the Alkanes Using the MCM Mechanism

An important policy application for the MCM mechanism has been the study of the reactivities of organic compounds under the conditions appropriate to the regional-scale transboundary formation and transport of ozone as observed in Europe. The reactivity scale developed using the MCM under these conditions is termed the Photochemical Ozone Creation Potential (POCP) scale (Derwent et al., 1998). This scale involves the calculation of the additional ozone formed in an air parcel following a five-day trajectory across Europe when the European emissions of a given organic compound are incremented over those in a base case. By indexing the additional amounts of ozone formed to those produced by the same mass of ethylene, a relative reactivity scale can be calculated. This is the POCP scale; the POCPs for a selected range of alkanes are presented in table VIII-C-2.

POCPs for the *n*-alkanes increase quickly with carbon number from ethane to *n*-butane and reach a maximum with 2,3-dimethylbutane and 3-methylpentane. They then steadily decline with increasing carbon number from *n*-hexane to *n*-dodecane. Cyclohexane shows the lowest POCP of any C_6 alkane, with about three-quarters of the reactivity of *n*-hexane. POCPs follow OH-reaction rate coefficients and increase from ethane to *n*-pentane. However, they decrease with increasing carbon number from C_5 onwards because of the increasing propensity of alkyl peroxy radicals to form organic nitrates. Hence, POCPs do not monotonically increase from C_5 onwards following OH-reaction rate coefficients because of these important mechanistic influences that inhibit ozone formation. This behavior is strongly reminiscent of that shown

Table VIII-C-2. Photochemical ozone creation potentials (POCPs) for selected alkanes determined using the MCM in an air parcel following a five-day trajectory across Europe

Alkane	POCP[a]	Alkane	POCP[a]
Ethane	10	2,3-Dimethylbutane	52
Propane	16	Heptane	38
Butane	32	2-Methylhexane	34
2-Methylpropane	31	3-Methylhexane	45
Pentane	40	Octane	36
2-Methylbutane	37	Nonane	36
2,2-Dimethylpropane	17	Decane	37
Hexane	41	Undecane	35
2-Methylpentane	40	Dodecane	34
3-Methylpentane	47	Cyclohexane	29
2,2-Dimethylbutane	23		

[a] POCPs are expressed as indices relative to ethylene = 100; POCPs are defined as follows for the *i*th organic compound:

$$\mathrm{POCP}_i = \frac{\text{Ozone increment for the } i\text{th compound}}{\text{Ozone increment for ethylene}} \times 100$$

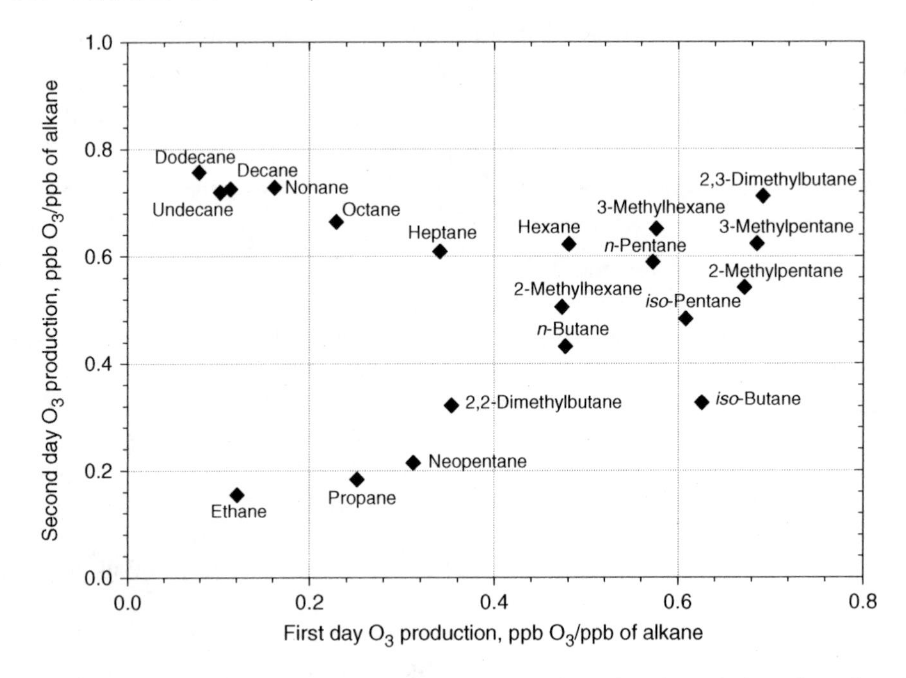

Figure VIII-C-3. First- and second-day ozone production following the emission of a spike of each alkane into an air parcel travelling along a five-day trajectory across Europe.

by the MIR scale in figure VIII-C-1 and is in stark contrast with that anticipated from the CBM mechanism where alkane reactivity on a per-carbon basis is assumed to be independent of carbon number.

On the POCP scale, ethane shows low reactivity but not the negligible reactivity found on the MIR scale because of the greater time for reaction available on the regional scale as compared with the urban scale. An important issue for the alkanes is therefore their ability to form ozone on multi-day time-scales. By emitting a spike of each organic compound at a fixed point within the five-day trajectory, Derwent et al. (2005) were able to study ozone formation from alkenes and carbonyl compounds during the first and subsequent days after the emission spike. Figure VIII-C-3 presents the corresponding results for the alkanes. First- and second-day ozone formation increases steadily with increasing carbon number, with both reaching their respective maxima with the C_6 alkanes: 2- and 3-methylpentane and 2,3-dimethylbutane. As carbon number increases beyond C_6, first-day reactivity collapses whilst second-day reactivity increases slightly. This behavior results directly from the increasing formation of alkyl nitrate from the reaction of organic peroxy radicals with NO and from the propensity of oxy radicals to undergo 1,5 H-atom isomerization reactions with carbon numbers beyond C_6, rather than decomposing to give two smaller chain-length ketone and aldehyde fragments. The detailed reaction pathways anticipated for the alkanes and represented in the MCM therefore exert an important influence on their reactivities on the regional scale and their propensities to form ozone on multi-day time-scales.

References

Abuin, E., and E.A. Lissi (1975), Arrhenius parameters for the photocleavage of butan-2-one triplets, *J. Photochem.*, *5*, 65–68.

Abuin, E.B., M.V. Encina, S. Díaz, and E.A. Lissi (1984), Cage reactions in the photolysis of 2,2,4,4-tetramethyl-3-pentanone, *Int. J. Chem. Kinet.*, *16*, 503–511.

Adachi, H., and N. Basco (1979a), Kinetic spectroscopy study of the reaction of CH_3O_2 with NO, *Chem. Phys. Lett.*, *63*, 490–492.

Adachi, H., and N. Basco (1979b), Kinetic spectroscopy study of the reaction of $C_2H_5O_2$ with NO, *Chem. Phys. Lett.*, *64*, 431–434.

Adachi, H., and N. Basco (1979c), The reaction of ethylperoxy radicals with NO_2, *Chem. Phys. Lett.*, *67*, 324–328.

Adachi, H., and N. Basco (1980), Reaction of CH_3O_2 with NO_2, *Int. J. Chem. Kinet.*, *12*, 1–16.

Adachi, H., and N. Basco (1982), Reactions of isopropylperoxy radicals with NO and NO_2, *Int. J. Chem. Kinet.*, *14*, 1243–1251.

Adachi, H., N. Basco, and D.G.L. James (1980), Mutual interactions of the methyl and methylperoxy radicals studied by flash photolysis and kinetic spectroscopy, *Int. J. Chem. Kinet.*, *12*, 949–977.

Albaladejo, J., B. Ballesteros, E. Jiménez, P. Martin, and E. Martinez (2002), A PLP–LIF kinetic study of the atmospheric reactivity of a series of C_4–C_7 saturated and unsaturated aliphatic aldehydes with OH, *Atmos. Environ.*, *36*, 3231–3239.

Alcock, W.G., and B. Mile (1975), The gas-phase reactions of alkylperoxy radicals generated by a photochemical technique, *Combust. Flame*, *24*, 125–128.

Alcock, W.G., and B. Mile (1977), Gas-phase reactions of alkylperoxy and alkoxy radicals. Part I. The photoinitiated oxidation of 2,3-dimethylbutane, *Combust. Flame*, *29*, 133–144.

Allan, W., D.C. Lowe, A.J. Gomez, H. Struthers, and G.W. Brailsford (2005), Interannual variation of ^{13}C in tropospheric methane: Implications for a possible atomic chlorine sink in the marine boundary layer, *J. Geophys. Res.*, *110*, D11306, doi:10.1029/2004JD005650.

Alumbaugh, R.L., G.O. Pritchard, and B. Rickborn (1965), Nonstereospecific mechanisms in the photolysis of cyclic ketones, *J. Phys. Chem.*, *69*, 3225–3231.

Alvárez-Idaboy, J.R., N. Mora-Diez, R.J. Boyd, and A. Vivier-Bunge (2001), On the importance of prereactive complexes in molecule-radical reactions: Hydrogen abstraction from aldehydes by OH, *J. Am. Chem. Soc.*, *123*, 2018–2024.

Amphlett, J.C., and E. Whittle (1967), Photolysis of halogens in presence of trifluoroacetaldehyde, *Trans. Faraday Soc.*, *63*, 80–90.

Anastasi, C., I.W.M. Smith, and D.A. Parkes (1978), Flash-photolysis study of spectra of CH_3O_2 and $C(CH_3)_3O_2$ radicals and kinetics of their mutual reactions and with NO, *J. Chem. Soc. Faraday Trans. I*, *74*, 1693–1701.

Anastasi, C., D.J. Waddington, and A. Woolley (1983), Reactions of oxygenated radicals in the gas phase, Part 10. Self-reactions of ethylperoxy radicals, *J. Chem. Soc., Faraday Trans. I*, *79*, 505–516.

Andersen, M.P.S., O.J. Nielsen, M.D. Hurley, J.C. Ball, T.J. Wallington, J. Stevens, J.W. Martin, D.A. Ellis, and S.A. Mabury (2004a), Atmospheric chemistry of n-$C_xF_{2x+1}CHO$ ($x = 1,3,4$): Reaction with Cl atoms, OH radicals; IR spectra of $C_xF_{2x+1}C(O)O_2NO_2$, *J. Phys. Chem. A*, *108*, 5189–5196.

Andersen, M.P.S., O.J. Nielsen, T.J. Wallington, M.D. Hurley, and W.B. DeMore (2005), Atmospheric chemistry of $CF_3OCF_2CF_2H$ and $CF_3OC(CF_3)_2H$: Reaction with Cl atoms and OH radicals, degradation mechanism, global warming potentials, and empirical relationship between k(OH) and k(Cl) for organic compounds, *J. Phys. Chem. A.*, *109*, 3926–3934.

Andersen, M.P.S., C. Stensby, O.J. Nielsen, M.D. Hurley, J.C. Ball, T.J. Wallington, J.W. Martin, D.A. Ellis, and S.A. Mabury (2004b), Atmospheric chemistry of n-$C_xF_{2x+1}CHO$ ($x = 1,3,4$): mechanism of the $C_xF_{2x+1}C(O)O_2 + HO_2$ reaction, *J. Phys. Chem. A*, *108*, 6325–6330.

Andersen, M.P.S., A. Toft, O.J. Nielsen, M.D. Hurley, T.J. Wallington, H. Chishima, K. Tonokura, S.A. Mabury, J.W. Martin, and D.A. Ellis (2006), Atmospheric chemistry of perfluorinated aldehyde hydrates (n-$C_xF_{2x+1}CH(OH)_2$, $x = 1,3,4$): hydration, dehydration, and kinetics and mechanism of Cl atom and OH radical initiated oxidation, *J. Phys. Chem. A*, *110*, 9854–9860.

Anderson, R.S., L. Huang, R. Iannone, and J. Rudolph (2007), Measurements of the $^{12}C/^{13}C$ kinetic isotope effects in the gas-phase reactions of light alkanes with chlorine atoms, *J. Phys. Chem. A*, *111*, 495–504.

Anderson, R.S., L. Huang, R. Iannone, A.E. Thompson, and J. Rudolph (2004), Carbon kinetic isotope effects in the gas phase reactions of light alkanes and ethene with the OH radical at 296 $\pm$ 4 K, *J. Phys. Chem. A.*, *108*, 11537–11544.

Archer, A.S., R.B. Cundall, G.B. Evans, and T.F. Palmer (1973a), The effect of temperature, pressure, and excitation wavelength on the photoluminescence of acetaldehyde vapour, *Proc. R. Soc. Lond., A*, *333*, 385–402.

Archer, A.S., R.B. Cundall, and T.F. Palmer (1973b), The role of excited states in the gas-phase photolysis of acetaldehyde, *Proc. R. Soc. Lond., A*, *334*, 411–426.

Arey, J., S.M. Aschmann, E.S.C. Kwok, and R. Atkinson (2001), Alkyl nitrate, hydroxyalkyl nitrate, and hydroxycarbonyl formation from the NO_x–air photooxidations of C_5–C_8 n-alkanes, *J. Phys. Chem. A.*, *105*, 1020–1027.

Ariya, P.A., V. Catoire, R. Sander, H. Niki, and G.W. Harris (1997), Trichloroethene and tetrachloroethene. Tropospheric probes for Cl- and Br-atom reactions during the polar sunrise, *Tellus, Ser. B*, *49*, 583–591.

Arnold, S.R., M.P. Chipperfield, and M.A. Blitz (2005), A three-dimensional model study of the effect of new temperature-dependent quantum yields for acetone photolysis. *J. Geophys. Res.*, *110*, D22305, doi:10.1029/2005JD005998.

Arnts, R.R., R.L. Seila, and J.J. Bufalini (1989), Determination of room temperature OH rate constants for acetylene, ethylene dichloride, ethylene dibromide, *p*-dichlorobenzene and carbon disulfide, *J. Air Pollut. Control Assoc.*, *39*, 453–460.

Arunachalam, S., R. Mathur, A. Holland, M.R. Lee, D. Olerud, and H. Jeffries (2003), Final report: Investigation of VOC reactivity. Assessment with comprehensive air quality modelling. Carolina Environmental Program, University of North Carolina at Chapel Hill, Chapel Hill, North Carolina.

Aschmann, S., and R. Atkinson (1995a), Rate constants for the gas-phase reactions of alkanes with Cl atoms at 296 ± 2 K, *Int. J. Chem. Kinet.*, *27*, 613–622.

Aschmann, S. M., and R. Atkinson (1995b), Rate constants for the reactions of the NO_3 radical with alkanes at 296 ± 2 K, *Atmos. Environ.*, *29*, 2311–2316.

Aschmann, S.M., and R. Atkinson (1998), Kinetics of the gas-phase reactions of the OH radical with selected glycol ethers, glycols, and alcohols, *Int. J. Chem. Kinet.*, *30*, 533–540.

Aschmann, S.M., J. Arey, and R. Atkinson (2000), Atmospheric chemistry of selected hydroxycarbonyls, *J. Phys. Chem. A*, *104*, 3998–4003.

Aschmann, S.M., J. Arey, and R. Atkinson (2001), Atmospheric chemistry of three C_{10} alkanes, *J. Phys. Chem. A*, *105*, 7598–7606.

Aschmann, S.M., J. Arey, and R. Atkinson (2002), Products and mechanism of the reaction of OH radicals with 2,2,4-trimethylpentane in the presence of NO, *Environ. Sci. Technol.*, *36*, 625–632.

Aschmann, S.M., J. Arey, and R. Atkinson (2003), Kinetics and products of the gas-phase reaction of OH radicals with 5-hydroxy-2-pentanone at 296 ± 2K, *J. Atmos. Chem.*, *45*, 289–299.

Aschmann, S.M., J. Arey, and R. Atkinson (2004), Products and mechanism of the reaction of OH radicals with 2,3,4-trimethylpentane in the presence of NO, *Environ., Sci. Technol.*, *38*, 5038–5046.

Aschmann, S.M., A.A. Chew, J. Arey, and R. Atkinson, (1997), Products of the gas-phase reaction of OH radicals with cyclohexane: Reactions of the cyclohexoxy radical, *J. Phys. Chem. A.*, *101*, 8042–8048.

Atkinson, R. (1986a), Kinetics and mechanisms of the gas-phase reactions of the hydroxyl radical with organic compounds under atmospheric conditions, *Chem. Rev.*, *86*, 69–201.

Atkinson, R. (1986b), Estimations of OH radical rate constants from H-atom abstraction from C—H and O—H bonds over the temperature range 250–1000 K, *Int. J. Chem. Kinet.*, *18*, 555–568.

Atkinson R. (1987), A structure–reactivity relationship for the estimation of rate constants for the gas-phase reactions of OH radicals with organic compounds, *Int. J. Chem. Kinet.*, *19*, 799–828.

Atkinson, R., (1989), Kinetics and mechanism of the gas-phase reactions of the hydroxyl radical with organic compounds, *J. Phys. Chem. Ref. Data, Monograph 1.*

Atkinson, R. (1990), Gas-phase tropospheric chemistry of organic compounds. A review, *Atmos. Environ.*, *24A*, 1–41.

Atkinson, R. (1991), Kinetics and mechanisms of the gas-phase reactions of the NO_3 radical with organic compounds, *J. Phys. Chem. Ref. Data*, *20*, 459–507.

Atkinson, R. (1994), Gas-phase tropospheric chemistry of organic compounds, *J. Phys. Chem. Ref. Data, Monograph 2*, 1–216.

Atkinson, R. (1995), Gas phase tropospheric chemistry of organic compounds, Issues, *Environ. Sci. and Technol.*, *4*, 65–89.

Atkinson, R. (1997a), Gas-phase tropospheric chemistry of volatile organic compounds: 1. Alkanes and alkenes, *J. Phys. Chem. Ref. Data*, *26*, 215–290.

Atkinson, R. (1997b), Atmospheric reactions of alkoxy and β-hydroxyalkoxy radicals, *Int. J. Chem. Kinet.*, *29*, 99–111.

Atkinson, R. (2003), Kinetics of the gas-phase reactions of OH radicals with alkanes and cycloalkanes, *Atmos. Chem. Phys.*, *3*, 2233–2307.

Atkinson, R., and J. Arey (2003), Atmospheric degradation of volatile organic compounds, *Chem. Rev.*, *103*, 4605–4638.

Atkinson, R., and S.M. Aschmann (1984), Rate constants for the reaction of OH radicals with a series of alkenes and dialkenes at 295 ± 1 K, *Int. J. Chem. Kinet.*, *16*, 1175–1186.

Atkinson, R., and S.M. Aschmann (1985), Kinetics of the gas-phase reaction of Cl atoms with a series of organics at 296 ± 2 K and atmospheric pressure, *Int. J. Chem. Kinet.*, *17*, 33–41.
Atkinson, R., and S.M. Aschmann (1987), Kinetics of the gas-phase reactions of Cl atoms with chloroethenes at 298 ± 2 K and atmospheric pressure, *Int. J. Chem. Kinet.*, *19*, 1097–1105.
Atkinson, R., and S.M. Aschmann (1988a), Rate constants for the reaction of OH radical with isopropylcyclopropane at 298 ± 2 K: Effects of ring strain on substituted cycloalkanes, *Int. J. Chem. Kinet.*, *20*, 339–342.
Atkinson, R., and S.M. Aschmann (1988b), Comment on "Flash photolysis resonance fluorescence investigation of the gas-phase reactions of OH radicals with a series of aliphatic ketones over the temperature range 240–440 K", *J. Phys. Chem.*, *92*, 4008–4008.
Atkinson, R., and S.M. Aschmann (1989), Rate constants for the reactions of the OH radical with the propyl and butyl nitrates and 1-nitrobutane at 298 ± 2 K, *Int. J. Chem. Kinet.*, *21*, 1123–1129.
Atkinson, R., and S.M. Aschmann (1992), OH radical reaction rate constants for polycyclic alkanes: Effects of ring strain and consequences for estimation methods, *Int. J. Chem. Kinet.*, *24*, 983–989.
Atkinson, R., and S.M. Aschmann (1995), Alkoxy radical isomerization products from the gas-phase OH radical-initiated reactions of 2,4-dimethyl-2-pentanol and 3,5-dimethyl-3-hexanol, *Environ. Sci. Tech.*, *29*, 528–536.
Atkinson, R., and W.P.L. Carter (1984), Kinetics and mechanisms of the gas-phase reactions of ozone with organic compounds under atmospheric conditions, *Chem. Rev.*, *84*, 437–470.
Atkinson, R., and W.P.L. Carter (1991), Reactions of alkoxy radicals under atmospheric conditions: The relative importance of decomposition versus reaction with O_2, *J. Atmos. Chem.*, *13*, 195–210.
Atkinson, R., and J.N. Pitts, Jr. (1978), Kinetics of the reactions of the OH radical with HCHO and CH_3CHO over the temperature range 299–426 K, *J. Chem. Phys.*, *68*, 3581–3584.
Atkinson, R., S.M. Aschmann, J. Arey, and B. Shorees (1992a), Formation of OH radicals in the gas phase reaction of O_3 with a series of terpenes, *J. Geophys. Res.*, *97*, 6065–6073.
Atkinson, R., S.M. Aschmann, and W.P.L. Carter (1983a), Kinetics of the reactions of O_3 and OH radicals with furan and thiophene at 298 ± 2 K, *Int. J. Chem. Kinet.*, *15*, 51–61.
Atkinson, R., S.M. Aschmann, and W.P.L. Carter (1983b), Rate constants for the gas-phase reactions of OH radicals with a series of bi- and tricycloalkanes at 299 ± 2 K: effects of ring strain, *Int. J. Chem. Kinet.*, *15*, 37–50.
Atkinson, R., S.M. Aschmann, W.P.L. Carter, and J.N. Pitts, Jr. (1982a), Rate constants for the gas-phase reaction of OH with a series of ketones at 299 ± 2 K, *Int. J. Chem. Kinet.*, *14*, 839–847.
Atkinson, R., S.M. Aschmann, W.P.L. Carter, and A.M. Winer (1982b), Kinetics of the gas-phase reactions of OH radicals with alkyl nitrates at 299 ± 2 K, *Int. J. Chem. Kinet.*, *14*, 919–926.
Atkinson, R., S.M. Aschmann, W.P.L. Carter, A.M. Winer, and J.N. Pitts, Jr. (1982c), Kinetics of the reactions of OH radicals with *n*-alkanes at 299 ± 2 K, *Int. J. Chem. Kinet.*, *14*, 781–788.
Atkinson, R., S.M. Aschmann, W.P.L. Carter, A.M. Winer, and J.N. Pitts, Jr. (1982d), Alkyl nitrate formation from the NO_x–air photooxidation of C_2–C_8 *n*-alkanes, *J. Phys. Chem.*, *86*, 4563–4569.
Atkinson, R., S.M. Aschmann, W.P.L. Carter, A.M. Winer, and J.N. Pitts, Jr. (1984a), Formation of alkyl nitrates from the reaction of branched and cyclic alkyl peroxy radicals with NO, *Int. J. Chem. Kinet.*, *16*, 1085–1101.
Atkinson, R., S.M. Aschmann, and J.N. Pitts, Jr. (1988), Rate constants for the reactions of the NO_3 radical with a series of organic compounds at 296 ± 2 K, *J. Phys. Chem.*, *92*, 3454–3457.
Atkinson, R., S.M. Aschmann, and A.M. Winer (1987), Alkyl nitrate formation from the reaction of a series of branched RO_2 radicals with NO as a function of temperature and pressure, *J. Atmos. Chem.*, *5*, 91–102.

Atkinson, R., S.M. Aschmann, A.M. Winer, and J.N. Pitts, Jr. (1981), Rate constants for the gas-phase reactions of O_3 with a series of carbonyls at 296 K, *Int. J. Chem. Kinet.*, *13*, 1133–1142.

Atkinson, R., S.M. Aschmann, A.M. Winer, and J.N. Pitts, Jr. (1982e), Rate constants for the reaction of OH radicals with a series of alkanes and alkenes at 299 ± 2 K, *Int. J. Chem. Kinet.*, *14*, 507–516.

Atkinson. R., D.L. Baulch, R.A. Cox, J.N. Crowley, R.F. Hampson, Jr., R.G. Hynes, M.E. Jenkin, J.A. Kerr, M.J. Rossi, and J. Troe (2004), Summary of evaluated kinetic and photochemical data for atmospheric chemistry, IUPAC, Web version, July.

Atkinson, R., D.L. Baulch, R.A. Cox, J.N. Crowley, R.F. Hampson, Jr., R.G. Hynes, M.E. Jenkin, J.A. Kerr, M.J. Rossi, and J. Troe (2005a), Summary of evaluated kinetic and photochemical data for atmospheric chemistry, IUPAC, Web version, March.

Atkinson, R., D.L. Baulch, R.A. Cox, J.N. Crowley, R.F. Hampson, R.G. Hynes, M.E. Jenkin, M.J. Rossi, and J. Troe (2005b), Evaluated kinetic and photochemical data for atmospheric chemistry: Volume II—reactions of organic species, *Atmos. Chem. Phys. Discuss.*, *5*, 6295–7168.

Atkinson, R., D.L. Baulch, R.A. Cox, R.F. Hampson, Jr., J.A. Kerr, and J. Troe (1992b), Evaluated kinetic and photochemical data for atmospheric chemistry, Supplement IV. IUPAC Subcommittee on Gas Kinetic Data Evaluation for Atmospheric Chemistry, *J. Phys. Chem. Ref. Data*, *21*, 1125–1568.

Atkinson, R., D.L. Baulch, R.A. Cox, R.F. Hampson, Jr., J.A. Kerr, M.J. Rossi, and J. Troe (1997), Evaluated kinetic and photochemical and heterogeneous data for atmospheric chemistry, Supplement V. IUPAC Subcommittee on Gas Kinetic Data Evaluation for Atmospheric Chemistry, *J. Phys. Chem., Ref. Data*, *26*, 521–1011.

Atkinson, R., D.L. Baulch, R.A. Cox, R.F. Hampson, J.A. Kerr, M.J. Rossi, and J. Troe, (1999), Evaluated kinetic and photochemical data for atmospheric chemistry. Supplement VII—IUPAC sub-committee on gas kinetic data evaluation for atmospheric chemistry, *J. Phys. Chem. Ref. Data*, *28*, 191–303.

Atkinson, R., G.M. Breuer, J.N. Pitts, Jr., and H.J. Sandoval (1976), Tropospheric and stratospheric sinks for halocarbons: photooxidation, $O(^1D)$ atom, and OH radical reactions, *J. Geophys. Res.*, *81*, 5765–5770.

Atkinson, R., W.P.L. Carter, S.M. Aschmann, A.M. Winer, and J.N. Pitts, Jr. (1984b), Kinetics of the reaction of OH radicals with a series of branched alkanes at 297 ± 2 K, *Int. J. Chem. Kinet.*, *16*, 469–481.

Atkinson, R., W.P.L. Carter, and A.M. Winer (1983c), Effects of temperature and pressure on alkyl nitrate yields in the NO_x photooxidation of *n*-pentane and *n*-hexane, *J. Phys. Chem.*, *87*, 2012–2018.

Atkinson, R., D.A. Hansen, and J.N. Pitts, Jr. (1975), Rate constants for the reaction of OH radicals with CHF_2Cl, CF_2Cl_2, $CFCl_3$, and H_2 over the temperature range 297–434K, *J. Chem. Phys.*, *63*, 1703–1706.

Atkinson, R., E.S.C. Kwok, J.Arey, and S.M. Aschmann (1995), Reactions of alkoxy radicals in the atmosphere, *Faraday Discuss.*, *100*, 23–37.

Atkinson, R., C.N. Plum, W.P.L. Carter, A.M. Winer, and J.N. Pitts, Jr. (1984c), Kinetics of the gas-phase reactions of NO_3 radicals with a series of alkanes at 296 ± 2 K, *J. Phys. Chem.*, *88*, 2361–2364.

Atkinson, R., C.N. Plum, W.P.L. Carter, A.M. Winer, and J.N. Pitts, Jr. (1984d), Rate constants for the gas-phase reactions of nitrate radicals with a series of organics in air at 298 ± 1 K, *J. Phys. Chem.*, *88*, 1210–1215.

Atkinson, R., E.C. Tuazon, and S.M. Aschmann (2000), Atmospheric chemistry of 2-pentanone and 2-heptanone, *Environ. Sci. Technol.*, *34*, 623–631.

Audley, G.J., D.L. Baulch, and I.M. Campbell (1981), Gas-phase reactions of hydroxyl radicals with aldehydes in flowing $H_2O_2 + NO_2 + CO$ mixtures, *J. Chem. Soc., Faraday Trans. 1*, *77*, 2541–2549.

Aumont, B., S. Madronich, I. Bey, and G.S. Tyndall (2000), Contribution of secondary VOC to the composition of aqueous atmospheric particles: A modelling approach, *J. Atmos. Chem.*, *35*, 59–75.

Aumont, B., S. Szopa, and S. Madronich (2005), Modelling the evolution of organic carbon during its gas-phase tropospheric oxidation: development of an explicit model based on a self-generating approach, *Atmos. Chem. Phys.*, *5*, 2497–2517.

Ausloos, P. (1961). Intramolecular rearrangements. II. Photolysis and radiolysis of 4-methyl-2-hexanone, *J. Phys. Chem.*, *65*, 1616–1618.

Ausloos, P., and E. Murad (1961), The fluorescence and phosphorescence of trifluoroacetone vapor, *J. Phys. Chem.*, *65*, 1519–1521.

Ausloos, P., and R.E. Rebbert (1961), Molecular rearrangements. III. Formation of 1-methylcyclobutanol in the photolyis of 2-pentanone, *J. Am. Chem. Soc.*, *83*, 4897–4899.

Ausloos, P., and R.E. Rebbert (1964), Photoelimination of ethylene from 2-pentanone, *J. Am. Chem. Soc.*, *86*, 4512–4513.

Ayhens, Y.V., J.M. Nicovich, M.L. McKee, and P.H. Wine (1997), Kinetic and mechanistic study of the reaction of atomic chlorine with methyl iodide over the temperature range 218–694 K, *J. Phys. Chem. A*, *101*, 9382–9390.

Ayscough, P.B., and E.W.R. Steacie (1956), The photolysis of hexafluoroacetone, *Proc. R. Soc.*, *A234*, 476–488.

Azad, K., and J.M. Andino (1999), Products of the gas-phase photooxidation reactions of 1-propanol with OH radicals, *Int. J. Chem. Kinet.*, *31*, 810–818.

Baba, M., and I. Hanazaki (1984), The $S_1(n,\pi^*)$ states of cyclopentanone and cyclobutanone in a supersonic nozzle beam, *J. Chem. Phys.*, *81*, 5426–5433.

Bacak, A., M.W. Bardwell, M.T. Raventos, C.J. Percival, G. Sanchez-Reyna, and D.E. Shallcross (2004), Kinetics of the reaction of CH_3O_2 + NO: A temperature and pressure dependence study with chemical ionization mass spectrometry, *J. Phys. Chem. A*, *108*, 10681–10687.

Bacher, C., G.S. Tyndall, and J.J. Orlando (2001), The atmospheric chemistry of glycolaldehyde, *J. Atmos. Chem.*, *39*, 171–189.

Badcock, C.C., M.J. Perona, G.O. Pritchard, and B. Rickborn (1969), Photolysis of 2-methylcyclohexanone, *J. Am. Chem. Soc.*, *91*, 543–546.

Baghal-Vayjooee, M.H., and S.W. Benson (1979), Kinetics and thermochemistry of the reaction Cl + cyclopropane = HCl + cyclopropyl. Heat of formation of the cyclopropyl radical, *J. Am. Chem. Soc.*, *101*, 2838–2840.

Baghal-Vayjooee, M.H., A.J. Colussi, and S.W. Benson (1979), The very-low-pressure study of the kinetics and equilibrium: Cl + CH_4 = CH_3 + HCl at 298 K. The heat of formation of the CH_3 radical, *Int. J. Chem. Kinet.*, *11*, 147–154.

Bagley, J.A., C. Canosa-Mas, M.R. Little, A.D. Parr, S.J. Smith, S.J. Waygood, and R.P. Wayne (1990), Temperature dependence of reactions of the nitrate radical with alkanes, *J. Chem. Soc. Faraday Trans.*, *86*, 2109–2114.

Baker, J., J. Arey, and R. Atkinson (2004), Rate constants for the gas-phase reactions of OH radicals with a series of hydroxyaldehydes at 296 ± 2 K, *J. Phys. Chem. A*, *108*, 7032–7037.

Baker, J., J. Arey, and R. Atkinson (2005), Rate constants for the reactions of OH radicals with a series of 1,4-hydroxyketones, *J. Photochem. Photobiol. A*, *176*, 143–148.

Baker, R.R., R.R. Baldwin, and R.W. Walker (1976), Addition of neopentane to slowly reacting mixtures of H_2 + O_2 at 480°C. 2. Addition of primary products from neopentane, and rate constants for H and OH attack on neopentane, *Combust. Flame*, *27*, 147–161.

Baldwin, R.R., and R.W. Walker (1979), Rate constants for hydrogen + oxygen system, and for H atoms and OH radicals + alkanes, *J. Chem. Soc. Faraday Trans. 1*, *75*, 140–154.

Baldwin, R.R., J.R. Barker, D.M. Golden, and D.G. Hendry (1977), Photochemical smog. Rate parameter estimates and computer simulations, *J. Phys. Chem.*, *81*, 2483–2492.

Baldwin, R.R., R.W. Walker, and R.W. Walker (1979), Addition of 2,2,3,3-tetramethylbutane to slowly reacting mixtures of hydrogen and oxygen, *J. Chem. Soc. Faraday Trans.*, *1*, 75, 1447–1457.

Baldwin, R.R., R.W. Walker, and R.W. Walker (1981), Addition of 2,2,3-trimethylbutane to slowly reacting mixtures of hydrogen and oxygen at 480° C, *J. Chem. Soc. Faraday Trans. 1*, *77*, 2157–2173.

Balestra-Garcia, C., G. Le Bras, and H. Mac Léod (1992), Kinetic study of the reactions OH + mono-, di-, and trichloroacetaldehyde and acetaldehyde by laser photolysis-resonance fluorescence at 298 K, *J. Phys. Chem.*, *96*, 3312–3316.

Balla, R.J., H.H. Nelson, and J.R. McDonald (1985), Kinetics of the reactions of isopropoxy radicals with NO, NO_2 and O_2, *Chem. Phys.*, *99*, 323–335.

Bamford, C.H., and R.G.W. Norrish (1935), Primary photochemical reactions. Part VI. Photochemical decomposition of *iso*-valeraldehyde and di-*n*-propyl ketone, *J. Chem. Soc.*, *1935*, 1504–1511.

Bamford, C.H., and R.G.W. Norrish (1938), Primary photochemical reactions. Part XII. The effect of temperature on the quantum yield of the decomposition of di-*n*-propyl ketone in the vapour phase and in solution, *J. Chem. Soc.*, 1544–1554.

Bamford, D.J., S.V. Filseth, M.F. Foltz, J.W. Hepburn, and C.B. Moore (1985), Photofragmentation dynamics of formaldehyde: $CO(v, j)$ distributions as function of initial rovibronic state and isotopic substitution, *J. Chem. Phys.*, *82*, 3032–3041.

Bardwell, M.W., A. Bacak, M.T. Raventós, C.J. Percival, G. Sanchez-Reyna, and D.E Shallcross (2005), Kinetics and mechanism of the $C_2H_5O_2 + NO$ reaction: A temperature and pressure dependence study using chemical ionization mass spectrometry, *Int. J. Chem. Kinet.*, *37*, 253–260.

Barker, J.R., S.W. Benson, and D.M. Golden (1977), The decomposition of dimethyl peroxide and the rate constant for $CH_3O + O_2 \rightarrow CH_2O + HO_2$, *Int. J. Chem. Kinet.*, *9*, 31–53.

Barker, J.R., L.L. Lohr, R.M. Shroll, and S. Reading (2003), Modeling the organic nitrate yields in the reaction of alkyl peroxy radicals with nitric oxide. 2. Reaction simulations, *J. Phys. Chem. A*, *107*, 7434–7444.

Barletta, B., S. Meinardi, I.J. Simpson, H.A. Khwaja, D.R. Blake, and F.S. Rowland (2002), Mixing ratios of volatile organic compounds (VOCs) in the atmosphere of Karachi, Pakistan, *Atmos. Environ.*, *36*, 3429–3443.

Barnes, I., V. Bastian, and K.H. Becker (1990), Reactions of IO radicals with sulfur-containing compounds, *Phys. Chem. Behav. Atmos. Pollut.*, Kluwer, Dordrecht, The Netherlands, 166–171.

Barnes, I., V. Bastian, K.H. Becker, E.H. Fink, and F. Zabel (1982), Reactivity studies of organic substances towards hydroxyl radicals under atmospheric conditions, *Atmos. Environ.*, *16*, 545–550.

Barrie, L.A., J.W. Bottenheim, R.C. Schnell, P.J. Crutzen, and R.A. Rasmussen (1988), Ozone destruction and photochemical reactions at polar sunrise in the lower Arctic atmosphere, *Nature*, *334*, 138–141.

Barry, J., G. Locke, D. Scollard, H. Sidebottom, J. Treacy, C. Clerbaux, R. Colin, and J. Franklin (1997), 1,1,1,3,3,-pentafluorobutane (HFC-365mfc): atmospheric degradation and contribution to radiative forcing, *Int. J. Chem. Kinet.*, *29*, 607–617.

Barry, J., H. Sidebottom, J. Treacy, and J. Franklin (1995), Kinetics and mechanism for the atmospheric oxidation of 1,1,2-trifluoroethane (HFC-143), *Int. J. Chem. Kinet.*, *27*, 27–36.

Barry, M.D., and P.A. Gorry (1984), Photofragmentation dynamics of CH_3I at 248 nm, *Mol. Phys.*, *52*, 461–473.

Bartolotti, L.J., and E.O. Edney (1994), Investigation of the correlation between the energy of the highest occupied molecular orbital (HOMO) and the logarithm of the OH rate constant of hydrofluorocarbons and hydrofluoroethers, *Int. J. Chem. Kinet.*, *26*, 913–920.

Batt, L. (1979), The gas-phase decomposition of alkoxy radicals, *Int. J. Chem. Kinet.*, *11*, 977–993.

Batt, L., and R.D. McCulloch (1976), The gas-phase pyrolysis of alkyl nitrites. II. *s*-Butyl nitrite, *Int. J. Chem. Kinet.*, *8*, 911–933.

Batt, L, and R.T. Milne (1976), The gas-phase pyrolysis of alkyl nitrites. I. *t*-Butyl nitrite, *Int. J. Chem. Kinet.*, *8*, 59–84.

Batt, L., and R.T. Milne (1977a), The gas-phase pyrolysis of alkyl nitrites. IV. Ethyl nitrite, *Int. J. Chem. Kinet.*, *9*, 549–565.

Batt, L., and R.T. Milne (1977b), The gas-phase pyrolysis of alkyl nitrites. III. Isopropyl nitrite, *Int. J. Chem. Kinet.*, *9*, 141–156.

Batt, L., and G.N. Robinson (1979), Reaction of methoxy radicals with oxygen. I. Using dimethyl peroxide as a thermal source of methoxy radicals, *Int. J. Chem. Kinet.*, *11*, 1045–1053.

Batt, L., and G.N. Robinson (1982), Arrhenius parameters for the decomposition of the *t*-butoxy radical, *Int. J. Chem. Kinet.*, *14*, 1053–1055.

Batt, L., M.W.M. Hisham, and M. MacKay (1989), Decomposition of the *t*-butoxy radical: II. Studies over the temperature range 303–393 K, *Int. J. Chem. Kinet.*, *21*, 535–546.

Batt, L., T.S.A. Islam, and G.N. Rattray (1978), The gas-phase pyrolysis of alkyl nitrites. VI. *t*-Amyl nitrite, *Int. J. Chem. Kinet.*, *10*, 931–943.

Baughcum, S., and S.R. Leone (1980), Photofragmentation infrared emission studies of vibrationally excited free radicals CH_3 and CH_2I, *J. Chem. Phys.*, *72*, 6531–6545.

Baugues, K. (1990), Preliminary planning information for updating the ozone regulatory impact analysis version of EKMA. Source Receptor Analysis Branch, Technical Support Division, U.S. Environmental Protection Agency, Research Triangle Park, North Carolina, USA.

Baulch, D.L., C.J. Cobos, R.A. Cox, C. Esser, P. Frank, T. Just, J.A. Kerr, M.J. Pilling, J. Troe, R.W. Walker, and J. Warnatz (1992), Evaluated kinetic data for combustion modelling, *J. Phys. Chem. Ref. Data*, *21*, 411–734.

Baxley, J.S., and J.R. Wells (1998), The hydroxyl radical reaction rate constant and atmospheric transformation products of 2-butanol and 2-pentanol, *Int. J. Chem. Kinet.*, *30*, 745–752.

Bayes, K.D., D.W. Toohey, R.R. Friedl, and S.P. Sander (2003), Measurements of quantum yields of bromine atoms in the photolysis of bromoform from 266 to 324 nm, *Geophys. Res.*, *108*, 4095, doi:10.1029/2002JD002877.

Beavan, S.W., H. Inoue, D. Phillips, and P.A. Hackett (1978), Collisional effects on the phosphorescence of 1,1,1-trifluoroacetone vapour, *J. Photochem., 8, 247–261.*

Beavan, S.W., D. Phillips, and R.G. Brown (1975), Phosphorescence of 1,1,1-trifluoroacetone using pulsed nitrogen laser excitation, *Chem. Phys. Lett.*, *56*, 542–544.

Becerra, R., and H.M. Frey (1988), Photolysis of cyclopentanone in the gas phase, *J. Chem. Soc., Faraday Trans.*, 2, 1941–1949.

Becker, K.H., and K. Wirtz (1989), Gas phase reactions of alkyl nitrates with hydroxyl radicals under tropospheric conditions in comparison with photolysis, *J. Atmos. Chem.*, *9*, 419–433.

Bednarek, G., M. Breil, A. Hoffmann, J.P. Kohlmann,V. Mörs, and R. Zellner (1996), Rate and mechanism of the atmospheric degradation of 1,1,1,2-tetrafluoroethane (HFC-134a), *Ber. Bunsenges. Phys. Chem.*, *100*, 528–539.

Behnke, W., W. Holländer, W. Koch, F. Nolting, and C. Zetzsch (1988), A smog chamber for studies of the photochemical degradation of chemicals in the presence of aerosols, *Atmos. Environ.*, *22*, 1113–1120.

Behnke, W., F. Nolting, and C. Zetzsch (1987), A smog chamber study of the impact of aerosols on the photodegradation of chemicals in the troposphere, *J. Aeros. Sci.*, *18*, 65–71.

Beichert, P., L. Wingen, J. Lee, R. Vogt, M.J. Ezell, M. Ragains, R. Neavyn, and B.J. Finlayson-Pitts (1995), Rate constants for the reactions of chlorine atoms with some simple alkanes at 298 K; measurement of a self-consistent set using both absolute and relative rate methods, *J. Phys. Chem.*, *99*, 13156–13162.

Benson, S.W., and G.B. Kistiakowsky (1942), The photochemical decomposition of cyclic ketones, *J. Am. Chem. Soc.*, *64*, 80–86.

Bera, R.K., and R.J. Hanrahan (1988), Investigation of gas phase reactions of hydroxyl radicals with fluoromethane and difluoromethane using argon-sensitized pulse radiolysis, *Radiat. Phys. Chem.*, *32*, 579–584.

Berntsen, T.K., and I.S.A. Isaksen (1997), A global three-dimensional chemical transport model for the troposphere. 1. Model description and CO and ozone results, *J. Geophys. Res.*, *102*, 21229–21280.

Berry, R.J., J. Yuan, A. Misra, and P. Marshall (1998), Experimental and computational investigations of the reaction of OH with CF_3I and the enthalpy of formation of HOI, *J. Phys. Chem. A*, *102*, 5182–5188.

Bethel, H.L., R. Atkinson, and J. Arey (2001), Kinetics and products of the reactions of selected diols with the OH radical, *Int. J. Chem. Kinet.*, *33*, 310–316.

Bethel, H.L., R. Atkinson, and J. Arey (2003), Kinetics and products of the reactions of OH radicals with 2-methyl-2-pentanol and 4-methyl-2-pentanol, *J. Photochem. Photobiol. A.*, *157*, 257–267.

Betterton, E.A., and M.R. Hoffmann (1988), Henry's Law constants of some environmentally important aldehydes, *Environ. Sci. Technol.*, *22*, 1415–1418.

Bevilacqua, T.J., D.R. Hanson, and C.J. Howard (1993), Chemical ionization mass spectrometric studies of the gas-phase reactions CF_3O_2 + NO, CF_3O + NO, and CF_3O + RH, *J. Phys. Chem.*, *97*, 3750–3757.

Bey, I., D.J. Jacob, R.M. Yantosca, J.A. Logan, B.D. Field, A.M. Fiore, Q. Li, H.Y. Liu, L.J. Mickley, and M.G. Schultz (2003), Global modelling of tropospheric chemistry with assimilated meteorology: Model description and evaluation, *J. Geophys. Res. 106*, 23073–23095.

Bhatnagar, A., and R.W. Carr (1994), Flash photolysis time-resolved mass spectrometric investigations of the reactions of CF_3O_2 and CF_3O radicals with NO, *Chem. Phys. Lett.*, *231*, 454–459.

Bhatnagar, A., and R.W. Carr (1995a), A kinetic study of the CF_3CFHO_2 + NO reaction and the dissociation of the CF_3CFHO radical, *Chem. Phys. Lett.*, *238*, 9–15.

Bhatnagar, A., and R.W. Carr (1995b), Temperature dependence of the reaction of CF_3CFClO_2 radicals with NO and the unimolecular decomposition of the CF_3CFClO radical, *J. Phys. Chem.*, *99*, 17573–17577.

Bhatnagar, A., and R.W. Carr (1996), HCFC-31: the $CHClFO_2$ + NO → CHClFO + NO_2 reaction and Cl atom elimination from CHClFO, *Chem. Phys. Lett.*, *258*, 651–656.

Bilde, M., and T.J. Wallington (1998), Atmospheric chemistry of CH_3I: reaction with atomic chlorine at 1–700 Torr total pressure and 295 K, *J. Phys. Chem. A*, *102*, 1550–1555.

Bilde, M., J.J. Orlando, G.S. Tyndall, T.J. Wallington, M.D. Hurley, and E.W. Kaiser (1999), FT-IR product studies of the Cl-initiated oxidation of CH_3Cl in the presence of NO, *J. Phys. Chem. A*, 103, 3963–3968.

Bilde, M., J. Sehested, T.J. Wallington, and O.J. Nielsen (1997), Atmospheric chemistry of CH_2BrCl: kinetics and mechanism of the reaction of F atoms with CH_2BrCl and fate of the CHBrClO radical, *J. Phys. Chem. A*, *101*, 5477–5488.

Bilde, M., T.J. Wallington, C. Ferronato, J.J. Orlando, G.S. Tyndall, E. Estupiñan, and S. Haberkorn (1998), Atmospheric chemistry of CH_2BrCl, $CHBrCl_2$, $CHBr_2Cl$, $CF_3CHBrCl$ and CBr_2Cl_2, *J. Phys. Chem. A*, *102*, 1976–1986.

Blacet, F.E. (1952), Photochemistry in the lower atmosphere, *Ind. Eng. Chem.*, *44*, 1339–1342.

Blacet, F.E., and J.G. Calvert (1951a), The photolysis of the aliphatic aldehydes. XIV. The butyraldehydes, *J. Am. Chem. Soc.*, *73*, 661–667.

Blacet, F.E., and J.G. Calvert (1951b), The photolysis of aliphatic aldehydes. XV. The butyraldehydes with iodine vapor, *J. Am. Chem. Soc.*, *73*, 667–674,

Blacet, F.E., and R.A. Crane (1954), The photolysis of aliphatic aldehydes. XVII. Propionaldehyde, n-butyraldehyde and isobutyraldehyde at 2380 and 1870 Å, *J. Am. Chem. Soc.*, *76*, 5337–5340.

Blacet, F.E., and J.D. Heldman (1942), The photolysis of the aliphatic aldehydes. X. Acetaldehyde and iodine mixtures, *J. Am. Chem. Soc.*, *64*, 889–893.

Blacet, F.E., and D.E. Loeffler (1942), The photolysis of the aliphatic aldehydes. XI. Acetaldehyde and iodine mixtures, *J. Am. Chem. Soc.*, *64*, 893–896.

Blacet, F.E., and A. Miller (1957), The photochemical decomposition of cyclohexanone, cyclopentanone and cyclobutanone, *J. Am. Chem. Soc.*, *79*, 4327–4329.

Blacet, F.E., and J.N. Pitts, Jr. (1952), The photolysis of the aliphatic aldehydes. XVI. Propionaldehyde and propionaldehyde–iodine mixtures, *J. Am. Chem. Soc.*, *74*, 3382–3388.

Blacet, F.E., and J.G. Roof (1936), The photolysis of the aliphatic aldehydes. III. Hydrogen from acetaldehyde, *J. Am. Chem. Soc.*, *58*, 278–280.

Blacet, F.E., and D. Volman (1938), The photolysis of the aliphatic aldehydes. VI. Acetaldehyde, *J. Am. Chem. Soc.*, *60*, 1243–1247.

Blades, A.T. (1970), Photolysis of cyclopentanone, *Can. J. Chem.*, *48*, 2269–2273.

Blake, D.R., and F.S. Rowland (1995), Urban leakage of liquefied petroleum gas and its impact on Mexico City air quality, *Science*, *269*, 953–956.

Blitz, M.A., D.E. Heard, and M.J. Pilling (2004), Pressure and temperature-dependent quantum yields for the photodissociation of acetone between 279 and 327.5 nm, *Geophys. Res. Lett.*, *31*, L06111, doi:10.1029/2003GL018793.

Blitz, M.A., D.E. Heard, and M.J. Pilling (2005), Wavelength dependent photodissociation of CH_3OOH: Quantum yields for CH_3O and OH, and measurement of the OH + CH_3OOH rate coefficient, *J. Photochem. Photobiol. A*, *176*, 107–113.

Blitz, M., M.J. Pilling, S.H. Robertson, and P.W. Seakins (1999), Direct studies on the decomposition of the *tert*-butoxy radical and its reaction with NO, *Phys. Chem. Chem. Phys.*, *1*, 73–80.

Bloss, C., V. Wagner, A. Bonzanini, M.E. Jenkin, K. Wirtz, M. Martin-Revejo, and M.J. Pilling (2005a), Evaluation of detailed aromatic mechanisms (MCM v3 and MCM v3.1) against environmental chamber data, *Atmos. Chem. Phys.*, *5*, 623–639.

Bloss, C., V. Wagner, M.E. Jenkin, R. Volamer, W.J. Bloss, J.D. Lee, D.E. Heard, K. Wirtz, M. Martin-Revejo, G. Rea, J.C. Wenger, and M.J. Pilling (2005b), Development of a detailed chemical mechanism (MCMv3.1) for the atmospheric oxidation of aromatic hydrocarbons. *Atmos. Chem. Phys.*, *5*, 641–644.

Bloss, W.J., D.M. Rowley, R.A. Cox, and R.L. Jones (2002), Rate coefficient for the BrO + HO_2 reaction at 298 K, *Phys. Chem. Chem. Phys.*, *4*, 3639–3647.

Bofill, J.M., S. Olivella, A. Solé, and J.M. Anglada (1999), The mechanism of methoxy radical oxidation by O_2 in the gas phase. Computational evidence for direct H atom transfer assisted by an intermolecular noncovalent O· · ·O bonding interaction, *J. Am. Chem. Soc.*, *121*, 1337–1347.

Bonard, A., V. Daële, J.L. Delfau, and C. Vovelle (2002), Kinetics of OH radical reactions with methane in the temperature range 295–660 K and with dimethyl ether and methyl-*tert*-butyl ether in the temperature range 295–618 K, *J. Phys. Chem. A*, *106*, 4384–4389.

Boone, G.D, F. Agyin, D.J. Robichaud, F.-M. Tao, and S.A. Hewitt (2001), Rate constants for the reactions of chlorine atoms with deuterated methanes: Experiment and theory, *J. Phys. Chem. A*, *105*, 1456–1464.

Borkowski, R.P., and P. Ausloos (1961), Intramolecular rearrangements. IV. Photolysis of 2-pentanone-4,5,5-d_3, *J. Phys. Chem.*, *65*, 2257–2260.

Bott, J.F., and N. Cohen (1989), A shock tube study of the reaction of the hydroxyl radical with H_2, CH_4, c-C_5H_{10}, and i-C_4H_{10}, *Int. J. Chem. Kinet.*, *21*, 485–498.

Bott, J.F., and N. Cohen (1991), A shock tube study of the reactions of the hydroxyl radical with several combustion species, *Int. J. Chem. Kinet.*, *23*, 1075–1094.

Boudries, H., G. Toupance, and A.L. Dutot (1994), Seasonal variation of atmospheric non-methane hydrocarbons on the western coast of Brittany, France, *Atmos. Environ.*, *28*, 1095–1112.

Bourbon, C., M. Brioukov, B. Hanoune, J.P. Sawerysyn, and P. Devolder (1996), Flow tube/LIF measurements of the rate constants for the reactions with NO of CF_3O and CF_3O_2 radicals, *Chem. Phys. Lett.*, *254*, 203–212.

Bourmada, N., C. Lafage, and P. Devolder (1987), Absolute rate constants of the reactions of OH with cyclohexane and ethane at 296 ± 2 K by the discharge flow method, *Chem. Phys. Lett.*, *136*, 209–214.

Bowman, J.H., D.J. Barket, Jr., and P.B. Shepson (2003), Atmospheric chemistry of nonanal, *Environ. Sci. Technol.*, *37*, 2218–2225.

Boyd, A.A., C.E. Canosa-Mas, D. King, R.P. Wayne, and M.R. Wilson (1991), Use of a stopped-flow technique to measure the rate constants at room temperature for reactions between the nitrate radical and various organic species, *J. Chem. Soc. Faraday Trans.*, *87*, 2913–2919.

Boyd, A.A., P.-M. Flaud, N. Daugey, and R. Lesclaux (2003), Rate constants for RO_2 + HO_2 reactions measured under a large excess of HO_2, *J. Phys. Chem. A.*, *107*, 818–821.

Boylan, J.W., M.T. Odman, J.G. Wilkinson, A.G. Russell, K.G. Doty, W.B. Norris, and R.T. McNider (2002), Development of a comprehensive, multiscale "one-atmosphere" modeling system: application to the Southern Appalachian Mountains. *Atmos. Environ.* *36*, 3721–3734.

Braslavsky, S., and J. Heicklen (1976), The gas phase reaction of O_3 with H_2CO, *Int. J. Chem. Kinet.*, *8*, 801–808.

Bravo-Pérez, G., J.R. Alvarez-Idaboy, A. Cruz-Torres, and M.E. Ruiz (2002), Quantum chemical and conventional transition-state theory calculations of rate constants for the NO_3 + alkane reaction, *J. Phys. Chem. A*, *106*, 4645–4650.

Bridier, I., R. Lesclaux, and B.Veyret (1992), Flash photolysis kinetic study of the equilibrium $CH_3O_2 + NO_2 \leftrightarrow CH_3O_2NO_2$, *Chem. Phys. Lett.*, *191*, 259–263.

Brown, A.C., C.E. Canosa-Mas, A.D. Parr, J.M.T. Pierce, and R.P.Wayne (1989), Tropospheric lifetimes of halogenated anaesthetics, *Nature, 341*, 635–637.

Brown, A.C., C.E. Canosa-Mas, A.D. Parr, K. Rothwell, and R.P. Wayne (1990a), Tropospheric lifetimes of three compounds for possible replacement of CFC and halons, *Nature, 347*, 541–543.

Brown, A.C., C.E. Canosa-Mas, A.D. Parr, and R.P. Wayne (1990b), Laboratory studies of some halogenated ethanes and ethers: measurements of rates of reaction with OH and of infrared absorption cross-sections, *Atmos. Environ.*, *24*, 2499–2511.

Brown, A.C., C.E. Canosa-Mas, and R.P. Wayne (1990c), A kinetic study of the reactions of OH with CH_3I and CF_3I, *Atmos. Environ.*, *24*, 361–367.

Brunet, V., and W.A. Noyes, Jr. (1958), Photochimie de l'Hexanone-2 en phase gazeuse, *Bull. Soc. Chim. France*, 121–123.

Bryukov, M.G., I.R. Slagle, and V.D. Knyazev (2002), Kinetics of reactions of Cl atoms with methane and chlorinated methanes, *J. Phys. Chem. A*, *106*, 10532–10542.

Bryukov, M.G., I.R. Slagle, and V.D. Knyazev (2003), Kinetics of reactions of Cl atoms with ethane, chloroethane, and 1,1-dichloroethane, *J. Phys. Chem. A*, *107*, 6565–6573.

Burkholder, J.B., M.K. Gilles, T. Gierczak, and A.R. Ravishankara (2002), The atmospheric degradation of 1-bromopropane ($CH_3CH_2CH_2Br$): The photochemistry of bromoacetone, *Geophys. Res. Lett.*, *29*, 1822, doi:10.1029/2002GL014712.

Burkholder, J.B., R.R. Wilson, T. Gierczak, R. Talukdar, S.A. McKeen, J.J. Orlando, G.L. Vaghjiani, and A.R. Ravishankara (1991), Atmospheric fate of CF_3Br, CF_2Br_2, CF_2ClBr, and CF_2BrCF_2Br, *J. Geophys. Res. 96*, 5025–5043.

Burrows, J.P., G.S. Tyndall, and G.K. Moortgat (1985), Absorption spectrum of NO_3 and kinetics of the reactions of NO_3 with NO_2, Cl, and several stable atmopsheric species at 298 K, *J. Phys. Chem.*, *89*, 4848–4856.

Butkovskaya, N.I., A. Kukui, and G. Le Bras (2004), Branching fractions for H_2O forming channels of the reaction of OH radicals with acetaldehyde, *J. Phys. Chem. A*, *108*, 1160–1168.

Butler, L.J., E.J. Hintsa, S.F. Shane, and Y.T. Lee (1986), The electronic state-selective photodissociation of CH_2BrI at 249, 210, and 193 nm, *J. Chem. Phys.*, *88*, 2051–2074.

Bytautas, L., and D.J. Klein (1998), Chemical combinatories for alkane-isomer enumeration and more, *J. Chem. Inf. Comput. Sci.*, *38*, 1063–1078.

Cabañas, B., P. Martín, S. Salgado, B. Ballesteros, and E. Martínez (2001), An experimental study on the temperature dependence for the gas-phase reactions of NO_3 radical with a series of aliphatic aldehydes, *J. Atmos. Chem.*, *40*, 23–39.

Cabañas, B., S. Salgado, P. Martín, M.T. Baeza, J. Albaladejo, and E. Martínez (2003), Gas-phase rate coefficients and activation energies for the reaction of NO_3 radicals with selected branched aliphatic aldehydes, *Phys. Chem. Chem. Phys.*, *5*, 112–116.

Cadman, P., A.W. Kirk, and A.F. Trotman-Dickenson (1976), Reactions of chlorine atoms with ethane, propane, isobutane, fluoroethane, 1,1-difluoroethane, 1,1,1-trifluoroethane and cyclopropane, *J. Chem. Soc. Faraday Trans.*, *72*, 1027–1032.

Calvert, J.G., and S.E. Lindberg (2003), A modeling study of the mechanism of the halogen–ozone–mercury homogeneous reactions in the troposphere during the polar spring, *Atmos. Environ.*, *37*, 4467–4481.

Calvert, J.G., and S.E. Lindberg (2004a), Potential influence of iodine-containing compounds on the chemistry of the troposphere in the polar spring. I. Ozone depletion, *Atmos. Environ.*, *38*, 5087–5104.

Calvert, J.G., and S.E. Lindberg (2004b), The potential influence of iodine-containing compounds on the chemistry of the troposphere in the polar spring. II. Mercury depletion, *Atmos. Environ.*, *38*, 5105–5116.

Calvert, J.G., R. Atkinson, K.H. Becker, R.M. Kamens, J.H. Seinfeld, T.J. Wallington, and G. Yarwood (2002), *The Mechanisms of Atmospheric Oxidation of Aromatic Hydrocarbons*, Oxford University Press, New York.

Calvert, J.G., R. Atkinson, J.A. Kerr, S. Madronich, G.K. Moortgat, T.J. Wallington, and G. Yarwood (2000), *The Mechanisms of Atmospheric Oxidation of the Alkenes*, Oxford University Press, New York.

Cameron, M., V. Sivakumaran, T.J. Dillon, and J.N. Crowley (2002), Reaction between OH and CH_3CHO: Part 1, Primary product yields of CH_3 (296 K), CH_3CO (296 K), and H (237–296 K), *Phys. Chem. Chem. Phys.*, *4*, 3628–3638.

Campbell, I.M., D.F. McLaughlin, and B.J. Handy (1976), Rate constants for reactions of hydroxyl radicals with alcohol vapours at 292 K, *Chem. Phys. Lett.*, *38*, 362–364.

Campbell, R.J., and E.W. Schlag (1967), Energy distribution in reaction products. The photolysis of cyclobutanone, *J. Am. Chem. Soc.*, *89*, 5103–5106.

Campbell, R.J., E.W. Schlag, and B.W. Ristow (1967), Mechanistic consequences of curved Stern–Volmer plots. The photolysis of cyclobutanone, *J. Am. Chem. Soc.*, *89*, 5098–5102.

Canosa-Mas, C.E., S. Carr, M.D. King, D.E. Shallcross, K.C. Thompson, and R.P. Wayne (1999), A kinetic study of the reactions of NO_3 with methyl vinyl ketone, methacrolein, acrolein, methyl acrylate and methyl methacrylate, *Phys. Chem. Chem. Phys.*, *1*, 4195–4202.

Canosa-Mas, C.E., S.J. Smith, S. Toby, and R.P. Wayne (1989), Laboratory studies of the reactions of the nitrate radical with chloroform, methanol, hydrogen chloride and hydrogen bromide, *J. Chem. Soc. Faraday Trans. II*, *85*, 709–725.

Cantrell, C.A., J.A. Davidson, K.L. Busarow, and J.G. Calvert (1986), The $CH_3CHO–NO_3$ reaction and possible nighttime PAN generation, *J. Geophys. Res.*, *91*, 5347–5353.

Cantrell, C.A., J.A. Davidson, A.H. McDaniel, R.E. Shetter, and J.G. Calvert (1990a), Temperature-dependent formaldehyde cross sections in the near-ultraviolet spectral region, *J. Phys. Chem.*, *94*, 3902–3908.

Cantrell, C.A., J.A. Davidson, R.E. Shetter, B.A. Anderson, and J.G. Calvert (1987), Reactions of NO_3 and N_2O_5 with molecular species of possible atmospheric interest, *J. Phys. Chem.*, *91*, 6017–6021.

Cantrell, C.A., R.E. Shetter, J.A. Lind, A.H. McDaniel, J.G. Calvert, D.P. Parrish, F.C. Fehsenfeld, M.P. Buhr, and M. Trainer (1993), An improved chemical amplifier technique for peroxy radical measusrements, *J. Geophys. Res.*, *98*, 2897–2909.

Cantrell, C.A., R.E. Shetter, A.H. McDaniel, J.G. Calvert, J.A. Davidson, D.C. Lowe, S.C. Tyler, R.J. Cicerone, and J.P. Greenberg (1990b), Carbon kinetic isotope effect in the oxidation of methane by hydroxyl radicals, *J. Geophys. Res.*, *95*, 22455–22462.

Cantrell, C.A., D.H. Stedman, and G. Wendel (1984), The measurement of atmospheric peroxy-radicals by chemical amplification, *J. Anal. Chem.*, *56*, 1496–1502.

Cantrell, C.A., W.R. Stockwell, L.G. Anderson, K.L. Busarow, D. Perner, A. Schmeltekopf, J.G. Calvert, and H.S. Johnston (1985), Kinetic study of the $NO_3–CH_2O$ reaction and its possible role in nighttime tropospheric chemistry, *J. Phys. Chem.*, *89*, 139–146.

Caralp, F., and W. Forst (2003), Chemical activation in unimolecular reactions of some unsubstituted alkoxy radicals, *Phys. Chem. Chem. Phys.*, *5*, 4653–4655.

Caralp, F., P. Devolder, Ch. Fittschen, N. Gomez, H. Hippler, R, Méreau, M.-T. Rayez, F. Striebel, and B. Viskolcz (1999), The thermal unimolecular decomposition rate constants of ethoxy radicals, *Phys. Chem. Chem. Phys.*, *1*, 2935–2944.

Caralp, F., R. Lesclaux, M.T. Rayez, J.C. Rayez, and W. Forst (1988), Kinetics of the combination reactions of chlorofluoromethylperoxy radicals with nitrogen dioxide in the temperature range 233–373 K, *J. Chem. Soc. Faraday Trans. 2*, *84*, 569–585.

CARB (1993), Proposed regulations for low-emissions vehicles and clean fuels. Staff Report and Technical Support Document. California Air Resources Board, Sacramento, California, USA.

Carl, S.A., and J.N. Crowley (2001), 298 K rate coefficients for the reaction of OH radicals with i-C_3H_7I, n-C_3H_7I and C_3H_8, *Atmos. Chem. Phys.*, *1*, 1–7.

Carless, H.A.J., and E.K.C. Lee (1970), Stereochemistry of the thermal and photolytic decomposition of *cis*-and *trans*-2,3-dimethylcyclobutanone in the gas phase, *J. Am. Chem. Soc.*, *92*, 4482.

Carless, H.A.J., and E.K.C. Lee (1972), Photolysis and pyrolysis of 2-*n*-propylcyclobutanone in the gas phase, *J. Am. Chem. Soc.*, *94*, 1–6.

Carless, H.A.J., J. Metcalfe, and E.K.C. Lee (1972), Photolytic and triplet benzene-sensitized decomposition of *cis*- and *trans*-2,3- and -2,4-dimethylcyclobutanones, *J. Am. Chem. Soc.*, *94*, 7221–7235.

Carr, Jr., R.W., D.G. Peterson, and F.K. Smith (1986), Flash photolysis of 1,3-dichlorotetrafluoroacetone in the presence of oxygen. Kinetics and mechanism of the oxidation of the chlorodifluoromethyl radicals, *J. Phys. Chem.*, *90*, 607–614.

Carslaw, N., D.J. Creasey, D.E. Heard, P.J. Jacobs, J.D. Lee, A.C. Lewis, J.B. McQuaid, M.J. Pilling, S. Bauguitte, S.A. Penkett, P.S. Monks, and G. Salisbury (2002), Eastern Atlantic Spring Experiment 1997 (EASE 97): 2. Comparisons of model concentrations of OH, HO_2 and RO_2 with measurements, *J. Geophys. Res., 107*, 4190, doi 10.1029/2001JD001568.

Carter, W.P.L. (1990), A detailed mechanism for the gas-phase atmospheric reactions of organic compounds, *Atmos. Environ.*, *24A*, 481–518.

Carter, W.P.L. (1994), Development of ozone reactivity scales for volatile organic compounds. *J. Air Waste Manage. Assoc.*, *44*, 881–899.

Carter, W.P.L. (1995), Computer modelling of environmental chamber measurements of maximum incremental reactivities of volatile organic compounds, *Atmos. Environ.*, *29*, 2513–2517.

Carter, W.P.L. (1998), Updated maximum incremental reactivity scale for regulatory applications. Preliminary Report to California Air Resource Board. University of California, Riverside, California, USA.

Carter, W.P.L. (2000), Documentation of the SAPRC-99 chemical mechanism for VOC reactivity assessment. Final Report to California Air Resource Board. Air Pollution Research Center and College of Engineering, Center for Environmental Research and Technology, University of California, Riverside, California, USA.

Carter, W.P.L., and R. Atkinson (1985), Atmospheric chemistry of alkanes, *J. Atmos. Chem.*, *3*, 377–405.

Carter, W.P.L., and R. Atkinson (1989), Alkyl nitrate formation from the atmospheric photooxidation of alkanes; a revised estimation method, *J. Atmos. Chem.*, *8*, 165–173.

Carter, W.P.L., K.R. Darnall, A.C. Lloyd, A.M. Winer, and J.N. Pitts, Jr. (1976), Evidence for alkoxy radical isomerization in photooxidations of C_4–C_6 alkanes under simulated atmospheric conditions, *Chem. Phys. Lett.*, *42*, 22–27.

Carter, W.P.L., A.C. Lloyd, J.L. Sprung, and J.N. Pitts, Jr. (1979), Computer modeling of smog chamber data: Progress in validating of a detailed mechanism for the photooxidation of propene and *n*-butane in photochemical smog, *Int. J. Chem. Kinet.*, *11*, 45–101.

Carter, W.P.L., D. Luo, and L.L. Malkina (1997), Environmental chamber studies for development of an updated photochemical mechanism for VOC reactivity assessment, Final Report to the California Air Resources Board, the Coordinating Research Council and the National Renewable Energy Laboratory. Air Pollution Research Center and College of Engineering, Center for Environmental Research and Technology, University of California, Riverside, California, USA.

Carter, W.P.L., D. Luo, L.L. Malkina, and D. Fitz (1995), The University of California, Riverside Experimental Chamber Data Base for Evaluating Oxidant Mechanisms, Volumes 1 and 2. University of California, Riverside, California, USA.

Carter, W.P.L., G. Tonnesen, and G. Yarwood (2003), Investigation of VOC reactivity effects using existing regional air quality models, Center for Environmental Research and Technology, University of California, Riverside, California.

Cassanelli, P., R.A. Cox, J.J. Orlando, and G.S. Tyndall (2006), An FTIR study of the isomerization of 1-butoxy radicals under atmospheric conditions, *J. Photochem. Photobiol.*, *177*, 109–115.

Cassanelli, P., D. Johnson, and R.A. Cox (2005), A temperature-dependent relative-rate study of the OH initiated oxidation of *n*-butane: The kinetics of the reactions of the 1-and 2-butoxy radicals, *Phys. Chem. Chem. Phys.*, *7*, 3702–3710.

Catoire, V., P.A. Ariya, H. Niki, and G.W. Harris (1997), FTIR study of the Cl- and Br-atom initiated oxidation of trichloroethylene, *Int. J. Chem. Kinet.*, *29*, 695–704.

Catoire, V., R. Lesclaux, P.D. Lightfoot, and M.T. Rayez (1994), Kinetic study of the reactions of CH_2ClO_2 with itself and with HO_2, and theoretical study of the reactions of CH_2ClO, between 251 and 600 K, *J. Phys. Chem.*, *98*, 2889–2898.

Catoire, V., R. Lesclaux, W.F. Schneider, and T.J. Wallington (1996), Kinetics and mechanisms of the self reactions of CCl_3O_2 and $CHCl_2O_2$ radicals and their reactions with HO_2, *J. Phys. Chem.*, *100*, 14356–14371.

Cattell, F.C., J. Cavanagh, R.A. Cox, and M.E. Jenkin (1986), A kinetics study of reactions of HO_2 and $C_2H_5O_2$ using diode laser absorption spectroscopy, *J. Chem. Soc. Faraday II, 82*, 1999–2018.

Cautreels, W., and K. van Cauwenberghe (1978), Experiments on distribution of organic pollutants between airborne particulate matter and corresponding gas-phase, *Atmos. Environ.*, *12*, 1133–1141.

Cavalli, F., I. Barnes, and K.H. Becker (2000), FT-IR kinetic and product study of the OH radical-initiated oxidation of 1-pentanol, *Environ. Sci. Technol.*, *34*, 4111–4116.

Cavalli, F., H. Geiger, I. Barnes, and K.H. Becker (2002), FTIR kinetic, product, and modeling study of the OH-initiated oxidation of 1-butanol in air, *Environ. Sci. Technol.*, *36*, 1263–1270.

Cavalli, F., M. Glasius, J. Hjorth, B. Rindone, and N.R. Jensen (1998), Atmospheric lifetimes, infrared spectra and degradation products of a series of hydrofluoroethers, *Atmos. Environ.*, *32*, 3767–3773.

Chang, J.S., and F. Kaufman (1977), Kinetics of the reactions of hydroxyl radicals with some halocarbons: $CHFCl_2$, CHF_2Cl, CH_3CCl_3, C_2HCl_3, and C_2Cl_4, *J. Chem. Phys.*, *66*, 4989–4994.

Chang, S., E. McDonald-Buller, Y. Kimura, G. Yarwood, J. Neece, M. Russell, P. Tanaka, and D. Allen (2002), Sensitivity of urban ozone formation to chlorine emission estimates, *Atmos. Environ.*, *36*, 4991–5003.

Cheema, S.A., K.A. Holbrook, G.A. Oldershaw, and R.W. Walker (2002), Kinetics and mechanism associated with the reactions of hydroxyl radicals and of chlorine atoms with 1-propanol under near-tropospheric conditions between 273 and 343 K, *Int. J. Chem. Kinet.*, *34*, 110–121.

Chen, J., V. Catoire, and H. Niki (1995), Mechanistic study of the $BrCH_2O$ radical degradation in 700 Torr air, *Chem. Phys. Lett.*, *245*, 519–528.

Chen, J., V. Young, H. Niki, and H. Magid (1997), Kinetic and mechanistic studies for reactions of $CF_3CH_2CHF_2$ (HFC-245fa) initiated by H-atom abstraction using atomic chlorine, *J. Phys. Chem. A*, *101*, 2648–2653.

Chen, L., S. Kutsuna, K. Tokuhashi, and A. Sekiya (2003), New technique for generating high concentrations of gaseous OH radicals in relative rate measurements, *Int. J. Chem. Kinet.*, *35*, 317–325.

Chen, L., K. Tokuhashi, S. Kutsuna, and A. Sekiya (2004), Rate constants for the gas-phase reaction of $CF_3CF_2CF_2CF_2CF_2CHF_2$ with OH radicals at 250–430 K, *Int. J. Chem. Kinet.*, *36*, 26–33.

Chen, T.-Y., I.J. Simpson, D.R. Blake, and F.S. Rowland (2001), Impact of the leakage of liquefied petroleum gas (LPG) on Santiago air quality, *Geophys. Res. Lett.*, *28*, 2193–2196.

Chen, X.-B., and W.-H. Fang (2002), Norrish I vs. II reactions of butanal: a combined CASSCF, DFT and MP2 study, *Chem. Phys. Lett.*, *361*, 473–482.

Chen, Y., and S. Ye (2004), Photochemical reaction mechanism of cyclobutanone: CASSCF study, *J. Quantum Chem.*, *97*, 725–735.

Chen, Y., and L. Zhu (2001), The wavelength dependence of the photodissociation of propionaldehyde in the 290–330 nm region, *J. Phys. Chem. A*, *105*, 9689–9696.

Chen, Y., L. Zhu, and J.S. Francisco (2002), Wavelength-dependent photolysis of *n*-butyraldehyde and *i*-butyraldehyde in the 280–330 nm region, *J. Phys. Chem. A*, *106*, 7755–7763.

Chew, A.A., and R. Atkinson (1996), OH radical formation yields from the gas-phase reactions of O_3 with alkenes and monoterpenes, *J. Geophys. Res.*, *101*, 28649–28653.

Chew, A.A., R. Atkinson, and S.M. Aschmann (1998), Kinetics of the gas-phase reactions of NO_3 radicals with a series of alcohols, glycol ethers, ethers and chloroalkenes, *J. Chem. Soc. Faraday Trans.*, *94*, 1083–1089.

Chiappero, M.S., F.E. Malanca, A. G.A. Argüello, S.T. Wooldridge, M.D. Hurley, J.C. Ball, T.J. Wallington, R.L.Waterland, and R.C. Buck (2006), Atmospheric chemistry of perfluoraldehyde ($C_xF_{2x+1}CHO$) and fluorotelomer aldehydes ($C_xF_{2x+1}CH_2CHO$): Quantification of the important role of photolysis, *J. Phys. Chem. A*, *110*, 11944–11953.

Chichinin, A., S. Téton, G. LeBras, and G. Poulet (1994), Kinetic investigation of the OH + CH_3Br reaction between 248 and 390 K, *J. Atmos. Chem.*, *18*, 239–245.

Chiltz, G., R. Eckling, P. Goldfinger, G. Huybrechts, H.S. Johnston, L. Meyers, and G. Verbeke (1963), Kinetic isotope effect in photochlorination of H_2, CH_4, $CHCl_3$, and C_2H_6, *J. Chem. Phys.*, *38*, 1053–1061.

Chiorboli, C., C.A. Bignozzi, A. Maldotti, P.F. Giardini, A. Rossi, and V. Carassiti (1983), Rate constants for the gas-phase reactions of OH radicals with β-dimethylstyrene and acetone. Mechanism of β-dimethylstyrene NO_x–air photooxidation, *Int. J. Chem. Kinet.*, *15*, 579–586.

Choo, K.Y., and S.W. Benson (1981), Arrhenius parameters for the alkoxy radical decomposition reactions. Part 1. Quenching kinetics of OH $A^2\Sigma$ and rate constants for reactions of OH $X^2\Pi$ with CH_3CCl_3 and CO, *Int. J. Chem. Kinet.*, *13*, 833–844.

Chow, J.M., A.M. Miller, and M.J. Elrod (2003), Kinetics of the $C_3H_7O_2$ + NO reaction: Temperature dependence of the overall rate constant and the i-$C_3H_7ONO_2$ branching channel, *J. Phys. Chem. A.*, *107*, 3040–3047.

Chowdhury, P.K., H.P. Upadhyaya, P.D. Naik, and J.P. Mittal (2002a), ArF laser photodissociation dynamics of hydroxyacetone: LIF observation of OH and its reaction rate with the parent, *Chem. Phys. Lett.*, *351*, 201–207.

Chowdhury, P.K., H.P. Upadhyaya, P.D. Naik, and J.P. Mittal (2002b), Direct observation of OH photofragment from triplet hydroxyacetone, *Chem. Phys. Lett.*, *356*, 476–482.

Chuang, C., M.F. Foltz, and C.B. Moore (1987), T_1 barrier height, S_1–T_1 intersystem crossing rate, and S_0 radical dissociation threshold for H_2CO, D_2CO and HDCO, *J. Chem. Phys.*, *87*, 3855–3864.

Clark, J.H., Moore, C.B., and N.S. Nagar (1978), Photodissociation of formaldehyde—Absolute quantum yields, radical reactions and NO reactions, *J. Chem. Phys.*, *68*, 1264–1271.

Clark, R.H., D. Husain, and J.Y. Jezequel (1982), The reaction of chlorine atoms, $Cl(3^2P_j)$, with nitric acid in the gas phase, *J. Photochem.*, *18*, 39–46.

Clarke, J.S., J.H. Kroll, N.M. Donahue, and J.G. Anderson (1998), Testing frontier orbital control: Kinetics of OH with ethane, propane, and cyclopropane from 180 to 360 K, *J. Phys. Chem. A*, *102*, 9847–9857.

Clemitshaw, K.C., and J.R. Sodeau (1987), Atmospheric chemistry at 4.2 K: a matrix isolation study of the reaction between CF_3O_2 radicals and NO, *J. Phys. Chem.*, *91*, 3650–3653.

Clyne, M.A.A., and P.M. Holt (1979a), Reaction kinetics involving ground $X^2\Pi$ and excited $A^2\Sigma^+$ hydroxyl radicals. Part 1. Quenching kinetics of hydroxyl $A^2\Sigma^+$ and rate constants for reactions of hydroxyl $X^2\Pi$ with trichloroethane and carbon monoxide., *J. Chem. Soc. Faraday Trans.*, *75*, 569–581.

Clyne, M.A.A., and P.M. Holt (1979b), Reaction kinetics involving ground $X^2\Pi$ and excited $A^2\Sigma^+$ hydroxyl radicals. Part 2—Rate constants for reactions of OH $X^2\Pi$ with halogenomethanes and halogenoethanes, *J. Chem. Soc. Faraday Trans. II*, *75*, 582–591.

Clyne, M.A.A., and R.F. Walker (1973), Absolute rate constants for elementary reactions in the chlorination of CH_4, CD_4, CH_3Cl, CH_2Cl_2, $CHCl_3$, $CDCl_3$ and $CBrCl_3$, *J. Chem. Soc. Faraday Trans. 1*, *69*, 1547–1567.

Coeur, C., V. Jacob, and P. Foster (1999), Gas-phase reaction of hydroxyl radical with the natural hydrocarbon bornyl acetate, *Phys. Chem. Earth (C)*, *24*, 537–539.

Coeur, C., V. Jacob, P. Foster, and P. Baussand (1998), Rate constant for the gas-phase reaction of hydroxyl radicals with the natural hydrocarbon bornyl acetate, *Int. J. Chem. Kinet.*, *30*, 497–502.

Cohen, N., and K.R. Westberg (1986), The use of transition state theory to extrapolate rate coefficients for reaction of O atoms with the alkanes, *Int. J. Chem. Kinet.*, *18*, 99–140.

Cohen, N., and K.R. Westberg (1991), Chemical kinetic data sheets for high-temperature reactions. Part II, *J. Phys. Chem. Ref. Data*, *20*, 1211–1311.

Collins, W.J., R.G. Derwent, C.E. Johnson, and D.S. Stevenson (2002), The oxidation of organic compounds in the troposphere and their global warming potentials, *Clim. Change*, *52*, 453–479.

Collins, W.J., D.S. Stevenson, C.E. Johnson, and R.G. Derwent (1997), Tropospheric ozone in a global-scale 3-D Lagrangian model and its response to NO_x emission controls, *J. Atmos. Chem.*, *26*, 223–274.

Commission of the European Communities (1991), Council directive amending directive 70/220/EEC on the approximation of the laws of member states relating to the measures to be taken against air pollution by emissions from motor vehicles. 91/441/EEC. Official Journal of the European Communities, L242/1-L242/106..

Commission of the European Communities (1999), The limitation of emissions of volatile organic compounds due to the use of organic solvents in certain activities and installations. 1999/13/EC. http://europa.eu.int/eur-lex/en/consleg/pdf/1999/en_1999L0013_do_001.pdf. Brussels, Belgium.

CONCAWE (1990), Closing the gasoline system—Control of gasoline emissions from the distribution system and vehicles. CONCAWE Report no. 3/90, Brussels, Belgium.

Cooper, D.L., N.L. Allan, and A. McCulloch (1990), Reactions of hydrofluorocarbons and hydrochlorofluorocarbons with the hydroxyl radical, *Atmos. Environ.*, *24A*, 2417–2419.

Cooper, D.L., T.P. Cunningham, N.L. Allan, and A. McCulloch (1992), Tropospheric lifetimes of potential CFC replacements: rate coefficients for reaction with the hydroxyl radical, *Atmos. Environ.*, *26A*, 1331–1334.

Costela, A., M.T. Crespo, and J.M. Figuera (1986), Laser photolysis of acetone at 308 nm, *J. Photochem.*, *34*, 165–173.

Cotter, E.S.N., N.J. Booth, C,E. Canosa-Mas, and R.P. Wayne (2001), Release of iodine in the atmospheric oxidation of alkyl iodides and the fates of iodinated alkoxy radicals, *Atmos. Environ.*, *35*, 2169–2178.

Cotter, E.S.N., C.E. Canosa-Mas, C.R. Manners, R.P. Wayne, and D.E. Shallcross (2003), Kinetic study of the reactions of OH with the simple alkyl iodides: CH_3I, C_2H_5I, 1-C_3H_7I and 2-C_3H_7I, *Atmos. Environ.*, *37*, 1125–1133.

Coulson, D.R., and N.C. Yang (1966), Deuterium isotope effects in the photochemistry of 2-hexanone, *J. Am. Chem. Soc.*, *88*, 4511–4513.

Cox, R.A., R.G. Derwent, A.E.J. Eggleton, and J.E. Lovelock (1976a), Photochemical oxidation of halocarbons in the troposphere, *Atmos. Environ.*, *10*, 305–308.

Cox, R.A., R.G. Derwent, P.M. Holt, and J.A. Kerr (1976b), Photo-oxidation of methane in the presence of NO and NO_2, *J. Chem. Soc. Faraday Trans. 1*, *72*, 2044–2060.

Cox, R.A., R.G. Derwent, S.V. Kearsey, L. Batt, and K.G. Patrick (1980a), Photolysis of methyl nitrite: kinetics of the reaction of the methoxy radical with O_2, *J. Photochem.*, *13*, 149–163.

Cox, R.A., R.G. Derwent, and M.R. Williams (1980b), Atmospheric photooxidation reactions. Rates, reactivity, and mechanisms for reaction of organic compounds with hydroxyl radicals, *Environ. Sci. Technol.*, *14*, 57–61.

Cox, R.A., K.F. Patrick, and S.A. Chant (1981), Mechanism of atmospheric photooxidation of organic compounds. Reactions of alkoxy radicals in oxidation of *n*-butane and simple ketones, *Environ. Sci. Technol.*, *15*, 587–592.

Cox, R.A., and G.S. Tyndall (1979), Rate constants for reactions of CH_3O_2 in the gas phase, *Chem. Phys. Lett.*, *65*, 357–360.

Cox, R.A., and G.S. Tyndall (1980), Rate constants for the reactions of CH_3O_2 with HO_2, NO and NO_2 using molecular modulation spectrometry, *J. Chem. Soc. Faraday Trans. II*, *76*, 153–163.

Crawford, M.A., J.J. Szente, M.M. Maricq, and J.S. Francisco (1997), Kinetics of the reaction between cyclopentylperoxy readicals and HO_2, *J. Phys. Chem. A*, *101*, 5337–5343.

Cronin, T.J., and L. Zhu (1998), Dye laser photolysis of *n*-pentanal from 280 to 330 nm, *J. Phys. Chem.*, *102*, 10274–10279.

Crowley, J.N., G. Saueressig, P. Bergamaschi, H. Fischer, and G.W. Harris (1999), Carbon kinetic isotope effect in the reaction CH_4 + Cl: A relative rate study using FTIR spectroscopy, *Chem. Phys. Letters*, *303*, 268–274.

Cundall, R.B., and A.S. Davies (1966), The mechanism of the gas phase photolysis of acetone, *Proc. R. Soc., Lond.*, *A290*, 563–582.

Cundall, R.B., and A.S. Davies (1967), Primary processes in the gas phase photochemistry of carbonyl compounds, *Prog. React. Kinet.*, *4*, 149–213.

Daële, V., A. Ray, I. Vassalli, G. Poulet, and G. Le Bras (1995), Kinetic study of reactions of C_2H_5O and $C_2H_5O_2$ with NO at 298 K and 0.55–2 Torr, *Int. J. Chem. Kinet.*, *28*, 1121–1133.

Dagaut, P., T.J. Wallington, and M.J. Kurylo (1988a), Temperature dependence of the rate constant for the HO_2 + CH_3O_2 gas-phase reaction, *J. Phys. Chem.*, *92*, 3833–3836.

Dagaut, P., T.J. Wallington, and M.J. Kurylo (1988b), Flash photolysis kinetic absorption spectroscopy study of the gas-phase reaction HO_2 + $C_2H_5O_2$ over the temperature range 228–380 K, *J. Phys. Chem.*, *92*, 3836–3839.

D'Anna, B., and C.J. Nielsen (1997), Kinetic study of the vapour-phase reaction between aliphatic aldehydes and the nitrate radical, *J. Chem. Soc. Faraday Trans.*, *93*, 3479–3483.

D'Anna, B., Ø. Andresen, Z. Gefen, and C.J. Nielsen (2001a), Kinetic study of OH and NO_3 radical reactions with 14 aliphatic aldehydes, *Phys. Chem. Chem. Phys.*, *3*, 3057–3063.

D'Anna, B., V. Bakken, J.A. Beukes, C.J. Nielsen, K. Brudnik, and J.T. Jodkowski (2003), Experimental and theoretical studies of gas phase NO_3 and OH radical reactions with formaldehyde, acetaldehyde and their isotopomers, *Phys. Chem. Chem. Phys.*, *5*, 1790–1805.

D'Anna, B., S. Langer, E. Ljungstrom, C.J. Nielsen, and M Ullerstam (2001b), Rate coefficients and Arrhenius parameters for the reaction of the NO_3 radical with acetaldehyde and acetaldehyde-1d, *Phys. Chem. Chem. Phys.*, *3*, 1631–1637.

D'Anna, B., S.R. Sellevåg, K. Wirtz, and C.J. Nielsen (2005), Photolysis study of perfluoro-2-methyl-3-pentanone under natural sunlight conditions, *Environ. Sci. Technol.*, *39*, 8708–8711.

Darnall, K.R., R. Atkinson, and J.N. Pitts, Jr. (1978), Rate constants for the reactions of the OH radical with selected alkanes at 300 K, *J. Phys. Chem.*, *82*, 1581–1584.

Darnall, K.R., W.P.L. Carter, A.M. Winer, A.C. Lloyd, and J.N. Pitts, Jr. (1976), Importance of RO_2 + NO in alkyl nitrate formation from C_4–C_6 alkane photooxidations under simulated atmospheric conditions, *J. Phys. Chem.*, *80*, 1948–1950.

Davidson, J.A., C.A. Cantrell, S.C. Tyler, R.E. Shetter, R.J. Cicerone, and J.G. Calvert (1987), Carbon kinetic isotope effect in the reaction of CH_4 with OH, *J. Geophys. Res.*, *92*, 2195–2199.

Davies, R.E., and P.J. Freyd (1989), $C_{167}H_{336}$ is the smallest alkane with more realizable isomers than the observed universe has "Particles", *J. Chem. Ed.*, *66*, 278–281.

Davis, D.D., W. Braun, and A.M. Bass (1970), Reactions of $Cl^2P_{3/2}$: Absolute rate constants for reaction with H_2, CH_4, C_2H_6, CH_2Cl_2, C_2Cl_4, and *c*-C_6H_{12}, *Int. J. Chem. Kinet.*, *2*, 101–114.

Davis, D.D., G. Machado, B. Conaway, Y. Oh, and R. Watson (1976), A temperature dependent kinetics study of the reaction of OH with CH_3Cl, CH_2Cl_2, $CHCl_3$, and CH_3Br, *J. Chem. Phys.*, *65*, 1268–1274.

Davis, M.E., W. Drake, D. Vimal, and P.S. Stevens (2005), Experimental and theoretical studies of the kinetics of the reactions of OH and OD with acetone and acetone d_6 at low pressure, *J. Photochem. Photobiol. A*, *176*, 162–171.

Davis, W., Jr. (1948), The photochemical decomposition of diethyl ketone at 3130 Å, *J. Am. Chem. Soc.*, *70*, 1868–1869.

Davis, W., Jr., and W.A. Noyes, Jr. (1947), Photochemical studies. XXXVIII. A further study of the photochemistry of methyl *n*-butyl ketone, *J. Am. Chem. Soc.*, *69*, 2153–2158.

Dawidowicz, E.A., and C.R. Patrick (1964), The photolysis of trifluoroacetone, *J. Chem. Soc.*, 4250–4254.

DeBruyn, W.J., S.X. Duan, X.Q. Shi, P. Davidovits, D.R. Worsnop, M.S. Zahniser, and C.E. Kolb (1992), Tropospheric heterogeneous chemistry of haloacetyl and carbonyl halides, *Geophys. Res. Lett.*, *19*, 1939–1942.

DeMore, W.B. (1992), Relative rate constants for the reactions of OH with methane and methyl chloroform, *Geophys. Res. Lett.*, *19*, 1367–1370.

DeMore, W.B. (1993a), Rate constants for the reactions of hydroxyl with HFC-134a (CF_3CH_2F) and HFC-134 (CHF_2CHF_2), *Geophys. Res. Lett.*, *20*, 1359–1362.

DeMore, W.B. (1993b), Rate constant ratio for the reactions of OH with CH_3D and CH_4, *J. Phys. Chem.*, *97*, 8564–8566.

DeMore, W.B. (1993c), Rates of hydroxyl radical reactions with some HFCs, *Proc. SPIE—Int. Soc. Opt. Eng.*, *1715*, 72–77.

DeMore,W.B. (1996), Experimental and estimated rate constants for the reactions of hydroxyl radicals with several halocarbons, *J. Phys. Chem.*, *100*, 5813–5820.

DeMore, W.B. (2005), Regularities in Arrhenius parameters for rate constants of abstraction reactions of hydroxyl radical with C—H bonds, *J. Photochem. Photobiol. A*, *176*, 129–135.

DeMore, W.B., and K.D. Bayes (1999), Rate constants for the reactions of hydroxyl radical with several alkanes, cycloalkanes, and dimethyl ether, *J. Phys. Chem. A.*, *103*, 2649–2654.

DeMore, W.B., S.P. Sander, D.M Golden, R.F. Hampson, M.J. Kurylo, C.J. Howard, A.R. Ravishankara, and C.E. Kolb (1994), Chemical kinetics and photochemical data for use in stratospheric modeling, Evaluation Number 11, JPL Publication 94–26.

DeMore, W.B., S.P. Sander, D.M Golden, R.F. Hampson, M.J. Kurylo, C.J. Howard, A.R. Ravishankara, C.E. Kolb, and M.J. Molina (1997), Chemical kinetics and photochemical data for use in stratospheric modeling, Evaluation Number 12, JPL Publication 97–4.

Deng, W., A.J. Davis, L. Zhang, D.R. Katz, and T.S. Dibble (2001), Direct kinetic studies of reactions of 3-pentoxy radicals wtih NO and O_2, *J. Phys. Chem. A*, *105*, 8985–8990.

Deng, W., C. Wang, D.R. Katz, G.R. Gawinski, A.J. Davis, and T.S. Dibble (2000), Direct kinetic studies of the reactions of 2-butoxy radicals wtih NO and O_2, *Chem. Phys. Lett.*, *330*, 541–546.

Denschlag, H.O., and E.K.C. Lee (1968), Benzene photosensitization and direct photolysis of cyclobutanone and cyclobutanone-2-*t* in the gas phase, *J. Am. Chem. Soc.*, *90*, 3628–3638.

Derwent, R.G. (1990), Evaluation of a number of chemical mechanisms for their application in models describing the formation of photochemical ozone in Europe, *Atmos. Environ.*, *24A*, 2615–2624.

Derwent, R.G., and M.E. Jenkin (1991), Hydrocarbons and the long-range transport of ozone and PAN across Europe, *Atmos. Environ.*, *25A*, 1661–1678.

Derwent, R.G., P. Grennfelt, and O. Hov (1991), Photochemical oxidants in the atmosphere, Nordic Council of Ministers, Status Report Nord 1991:7, Copenhagen, Denmark.

Derwent, R.G., M.E. Jenkin, S.M. Saunders, and M.J. Pilling (1998), Photochemical ozone creation potentials for organic compounds in North West Europe calculated with a master chemical mechanism, *Atmos. Environ.*, *32*, 2429–2441.

Derwent, R.G., M.E. Jenkin, S.M. Saunders, and M.J. Pilling (2001), Characterisation of the reactivities of volatile organic compounds using a master chemical mechanism. *J. Air Waste Manage. Assoc.*, *51*, 699–707.

Derwent, R.G., M.E. Jenkin, S.M. Saunders, M.J. Pilling, and N.R. Passant (2005), Multi-day ozone formation for alkanes and carbonyls investigated with a master chemical mechanism under European conditions, *Atmos. Environ.*, *39*, 627–635.

Derwent, R.G., M.E. Jenkin, S.M. Saunders, M.J. Pilling, P.G. Simmonds, N.R. Passant, G.J. Dollard, P. Dumitrean, and A. Kent (2003), Photochemical ozone formation in north west Europe and its control, *Atmos. Environ.*, *37*, 1983–1991.

Desai, J., J. Heicklen, A. Bahta, and R. Simonaitis (1986), The photo-oxidation of *i*-C_3H_7CHO vapour, *J. Photochem*, *34*, 137–164.

DeSain, J.D., S.J. Klippenstein, C.A. Taatjes, M.D. Hurley, and T.J. Wallington (2003), Product formation in the Cl-initiated oxidation of cyclopropane, *J. Phys. Chem. A*, *107*, 1992–2002.

Deshmukh, S., and W.P. Hess (1994), Photodissociation of acetyl chloride: Cl and CH_3 quantum yields and energy distributions, *J. Chem. Phys.*, *100*, 6429–6433,

Destriau, M., and J. Troe (1990), Thermal dissociation and recombination of alkyl and haloalkyl peroxynitrates: an SACM modeling study, *Int. J. Chem. Kinet.*, *22*, 915–934.

Devolder, P. (2003), Atmospheric fate of small alkoxy radicals: recent experimental and theoretical advances, *J. Photochem. Photobiol. A. Chem.*, *157*, 137–147.

Devolder, P., Ch. Fittschen, A. Frenzel, H. Hippler, G. Poskrebyshev, F. Striebel, and B. Viskolcz (1999), Complete falloff curves for the unimolecular decomposition of *i*-propoxy radicals between 330 and 408 K, *Phys. Chem. Chem. Phys.*, *1*, 675–682.

Dhanya, S. and R.D. Saini (1997), Rate constants of OH radical reactions in the gas phase with some fluorinated compounds: a correlation with molecular parameters, *Int. J. Chem. Kinet.*, *29*, 187–194.

Diau, E.W.-G., C. Kötting, and A.H. Zewail (2001), Femtochemistry of Norrish Type-I reactions: III. The anomalous predissociation dynamics of cyclobutanone on the S_1 surface, *Chemphyschem*, *2*, 294–309.

Dillemuth, F.J., D.R. Skidmore, and C.C. Schubert (1960), The reaction of ozone with methane, *J. Phys. Chem.*, *64*, 1496–1499.

Dillon, T.J., D. Hölscher, V. Sivakumaran, A. Horowitz, and J.N. Crowley (2005), Kinetics of the reactions of HO with methanol (210–351 K) and with ethanol (216–368 K), *Phys. Chem. Chem. Phys.*, *7*, 349–355.

Dimitriades, B. (1996), Scientific basis for the VOC reactivity issues raised by Section 183(e) of the Clean Air Act Amendments of 1990, *J. Air Waste Manage. Assoc.*, *46*, 963–970.

Dimitriades, B. (1999), Scientific basis of an improved EPA policy on control of organic emissions for ambient ozone reduction, *J. Air Waste Manage. Assoc.*, *49*, 831–838.

Ding, H., A.J. Orr-Ewing, and R.N. Dixon (1999), Rotational structure in the $\tilde{A}\ ^1A''$—X^1A' spectrum of formyl chloride, *Phys. Chem. Chem. Phys.*, *1*, 4181–4185.

Dlugokencky, E.J., and C.J. Howard (1989), Studies of NO_3 radical reactions with some atmospheric organic compounds at low pressures, *J. Phys. Chem.*, *93*, 1091–1096.

Dóbé, S., T. Bérces, and F. Márta (1986), Gas phase decomposition and isomerization reactions of 2-pentoxy radicals, *Int. J. Chem. Kinet.*, *18*, 329–344.

Dóbé, S., L.A. Khachatryan, and T. Bérces (1989), Kinetics of reactions of hydroxyl radicals with a series of aliphatic aldehydes, *Ber. Bunsenges. Phys. Chem.*, *93*, 847–852.

Dóbé, S., T. Turányi, T. Bérces, and F. Márta (1991), The kinetics of hydroxyl radical with cyclopropane and cyclobutane, *Proc Indian Acad. Sci. (Chem. Sci.)*, *103*, 499–503.

Dóbé, S., T. Turányi, A.A. Iogansen, and T. Bérces (1992), Rate constants of the reactions of OH radicals with cyclopropane and cyclobutane, *Int. J. Chem. Kinet.*, *24*, 191–198.

Dobis, O., and S.W. Benson (1987), Analysis of flow dynamics in a new, very low pressure reactor. Application to the reaction: $Cl + CH_4 \leftrightarrow HCl + CH_3$, *Int. J. Chem. Kinet.*, *19*, 691–708.

Dobis, O., and S.W. Benson (1990), Kinetics of Cl atom reactions with C_2H_6, C_2H_5, and C_2H_4, Rates of disproportionation and recombination of ethyl radicals, *J. Am. Chem. Soc.*, *112*, 1023–1029.

Dobis, O., and S.W. Benson (1991), Temperature coefficients of the rates of Cl atom reactions with C_2H_6, C_2H_5, and C_2H_4. The rates of disproportionation and recombination of ethyl radicals, *J. Am. Chem. Soc.*, *113*, 6377–6386.

Dobis, O., S.W. Benson, and T.J. Mitchell (1994), Kinetics of the reactions of Cl atoms with C_2D_6 and C_2D_5 and the disproportionation of 2 C_2D_5. The deuterium isotope effect, *J. Phys. Chem.*, *98*, 12284–12293.

Dodd, R.E., and J.W. Smith (1957), The photolysis of trifluoracetaldehyde, *J. Chem. Soc.*, *282*, 1465–1473.

Dodge, M.C. (2000), Chemical oxidant mechanisms for air quality modeling: critical review, *Atmos. Environ.*, *34*, 2103–2130.

Dognon, A.M., F. Caralp, and R. Lesclaux (1985), Reactions of chlorofluoromethyl peroxy radicals with nitric oxide: a kinetic study in the temperature range 230–430 K, *J. Chim. Phys. Phys.-Chim. Biol.*, *82*, 349–352.

Dollard, G.J., P. Dumitrean, S. Telling, J. Dixon, and R.G. Derwent (2007), Observed trends in ambient concentrations of C_2–C_8 hydrocarbons in the United Kingdom over the period from 1993 to 2004, *Atmos. Environ.*, *41*, 2559–2569.

Donaghy, T., I. Shanahan, M. Hande, and S. Fitzpatrick (1993), Rate constants and atmospheric lifetimes for the reactions of OH radicals and Cl atoms with haloalkanes, *Int. J. Chem. Kinet.*, *25*, 273–284.

Donahue, N.M. (2003), Reaction barriers: origin and evolution, *Chem. Rev.*, *103*, 4593–4604.

Donahue, N.M., J.G. Anderson, and K.L. Demerjian (1998), New rate constants for ten OH–alkane reactions from 300 to 400 K: An assessment of accuracy, *J. Phys. Chem. A*, *102*, 3121–3126.

Dorfman, L.M., and Z.D. Sheldon (1949), The mechanism of the photo-chemical decomposition of diethyl ketone, *J. Chem. Phys.*, *17*, 511–516.

Doussin, J.F., B. Picquet-Varrault, R. Durand-Jolibois, H. Loirat, and P. Carlier (2003), A visible and FTIR spectrometric study of the nighttime chemistry of acetaldehyde and PAN under simulated atmospheric conditions, *J. Photochem. Photobiol. A.*, *157*, 283–293.

Drew, R.M., J.A. Kerr, and J. Olive (1985), Relative rate constants of the gas-phase decomposition reactions of the *s*-butoxy radical, *Int. J. Chem. Kinet.*, *17*, 167–176.

Droege, A.T., and F.P. Tully (1986a), Hydrogen-atom abstraction from alkanes by OH. 3. Propane, *J. Phys. Chem.*, *90*, 1949–1954.

Droege, A.T., and F.P. Tully (1986b), Hydrogen-atom abstraction from alkanes by OH. 5. *n*-Butane, *J. Phys. Chem.*, *90*, 5937–5941.

Droege, A.T., and F.P. Tully (1987), Hydrogen-atom abstraction from alkanes by OH. 6. Cyclopentane and cyclohexane, *J. Phys. Chem.*, *91*, 1222–1225.

Duffy, B.L., P.F. Nelson, Y. Ye, and I.A. Weeks (1999), Speciated hydrocarbon profiles and calculated reactivities of exhaust and evaporative emissions from 82 in-use light-duty Australian vehicles, *Atmos. Environ.*, *33*, 291–307.

Dunlop, J.R., and F.P. Tully (1993a), A kinetic study of OH radical reactions with methane and perdeuterated methanes, *J. Phys. Chem.*, *97*, 11148–11150.

Dunlop, J.R., and F.P. Tully (1993b), Catalytic dehydration of alcohols by OH. 2-Propanol: an intermediate case, *J. Phys. Chem.*, *97*, 6457–6464.

Dunn, J.R., and K.O. Kutschke (1958), Photooxidation of ketones in an O^{18} enriched environment, *Can. J. Chem.*, *36*, 421–424.

Durham, R.W.. G.R. Martin, and H.C. Sutton (1949), A radioactive tracer method for the analysis of mixtures of short-lived free radicals, *Nature*, *164*, 1052–1053.

Dzvonik, M.K., S. Yang and R. Bersohn (1974), Photodissociation of molecular-beams of aryl halides, *J. Chem. Phys.*, *61*, 4408–4421.

Eberhard, J., and C.J. Howard (1996), Temperature-dependent kinetics studies of the reaction of $C_2H_5O_2$ and n-$C_3H_7O_2$ radicals with NO, *Int. J. Chem. Kinet.*, *28*, 731–740.

Eberhard, J., and C.J. Howard (1997), Rate coefficients for the reactions of some C_3 to C_5 hydrocarbon peroxy radicals with NO, *J. Phys. Chem. A*, *101*, 3360–3366.

Eberhard, J., C. Müller, D.W. Stocker, and J.A. Kerr (1995), Isomerization of alkoxy radicals under atmospheric conditions, *Environ. Sci. Technol*, *29*, 232–241.

Eberhard, J., P.W. Villalta, and C.J. Howard (1996), Reaction of isopropyl peroxy radicals with NO over the temperature range 201–401 K, *J. Phys. Chem.*, *100*, 993–997.

Edney, E.O., and D.J. Driscoll (1992), Chlorine initiated photooxidation studies of hydrochlorofluorocarbons (HCFCs) and hydrofluorocarbons (HFCs): results for HCFC-22 ($CHClF_2$); HFC-41 (CH_3F); HCFC-124 ($CClFHCF_3$); HFC-125 (CF_3CHF_2); HFC-134a (CF_3CH_2F); HCFC-142b ($CClF_2CH_3$); and HFC-152a (CHF_2CH_3), *Int. J. Chem. Kinet.*, *24*, 1067–1081.

Edney, E.O., B.W. Gay, Jr., and D.J. Driscoll (1991), Chlorine-initiated oxidation studies of hydrochlorofluorocarbons: results for HCFC-123 (CF_3CHCl_2) and HCFC-141b ($CFCl_2CH_3$), *J. Atmos. Chem.*, *12*, 105–120.

Edney, E.O., T.E. Kleindienst, and E.W. Corse (1986), Room temperature rate constants for the reaction of OH with selected chlorinated and oxygenated hydrocarbons, *Int. J. Chem. Kinet.*, *18*, 1355–1371.

Eichholtz, M., A. Schneider, J.-T. Vollmer, and H.Gg. Wagner (1997), Kinetic investigations of the reactions of tetramethylgermane, tetraethylgermane, tetramethyloxygermane and 3,4-diethylpentane with $O(^3P)$ atoms, *Z. Phys. Chem. (Munich)*, *199*, 267–274.

Elfers, G., F. Zabel, and K.H. Becker (1990), Determination of the rate constant ratio for the reactions of the ethylperoxy radical with NO and NO_2, *Chem. Phys. Lett.*, *168*, 14–19.

Ellis, D.A., J.W. Martin, A.O. De Silva, S.A. Mabury, M.D. Hurley, M.P.S. Andersen, and T.J. Wallington (2004), Degradation of fluorotelomer alcohols: A likely atmospheric source of perfluorinated carboxylic acids, *Environ. Sci. Tech.*, *38*, 3316–3321.

Ells, V.R., and W.A. Noyes, Jr. (1938), Photochemical studies. XXVIII. The photochemical decomposition of ethyl methyl ketone by wave lengths from 1850–2000 Å, *J. Am. Chem. Soc.*, *60*, 2031–2036.

Elrod, M.J., D.L. Ranschaert, and N.J. Schneider (2001), Direct kinetics study of the temperature dependence of the CH_2O branching channel for the $CH_3O_2 + HO_2$ reaction, *Int. J. Chem. Kinet.*, *33*, 363–376.

Elterman, L. (1968), UV, visible and IR attenuation for altitudes to 50 km, AFCRL-68-0153, No. 285, Air Force Cambridge Research Laboratories, Bedford, MA.

Emrich, M., and P. Warneck (2000), Photodissociation of acetone in air: Dependence on pressure and wavelength of the excited singlet state, *J. Phys. Chem. A*, *104*, 9436–9442.

Enami, S., S. Hashimoto, M. Kawasaki, Y. Nakano, T. Ishiwata, and T.J. Wallington (2005), Observation of adducts in the reaction of Cl atoms with XCH_2I (X = H, CH_3, Cl, Br, I) using cavity ring-down spectroscopy, *J. Phys. Chem. A*, *109*, 1587–1593.

Enami, S., J. Ueda, M. Goto, Y. Nakano, S. Aloisio, S. Hashimoto, and M. Kawasaki (2004), Formation of iodine monoxide radical from the reaction of CH_2I with O_2, *J. Phys. Chem.*, *108*, 6347–6350.

Encina, M.V., A. Nogales, and E.A. Lissi (1975), Temperature dependence of 4-methyl-2-pentanone photochemistry, *J. Photochem.*, *4*, 75–89.

Eskola, A.J., D. Wojcik-Pastuszka, E. Ratajczak, and R.S. Timonen (2006), Kinetics of the reactions of CH_2Br and CH_2I radicals with molecular oxygen at atmospheric temperatures, *Phys. Chem. Chem. Phys.*, *8*, 1416–1424.

Espada, C., J. Grossenbacher, K. Ford, T. Couch, and P.B. Shepson (2005), The production of organic nitrates from various anthropogenic volatile organic compounds, *Int. J. Chem. Kinet.*, *37*, 675–685.

Etheridge, D.M., L.P. Steel, R.J. Francey, and R.L. Langenfelds (1998), Atmospheric methane between A.D. 1000 and present: Evidence for anthropogenic emissions and climate variability, *J. Geophys. Res. 103*, 15979–15993.

Falgayrac, G., F. Caralp, N. Sokolowski-Gomez, P. Devolder, and Ch. Fittschen (2004), Rate constants for the decomposition of 2-butoxy radicals and their reaction with NO and O_2, *Phys. Chem. Chem. Phys.*, *6*, 4127–4132.

Fang, T.D., P.H. Taylor, and R.J. Berry (1999), Kinetics of the reaction of OH radicals with CH_2ClCF_2Cl and CH_2ClCF_3 over an extended temperature range, *J. Phys. Chem. A*, *103*, 2700–2704.

Fang, T.D., P.H. Taylor, and B. Dellinger (1996), Absolute rate measurements of the reaction of OH radicals with HCFC-21 ($CHFCl_2$) and HCFC-22 (CHF_2Cl) over an extended temperature range, *J. Phys. Chem.*, *100*, 4048–4054.

Fang, T.D., P.H. Taylor, B. Dellinger, C.J. Ehlers, and R.J. Berry (1997), Kinetics of the OH + CH_3CF_2Cl reaction over an extended temperature range, *J. Phys. Chem. A*, *101*, 5758–5764.

Fang, W.-H., and R.-Z. Liu (2001), *Ab initio* studies of dissociation pathways on the ground- and excited-state potential energy surfaces for formyl chloride (HClCO), *J. Chem. Phys.*, *115*, 10431–10437.

Feilberg, K.L., B. D'Anna, M.S. Johnson, and C.J. Nielsen (2005a), Relative tropospheric photolysis rates of HCHO, $H^{13}CHO$, $HCH^{18}O$, and DCDO measured at the European photoreactor facility, *J. Phys. Chem. A*, *109*, 8314–8319.

Feilberg, K.L., D.W.T. Griffith, M.S. Johnson, and C.J. Nielsen (2005b), The ^{13}C and D kinetic isotope effects in the reaction of CH_4 with Cl, *Int. J. Chem. Kinet.*, *37*, 110–118.

Feilberg, K.L., M.S. Johnson, and C.J. Nielsen (2004), Relative reaction rates of HCHO, HCDO, DCDO, $H^{13}CHO$, and $HCH^{18}O$ with OH, Cl, Br, and NO_3 radicals, *J. Phys. Chem. A*, *108*, 7393–7398.

Fenter, F.F., V. Catoire, R. Lesclaux, and P.D. Lightfoot (1993), The ethylperoxy radical—Its ultraviolet spectrum, self-reaction, and reaction with HO_2, each studied as a function of temperature, *J. Phys. Chem.*, *97*, 3530–3538.

Ferenac, M.A., A.J. Davis, A.S. Holloway, and T.S. Dibble (2003), Isomerization and decomposition reactions of primary alkoxy radicals derived from oxygenated solvents, *J. Phys. Chem. A*, *107*, 63–72.

Ferrari, C., A. Roche, V. Jacob, P. Foster, and P. Baussand (1996), Kinetics of the reactions of OH radicals with a series of esters under simulated conditions at 295 K, *Int. J. Chem. Kinet.*, *28*, 609–614.

Finklestein, A., and W.A. Noyes, Jr. (1953), The reactions of radicals from diethyl ketone with oxygen and general discussion, *Faraday Soc. Discuss.*, *14*, 2–16.

Finlayson-Pitts, B.J. (1993), Chlorine atoms as a potential tropospheric oxidant in the marine boundary layer, *Res. Chem. Intermed.*, *19*, 235–249.

Finlayson-Pitts, B.J., M.J. Ezell, T.M. Jayaweera, H.N. Berko, and C.C. Lai (1992), Kinetics of the reactions of OH with methyl chloroform and methane: implications for global tropospheric OH and the methane budget, *Geophys. Res. Lett.*, *19*, 1371–1374.

Finlayson-Pitts, B.J., M.J. Ezell, and J.N. Pitts (1989), Formation of chemically active chlorine compounds by reactions of atmospheric NaCl particles with gaseous N_2O_5 and $ClONO_2$, *Nature*, *337*, 241–244.

Fittschen, Ch., A. Frenzel, K. Imrik, and P. Devolder (1999), Rate constants for the reactions of C_2H_5O, *i*-C_3H_7O, and *n*-C_3H_7 with NO and O_2 as a function of temperature, *Int. J. Chem. Kinet.*, *31*, 860–866.

Fittschen, Ch., H. Hippler, and B. Viskolcz (2000), The β C—C bond scission in alkoxy radicals: Thermal unimolecular decomposition of *t*-butoxy radicals, *Phys. Chem. Chem. Phys.*, *2*, 1677–1683.

Flocke, F., et al. (2006), unpublished personal communication to the authors.

Flowers, B.A., M.E. Angerhofer, W.R. Simpson, T. Nakayama, and Y. Matsumi (2005), Nitrate radical quantum yield from peroxyacetyl nitrate photolysis, *J. Phys. Chem. A*, *109*, 2552–2558.

Folberth, G.A., D.A. Hauglustaine, J. Lathiere, and F. Brocheton (2005), Impact of biogenic hydrocarbons on tropospheric chemistry: results from a global chemistry–climate model, *Atmos. Chem. Phys. Discuss* *5*, 10517–10612.

Ford, H.W., and N. Endow (1957), Rate constants at low concentrations. IV. Reactions of atomic oxygen with various hydrocarbons, *J. Chem. Phys.*, *27*, 1277–1279.

Förgeteg, S., T. Bérces, and S. Dóbé (1979), The kinetics and mechanism of *n*-butyraldehyde photolysis in the vapor phase at 313 nm, *Int. J. Chem. Kinet.*, *11*, 219–237.

Förgeteg, S, S. Dóbé, and T. Bérces (1978), Effect of pressure on the primary photo-chemical processes of *n*-butyraldehyde, *React. Kinet. Catal. Lett.*, *9*, 331–336.

Francisco, J.S. (1992), Gas-phase hydrolysis of trifluoromethyl carbonyl halides to trifluoroacetic acid, *J. Phys. Chem.*, *96*, 8894–8899.

Francisco, J.S., and Z. Li (1989), Dissociation pathways of carbonyl halides, *J. Phys. Chem.*, *93*, 8118–8122.

Francisco, J.S., and Y. Zhao (1992), *Ab initio* studies of dissociation pathways on the ground state potential energy surface for HFCO and HClCO, *J. Chem. Phys.*, *96*, 7587–7596.

Gaffney, J.S. and S.Z. Levine (1979), Predicting gas phase organic molecule reaction rates using linear free energy correlations. I. $O(^3P)$ and OH addition and abstraction reactions, *Int. J. Chem. Kinet.*, *11*, 1197–1209.

Gaffney, J.S., R. Fajer, G.I. Senum, and J.H. Lee (1986), Measurement of the reactivity of OH with methyl nitrate: Implications for prediction of alkyl nitrate–OH reaction rates, *Int. J. Chem. Kinet.*, *18*, 399–407.

Gandini, A., and P.A. Hackett (1976/1977), The phosphorescence of 1,1,1-trifluoroacetone, *J. Photochem.*, *6*, 75–76.

Gandini, A., and P.A. Hackett (1977a), Electronic relaxation processes in acetone and 1,1,1-trifluoroacetone vapor and the gas phase recombination of the acetyl radical at 22°C, *J. Am. Chem. Soc.*, *99*, 6195–6205.

Gandini, A., and P.A. Hackett (1977b), Vibrational relaxation of excited acetaldehyde vapor. A search for unquenchable species, *Chem. Phys. Lett.*, *52*, 107–110.

Gandini, A., and K.O. Kutschke (1968), The primary process in the photolysis of hexafluoroacetone vapour. II. The fluorescence and phosphorescence, *Proc. Roy. Soc.*, *A306*, 511–528.

Gandini, A., D.A. Whytock, and K.O. Kutschke (1968a), The primary process in the photolysis of hexafluoroacetone vapour. III. The efficiency of intersystem crossing, *Proc. R. Soc., A306*, 529–536.

Gandini, A., D.A. Whytock, and K.O. Kutschke (1968b), The primary process in the photolysis of hexafluoroacetone vapour. IV. Identification of singlet quenching addends, *Proc. R. Soc., A306*, 537–540.

Gandini, A., D.A. Whytock, and K.O. Kutschke (1968c), The primary process in the photolysis of hexafluoroacetone vapour. V. Mechanistic considerations, *Proc. R. Soc., A306*, 541–551.

Gardner, E.P., R.D. Wijayaratne, and J.G. Calvert (1984), Primary quantum yields of photodecomposition of acetone in air under tropospheric conditions, *J. Phys. Chem., 88*, 5069–5076.

Garland, N.L., and H.H. Nelson (1996), Temperature dependent kinetics of the reaction OH + $CF_3CH_2CF_3$, *Chem. Phys. Lett., 248*, 296–300.

Garland, N.L., L.J. Medhurst, and H.H. Nelson (1993), Potential chlorofluorocarbon replacements: OH reaction rate constants between 250 and 315 K and infrared absorption spectra, *J. Geophys. Res., 98*, 23107–23111.

Garraway, J., and R.J. Donovan (1979), Gas-phase reaction of OH with alkyl iodides, *J. Chem. Soc. Chem. Commun.*, 1108.

Geiger, H., I. Barnes, K.H. Becker, B. Bohn, T. Brauers, B. Donner, H.P. Dorn, M. Elend, C.M.F. Dinis, D. Grossmann, H. Hass, H. Hein, A. Hoffmann, L. Hoppe, F. Hulsemann, D. Kley, B. Klotz, H.G. Libuda, T. Maurer, D. Mihelcic, G.K. Moortgat, R. Olariu, P. Neeb, D. Poppe, L. Ruppert, C.G. Sauer, O. Shestakov, H. Somnitz, W.R. Stockwell, L.P. Thuner, A. Wahner, P. Weisen, F. Zabel, R. Zellner, and C. Zetzsch (2002), Chemical mechanism development: Laboratory studies and model applications, *J. Atmos. Chem., 42*, 323–357.

Gery, M.W., G.Z. Whitten, J.P. Killus, and M.C. Dodge (1989), A photochemical kinetics mechanism for urban and regional scale computer modeling, *J. Geophys. Res., 94*, 12925–12956.

Gherman, G.F., R.A. Friesner, T.-H. Wong, Z. Min, and R. Bersohn (2001), Photodissociation of acetaldehyde: The CH_4 + CO channel, *J. Chem. Phys., 114*, 6128–6133.

Gierczak, T., J.B. Burkholder, S. Bauerle, and A.R. Ravishankara (1998), Photochemistry of acetone under tropospheric conditions, *Chem. Phys., 231*, 229–244.

Gierczak, T., M.K. Gilles, S. Bauerle, and A.R. Ravishankara (2003), Reaction of hydroxyl radical with acetone. 1. Kinetics of the reactions of OH, OD, and ^{18}OH with acetone and acetone-d_6, *J. Phys. Chem. A, 107*, 5014–5020.

Gierczak, T., R.K. Talukdar, J.B. Burkholder, R.W. Portmann, J.S. Daniel, S. Solomon, and A.R. Ravishankara (1996), Atmospheric fate and greenhouse warming potentials of HFC 236fa and HFC 236ea, *J. Geophys. Res., 101*, 12905–12911.

Gierczak, T., R.K. Talukdar, S.C. Herndon, G.L. Vaghjiani, and A.R. Ravishankara (1997), Rate coefficients for the reactions of hydroxyl radicals with methane and deuterated methanes, *J. Phys. Chem. A, 101*, 3125–3134.

Gierczak, T., R. Talukdar, G.L. Vaghjiani, E.R. Lovejoy, and A.R. Ravishankara (1991), Atmospheric fate of hydrofluoroethanes and hydrofluorochloroethanes: 1. Rate coefficients for reactions with OH, *J. Geophys. Res., 96*, 5001–5011.

Giessing, A.M.B., A. Feilberg, T.E. Møgelberg, J. Sehested, M. Bilde, T.J. Wallington, and O.J. Nielsen (1996), Atmospheric chemistry of HFC-227ca: spectrokinetic investigation of the $CF_3CF_2CF_2O_2$ radical, its reactions with NO and NO_2 and the atmospheric fate of the $CF_3CF_2CF_2O$ radical, *J. Phys. Chem., 100*, 6572–6579.

Gilbert, R.G., K. Luther, and J. Troe (1983), Theory of thermal unimolecular reactions in the fall-off range. II. Weak collision rate constants, *Ber. Bunsen-Ges. Phys. Chem., 87*, 169–177.

Gill, R.J., and G.H. Atkinson (1979), Wavelength dependence of HCO formation in the photolysis of acetaldehyde, *Chem. Phys. Lett., 64*, 426–430.

Gilles, M.K., J.B. Burkholder, T. Gierczak, P. Marshall, and A.R. Ravishankara (2002), Rate coefficient and product branching measurements for the reaction OH + bromopropane from 230 to 360 K, *J. Phys. Chem. A, 106*, 5358–5366.

Gilles, M.K., R.K. Talukdar, and A.R. Ravishankara (2000), Rate coefficients for the OH + CF_3I reaction between 271 and 370 K, *J. Phys. Chem. A, 104*, 8945–8950.

Gilles, M. K., A.A. Turnipseed, R.K. Talukdar, Y. Rudich, P.W. Villalta, L.G. Huey, J.B. Burkholder, and A.R. Ravishankara (1996), Reactions of $O(^3P)$ with alkyl iodides: rate coefficients and reaction product, *J. Phys. Chem.*, *100*, 14005–14015.

Gipson, G.L., and J.O. Young (1999), Chapter 8, Gas phase chemistry. Science algorithms of the EPA Models—3 Community Multiscale Air Quality modelling system. http://www.epa.gov/asmdnerl/CMAQ/ch08.pdf. CMAS, North Carolina.

Glasius, M., A. Calogirou, N.R. Jensen, J. Hjorth, and C.J. Nielsen (1997), Kinetic study of gas-phase reactions of pinonaldehyde and structurally related compounds, *Int. J. Chem. Kinet.*, *29*, 527–533.

Glasson, W.A. (1975), Methoxy radical reactions in atmospheric chemistry, *Environ. Sci. Technol.*, *9*, 1048–1053.

Glicker, S., and L.J. Stief (1971), Photolysis of formaldehyde at 1470 and 1236 Å, *J. Chem. Phys.*, *54*, 2852–2857.

Goddard, J.D., and H.F. Schaefer, III (1979), The photodissociation of formaldehyde: Potential energy surface features, *J. Chem. Phys.*, *70*, 5117–5134.

Godwin, F.G., C. Paterson, and P.A. Gory (1987), Photofragmentation dynamics of n-C_3H_7I and i-C_3H_7I at 248 nm, *Mol. Phys.*, *61*, 827–848.

Golemba, F.J., and J.E. Guillet (1972), Photochemistry of ketone polymers. V. Photochemistry of the linear 2-alkanones in solution, *Macromolecules*, *5*, 63–68.

Gong, H., A. Matsunaga, and P. J. Ziemann (2005), Products and mechanism of secondary organic aerosol formation from reactions of linear alkenes with NO_3 radicals, *J. Phys. Chem. A*, *109*, 4312–4324.

Good, D.A., and J.S. Francisco (2003), Atmospheric chemistry of alternative fuels and alternative chlorofluorocarbons, *Chem. Rev.*, *103*, 4999–5023.

Gordon, S., and W.A. Mulac (1975), Reaction of the $OH(X^2\pi)$ radical produced by the pulse radiolysis of water-vapor, *Int. J. Chem. Kinet., Symposium No. 1*, 289–299.

Gorse, R.A., and D.H. Volman (1974), Photochemistry of the gaseous hydrogen peroxide–carbon monoxide system. II. Rate constants for hydroxyl radical reactions with hydrocarbons and for hydrogen atom reaction with hydrogen peroxide, *J. Photochem.*, *3*, 115–122.

Green, W.H., C.B. Moore, and W.F. Polik (1992), Transition-state and rate constants for unimolecular reactions, *Ann. Rev. Phys. Chem.*, *43*, 591–626.

Greenhill, P.G., and B.V. O'Grady (1986), The rate-constant of the reaction of hydroxyl radicals with methanol, ethanol and (D_3)methanol, *Aust. J. Chem.*, *39*, 1775–1787.

Greiner, N.R. (1970), Hydroxyl radical kinetics by kinetic spectroscopy. VI. Reactions with alkanes in the range 300–500°K, *J. Chem. Phys.*, *53*, 1070–1076.

Grosjean, E., R.A. Rasmussen, and D. Grosjean (1998), Ambient levels of gas phase pollutants in Porto Alegre, Brazil, *Atmos. Environ.*, *32*, 3371–3379.

Gruver, J.T., and J.G. Calvert (1956), The vapor phase photolysis of 2-methylbutanal at wavelength 3130 Å, *J. Am. Chem. Soc.*, *78*, 5208–5211.

Gruver, J.T., and J.G. Calvert (1958), The vapor phase photolysis of (+)2–methylbutanal-iodine mixtures at 3130 Å, *J. Am. Chem. Soc.*, *80*, 3524–3527.

Guicherit, R., and H. van Dop (1977), Photochemical production of ozone in western Europe and its relation to meteorology, *Atmos. Environ.*, *11*, 145–155.

Guo, H., and G.C. Schatz (1990), Time-dependent dynamics of methyl iodide photodissociation in the first continuum, *J. Chem. Phys.*, *93*, 393–402.

Gutman, D., N. Sanders, and J.E. Butler (1982), Kinetics of the reactions of methoxy and ethoxy radicals with oxygen, *J. Phys. Chem.*, *86*, 66–70.

Haagen-Smit, A.J. (1952), Chemistry and physiology of Los Angeles smog, *Ind. Eng. Chem.*, *44*, 1342–1346.

Hack, W., K. Hoyermann, C. Kersten, M. Olzmann, and B. Viskolcz (2001), Mechanism of the 1-C_4H_9 + O reaction and the kinetics of the intermediate 1-C_4H_9O radical, *Phys. Chem. Chem. Phys.*, *3*, 2365–2371.

Hackett, P.A., and K.O. Kutschke (1977), Photochemistry and photophysics of hexafluoroacetone vapor at low pressures, *J. Phys. Chem.*, *81*, 1245–1252.

Hackett, P.A., and D. Phillips (1974a), Photochemistry of halogenated acetones. I. Spectroscopic studies, *J. Phys. Chem.*, *78*, 665–671.

Hackett, P.A., and D. Phillips (1974b), Photochemistry of halogenated acetones. II. Rate constant measurements, *J. Phys. Chem.*, *78*, 671–678.

Hackett, P.A., and D. Phillips (1974c), Photochemistry of halogenated acetones. III. Vibrational relaxation in singlet states, *J. Phys. Chem.*, *78*, 679–682.

Hackett, P.A., and D. Phillips (1974d), Photochemistry of halogenated acetones. IV. Quenching of the excited states, *J. Phys. Chem.*, *78*, 682–686.

Hägele, J., K. Lorenz, D. Rhäsa, and R. Zellner (1983), Rate constants and CH_3O product yield of the reaction $OH + CH_3OH \rightarrow$ products, *Ber. Bunsenges. Phys. Chem.*, *87*, 1023–1026.

Hakami, A., M. Arhami, and A.G. Russell (2004a), Further analysis of VOC reactivity metrics and scales. Department of Civil Engineering, Georgia Institute of Technology, Atlanta, Georgia, USA.

Hakami, A., M.S. Bergin, and A.G. Russell (2003), Assessment of the ozone and aerosol formation potentials (reactivities) of organic compounds over the eastern United States. Department of Civil and Environmental Engineering, Georgia Institute of Technology, Atlanta, Georgia, USA.

Hakami, A., M.S. Bergin, and A.G. Russell (2004b), Ozone formation potential of organic compounds in the eastern United States. A comparison of episodes, inventories, and domains, *Environ. Sci. Technol.*, *38*, 6748–6759.

Hall, G.E., H.W. Metzier, J.T. Muckerman, J.M. Preses, and R.E. Weston, Jr. (1995), Studies of the 193 nm photolysis of diethyl ketone and acetone using time-resolved Fourier transform emission spectroscopy, *J. Chem. Phys.*, *102*, 6660–6668.

Halpern, A.M., and W.R. Ware (1970), Luminescent properties of hexafluoroacetone. III. Vibrational relaxation and radiationless processes in the gas phase, *J. Chem. Phys.*, *53*, 1969–1977.

Halpern, A.M., and W.R. Ware (1971), Excited singlet state radiative and nonradiative transition probabilities for acetone, acetone-d_6, and hexafluoroacetone in the gas phase, in solution, and in the neat liquid, *J. Chem. Phys.*, *54*, 1271–1276.

Handwerk, V., and R. Zellner (1978), Kinetics of the reactions of OH radicals with some halocarbons ($CHClF_2$, CH_2ClF, CH_2ClCF_3, CH_3CClF_2, CH_3CHF_2) in the temperature range 260–370 K, *Ber. Bunsenges. Phys. Chem.*, *82*, 1161–1166.

Hanson, D., J. Orlando, B. Nozière, and E. Kosciuch (2004), Proton transfer mass spectrometry studies of peroxy radicals, *Int. J. Mass Spectrom.*, *239*, 147–159.

Hanst, P.L., and B.W. Gay, Jr. (1983), Atmospheric oxidation of hydrocarbons: Formation of hydroperoxides and peroxyacids, *Atmos. Environ.*, *17*, 2259–2265.

Harris, S.J., and J.A. Kerr (1988), Relative rate measurements of some reactions of hydroxyl radicals with alkanes studied under atmospheric conditions, *Int. J. Chem. Kinet.*, *20*, 939–955.

Harris, S.J., and J.A. Kerr (1989), A kinetic and mechanistic study of the formation of alkyl nitrates in the photo-oxidation of *n*-heptane studied under atmospheric conditions, *Int. J. Chem. Kinet.*, *21*, 207–218.

Harrison, A.J., B.J. Cederholm, and M.A. Terwilliger (1959), Absorption of acyclic oxygen compounds in the vacuum ultraviolet. I. Alcohols, *J. Chem. Phys.*, *30*, 355–356.

Hartley, G.H., and J.E. Guillet (1968a), Photochemistry of ketone polymers. I. Studies of ethylene–carbon monoxide copolymers, *Macromolecules*, *1*, 165–170.

Hartley, G.H., and J.E. Guillet (1968b), Photochemistry of ketone polymers. II. Studies of model compounds, *Macromolecules*, *1*, 413–417.

Hartmann, D., J. Karthäuser, J.P. Sawerysyn, and R. Zellner (1990), Kinetics and HO_2 product yield of the reaction $C_2H_5O + O_2$ between 295 and 411 K, *Ber. Bunsenges. Phys. Chem.*, *94*, 639–645.

Harwood, M.H., J.M. Roberts, G.J. Frost, A.R. Ravishankara, and J.B. Burkholder (2003), Photochemical studies of $CH_3C(O)OONO_2$ (PAN) and $CH_3CH_2C(O)OONO_2$ (PPN): NO_3 quantum yields, *J. Phys. Chem. A*, *107*, 1148–1154.

Hashikawa, Y., M. Kawasaki, R.L. Waterland, M.D. Hurley, J.C. Ball, T.J. Wallington, M.P. Sulback Andersen, and O.J. Nielsen (2004), Gas phase UV and IR absorption spectra of $C_xF_{2x+1}CHO$ (x = 1–4), *J. Fluorine Chem.*, *125*, 1925–1932.

Hasson, A.S., and I.W.M. Smith (1999), Chlorine atom initiated oxidation of chlorinated ethenes: results for 1,1-dichloroethene ($H_2C{=}CCl_2$), 1,2-dichloroethene ($HClC{=}CClH$), trichloroethene ($HClC{=}CCl_2$), and tetrachloroethene ($Cl_2C{=}CCl_2$), *J. Phys. Chem. A*, *103*, 2031–2043.

Hasson, A.S., C.M. Moore, and I.W.M. Smith (1998), The fluorine atom initiated oxidation of CF_3CFH_2 (HFC-134a) studied by FTIR spectroscopy, *Int. J. Chem. Kinet.*, *30*, 541–554.

Hasson, A.S., G.S. Tyndall, and J.J. Orlando (2004), A product yield study of the reaction of HO_2 radicals with ethyl peroxy, acetyl peroxy, and acetonyl peroxy radicals, *J. Phys. Chem. A*, *108*, 5979–5989.

Hayman, G.D., and F. Battin-Leclerc (1995), Kinetics of the reactions of the HO_2 radical with peroxyl radicals derived from hydrochlorofluorocarbons and hydrofluorocarbons, *J. Chem. Soc. Faraday Trans.*, *91*, 1313–1323.

Hayman, G.D., M.E. Jenkin, T.P. Murrells, and C.E. Johnson (1994), Tropospheric degradation chemistry of HCFC-123 (CF_3CHCl_2): a proposed replacement chlorofluorocarbon, *Atmos. Environ.*, *28*, 421–437.

Heicklen, J. (1965), The photolysis of phosgene–ethylene mixtures, *J. Am. Chem. Soc.*, *87*, 445–453.

Heicklen, J., J. Desai, A. Bahta, C. Harper, and R. Simonaitis (1986), The temperature and wavelength dependence of the photooxidation of propionaldehyde, *J. Photochem.*, *34*, 117–135.

Heimann, G., and P. Warneck (1992), OH radical induced oxidation of 2,3-dimethylbutane in air, *J. Phys. Chem.*, *96*, 8403–8409.

Heimann, G., H.-J. Benkelberg, O. Boge, and P. Warneck (2002), Photodecomposition of iodopentanes in air: Product distributions from the self-reactions of *n*-pentyl peroxyl radicals, *Int. J. Chem. Kinet.*, *34*, 126–138.

Hein, H., A. Hoffmann, and R. Zellner (1998), Direct investigations of reactions of 2-butoxy radicals using laser pulse initiated oxidation. Reaction with O_2 and unimolecular decomposition at 293 K and 50 mbar, *Ber. Bunsen-Ges. Phys. Chem.*, *102*, 1840–1849.

Hein, H., A. Hoffmann, and R. Zellner (1999), Direct investigations of reactions of 1-butoxy and 1-pentoxy radicals using laser pulse inititated oxidation: Reaction with O_2 and isomerisation at 293 K and 50 mbar, *Phys. Chem. Chem. Phys.*, *1*, 3743–3752.

Hein, H., H. Somnitz, A. Hoffmann, and R. Zellner (2000), A combined experimental and theoretical investigation of the reactions of 3-pentoxy radicals: Reaction with O_2 and unimolecular decomposition, *Z. Phys. Chem.*, *214*, 449–471.

Heiss, A., and K. Sahetchian (1996), Isomerization reactions of the n-C_4H_9O and n-OOC_4H_8OH radicals in oxygen, *Int. J. Chem. Kinet.*, *28*, 531–544.

Helleis, F., G.K. Moortgat, and J.N. Crowley (1993), Kinetic investigation of the reactions of CD_3O_2 with NO and NO_3 at 298 K, *J. Phys. Chem.*, *100*, 17846–17854.

Heneghan, S.P., P.A. Knoot, and S.W. Benson (1981), The temperature coefficient of the rates in the system $Cl + CH_4 \leftrightarrow CH_3 + HCl$, thermochemistry of the methyl radical, *Int. J. Chem. Kinet.*, *13*, 677–691.

Herndon, S.C., T. Gierczak, R.K. Talukdar, and A.R. Ravishankara (2001), Kinetics of the reactions of HO with several alkyl halides, *Phys. Chem. Chem. Phys.*, *3*, 4529–4535.

Herron, J.T. (1988), Evaluated chemical kinetic data for the reactions of atomic oxygen $O(^3P)$ with saturated organic compounds in the gas phase, *J. Phys. Chem. Ref. Data 17*, 967–1026.

Herron, J.T., and R.E. Huie (1969), Rates of reaction of atomic oxygen. II. Some C_2 to C_8 alkanes, *J. Phys. Chem.*, *73*, 3327–3337.

Hess, W.P., and F.P. Tully (1988), Catalytic conversion of alcohols to alkenes by OH, *Chem. Phys. Lett.*, *152*, 183–189.

Hess, W.P., and F.P. Tully (1989), Hydrogen-atom abstraction from methanol by OH, *J. Phys. Chem.*, *93*, 1944–1947.

Hess, W.P., S.J. Kohler, H.K. Haugen, and S.R. Leone (1986), Application of an INGAASP diode-laser to probe photodissociation dynamics—I* quantum yields from n-C_3F_7I and i-C_3F_7I and CH_3I by laser gain vs. absorption-spectroscopy, *J. Chem. Phys.*, *84*, 2143–2149.

Hess, W.P., R. Naaman, and S.R. Leone (1987), Nonunity I* quantum yield in the 193 nm photodissociation of methyl iodide, *J. Phys. Chem.*, *91*, 6085–6087.

Hickson, K.M., and L.F. Keyser (2004), Kinetics of the $Cl(^2P_J)$ + C_2H_6 reaction between 177 and 353 K, *J. Phys. Chem. A*, *108*, 1150–1159.

Highwood, E.J., K.P. Shine, M.D. Hurley, and T.J. Wallington (1999), Estimation of direct radiative forcing due to non-methane hydrocarbons, *Atmos. Environ.*, *33*, 759–767.

Hitsuda, K., K. Takahashi, Y. Matsumi, and T.J. Wallington (2001a), Kinetics of the reactions of $Cl(^2P_{1/2})$ and $Cl(^2P_{3/2})$ atoms with C_3H_8, C_3D_8, n-C_4H_{10}, and i-C_4H_{10} at 298 K, *Chem. Phys. Lett.*, *346*, 16–22

Hitsuda, K., K. Takahashi, Y. Matsumi, and T.J. Wallington (2001b), Kinetics of the reactions of $Cl(^2P_{1/2})$ and $Cl(^2P_{3/2})$ atoms with C_2H_6, C_2D_6, CH_3F, C_2H_5F, and CH_3CF_3 at 298 K, *J. Phys. Chem. A*, *105*, 5131–5136.

Hjorth, J., G. Ottobrini, and G. Restelli (1988), Reaction between NO_3 and CH_2O in air: a determination of the rate constant at 295 ± 2 K, *J. Phys. Chem.*, *92*, 2669–2672.

Ho, P., and A.V. Smith (1982), Rotationally excited CO from formaldehyde photodissociation, *Chem. Phys. Lett.*, *90*, 407–411.

Hochanadel, C.J., J.A. Ghormley, J.W. Boyle, and P.J. Ogren (1977), Absorption spectrum and rates of formation and decay of the CH_3O_2 radical, *J. Phys. Chem.*, *81*, 3–7.

Holt, T., R. Atkinson and J. Arey (2005), Effect of water vapor concentration on the conversion of a series of 1,4-hydroxycarbonyls to dihydrofurans, *J. Photochem. Photobiol. A: Chemistry*, *176*, 231–237.

Hooshiyar, P.A., and H. Niki (1995), Rate constants for the gas-phase reactions of Cl- atoms with C_2–C_8 alkanes at $T = 296 \pm 2$ K, *Int. J. Chem. Kinet.*, *27*, 1197–1206.

Hopkinson, A.C., E. Lee-Ruff, and M.H. Lien (1988), Molecular orbital calculations of excited state cyclobutanone and its photocarbene, *Tetrahedron*, *44*, 6815–6820.

Horie, O., J.N. Crowley, and G.K. Moortgat (1990), Methylperoxy self-reaction: Product and branching ratio between 223 and 333 K, *J. Phys. Chem.*, *94*, 8198–8203.

Horowitz, A. (1986), Use of hydrogen chloride as a probe for the determination of primary processes in acetaldehyde photolysis, *J. Phys. Chem.*, *90*, 4393–4397.

Horowitz, A. (1991), Wavelength dependence of the primary photodissociation processes in acetone photolysis, *J. Phys. Chem.*, *95*, 10816–10823.

Horowitz, A., and J.G. Calvert (1978a), The quantum efficiencies of the primary processes in formaldehyde photolysis at 3130 Å and 25°C, *Int. J. Chem. Kinet.*, *10*, 713–732.

Horowitz, A., and J.G. Calvert (1978b), Wavelength dependence of the quantum efficiencies of the primary processes in formaldehyde photolysis at 25°C, *Int. J. Chem. Kinet.*, *10*, 805–819.

Horowitz, A., and J.G. Calvert (1982), Wavelength dependence of the primary processes in acetaldehyde photolysis, *J. Phys. Chem.*, *86*, 3105–3114.

Horowitz, A., C.J. Kershner, and J.G. Calvert (1982), Primary processes in the photolysis of acetaldehyde at 3000 Å and 25°C, *J. Phys. Chem.*, *86*, 3094–3105.

Horowitz, L.W., J. Liang, G.M. Gardner, and D.J. Jacob (1998), Export of reactive nitrogen from North America during summertime: Sensitivity to hydrocarbon chemistry, *J. Geophys. Res.* *103*, 13451–13476.

Horowitz, L.W., S. Walters, D.L. Mauzerall, L.K. Emmons, P.J. Rasch, C. Granier, X. Tie, J.-F. Lamarque, M.G. Schultz, G.S. Tyndall, J.J. Orlando, and G.P. Brasseur (2003), A global simulation of tropospheric ozone and related tracers: Description and evaluation of MOZART, version 2, *J. Geophys. Res.*, *108*, 4784, doi:10.1029/2002JD002853.

Hou, H., and B. Wang (2005), A systematic computational study on the reactions of HO_2 with RO_2: the HO_2 + CH_3O_2 (CD_3O_2) and HO_2 + CH_2FO_2 reactions, *J. Phys. Chem. A*, *109*, 451–460.

Hou, H., L. Deng, J. Li, and B. Wang (2005), A systematic computational study of the reactions of HO_2 with RO_2: the HO_2 + CH_2ClO_2, $CHCl_2O_2$, and CCl_3O_2 reactions, *J. Phys. Chem. A*, *109*, 9299–9309.

Hou, H., B. Wang, and Y. Gu (1999), Theoretical investigation of the $O(^3P)$ + CHX_2 (X = F, Cl) reactions, *J. Phys. Chem. A*, *103*, 8075–8081.

Hou, H., B. Wang, and Y. Gu (2000a), Ab initio and RRKM studies of the unimolecular reactions of CH_2XCHFO (X = H, F) radicals, *Phys. Chem. Phys. Chem.* *2*, 61–65.

Hou, H., B. Wang, and Y. Gu (2000b), Decomposition and isomerization of the CH_3CHClO radical: ab initio and RRKM study, *J. Phys. Chem. A*, *104*, 1570–1575.

Houston, P.L., and C.B. Moore (1976), Formaldehyde photochemistry: Appearance rate, vibrational relaxation, and energy distribution of the CO product, *J. Chem. Phys.*, *65*, 757–770.

Houweling, S., F. Dentener, and J. Lelieveld (1998), The impact of nonmethane hydrocarbon compounds on tropospheric chemistry, *J. Geophys. Res.* *103*, 10673–10696.

Howard, C.J., and K.M. Evenson (1976), Rate constants for the reactions of OH with CH_4 and fluorine, chlorine, and bromine substituted methanes at 296 K, *J. Chem. Phys.*, *64*, 197–202.

Hsu, K.-J., and W.B. DeMore (1994), Rate constants for the reactions of OH with CH_3Cl, CH_2Cl_2, $CHCl_3$, and CH_3Br, *Geophys. Res. Lett.*, *21*, 805–808.

Hsu, K.-J., and W.B. DeMore (1995), Rate constants and temperature dependences for the reactions of hydroxyl radical with several halogenated methanes, ethanes, and propanes by relative rate measurements, *J. Phys. Chem.*, *99*, 1235–1244.

Hucknall, D.J., D. Booth, and R.J. Sampson (1975), Reactions of hydroxyl radicals with alkanes, *Int. J. Chem. Kinet., Symposium 1*, 301–316.

Huder, K., and W.B. DeMore (1993), Rate constant for the reaction of OH with CH_3CCl_2F (HCFC-141b) determined by relative rate measurements with CH_4 and CH_3CCl_3, *Geophys. Res. Lett.*, *20*, 1575–1577.

Hunter, T.F., S. Lunt, and K.S. Kristjansson (1983), Photofragmentation of CH_3I, CD_3I and CF_3I. Formation of $I(^2P_{1/2})$ as a function of wavelength, *J. Chem. Soc. Faraday Trans. 2*, *79*, 303–316.

Hurley, M.D., J.C. Ball, T.J. Wallington, M.P.S. Andersen, O.J. Nielsen, D.A. Ellis, J.W. Martin, and S.A. Mabury (2006), Atmospheric chemistry of n-$C_xF_{2x+1}CHO$ (x = 1, 2, 3, 4): Fate of n-$C_xF_{2x+1}C(O)$ radicals, *J. Phys. Chem. A*, *110*, 12443–12447.

Hurley, M.D., W.F. Schneider, T.J. Wallington, D.J. Mann, J.D. DeSain, and C.A. Taatjes (2003), Kinetics of elementary reactions in the chain chlorination of cyclopropane, *J. Phys. Chem. A*, *107*, 2003–2010.

Hurley, M.D., T.J. Wallington, G.A. Buchanan, L.K. Gohar, G. Marston, and K.P. Shine (2005), IR spectrum and radiative forcing of CF_4 revisited, *J. Geophys. Res.*, *110*, D02102, doi:10.1029/2004JD005201.

Huybrechts, G., J. Olbregts, and K. Thomas (1967), Gas-phase chlorine-photosensitized oxidation and oxygen-inhibited photochlorination of tetrachloroethylene and pentachloroethane, *Trans. Faraday Soc.*, *63*, 1647–1655.

Hynes, R.G., D.E. Angove, S.M. Saunders, V. Haverd, and M. Azzi (2005), Evaluation of two MCMv3.1 alkene mechanisms using indoor environmental chamber data, *Atmos. Environ.*, *39*, 7251–7262.

Iannone, R., R.S. Anderson, A. Vogel, P.S. Eby, M.J. Whiticar, and J. Rudolph (2005), The hydrogen kinetics isotope effects of the reactions of n-alkanes with chlorine atoms in the gas phase, *J. Atmos. Chem.*, *50*, 121–138.

Iannone, R., R.S. Anderson, A. Vogel, J. Rudolph, P. Eby, and M.J. Whiticar (2004), Laboratory studies of the hydrogen kinetic isotope effects (KIES) of the reaction of non-methane hydrocarbons with the OH radical in the gas phase, *J. Atmos. Chem.*, *47*, 191–208.

IPCC (1990), Intergovernmental Panel on Climate Change; Climate change: The IPCC scientific assessment, Cambridge University Press, Cambridge, United Kingdom.

IPCC (2001), Intergovernmental Panel on Climate Change; Climate change 2001: The scientific basis, Cambridge University Press, New York.

IPCC (2005), Intergovernmental Panel on Climate Change; IPCC/TEAP special report on safeguarding the ozone layer and the global climate system, Cambridge University Press, New York.

Ishakawa, H., and W.A. Noyes, Jr. (1962a), The triplet state of benzene, *J. Am. Chem. Soc.*, *84*, 1502–1503.

Ishakawa, H., and W.A. Noyes, Jr., (1962b), Photosensitization by benzene vapor: Biacetyl. The triplet state of benzene, *J. Chem. Phys.*, *37*, 583–591.

IUPAC (2005), Subcommittee for gas kinetic data evaluation, Evaluated gas kinetic data for atmospheric chemistry: www.iupac-kinetic.ch.cam.ac.uk.

Iwasaki, E., Y. Matsumi, K. Takahashi, T.J. Wallington, M.D. Hurley, J.J. Orlando, E.W. Kaiser, and J.G. Calvert (2008), Atmospheric chemistry of cyclohexanone: UV spectrum and kinetics of reaction with chlorine atoms, *Int. J. Chem. Kinet.*, *40*, 323–329.

Jäger, M., H. Heydtmann, and C. Zetzsch (1996), Vacuum ultraviolet spectrum and quantum yield of the 193 nm photolysis of phosgene, *Chem. Phys. Lett.*, *263*, 817–821.

Japar, S.M., T.J. Wallington, J.F.O. Richert, and J.C. Ball (1990), The atmospheric chemistry of oxygenated fuel additives: *t*-butyl alcohol, dimethyl ether, and methyl *t*-butyl ether, *Int. J. Chem. Kinet.*, *22*, 1257–1269.

Jayanty, R.K.M., R. Simonaitis, and J. Heicklen (1975), The photolysis of chlorofluoromethanes in the presence of O_2 or O_3 at 213.9 nm and their reaction with $O(^1D)$, *J. Photochem.*, *4*, 381–398.

Jeffries, H.E. (1991), UNC solar radiation models, Report for EPA Cooperative Agreements CR813107, CR813964, CR815779, University of North Carolina, USA.

Jemi-Alade, A.A., P.D. Lightfoot, and R. Lesclaux (1991), Measurements of the UV absorption cross sections of $CF_3CCl_2O_2$, CF_3CHFO_2, $CFCl_2CH_2O_2$, and $CF_2ClCH_2O_2$ in the gas phase, *Chem. Phys. Lett.*, *179*, 119–124.

Jenkin, M.E. (2004), Modelling the formation and composition of secondary organic aerosol from α- and β-pinene ozonolysis using MCM v3, *Atmos. Chem. Phys.*, *4*, 1741–1757.

Jenkin, M.E., and R.A. Cox (1991), Kinetics of reactions of CH_3O_2 and $HOCH_2CH_2O_2$ radicals produced by photolysis of iodomethane and 2-iodoethanol, *J. Phys. Chem.*, *95*, 3229–3237.

Jenkin, M.E., R.A. Cox, G.D. Hayman, and L.J. Whyte (1988), Kinetic study of the reactions $CH_3O_2 + CH_3O_2$ and $CH_3O_2 + HO_2$ using molecular modulation spectroscopy, *J. Chem. Soc. Faraday Trans. 2*, *84*, 913–930.

Jenkin, M.E., S.M. Saunders, and M.J. Pilling (1997), The tropospheric degradation of volatile organic compounds: a protocol for mechanism development, *Atmos. Environ.*, *31*, 81–104.

Jenkin, M.E., S.M. Saunders, V. Wagner, and M.J. Pilling (2003), Protocol for the development of the Master Chemical Mechanism, MCM v3 (Part B): tropospheric degradation of aromatic volatile organic compounds, *Atmos. Chem. Phys.*, *3*, 181–193.

Jeong, K.-M., and F. Kaufman (1979), Rates of the reactions of 1,1,1-trichloroethane (methyl chloroform) and 1,1,2-trichloroethane with OH, *Geophys. Res. Lett.*, *6*, 757–759.

Jeong, K-M., and F. Kaufman (1982), Kinetics of the reaction of hydroxyl radical with methane and with nine Cl- and F-substituted methanes. 1. Experimental results, comparisons, and applications, *J. Phys. Chem.*, *86*, 1808–1815.

Jeong, K.-M., K.-J. Hsu, J.B. Jeffries, and F. Kaufman (1984), Kinetics of the reactions of OH with C_2H_6, CH_3CCl_3, $CH_2ClCHCl_2$, $CH_2ClCClF_2$, and CH_2FCF_3, *J. Phys. Chem.*, *88*, 1222–1226.

Jiang, Z., P.H. Taylor, and B. Dellinger (1992), Laser photolysis/laser-induced fluorescence studies of the reaction of OH with 1,1-dichloroethane over an extended temperature range, *J. Phys. Chem.*, *96*, 8964–8966.

Jiang, Z., P.H. Taylor, and B. Dellinger (1993), Laser photolysis/laser-induced fluorescence studies of the reaction of OH with 1,1,1,2- and 1,1,2,2-tetrachloroethane over an extended temperature range, *J. Phys. Chem.*, *97*, 5050–5053.

Jiménez, E., B. Ballesteros, E. Martinez, and J. Albaladejo (2005a), Tropospheric reaction of OH with selected linear ketones: Kinetic studies between 228 and 405 K, *Environ. Sci. Tech.*, *39*, 814–820.

Jiménez, E., M.K. Gilles, and A.R. Ravishankara (2003), Kinetics of the reactions of the hydroxyl radical with CH_3OH and C_2H_5OH between 235 and 360 K, *J. Photochem. Photobiol. A*, *157*, 237–245.

Jiménez, E., B. Lanza, A. Garzón, B. Ballesteros, and J. Albaladejo (2005b), Atmospheric degradation of 2-butanol, 2-methyl-2-butanol, and 2,3-dimethyl-2-butanol: OH kinetics and UV absorption cross sections, *J. Phys. Chem. A*, *109*, 10903–10909.

Jobson, B.T., H. Niki, Y. Yokouchi, J. Bottenheim, F. Hopper, and R. Leaitch (1994), Measurements of C2–C6 hydrocarbons during the Polar Sunrise 1992 Experiment: Evidence for Cl atom and Br atom chemistry, *J. Geophys. Res.*, *99*, 25355–25368.

Johnson, D., S. Carr, and R.A. Cox (2005), The kinetics of the gas-phase decomposition of the 2-methyl-2-butoxyl and 2-methyl-2-pentoxyl radicals, *Phys. Chem. Chem. Phys.*, *7*, 2182–2190.

Johnson, D., P. Cassanelli, and R.A. Cox (2004a), Isomerization of simple alkoxyl radicals: New temperature-dependent rate data and structure activity relationship, *J. Phys. Chem. A*, *108*, 519–523.

Johnson, D., P. Cassanelli, and R.A. Cox (2004b), Correlation-type structure activity relationships for the kinetics of the decomposition of simple and β-substituted alkoxyl radicals, *Atmos. Environ.*, *38*, 1755–1765.

Jolley, J.E. (1957), The photooxidation of diethyl ketone, *J. Am. Chem. Soc.*, *79*, 1537–1542.

Jolly, G.S., G. Paraskevolpoulos, and D.L. Singleton (1985), Rate of OH radical reactions. XII. The reactions of OH with c-C_3H_6, c-C_5H_{10}, and c-C_7H_{14}. Correlation of hydroxyl rate constants with bond dissociation energies, *Int. J. Chem. Kinet.*, *17*, 1–10.

Judge, R.H., and D.C. Moule (1985), The $\tilde{A}^1A'' \leftarrow X^1A'$ electronic transition in formyl chloride, CHClO, *J. Mol. Spectros.*, *113*, 302–309.

Jungkamp, T.P.W., J.N. Smith, and J.H. Seinfeld (1997), Atmospheric oxidation mechanism of n-butane: the fate of alkoxy radicals, *J. Phys. Chem.*, *101*, 4392–4401.

Kaiser, E.W., and T.J. Wallington (1994), FTIR product study of the Cl initiated oxidation of CH_3Cl: evidence for HCl elimination from the chloromethoxy radical, *J. Phys. Chem.*, *98*, 5679–5685.

Kaiser, E.W., and T.J. Wallington (1995), CH_3CO reactions with Cl_2 and O_2: more evidence for HCl elimination from the CH_3CHClO radical, *J. Phys. Chem.*, *99*, 8669–8672.

Kaiser, E.W., L. Rimai, E. Schwab, and E.C. Lim (1992), Application of time-resolved infrared spectroscopy to the determination of absolute rate constants for chlorine atom + ethane and chlorine atom + chloroethane, *J. Phys. Chem.*, *96*, 303–306.

Kaiser, E.W., T.J. Wallington, Y. Hashikawa, and M. Kawasaki (2004), The rate constant ratio $k_1(Cl + C_2H_6)/k_2\ (Cl + CH_4)$ from 250 to 700 K, *J. Phys. Chem. A*, *108*, 10141–10146.

Kakesu, M., H. Bandow, N. Takenaka, Y. Maeda, and N. Washida (1997), Kinetic measurements of methyl and ethyl nitrate reactions with OH radicals, *Int. J. Chem. Kinet.*, *29*, 933–941.

Kalabokas, P.D., L.G. Viras, J.G. Bartzis, and C.C. Repapis (2000), Mediterranean rural ozone characteristics around the urban area of Athens, *Atmos. Environ.* *34*, 5199–5208.

Kambanis, K.G., Y.G. Lazarou, and P. Papagiannakopoulos (1995), Absolute reaction rate of chlorine atoms with neopentane, *Int. J. Chem. Kinet.*, *27*, 343–350.

Kamboures, M.A., J.C. Hansen, and J.S. Francisco (2002), A study of the kinetics and mechanisms involved in the atmospheric degradation of bromoform by atomic chlorine, *Chem. Phys. Lett.*, *353*, 335–344.

Kamiya, K., and K. Morokuma (1991), Potential energy surface for unimolecular dissociation and rearrangement reactions of the ground electronic state of HFCO, *J. Chem. Phys.*, *94*, 7287–7298.

Kan, C.S., J.G. Calvert, and J.H. Shaw (1980), Reactive channels of the CH_3O_2–CH_3O_2 reaction, *J. Phys. Chem.*, *84*, 3411–3417.

Kan, C.S., R.D. McQuigg, M.R. Whitbeck, and J.G. Calvert (1979), Kinetic flash spectroscopic study of the CH_3O_2–CH_3O_2 and CH_3O_2–SO_2 reactions, *Int. J. Chem. Kinet.*, *11*, 921–933.

Kash, P.W., G.C.G. Waschewsky, R.E. Morss, L.J. Butler, and M.M. Francl (1994), Competing C—Br and C—C bond fission following $^1[n(O),\pi^*(C{=}O)]$ excitation in bromoacetone: Conformation dependence of nonadiabaticity at a conical intersection, *J. Chem. Phys.*, *100*, 3463–3475.

Kasner, J.H., P.H. Taylor, and B. Dellinger (1990), Laser photolysis/laser induced fluorescence study of OH–C_2H_5Cl rate constants from 294 to 789 K, *J. Phys. Chem.*, *94*, 3250–3253.

Kawasaki, M., S.J. Lee, and R. Bersohn (1975), Photodissociation of molecular-beams of methylene iodide and iodoform, *J. Chem. Phys.*, *63*, 809–814.

Keller-Rudek, H., and G.K. Moortgat (2005), MPI-Mainz-UV-VIS *Spectral Atlas of Gaseous Molecules*, catalogue of spectra available online: http://www.atmosphere.mpg.de/spectral-atlas-mainz

Kemper, M.J.H., J.M.F. van Dijk, and H.M. Buck (1978), Ab initio calculation of the photochemistry of formaldehyde. The search for a hydroxycarbene intermediate, *J. Am. Chem. Soc.*, *100*, 7841–7846.

Kenner, R.D., K.R. Ryan, and I.C. Plumb (1993), Kinetics of the reaction of CH_3O_2 with ClO at 293 K, *Geophys. Res. Lett.*, *20*, 1571–1574.

Kerr, J.A., and D.W. Sheppard (1981), Kinetics of the reactions of hydroxyl radicals with aldehydes studied under atmospheric conditions, *Environ. Sci. Tech.*, *15*, 960–963.

Kerr, J.A., and D.W. Stocker (1986), Kinetics of the reactions of hydroxyl radicals with alkyl nitrates and with some oxygen-containing organic compounds studied under simulated atmospheric conditions, *J. Atmos. Chem.*, *4*, 253–262.

Kerr, J.A., and D.W. Stocker (2001–2002), Strengths of chemical bonds, in *CRC Handbook of Chemistry and Physics*, 82nd edition, edited by D.R. Lide, pp. 9-51–9-74, CRC Press, Boca Raton.

Keyser, L. F. (1978), Absolute rate and temperature dependence of the reaction between chlorine (^2P) atoms and methane, *J. Chem. Phys.*, *69*, 214–218.

Khalil, M.A.K., and Rasmussen, R.A. (1999), Atmospheric methyl chloride, *Atmos. Environ.*, *33*, 1305–1321.

Kim, Y.S., W.K. Kang, D.-C. Kim, and K.-H. Jung (1997), Photodissociation of *tert*-butyl iodide at 277 and 304 nm: Evidence for direct and indirect dissociation in A-band photolysis of alkyl iodide, *J. Phys. Chem. A*, *101*, 7576–7581.

King, R.A., W.D. Allen, and H.F. Schaefer, III (2000), On apparent quantized transition-state thresholds in the photofragmentation of acetaldehyde, *J. Chem. Phys.*, *112*, 5585–5592.

Kirk, A.D., and G.B. Porter (1962), Kinetics of excited molecules. III. Photooxidation of acetone, *J. Phys. Chem.*, *66*, 556–557.

Kirsch, L.J., and D.A. Parkes (1981), Recombination of *tertiary*-butyl peroxy-radicals. 1. Product yields between 298 K and 373 K, *J. Chem. Soc. Faraday Trans. I*, *77*, 293–307.

Kirsch, L.J., D.A. Parkes, D.J. Waddington, and A. Woolley (1979), Reactions of oxygenated radicals in the gas phase. Part 6. Reactions of isopropylperoxy and isopropoxy radicals, *J. Chem. Soc. Faraday Trans. 1*, *75*, 2678–2687.

Klamt, A. (1993), Estimation of gas-phase hydroxyl radical rate constants of organic compounds from molecular orbital calculations, *Chemosphere*, *27*, 1273–1289.

Klein, T., I. Barnes, K.H. Becker, E.H. Fink, and F. Zabel (1984), Pressure dependence of the rate constants for the reactions of C_2H_4 and C_3H_6 with OH radicals at 295 K, *J. Phys. Chem.*, *88*, 5020–5025.

Kleindienst, T.E., P.B. Shepson, and C.M. Nero (1989), The production of chlorine atoms from the reaction of hydroxyl radical with chlorinated hydrocarbons, *Int. J. Chem. Kinet.*, *21*, 863–884.

Klemm, O., I.C. Ziomas, D. Balis, P. Suppan, J. Slemr, R. Romero, and L.G. Vyras (1998), A summer air-pollution study in Athens, Greece, *Atmos. Environ.*, *32*, 2071–2087.

Klemm, R.G., D.N. Morrison, P. Gilderson, and A.T. Blades (1965), The photolysis of cyclopentanone and cyclobutanone, *Can. J. Chem.*, *43*, 1934–1941.

Klöpffer, V.W., R. Frank, E. Kohl, and F. Haag (1986), Quantitative Erfassung der photochemischen Transformationsprozesse in der Troposphäre, *Chemiker-Zeitung*, *110*, 57–61.

Knee, J.L., L.R. Khundkar, and A.H. Zewail (1985), Picosecond monitoring of a chemical reaction in molecular-beams-photofragmentation of $R—I \rightarrow R^{\ddagger} + I$, *J. Chem. Phys.*, *83*, 1996–1997.

Knox, J. H., and R.L. Nelson (1959), Competitive chlorination reactions in the gas phase: Hydrogen and C_1–C_5 saturated hydrocarbons, *Trans. Faraday. Soc.*, *55*, 937–946.

Kockarts, G. (1994), Penetration of solar radiation in the Schuman–Runge bands of molecular oxygen: a robust approximation, *Ann. Geophysicae*, *12*, 1207–1217.

Koffend, J.B., and N. Cohen (1996), Shock tube study of OH reactions with linear hydrocarbons near 1100 K, *Int. J. Chem. Kinet.*, *28*, 79–88.

Koffend, J.B., and S.R. Leone (1981), Tunable laser photodissociation: Quantum yield of $I^*(^2P_{1/2})$ for CH_2I_2, *Chem. Phys. Lett.*, *81*, 136–141.

Kowalczyk, J., A. Jowko, and M. Symanowicz (1998), Kinetics of radical reactions in freons, *J. Radioanal. Nucl. Chem.*, *232*, 75–78.

Kozlov, S.N., V.L. Orkin, R.E. Huie, and M.J. Kurylo (2003a), OH reactivity and UV spectra of propane, *n*-propyl bromide, and isopropyl bromide, *J. Phys. Chem. A*, *107*, 1333–1338.

Kozlov, S.N., V.L. Orkin, and M.J. Kurylo (2003b), An investigation of the reactivity of OH with fluoroethanes: CH_3CH_2F (HFC-161), CH_2FCH_2F (HFC-152), and CH_3CHF_2 (HFC-152a), *J. Phys. Chem. A*, *107*, 2239–2246.

Kramp, F., and S.E. Paulson (1998), On the uncertainties in the rate coefficients for OH reactions with hydrocarbons, and the rate coefficients of the 1,3,5-trimethylbenzene and *m*-xylene reactions with OH radicals in the gas phase, *J. Phys. Chem. A*, *102*, 2685–2690.

Kraus, J.W., and J.G. Calvert (1957), The disproportionation and combination reactions of butyl free radicals, *J. Am. Chem. Soc.*, *79*, 5921–5926.

Kukui, A., and G. Le Bras (2001), Theoretical study of the thermal decomposition of several β-chloroalkoxy radicals, *Phys. Chem. Chem. Phys.*, *3*, 175–178.

Kurosaki, Y., and K. Yokoyama (2002), Photodissociation of acetaldehyde, $CH_3CHO \rightarrow CH_4 + CO$: Direct ab initio dynamics study, *J. Phys. Chem.*, *106*, 11415–11421.

Kurylo, M.J., and T.J. Wallington (1987), The temperature dependence of the rate constant for the gas phase disproportionation reaction of CH_3O_2 radicals, *Chem. Phys. Lett.*, *138*, 543–547.

Kurylo, M.J., P.C. Anderson, and O. Klais (1979), A flash photolysis resonance fluorescence investigation of the reaction $OH + CH_3CCl_3 = H_2O + CH_2CCl_3$, *Geophys. Res. Lett.*, *6*, 760–762.

Kurylo, M.J., P. Dagaut, T.J. Wallington, and D.M. Neuman (1987), Kinetic measurements of the gas phase $HO_2 + CH_3O_2$ cross-disproportionation reaction at 298 K, *Chem. Phys. Lett.*, *139*, 513–518.

Kutschke, K.O., M.H.J. Wijnen, and E.W.R. Steacie (1952), Mechanism of the photolysis of diethyl ketone, *J. Am. Chem. Soc.*, *74*, 714–718.

Kwok, E.S.C., and R. Atkinson (1995), Estimation of hydroxyl radical reaction rate constants for gas-phase organic compounds using a structure–reactivity relationship: An update, *Atmos. Environ.*, *29*, 1685–1695.

Kwok, E.S.C., J. Arey, and R. Atkinson (1996), Alkoxy radical isomerization in the OH radical-initiated reactions of C_4–C_8 *n*-alkanes, *J. Phys. Chem.*, *100*, 214–219.

Lancar, I., G. Le Bras, and G. Poulet (1993), Oxidation of CH_3CCl_3 and CH_3CFCl_2 in the atmosphere: kinetic studies of OH reactions, *J. Chim. Phys.*, *90*, 1897–1908.

Langer, S., and E. Ljungström (1995), Rates of reaction between the nitrate radical and some aliphatic alcohols, *J. Chem. Soc. Faraday Trans.*, *91*, 405–410.

Langer, S., E. Ljungstrom, and I. Wangberg (1993), Rates of reaction between the nitrate radical and some aliphatic esters, *J. Chem. Soc. Faraday Trans.*, *89*, 425–431.

Lantz, K.O., R.E. Shetter, C.A. Cantrell, S.J. Flocke, J.G. Calvert, and S. Madronich (1996), Theoretical, actinometric, and radiometric determination of the photolysis rate coefficient of NO_2 during the Mauna Loa Observatory Photochemistry Experiment 2, *J. Geophys. Res.*, *101*, 14513–14629.

Latella, A., G. Stani, L. Cobelli, M. Duane, H. Junninen, C. Astorga, and B.R. Larsen (2005), Semicontinuous GC analysis and receptor modelling for source apportionment of ozone precursor hydrocarbons in Bresso, Milan, 2003, *J. Chromatogr. A*, *1071*, 29–39.

Laval-Szopa, S. (2003), Développement d'une chaîne automatique d'écriture de schémas chimiques explicites et réduits adaptés à l'étude de la pollution photooxydante aux différentes échelles, PhD Thesis, University of Paris XII.

Law, K.S., P.H. Plantevin, D.E. Shallcross, H.L. Rogers, J.A. Pyle, C. Grouhel, V. Thouret, and A. Marenco (1998), Evaluation of modeled O_3 using measurement of ozone by Airbus in-service aircraft (MOZAIC) data, *J. Geophys. Res.*, *103*, 25721–25737.

Lazarou, Y.G., C. Michael, and P. Papagiannakopoulos (1992), Kinetics of the reaction of chlorine atoms with dimethylnitramine, *J. Phys. Chem.*, *96*, 1705–1708.

Le Calvé, S., D. Hitier, G. Le Bras, and A. Mellouki (1998), Kinetic studies of OH reactions with a series of ketones, *J. Phys. Chem. A*, *102*, 4579–4584.

Le Calvé, S., G. Le Bras, and A. Mellouki (1997), Kinetic studies of OH reactions with a series of methyl esters, *J. Phys. Chem. A*, *101*, 9137–9141.

Le Crâne, J.P., E. Villenave, M.D. Hurley, T.J. Wallington, and J.C. Ball (2005), Atmospheric chemistry of propionaldehyde: Kinetics and mechanisms of reactions with OH radicals and Cl atoms, UV spectrum, and self-reaction kinetics of $CH_3CH_2C(O)O_2$ radicals at 298 K, *J. Phys. Chem. A*, *109*, 11837–11850.

Lee, E.K.C. (1967), Benzene photosensitizaton of cyclopentanone and cyclopentanone-2-*t*, *J. Phys. Chem.*, *71*, 2804–2813.

Lee, E.K.C., and R.S. Lewis (1980), Photochemistry of simple aldehydes and ketones in the gas phase, *Adv. Photochem.*, *12*, 1–96.

Lee, F.S.C., and F.S. Rowland (1977), Competitive radiotracer evaluation of relative rate constants at stratospheric temperatures for reactions of ^{38}Cl with CH_4, and C_2H_6 vs. $CH_2{=}CHBr$, *J. Phys. Chem.*, *81*, 86–87.

Lee, S.J., and R. Bersohn (1982), Photo-dissociation of a molecule with 2 chromophores—CH_2IBr, *J. Phys. Chem.*, *86*, 728–730.

Leighton, P.A. (1961), *Photochemistry of Air Pollution*, Academic Press, New York.

Leighton, P.A., and F.E. Blacet (1933), The photolysis of the aliphatic aldehydes. II. Acetaldehyde, *J. Am. Chem. Soc.*, *55*, 1766–1774.

Lendvay, G., and B. Viskolcz (1998), Ab intio studies of the isomerization and decomposition reactions of the 1-butoxy radical, *J. Phys. Chem. A*, *102*, 10777–10780.

Lesclaux, R., and F. Caralp (1984), Determination of the rate constants for the reactions of dichlorofluoromethylperoxy radical with nitrogen oxides (NO and NO_2) by laser photolysis and time resolved mass spectrometry, *Int. J. Chem. Kinet.*, *16*, 1117–1128.

Lesclaux, R., A.M. Dognon, and F. Caralp (1987), Photooxidation of halomethanes at low temperature: the decomposition rate of CCl_3O and $CFCl_2O$ radicals, *J. Photochem. Photobiol.*, *41*, 1–11.

Letokhov, V.S. (1972), Selective laser photochemical reactions by means of photopredissociation of molecules, *Chem. Phys. Lett.*, *15*, 221.

Leu, G-H., and Y.-P. Lee (1994), Temperature dependence of the rate coefficient of the reaction $OH + CF_3CH_2F$ over the range 255–424 K, *J. Chin. Chem. Soc. (Taipei)*, *41*, 645–649.

Leu, M.-T., and W.B. DeMore (1976), Rate constants at 295 K for the reactions of atomic chlorine with H_2O_2, HO_2, O_3, CH_4 and HNO_3, *Chem. Phys. Lett.*, *41*, 121–124.

Lewis, R.S., S.P. Sander, S. Wagner, and R.T. Watson (1980), Temperature-dependent rate constants for the reaction of ground-state chlorine with simple alkanes, *J. Phys. Chem.*, *84*, 2009–2015.

Li, Z. (2004), Kinetic study of OH radical reactions with volatile organic compounds using relative rate/discharge fast flow/mass spectrometer technique, *Chem. Phys. Lett.*, *383*, 592–600.

Li, Z., and J.S. Francisco (1989), Dissociation dynamics of perhaloalkoxy radicals, *J. Am. Chem. Soc.*, *111*, 5660–5667.

Liberles, A., S. Kang, and A. Greenberg (1973), Semiempirical calculations on the ring opening of substituted cyclopropanones, *J. Org. Chem.*, *38*, 1922–1924.

Libuda, H.G. (1992), Spektroskopische und kinetische Untersuchengen an halogenierten Carbonylverbindungen von atmosphärischen Interesse, Ph.D. Thesis, University of Wuppertal, Germany.

Libuda, H.G., O. Shestakov, J. Theloke, and F. Zabel (2002), Relative-rate study of thermal decomposition of the 2-butoxyl radical in the temperature range 280–313 K, *Phys. Chem. Chem. Phys.*, *4*, 2579–2586.

Lide, D.R. (2001–2002), *CRC Handbook of Chemistry and Physics*, 82nd Edition, CRC Press, Boca Raton, Florida.

Lightfoot, P.D., R.A. Cox, J.N. Crowley, M. Destriau, G.D. Hayman, M.E. Jenkin, G.K. Moortgat, and F. Zabel (1992), Organic peroxy radicals: kinetics, spectroscopy and tropospheric chemistry, *Atmos. Environ. Part A*, *26*, 1805–1961.

Lightfoot, P.D., S.P. Kirwan, and M.J. Pilling (1988), Photolysis of acetone at 193.3 nm, *J. Phys. Chem.*, *92*, 4938–4946.

Lightfoot, P.D., R. Lesclaux, and B. Veyret (1990a), Flash photolysis study of the $CH_3O_2 + CH_3O_2$ reaction: Rate constants and branching ratios from 248 to 573 K, *J. Phys. Chem.*, *94*, 700–707.

Lightfoot, P.D., P. Roussel, F. Caralp, and R. Lesclaux (1991), Flash photolysis study of the CH_3O_2 + CH_3O_2 and CH_3O_2 + HO_2 reactions between 600 and 719 K: Unimolecular decomposition of methylhydroperoxide, *J. Chem. Soc. Faraday Trans.*, *87*, 3213–3220.

Lightfoot, P.D., P. Roussel, B.Veyret, and R. Lesclaux (1990b), Flash photolysis study of the spectra and self-reactions of neopentylperoxy and *t*-butylperoxy radicals, *J. Chem. Soc. Faraday Trans.*, *86*, 2927–2936.

Lightfoot, P.D., B. Veyret, and R. Lesclaux (1990c), Flash photolysis study of the CH_3O_2 + HO_2 reaction between 248 and 573 K, *J. Phys. Chem.*, *94*, 708–714.

Lim, Y.B., and P.J. Ziemann (2005), Products and mechanism of secondary organic aerosol formation from reactions of *n*-alkanes with OH radicals in the presence of NO_x, *Environ. Sci. Technol.*, *39*, 9229–9236.

Lin, C.L., M.T. Leu and W.B. DeMore (1978), Rate constant for the reaction of atomic chlorine with methane, *J. Phys. Chem.*, *82*, 1772–1777.

Lin, C.-Y., and J.-J. Ho (2002), Theoretical studies of isomerization reactions of 2-pentoxy radical and its derivatives including the unsaturated alkoxy radicals, *J. Phys. Chem. A*, *106*, 4137–4144.

Lissi, E.A., E. Abuin, and M.V. Encina (1973/1974), Photochemistry of butanone and methyl butanone, *J. Photochem.*, *2*, 377–392.

Liu, R., R.E. Huie, and M.J. Kurylo (1990), Rate constants for the reactions of the OH radical with some hydrochlorofluorocarbons over the temperature range 270–400 K, *J. Phys. Chem.*, *94*, 3247–3249.

Lloyd, A.C., K.R. Darnall, A.M. Winer, and J.N. Pitts, Jr. (1976a), Relative rate constants for reactions of the hydroxyl radical with a series of alkanes, alkenes, and aromatic hydrocarbons, *J. Phys. Chem.*, *80*, 789–794.

Lloyd, A.C., K.R. Darnall, A.M. Winer, and J.N. Pitts, Jr. (1976b), Relative rate constants for the reactions of OH radicals with isopropyl alcohol, diethyl and di-*n*-propyl ether at 305 ± 2 K, *Chem. Phys. Lett.*, *42*, 205–209.

Locht, R., B. Leyh, A. Hoxha, D. Dehareng, H.W. Jochims, and H. Baumgärtel (2000), About the vacuum UV photoabsorption spectrum of methyl fluoride (CH_3F): the fine structure and its vibrational analysis, *Chem. Phys.*, *257*, 283–299.

Lohr, L.L., J.R. Barker, and R.M. Shroll (2003), Modeling the organic nitrate yields in the reaction of alkyl peroxy radicals with nitric oxide. 1. Electronic structure calculations and thermochemistry, *J. Phys. Chem. A.*, *107*, 7429–7433.

Lorenz, K., D. Rhäsa, R. Zellner, and B. Fritz (1985), Laser photolysis–LIF kinetic studies of the reactions of CH_3O and CH_2CHO with O_2 between 300 and 500 K, *Ber Bunsen. Phys. Chem. Chem. Phys.*, *89*, 341–342.

Lotz, Ch., and R. Zellner, unpublished results quoted in Orlando et al. (2003b).

Louis, F., D.R. Burgess, M.-T. Rayez, and J.-P. Sawerysyn (1999), Kinetic study of the reactions of CF_3O_2 radicals with Cl and NO, *Phys. Chem. Chem. Phys.*, *1*, 5087–5096.

Luke, W.T., and R.R. Dickerson (1988), Direct measurements of the photolysis rate coefficient of ethyl nitrate, *Geophys. Res. Lett.*, *15*, 1181–1184.

Luke, W.T., R.R. Dickerson, and L.J. Nunnermacker (1989), Direct measurements of the photolysis rate coefficients and Henry's law constants of several alkyl nitrates, *J. Geophys. Res.*, *94*, 14905–14921.

Madronich, S., and J.G. Calvert (1990), Permutation reactions of organic peroxy radicals in the troposphere, *J. Geophys. Res.*, *95*, 5697–5715.

Madronich, S., and S.J. Flocke (1998), The role of solar radiation in atmospheric chemistry, in *Handbook of Environmental Chemistry*, Vol. II, Part L, *Reactions and Processes*, edited by P. Boule, pp. 1–26, Springer-Verlag, Heidelberg.

Madronich, S., and G. Weller (1990), Numerical integration errors in calculated tropospheric photodissociation rate coefficients, *J. Atmos. Chem.*, *10*, 289–300.

Magneron, I., V. Bossoutrot, A. Mellouki, G. Laverdet, and G. Le Bras (2003), The OH-initiated oxidation of hexylene glycol and diacetone alcohol, *Environ. Sci. Tech.*, *37*, 4170–4181.

Magneron, I., A. Mellouki, G. Le Bras, G.K. Moortgat, A. Horowitz, and K. Wirtz (2005), Photolysis and OH-initiated oxidation of glycolaldehyde under atmospheric conditions, *J. Phys. Chem. A*, *109*, 4552–4561.

Mahmud, K., P. Marshall, and A. Fontijn (1988), The reaction of $O(^3P)$ atoms with ethane: An HTP kinetics study from 300 to 1270 K, *J. Chem. Phys.*, *88*, 2393–2397.

Majer, J.R., and Z.Y. Al-Saigh (1984), Photochemistry of 1,1,1-trifluoro-3-bromoacetone. 2. Dual fluorescence in the gas phase, *J. Phys. Chem.*, *88*, 1157–1159.

Majer, J.R., and J.P. Simons (1964), Photochemical processes in halogenated compounds, *Adv. Photochem.*, *1*, 137–181.

Majer, J.R., C. Olavesen, and J.C. Robb (1971), Wavelength effect in the photolysis of halogenated ketones, *J. Chem. Soc. (B)*, 48–55.

Majer, J.R., J.C. Robb, and Z.Y. Al-Saigh (1976), Photochemistry of 1,1,1-trifluoro-3-bromoacetone, *J. Chem. Soc.*, *72*, 1697–1706.

Mannik, L., G.M. Keyser, and K.B. Woodall (1979), Deuterium isotope separation by tunable-laser predissociation of formaldehyde, *Chem. Phys. Lett.*, *65*, 1231–1233.

Manning, A.J., D.B. Ryall, R.G. Derwent, P.G. Simmonds, and S. O'Doherty (2003), Estimating European emissions of ozone-depleting and greenhouse gases using observations and a back-attribution technique, *J. Geophys. Res.*, *108*, D4405, doi:10.1029/2002JD002312.

Manning, R.G., and M.J. Kurylo (1977), Flash photolysis resonance fluorescence investigation of the temperature dependencies of the reactions of $Cl(^2P)$ atoms with CH_4, CH_3Cl, CH_3F, $CH_3F^{\dagger}$, and C_2H_6, *J. Phys. Chem.*, *81*, 291–296.

Maricq, M.M., and J.J. Szente (1992), Flash photolysis–time-resolved UV spectroscopy of the 1,2,2,2-tetrafluoroethylperoxy radical self-reaction, *J. Phys. Chem.*, *96*, 10862–10868.

Maricq, M.M, and J.J. Szente (1994), A kinetic study of the reaction between ethylperoxy and HO_2, *J. Phys. Chem.*, *98*, 2078–2082.

Maricq, M.M., and J.J. Szente (1995), The 193 and 248 nm photodissociation of $CF_3C(O)Cl$, *J. Phys. Chem.*, *99*, 4554–4557.

Maricq, M.M., and J.J. Szente (1996), Kinetics of the reaction between ethylperoxy radicals and nitric oxide, *J. Phys. Chem.*, *100*, 12374–12379.

Maricq, M.M., J. Shi, J.J. Szente, L. Rimai, and E.W. Kaiser (1993), Evidence for the three-center elimination of hydrogen chloride from 1-chloroethoxy, *J. Phys. Chem.*, *97*, 9686–9694.

Maricq, M.M., J.J. Szente, M.D. Hurley, and T.J. Wallington (1994), Atmospheric chemistry of HFC-134a: kinetic and mechanistic study of the $CF_3CFHO_2 + HO_2$ reaction, *J. Phys. Chem.*, *98*, 8962–8970.

Maricq, M.M., J.J. Szente, G.A. Khitrov, T.S. Dibble, and J.S. Francisco (1995), CF_3CO dissociation kinetics, *J. Phys. Chem.*, *99*, 11875–11882.

Maricq, M.M, J.J. Szente, G.A. Khitrov, and J.S. Francisco (1996), The $CF_3C(O)O_2$ radical. 2. Its UV spectrum, self-reaction kinetics, and reaction with NO, *J. Phys. Chem.*, *100*, 4514–4520.

Markert, F., and O.J. Nielsen (1992), The reactions of OH radicals with chloroalkanes in the temperature range 295–360 K, *Chem. Phys. Lett.*, *194*, 123–127.

Marling, J. (1977), Isotope separation of oxygen-17, oxygen-18, carbon-13, and deuterium by ion laser induced formaldehyde photopredissociation, *J. Chem. Phys.*, *66*, 4200–4225.

Martell, J.M., and R.J. Boyd (1995), *Ab Initio* studies of reactions of hydroxyl radicals with fluorinated ethanes, *J. Phys. Chem.*, *99*, 13402–13411.

Martell, J.M., H. Yu, and J.D. Goddard (1997), Molecular decomposition of acetaldehyde and formamide: theoretical studies using Hartree–Fock, Moller–Plesset and density functional theories, *Mol. Phys.*, *82*, 497–502.

Martin, G.R., and H.C. Sutton (1952), Radioactive tracer studies of free radical mechanisms. Part 1. The photolysis of acetone + iodine mixtures, *Trans. Faraday Soc.*, *48*, 812–823.

Martin, J.-P., and G. Paraskevopoulos (1983), A kinetic study of the reactions of OH radicals with fluoroethanes. Estimates of C—H bond strengths in fluoroalkanes, *Can. J. Chem.*, *61*, 861–865.

Martin, P., E.C. Tuazon, S.M. Aschmann, J. Arey, and R. Atkinson (2002), Formation and atmospheric reactions of 4,5-dihydro-2-methylfuran, *J. Phys. Chem. A*, *106*, 11492–11501.

Martin, T.W., and J.N. Pitts, Jr. (1955), Structure and reactivity in the vapor phase photolysis of ketones. II. Methyl neopentyl ketone, *J. Am. Chem. Soc.*, *77*, 5465–5468.

Martinez, R.D., A.A. Buitrage, N.W. Howell, C.H. Hern, and J.A. Joens (1992), The near ultraviolet U.V. absorption spectra of several aliphatic aldehydes and ketones at 300 K, *Atmos. Environ.*, *26A*, 785–792.

Martino, P.C., P.B. Shevlin, and S.D. Worley (1977), The electronic structure of cyclopropanone, *J. Am. Chem. Soc.*, *99*, 8003–8006.

Masaki, A., S. Tsunashima, and N. Washida (1994), Rate constant for the reaction of CH_3O_2 with NO, *Chem. Phys. Lett.*, *218*, 523–528.

Masaki, A., S. Tsunashima, and N. Washida (1995), Rate constants for reactions of substituted methyl radicals (CH_2OCH_3, CH_2NH_2, CH_2I, and CH_2CN) with O_2, *J. Phys. Chem.*, *99*, 13126–13131.

Mashino, M., Y. Ninomiya, M. Kawasaki, T.J. Wallington, and M.D. Hurley (2000), Atmospheric chemistry of $CF_3CF{=}CF_2$: kinetics and mechanism of its reactions with OH radicals, Cl atoms, and ozone, *J. Phys. Chem. A*, *104*, 7255–7260.

Masson, C.R. (1952), The photolysis of di-*n*-propyl ketone, *J. Am. Chem. Soc.*, *74*, 4731–4735.

Mathias, E., E. Sanhueza, I.C. Hisatsune, and J. Heicklen (1974), Chlorine atom sensitized oxidation and the ozonolysis of tetrachloroethylene, *Can. J. Chem.*, *52*, 3852–3862.

Matsumi, Y., K. Izumi, V. Skorokhodov, M. Kawasaki, and N. Tanaka (1997), Reaction and quenching of $Cl(^2P_j)$ atoms in collisions with methane and deuterated methanes, *J. Phys. Chem. A*, *101*, 1216–1221.

Maul, C., T. Haas, K.-H. Gericke, and F.J. Comes (1995), Spin selectivity in the ultraviolet photodissociation of phosgene, *J. Chem. Phys.*, *102*, 3238–3247.

Mazely, T.L., R.R. Friedl, and S.P. Sander (1995), Production of NO_2 from photolysis of peroxyacetyl nitrate, *J. Phys. Chem.*, *99*, 8162–8169.

Mazely, T.L., R.R. Friedl, and S.P. Sander (1997), Quantum yield of NO_3 from peroxyacetyl nitrate photolysis, *J. Phys. Chem. A*, *101*, 7090–7097.

McAdam, K., B. Veyret, and R. Lesclaux (1987), UV absorption spectra of HO_2 and CH_3O_2 radicals and the kinetics of their mutual reactions at 298 K, *Chem. Phys. Lett.*, *133*, 39–44.

McCarthy, R.L., F.A. Bower, and J.P. Jesson (1977), The fluorocarbon–ozone theory—1. Production and release—World production and release of CCl_3F and CCl_2F_2 (Fluorocarbons 11 and 12) through 1975, *Atmos. Environ.*, *11*, 491–497.

McCaulley, J.A., N. Kelly, M.F. Golde, and F. Kaufman (1989), Kinetic studies of the reactions of F and OH with CH_3OH, *J. Phys. Chem.*, *93*, 1014–1018.

McGee, T.H. (1968), The photolysis of cyclobutanone, *J. Phys. Chem.*, *72*, 1621–1625.

McGivern, W.S., J.S. Francisco, and S.W. North (2002), Investigation of the atmospheric oxidation pathways of bromoform: initiation via OH/Cl reactions, *J. Phys. Chem. A*, *106*, 6395–6400.

McGivern, W.S., H. Kim, and J.S. Francisco (2004), Investigation of the atmospheric oxidation pathways of bromoform and dibromomethane: initiation via UV photolysis and hydrogen abstraction, *J. Phys. Chem. A*, *108*, 7247–7252.

McGivern, W.S., O. Sorkhabi, A.G. Suits, A. Derecskei-Kovacs, and S.W. North (2000), Primary and secondary processes in the photodissociation of $CHBr_3$, *J. Phys. Chem. A*, *104*, 10085–10091.

McKeen, S.A., T. Gierczak, J.B. Burkholder, P.O. Wennberg, T.F. Hanisco, E.R. Keim, R.-S. Gao, S.C. Liu, A.R. Ravishankara, and D.W. Fahey (1997), The photochemistry of acetone in the upper troposphere: A source of odd-hydrogen radicals, *Geophys. Res. Lett.*, *24*, 3177–3180.

McLoughlin, P., R. Kane, and I. Shanahan (1993), A relative rate study of the reaction of chlorine atoms (Cl) and hydroxyl radicals (OH) with a series of ethers, *Int. J. Chem. Kinet.*, *25*, 137–149.

MCM (2005), *The Master Chemical Mechanism*, available at: http://mcm.leeds.ac.uk/MCM.

McMillan, G.R., J.G. Calvert, and J.N. Pitts, Jr. (1964), Detection and lifetime of enol-acetone in the photolysis of 2-pentanone vapor, *J. Am. Chem. Soc.*, *86*, 3602–3605.

McNesby, J.R., and A.S. Gordon (1958), Photolysis and pyrolysis of 2-pentanone-1,1,1-3,3-d_5, *J. Am. Chem. Soc.*, *80*, 261–264.

McQuigg, R.D., and J.G. Calvert (1969), The photodecompostion of CH_2O, CD_2O, CHDO and CH_2O–CD_2O mixtures at xenon flash lamp intensities, *J. Am. Chem. Soc.*, *91*, 1590–1599.

Meagher, J.F., E.B. Cowling, F.C. Fehsenfeld, and W.J. Parkhurst (1998), Ozone formation and transport in southwestern United States: Overview of the SOS Nashville/Middle Tennessee ozone study, *J. Geophys. Res.*, *103*, 22213–22223.

Medhurst, L.J., J. Fleming, and H.H. Nelson (1997), Reaction rate constants of $OH + CHF_3 \rightarrow$ products and $O(^3P) + CHF_3 \rightarrow OH + CF_3$ at 500–750 K, *Chem. Phys. Lett.*, *266*, 607–611.

Meier, U., H.H. Grotheer, and Th. Just (1984), Temperature dependence and branching ratio of the $CH_3OH + OH$ reaction, *Chem. Phys. Lett.*, *106*, 97–101.

Meier, U., H.H. Grotheer, G. Riekert, and Th. Just (1985a), Study of hydroxyl reactions with methanol and ethanol by laser-induced fluorescence, *Ber. Bunsenges. Phys. Chem.*, *89*, 325–327.

Meier, U., H.H. Grotheer, G. Riekert, and Th. Just (1985b), Temperature dependence and branching ratio of the $C_2H_5OH + OH$ reaction, *Chem. Phys. Lett.*, *115*, 221–225.

Meier, U., H.H. Grotheer, G. Riekert, and Th. Just (1987), Reactions in a non-uniform flow tube temperature profile: effect on the rate coefficient for the reaction $C_2H_5OH + OH$, *Chem. Phys. Lett.*, *133*, 162–164.

Meller, R. and G.K. Moortgat (1997), $CF_3C(O)Cl$: Temperature-dependent (223–298 K) absorption cross-sections and quantum yields at 254 nm, *J. Photochem. Photobiol. A*, *108*, 105–116.

Meller, R., and G.K. Moortgat (2000), Temperature dependence of the absorption cross-sections of formaldehyde between 223–323 K in the wavelength range 225–375 nm, *J. Geophys. Res.*, *105*, 7089–7101.

Meller, R., D. Boglu, and G.K. Moortgat (1994), EUR 16171 EN, Editor K. H. Becker, *Tropospheric Oxidation Mechanisms* (Joint EC/EuroTrac/GDCU Workshop, LACTOZ–HALIPP, Leipzig, Sept. 20–22, 1994), pp. 183–190.

Mellouki, A. (1998), Kinetic studies of Cl atom reactions with series of alkanes using the pulsed laser photolyis–resonance fluorescence method, *J. Chim. Phys.*, *95*, 513–522.

Mellouki, A., G. Le Bras, and H. Sidebottom (2003), Kinetics and mechanisms of the oxidation of oxygenated organic compounds in the gas phase, *Chem. Rev.*, *103*, 5077–5096.

Mellouki, A., F. Oussar, X. Lun, and A. Chakir (2004), Kinetics of the reactions of the OH radical with 2-methyl-1-propanol, 3-methyl-1-butanol and 3-methyl-2-butanol between 241 and 373 K, *Phys. Chem. Chem. Phys.*, *6*, 2951–2955.

Mellouki, A., R.K. Talukdar, A.-M. Schmoltner, T. Gierczak, M.J. Mills, S. Solomon, and A.R. Ravishankara (1992), Atmospheric lifetimes and ozone depletion potentials of methyl bromide (CH_3Br) and dibromomethane (CH_2Br_2), *Geophys. Res. Lett.*, *19*, 2059–2062.

Mellouki, A., S. Téton, G. Laverdet, A. Quilgars, and G. Le Bras (1994), Kinetic studies of OH reactions with H_2O_2, C_3H_8 and CH_4 using the pulsed laser photolysis–laser induced fluorescence method, *J. Chim. Phys.*, *91*, 473–487.

Mellouki, A., S. Téton, and G. LeBras (1995), Rate constant for the reaction of OH radical with HFC-365mfc ($CF_3CH_2CF_2CH_3$), *Geophys. Res. Lett.*, *22*, 389–392.

Mendenhall, G.D., D.M. Golden, and S.W. Benson (1975), The very-low-pressure pyrolysis (VLPP) of *n*-propyl nitrate, *tert*-butyl nitrite, and methyl nitrite. Rate constants for some alkoxy radical reactions, *Int. J. Chem. Kinet.*, *7*, 725–737.

Méreau, R., M.-T. Rayez, F. Caralp, and J.-C. Rayez (2000a), Theoretical study on the comparative fate of the 1-butoxy and β-hydroxy-1-butoxy radicals, *Phys. Chem. Chem. Phys.*, *2*, 1919–1928.

Méreau, R., M.-T. Rayez, F. Caralp, and J.-C. Rayez (2000b), Theoretical study of alkoxyl radical decomposition reactions: Structure–activity relationships, *Phys. Chem. Chem. Phys.*, *2*, 3765–3772.

Méreau, R., M.T. Rayez, F. Caralp, and J.C. Rayez (2003), Isomerisation reactions of alkoxy radicals: theoretical study and structure–activity relationships, *Phys. Chem. Chem. Phys.*, *5*, 4828–4833.

Méreau, R., M.-T. Rayez, J.-C. Rayez, F. Caralp, and R. Lesclaux (2001), Theoretical study on the atmospheric fate of carbonyl radicals: kinetics of decomposition reactions, *Phys. Chem. Chem. Phys.*, *3*, 4712–4717.

Metcalfe, S.E., and R.G. Derwent (2005), *Atmospheric Pollution and Environmental Change. Key Issues in Environmental Change*, Hodder Arnold, London.

Metha, G.F., A.C. Terentis, and S.H. Kable (2002), Near threshold photochemistry of propanal, barrier height, transition state structure, and product state distributions for the HCO channel, *J. Phys. Chem. A*, *106*, 5817–5827.

Meunier, N., J.F. Doussin, E. Chevallier, R. Durand-Jolibois, B. Picquet-Varrault, and P. Carlier (2003), Atmospheric fate of alkoxy radicals: branching ratio of evolution pathways for 1-propoxy, 2-propoxy, 2-butoxy, and 3-pentoxy radicals, *Phys. Chem. Chem. Phys.*, *5*, 4834–4839.

Meyrahn, H., G.K. Moortgat, and P. Warneck (1982), The photolysis of acetaldehyde under atmospheric conditions, in *Atmospheric Trace Constituents*, edited by F. Herbert Friedr. pp. 04–71, Vieweg & Sohn, Braunschweig/Wiesbaden, and data tables provided to the authors by Dr. Moortgat.

Meyrahn, H., J. Pauly, W. Schneider, and P. Warneck (1986), Quantum yields for the photodissociation of acetone in air and an estimation for the life time of acetone in the lower troposphere, *J. Atmos. Chem.*, *4*, 277–291.

Michael, J.L., and W.A. Noyes, Jr. (1963), The photochemistry of mixtures of 2-pentanone and 2-hexanone with biacetyl, *J. Am. Chem. Soc.*, *85*, 1027–1032.

Michael, J.V., and J.H. Lee (1977), Selected rate constants for H, O, N, and Cl atoms with substrates at room temperatures, *Chem. Phys. Lett.*, *51*, 303–306.

Michael, J.V., D.G. Keil, and R.B. Klemm (1982), A resonance fluorescence kinetic study of oxygen atom + hydrocarbon reactions. V. $O(^3P)$ + neopentane (415–922 K), 19th *Symp. (Int.) Combust. [Proc.]*, 39–50.

Michael, J.V., D.G. Keil, and R.B. Klemm (1985), Rate constants for the reaction of hydroxyl radicals with acetaldehyde from 244–528 K, *J. Chem. Phys.*, *83*, 1630–1636.

Miller, C.E., J.I. Lynton, D.M. Keevil, and J.S. Francisco (1999), Dissociation pathways of peroxyacetyl nitrate (PAN), *J. Phys. Chem. A*, *103*, 11451–11459.

Miyoshi, A., K. Ohmori, K. Tsuchiya, and H. Matsui (1993), Reaction rates of atomic oxygen with straight chain alkanes and fluoromethanes at high temperatures, *Chem. Phys. Lett.*, *204*, 241–247.

Miyoshi, A., K. Tsuchiya, N. Yamauchi, and H. Matsui (1994), Reaction of atomic oxygen (^3P) with selected alkanes, *J. Phys. Chem.*, *98*, 11452–11458.

Møgelberg, T.E., M. Bilde, J. Sehested, T.J. Wallington, and O.J. Nielsen (1996a), Atmospheric chemistry of 1,1,1,2-tetrachloroethane (CCl_3CH_2Cl): Spectrokinetic investigation of the CCl_3CClHO_2 radical, its reactions with NO and NO_2, and atmospheric fate of the CCl_3CClHO radical, *J. Phys. Chem.*, *100*, 18399–18407.

Møgelberg, T.E., A. Feilberg, A.M.B. Giessing, J. Sehested, M. Bilde, T.J. Wallington, and O.J. Nielsen (1995a), Atmospheric chemistry of HFC-236cb: spectrokinetic investigation of the $CF_3CF_2CFHO_2$ radical, its reaction with NO and NO_2 and fate of the CF_3CF_2CFHO radical, *J. Phys. Chem.*, *99*, 17386–17393.

Møgelberg, T.E., O.J. Nielsen, J. Sehested, and T.J. Wallington (1995b), Atmospheric chemistry of HCFC-133a: the UV absorption spectra of CF_3CClH and CF_3CClHO_2 radicals, reactions of CF_3CClHO_2 with NO and NO_2, and fate of CF_3CClHO radicals, *J. Phys. Chem.*, *99*, 13437–13444.

Møgelberg, T.E., O.J. Nielsen, J. Sehested, T.J. Wallington, and M.D. Hurley (1994a), Atmospheric chemistry of CF_3COOH. Kinetics of the reaction with OH radicals, *Chem. Phys. Lett.*, *226*, 171–177.

Møgelberg, T.E., O.J. Nielsen, J. Sehested, T.J. Wallington, and M.D. Hurley (1995c), Atmospheric chemistry of HFC-272ca: spectrokinetic investigation of the $CH_3CF_2CH_2O_2$ radical, its reactions with NO and NO_2, and the fate of the $CH_3CF_2CH_2O$ radical, *J. Phys. Chem.*, *99*, 1995–2001.

Møgelberg, T.E., O.J. Nielsen, J. Sehested, T.J. Wallington, M.D. Hurley, and W.F. Schneider (1994b), Atmospheric chemistry of HFC-134a. Kinetic and mechanistic study of the CF_3CFHO_2+NO_2 reaction, *Chem. Phys. Lett.*, *225*, 375–380.

Møgelberg, T.E., J. Platz, O.J. Nielsen, J. Sehested, and T.J. Wallington (1995d), Atmospheric chemistry of HFC-236fa: spectrokinetic investigation of the $CF_3CHO_2CF_3$ radical, its reactions with NO and the fate of the CF_3CHOCF_3 radical, *J. Phys. Chem.*, *99*, 5373–5378.

Møgelberg, T.E., J. Sehested, M. Bilde, T.J. Wallington, and O.J. Nielsen (1996b), Atmospheric chemistry of HFC-227ea: spectrokinetic investigation of the $CF_3CFO_2(\bullet)CF_3$ radical, its reactions with NO and NO_2 and the atmospheric fate of the $CF_3CFO(\bullet)CF_3$ radical, *J. Phys. Chem.*, *100*, 8882–8889.

Møgelberg, T.E., J. Sehested, O.J. Nielsen, and T.J. Wallington (1995e), Atmospheric chemistry of pentachloroethane: absorption spectra for CCl_3CCl_2 and $CCl_3CCl_2O_2$ radicals, kinetics of the $CCl_3CCl_2O_2$ + NO reaction and fate of the CCl_3CCl_2O radical, *J. Phys. Chem.*, *99*, 16932–16938.

Møgelberg, T.E., J. Sehested, G.S. Tyndall, J.J. Orlando, J.M. Fracheboud, and T.J. Wallington (1997a), Atmospheric chemistry of HFC-236cb: fate of the alkoxy radical CF_3CF_2CFHO, *J. Phys. Chem. A*, *101*, 2828–2832.

Møgelberg, T.E., J. Sehested, T.J. Wallington, and O.J. Nielsen (1997b), Atmospheric chemistry of HFC-134a: kinetics of the decomposition of the alkoxy radical CF_3CFHO, *Int. J. Chem. Kinet.*, *29*, 209–217.

Mohan-Rao, A.M., G.G. Pandit, P. Sain, S. Sharma, T.M. Krishnamoorthy, and K.S.V. Nambi (1997), Non-methane hydrocarbons in industrial locations of Bombay, *Atmos. Environ.*, *31*, 1077–1085.

Molina, L.T., and M.J. Molina (1986), Absolute absorption cross sections of ozone in the 185–150 nm wavelength range, *J. Geophys. Res.*, *91*, 14501–14508.

Molina, M.J., and F.S. Rowland (1974), Stratospheric sink for chlorofluoromethanes: chlorine-atom catalyzed destruction of ozone, *Nature*, *249*, 810–814.

Moore, S.B., and R.W. Carr (1990), Kinetics of the reactions of chlorodifluoromethyldioxy radicals with nitrogen dioxide, *J. Phys. Chem.*, *94*, 1393–1400.

Moortgat, G.K., and P. Warneck (1979), CO and H_2 quantum yields in the photodecomposition of formaldehyde in air, *J. Chem. Phys.*, *70*, 3639–3651.

Moortgat, G.K., R.A. Cox, G. Schuster, J.P. Burrows, and G.S. Tyndall (1989), Peroxy radical reactions in the photo-oxidation of CH_3CHO, *J. Chem. Soc. Faraday Trans. 2*, *85*, 809–829.

Moortgat, G.K., W. Seiler, and P. Warneck (1983), Photodissociation of HCHO in air: CO and H_2 quantum yields at 220 and 300 K, *J. Chem. Phys.*, *78*, 1185–1190.

Morabito, P., and J. Heicklen (1987), The reactions of alkoxyl radicals with O_2. IV. n-C_4H_9O radicals, *Bull. Chem. Soc. Jpn.*, *60*, 2641–2650.

Morris, E.D., Jr., and H. Niki (1971a), Mass spectrometric study of the reaction of hydroxyl radical with formaldehyde, *J. Chem. Phys.*, *55*, 1991–1992.

Morris, E.D., Jr., and H. Niki (1971b), Reactivity of hydroxyl radicals with olefins, *J. Phys. Chem.*, *75*, 3640–3641.

Morris, E.D., Jr., and H. Niki (1974), Reaction of the nitrate radical with acetaldehyde and propylene, *J. Phys. Chem.*, *78*, 1337–1338.

Morris, E.D., Jr., D.H. Stedman, and H. Niki (1971), Mass spectrometric study of the reactions of the hydroxyl radical with ethylene, propylene, and acetaldehyde in a discharge-flow system, *J. Amer. Chem. Soc*, *93*, 3570–3572.

Morris, E.R., and J.C.J. Thynne (1968), Intramolecular elimination reactions in the photolysis of fluoroaldehydes, *J. Phys. Chem.*, *72*, 3351–3352.

Morrison, B.M., and J. Heicklen (1980), The reactions of HO with CH_2O and of HCO with NO_2, *J. Photochem.*, *13*, 189–199.

Morrissey, R.J., and C.C. Schubert (1963), The reactions of ozone with propane and ethane, *Combust. Flame*, *7*, 263–268.

Mörs, V., A. Hoffmann, W. Malms, and R. Zellner (1996), Time resolved studies of intermediate products in the oxidation of HCFC 141b ($CFCl_2CH_3$) and HCFC 142b (CF_2ClCH_3), *Ber. Bunsenges. Phys. Chem.*, *100*, 540–552.

Moschonas, N., and S. Glavas (1996), C_3–C_{10} hydrocarbons in the atmosphere of Athens, Greece, *Atmos. Environ.*, *30*, 2769–2772.

Müller-Remmers, P.L., P.C. Mishra, and K. Jug (1984), A SINDO1 study of photoisomerization and photofragmentation of cyclopentanone, *J. Am. Chem. Soc.*, *108*, 2538–2543.

Mund, Ch., Ch. Fockenberg, and R. Zellner (1998), LIF spectra of n-propoxy and i-propoxy radicals and kinetics of their reactions with O_2 and NO_2, *Ber. Bunsen-Ges. Phys. Chem.*, *102*, 709–715.

Mund, Ch., Ch. Fockenberg, and R. Zellner (1999), Correction to LIF spectra of *n*-propoxy and *i*-propoxy radicals and kinetics of their reactions with O_2 and NO_2, *Phys. Chem. Chem. Phys.*, *1*, 2037.

Nakashima, K., K. Uchida-Kai, M. Koyanogi, and Y. Kanda (1982), Solvent effects on the intensities of forbidden bands of molecules. Absorption spectra of acetone and cyclopentanone, *Bull. Chem. Soc. Jpn.*, *55*, 415–419.

Nau, W.M., and J.C. Scaiano (1996), Oxygen quenching of excited aliphatic ketones and diketones, *J. Phys. Chem.*, *100*, 11360–11367.

Neckel, H., and D. Labs (1984), The solar radiation between 3300 and 12500 Å, *Solar Phys.*, *90*, 205–258.

Neeb, P. (2000), Structure–reactivity based estimation of the rate constants for hydroxyl radical reactions with hydrocarbons, *J. Atmos. Chem.*, *35*, 295–315.

Nelson, D.D., Jr., J.C. Wormhoudt, M.S. Zahniser, C.E. Kolb, M.K.W. Ko, and D.K. Weisenstein (1997), OH reaction kinetics and atmospheric impact of 1-bromopropane, *J. Phys. Chem. A*, *101*, 4987–4990.

Nelson, D.D., Jr., M.S. Zahniser, and C.E. Kolb (1992), Chemical kinetics of the reactions of the OH radical with several hydrochlorofluorocarbons, *J. Phys. Chem.*, *96*, 249–253.

Nelson, D.D., Jr., M.S. Zahniser, and C.E. Kolb (1993), OH reaction kinetics and atmospheric lifetimes of CF_3CFHCF_3 and CF_3CH_2Br, *Geophys. Res. Lett.*, *20*, 197–200.

Nelson, D.D., Jr., M.S. Zahniser, C.E. Kolb, and H. Magid (1995), OH reaction kinetics and atmospheric lifetime estimates for several hydrofluorocarbons, *J. Phys. Chem.*, *99*, 16301–16306.

Nelson, L., O. Rattigan, R. Neavyn, H. Sidebottom, J. Treacy, and O.J. Nielsen (1990a), Absolute and relative rate constants for the reactions of hydoxyl radicals and chlorine atoms with a series of aliphatic alcohols and ethers at 298 K, *Int. J. Chem. Kinet.*, *22*, 1111–1126.

Nelson, L., I. Shanahan, H.W. Sidebottom, J. Treacy, and O.J. Nielsen (1990b), Kinetics and mechanism for the oxidation of 1,1,1-trichloroethane, *Int. J. Chem. Kinet. 22*, 577–590.

Nelson, L., J.J. Treacy, and H.W. Sidebottom (1984), Oxidation of methylchloroform, *Phys. Chem. Behav. Atmos. Pollut. Proc. Eur. Symp.*, 258–263.

Nesbitt, D.J., and S.R. Leone (1982), Laser-initiated Cl_2/hydrocarbon chain reactions: Time-resolved infrared emission spectra of product vibrational excitation, *J. Phys. Chem.*, *86*, 4962–4973.

Newell, R.E., J.W. Kidson, D.G. Vincent, and G.J. Boer (1972), *The General Circulation of the Tropical Atmosphere*, MIT Press, Cambridge, MA.

Nicholson, A.J.C. (1954), The photochemical decomposition of the aliphatic methyl ketones, *Trans. Faraday Soc.*, *50*, 1067–1073.

Nicol, C.H., and J.G. Calvert (1967), Relations between photodecomposition modes and molecular structure in the series of carbonyl compounds, *n*-C_3H_7COR, *J. Am. Chem. Soc.*, *89*, 1790–1798.

Nicolet, M., (1977), *Etude des reactions chimiques de l'ozone dans la stratosphere*, Institut Royal Météorologique de Belgique, Brussels.

Nielsen, O.J. (1991), Rate constants for the gas-phase reactions of OH radicals with CH_3CHF_2 and $CHCl_2CF_3$ over the temperature range 295–388 K, *Chem. Phys. Lett.*, *187*, 286–290.

Nielsen, O.J., T. Ellermann, E. Bartkiewicz, T.J. Wallington, and M.D. Hurley (1992a), UV absorption spectra, kinetics, and mechanisms of the self reaction of CHF_2O_2 radicals in the gas phase at 298 K, *Chem. Phys. Lett.*, *192*, 82–88.

Nielsen, O.J., T. Ellermann, J. Sehested, E. Bartkiewicz, T.J. Wallington and M.D. Hurley (1992b), UV absorption spectra, kinetics and mechanisms of the self reaction of CF_3O_2 radicals in the gas phase at 298 K, *Int. J. Chem. Kinet.*, *24*, 1009–1021.

Nielsen, O.J., T. Ellermann, J. Sehested, and T.J. Wallington (1992c), UV absorption spectrum, and kinetics and mechanism of the self reaction of $CHF_2CF_2O_2$ radicals in the gas phase at 298 K, *J. Phys. Chem.*, *96*, 10875–10879.

Nielsen, O.J., E. Gamborg, J. Sehested, T.J. Wallington, and M.D. Hurley (1994), Atmospheric chemistry of HFC-143a: spectrokinetic investigation of the CF_3CH_2O radical, its reactions with NO and NO_2, and the fate of CF_3CH_2O, *J. Phys. Chem.*, *98*, 9518–9521.

Nielsen, O.J., J. Munk, P. Pagsberg, and A. Sillesen (1986), Absolute rate constants for the gas-phase reaction of OH radicals with cyclohexane and ethane at 295 K, *Chem. Phys. Lett.*, *128*, 168–171.

Nielsen, O.J., H.W. Sidebottom, M. Donlon, and J. Treacy (1991), An absolute- and relative-rate study of the gas-phase reaction of OH radicals and Cl atoms with *n*-alkyl nitrates, *Chem. Phys. Lett.*, *178*, 163–170.

Niki, H., P.D. Maker, C.M. Savage, and L.P. Breitenbach (1978), Relative rate constants for the reaction of hydroxyl radical with aldehydes, *J. Phys. Chem.*, *82*, 132–134.

Niki, H., P.D. Maker, C.M. Savage, and L.P. Breitenbach (1980), An FTIR (Fourier transform IR) study of the chlorine atom-initiated oxidation of dichloromethane and chloromethane, *Int. J. Chem. Kinet.*, *12*, 1001–1012.

Niki, H., P.D. Maker, C.M. Savage, and L.P. Breitenbach (1981a), Fourier transform infrared studies of the self-reaction of CH_3O_2 radicals, *J. Phys. Chem.*, *85*, 877–881.

Niki, H., P.D. Maker, C.M. Savage, and L.P. Breitenbach (1981b), An FT IR study of the isomerization and O_2 reaction of *n*-butoxy radicals, *J. Phys. Chem.*, *85*, 2698–2700.

Niki, H., P.D. Maker, C.M. Savage, and L.P. Breitenbach (1982), Fourier transform infrared studies of the self-reaction of $C_2H_5O_2$ radicals, *J. Phys. Chem.*, *86*, 3825–3829.

Niki, H., P.D. Maker, C.M. Savage, and L.P. Breitenbach (1983), A Fourier transform infrared study of the kinetics and mechanism for the reaction HO + CH_3OOH, *J. Phys. Chem.*, *87*, 2190–2193.

Niki, H., P.D. Maker, C.M. Savage, and L.P. Breitenbach (1984), Fourier transform infrared study of the kinetics and mechanisms for the reaction of hydroxyl radical with formaldehyde, *J. Phys. Chem.*, *88*, 5342–5344.

Nip, W.S., D.L. Singleton, R. Overend, and G. Paraskevopoulos (1979), Rates of hydroxyl radical reactions. 5. Reactions with fluoromethane, difluoromethane, trifluoromethane, fluoroethane, and 1,1-difluoroethane at 297 K, *J. Phys. Chem.*, *83*, 2440–2443.

Nishida, S., K. Takahashi, Y. Matsumi, M. Chiappero, G. Argüello, T.J. Wallington, M.D. Hurley, and J.C. Ball (2004), CF_3ONO_2 yield in the gas phase reaction of CF_3O_2 radicals with NO in 20–700 Torr of air at 296 K, *Chem. Phys. Lett.*, *388*, 242–247.

NIST, Chemical Kinetics Database: Version 2Q98 [http://kinetics.nist.gov].

Nizkorodov, S.A., J.D. Crounse, J.L. Fry, C.M. Roehl, and P.O. Wennberg (2005), Near-IR photodissociation of peroxy acetyl nitrate, *Atm. Chem. Phys.*, *5*, 385–392.

Noble, M., and E.K.C. Lee (1984a), Mode specificity and energy-dependence of the dynamics of single rovibrational levels of S_1 acetaldehyde in a supersonic jet, *J. Chem. Phys.*, *80*, 134–139.

Noble, M., and E.K.C. Lee (1984b), The jet-cooled electronic-spectrum of acetaldehyde and deuterated derivatives at rotational resolution, *J. Chem. Phys.*, *81*, 1632–1642.

Nolting, F., W. Behnke, and C. Zetzsch (1988), A smog chamber for studies of the reactions of terpenes and alkanes with ozone and OH, *J. Atmos. Chem.*, *6*, 47–59.

Norrish, R.G.W., and M.E.S. Appleyard (1934), Primary photochemical reactions. Part IV. Decomposition of methyl ethyl ketone and methyl butyl ketone, *J. Chem. Soc.*, 874–880.

North, S.W., D.A. Blank, J.D. Gezelter, C.A. Longfellow, and Y.T. Lee (1995), Evidence for stepwise dissociation dynamics in acetone at 248 and 193 nm, *J. Chem. Phys.*, *102*, 4447–4460.

Noyes, W.A., Jr., and I. Unger (1966), Singlet and triplet states: Benzene and simple aromatic compounds, *Adv. Photochem.*, *4*, 49–79.

Noyes, W.A., Jr., G.B. Porter, and J.E. Jolley (1956), The primary photochemical process in simple ketones, *Chem. Rev.*, *56*, 49–94.

Odman, T., and C.L. Ingram (1996), Multi-scale Air Quality Simulation Platform (MAQSIP): Source code, documentation and validation, MCNC Technical Report ENV-96TR002-v1.0, Research Triangle Park, North Carolina.

Oh, S., and J.M. Andino (2000), Effects of ammonium sulfate aerosols on the gas-phase reactions of the hydroxyl radical with organic compounds, *Atmos. Environ.*, *34*, 2901–2908.

Oh, S., and J.M. Andino (2001), Kinetics of the gas-phase reactions of hydroxyl radicals with C_1–C_6 aliphatic alcohols in the presence of ammonium sulfate aerosols, *Int. J. Chem. Kinet.*, *33*, 422–430.

Oh, S., and J.M. Andino (2002), Laboratory studies of the impact of aerosol composition on the heterogeneous oxidation of 1-propanol, *Atmos. Environ.*, *36*, 149–156.

Ohta, T., and I. Mizoguchi (1980), Determination of the mechanism of the oxidation of the trichloromethyl radical by the photolysis of chloral in the presence of oxygen, *Int. J. Chem. Kinet.*, *12*, 717–727.

Okabe, H. (1977), Photodissociation of thiophosgene, *J. Chem. Phys.*, *66*, 2058–2062.

Okabe, H., and D.A. Becker (1963), Vacuum ultraviolet photochemistry. VII. Photolysis of *n*-butane, *J. Chem. Phys.*, *39*, 2549–2555.

Olkhov, R.V., and I.W.M. Smith (2003), Time-resolved experiments on the atmospheric oxidation of C_2H_6 and some C_2 hydrofluorocarbons, *Phys. Chem. Chem. Phys.*, *5*, 3436–3442.

Olsen, J.F., S. Kang, and L. Burnelle (1971), The question of an open-ring form of cyclopropanone, *J. Mol. Struct.*, *9*, 305–313.

O'rji, L.N., and D.A. Stone (1992), Relative rate constant measurements for the gas-phase reactions of hydroxyl radicals with 4-methyl-2-pentanone, *trans*-4-octene, and *trans*-2-heptene, *Int. J. Chem. Kinet.*, *24*, 703–710.

Orkin, V.L., and V.G. Khamaganov (1993a), Determination of rate constants for reactions of some hydrohaloalkanes with OH radicals and their atmospheric lifetimes, *J. Atmos. Chem.*, *16*, 157–167.

Orkin, V.L., and V.G. Khamaganov (1993b), Rate constants for reactions of OH radicals with some Br-containing haloalkanes, *J. Atmos. Chem.*, *16*, 169–178.

Orkin, V.L., R.E. Huie, and M.J. Kurylo (1996), Atmospheric lifetimes of HFC-143a and HFC-245fa: flash photolysis resonance fluorescence measurements of the OH reaction rate constants, *J. Phys. Chem.*, *100*, 8907–8912.

Orkin, V.L., V.G. Khamaganov, A.G. Guschin, R.E. Huie, and M.J. Kurylo (1997), Atmospheric fate of chlorobromomethane: rate constant for the reaction with OH, UV spectrum, and water solubility, *J. Phys. Chem. A*, *101*, 174–178.

Orlando, J.J. (2003), Atmospheric chemistry of organic bromine and iodine compounds, in *The Handbook of Environmental Chemistry*, Volume 3R, edited by Alasdair H. Neilson, pp. 253–299, Springer-Verlag GmbH.

Orlando, J.J., and G.S. Tyndall (2002a), Oxidation mechanisms for ethyl chloride and ethyl bromide under atmospheric conditions, *J. Phys.Chem. A*, *106*, 312–319.

Orlando, J.J., and G.S. Tyndall (2002b), Mechanisms for the reactions of OH with two unsaturated aldehydes: crotonaldehyde and acrolein, *J. Phys. Chem. A*, *106*, 12252–12259.

Orlando, J.J., L.T. Iraci, and G.S. Tyndall (2000), Chemistry of the cyclopentoxy and cyclohexoxy radicals at subambient temperatures, *J. Phys. Chem. A*, *104*, 5072–5079.

Orlando, J.J., C.A. Piety, J.M. Nicovich, M.L. McKee, and P.H. Wine (2005), Rates and mechanisms for the reactions of chlorine atoms with iodoethane and 2-iodopropane, *J. Phys. Chem. A*, *109*, 6659–6675.

Orlando, J.J., G.S. Tyndall, E.C. Apel, D.D. Riemer, and S.E. Paulson (2003a), Rate coefficients and mechanisms of the reaction of Cl-atoms with a series of unsaturated hydrocarbons under atmospheric conditions, *Int. J. Chem. Kinet.*, *35*, 334–353.

Orlando, J.J., G.S. Tyndall, M. Bilde, C. Ferronato, T.J. Wallington, L. Vereecken, and J. Peeters (1998), Laboratory study of the mechanism of OH- and Cl-initiated oxidation of ethylene, *J. Phys. Chem. A*, *102*, 8116–8123.

Orlando, J.J., G.S. Tyndall, J.M. Fracheboud, E.G. Estupiñan, S. Haberkorn, and A. Zimmer (1999a), The rate and mechanism of the gas-phase oxidation of hydroxyacetone, *Atmos. Environ.*, *33*, 1621–1629.

Orlando, J.J., G.S. Tyndall, and S.E. Paulson (1999b), Mechanism of the OH-initiated oxidation of methacrolein, *Geophys. Res. Lett.*, *26*, 2191–2194.

Orlando, J.J., G.S. Tyndall, and T.J. Wallington (1996a), The atmospheric oxidation of CH_3Br: chemistry of the CH_2BrO radical, *J. Phys. Chem.*, *100*, 7026–7033.

Orlando, J.J., G.S. Tyndall, and T.J. Wallington (2003b), The atmospheric chemistry of alkoxy radicals, *Chem. Rev.*, *103*, 4657–4689.

Orlando, J.J., G.S. Tyndall, T.J. Wallington, and M. Dill (1996b), Atmospheric chemistry of CH_2Br_2: rate coefficients for its reaction with Cl atoms and OH radicals and the chemistry of the $CHBr_2O$ radical, *Int. J. Chem. Kinet.*, *28*, 433–442.

Overend, R., and G. Paraskevopoulos (1978), Rates of OH radical reactions. 4. Reactions with methanol, ethanol, 1-propanol, and 2-propanol at 296 K, *J. Phys. Chem*, *82*, 1329–1333.

Pagsberg, P., J. Munk, A. Sillesen, and C. Anastasi (1988), UV spectrum and kinetics of hydroxymethyl radicals, *Chem. Phys. Lett.*, *146*, 375–381.

Papagni, C., J. Arey, and R. Atkinson (2000), Rate constants for the gas-phase reactions of a series of C_3–C_6 aldehydes with OH and NO_3 radicals, *Int. J. Chem. Kinet.*, *32*, 79–84.

Paraskevopoulos, G., D.L. Singleton, and R.S. Irwin (1981), Rates of OH radical reactions. 8. Reactions with CH_2FCl, CHF_2Cl, $CHFCl_2$, CH_3CF_2Cl, CH_3Cl, and C_2H_5Cl at 297 K, *J. Phys. Chem.*, *85*, 561–564.

Parkes, D.A. (1974), The roles of alkylperoxy and alkoxy radicals in alkyl radical oxidation at room temperature, in *Proceedings of the 15th International Symposium on Combustion*, pp. 795–804, The Combustion Institute, Pittsburgh, PA.

Parkes, D.A. (1977), The oxidation of methyl radicals at room temperature, *Int. J. Chem. Kinet.*, *9*, 451–469.

Parkes, D.A., D.M. Paul, C.P. Quinn, and R.C. Robson (1973), The ultraviolet absorption by alkylperoxy radicals and their mutual reactions, *Chem. Phys. Lett.*, *23*, 425–429.

Parmar, S.S., and S.W. Benson (1989), Kinetics and thermochemistry of the reaction $C_2D_6 + Cl \leftrightarrow C_2D_5 + DCl$/ The heat of formation of the C_2D_5 and C_2H_5 radicals, *J. Am. Chem. Soc.*, *111*, 57–61.

Parmenter, C.S., and W.A. Noyes, Jr. (1963), Energy dissipation from excited acetaldehyde molecules, *J. Am. Chem. Soc.*, *85*, 416–421.

Parrish, D.D., M. Trainer, V. Young, P.D. Goldan, W.C. Kuster, B.T. Jobson, F.C. Fehsenfeld, W.A. Lonnemann, R.D. Zika, C.T. Farmer, D.D. Riemer, and M.O Rodgers (1998), Internal consistency tests for evaluation of measurements of anthropogenic hydrocarbons in the troposphere, *J. Geophys. Res.*, *103*, 22239–22359.

Passant, N.R. (2002), Speciation of UK emissions of non-methane volatile organic compounds, AEA Technology Report AEAT/ENV/R/0545 Issue 1, Culham, UK.

Pate, C.T., R. Atkinson, and J.N. Pitts, Jr. (1976), Rate constants for the gas phase reaction of peroxyacetyl nitrate with selected atmospheric constituents, *J. Environ. Sci. Health, Part A*, *11*, 19–31.

Pearce, C., and D.A. Whytock (1971), The photolysis of trifluoroacetaldehyde at 313 nm, *J. Chem. Soc., Chem. Commun.*, 1464–1466.

Pearson, G.S. (1963), The photooxidation of acetone, *J. Phys. Chem.*, *67*, 1686–1692.

Peeters, J., and V. Pultau (1992), Rate constant of CF_3CFHO_2 + NO, *Proceedings of CEC/EUROTRAC Workshop on Chemical Mechanisms Describing Tropospheric Processes*, September 1992, edited by J. Peeters, pp. 225–230.

Peeters, J., and V. Pultau (1993), Reactions of hydrofluorocarbons and hydrochlorofluorocarbons derived peroxy radicals with nitric oxide. Results for CF_3CH_2F (HFC 134a), CF_3CH_3 (HFC 143a), CF_2ClCH_3 (HCFC 142b), CF_3CCl_2H (HCFC 123) and CF_3CHFCl (HCFC 124), *Proceedings of the Sixth European Symposium on Physico-chemical Behaviour of Atmospheric Pollutants, Varese (Italy)* Oct. 1993, edited by G. Angeletti and G. Restelli, European Commission Publications, pp. 372–378.

Peeters, J., G. Fantechi, and L.J. Vereecken (2004), A generalized structure–activity relationship for the decomposition of (substituted) alkoxy radicals, *J. Atm. Chem.*, *48*, 59–80.

Peeters, J., L.Vereecken, and G. Fantechi (2001), The detailed mechanism of the OH-initiated atmospheric oxidation of α-pinene: a theoretical study, *Phys. Chem. Chem. Phys.*, *3*, 5489–5504.

Peeters, J., J. Vertommen, and I. Langhans (1992), Rate constants of the reactions of CF_3O_2, i-$C_3H_7O_2$, and t-$C_4H_9O_2$ with NO, *Ber. Bunsenges. Physik.*, *96*, 431–436.

Percival, C.J., G. Marston, and R.P. Wayne (1995), Correlations between rate parameters and calculated molecular properties in the reactions of the hydroxyl radical with hydrofluorocarbons, *Atmos. Environ.*, *29*, 305–311.

Perry, R.A., R. Atkinson, and J.N. Pitts, Jr. (1976), Rate constants for the reaction of OH radicals with $CHFCl_2$ and CH_3Cl over the temperature range 298–423 K, and with CH_2Cl_2 at 298°K, *J. Chem. Phys.*, *64*, 1618–1620.

Person, M.D., P.W. Kash, and L.J. Butler (1992a), A new class of Norrish type I processes: α-bond cleavage upon $[n,\pi^*$ (C=O)] excitation in the acid halides, *J. Phys. Chem.*, *96*, 2021–2023.

Person, M.D., P.W. Kash, and L.J. Butler (1992b), Nonadiabaticity and the competition between alpha and beta bond fission upon $^1[n,\pi^*(C{=}O)]$ excitation in acetyl and bromoacetyl chloride, *J. Chem. Phys.*, *97*, 355–373.

Picquet, B., S. Heroux, A. Chebbi, J.-F. Doussin, R. Durand-Jolibois, A. Monod, H. Loirat, and P. Carlier (1998), Kinetics of the reactions of OH radicals with some oxygenated volatile organic compounds under simulated atmospheric conditions, *Int. J. Chem. Kinet.*, *30*, 839–847.

Piety, C.A., R. Soller, J.M. Nicovich, M.L. McKee, and P.H. Wine (1998), Kinetic and mechanistic study of the reaction of atomic chlorine with methyl bromide over an extended temperature range, *Chem. Phys.*, *231*, 155–169.

Pilgrim, J. S., A. McIlroy, and C. A. Taatjes (1997), Kinetics of Cl atom reactions with methane, ethane, and propane from 292 to 800 K, *J. Phys. Chem. A*, *101*, 1873–1880.

Pinho, P.G., C.A. Pio, and M.E. Jenkin (2005), Evaluation of isoprene degradation in the detailed tropospheric chemical mechanism, MCM v3, using environmental chamber data, *Atmos. Environ.*, *39*, 1303–1322.

Pitari, G., E. Mancini, V. Rizi, and D.T. Shindell (2002), Impact of future climate and emissions changes on stratospheric aerosols and ozone, *J. Atmos. Sci.*, *59*, 414–440.

Pitts, J.N., Jr., and F.E. Blacet (1950), Methyl ethyl ketone photochemical processes, *J. Am. Chem. Soc.*, *72*, 2810–2811.

Pitts, J.N., Jr., and F.E. Blacet (1952), The vapor-phase photolysis of acetone–iodine mixtures, *J. Am. Chem. Soc.*, *74*, 455–457.

Pitts, J.N., Jr., and A.D. Osborne (1961), A comparative study of radiolysis, photolysis and electron impact data of ketones, Chapter 8 in *Chemical Reactions in the Lower Atmosphere*, pp. 129–138, Interscience, New York.

Pitts, J.N., Jr., and J.K.S. Wan (1966), Photochemistry of ketones and aldehydes, Chapter 16 in *Chemistry of the Carbonyl Group*, edited by S. Patai, pp. 823–915, Interscience, New York.

Pitts, J.N., Jr., D.R. Burley, J.C. Mani, and A.D. Broadbent (1968), Molecular structure and photochemical reactivity. VIII. Type II photoelimination of alkenes from alkyl phenyl ketones, *J. Am. Chem. Soc.*, *90*, 5900–5902.

Plagens, H., R. Bröske, M. Splittler, L. Ruppert, I. Barnes, and K.H. Becker (1998), Atmospheric loss processes of hexanal. Photolysis and reaction with OH and Cl radicals, in *Proceedings of the Second Workshop of the Eurotrac-2 Subproject CMD, Chemical Mechanism Development*, 23–25 September, Karlsruhe.

Platz, J., O.J. Nielsen, J. Sehested, and T.J. Wallington (1995), Atmospheric chemistry of 1,1,1-trichloroethane: UV spectra and self-reaction kinetics of CCl_3CH_2 and $CCl_3CH_2O_2$ radicals, kinetics of the reaction of $CCl_3CH_2O_2$ radicals with NO and NO_2, and the fate of the alkoxy radical CCl_3CH_2O, *J. Phys. Chem.*, *99*, 6570–6579.

Platz, J., J. Sehested, O.J. Nielsen, and T.J. Wallington (1999), Atmospheric chemistry of cyclohexane: UV spectra of c-C_6H_{11} and $(c$-$C_6H_{11})O_2$ radicals, kinetics of the reactions of $(c$-$C_6H_{11})O_2$ radicals with NO and NO_2, and the fate of the alkoxy radical $(c$-$C_6H_{11})O$, *J. Phys. Chem. A*, *103*, 2688–2695.

Plumb, I.C., and K.R. Ryan (1982), Kinetics of the reaction of CF_3O_2 with NO, *Chem. Phys. Lett.*, *92*, 236–238.

Plumb, I.C., K.R. Ryan, J.R. Steven, and M.F.R. Mulcahy (1979), Kinetics of the reaction of CH_3O_2 with NO, *Chem. Phys. Lett.*, *63*, 255–258.

Plumb, I.C., K.R. Ryan, J.R. Steven, and M.F.R. Mulcahy (1981), Rate coefficient for the reaction of CH_3O_2 with NO at 295 K, *J. Phys. Chem.*, *85*, 3136–3138.

Plumb, I.C., K.R. Ryan, J.R. Steven, and M.F.R. Mulcahy (1982), Kinetics of the reaction of $C_2H_5O_2$ with NO at 295 K, *Int. J. Chem. Kinet.*, *14*, 183–194.

Pope, F.D., C.A. Smith, M.N.R. Ashfold, and A.J. Orr-Ewing (2005a), High-resolution absorption cross sections of formaldehyde at wavelengths 313 to 320 nm, *Phys. Chem. Chem. Phys.*, *7*, 79–84.

Pope, F.D., C.A. Smith, P.R. Davis, D.E. Shallcross, M.N.R. Ashfold, and A.J. Orr-Ewing (2005b), Photochemistry of formaldehyde under tropospheric conditions, *Faraday Disc.*, *130*, 59–72.

Poulet, G., G. Le Bras, and J. Combourieu (1974), Etude cinétique des réactions du chlore atomique et du radical ClO avec le méthane par la technique du réacteur à écoulement rapide, couplé à un spectrométre de masse, *J. Chim. Phys.*, *71*, 101–106.

Prather, M.J. (2002), Lifetimes of atmospheric species: Integrating environmental impacts, *Geophys. Res. Lett.*, *29*, 2063, doi:10.1029/2002GL016299.

Pritchard, H.O., J.B. Pyke, and A.F. Trotman-Dickenson (1954), A method for the study of chlorine atom reactions. The reaction $Cl + CH_4 \rightarrow CH_3 + HCl$, *J. Am. Chem. Soc.*, *76*, 1201–1202.

Pritchard, H.O., J.B. Pyke, and A.F. Trotman-Dickenson (1955), The study of chlorine atom reactions in the gas phase, *J. Am. Chem. Soc.*, *77*, 2629–2633.

Qi, B., Y.H. Zhang, Z.M. Chen, K.S. Shao, X.Y. Tang, and M. Hu (1999), Identification of organic peroxides in the oxidation of C_1–C_3 alkanes, *Chemosphere*, *38*, 1213–1221.

Qian, H.-B., D. Turton, P.W. Seakins, and M.J. Pilling (2002), A laser flash photolysis/IR diode laser absorption study of the reaction of chlorine atoms with selected alkanes, *Int. J. Chem. Kinet.*, *34*, 86–94.

Qiu, L.X., S.-H. Shi, S.-B. Xing, and X.G. Chen (1992), Rate constants for the reactions of OH with five halogen-substituted ethanes from 292 to 366 K, *J. Phys. Chem.*, *96*, 685–689.

Raber, W.H. (1992), Ph.D. Thesis, Johannes Gutenberg-Universität, Mainz, Germany.

Raber, W.H., and G.K. Moortgat (1996), Photooxidation of selected carbonyl compounds in air: methyl ethyl ketone, methyl vinyl ketone, and methacrolein and methylglyoxal, Chapter 9 in *Progress and Problems in Atmospheric Chemistry*, edited by J. Barker, pp. 328–373, World Scientific, Singapore.

Raff, J.D., P.S. Stevens, and R.A. Hites (2005), Relative rate and product studies of the OH–acetone reaction, *J. Phys. Chem. A*, *109*, 4728–4735.

Rajakumar, B., R.W. Portmann, J.B. Burkholder, and A.R. Ravishankara (2006), Rate coefficients for the reactions of OH with $CF_3CH_2CH_3$, CF_3CHFCH_2F and $CHF_2CHFCHF_2$ between 238 and 375 K, *J. Phys. Chem. A*, *110*, 6724–6731.

Rakovski, S., and D. Cherneva (1990), Kinetics and mechanism of the reaction of ozone with aliphatic alcohols, *Int. J. Chem. Kinet.*, *22*, 321–329.

Ramacher, B., J.J. Orlando, and G.S. Tyndall (2001), Temperature-dependent rate coefficient measurements for the reaction of bromine atoms with trichloroethene, ethene, acetylene, and tetrachloroethene in air, *Int. J. Chem. Kinet.*, *33*, 198–211.

Ranschaert, D.L., N.J. Schneider, and M.J. Elrod (2000), Kinetics of the $C_2H_5O_2 + NO_x$ reactions: Temperature dependence of the overall rate constant and the $C_2H_5ONO_2$ branching channel of $C_2H_5O_2 + NO$, *J. Phys. Chem. A*, *104*, 5758–5765.

Rattigan, O., E. Lutman, R.L. Jones, R.A. Cox, K. Clemitshaw, and J. Williams (1992), Temperature-dependent absorption cross-sections of gaseous nitric acid and methyl nitrate, *J. Photochem. Photobiol. A: Chemistry*, *66*, 313–326.

Rattigan, O.V., D.M. Rowley, O. Wild, R.L. Jones, and R.A. Cox (1994), Mechanism of atmospheric oxidation of 1,1,1,2-tetrafluoroethane, *J. Chem. Soc. Faraday Trans.*, *90*, 1819–1829.

Rattigan, O.V., D.E. Shallcross, and R.A. Cox (1997), UV absorption cross-sections and atmospheric photolysis rates of CF_3I, CH_3I, C_2H_5I and CH_2ICl, *J. Chem. Soc., Faraday Trans.*, *93*, 2839–2846.

Rattigan, O.V., O. Wild, and R.A. Cox (1998), UV absorption cross-sections and atmospheric photolysis lifetimes of halogenated aldehydes, *J. Photochem. Photobiol. A: Chemistry*, *112*, 1–7.

Ravishankara, A.R., and D.D. Davis (1978), Kinetic rate constants for the reaction of OH with methanol, ethanol, and tetrahydrofuran at 298 K, *J. Phys. Chem.*, *82*, 2852–2853.

Ravishankara, A.R., and P.H. Wine (1980), A laser flash photolysis–resonance fluorescence kinetics study of the reaction $Cl(^2P) + CH_4 \rightarrow CH_3 + HCl$, *J. Chem. Phys.*, *72*, 25–30.

Ravishankara, A.R., F.L. Eisele, N.M. Kreutter, and P.H. Wine (1981), Kinetics of the reaction of CH_3O_2 with NO, *J. Chem. Phys.*, *74*, 2267–2274.

Ravishankara, A.R., F.L. Eisele, and P.H. Wine (1980), Pulsed laser photoysis–long path laser absorption kinetics study of the reaction of methylperoxy radcials with NO_2, *J. Chem. Phys.*, *73*, 3743–3749.

Ravishankara, A.R., S. Solomon, A.A.Turnipseed, and R.F. Warren (1993), Atmospheric lifetimes of long-lived halogenated species, *Science*, *259*, 194–199.

Ravishankara, A.R., A.A. Turnipseed, N.R. Jensen, S. Barone, M. Mills, C.J. Howard, and S. Solomon (1994), Do hydrofluorocarbons destroy stratospheric ozone?, *Science*, *263*, 71–75.

Ray, G.W., L.F. Keyser, and R.T. Watson (1980), Kinetics study of the Cl (^2P) + $Cl_2O \rightarrow Cl_2$ + ClO reaction at 298 K, *J. Phys. Chem.*, *84*, 1674–1681.

Rayez, M.T., J.C. Rayez, T. Berces, and G. Lendvay (1993), Theoretical study of the reactions of OH radicals with substituted acetaldehydes, *J. Phys. Chem.*, *97*, 5570–5576.

Rebbert, R.E., and P. Ausloos (1967), Comparison of the direct and sensitized photolysis of 3-methylpentanal in the vapor phase, *J. Am. Chem. Soc.*, *89*, 1573–1579.

Reisen, F., S.M. Aschmann, R. Atkinson, and J. Arey (2005), 1,4-Hydroxycarbonyl products of the OH radical initiated reactions of C_5–C_8 *n*-alkanes in the presence of NO, *Environ. Sci. Technol.*, *39*, 4447–4453.

Richter, H.R., J.R. Sodeau, and I. Barnes (1993), The photolysis and chlorine-initiated photo-oxidation of trifluoroacetaldehyde, STEP-HALOLSIDE/AFEAS Workshop, Dublin, 23–25 March, pp. 182–188.

Riley, S.J., and K.R. Wilson (1972), Excited fragments from excited molecules: Energy partitioning in the photodissociation of alkyl iodides, *Faraday Discuss. Chem. Soc.*, *53*, 132–146.

Rinsland, C.P., R. Zander, L.R. Brown, C.B. Farmer, J.H. Park, R.H. Norton, J.M. Russell III, and O.F. Raper (1986), Detection of carbonyl fluoride in the stratosphere, *Geophys. Res. Lett.*, *13*, 769–772.

Roberts, J.M. (1990), The atmospheric chemistry of organic nitrates, *Atmos. Environ.*, *24A*, 243–287.

Roberts, J.M., and R.W. Fajer (1989), UV absorption cross sections of organic nitrates of potential atmospheric importance and estimation of atmospheric lifetimes, *Environ. Sci. Technol.*, *23*, 945–951.

Rodriguez, H.J., J.-C. Chang, and T.F. Thomas (1976), Thermal, photochemical, and photophysical processes in cyclopropanone vapor, *J. Am. Chem. Soc.*, *98*, 2027–2041.

Roehl, C.M., J.B. Burkholder, G.K. Moortgat, A.R. Ravishankara, and P.J. Crutzen (1997), Temperature dependence of UV absorption cross sections and atmospheric implications of several alkyl iodides, *J. Geophys. Res.*, *102*, 12819–12829.

Roehl, C.M., J.J. Orlando, G.S. Tyndall, R.E. Shetter, G.J. Vázquez, C.A. Cantrell, and J.G. Calvert (1994), Temperature dependence of the quantum yields for the photolysis of NO_2 near the dissociation limit, *J. Phys. Chem.*, *98*, 7837–7843.

Rogers, J.D. (1990), Ultraviolet absorption cross sections and atmospheric photodissociation rate constants for formaldehyde, *J. Phys. Chem.*, *94*, 4011–4015.

Rotman, D.A., C.S. Atherton, D.J. Bergmann, P.J. Cameron-Smith, C.C. Chuang, P.S. Connell, J.E. Dignon, A. Franz, K.E. Grant, D.E. Kinnison, C.R. Molenkamp, D.D. Proctor, and J.R. Tannahill (2004), IMPACT, the LLNL 3-D global atmospheric chemical transport model for the combined troposphere and stratosphere: Model description and analysis of ozone and other trace gases, *J. Geophys. Res.*, *109*, D04303, doi:10.1029/2002JD003155.

Rowley, D.M., R. Lesclaux, P.D. Lightfoot, K. Hughes, M.D. Hurley, S. Rudy, and T.J. Wallington (1992a), A kinetic and mechanistic study of the reaction of neopentylperoxy radicals with HO_2, *J. Phys. Chem.*, *96*, 7043–7048.

Rowley, D.M., R. Lesclaux, P.D. Lightfoot, B. Noziere, T.J. Wallington, and M.D. Hurley (1992b), Kinetic and mechanistic studies of the reactions of cyclopentylperoxy and cyclohexylperoxy radicals with HO_2, *J. Phys. Chem.*, *96*, 4889–4894.

Rowley, D.M., P.D. Lightfoot, R. Lesclaux, and T.J. Wallington (1991), UV absorption spectrum and self-reactions of cyclohexylperoxy radicals, *J. Chem. Soc. Faraday Trans.*, *87*, 3221–3226.

Rowley, D.M., P.D. Lightfoot, R. Lesclaux, and T.J. Wallington (1992c), Ultraviolet absorption spectrum and self-reaction of cyclopentylperoxy radicals, *J. Chem. Soc. Faraday Trans.*, *88*, 1369–1376.

RRWG (1999), VOC reactivity science assessment: Executive Summary. Reactivity Research Working Group, http://www.cgenv.com/narsto/reactinfo.html.

Rubin, J.I., A.J. Kean, R.A. Harley, D.B. Millet, and A.H. Goldstein (2006), Temperature dependence of volatile organic compound evaporative emissions from motor vehicles, *J. Geophys. Res.*, *111*, D03305, doi:10.1029/2005JD006458.

Rudolph, J., R. Koppmann, and Ch. Plass-Dülmer (1996), The budgets of ethane and tetrachloroethene: Is there evidence for an impact of reactions with chlorine atoms in the troposphere?, *Atmos. Environ.*, *30*, 1887–1894.

Russell, G.A. (1957), Deuterium-isotope effects in the autoxidation of aralkyl hydrocarbons. Mechanism of the interaction of peroxy radicals, *J. Am. Chem. Soc.*, *79*, 3871–3877.

Rust, F., and C.M. Stevens (1980), Carbon kinetic isotope effect in the oxidation of methane by hydroxyl, *Int. J. Chem. Kinet.*, *12*, 371–377.

Ryan, K.R., and I.C. Plumb (1984), Kinetics of the reactions of CCl_3 with O and O_2 and of CCl_3O_2 with NO at 295 K, *Int. J. Chem. Kinet.*, *16*, 591–602.

Sakurai, H., and S. Kato (1999), A theoretical study of the Norrish type I reaction of acetone, *J. Molec. Struct. (Theochem), 461–462*, 145–152.

Sander, S.P., and R.T. Watson (1980), Kinetics studies of the reactions of CH_3O_2 with NO, NO_2 and CH_3O_2 at 298 K, *J. Phys. Chem.*, *84*, 1664–1674.

Sander, S.P., and R.T. Watson (1981), Temperature dependence of the self-reaction of CH_3O_2 radicals, *J. Phys. Chem.*, *85*, 2960–2964.

Sander, S.P., R.R. Friedl, D.M. Golden, M.J. Kurylo, R.E. Huie, V.L. Orkin, G.K. Moortgat, A.R. Ravishankara, C.E. Kolb, M.J. Molina, and B.J. Finlayson-Pitts (2003), *Chemical Kinetics and Data for Use in Atmospheric Studies*, Evaluation Number 14, NASA, Pasadena, CA; http://jpldataeval.jpl.nasa.gov.

Sander, S.P., M.J. Kurylo, V.L. Orkin, D.M. Golden, R.E. Huie, B.J. Finlayson-Pitts, C.E. Kolb, M.J. Molina, R.R. Friedl, A.R. Ravishankara, and G.K. Moortgat (2005), NASA–JPL Data Evaluation [http://jpldataeval.jpl.nasa.gov/].

Sanders, N., J.E. Butler, L.R. Pasternack, and J.R. McDonald (1980a), $CH_3O(X^2E)$ production from 266 nm photolyis of methyl nitrite and reaction with NO, *Chem. Phys.*, *49*, 17–22.

Sanders, N., J.E. Butler, L.R. Pasternack, and J.R. McDonald (1980b), $CH_3O(X^2E)$ production from 266 nm photolyis of methyl nitrite and reaction with NO, *Chem. Phys.*, *48*, 203–208.

Sanhueza, E. (1977), The chlorine atom sensitized oxidation of $HCCl_3$, HCF_2Cl, and HCF_3, *J. Photochem.*, *7*, 325–334.

Sanhueza, E., and J. Heicklen (1975a), Chlorine-atom sensitized oxidation of dichloromethane and chloromethane, *J. Phys. Chem.*, *79*, 7–11.

Sanhueza, E., and J. Heicklen (1975b), Oxidation of *cis*- and *trans*-1,2-dichloroethene, *Int. J. Chem. Kinet.*, *7*, 589–604.

Sanhueza, E., R. Simonaitis, and J. Heicklen (1979), The reaction of CH_3O_2 with SO_2, *Int. J. Chem. Kinet.*, *11*, 907–914.

Sarzynski, D., and B. Sztuba (2002), Gas-phase reactions of Cl atoms with propane, *n*-butane, and isobutane, *Int. J. Chem. Kinet.*, *34*, 651–658.

Saueressig, G., P. Bergamaschi, J.N. Crowley, H. Fischer, and G.W. Harris (1995), Carbon kinetic isotope effect in the reaction of CH_4 with Cl atoms, *Geophys. Res. Lett.*, *22*, 1225–1228.

Saueressig, G., P. Bergamaschi, J.N. Crowley, H. Fischer, and G.W. Harris (1996), D/H kinetic isotope effect in the reaction $CH_4 + Cl$, *Geophys. Res. Lett.*, *23*, 3619–3622.

Saueressig, G., J.N. Crowley, P. Bergamaschi, C. Bruhl, C.A.M. Brenninkmeijer, and H. Fischer (2001), Carbon 13 and D kinetic isotope effects in the reaction of CH_4 and $O(^1D)$ and OH: New laboratory measurements and their implications for the isotopic composition of stratospheric methane, *J. Geophys. Res.*, *106*, 23127–23138.

Saunders, S.M., D.L. Baulch, K.M. Cooke, M.J. Pilling, and P.I. Smurthwaite (1994), Kinetics and mechanisms of the reactions of OH with some oxygenated compounds of importance in tropospheric chemistry, *Int. J. Chem. Kinet.*, *26*, 113–130.

Saunders, S.M., M.E. Jenkin, R.G. Derwent, and M.J. Pilling (2003), Protocol for the development of the Master Chemical Mechanism, MCM v3 (Part A): tropospheric degradation of non-aromatic volatile organic compounds, *Atmos. Chem. Phys.*, *3*, 161–180.

Savoie, D.L., and J.M. Prospero (1982), Particle size distribution of nitrate and sulfate in the marine atmosphere, *Geophys. Res. Lett.*, *9*, 1207–1210.

Sawerysyn, J.-P., C. Lafage, B. Mériaux, and A. Tighezza (1987), Cinétique de la réaction des atomes de chlore avec le méthane à 294±1 K: Une nouvelle étude par la technique du réacteur à écoulement rapide et à décharge couplé à un spectromètre de masse, *J. Chim. Phys.*, *84*, 1187–1193.

Scala, A.A., and D.G. Ballan (1972), The vacuum ultraviolet photolysis of cyclopentanone, *Can. J. Chem.*, *50*, 3938–3943.

Schiffmann, A., D.D. Nelson, Jr., M.S. Robinson, and D.J. Nesbitt (1991), High-resolution infrared flash kinetic spectroscopy of OH radicals, *J. Phys. Chem.*, *95*, 2629–2636.

Schlyer, D.F., A.P. Wolf, and P.P. Gaspar (1978), Rate constants for the reactions of atomic chlorine with group 4 and group 5 hydrides, *J. Phys. Chem.*, *82*, 2633–2637.

Schmitt, G., and F.J. Comes (1980), Photolysis of CH_2I_2 and 1,1-$C_2H_4I_2$ at 300 nm, *J. Photochem.*, *14*, 107–123.

Schmitt, G., and F.J. Comes (1987), Competitive photodecomposition reactions of chloroiodomethane, *J. Photochem. Photobiol. A., Chemistry*, *41*, 13–20.

Schmoltner, A.M., R.K. Talukdar, R.F. Warren, A. Mellouki, L. Goldfarb, T. Gierczak, S.A. McKeen, and A.R. Ravishankara (1993), Rate coefficients for reactions of several hydrofluorocarbons with OH and $O(^1D)$ and their atmospheric lifetimes, *J. Phys. Chem.*, *97*, 8976–8982.

Schneider, W.F., T.J. Wallington, J.R. Barker, and E.A. Stahlberg (1998), CF_3CFHO radical: decomposition vs. reaction with O_2, *Ber. Bunsen-Ges.*, *102*, 1850–1856.

Scholtens, K.W., B.M. Messer, C.D. Cappa, and M.J. Elrod (1999), Kinetics of the CH_3O_2 + NO reaction: Temperature dependence of the overall rate constant and an improved upper limit for the CH_3ONO_2 branching ratio, *J. Phys. Chem. A*, *103*, 4378–4384.

Schroeder, W.H., K.G. Anlauf, L.A. Barrie, J.Y. Lu, A. Steffen, D.R. Schneeberger, and T. Berg (1998), Arctic springtime depletion of mercury, *Nature*, *394*, 331–332.

Schubert, C.C., and R.N. Pease (1956a), The oxidation of lower paraffin hydrocarbons. I. Room temperature reaction of methane, propane, *n*-butane and isobutane with ozonized oxygen, *J. Am. Chem. Soc.*, *78*, 2044–2048.

Schubert, C.C., and R.N. Pease (1956b), The oxidation of lower paraffin hydrocarbons. II. Observations on the role of ozone in the slow combustion of isobutane, *J. Am. Chem. Soc.*, *78*, 5553–5556.

Scollard, D.J., J.J. Treacy, H.W. Sidebottom, C. Balestria-Garcia, G. Laverdet, G. Le Bras, H. MacLéod, and S. Téton (1993), Rate constants for the reactions of hydroxyl radicals and chlorine atoms with halogenated aldehydes, *J. Phys. Chem.*, *97*, 4683–4688.

Seeley, J.V., J.T. Jayne, and M.J. Molina (1996), Kinetic studies of chlorine atom reactions using the turbulent flow tube technique, *J. Phys. Chem.*, *100*, 4019–4025.

Sehested, J., and T.J. Wallington (1993), Atmospheric chemistry of hydrofluorocarbon 134a. Fate of the alkoxy radical CF_3O, *Environ. Sci. Technol.*, *27*, 146–152.

Sehested, J., and O.J. Nielsen (1993), Absolute rate constants for the reaction of trifluoromethylperoxy (CF_3O_2) and trifluoromethoxy (CF_3O) radicals with nitrogen monoxide at 295 K, *Chem. Phys. Lett.*, *206*, 369–375.

Sehested, J., T. Ellermann, O.J. Nielsen, T.J. Wallington, and M.D. Hurley (1993a), UV absorption spectrum, and kinetics and mechanism of the self reaction of $CF_3CF_2O_2$ radicals in the gas phase at 295 K, *Int. J. Chem. Kinet.*, *25*, 701–717.

Sehested, J., T.E. Møgelberg, K. Fagerström, G. Mahmoud, and T.J. Wallington (1997), Absolute rate constants for the self reaction of HO_2, CF_3CFHO_2, and CF_3O_2 radicals and the cross reactions of HO_2 with FO_2, HO_2 with CF_3CFHO_2, and HO_2 with CF_3O_2 at 295 K, *Int. J. Chem. Kinet.*, *29*, 673–676.

Sehested, J., O.J. Nielsen, and T.J. Wallington (1993b), Absolute rate constants for the reaction of NO with a series of peroxy radicals in the gas phase at 295 K, *Chem. Phys. Lett.*, *213*, 457–464.

Selby, K., and D.J. Waddington (1979), Reaction of oxygenated radicals in the gas phase. Part 4. Reactions of methylperoxyl and methoxyl radicals, *J. Chem. Soc. Perkin Trans. 2*, *9*, 1259–1263.

Sellevåg, S.R., T. Kelly, H. Sidebottom, and C.J. Nielsen (2004), A study of the IR and UV–Vis absorption cross-sections, photolysis and OH-initiated oxidation of CF_3CHO and CF_3CH_2CHO, *Phys. Chem. Chem. Phys.*, *6*, 1243–1252.

Sellevåg, S.R., V. Stenstrøm, T. Helgaker, and C.J. Nielsen (2005), Atmospheric chemistry of CHF_2CHO: Study of the IR and UV–Vis absorption cross sections, photolysis and OH-, Cl-, and NO_3-initiated oxidation, *J. Phys. Chem. A*, *109*, 3652–3662.

Semmes, D.H., A.R. Ravishankara, C.A. Gump-Perkins, and P.H. Wine (1985), Kinetics of the reactions of hydroxyl radical with aliphatic aldehydes, *Int. J. Chem. Kinet.*, *17*, 303–313.

Senapati, D., K. Kavia, and P.K. Das (2002), Photodissociation dynamics of CH_2ICl at 222, 236, 266, 280, and ~304 nm, *J. Phys. Chem. A*, *106*, 8479–8482.

Setokuchi, O., and M. Sato (2002), Direct dynamics of an alkoxy radical (CH_3O, C_2H_5O, and i-C_3H_7O) reactions with an oxygen molecule, *J. Phys. Chem. A*, *106*, 8124–8132.

Sevin, A., E. Fazilleau, and P. Ghaquin (1981), Etude théorique des ruptures thermique et photochimique de la cyclopropanone, *Tetrahedron*, *37*, 3831–3837.

Shallcross, D.E., P. Biggs, C.E. Canosa-Mas, K.C. Clemitshaw, M.G. Harrison, M. Reyes Lópes Alañon, J.A. Pyle, A. Vipond, and R.P. Wayne (1997), Rate constants for the reaction between OH and CH_3ONO_2 and $C_2H_5ONO_2$ over a range of pressure and temperature, *J. Chem. Soc. Faraday Trans.*, *93*, 2807–2811.

Shepson, P.B., and J. Heicklen (1981), Photooxidation of ethyl iodide at 22 °C, *J. Phys. Chem.*, *85*, 2691–2694.

Shepson, P.B., and J. Heicklen (1982a), The photo-oxidation of propionaldehyde, *J. Photochem.*, *18*, 169–184.

Shepson, P.B., and J. Heicklen (1982b), The wavelength and pressure dependence of the photolysis of propionaldehyde in air, *J. Photochem.*, *19*, 215–227.

Shi, J., T.J. Wallington, and E.W. Kaiser (1993), FTIR product study of the Cl initiated oxidation of C_2H_5Cl: atmospheric fate of the alkoxy radical CH_3CHClO, *J. Phys. Chem.*, *97*, 6184–6192.

Shindell, D.T., G. Faluvegi, and N. Bell (2003), Pre-industrial-to-present-day radiative forcing by tropospheric ozone from improved simulations with the GISS chemistry-climate GCM, *Atmos. Chem. Phys.*, *3*, 1675–1702.

Shindell, D.T., G. Faluvegi, D.S. Stevenson, M.C. Krol, L.K. Emmons, J.-F. Lamarque, G. Pétron, F.J. Dentener, K. Ellingsen, M.G. Schultz, O. Wild, M. Amann, C.S. Atherton, D.J. Bergmann, I. Bey, T. Butler, J. Cofala, W.J. Collins, R.G. Derwent, R.M. Doherty, J. Drevet, H.J. Eskes, A.M. Fiore, M. Gauss, D.A. Hauglustaine, L.W. Horowitz, I.S.A. Isaksen, M.C. Krol, M.G. Lawrence, V. Montanaro, J.-F. Muller, G. Pitari, M.J. Prather, J.A. Pyle, S. Rast, J.M. Rodriguez, M.G. Sanderson, N.H. Savage, S.E. Strahan, K. Sudo, S. Szopa, N. Unger, T.P.C. van Noije, and G. Zeng (2006), Multimodel simulations of carbon monoxide: Comparison with observations and projected near-future changes. *J. Geophys. Res.*, *111*, D19306, doi: 10:1029/2006JD007100.

Shortridge, R.G., Jr., and E.K.C. Lee (1970), Benzene photosensitization and direct photolysis of cyclohexanone and cyclohexanone-*2-t* in the gas phase, *J. Am. Chem. Soc.*, *92*, 2228–2236.

Shortridge, R.G., Jr., C.F. Rusbult, and E.K.C. Lee (1971), Fluorescence excitation study of cyclobutanone, cyclopentanone, and cyclohexanone in the gas phase, *J. Am. Chem. Soc.*, *93*, 1863–1867.

Sieger, R.A. (1954),The vapor phase photolysis of trifluoroacetone at wavelength 3130 Å, M.Sc. Thesis, Ohio State University, Columbus, Ohio.

Sieger, R.A., and J.G. Calvert (1954), The vapor phase photolysis of trifluoroacetone at wave length 3130 Å, *J. Am. Chem. Soc.*, *76*, 5197–5201.

Simon, F.-G., W. Schneider, and G.K. Moortgat (1990), UV-absorption spectrum of the methylperoxy radical and the kinetics of its disproportionation reaction at 300 K, *Int. J. Chem. Kinet.*, *22*, 791–813.

Simonaitis, R., and J. Heicklen (1979), Mechanism of SO_2 oxidation by CH_3O_2 radicals—Rate coefficients for the reactions of CH_3O_2 with $SO_2$and NO, *Chem. Phys. Lett.*, *65*, 361–365.

Simonaitis, R., and J. Heicklen (1981), Rate coefficient for the reaction of CH_3O_2 with NO from 218 K to 365 K, *J. Phys. Chem.*, *85*, 2946–2949.

Simonaitis, R., and J. Heicklen (1983), A flash photolysis study of the photo-oxidation of acetaldehyde at room temperature, *J. Photochem.*, *23*, 299–309.

Singh, H.B. (1976), Phosgene in the ambient air, *Nature*, *264*, 23–24.

Singh, H.B., and J.F. Kasting (1988), Chlorine–hydrocarbon photochemistry in the marine troposphere and lower stratosphere, *J. Atmos. Chem.*, *7*, 261–285.

Singh, H.B., A.N. Thakur, Y.E. Chen, and M. Kanakidou (1996), Tetrachloroethylene as an indicator of low Cl atom concentrations in the troposphere, *Geophys. Res. Lett.*, *23*, 1529–1532.

Singleton, D.L., G. Paraskevopoulos, and R.S. Irwin (1980), Reaction of OH with CH_3CH_2F: the extent of H abstraction from the α and β positions, *J. Phys. Chem.*, *84*, 2339–2343.

Sivakumaran, V., and J. Crowley (2003), Reaction between OH and CH_3CHO, Part 2, Temperature dependent rate coefficients (201–348 K), *Phys. Chem. Chem. Phys.*, *5*, 106–111.

Sivakumaran, V., D. Hölscher, T.J. Dillon, and J. Crowley (2003), Reaction between OH and HCHO: Temperature dependent rate coefficients (202–399 K) and product pathways (298 K), *Phys. Chem. Chem. Phys.*, *5*, 4821–4827.

Smith, B.J., M.T. Nguyen, W.J. Bouma, and L. Radom (1991), Unimolecular rearrangements connecting hydroxyethylidene (CH_3COH), acetaldehyde ($CH_3CH{=}O$), and vinyl alcohol ($CH_2{=}CHOH$), *J. Am. Chem. Soc.*, *113*, 6452–6458.

Smith, C.A., L.T. Molina, J.J. Lamb, and M.J. Molina (1984), Kinetics of the reaction of OH with pernitric and nitric acids, *Int. J. Chem. Kinet.*, *16*, 41–55.

Smith, C.A., F.D. Pope, B. Cronin, C.B. Parkes, and A.J. Orr-Ewing (2006), Absorption cross sections of formaldehyde at wavelengths from 300 to 340 nm at 294 and 245 K, *J. Phys. Chem. A*, *110*, 11645–11653.

Smith, G.D., L.T. Molina, and M.J. Molina (2002a), Measurement of radical quantum yields from formaldehyde photolysis between 269 and 339 nm, *J. Phys. Chem. A*, *106*, 1233–1240.

Smith, I.W.M., and A.R. Ravishankara (2002), Role of hydrogen-bonded intermediates in the bimolecular reactions of the hydroxyl radical, *J. Phys. Chem. A*, *106*, 4798–4807.

Smith, J.D., J.D. DeSain, and C.A. Taatjes (2002b), Infrared laser absorption measurements of HCl($v = 1$) production in reactions of Cl atoms wtih isobutane, methanol, acetaldehyde, and toluene at 295 K, *Chem. Phys. Lett.*, *366*, 417–425.

Smith, R.H. (1978), Rate constant and activation energy for the gaseous reaction between hydroxyl and formaldehyde, *Int. J. Chem. Kinet.*, *10*, 519–528.

Sodeau, J.R., and E.K.C. Lee (1978), Intermediacy of hydroxymethylene (HCOH) in the low temperature matrix photochemistry of formaldehyde, *Chem. Phys. Lett.*, *57*, 71–74.

Solignac, G., A. Mellouki, G. LeBras, I. Barnes, and Th. Benter (2006), Reaction of Cl atoms with $C_6F_{13}CH_3OH$, $C_6F_{13}CHO$, and C_3F_7CHO, *J. Phys. Chem. A*, *110*, 4450–4457.

Somnitz, H., and R. Zellner (2000a), Theoretical studies of unimolecular reactions of C_2–C_5 alkoxy radicals. Part I. Ab initio molecular orbital calculations, *Phys. Chem. Chem. Phys.*, *2*, 1899–1905.

Somnitz, H., and R. Zellner (2000b), Theoretical studies of unimolecular reactions of C_2–C_5 alkoxy radicals. Part II. RRKM dynamical calculations, *Phys. Chem. Chem. Phys. A*, *2*, 1907–1918.

Somnitz, H., and R. Zellner (2000c), Theoretical studies of unimolecular reactions of C_2–C_5 alkoxy radicals. Part III. A microscopic structure activity relationship (SAR), *Phys. Chem. Chem. Phys.*, *2*, 4319–4325.

Somnitz, H., and R. Zellner (2001), Theoretical studies of the thermal and chemically activated decomposition of CF_3CY_2O radicals (Y = F, H) radicals, *Phys. Chem. Chem. Phys*, *3*, 2352–2364.

Somnitz, H., M. Fida, T. Ufer, and R. Zellner (2005), Pressure dependence for the CO quantum yield in the photolysis of acetone at 248 nm: A combined experimental and theoretical study, *Phys. Chem. Chem. Phys.*, *7*, 3342–3352.

Sørensen, M., M.D. Hurley, T.J. Wallington, T.S. Dibble, and O.J. Nielsen (2002), Do aerosols act as catalysts in the OH radical initiated atmospheric oxidation of volatile organic compounds?, *Atmos. Environ.*, *36*, 5947–5952.

Sørensen, M., E.W. Kaiser, M.D. Hurley, T.J. Wallington, and O.J. Nielsen (2003), Kinetics of the reaction of OH radicals with acetylene in 25–7750 Torr of air at 296 K, *Int. J. Chem. Kinet.*, *35*, 191–197.

Soto, M.R., and M. Page (1990), Features of the potential energy surface for reactions of hydroxyl with formaldehyde, *J. Phys. Chem.*, *94*, 3242–3246.

Spicer, C.W., E.G. Chapman, B.J. Finlayson-Pitts, R.A. Plastridge, J.M. Hubbe, J.D. Fast, and C.M. Berkowitz (1998), Unexpectedly high concentrations of molecular chlorine in coastal air, *Nature*, *394*, 353–356.

Spittler, M., I. Barnes, K.H. Becker, and T.J. Wallington (2000), Product study of the $C_2H_5O_2$ + HO_2 reaction in 760 Torr of air at 284–312 K, *Chem. Phys. Lett.*, *321*, 57–61.

Srinivasan, R. (1959), The photochemical type II process in 2-hexanone-5,5-d_2 and 2-hexanone, *J. Am. Chem. Soc.*, *81*, 5061–5065.

Srinivasan, R. (1961), Photochemistry of cyclopentanone. I. Details of the primary process, *J. Am. Chem. Soc.*, *83*, 4344–4347.

Srinivasan, R. (1964), Photochemistry of the cyclic ketones, *Adv. Photochem.*, *1*, 83–113.

Stachnik, R.A., L.T. Molina, and M.J. Molina (1986), Pressure and temperature dependence of the reaction of OH with nitric acid, *J. Phys. Chem.*, *90*, 2777–2780.

Stedman, D.H., and H. Niki (1973), Ozonolysis rates of some atmospheric gases, *Environ. Lett.*, *4*, 303–310.

Stemmler, K., W. Mengon, and J.A. Kerr (1997), Hydroxyl-radical-initiated oxidation of isobutyl isopropyl ether under laboratory conditions related to the troposphere, *J. Chem. Soc. Faraday Trans.*, *93*, 2865–2875.

Stevenson, D.S., F.J. Dentener, M.G. Schultz, K. Ellingson, T.P.C. van Noije, O. Wild, G. Zeng, M. Amann, C.S. Atherton, N. Bell, D.J. Bergmann, I. Bey, T. Butler, J. Cofala, W.J. Collins, R.G. Derwent, R.M. Doherty, J. Drevet, H.J. Eskes, A.M. Fiore, M. Gauss, D.A. Hauglustaine, L.W. Horowitz, I.S.A. Isaksen, M.C. Krol, J.-F. Lamarque, M.G. Lawrence, V. Montanaro, J.-F. Muller, G. Pitari, M.J. Prather, J.A. Pyle, S. Rast, J.M. Rodriguez, M.G. Sanderson, N.H. Savage, D.T. Shindell, S.E. Strahan, K. Sudo, and S. Szopa (2006), Multi-model ensemble simulations of present-day and near-future tropospheric ozone, *J. Geophys. Res.*, *111*, D08301, doi: 10.1029/2005JD006338.

Stevenson, D.S., R.M. Doherty, M.G. Sanderson, W.J. Collins, C.E. Johnson, and R.G. Derwent (2004), Radiative forcing from aircraft NO_x emissions: mechanisms and seasonal dependence, *J. Geophys. Res.*, *109*, D17307, doi:10.1019/2004JD004759.

Stickel, R.E., J.M. Nicovich, S. Wang, Z. Zhao, and P.H. Wine (1992), Kinetic and mechanistic study of the reaction of atomic chlorine with dimethyl sulfide, *J. Phys. Chem.*, *96*, 9875–9883.

Stief, L.J., D.F. Nava, W.A. Payne, and J.V. Michael (1980), Rate constant for the reaction of hydroxyl radical with formaldehyde over the temperature range 228–362 K, *J. Chem. Phys.*, *73*, 2254–2258.

Stockwell, W.R., F. Kirchner, M. Kuhn, and S. Seefeld (1997), A new mechanism for regional atmospheric chemistry modelling, *J. Geophys. Res. 102*, 25847–25879.

Stockwell, W.R., P. Middleton, J.S. Chang, and X. Tang (1990), The second generation regional acid deposition model chemical mechanism for regional air quality modeling, *J. Geophys. Res.*, *95*, 16343–16367.

Strachan, A.N., and F.E. Blacet (1955), The photolysis of chloroacetone at 3130 Å, *J. Am. Chem. Soc.*, *77*, 5254–5257.

Strunin, V.P., N.K. Serdyuk, E.N. Chesnokov, and V.N. Panfilov (1975), Kinetic isotope effects in the chlorination of methane, *React. Kinet. Catal. Lett.*, *3*, 97–104.

Stutz, J., M.J. Ezell, A.A. Ezell, and B.J. Finlayson-Pitts (1998), Rate constants and kinetic isotope effects in the reactions of atomic chlorine with *n*-butane and simple alkenes at room temperature, *J. Phys. Chem. A*, *102*, 8510–8519.

Sudo, K., M. Takahashi, J. Kurokawa, and H. Akimoto (2002), CHASE: A global chemical model of the troposphere 1. Model description, *J. Geophys. Res. 107*, D4339, doi:10.1029/2001JD001113.

Sumathi, R., and A.K. Chandra (1993), Photodecomposition of acetyl chloride on the excited singlet state surface, *J. Chem. Phys.*, *99*, 6531–6536.

Suong, J.Y., and R.W. Carr (1982), The photo-oxidation of 1,3-dichlorotetrafluoroacetone: mechanism of the reaction of CF_2Cl with oxygen, *J. Photochem.*, *19*, 295–302.

Szilágyi, I., S. Dóbé, and T. Bérces (2000), Rate constant for the reaction of the OH-radical with CH_2F_2, *React. Kinet. Catal. Lett.*, *70*, 319–324.

Szilágyi, I., S. Dóbé, T. Bérces, F. Márta, and B. Viskolcz (2004), Direct kinetic study of reactions of hydroxyl radicals with alkyl formates, *Z. Phys. Chem.*, *218*, 479–492.

Szopa, S., B. Aumont, and S. Madronich (2005), Assessment of the reduction methods used to develop chemical schemes: building of a new chemical scheme for VOC oxidation suited to three-dimensional multi-scale HO_x–NO_x–VOC chemistry simulations, *Atmos. Chem. Phys.*, *5*, 2519–2538.

Tadić, J., I. Juranić, and G.K. Moortgat (2001a), Pressure dependence of the photooxidation of selected carbonyl compounds in air: *n*-butanal and *n*-pentanal, *J. Photochem. Photobiol. A*, *143*, 168–179.

Tadić, J., I. Juranić, and G.K. Moortgat (2001b), Photooxidation of *n*-hexanal in air, *Molecules*, *6*, 287–299.

Tadić, J.M., I.O. Juranić and G.K. Moortgat (2002), Photooxidation of *n*-heptanal in air: Norrish type I and II processes and quantum yield total pressure dependency, *J. Chem. Soc., Perkins Trans.*, *2*, 135–140.

Takagi, H., N. Washida, H. Bandow, H. Akimoto, and M. Okuda (1981), Photooxidation of C_5–C_7 cycloalkanes in the NO–H_2O–air system, *J. Phys. Chem.*, *85*, 2701–2705.

Takahashi, K., T. Nakayama, and Y. Matsumi (2004), Hydrogen atom formation in the photolysis of acetone at 193 nm, *J. Phys. Chem. A*, *109*, 8002–8008.

Takekawa, H., H. Minoura, and S. Yamazaki (2003), Temperature dependence of secondary organic aerosol formation by photo-oxidation of hydrocarbons, *Atmos. Environ.*, *37*, 3413–3424.

Taketani, F., T. Nakayama, K. Takahashi, Y. Matsumi, M.D. Hurley, T.J. Wallington, A.Toft, and M.P.S. Andersen (2005), Atmospheric Chemistry of CH_3CHF_2 (HFC-152a): Kinetics, mechanisms, and products of Cl atom and OH radical initiated oxidation in presence and absence of NO_x, *J. Phys. Chem. A*, *109*, 9061–9069.

Talukdar, R.K., J.B. Burkholder, M. Hunter, M.K. Gilles, J.M. Roberts, and A.R. Ravishankara (1997a), Atmospheric fate of several alkyl nitrates. 2. UV absorption cross-sections and photodissociation quantum yields, *J. Chem. Soc. Faraday Trans.*, *93*, 2797–2805.

Talukdar, R.K., J.B. Burkholder, A.-M. Schmoltner, J.M. Roberts, R.R. Wilson, and A.R. Ravishankara (1995), Investigation of the loss processes for peroxyacetyl nitrate in the atmosphere: UV photolysis and reaction with OH, *J. Geophys. Res.*, *100*, 14163–14174.

Talukdar, R.K., T. Gierczak, D.C. McCabe, and A.R. Ravishankara (2003), Reaction of hydroxyl radical with acetone. 2. Products and reaction mechanism, *J. Phys. Chem. A*, *107*, 5021–5032.

Talukdar, R.K., S.C. Herndon, J.B. Burkholder, J.M. Roberts, and A.R. Ravishankara (1997b), Atmospheric fate of several alkyl nitrates. Part 1. Rate coefficients of the reactions of alkyl nitrates with isotopically labelled hydroxyl radicals, *J. Chem. Soc. Faraday Trans.*, *93*, 2787–2796.

Talukdar, R.K., A. Mellouki, J.B. Burkholder, M.K. Gilles, G. Le Bras, and A.R. Ravishankara (2001), Quantification of the tropospheric removal of chloral (CCl_3CHO): Rate coefficient for the reaction of OH, UV absorption cross sections, and quantum yields, *J. Phys. Chem., A*, *105*, 5188–5196.

Talukdar, R., A. Mellouki, T. Gierczak, S. Barone, S.-Y. Chiang, and A.R. Ravishankara (1994), Kinetics of the reactions of OH with alkanes, *Int. J. Chem. Kinet.*, *26*, 973–990.

Talukdar, R., A. Mellouki, T. Gierczak, J.B. Burkholder, S.A. McKeen, and A.R. Ravishankara (1991a), Atmospheric fate of CF_2H_2, CH_3CF_3, CHF_2CF_3, and CH_3CFCl_2: rate coefficients for reactions with OH and UV absorption cross sections of CH_3CFCl_2, *J. Phys. Chem.*, *95*, 5815–5821.

Talukdar, R., A. Mellouki, T. Gierczak, J.B. Burkholder, S.A. McKeen, and A.R. Ravishankara (1991b), Atmospheric lifetime of CHF_2Br, a proposed substitute for halons, *Science*, *252*, 693–695.

Talukdar, R., A. Mellouki, A-M. Schmoltner, T. Watson, S. Montzka, and A.R. Ravishankara (1992), Kinetics of the OH reaction with methyl chloroform and its atmospheric implications, *Science*, *257*, 227–230.

Tanaka, P.L., D.D. Riemer, S. Chang, G. Yarwood, E.C. McDonald-Buller, E.C. Apel, J.J. Orlando, D.J. Neece, C.B. Mullins, and D.T. Allen (2003), Direct evidence for chlorine-enhanced urban ozone formation in Houston, Texas, *Atmos. Environ.*, *37*, 1393–1400.

Tang, K.Y., and E.K.C. Lee (1976), Laser photolysis of cyclobutanone photodecomposition from selected vibronic levels at long wavelengths, *J. Phys. Chem.*, *80*, 1833–1836.

Tang, Y. and L. Zhu (2004), Wavelength dependent photolysis of *n*-hexanal and *n*-heptanal in the 280–340 nm region, *J. Phys. Chem. A*, *108*, 8307–8316.

Taniguchi, N., T.J. Wallington, M.D. Hurley, A.G. Guschin, L.T. Molina, and M.J. Molina (2003), Atmospheric chemistry of $C_2F_5C(O)CF(CF_3)_2$: Photolysis and reaction with Cl atoms, OH radicals, and ozone, *J. Phys. Chem. A*, *107*, 2674–2679.

Taylor, P.H., Z. Jiang, and B. Dellinger (1992), Laser photolysis/laser-induced fluorescence studies of the reaction of hydroxyl with 1,1,2-trichloroethane over an extended temperature range, *J. Phys. Chem.*, *96*, 1293–1296.

Taylor, P.H., Z. Jiang, and B. Dellinger (1993), Determination of the gas-phase reactivity of hydroxyl with chlorinated methanes at high temperature: effects of laser/thermal photochemistry, *Int. J. Chem. Kinet.*, *25*, 9–23.

Taylor, P.H., S. McCarron, and B. Dellinger (1991), Investigation of 1,2-dichloroethane–hydroxyl kinetics over an extended temperature range: effect of chlorine substitution, *Chem. Phys. Lett.*, *177*, 27–32.

Taylor, P.H., M.S. Rahman, M. Arif, B. Dellinger, and P. Marshall (1996), Kinetic and mechanistic studies of the reaction of hydroxyl radicals with acetaldehyde over an extended temperature range, *Twenty-sixth Symposium (International) on Combustion*, The Combustion Institute, Pittsburgh, PA, 497–504.

Taylor, P.H., T. Yamada, and P. Marshall (2006), The reaction of OH with acetaldehyde and deuterated acetaldehyde: further insight into the reaction mechanism at both low and elevated temperatures, *Int. J. Chem. Kinet.*, *38*, 489–495.

Taylor, R.P., and F.E. Blacet (1956), The photolysis of chloroketones, *J. Am. Chem. Soc.*, *78*, 706–707.

Taylor, W.D., T.D. Allston, M.J. Mascato, G.B. Fazekas. R. Kozlowski, and G.A. Takacs (1980), Atmospheric photodissociation lifetimes for nitromethane, methyl nitrite, and methyl nitrate, *Int. J. Chem. Kinet.*, *12*, 231–240.

Temme, C., J.W. Einax, R. Ebinghaus, and W.H. Schroeder (2003), Measurements of atmospheric mercury species at a coastal site in the Antarctic and over the south Atlantic Ocean during polar summer, *Environ. Sci. Technol.*, *37*, 22–31.

Temps, F., and H.Gg. Wagner (1984), Rate constants for the reactions of OH-radicals with CH_2O and HCO, *Ber. Bunsenges. Phys. Chem.*, *88*, 415–418.

Terentis, A.C., P.T. Knepp, and S.H. Kable (1995), Nascent state distribution of the HCO photoproduct arising from the 309 nm photolysis of propionaldehyde, *J. Phys. Chem.*, *99*, 12704–12710.

Téton, S., A. El Boudali, and A. Mellouki (1996a), Rate constants for the reactions of OH radicals with 1- and 2-bromopropane, *J. Chim. Phys.*, *93*, 274–282.

Téton, S., A. Mellouki, and G. Le Bras (1996b), Rate constants for reactions of OH radicals with a series of asymmetrical ethers and *tert*-butyl alcohol, *Int. J. Chem. Kinet.*, *28*, 291–297.

Thelen, M.-A., P. Felder, and J.R. Huber (1993), The photofragmentation of methyl hydroperoxide CH_3OOH at 193 and 248 nm in a cold molecular beam, *Chem. Phys. Lett.*, *213*, 275–281.

Thévenet, R., A. Mellouki, and G. Le Bras (2000), Kinetics of OH and Cl reactions with a series of aldehydes, *Int. J. Chem. Kinet.*, *32*, 676–685.

Thomas, S.S., and J.G. Calvert (1962), Photooxidation of 2,2′-azoisobutane at 25°, *J. Am. Chem. Soc.*, *84*, 4207–4212.

Thomas, T.F., and H.J. Rodriguez (1971), Photochemistry of cyclopropanone, *J. Am. Chem. Soc.*, *93*, 5918–5920.

Thüner, L.P., I. Barnes, K.H. Becker, T.J. Wallington, L.K. Christensen, J.J. Orlando, and B. Ramacher (1999), Atmospheric chemistry of tetrachloroethene ($Cl_2C{=}CCl_2$): Products of chlorine atom initiated oxidation, *J. Phys. Chem. A*, *103*, 8657–8663.

Tokuhashi, K., L. Chen, S. Kutsuna, T. Uchimaru, M. Sugie, and A. Sekiya (2004), Environmental assessment of CFC alternatives: Rate constants for the reactions of OH radicals with fluorinated compounds, *J. Fluorine Chem.*, *125*, 1801–1807.

Tomas, A., F. Caralp, and R. Lesclaux (2000), Decomposition of the CF_3CO radical: Pressure and temperature dependencies of the rate constant, *Z. Physik. Chem.*, *Int. J. Res. Phys. Chem. Chem. Phys.*, *214*, 1349–1365.

Trentelman, K.A., S.H. Kable, B.D. Moss, and P.L. Houston (1989), Photodissociation dynamics of acetone at 193 nm—Photofragment internal and translational energy-distributions, *J. Chem. Phys.*, *91*, 7498–7513.

Treves, K., and Y. Rudich (2003), The atmospheric fate of C_3–C_6 hydroxyalkyl nitrates, *J. Phys. Chem. A*, *107*, 7809–7817.

Troe, J. (1969), Ultraviolettspektrum und Reaktionen des HO_2-Radikals im thermischen Zerfall von H_2O_2, *Ber. Bunsenges. Phys. Chem.*, *73*, 946–952.

Troe, J. (1983), Theory of thermal unimolecular reactions in the fall-off range. I. Strong collision rate constants, *Ber. Bunsenges. Phys. Chem.*, *87*, 161–169.

Tsalkani, N., A. Mellouki, G. Poulet, G. Toupance, and G. Le Bras (1988), Rate constant measurements for the reactions of OH and Cl with peroxyacetyl nitrate at 298 K, *J. Atmos. Chem.*, *7*, 409–419.

Tschuikow-Roux, E., J. Niedzielski, and F. Faraji (1985a), Competitive photochlorination and kinetic isotope effects for hydrogen/deuterium abstraction from the methyl group in C_2H_6, C_2D_6, CH_3CHCl_2, CD_3CHCl_2, CH_3CCl_3, and CD_3CCl_3, *Can. J. Chem.*, *63*, 1093–1099.

Tschuikow-Roux, E., T. Yano, and J. Niedzielski (1985b), Reactions of ground state chlorine atoms with fluorinated methanes and ethanes, *J. Chem. Phys.*, *82*, 65–74.

Tsunashima, S., O. Ohsawa, C. Takahashi, and S. Sato (1973), The photolysis of cyclopentanone, *Bull. Chem. Soc. Japan*, *46*, 83–86.

Tuazon, E.C., and R. Atkinson (1993a), Tropospheric transformation products of a series of hydrofluorocarbons and hydrochlorofluorocarbons, *J. Atmos. Chem.*, *17*, 179–199.

Tuazon, E.C., and R. Atkinson (1993b), Tropospheric degradation products of 1,1,1,2-tetrafluoroethane (HFC-134a), *J. Atmos. Chem.*, *16*, 301–312.

Tuazon, E.C., and R. Atkinson (1994), Tropospheric reaction products and mechanisms of the hydrochlorofluorocarbons 141b, 142b, 225ca, and 225cb, *Environ. Sci. Technol.*, *28*, 2306–2313.

Tuazon, E.C., S.M. Aschmann, M.V. Nguyen, and R. Atkinson (2003), H-atom abstraction from selected C—H bonds in 2,3-dimethylpentanal, 1,4-cyclohexadiene, and 1,3,5-cycloheptatriene, *Int. J. Chem. Kinet.*, *35*, 415–426.

Tuazon, E.C., R. Atkinson, S.M. Aschmann, J. Arey, A.M. Winer, and J.N. Pitts Jr. (1986), Atmospheric loss processes of 1,2-dibromo-3-chloropropane and trimethyl phosphate, *Environ. Sci. Technol.*, *20*, 1043–1046.

Tuazon, E.C., W.P.L. Carter, R. Atkinson, and J.N. Pitts, Jr. (1983), The gas-phase reaction of hydrazine and ozone: A nonphotolytic source of OH radicals for measurement of relative OH radical rate constants, *Int. J. Chem. Kinet.*, *15*, 619–629.

Tully, F.P., A.T. Droege, M.L. Koszykowski, and C.F. Melius (1986a), Hydrogen-atom abstraction from alkanes by OH. 2. Ethane, *J. Phys. Chem.*, *90*, 691–698.

Tully, F.P., J.E.M. Goldsmith, and A.T. Droege (1986b), Hydrogen-atom abstraction from alkanes by OH. 4. Isobutane, *J. Phys. Chem.*, *90*, 5932–5937.

Tully, F.P., M.L. Koszykowski, and J.S. Binkley (1985), Hydrogen-atom abstraction from alkanes by OH. 1. Neopentane and neooctane, *20th International Symposium on Combustion, 1984*, The Combustion Institute, Pittsburgh, PA. pp. 715–721.

Turpin, E., C. Fittschen, A. Tomas, and P. Devolder (2003), Reaction of OH radicals with acetone: Determination of the branching ratio for the abstraction pathway at 298 K and 1 Torr, *J. Atmos. Chem.*, *46*, 1–13.

Turnipseed, A.A., S.B. Barone, and A.R. Ravishankara (1994), Kinetics of the reactions of CF_3O_x with NO, O_3, and O_2, *J. Phys. Chem.*, *98*, 4594–4601.

Tyler, S.C., H.O. Ajie, A.L. Rice, R.J. Cicerone, and E.C. Tuazon (2000), Experimentally determined kinetic isotope effects in the reaction of CH_4 with Cl: Implications for atmospheric CH_4, *Geophys. Res. Lett.*, *27*, 1715–1718.

Tyndall, G.S., R.A. Cox, C. Granier, R. Lesclaux, G.K. Moortgat, M.J. Pilling, A.R. Ravishankara, and T.J. Wallington (2001), Atmospheric chemistry of small organic peroxy radicals, *J. Geophys. Res.*, *106*, 12157–12182.

Tyndall, G.S., J.J. Orlando, T.J. Wallington, M. Dill, and E.W. Kaiser (1997a), Kinetics and mechanisms of the reactions of chlorine atoms with ethane, propane, and *n*-butane, *Int. J. Chem. Kinet.*, *29*, 43–55.

Tyndall, G.S., J.J. Orlando, T.J. Wallington, and M.D. Hurley (1997b), Pressure dependence of the rate coefficients and product yields for the reaction of CH_3CO radicals with O_2, *Int. J. Chem. Kinet.*, *29*, 655–663.

Tyndall, G.S., J.J. Orlando, T.J. Wallington, M.D. Hurley, M. Goto, and M. Kawasaki (2002), Mechanism of the reaction of OH radicals with acetone and actaldehyde at 251 and 296 K, *Phys. Chem. Chem. Phys.*, *4*, 2189–2193.

Tyndall, G.S., T.A. Staffelbach, J.J. Orlando, and J.G. Calvert (1995), Rate coefficients for the reactions of OH radicals with methylgloxal and acetaldehyde, *Int. J. Chem. Kinet.*, *27*, 1009–1020.

Tyndall, G.S., T.J. Wallington, and J.C. Ball (1998), FTIR product study of the reactions $CH_3O_2 + CH_3O_2$ and $CH_3O_2 + O_3$, *J. Phys. Chem. A*, *102*, 2547–2554.

Ullerstam, M., S. Langer, and E. Ljungstrom (2000), Gas phase rate coefficients and activation energies for the reaction of butanal and 2-methyl-propanal with nitrate radicals, *Int. J. Chem. Kinet.*, *32*, 294–303.

UN ECE (2005), Protocol concerning the control of emissions of volatile organic compounds, United Nations Economic Commission for Europe, http://www.unece.org/env/lrtap/vola_h1.htm, Geneva, Switzerland.

U.S. Standard Atmosphere (1976), National Oceanic and Atmospheric Administration, National Aeronautics and Space Administration, United States Air Force, Washington, DC, October.

Vaghjiani, G.L., and A.R. Ravishankara (1989a), Kinetics and mechanism of OH reaction with CH_3OOH, *J. Phys. Chem.*, *93*, 1948–1959.

Vaghjiani, G.L., and A.R. Ravishankara (1989b), Absorption cross-sections of CH_3OOH, H_2O_2, and D_2O_2 vapors between 210 and 365 nm at 297 K, *J. Geophys. Res.*, *94*, 3487–3492.

Vaghjiani, G.L., and A.R. Ravishankara (1990), Photodissociation of H_2O_2 and CH_3OOH at 248 nm and 298 K: Quantum yields for OH, $O(^3P)$ and $H(^2S)$, *J. Chem. Phys.*, *92*, 996–1003.

Vaghjiani, G.L., and A.R. Ravishankara (1991), New measurements of the rate coefficient for the reaction of OH with methane, *Nature*, *350*, 406–409.

Vandenberk, S., and J. Peeters (2003), The reaction of acetaldehyde and propionaldehyde with hydroxyl radicals: Experimental determination of the primary H_2O yield at room temperature, *J. Photochem. Photobiol. A*, *157*, 269–274.

Vandenberk, S., L. Vereecken, and J. Peeters (2002), The acetic acid forming channel in the acetone + OH reaction: A combined experimental and theoretical investigation, *Phys. Chem. Chem. Phys.*, *4*, 461–466.

Van Hoosier, M.E., J.-D.F. Bartoe, G.E. Brueckner, and K.D. Prinz (1988), Absolute solar spectral irradiance 120 nm–400 nm (results from the Solar Ultraviolet Spectral Irradiance Monitor—SUSIM—Experiment on board Spacelab 2), *Astro. Lett. Commun.*, *27*, 163–168.

Van Veen, G.N.A., T. Baller, A.E. De Vries, and N.J.A. Van Veen (1984), The excitation of the umbrella mode of CH_3 and CD_3 formed from photodissociation of CH_3I and CD_3I at 248 nm, *Chem. Phys.*, *87*, 405–417.

Vasvári, G., I. Szilágyi, A. Bendsura, S. Dóbé, T. Bérces, E. Henon, S. Canneaux, and F. Bohr (2001), Reaction and complex formation between OH radical and acetone, *Phys. Chem. Chem. Phys.*, *3*, 551–555.

Vereecken, L., and J. Peeters (2003), The 1,5-H-shift in 1-butoxy: a case study in the rigorous implementation of transition state theory for a multirotamer system, *J. Chem. Phys.*, *119*, 5159–5170.

Vésine, E., V. Bossoutrot, A. Mellouki, G. Le Bras, J. Wenger, and H. Sidebottom (2000), Kinetic and mechanistic study of OH- and Cl-initiated oxidation of two unsaturated HFCs: $C_4F_9CH{=}CH_2$ and $C_6F_{13}CH{=}CH_2$, *J. Phys. Chem. A*, *104*, 8512–8520.

Villalta, P.W., L.G. Huey, and C.J. Howard (1995), A temperature-dependent kinetics study of the CH_3O_2 + NO reaction using chemical ionization mass spectrometry, *J. Phys. Chem.*, *99*, 12829–12834.

Villenave, E., and R. Lesclaux (1995), The UV absorption spectra of CH_2Br and CH_2BrO_2 and the reaction kinetics of CH_2BrO_2 with itself and with HO_2 at 298 K, *Chem. Phys. Lett.*, *236*, 376–384.

Villenave, E., and R. Lesclaux (1996), Kinetics of the cross reactions of CH_3O_2 and $C_2H_5O_2$ radicals with selected peroxy radicals, *J. Phys. Chem.*, *100*, 14372–14382.

Villenave, E., V.L. Orkin, R.E. Huie, and M.J. Kurylo (1997), Rate constant for the reaction of OH radicals with dichloromethane, *J. Phys. Chem. A*, *101*, 8513–8517.

Viskolcz, B., and T. Bérces (2000), Enthalpy of formation of selected carbonyl radicals from theory and comparison with experiment, *Phys. Chem. Chem. Phys.*, *2*, 5430–5436.

Voicu, I., I. Barnes, K.H. Becker, T.J. Wallington, Y. Inoue, and M. Kawasaki (2001), Kinetic and product study of the Cl-initiated oxidation of 1,2,3-trichloropropane ($CH_2ClCHClCH_2Cl$), *J. Phys. Chem. A*, *105*, 5123–5130.

Vollenweider, J.K., and H. Paul (1986), On the rates of decarbonylation of hydroxyacetyl and other acyl radicals, *Int. J. Chem. Kinet.*, *18*, 791–800.

Volz-Thomas, A., H. Geiss, A. Hofzumahaus, and K.H. Becker (2003), Introduction to special section: Photochemistry experiment in BERLIOZ, *J. Geophys. Res.*,*108*, 8252, doi:10.1029/2001JD002029.

Von Kuhlmann, R., M.G. Lawrence, P.J. Crutzen, and P.J. Rasch (2003), A model for studies of tropospheric ozone and nonmethane hydrocarbons: Model description and ozone results, *J. Geophys. Res.*, *108, 4294*, doi:10.1029/2002JD002893.

Wagner, P.J., and G.S. Hammond (1968), Properties and reactions of organic molecules in their triplet states, *Adv. Photochem.*, *5*, 21–156.

Wagner, V., M.E. Jenkin, S.M. Saunders, J. Stanton, K. Wirtz, and M.J. Pilling (2003), Modelling of the photooxidation of toluene: Conceptual ideas for validating detailed mechanisms, *Atmos. Chem. Phys.*, *3*, 89–106.

Wallington, T.J. (1991), Fourier-transform infrared product study of the reaction of CH_3O_2 + HO_2 over the pressure range 15–700 torr at 295 K, *J. Chem. Soc. Faraday Trans.*, *87*, 2379–2382.

Wallington, T.J., and M.D. Hurley (1992a), A kinetic study of the reaction of chlorine atoms with CF_3CHCl_2, CF_3CH_2F, $CFCl_2CH_3$, CF_2ClCH_3, CHF_2CH_3, CH_3D, CH_2D_2, CHD_3, CF_4 and CD_3Cl, *Chem. Phys. Lett.*, *189*, 437–442.

Wallington, T.J., and M.D. Hurley (1992b), Are anomalies in the kinetics of halogenation caused by hot atom effects? An experimental investigation, *Chem. Phys. Lett.*, *194*, 309–312.

Wallington, T.J., and M.D. Hurley (1992c), FTIR product study of the reaction of CD_3O_2 + HO_2, *Chem. Phys. Lett.*, *193*, 84–88.

Wallington, T.J., and M.D. Hurley (1993), Atmospheric chemistry of HC(O)F: reaction with OH radicals, *Environ. Sci. Technol.*, *27*, 1448–1452.

Wallington, T.J., and S.M. Japar (1990a), FTIR product study of the reaction of $C_2H_5O_2$ + HO_2 in air at 295 K, *Chem. Phys. Lett.*, *166*, 495–499.

Wallington, T.J., and S.M. Japar (1990b), Reaction of CH_3O_2 + HO_2 in air at 295 K: A product study, *Chem. Phys. Lett.*, *167*, 513–518.

Wallington, T.J., and E.W. Kaiser (1999), Comment on the "The fluorine atom initiated oxidation of CF_3CFH_2 (HFC-134a) studied by FTIR spectroscopy" by Hasson et al., *Int. J. Chem. Kinet.*, *31*, 397–398.

Wallington, T.J., and M.J. Kurylo (1987a), Flash photolysis resonance fluorescence investigation of the gas-phase reactions of OH radicals with a series of aliphatic ketones over the temperature range 240–440 K, *J. Phys. Chem.*, *91*, 5050–5054.

Wallington, T.J., and M.J. Kurylo (1987b), The gas phase reactions of hydroxyl radicals with a series of aliphatic alcohols over the temperature range 240–440 K, *Int. J. Chem. Kinet.*, *19*, 1015–1023.

Wallington, T.J., and O.J. Nielsen (1991), Pulse radiolysis study of 1,2,2,2-tetrafluoroethylperoxy (CF_3CFHO_2) radicals in the gas phase at 298 K, *Chem. Phys. Lett.*, *187*, 33–39.

Wallington, T.J., and W.F. Schneider (1994), The stratospheric fate of CF_3OH, *Environ. Sci. Tech.*, *28*, 1198–1200.

Wallington, T.J., J.M. Andino, and S.M. Japar (1990), FTIR product study of the self reaction of $CH_2ClCH_2O_2$ radicals in air at 295 K, *Chem. Phys. Lett.*, *165*, 189–194.

Wallington, T.J., J.M. Andino, A.R. Potts, and O.J. Nielsen (1992a), Pulse radiolysis and Fourier transform infrared study of neopentyl peroxy radicals in the gas phase at 297 K, *Int. J. Chem. Kinet.*, *24*, 649–663.

Wallington, T.J., R. Atkinson, and A.M. Winer (1984), Rate constant for the gas phase reaction of OH radicals with peroxyacetyl nitrate (PAN) at 273 and 297 K, *Geophys. Res. Lett.*, *11*, 861–864.

Wallington, T.J., R. Atkinson, A.M. Winer, and J.N. Pitts, Jr. (1986), Absolute rate constants for the gas-phase reactions of the NO_3 radical with CH_3SCH_3, NO_2, CO, and a series of alkanes at 298 ± 2 K, *J. Phys. Chem.*,*90*, 4640–4644.

Wallington, T.J., R. Atkinson, A.M. Winer, and J.N. Pitts, Jr. (1987a), A study of the reaction $NO_3 + NO_2 + M \rightarrow N_2O_5 + M$ ($M = N_2, O_2$), *Int. J. Chem. Kinet.*, *19*, 243–249.

Wallington, T.J., J.C. Ball, O.J. Nielsen, and E. Bartkiewicz (1992b), Spectroscopic, kinetic and mechanistic study of CH_2FO_2 radicals in the gas phase at 298 K, *J. Phys. Chem.*, *96*, 1241–1246.

Wallington, T.J., M. Bilde, T.E. Møgelberg, J. Sehested, and O.J. Nielsen (1996a), Atmospheric chemistry of 1,2-dichloroethane: UV spectra of $CH_2ClCHCl$ and $CH_2ClCHClO_2$ radicals, kinetics of the reactions of $CH_2ClCHCl$ radicals with O_2 and $CH_2ClCHClO_2$ radicals with NO and NO_2, and fate of the alkoxy radical $CH_2ClCHClO$, *J. Phys. Chem.*, *100*, 5751–5760.

Wallington, T.J., P. Dagaut, and M.J. Kurylo (1992c), Ultraviolet absorption cross sections and reaction kinetics and mechanisms for peroxy radicals in the gas phase, *Chem. Rev.*, *92*, 667–710.

Wallington, T.J., P. Dagaut, R. Liu, and M.J. Kurylo (1988a), Gas-phase reactions of hydroxyl radicals with the fuel additives methyl *tert*-butyl ether and *tert*-butyl alcohol over the temperature range 240–440 K, *Environ. Sci. Tech.*, *22*, 842–844.

Wallington, T.J., P. Dagaut, R. Liu, and M.J. Kurylo (1988b), Rate constants for the gas phase reactions of OH with C_5 through C_7 aliphatic alcohols and ethers: Predicted and experimental values, *Int. J. Chem. Kinet.*, *20*, 541–547.

Wallington, T.J., T. Ellermann, O.J. Nielsen, and J. Sehested (1994a), Atmospheric chemistry of FCO_x radicals: UV spectra and self-reaction kinetics of FCO and $FC(O)O_2$ and kinetics of some reactions of FCO_x with O_2, O_3, and NO at 296 K, *J. Phys. Chem.*, *98*, 2346–2356.

Wallington T.J., C.A. Gierczak, J.C. Ball, and S.M. Japar (1989a), Fourier transform infrared study of the self reaction of $C_2H_5O_2$ radicals in air at 295 K, *Int. J. Chem. Kinet.*, *21*, 1077–1089.

Wallington, T.J., M.D. Hurley, J.C. Ball, T. Ellermann, O.J. Nielsen, and J. Sehested (1994b), Atmospheric chemistry of HFC-152: UV absorption spectrum of CH_2FCFHO_2 radicals, kinetics of the reaction $CH_2FCFHO_2 + NO \rightarrow CH_2FCHFO + NO_2$, and fate of the alkoxy radical CH_2FCFHO, *J. Phys. Chem.*, *98*, 5435–5440.

Wallington, T.J., M.D. Hurley, J.C. Ball and E.W. Kaiser (1992d), Atmospheric chemistry of hydrofluorocarbon 134a; fate of the alkoxy radical CF_3CFHO, *Environ. Sci. Technol.*, *26*, 1318–1324.

Wallington, T.J., M.D. Hurley, J.M. Fracheboud, J.J. Orlando, G.S. Tyndall, J. Sehested, T.E. Møgelberg, and O.J. Nielsen (1996b), Role of excited CF_3CFHO radicals in the atmospheric chemistry of HFC-134a, *J. Phys. Chem.*, *100*, 18116–18122.

Wallington, T.J., M.D. Hurley, O.J. Nielsen, and J. Sehested (1994c), Atmospheric chemistry of CF_3CO_x radicals: fate of CF_3CO radicals, the UV absorption spectrum of $CF_3C(O)O_2$ radicals, and kinetics of the reaction $CF_3C(O)O_2 + NO \rightarrow CF_3C(O)O + NO_2$, *J. Phys. Chem.*, *98*, 5686–5694.

Wallington, T.J., M.D. Hurley, and W.F. Schneider (1996c), Atmospheric chemistry of CH_3Cl: mechanistic study of the reaction of CH_2ClO_2 radicals with HO_2, *Chem. Phys. Lett.*, *251*, 164–173.

Wallington, T.J., M.D. Hurley, W.F. Schneider, J. Sehested, and O.J. Nielsen (1994d), Mechanistic study of the gas phase reaction of CH_2FO_2 radicals with HO_2, *Chem. Phys. Lett.*, *218*, 34–42.

Wallington, T.J., M.D. Hurley, J. Xia, D.J. Wuebbles, S. Sillman, A. Ito, J.E. Penner, D.A. Ellis, J. Martin, S.A. Mabury, O.J. Nielsen, and M.P.S. Andersen (2006), Formation of $C_8F_{17}COOH$ (PFNA), $C_7F_{15}COOH$ (PFOA), and other perfluorocarboxylic acids (PFCAs) during the atmospheric oxidation of 8:2 fluorotelomer alcohol (n-$C_8F_{17}CH_2CH_2OH$), *Environ. Sci. Technol.*, *40*, 924–930.

Wallington, T.J., D.M. Neuman, and M.J. Kurylo (1987b), Kinetics of the gas phase reactions of hydroxyl radicals with ethane, benzene, and a series of halogenated benzenes over the temperature range 234–438 K, *Int. J. Chem. Kinet.*, *19*, 725–739.

Wallington, T.J., O.J. Nielsen, and J. Sehested (1997), Reactions of organic peroxy radicals in the gas phase, in *Peroxyl Radicals*, edited by Z. Alfassi, pp. 113–172, John Wiley, New York, ISBN 0-471-97065-4.

Wallington, T.J., O.J. Nielsen, and K. Sehested (1999), Kinetics of the reaction of CH_3O_2 with NO_2, *Chem. Phys. Lett.*, *313*, 456–460.

Wallington, T.J., J.J. Orlando, and G.S. Tyndall (1995a), Atmospheric chemistry of chloroalkanes: intramolecular elimination of HCl over the temperature range 264–336 K, *J. Phys. Chem.*, *99*, 9437–9442.

Wallington, T.J., W.F. Schneider, T.E. Møgelberg, O.J. Nielsen, and J. Sehested (1995b), Atmospheric chemistry of FCO_x radicals: kinetic and mechanistic study of the $FC(O)O_2 + NO_2$ reaction, *Int. J. Chem. Kinet.*, *27*, 391–402.

Wallington, T.J., W.F. Schneider, D.R. Worsnop, O.J. Nielsen, J. Sehested, W. DeBruyn, and J.A. Shorter (1994e), Environmental impact of CFC replacements: HFCs and HCFCs, *Environ. Sci. Technol.*, *28*, 320A–326A.

Wallington, T.J., J. Sehested, and O.J. Nielsen (1994f), Atmospheric chemistry of $CF_3C(O)O_2$ radicals. Kinetics of their reaction with NO_2 and kinetics of the thermal decomposition of the product $CF_3C(O)O_2NO_2$, *Chem. Phys. Lett.*, *226*, 563–569.

Wallington, T.J., L.M. Skewes, and W.O. Siegl (1989b), A relative rate study of the reaction of Cl with a series of chloroalkanes at 295 K, *J. Phys. Chem.*, *93*, 3649–3651.

Wallington, T.J., L.M. Skewes, W.O. Siegl, C.-H. Wu, and S.M. Japar (1988c), Gas phase reaction of Cl atoms with a series of oxygenated organic species at 295 K, *Int. J. Chem. Kinet.*, *20*, 867–875.

Wang, B., H. Hou, and Y. Gu (1999), Ab initio studies of the reaction of $O(^3P)$ with CHFCl radical, *J. Phys. Chem. A*, *103*, 5075–5079.

Wang, J., H. Chen., G.P. Glass, and R.F. Curl (2003), Kinetic study of the reaction of acetaldehyde with OH, *J. Phys. Chem. A*, *107*, 10834–10844.

Wang, J.J., and L.F. Keyser (1999), Kinetics of the $Cl(^2P_J) + CH_4$ reaction: Effects of secondary chemistry below 300 K, *J. Phys. Chem. A*, *103*, 7460–7469.

Wang, S.C., S.E. Paulson, D. Grosjean, R.C. Flagan, and J.H. Seinfeld (1992), Aerosol formation and growth in atmospheric organic/NO_x systems—1. Outdoor smog chamber studies of C_7- and C_8-hydrocarbons, *Atmos. Environ.*, *26A*, 403–420.

Wantuck, P.J., R.C. Oldenborg, S.L. Baughcum, and K.R. Winn (1987), Removal rate constant measurement for CH_3O by O_2 over the 298–973 K range, *J. Phys. Chem.*, *91*, 4653–4655.

Warneck, P. (2001), Photodissociation of acetone in the troposphere: an algorithm for the quantum yield, *Atmos. Environ.*, *35*, 5773–5777.

Waschewsky, G.C.G., P.W. Kash, T.L. Myers, D.C. Kitchen, and L.J. Butler (1994), What Woodward and Hoffmann didn't tell us: The failure of the Born–Oppenheimer approximation in competing reaction pathways, *J. Chem. Soc. Faraday Trans.*, *90*, 1581–1598.

Waterland, R.L., and K.D. Dobbs (2006), *Proceedings of 19th International Symposium on Gas Kinetics*, edited by P. Dagaut and A. Mellouki, pp. 111–112, Orleans, France, July.

Watson, R. (1977), Rate constants for reactions of ClO_x of atmospheric interest, *J. Phys. Chem. Ref. Data*, *6*, 871–917.

Watson, R.T., G. Machado, B. Conaway, S. Wagner, and D.D. Davis (1977), A temperature dependent kinetics study of the reaction of OH with CH_2ClF, $CHCl_2F$, $CHClF_2$, CH_3CCl_3, CH_3CF_2Cl, and $CF_2ClCFCl_2$, *J. Phys. Chem.*, *81*, 256–262.

Watson, R., G. Machado, S. Fischer, and D.D. Davis (1976), A temperature dependence kinetics study of the reactions of Cl ($^2P_{3/2}$) with O_3, CH_4, and H_2O_2, *J. Chem. Phys.*, *65*, 2126–2138.

Watson, R.T., A.R. Ravishankara, G. Machado, S. Wagner, and D.D. Davis (1979), A kinetics study of the temperature dependence of the reactions of OH($^2\Pi$) with CF_3CHCl_2, CF_3CHClF, and CF_2ClCH_2Cl, *Int. J. Chem. Kinet.*, *11*, 187–197.

Wayne, R.P., I. Barnes, P. Biggs, J.P. Burrows, C.E. Canosa-Mas, J. Hjorth, G. Le Bras, G.K. Moortgat, D. Perner, G. Poulet, G. Restelli, and H. Sidebottom (1991), The nitrate radical: Physics, chemistry, and the atmosphere, *Atmos. Environ. A*, *25*, 1–203.

Weaver, J., J. Meagher, and J. Heicklen (1976/1977), Photo-oxidation of CH_3CHO vapor at 3130 Å, *J. Photochem.*, *6*, 111–126.

Weaver, J., R. Shortridge, J. Meagher, and J. Heicklen (1975), The photooxidation of $CD_3N_2CD_3$, *J. Photochem.*, *4*, 109–120.

Weibel, D.E., G.A. Arguello, E.R. de Staricco, and E.H. Staricco (1995), Quantum yield in the gas phase photolysis of perfluoroacetyl chloride: a comparison with related compounds, *J. Photochem. Photobiol. A*, *86*, 27–31.

Weir, D.S. (1961), The photolysis and fluorescence of diethyl ketone and diethyl ketone–biacetyl mixtures at 3130 and 2537 Å, *J. Am. Chem. Soc.*, *83*, 2629–2033.

Weir, D.S. (1962), Energy exchange in the 3-methyl–2-butanone–biacetyl system at 3130 Å, *J. Am. Chem. Soc.*, *84*, 4039–4040.

Weston, R.E. (1996), Possible greenhouse effects of tetrafluoromethane and carbon dioxide emitted from aluminum production, *Atmos. Environ.*, *30*, 2901–2910.

Wettack, F.S., and W.A. Noyes, Jr. (1968), The direct and sensitized photolyses of 2-pentanone, *J. Am. Chem. Soc.*, *90*, 3901–3906.

Whiteway, S.G., and C.R. Masson (1955), The carbon monoxide quantum yield in the photolysis of diisopropyl ketone, *J. Am. Chem. Soc.*, *77*, 1508–1509.

WHO (1987), *Air Quality Guidelines for Europe*, World Health Organisation Regional Publications, European series No.23, WHO Regional Office for Europe, Copenhagen, Denmark.

Whytock, D.A., and K.O. Kutschke (1968),The primary process in the photolysis of hexafluoroacetone vapour. I. Photodecomposition studies, *Proc. R. Soc.*, *A306*, 503–510.

Whytock, D.A., J.H. Lee, J.V. Michael, W.A. Payne, and L.J. Stief (1977), Absolute rate of the reaction of Cl(^2P) with methane from 200–500 K, *J. Chem. Phys.*, *66*, 2690–2695.

Wiebe, H.A., A. Villa, T.M. Hellman, and J. Heicklen (1973), Photolysis of methyl nitrite in the presence of nitric oxide, nitrogen dioxide, and oxygen, *J. Am. Chem. Soc.*, *95*, 7–13.

Wijnen, M.H.J. (1961), Photolysis of phosgene in the presence of ethylene, *J. Am. Chem. Soc.*, *83*, 3014–3017.

Wijnen, M.H.J. (1962), Initial stages of the chlorine atom induced polymerization of acetylene, *J. Chem. Phys.*, *36*, 1672–1675.

Wild, O., and M.J. Prather (2000), Excitation of the primary tropospheric chemical mode in a global three-dimensional model, *J. Geophys. Res.*, *105*, 24647–24660.

Wilson, E.W., Jr., W.A. Hamilton, H.R. Kennington, B. Evans III, N.W. Scott, and W.B. DeMore (2006), Measurement and estimation of rate constants for the reactions of hydroxyl radical with several alkanes and cycloalkanes, *J. Phys. Chem. A.*, *110*, 3593–3604.

Wilson, E.W, Jr., A.M. Jacoby, S.J. Kukta, L.E. Gilbert, and W.B. DeMore (2003), Rate constants for reaction of CH_2FCH_2F (HFC-152) and CH_3CHF_2 (HFC-152a) with hydroxyl radicals, *J. Phys. Chem. A*, *107*, 9357–9361.

Wilson, E.W., Jr., A.A. Sawyer, and H.A. Sawyer (2001), Rates of reactions for cyclopropane and difluoromethoxydifluoromethane with hydroxyl radicals, *J. Phys. Chem. A*, *105*, 1445–1448.

Wilson, J.E., and W.A. Noyes, Jr. (1943), Photochemical studies. XXXVI. Quantum yields during the photochemical decomposition of *n*-butyl methyl ketone, *J. Am. Chem. Soc.*, *65*, 1547–1550.

Winer, A.M., A.C. Lloyd, K.R. Darnall, R. Atkinson, and J.N. Pitts, Jr. (1977), Rate constants for the reaction of OH radicals with *n*-propyl acetate, *sec*-butyl acetate, tetrahydrofuran, and peroxyacetyl nitrate, *Chem. Phys. Lett.*, *51*, 221–226.

Winer, A.M., A.C. Lloyd, K.R. Darnall, and J.N. Pitts, Jr. (1976), Relative rate constants for the reaction of the hydroxyl radical with selected ketones, chloroethenes, and monoterpene hydrocarbons, *J. Phys. Chem.*, *80*, 1635–1639.

Wingenter, O.W., M.K. Kubo, N.J. Blake, T.W., Smith Jr., D.R. Blake, and F.S. Rowland (1996), Hydrocarbon and halocarbon measurements as photochemical and dynamical indicators of atmospheric hydroxyl, atomic chlorine, and vertical mixing obtained during Lagrangian flights, *J. Geophys. Res.*, *101*, 4331–4340.

Wingenter, O.W., B.C. Sive, N.J. Blake, D.R. Blake, and F.S. Rowland (2005), Atomic chlorine concentrations derived from ethane and hydroxyl measurements over the equatorial Pacific Ocean: Implication for dimethyl sulfide and bromine monoxide, *J. Geophys. Res.*, *110*, D20308, doi:10.1029/2005JD005875.

WMO (1986), *Atmospheric Ozone 1985*, Assessment of our understanding of the processes controlling its present distribution and change, World Meteorological Organization Global Ozone Research and Monitoring Project—Report No. 16, Geneva, Switzerland.

WMO (1989), *Scientific Assessment of Stratospheric Ozone: 1989*, World Meteorological Organization Global Ozone Research and Monitoring Project—Report No. 20, Geneva, Switzerland.

WMO (2003), *Scientific Assessment of Ozone Depletion: 2002*, World Meteorological Organization Global Ozone Research and Monitoring Project—Report No. 47, Geneva, Switzerland.

Wollenhaupt, M., S.A. Carl, A. Horowitz, and J.N. Crowley (2000), Rate coefficients for reaction of OH with acetone between 202 and 395 K, *J. Phys. Chem. A*, *104*, 2695–2705.

Wollenhaupt, M., and J.N. Crowley (2000), Kinetic studies of the reactions $CH_3 + NO_2 \rightarrow$ products, $CH_3O + NO_2 \rightarrow$ products, and $OH + CH_3C(O)CH_3 \rightarrow CH_3C(O)OH + CH_3$, over a range of temperature and pressure, *J. Phys. Chem. A*, *104*, 6429–6438.

Woodbridge, E.I., T.R. Fletcher, and S.R. Leone (1988), Photofragmentation of acetone at 193 nm: Rotational- and vibrational-state distributions of the CO fragment by time-resolved FTIR emission spectroscopy, *J. Phys. Chem.*, *92*, 5387–5393.

Wu, F., and R.W. Carr (1991), An investigation of temperature and pressure dependence of the reaction of CF_2ClO_2 radicals with nitrogen dioxide by flash photolysis and time resolved mass spectrometry, *Int. J. Chem. Kinet.*, *23*, 701–715.

Wu, F., and R.W. Carr (1992), Time-resolved observation of the formation of CF_2O and CFClO in the $CF_2Cl + O_2$ and $CFCl_2 + O_2$ reactions: The unimolecular elimination of chlorine atoms from CF_2ClO and $CFCl_2O$ radicals, *J. Phys. Chem.*, *96*, 1743–1748.

Wu, F., and R.W. Carr (1995), Observation of the $CFCl_2CH_2O$ radical in the flash photolysis of $CFCl_2CH_3$ in the presence of O_2. Kinetics of the reactions of Cl and ClO with $CFCl_2CH_2O_2$ and CH_2CFClO_2, *J. Phys. Chem.*, *99*, 3128–3136.

Wu, F., and R.W. Carr (1996a), Temperature dependence of the reaction of NO with $CFCl_2CH_2O_2$ and CH_3CFClO_2 radicals, *Int. J. Chem. Kinet.*, *28*, 9–19.

Wu, F., and R.W. Carr (1996b), Kinetic study of the reaction of the $CFCl_2CH_2O$ radical with O_2, *J. Phys. Chem. A*, *100*, 9352–9359.

Wu, F., and R.W. Carr (1999), The chloromethoxy radical: kinetics of the reaction with O_2 and the unimolecular elimination of HCl at 306 K, *Chem. Phys. Lett.*, *305*, 44–50.

Wu, F., and R.W. Carr (2001), Kinetics of CH_2ClO radical reactions with O_2 and NO, and the unimolecular elimination of HCl, *J. Phys. Chem. A*, *105*, 1423–1432.

Wu, F., and R.W. Carr (2003), Reactivity of the CF_3CFHO radical: thermal decomposition and reaction with O_2, *J. Phys. Chem. A*, *107*, 10733–10742.

Wu, H., Y. Mu, X. Zhang, and G. Jiang (2003), Relative rate constants for the reactions of hydroxyl radicals and chlorine atoms with a series of aliphatic alcohols, *Int. J. Chem. Kinet.*, *35*, 81–87.

Wuebbles, D.J. (1983), Chlorocarbon emission scenarios: Potential impact on stratospheric ozone, *J. Geophys. Res.*, *88*, 1433–1443.

Xing, J.-H., and A. Miyoshi (2005), Rate constants for the reactions of a series of alkylperoxy radicals with NO, *J. Phys. Chem. A*, *109*, 4095–4101.

Xing, J.-H., Y. Nagai, M. Kusuhara, and A. Miyoshi (2004), Reactions of methyl- and ethylperoxy radicals with NO studied by time-resolved negative ionization mass spectrometry, *J. Phys. Chem. A*, *108*, 10458–10463.

Xing, S-B., S-H. Shi, and L-X. Qiu (1992), Kinetics studies of reactions of OH radicals with four haloethanes. Part I. Experiment and BEBO calculation, *Int. J. Chem. Kinet.*, *24*, 1–10.

Xu, D., J.S. Francisco, J. Huang, and W.M. Jackson (2002a), Ultraviolet photodissociation of bromoform at 234 and 267 nm by means of ion velocity imaging, *J. Chem. Phys.*, *117*, 2578–2585.

Xu, D., J. Huang, J.S. Francisco, J.C. Hansen, and W.M. Jackson (2002b), Photodissociation of carbonic dibromide at 267 nm: Observation of three-body dissociation and molecular elimination of Br_2, *J. Chem. Phys.*, *117*, 7483–7490.

Yadav, J.S., and J.D. Goddard (1986), Acetaldehyde photochemistry: The radical and molecular dissociations, *J. Chem. Phys.*, *84*, 2682–2690.

Yamabe, S., T. Minato and Y. Osamura (1979), A theoretical study on the paths of photodissociation: cyclopropanone $\rightarrow CO + C_2H_4$, *J. Am. Chem. Soc.*, *101*, 4525–4531.

Yamada, T., T.D. Fang, P.H. Taylor, and R.J. Berry (2000), Kinetics and thermochemistry of the OH radical reaction with CF_3CCl_2H and CF_3CFClH, *J. Phys. Chem. A*, *104*, 5013–5022.

Yamada, T., P.H. Taylor, A. Goumri, and P. Marshall (2003), The reaction of OH with acetone and acetone-d_6 from 298 to 832 K: Rate coefficients and mechanism, *J. Chem. Phys.*, *119*, 10600–10606.

Yamaguchi, Y., S.S. Wesolowski, T.J. Van Huis, and F.F. Schaefer (1998), The unimolecular dissociation of H_2CO on the lowest triplet potential-energy surface, *J. Chem. Phys.*, *108*, 5281–5288.

Yang, X., P. Felder, and J.R. Huber (1993), Photodissociation of methyl nitrate in a molecular beam, *J. Phys. Chem.*, *97*, 10903–10910.

Yarwood, G., N. Peng, and H. Niki (1992), FTIR spectroscopic study of the chlorine- and bromine-atom-initiated oxidation of ethane, *Int. J. Chem. Kinet.*,*24*, 369–384.

Yetter, R.A., H. Rabitz, F.L. Dryer, R.G. Maki, and R.B. Klemm (1989), Evaluation of the rate constant for the reaction $OH+H_2CO$: Application of modeling and sensitivity analysis techniques for determination of the product branching ratio, *J. Chem. Phys.*, *91*, 4088–4097.

Yeung, E.S., and C.B. Moore (1972), Isotopic separation by photopredissociation, *Appl. Phys. Lett.*, *21*, 109–110.

Yujing, M., and A. Mellouki (2000), The near-UV absorption cross sections for several ketones, *J. Photochem. Photobiol. A: Chemistry*, *134*, 31–36.

Yujing M., and A. Mellouki (2001), Temperature dependence for the rate constants of the reaction of OH radicals with selected alcohols, *Chem. Phys. Lett.*, *333*, 63–68.

Yujing, M., Y. Wenxiang, P. Yuexiang, and Q. Lianxiong (1993), Chemical kinetics of OH radical with HCFC-22 and the residual life time in the atmosphere, *J. Environ. Sci.*, *5*, 481–490.

Zabarnick, S., and J. Heicklen (1985a), Reactions of alkoxy radicals with O_2. I. C_2H_5O radicals, *Int. J. Chem. Kinet.*, *17*, 455–476.

Zabarnick, S., and J. Heicklen (1985b), Reactions of alkoxy radicals with O_2. II. n-C_3H_7O radicals, *Int. J. Chem. Kinet.*, *17*, 477–501.

Zabarnick, S., and J. Heicklen (1985c), Reactions of alkoxy radicals with O_2. III. i-C_4H_9O radicals, *Int. J. Chem. Kinet.*, *17*, 503–524.

Zabarnick, S., J.W. Fleming, and M.C. Lin (1988), Kinetics of hydroxyl radical reactions with formaldehyde and 1,3,5-trioxane between 290 and 600 K, *Int. J. Chem. Kinet.*, *20*, 117–129.

Zabel, F. (1995), Unimolecular decomposition of peroxynitrates, *Z. Phys. Chem.*, *188*, 119–142.

Zabel, F. (1999), Reaktionswege von Alkoxyradikalen unter atmosphärishen Bedingungen and UV-Absorptionspektrum von Carbonylverbindungen, in *Annual Report 1998 on BMBF-Project TFS/LT3*, coordinated by K.H. Becker, Bergische Universität Wuppertal, Germany, FKZ07TFS30, p. 69.

Zabel, F., F. Kirchner, and K.H. Becker (1994), Thermal decomposition of $CF_3C(O)O_2NO_2$, $CClF_2C(O)O_2NO_2$, $CCl_2FC(O)O_2NO_2$, and $CCl_3C(O)O_2NO_2$, *Int. J. Chem. Kinet.*, *26*, 827–845.

Zabel, F., A. Reimer, K.H. Becker, and E.H. Fink (1989), Thermal decomposition of alkyl peroxynitrates, *J. Phys. Chem.*, *93*, 5500–5507.

Zador, J., V. Wagner, K. Wirtz, and M.J. Pilling, (2005), Quantitative assessment of uncertainties for a model of tropospheric ethene oxidation using the European Photoreactor (EUPHORE). *Atmos. Environ.*, *39*, 2805–2817.

Zahniser, M.S., B.M. Berquist, and F. Kaufman (1978), Kinetics of the reaction $Cl + CH_4 \rightarrow CH_3 + HCl$ from 200° to 500° K, *Int. J. Chem. Kinet.*, *10*, 15–29.

Zahra, A., and W.A. Noyes, Jr. (1965), The photochemistry of methyl isopropyl ketone, *J. Phys. Chem.*, *69*, 943–948.

Zellner, R. (1987), Recent advances in free radical kinetics of oxygenated hydrocarbon radicals, *J. Chim. Phys.*, *84*, 403–407.

Zellner, R., G. Bednarek, A. Hoffmann, J.P. Kohlmann, V. Mors, and H. Saathoff (1994), Rate and mechanism of the atmospheric degradation of 2H-heptafluoropropane (HFC-227), *Ber. Bunsenges. Phys. Chem.*, *98*, 141–146.

Zellner, R., B. Fritz, and K. Lorenz (1986), Methoxy formation in the reaction of CH_3O_2 radicals with NO, *J. Atmos. Chem.*, *4*, 241–251.

Zeng, G., and J.A. Pyle (2003), Changes in tropospheric ozone between 2000 and 2100 modeled in a chemistry-climate model, *Geophys. Res. Lett.*, *30*, 1392, doi:10.1029/2002GL016708.

Zhang, D., J. Zhong, and L. Qiu (1997), Kinetics of the reaction of hydroxyl radicals with CH_2Br_2 and its implications in the atmosphere, *J. Atmos. Chem.*, *27*, 209–215.

Zhang, J.Y., T. Dransfield, and N.M. Donahue (2004a), On the mechanism for nitrate formation via the peroxy radical plus NO reaction, *J. Phys. Chem. A*, *108*, 9082–9095.

Zhang, L., K.M. Callahan, D. Derbyshire, and T.S. Dibble (2005), Laser-induced fluorescence spectra of 4-methylcyclohexoxy radical and perdeuterated cyclohexoxy radical and direct kinetic studies of their reactions with O_2, *J. Phys. Chem.*, *109*, 9232–9240.

Zhang, L., K.A. Kitney, M.A. Ferenac, W. Deng, and T.S. Dibble (2004b), LIF spectra of cyclohexoxy radical and direct kinetic studies of its reaction with O_2, *J. Phys. Chem. A*, *108*, 447–454.

Zhang, Z., R.E. Huie, and M.J. Kurylo (1992a), Rate constants for the reactions of OH with CH_3CFCl_2 (HCFC-141b), CH_3CF_2Cl (HCFC-142b), and CH_2FCF_3 (HFC-134a), *J. Phys. Chem.*, *96*, 1533–1535.

Zhang, Z., R. Liu, R.E. Huie, and M.J. Kurylo (1991), Rate constants for the gas phase reactions of the OH radical with $CF_3CF_2CHCl_2$ (HCFC-225ca) and CF_2ClCF_2CHClF (HCFC-225cb), *Geophys. Res. Lett.*, *18*, 5–7.

Zhang, Z., S. Padmaja, R.D. Saini, R.E. Huie, and M.J. Kurylo (1994), Reactions of hydroxyl radicals with several hydrofluorocarbons: the temperature dependencies of the rate constants for $CHF_2CF_2CH_2F$ (HFC-245ca), $CF_3CHFCHF_2$ (HFC-236ea), CF_3CHFCF_3 (HFC-227ea), and $CF_3CH_2CH_2CF_3$ (HFC-356ffa), *J. Phys. Chem.*, *98*, 4312–4315.

Zhang, Z., R.D. Saini, M.J. Kurylo, and R.E. Huie (1992b), Rate constants for the reactions of the hydroxyl radical with $CHF_2CF_2CF_2CHF_2$ and $CF_3CHFCHFCF_2CF_3$, *Chem. Phys. Lett.*, *200*, 230–234.

Zhang, Z., R.D. Saini, M.J. Kurylo, and R.E. Huie (1992c), A temperature dependent kinetic study of the reaction of the hydroxyl radical with CH_3Br, *Geophys. Res. Lett.*, *19*, 2413–2416.

Zhu, L., and J.W. Bozzelli (2002), Structures, rotational barriers, and thermochemical properties of chlorinated aldehydes and corresponding acetyl ($CC^{\bullet}{=}O$) and formyl methyl radicals ($C^{\bullet}C{=}O$) and additivity groups, *J. Phys. Chem. A*, *106*, 345–355.

Zhu, L., and C.-F. Ding (1997), Temperature dependence of the near UV absorption spectra and photolysis products of ethyl nitrate, *Chem. Phys. Lett.*, *265*, 177–184.

Zhu, L., T. Cronin, and A. Narang (1999), Wavelength-dependent photolysis of *i*-pentanal and *t*-pentanal from 380–330 nm, *J. Phys. Chem.*, *103*, 7248–7253.

Zhu, Q.J., J.R. Cao, Y. Wen, J.M. Zhang, X.A. Huang, W.Q. Fang, and X.J. Wu (1988), Photodissociation channels and energy partitioning in the photofragmentation of alkyl iodides, *Chem. Phys. Lett.*, *144*, 486–492.

Zou, P., J. Shu, T.J. Sears, G.E. Hall, and S.W. North (2004), Photodissociation of bromoform at 248 nm: Single and multiphoton processes, *J. Phys. Chem. A*, 108, 1482–1488.

Zuckermann, H., B. Schmitz, and Y. Haas (1988), Dissociation energy of an isolated triplet acetone molecule, *J. Phys. Chem.*, *92*, 4835–4837.

Appendix A

Structures of Some Common Alkanes

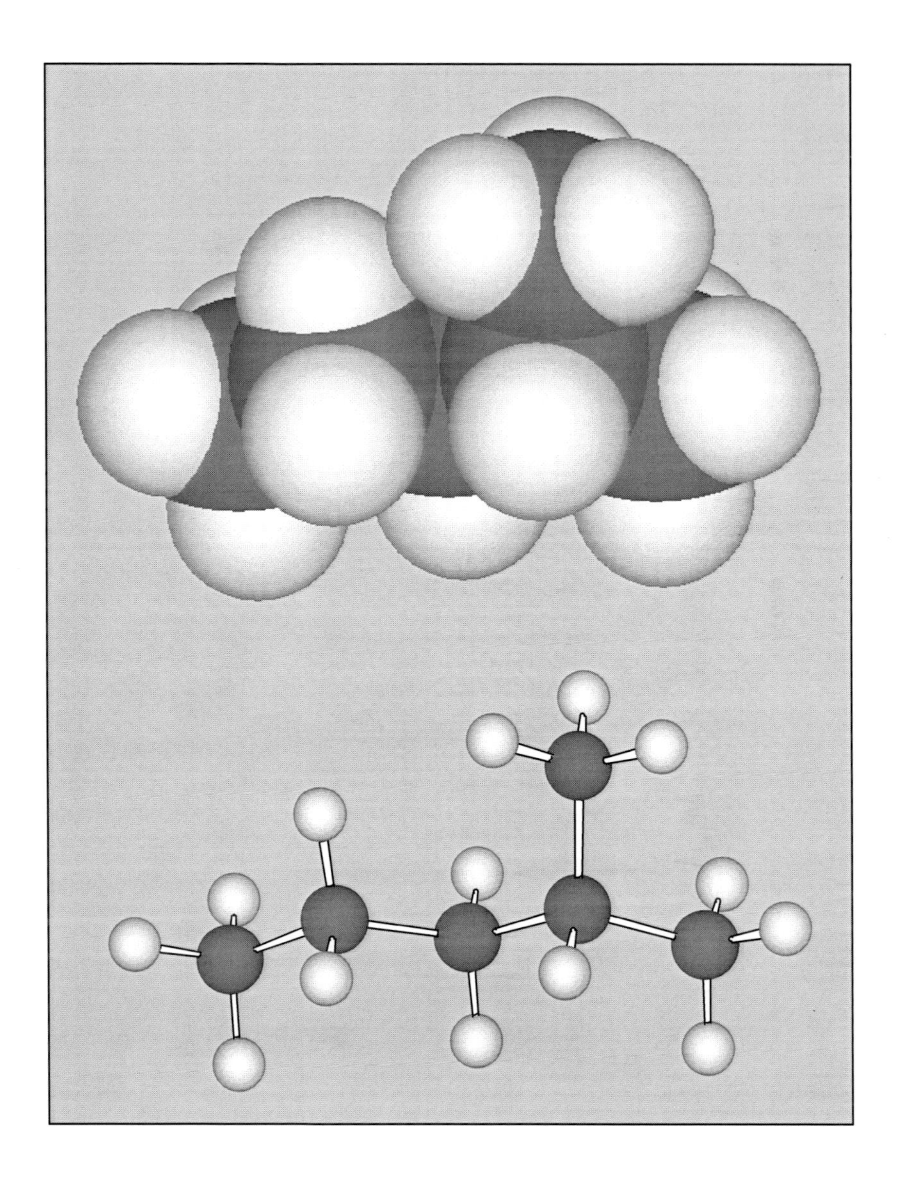

1. Acyclic (Open-Chain) Alkanes

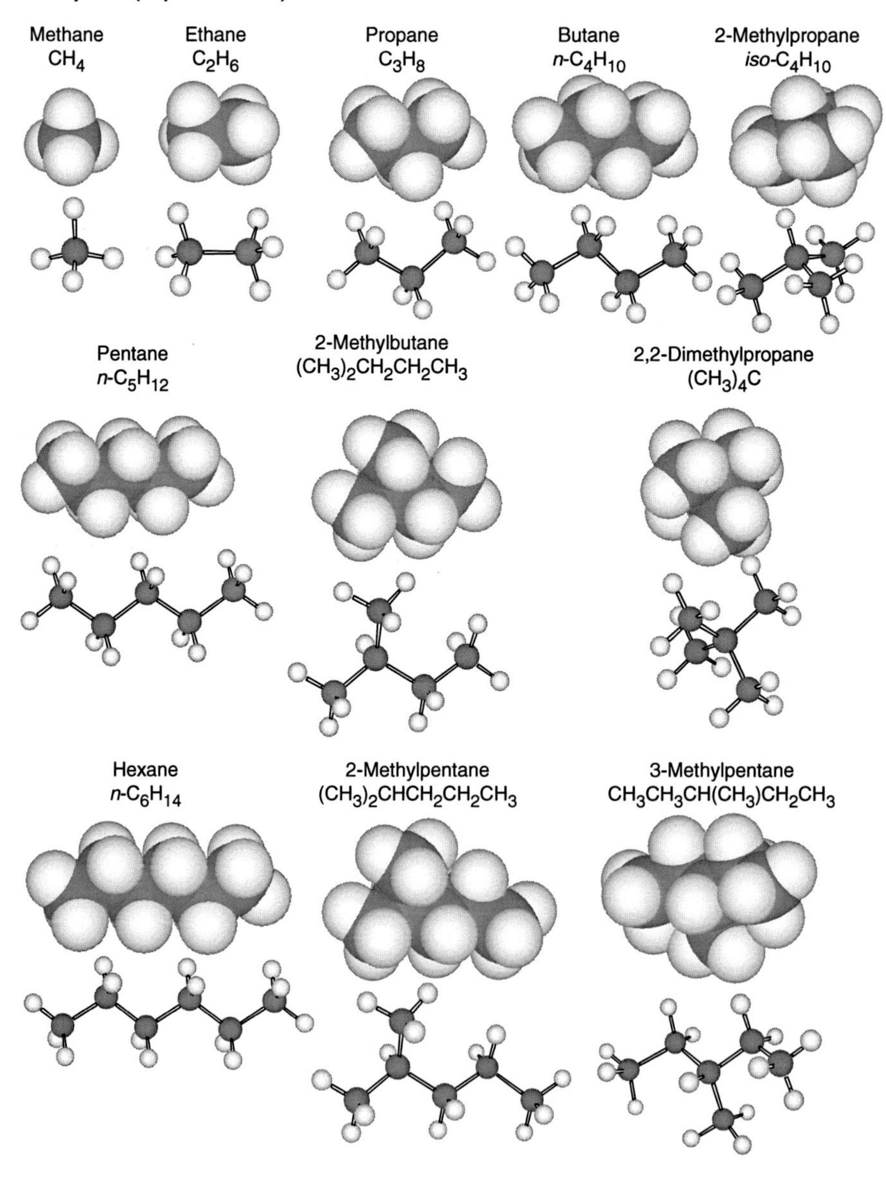

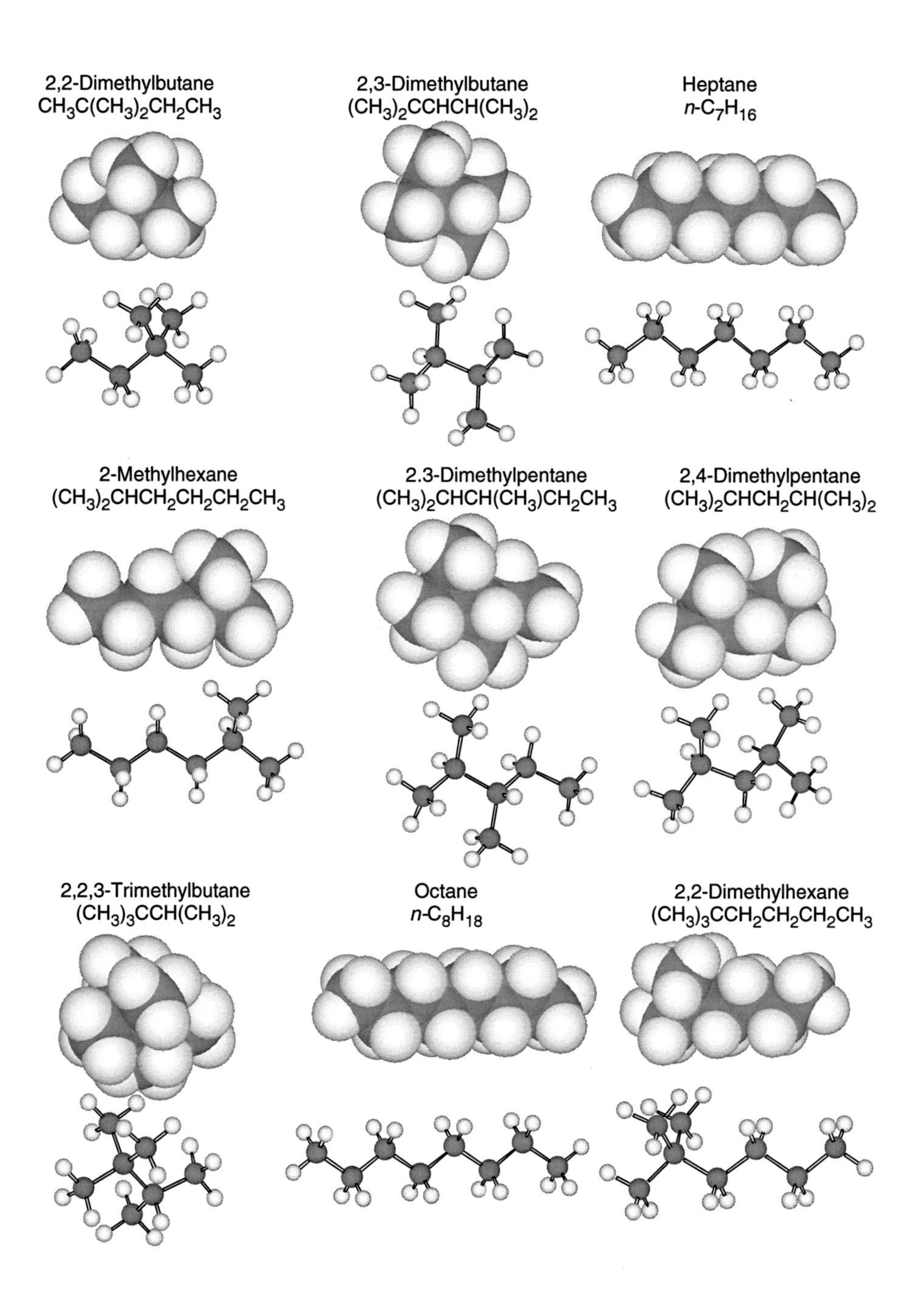
2,2-Dimethylbutane
$CH_3C(CH_3)_2CH_2CH_3$
2,3-Dimethylbutane
$(CH_3)_2CCHCH(CH_3)_2$
Heptane
n-C_7H_{16}
2-Methylhexane
$(CH_3)_2CHCH_2CH_2CH_2CH_3$
2.3-Dimethylpentane
$(CH_3)_2CHCH(CH_3)CH_2CH_3$
2,4-Dimethylpentane
$(CH_3)_2CHCH_2CH(CH_3)_2$
2,2,3-Trimethylbutane
$(CH_3)_3CCH(CH_3)_2$
Octane
n-C_8H_{18}
2,2-Dimethylhexane
$(CH_3)_3CCH_2CH_2CH_2CH_3$

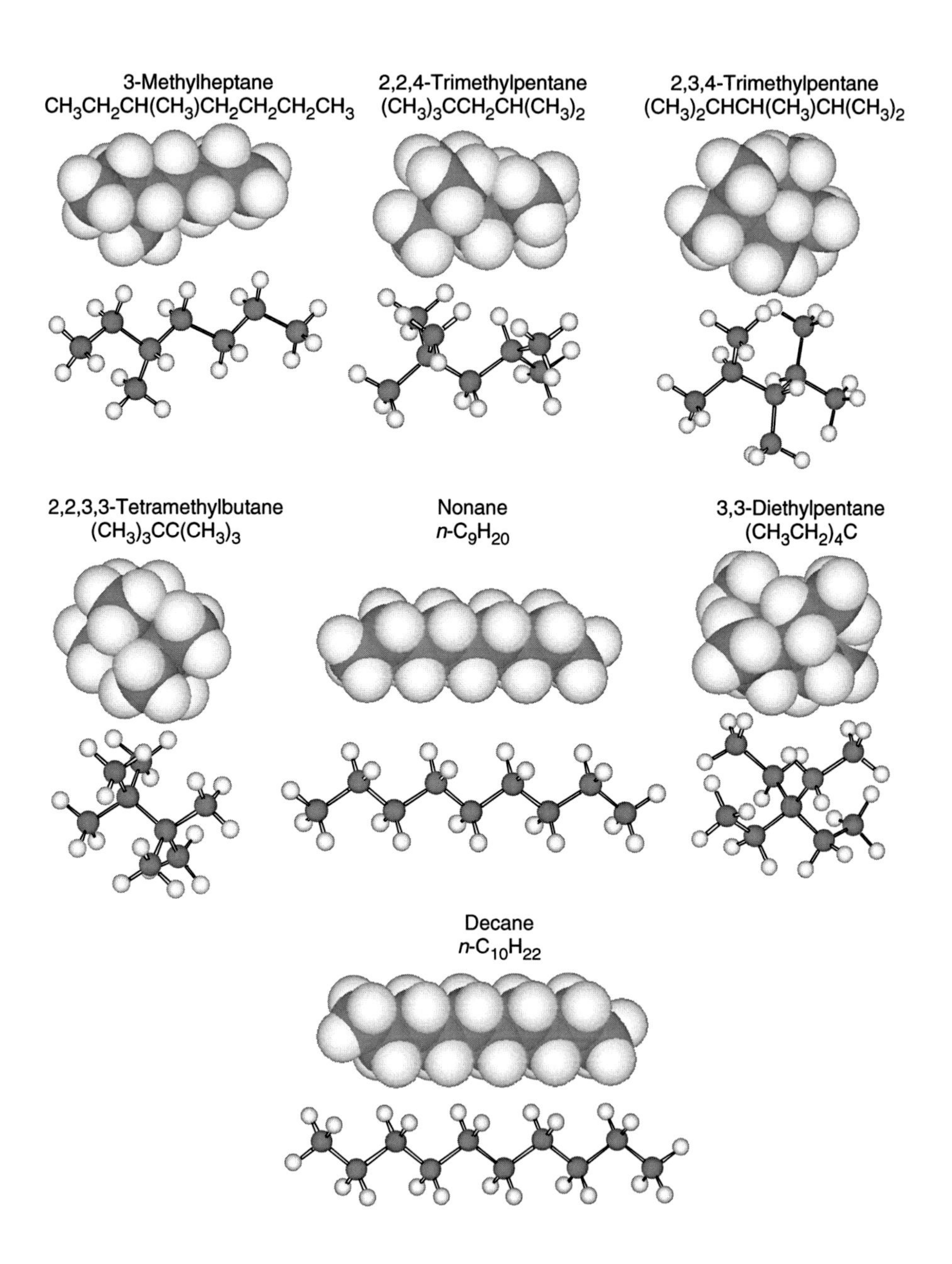
3-Methylheptane
$CH_3CH_2CH(CH_3)CH_2CH_2CH_2CH_3$
2,2,4-Trimethylpentane
$(CH_3)_3CCH_2CH(CH_3)_2$
2,3,4-Trimethylpentane
$(CH_3)_2CHCH(CH_3)CH(CH_3)_2$
2,2,3,3-Tetramethylbutane
$(CH_3)_3CC(CH_3)_3$
Nonane
n-C_9H_{20}
3,3-Diethylpentane
$(CH_3CH_2)_4C$
Decane
n-$C_{10}H_{22}$

2. Cyclic Alkanes

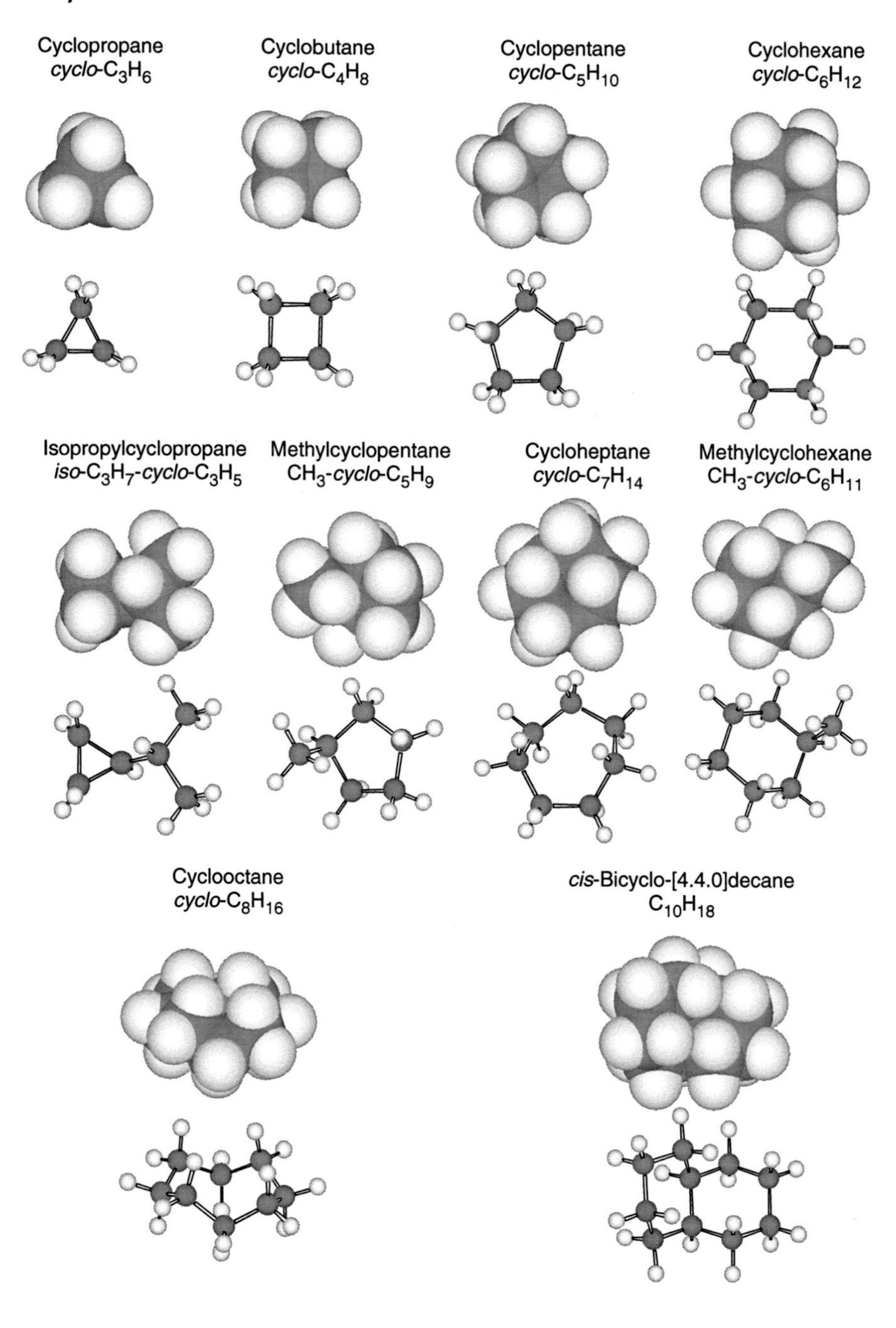

Appendix B

Structures of Some Common Haloalkanes

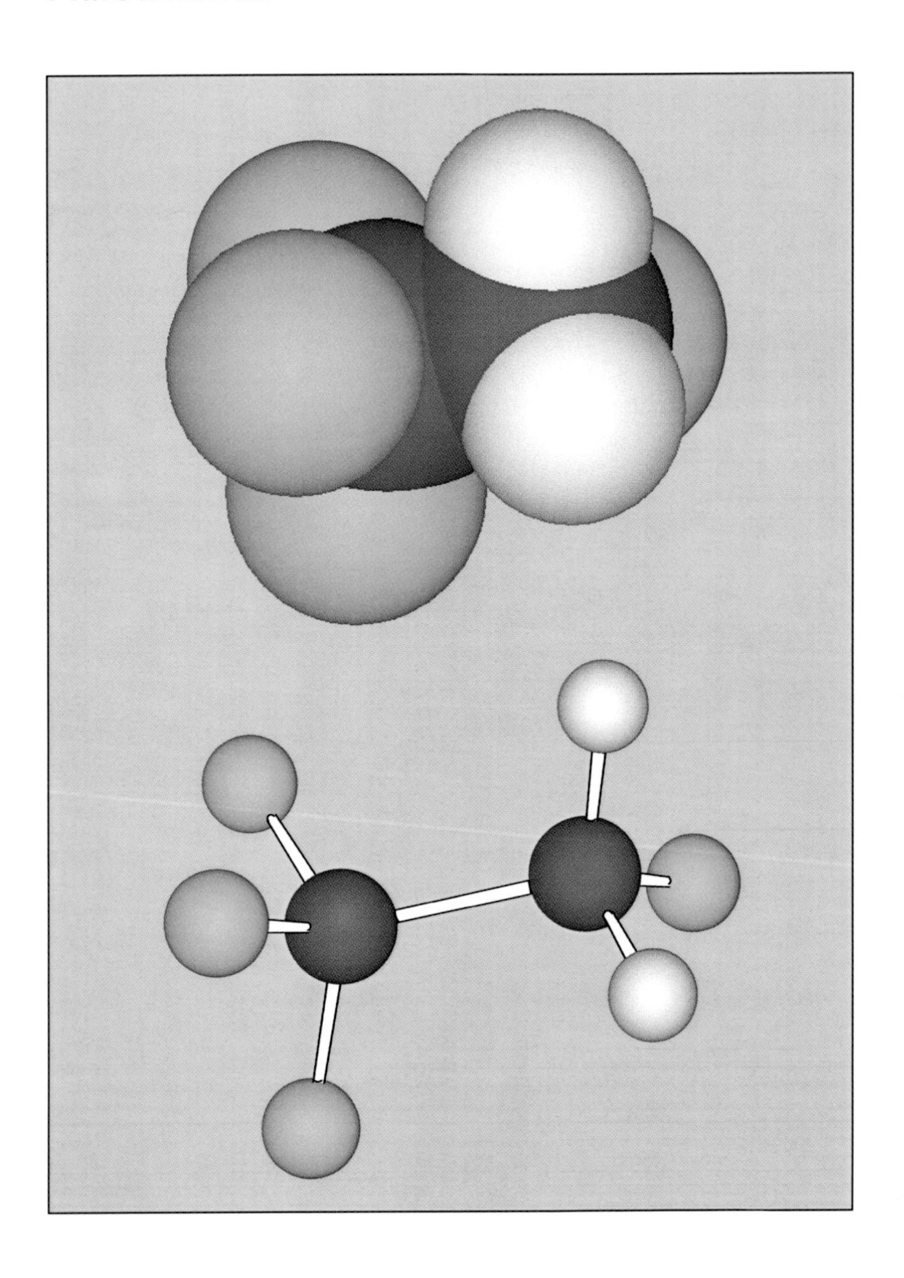

I. Halogen-Atom-Substituted Methanes

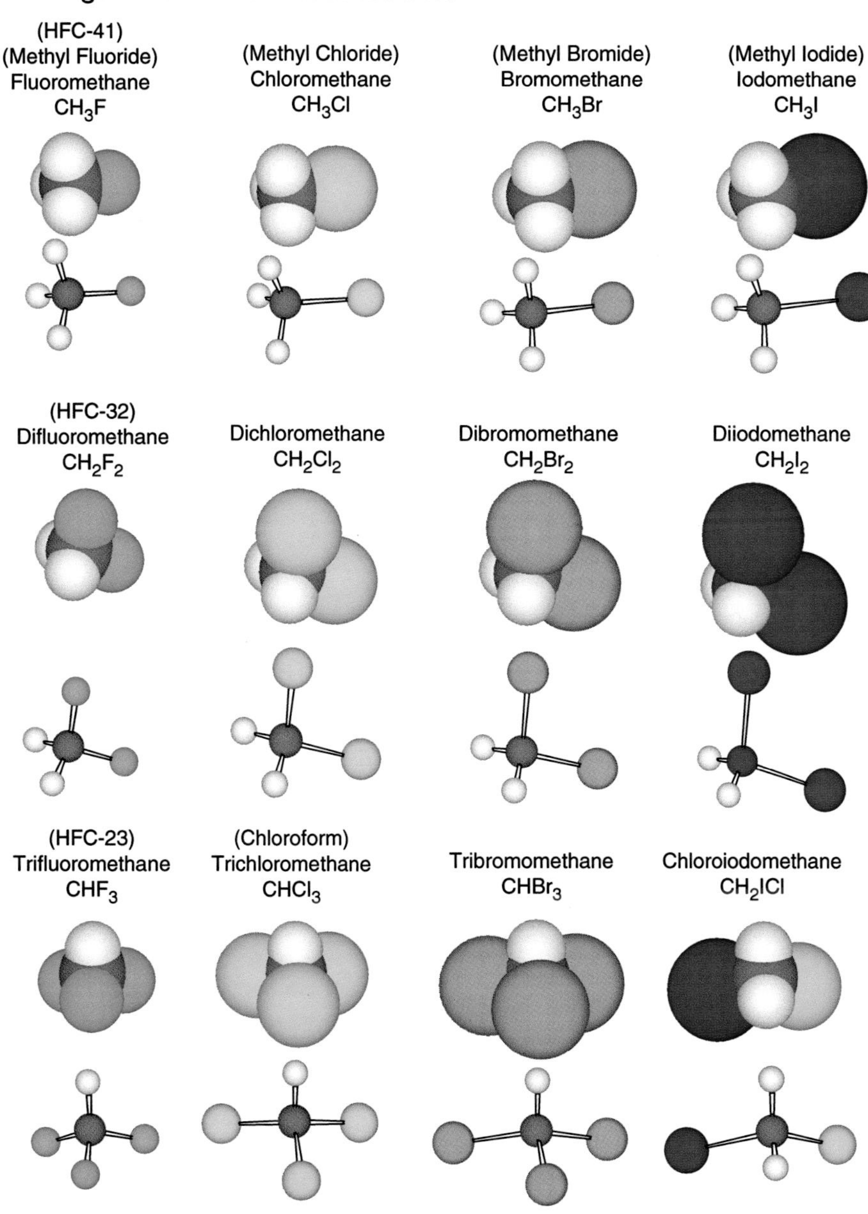

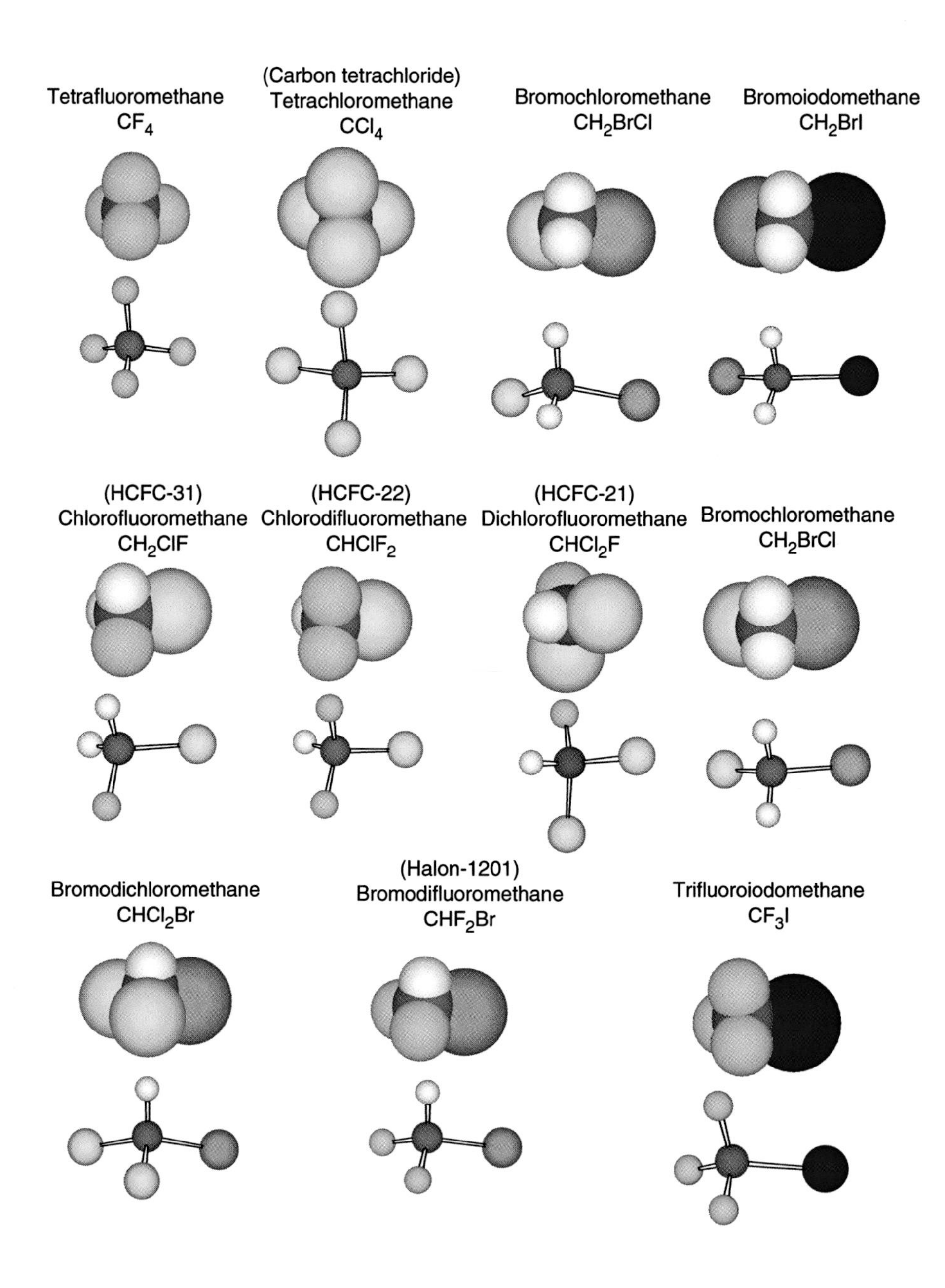
Tetrafluoromethane
CF4
(Carbon tetrachloride)
Tetrachloromethane
CCl4
Bromochloromethane
CH2BrCl
Bromoiodomethane
CH2BrI
(HCFC-31)
Chlorofluoromethane
CH2ClF
(HCFC-22)
Chlorodifluoromethane
CHClF2
(HCFC-21)
Dichlorofluoromethane
CHCl2F
Bromochloromethane
CH2BrCl
Bromodichloromethane
CHCl2Br
(Halon-1201)
Bromodifluoromethane
CHF2Br
Trifluoroiodomethane
CF3I

2. F-Substituted Ethanes

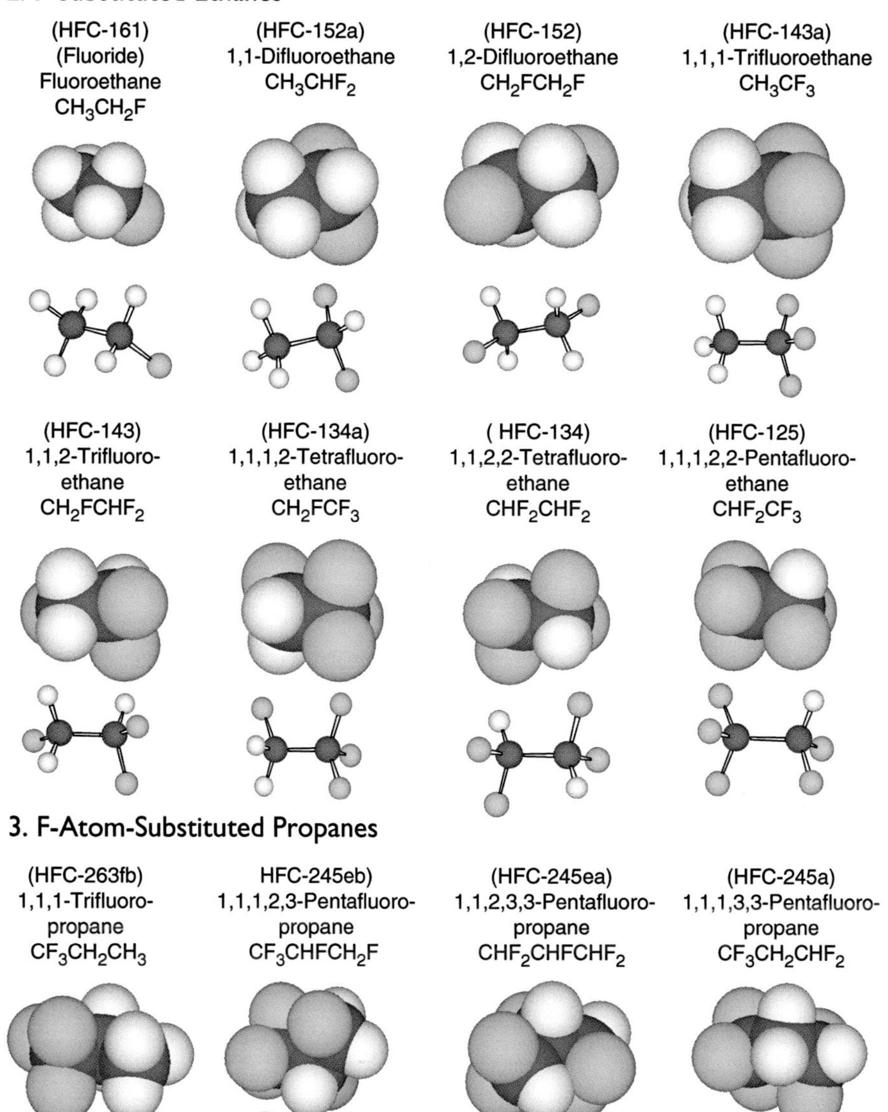

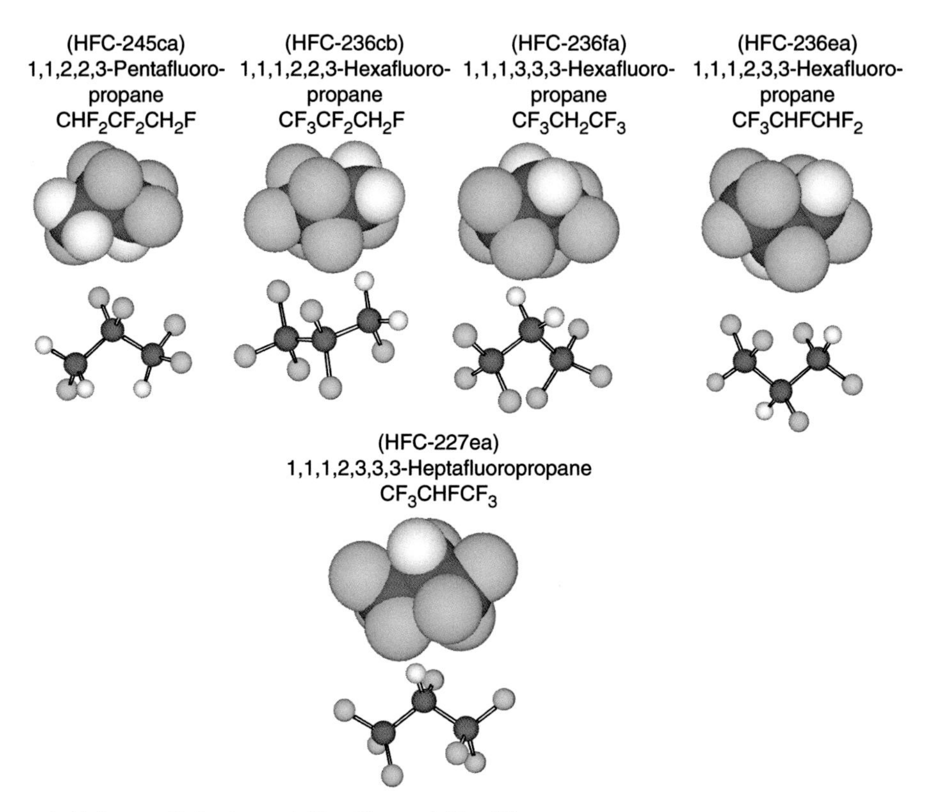

4. F-Atom-Substituted C_4, C_5, and C_6 Alkanes

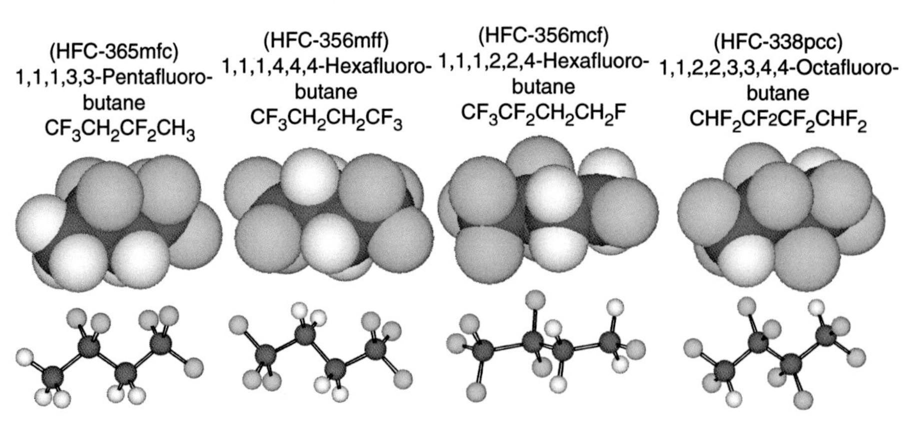

(HFC-458mfcf)
1,1,1,3,3,5,5,5-Octafluoropentane
$CF_3CH_2CF_2CH_2CF_3$

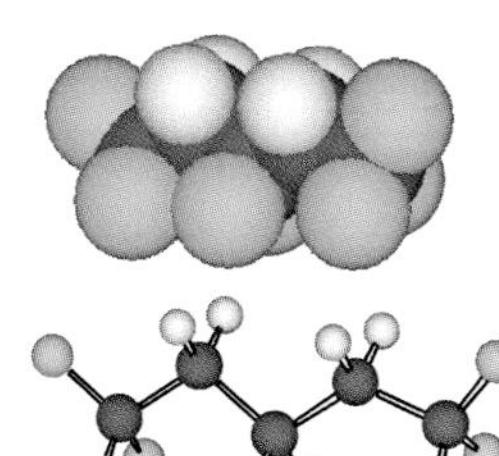

(HFC-43-10mee)
1,1,1,2,2,3,4,5,5,5-Decafluoropentane
$CF_3CF_2CHFCHFCF_3$

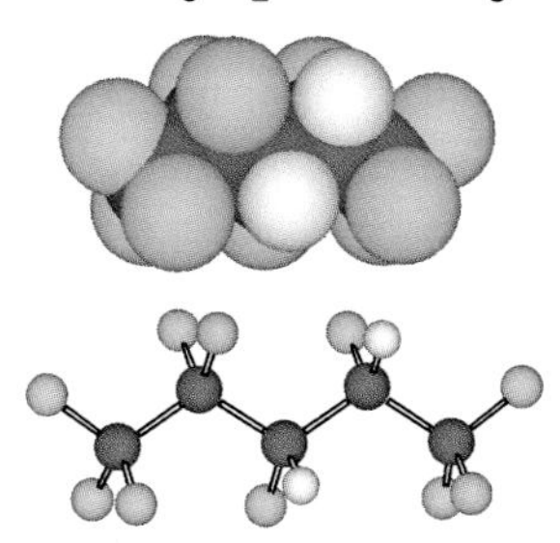

(HFC-55-10mcff)
1,1,1,2,2,5,5,6,6,6-Decafluorohexane
$CF_3CF_2CH_2CH_2CF_2CF_3$

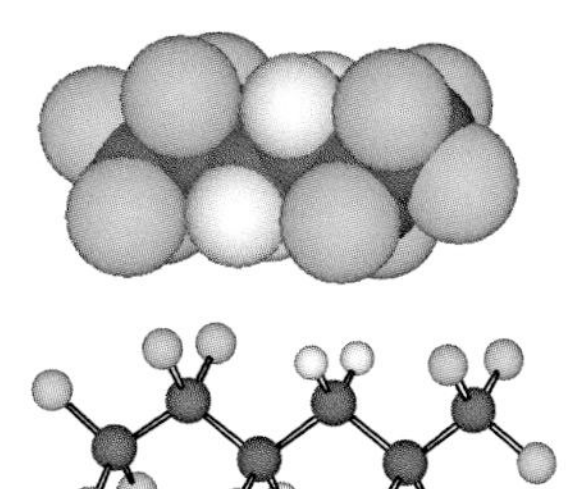

(HFC-52-13p)
1,1,1,2,2,3,3,4,4,5,5,6,6-Tridecafluorohexane
$CHF_2CF_2CF_2CF_2CF_2CF_3$

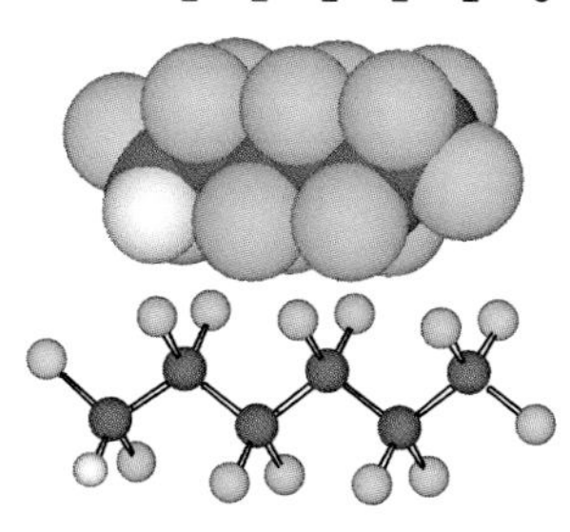

5. Structures of Some Cl-Atom-Substituted Ethanes

(Ethyl chloride)
Chloroethane
CH_3CH_2Cl

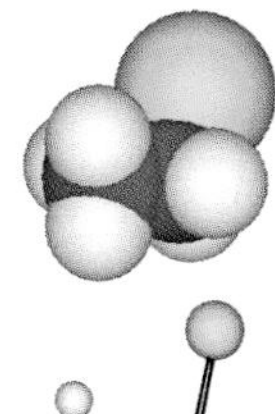

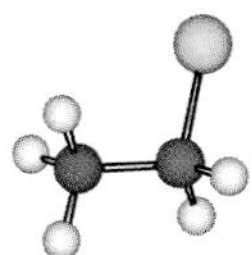

(Ethylene dichloride)
1,2-Dichloroethane
CH_2ClCH_2Cl

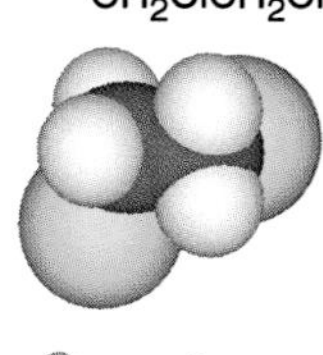

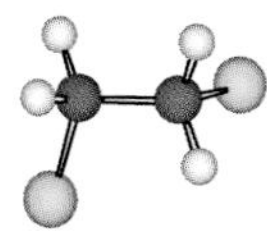

1,1-Dichloroethane
CH_3CHCl_2

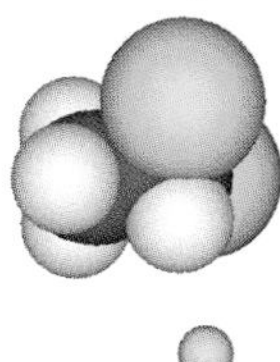

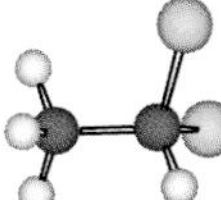

1,1,2-Trichloroethane
$CH_2ClCHCl_2$

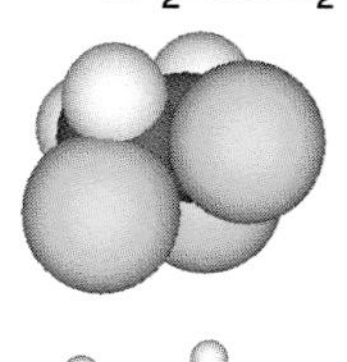

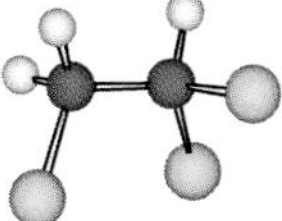

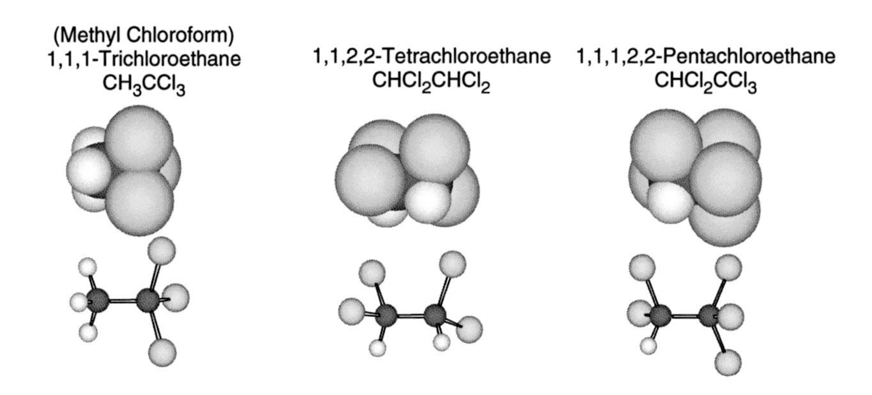

6. Cl-Atom-Substituted C_3 and C_4 Alkanes

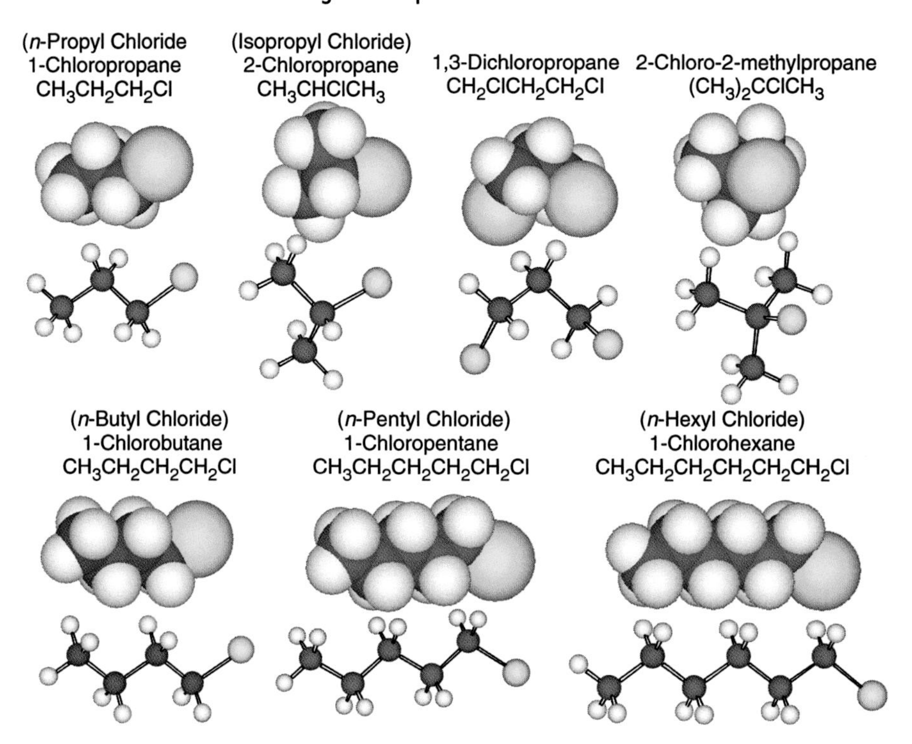

7. Br- and I-Atom-Substituted C_2–C_6 Alkanes

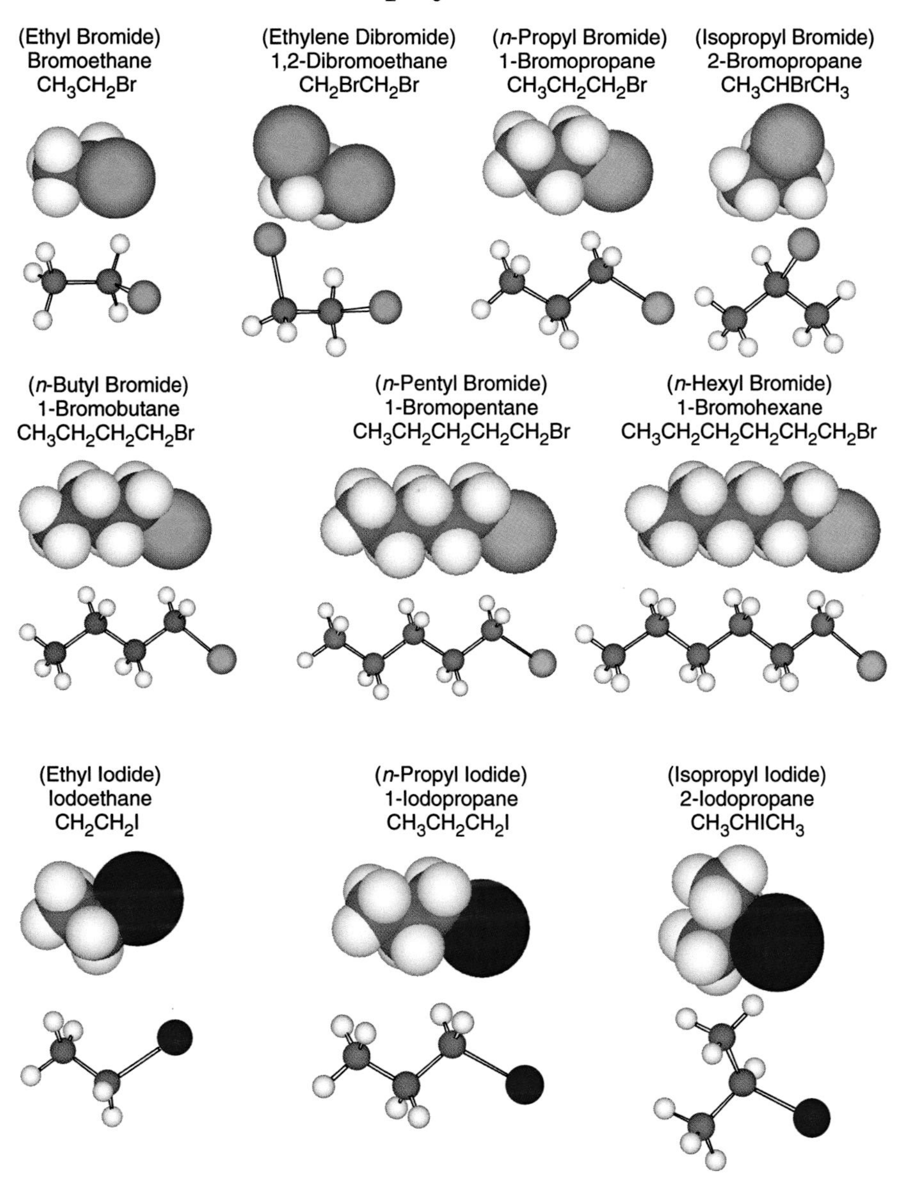

8. Some Cl- and F-Atom-Substituted C_2, C_3, and C_4 Alkanes

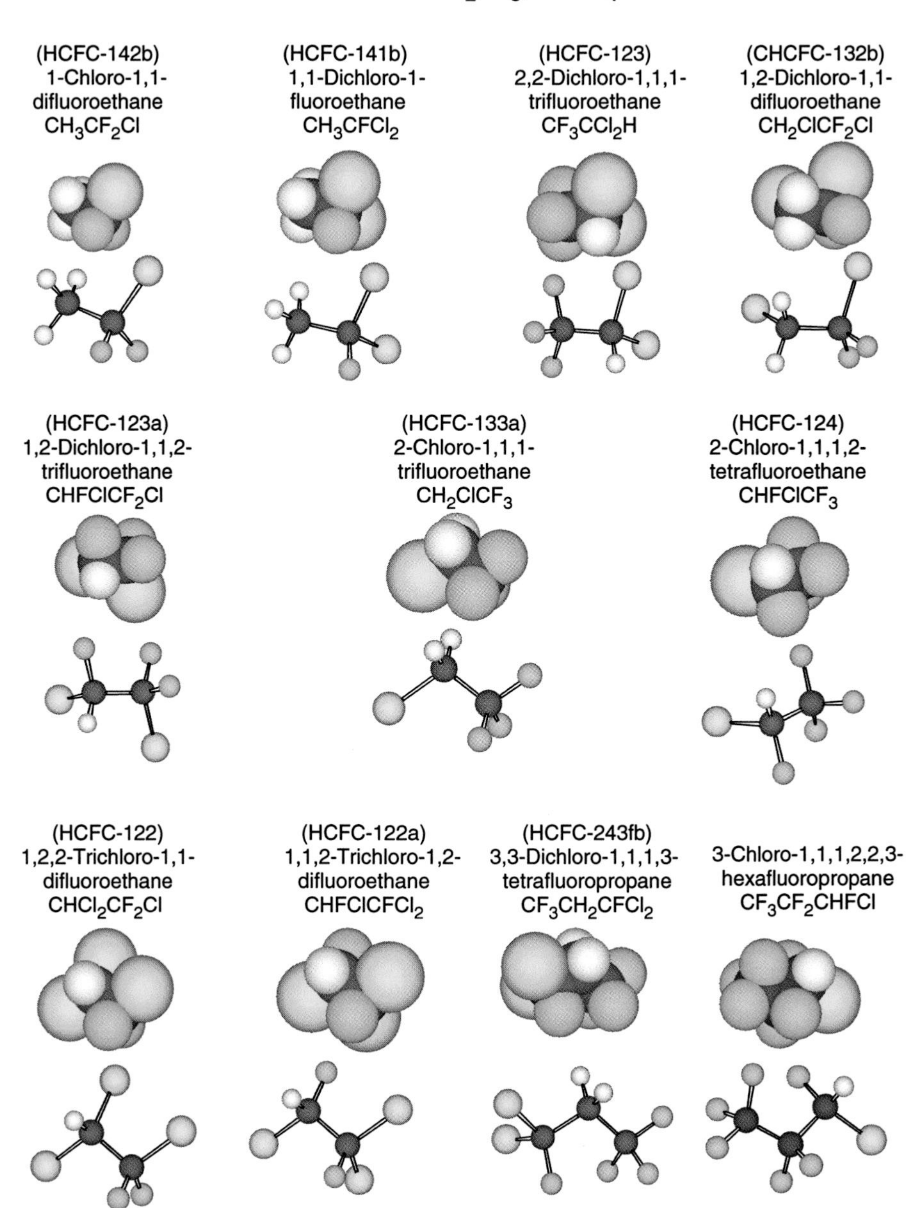

(HCFC-243cc)
1,1-Dichloro-1,2,2-trifluoropropane
$CH_3CF_2CFCl_2$

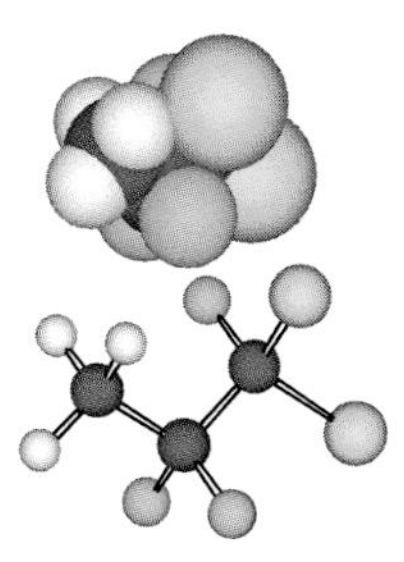

(HCFC-225ca)
3,3-Dichloro-1,1,1,2,2-pentafluoropropane
$CF_3CF_2CCl_2H$

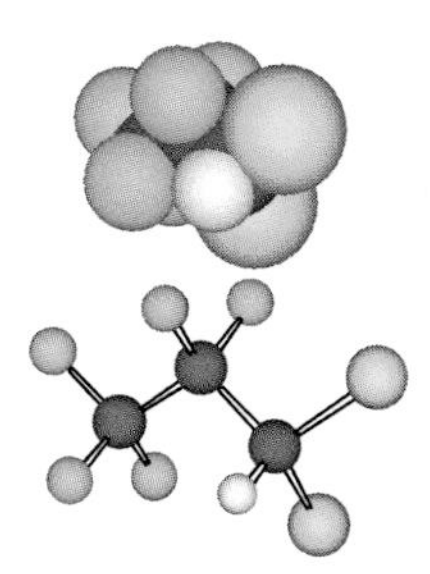

(HCFC-225cb)
1,3-Dichloro-1,1,2,2,3-pentafluoropropane
CF_2ClCF_2CHFCl

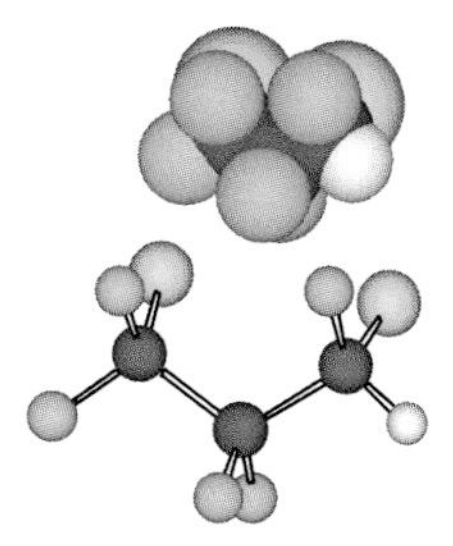

9. Some F-, Cl-, and Br-, Substituted C2 and C3 Alkanes

(Halon-2301)
2-Bromo-1,1,1-trifluoroethane
CF_3CH_2Br

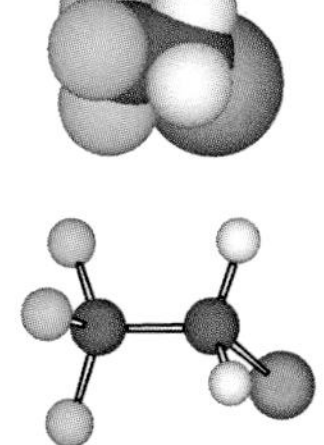

(Halon-2401)
2-Bromo-1,1,1,2-tetrafluoroethane
CF_3CHFBr

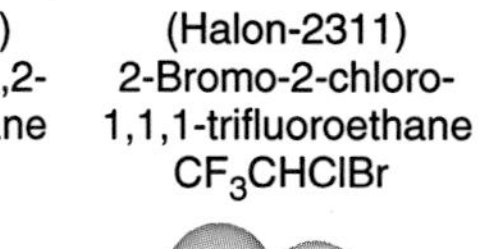

(Halon-2311)
2-Bromo-2-chloro-1,1,1-trifluoroethane
$CF_3CHClBr$

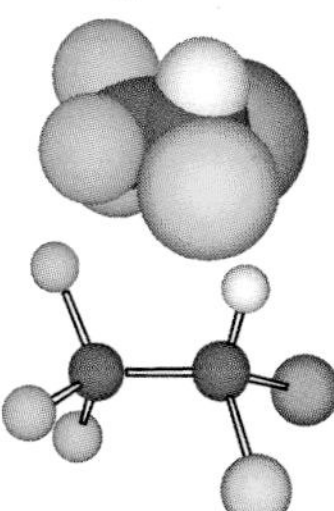

1-Bromo-2-chloro-1,1,2-trifluoroethane
$CF_2BrCHFCl$

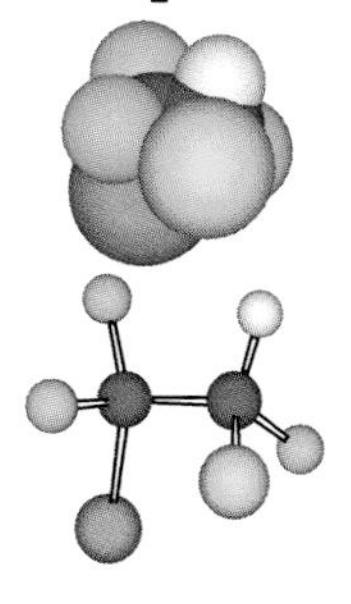

1,2-Dibromo-3-chloropropane
$ClCH_2CHBrCH_2Br$

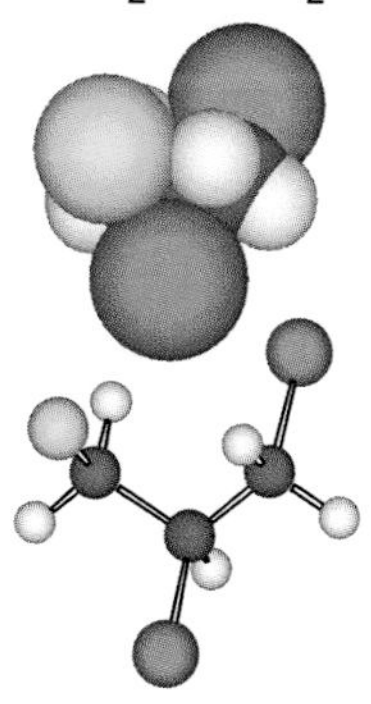

Appendix C

Structures of Some Common Atmospheric Oxidation Products of the Alkanes

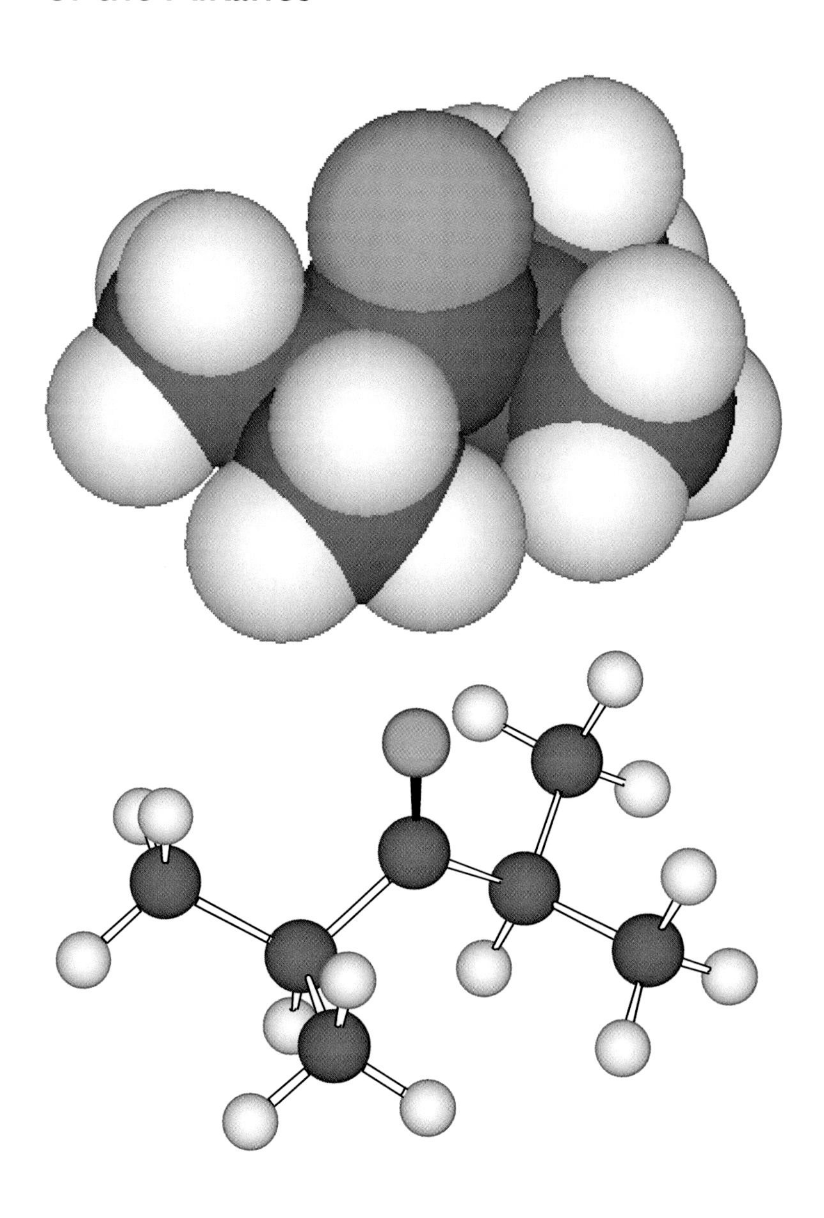

1. Aldehydes

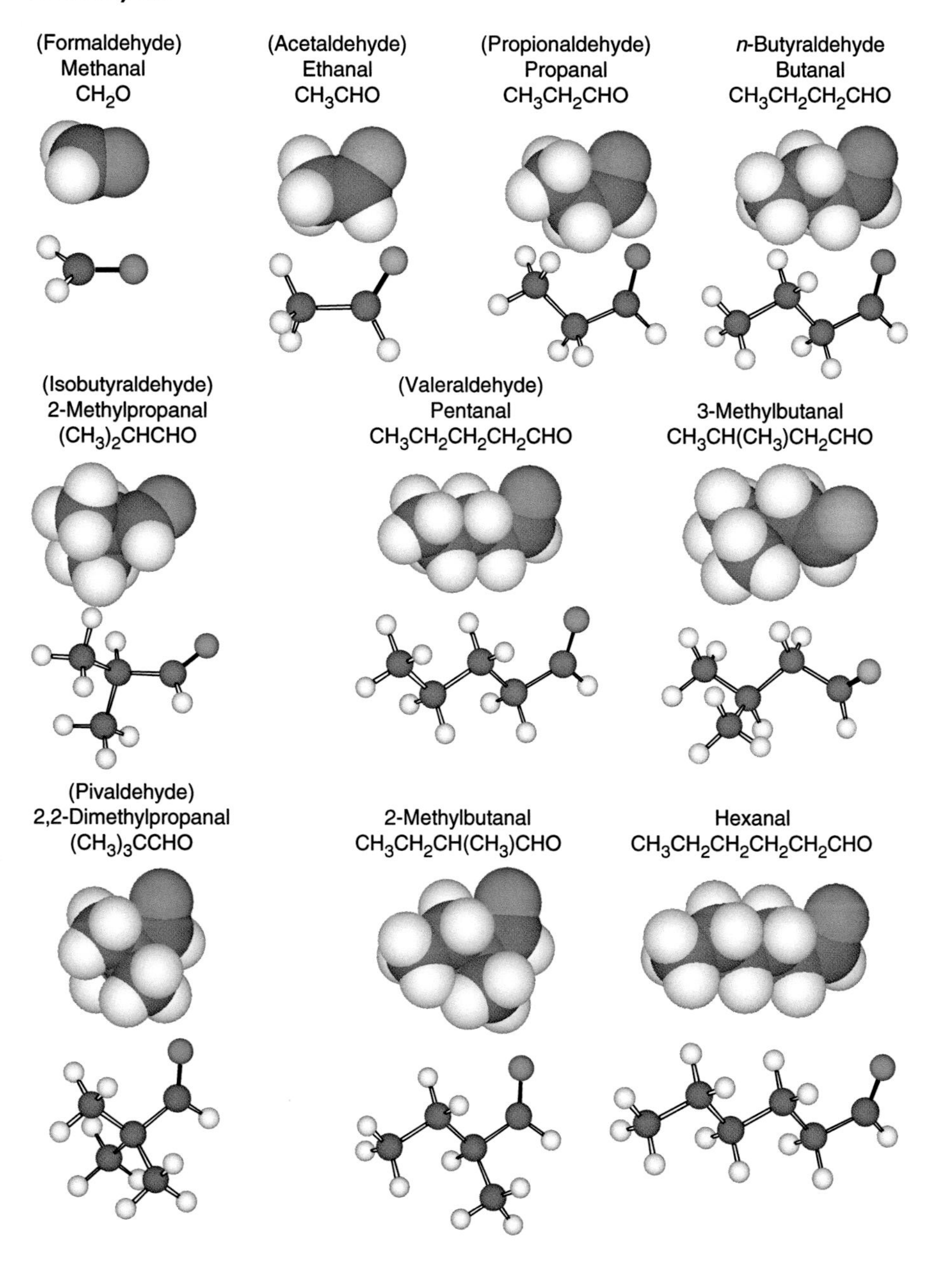
(Formaldehyde)
Methanal
CH_2O
(Acetaldehyde)
Ethanal
CH_3CHO
(Propionaldehyde)
Propanal
CH_3CH_2CHO
n-Butyraldehyde
Butanal
$CH_3CH_2CH_2CHO$
(Isobutyraldehyde)
2-Methylpropanal
$(CH_3)_2CHCHO$
(Valeraldehyde)
Pentanal
$CH_3CH_2CH_2CH_2CHO$
3-Methylbutanal
$CH_3CH(CH_3)CH_2CHO$
(Pivaldehyde)
2,2-Dimethylpropanal
$(CH_3)_3CCHO$
2-Methylbutanal
$CH_3CH_2CH(CH_3)CHO$
Hexanal
$CH_3CH_2CH_2CH_2CH_2CHO$

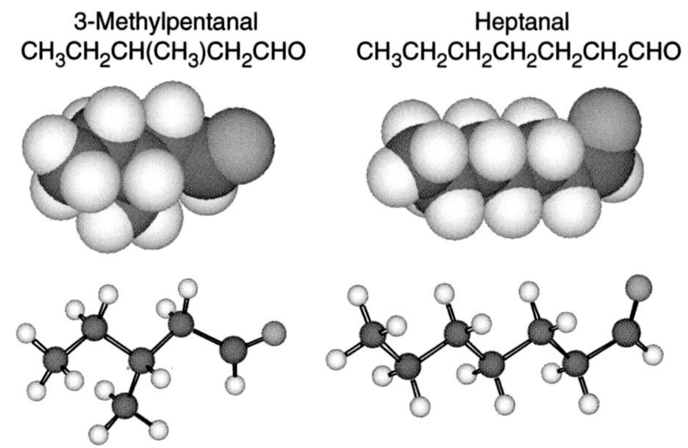
3-Methylpentanal
CH3CH2CH(CH3)CH2CHO
Heptanal
CH3CH2CH2CH2CH2CH2CHO

2. Ketones

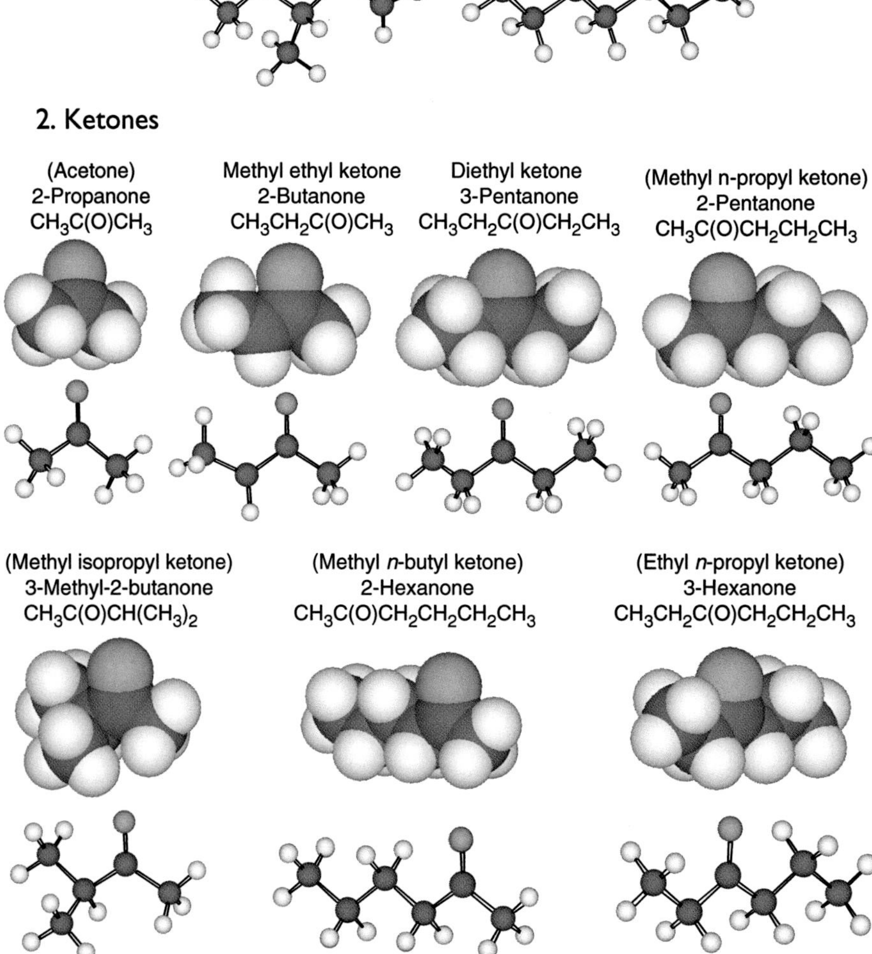
(Acetone)
2-Propanone
CH3C(O)CH3
Methyl ethyl ketone
2-Butanone
CH3CH2C(O)CH3
Diethyl ketone
3-Pentanone
CH3CH2C(O)CH2CH3
(Methyl n-propyl ketone)
2-Pentanone
CH3C(O)CH2CH2CH3
(Methyl isopropyl ketone)
3-Methyl-2-butanone
CH3C(O)CH(CH3)2
(Methyl n-butyl ketone)
2-Hexanone
CH3C(O)CH2CH2CH2CH3
(Ethyl n-propyl ketone)
3-Hexanone
CH3CH2C(O)CH2CH2CH3

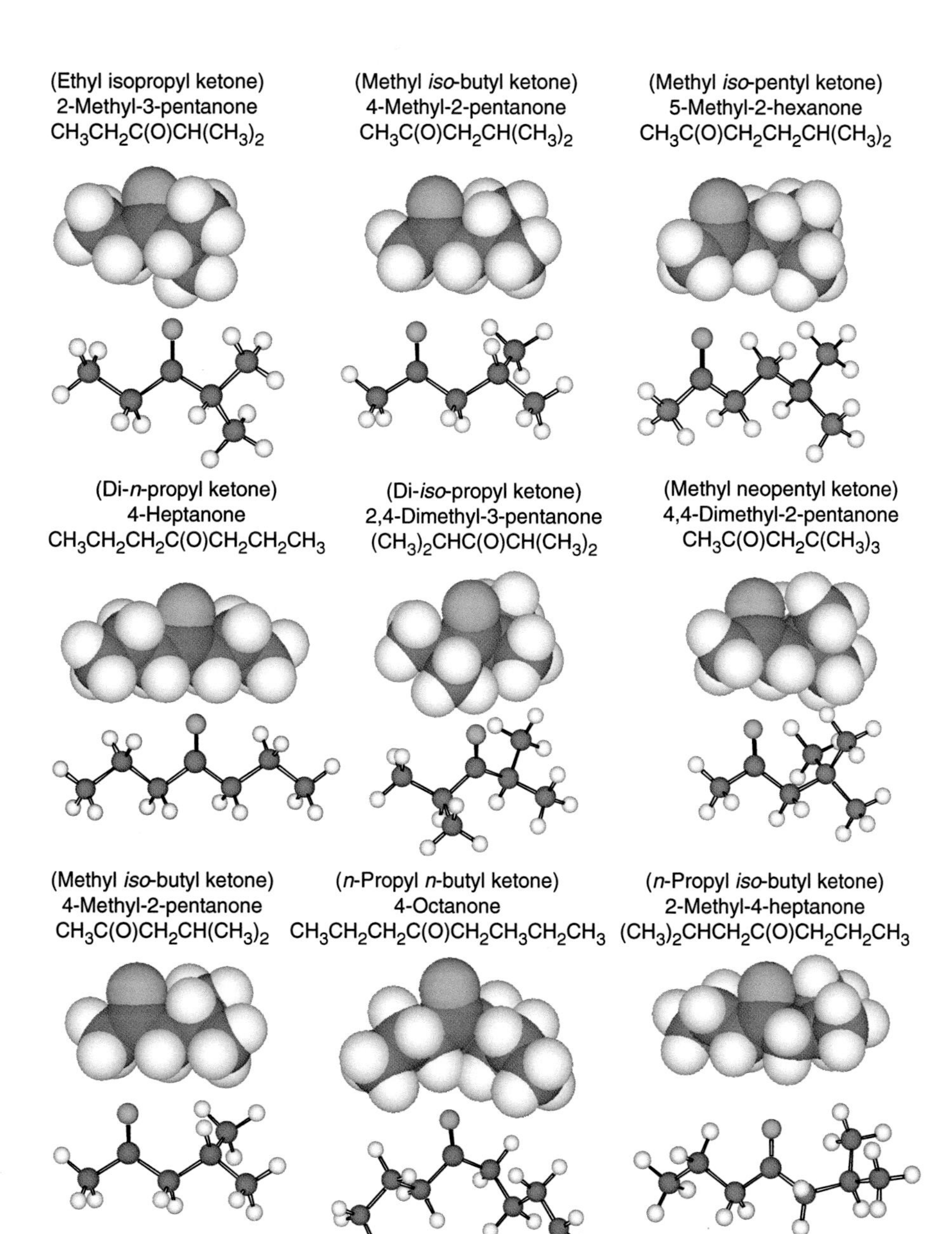
(Ethyl isopropyl ketone)
2-Methyl-3-pentanone
$CH_3CH_2C(O)CH(CH_3)_2$
(Methyl *iso*-butyl ketone)
4-Methyl-2-pentanone
$CH_3C(O)CH_2CH(CH_3)_2$
(Methyl *iso*-pentyl ketone)
5-Methyl-2-hexanone
$CH_3C(O)CH_2CH_2CH(CH_3)_2$
(Di-*n*-propyl ketone)
4-Heptanone
$CH_3CH_2CH_2C(O)CH_2CH_2CH_3$
(Di-*iso*-propyl ketone)
2,4-Dimethyl-3-pentanone
$(CH_3)_2CHC(O)CH(CH_3)_2$
(Methyl neopentyl ketone)
4,4-Dimethyl-2-pentanone
$CH_3C(O)CH_2C(CH_3)_3$
(Methyl *iso*-butyl ketone)
4-Methyl-2-pentanone
$CH_3C(O)CH_2CH(CH_3)_2$
(*n*-Propyl *n*-butyl ketone)
4-Octanone
$CH_3CH_2CH_2C(O)CH_2CH_3CH_2CH_3$
(*n*-Propyl *iso*-butyl ketone)
2-Methyl-4-heptanone
$(CH_3)_2CHCH_2C(O)CH_2CH_2CH_3$

(*n*-Propyl *sec*-butyl ketone)
3-Methyl-4-heptanone
$CH_3CH_2CH_2C(O)CH(CH_3)CH_2CH_3$

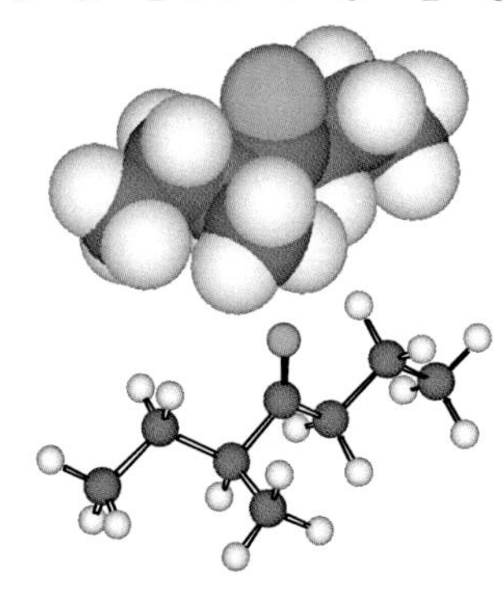

(*n*-Propyl *tert*-butyl ketone)
2,2-Dimethyl-3-hexanone
$(CH_3)_2CC(O)CH_2CH_2CH_3$

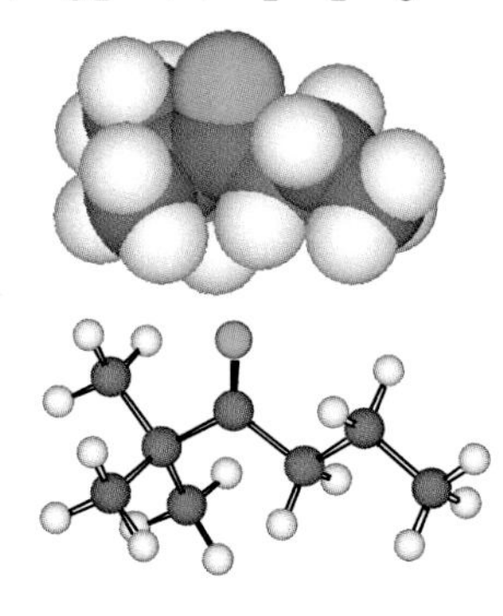

(Di-*iso*-butyl ketone)
2,6-Dimethyl-4-heptanone
$(CH_3)_2CHCH_2C(O)CH_2CH_2(CH_3)_2$

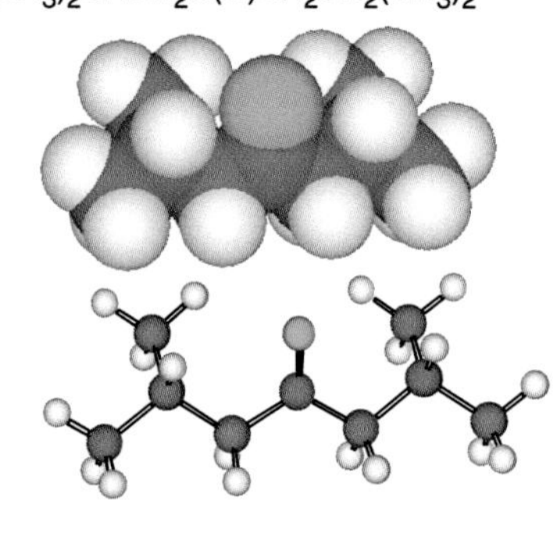

(Di--*n*-butyl ketone)
5-Nonanone
$CH_3CH_2CH_2CH_2C(O)CH_2CH_2CH_2CH_3$

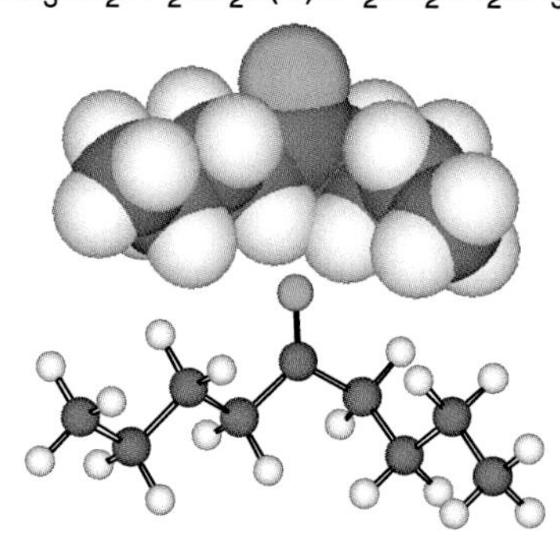

(Di-*sec*-butyl ketone)
3,5-Dimethyl-4-heptanone
$CH_3CH_2CH(CH_3)C(O)CH(CH_3)CH_2CH_3$

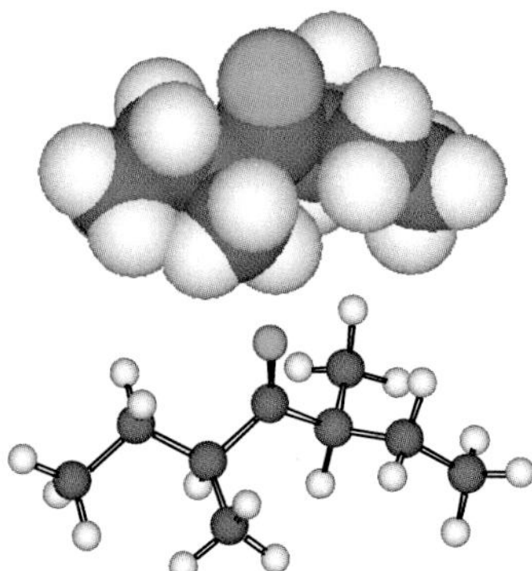

(Di-*tert*-butyl ketone)
2,2,4,4-Tetramethyl-3-pentanone
$(CH_3)_3CC(O)C(CH_3)_3$

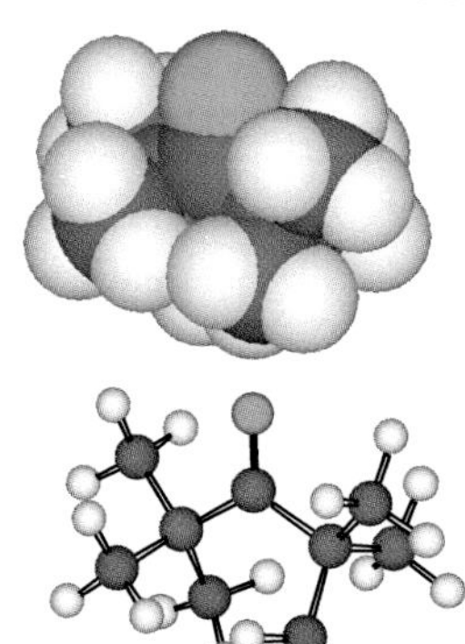

3. Cyclic Ketones

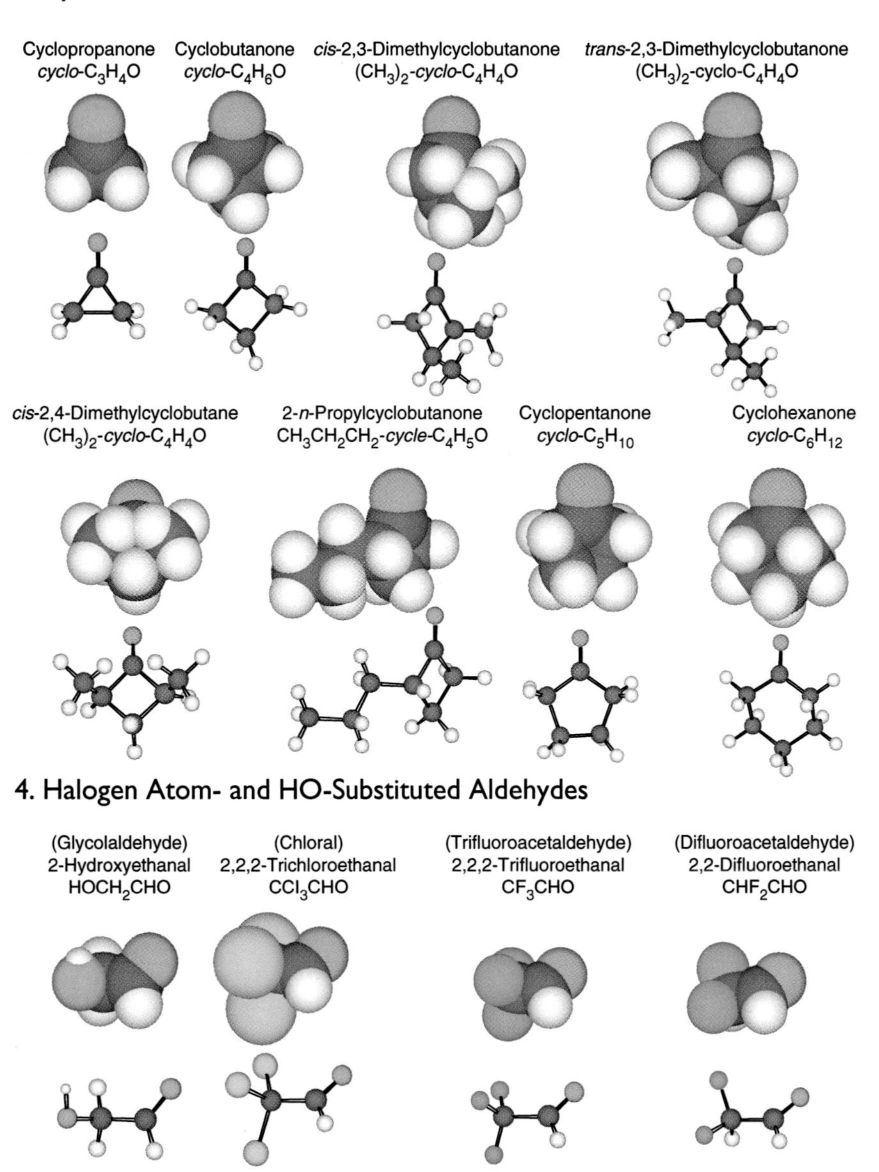

4. Halogen Atom- and HO-Substituted Aldehydes

5. Halogen Atom- and HO-Substituted Ketones

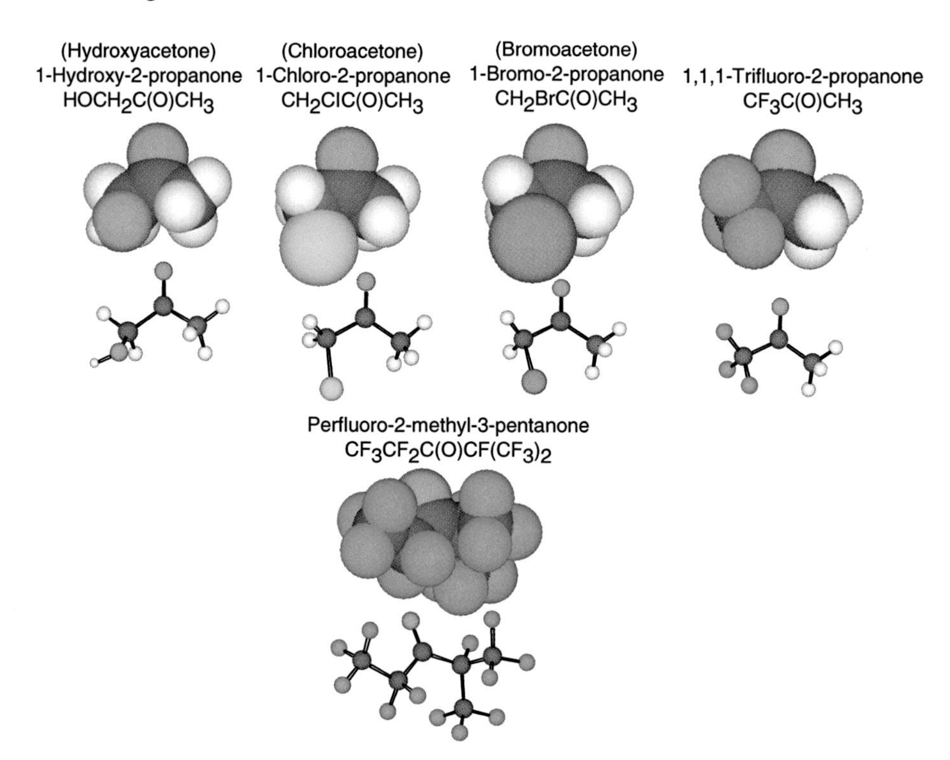

6. Formyl Halides and Carbonyl Halides

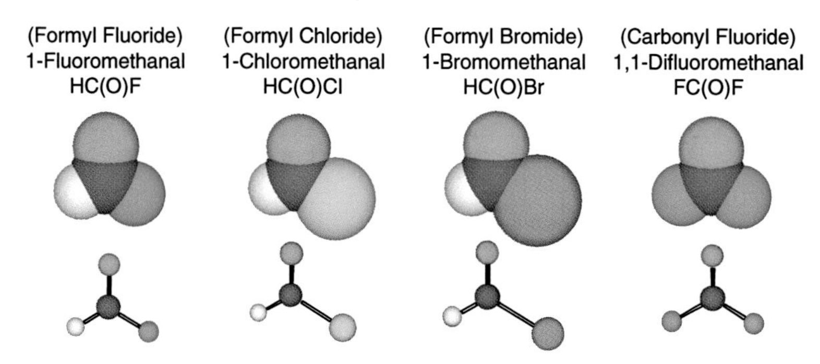

(Phosgene)
(Carbonyl Chloride)
1,1- Dichloromethanal
ClC(O)Cl

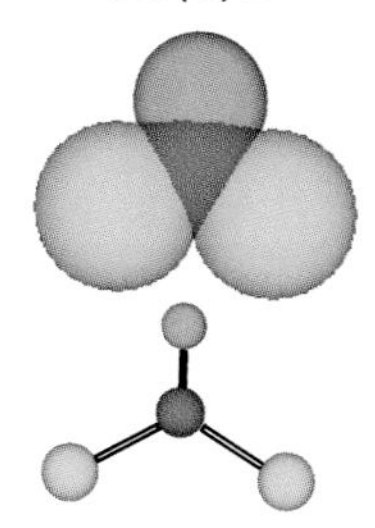

Carbonyl Bromide
1,1-Dibromomethanal
BrC(O)Br

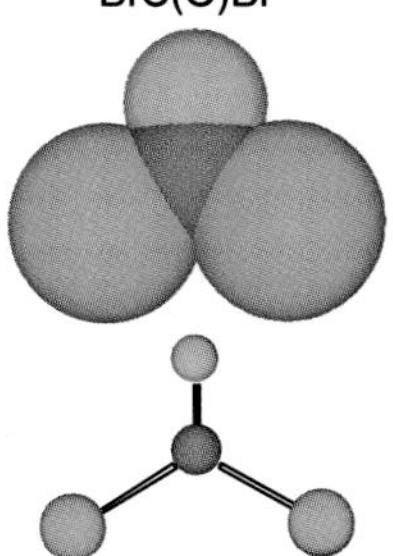

7. Acetyl Halides

(Acetyl Chloride)
1-Chloroethanal
$CH_3C(O)Cl$

Chloroacetyl Chloride
1,2-Dichloroethanal
$CH_2ClC(O)Cl$

(Dichloroacetyl Chloride)
1.2.2-Trichloroethanal
$CHCl_2(O)Cl$

Trichloroacetyl Chloride
1,2,2,2-Tetrachloroethanal
$CCl_3C(O)Cl$

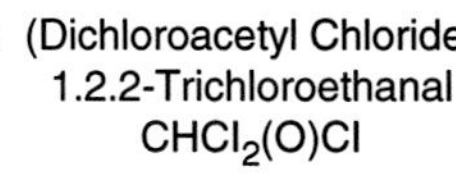

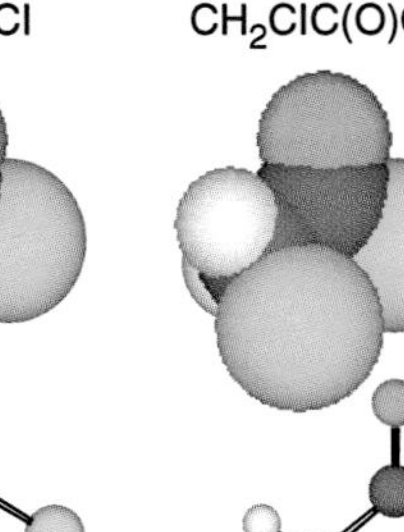

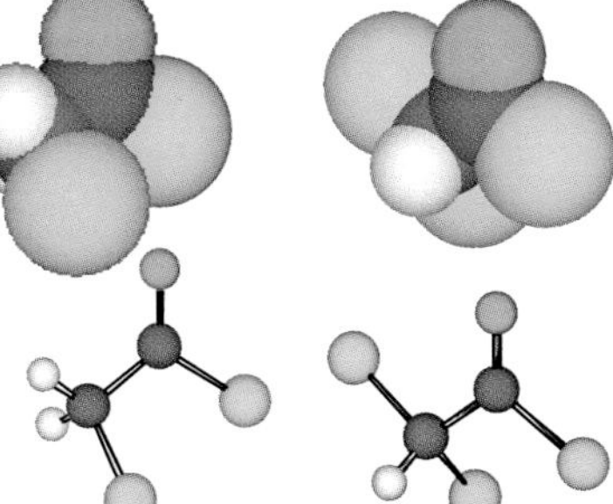

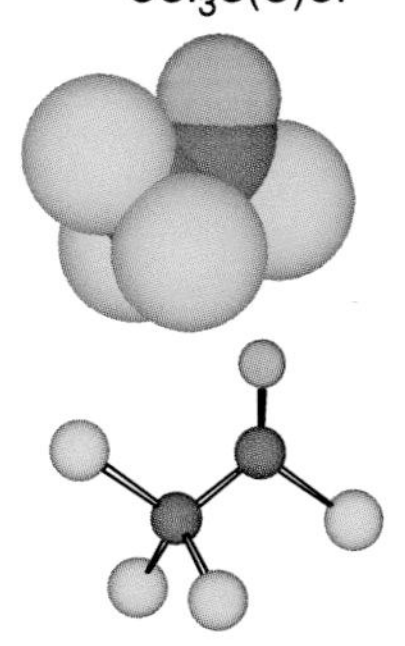

(Trifluoroacetyl Chloride)
1-Chloro-2,2,2-trifluoroethanal
$CF_3C(O)Cl$

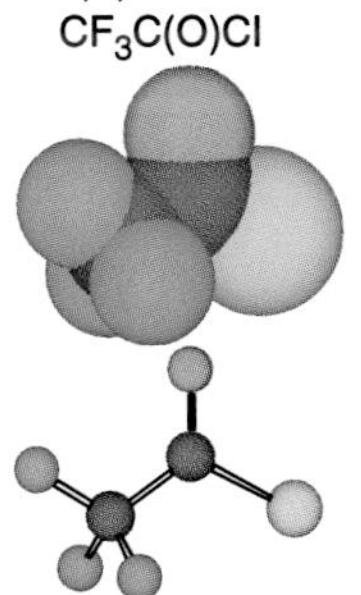

8. Acyclic Alcohols

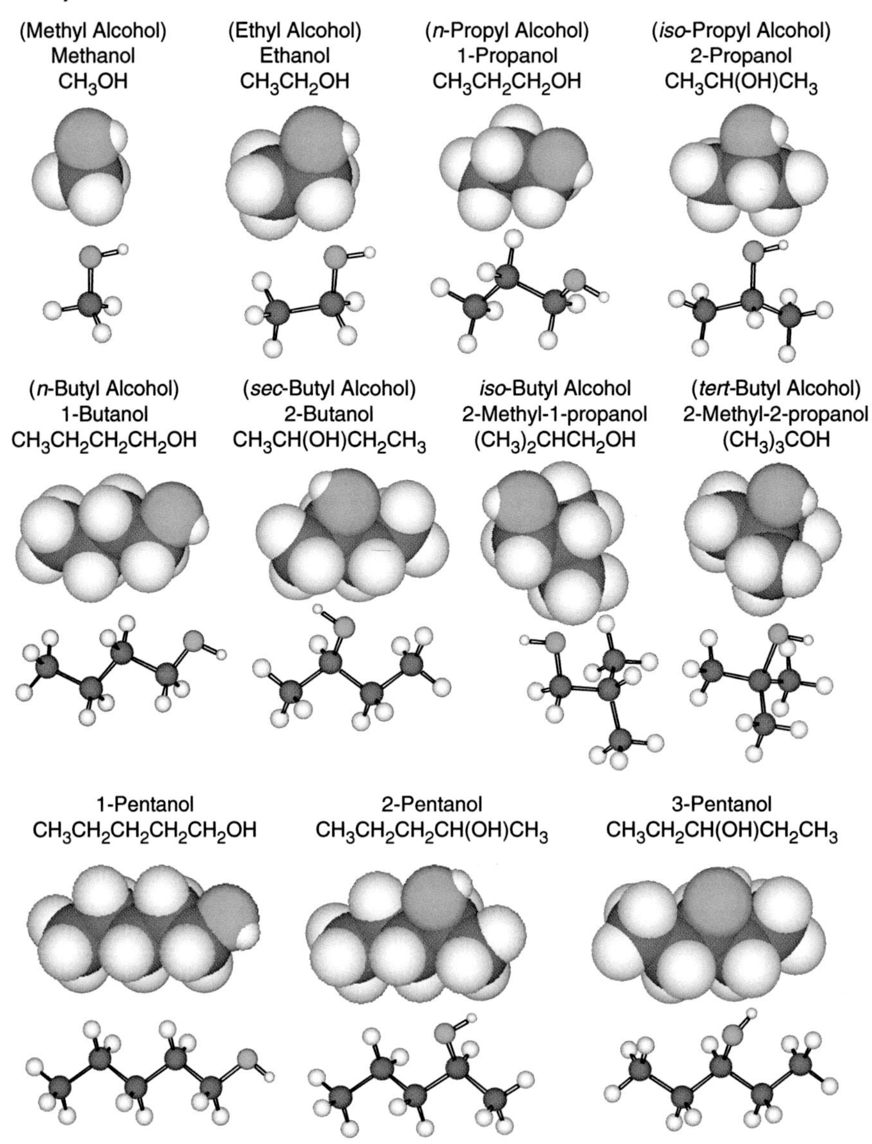

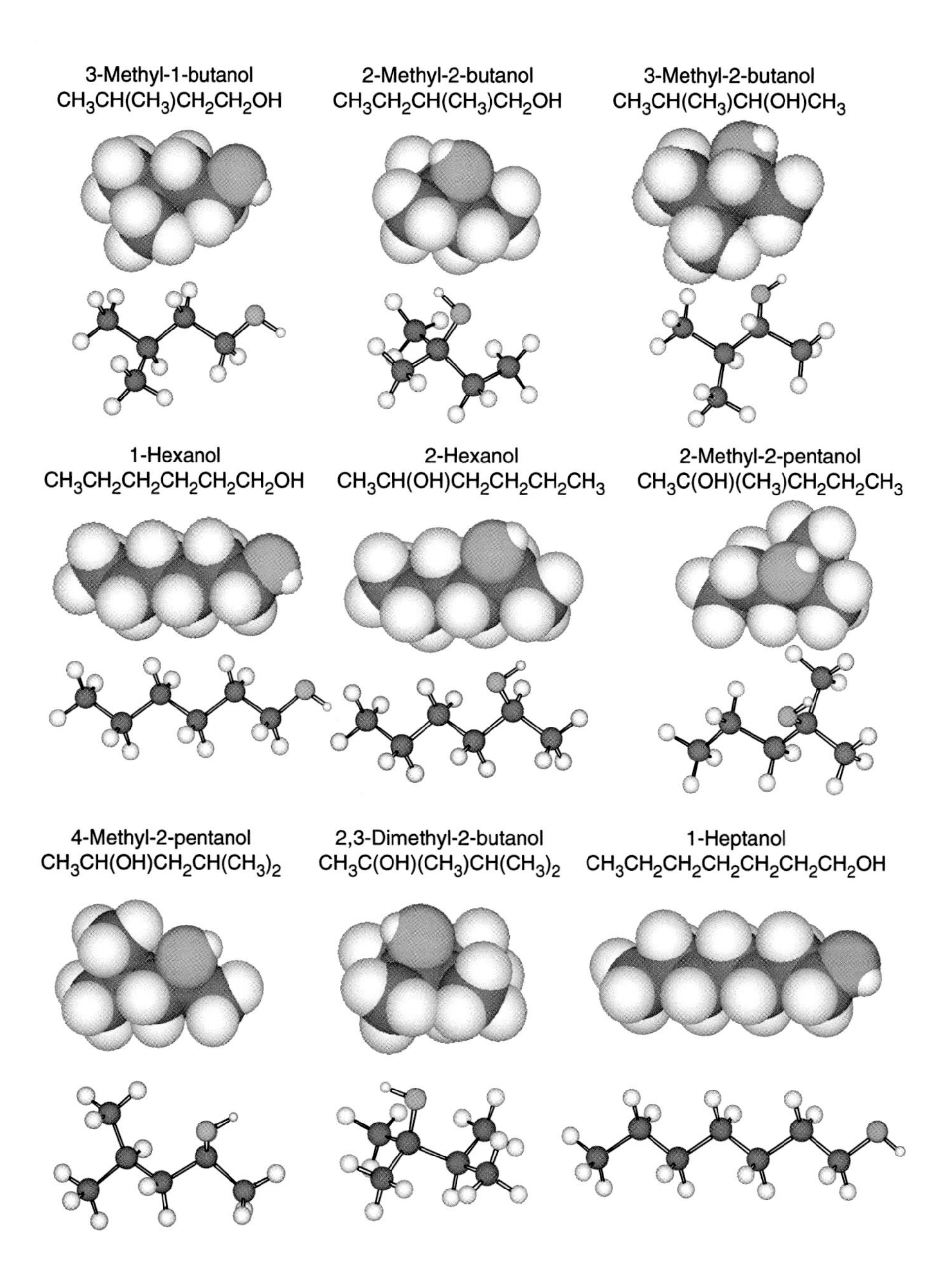

3-Methyl-1-butanol
$CH_3CH(CH_3)CH_2CH_2OH$
2-Methyl-2-butanol
$CH_3CH_2CH(CH_3)CH_2OH$
3-Methyl-2-butanol
$CH_3CH(CH_3)CH(OH)CH_3$
1-Hexanol
$CH_3CH_2CH_2CH_2CH_2CH_2OH$
2-Hexanol
$CH_3CH(OH)CH_2CH_2CH_2CH_3$
2-Methyl-2-pentanol
$CH_3C(OH)(CH_3)CH_2CH_2CH_3$
4-Methyl-2-pentanol
$CH_3CH(OH)CH_2CH(CH_3)_2$
2,3-Dimethyl-2-butanol
$CH_3C(OH)(CH_3)CH(CH_3)_2$
1-Heptanol
$CH_3CH_2CH_2CH_2CH_2CH_2CH_2OH$

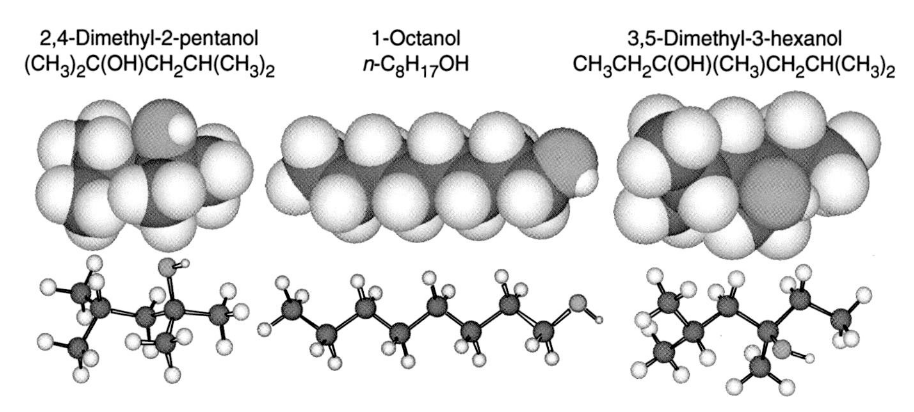

9. Alkyl Hydroperoxides

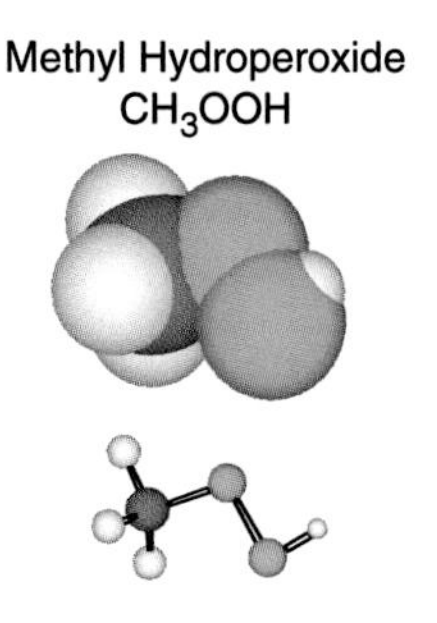

10. Alkyl Nitrates

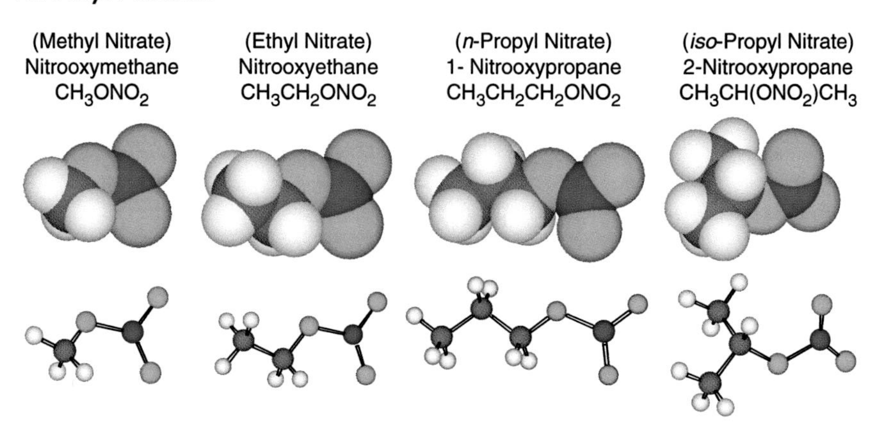

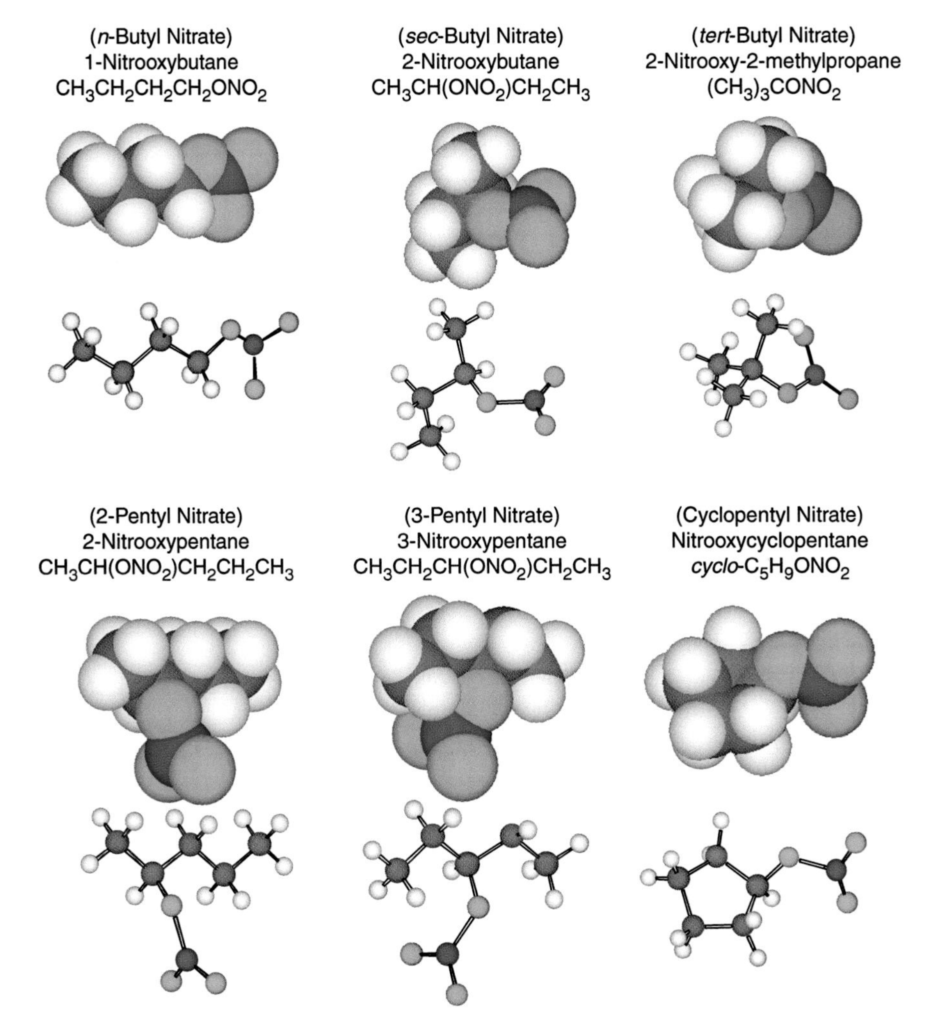
(*n*-Butyl Nitrate)
1-Nitrooxybutane
$CH_3CH_2CH_2CH_2ONO_2$
(*sec*-Butyl Nitrate)
2-Nitrooxybutane
$CH_3CH(ONO_2)CH_2CH_3$
(*tert*-Butyl Nitrate)
2-Nitrooxy-2-methylpropane
$(CH_3)_3CONO_2$
(2-Pentyl Nitrate)
2-Nitrooxypentane
$CH_3CH(ONO_2)CH_2CH_2CH_3$
(3-Pentyl Nitrate)
3-Nitrooxypentane
$CH_3CH_2CH(ONO_2)CH_2CH_3$
(Cyclopentyl Nitrate)
Nitrooxycyclopentane
cyclo-$C_5H_9ONO_2$

11. Peroxyacyl Nitrates

[Peroxyacetyl Nitrate (PAN)]
(Ethaneperoxoic Nitric Anhydride)
1-Nitroperoxyethanal
$CH_3C(O)OONO_2$

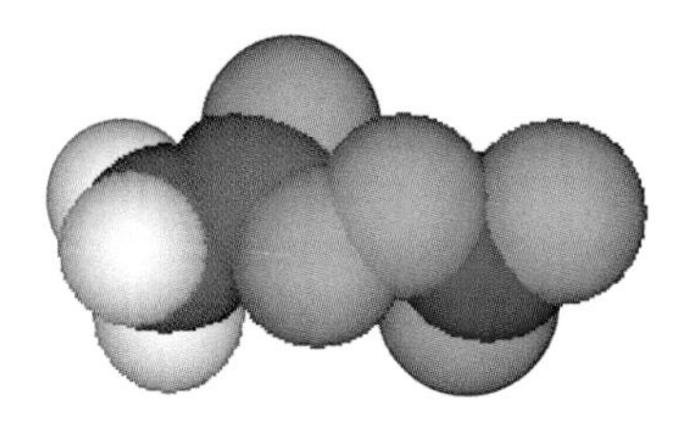

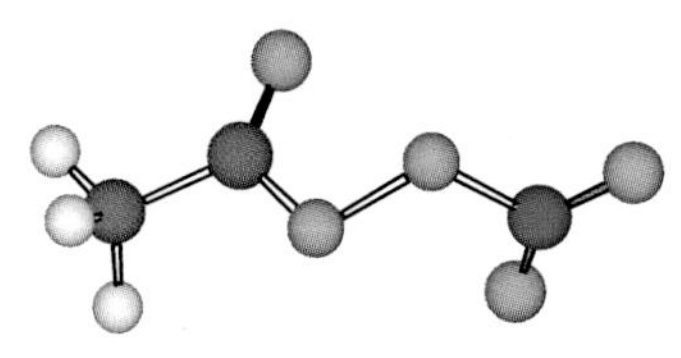

[Peroxypropionyl Nitrate (PPN)]
(Propaneperoxoic Nitric Anhydride)
1-Nitroperoxypropanal
$CH_3CH_2C(O)OONO_2$

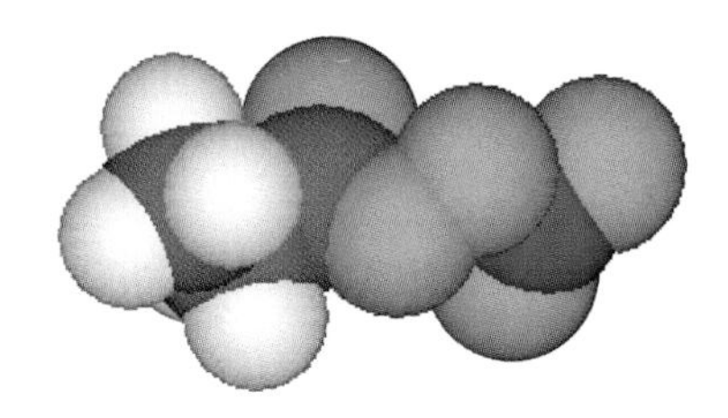

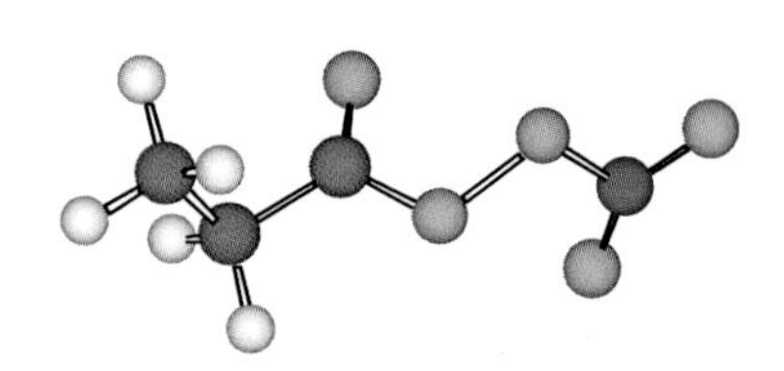

Author Index

Subject Index